MW01629201

MODERN ELECTROPLATING

MODERN ELECTROPLATING

FOURTH EDITION

Edited by

Mordechay Schlesinger
University of Windsor, Windsor Ontario, Canada

Milan Paunovic
IBM T.J. Watson Research Center, Yorktown Heights, NY

Sponsored by

THE ELECTROCHEMICAL SOCIETY, INC. *Pennington, New Jersey*

A WILEY-INTERSCIENCE PUBLICATION

JOHN WILEY & SONS, INC.

New York • Chichester • Weinheim • Brisbane • Singapore • Toronto

Published simultaneously in Canada.

For ordering and customer service, call 1-800-CALL WILEY

Library of Congress Cataloging in Publication Data is available.

Schlesinger, Mordechay
Modern Electroplating, Fourth Edition/Mordechay Schlesinger and Milan Paunovic.

ISBN 0-471-16824-6

Printed in the United States of America.

10 9 8 7 6 5 4 3 2 1

CONTRIBUTORS

JOSEPH A. ABYS, Lucent Technologies, Electroplating Chemicals and Services, Staten Island, NY

GEORGE A. DI BARI, INCO, Saddle Brook, NJ

JACK W. DINI, Lawrence Livermore National Laboratory, Livermore, CA

TAKAYUKI HOMMA, Department of Applied Chemistry, Waseda University, Tokyo, Japan

MASANOBU IZAKI, Department of Inorganic Chemistry, Osaka Municipal Technical Research Institute, Osaka, Japan

MANFRED JORDAN, Dr. Ing. Mas Schlötter GmbH & Co. KG, Galvanotechnik, D-73304 Geisling/Steige, Germany

MASARU KATO, Central Research Laboratory, Kanto Chemical Company, Soka Saitama-ken, Japan

PAUL A. KOHL, Georgia Institute of Technology, School of Chemical Engineering, Atlanta, GA

SHINICHI KOMABA, Department of Applied Chemistry, Waseda University, Tokyo, Japan

N. V. MANDICH, HBM Electrochemical Company, Lansing, IL

TOSHIYUKI MOMMA, Department of Applied Chemistry, Waseda University, Tokyo, Japan

IZUMI OHNO, Tokyo Institute of Technology, Department of Metallurgical Engineering, Tokyo, Japan

YUTAKA OKINAKA, Advanced Research Center for Science and Engineering, Waseda University, Tokyo, Japan

TETSUYA OSAKA, Department of Applied Chemistry, Waseda University, Tokyo, Japan

MILAN PAUNOVIC, IBM T.J. Watson Research Center, Yorktown Heights, NY

MORDECHAY SCHLESINGER, Department of Physics, University of Windsor, Windsor Ontario, Canada

T. E. SCHLESINGER, Department of Electrical and Computer Engineering, Carnegie Mellon University, Pittsburgh, PA

DEXTER D. SNYDER, General Motors Research and Development Center, Warren, MI

DONALD L. SNYDER, ATO Tech, Cleveland, OH

DENNIS R. TURNER, 59 Susan Drive, Chatham, NJ

MICHA TOMKIEWICZ, Department of Physics, Brooklyn College of SUNY, Brooklyn, NY

ROLF WEIL, 47 Carteret Street, West Orange, NJ

RENÉ WINAND, Department of Metallurgy and Electrochemistry, University of Bruxelles, Bruxelles, Belgium

TOKIHIKO YOKOSHIMA, Department of Applied Chemistry, Waseda University, Tokyo, Japan

YUN ZHANG, Lucent Technologies, Electroplating Chemicals and Services, Staten Island, NY

CONTENTS

PREFACE

In planning this new edition of *Modern Electroplating*, we have realized from the start that it would be impossible to include in one volume both the fundamental aspects and the technology itself. For this reasons we have decided to publish the recent developments in the science of deposition in a separate volume titled *Fundamentals of Electrochemical Deposition*. That volume was published in November 1998. Therefore, the present volume includes only a brief summary of fundamental technological advancements, and this is presented in the first, introductory chapter.

Since the last edition of *Modern Electroplating* in 1975, electrochemical deposition has evolved from an ill-defined area, as the Preface to the previous edition calls it, into an exact science. This development is, in the first place, seen as responsible for the ever-increasing number and widening types of applications of this branch of practical science and engineering.

The most significant developments in any field of science or technology in general, and in electrochemistry in particular, are made by those only who possess a good understanding of the fundamental aspects of the discipline, which in this case is electrochemical deposition. We, the editors, found it necessary and highly desirable to seek and present to the reader a companion volume that, for all intents and purposes, makes essentially a completely new contribution and not just a revised version of the earlier editions. Thus, for the sake of illustration, the present edition includes a chapter devoted to the electrodeposition of semiconductors. Another deals with environmental issues. Last, but not least, in this connection, neither of the editors nor the vast majority of the contributors were associated with any of the earlier editions.

Technological areas in which the possession of technical knowledge of electroplating is found to be essential include all aspects of electronics; macro-, micro-, and nano-optics; opto-electronics; and sensors of most types. In addition, a number of key industries, such as the automotive industry, employ methods of electroplating. This is so even when other methods such as evaporation and sputtering CVD (chemical vapor deposition) are an option. Electroplating is therefore often used for reasons of economy and/or convenience.

This volume is divided into 26 chapters. After a three-part introductory chapter by Paunovic, Schlesinger, and Weil come 13 chapters dealing with the electrodeposition of copper (Dini), nickel (DiBari), gold (Kohl), silver (Schlesinger), tin (Abys et al.), chromium (Snyder et al.), lead and alloys (Jordan), tin-lead alloys (Jordan), zinc and alloys (Winand), iron and alloys (Izaki), palladium and alloys (Abys et al.), nickel and cobalt alloys (DiBari), and semiconductors (T. E. Schlesinger). Closing this series of chapters is one on deposition on nonconductors (Schlesinger), and conductive polymers (Osaka et al.). Next come 6 chapters dealing with electroless deposition of copper (Paunovic), nickel (Schlesinger), cobalt (Osaka), palladium and platinum (Ohno), gold (Okinaka), and electroless alloys

(Ohno). Finally, 4 chapters close the book, and these are on preparation for deposition (Dexter Snyder), manufacturing technologies (Turner), manufacturing control (Turner), and environmental considerations (Tomkiewicz).

In the preface to *Fundamentals of Electrochemical Deposition* we stated that it may be considered a lucky coincidence that this volume is published close to the time that copper interconnection technology is introduced in the microelectronic industry. This is still the case. There has been a truly revolutionary change from physical to electrochemical techniques in the production of microconductors on silicon, and developments in electrochemical deposition are bound to generate and maintain in the twenty-first century an increased interest and urgent need for up-to-date information regarding the technology. The present volume together with the *Fundamentals* volume should be of great help in understanding these advancements.

The chapters were written by different authors and so differences in style and approach will be evident. We the editors have tried to smooth those differences without changing the basic message present in each chapter. We also intend this volume to be a useful reference for practitioners of deposition as well as for individuals who are about to enter this modern ever-evolving field of practical knowledge. For this reason each chapter is complete and may be read and consulted separately, and certainly the book can be read in any order.

Our thanks and heartfelt gratitude go to many members of the Electrochemical Society and in particular to those of the Electrodeposition Division. Our thanks also go to our respective families for their patience and understanding during the hectic long hours we spent in preparing this volume.

MORDECHAY SCHLESINGER

Windsor Ontario, Canada

MILAN PAUNOVIC

Yorktown Heights, New York

Conversion Factors

1 centimeter (cm) = 0.934 inch (in.)
1 millimeter (mm) = 0.0394 inch (in.)
1 micrometer (μm, micron) = 0.0394 mil = 39.37 microinch (μ in.)
1 square decimeter (dm^2) = 15.5 square inch ($in.^2$) = 0.1076 square foot (ft^2)
1 square centimeter (cm^2) = 0.155 square inch ($in.^2$)
1 square millimeter (mm^2) = 0.00155 square inch ($in.^2$)
1 kilogram (kg) = 2.205 pound (lb)
1 gram (g) = 0.0353 ounce (oz) avoirdupois = 0.0321 ounce Troy
1 liter (l) = 0.264 gallon U.S. (gal) = 0.220 gallon British
1 ampere per square decimeter (A/dm^2) = 9.29 ampere per square foot (A/ft^2) – see chart
1 gram per liter (g/l) = 0.133 ounce per gallon, U.S. (oz/gal) – see chart
1 kilogram per square millimeter (kg/mm^2) = 1.422 pounds per square inch ($lb/in.^2$, psi) – see chart; strictly, unit should be kilogram-force, or kgf/mm^2; in S.I. units, 1 $kgf/mm^2 = 9.806 \times 10^6$ newton/square meter (N/m^2) = 9.806 MN/m^2; 1 N/m^2 = 1 Pa (pascal)

Approximate Conversion Factors for Mental Calculation (accurate to 10% or better)

To convert from	To	Do this
A/dm^2	A/ft^2	Multiply by 10
A/dm^2	$A/in.^2$	Divide by 15 (multiply by 2, divide by 30)
Celsius (Centigrade)*	Fahrenheit, F	Multiply by 9/5 (1.8) and add 32
g/l	oz/gal	Divide by 7.5 (multiply by 4, divide by 30)
kg/mm^2	$lb/in.^2$(psi)	Multiply by 1500
mm	inch	Divide by 25 (multiply by 4, divide by 100)
micrometers (m)	mil	Divide by 25 (multiply by 4, divide by 100)

*Exact.

Graphical Conversions

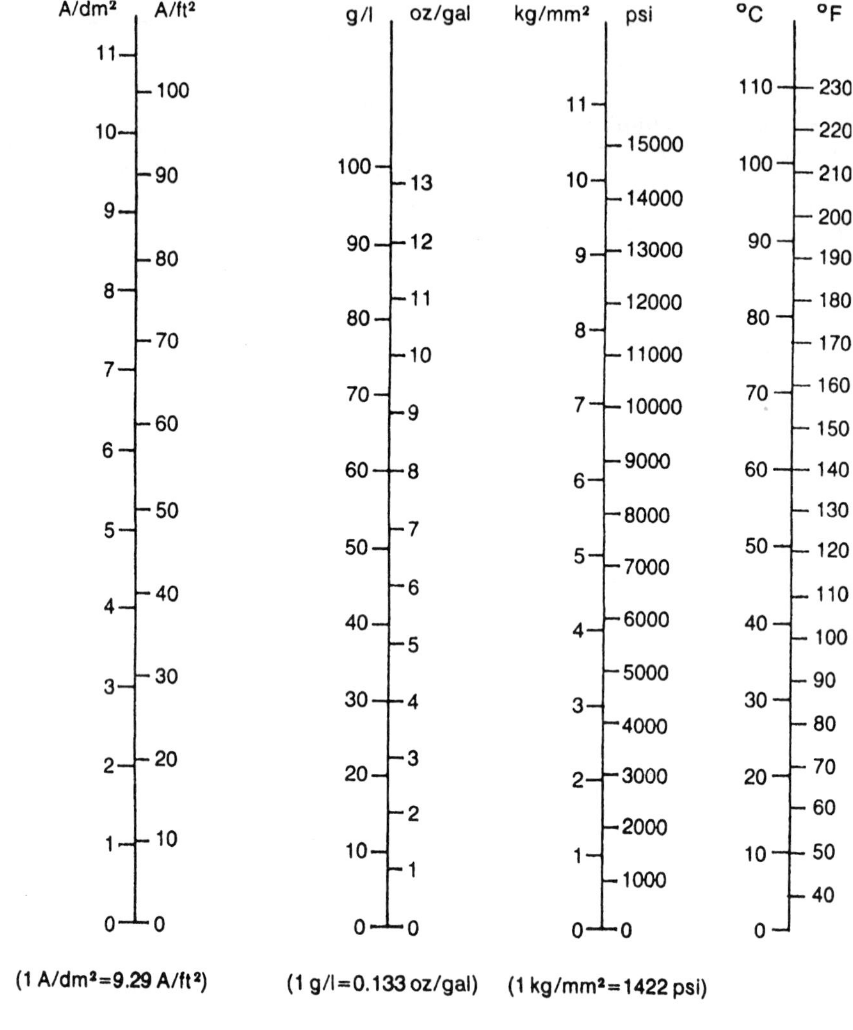

(1 A/dm² = 9.29 A/ft²) (1 g/l = 0.133 oz/gal) (1 kg/mm² = 1422 psi)

The Electrochemical Society Series

Corrosion Handbook
Edited by Herbert H. Uhlig

Modern Electroplating, Third Edition
Edited by Frederick A. Lowenheim

Modern Electroplating, Fourth Edition
Edited by Mordechay Schlesinger and Milan Paunovic

The Electron Microprobe
Edited by T. D. McKinley, K. F. J. Heinrich, and D. B. Wittry

Chemical Physics of Ionic Solutions
Edited by B. E. Conway and R. G. Barradas

High-Temperature Materials and Technology
Edited by Ivor E. Campbell and Edwin M. Sherwood

Alkaline Storage Batteries
S. Uno Falk and Alvin J. Salkind

The Primary Battery (*in Two Volumes*)
Volume I Edited by George W. Heise and N. Corey Cahoon
Volume II Edited by N. Corey Cahoon and George W. Heise

Zinc-Silver Oxide Batteries
Edited by Arthur Fleischer and J. J. Lander

Lead-Acid Batteries
Hans Bode
Translated by R. J. Brodd and Karl V. Kordesch

Thin Films-Interdiffusion and Reactions
Edited by J. M. Poate, M. N. Tu, and J. W. Mayer

Lithium Battery Technology
Edited by H. V. Venkatasetty

Quality and Reliability Methods for Primary Batteries
P. Bro and S. C. Levy

Techniques for Characterization of Electrodes and Electrochemical Processes
Edited by Ravi Varma and J. R. Selman

Electrochemical Oxygen Technology
Kim Kinoshita

Synthetic Diamond: Emerging CVD Science and Technology
Edited by Kari E. Spear and John P. Dismukes

Corrosion of Stainless Steels
A. John Sedriks

Fundamentals of Electrochemical Deposition
Milan Paunovic and Mordechay Schlesinger

Semiconductor Wafer Bonding: Science and Technology
Q.-Y. Tong and U. Göscle

Uhlig's Corrosion Handbook, Second Edition
Edited by R. Winston Revie

The Electrochemical Society
65 South Main Street
Pennington, NJ 08534-2839 USA
http://www.electrochem.org

MODERN ELECTROPLATING

1 Fundamental Considerations

INTRODUCTION

In the preparation of the fourth edition of *Modern Electroplating*, we realized that the first chapter in the third edition (1974), needed to be enlarged considerably in order to cover all the significant progress made since 1974. As the prospect of adding material to the chapter on fundamentals began to suggest an inbalance in the new edition, we chose to publish a separate volume titled *Fundamentals of Electrochemical Deposition* (*Fundamentals* in further text) that would treat the basic aspects of electrochemical deposition [1]. For this reason we provide in this fourth edition only a brief review of these fundamentals. The number of references in this chapter is also limited, and the reader is urged to consult the more extensive list of references given in *Fundamentals*.

This chapter is divided into three parts. Part A treats electrochemical aspects, part B treats physical aspects, and part C treats material science. Our objective is for this presentation of the fundamentals to provide basis for understanding only the electrchemical deposition processes treated in this volume. Information on a higher level is presented in the *Funamentals* volume.

Modern Electroplating, Fourth Edition, Edited by Mordechay Schlesinger and Milan Paunovic.
ISBN 0-471-16824-6

Part A
Electrochemical Aspects

MILAN PAUNOVIC

1 INTRODUCTION

In part A we discuss (1) electrode potential, (2) kinetics and mechanism of electrodeposition, (3) growth mechanism, and (4) electroless and displacement depositions. All four topics are presented in a concise manner that emphasizes the most important points. Most concepts are clarified by solutions of numerical examples. These examples are usefull for this chapter and for the chapters that follow.

2 ELECTRODE POTENTIAL

When a metal M is immersed in an aqueous solution containing ions of that metal M^{z+} (e.g., salt MA) there will be an exchange of metal ions M^{z+} between two phases, the metal and the solution. Some M^{z+} ions from crystal lattice enter the solution, and some ions from the solution enter the crystal lattice. Initially one of these reactions may occur faster than the other. Let us assume that conditions are such that more M^{z+} ions leave than enter the crystal lattice. In this case there is an excess of electrons on the metal and the metal acquires negative charge, q_M^- (charge on the metal per unit area). In response to the charging of the metal side of the interphase, there is also a rearrangement of charges on the solution side of the interphase. The negative charge on the metal attracts positively charged M^{z+} ions from the solution and repels negatively charged A^{z-} ions. The result of this is an excess of positive M^{z+} ions in the solution in the vicinity of the metal interphase. Thus, in this case, the solution side of the interphase acquires opposite and equal charge, q_S^+ (the charge per unit area on the solution side of the interphase). This positive charge at the solution side of the interphase slows down the rate of M^{z+} ions leaving the crystal lattice (due to repulsion) and accelerates the rate of ions entering the crystal lattice. After a certain period of time a dynamic equilibrium between the metal M and its ions in the solution will result, Equation (1a):

$$M^{z+} + ze \leftrightarrows M \tag{1a}$$

where z is the number of electrons involved in the reaction. Reaction from left to the right consumes electrons and is called reduction. Reaction from right to the left liberates electrons and is called oxidation. At the dynamic equilibrium the same number of M^{z+} ions enter, $\overleftarrow{n}$, and the same number of M^{z+} ions leave crystal lattice, $\overrightarrow{n}$, Figure 1 and equation (1b):

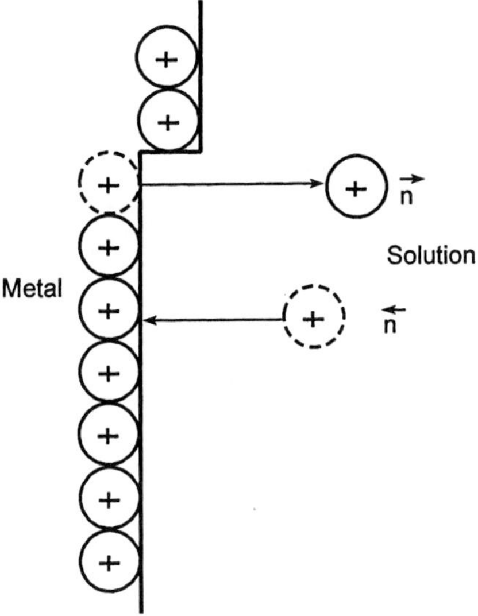

Figure 1 Formation of metal-solution interphase; equlibrium state: $\vec{n} = \vec{n}$.

$$\vec{n} = \overleftarrow{n} \tag{1b}$$

The interphase region is neutral at equilibrium:

$$q_M = -q_S \tag{1c}$$

The result of the charging of the interphase is the potential difference, $\Delta\phi$ (M, S), between the potentials of the metal, ϕ_M, and the solution, ϕ_S:

$$\Delta\phi(M, S) = \phi_M - \phi_S \tag{2}$$

In order to measure the potential difference of an interphase, one must connect it to another one and thus form an electrochemical cell. Potential difference across this electrochemical cell can be measured.

For example, consider the cell shown in Figure 2. This cell may be schematically represented in the following way:

$$\mathrm{Pt, H_2}\,(p = 1)\,|\,\mathrm{H^+}(a = 1)\,||\,\mathrm{Cu^{2+}}(a = 1)\,|\,\mathrm{Cu}\,|\,\mathrm{Pt}$$

where the left-hand electrode is the normal hydrogen reference electrode; a stands for activity and p for the pressure of H_2. In case when $p = 1$ atm and the activity of H^+ ions is 1 the hydrogen electrode is called the standard hydrogen electrode and its potential is zero by convention. The measured value of the potential difference of this cell is +0.337 V at 250°C. This measured cell potential difference, +0.337 V, is called the relative standard electrode potential of Cu and is denoted E^0. The standard electrode potential of other electrodes is obtained in a similar way, by forming a cell consisting of the standard hydrogen electrode (SHE) and the electrode under investigation. Standard electrode potentials at 25°C are listed in Table 1.

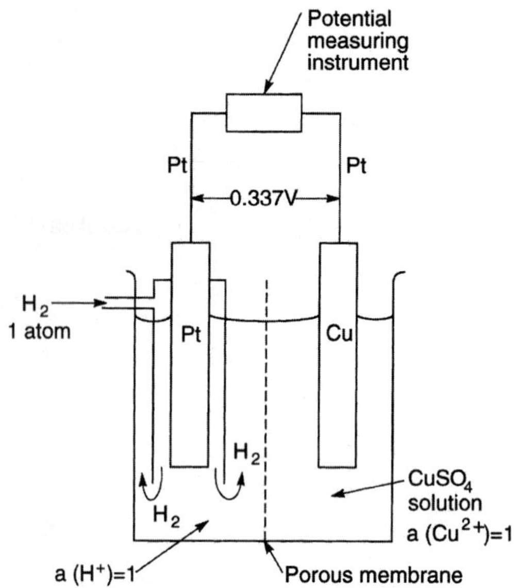

Figure 2 Relative standard electrode potential E^0 of a Cu/Cu^{2+} electrode.

TABLE 1 Standard Electrode Potentials

Metal/Metal-Ion Couple	Electrode Reaction	Standard Value (V)
Au/Au^{+}	$Au^{+} + e \Leftrightarrow Au$	1.692
Au/Au^{3+}	$Au^{3+} + 3e \Leftrightarrow Au$	1.498
Pd/Pd^{2+}	$Pd^{2+} + 2 \Leftrightarrow Pd$	0.951
Cu/Cu^{+}	$Cu^{+} + e \Leftrightarrow Cu$	0.521
Cu/Cu^{2+}	$Cu^{2+} + 2e \Leftrightarrow Cu$	0.3419
Fe/Fe^{3+}	$Fe^{3+} + 3e \Leftrightarrow Fe$	−0.037
Pb/Pb^{2+}	$Pb^{2+} + 2e \Leftrightarrow Pb$	−0.1262
Ni/Ni^{2+}	$Ni^{2+} + 2e \Leftrightarrow Ni$	−0.257
Co/Co^{2+}	$Co^{2+} + 2e \Leftrightarrow Co$	−0.28
Fe/Fe^{2+}	$Fe^{2+} + 2e \Leftrightarrow Fe$	−0.447
Zn/Zn^{2+}	$Zn^{2+} + 2e \Leftrightarrow Zn$	−0.7618
Al/Al^{3+}	$Al^{3+} + 3e \Leftrightarrow Al$	−1.662
Na/Na^{+}	$Na^{+} + e \Leftrightarrow Na$	−2.71

Source: G. Millazzo and S. Caroli, *Tables of Standard Electrode Potentials*, Wiley, New York, 1978.

The potential E of the M^{z+}/M electrode is a function of the activity (see Eq. (5) below) of metal ions in the solution according to the Nernst equation

$$E = E^0 + \frac{RT}{zF} \ln a(M^{z+}) \tag{3}$$

or, converting the natural logarithm into the decimal logarithm yields

$$E = E^0 + 2.303 \frac{RT}{zF} \log a(M^{z+}) \tag{4}$$

where R, T, z, and F are the gas constant, absolute temperature, number of electrons involved in the reaction (la), and Faraday's constant (96,500 coulombs), respectively. The activity of the ion, $a(M^{z+})$, is defined by

$$a(M^{z+}) = \gamma c(M^{z+}) \tag{5}$$

where $c(M^{z+})$ is the concentration of M^{z+} in moles per liter and $\gamma(M^{z+})$ is the activity coefficient of M^{z+}; when the concentration of a solution is low, such as 0.001 molar or lower, the activity may be replaced by concentration in moles per liter.

The activity coefficient γ is a dimensionless quantity which depends on the concentration of all ions present in the solution (ionic strength). The individual activity coefficients of the specific ionic species cannot be measured experimentally, but it can be calculated. The experimentally measurable quantity is the mean total ionic activity $\gamma_\pm$:

$$\gamma_\pm = \sqrt{\gamma_+\gamma_-} \tag{6}$$

which is the geometric mean (the square root of the product) of the activity coefficients of the individual ionic species [4].

When the activity of M^{z+} in the solution is equal to 1, $a(M^{z+})=1$, then by Eqs. (3) and (4), since $\ln 1=0$,

$$E = E^0 \tag{7}$$

where E^0 is the relative standard electrode potential of the M^{z+}/M electrode. The quantity RT/F has the dimension of voltage and at 298 K (25°C) has the value of 0.0257 V and $2.303\,(RT/F)=0.0592$ V. With these values Eq. (4) reads

$$E = E^0 + \frac{0.0592}{z} \log a(M^{z+}) \tag{8}$$

The use of Eq. (8) is illustrated in Examples 1 and 2.

Example 1

Calculate the reversible electrode potential of a Cu electrode immersed in a $CuSO_4$ aqueous solution with concentrations 1.0, 0.1, 0.01, and 0.001 mol liter^{-1} at 25°C. The standard electrode potential for a Cu/Cu^{2+} electrode is 0.337 V. Use concentrations in Eq. (8) instead of activities in an approximate calculation.

From Eq. (8), for $z=2$, $E^0=0.337$, and for the concentration 1.0 mol liter^{-1} solution, we obtain $E=0.337+(0.0592/2)\log 1=0.337$ V, since $\log 1=0$. For the 0.1 mol liter^{-1} solution, we obtain $E=0.337+(0.0592/2)\log 0.1=0.37$ V, since $\log 0.1$ is -1. Using the same procedure for 0.01 and 0.001 mol liter^{-1} solutions, we find that $E=0.278$ and 0.248 V, respectively.

Example 2

Now calculate the reversible electrode potential of a Cu electrode for the conditions given in the Example 1, but use activities in Eq. (8) instead of concentrations. The mean activity coefficients of the above solutions, 1.0, 0.1, 0.01, and 0.001 are 0.043, 0.158, 0.387, and 0.700, respectively. Activities of these solutions are calculated using Eq. (5). For the 1.0 mol liter^{-1} solution, and

TABLE 2 Reversible Electrode Potential *E* of a Cu Electrode Immersed in a $CuSO_4$ Aqueous Solution

$CuSO_4$ Concentration, *c* (mol/L)	*E*, *V* Calculated Using *c*	Activity, *a*	*E*, *V* Calculated Using *a*	ΔE, *V*
1.0	0.337	4.3×10^{-2}	0.297	0.040
0.1	0.307	1.58×10^{-2}	0.284	0.023
0.01	0.278	3.87×10^{-3}	0.266	0.012
0.001	0.248	7.00×10^{-4}	0.244	0.004

Note: See Example 2.

$\gamma = 0.043$, we find that the activity of this solution is $a_{1.00} = \gamma\ c(Cu^{2+}) = 0.043 \times 1 = 0.043$. Activities for solutions 0.1, 0.01, and 0.001 mol litre^{-1} are 1.58×10^{-2}, 3.87×10^{-3}, and 7.00×10^{-4}, respectively.

Using Eq. (8) for the 1.00 mol liter^{-1} solution, the reversible electrode potential, at 25°C is $E = 0.337 + (0.0592/2)\log 0.043 = 0.337 - 0.0400 = 0.297$ V. For 0.1, 0.01, and 0.001 mol liter^{-1} solutions, the reversible electrode potentials, at 25°C are 0.284, 0.266, and 0.244 V, respectively.

Hence Examples 1 and 2 illustrate that the effect of considering the activity coefficient in calculating electrode potential values is not substantial, and its effect decreases with decrease in concentration, as seen in column 5 Table 2. The difference ΔE in Table 2 is due to ion–ion interactions in the solution (1). The ion–ion interactions include interactions of the hydrated Cu^{2+} ions with one another and with SO_4^{2-} anions. In using concentration in Eq. (8) instead of activity, one thus neglects the ion–ion interactions.

In deposition of alloys it is frequently necessary to complex metal ions, as will be shown in part B of this chapter. In this case the concentration or activity of metal ions in solution and the reversible electrode potential are calculated using the stability constant of the complex.

3 KINETICS AND MECHANISM OF ELECTRODEPOSITION

3.1 Relationship between Current and Potential

When an electrode is made a part of an electrochemical cell through which current is flowing, its potenatial will differ from the equilibrium potential. If the equilibrium potential of the electrode (potential in the absence of current) is *E* and potential of the same electrode as a result of current flowing is *E*(*I*), then the difference η between these two potentials

$$\eta = E(I) - E \tag{9}$$

is called *overpotential*.

An example of an electrochemical cell of industrial importance is shown in Figure 3. This figure shows how the potential *E*(*I*) of a copper cathode in the electrodeposition of copper can be measured using the three-electrode cell. The

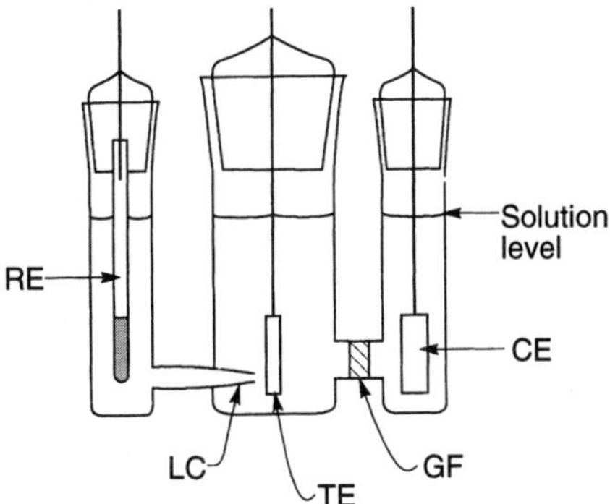

Figure 3 Three-component, three-electrode, electrochemical cell; RE, reference electrode; LC, Lugin capilary; TE, test electrode; GF, glass frit; CE, counterelectrode; P, power supply; V, voltmeter.

potential $E(I)$ of a Cu cathode, the test electrode, is measured versus a reference electrode (e.g., saturated calomel electrode), a Lugin capilary, and a high-input impedance device (voltmeter) so that a negligible current is drawn through the reference electrode.

For large negative values of overpotentials ($\eta \geq 100\,\text{mV}$) the current density i ($i = I/S$, where S is the surface area of the electrode) increases exponentially with the overpotential η according to the Eq. (1):

$$i = -i_0\, e^{-\alpha z f \eta} \tag{10}$$

and for large positive values of overpotential (anodic processes) according to the equation

$$i = -i_0\, e^{(1-\alpha) z f \eta} \tag{11}$$

where i_0 is the exchange current density ($i_0 = i$ when $\eta = 0$), α the transfer coefficient, F the faraday constant, R the gas constant, T the absolute temperature, and

$$f = \frac{F}{RT} \tag{12}$$

From Eq. (10) and (11) follows that for $\eta = 0$, $i = i_0$. Thus, when an electrode is at equilibrium, there is a constant exchange of charge carriers (electrons or ions) across the metal-solution interphase (Eq. 1b).

At 25°C,

$$f = (96487\text{ C mol}^{-1})/(8.3144\text{ JK}^{-1}\text{ mol}^{-1} \times 298\text{ K}) = 38.94\text{ V} \tag{13}$$

since the Faraday constant $F = 96{,}487.0\text{ C mol}^{-1}$, the gas constant $R = 8.3144$ joule mol^{-1} deg K^{-1}, and joule = volts×coulomb. These exponential relationships show that even small changes in η produce large changes in the current density, as seen from Figure 4. Taking the logarithm of Eqs. (10) and (11) and solving the resulting equations for η^{one} obtains the Tafel equation:

$$\eta = a \pm b\, \log |i| \tag{14}$$

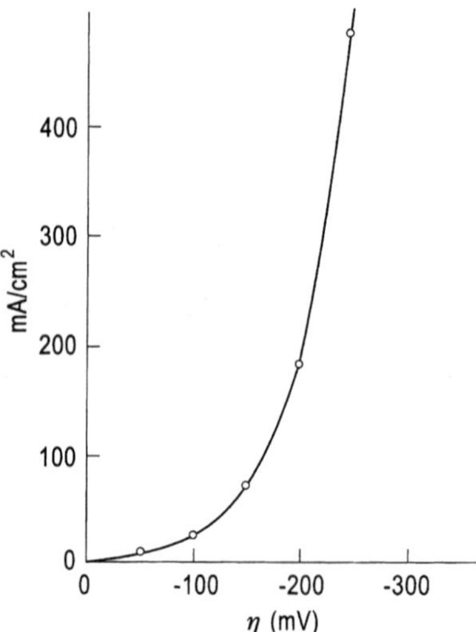

Figure 4 Exponential relationship between current density and overpotential for electro-deposition of copper from an aqueous solution of 0.15 N $CuSO_4$ and 1.0 N H_2SO_4.

where a and b are constants and $|i|$ is the absolute value of the current density. The $\pm$ sign holds for anodic and cathodic process, respectively. The theoretical value of the constant a, for the cathodic process (a_c), is

$$a_c = \frac{2.303RT}{\alpha zF} \log i_0 \tag{15}$$

and that of b_c is

$$b_c = \frac{2.303RT}{\alpha zF} \tag{16}$$

The use of Eq. (10) and (14) is illustrated in Examples 3 and 4.

Example 3

Matson and Bockris [3] studied the electrodeposition of copper from the aqueous solution 0.15 N $CuSO_4$ and 1.0 N H_2SO_4 and determined that the exchange current density for this process is 3.7 mA cm^{-2} (3.7×10^{-3} A cm^{-2}). Using Eq. (10), we calculate the cathodic current densities i as a function of the overpotential η, for $\eta = -50, -100, -150, -200$, and -250 mV. For α we take the most frequent value, $\alpha = 0.5$, and for temperature, 25°C. Next we plot $i = f(\eta)$.

Using Eq. (10), we see that i (in A Cm^{-2}) for $\eta = -50$ mV (-0.050 V) is

$$i = -i_0\, e^{-\alpha f\eta} = -3.7 \times 10^{-3}\, e^{0.9737} = -9.80 \times 10^{-3}\ \text{A cm}^{-2}$$

since we have $-\alpha f\eta = -0.5 \times 38.94 \times (-0.050) = 0.9737$ and $e = 2.7183$. For $\eta = -$ 100 mV(0.100 V),

$$i = -i_0\, e^{-\alpha f\eta} = -3.7 \times 10^{-3}\, e^{1.947} = -25.9 \times 10^{-3}\ \text{A cm}^{-2}$$

since we now have $-\ \alpha f\eta = -0.5 \times 38.94 \times (-0.100) = 1.947$.

In the same way we find i for the other η values. These are shown in Table 3.

The calculated values of η are used to plot the $i = f(\eta)$ function corresponding to this example, Figure 4. It again illustrates that even small increases in η result in large changes in the current density i.

Example 4

Using the calculated values of i for the given values of η from Table 3 plot the corresponding Tafel equation, $\eta = f(\log i)$.

The solution is given in Figure 5. It shows that for large values of η ($\eta \geq 100$ mV), the function $\eta = f(\log i)$ is a straight line. The experimentally determined cathodic and anodic Tafel lines for electrodeposition of copper in acid copper sulfate solution are shown in Figure 6. The figure shows that the transfer coefficients for the anodic and the cathodic process are different and that extrapolated value of i at $\eta = 0$ gives the exchange currenty density. The transfer coefficient α_c for the cathodic process (deposition of Cu^{2+}) is obtained from the slope $d\eta/d(\log i)$ of the cathodic Tafel line. That for the anodic process, α_a, is obtained from the slope of the Tafel line for the anodic process.

TABLE 3 Current and Overpotential for Electrodeposition of Copper from Acid Copper Sulfate Solution

Overpotential $-\eta$ (mV)	Current Density, i (mA/cm^2)	Log i
50	9.80	0.991
100	25.9	1.41
150	68.7	1.84
200	182	2.26

Note: See Example 3.

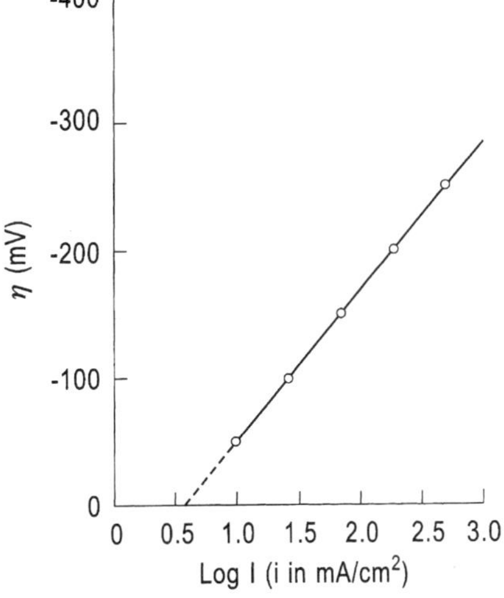

Figure 5 Tafel plot, $\eta = f(\log i)$, for electrodeposition of copper from the aqueous solution of 0.15 $CuSO_4$ and 1.0 N H_2SO_4.

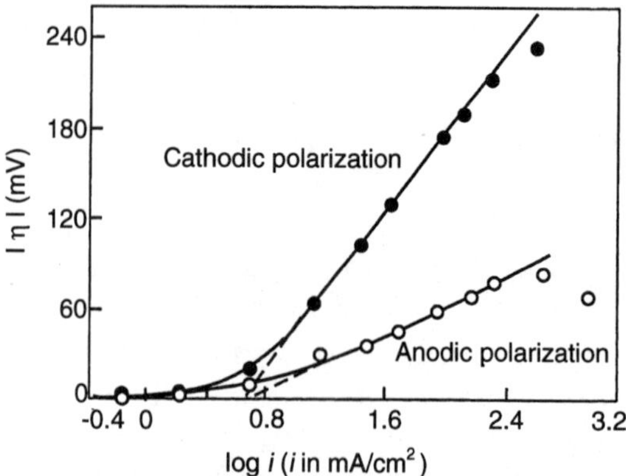

Figure 6 Current-potential relationship for the electrodeposition of copper from an acid $CuSO_4$ solution. (From J. O'M. Bockris, *Transactions of the Symposium on Electrode Processes*, E. Yeager, ed., Wiley, New York, 1961; with permission from the Electrochemical Society)

3.2 Influence of Mass Transport on Electrode Kinetics

The current-potential relationship defined by Eqs. (10), (11), and (14) is valid for the case where the charge transfer, Eq. (1), is the slow process (rate-determining step). This relationship has a limit where the rate of deposition reaction is limited by transport of M^{z+} ions. A general current-potential relationship is shown in Figure 7. The limiting, or the maximum, current density is given by [2]

$$i_{\mathrm{L}} = \frac{nFD}{\delta} c_{\mathrm{b}} \tag{17}$$

where D is the diffusion coefficient of the depositing species M^{z+}, c_{b} is the bulk concentration of M^{z+} ions in the solution, δ is the diffusion layer thickness, n the number of electrons involved in the reaction, and F is the Faraday constant. The

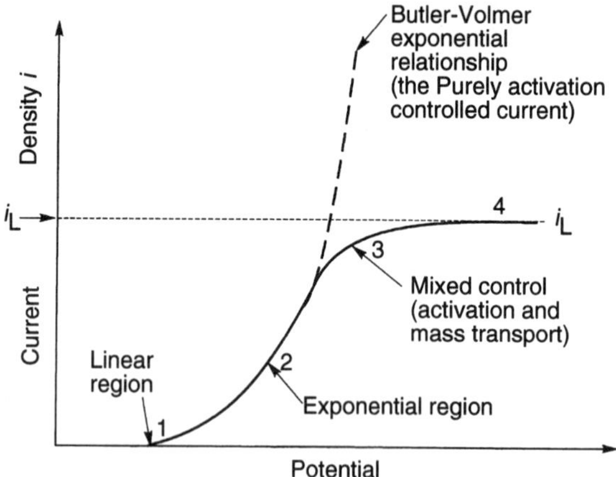

Figure 7 Four regions in the general current-overpotential relationship: 1, linear; 2, exponential; 3, mixed control; 4, limiting current density region.

diffusion layer thickness δ is defined by the Nemst diffusion-layer model illustrated in Figure 8. This model assumes that the concentration of M^{z+} ions has a bulk concentration c_b up to a distance δ from the electrode surface and then falls off linearly to $c_{x=0}$ at the electrode surface. In this model it is assumed that the liquid layer of thickness δ is practically stationary (quiescent). At a distance greater than δ from the surface, the concentration of the reactant M^{z+} is assumed to be equal to that in the bulk. At these distances, $x > \delta$, stirring is efficient. Ions M^{z+} must diffuse through the diffusion layer to reach the electrode surface.

At the values of the limiting (maximum) current density the species M^{z+} are reduced as soon as they reach electrode. At these conditions the concentration of the reactant M^{z+} at the electrode is nil, and the rate of deposition reaction is controlled by the rate of transport of the reactant M^{z+} to the electrode. If an external current greater than the limiting current i_L is forced through the electrode, the double layer is further charged, and the potential of the electrode will change until some other process, other than reduction of M^{z+}, can occur. It will be shown later that the limiting current density is of great practical importance in metal deposition, since the type and quality of metal deposits depend on the relative values of the deposition current and the limiting current. One extreme example is shown in Figure 9.

Example 5

Calculate the diffusion limiting current density i_L for the deposition of a metal ion M^{2+}at a cathode in a quiescent (unstirred) solution assuming the diffusion layer thickness δ is 0.05 cm. The concentration of M^{2+} ions in the bulk (c_b) is

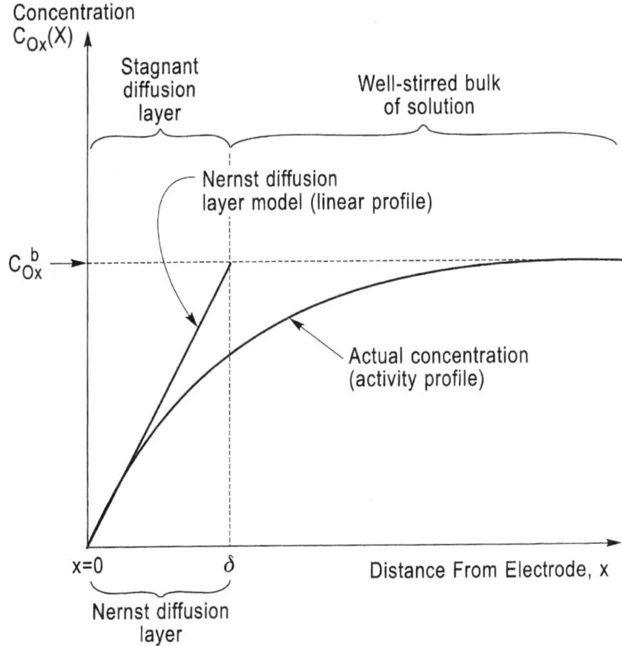

Figure 8 Variation of the concentration of the reactant during non-steady-state electrolysis; c_{Ox}^{b} is the concentration in the bulk; $c_{Ox}(x)$ is the concentration at the surface.

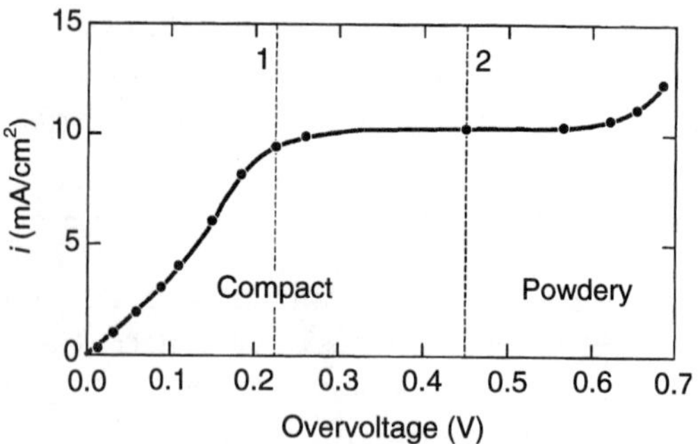

Figure 9 Overpoential characteristic of transition from compact to powdery deposit in electrodeposition of Cu from $CuSO_4$ (0.1 M) + H_2SO_4 (0.5 M) solution. (From N. Ibl, in *Advances in Electrochemistry and Electrochemical Engineering*, vol. 2, C. W. Tobias, ed., Wiley, New York, 1962; with permission from Wiley)

10^{-2} mol liter^{-1} (10^{-5} mol cm^{-3}) and the diffusion coefficient D of M^{2+}, in unstirred solution, is $2 \times 10^{-5}\,\text{cm}^2\,\text{s}^{-1}$.

Using Eq. (17), we calculate that the limiting diffusion current density for this case is

$$i_L = \frac{nFD}{\delta} c_b = \frac{2 \times 96487 \times 2 \times 10^{-5} \times 10^{-5}}{0.05} = 7.72 \times 10^{-4}\,\text{A cm}^{-2} = 0.72\,\text{mA cm}^{-2}$$

In the same way we find that for $c_b = 10^{-1}$ mol liter^{-1} (10^{-4} mol cm^{-3}) $i_L = 7.20\,\text{mA cm}^{-2}$.

In the stirred electrolyte solution the diffusion layer thickness δ and the limiting current density i_L depend on the nature of stirring.

For a rotating electrode the diffusion layer thickness depends on the angulat speed of rotation ω according to

$$\delta = \frac{1.61 D^{1/3} \nu^{1/6}}{\sqrt{\omega}} \tag{18}$$

and the limiting current density depends on ω according to

$$i_L = \frac{0.62 nFaD^{2/3} c}{\nu^{1/6}} \sqrt{\omega} \tag{19}$$

where D is the diffusion coefficient, ν the kinematic viscosity (the coefficient of viscosity/density of the liquid), ω rotation angular speed (in radians s^{-1}), c the concentration of the solution, and a is the disc surface area [2]. We use these equations in the next two examples.

Example 6

Determine diffusion layer thickness for a rotating electrode at 60, 240, and 360 rpm. For kinematic viscosity ν use $10^{-2}\,\text{cm}^2\,\text{s}^{-1}$ and for the diffusion coefficient D use $10^{-5}\,\text{cm}^2\,\text{s}^{-1}$.

For the above given values for ν and D, Eq. (18) is

$$\delta = \frac{1.61 \times 10^{-2}}{\sqrt{\omega}} = \frac{1.61 \times 10^{-2}}{\sqrt{2\pi N}}$$

In the above equation the rotation speed ω in radians s^{-1} is replaced by $\omega = 2\pi N$, where N is the number of rotations per second, rps.

Applying the above equation for $N = 1$ rps (60 rpm), one gets $\delta = 64\,\mu m$. For $N = 4$ rps (240 rpm), $\delta = 32\,\mu m$ and for $N = 6$ rps (360 rpm), $\delta = 26\,\mu m$.

Example 7

Calculate the limiting current density i_L for the deposition of a metal ion M^{2+} at a rotating disc cathode with a surface area $a = 1\ cm^2$ and the rotating speed 300 rpm (5 rps) using Eq. (19) assuming that the concentration of M^{2+} is $10^{-5}\,mol\,cm^{-3}$ ($10^{-2}\,mol\,liter^{-1}$). The diffusion coefficient D of M^{2+} is $2 \times 10^{-5}\,cm^2$, $\nu = 10^{-2}\,cm^2\,s^{-1}$.

Substitution of the above given values for a, D, and ν into Eq. (19) yields

$$i_L = 1.91 \times 10^2\ c\sqrt{\omega}$$

For $c = 10^{-2}\,mol\,liter^{-1} = 10^{-5}\,mol\,cm^{-3}$, and $\omega = 2\pi N = 31.41$

$$i_L = 1.07 \times 10^{-2}\,A\,cm^{-2} = 10.7\,mA\,cm^{-2}$$

This value should be compared with the i_L value for a quiescent (unstirred) solution in Example 5.

3.3 Faraday's Law

Faraday's law states that the amount of electrochemical reaction that occurs at an electrode is proportional to the quantity of electric charge Q passed through an electrochemical cell. Thus, if the weight of a product of electrolysis is w, then Faraday's law states that

$$w = ZQ \tag{20}$$

where Z is *the electrochemical equivalent*, the constant of proportionality. Since Q is the product of the current I, in ampers, and the elapsed time t, in seconds,

$$Q = It \tag{21}$$

$$w = ZIt \tag{22}$$

According to the Faraday's law the production of one gram equivalent of a product at the electrode, W_{eq}, in a cell requires 96,487 coulombs. The constant 96,487 is termed the Faraday *constant F*. The coulomb is the quantity of electricity transported by the flow of one ampere for one second.

The Faraday constant represents one mole of electrons and its value can be calculated from

$$F = N_A\, e \tag{23}$$

where N_A is Avogadro's number (6.0225×10^{23} molecules mol^{-1}) and e is the charge of a single electron (1.6021×10^{-19} coulombs, C)

$$F = (6.0225 \times 10^{23})(1.6021 \times 10^{-19}) = 96,487\,\mathrm{C\,mol^{-1}} \tag{24}$$

One equivalent, w_{eq}, is that fraction of a molar (atomic) unit of reaction that corresponds to the transfer of one electron. For example, w_{eq} for silver is the gram atomic weight of silver, since the reduction of Ag^+ requires one electron. The deposition of copper from a Cu^{2+} salt involves two electrons, and the w_{eq} for Cu is (gram atomic weight of Cu)/2. In general,

$$w_{eq} = \frac{A_{wt}}{n} \tag{25}$$

where A_{wt}, is the atomic wei⌐ht of of metal deposited on the cathode, and n number of electrons involved in the :leposition reaction.

From Eqs. (20) and (22) it follows that when $Q = 1$ coulomb, or $Q = 1$ ampere second, then

$$w_{Q=1} = Z \tag{26}$$

Thus the electrochemical equivalent of a metal M, $Z(M)$, is the weight in grams produced, or consumed, by one coulomb (one ampere second). The combination of Eqs. (20) and (26) yields

$$w = w_{Q=1}\,Q \tag{27}$$

The value of Z, or $w_{Q=1}$, can be evaluated in the following way. Since 96,487 coulombs are required for the deposition of an equivalent of a metal, w_{eq}, from Eq. (20) it follows that

$$w_{eq} = 96,487\,Z \tag{28}$$

and

$$Z = w_{Q=1} = \frac{w_{eq}}{96,487} = \frac{w_{eq}}{F} \tag{29}$$

Since $w_{eq} = A_{wt}/n$, Eq. (25),

$$Z = \frac{A_{wt}}{nF} \tag{30}$$

Finally, from Eqs. (20) and (30)

$$w = ZQ = \frac{A_{wt}}{nF}Q \tag{31}$$

Example 8

Determine the electrochemical equivalent of Cu for the case of electrodeposition of Cu from Cu^{2+} solutions. The atomic weight of Cu is 63.55.

From Eq. (29),

$$Z(Cu^{2+}) = \frac{A_{wt}Cu}{nF} = \frac{63.55}{2 \times 96,487} = 3.293 \times 10^{-4}\,\mathrm{g\,C^{-1}}$$

Example 9

A current of 300 mA was passed for 20 minutes through an electrochemical cell containing copper electrodes in H_2SO_4 acidified $CuSO_4$ aqueous solution. Calculate the amount of copper deposited at the cathode. The gram atomic weight of Cu is 63.55.

In order to calculate w using Eq. (31), we find first the number of coulombs Q, the quantity of electricity passed during the electrolysis

$$Q = It = 0.300\,\text{A} \times 1200\,\text{s} = 360\,\text{As} = 360\,\text{C}$$

The number of electrons n involved in the Cu deposition reaction

$$\text{Cu}^{2+} + 2e = \text{Cu}$$

is 2. Substituting these values into Eq. (31), we get the amount of Cu deposited at the electrode

$$w = \frac{63.55 \times 360}{2 \times 96,487} = 0.118\,\text{g}$$

Alternatively, using the value of the electrochemical equivalent for Cu, $Z(\text{Cu}^{2+}) = 3.293 \times 10^{-4}\,\text{g}\,\text{C}^{-1}$, and Eq. (31), we get

$$w = ZQ = 3.293 \times 10^{-4} \times 0.300 \times 1200 = 0.118\,\text{g}$$

3.4 Current Efficiency

When two or more reactions occur simultaneously at an electrode, the number of coulombs of electricity passed corresponds to the sum of the number of equivalents of each reaction. For example, during deposition of Cu from a solution of cupric nitrate in dilute nitric acid, three cathodic reactions occur: the deposition of Cu (the reduction of cupric ions) and the reduction of both nitrate and hydrogen ions.

The current efficiency CE of the jth, process, namely of any one of the simultaneous reactions, is defined as the number of coulombs required for that reaction, Q_j, divided by the total number of coulombs passed, Q_{total}

$$\text{CE} = \frac{Q_j}{Q_{\text{total}}} \tag{32}$$

An alternative equation defining current efficiency is

$$\text{CE} = \frac{w_j}{w_{\text{total}}} \tag{33}$$

where w_j is the weight of metal j actually deposited and w_{total} is that which would have been deposited if all the current had been used for depositing the metal j.

Thus, in general, at a current efficiency under 100%, the remainder of the current is used in side processes, such as the reduction of hydrogen and nitrate ions in the example above.

Example 10

When a current of 3 A flows for 8 minutes through a cell composed of two Pt electrodes in a solution of $Cu(NO_3)_2$ in dilute HNO_3 acid, 0.36 g of Cu is deposited on the cathode. Calculate the current efficiency for the deposition of copper.

The number of coulombs required for deposition of 0.36 g of Cu is obtained from Eq. (31):

$$Q_j = \frac{\text{wn}\,96{,}487}{A_{\text{wt}}} = \frac{0.36 \times 2 \times 96{,}487}{63.55} = 1093$$

$$Q_{\text{total}} = It = 3 \times 480 = 1440$$

$$\text{and} \quad \text{CE} = \frac{Q_j}{Q_{\text{total}}} = \frac{1093}{1440} = 0.759, \text{or CE} = 75.9\%$$

3.5 Deposit Thickness

The deposit thickness may be evaluated by considering the volume of the deposit. Since the volume of the deposit V is the product of the plated surface area a, and the thickness (height) h, it follows that $h = V/a$. The volume of the deposite is related to the weight of the deposit w and the density of the deposit d, by the relationship defining the density, $d = w/V$. Thus

$$h = \frac{V}{a} = \frac{w}{ad} \tag{34}$$

In the case where it is necessary to calculate the time t (second, s) required to obtain the desired deposit thickness h, at a given current density, we introduce the Faraday's law, Eq. (31), into Eq. (34) and obtain

$$h = \frac{w}{ad} = \frac{ZQ}{ad} = \frac{ZIt}{ad} \text{ cm} \tag{35}$$

$$t = \frac{had}{ZI} \text{ s} \tag{36}$$

Example 11

The weight of a Co deposit is 0.00208 g on a substrate with surface area of 7.5 cm^2. What is the thickness of the Co deposit. Remember that the density of Co is 8.71 $g\,cm^{-3}$.

Using Eq. (34), we have that

$$h = \frac{0.00208}{7.5 \times 8.71} = 3.184 \times 10^{-5}\,\text{cm} = 3184\,\text{Å}.$$

Example 12

Determine the time required to obtain a Cu deposit of thickness 1 μm (10^{-4} cm) when electrodeposition is done at 4 and 6 A. The surface area a of the substrate is 314 cm^2.

From Eq. (36), for $I = 4$ A,

$$t = \frac{had}{ZI} = \frac{10^{-4} \times 314 \times 8.93}{3.29 \times 10^{-4} \times 4} = 213\,\text{s} = 3\,\text{m}\,33\,\text{s},$$

for $I = 6$ A, in the same way, one obtains that $t = 142$ s = 2 m 22 s.

3.6 Atomistic Aspects of Electrodeposition

In the electrodeposition of metals, generally a metal ion M^{z+} is transferred from the solution into the ionic metal lattice. A simplified atomistic representation of this process is

$$M^{z+}\ (\text{solution}) \rightarrow M^{z+}\ (\text{lattice}) \tag{37}$$

This reaction is accompanied by the transfer of z electrons from an external electron source (e.g., power supply) to the electron gas in the metal M.

Before discussing the individual atomistic processes that make up the overall electrodeposition process, Eq. (37), it is necessary to consider the basic characteristics of the bulk and the surface structures of metals [1]. A metal may be considered to be a fixed lattice of positive ions permeated by a gas of free electrons. Positive ions are the atomic cores, while the negative charges are the valence electrons. For example, the copper atom has a configuration (electronic structure) $1s^2 2s^2 2p^6 3s^2 3p^6 3d^{10} 4s^1$ (the superscripts indicate the number of electrons in the orbit-configuration) with a single valence electron (4s). The atomic core of Cu^+ has the set of configurations given above, less the one valence electron $4s^1$. The free electrons form what is known as the *electron gas* in the metal, and they move nearly freely through the volume of the metal. Each metal atom thus contributes its single valence electron to the electron gas in the metal. Interactions between the free electrons and the metal ions are largely responsible for the metallic bond.

Surfaces may be divided into ideal and real. Ideal surfaces exhibit no surface lattice defects (vacancies, impurities, grain boundaries, dislocations, etc.). Real surfaces have a variety of defects. For example, the density of metal surface atoms is about $10^{15}\,cm^{-2}$, while the density of dislocations is of the order of $10^8\,cm^{-2}$. The structure of real surfaces differs from those of ideal surfaces by surface roughness. While an ideal surface is atomically smooth, a real surface may have defects, steps, kinks, vacancies, and clusters of adatoms (Fig. 10).

The atomic processes that make up the electrodeposition process, Eq. (37), can be viewed considering the structure of the initial, M^{z+} (solution), and the final state, M^{z+} (lattice). Since metal ions in an aqueous solution are hydrated the structure of the initial state in Eq. (37) should be represented by $[M(H_2O)_x]^{z+}$. The structure of the final state is an M adion (adatom; adsorbed ion, atom) at a kink site (Fig. 11), since it is generally assumed that atoms (ions) are attached to a crystal via a kink site [1]. Thus the final step of the overall reaction, Eq. (37), is the incorporation of the

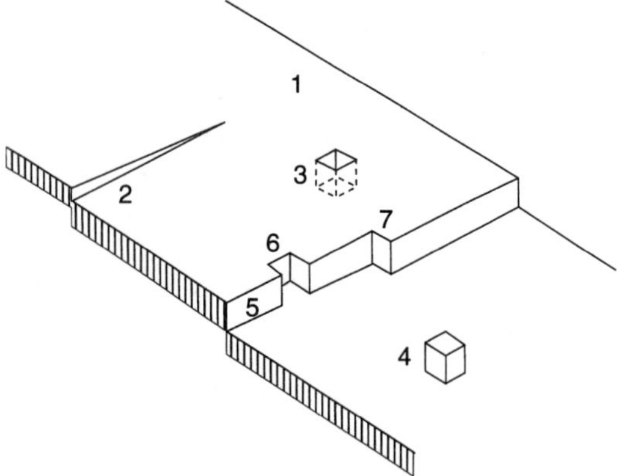

Figure 10 Some simple defects found on a low-index crystal face: 1, perfect flat face, terrace; 2, an emerging screw dislocation; 3, a vacancy in the terrace; 4, an adatom on the terrace; 5, a monatomic step in the surface, a ledge; 6, a vacancy in the ledge; 7, a kink, a step in the ledge.

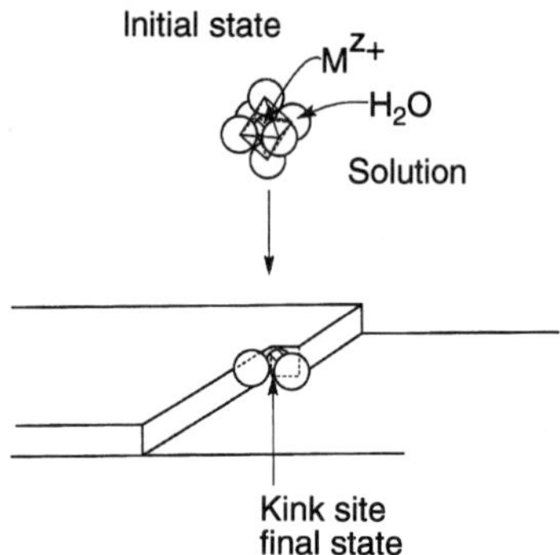

Figure 11 Initial and final states in metal deposition.

M^{z+} adion into the kink site. Because of surface inhomogeneity the transition from the initial state $[M(H_2O)_x]^{z-}$ (solution) to the final state M^{z+} (kink),

$$[M(H_2O)_x]^{z+} \text{ (solution)} \rightarrow M^{z+} \text{ (kink)} \tag{38}$$

may proceed via either of the two mechanisms: (1) step-edge site ion-transfer or (2) terrace site ion-transfer.

Step-Edge Ion-Transfer Mechanism The step-edge site ion-transfer, or direct transfer mechanism, is illustrated in Figure 12. As the figure shows, this mechanism ion-transfer from the solution takes place on a kink site of a step edge or on any other site on the step edge. In both cases the result of the ion-transfer is an M adion in the metal crystal lattice. In the first case of a direct transfer to the kink site, the M adion is in the half-crystal position, where it is bonded to the crystal lattice with one-half of the bonding energy of the bulk ion. Thus M adion belongs to the bulk crystal. However, it still has some water of hydration (Fig. 12). In the second case of a direct

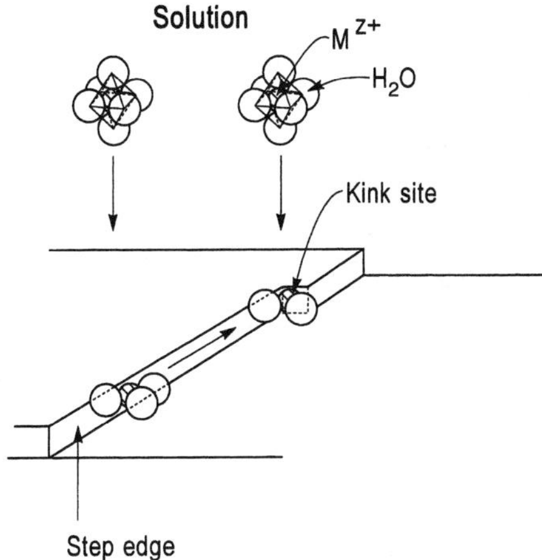

Figure 12 Step-edge ion-transfer mechanism.

transfer to the step-edge site other than a kink, the transfed metal ion diffuses along the step edge until it finds a kink site (Fig. 12). Thus, in a step-edge site transfer mechanism, there are two possible paths: direct transfer to a kink site and the step-edge diffusion path.

Terrace Ion-Transfer Mechanism In the terrace site transfer mechanism a metal ion is transferred from the solution to the flat face of the terrace region (Fig. 13). At this position the metal ion is in the adion (adsorbed-like) state having most of the water of hydration. It is weakly bound to the crystal lattice. From this position it diffuses on the surface, seeking a position of lower energy. The final position is a kink site.

In view of these two mechanisms, the step-edge and terrace ion-transfer, the overall current density i is considered to be composed of two components:

$$i = i_{se} + i_{te} \tag{39}$$

where i_{se} and i_{te} are the step-edge and terrace site current density components, respectively.

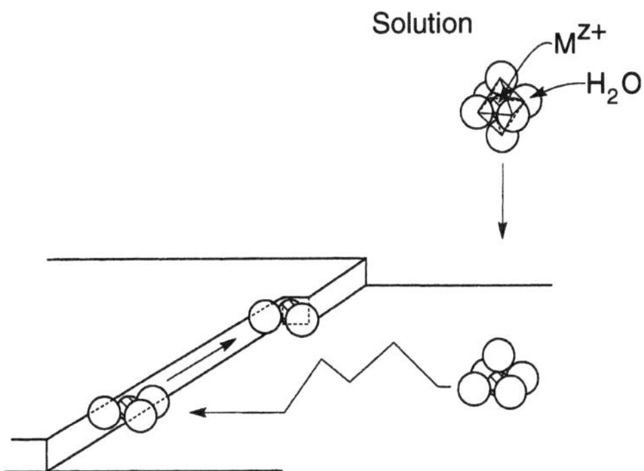

Figure 13 Ion transfer to a terrace site, surface diffusion, and incorporation at kink site.

The initial theoretical treatment of these mechanisms of deposition was given by Lorenz [5–8]. The initial experimental studies on surface diffusion were published by Mehl and Bockris [9, 10]. Conway and Bockris [11, 12] calculated the activation energies for the ion-transfer process at various surface sites. Simulation of crystal growth via surface diffusion was discussed by Gilmer and Bennema [13].

3.7 Pulse Deposition Techniques

One important effect in pulse deposition techniques is a modification of the diffusion layer [14]. The Nernst diffusion-layer model is illustrated in Figure 8 and described in the Section 3.2. Under pulse deposition conditions the Nernst diffusion-layer is split into two diffusion-layers as is schematically shown in Figure 14. In the pulsating diffusion-layer, which is in the immediate vicinity of the cathode, the metal ion concentration pulsates with the frequency of the pulsating current [14]. Pulse deposition techniques are used mostly to improve the distribution of the deposit (current distribution), the leveling, and the brightness of the deposit [15].

The four major waveforms that are used in pulse deposition techniques are shown in Figure 15. In the *rectangular-pulse deposition* technique the waveform consists of pulses of a current or potential of a rectangular shape separated by intervals of zero current or potential. The shape of this waveform is shown in figure 15*a* where I or E is the maguitude of the applied current or potential, t_{on} is the on period of a pulse, t_{off} is the off period, and λ is the cycle time.The waveform of *periodic reverse deposition* is shown in Figure 15*b*. It is seen that in this technique the applied current or potential is periodically switched from cathodic to anodic polarization; t_c is the cathodic pulse period, t_a the anodic pulse period, and λ the cycle time.

A *superimposed sinusoidal deposition* waveform is shown in Figure 15*c*. This waveform is the sum of a sinusoidal alternating (ac) wave, current or potential, and a direct cathodic current (dc). If the amplitude of the sine wave is greater than the dc offset, then the wave form consists of both a cathodic and an anodic portion.

There is now increased interest in pulse deposition techniques because of recently published data on the beneficial use of pulse plating in the fabrication process of integrated circuits [16].

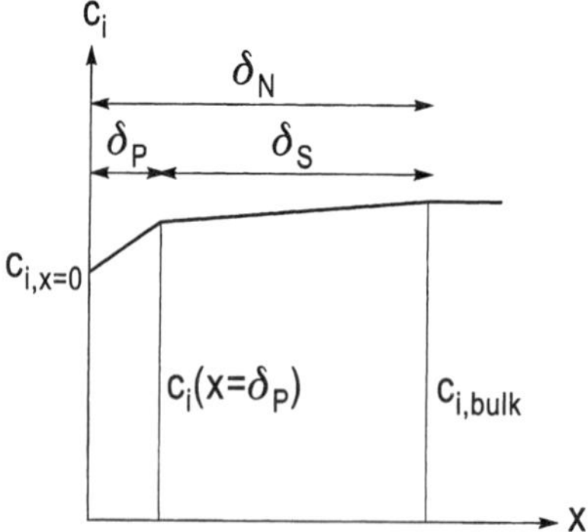

Figure 14 Schematic concentration profile at the cathode for pulse plating conditions; δ_p, pulsating diffusion layer thickness; δ_s, stationary diffusion layer thickness; δ_N, Nernst diffusion layer thickness.

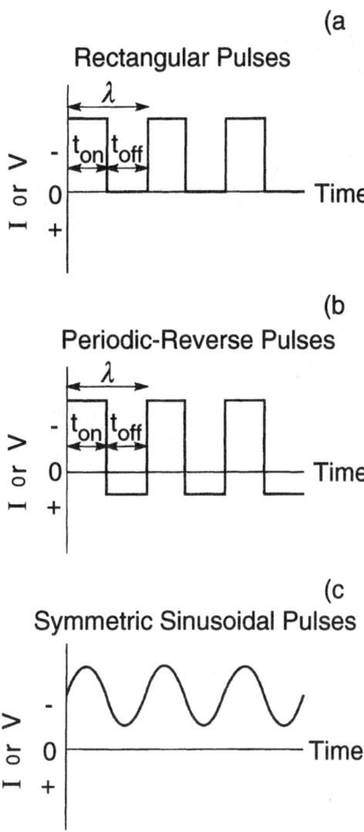

Figure 15 Three major wave forms used in pulse deposition; t_{on}, the on period of a pulse; t_{off}, the off period; λ, the cycle time; t_c, the cathodic pulse period; t_a, the anodic pulse period.

4 GROWTH MECHANISM

There are two basic mechanisms for formation of a coherent deposit [1, 17]: layer growth and three-dimensional (3D) crystallites growth (or nucleation-coalescence growth). A schematic illustration of these two mechanisms is given in Figure 16.

In the layer growth mechanism a crystal enlarges by a spreading of discrete layers (steps), one after another across the surface. In this case a growth layer, a step, is a structure component of a coherent deposit. Steps, or growth layers, are the structural components for the construction of a variety of growth forms in the electrodeposition of metals (e.g., columnar crystals, whiskers, and fiber texture). We can distinguish among monoatomic steps, polyatomic microsteps, and polyatomic macrosteps. In general, there is a tendency for a large number of thin steps to bunch into a system of a few thick steps. Many monoatomic steps can unite (bunch, coalesce) to form a polyatomic step.

In the 3D crystallites growth mechanism the structural components are 3D crystallites, and a coherent deposit is built up as a result of coalescence (joining) of these crystallites. The growth sequence of electrodeposition via nucleation-coalescence consists of four stages: (1) formation of isolated nuclei and their growth to TDC (3D crystallites), (2) coalescence of TDC, (3) formation of linked network, and (4) formation of a continuous deposit.

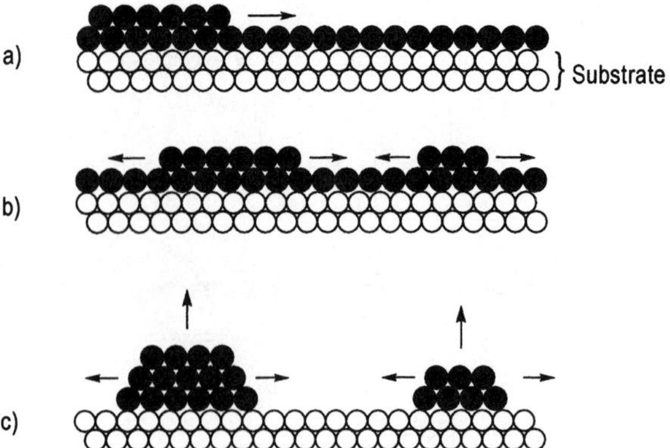

Figure 16 Schematic representation of layer growth (*a*, *b*) and the nucleation-coalescence mechanism (*c*).

Development of Columnar Microstructure The columnar microstructure is perpendicular to the substrate surface, is shown schematically in Figure 17. This microstructure is composed of relatively fine grains near the substrate but then changes to a columnar microstructure with much coarser grains at greater distances from the substrate. The development of the columnar microstructure may be interpreted as the result of growth competition among adjacent grains. The low-surface-energy grains grow faster than the high-energy ones. This rapid growth of the low-surface-energy grains at the expense of the high-energy grains results in an increase in mean grain size with increased thickness of deposit and the transition from a fine grain size near the substrate to a coarse, columnar grain size [1, 18].

Overpotential Dependence of Growth Forms The development of growth forms on overpotentials stems from the potential dependence of the nucleation and growth processes. Competition between nucleation and growth processes is strongly influenced by the potential of the cathode [1, 19].

4.1 Effect of Additives

Additives affect deposition and crystal building processes as adsorbates (adsorbed substances) at the surface of the cathode. There are two basic types of adsorption: chemisorption (an abreviation of *chemical adsorption*) and physisorption (an abreviation of *physical adsorption*). In chemisorption the chemical attractive forces of adsorption act between the surface and the absorbate (usually these are covalent

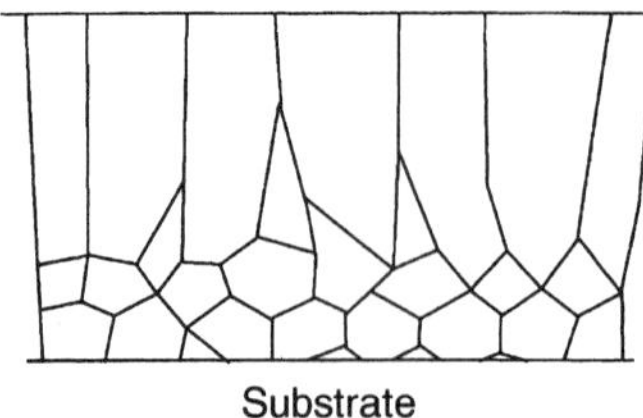

Figure 17 Schematic cross section (perpendicular to the substrate) of the columnar deposit.

bonds). Thus there is a chemical combination between the substrate and the adsorbate where electrons are shared and/or transferred. New electronic configurations may be formed through this sharing of electrons. In physisorption the physical forces of adsorption, Van der Waals or electrostatic forces, act between the surface and the adsorbate; there is no electron transfer and no electron sharing.

The adsorption energy for chemisorbed species is greater than that for physisorbed species. Typical values for chemisorption are in the range of 20 to $100\,\mathrm{kcal\,mol^{-1}}$ and for physisorption, around $5\,\mathrm{kcal\,mol^{-1}}$.

4.2 Effect of Additives on Nucleation and Growth

Adsorbed additives affect the kinetics of electrodeposition and the growth mechanism by changing the concentration of growth sites on a surface, the concentration of adions on the surface, the diffusion coefficient D, and the activation energy of surface diffusion of adions. In the presence of adsorbed additives the mean free path for lateral diffusion of adions is diminished, which is equivalent to a decrease in the diffusion coefficient D (diffusivity) of adions. This decrease in D may result in an increase in adion concentration at steady state and thus an increase in the frequency of the two-dimensional nucleation between diffusing adions.

Additives can also influence the propagation of microsteps and cause bunching and the formation of macrosteps. The type of deposit obtained at constant current density may depend on the surface coverage of additives.

4.3 Leveling

Leveling is defined as the progressive reduction of the surface roughness during deposition. Surface roughness may be the result of mechanical polishing. In this case scratches on the cathode represent the initial roughness and the result of cathodic leveling is a smooth (flat) deposit or a deposit of reduced roughness. During this type of leveling more metal is deposited in recessed areas than on peaks (bumps). This leveling is of great value in the metal finishing industry. Leveling can be achieved in solutions either in the absence of addition agents or in the presence of leveling agents. Thus there are two types of leveling processes: (1) geometric leveling corresponding to leveling in the absence of specific agents and (2) true or electrochemical leveling corresponding to leveling in the presence of leveling agents.

Dukovic and Tobias [20] and Madore and Landolt [21] developed theoretical models of leveling during electrodeposition in both the absence and presence of additives. Both models indicate that a significant leveling of semicircular and triangular grooves by uniform current distribution (geometric leveling) is achieved when the thickness of deposit is at least equal to to the depth of the groove. Leveling in the presence of leveling agents (true leveling) is achieved at much lower deposit thickness.

Theories of leveling by additives are based on (1) the correlation between an increase in the polarization produced by the leveling agents, Leidheiser [22], and (2) preferential adsorption of a leveling agent on the high point (peaks of the substrate or flat surfaces) and by this inhibition (slowing down) of deposition at these points [1, 23].

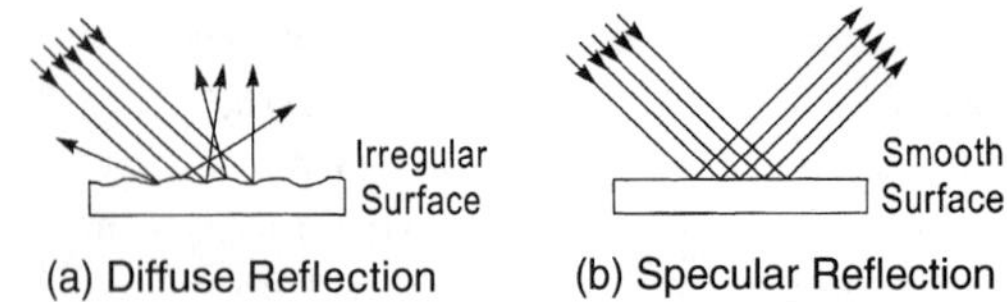

Figure 18 Diffuse and specular reflection. The surface of the object that has irregularities diverges out an initially parallel beam of light in all directions to produce diffusion reflection (*a*). A smooth, bright, surface specularly reflects a parralel beam of light in one direction only (*b*).

4.4 Brightening

The brightness of a surface is defined as the optical reflecting power of the surface. It is measured by the amount of light specularly reflected (Fig. 18). Weil and Paquin [24] showed a linear relationship between the logarithm of the amount of light specularly reflected by a surface and the fraction of the surface having roughness of less than 1500 Å. Kardos and Foulke [23] distinguish three possible mechanisms for bright deposition: (1) diffusion-controlled leveling, (2) grain refining, and (3) randomization of crystal growth. Some fundamental aspects of brightening and leveling are reviewed by Onicin and Muresa [25].

4.5 Consumption of Additives

An additive can be consumed at the cathode by incorporation into the deposit and/or by the electrochemical reaction at the cathode or anode. Consumption of coumarin in the deposition of nickel from the Watts-type solution was studied extensively. For example, Rogers and Taylor [26] found that the coumarin concentration decreases linearly with time and that the rate of coumarin consumption is a function of concentration. The rate of consumption increases with increase in the bulk concentration of the coumarin. Thus, in the electrodeposition of nickel from a Watts-type solution, the total current density is the sum of the current densities for nickel deposition, hydrogen evolution, and additive reduction.

In general, the control of an electrodeposition solution involves monitoring the concentration of additives and by-products of the reaction of additives at the electrodes.

5 ELECTROLESS AND DISPLACEMENT DEPOSITION

Electroless and displacement depositions have one basic characteristic in common: no power supply is necessary to drive the deposition reaction

$$M^{z+}_{\text{solution}} + ze \xrightarrow{\text{electrode}} M_{\text{lattice}} \tag{40}$$

In the electroless deposition process z electrons are supplied by a reducing agent, Red, present in the solution. The electron-donating species Red donates electrons to the catalytic surface. In the displacement deposition process z electrons are supplied by the substrate; the substrate S is the electron-donating species.

In general, an electrode with lower electrode potential in Table 1 will reduce ions of an electrode with higher electrode potential (Fig. 19).

5.1 Electroless Deposition

The overall reaction of electroless metal deposition is

$$M^{z+}_{\text{solution}} + \text{Red}_{\text{solution}} \xrightarrow{\text{catalytic surface}} M_{\text{lattice}} + \text{Ox}_{\text{solution}} \tag{41}$$

where Ox is the oxidation product of the reducing agent Red. The catalytic surface may be the substrate S itself or catalytic nuclei of the metal M dispersed on a noncatalytic substrate surface. The reaction represented by Eq. (41) must be conducted in a way such that a homogeneous reaction between M^{z+} and Red, in the bulk of the solution, is suppressed.

According to the mixed-potential theory of electroless deposition [27–29] the overall reaction given by Eq. (41) can be decomposed into one reduction reaction, *the cathodic partial reaction*

$$M^{z+}_{\text{solution}} + ze \xrightarrow{\text{catalytic surface}} M_{\text{lattice}} \tag{42}$$

and one oxidation reaction, *the anodic partial reaction*

$$\text{Red}_{\text{solution}} \xrightarrow{\text{catalytic surface}} \text{Ox}_{\text{solution}} + me \tag{43}$$

Thus overall reaction, Eq. (41), is the result of the combination of two different partial reactions, Eqs. (42) and (43). These two partial reactions, however, occur at one and the same electrode, the same metal-solution interphase. In order that the overall electroless deposition reaction may proceed, the equilibrium (rest) potential of the reducing agent, $E_{\text{eq, Red}}$ Eq. (43), must be more negative than that of the metal

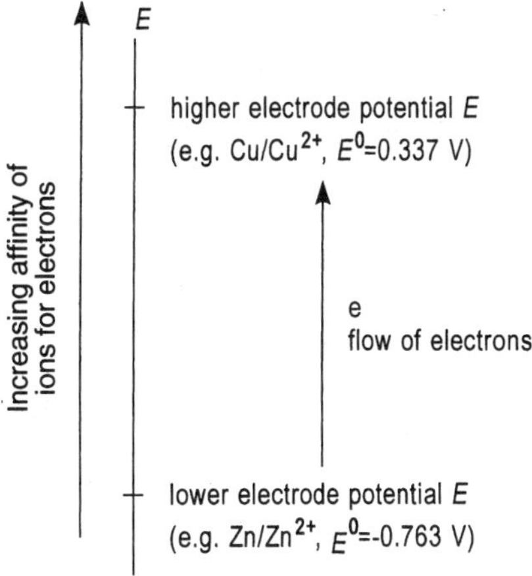

Figure 19 An electrode with lower electrode potential will reduce ions of an electrode with higher electrode potential.

electrode, $E_{eq,M}$ Eq. (42), so that the reducing agent Red can function as an electron donor and M^{z+} as an electron acceptor. This is in accord with the discussion in Section 5.1.

Evans Diagram According to the mixed-potential theory, the overall reaction of the electroless deposition can be described electrochemically in terms of two current-potential (i-E) curves, as shown schematically in Figure 20. This figure shows a general Evans diagram with current-potential functions $i=f(E)$ for the individual electrode processes, Eqs. (42) and (43). In this the sign of the current density is supressed. According to this presentation of the mixed-potential theory, the current-potential curves for the individual processes $i_{cathodic} = i_M = f(E)$ and $i_{anodic} = i_{Red} = f(E)$, intersect. The coordinates of this intersection have the following meaning: (1) the abscisa, the current density of the intersection, is the deposition current density i_{dep} (i.e., log i_{dep}), that is, the rate of electroless deposition in terms of mA/cm^2, and (2) the ordinate, the potential of the intersection, is the mixed potential, E_{mp}.

Mixed Potential, E_{mp} When a catalytic surface S is introduced into an aqueous solution containing M^{z+} ions and a reducing agent, the partial reaction of reduction, Eq. (42), and the partial reaction of oxidation, Eq. (43), occur simultaneously. Each of these partial reactions strives to establish its own equilibrium potential, E_{eq}. The result of this "competition" is the creation of a steady state with the compromised potential called the *steady-state mixed potential*, E_{mp}. The result of this mixed potential is that the potential of the redox couple Red/Ox Eq. (43) is raised anodically from the reversible value $E_{eq,Red}$ (Fig. 20), and the potential of the metal

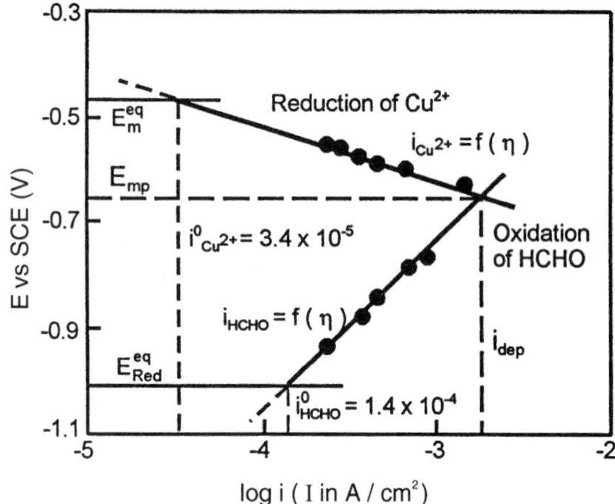

Figure 20 Current-potential curves for reduction of Cu^{2+} ions and for oxidation of reducing agent Red, formaldehyde, combined into one graph (Evans diagram). Solution for the Tafel line for the reduction of Cu^{2+} ions: 0.1 M $CuSO_4$, 0.175 M EDTA, pH 12.50, E_{eq} (Cu/Cu^{2+}) = 0.47 V versus SCE; for the oxidation of formaldehyde: 0.05 M HCHO and 0.075 M EDTA, pH 12.50, E_{eq} (HCHO) = 1.0 V versus SCE; temperature 25°C (±0.5°C). (From M. Paunovic, *Plating*, **55**, 1161 (1968), with permission from American Electroplaters and Metal Finishers Society).

electrode M/M^{z+} Eq. (42) is depressed cathodically from its reversible value $E_{eq,M}$, down to the mixed potential E_{mp} (Fig. 20).

Mixed-potential theory was tested experimentally and verified extensivelly for the case of electroless deposition of Cu, Au, and Ni [1]. Its significance is in the fact that the kinetics and mechanism of electroless deposition can be studied and interpreted by examining partial electrode processes [29].

5.2 Displacement Deposition

It is in Section 5.1 mentioned that in the displacement deposition process z electrons in the deposition reaction, Eq. (40), are supplied by the substrate. The substrate is the electron-donating species. We describe here example of displacement deposition of Cu on Zn,. which occurs when a strip of Zn is placed in a solution of $CuSO_4$. In order to find out what reactions will occur in this system, is necessary to consider standard electrode potentials (Table 1). The standard electrode potential E^0 of Cu/Cu^{2+} is 0.337 V and that for Zn/Zn^{2+} is −0.763 V. Since Zn/Zn^{2+} has a lower electrode potential than the Cu/Cu^{2+} system, Zn will reduce Cu^{2+} ions in the solution. Thus there are two partial reactions in this system, as in an electroless deposition system (Fig. 21). In displacement deposition of Cu on Zn, electrons are supplied in the oxidation reaction of Zn

$$Zn \rightarrow Zn^{2+} + 2e \tag{44}$$

where Zn from the substrate dissolves into the solution and thus supplies electrons necessary for the reduction-deposition reaction

$$Cu^{2+} + 2e \rightarrow Cu \tag{45}$$

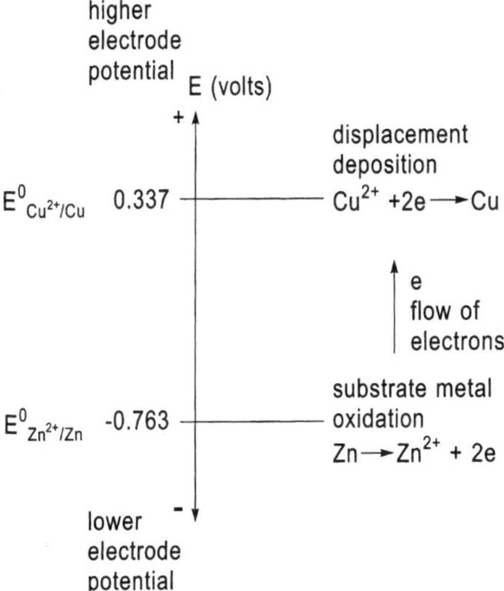

Figure 21 Relationship between partial reactions in displacement deposition of Cu on Zn, $Zn + Cu^{2+} = Zn^{2+} + Cu$.

The overall displacement deposition reaction is

$$Zn + Cu^{2+} \rightarrow Zn^{2+} + Cu \qquad (46)$$

It is obtained by combination of the two partial reactions, oxidation and reduction, reactions (44) and (45), respectively.

In general, Ox/Red (M^{z+}/M) couples with high standard electrode potentials are reduced by Ox/Red (M^{z+}/M) couples with low standard electrode potentials. Or, in other words, low-potential couples reduce high-potential couples (see Table 1 and Fig. 19). The thickness of the deposited metal in this case is self-limiting since the displacement depsotion process needs exposed (free) substrate surface in order to proceed.

Part B
Physical Considerations

MORDECHAY SCHLESINGER

6 ELECTRODEPOSITION OF ALLOYS

6.1 General

Alloy deposition is as old an art and science as the electrodeposition of individual metals (e.g., brass, which is an alloy of copper and zinc, deposition was invented circa 1840). As can be expected, alloy deposition is subject to the same principles as single-metal plating. Indeed progress in both types of plating has depended on similar advances in electrodeposition science and/or technology. The subject of alloy electroplating is being dealt with by an ever-increasing number of scientific and technical publications. The reason for this is the vastness of the number of possible alloy combinations and the concomitant possible practical applications.

Properties of alloy deposits superior to those of single-metal electroplates are common place and are widely described in the literature. In other words, it is recognized that alloy deposition often provides deposits with properties not obtained by employing electrodeposition of single metals. It is further the case that, relative to the single component metals involved, alloy deposits can have different properties in certain composition ranges. They can be denser, harder, more corrosion resistant, more protective of the underlying basis metal, tougher and stronger, more wear resistant, different (better) in magnetic properties, more suitable for subsequent electroplate overlays and conversion chemical treatments, and superior in antifriction applications.

Electrodeposited binary alloys may or may not be the same in phase structure as those formed metallurgically. By way of illustration, we note that in the case of brass (Cu–Zn alloy), X-ray examination reveals that, apart from the superstructure of β-brass, virtually, the same phases occur in the alloys deposited electrolytically as those formed in the melt. Phase limits of electrodeposited alloys closely agree with those in the bulk form. Somewhat opposite is the case of the Ag–Pb alloy. In the cast alloy form it contains "large" silver crystals with lead present in the grain boundaries as dendrites. In the electrodeposited form the alloy contains exceedingly fine grains exhibiting no segregation of lead.

By choosing specific metal combinations, electrodeposited alloys can be made to exhibit hardening as a result of heat treatment subsequent to deposition. Such treatment, it should be noted, causes solid precipitation. When alloys (e.g., Cu–Ag, Cu–Pb, and Cu–Ni) are coelectrodeposited within the limits of diffusion currents, equilibrium solutions or supersaturated solid solutions are in evidence as is observed by X-ray analysis. The actual type of deposit can, for instance, be determined by the work value of nucleus formation under the overpotential conditions of the more electronegative metal. When the metals are codeposited at low polarization values, the formation of solid solutions or of supersaturated solid solutions results. This is so even when the metals are not mutually soluble in the solid state according to the corresponding phase diagram. As a rule, codeposition at high polarization values, on the other hand, results in two phase alloys even with systems capable of forming continuous series of solid solutions.

6.2 Principles

The electrodeposition of an alloy requires, by definition, the codeposition of two or more metals. In other words, their ions must be present in an electrolyte that provides a "cathode film" where the individual deposition potentials can be made to be close or even the same.

Three main stages in the cathodic deposition of alloys (or single metals) are to be recognized:

1. *Ionic migration.* The hydrated ion(s) in the electrolyte migrate(s) toward the cathode under the influence of the applied potential as well as through diffusion and/or convection.
2. *Electron transfer.* At the cathode surface area, the hydrated metal ion(s) enter the diffusion double layer where the water molecules of the hydrated ion are aligned by the field present in this layer. Subsequently the metal ion(s) enter(s) the fixed double layer where, because of the higher field present, the hydrated shell is lost. Then on the cathode surface, the individual ion may be neutralized and is adsorbed.
3. *Incorporation.* The adsorbed atom wanders to a growth point on the cathode and is incorporated in the growing lattice.

6.3 Deposition

Now the deposition potential, which is the potential at which deposition occurs for a given metal, is determined by two major quantities, as was discussed in the first part of this chapter (see Eqs. (3) and (4) above). Specifically, one quantity is the standard electrode potential, E^0 (it is defined as the potential that exists when the metal is immersed in a solution of its ions at unit activity) which is a characteristic of the metal in question. The other quantity is a function of the ion's activity value in the electrolyte. This value is in turn proportional to the ionic concentration as long as the latter is of moderate value; otherwise, it is a complicated function of all the units in the solution. For two metals to be codeposited and thus produce an alloy, they must both be present in the electrolyte, and their individual deposition potential should be the same or nearly the same. It is clear that for two metals whose standard electrode potentials differ by a great amount, the only way to achieve these deposition potentials is by controlling the value of the activity, that is, changing the respective concentrations. The deposition potentials can thus be brought into harmony. Detailed calculations (see Fundamentals), for instance, indicate that copper ion concentration can be brought into harmony with zinc ion concentration in certain solutions and thus used in Zn–Cu alloy plating.

Using an example, we wish to stress the point that in a solution containing "simple" salts of zinc and copper where ion concentration, and so the activity, values are close together, an alloy deposition is virtually impossible. That is due to the large difference between the standard electrode potentials E^0 for copper (−0.345 at 25°C) and that for zinc (−0.762 at 25°C).

In a mixed copper-zinc solution of complex cyanide, however, the Cu^+ ion concentration can be reduced to the order of 10^{-18} mol liter^{-1} and the concentration ratio (zinc ion)/(copper ion) will be made very large. This will compensate for the large difference in their standard electrode potentials. A detailed treatment for this case shows, that copper cyanide complex is of the form $Cu(CN)_3^{2-}$ for which the dissociation constant value is known. The dissociation constant for the zinc cyanide complex, $Zn(CN)_4^{2-}$ is also well known. Calculations using these values make it clear that their respective deposition electrode potentials can be brought close enough together.

To express the above in a different, more specific way, we state that codeposition of two or more metals is possible under suitable conditions of potential and polarization. The necessary condition for the simultaneous deposition of two or more metals is that the cathode potential versus current density curves (polarization curves) be similar and close together. Figure 22 depicts a typical deposition potential plotted against current density for two different individual metals. The curves indicate that in a bath containing both metals, M_1 and M_2, may codeposit at, say, potential V_2 in the ratio I_4/I_2.

It should be understood that the plating bath may be depleted of one metal ion faster than of another. In order to keep deposition conditions under control, (uniform), metal ions must be replenished in direct proportion to their rates of deposition dictated by the specific alloy. Clearly, the ideal case is where the polarization curves of the component metals being codeposited are identical. In practice, however, it is next to impossible to realize this condition.

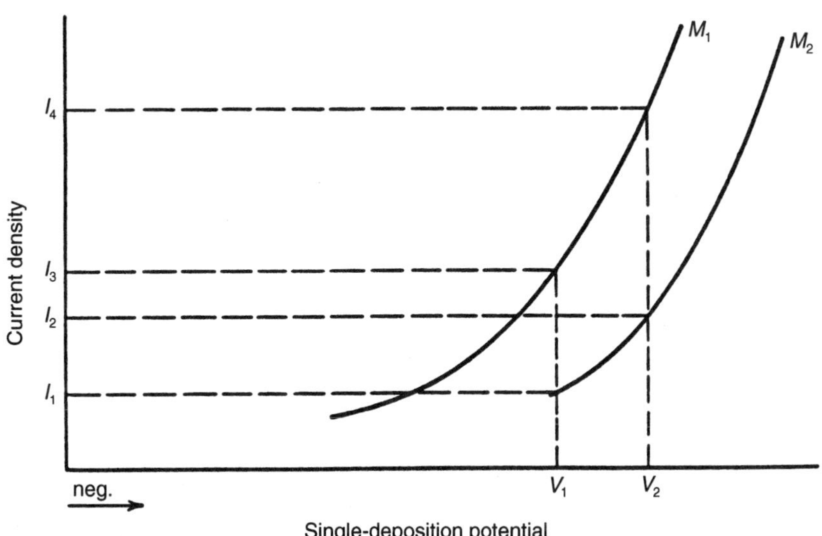

Figure 22 Single-electrode potentials of metals M_1 and M_2 versus current density in the same type of bath.

Electrochemically viewed, even when a single-metal deposition is accompanied by hydrogen evolution, it may well be said that one deals with alloy plating in which hydrogen is the codepositing element. This is equally so when hydrogen is discharged as gas, since the conditions for codeposition are met. Alloy plating of metals makes it into a process of production of hydrogen and the two or more metals that are being alloyed.

Another path to alloy deposition is that via interdiffusion in electrodeposited bilayers or multilayers. In this case different coatings are deposited alternately, using two different plating baths or even one bath where the deposition potential is varied periodically and then heat treatment is applied to promote mutual diffusion thus ending up with an alloy. For example, an alloy of 80% Ni and 20% Cr can be produced by the deposition of alternating layers of 19 μm (micrometer) thick Ni and 6 μ thick Cr. Subsequent heating to 1000°C for four to five hours produces completely diffused alloys of rather high quality in terms of their corrosion properties. Brass can also be produced by interdiffusion of Cu and Zn under suitable conditions.

7 STRUCTURE AND PROPERTIES OF DEPOSITS

7.1 General

In their solid state, atoms (ions) are arranged in a regular pattern that may be described by the three-dimensional repetition of a certain pattern unit. The structure therefore is said to be periodic. When the periodicity of the pattern extends throughout a piece of the material we refer to it as a *single crystal.* In polycrystalline material the periodicity is interrupted at so-called grain boundaries. The size of grains through which the structure is periodic varies markedly as discussed below. When the size of the grains or crystallites becomes comparable to the size of the pattern unit, the substance is called *amorphous.*

Grains are understood here to be individual crystallites in a polycrystalline body of material. This definition, however, is not common to all authors. Thus some authors refer to clumps of crystallites as "grains" while others call such clumps "islands".

Crystalline material, as stated above (whether single-crystal, polycrystalline or even nanocrystalline-amorphous) is made such that its component (constituent-base) ions,atoms, or molecules are arranged on a three-dimensional regular, repetitive pattern called a *lattice.* In metals the constituent bases are, singly or multiply, ionized metal atoms. The commonest form among (plated) metals is the face-centered-cubic (fcc) lattice, which is the lattice structure of Ag, Al, Au, Cu, Ni, and others. Here fcc refers to a lattice arrangement of metal ions at each of the corners of a cube plus one atom each in the center of every cube face.

Next in frequency is the hexagonal-close-packed (hcp) lattice, which is assumed by Co, Zn, and some others. Here hcp refers to an arrangement of metal ions with planes of ions placed at the corners of hexagons which are separated by planes of ions grouped in sets of three between hexagons in the adjacent planes. This way one of the two close packing arrangements of hard spheres is realized. The other way of close packing of hard spheres is realized by the fcc arrangement. That arrangement is therefore sometimes referred to as *cubic close packed.*

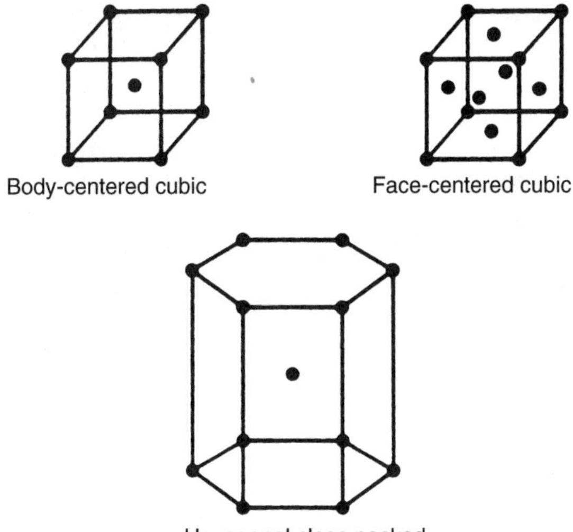

Figure 23 Unit cells of the three most important lattices. Reproduced with permission.

The body-centered cubic (bcc) is the lattice symmetry of Fe, for instance. Here bcc refers to a crystal arrangement of ions at the corners of a cube, and one ions in the centre of the cube equidistant from each face. Figure 23 depicts the unit cells corresponding to the above-mentioned structures. In the last analysis, crystalline structure exhibited as a product of electrodeposition depends on a "competition" between rates of new crystallite formation and existing crystal growth.

It ought to be stressed that actually a large number of variables in the plating process have bearing on structure. Among these are metal ion concentration, additives, current density, temperature, agitation, and polarization. It is outside the scope of this chapter to discuss these effects in more detail.

7.2 Substrate and Environment

Another factor that has substantial influence on electrodeposits, their structure, and their properties is the nature of the substrate upon which the plating occurs. Two phenomena are important in this context: *epitaxy* and *pseudomorphism* Epitaxy refers to the systematic structural kinship between the atomic lattices of the substrate and the deposit at or near the interface. In other words, it concerns the induced continuation of the morphology/or structure of the substrate material into a coating applied on to it. An important parameter of epitaxial growth is the substrate's temperature. For a given material being deposited on a substrate, all other conditions being fixed, there exists an *epitaxial temperature*. That is a temperature above which epitaxial growth is possible; below this temperature no epitaxial growth can occur. For example, for silver evaporated on NaCl, the temperature is 150°C. Pseudomorphism is the continuation of grain boundaries and similar geometric features of the substrate into the deposit. An alternative definition states that deposits that are stressed to fit upon a substrate are said to be pseudomorphic. A working rule is that pseudomorphism persists deeper (up to 10 nm) into the deposit than epitaxy.

Prior cleaning, or the cleanliness state of the substrate, also greatly influences the structure and adhesion of the deposit. In general, both mechanical and chemical cleaning methods may be required.

Metal surfaces, exposed to certain types of gas will generally undergo transformation, and their transformation can be accompanied by significant changes in their properties. The rate and the extent of change are dependent on the metal, the gases, and the new phase/product that will form at the interface between the two original phases. We mention here the case of Si oxidizing to form SiO_2 which undergoes expansion thus creating strong (compressive) stress on the interfacial surface.

An example of high practical significance is the copper deposits used in microelectronics, mirrors, and other optical applications. These deposits have been observed to soften over time even when stored at room temperature for as short as four to six weeks. The surfaces of mirrors and other precision objects made of copper will deform after a few months. This kind of degradation can be counterweighed by a suitable metal overcoating. Another, though not always practical, way is to subject the object to heat treatment at about 300°C. These degradation phenomena are the direct results of microstructural instabilities, often referred to as *re-crystallization* in the copper. It is worth stressing that re-crystallization is not limited to copper. "As deposited" electroless Ni–P is subject to a similar process.

Crystals grown via electroplating may be oriented every which way. That is to say, the direction axes of individual crystallites may be randomly distributed. The case, however, where one axis is oriented or fixed in *nearly* one direction, is said to be a single texture. Where two axes are fixed or oriented, it is said to be double texture. Monocrystalline orientation refers to the case where there are three such nearly oriented axes. This includes, for instance, epitaxial films. Orientation is viewed with respect to any fixed (in space) frame of reference.

Electrochemical parameters, and not substrate properties, are the main deciding factors in the texture of deposits. This is particularly evident if the deposit's thickness is 1 μm or more. In deposits of lesser thickness, the substrate plays an important role as well (see above). Another nonelectrochemical factor may be the codeposition of particulate matter with some metal deposits. To summarize, we note that texture is mostly influenced by deposition current density, itself being a function of bath parameters. Not surprising then is the fact that in the case of physical vapor deposition (PVD) the deposition rate, which serves as the effective current density, is the determining factor in setting the texture of the coating.

7.3 Properties

The functional relationship between structure and property is complex. To be precise, there are different microscopic and macroscopic interactions involved in materials. There are both quantum mechanical interactions between constituent atoms or molecules and classical-type interactions between grains and groups of grains. Electrodeposition adds some measure of complication, since it is inherently a nonequilibrium process. The product films thus exhibit many imperfections that vary in number and nature from grain to grain. In this connection it is also important to remember that the most basic imperfection of a crystal lattice is the surface itself where the periodicity, the hallmark of bulk material, is cut off.

For illustration of the complexity discussed above, we note the following few practical cases:

1. Gold, when electrodeposited with about 0.5% cobalt; it has a hardness of about four to five times that of annealed electrodeposited gold. This degree of hardness cannot be achieved using any of the known metallurgical methods. It is widely believed that grain sizes of 20 to 30 nm in this material are responsible for this high degree of hardness. Cobalt-hardened gold films have found a large number of applications in electronics and other areas.
2. Texture has a marked influence on the properties of a given deposit. Seemingly unrelated parameters (properties) such as corrosion resistance, hardness, magnetic properties, porosity, and contact resistance are all texture dependent. The specific texture of a copper substrate upon which a nano-multilayer system such as Cu–Ni layer pairs are deposited has direct bearing on the deposit's texture and consequently magnetic characteristics.
3. Different nickel deposits show a great variety of contact resistance values. This is particularly so after the deposits have been exposed to the atmosphere for extended periods of time. The differences in these values may be best explained in terms of variations in plated texture. Nickel electrodeposits with polycrystalline nature have been observed to behave as single crystals if their grains are oriented such that the (100) planes are parallel to the surface. Not surprisingly the oxidation rate in (100) oriented single crystals is self-limiting at ambient temperature.
4. The optical properties of some thin film light polarizers owe their ability to polarize light to the specific texture of their structure.

7.4 Impurities

Electroplated films almost always contain various types of inclusions or impurities. These additives may be from one or more of the following origins:

1. Added chemicals (levelers, brighteners, etc.)
2. Added particles (for composite coating)
3. Cathodic products (complex metal ions)
4. Hydroxides (of the depositing metals)
5. Bubbles (hydrogen gas, etc.)

In general, it may be stated that deposits produced using low-current densities possess higher impurity content than deposits produced using high-current densities (see Fundamentals).

Small amounts (in ppm) of impurities can also influence material strength. For example, as noted earlier, a small amount of cobalt in electrodeposited gold enhances its hardness. Sulfur impurities can, on the otherhand, be detrimental to nickel deposits. An increase in sulfur content is known to reduce the fracture resistance of electroformed nickel. If no other impurities are present in the electrodeposited nickel, hardness alone can be used as an indicator of sulfur impurity content.

From the foregoing discussion it should be clear that in practice, it is virtually impossible to obtain and maintain electrodepositing baths free from impurities. This is a big consideration in the research laboratory. In research work on electrode kinetics, for example, careful and often complex purification procedures are followed in order to remove some key contaminants. Otherwise, no reproducible or reliable results can be obtained. The plating shop worker, on the other hand, deals with highly contaminated (from the point of view of the researcher) solutions from a variety, and by and large, unknown, sources. It is for this reason that technical practice must and does rely on empirical observations.

In some deposits, notably those of nickel, electrical resistance follows current density at low temperatures in the sense that films deposited at a low-current density (e.g., $10\,mA\,cm^{-2}$) show lower resistance than those deposited at higher density. This effect of low resistance occurs in the low-temperature range of 4 to 40°K but disappears at closer to room temperature.

A special type of impurity is hydrogen. Hydrogen is codeposited with most metals. Because of its low atomic number, it can be expected to be readily adsorbed by the basis metal. The source of the hydrogen may vary from the associated preparatory processes such as electro cleaning to the specific chemical reactions associated with the plating process itself. Regardless of its origin the presence of hydrogen may result in embrittlement, which means a substantial reduction in ductility. Hydrogen embrittlement (HE) is an expression used to describe a large variety of fracture phenomena having in common the presence of hydrogen in the metal or alloy as a solute.

Many types of mechanisms have been suggested in the attempts to explain HE in different systems. No consensus has been reached to date as to the validity of these. The absorbed hydrogen in steel, for instance, makes it most susceptible to embrittlement. At least in part, this is due to the hydrogen interfering with the normal flow or slip of the lattice planes under stress. If, as often is the case, a deposit contains voids on the microscopic scale, hydrogen may accumulate in molecular form possibly developing pressures exceeding the tensile strength of the basis metal leading to the development of blisters.

Finally, in physical vapor-deposited as well as sputter-deposited films, incorporated gases can increase stress and raise annealing temperatures. Similar effects are present in electron beam evaporated films.

8 MULTILAYERED AND COMPOSITE FILMS

8.1 General

With the introduction of many different reliable ways,including electrodeposition, to produce systems with (nanometer) nm-scale structural and composition variations, there has came the ability for greater control of material properties, which has led to the design of new materials.

Nano structural materials are divided into three main types: one-dimensional (commonly known as multilayered) structures made of alternate thin layers of different composition; two-dimensional structures (commonly known as wire-type elements suspended within a three-dimensional matrix), and three-dimensional constructs which may be made of a distribution of fine particles suspended within a matrix (either in periodic or random fashion) or an aggregate of two or more phases with a nanometric grain size. Figure 24 depicts the above mentioned constructs.

The semiconductor community has made band gap engineering a reality by exploiting the ability to produce multiple quantum well material. Here the quantum well refers to a potential well with dimensions such that quantum mechanical effects are taken into account. In other words, one thinks of electrons/holes as being trapped in the well, having distinct energy levels rather than a continuum. The progress from one-dimensional nano structural materials to two- and three-dimensional ones (i.e., to quantum wires and quantum dots; see Fig. 24) has provided enhanced carrier (electrons and/or holes) confinement by their presence which permit more control of the energy band structure.

Accurate control of a micro structure on nano metric scale makes it possible to control magnetic and mechanical properties to a hitherto unattainable degree. In particular, magnetic nano structures have recently become the subject of an increasing number of experimental and theoretical studies. The materials are made of alternating layers, around 10 Å thick, of magnetic (e.g., cobalt) and nonmagnetic metals (e.g., copper).

The magnetic layers are exchange-coupled to one another with a sign that oscillates with the thickness of the spacer (nonmagnetic) layer. This arrangement can be understood if one considers the multilayer systems as magnetic layers separated by nonmagnetic layers. The number of free conduction electrons is dependent on the number of nonmagnetic atoms, that is, the thickness of the nonmagnetic layer. As this number increases, subsequent orbitals in the magnetic atoms will experience alternating electron surpluses and deficits, causing a sinusoidal variation in the average magneton number per atom. Thus the magnitude of the exchange interaction varies not only in the usual fashion, inversely with the square of the distance between magnetic layers but also in a sinusoidal manner. These results were known recently for sputtered, evaporated, and molecular beam epitaxy (MBE) grown samples.

Electrodeposition of composition-modulated films was first performed by Brenner in 1939 [30] who employed two separate baths for the two components and a

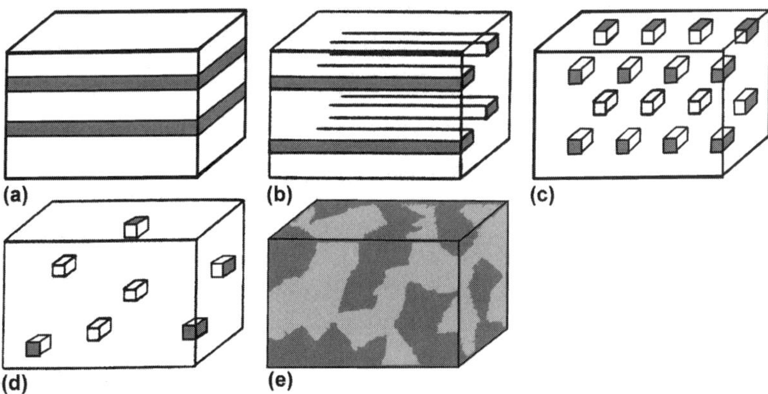

Figure 24 (*a*) Quantum wells—multilayers; (*b*) quantum wires; (*c*) ordered dots; (*d*) random dots; (*e*) nanometer-sizes dots.

"periodic" immersion of the deposit in the two baths. This method proved too cumbersome to be adopted in practice, however. Deposition from a single bath with the presence of salts of the two components of the multilayer is what has evolved.

A serious problem was found in the deposition of two metals from one bath. Specifically, while a layer of the more noble member could, be deposited by choosing the potential to be between the reduction potentials of the two metals, the potential had to be set to a value appropriate for the reduction of the less noble member so that both would be deposited and result in an alloy layer rather than pure metal.

8.2 Electrodeposition of Nanostructures

In 1986 Yahalom and Zadok [31] pointed to methods to produce composition-modulated alloys by electrodeposition, initially for the Cu–Ni couple. Modulation was obtained to thicknesses down to 8 Å. The principle of their method is as follows:

Traces of metal A ions are introduced into a concentrated solution of metal B, assuming that metal A is nobler than B. At a low enough polarization potential the rate of reduction of metal B is high because it is determined by its activation polarization. The rate of reduction of metal A is slow and controlled by diffusion. At a predetermined considerably less negative polarization potential only metal A is reduced. The potential is simply switched between these two potential values, forming a modulated structure composed of pure A layers and layers of B with traces of A in the B layers. The actual deposition of the multilayered composite can be carried out by either current or potential control.

The method is suitable for a considerable number of metal couples. The main caveat, however, is that both metals can be deposited from similar baths. Another is that they differ sufficiently in their degree of nobility.

The method was first tried using the Cu–Ni system, which is still the most-studied system by far when it comes to electrochemically produced layered structures. The reasons for this are quite compelling. There is a sufficient difference in reduction potentials of the two metals. There is the similarity in crystal structure (fcc) and proximity of lattice parameters, ensuring a good coherency between the layers. Further considerable data exist on the magnetic, mechanical, and other properties of the Cu–Ni system deposited by other than electrodeposition, enabling comparison of the modulated layers produced by the chemical method.

8.3 Analysis of the Deposit

A standard method for confirming the coherence of the layers is via X-ray diffraction spectra. If the layers are coherent, and there are a sufficient number of them to provide a relatively strong Bragg diffraction pattern, then satellites due to super lattice formation should appear on each side of the Bragg diffraction peak (super lattice refers here to the periodicity of the layered structure).

The mathematical expression for the difference between the sine values of the two satellites is

$$\sin\theta_s^+ - \sin\theta_s^- = \frac{\lambda}{\Lambda}$$

where λ is the X-ray wave length and where Λ is the super lattice repetitive period. The expression is a consequence of an analogous formula for light diffraction by optical grating.

In general, one can expect systems like these to have properties different from the bulk. In bulk materials, most ions are in a crystalline environment; that is, they do not "see" the surface. In a crystalline super lattice most atoms "know" that are near an interface, that is, in a noncrystalline/ nonperiodic environment. For an extended treatment of the mechanica, magnetic, and other properties of multilayered thin film systems, the reader should consult "Fundamentals" (chapter 17).

8.4 Conclusion

All the possible methods of deposition have inherent advantages and disadvantages with regard to the quality of the multilayers they create and the ease of their production. However, electrodeposition seems best to fulfill imposed financial and temporal restrictions. Specifically, vapor deposition is expensive, creates quasi-amorphous interfaces, and is time-consuming in terms of controlling the alternation of deposition material. Electroless deposition may suffer from the same drawback with regard to mechanical switching between solutions for alternating layer materials. Finally, sputtering requires an expensive vacuum system and cannot easily be extended to industrial usage. For these reasons electrodeposition is now the production method of choice for most practical needs.

The constructs thus obtained provide useful paradigm for fundamental studies as well as for a plethora of applications. Some applications are in magnetically operated read/write heads, others in sensors, but these are just a precious few. Recently (January 1998) IBM became the first company to market giant magneto resistance (GMR) based multilayer read/write heads in a family of disc drive products designated Deskstar 16 GP. The use of such technology allows the reading and writing of extremely small magnetic bits of information that facilitates high-density storage. This provides for the ability to store 3.2 gigabytes of data on a 95 mm diameter disc.

We must again emphasize that electrodeposition presents, in principle, several advantages for the investigation and the production of layered alloys. Among these characteristics are the tendency of electrodeposited materials to grow epitaxial and thus to form materials with a texture influenced by the substrate. Electrodeposition thus can be used in systems that do not lend themselves to vacuum deposition. The electrodeposition process is inexpensive, and it can be scaled up with relative ease for use on large parts; further it is a room temperature technology. This last point is important for systems in which undesirable interdiffusion between the adjacent layers may readily occur.

The studies so far of electrodeposited multilayer materials clearly show that electrodeposition is a viable technique for the production of thin multilayered materials in systems that, from the electrochemical standpoint, are adaptable to the pulsed deposition technique.

Recent results have been able to demonstrate that the coatings produced by electrodeposition display the same coherency and layer thickness uniformity as those composition modulated alloys produced by vacuum evaporation or sputter deposition. Demonstration of the capability of electrodeposition to produce materials with predesignable, variable, and controllable composition down to practically the atomic scale constitutes an important step toward the realization of custom-tailored materials. On the theoretical side, the lack, at the time of writing, of a single satisfactory theory for the possible explanation of the ever-growing amount of empirical results is somewhat disappointing.

9 INTERDIFFUSION IN THIN FILMS

9.1 General

Modern integrated circuits are made of layered thin film structures which are subject to interdiffusion during the thermal processing stage in fabrication. Diffusion, in general and not only in the case of thin films, is a thermodynamically irreversible self-driven process. It is best defined in simple terms such as the tendency of two different gases to mix when separated by a porous partition. It drives toward an equilibrium, maximum entropy state of a system. It does so by eliminating concentration gradients of, for instance, impurity atoms or vacancies in a solid or between physically connected thin films. In case of two gases separated by a porous partition, it leads to the eventual perfect mixing of the two.

In equilibrium in a solid, the impurities or vacancies will be distributed uniformly. Similarly in case of two gases, as above, once a thorough mixture has been formed on both sides of the partition, the diffusion process is complete. As well at that stage, the entropy of the system has reached it's maximum value by virtue of the fact that the *information* regarding the whereabouts of the two gases has been minimized. In general, it ought to be remembered that entropy of a system is a measure of the information available about the system. Thus the constant increase of entropy in the universe, it is argued, should lead eventually to an absolute chaotic state where no information is available at all.

The diffusion process in general may be viewed as the model for specific, well-defined, transport problems. In particle diffusion, one is concerned with the transport of particles through systems of particles in a direction perpendicular to surfaces of constant concentration. In a viscous fluid flow, one is concerned with the transport of momentum by particles in a direction perpendicular to the flow. In electrical conductivity it is with the transport of charges by particles in a direction perpendicular to equal potential surfaces. In solids, for instance, the chemical potential is identified with the Fermi energy level. When two solids or thin films are brought into contact, such as in the case of a p-n junction, charged particles will undergo interdiffusion such that the chemical potentials or Fermi levels will be balanced, that is, reach the same level.

Let us assume now that particle diffusion occurs as a result of difference in concentrations as is the case in a deposition bath. Then a relation known as *Fick's law* is in effect, and it is written as

$$\text{particle flux} = (\text{diffusion coefficient}) \times (\text{driving force})$$

or using the language of mathematics,

$$J_n = -D \text{ grad } c \tag{47}$$

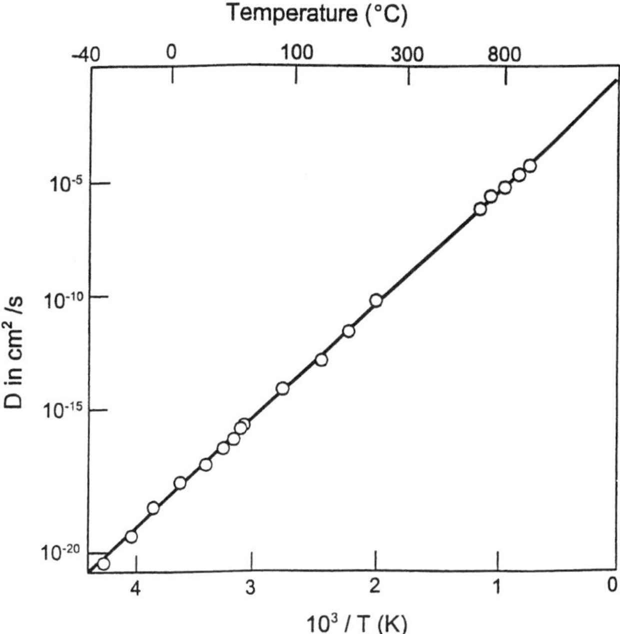

Figure 25 Diffusion coefficient of carbon in alpha iron. From "Introduction to solid state physics, VII" by C. Kittel, with permission from John Wiley & Sons, Inc.

where c is the concentration of the diffusing species (ions or particles). The concentration gradient is acting as the driving force.

The term "flux" here refers to the amount of the diffusing substance passing through a cross section of unit area in unit time. The term "gradient" refers to the change in substance concentration as a function of distance. Both quantities, since they have directionality in addition to a numeric value, are viewed as vectors quantities. It has been found experimentally that the diffusion constant/coefficient, **D** in Eq. (47), depends on temperature according to an exponential type expression, (see Fig. 25). This type of temperature dependency is indeed typical of activation energy driven processes, in general.

To actually diffuse, an atom-ion must overcome a potential energy barrier due to neighbors. If, as above, the potential energy barrier height is E, then the statistical mechanical considerations indicate that the atom will have sufficient thermal energy to pass over the barrier a fraction $\exp(-E/kT)$ of the time. Here T stands for the absolute, Kelvin, temperature while k is the Boltzman's constant. If f is a characteristic atomic vibrational frequency, then the probability p that during unit time the atom will pass the potential energy barrier is given by

$$p \cong f \exp(-E/kT) \tag{48}$$

In other words, in unit time the atom makes f (of the order of 10^{14}/second) attempts to surmount the barrier with the probability each time of the exponent. Thus the quantity p is also known as *jump frequency*.

9.2 Diffusion in Electrodeposits

Diffusion is liable to corrupt the properties of a deposit and defeat the purpose for which the electrodeposition was performed in the first place. This may particularly be so at the basis metal–film interface. Thus, for instance, in deposits for decorative purposes, diffusion of the coated underlayer (metal) to the surface will debase the intended appearance. Another example is gold plating of electronic contacts, which is often practised in order to avoid corrosion of the contact areas. In gold plating the underlayer is often copper. The copper can diffuse to the surface of the gold. As a result of the copper oxidation the contact resistance will be altered, and markedly for the worse. As a practical fact we note here that a 3 μm thick gold film deposit will be covered with the oxide of its underlying copper, if exposed to 300°C, within one month. If exposed to 500°C a gold layer of 30 μm thickness will be diffused through in four to five days.

In some instances, to improve solder ability, tin is deposited on nickel surfaces. In a short time, however, interdiffusion of the two metals results in the growth of an intermetallic $NiSn_3$ compound that is much less amenable to soldering. In the case of tin electrolessly deposited over nickel surfaces, the interdiffusion results in pores in both films. Pores are to be avoided, of course, if conductivity or contact resistance is an issue.

Earlier in this chapter mention was made of hydrogen embrittlement. It was indicated that the presence of hydrogen in, say, steel will result in the same effect. That predicament concerns an interstitial solid solution. The solute atoms (hydrogen in this case) may move along the interstices between the solvent atoms (iron in this case) without having to displace them. This state of affairs is facilitated by the hydrogen atoms being much smaller then the interionic space in the lattice of the solvent.

Diffusion must not, however, always be viewed as a harmful phenomenon. In some cases it is desirable if not essential, as in welding where diffusion ensures joining of the welded parts. Steel is often coated with tin to protect it from corrosion. In this case the formation, via interdiffusion, of the intermetallic $FeSn_2$ is the key for effective protection.

Yet another positive aspect of the diffusion phenomenon is the creation of alloys by first depositing alternate layers of different coatings, and then an alloy is created by heating to promote interdiffusion to produce an alloy. Specifically brass deposits can be produced by first depositing copper and zinc layers alternately. Subsequent heating produces the required brass. This type of approach obviates the undesired direct method of brass deposition via cyanide process.

In the case of jet engine parts which are routinely subject to temperatures close to 500°C, diffusion alloyed nickel-cadmium is found to serve as effective corrosion protective agent.

9.3 Void Formation

Whenever the interdiffusion of two metals is uneven, it can create vacancies or voids. The uneven diffusion is the result of unequal mobilities between a metal couple. The voids occur individually near the common interface. Voids like bubbles coalesce, resulting in porosity and loss of strength. Many thin film couples exhibit this phenomenon, which is referred to as *Kirkendall void creation*. Al–Au, Cu–Pt, and Cu–Au are just a few examples. To be specific, it has been found, for instance, that in Au–Ni about five times more Ni atoms diffuse into Au than Au atoms diffuse into Ni.

Kirkendall void formation can, however, be prevented from occurring by choosing the "right" metal species. For instance, while platinum coating upon copper is subject to the Kirkendall void creation process, the same coating upon electrodeposited nickel is free of it, even if the surface is heated to as high as 600°C for many hours (more than 10 hours).

Again, the effect has useful aspects as well. It can help in controlling porosity, say, of an electroformed object requiring cooling. Indeed, this is a consideration in a number of electronics applications.

9.4 Diffusion Barriers

Diffusion barriers are coatings that serve to protect against undesirable diffusion. One of the best examples is that of a 100 μm thick electrodeposited copper layer. That serves as an effective barrier against the diffusion of carbon. Another example is nickel and nickel alloys (notably electrolessly deposited Ni–P) that block diffusion of copper into and through gold over plate. This is achieved by the deposition of a relatively thin Ni–P layer (less than 1 μm) between the copper and its overlayer. Naturally the effectiveness of the diffusion barrier increases with its thickness. Another factor in the effectiveness of a diffusion barrier is its crystallographic properties such as grain size and preferred crystalline orientation.

In electronic applications it is usual to deposit copper and/or copper alloy and tin in sequence. With a nickel diffusion barrier layer, 0.5 μm thick, between the layers present, no failure occurs. Without the nickel layers between bronze, copper, or tin layers, for instance, intermetallic brittle layer(s) and Kirkendall voids are formed, leading eventually to separation of the coated system and substrate.

Also, when tin-containing solder connections are made to copper, intermetallic materials are formed. These materials keep growing to render weak surfaces. Again, a nickel layer between substrate and solder provides a solution to the problem.

9.5 Diffusion Welding

Here one utilizes diffusion to bring about joints of high quality. In other words, here again, we have practical and useful aspects of thin film interdiffusion. In practice, clean (very clean) cleaved or otherwise smoothened metal surfaces must be made for these to be in a firm mechanical contact while using a strong force that is still insufficient to cause macroscopic deformation even at an elevated temperature. This contact is usually done in vacuum or at least in an inert atmosphere. The problems of hard to reach joints and objectionable thermal conditions, and the resultant undesired microstructures such as Kirkendall voids, are minimized if not eliminated all together. Thus there can be achieved good quality, distortion-free joints that require no additional machining or other post treatment.

Often before welding can be performed, a special layer must be applied in the form of a thin film coating or a thin foil in order to promote joining. Such coating can be applied by electroless or electrodeposition methods.

Five essential process variables common to all diffusion bonding techniques are to be considered: (1) temperature, (2) pressure, (3) time, (4) surface condition, and (5) process atmosphere.

Process temperature is usually 1/2 to $\frac{3}{4}$ of the melting point of the lower melting point metal in the intended weld or bond. It is the purpose of attaining an elevated temperature to promote or accelerate the interdiffusion of atoms at the joint interface and also to provide some metal softening, which in turn aids in surface deformation. The pressure application establishes a firm and robust mechanical contact of the surfaces and further breaks up surface oxides (hence the frequent use of silver). As a result the surface is clean for bonding. The period of time at which the elevated temperature is maintained depends on metallurgical and other considerations.

9.6 Electromigration

In an electrical cord or wire, electricity is conducted without transport of atoms in the conductor. The common, free-electron model of electric conductivity makes a basic assumption that electrons are free to move in and through a metal lattice constrained only by scattering events. Scattering is the cause of electrical resistance and what is termed Joule heating. Scattening, however, does not cause displacement of atoms or ions as long as the current density is moderate. At high-current densities (of about 10^4 amp cm^{-2}) the current can displace ions and thus cause transport of mass. The mass transport caused by the electric field and the charge carriers is known as *electromigration*. It is present in interconnecting lines in microelectronic devices, since in these lines the current density values are high. By way of example, a current of 2 mA applied to a 1 μ wide aluminum line of 0.2 μ thickness, represents a current density value of 10^5 amp cm^{-2}; this current density will cause mass transport in the line even at room temperature. It constitutes a reliability failure endemic to thin film circuits. As modern electronic circuitry becomes smaller and smaller, the current densities will become larger and the probability of circuit failure due to electromigration will be more of a problem. Both voids and extrusions can result. A line of aluminum that is subject to an electric field and current density of sufficient magnitude can cause electromigration. The line can be expected to undergo morphological change such that depletion occurs at the negative (cathode) end while extrusion is present at the positive (anode) side. This means that the material migration is in the direction of the movement of the charge-carrying electrons. The driving force behind electromigration consists of two parts. The first is the direct action of the electrostatic field on the diffusing atoms. The second is the momentum exchange of the moving charge carriers with the diffusing atoms. These forces are referred to as the*direct* force and the *electron wind* force, respectively. The electron wind force is usually far greater than the direct force. There have been attempts to quantitatively, using quantum mechanical methods, explain and estimate the electron wind force. None is widely accepted, and the common practice is to adhere to semiclassical treatments. Remedies include lowering the operating temperature or using other metals (than Al) that have a higher activation energy for diffusion where electromigration is less of a problem.

Part C
Material Science

ROLF WEIL

In previous editions this part was a chapter called "Metallurgical Principles". However, since metals are no longer the only electrodeposited materials, the new name seems to be more appropriate. The title also implies that electrodeposition has become more scientific for which its greater use in electronic application is partially responsible. The topics of material science that are relevant to electrodeposits are the structure, characterization, the various classes of plated materials, and the properties and internal stress.

10 STRUCTURE

By structure we mean the arrangement of the atoms or atom groups. In the majority of electrodeposited materials, the atoms are arranged in a uniform, three-dimensional array. Actually ions form the array, and the valence electrons participate in the bonding. The volume over which this arrangement extends uniformly is called a crystal. When many crystals form a solid material, they are called grains. If the array of atoms is random, the material is amorphous. However, in materials considered to be amorphous, there are generally still very small groups of atoms that possess the same arrangement as crystals [32].

Most electrodeposits exist in one of three crystal habits. The most common one is face-centered cubic (fcc) in which atoms or atom groups are located at the corners of a cube and in the center of its faces. Another common crystal structure is body-centered cubic (bcc) in which atoms or atom groups are located at the corners of a cube and in its center. Materials less often have the hexagonal structure. The atomic arrangement of the basal plane of the hexagonal-close-packed (hcp) structure is the same as that on the one perpendicular to the cube diagonal of the face-centered-cubic one. The habits only differ in the third atom layer.

In many electrodeposits there is a crystal direction that grows faster toward the anode than the other ones. Grains possessing this direction can also grow sideways and cover the less favorably oriented ones. They can grow sideways until they encounter grains of the same orientation. In this way the deposit consists mostly of the grains possessing the favorable growth direction.

If the grains are not randomly oriented, the condition is called a texture. In the case of electrodeposits the texture is a fiber axis because just as in a wire drawn through a die, the directions perpendicular to the preferred orientation are randomly oriented. When electrodeposits are annealed, they generally recrystallize; that is, new grains form. When deposits recrystallize, the texture often changes.

There are defects in the crystalline arrangement. At the boundaries of grains there has to be a departure from the regular crystalline arrangement to accommodate different orientations or different crystal structures. Other defects in electrodeposits are primarily dislocations, twins, and codeposited foreign atoms or molecular groups. A dislocation is a line defect that can be thought of as being the boundary between a portion of a crystal plane that has been displaced by an applied force (i.e., slipped) and that which has not. However, dislocations can originate due to factors other than slip.

The density of dislocations in electrodeposits often approaches that of heavily, plastically deformed metals. Twins are a special type of grain that possesses two boundaries; the crystal arrangement on one side is the mirror image of that on the other side. Codeposited foreign atoms or molecular groups originate from intentional additions to the plating solutions or impurities in it.

10.1 Characterization of the Structure

The most frequently used instrument to reveal the structure of electrodeposits is the scanning electron microscope (SEM). Its advantage over the optical microscope is its greater depth of field allowing a rough topography to be completely in focus. Surfaces that are electrically conducting can be examined with essentially no preparation. The SEM also has better resolution than the optical one. Most scanning electron microscopes have an attachment for chemical analysis by X-ray fluorescence. Because the electron beam can be focussed on a small area, individual structural components can be chemically analyzed. Characteristic X-rays can also be imaged so that the distribution of a particular chemical element can be seen.

The main disadvantage of the SEM is that is is not possible to produce electron diffraction patterns. Thus the crystal structure and orientation of grains cannot be determined. Great care has to be taken when the SEM is used to determine the grain size.

When secondary electrons are used for imaging, nodules consisting of many crystallites can have the appearance of grains and be easily mistaken for them. A scanning electron micrograph of such nodules using secondary electrons for imaging is shown in Figure 26. It is possible to use backscattered electrons to image differences in orientation in the specimen and thus between grains. However, if there is a texture, which is frequently the case in electrodeposits, such a determination of the grain size is difficult. The SEM can also be operated to detect shorts in electrical circuits.

The transmission electron microscope (TEM) has better resolution than the SEM and the capability to obtain electron diffraction patterns. By special techniques it is possible to obtain atomic resolution in the TEM. The main disadvantage of the TEM is that only very thin specimens can be examined. Thus extensive specimen preparation is required for most deposits.

With the TEM it is possible to unambigeously determine the grain size. The determination is performed by placing the field-limiting aperature around an area believed to be a grain. Care must be taken that surrounding areas are not included by the aperature. An electron diffraction pattern is then obtained from the area defined by the aperature. If the pattern is that of a single crystal, the area is a grain. The orientation of the grain as well as directions in it can be determined from the diffraction pattern.

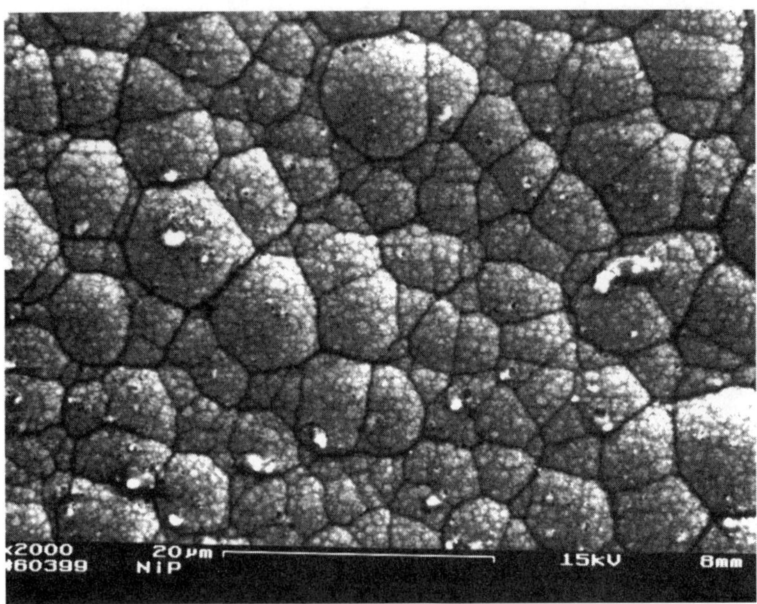

Figure 26 Scanning electron micrograph taken with secondary electrons of nodules in electroless nickel which can be mistaken for grains.

Several features of interest in electrodeposits seen in the TEM are shown in Figure 27. Twins, grain boundaries, and some dislocations are shown. The twins are the areas bounded by two parallel lines. Dislocations are better shown in Figure 28. The location of codeposited molecules or micropores can be imaged by the defocusing method [33]. Molecular groups appear as white dots in Figure 27. Focused ion beam technique can also be used to determine the grain size.

Recently a new class of microscopes has came into use that is particularly suited for the study of the structure of electrodeposits. They are the scanning-probe microscopes. Their main advantage for the study of electrodeposits is that they can operate in liquids. They can therefore be used for in situ studies. However, the scanning-probe microscopes cannot be used to observe continualy changing processes because it takes a relatively long time to form an image. They can have resolutions comparable to the TEM.

Three types of instruments most relevant for studies of electrodeposits are the atomic force microscope (AFM), the scanning tunneling microscope (STM), and the scanning electrochemical microscope (SECM). In the AFM and the STM, a very fine tip only a few atoms in diameter scans the surface and remains at a constant, very small distance above it. In the AFM this distance is controlled by the repulsive force which is present when two objects are only a few angstroms apart.

The tunneling current, which can also be present under these circumstances, controls the distance between the tip and the specimen surface in the STM. A precisely controlled feedback system keeps the distance between the tip and the specimen surface constant. In this way the tip follows the surface topography. The tip is embedded in a cantilever beam. A laser measures the deflection of the beam and thereby the movement of the tip.

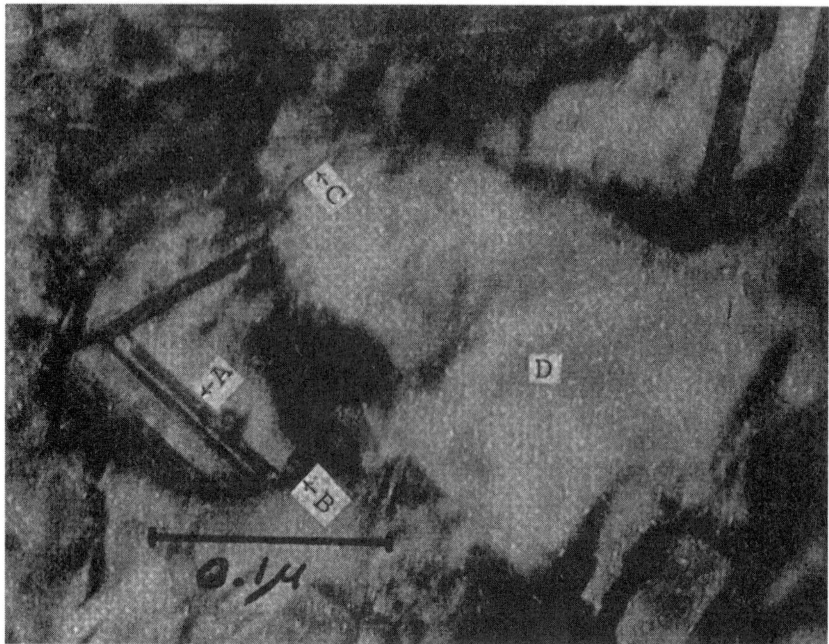

Figure 27 Transmission electron micrograph of semibright nickel showing twin (*A*), grain boundary (*B*), dislocation (*C*), and sites (white dots) of codeposited foreign material (*D*).

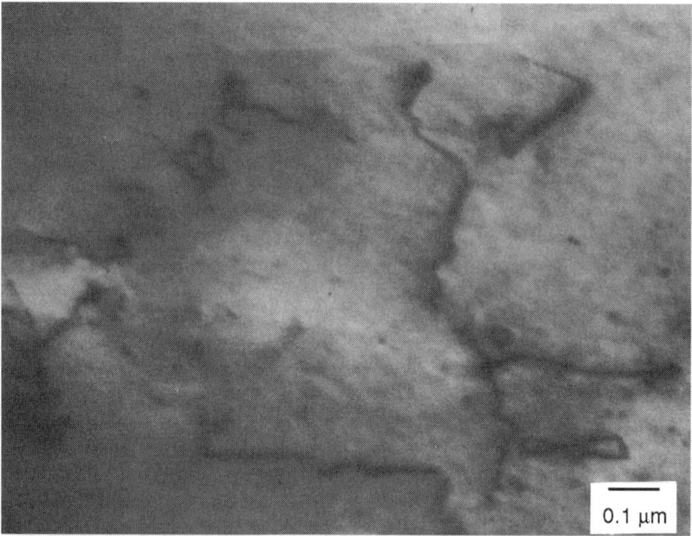

Figure 28 Transmission electron micrograph showing dislocations in electroplated copper.

The STM has two disadvantages for studying electrodeposits. The tunneling current can affect the structure of an electrodeposit and all studied materials have to be electrical conductors. Both the AFM [34] and the STM [35] have been used for in situ studies of electrodeposition. An atomic force electron micrograph of monoatomic terraces on a copper surface taken in situ is provided in Figure 29.

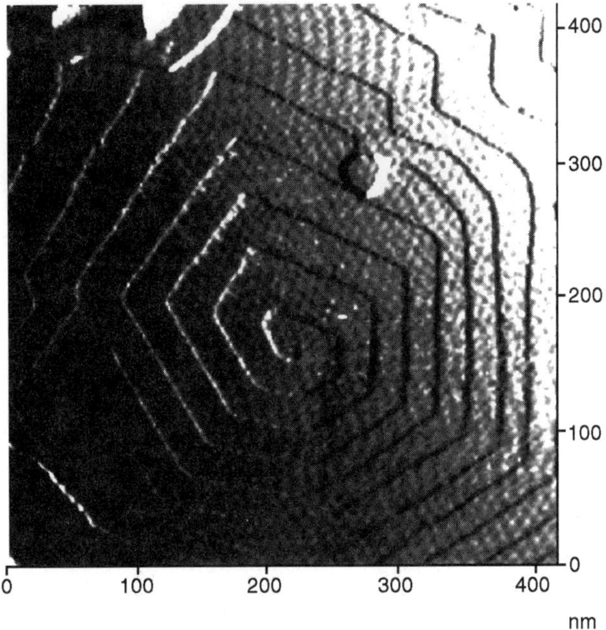

Figure 29 Atomic force micrograph taken in situ of monotomic steps on an electroplated copper surface (from [34]).

In the SECM the probe is a microelectrode. The probe hovers over the surface at a constant distance and maps the electrochemical activity of the surface [36].

The fiber axis is generally determined by X-ray diffraction. The crystal direction perpendicular to the planes that result in the highest-intensity diffraction lines is usually taken as the fiber axis. Care must be taken to ensure that the preferred orientation is not a direction perpendicular to crystal planes which do not diffract as for example the {211} planes of face-centered-cubic metals.

10.2 Classes of Plated Materials

The materials that have been electrodeposited are relatively pure metals, alloys, ceramics, conducting polymers, semiconductors, and superconductors. The majority of plated materials are the relatively pure metals. The most common ones are copper, nickel, gold, zinc, chromium, and tin.

Alloys are mixtures of atomic species of which at least one is a metal. The basis of alloy behavior is the equilibrium or phase diagram, which shows the phases to be expected at a given temperature for a particular composition. Although electrodeposited alloys mostly exhibit the phases predicted by the equilibrium diagram, there are exceptions.

Alloys that should consist of two phases are sometimes plated as supersaturated solid solutions. Examples are Ni–P alloys [32]. According to the equilibrium diagram, phosphorus is essentially insoluble in nickel at ambient temperatures. The alloys therefore should consist of essentially pure nickel and the phase Ni_3P. However, alloys containing as much as 13% P by weight are essentially solid solutions [32]. Sometimes phases not shown on equilibrium diagrams form in electrodeposits. An example is the metastable phase NiSn. The semiconductors that can be electroplated are described in a handbook [37].

The effect of the structure of the substrate on that of the deposit depends on several factors. If the interatomic distance in a particular crystal plane of the substrate closely matches that of one of the deposit, the structure of the former may be continued into the latter. Such a condition is called *epitaxy*.

Oxides and other films on the substrate surface, and distortions due to plastic flow as may result from mechanical polishing, hinder the development of epitaxy. Plating conditions that result in high overvoltages such as high current densities or certain additions to the plating solution are favorable to the formation of three-dimensional nuclei, that is, the start of new grains. Then the relationship between the substrate and the deposit tends to be lost. On the other hand, elevated temperatures of the plating solution and low current densities, which permit the migration of atoms to sites where they can be incorporated into the existing structure, favor epitaxial growth.

Even in the absence of epitaxy, the morphology of the substrate can affect the structure of the deposit. Despite being cathodic, the substrate surface may dissolve in the plating solution and then redeposit with the plated metal as an alloy. Such is the case in some trivalent chromium plating solutions [38]. Etched substrate surfaces have been observed [39] to cause twins to form in the deposit.

The shape and size of the grains is determined by the plating conditions and the composition of the solution. Simple, pure plating solutions such as acidified copper sulfate produce large, columnar grains that generally exhibit a fiber axis. Solutions based on complex compounds such as copper cyanide or solutions containing active addition agents tend to yield fine-grained deposits that usually do not exhibit a texture.

Some deposits such as bright nickel show the so-called banded structure [40]. It manifests itself by a series of closely spaced, parallel lines seen in the cross section. It is caused by periodic variations of the overpotential. When the magnitude of the overpotential reaches its maximum value, growth is interupted. This growth interruption appears as a line in the cross section. The periodicity of the interruptions results in the banded structure.

Electrolytic deposition is uniquely suited to produce the wiring of printed circuits and the interconnections of multilayer microchips. By the use of photoresist, the areas to be plated can be precisely defined. Autocatalytic, known as electroless plating, permits a conducting layer to be deposited on insulating materials. Thin layers of solder can be deposited for making connections. Some unique materials for use in electronic applictions have been developed. Of special importance are materials having structures on the nanometer scale.

11 HARDNESS

The most frequently conducted test of the properties of electrodeposits that may be indicative of the strength and ductility is that of the hardness. Hardness readings are only of value when they have been experimentally related to the strength or ductility of a specific material. They can also be useful as an indication of wear resistance.

For electrodeposits even the qualitiative relationships commonly observed among hardness, tensile strength, and ductility do not always prevail. For example, it would be expected that the hardness increases with the tensile strength and decreases with the ductility. However, the reverse effect is common among electrodeposits. It is particularly important, therefore, in dealing with deposits to be sure of the significance of the hardness values.

The most commonly performed hardness test is the indention test. It consists of pushing an indentor by an applied load into the material. The diameter or diagonal of the indent is measured with a microscope having a filar eyepiece. The indention is generally made on the cross section because the thickness of the coating has to be at least 14 times greater than the depth of the indent. Failing this requirement, the substrate affects the hardness value by what is commonly called the *anvil effect*.

The largest indent on the cross section for the greatest accuracy is obtained with the Knoop indentor, which is an assymmetric, pyramidal, diamond point. It produces an indent whose length is about 7 times its width and 30 times its depth. The length of the indent should be, of course, parallel to the substrate-deposit interface. The Vickers indentor, which produces an indent whose length and width are equal, is also used for deposits. The nanohardness tester produces such a small indent that it can be used on the surface without causing the anvil effect. Since the hardness depends on the applied load, it should always be specified.

12 ADHESION

The usual meaning of adhesion is the extent to which one item sticks to another. If an electrodeposit separates from the substrate, the adhesion is poor. Adhesion of metallic substrates is generally better than is required for normal service. Plating on plastics can present a serious adhesion problem. The Jaquet test described in Chapter 28 is most often used with deposits on plastics.

There is no nondestructive test for adhesion. The destructive tests are at best only semiquantitive. Adhesion depends on a bond between the coating and the substrate. The strength of such a bond should equal or exceed the cohesive strength of the weaker of the two material involved.

Adhesion can be regarded as a measure of the degree to which a bond has been developed from place to place at the coating-substrate interface. For example, if only half of an interface area is clean so that a bond can be established, it can be said that 50% adhesion was achieved. Generally, adhesion is tested by pulling off the coating with pliers. The test should be performed with production items rather than test specimens.

Good adhesion is of no value unless the bond strength is satisfactory. The Ollard test for bonding is illustrated in Figure 30. It consists of plating a metal cylinder to a thickness of about 2.5 mm and then machining the deposit so as to leave a shoulder that is supported by a die. A load is applied to separate the deposit from the substrate and is indicative of the bond strength. Roehl [41] refined Ollard's method and extensively studied nickel on steel. His work and other investigations indicated that the bond strength equals or exceeds the cohesive strength of the weaker of the two involved metals.

Fracture rarely occurs at the interface if a good plating technique was employed. If the load is applied perpendicular to the deposit-substrate interface, the value of the bond strength is more quantitative. To apply the load in this way, a nodule shaped by an electroforming technique is deposited on the substrate and subsequently pulled

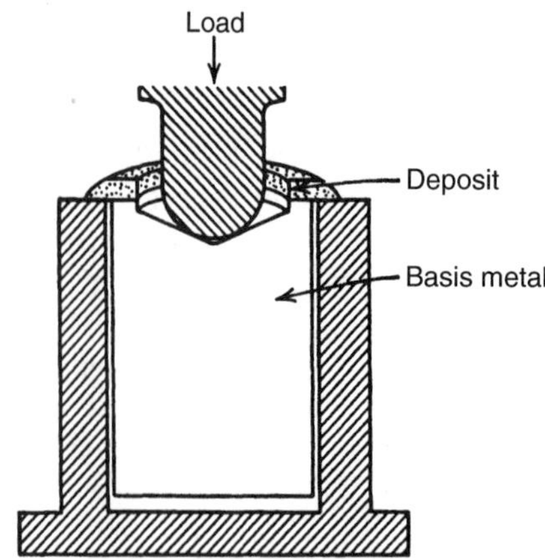

Figure 30 Ollard test specimen in testing fixture.

off. The bond strength is then the load needed to pull off the nodule divided by its area.

The main factors that are detrimental to adhesion are brittle layers, which can form by diffusion between the deposit and substrate. Brittle layers form especially if annealing is involved. An example is copper plated on zinc die castings which forms blisters during the baking of enamels applied over the coating.

13 MECHANICAL PROPERTIES

The relevant mechanical properties are the modulus of elasticity, the yield, tensile and fatigue strengths, and the ductility. They are illustrated in Figure 31 [42] which gives the stress-strain curve of a ductile material. The slope in the elastic portion is the modulus of elasticity. The yield strength is the stress to produce a small amount of plastic deformation called the *offset*, which in Figure 31 is 0.2%. The tensile strength is the maximum stress. The ductility is the strain to fracture. The fatigure strength cannot be determined from a stress-strain curve. The mechanical properties are of particular importance for electroforms and printed circuits.

The two methods for determining the mechanical properties are the tensile and the bulge tests. Tensile testing consists of applying a load continuously to a so-called dog-bone-shaped specimen until it fractures. The normally used tensile specimens are unsuitable for most electrodeposits. The two-inch-wide reduced section of the standard tensile specimen has a width-to-thickness ratio that results in a nonuniform strain and therefore low values of the ductility [43].

Using a width-to-thickness ratio typical of the wiring of printed circuits and then scaling the other dimensions results in a tiny tensile specimen. Several machines to test such small specimens are in use [44]. The bulge test consists of clamping a foil over an orifice and introducing a fluid to cause a hemispherical bulge. A stress-strain curve can be constructed using measurements of the fluid pressure and the bulge height. The wiring of printed circuits, which are thermally cycled, can fail due to stresses that develop because of differences in the coefficients of thermal expansion

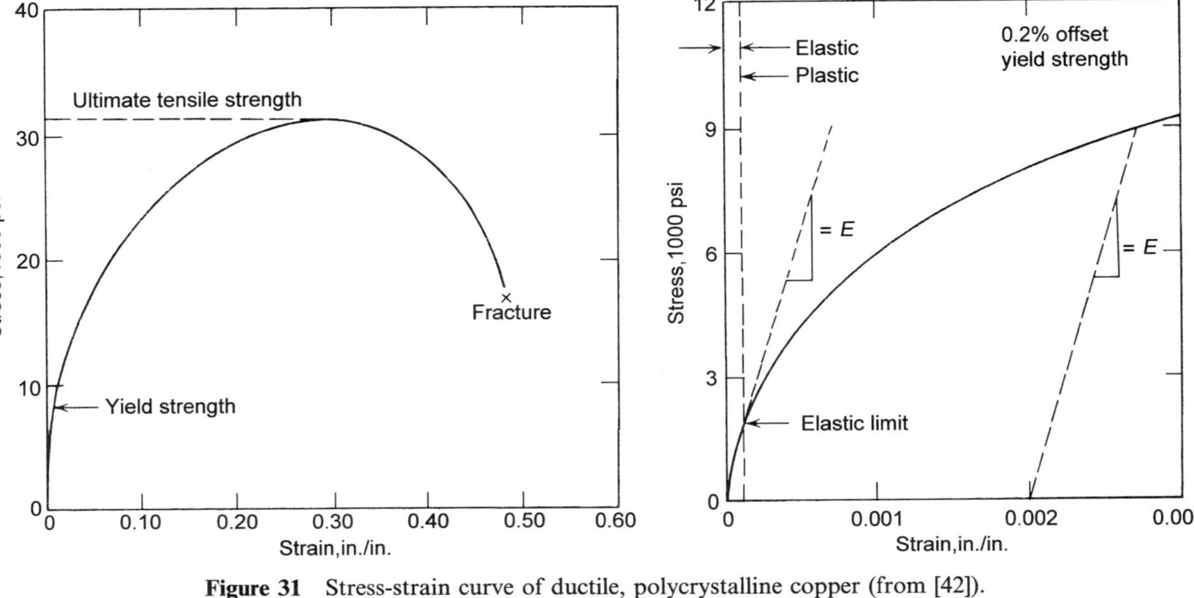

Figure 31 Stress-strain curve of ductile, polycrystalline copper (from [42]).

between deposit and substrate. Thermal-cycling tests are used to test for such failures.

The moduli of elasticity of electrodeposits are generally smaller than those of the same metal formed in other ways. The moduli should all be the same. Possible reasons for the smaller values are the difficulty of obtaining accurate values and the possibility that deposits do not behave elastically.

The primary sources of strength are impediments to the movement of dislocations. In electrodeposits the main impediments are grain boundaries. Addition agents strengthen deposits primarily by refining the grains. Codeposited foreign substances can also increase the dislocation density. Dislocations can hinder the movement of others and thus strengthen materials.

Ductility is the amount of plastic deformation that can occur prior to fracture. Some deposits are inherently brittle because they contain cracks that readily lead to fracture. The cracks can be caused by the internal stresses, which will be subsequently discussed, and by stress corrosion.

Fine-grained deposits tend to be brittle because plastic deformation which occurs primarily by dislocation motion is impeded by the grain boundaries. Some deposits appear to be brittle but are really quite ductile. The reason for this behavior is a phenomenom called *necking*. Here the plastic deformation prior to fracture occurs in a very small volume. The overall plastic deformation is then quite small and thus indicative of poor ductility. However, the ductility, as indicated by the reduction in the cross-sectional area prior to fracture, may be quite good. When the deposit adheres well to the substrate, necking is essentially prevented.

Annealing of electrodeposits is mostly a low-temperature heat treatments to remove hydrogen, which can be a source of embrittlement. Electroforms may be heat treated so as to cause recrystallization. The recrystallization temperature of electrodeposits is generally lower than that of the same wrought metals. Some fine-grained copper electrodeposits have been observed by the author to recrystallize at ambient temperatures.

Precipitation hardening is possible for some supersaturated solid solution. Electroless nickel is an example of such an alloy that can be heat treated to cause the precipitation of Ni_3P or Ni_3B and thereby substantially increase the hardness and wear resistance [32].

Electrodeposition is well suited for the production of composites. The incorporation of hard material such as diamonds or ceramic particles can markedly improve the wear resistance. Composites consisting of thin alternate layers of two different materials can possess better strength properties than those of the individual ones. Composites have been produced consisting of alternate layers of two different metals such as nickel and copper, layers of alloys of different composition, or layers of different phases such as alpha and beta brass. These composites have been produced in one solution by alternately changing the plating conditions rather than by plating in two different solutions, which is impractical.

14 MAGNETIC PROPERTIES

There are applications of electrodeposits where ferromagnetic properties, the only ones discussed here, are important. The magnetic properties are illustrated in terms of the hysteresis loop shown in Figure 32 which gives a graph of the magnetic field on the abscissa and the magnitization on the ordinate. When a magnetic field is imposed on a material, a magnetization results. The remanences $+\mathbf{B}_r$ and $-\mathbf{B}_r$ are the magnitizations remaining after the field has been removed. Thus a magnetic element used for storing information is either in the positive or the negative remanence state, which constitutes its memory.

In digital computers, for example, the two states represent 0 and 1. The coercive force $\mathbf{H}_r$ is the field required to maintain zero magnetization and therefore is the force needed to switch from $+\mathbf{B}_r$ to $-\mathbf{B}_r$, and vice versa. The saturation magnetization $\mathbf{B}_m$ is a measure of the total available magnetization. The squareness of the hysteresis loop is defined as the ratio $\mathbf{B}_r/\mathbf{B}_m$.

The principal structural element in magnetism is the domain. It is a small region that has a spontaneous magnetization. The sizes of domains vary greatly. For information storage, the domains should be as small as possible; their size is then limited by the width of their walls, which are less than one micrometer.

At the other extreme, a whole magnet can be one domain. If the magnetization directions of all domains are randomly oriented, there is no net bulk magnetization. The effect of an imposed field is to align domains along its direction.

Domains already aligned grow at the expense of less favorably oriented ones by moving their walls. At the saturation magnetization, the only existing domains are those oriented in the proper direction. Since domains do not completely revert to the random orientation when the field is removed, there is remanence. The coercive force is that required to cause a random oriention of the magnetization directions of the various domains. The magnitude of the magnetization decreases drastically when the material is heated to the Curie temperature.

Magnetic materials are classified as hard or soft depending on the value of the coercive force. Materials with coercive forces exceeding 200 *oersteds* are considered magnetically hard. They are used, for example, in disc elements for information storage. Electroless Co–P (cobalt-phosphorus) alloys are magnetically hard materials. When it is desired that relatively low fields are capable of switching from one remanence state to another, for example, as in plated-wire memory elements, magnetically soft materials are used. Examples of magnetically soft material are Fe–Ni (iron-nickel) alloys. A high remanence and a square hysteresis loop are desirable for both types of materials.

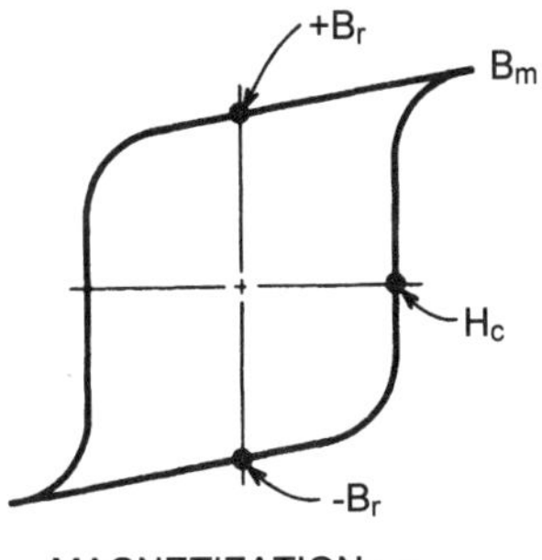

MAGNETIZATION →

Figure 32 Hysteresis loop.

The microstructure of electrodeposited materials affects the magnetic properties in several ways. The previously discussed defects, namely dislocations, grain and twin boundaries, and some codeposited materials, make it more difficult for domain walls to move, and this results in higher coercive forces. The ease of moving domains also depends on the crystallographic orientations. Since each magnetic material possesses a particular oriention in which it is most easily magnetized, the texture is in this case of practical importance.

When a material is magnetized, its shape and volume change. The change, which is called *magnetostriction*, is due to a strain resulting from the magnitization. Magnetostriction can also be a change in the magnitization due to strains. When deposition occurs in a magnetic field, the structure, texture, and throwing power of both magnetic and nonmagnetic materials can be affected.

15 INTERNAL STRESS

Electrodeposits are often laid down in a state of stress. Such a stress is named internal, or residual, because all or part of it remains in the deposit. Electroplaters are familiar with one type, *macrostress* as it manifests itself in the bending of parts plated on one side. If one end of a strip is clamped and the other one bends toward where the anode was located, the macrostress is tensile. If the strip bends in the opposite way the macrostress is compressive. Macrostresses can cause distortions, cracking of the deposit, loss of adhesion to the substrate and increased corrosion. Tensile macrostresses can also deteriorate the fatigue properties. The other type of stress, *microstress*, manifests itself primarily in an increase in the hardness.

The most commonly commercially used macrostress-measuring device is the spiral contractometer. The problem with devices such as the contractometer in which the specimen is allowed to deform is that some of the stress is relieved. There are devices, used in research in which an opposing force is applied so as not to allow the specimen to deform. Microstresses can be determined only from the broadening of X-ray diffraction lines. X-ray diffraction can also be used to measure macrostresses. However, it usually does not yield accurate values for electrodeposits because of the broadening of the diffraction lines due to microstresses and their assymmetry due to twinning.

The structural causes of internal macrostresses are not fully understood. There is evidence that the coalescence of grains or parts of them growing laterally from different nucleation centers is a cause. The stress fields around oriented arrays of dislocation produced by coalescence or other growth processes can add in a way to result in macrostresses. The sign of the macrostress depends on the orientation of the array. The codeposition of hydrogen has been identified as a cause of macrostress.

If hydrogen diffuses out of a layer of the deposit allowing it to shrink, a tensile macrostress can develop. A tensile macrostress can also develop if hydrogen diffuses onto the substrate or previously laid-down layers causing them to expand. If micropores form during deposition and hydrogen then diffuses into them causing them to expand, a compressive macrostress can be caused.

ACKNOWLEDGMENTS

Figure 32 was taken by Mr. Tseng-Ming Chou of Stevens Institute of Technolgy. Figure 35 was supplied by Professor Dale P. Barkey of the University of New Hampshire; it previously appeared in Ref. [3].

GLOSSARY

alloy A body of material composed of two or more elements of which at least one is a metal

anvil effect An error in hardness measurements resulting when the depth of the indent is less than 14 times the thickness

backscattered electrons Electrons scattered by the near-surface region into the detector of a scanning electron microscope

body-centered cubic (bcc) A crystal arrangement of atoms or atom groups located at the corners of a cube and in its center

coalescence The joining of groups of atoms often caused by surface tension

coercive force Magnetic field required to reduce magnetization to zero

cohesive strength The force required on a bulk basis to break the atomic bonds in the body of a material

crystal A body of material having an orderly, repetitious arrangement of atoms that may or may not be indicated by the external shape

crystal habit Type of atomic arrangement in a crystal

curie temperature Temperature above which a material is no longer ferromagnetic

dislocation Line defect in crystals which can be visualized as being the boundary between the portion of a plane which has slipped and which has not

dog bone Shape of a tensile specimen having a narrower central portion and wider flanges

domain A region in a ferromagnetic material in which all magnetic moments are aligned

ductility The strain which a body can undergo before fracturing

electroless Autocatalytic deposition without an externally applied current

epitaxy Induced continuation of the structure of the substrate into the deposit

face-centered cubic (fcc) Crystal arrangement of atoms or atom groups at the corners of a cube and in the centers of the faces

fatigue Tensile stress on a body which varies cyclically with time to a lower tensile value or a compressive one

fiber axis Texture observed in wires and electrodeposits in which there is only a preferentially oriented direction

filar eyepiece Eyepiece of an optical microscope containing a cross hair that can be moved by a screw for making measurements

grain An individual crystal in a polycrystalline body of material

hexagonal close packed (hcp) A crystal arrangement wherein planes of atoms located at the corners of hexagons are separated by planes of atoms grouped in sets of three between hexagons in adjacent planes. The atomic arrangement on the basal plane is the same as that of the cube diagonal of fcc

hysteresis loop A graph of magnetic field versus magnetization

macrostress Internal stress either tensile or compressive that extends with the same sign over the entire part

magnetostriction Strain resulting from externally applied magnetic fields or a change in magnetization resulting from changes in the macrostress

microstress Internal stress that changes from tensile to compressive over microscopic regions

modulus of elasticity Linear ratio of stress to strain in the initial portion of a stress-strain diagram

necking A condition when most of the strain is concentrated in a small portion of a specimen

phase A mechanically separable, homogeneous portion of a heterogeneous system

preferred orientation An arrangement of grains wherein all or many grains are oriented so that in them some crystallographic feature bears a constant directional relationship to some reference feature such as a sheet surface or a wire axis

recrystallization The formation, usually by heat treatment, of new grains from strained or extremely small ones

remanence Magnetization remaining after the field is reduced to zero

resolution The smallest detail that a magnifying instrument can reveal in an object

reduction in area The strain calculated from the ratio of the change in the cross-sectional areas to the origional one

secondary electrons Electrons produced in the specimen by electron bombardment in the scanning electron microscope. They are less energetic than the back-scattered ones and are used to reveal the topography

slip The sliding within a crystal of a block of atoms over another one by some integral number of interatomic distances

strain Deformation resulting from the application of a stress

stress Load or deforming force per unit area

stress, internal Stress in a part not caused by externally applied loads

tensile strength The maximum stress that a body can support without breaking without regard to the deformation that occurs

texture The presence of preferred orientations

topography Relief features of an area

tunneling Flow of an electrical current between bodies separated by only a few angstroms

twin Portion of a grain wherein the atomic arragement is the mirror image of the surroundings

yield strength Stress that causes a very small amount of plastic deformation, usually 0.02%

REFERENCES

1. M. Paunovic and M. Schlesinger, *Fundamentals of Electrochemical Deposition*, Wiley, New York, 1998.
2. A. J. Bard and L. R. Faulkner, *Electrochemical Methods*, Wiley, New York, 1980.
3. E. Mattsson and J. O'M. Bockris, *Trans. Faraday Soc.*, **55**, 1586 (1959).
4. J. O'M. Bockris and A. K. N. Reddy, *Modern Electrchemistry*, vol. 1, Plenum Press, New York, 1998.
5. W. Lorenz, *Z. Phys. Chem.*, **202B**, 275 (1953).
6. W. Lorenz, *Z. Elektrochem.*, **57**, 382 (1953).
7. W. Lorenz, *Naturwissenschaften*, **40**, 576 (1953).
8. W. J. Lorenz, *Z. Naturforsch*, **9a**, 716 (1954).
9. W. Mehl and J. O'M. Bockris, *J. Chem. Phys.*, **27**, 817 (1957).
10. W. Mehl and J. O'M. Bockris, *Can. J. Chem.*, **37**, 190 (1959).
11. B. E. Conway and J. O'M. Bockris, *Proc. Roy. Soc. London*, **A248**, 394 (1958).
12. B. E. Conway and J. O'M. Bockris, *Electrochim. Acta*, **3**, 340 (1961).
13. G. H. Gilmer and P. Bennema, *J. Appl. Phys.*, **43**, 1347 (1972).
14. N. Ibl, *Surf. Technol.*, **10**, 81 (1980).
15. A. M. Pesco and H. Y. Cheh, in *Modern Aspects of Electrochemistry*, no. 19, B. E. Conway, J. O'M. Bockris, and R. E. White, eds., Plenum Press, New York, 1989.
16. C. H. Ting, V. Dubin, and R. Cheung, in *Fundamental Aspects of Electrochemical Deposition and Dissolution Including Modeling*, M. Paunovic, M. Datta, M. Matlosz, T. Osaka, and J. B. Talbot, eds., *Proceedings*, vol. 97-27, Electrochemical Society, Pennington, NJ, 1997, p. 321.
17. E. Budevski, G. Staikov, and W. J. Lorenz, *Electrochemical Phase Formation and Growth*, VCH Publishers, New York, 1996.
18. D. J. Srolovitz, A. Mazor, and G. G. Bukiet, *J. Vac. Sci. Technol.*, **A6**, 2371 (1988).
19. A. Damjanovic, M. Paunovic, and J. O'M. Bockris, *J. Electroanal. Chem.*, **9**, 93 (1965).
20. J. O. Dukovic and C. Tobias, *J. Electrochem. Soc.*, **137**, 3748 (1990).
21. C. Madore and D. Landolt, *J. Electrochem. Soc.*, **143**, 3936 (1996).
22. H. Leidheiser Jr., *Z. Elektrochemie*, **59**, 756 (1955).
23. O. Kardos and D. G. Foulke, in *Advances in Electrochemistry and Electrochemical Engineering*, vol. 2, C. W. Tobia, ed., Wiley-Interscience Publishers, New York, 1962.
24. R. Weil and R. Paquin, *J. Electrochem. Soc.*, **107**, 87 (1960).
25. L. Oniciu and J. Muresan, *J. Appl. Electrochem.*, **21**, 565 (1991).
26. G. T. Rogers and K. J. Taylor, *Electrochim. Acta*, **8**, 887 (1963); **11**, 1685 (1966); **13**, 109 (1968).
27. M. Paunovic, *Plating*, **55**, 1161 (1968).
28. M. Saito, *J. Met. Fin. Soc. Jpn.*, **17**, 14 (1966).
29. M. Paunovic, in *Electrochemistry in Transition*, O. J. Murphy, S. Srinivasan, and B. E. Conway, eds., Plenum Press, New York, 1992, p. 479.
30. A. Brenner, Ph. D. thesis University of Maryland, 1939.
31. J. Yahalom and O. Zadok, *J. Materials Sci.*, **22**, 499 (1987).
32. R. Weil and K. Parker, in *Electroless Plating*, G. O. Mallory and J. B. Hajdu, eds., American Electroplaters and Surface Finishers Society, Orlando, FL, 1990, pp. 111–138.
33. L. Albert, R. Schneider, and H. Fischer, *Z. Naturwiss.*, **19**, 1120 (1964).

34. Q. Wu and D. Barkey, *J. Electrochem. Soc.*, **144**, L261–L262 (1997).
35. C. J. Weber, H. W. Pickering, and K. G. Weil, *J. Electrochem. Soc.*, **144**, 2364–2369 (1997).
36. A. J. Bard, F.-R. F. Fan, J. Kwak, and O. Lev, *Anal. Chem.*, **6**, 132 (1989).
37. R. K. Parker, S. N. Sahu, and S. Chandra, *Handbook of Semiconductor Electrodeposition*, Marcel Dekker, New York, 1996.
38. C. Sheu and R. Weil, *J. Electrochem. Soc.*, **137**, 2052–2055 (1990).
39. E. C. Felder, S. Nakahara, and R. Weil, *Thin Solid Films*, **84**, 197–203 (1981).
40. C. C. Nee and R. Weil, *Surf. Technol.*, **25**, 7–15 (1985).
41. E. J. Roehl, *Iron Age*, **146** (13), 17 (Sept. 26 1940); **146** (14), 30 (Oct. 3 1940).
42. H. W. Hayden, W. G. Moffatt, and J. Wulff, *The Structure and Properties of Materials*, vol. 3, Wiley, New York, 1965, 4.
43. K. Lin, I. Kim, and R. Weil, *Plating Surf. Finish.*, **75**, 52–56 (1988).
44. I. Kim and R. Weil, *Testing of Metallic and Inorganic Coatings*, **ASTM STP 947**, W. B. Harding and G. DiBari, eds., American Society for Testing and Materials, Philadelphia, 1987, pp. 11–18.

2 Electrodeposition of Copper

JACK W. DINI

INTRODUCTION

Copper is the most common metal plated, exclusive of continuous strip plating and nickel [1]. Major uses of electroplated copper are plating on plastics, printed wiring boards, zinc die castings, automotive bumpers, rotogravure rolls, electrorefining, and electroforming [2]. Electroplated copper is playing a major role in the change from aluminum to copper in semiconductor interconnect technology. This materials change has been heralded as a "major breakthrough" in *The New York Times* (September 22, 1997) and a "dazzling technical advance" by *Time* magazine (October 10, 1997) [2a]. It signals one of the most important changes in materials that the semiconductor industry has experienced since its creation [2b]. Copper is electrodeposited for numerous engineering and decorative applications requiring a wide range of mechanical and physical properties. The range extends from properties superior to full-hard wrought copper to properties equivalent to annealed pure copper [3].

Copper is an excellent choice for an underplate, since it often covers minor imperfections in the base metal. It is relatively inert in most plating solutions of other common metals, it has a very high plating efficiency, resulting in excellent coverage even on difficult to plate parts, and lastly, it is highly conductive, making it an excellent coating for printed wiring boards or as a coating on steel wire used to conduct electricity [1].

Copper deposits also act as thermal expansion barriers by absorbing the stress produced when metals with different thermal expansion coefficients undergo temperature changes, and this is particularly helpful with plastic substrates. The leveling and brightness properties of copper deposits can be further enhanced by buffing, and since copper is much softer than steel or nickel, it is easy and relatively inexpensive to buff [4].

Of the plating systems that have been studied, only relatively few have revealed a stage of commercial importance for electrodeposition of copper. These are the alkaline cyanide and pyrophosphate complex ion systems and the acid sulfate and fluoborate simple ion systems. Other types of solutions have been too unstable or lacking in good deposit characteristics over sufficiently wide current density ranges [5]. In recent years some alkaline noncyanide systems have been developed for replacing cyanide.

Modern Electroplating, Fourth Edition, Edited by Mordechay Schlesinger and Milan Paunovic.
ISBN 0-471-16824-6

The areas of application of the various copper plating solutions overlap somewhat, but each has its fairly well-defined area of usefulness. Clearly, the most heavily used solutions are the acid copper sulfate. Open literature publications and patents for sulfate solutions since the previous edition of this book in 1974 [6] far outnumber those of all the other solutions combined. Deposits produced in cyanide solutions are typically thin (<12.5 μm) and are used as an undercoating for nickel and chromium, as a heat treatment stop-off for selective hardening of ferrous parts, or as an intermediate step prior to additional plating. For example, a copper cyanide deposit is typically a key part of the activation cycle for preparing aluminum, beryllium, and zinc die castings for plating.

Copper deposits from cyanide solutions are not generally suitable for deposition of relatively thick deposits for electroforming and similar applications. Cyanide solutions are finding less and less favor because of their toxicity and waste treatment problems and are being replaced by noncyanide solutions. Pyrophosphate solutions, once used heavily for plating through holes on printed wiring boards have been almost completely replaced by high-throw acid sulfate solutions. Fluoborate solutions have been advertised for many years as having the capability to deposit copper at very high current densities. However, present commercial usage of this type of solution is minimal simply because other solutions such as those with sulfate ions can do the same job and are less expensive, easier to control and less susceptible to impurities. The chemical cost of acid fluoborate electrolyte is approximately twice that of acid sulfate per gallon, which is one strong reason why fluoborate copper has not gained a significant share of the through hole plating market [7]. Continuous copper plating from fluoborate solutions on electroless coated plastic circuits has been reported [8].

Part A
Acid Copper*

1 HISTORY AND DEVELOPMENT

Acid copper deposition was referred to as early as 1810 [9]. In 1831 Bessemer [10] copper-plated steel castings of frogs, insects, and plants by immersion in copper sulfate solution on a zinc tray. In 1836 the Daniel cell was first used for electrodepositing copper [11] and a technical report was published by De la Rue [12]. In 1839 Jacobi made Russian bank notes using copper-plated electrotypes, and in 1840 he received the first patent for making electrotypes [13]. Smee discussed commercial copper-plating processes in 1843 [9]. During the next 70 years, progress was directed principally toward developing specific applications for acid copper electrodeposition. Most of the efforts dealt with copper^{2+} sulfate-sulfuric acid solutions, but oxalate [14], nitrate [15], acetate [15], fluosilicate [16, 17], and copper $^{1+}$ chloride [16] solutions were also investigated.

In more recent times the following solutions have been evaluated: sulfate-oxalate-boric acid [18], sulfate-oxalate [19, 20], copper^{1+} chloride [21], copper^{1+} chloride-sodium thiosulfate [22], benzene disulfonic acid [23], copper^{1+} iodide and bromide [24], iodide and chloride [25], fluoborate [26, 27], alkane sulfonic acid [28, 29], sulfamic acid [30, 31], copper^{2+} formate with ammonium salts [32], phosphate-sulfate [33], fluosilicate [34], fluosilicate-silicic acid [35], copper^{2+} glycolate, lactate, malate, and tartrate [36].

At the present time only the sulfate and fluoborate solutions are commercially used. Passal reviewed the first 50 years of AES (American Electroplaters' Society) copper-plating history in 1959 [5] and Van Tilburg covered the 75 year history in 1984 [11]. Other reviews of acid copper plating can be found in refs. [6] and [37–41].

2 APPLICATIONS

Electrodeposition of copper from acid solutions is extensively used for electroforming, electrorefining, and electroplating. Refiners and electroformers, in particular, employ acid solutions because costs of chemicals and power are low and because the solutions are simple and easy to control. In the electrowinning and electrorefining industries, acid solutions are employed exclusively. More than 80% of the domestic production of primary copper is refined electrolytically.

Acid copper sulfate solutions are widely used for plating of printed wiring boards and for semiconductor interconnect technology. Electroformed copper articles include band instruments, heat exchangers, reflectors, and a variety of articles for military and aerospace applications. All three main types of printing processes

*Revision of section by W. H. Safranek in the 1974 edition.

(electrotyping, rotogravure work, and lithography) use copper, and sometimes nickel and chromium [42].

Acid copper solutions containing organic brightening and leveling agents are used extensively to deposit smooth copper on rough steel and etched plastics. Zinc die castings are plated with approximately 15 μm of leveling acid copper [43, 44], before nickel and chromium plating, to eliminate buffing before plating. Because of the excellent microthrowing power of the acid copper, pits, pores, or crevices in either steel or zinc surfaces are well filled with copper, and this improves resistance to corrosion or blistering [45–47]. Acid copper deposits are one part of the sequence for decorative plating of aluminum wheels for automotive applications. In some cases the copper is buffed to add luster to the low current density areas of the wheel and to flow some of the copper over small pits and voids [48]. Coatings on plastics need to be bright and ductile and have the capability to expand and contract with the thermal expansion of the plastic without cracking, blistering, or peeling. Bright decorative acid copper deposits meet these requirements [4, 49].

One of the most important steps in the production of plated wire for memory use is copper plating in acid sulfate solution [50]. Many kilometers of steel wire are given a copper cyanide strike and plated with copper in acid solutions to produce a high strength electrical cable. Thick deposits (200 μm) of copper are applied to steel rolls and then engraved for use in printing and marking papers and textiles. Stainless steel cooking vessels are copper plated in acid solutions to improve the heat diffusion characteristics of outer surfaces and avoid local hot spots. Copper plating for stopping off carburizing on selected areas is accomplished by striking in a cyanide solution followed by plating in acid solutions. Acid copper plating is sometimes used for building up worn or overmachined parts, especially when copper surfaces are desired for protection against fretting corrosion. Optical surfaces can be produced on parts by single point diamond turning specific acid copper deposits [51–54].

Metal powders produced by deposition in acid solutions are used for making sintered compacts and pigments. The powder is deposited from dilute solutions at high temperatures, brushed off the cathodes, filtered, washed, ground into fine particles, screened, and blended for use in manufacturing powder compacts [55, 56]. Copper sulfate solutions have also been recommended for plating "low-density" powder compacts, to fill porosity near the surface [57].

3 PRINCIPLES

The copper^{2+} salts in either the sulfate or the fluoborate solution are highly ionized except for small amounts of less ionized complex salts formed with certain addition agents. The addition of sulfuric acid to the sulfate solution, or fluoboric acid to the fluoborate solution, is necessary for obtaining acceptable deposits. Because of the high conductivity of commercial solutions and because anode and cathode polarizations are small, voltages required for depositing copper are less for acid than for alkaline solutions. Electrorefining plants employ the copper sulfate solution largely for this reason. Tank voltages for refining copper are frequently as low as 0.2 V for a cathode and anode current density of 1.6 to 2.2 $\mathrm{A\,dm^{-2}}$.

Anode and cathode polarizations are nearly negligible in purified solutions used at low current densities. Even at the high cathode current density of 21.5 $\mathrm{A\,dm^{-2}}$, a 6 V current source is ample when the solution is efficiently agitated. Excessive polarization of the anodes in the sulfate solution may occur when the anode current density exceeds about 5 $A\,dm^{-2}$. With the fluoborate solution, the anode current density can be at least 40 $A\,dm^{-2}$ without encountering excessive anode polarization. Anode and cathode efficiencies are nearly 100% at all practical current densities. The rate of deposition obtainable depends chiefly on the efficiency of agitation in preventing excessive polarization. Acceptable deposits were reported at a current density as high as 260 $A\,dm^{-2}$, equivalent to 3.63 $mm\,h^{-1}$ [58], with violent agitation.

The first stages of the formation of a copper deposit depend on the deposition rate, the substrate surface nature, and the deposition technique [59]. The final stage in the growth involves an equilibrium of copper electrochemically dissolving and precipitating [60]. The character of copper deposits is influenced by the concentrations of copper salts, additives, free acid, temperature, cathode current density, and the nature and degree of agitation. At potentials between −60 and −30 mV, the growth of bulk copper proceeds in cycles of nucleation, agglomeration, and crystallization.

The concentration profile of copper in an acid electrolyte has been measured in an attempt to understand the effects of the supporting electrolyte on the rate of mass transfer of Cu^{2+} ion toward the cathode surface [61]. Overpotential relaxation experiments have shown that calculated mass-transfer boundary layer thicknesses increased with increasing electrode length and solution viscosity and decreased with increasing current density [62]. Impedance behavior of a copper cathode in an acid copper electrolyte has also been studied [63].

4 FUNCTIONS OF SOLUTION CONSTITUENTS

4.1 Copper and Sulfuric Acid

Copper^{2+} sulfate ($CuSO_4 \cdot 5H_2O$) and sulfuric acid, or copper^{2+} fluoborate [$Cu(BF_4)_2$] and fluoboric acid, are the primary constituents of the sulfate and fluoborate solutions, respectively. The copper salts furnish the metal ions in solutions such as those given in Table 1, which contains formulations for conventional and high-throw solutions. The latter are used for plating printed wiring boards and are discussed in detail in a subsequent section. Copper can be deposited at very low cathode current densities from the acid free aqueous solutions of the salts [64, 65], but at higher current densities the deposits from the sulfate solution are spongy and contain occluded salts. Plate characteristics are improved, solution conductivity is increased, and anode and cathode polarizations are greatly reduced when free acid is added to either solution [26, 66]. The acid also prevents the precipitation of basic salts.

The concentration of copper sulfate is not particularly critical, although the resistivity of the solution is greater when the concentration is increased [67]. Cathode polarization increases slightly at copper sulfate concentrations above 1 M (250 $g\,liter^{-1}$) [68]. A concentration of less than 60 $g\,liter^{-1}$ copper sulfate results in a decreased cathode efficiency [69]. Changes in the concentration of copper sulfate have little effect on grain size, but some grain refinement occurs as a result of increasing the sulfuric acid concentration to 1.5 N (72 $g\,liter^{-1}$) [70]. When very high

TABLE 1 Formulations of Acid Copper Solutions

Copper Sulfate Solutions	Conventional Solutions	High-Throw Solutions
Copper sulfate, $CuSO_4 \cdot 5H_2O$, g liter^{-1}	200–250	60–100
Sulfuric acid, H_2SO_4, g liter^{-1}	45–90	180–270
Chloride, mg liter^{-1}	—	50–100
Copper Fluoborate Solutions	**Low-Concentration Solutions**	**High-Concentration Solutions**
Copper fluoborate, $Cu(BF_4)_2$, g liter^{-1}	225	450
Fluoboric acid, HBF_4, g liter^{-1}	15	30
Boric acid, H_3BO_3, g liter^{-1}	15	30

cathode current densities are used, a high concentration of copper sulfate, within the limits given in Table 1, is recommended. The solubility of copper sulfate is decreased when the sulfuric acid concentration is increased [66].

Changes in sulfuric acid concentration have more influence than changes in copper sulfate concentration on anode and cathode polarization, and on solution conductivity. Cathode polarization decreases when a small amount of sulfuric acid is added to a solution of copper sulfate, reaches a minimum at about 0.4 M (38 g liter^{-1}), and increases with a further increase in sulfuric acid concentration [68]. Throwing power increases dramatically in low copper (15 g liter^{-1}), high sulfuric acid solutions [71]. Excess sulfuric acid drastically increases the cathodic overpotential and introduces a smaller ratio of level plane of electrodeposits resulting in nodular precipitates [72]. It also drastically changes the X-ray diffraction pattern by increasing the surface overpotential and inducing three-dimensional nucleation at the higher overpotential [73]. Specific conductivity is nearly doubled when the concentration of sulfuric acid is raised from 50 to 100 g liter^{-1} [74]. An additional increase in free acid concentration to 200 g liter^{-1}, which is the level widely used in electrorefining, reduces the resistivity from 4.2–4.3 to 1.6–1.9 microhm-cm [75].

4.2 Chloride

Chloride ion, in bright and high-throw acid sulfate solutions, reduces anode polarization [76] and eliminates striated deposits in high current density areas [40]. It is important to control the chloride ion at 60 to 80 ppm. Below 30 ppm, deposits will be dull, striated, coarse, and step-plated. Above 120 ppm, deposits will be coarse grained and dull, and the anodes will polarize, causing plating to stop [41]. Cl^- affects the surface appearance, structure, microhardness, crystallographic orientation, and internal stress of the deposits [77–79]. A minimal amount of chloride was essential to the ductility of deposits from two different proprietary copper-plating solutions. The elongation in each case was found to rise dramatically for chloride additions in the 10 mg liter^{-1} range [80].

Among the halides, Cl^- is the most effective over a wide range of concentrations (40 to 150 mg liter^{-1}) in keeping stress to a null value, and this is particularly important considering the synergistic effect of Cl^- over some brightening additives normally used [81]. The presence of about 50 mg liter^{-1} chloride is optimum for permitting an increase in microhardness without raising internal stress [82]. Cl^- exerts no influence on throwing power [82, 83].

In the presence of Cl^- and additives, such as thiourea or a commercial proprietary brightener, it is suggested that in addition to CuCl surface films, adsorbed Cu(I) or Cu(I) complex-Cl bridge films inhibit surface diffusion of adsorbed Cu atoms, and this becomes the rate determining step in the deposition mechanism [84–86]. With solutions containing polyethylene glycols (PEG), the absence, or reduced quantities of chloride alters the suppression effect of PEG, adversely affecting deposit morphology. In the presence of chloride, PEG fractions inhibited currents much more strongly than in the presence of PEG alone [87].

Small amounts of chloride ion are known to have an accelerating effect on the deposition of copper. This supports the hypothesis that chloride ions act as binding sites for surfactants such as polyethylene glycol to the electrode surface [88–90]. Excess chloride can produce insoluble copper chlorides at the anode surface, hindering the deposition process [91]. Although chloride ion is adsorbed on depositing copper, up to 10^{-3} M chloride ion has little effect on the current-voltage curve for acid sulfate solutions containing no brighteners or additives [92].

4.3 Fluoborate

Copper fluoborate is more soluble than copper sulfate. Metal ion concentration can be more than double in the fluoborate solution in comparison with a copper sulfate solution containing 50 to 75 g liter^{-1} of sulfuric acid. If the acid concentration in the fluoborate solution is too low (pH more than 1.7), deposits are dull, dark, and brittle. Boric acid is added to stabilize the solution and prevent decomposition of copper fluoborate; this increases resistivity slightly. In a solution with a concentration of more than 15 g liter^{-1} fluoboric acid or 220 g liter^{-1} copper fluoborate, an increase in the concentration of either the salt or the acid lowers the resistivity.

5 ADDITION AGENTS

Addition agents for brightening, hardening, grain refining, surface smoothing, increasing the limiting current density, and reducing trees are frequently added to acid copper sulfate solutions. An extensive list of additives used in acid copper plating prior to 1959 can be found in [5] and in the acid copper chapter in the previous edition of this book [6]. Additives covered in patents granted in recent years are listed in Table 2. Materials that have been reported in recent technical literature publications include benzotriazole [84, 85, 93, 94], cadmium [95], casein [96], cobalt [97], dextrin [96], dimethylamino derivatives [98], disulphides [98, 99], 1,8-disulphonic acid [93], disodic 3,3-dithiobispropanesulphonate [100], 4,5-dithiaoctane-1,8 disulphonic acid [93], dithiothreitol [101], ethylene oxide [100], gelatin [102], glue [76, 96], gulac [76], lactose benzoylhydrazone [103], 2-mercaptoethanol [104], molasses [96], sulphonated petroleum [76], *o*-phenanthroline [95, 105], polyethoxy ether [106], poly-

TABLE 2 Additives that have been Patented for Copper Plating

Patent Reference	Date of Issue	Investigator	Additives
		Acid Copper Sulfate	
U.S.5,433,840	7/18/95	Dahms et al.	Polyalkylene glycol
U.S.5,431,803	7/11/95	DiFranco et al.	Animal glue
U.S.5,417,841	5/23/95	Frisby	Alkoxythio and sulfonated compounds
U.S.5,403,465	4/4/95	Apperson et al.	Animal glue and an impurity
U.S.5,328,589	7/12/94	Martin	Polymers comprising ether groups
U.S.5,252,196	10/12/93	Sonnenberg et al.	Numerous additives
U.S.5,215,645	6/1/93	DiFranco et al.	High-protein polymers of amino acids
U.S.5,174,886	12/29/92	King et al.	Polyalkylene glycols
U.S.5,151,170	8/29/92	Montgomery et al.	Peroxide oxidation; product of a dialkylaminothioxomethyl-thioalkane sulfonic acid
U.S.5,145,572	8/8/92	Hupe et al.	Hydroquinone or ethoxylated alkylphenols
U.S.5,068,013	11/26/91	Bernards et al.	Polyethylene oxides
U.S.5,051,154	9/24/91	Bernards et al.	Polyethylene oxides, glycols, and amines
U.S.5,024,736	6/18/91	Clauss et al.	Disubstituted ethane sulfonic compounds
U.S.4,990,224	2/5/91	Mahmoud	Urea, sodium lauryl sulfate, and tosyl or mesyl sulfonic acid
U.S.4,975,159	12/4/90	Dahms	Alkoxylated lactam amides
U.S.4,954,226	9/4/90	Mahmoud	Urea and glycerin
U.S.4,948,474	8/14/90	Miljkovic	Alkylarylene
U.S.4,897,165	1/30/90	Bernards et al.	Copper/sulfuric acid ratios
U.S.4,781,801	11/1/88	Frisby	Polyether surfactants+ sulfurized benzene+ grain refiner
U.S.4,673,467	6/16/87	Nee	Polyethers+mercapto-imidazole+sulfurized benzene
U.S.4,555,315	11/26/85	Barbieri et al.	Polyethers+organic divalent sulfur compound+ tertiary alkyl amine with polyepichlorohydrin
U.S.4,551,212	11/5/85	Rao et al.	Phenazine dyestuffs
U.S.4,540,473	11/22/83	Bindra et al.	S containing anions other than SO_4
U.S.4,521,282	7/11/84	Tremmel	Sulfamic acid
U.S.4,490,220	12/25/84	Houman	Nitrogen-carbon-sulfur compound
U.S.4,430,173	2/7/84	Boudot et al.	Sodium salt of W-sulfo-n-propyl N,N-diethyldithiocarbamate, polyethylene glycol, and crystal violet
U.S.4,376,685	6/24/81	Watson	Alkylated polyalkyleneimine

TABLE 2 *(Continued)*

Patent Reference	Date of Issue	Investigator	Additives
U.S.4,347,108	8/31/82	Willis	Nitrogen and sulfur containing compounds
U.S.4,336,114	6/22/82	Mayer et al.	Phthalocyanine, tertiary alkyl amine with polyepichlorohydrin, polyethyleneimine
U.S.4,334,966	12/2/81	Beach et al.	Polyether, mercaptoimidazole, sulfurized benzene
U.S.4,310,392	1/12/82	Kohl	Phenolphthalein
U.S.4,272,335	2/19/80	Combs	Substituted phthalocyanine radical
U.S.4,242,181	12/30/80	Malak	Regular coffee
U.S.4,134,803	1/16/79	Eckles et al.	Disulfides, sulfonic acids, and aliphatic aldehyde
U.S.4,110,176	8/29/78	Creutz et al.	Alkoxylated polyalkyleneimine
U.S.4,038,161	7/26/77	Eckles et al.	Epihalohydrins
U.S.4,036,710	7/19/77	Kardos et al.	Di or triaminotriphenylmethane dyes and sulfoalkylsulfides
U.S.4,014,760	3/29/77	Kardos et al.	Aryl and sulfoalkyl sulfide compounds
U.S.4,009,087	2/22/77	Kardos et al.	Heteroaromatic, sulfoalkylsulfide and sulfoarylsulfide compounds
U.S.3,966,565	6/29/76	Kardos et al.	Aryl amine and sulfoalkyl sulfide compounds
U.S.3,956,120	5/11/76	Kardos et al.	Amines and sulfoalkyl sulfide compounds
U.S.3,956,084	11/21/74	Kardos et al.	Aryl amine and sulfoalkyl sulfide compounds
U.S.3,956,078	5/11/76	Kardos et al.	Amines and sulfoalkyl sulfide compounds
U.S.3,956,079	5/11/76	Kardos et al.	Aryl amine and sulfoalkyl sulfide compounds
U.S.3,940,320	2/24/76	Kardos et al.	Aryl N-heteroaromatic and sulfoalkyl sulfide compounds
U.S.3,923,613	8/27/74	Immel	Urea
U.S.3,841,979	10/15/74	Arcilesi	Polyethers
U.S.3,804,729	4/16/74	Kardos et al.	Polysulfides, heterocyclic sulfur and polyether compounds
U.S.3,798,138	3/19/74	Ostrow et al.	2-thia- or 2-imidazolidinethiones, aldehydes, and carbon sulfur groups
U.S.3,778,354	12/11/73	Toledo	cobalt
U.S.3,775,265	8/25/70	Bharucha	Amine for plating on Al
U.S.3,775,264	3/9/72	Bharucha	Amine and ammonia for plating on aluminum
U.S.3,770,598	11/6/73	Creutz	Polyethyleneimine

TABLE 2 *(Continued)*

Patent Reference	Date of Issue	Investigator	Additives
U.S.3,770,597	3/16/71	Tixier	Formaldehyde and thiourea
U.S.3,769,179	1/19/72	DuRose	High-acid, low-copper+ grain refiners
U.S.3,715,289	2/8/71	Pope Jr.	Ethylene oxide adduct+ 2 mercaptopyridine
U.S.3,682,788	8/8/72	Kardos et al.	Polysulfides, thiourea, and polyethers
U.S.3,767,539	10/23/73	Clauss et al.	Selenium compounds
U.S.3,751,289	8/7/73	Arcilesi	Polyethers
U.S.3,743,584	7/3/73	Clauss et al.	Polymeric phenazonium compounds
U.S.3,732,151	5/8/73	Abbott	Triarylmethane and sulfurized sulfonated aromatics
U.S.3,725,220	4/3/73	Kessler et al.	Sulfonium compounds
		Pyrophosphate Copper	
U.S.5,100,517	3/31/92	Starinshak et al.	Plating of wire with insoluble anodes
U.S.3,928,148	12/23/75	Lerner	Pyro+cyanide
U.S.3,784,454	1/8/74	Lyde	Mercaptothiadiazoles, aliphatic dicarboxylic acids, and hydroxyethyl cellulose
U.S.3,729,393	4/24/73	Lyde	Mercaptothiazoles, thiazoles, or pyrimidines+ alkaryl-sulfonic acids
U.S.3,775,268	12/30/71	Fino et al.	Lead
U.S.3,674,660	7/4/72	Lyde	Iminodiacetic, cinammic, aliphatic, carboxylic acids, and hydroxyethylcellulose
		Cyanide Copper	
U.S.3,790,451	2/5/74	Weisenberger et al.	Acetylenic alcohol+complexing agent+hydroxy acid
		Cyanide-Free Copper	
U.S.4,933,051	6/12/90	Kline	Organophosphonates
U.S.4,521,282	6/4/85	Tremmel	Sulfamic acid
U.S.4,469,569	9/4/84	Tomaszewski et al.	pH 7.5–10.5+a Cu soluble anode and a ferrite insoluble anode
U.S.4,462,874	7/31/84	Tomaszewski et al.	pH 6–10.5+a Cu soluble anode and a Ni–Fe alloy insoluble anode
U.S.4,389,286	8/28/82	McCoy	Copper, tin, lead, and glucoheptonic acid

TABLE 2 ***(Continued)***

Patent Reference	Date of Issue	Investigator	Additives
U.S.3,928,147	10/9/73	Kowalski	Phosphonates for zinc die castings
U.S.3,833,486	3/26/73	Nobel et al.	Phosphonates
U.S.3,706,635	11/15/71	Kowalski	Phosphonates
U.S.3,706,634	11/15/71	Kowalski	Phosphonates
U.S.3,475,293	10/28/69	Haynes et al.	Phosphonates

ethylene glycol [87, 107, 108], polyethylene imine [107], poly N,N'-diethylsaphranin [100], polypropylene ether [109, 110], propylene oxide [100], sugar [96], thiocarbamoyl-thio-alkane sulfonates [111], and thiourea [76, 81, 84, 85, 96, 112–114]. Chloride, which can also be considered an additive, is discussed in a separate section.

Many of the present-day, commercially available additives contain three components designated as carrier, leveler, and brightener. Reid [115] reports that "Carriers are typically polyalkylene glycol type polymers with a molecular weight around 2000, levelers are typically alkane surfactants containing sulfonic acids and amine or amide functionalities, and brighteners are typically propane sulfonic acids which are derivatized with surface active groups containing pendant sulfur atoms."

The use of a particular additive must be evaluated for each application because undesirable characteristics can then be avoided. For example, many of the addition agents proposed result in embrittlement of the plate. Deposition potentials are generally higher when addition agents are added. Cathode polarization is greatly increased by adding gelatin (0.2 g liter^{-1}) [70, 116] or glue [68, 117]. These additions result in grain refinement, but this is chiefly unidirectional because the structure remains columnar and becomes more fibrous [70]. Gelatin additions to the sulfate solution introduce porosity, organic inclusions [64, 118], or both.

Phenolsulfonic acid is used in the electrotyping industry, but results with it depend on the sulfonation and purification procedures [119]. Deposits become harder and smoother after a solution has been electrolyzed or dummied for a short time following an addition of phenol or phenolsulfonic acid [120].

The smoothing and grain refining tendencies of addition agents are sometimes associated with the formation of complex ions with copper or of colloids at or near the cathode interface. Gelatin or glycine, for example, forms complex ions with copper [121–123] and also exists in colloidal form [124]. Particles of colloids arising from additions of selenious and arsenic oxides have been observed by ultramicroscopic examination to concentrate at the cathode [125].

Nodulation prevention in refineries is of high priority, and it is the major concern in overall cathode quality [114]. Nodulation is suppressed by proper selection of solution operating conditions but, most important, by proper choice of addition agents. In most refineries operating presently, addition agents consist of animal glue, thiourea, chloride ion, and sometimes a sulphonated hydrocarbon. It has recently been shown that thiourea, which is added to give a smooth copper surface, can initiate nodulation, and this effect is always associated with a large increase in overpotential, >100 mV [114].

6 OPERATING CONDITIONS

6.1 Temperature

Temperatures may vary from 18° to 60°C; however, a temperature between 32° and 43°C is common, since it can be maintained economically with little or no heating or cooling. An increase in the temperature results in a higher conductivity and reduced anode and cathode polarization [70]. A temperature below 30°C is recommended for plating bright copper in acid solutions to maintain good leveling power. These solutions are customarily agitated with air.

6.2 Current Density/Agitation

An increase in current density in either the sulfate or the fluoborate solution results in increased cathode polarization (but not to the extent noted for many other solutions). Cathode films become more depleted in copper (II) ion and more concentrated in sulfate ion when the current density is increased [126]. Clear evidence has been reported of grain refinement produced by increasing the current density [127–130]. For example, an increase in current density from 1 to 7 A dm^{-2} reduced grain size by about one-third [130].

Current density and agitation must be balanced in order to obtain deposits having the desired properties. For producing electrotypes, cathode current densities of 16 to 22 A dm^{-2} are generally employed when using the sulfate solution agitated with air. Fast moving, endless wire can be plated at 50 A dm^{-2} [131]. Still higher current densities are used when sufficient agitation can be supplied. When movement of the work is impractical or when air agitation fails to provide good mixing at all significant surfaces, the current density is usually kept at 3.7 to 5.4 A dm^{-2}. It is claimed that higher current densities are practical with the fluoborate solution when agitation is the same as for the sulfate solution [26, 27].

6.3 Ultrasonic Agitation

Although results with ultrasonic agitation have shown some improvements, practical applications are still limited to surface cleaning processes [132, 133]. Researchers have shown that ultrasonic agitation can result in an increase in limiting current density, current efficiency, and a decrease in concentration polarization in acid sulfate solutions [134, 135] as well as pyrophosphate [134] and EDTA solutions [136]. It can also increase the hardness of deposits. This change is believed to be due to a substantial reduction in porosity and to work hardening caused by the repeated impacts of electrolyte jets on the surface of the deposit during the collapse of cavities at high intensities of 13kHz [137, 138]. Ultrasonic agitation is dependent on the location of the transducer in the plating tank.

6.4 Other Forms of Agitation

A reciprocating paddle cell has been used to deposit coatings of uniform thickness on large surface areas [139]. The paddle is a pair of confronting, separated triangular blocks aligned parallel to the cathode and was designed in this manner to provide uniform laminar flow. Mass-transfer characteristics of the paddle cell, which induces a nearly periodic flow, have been evaluated for acid copper solutions. Of importance are the paddle length scales and spatial location of the cathode relative to the paddle stroke boundary [140]. Also the paddle height above the cathode significantly affects deposition uniformity [141].

6.5 Filtration and Purification

Filtration requirements depend on the dirt load of air, any dirt brought in by the work, and the amount of detached anode sludge particles. If the cathodes are steel and if they are incompletely protected with a copper or nickel strike, particles that become detached from immersion deposits will also contribute to the filtration requirements. Although it is possible to maintain acceptably smooth deposits with only occasional batch filtration, continuous filtration is usually preferred. Cellulosic filter aids are satisfactory, but siliceous filter aids should be avoided when the flouborate solution is filtered; filter papers are satisfactory.

Potassium permanganate is often added to solutions used for plating of printed wiring boards to oxidize contaminants to species that are more easily removed by carbon treatment. Manganese, which accumulates in the solutions as a result of this procedure, exerts a detrimental effect on deposit tensile properties and interferes with cyclic voltammetric stripping analysis for the additive. Use of hydrogen peroxide at slightly elevated temperatures produces comparable purification results without the disadvantages associated with the use of permanganate [142].

Up to the early to mid-1980s many of the bright leveling acid copper processes contained some hazardous organic additives. By contrast, acid copper processes available today can produce the same leveling and fine-grained amorphous deposits as older processes, but without the concern for the safety of workers and the environment due to the additives. Waste treatment is easily accomplished by increasing the pH to about 9 to precipitate the copper. Once the precipitated copper is filtered out, the remaining solution does not contain any controlled material [4].

6.6 Equipment

Steel tanks with rubber or plastic are preferred for large acid copper solutions, but glass fiber reinforced plastic tanks are used for small volumes of solution. Lining materials that are generally suitable for either the sulfate or the fluoborate solution are properly formulated natural hard rubber, neoprene rubber, polyethylene, or plasticized vinyl chloride polymers. Air lines can be made of hard rubber or polymerized vinylidine chloride. Special grades of carbon pipe and tubing make efficient heat exchangers or cooling coils. Lead is, however, satisfactory in the sulfate solution and is less expensive. Rubber or rubber-lined filters are used for continuous filtration, but stainless steel is satisfactory for short periods.

Graphite is recommended for use as a heat exchanger for copper fluoborate solutions [143]. For further details, see [144].

6.7 Anodes

Rolled and cast bars and electrolytic copper sheets have been employed as anodes in sulfate and fluoborate solutions. Another choice is high-purity, oxygen-free anodes, which are commercially available in several shapes and sizes. The benefit of oxygen-free anodes is that anode sludge is decreased [145]. On the other hand, the tenacity of the anode film is improved, and the number of particles that become detached from anode surfaces in air agitated solutions is decreased by adding 0.02% to 0.04% phosphorus to cast copper [146, 147]. The films on phosphorized copper anodes are responsible for the slight polarization of about 0.5V [145]. Rolled copper anodes containing at least 0.004% phosphorus are customarily recommended by the vendors of brighteners for plating bright copper in copper sulfate solutions. Chunks of polarized copper are frequently used in titanium anode baskets. Bagged copper anodes of phosphorized copper are frequently used in acid copper plating of printed wiring boards to minimize sludge formation and to produce better anode corrosion. The black film formed on copper anodes containing phosphorus contains Cu^{+1}, Cl, and P and is envisioned to be a porous like CuCl-like matrix that is laden with aqueous copper sulfate solution. The beneficial effect of P in copper anodes is to inhibit disproportionation of Cu^{+1} [148].

Anode film particles often become detached from the anodes. Air agitation promotes the detachment, causing some particles to be dissolved by the free acid. If the work is racked so that cathode shelves lie in a horizontal plane, particles will settle out on these areas and roughen the plate. In such cases the anode sludge can sometimes be decreased and the deposits made smoother by raising the solution temperature or increasing the acid concentration. Fine copper particles can be prevented from reaching the cathode by bagging anodes with woven Dynel or polypropylene. To allow good mixing of the solution adjacent to the anodes, bags can be made in the form of envelopes, enclosing several anodes placed edgewise to the cathodes.

To avoid excessive polarization at any anode in the copper sulfate solution, the anode current density should not be more than about $5\,A\,dm^{-2}$ in unagitated solutions. With vigorous agitation, the limiting anode current density is more than $17\,A\,dm^{-2}$. The anode current density in an unagitated fluoborate solution can be as high as $40\,A\,dm^{-2}$, and with air agitation, it can be increased to $55\,A\,dm^{-2}$.

Small amounts of silver, sulfur, lead, tin, nickel, and other elements are common impurities in rolled, cast, and electrolytic copper [145]. Silver as an impurity in the anodes employed in sulfate and fluoborate solutions is of much less consequence than it is in cyanide solutions. Oxygen-free high-conductivity and electrolytic copper were found to have a lower impurity content than rolled or ordinary cast anodes. Two batches of high-purity anodes with average grain sizes of 10 and 0.01 mm performed similarly [145]. The anodic behavior of copper anodes containing arsenic or antimony as impurities has been investigated for electrorefining operations [149]. Lead containing 3% tin and 3% antimony has been proposed as an insoluble anode for facilitating the plating of printing rolls [150]. An insoluble anode that has been used in copper electrowinning was made of an alloy of copper, silicon, iron, and lead [69]. Graphite is the only electrically conductive material known to be insoluble as an anode in the fluoborate solution. When used as an anode, graphite produces a sludge of finely divided carbon particles. Recent reviews on anode reactions can be found in [151–154].

6.8 Specifications

Information on standards, specifications, and recommended uses for copper plating can be found in [39] and [155].

7 EFFECTS OF IMPURITIES IN PLATING SOLUTIONS

Acid copper solutions are more tolerant of ionic impurities than many other plating solutions. Many metallic ions introduced regularly by carryover with the work, by dissolution of impurities in the anode, or by dissolution of the basis metal (e.g., iron, nickel, or zinc) can be expected to accumulate in the solution because conditions are usually not satisfied for effecting codeposition of such metallic impurities (nickel, cobalt, zinc, iron) with copper. For example, less than 2 ppm nickel was codeposited with copper in a sulfate solution containing 15 g $liter^{-1}$ nickel [156].

Nickel and iron reduce the conductivity of the sulfate solution to the same degree as an equivalent increase in copper. The deposition potentials of arsenic and antimony are apparently near that of copper in the sulfate solution, since codeposition was reported to occur [157]. Arsenic and antimony in concentrations of 10 to 80 g $liter^{-1}$ and 0.02 to 0.1 g $liter^{-1}$, respectively, embrittled deposits and roughened surfaces. Addition of gelatin or tannin, however, effectively inhibited codeposition of these impurities and prevented roughness and embrittlement caused by them. Antimony readily codeposits with copper, but only small amounts of arsenic could be detected [158]. Bismuth, like arsenic and antimony, caused granular deposits [159]. Antimony and bismuth are believed to form insoluble complex compounds with arsenate [160].

Small concentrations of alkali metal and alkaline earth salts were found to smooth copper deposits [16]. Tin salts were also reported to smooth deposits and were, at one time, purposely added to the sulfate solution for this reason [15]. Lead is completely precipitated as partly sulfate and partly silver. If silver is a contaminant, a small amount will be codeposited with copper. A high current density favors the codeposition of silver [156]. Silver chloride caused pitting in semibright copper plate [161].

Nitrates are reduced to ammonia at the cathode in copper sulfate solutions [162]. A reduction product of the sulfate ion, which is said to have a grain-coarsening effect, can be removed by heating the solution and adding an oxidizing agent [163], but oxidizing agents such as hydrogen peroxide or potassium permanganate are said to reduce throwing power [161].

In the fluoborate solution, lead is the only metallic impurity known to interfere with the deposition of ductile plates; it can be precipitated by adding sulfuric acid. Besides lead, metals like silver, gold, arsenic, and antimony might be codeposited with copper, but the effects of such impurities are not known.

Organic impurities originating from decomposition of addition agents or leaching of elastomeric tank linings sometimes embrittle deposits, but they can be removed by treating the solution with an appropriate activated carbon, followed by filtration. Treatment with activated carbon is desirable when a new solution is being prepared—especially the fluoborate solution, which may be contaminated with impurities leached from rubber shipping drums.

8 ANALYTICAL METHODS

The concentration of copper sulfate or copper fluoborate can be approximated by specific gravity measurements, but the contribution of sulfuric acid to the specific gravity of the sulfate solution must be taken into account [164]. The acid concentration of fluoborate solutions is controlled by measuring pH, using colorimetric pH papers. Analytical methods for determining copper and sulfate concentration are described by Langford [165], who also presents procedures for determining copper and fluoboric acid concentration. Methods for determining trace amounts of lead [166], nickel [167], chromium [168], and chloride [77, 169] have been reported.

The concentration of addition agents is controlled by many techniques: empirical methods, such as evaluating the appearance of deposits on special bent or sloping cathodes immersed in beaker samples of the plating solution, Hull cell tests [170–172], and a variety of sophisticated analytical procedures.

Tench and Ogden and their colleagues pioneered in the use of cyclic voltammetry stripping (CVS) [173–177]. They were also instrumental in using CVS with copper pyrophosphate solutions and this is covered later in this chapter. Others who have used CVS with acid copper solutions can be found in [111, 112, 178–183]. One difficulty with CVS stems from the tendency of most acid copper solutions to "age" during operation by forming additive by-products [183]. Therefore a standard curve of stripping charge versus additive concentration for a fresh solution cannot be used for an aged one. Tench and White [184] modified the technique to step the electrode potential over a range to include plating, stripping, cleaning, and equilibration and called this technique cyclic pulse voltammetric stripping (CPVS). These researchers and others reported that CPVS was highly effective in mitigating the effects of additive breakdown products and other contaminants [181, 183–185].

Other methods that have been used with acid copper solutions include chromatography [102, 186], ion chromatography [180], high-performance liquid chromatography [180, 187–190], differential pulse polarography [191], and spectrophotometry [192–194]. Kanazawa et al. [195] used a quartz crystal microbalance to measure efficiency of an acid copper solution, and Mansfeld [196] developed a copper plating solution unit that operates by monitoring the polarization of a copper cathode at a constant applied direct current (dc). This unit is used to detect the presence of excess amounts of organic contaminant in the solutions, and it has worked well in production situations. Another technique for testing the influence of additives is a modified rotating-cone electrode, which is recommended because of reproducible and controlled mass-transfer performance at fixed rotation speeds [197]. Troubleshooting has been covered by Mohler [198] and Rudolph [199].

9 PROPERTIES AND STRUCTURE

Copper is electrodeposited for numerous engineering and decorative applications requiring a wide range of mechanical and physical properties. This range extends from properties superior to full-hard wrought copper to properties equivalent of annealed pure copper and is summarized in Table 3. This subject is extremely broad, and no attempt will be made to provide comprehensive coverage on the matter since there already exists a number of excellent reviews on the topic. The best source is Safranek's book on the properties of electrodeposits [3]. An extensive research project sponsored by the *American Electroplaters and Surface Finishers Society* showed that the mechanical properties of electroplated copper varied widely depending

TABLE 3 Ranges of Properties of Copper Electrodeposits

Outstanding Characteristic	Tensile Strength		Elongation, % (2 in.)	Internal Stress[a]		Hardness (VHN_{200}),	Electrical Resistivity,
	$kg\ mm^{-2}$	kpsi		$kg\ mm^{-2}$	psi	$kg\ mm^{-2}$	Microhm, cm
High strength	45–63	64–90	4–18	−4.2 or 3.6 to 5.5	−6000 or 5000 to 7800	131–159	1.75–2.02
Hardness	3.5–55	5–79	0–10	−4.2 or 3.5	−6000 to 5000	193–350	1.96–4.60
Low electrical resistivity	18–27	26–38	15–41	−0.5 to 1.6	−700 to 2200	48–64	1.70–1.73
Near zero stress	14–23	20–33	8–24	−0.08 to 0.06	−110 to 80	56–57	1.71–1.72
Good leveling	36	51	14–19	2.0	2900	128–137	1.82
Thermal stability[b]	22.5–30	32–43	26–39	0.5 to 2.9	700 to 4100	55–106	1.73–1.76

Source: From Safranek [3]. Reprinted with permission of the American Electroplaters & Surface Finishers Soc., Orlando, FL.

[a] Negative values indicate a compressive stress.

[b] Deposits that changed <0.02% in length after heating to 400°C.

on factors such as solution composition, current density, temperature, impurities, and addition agents [200–202].

The grain structures of copper deposits usually correspond to one of four types, shown in Figure 1. Deposits from the sulfate solutions containing no addition agents are columnar (Figure 1*a*). Additions of gelatin, phenolsulfonic acid, and many other agents cause the fibrous structure shown in Figure 1*b*. Such copper usually is 15% to 20% harder than columnar copper. The fine-grained deposit in Figure 1*c* is characteristic of copper deposited in copper cyanide solutions at high temperatures (70°–80°C), copper pyrophosphate solutions, copper sulfate solutions containing triisopropanolamine or other amines, and copper sulfate solutions containing popularly used brighteners. The banded copper in Figure 1*d* is characteristic of copper deposited with periodic current reversal from cyanide solutions containing brighteners. Copper is deposited from the sulfate solution as fcc crystals that are randomly oriented [203], unless deposited at less than 1.5 A dm^{-2}, when the basis metal can exert an influence on the structure so that crystals in the basis metal and plate are oriented similarly [204].

Grain continuity from the basis metal into sulfate copper deposits has been detected by many investigators [129, 205, 206]. Thus the structure of the deposit is often influenced by that of the basis metal. The basis metal structure was reproduced in the copper plate when the copper substrate was cleaned and immersed in an acid solution before plating [207, 208], but was not reproduced when the basis metal was cleaned but not dipped in acid solution before plating [208]. The grain size of the substrate appeared to influence the reproduction of the basis metal structure in the copper deposit [208]. The structure of copper, electrodeposited from sulfate solutions, was not influenced by dislocations such as etch pits in the substrate surface, unless the etch pits contained oxides or sulfides [209]. Oxide and sulfide particles in the surface layer of the substrate behaved as nucleation sites.

Tensile strength, in general, is inversely proportional to the square root of the grain diameter as shown in Figure 2 for a variety of copper deposits and wrought copper [210]. Loss of tensile properties such as reduced elongation and tensile strength of electrodeposited copper has been noted by many researchers. It affects copper in a number of applications: printed wiring boards that exhibit cracking in deposits as a result of soldering operations, optics parts for physics experiments that require high temperature stability of the copper to maintain precise stability during usage, and shaped charge parts which undergo deformation at very high strain rates. It is postulated that segregation at grain boundaries is driven by a reduction in grain boundary interfacial energy which, in turn, leads to a reduction in the grain boundary cohesive strength [211–213].

Zakraysek [214–217] observed a thermally induced ductile-brittle transition in electrodeposited copper with embrittlement occurring as a result of fracture along grain boundaries. Lin and Sheppard [218] noted an embrittlement effect when electroplated copper having a columnar grain structure was tensile tested at temperatures between 150° and 300°C. The effect was also strain-rate dependent at the lower temperature. This was not noted for deposits annealed without straining at the same temperatures and then tested at room temperature. The behavior is believed to be associated with inhibition of grain growth and grain boundary weakening, possible as the result of vacancy effects at the grain boundaries resulting from chloride in the deposit [218].

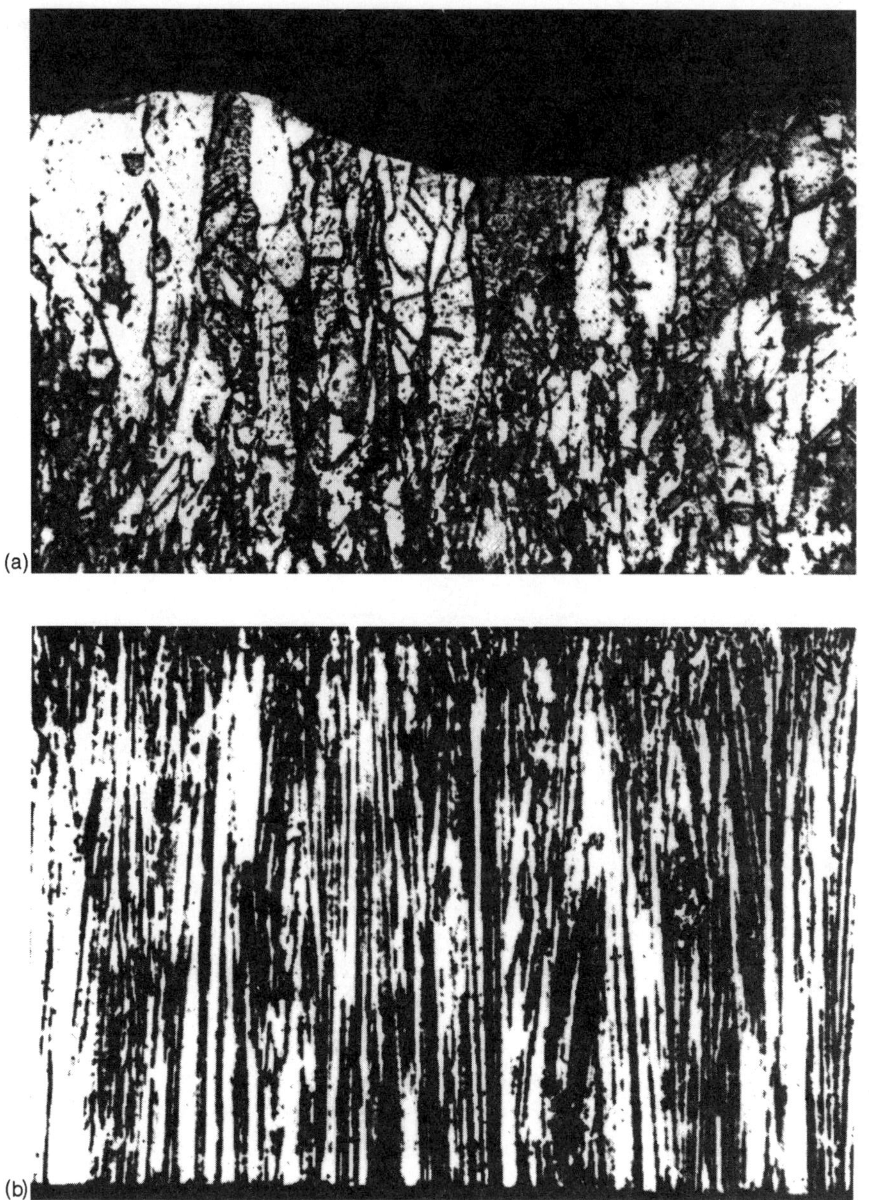

Figure 1 Structure of copper deposits: Cross-sectional views (~500×) after etching with ferric chloride reagent. (*a*) columnar; (*b*) fibrous; (*c*) fine-grained; (*d*) banded. From Lowenheim [6]. Reprinted with permission of John Wiley & Sons, Inc.

The surface mobilities of vacancies in copper can be increased by as much as four orders of magnitude by the presence of monolayers of adsorbed halides [219]; chloride surface contamination can induce cracking in copper by the vacancy mechanism at temperatures as low as 200°C [220]. Since small quantities of chloride are used in acid copper sulfate solutions, some chloride can be incorporated in the deposit [218]. Thin copper chloride crystals have been epitaxially absorbed onto copper single crystals from a sulfate solution containing only 7 ppm chloride [221].

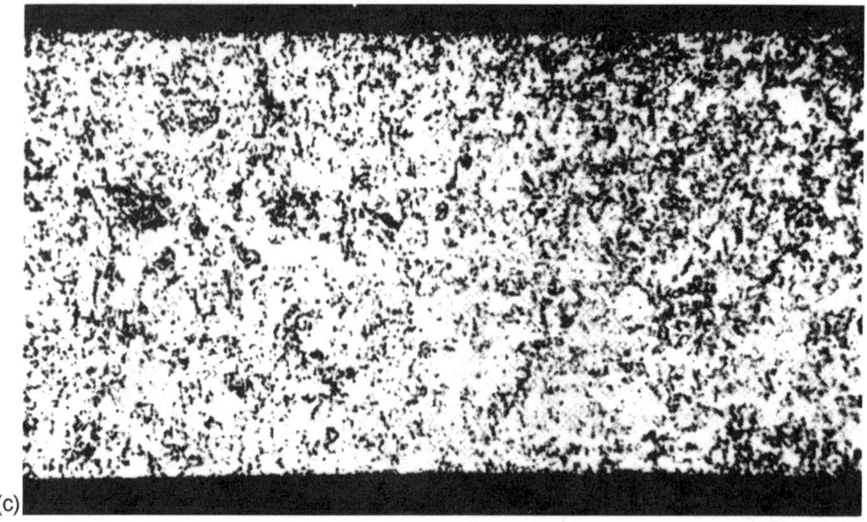
(c)

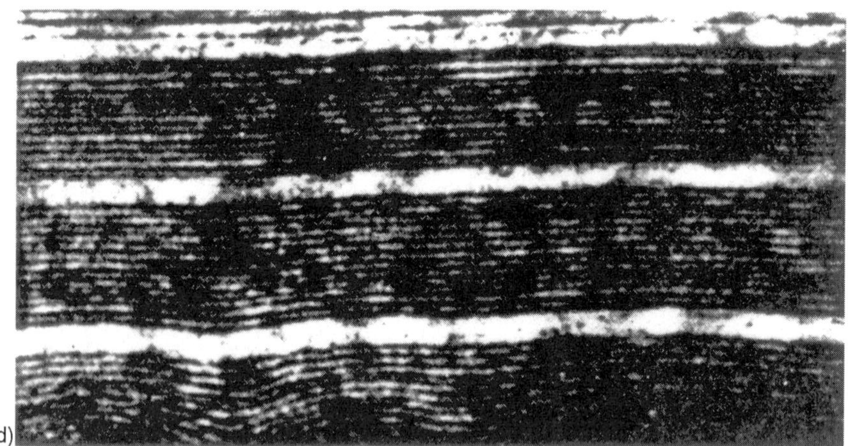
(d)

Figure 1 (*Continued*).

Others noted poor performance of electrodeposited copper in shaped charge applications [211, 222]. In these cases, strain rates were very high (10^4 and higher) compared to rates around 10^{-3} for typical tensile tests. Despite the radical difference in deformation history, Lassila [211] postulates that the fundamental material characteristic of high temperature embrittlement is the primary cause of the poor performance. The presence of segregated impurities is believed to be the primary cause of brittle fracture and particulation. Support of this hypothesis was provided by Sole and Szendrei [223] who electroformed shaped charge liners that exhibited good performance by using additive-free acid sulfate solutions with pulse plating. In a similar vein, Woodman et al. [210] suggested that use of pulse plating can provide high strengths without additives, thereby avoiding embrittlement.

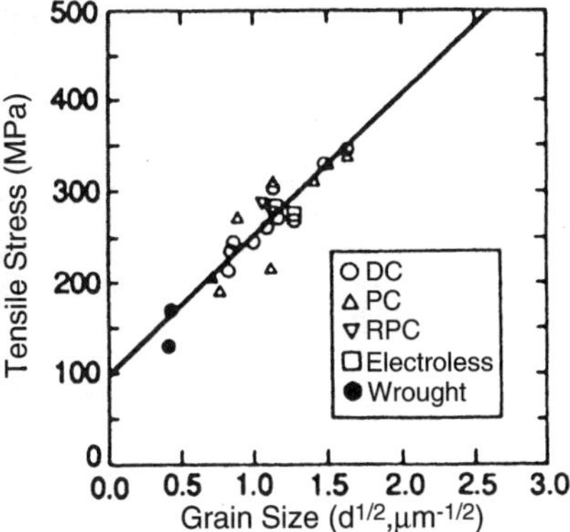

Figure 2 Plot of tensile strength versus inverse square-root of grain size. From Woodman [210]. Reproduced by permission of the Electrochemical Society, Inc.

Merchant [224, 225] characterized thermal response of electrodeposited copper by monitoring changes in microstructure, tensile strength, elongation, and microhardness following 30 minute isothermal annealing at temperatures between 23° and 400°C. By judicious control of additives to the electrolyte, considerable enhancements of strength and hardness were obtained. The annealing gradually removed embrittlement with the removal rate dependent on the time/temperature parameters of the thermal exposure [225]. Embrittlement at 180°C with increasing deposit thickness was attributed to developing low-density regions in the morphological boundaries [226]. Others who have investigated thermal response of acid copper coatings can be found in [227] and [228].

10 CURRENT MODULATION TECHNIQUES

Current modulation techniques such as periodic reverse, pulse plating, and asymmetric alternating current plating have been used to improve deposit properties. For overviews of these techniques, see [229–235].

With periodic reverse plating (PR), parts are plated in a conventional manner for a selected time and are then deplated for a shorter period by reversing the current. Malone [236] used PR to enhance the uniformity and grain structure, regardless of deposit thickness. A current density of about 5 A dm^{-2} and a PR cycle with a cathodic to anodic ratio of 2:1 in a solution containing no additives provided consistent mechanical properties and excellent thermal stability. Wan [237] observed that the hardness of deposits increased when PR or pulse plating were used instead of dc.

Pulsed current techniques involve application of a forward current for a certain time interval with a short, high-energy reverse pulse periodically interposed. The main difference between dc and pulsed current is that with dc plating only voltage (or current) can be controlled, while with pulse plating three parameters—on-time, off-time, and peak current density—can be varied independently [238]. These variables are believed by many to create a mass transport situation, an electrocrystallization condition, and adsorption and desorption phenomena which are not otherwise possible [229, 238, 239].

In acid copper deposition, pulsed current has been used for improvement of mechanical and physical properties of deposits [210, 234, 240]. Benefits include finer grain structure [210, 241–245], increased hardness [237, 246, 247], reduced stress [248], reduced surface roughness [238, 249, 250], reduced porosity [246], and improved leveling [251, 252] and throwing power [253]. Use of pulsed reversed current in acid copper sulfate solutions containing a combination of polyethers, sulphopropyl sulphides, and chloride ions can provide enhanced throwing power because of a change in the polarization characteristics of the electrolyte [239, 254]. Enhanced throwing power was not obtained when single additives were present. Best results were obtained with an anodic/cathodic ratio of 3:1 and a cathodic/anodic ratio of 20:1 [254]. Macrothrowing power was poor for pulse plating [251].

Benefits claimed for printed wiring board plating include: improving the through hole distribution of solutions used for plating printed wiring boards [238, 254, 255–259], a reduction in average plating time by 30% to 50% [259], a reduction in average track height of 50% [259], and a finer grained, more ductile deposit with less stress [248]. Holmbom and Jacobsson [260] observed that by pulse plating at high frequencies (above 1 kHz), through hole plating uniformity was improved compared with that of dc deposits. At lower current frequencies, the uniformity was worse than that of dc plating. Deposits prepared by periodic reverse exhibited a similar uniformity to pulse plated deposits. However, appearance was completely dull due to loss of brighteners which were desorbed during the reversal part of the plating cycle. In addition, chlorine evolution occurred at the cathodes during the reversed pulses. A computer model that predicts the effect of pulse plating copper in small holes was developed by Ng et al. [253]. Leisner et al. [261] investigated throwing power under pulsed current conditions in an attempt to optimize through hole plating. However, as subsequently mentioned in the section on printed wiring boards [262], although many claims have been made for pulse plating, no manufacturing process based on this technology other than that described by Engelhaupt [258] has been reported.

Deposits produced by using pulse plating to selectively deposit copper (spot plating) were brighter and contained larger crystallites than those applied by using direct current [263]. Puippe and Ibl [245] noted that an increase in off-time was accompanied by an increase in grain size, similar to the way this system reacts with dc. For optics applications, a dendritic crystal structure is deposited using a bipolar pulse plating technique. This surface is then oxidized to provide a highly antireflective surface [264]. Wan et al. [265] reported that pulse time exerted the greatest effect on current efficiency, and Popov et al. [266] showed that the effective current density is dependent not only on values of the effective overpotential but also on the ratio of pulse-to-pause duration and on the frequency of the input pulsating potential. Rudder [267] compared pulse plating of acid copper against cyanide copper and found that the leveling and thickness performance of the pulse-plated cyanide copper was better than that of acid copper.

Reverse pulse plating, was investigated by a number of researchers. White and Galasco [268] obtained surface distributions on variably spaced, 50 μm wide lines and deposit properties typical for dc by employing reverse pulsed current in additive-free solutions. Current efficiency for reverse pulse plating was lower than that for pulse plating or for dc plating [269]. Mann [270] noted that when using high reversal frequencies, plating speed, throwing power, and surface quality were improved in comparison with dc plating. By combining pulse current and pulse reverse current,

Zhou et al. [271] developed an additive-free solution that produced deposits comparable to commercial additive solutions. A benefit claimed for this approach is that waste disposal cost can be greatly reduced compared to the current process due to the possible in-process recycling of rinse water and plating solution in the additive-free solution.

Morphological changes in copper deposits caused by use of superimposed alternating current (ac) were correlated with reduction in grain size and smooth deposits. Hexagonal pryamidal growth obtained with dc on the [111] copper plane transformed to layers, triangular pyramids, and polycrystalline deposits under the influence of superimposed ac [272].

Additional information on pulse plating form acid copper solutions can be found in [273–276]. Pulsed current techniques have been used by researchers at Battelle and this is discussed in more detail in the section on high speed electrodeposition.

11 PLATING ON STEEL, ZINC, PLASTICS, AND ALUMINUM

Historically, buffed acid copper deposits have been widely used as an undercoat for nickel-chromium deposits. The development of reliable processes capable of producing fully bright, ductile, high leveling deposits that did not require buffing served to promote even wider use of acid copper solutions [277].

The fact that copper is much cheaper than nickel suggests that it might be considered an economical replacement for a large proportion of the total thickness of electroplated coatings. However, an equal thickness of copper undercoat is not comparable with nickel. It has been proved fairly conclusively that inferior corrosion resistance results when part of the bright nickel layer in a bright nickel plus decorative chromium coating is replaced by an equal thickness of copper [278].

In terms of corrosion performance, copper underlayers are generally not detrimental when there is adequate thicknesses of double-layer nickel. Under microdiscontinuous chromium and double-layer nickel, copper layers are neither detrimental nor beneficial. In essence, the nickel and chromium layers are so protective that parts are obsolete before the copper would even start to corrode [279]. These general characteristics have been confirmed in panel exposure programs involving steel, zinc, plastics, and aluminium [280–281]. Jonkind [282] compared various processes for Cu–Ni–Cr plating of aluminum bumper bar stock (Alloy X-7046) with Ni–Cr–plated steel bumpers. The aluminum-plated parts showed at least equal corrosion test results with the steel-plated parts in CASS, neutral salt spray, and atmospheric tests.

Acid copper deposits are used over a cyanide copper strike for plating zinc die castings [283]. The ductility of copper deposits is one of the most important parameters in the coating of plastics [284]. It is suggested that elongation should be at least 15% [285]. Typical bright copper solutions for plating on plastics contain around 60 g liter^{-1} copper and 60 g liter^{-1} sulfuric acid. Lower copper concentrations (10–20 g liter^{-1}) and higher contents of sulfuric acid (200 g liter^{-1}) are also used because of their excellent leveling capability and throwing power [284]. A high-throw acid copper strike is very effective in improving quality by increasing the copper thickness in low current density areas while also reducing cost and waste treatment requirements when compared to a nickel or pyrophosphate copper strike [49].

12 PLATING OF PRINTED WIRING BOARDS

When printed conductors are used on both sides of a printed wiring board, electrical connections between the two sides of the board are most reliably made via plated through holes in the board. Because copper is widely used as the base conductor metal in a printed circuit, it is natural that copper also be used for plating the through holes. A historical summary of the evolution of acid sulfate copper plating processes and their benefits as they relate to printed wiring board market needs is presented in Table 4.

The precise technical requirements of electronic products and the demands of environmental safety compliance have been the driving forces exerting major influence on plating practices [41]. The preferred plating process is acid copper sulfate with organic additives. Copper pyrophosphate deposits, once the standard of the industry, have been almost entirely replaced by acid copper, except for some military and special applications [41, 262].

The success of the acid copper sulfate solutions is attributed to their good throwing power, ease of control, good mechanical and physical properties, and their favorable properties regarding waste disposal and treatment compared to pyrophosphate and cyanide solutions. The major factors that affect throwing power, the ability of a plating solution to produce a relatively uniform distribution of metal upon a cathode of irregular shape, are cathode polarization, cathode efficiency, and solution conductivity. Rothschild [71] compared throwing power for a variety

TABLE 4 Historical Summary of Acid Copper Evolution Related To Printed Wiring Plating

Year	Innovation	Benefits
1970	Original high-throw process	Improved throwing power when compared to conventionally used pyrophosphate copper
1973	Ductile high-throw process	Improved ductility agents to consistently pass thermal shock testing
1976	Process for high aspect ratio MLB	Improved ductility for MLBs with high aspect ratios
1979	High temperature and high throw	Improved distribution without need for chillers
1983	Increased current density with high throw (4.3 to 6.5 $A\,dm^{-2}$)	Improved productivity and lower operational cost
1984	Process control/additive analysis	Elimination of subjective readings and lower operational costs
1984	High current density and high throw (10.8 $A\,dm^{-2}$)	Improved productivity and lower operational cost

Source: From Mayer and Barbieri, [178].

of copper-plating solutions by using a Haring cell [286]. This work, summarized in Table 5, clearly shows the benefit obtained with the low copper–high acid formulations which have become the standard of the industry for plating printed wiring boards. Some chloride is also essential in these solutions and this is discussed in the section on functions of solution constituents. Turner [287] and Walker and Cook [82] obtained results similar to those of Rothschild [71], namely through hole plating ability improved with decreasing copper concentration and increasing acid content. A noncyanide formulation provided even better throwing power than the low copper–high acid formulation.

Fundamental studies on through hole electroplating [288] and a thorough review of plating into through holes and blind holes were published by Yung et al. [262]. A mathematical model for through hole plating was developed by Hazlebeck and Talbot [289]. Kessler and Alkire outlined procedures for investigating complex electrochemical phenomena appearing during plating of printed wiring boards [290] and developed a model for predicting copper thickness distribution on multilayer printed wiring board through holes [291]. References [292–295] provide additional information on plating high aspect ratio holes.

Innovative agitation schemes such as jet plating for high speed plating ($>10\,\mathrm{A\,dm^{-2}}$) of printed wiring boards have been implemented in production lines [262, 288, 296, 297]. Other techniques that have been used include high-speed additives [41], vibratory agitation [298], and forced solution flooding [299]. Deposit distribution on a printed wiring board was not improved when ultrasonic agitation was compared with mechanical agitation [300]. Hewlett-Packard utilized a conveyorized impingement-agitated plating system where copper plating was done at $7.5\,\mathrm{amp\,dm^{-2}}$, which is at least twice the current density usually used for plating printed wiring boards. The significant agitation in the process was accomplished by use of hydraulic injection. Agitation via panel motion or entrained air bubbles was insignificant [301].

Most additive systems for acid copper sulfate solutions used for plating printed wiring boards are based on some combination of a polymer surfactant of a high

TABLE 5 Values of Throwing Power (Haring Cell 11:1 Ratio)

Solution	Throwing Power
Conventional acid sulfate copper[a]	14
Conventional acid fluoborate copper[b]	14
Cyanide copper	50
Low copper–high fluoboric acid copper[c]	58
Pyrophosphate copper	62
Low copper–high sulfuric acid copper[d]	87
Alkaline noncyanide copper[e]	95

Sources: From Rothschild, (B. F. Rothschild, *Plating and Surface Finishing*, 66, 70 (May 1979). Ratio from Haring and Blum, ref (H. E. Haring and W. Blum, *Trans. Am. Electrochem. Soc.*, 44, 313 (1923).

[a] Copper, $48\,\mathrm{g\,liter^{-1}}$, sulfuric acid $75\,\mathrm{g\,liter^{-1}}$.

[b] Copper $120\,\mathrm{g\,liter^{-1}}$, fluoboric acid $30\,\mathrm{g\,liter^{-1}}$.

[c] Copper $15\,\mathrm{g\,liter^{-1}}$, fluoboric acid $340\,\mathrm{g\,liter^{-1}}$,

[d] Copper $15\,\mathrm{g\,liter^{-1}}$, sulfuric acid $210\,\mathrm{g\,liter^{-1}}$, $30\,\mathrm{mg\,liter^{-1}}$ chloride.

[e] From L.C. Tomaszewski and R.A. Tremmel, Proc. AESF SUR/FIN 85, Session D, Detroit, MI (1985).

molecular weight such as polyethylene glycol which apparently inhibits copper deposition by forming an adsorbed film on the cathode that mediates transport of species from the solution and a sulfur-containing and/or nitrogen-containing organic species such as disulfide or organic sulphonate, which induces leveling [80]. Some single blend additives can be comprised of three components: (1) carrier agent, (2) leveling agent, and (3) ductility agent. Carrier agents are the primary grain refiners and impart brightness to the deposit. Leveling agents level out drilling imperfections in plated through holes and also prevent fault planes and foldovers in multilayer board plating. Ductility agents create the equiaxed grain structure that is best suited to withstand thermal shocks [178, 302].

In a study of eight commercial acid copper solutions, all yielded sufficiently ductile deposits for multilayer circuit board applications, but some of the solutions exhibited a relatively strong dependence on solution agitation [303]. In one production situation, a 40% annual savings was realized by operating a solution with additives at 6.4 amp dm^{-2} versus 3.2 amp dm^{-2} [178]. In additive-free solutions, increased levels of agitation improved deposit quality dramatically up to the point of eliminating ion concentration gradients causing mass transport-limited plating conditions. With additives, the plating solutions were considerably less sensitive to changes in plating parameters such as the current density and level of agitation [304]. Byle and Bratin [305] also studied the effects of additives.

Yung et al. [288] utilized a gap cell, which provided accurate nondestructive profile measurements without tedious cross sectioning to show that agitation, which is necessary inside the holes to replenish the plating solution, is also necessary to provide a certain amount of agitation on the surface of the board.

Pulsed current techniques (discussed in detail in Section 10 on current modulation techniques) are claimed to offer benefits for plating of printed wiring boards. However, aside from Engelhaupt [258], no other manufacturing process based on this technology has been reported [262].

Nonuniform copper plating on printed wiring boards may be attributed to nonuniform anode currents due to high contact resistances between anode hooks and anode rods. Turner [306] developed an anode hook design to make a stable low-resistance contact by clamping the hook to the anode rod. Use of titanium for the hook and rod resulted in a low maintenance contact system.

Optimizing the performance of a plating process does not need to involve extensive, tedious testing. The traditional approach of examining one variable at a time while holding all others constant not only requires repeated tedious experiments, but allows bias error in each measurement [307]. Use of factorial, screening, response-surface designs, or Taguchi techniques allows for accurate solutions from a minimum of test runs [308]. Any of these approaches in studying the effects of many variables is highly recommended for both research and production activities.

Barringer and Carano [309] used the Taguchi technique to optimize the leveling of an acid copper plating solution for printed wiring boards. The effects of temperature, leveler (one component included in the addition agent), and ratio of sulfuric acid to copper metal were investigated. The best leveling was obtained at the low temperature, increasing the ratio of sulfuric acid to copper resulted in some slight improvement of leveling, and additions of leveler produced no improvements. Taguchi techniques were applied by Wan and McCaskie [307] to elucidate the relative importance of various process parameters on dendrite formation in an acid copper system. Chloride and arsenic concentration were two critical factors that affected dendritic growth, while lead and temperature exerted little effect. Elbs and Rasmussen [310] utilized Taguchi techniques for optimizing the current efficiency of a cyanide copper solution.

An important requirement of copper deposits is that they have sufficient strength to withstand subsequent soldering of the circuits. During this operation, high forces are generated by differential expansion of the board material (usually epoxy glass) and the copper. Therefore the copper must be either strong enough to contain and deform the epoxy glass or have sufficient elongation to be stretched without fracture [311]. With the proper additives, fine-grained deposits with tensile strengths around 345 MPa and a minimum elongation of 10% are routinely produced [41].

Most of the studies on the ductility of copper deposits for printed circuit boards (PCB) have been done under a uniaxial stress state. However, the stress state of electrodeposited copper in through holes of a PCB is definitely biaxial. Therefore Ye et al. [312] correlated cathodic overpotential, solution composition, and aging with characteristics such as crystallographic texture, roughness, and ductility. Copper foils with a low [220] preferred crystallographic orientation and a smooth surface were obtained when deposition was done at 87 to 113 mV cathodic overpotential from solutions with a low chloride content (<40 ppm). Highest ductilities were achieved under these plating conditions.

Kang et al. [313] studied surface morphologies of foils with different textures. Increasing solution temperature and decreasing current density changed the texture from <111> to <110>. The mechanical properties at elevated temperature of a proprietary formulation for multilayer board plating was evaluated in detail by Fox [314]. The deposit was found to exhibit adequate localized ductility, although, as is generally the case for electroplated copper, its final elongation after fracture was lower than that for wrought, annealed copper. Merchant reported on the defect structure, isothermal anneal kinetics, and thermal response of electrodeposited copper [224, 225, 315, 316].

13 PATTERNED ELECTRODEPOSITION FOR MICROELECTRONICS

Computers, microprocessors, and other microelectronic devices could not exist without the technology of depositing thin metal or alloy films with fine lithographic patterns [317]. Romankiw et al. [318–321] have demonstrated the capabilities of resist patterned electroplating for the past two decades. Dukovic [317] suggests that there are three trends that appear to be propelling patterned electrodeposition into a phase of major expansion. First, high-speed machines require the high conductivity of copper, and acid copper plating is simple and capable of reaching high rates. Second, the pattern replication powers of electrodeposition are ideally suited to the level of miniaturization required for wiring structures. Third, demands for cost reduction have increased the importance of reducing the capital and operating expenses associated with metal deposition processes, and it is therefore likely that electroplated copper thin film, which is important today, will continue to play a central role in the future.

IBM researchers have successfully implemented copper electroplating technology for the fabrication of chip interconnect structures [2a, 2b, 321a]. The process, termed *damascene copper electroplating*, meets the challenges of filling trenches and vias with copper without creating a void or seam. Under proper conditions, electroplating inside trenches occurs preferentially in the bottom, leading to void-free deposits, a phenomenon referred to as *superfilling*. Proprietary additives are used in the plating solution to produce the superfilling properties.

Electrodeposited bonding bumps are indispensable microconnectors for high-density interconnection in recent microelectronics applications. The bumps generally have mushroom shape, and are deposited onto dot-shaped cavities of several 10 to 200 μm in diameter formed by photoresist. Kondo et al. [322] discussed the shape evolution of copper bumps, and Dukovic [323] developed a two-dimensional numerical computation of tertiary current distribution.

High-speed selective jet electrodeposition has also been used to produce deposits in a selective manner without the need for masking [324]. The principle is that a nonsubmersed, free-standing electrolyte jet impinges onto a substrate and deposition occurs within the impingement region, with little or no deposition occurring in the surrounding areas. Bocking [324] determined the effects of temperature, velocity, and concentration during high-speed selective jet plating in an acid copper sulfate solution without additives.

The development of laser enhanced electroplating processes offers a promising technique for high-speed and maskless selective plating and/or as a repair and engineering design-change scheme for microcircuits [300, 325–328]. For this, temperature is used to modify the position of the equilibrium potential in a localized region so that electrodeposition is driven by the potential difference between this region and the nonirradiated regions. Use of a focused argon laser beam (488 nm) in an acid copper solution provided plating rates as high as 25 $\mu m\,s^{-1}$ [300]. Bindra et al. [327, 328] discussed the mechanism of laser enhanced acid copper plating and Paatsch et al. [329] reported on laser induced deposition of copper on p-type silicon. It was demonstrated that the increase in the plating rate under laser illumination results principally from photo-induced heating of the electrode surface [328].

14 ELECTROFORMING

Copper finds extensive use in electroforming [330, 331]. Acid sulfate with periodic reversal of current [236, 332], acid sulfate with oxygen reduction additive, and pyrophosphate solutions provide excellent properties for aerospace applications [236]. However, although acceptable properties are attainable via pyrophosphate deposition, very little electroforming is done with these solutions. Electroforming with acid copper or nickel is used as one of the primary methods of fabricating the outer shells of regeneratively cooled thrust chambers for advanced design rockets and other applications.

Typically the process involves filling machined (or etched) channels with wax, making the wax conductive with silver powder, plating thick copper to seal over the channels, and then removing the wax. Some thrust chamber requirements for copper require that the deposits retain useful mechanical properties and metallurgical structure at temperatures up to 400°C. The electroformed copper deposits often contain some oxygen, and this can be deleterious at high temperatures because hydrogen can combine with the oxygen and produce water.

With the high temperatures involved, steam pressure generated by the reaction often exceeds the strength of the copper and causes plastic deformation and/or tearing, frequently manifesting itself by grain boundary cracking or cavities [333, 334]. This problem is avoided by using an oxygen control additive such as sugar [335, 336] or potassium aluminum sulfate [337].

All of the pentoses are suitable for use in the copper sulfate solution, such as xylose, arabinose, ribose, and lyxose. These materials act as oxygen scavengers in the solution by picking up oxygen and thereby preventing the anodes from being oxidized [172]. Table 6 includes gas and carbon content of a variety of copper deposits and clearly shows the high purity attainable with an oxygen control additive such as d-xylose [338].

Farmer et al. [339], in evaluating copper for a physics accelerator application, noted that copper deposited from acid sulfate solution containing one set of proprietary additives was unstable during heat treatment and exhibited void formation around 450°C. By contrast, another acid copper solution with different proprietary additives formed no voids when heated at 1000°C. Deposits from a cyanide solution and an acid copper solution containing d-xylose also showed no void formation after heating at 1000°C, giving further proof that deposits with low carbon and gas contents (Table 6) are stable at high temperatures.

Electroformed targets for a 14 MeV neutron source facility did not require the high-temperature stability just discussed. However, they did require a deposit elongation around 20% because of a forming operation after plating. This was obtained by using a medium copper (150 g liter^{-1} copper sulfate), high acid (188 g liter^{-1} sulfuric acid) formulation [340].

Copper deposits usually contain small amounts of other impurities besides oxygen and these can affect properties such as electrical resistivity [3]. Hydrogen is present in different bound states in the deposits and the total amount of hydrogen is often five to six orders of magnitude larger than the equilibrium solubility of hydrogen at room temperature [341]. It is most likely that the hydrogen is bound to crystal

TABLE 6 Carbon and Gas Contents of Various Copper Electrodeposits and OFHC Copper

	Content (wt. ppm)			
Plating Solution	C	H	O	N
Low copper–high acid proprietary	50	11	50	5
D-xylose[a]	12	1	11	2
Pyrophosphate proprietary	21	6	140	17
High copper–low acid proprietary	190	14	50	9
Cyanide	NA	3	53	6
OFHC (wrought)	90	4	50	<10

Source: From Dini [338].

[a] Oxygen control additive.

lattice defects such as vacancies and dislocations, or adsorbed at the grain boundaries, or present in microvoids. Deposits obtained at low current densities contain a larger amount of hydrogen than those obtained at high current densities.

Hydrogen is evolved from the deposits as a function of temperature with the major amount being eliminated at temperatures around 450°C [341]. Sulfate (0.071–0.099 wt%) and sulfide (0.040–0.069 wt%) have been found in deposits from acid copper solutions. These impurities decreased with the increase in plating temperature, with the increase in acidity of the plating solution, and with the addition of 0.1 g liter^{-1} gelatin [342].

Strongly enhanced ductility (up to 40%) and reflectivity were achieved in the presence of dimethylamino derivatives [98]. The ductility of the coatings in contrast to the tensile strength appeared to be correlated to the amounts of codeposited carbon and sulfur. An elongation of 15% and low stress were obtained from solutions containing polypropylene ether. These deposits were essentially sulfur free, with carbon and nitrogen contents of about 0.015 to 0.003 and 0.001 wt% [110]. A high-throw acid copper solution has been used with glue and phenol sulphonic acid for electroforming of bellows [343].

Although not electroforming, very thick deposits (around 5000 μm) were produced on the outer surface rotor of a 300 MVA superconducting generator. This was done in acid sulfate solution with a proprietary brightener. Deposit properties were; yield strength around 210 MPa, elongation 15%, and high electrical conductivity at cryogenic temperatures [344]. Radio-frequency quadrople accelerator components up to 4 m long were plated with 250 μm in an acid sulfate solution containing a brightener system [345].

15 HIGH-SPEED ELECTROPLATING

A current density of 300 A dm^{-2} or more is practical for depositing most metals if the solution flow rate is in the range of 1.0 to 1.2 m s^{-1}, depending on the topography of the surface, the solution viscosity, and other factors [346]. High current densities for achieving fast rates have been used in both copper sulfate and copper fluoborate solutions. Large variations in the copper and sulfuric acid concentrations had no significant effect on the appearance of deposits from the sulfate solution. The tensile strength of copper deposited with turbulent flow in a sulfate solution increased from 350 to 450 MPa as the current density was raised from 100 to 300 A dm^{-2}, which increased the deposition rate from 25 to 75 μm min^{-1}. High-speed copper deposited in a fluoborate solution exhibited a lower tensile strength, 280 to 350 MPa over the range of 100 to 300 A dm^{-2}. However, all of these values are higher than those reported for copper deposited at conventional rates.

High-speed copper plating has been developed for commercial applications such as electroforming of flexible printed circuits [347], in-mold plating of polymeric materials [348], and plating of round bars [349]. A process for electroforming copper circuit paths on a reusable metal substrate and transferring the circuitry to a flexible polymeric surface coated with adhesive has been advanced to production. The copper circuit paths are electroformed at rates up to 50 μm min^{-1} using fast rate electrodeposition technology. The process does not generate spent etching solutions or plating rinse water and costs can be 70% to 80% of those associated with the conventional etched circuit process for high volume production. Another benefit is that high-speed deposition conserves equipment and floor space [348].

A combination of high-speed plating with pulsed current (fast rate, interrupted current, FRIC) allows for deposition of copper with different structures at rates of 50 μm min^{-1} or higher [350]. By controlling the frequency and/or duty cycle with the high-speed deposition rate, it is possible to tailor the deposit structure to suit the application. For example, with applications that need structural strength particularly at corners, the FRIC technique at high frequency would be used to promote growth of mainly polycrystalline copper which would be equally strong in all directions. By contrast, for applications such as wire coatings that require maximum electrical conductivity, columnar structures produced at moderate frequency are preferred [350].

For agitation of the electrolyte, forced circulation across the cathode by pumping electrolyte is most commonly used, although ultrasonic agitation and gas sparging have also been investigated [351]. When plating wire and strip using direct current, the deposition rate can be increased by intensifying the electrolyte flow, by optimizing the gap between the cathode and anode, and by the use of high-performance electrical contacts [352].

16 DIAMOND TURNING

Diamond turning surface finishing uses single-point diamond tools on a precision lathe under precisely controlled machine and environmental conditions. Diamond tools are used when it is necessary to machine the smoothest possible surface texture. When the surface to be cut is one of the preferred diamond turnable materials such as copper, electroless nickel, or aluminum, surface textures as smooth as 1 nm can be obtained. Coatings offer significant advantages for diamond turning applications inasmuch as they can be applied to lightweight substrates such as aluminum or beryllium, and copper coatings produced in acid sulfate solutions have been used in many applications. A number of 1.3 m diameter molybdenum parts required form accuracy of approximately 250 Å, and this was obtained by coating the parts with 0.3 to 0.45 mm of copper prior to diamond turning [52]. A physical vapor deposition process was used to provide an initial, thin (6 μm) adherent copper layer, and this was followed by thick plating in an acid copper solution containing proprietary additives. An important requirement for these parts, as well as for many other diamond turning applications, is long-term microstructural stability. If the deposit undergoes recrystallization after machining, the precise surface finish is degraded. Copper deposits have been shown to markedly soften after storage at room temperature for 30 days [353]. For example, copper mirror surfaces for the Antares laser system changed over a six month period due to recrystallization [51]. The problem was eliminated by employing a low temperature heat treatment (1 h at 250°C). This same treatment provided stability for the proprietary copper used for the 1.3 m diameter molybdenum parts mentioned earlier [52].

Rao and Trager [354] reported that copper for diamond turning should have a grain size from about 200 to 500 Å and a hardness from about 250 to 320 Knoop (15 g load). An additive comprising a mixture of phenazine dyestuffs in the combination from about 30% to 40% by weight of a Janus Green B type dyestuff and about 70% to 60% by weight of Safranine T provided these properties.

By judicious choice of plating operating conditions, such as current density, pH, solution composition, and additive concentration, it is possible to deposit coatings that are in a stress-free state. A plot of current density versus stress was used to define conditions for depositing 1 mm thick copper with zero stress on glass substrates and on pyrex optics. This deposit was capable of withstanding single-point diamond turning to provide an optical surface and was able to withstand heating at 250°C for four hours without degradation of the metal-glass bond [53].

Occasionally, a phenomenon referred to as *black spots* has been obtained. These are small, localized defects noted on electroplated copper after single-point diamond turning. Analysis of these spots has revealed oxides, sulfates, and chlorides, materials which are components of the plating solution [355, 356]. It is speculated that the most likely cause of these defects is a void or pit exposed by the diamond cutting tool. In an effort to eliminate these minuscule pits Dini et al. [356] evaluated plating under reduced pressure (around 100 torr). Copper deposits produced under these conditions exhibited finer and denser grain structure and a reduced surface roughness.

17 MISCELLANEOUS

17.1 Magnetics

Effects of magnetic fields on electroplating of copper were investigated by Takeo et al. [357] who noted a reduction in energy required for deposition with increasing magnetic field as a consequence of reduced concentration overpotential. Changes in structure and hardness were also noted as the applied magnetic field strength was increased. Ismail and Fahidy [358] studied the effect of uniform magnetic fields on the morphological properties of fine mesh metal screen deposits. Popov et al. [359] provided some correlation between dendritic growth and copper powder formation.

17.2 Striations

Occasionally spiral patterns (striations) that follow the streamlines of the fluid are obtained during electrodeposition of copper, zinc, and silver [360]. This is accompanied by small or negative polarization resistance and is typically observed only below a certain current density. The range of current density depends on the electrolyte and its concentration. Studies with rotating hemispherical electrodes (RHE) revealed no striations in the deposition of copper from sulfate solution unless a certain amount of gelatin was added. With increasing additions of gelatin (10–20 ppm), the polarization curve became progressively steeper, vertical and finally S-shaped, while shifting to a more negative potential [360]. Spiral patterns were also observed in acid copper sulfate solutions containing polyethylene glycols and chloride [89].

17.3 Underpotential Deposition

Underpotential deposition (UPD) involves formation of up to a monolayer of atoms on foreign substrates at potentials that are more positive than the potential of the reversible deposition in the same solution, that is, before bulk deposition can occur. The fact that UPD exists means that the chemical potential of adatoms deposited at the underpotentials is different from that of the corresponding bulk metal. The adatoms are more strongly bound to the surface of the foreign substrate than to the surface of its own species [361, 362].

The likelihood that UPD plays an important role in electroplating processes has long been recognized [363]. An example wherein UPD plays an important role is in the electroplating of gold where ions of metal (or semimetal) serve as brightening agents for gold plated from cyanide and noncyanide solutions. Gold deposition proceeds in the region of UPD of the foreign metals, which are incorporated only at trace levels in the deposit. These foreign metals act as brighteners because of their strong influence on the nucleation process [363]. With copper, it's been demonstrated that the UPD of Pb or Sn on copper can be used to produce electroplated Cu–Pb and Cu–Sn alloys, with small amounts of alloyed Pb and Sn. This is done in acid solutions that do not contain complexants. These alloys are of interest as possible on-chip wiring for very large scale integration where the content of alloying agent must be kept small in order to maintain a low resistivity [363]. Some other studies on UPD include Cu on Au [364] and Cu and Ag on Pt [365].

REFERENCES

1. L. W. Flott, *Metal Finishing*, **94**, 55 (March 1996).
2. R. Sard, *Encyclopedia of Materials Science and Engineering*, vol. 2, M. B. Bever, ed., 1423 Wiley, New York, 1986.

2a. P. C. Andricacos, *Interface*, **7**, 23 (Spring 1998).

2b. P. C. Andricacos, *Interface*, **8**, 32 (Spring 1998).

3. W. H. Safranek, *The Properties of Electrodeposited Metals and Alloys*, 2nd ed., American Electroplaters and Surface Finishers Soc., Orlando, FL, 1986.
4. D. L. Snyder, *Met. Finish.*, **89**, 37 (April 1991).
5. F. Passal, *Plating*, **46**, 628 (1959).
6. F. A. Lowenheim, *Modern Electroplating*, 3rd ed., Wiley, New York, 1974.
7. M. Carano, *Electronics*, **32** (3), 29 (March 1986).
8. J. F. D'Amico and M. A. De Angelo, *J. Electrochem. Soc.*, **123**, 478 (1976).
9. A. Smee, *Elements of Electrometallurgy*, 2nd ed., Longmans, Green, London, 1843.
10. G. R. Strickland, *Prod. Finish.* (London), **24**, 24 (Sept. 1971).
11. G. C. Van Tilburg, *Plating Surf. Finish.*, **71**, 78 (April 1984).
12. W. De la Rue, *London and Edinborough Phil. Mag.*, **9**, 484 (1836).
13. W. H. Safranek, F. B. Dahle, and C. L. Faust, *Plating*, **35**, 39 (1948).
14. Langbein, *Electrochem. Metall.*, **3**, 310 (1903).
15. Beadle, *Electrochem. Metall.*, **1**, 163 (1901).
16. E. F. Kern, *Trans. Am. Electrochem. Soc.*, **15**, 441 (1909).
17. E. F. Kern, U.S. Patent 946,903 (1910).
18. C. G. Fink and C. Y. Wong, *Trans. Electrochem. Soc.*, **63**, 65 (1933).
19. A. I. Levin, S. M. Scherbakov, F. A. Zorin, V. M. Svitskii, and R. A. Gokhina, *J. Appl. Chem. (USSR)*, **13**, 686 (1940); *Chem. Abstr.*, **35**, 3536.

20. A. I. Levin, *J. Appl. Chem. (USSR)*, **14**, 78 (1941); *Chem. Abstr.*, **36**, 972.
21. N. P. Diev and A. J. Loshkarev, *J. Appl. Chem. (USSR)*, **12**, 585 (1939); *Chem. Abstr.*, **33**, 8504.
22. D. C. Gernes, G. A. Lorenz, and G. H. Montillon, *Trans. Electrochem. Soc.*, **77**, 177 (1940).
23. J. R. Stack, U.S. Patent 2,294,053 (1942).
24. M. Schlotter, J. Korpiun, and W. Nurmeister, *Z. Metall.*, **25**, 107 (1933).
25. D. Thomas and J. C. Silva, *J. Electrochem. Soc.*, **143**, 458 (1996).
26. C. Struyk and A. E. Carlson, *Mon. Rev. Am. Electroplat. Soc.*, **33**, 923 (1946).
27. K. S. Willson, A. H. DuRose, and D. G. Ellis, *Plating*, **35**, 252, 304 (1948).
28. C. L. Faust, B. Agruss, E. L. Combs, and W. A. Pruell, *Mon. Rev. Am. Electroplat. Soc.*, **34**, 541, 709 (1947).
29. W. A. Proell, U.S. Patent 2,525,943 (1950).
30. L. Cambi and R. Piontelli, *Rend. 1st. Lombardo Sci. Lett.*, **72** (1), 128 (1939); *Chem Abstr.*, **34**, 4995 (1940).
31. S. Venkatachalam and T. L. Rama Char, *Electroplat. Met. Finish.*, **14**, 3 (1961).
32. J. E. Stareck and F. Passal, U.S. Patent 2,383,895 (1945).
33. C. B. F. Young, *Met. Finish.*, **47**, 56 (November 1949).
34. D. Uceda, T. O'Keefe, and E. Cole, *J. Electrochem. Soc.*, **137**, 1397 (1990).
35. I. Hoshino and T. Fumoto, Japanese Patent 5306 (1955).
36. K. Hosokawa and T. Inui, *Kinzoku Hyomen Gijutsu*, **17** (4), 138 (1966); *Chem Abstr.*, **67**, 70009.
37. F. A. Lowenheim, *Electroplating*, McGraw-Hill, New York, 1978.
38. L. M. Weisenberger, "Copper Plating," in *Metals Handbook*, 9th ed., vol. 5, *Surface Cleaning, Finishing and Coating*, American Soc. for Metals, (1982).
39. L. M. Weisenberger and B. J. Durkin, "Copper Plating," in *ASM Handbook*, vol. 5, *Surface Engineering*, ASM International (1994).
40. A. Sato and R. Barauskas, *Metal Finishing, 64th Guidebook and Directory Issue*, vol. 94, no. 1A, 214 (1996).
41. E. F. Duffek, "Plating" in *Printed Circuits Handbook*, 4th ed., C. F. Coombs Jr., ed., McGraw-Hill, New York, 1996.
42. G. F. Bidmead, *Trans. Inst. Met. Finish.*, **59**, 129 (1981).
43. W. H. Safranek and H. R. Miller, *Plating*, **55**, 233 (1968); **56**, 11 (1969).
44. W. H. Safranek and J. G. Beach, *Tech. Rep. on Electroplating Bright, Leveling and Ductile Copper*, Copper Development Assoc., Inc. (Oct. 1968).
45. D. R. Millage, *Precision Metal Molding*, **23**, 38 (Feb. 1965).
46. H. G. Creutz, *Proc. Am. Electroplat. Soc.*, **51**, 82 (1964).
47. W. H. Safranek, Tech. Paper 105, Soc. Die Casting Engineers (1966).
48. L. Harrison, N. Martyak, J. McCaskie, M. McNeil, R. Schultz, and H. Smith, *Met. Finish.*, **92**, 11 (Dec. 1994).
49. D. J. Combs, *Proc. AESF SUR/FIN 80*, Session B, Boston (June 1981); *Plating Surf. Finish.*, **68**, 58 (July 1981).
50. R. L. Barauskas and A. E. Guttensohn, *Plating*, **60**, 355 (1973).
51. B. M. Hogan, *Proc. AESF SUR/FIN 84*, Session H, New York (July 1984).
52. T. G. Beat, W. K. Kelley, and J. W. Dini, *Plating Surf. Finish.*, **75**, 7 (Feb. 1988).

53. W. C. Cowden, T. G. Beat, T. A. Wash, and J. W. Dini, in *Metallized Plastics 1: Fundamental and Applied Aspects*, K. L. Mittal and J. R. Susko, eds., Plenum Press, New York, 93, 1989.
54. J. W. Dini, *Plating Surf. Finish.*, **79**, 121 (May 1992).
55. F. Wills and E. J. Clugston, *J. Electrochem. Soc.*, **106**, 362 (1959).
56. H. J. Modi and G. S. Tendolkar, *J. Sci. Ind. Res.*, **12B**, 431 (1953).
57. W. H. Safranek and M. Wirth, *Mater. Methods*, **44** (5), 104 (1956).
58. C. W. Bennett, *Trans. Am. Electrochem. Soc.*, **21**, 253 (1912).
59. P. Vanden Brande and R. Winand, *J. Appl. Electrochem.*, **23**, 1089 (1993).
60. J. E. T. Anderson, G. Bech-Nielson, and P. Moller, *Surf. Coatings Technol.*, **70**, 87 (1994).
61. Y. Fukunaka, K. Denpo, M. Iwata, K. Maruoka, and Y. Kondo, *J. Electrochem. Soc.*, **130**, 2492 (1983).
62. V. Anantharaman and P. N. Pintauro, *J. Electrochem. Soc.*, **136**, 1727 (1989).
63. J. D. Reid and A. P. David, *J. Electrochem. Soc.*, **134**, 1389 (1987).
64. R. Taft and H. E. Messmore, *J. Phys. Chem.*, **35**, 2585 (1931).
65. R. Taft and O. R. Bingham, *J. Phys. Chem.*, **36**, 2338 (1932).
66. H. M. Goodwin and W. H. Horsch, *Chem. Metall. Eng.*, **21**, 181 (1919).
67. E. F. Kern and M. Y. Chang, *Trans. Am. Electrochem. Soc.*, **41**, 181 (1922).
68. G. M. Kimber, D. H. Napier, and D. H. Smith, *J. Appl. Chem.*, **17**, 29 (1967).
69. C. W. Eichrodt, *Trans. Am. Electrochem. Soc.*, **45**, 381 (1924).
70. A. K. Graham, *Trans. Am. Electrochem. Soc.*, **52**, 157 (1927).
71. B. F. Rothschild, *Plating Surf. Finish.*, **66**, 70 (May 1979).
72. Y. Fukunaka, H. Doi, Y. Nakamura, and Y. Kondo, in *Electrochemical Technology in Electronics*, vol. 88-23, L. T. Romankiw and T. Osaka, eds., Electrochemical Soc., 83 (1988).
73. Y. Fukunaka, H. Doi, and Y. Kondo, *J. Electrochem. Soc.*, **137**, 88 (1990).
74. H. K. Richardson and F. D. Taylor, *Trans. Am. Electrochem. Soc.*, **20**, 179 (1911).
75. S. Skowronski and E. A. Reinsoso, *Trans. Am. Electrochem. Soc.*, **52**, 205 (1927).
76. G. Fabricius and G. Sundholm, *J. Appl. Electrochem.*, **15**, 797 (1985).
77. A. Suresh, *Plating Surf. Finish.*, **71**, 108 (May 1984).
78. N. Pradhan, P. G. Krishna, and S. C. Das, *Plating Surf. Finish.*, **83**, 56 (Mar. 1996).
79. J. W. Chang and R. Weil, *Proc. AESF SUR/FIN 86*, Session G, Philadelphia (June 1986).
80. D. Anderson, R. Haak, C. Ogden, D. Tench, and J. White, *J. Appl. Electrochem.*, **15**, 631 (1985).
81. R. E. Gana, M. G. Figueroa, and R. J. Larrain, *J. Appl. Electrochem.*, **9**, 465 (1979).
82. R. Walker and S. D. Cook, *Surf. Technol.*, **11**, 189 (1980).
83. W. A. Fairweather, *Trans. Inst. Met. Finish.*, **62**, 5 (1984).
84. S. Yoon, M. Schwartz, and K. Nobe, *Plating Surf. Finish.*, **81**, 65 (Dec. 1994).
85. S. Yoon, M. Schwartz, and K. Nobe, *Plating Surf. Finish.*, **82**, 64 (Feb. 1995).
86. Z. Nagy, J. P. Blaudeau, N. C. Hung, L. A. Curtiss, and D. J. Zurawski, *J. Electrochem. Soc.*, **142**, L87 (June 1995).
87. J. D. Reid and A. P. David, *Plating Surf. Finish.*, **74**, 66 (Jan. 1987).
88. L. S. Melnicki, *Proc. Electrochemical Technology in Electronics*, vol. 88-23, L. T. Romankiw and T. Osaka, eds., Electrochemical Society, 95, (1988).

89. M. Hill and G. T. Rogers, *J. Electroanal. Chem.*, **86**, 179 (1978).
90. M. Yokoi, S. Konishi, and T. Hayashi, *Denki Kagaku*, **51**, 460 (1983).
91. M. Goodenough and K. J. Whitlaw, *Trans. Inst. Met. Finish.*, **67**, 57 (1989).
92. M. R. H. Hill and G. T. Rogers, *J. Electroanal. Chem.*, **68**, 149 (1976).
93. E. E. Farndon, F. C. Walsh, and S. A. Campbell, *J. Appl. Electrochem.*, **25**, 574 (1995).
94. M. J. Armstrong and R. H. Muller, *J. Electrochem. Soc.*, **138**, 2303 (1991).
95. R. Weil and J.-W. Chang, *Plating Surf. Finish.*, **75**, 60 (Nov. 1988).
96. J. B. Mohler, *Met. Finish.*, **85**, 121 (June 1987).
97. J. E. Breen, E. Toledo, and V. White, Soc. of Automotive Engineers, Paper 710793, Los Angeles, CA, (Sept. 1971).
98. D. Stoychev, I. Vitanova, R. Buyukliev, N. Petkova, I Popova, and I. Pojarliev, *J. Appl. Electrochem.*, **22**, 987 (1992).
99. D. Stoychev, I. Vitanova, R. Buyukliev, N. Petkova, I. Popova, and I. Pojarliev, *J. Appl. Electrochem.*, **22**, 978 (1992).
100. R. Rashkov and C. Nanev, *J. Appl. Electrochem.*, **25**, 603 (1995).
101. Y. N. Sadana and S. Nageswar, *J. Appl. Electrochem.*, **14**, 489 (1984).
102. N. Koura, A. Tsutsumi, and K. Watanabe, *Plating Surf. Finish.*, **77**, 58 (Sept. 1990).
103. G. H. Awad, F. M. Abdel Halim, and M. I. El-Diehey, *Met. Finish.*, **78**, 65 (Sept. 1980).
104. R. Lakshmana Sarma and S. Nageswar, *J. Appl. Electrochem.*, **12**, 329 (1982).
105. J. W. Chang and R. Weil, *Proc. AESF SUR/FIN 85*, Session F, Detroit (July 1985).
106. K. S. Rajam and I. Rajagopal, *Met. Finish.*, **77**, 17 (April 1979).
107. G. A. Hope, G. M. Brown, D. P. Schweinsberg, K. Shimizu, and K. Kobayashi, *J. Appl. Electrochem.*, **25**, 890 (1995).
108. D. Stoychev and C. Tsvetanov, *J. Appl. Electrochem.*, **26**, 741 (1996).
109. E. Knaak, J. Hupe, and W. Metzger, *Proc. 12th World Congress on Surface Finishing*, Paris, France, 347 (Oct. 1988).
110. W. Metzger, E. Knaak, and J. Hupe, *Plating Surf. Finish.*, **75**, 64 (July 1988).
111. M. Grall, G. Durand, and J. M. Couret, *Plating Surf. Finish.*, **72**, 72 (Dec. 1985).
112. H. M. Wu, M. L. Lay, and C. H. Huang, *Plating Surf. Finish.*, **79**, 66 (Sept. 1992).
113. P. Malathy and B. A. Shenoi, *Met. Finish.*, **75**, 56 (Mar. 1977).
114. D. F. Suarez and F. A. Olson, *J. Appl. Electrochem.*, **22**, 1002 (1992).
115. J. Reid, *PC Fab*, **10**, 65 (Nov. 1987).
116. C. G. Fink and C. A. Philippi, *Trans. Am. Electrochem. Soc.*, **50**, 267 (1926).
117. E. F. Kern and R. W. Rowen, *Trans. Am. Electrochem. Soc.*, **41**, 379 (1929).
118. C. Marie and P. Jacquet, *J. Chim. Phys.*, **26**, 189 (1929); *Chem Abstr.*, **23**, 5091 (1929).
119. E. I. Peters, *Electrotypers and Stereotypers Mag.*, **44**, 48 (1958).
120. R. O. Hull and W. Blum, *J. Res. Natl. Bur. Stand.*, **5**, 767 (1930).
121. G. Fuseya and K. Murata, *Trans. Am. Electrochem. Soc.*, **50**, 235 (1926).
122. G. Fuseya and K. Murata, *Trans. Am. Electrochem. Soc.*, **52**, 249 (1927).
123. T. J. Barker, *Chem. News*, **97**, 37, 51 (1908).
124. P. K. Frolich, *Trans. Am. Electrochem. Soc.*, **46**, 67 (1924).
125. M. N. Polukarov, *J. Gen. Chem. (USSR)*, **18**, 1249 (1948).
126. A. Brenner, *Proc. Am. Electroplat. Soc.*, **28** (1941).
127. V. DeNora, *Met. Ital.*, **31**, 607 (1939); *Chem. Abstr.*, **35**, 5797 (1940).
128. L. L. Shreir and J. W. Smith, *J. Electrochem. Soc.*, **99**, 64 (1952).

129. L. L. Shreir and J. W. Smith, *J. Electrochem. Soc.*, **99**, 450 (1952).
130. A. Butts and V. DeNora, *Trans. Electrochem. Soc.*, **79**, 163 (1941).
131. O. E. Adler and M. J. Krinowitz, U.S. Patent 2,317,350 (1943).
132. M. C. Hsiao and C. C. Wan, *Plating Surf. Finish.*, **76**, 46 (Mar. 1989).
133. M. C. Hsiao and C. C. Wan, *Proc. AESF SUR/FIN 89*, Session J-2, Cleveland (June 1989).
134. R. Vasudevan, R. Devanathan, and K. G. Chidambaram, *Met. Finish.*, **90**, 23 (Oct. 1992).
135. W. R. Wolfe, H. Chessin, E. Yeager, and F. Hovorka, *J. Electrochem. Soc.*, **101**, 590 (1954).
136. A. Chiba and W. C. Wu, *Plating Surf. Finish.*, **79**, 62 (Dec. 1992).
137. R. Walker and C. T. Walker, *Nature*, **250**, 410 (Aug. 2, 1974).
138. C. T. Walker and R. Walker, *J. Electrochem. Soc.*, **124**, 661 (1977).
139. J. V. Powers and L. T. Romankiw, U.S. Patent 3,652,442 (1972).
140. D. T. Schwartz, B. G. Higgins, P. Stroeve, and D. Borowski, *J. Electrochem. Soc.*, **134**, 1639 (1987).
141. D. E. Rice, D. Sundstrom, M. F. McEachern, L. A. Klumb, and J. B. Talbot, *J. Electrochem. Soc.*, **135**, 2777 (1988).
142. D. Anderson, M. Hanna, C. Ogden, and D. Tench, *Plating Surf. Finish.*, **70**, 70 (Dec. 1983).
143. A. E. Carlson and C. Struyk, *Metal Finishing Guidebook-Directory*, Metals and Plastics Publications, Westwood, NJ, 1970, p. 276.
144. L. J. Durney, ed., *Electroplating Engineering Handbook*, 4th ed., Van Nostrand Reinhold, New York, 1984.
145. W. H. Safranek and C. L. Faust, *Plating*, **42**, 1541 (1955).
146. R. P. Nevers, R. L. Hungerford, and E. W. Palmer, *Plating*, **41**, 1301 (1954).
147. R. P. Nevers, R. L. Hungerford, and E. W. Palmer, *Iron Age*, **174**, 114 (Aug. 12, 1954).
148. G. S. Frankel, A. G. Schrott, H. S. Isaacs, J. Horkans, and P. C. Andricacos, *J. Electrochem. Soc.*, **140**, 961 (1993).
149. J. C. Minotas, H. Djellab, and E. Ghali, *J. Appl. Electrochem.*, **19**, 777 (1989).
150. E. H. Mundell, U.S. Patent 2,852,450 (1958).
151. C. T. Walker, *Met. Finish.*, **86**, 37 (Apr. 1988).
152. W. C. Walker, *Plating Surf. Finish.*, **77**, 16 (Oct. 1990).
153. D. R. Gabe, in *Oxides and Oxide Films*, vol. 6, A. K. Vijh, ed., Marcel Dekker, New York, 1981.
154. J. Aromaa, O. Forsen, J. Kukkonen, M. Tavi, and Y. Leppanen, *Proc. 12th World Congress on Surface Finishing*, Paris, France, 487 (Oct. 1988).
155. A. C. Hamilton Jr., *AESF Shop Guide*, 9th ed., **79** (1991).
156. S. J. Wallden, S. T. Henriksson, P. J. Arbstedt, and T. Mioen, *J. Met.*, **11** (8), 528 (1959).
157. C. Yu Wen and E. F. Kern, *Trans. Am. Electrochem. Soc.*, **20**, 121 (1911).
158. V. Hospadaruk and C. A. Winkler, *Trans. Am. Inst. Min. Met. Eng.*, **197**, 1375 (1953).
159. H. Bovet, *Pro-metal*, **7** (43), 452 (1955); Translation 3738, H. Butcher, Altadena, CA.
160. M. A. Mosher, *J. Electrochem. Soc.*, **107**, 7C (1960).
161. M. R. Marshall, *Met. Finish. J.*, **3**, 457 (1957).
162. R. Taft, *Trans. Am. Electrochem. Soc.*, **63**, 75 (1933).
163. L. L. Shreir and J. W. Smith, *Trans. Faraday Soc.*, **50**, 393 (1954).
164. J. B. Mohler, *Met. Finish.*, **85**, 47 (July 1987).

165. K. E. Langford, *Analysis of Electroplating and Related Solutions*, 4th ed., Robert Draper, Teddington, Middlesex, England, 1971.

166. E. J. Serfass, W. S. Levine, and M. H. Perry, *Plating*, **37**, 166 (1950).

167. E. J. Serfass, W. S. Levine, P. J. Prong, M. H. Perry, and R. B. Freeman, *Plating*, **37**, 495 (1950).

168. E. J. Serfass, M. H. Perry, and S. Sen, *Plating*, **37**, 389 (1950).

169. E. J. Serfass and M. H. Perry, *Plating*, **37**, 62 (1950).

170. R. O. Hull, U. S. Patent 2,149,344 (1939).

171. M. K. Sanicky, *Plating Surf. Finish.*, **72**, 20 (Oct. 1985).

172. J. W. Dini, Electrodeposition: The Materials Science of Coatings and Substrates, Noyes Publ., Park Ridge, NJ, 1993.

173. C. Ogden and D. Tench, *Proc. AES Second Design and Finishing of Printed Wiring and Hybrid Circuits Symp.*, San Francisco, CA (Jan. 1980).

174. R. Haak, C. Ogden, and D. Tench, *Plating Surf. Finish.*, **68**, 52 (Apr. 1981).

175. R. Haak, C. Ogden, and D. Tench, *Plating Surf. Finish.*, **69**, 62 (Mar. 1982).

176. W. O. Freitag, C. Ogden, D. Tench, and J. White, *Plating Surf. Finish.*, **70**, 55 (Oct. 1983).

177. C. Ogden and D. Tench in *Application of Polarization Measurements in the Control of Metal Deposition*, I. H. Warren, ed., Elsevier, New York, 1984.

178. L. J. Mayer and S. C. Barbieri, *Proc. AES 12th Plating in the Electronics Industry Symp.*, Kissimmee, FL (1985).

179. R. Gluzman, *Trans. Inst. Met. Finish.*, **63**, 134 (1985).

180. J. E. McCaskie, *Proc. AESF SUR/FIN 87*, Session M-3, Chicago (July 1987).

181. G. L. Fisher and P. J. Pellegrino, *Proc. AESF SUR/FIN 87*, Session M-7, Chicago (July 1987).

182. S. Yamato and M. Tadakoshi, *Met. Finish.*, **85**, 67 (Oct. 1987).

183. G. L. Fisher and P. J. Pellegrino, *Plating Surf. Finish.*, **75**, 88 (June 1988).

184. D. Tench and J. White, *J. Electrochem. Soc.*, **132**, 831 (1985).

185. D. Tench et al., Final Report, *Manufacturing Technology for Printed Wiring Board Electrodeposition Processes*, AFWAL Contract No. F33615-81-C-5108 (1985).

186. A. Gemmler, T. Bolch, R. De Donker, and J. Vanhumbeeck, *Proc. AESFSUR/FIN 87*, Session O-6, Chicago (July 1987).

187. B. Bressel, *Galvanotechnik*, **75**, 1488 (1984).

188. B. Bressel, *Galvanotechnik*, **76**, 1970 (1985).

189. J. D. Reid, *Plating Surf. Finish.*, **75**, 108 (May 1988).

190. J. D. Reid, *Proc. AESF SUR/FIN 90*, Session H, Boston (July 1990).

191. D. Engelhaupt and J. Canright, *Proc. AESF SUR/FIN 84*, Session J, New York (July 1984).

192. T. R. Mattoon, P. McSwiggen, and S. A. George, *Met. Finish.*, **83**, 83 (Feb. 1985).

193. S. A. George and T. R. Mattoon, *Met. Finish.*, **83**, 23 (March 1985).

194. S. A. George and T. R. Mattoon, *Met. Finish.*, **83**, 29 (April 1985).

195. K. K. Kanazawa, G. L. Borges, S. Doss, and C. Hildebrand, *Proc. AESF SUR/FIN 95*, Session B, Baltimore (June 1995).

196. F. Mansfeld, *Plating Surf. Finish.*, **65**, 60 (May 1978).

197. D. Reichenbach, *Plating Surf. Finish.*, **81**, 106 (May 1994).

198. J. B. Mohler, *Met. Finish.*, **85**, 49 (August 1987).

199. T. Rudolph, *Plating Surf. Finish.*, **64**, 34 (Dec. 1977).

200. V. A. Lamb and D. R. Valentine, *Plating*, **52**, 1289 (1965).
201. V. A. Lamb and D. R. Valentine, *Plating*, **53**, 86 (1966).
202. V. A. Lamb, C. E. Johnson, and D. R. Valentine, *J. Electrochem. Soc.*, **117**, 291C, 341C, 381C (1970).
203. J. W. Cuthbertson, *Trans. Electrochem. Soc.*, **77**, 157 (1940).
204. W. A. Wood, *Proc. Phys. Soc. (London)*, **43**, 138 (1931).
205. A. K. Huntington, *Trans. Faraday Soc.*, **1**, 324 (1905).
206. A. W. Hothersall, *Trans. Faraday Soc.*, **31**, 1242 (1935).
207. W. Blum and H. S. Rawdon, *Trans. Am. Electrochem. Soc.*, **44**, 305 (1923).
208. A. K. Graham, *Trans. Am. Electrochem. Soc.*, **44**, 427 (1923).
209. T. B. Vaughan and H. J. Pick, *Electrochim. Acta*, **2**, 179 (1960).
210. A. S. Woodman et al., *Proc. 2nd International Symp. on Electrochemical Technology Applications in Electronics*, L. T. Romankiw, M. Datta, T. Osaka, and Y. Yamazaki, eds., Electrochemical Society, 86 (1993).
211. D. H. Lassila, *Shock Wave and High Strain Rate Phenomena in Materials*, M. A. Meyers, L. E. Murr, and K. P. Staudhammer, eds., Marcel Dekker, New York 1992, 543.
212. D. McLean and H. R. Tipler, *Met. Sci. J.*, **4**, 103 (1970).
213. D. L. Wood, *Trans. TMS-AIME 209:209* (1957).
214. L. Zakraysek, *Fractography and Material Science*, ASTM STP 733, L. N. Gilbertson and R. D. Zipp, eds., ASTM, 428 (1981).
215. L. Zakraysek, *Circuits Manuf.*, **23**, 38 (Dec. 1983).
216. L. Zakraysek, *Met. Finish.*, **85**, 29 (Apr. 1987).
217. L. Zakraysek, *Circuit World*, **15**, no. 4, 9 (1989).
218. K. Lin and K. G. Sheppard, *Plating Surf. Finish.*, **80**, 40 (Aug. 1993).
219. F. Delamare and G. E. Rhead, *Surf. Sci.*, **28**, 267 (1971).
220. C. L. Bianchi and J. R. Galvele, *Corrosion Sci.*, **27**, 631 (1987).
221. J. W. Chang, Ph.D. thesis, Stevens Institute of Technology, Hoboken, NJ (1988).
222. J. W. Dini and W. H. Gourdin, *Plating Surf. Finish.*, **77**, 54 (Aug. 1990).
223. M. J. Sole and T. Szendrei, *Proc. AESF SUR/FIN 91*, Toronto, Session H, Toronto, 423 (1991).
224. H. D. Merchant, *Defect Structure, Morphology and Properties of Deposits*, H. D. Merchant, ed., Minerals, Metals and Materials Soc., Warrendale, PA, 301 (1995).
225. H. D. Merchant, *J. Electron. Matl.*, **22**, 631 (1993).
226. R. J. De Angelis, D. B. Knorr, and H. D. Merchant, *Defect Structure, Morphology and Properties of Deposits*, H. D. Merchant, ed., Minerals, Metals and Materials Soc., Warrendale, PA, 87 (1995).
227. D. S. Stoychev, I. V. Tomov, and I. B. Vitanova, *J. Appl. Electrochem.*, **15**, 879 (1985).
228. Sv. Surnev and I. Tomov, *J. Appl. Electrochem.*, **19**, 752 (1989).
229. J. Cl. Puippe and F. Leaman, eds., *Theory and Practice of Pulse Plating*, American Electroplaters and Surface Finishers Soc. (1986).
230. *Proc. AESF 1st Intl. Pulse Plating Symp.*, Boston, MA (Apr. 1979).
231. *Proc. AESF 2nd Intl. Pulse Plating Symp.*, Rosemont, IL (Oct. 1981).
232. *Proc. AESF 3rd Intl. Pulse Plating Symp.*, Washington, DC (Oct. 1986).
233. C. C. Wan, H. Y. Cheh, and H. B. Linford, *Plating*, **61**, 559 (1974).
234. G. Devaraj and S. Guruviah, *Materials Chem. Phys.*, **25**, 439 (1990).
235. J. W. Dini, *Met. Finish.*, **61**, 52 (July 1963).

236. G. A. Malone, "Investigation of Electroforming Techniques," NASA-CR-134776 (Apr. 1975), and NASA-CR-134959 (Dec. 1975).
237. C. C. Wan, in *Proc. AESF 1st Intl. Pulse Plating Symp.*, Boston, MA (Apr. 1979).
238. M. R. Kalantary and D. R. Gabe, *J. Materials Sci.*, **30**, 4515 (1995).
239. T. Pearson and J. K. Dennis, *J. Appl. Electrochem.*, **20**, 196 (1990).
240. A. J. Avila and M. J. Brown, *Plating*, **57**, 70 (1970).
241. A. Hickling and H. P. Rothbaum, *Trans. Inst. Met. Finish.*, **34**, 199 (1957).
242. N. Ibl, J. Cl. Puippe, and H. Angerer, *Surf. Tech.*, **6**, 287 (1978).
243. Y. M. Polukarov et al., *Electrokhim.*, **18**, 1224 (1982).
244. M. Yokoi and T. Hayashi, *Denki Kagaku*, **46**, 195 (1978).
245. J. Cl. Puippe and N. Ibl, *Plating Surf. Finish.*, **67**, 68 (June 1980).
246. G. Devaraj and S. K. Seshadri, *Plating Surf. Finish.*, **79**, 72 (Aug. 1992).
247. D. S. Stoychev and M. S. Aroyo, *Proc. SUR/FIN 96*, Cleveland, Session U, 851 (1996).
248. W. F. Hall and A. R. Chaudhuri, *Proc. AESF 10th Symp. on Plating in the Electronics Industry*, San Francisco (1983).
249. A. R. Despic and K. I. Popov, *J. Appl. Electrochem.*, **1**, 275 (1971).
250. K. I. Popov, M. D. Maksimovic, and D. C. Totovski, *Surf. Technol.*, **17**, 125 (1982).
251. H. Y. Cheh, H. B. Linford, and C. C. Warr, American Electroplaters Society Research Project 35 (1977).
252. K. I. Popov, D. C. Totovski, and M. D. Maksimovic, *Surf. Technol.*, **19**, 181 (1983).
253. H. K. Ng, A. C. C. Tseung, and D. B. Hibbart, *J. Electrochem. Soc.*, **127**, 1034 (1980).
254. T. Pearson, J. K. Dennis, J. F. Houlston, and M. Dorey, *Trans. Inst. Met. Finish.*, **69**, 9 (1991).
255. T. Pearson and J. K. Dennis, *Trans. Inst. Met. Finish.*, **69**, 75 (1991).
256. A. Montgomery, *Circuit World*, **15**, 33 (Jan. 1989).
257. M. R. Kalantary, D. R. Gabe, and M. Goodenough, *Met. Finish.*, **89**, 21 (Apr. 1991).
258. D. Engelhaupt, "Pulse Plating of Copper for Printed Wiring Boards," Martin Marietta Orlando Aerospace, Report OR 18,258 (March 1985).
259. S. Coughlan, *Trans. Inst. Met. Finish.*, **73**, 54 (1995).
260. G. Holmbom and B. E. Jacobsson, *Surface and Coatings Technology*, **35**, 333 (1988).
261. P. Leisner, G. Bech-Nielsen, and P. Moller, *Proc. AESF SUR/FIN 92*, Atlanta, Session S, 1011 (1992).
262. E. K. Yung, L. T. Romankiw, and R. C. Alkire, *Electrodeposition Technology, Theory and Practice*, Electrochemical Society, L. T. Romankiw and D. R. Turner, eds., 75 (1987); *J. Electrochem. Soc.*, **136**, 206 (1989).
263. N. R. K. Vilambi and D.-T. Chin, *Plating Surf. Finish.*, **75**, 67 (Jan. 1988).
264. D. E. Engelhaupt, U.S. Patent 5,326,454 (July 5, 1994).
265. C. C. Wan, H. Y. Cheh, and H. B. Linford, *J. Appl. Electrochem.*, **9**, 29 (1979).
266. K. I. Popov, M. V. Vojnovic, LJ. M. Vracar, and B. J. Lazarevic, *J. Appl. Electrochem.*, **4**, 267 (1974).
267. D. R. Rudder, *Proc. AESF SUR/FIN 85*, Detroit, Session F (July 1985).
268. J. R. White and R. T. Galasco, *Plating Surf. Finish.*, **75**, 122 (May 1988).
269. J. R. White, *J. Appl. Electrochem.*, **17**, 977 (1987).
270. J. Mann, *Trans. Inst. Metal Finish.*, **56**, 70 (1978).
271. C. D. Zhou, R. P. Renz, E. J. Taylor, E. C. Stortz, and B. Grant, *Proc. AESF SUR/FIN 96*, Cleveland, 517 (June 1996).

272. B. S. Sheshadi, *Plating Surf. Finish.*, **65**, 43 (July 1978).
273. N. Ibl, *Proc. 2nd Int. Pulse Plating Symp.*, Rosemont IL (Oct. 1981).
274. C. C. Wan, *Proc. 2nd Int. Pulse Plating Symp.*, Rosemont, IL (Oct. 1981).
275. D.-T. Chin, *J. Electrochem. Soc.*, **130**, 1657 (1983).
276. C. W. Yeow and D. B. Hibbert, *J. Electrochem. Soc.*, **130**, 786 (1983).
277. R. J. Clauss and N. C. Adamowicz, *Plating Surf. Finish.*, **57**, 236 (Mar. 1970).
278. J. K. Dennis and T. E. Such, *Nickel and Chromium Plating*, 3rd ed., Woodhead Publ. Cambridge, England, 1993.
279. G. A. DiBari, private communication, Sept. 1996.
280. G. A. DiBari, *Met. Finish.*, **75**, 17 (June 1977); **75**, 17 (July 1977).
281. G. A. DiBari, *Plating Surf. Finish.*, **64**, 68 (May 1977).
282. J. C. Jongkind, *Plating Surf. Finish.*, **62**, 1135 (1975).
283. S. K. Jalota, *Electroplating of Zinc Die Castings*, Bath Press, Bath, Avon (1987).
284. R. Suchentrunk, Metallizing of Plastics—A Handbook of Theory and Practice, R. Suchentrunk, ed., Finishing Publications, Orlando, FL, 1993.
285. W. Meyer, *Metalloberflache*, **34**, 489 (1980).
286. H. E. Haring and W. Blum, *Trans. Am. Electrochem. Soc.*, **44**, 313 (1923).
287. D. R. Turner, *Plating Surf. Finish.*, **66**, 32 (July 1979).
288. E. K. Yung and L. T. Romankiw, *Electrodeposition Technology, Theory and Practice*, L. T. Romankiw and D. R. Turner, eds., Electrochemical Society, 607 (1987); *J. Electrochem. Soc.*, **136**, 756 (1989).
289. D. A. Hazlebeck and J. B. Talbot, *Proc. Electrochemical Technology in Electronics*, vol. 88-23, L. T. Romankiw and T. Osaka, eds., Electrochemical Soc., 143 (1988).
290. T. Kessler and R. Alkire, *Plating Surf. Finish.*, **63**, 22 (Sept. 1976).
291. T. Kessler and R. Alkire, *J. Electrochem. Soc.*, **123**, 990 (1976).
292. L. Mayer and S. Barbieri, *Plating Surf. Finish.*, **68**, 46 (Mar. 1981).
293. S. Barbieri, L. Mayer, and J.-H. Shyu, *Trans. Inst. Met. Finish.*, **62**, 25 (1984).
294. G. L. Fisher, W. Sonnenberg, J. Bladon, and R. Bernards, *Trans. Inst. Met. Finish.*, **68**, 50 (1990).
295. S. Yamato, C. Kan, and A. Ohharada, *Proc. AESF SUR/FIN 92*, Atlanta, Session L, 679 (June 1992).
296. P. Davidson, *Proc. AESF SUR/FIN 86*, Philadelphia, Session I, 27 (June 1986).
297. M. Carano, in *Proc. 13th AESF Symp. on Plating in the Electronics and Printed Circuit Industry*, 14 (1986).
298. S. A. Amadi, D. R. Gabe, and M. R. Goodenough, *Met. Finish.*, **91**, 23 (May 1993).
299. E. J. Glantz, *PC Fab.*, **12**, 60 (Feb. 1989).
300. M. C. Hsiao and C. C. Wan, *J. Electrochem. Soc.*, **138**, 2273 (1991).
301. D. P. Davidson, *Circuits Manufacturing*, **27**, 36 (Aug. 1987).
302. L. J. Mayer and S. C. Barbieri, *Plating Surf. Finish.*, **68**, 46 (1981).
303. R. Haak, C. Ogden, and D. Tench, *Plating Surf. Finish.*, **68**, 59 (Oct. 1981).
304. W. Engelmaier and T. Kessler, *J. Electrochem. Soc.*, **125**, 36 (1978).
305. F. Byle and P. Bratin, *Proc. AESF SUR/FIN 90*, Session H, Boston, (July 1990).
306. D. R. Turner, *Plating Surf. Finish.*, **63**, 41 (Jan. 1976).
307. H. H. Wan and J. E. McCaskie, *Plating Surf. Finish.*, **78**, 80 (Nov. 1991).
308. J. W. Dini, *Plating Surf. Finish.*, **76**, 64 (Sept. 1989).

309. T. Barringer and M. Carano, *Plating Surf. Finish.*, **73**, 36 (Mar. 1986).
310. A. Elbs and J. Rasmussen, *Galvanotechnik*, **85**, 3998 (1994).
311. D. A. Luke, *Trans. Inst. Met. Finish.*, **51**, 113 (1973).
312. X. Ye, M. DeBonte, J. P. Celis, and J. R. Roos, *J. Electrochem. Soc.*, **139**, 1592 (1992).
313. S. Kang, J.-S. Yang, and D. N. Lee, *Plating Surf. Finish.*, **82**, 67 (Oct. 1995).
314. A. Fox, *J. Testing and Evaluation*, **4**, no. 1, 74 (Jan. 1976).
315. H. D. Merchant, *J. Electron. Mater.*, **24**, 919 (1995).
316. H. D. Merchant, *Defect Structure, Morphology and Properties of Deposits*, H. D. Merchant, ed., Minerals, Metals and Materials Soc., Warrendale, PA, 1 (1995).
317. J. O. Dukovic, *Advances in Electrochemical Science and Engineering*, H. Gerischer and C. W. Tobias, eds., VCH, Germany, 117 (1994).
318. L. T. Romankiw, "A Review of Plating through Polymeric Resist Masks," *Extended Abstracts of the Electrochemical Society*, 79-2, Abstract No. 462, 1165–1166, Electrochemical Soc., Pennington, NJ (1979).
319. L. T. Romankiw, *Oberflache-Surface*, **25**, 238 (1984).
320. L. T. Romankiw and T. A. Palumbo, *Electrodeposition Technology, Theory and Practice*, L. T. Romankiw and D. R. Turner, eds., Electrochemical Society, 13 (1987).
321. L. T. Romankiw, "Electrochemical Technology in The Electronics Industry and Its Future," *New Materials and New Processes*, JEC Press, Cleveland, OH, 1988, 3, 39.
321a. P. C. Andricacos, C. Uzoh, J. O. Dukovic, J. Horkans, and H. Deligianni, *IBM J. Res. Develop.*, **42**, 567 (1998).
322. K. Kondo, K. Fukui, K. Uno, and K. Shinohara, *J. Electrochem. Soc.*, **143**, 1880 (1996).
323. J. O. Dukovic, *IBM J. Res. Develop.*, **37**, 125 (Mar. 1993).
324. C. Bocking, *Trans. Inst. Met. Finish.*, **69**, 119 (1991).
325. R. J. von Gutfeld, R. E. Acosta, and L. T. Romankiw, *IBM J. Res. Develop.*, **26**, 136 (Mar. 1982).
326. J. Cl. Puippe, R. E. Acosta, and R. J. von Gutfeld, *J. Electrochem. Soc.*, **128**, 2539 (1981).
327. P. Bindra, G. V. Arbach, and U. Stimming, *J. Electrochem. Soc.*, **134**, 2893 (1987).
328. P. Bindra, D. Light, G. V. Arbach, and U. Stimming, *Proc. Electrochemical Technology in Electronics*, vol. 88-23, L. T. Romankiw and T. Osaka, eds., Electrochemical Soc., 341 (1988).
329. W. Paatsch, W. Kautck, and N. Sorg, *Proc. AESF SUR/FIN 90*, Session I, Boston (July 1990).
330. M. J. Sole, *J. Metals*, **46**, 29 (June 1994).
331. R. Suchentrunk, *Trans. Inst. Met. Finish.*, **64**, 19 (1986).
332. G. A. Malone, *Proc. AESF SUR/FIN 79*, Session H, Atlanta (June 1979).
333. M. R. Louthan Jr., "The Effect of Hydrogen on Metals," in *Corrosion Mechanisms*, F. Mansfeld, ed., Marcel Dekker, New York, 1987.
334. M. Koiwa, A. Yamanaka, M. Arita, and H. Numakura, *J. Inst. Metals*, **30**, 991 (1989).
335. J. R. Denchfield, U.S. Patent 3,616,330 (1971).
336. F. T. Schuler and F. Mansfeld, *J. Vac. Sci. Technol.*, **12**, 758 (July–Aug. 1975).
337. D. Engelhaupt, private communication, July 1996.
338. J. W. Dini, *Thin Solid Films*, **95**, 123 (1982).
339. J. C. Farmer et al., *Plating Surf. Finish.*, **75**, 48 (Mar. 1988).
340. W. K. Kelley, J. W. Dini, and C. M. Logan, *Plating Surf. Finish.*, **69**, 54 (Mar. 1982).

341. V. V. Gubin, L. T. Zhuravlev, B. M. Platonov, and Yu. M. Polukarov, *Soviet Electrochem.*, **20** (5), 671 (1984).

342. R. D. Srivastava and S. Kumar, *Trans. Inst. Met. Finish.*, **50**, 102 (1972).

343. N. V. Shanmugam and S. J. Pr. Thangavelu, *Trans. Met. Finish. Assoc. India*, **3** (3), 27 (July–Sept. 1994).

344. L. A. Doggrell and G. D. Hooper, *Proc. AES Int. Symp. on Electroforming/Deposition Techniques*, Los Angeles, CA (March 1983).

345. A. Mayer, J. L. Uher, and W. A. Wright, *Plating Surf. Finish.*, **70**, 29 (Oct. 1983).

346. W. H. Safranek, *Plating Surf. Finish.*, **69**, 48 (Apr. 1982).

347. G. R. Schaer and T. Wada, *Plating Surf. Finish.*, **68**, 52 (July 1981).

348. G. R. Schaer and J. R. Preston, U.S. Patent 4,597,836 (1986).

349. L. J. J. Janssen, *J. Appl. Electrochem.*, **18**, 339 (1988).

350. M. F. El-Shazly, J. L. White, and E. W. Brooman, *Proc. AESF 3rd Int. Pulse Plating Symp.*, American Electroplaters and Surface Finishers Soc., Orlando, FL, (Oct. 1986).

351. D. A. Uceda and T. J. O'Keefe, *J. Appl. Electrochem.*, **20**, 327 (1990).

352. P. Kutzschbach, W. Rempt, K.-D. Baum, and H. Liebscher, *Wire*, **45** (6), 336 (1995).

353. H. J. Wiesner and W. J. Frey, *Plating Surf. Finish.*, **66**, 51 (Feb. 1979).

354. S. T. Rao and L. Trager, U.S. Patent 4,551,212 (Nov. 1985).

355. M. S. Abrahams, S. T. Rao, C. J. Buiocchi, and L. Trager, *J. Electrochem. Soc.*, **133**, 1786 (1986).

356. J. W. Dini, T. G. Beat, W. C. Cowden, L. E. Ryan, and W. B. Hewitt, *Proc. SUR/FIN 92*, Session R, Atlanta (June 1992).

357. H. Takeo, C. Tam, and J. Dash, "Magnetic Effects on Electroplating of Copper," *Proc. AES 11th Plating in the Electronics Industry Symp.*, Orlando, FL (Feb. 1984).

358. M. I. Ismail and T. Z. Fahidy, *J. Appl. Electrochem.*, **11**, 543 (1981).

359. K. I. Popov, LJ. M. Djukic, M. G. Pavlovic, and M. D. Maksimovic, *J. Appl. Electrochem.*, **9**, 527 (1979).

360. J. Jorne and M. G. Lee, *J. Electrochem. Soc.*, **143**, 865 (1996).

361. R. R. Adzic, in *Advances in Electrochemistry and Electrochemical Engineering*, vol. 13, H. Gerischer, ed., Wiley, New York, 1984, 159.

362. D. M. Kolb, in *Advances in Electrochemistry and Electrochemical Engineering*, vol. 11, H. Gerischer and C. Tobias, eds., Wiley, New York, 1978, 125.

363. J. Horkans, I.-C. H. Chang, P. C. Andricacos, and H. Deligianni, *J. Electrochem. Soc.*, **142**, 2244 (1995).

364. J. C. Farmer, *J. Electrochem. Soc.*, **132**, 2640 (1985).

365. J. S. Hammond and N. Winograd, *J. Electrochem. Soc.*, **124**, 826 (1977).

Part B
Cyanide Copper*

18 HISTORY AND DEVELOPMENT

Before 1915 copper cyanide solutions were invariably prepared by dissolving copper carbonate in alkali cyanide. The use of tartrates is mentioned in early references [366, 367], but their value apparently was not fully recognized. An exception was the combination cleaning and plating solution described by Watts [368] in 1915. Deposits from these early solutions were relatively thin and were used largely as bases for oxidized and other decorative types of finishes.

The first commercially successful high-efficiency solution was introduced in 1938. It was used extensively throughout the automotive industry in subsequent years for plating zinc die castings as well as steel parts. High rates of deposition were achieved by operating at temperatures around 80°C. Alkali was added to improve anode corrosion and to increase the conductivity of the solution [369]. Alkali thiocyanate was used by Wernlund [370] in 1941 to brighten the deposits. Successful commercialization of the process required the use of specific surface active agents to sequester organic contaminants and prevent pitting [371]. This solution permitted deposition of smooth, bright deposits in the thickness range of 25 to 50 μm. The basic formulations of the proprietary high-efficiency solutions now in commercial operation are similar to these, except that thiocyanate and other active sulfur containing constituents are not recommended in the new processes.

The next major advance in high-efficiency cyanide copper solutions was achieved by employing periodic reverse plating cycles [372]. Other current manipulation techniques such as interrupted current are occasionally used to modify the properties of the coatings. A wide variety of addition agents have been developed and this is covered in a later section of this chapter. Reviews of cyanide copper plating can be found in [373–379].

19 APPLICATIONS

Cyanide copper plating is used throughout the metal finishing industry for many applications, although not as extensively today as in the 1970s because of environmental issues. Horner [380] estimates that present usage is less than 50% of that in the 1970s. Possible site contamination, worker safety considerations, and high waste treatment and reporting costs are some drawbacks [381].

Applications include as an undercoat for other deposits to protect basis metal or promote adhesion; for surface improvement in buffing, soldering, lubricity, rotogravure, printing rolls, decorative plating of zinc die castings [382–385]; in plating of steel parts with copper before nickel and chromium plating [384–385]; as a stopoff to prevent case hardening on selected areas of ferrous metal surfaces; and in

*Revision of section by R. R. Bair and A. K. Graham in the 1974 edition.

plating of coinage [386, 387]. A high-speed cyanide copper-plating process is used to produce heavy coatings of copper on steel wire for electrical use. For noncorrosive environments, aluminum is often plated with copper cyanide after zincating. This is followed by a decorative nickel-chromium coating [388].

20 FUNCTIONS OF MAJOR SOLUTION CONSTITUENTS

20.1 Copper Cyanide

Typical formulations for cyanide solutions are give in Table 7. Copper cyanide is insoluble in water but dissolves in aqueous solutions containing alkali metal cyanides to form four stable soluble complexes and free species in water: $CuCN_{aq}$, $[Cu(CN)_2]^-$, $[Cu(CN)_3]^{2-}$, $[Cu(CN)_4]^{3-}$ and Cu^+ and CN^-. In all four oxidation complexes, the oxidation number of copper is +1. A model of copper discharge from cyanide electrolyte was developed by Dudek and Fedkiw [389].

Deposition occurs from the $Cu(CN)_2^-$ and $Cu(Cn)_3^{2-}$ complexes [390]. Cuprous cyanide complexes shift the copper deposition potential to more negative values which avoid displacement deposition on less noble substrates. The predominant copper cyanide species discharged in copper strike plating is proposed to be $Cu(CN)_3^{2-}$; although $Cu(CN)_4^{3-}$ is at a higher concentration, it is not as electroactive. These kinetics give rise to the inherent good "macroscopic throwing power" observed in the cyanide solution [391]. Additional information on kinetics can be found in [392].

The Taguchi method was used to evaluate a number of parameters in a copper cyanide solution. An increase in copper concentration and a decrease in potassium cyanide resulted in good current efficiency over a wide current density range. Hydroxide and carbonate concentration were of secondary importance [393].

TABLE 7 Copper Cyanide Solution Formulations

	Strike ($g\,liter^{-1}$)		Rochelle ($g\,liter^{-1}$)		High Efficiency ($g\,liter^{-1}$)	
Constituent	Typical	Limits	Typical	Limits	Typical	Limits
CuCN	22	15–30	26	19–45	75	49–127
NaCN	33	23–48	35	26–68	102	62–154
or						
KCN	43	31–64	—		136	76–178
Na_2CO_3	15	0–15	30	15–16		
NaOH	—	—	—		15	22–37
or						
KOH	—	—	—		15	31–52
Rochelle salt ($KNaC_4H_4O_6 \cdot 4H_2O$)	15	15	45	30–60	—	—
By analysis						
Copper	16	11–21	18	13–32	55	34-89
Free cyanide	9	6–11	6	4–9	19	10–20

20.2 Free Cyanide

The alkali cyanide in excess of the tricomplex salt $Na_2[Cu(CN)_3]$ is termed *free cyanide*. The ratio of the free cyanide to the copper metal content in the strike and Rochelle solutions is somewhat higher than in the high efficiency solutions. This accounts, in part, for the lower cathode efficiency. Hydrogen is simultaneously liberated at the cathode in the strike and Rochelle solutions, and this hydrogen evolution provides additional cleaning. Strike and Rochelle solutions are also operated at a relatively high free cyanide-to-metal ratio in order to produce the desired type of deposit. The excellent adhesion of strike deposits can be attributed largely to the high free cyanide combined with low metal which minimizes any tendency to form nonadherent immersion deposits. Free cyanide is essential in all cyanide copper plating solutions in order to obtain good corrosion of the copper anodes. If it is too low, the anodes polarize and become coated with an insulating film. Free cyanide contributes to conductivity in strike and Rochelle solutions, but not significantly in high efficiency solutions that contain alkali and higher concentrations of the complex copper ions. The concentration of the free cyanide increases at higher temperatures, since lower complexes are formed and free cyanide is thereby liberated [390]. Free cyanide is normally determined by titrating a sample of the solution, at or below room temperature, with silver nitrate, using potassium iodide as an indicator.

20.3 Sodium or Potassium Hydroxide

Sodium or potassium hydroxide is added to high-efficiency solutions primarily to provide good electrical conductivity and to improve throwing power. Alkali is also essential in these solutions for good anode corrosion. The brightness of the deposits from high-efficiency solutions is also influenced by the alkali metal hydroxide. Usually only small amounts of caustic are added to strike or Rochelle solutions for adjusting or maintaining the pH within the proper range. Many users of high-efficiency cyanide copper solutions base their formulations on sodium cyanide to complex the cyanide copper and add potassium hydroxide as a means of introducing the potassium ion.

21 COMPARISON OF SODIUM AND POTASSIUM FORMULATIONS

The complex salts formed with sodium or potassium cyanide are similar in composition but the potassium salts are more soluble. Horner [394] compared the throwing power and efficiencies of sodium and potassium formulations and the effects of free cyanide, carbonate content, and other factors. He concluded that potassium salts offered no advantage for cathode efficiency over sodium salts but did show improved throwing power. Hatherley et al. [387] reported that sodium and potassium formulations each possess various advantages and disadvantages. Conductivity is higher in potassium solutions, especially at high concentrations of carbonate. Cathode polarization curves also favor potassium formulations over sodium. By contrast, sodium solutions exhibit less of a risk of anode passivation.

Juhos et al. [395] also found that sodium formulations favored the anode reactions. Hatherley et al. [387] suggest that commercial considerations might be used to specify optimum carbonate level where the advantage of high conductivity has to be set against reduced cathode efficiency.

21.1 Carbonate

Carbonate exerts a strong buffer action at a pH of 10.8 to 11.5 in strike and Rochelle solutions [396, 397] and facilitates pH control. It also reduces anode polarization in these solutions. No beneficial effects of carbonate have been observed in high-efficiency solutions that contain substantial amounts of caustic. Commercial high-efficiency cyanide copper solutions contain an average of about 50 g $liter^{-1}$ sodium carbonate. It is formed by oxidation of the cyanide radical at the surface of polarized anodes or insoluble anodic metal surfaces. Absorption of carbon dioxide from the air by the caustic alkali in the solution and hydrolysis of the cyanide are other sources of carbonates formed in the high-efficiency solutions.

The oxidation of cyanide should be avoided because it increases the cyanide consumption, causes a rapid buildup of carbonates, and complicates control of free cyanide. Concentrations up to about 90 g $liter^{-1}$ sodium carbonate or 120 g $liter^{-1}$ potassium carbonate have no significant effect in reducing the plating speed of high-efficiency solutions. If maximum plating speed is required, however, the concentrations should be controlled below these levels.

21.2 Tartrates

Tartrates are used primarily in the low-efficiency solutions. It is believed that the potassium-sodium tartrate or Rochelle salt ($KNaC_4H_4O_6 \cdot 4H_2O$) forms temporary complexes with copper by reacting with products of electrolysis produced in the anode film. Solutions containing tartrates may be operated with lower free cyanide and at higher current densities and efficiencies without impairing anode corrosion [396, 398]. Obtaining better quality deposits by the use of Rochelle salt is associated in part with the formation of complex salts which are probably present in the cathode film. Solutions are normally operated with about 45 g $liter^{-1}$ Rochelle salt. Wagner and Beckwith [398] claimed that the optimum concentration was about 22.5 g $liter^{-1}$, with somewhat lower efficiencies being obtained at higher values. The substitution of sodium-potassium citrate for Rochelle salt [399] has been investigated and several other proprietary compounds of a similar type are claimed to offer advantages.

22 ADDITION AGENTS

Additives for cyanide copper systems include compounds having active sulfur groups and/or containing metalloids such as selenium or tellurium. Other agents that have worked are organic amines or their reaction products with active sulfur containing compounds; inorganic compounds containing metals such as selenium, tellurium, lead, thallium, antimony, arsenic; nitrogen, and sulfur heterocyclic compounds; and unsaturated alcohols, saccharin, and polyethylene glycol. An extensive listing of additives used in acid and cyanide copper prior to 1959 can be found in [374] and more recent information in [400–411]. Of interest are data from Table 2 in the section on acid copper plating which lists only one U.S. patent for cyanide copper plating since the mid-1970s compared to over 60 for acid copper.

23 SOLUTION TYPES: STRIKE AND ROCHELLE SOLUTIONS

Copper cyanide solutions may be classified in three categories: strike, Rochelle, and high-efficiency. Strike solutions are used extensively to apply relatively thin coatings of copper as an undercoating for other metals. Rochelle solutions, in general, are not as sensitive to contamination as the high-efficiency solutions and are used in applications where an intermediate thickness of copper is applied. Deposits from these solutions vary from dull to semibright; the brightness or luster is not of major importance for most applications.

Copper strike solutions are used for applying a thin coating of copper on aluminum and zinc die castings before plating with other metals [384]. The adherent copper coating prevents chemical attack when these substrates are subsequently plated with heavier coatings from other solutions. The thickness of strike deposits normally varies from about 0.5 to 1 µm, although it may sometimes be as high as 2.5 µm. The heavier strike deposits must be used when the subsequent plating operation is from an acid electrolyte. For instance, the copper strike deposit on zinc die castings must be three to five times as heavy for subsequent acid copper plating as compared with cyanide copper plating.

The bond strength between electroless nickel and a variety of aluminum alloys is higher when a copper strike is used [412]. Besides showing good adhesion of subsequent electrodeposits, the copper strike from cyanide solutions is an additional cleaning operation. Furthermore a thin copper strike (1.3 µm) under lead coatings over steel has shown improved protective value in atmospheric exposure tests compared to lead coatings alone [413, 414]. Aluminum parts can be plated in a barrel if the free cyanide in the copper strike is high enough (15–20 g $liter^{-1}$) to prevent displacement of copper on zinc [415]. A good review of copper cyanide striking is provided by Mohler [416].

24 OPERATING CONDITIONS AND SOLUTION CHARACTERISTICS

The operating conditions of strike and Rochelle solutions are given in Table 8. Strike solutions are usually operated in the temperature range of 40° to 60°C. The cathode efficiencies are relatively low (10%–60%) in the normal plating range of 1.0 to 3.2 A dm^{-2}. The solution composition and operating characteristics vary widely in actual commercial practice. Many users prefer to add 15 to 30 g $liter^{-1}$ sodium carbonate. When steel parts are plated, 15 to 30 g $liter^{-1}$ sodium hydroxide may be added to improve the conductivity. Rochelle salt, or its equivalent, is sometimes used to improve anode and cathode characteristics.

Because of the higher cathode efficiency, heavier deposits can be more practically produced from Rochelle solutions than from strike solutions. These plating characteristics are achieved by operating at higher temperatures, controlled pH, and higher metal concentration, and adding substantial amounts of Rochelle salt to the

TABLE 8 Operating Conditions for Typical Cyanide Solutions

Conditions	Strike	Rochelle	High Efficiency
Temperature (°C)	40–60	55–75	60–80
Cathode current density ($A\,dm^{-2}$)	0.5–4	1.5–6.5	1–11
Anode current density($A\,dm^{-2}$)	0.5–1	0.8–3.3	1.5–5
Efficiency range			
Cathode (%)	10–60	30–90	70–99+
Anode (%)	95–100	50–70	—
Ratio of anode to cathode area	3:1	2:1	3:2
Agitation			
Cathode rod	Optional	Preferred	Either or both
Air			
Solution voltage (V)	6	6	0.75–4
Limiting thickness (μm)	2.5	13	>25

solution. The average cathode efficiency is approximately 50% higher than that obtained from strike solutions. The anode and cathode efficiencies vary with the temperature and degree of agitation [396]. Rochelle solutions are normally operated without metal-containing addition agents, but lead had been used as a grain refining and brightening agent [417].

Polarization phenomena in copper cyanide solutions have been extensively investigated [387, 390, 396, 418, 419]. Hatherley and coworkers [387] reported that potassium and sodium based solutions may both be operated at close to 100% cathode efficiencies, but that potassium solutions can sustain efficient copper deposition to a higher maximum current than can sodium. They also noted that increasing free cyanide diminishes the limited current density that can be sustained at maximum efficiency, and this relationship is linear for both sodium and potassium solutions. An increase in the carbonate or hydroxide content is also associated with a reduction in the maximum current for efficient copper deposition. Last, the efficiency of solutions cannot be restored by precipitation of carbonate with lime, nor by any of the other chemical agents that have been proposed in the past [387, 390].

Graham and Read [396, 418, 419] defined the limiting current density beyond which excessive polarization occurs for a given set of solution variables. Excessive polarization causes insulation of the anodes and usually necessitates their removal for cleaning. Variations in metal content and free cyanide have little effect on the limiting current density. If the carbonate content is too high, the copper may act as an insoluble anode without exhibiting excessive polarization. Reducing the pH from 12.8 to as low as 10.3 does not greatly alter the limiting current density. The presence of tartrates does not overcome the tendency of the anodes to polarize excessively, but the allowable current density is raised when both carbonate and tartrate are present. If insoluble iron anodes are used along with copper anodes, higher anode current densities may be used because the iron anodes depolarize the copper.

The throwing power of Rochelle solutions is superior to that of high-efficiency cyanide solutions. Most of the published data on throwing power are limited to current densities below 1.0 $A\,dm^{-2}$. The factors influencing the throwing power of Rochelle solutions when operated at more practical current densities and temperatures may be quite different. The greatest single factor contributing to good throwing power and metal distribution appears to be associated with a decrease in cathode efficiency with increasing current density. High efficiencies, in general, contribute to poor throwing power, but there are also other factors, as discussed later.

25 MAINTENANCE AND CONTROL

Copper strike solutions in commercial use vary in metal content but are usually maintained at a copper metal concentration of about 10 to 16 g liter^{-1}. The metal content increases during operation because the copper anode efficiency is higher than the cathode efficiency. It is common practice to use a combination of steel and copper anodes in the strike solutions to obtain an overall anode efficiency equal to the average cathode efficiency. The control of the critical free sodium cyanide is simplified, and a more stable solution composition is thereby attained. The free sodium cyanide should be controlled within the range of 6 to 11 g liter^{-1}.

Continuous filtration of the strike solutions through activated carbon is recommended for the best results. This prevents particle roughness on the shelf areas of the plated parts and minimizes drag in of organic contamination into subsequent plating solutions.

Control of the pH of Rochelle solutions is important for obtaining the best performance. Operation within the pH range of 12.2 to 12.8 is recommended [396]. The range 12.5 to 12.8 is preferred because the solution is buffered in this range by the carbonate. Operation at a pH in excess of 12.9 should be avoided, owing to decreased anode efficiencies and chemical attack of the solution on zinc die castings.

The pH of Rochelle copper solutions can be determined by a pH meter, by colorimetric methods, or by titrating with hydrochloric or sulfuric acid. The first two methods are rapid and simple; the titration method, however, has the advantage of indicating directly the approximate excess of caustic above the sulfo-orange sodium carbonate endpoint. The pH at the sodium carbonate buffer point is approximately 11.5; it can be raised to 12.7 by the addition of 0.75 g liter^{-1} caustic soda.

Controlling the free cyanide concentration of Rochelle solutions within the range of 4 to 9 g liter^{-1} is also important. Excessive anode polarization is encountered at very low free cyanide and the quality of the deposits is also adversely affected. Operating with a significantly higher free cyanide concentration than recommended results in lower cathode efficiencies and in the production of dull deposits that are difficult to buff.

26 HIGH-EFFICIENCY CYANIDE COPPER SOLUTIONS

Solutions that are formulated and operated under conditions that give essentially 100% anode and cathode efficiencies are classified in this category. The rates of deposition are appreciably higher than can be obtained from strike or Rochelle solutions.

Many of the high-efficiency cyanide copper solutions are used for plating of automotive bright trim parts. As pointed out earlier, Rochelle solutions may be used for plating zinc die castings. Most installations, however, use high-efficiency solutions in order to obtain the required thickness relatively quickly. The brightness of deposits from the high efficiency solutions is also an important factor.

Copper is also plated from high-efficiency solutions on selected surfaces of ferrous metal parts to prevent carbon penetration during case hardening. Thickness requirements vary from 13 to 25 μm, depending on the type of carburizing medium used, time of treatment in the carburizing solution, and type of surface finish on the steel. Another use of high-efficiency solutions is the plating of steel wire with a heavy coating of copper [420–422]. Part-to-part spread in deposit thickness for barrel plating is considerably less for alkaline solutions (cyanide, pyrophosphate, and amine) than for acid solutions (sulfate and fluoborate) [423].

27 OPERATING CONDITIONS AND SOLUTION CHARACTERISTICS

High-efficiency cyanide solutions are normally operated in the temperature range of 60° to 80°C, (see Table 8), with the higher temperature preferred for maximum plating speed. Many of the high-efficiency electrolytes are operated at temperatures as low as 60°C to fit specific needs and equipment. It is emphasized that a reduction in solution temperature can restrict the width of bright plating range and maximum usable current densities. The cathode current density range in commercial practice varies from about 1 to 11 A dm^{-2}, depending on the contour of the parts being plated. Automotive parts are generally plated at current densities in the range of 2 to 5.5 A dm^{-2}, whereas wire is plated from high metal solutions at current densities as high as 11 A dm^{-2}. Maximum plating speeds are achieved with potassium formulated solutions containing high metal and operated with vigorous solution circulation. The solutions must be operated at substantially 100% cathode efficiency in order to produce good quality deposits and to avoid hydrogen evolution at the cathode, which produces dull, burned deposits.

These high-efficiency solutions would normally be expected to have poor throwing power. However, metal distribution on plated parts is better than would be predicted from the efficiency factor alone. One reason for this is that the high concentration of salts, particularly alkali, and high operating temperatures greatly improve the conductivity and reduce the difference in current density on protruding and recessed areas. Rochelle salt solutions have good throwing power because the cathode efficiency decreases with increase in current density.

The anode current densities may vary from 1.5 to 5 A dm^{-2}. The anode efficiency under these conditions is essentially 100%. At very low anode current densities (0.2–0.6 A dm^{-2}), particles of copper may be formed or generated at the anode surface through intercrystalline corrosion. The particles settle on shelf areas of the parts and produce serious roughness. Certain types of proprietary organic addition agents are claimed to play an important role in improving anode corrosion [424]. They apparently form a thin organic film over the anode surface and thereby promote smooth anode corrosion and minimize any tendency to generate copper particles.

Solution circulation or movement of the electrolyte is also important in obtaining performance. In addition to increasing the plating rate, vigorous circulation is effective in minimizing pitting caused by hydrogen bubbles clinging to the surface of parts being plated. Circulation is accomplished commercially by pumping the electrolyte or by air agitation.

The plating characteristics of some of the high-efficiency solutions can be improved by utilizing current manipulation techniques (discussed in more detail in the section on acid copper). Parts are plated in the conventional manner for a selected time and are then deplated for a shorter period by reversing the current. This technique is referred to as periodic current reversal plating (PR). Direct plating cycles of 2 to 40 s with 0.5 to 10 s reverse (deplating) cycles were described by Jernstedt [372]. Longer cycles, in excess of 60 s direct with reverse cycles in excess of 12, were disclosed by Wernlund [425]. Ismail [426] reported that surface roughness increased with frequency and amplitude of the PR current used. Increasing the solution temperature led to brighter surfaces up to 50°C, and a PR ratio of two gave the best surface brightness. The PR plate was brighter and had a higher corrosion resistance than the corresponding direct current plated deposits. A number of production applications utilizing PR are discussed in [427].

Current interruption cycles are also used. For example, zinc die castings are frequently plated using a 10 s plating cycle followed by interrupting the current for 1 s and then repeating the cycle. This approach can provide excellent deposit brightness from bright plating solutions which are contaminated sufficiently that acceptable deposits cannot be produced by use of continuous direct current.

One of the advantages gained by employing periodic current reversal or interrupted cycles is leveling. The degree of leveling obtained is greatest with PR, particularly with relatively long reversal cycles. Deposits so plated show a laminar structure, whereas those plated with conventional direct current are columnar. The leveling obtained with current interruption is less than with current reversal, but it is adequate for covering minor surface marks such as buffing wheel lap marks on zinc die castings. The uniformity of distribution of the copper on irregularly shaped parts is also improved when current reversal is used. It prevents excessive buildup of metal on high current density areas and yields significant savings in anode consumption.

The choice between periodic current reversal and current interruption greatly favors the latter. Periodic reversal has a low net rate of deposition because on the reversal the copper is being depleted. As an example, a 60/20 PR cycle has only a 50% net deposition rate if the current applied is the same for the plate and deplete cycles. The use of PR has not had broad acceptance because of this factor. Rudder [428] compared pulse plating of cyanide copper versus acid copper and found that the leveling and thickness performance of the pulse plated cyanide copper was better than that of the acid copper.

28 MAINTENANCE AND CONTROL

High-quality copper deposits can be produced consistently from the high-efficiency solutions by following good plating practice. Analytical methods are available for determining the concentrations of the normal solution components [429–433]. Methods are also available for determining contaminants such as carbonates, ferrocyanide, and zinc. The Hull cell is used extensively for detecting organic contamination, controlling addition agents, and determining the effectiveness of various purification treatments [434–436].

The composition of the solutions in commercial operations varies widely. The copper metal content of the average solution is maintained at about 45 to 55 g liter^{-1}.

Solutions used for plating steel wire or rod, however, are usually maintained at a metal concentration of 60 to 90 g liter^{-1}. The free-sodium cyanide is normally maintained in the range 10 to 20 g liter^{-1}. The free cyanide can be lower when the metal content is low. Solutions containing high metal usually require a slight increase in free cyanide for optimum plating performance. An interesting exception to the usual composition and operating conditions for so-called stable copper cyanide plating solutions has been described by Dingley [437] who operated solutions under conditions to give high current efficiencies. Anode to cathode areas had to be greater than 4:1 to maintain solution stability.

Good cleaning and thorough rinsing of the parts before plating are very important, since high-efficiency solutions are sensitive to organic contamination, which produces dull deposits and sometimes pitting. Continuous carbon treatment of the copper strike is recommended as a precaution in minimizing drag-in of organic contaminants. The high-efficiency solutions are normally purified continuously by passing them through a filter packed with activated carbon. A periodic batch treatment with activated carbon may be required if excessive amounts of contaminants are present. The effectiveness of the purification can, in some cases, be improved by adding about 1 ml liter^{-1} of 30% hydrogen peroxide to the solution before adding the carbon. The peroxide can be added safely by first diluting it with 15 parts of water. Its effectiveness is attributed largely to its ability to convert some types of organic contaminants to a less soluble form by oxidation.

Excessive carbonate in high-efficiency solutions reduces the bright current density range and produces grainy deposits. Carbonate should be removed if the concentration exceeds about 90 g liter^{-1} sodium carbonate or 118 g liter^{-1} potassium carbonate. It can be precipitated by adding calcium hydroxide (hydrated lime). Excess sodium carbonate may also be crystallized out of sodium formulated plating solutions by cooling the solutions to about −30°C. This procedure is unsuitable for solutions maintained solely with potassium salts because of the greater solubility of potassium carbonate at all temperatures. Crain and others [438] discuss the causes of carbonate buildup in cyanide copper-plating solutions and control procedures, including a patented method applicable to both sodium and potassium solutions.

Chromate is harmful in the high-efficiency solutions even when present in very low concentrations (5–10 ppm). Its action is similar to that observed in other cyanide copper-plating solutions in that it produces dull, nonuniform deposits. In low concentrations it produces blotchy deposits at low current densities and this condition is extended to higher current densities as the concentration of chromate increases. Adverse effects of chromate can be eliminated by adding a chelating agent of the ethylenediaminetetraacetic acid type. Although chelating agents of this type do not function as reducing agents, the chromate is probably reduced to a Cr^{3+} compound by the Cu^{1+} ion in the presence of the chelating agent. Reducing agents such as sugar derivatives and sodium stannite are also used [439]. In the absence of Rochelle salt, the reduced chromium will gradually precipitate and is removed by filtration.

Hexavalent chromium in just a few ppm can cause dullness, blistering, skip plating, and reduced cathode efficiency. It also gives rise to a reduced bond strength between the copper deposit and steel substrate [440]. Proprietary compounds are available that reduce the hexavalent chromium to the trivalent state without any side effects.

Zinc, which contributes to step plating in low current density areas, codeposits with the copper to form brass deposits which are somewhat brittle. The brassy appearance of the deposits can be detected when zinc concentrations are as low as 1.5 to 2.25 g $liter^{-1}$. Excessive buildup of zinc in high-efficiency solutions can be avoided by removing die castings that fall off racks into the plating solution and by applying an adequate thickness of copper in the strike before the higher-speed copper plating. If the zinc builds up to an objectionable concentration, the excess can be removed by dummy electrolysis using an average cathode current density of 0.3 to 0.5 A dm^{-2}. Small amounts of sodium thiocyanate are effective in sequestering excessive zinc.

Iron contamination comes from salts, water, drag-in, and attack of steel in the cyanide solution. The iron can be complexed into a stable ferrocyanide salt that does not codeposit but does accumulate in the solution. If the iron level gets too high, the anode efficiency is lowered, thus causing carbonate buildup.

For a discussion of impurities in copper cyanide plating solutions, including their effects on the deposits and methods for their removal, see [382, 441, 442]. Horner [382] is an excellent source on troubleshooting of copper cyanide solutions.

29 ANODES

Cast ball, cast elliptical, or electrolytic copper sheet anodes are generally used with strike solutions. Steel anodes are sometimes used in conjunction with copper in a low efficiency solution such as a strike to safeguard against excessive anode polarization and as a means of controlling the metal content [396, 418]. Copper compounds formed at the anode are soluble in these solutions when the anodes are operated below the critical current density [418]. Excessive current densities usually cause polarization of the anodes by formation of an insulating film. Cast, rolled, or oxygen-free high-purity copper anodes are preferred for Rochelle and high-efficiency solutions.

Copper anodes for high-efficiency solutions should be substantially free of oxide inclusions. Copper oxide particles produce serious roughness on the plated parts because copper oxide is readily reduced to particles of copper at the cathode. Electrolytic sheet copper contains substantial amounts of oxide and other occlusions; it is therefore not recommended for high-efficiency solutions. Crystal structure of the copper is also an important factor in anode corrosion. A large grain structure, in general, is preferred. Several types of good quality copper anodes produced by melting, extrusion, or casting under an inert atmosphere are available.

30 MATERIALS OF CONSTRUCTION

Rubber or polyvinylchloride lined steel plating tanks are more satisfactory than plain steel for high-efficiency cyanide copper-plating solutions. Polypropylene tanks with adequate reinforcing may also be used, provided that the operating temperature is not excessive [378]. Leaching should be done on any new tanks or equipment coming in contact with the plating solution. This removes any material that may leach into the plating solution and cause poor quality deposits.

Unlined steel tanks are also widely used. However, under certain adverse conditions the steel may become susceptible to corrosion from the electrolyte.

Increasing the alkalinity of the solution by adding hydroxide and raising the free cyanide content serve to protect the steel and prevent its dissolution [382, 443].

The most satisfactory method of heating the electrolyte is to circulate it through an external heat exchanger and return it to the plating tank through a filter. Particles that may be generated in the heat exchanger are thereby removed. Brass or bronze fittings of any type are not recommended, since they are chemically attacked by the electrolyte. Lead-containing equipment parts must also be avoided, since lead can contaminate some types of high-efficiency processes.

31 ENVIRONMENTAL

Copper cyanide processes are relatively easy to operate and produce excellent deposits; however, they use extremely dangerous solutions and therefore must be monitored carefully because of the potential of poisoning the workers and the environment. Accidental spills of acid into the cyanide tank or cyanide spilled into a sewer system can result in a very dangerous condition. Cyanide's waste treatment must also be very carefully conducted in a separate tank under closely controlled conditions [444]. Cyanide-bearing solutions require oxidation of the cyanide with an oxidizing agent such as chlorine or hypochlorite, followed by precipitation of the heavy metals [378]. While it is unlikely to happen on a frequent basis, if the cyanide copper solution ever requires disposal, the cost is prohibitive, compared to alkaline non cyanide copper [445].

32 STRUCTURE AND PROPERTIES

The best source for property data is Safranek's [446] treatise on properties of electrodeposits. Another comprehensive reference is that of Lamb et al. [447, 448] who provide data on Young's modulus, fatigue strength, effects of cold rolling and annealing on mechanical properties, tensile and elongation data for temperatures from −78° to 325°C, thermal expansivity, texture analysis, and gas content for solutions containing no addition agents. The following paragraph summarizes information that appeared in Safranek's book; very little has appeared on properties of cyanide deposits since publication of his book.

> Copper deposited in low concentration cyanide solutions at low temperatures (40°C) was coarse grained. Fibrous structure, which is typical of cyanide deposited at 60°C was changed to a fine-grained, equiaxed structure by adding Rochelle salts or increasing the copper cyanide concentration to 75 g liter^{-1} and the temperature to 80°C. Strong and hard deposits with a tensile strength of at least 43 MPa were deposited with periodically reversed current in solutions at 80°C containing 75 g liter^{-1} copper cyanide [446].

Copper plated from cyanide solutions is highly embrittling to high strength steel substrates because of the inefficiency of the solutions and the fact that cyanide solutions are highly conducive to hydrogen entry into steel [449]. For waveguide applications, copper deposited from a cyanide solution formed no voids during heat treatment to 1000°C [450].

REFERENCES

366. Walenn, British Patent 1540 (1857).
367. A. Watt and A. Philip, *Electro-Plating and Electro-Refining of Metals*, 2nd ed., Van Nostrand, New York, 1911, 156.
368. O. P. Watts, *Trans. Am. Electrochem. Soc.*, **27**, 141 (1915).
369. C. J. Wernlund, H. L. Benner, and R. R. Bair, U.S. Patent 2,287,654 (1942).
370. C. J. Wernlund, U.S. Patent 2,347,448 (1944).
371. D. A. Holt, U.S. Patent 2,255,057 (1941).
372. G. W. Jernstedt, U.S. Patents 2,451,340; 2,451,341 (1948).
373. F. A. Lowenheim, *Modern Electroplating*, 3rd ed., Wiley, New York, 1974.
374. F. Passal, *Plating*, **46**, 628 (1959).
375. G. C. Van Tilburg, *Plating Surf. Finish.*, **71**, 78 (Apr. 1984).
376. F. A. Lowenheim, *Electroplating*, McGraw-Hill, New York, 1978.
377. L. M. Weisenberger, "Copper Plating," in *Metals Handbook*, 9th ed., vol. 5, *Surface Cleaning, Finishing and Coating*, American Soc. for Metals, 1982.
378. L. M. Weisenberger and B. J. Durkin, "Copper Plating," in *ASM Handbook*, vol. 5, *Surface Engineering*, ASM International, 1994.
379. A. Sato and R. Barauskas, *Metal Finishing*, 64th Guidebook and Directory Issue, vol. 94, no. 1A, 214 (1996).
380. J. Horner, private communication, May 1996.
381. B. A. Smith, W. S. Rapacki, and T. S. Davidson, *Plating Surf. Finish.*, **79**, 11 (Aug. 1992).
382. J. Horner, *Plating Surf. Finish.*, **74**, 34 (Mar. 1987).
383. S. K. Jalota, *Electroplating of Zinc Die Castings*, Bath Press, Bath, Avon, 1987.
384. J. K. Dennis and T. E. Such, *Nickel and Chromium Plating*, 3rd ed., Woodhead Publ., Cambridge, England, 1993.
385. G. A. DiBari, *Met. Finish.*, **75**, 17 (June 1977); **75**, 17 (July 1977).
386. Staff Report, *Plating Surf. Finish.*, **68**, 36 (July 1981).
387. P. G. Hatherley, K. G. Watkins, and M. McMahon, *Trans. Inst. Met. Finish.*, **70**, 177 (1992).
388. D. S. Lashmore, *Plating Surf. Finish.*, **67**, 36 (Jan. 1980).
389. D. A. Dudek and P. S. Fedkiw, *Proc. AESF SUR/FIN 96*, Cleveland, OH, Session T (June 1996).
390. P. G. Hatherley and P. J. Carpenter, *Trans. Inst. Met. Finish.*, **73**, 85 (1995).
391. D. Chu and P. S. Fedkiw, *J. Electronanal. Chem.*, **345**, 107 (1993).
392. R. E. Sinitski, V. Srinivasan, and R. Haynes, *J. Electrochem. Soc.*, **127**, 47 (1980).
393. A. Elbs and J. Rasmussen, *Galvanotechnik*, **85**, 3998 (1994).
394. J. Horner, *Proc. Am. Electroplaters' Soc.*, **51**, 71 (1964).
395. Cs. Juhos, Sf. Gheorghe, E. Gruenwald, and Cs. Varhely, *Galvanotechnik*, **81**, 857 (1990).
396. A. K. Graham and H. J. Read, *Met. Ind. (NY)*, **35**, 559, 617 (1937); **36**, 15, 77, 120, 169 (1938).
397. A. K. Graham, *Met. Ind. (NY)*, **36**, 279 (1938).
398. R. M. Wagner and M. M. Beckwith, *Proc. Am. Electroplat. Soc.*, **25**, 147 (1938).
399. C. W. Smith and C. B. Munton, *Met. Finish.*, **39**, 415 (1941).
400. B. Ostrow and F. Nobel, U.S. Patent 3,021,266 (1962).
401. F. Passal, U.S. Patent 3,030,282 (1962).

402. F. Passal, U.S. Patent 3,111,465 (1963).
403. P. Leenders and H. Creutz, U.S. Patent 3,084,112 (1963).
404. E. Hadley, U.S. Patent 3,179,577 (1965).
405. P. Leenders, U.S. Patent 3,219,560 (1965).
406. A. Debo, U.S. Patent 3,216,913 (1965).
407. A. H. DuRose, U.S. Patent 3,269,925 (1966).
408. B. D. Ostrow and F. I. Nobel, U.S. Patent 3,309,292 (1967).
409. J. R. Crain, U.S. Patent 3,296,101 (1967).
410. L. M. Weisenberger et al., U.S. Patent 3,790,451 (Feb. 1974).
411. M. Pushpavanam, *Met. Finish.*, **84**, 53 (Apr. 1986).
412. J. W. Dini and H. R. Johnson, *Proc. Electroless Nickel Conf. III, Products Finishing Magazine* (Mar. 1983).
413. A. H. DuRose, in *ASTM Special Tech. Publ.*, **197** (1956).
414. International Lead-Zinc Research Organization, Inc., Project LE-36, Rep. 28 (Dec. 1968).
415. G. Schaer, *Plating Surf. Finish.*, **68**, 51 (Mar. 1981).
416. J. B. Mohler, *Met. Finish.*, **82**, 41 (May 1984).
417. H. L. Benner and C. J. Wernlund, *Trans. Electrochem. Soc.*, **80**, 355 (1941).
418. H. J. Read and A. K. Graham, *Trans. Electrochem. Soc.*, **74**, 411 (1938).
419. A. K. Graham, Plating and Metal Finishing Guidebook, *Met. Ind. (NY)*, **26** (1940).
420. *Iron Age*, **184** (2), 98 (July 9, 1959).
421. H. Kenmore and Manson, U.S. Patent 2,680,710 (1954).
422. E. D. Boelter, U.S. Patent 2,854,389 (1948).
423. S. E. Craig Jr., R. E. Harr, and S. Y. Wu, *Plating*, **60**, 1239 (1973).
424. C. J. Wernlund, U.S. Patent 2,774,728 (1956).
425. C. J. Wernlund, U.S. Patent 2,701,234 (1955).
426. M. I. Ismail, *J. Appl. Electrochem.*, **9**, 407 (1979).
427. J. W. Dini, *Met. Finish.*, **61**, 52 (July 1963).
428. D. R. Rudder, *Proc. AESF SUR/FIN 85*, Detroit, MI, Session L (1985).
429. K. E. Langford, *Analysis of Electroplating and Related Solutions*, 4th ed. Robert Draper, Teddington, Middlesex, England, 1971.
430. S. Sriveeraraghaven and S. R. Natarajan, *Met. Finish.*, **73**, 37 (Aug. 1975).
431. J. D. Kostura, *Plating Surf. Finish.*, **70**, 70 (June 1983).
432. L. J. Durney, ed., *Electroplating Engineering Handbook*, 4th ed., Van Nostrand Reinhold, New York, 1984.
433. J.-Y. Hwang, Y.-Y. Wang, and C.-C. Wan, *Plating Surf. Finish.*, **74**, 56 (Apr. 1987).
434. R. O. Hull, U.S. Patent 2,149,344 (1939).
435. M. K. Sanicky, *Plating Surf. Finish.*, **72**, 20 (Oct. 1985).
436. J. W. Dini, *Electrodeposition: The Materials Science of Coatings and Substrates*, Noyes Publs, Park Ridge, NJ, 1993.
437. W. Dingley, J. Bednar, and R. R. Rogers, *Plating*, **53**, 602 (1966); Discussion, **54**, 397 (1967).
438. J. R. Crain, *Plating*, **51**, 31 (1964); W. L. Bohman and M. Ceresa, U.S. Patent 2,858,257 (1958); M. Ceresa and J. Crain, U.S. Patent 2,861,927 (1958); W. L. Bohman and M. Ceresa, U.S. Patent 2,861,928 (1958).
439. H. G. McLeod and D. A. Swalheim, U.S. Patent 2,885,331 (1959).

440. N. W. Phasey, *Trans. Inst. Met. Finish.*, **51**, 77 (1973).
441. F. I. Nobel and B. D. Ostrow, *Plating*, **50**, 823 (1963).
442. D. A. Swalheim, *Plating Surf. Finish.*, **64**, 36 (Apr. 1977).
443. K. G. Watkins, P. J. Carpenter, and P. G. Hatherley, *Trans. Inst. Metal. Finish.*, **71**, 85 (1993).
444. D. L. Snyder, *Met. Finish.*, **89**, 37 (Apr. 1991).
445. F. Altmayer, *Plating Surf. Finish.*, **80**, 42 (Feb. 1993).
446. W. H. Safranek, *The Properties of Electrodeposited Metals and Alloys*, 2nd ed., American Electroplaters and Surface Finishers Soc., Orlando, FL, 1986.
447. V. A. Lamb and D. R. Valentine, *Plating*, **52**, 1289 (1965); *Plating*, **53**, 86 (1966).
448. V. A. Lamb, C. E. Johnson, and D. R. Valentine, *J. Electrochem. Soc.*, **117**, 291C, 341C, 381C (1970).
449. N. V. Parthasaradhy, *Met. Finish.*, **72**, 36 (Aug. 1974).
450. J. C. Farmer, H. R. Johnson, H. A. Johnsen, J. W. Dini, D. Hopkins, and C. P. Steffani, *Plating Surf. Finish.*, **75**, 48 (Mar. 1988).

Part C
Alkaline Noncyanide Copper

Alkaline noncyanide copper-plating solutions have found increasing popularity since the mid-1980s because of environmental issues. The cost of using and disposing of cyanides and associated environmental concerns have led to efforts to replace cyanides. Besides the obvious plus of elimination of cyanide from the wastewater stream, these new solutions are safe to work with and are easily waste treated with lime in the same treatment process used for nickel, acid copper, and acids [451]. Other benefits include faster barrel plating speed, lower sludge volume generation because of lower metal concentrations, simplified wastewater treatment, no trouble with carbonate buildup, and lower OSHA safety concerns. Disadvantages include higher operating costs, difficulty in using the process on zinc die castings, greater sensitivity to impurities, and a chemistry more difficult to control [451]. Copper concentration limits and operating conditions for alkaline noncyanide solutions are summarized in Table 9.

Typically the throwing power of the noncyanide solutions is very good as can be seen from Table 5 in Part A of this chapter. An alkaline noncyanide solution provided even better throwing power than the high-throw acid copper formulation used for plating printed wiring boards. Since the noncyanide processes use cupric copper ions, while the cyanide processes contains monovalent copper, the latter provide for faster plating at the same current density. However, the noncyanide processes can operate at higher current densities thus yielding faster plating overall [452]. Most noncyanide processes use air agitation, while at least one, (phosphonate) uses a purification cell, with proprietary anodes to prevent build-up of too much cuprous copper [453]. The noncyanide processes require excellent cleaning prior to plating and are not as tolerant of poor cleaning practices as cyanide processes. However, adhesion can be as good as that obtained with cyanide. For example, excellent results have been reported for aluminum prepared for plating by substituting a pyrophosphate strike for cyanide [454, 455]. Similar good results were obtained with 1018 steel and cast iron. The deposits have good mechanical properties including higher ductility than cyanide deposits, as well as outstanding metal

TABLE 9 Concentration Limits and Operating Conditions of Alkaline Noncyanide Copper-Plating Solutions

Constituent or Condition	Range
Copper metal (g liter^{-1})	5–14
pH	9–10.5
Temperature (°C)	38–65
Cathode current density (A dm^{-2})	0.5–3.0
Tank voltage	2–12
Anode : cathode ratio	1.5 : 1
Copper anodes	OFHC or EPT 110 copper

distribution and very good tolerance to common impurities [456]. Safranek and Miller [457, 458] reported that a cyanide strike can be replaced with a pyrophosphate strike if ultrasonic agitation is used during the striking process. Applications, where the copper deposit is used as a heat treat masking barrier prior to carburizing, nitriding, or through hardening, have proved successful [459, 460]. A number of proprietary formulations are available and these are based on a variety of chelating ligands, most commonly carboxylic acids, amines, and phosphonates [460–463]. Table 2 in Part A on acid copper plating lists a number of U.S. patents. Nonproprietary formulations that have been evaluated include solutions containing amine-ammonia [464], chloride [465], citrate [465, 466], EDTA [467, 468], glycerolate [469], phosphate [470], pyrophosphate [454, 455, 471], tartrate [472, 473], and triethanolamine [474].

REFERENCES

451. D. L. Snyder, *Met. Finish.*, **89**, 37 (Apr. 1991).
452. F. Altmayer, *Plating Surf. Finish.*, **80**, 40 (Feb. 1993).
453. T. W. Bleeks and T. S. Davidson, "An Alternative to Cyanide Copper," *Proc. 13th AESF/EPA Conf. on Environmental Control for the Surface Finishing Industry*, Orlando, FL, 179 (Jan. 1992).
454. H. R. Johnson, W. D. Bonivert, and J. Hachman, "Alternatives for Cyanide Copper Plating," Spring Meeting, Materials Research Society (MRS), Symposium W, Environmentally Conscious Materials Processing (May 1991).
455. C. P. Steffani and J. W. Dini, *Int. J. Environmentally Conscious Design Manuf.*, **3**, no. 2, 43 (1994).
456. L. C. Tomaszewski and R. A. Tremmel, *Proc. AESF SUR/FIN 85*, Session D, Detroit, MI (1985).
457. W. H. Safranek and H. R. Miller, *Plating*, **55**, 233 (1968).
458. W. H. Safranek, *Plating*, **56**, 11 (1969).
459. B. A. Smith, S. Papacki, and T. S. Davidson, *Plating Surf. Finish.*, **79**, 11 (Aug. 1992).
460. L. M. Weisenberger and B. J. Durkin, "Copper Plating," *ASM Handbook*, vol. 5, *Surface Engineering*, ASM International, 1994.
461. R. T. Haynes, R. R. Irani, and R. P. Langguth, U.S. Patent 3,475,293 (1969).
462. J. Szotek, *Proc. AESF SUR/FIN 93*, Session F, Anaheim CA, (1993).
463. D. F. Maule and B. Srinivasan, *Proc. AESF SUR/FIN 93*, Session F, Anaheim CA, (1993).
464. M. B. I. Janjua, J. Yernaux, P. J. Nicoll, and N. R. Bharucha, *Plating*, **60**, 1124 (1973).
465. G. Carneval and J. Bebczuk de Cusminsky, *J. Electrochem. Soc.*, **128**, 1215 (1981).
466. E. Chassaing, K. Vu Quang, and R. Wiart, *J. Appl. Electrochem.*, **16**, 591 (1986).
467. A. Chiba and W. C. Wu, *Plating Surf. Finish.*, **79**, 62 (Dec. 1992).
468. R. M. Krishnan, M. Kanagasabapathy, S. Jayakrishnan, S. Sriveeraraghavan, R. Anantharam, and S. R. Natarajan, *Plating Surf. Finish.*, **82**, 56 (July 1995).
469. O. Trost and B. Pihlar, *Met. Finish.*, **90**, 125 (June 1992).
470. F. M. Al-Kharafi and Y. A. El-Tantawy, *J. Electrochem. Soc.*, **128**, 2073 (1981).
471. Y. Gan, *Plating Surf. Finish.*, **79**, 81 (June 1992).

472. P. L. Cavallotti, D. Colombo, E. Galbiati, A. Piotti, and F. Kruger, *Plating Surf. Finish.*, **75**, 78 (Apr. 1988).

473. S. Bharathi, S. Rajendran, V. N. Loganathan, C. Krishna, and K. R. Ananndakumar-annair, *Proc. AESF SUR/FIN 96*, Session G, Cleveland, OH, (1996).

474. S. M. Mayanna and B. N. Maruthi, *Met. Finish.*, **94**, 42 (Mar. 1996).

Part D
Pyrophosphate Copper*

33 HISTORY AND DEVELOPMENT

The earliest published reference to copper pyrophosphate deposition was by Roseleur in 1847 [475]. In 1883 Gutensohn [476] was granted a patent for copper pyrophosphate plating on a number of substrates. Others associated with the early state of the art included Brand [477], Delval [478], Royer [479], and Regelsberger [480]. Pioneering work by Stareck [481, 482] led to the development of a commercial copper pyrophosphate in 1941. The same year Coyle [483] described the main aspects of the process. Reviews of copper pyrophosphate plating can be found in [484–490].

34 APPLICATIONS

The production of electroformed objects such as waveguides, paint spray masks [491], helical antennae [492], heat exchangers [493], molds for making toys [491, 494], and hard, high strength [495], wear-resistant [496] deposits are some uses of this solution. Pyrophosphate deposits are also used on steel [491, 497, 498] and aluminum parts [499–504], and in some cases, they serve as a replacement for cyanide deposits [505, 506]. Other applications include plating zinc die castings [507, 508] before bright nickel and chromium plating, as a lubricant for wire deep drawing operations [485, 491], as a stopoff on steel for selective hardening operations such as nitriding and carburizing [485, 509–512], in roll plating [485], in minimizing hydrogen embrittlement [513, 514], and in the manufacture of plated steel cord for radial tires and high pressure plastic pipes [515, 516]. Pyrophosphate solutions are also used for plating through holes on printed wiring boards although in most cases this application has been replaced by high-throw acid copper sulfate formulations [490, 517–519]. Use of pyrophosphate processes for automotive decorative plating has been completely eliminated for many years [520]. Pyrophosphate solutions have been used for plating on plastics; however, these solutions are considerably more expensive to make up than high-throw acid copper formulations [521].

35 BASIC CHEMISTRY

The copper pyrophosphate plating solution contains copper pyrophosphate ($Cu_2P_2O_7 \cdot 3H_2O$), either potassium pyrophosphate ($K_4P_2O_7$) or the corresponding sodium salt ($Na_4P_2O_7$), nitrate (NO_3), and ammonia (NH_3). The pyrophosphate salts react in aqueous solution to form alkaline complex anions, the major complex $Cu(P_2O_7)_2^{6-}$ and also $(Cu_2P_2O_7)^{2-}$. The solution is of intermediate stability [522]. Between pH values of 7 and 11, the stability of the complex anion is evidenced by the

*Revision of Section by J. W. Dini in the 1974 edition.

slow but sure hydrolysis of $(P_2O_7)^{4-}$ to $(PO_4)^{3-}$. At pH values above 11, $Cu(OH)_2$ precipitates; at pH values noticeably below 7 either $CuH_2P_2O_7$ or $Cu_2P_2O_7$ precipitates. Acidification below pH 7 converts the $(P_2O_7)^{4-}$ to either $(H_2P_2O_7)^{2-}$ or $(HPO_4)^{2-}$ and thus destroys the complex anion [523]. Additional information on the chemistry of copper pyrophosphate complexes may be found in [486–488, 524–534].

36 CONSTITUENTS

36.1 Copper and Pyrophosphate

Typical formulations are included in Table 10. The copper and pyrophosphate contents of the plating solution are critical in terms of the ratio of one to the other. To promote anode corrosion and to increase electrical conductivity, the solution must contain excess complexing compound. Although either may be used, potassium pyrophosphate is preferred over the sodium salt because it is more soluble, and a potassium solution has a higher electrical conductivity because of the higher mobility of potassium ions.

36.2 Nitrate

The presence of a nitrate results in a higher maximum current density because, especially at current densities above 3.2 $A\,dm^2$, the nitrate ion reduces cathode polarization by acting as a hydrogen acceptor according to the equation: [485]

$$NO_3^- + 10H^+ + 8e = NH_4^+ + 3H_2O$$

As the solution is used extensively, reduction of the nitrate ion also leads to a buildup of nitrite [535].

36.3 Ammonia

A small amount of ammonia is used to produce more uniform and lustrous deposits and to improve anode corrosion. Excess ammonia can cause cuprous oxide to form and this can hinder adhesion [522].

Because ammonia evaporates from the plating solution, it is usually added daily in quantities either dependent on the size of the tank, for example, 140 g of ammonium hydroxide per square meter of exposed surface per day [536] or as determined by chemical analysis of the solution. In some installations, gaseous ammonia is used [537]. Kojima [538] disclosed that ammonia consumption increased with

TABLE 10 Optimum Range of Solution Constituents

Analytical Constitutent	Composition ($g\,liter^{-1}$)
Copper, Cu^{2+}	22–38
Pyrophosphate, $(P_2O_7)^{4-}$	150–250
Nitrate, NO_3^-	5–10
Ammonia, NH_3	1–3
Orthophosphate, $(HPO_4)^{2-}$	No greater than 113
Organic additives	As required

increasing temperature, pH, agitation, and ammonia content in the solution, while current had little effect on consumption.

36.4 Orthophosphate

Orthophosphate formed by the hydrolysis of pyrophosphate is beneficial in that it promotes anode corrosion and acts as a buffer. At all concentrations it affects the properties of the plating solutions and deposits. Increasing orthophosphate concentration decreases the throwing power and efficiency of the solution and causes a reduction in the ductility of deposits [539].

Low pH (less than 7), increasing P_2O_7 concentration, high P_2O_7/Cu ratio, high temperatures (greater than 60°C) in the plating solution or heat exchanger, or local overheating will cause orthophosphate to build up. Orthophosphate cannot be removed chemically from a pyrophosphate solution. The concentration can be reduced only by either discarding the solution altogether or by diluting and rebuilding it.

36.5 Additives

Additives have more of an influence on deposit properties than any other plating variable as will be shown in the section on properties and structure. When used at controlled, limited concentrations, organic additives refine the grain structure, provide desired tensile and ductility properties, impart leveling characteristics to the plating solution, and act as brighteners. However, decomposition products from an excessive additive concentration can cause brittle copper deposits. Thus, for optimum quality deposits, no more additive should be added to a solution than is necessary to replace losses by consumption.

Organic additives covered in the patent literature as brighteners include mercaptothiadiazoles [540–550], mercaptothiazoles [551–556], mercaptobenzimidazole [557], and pyrimidines [555, 558, 559]. Other organic and inorganic materials, such as glycerol [560, 561], triethanolamine [560, 561], trioxyglutaric acid [525, 562], diphenylamine sulfonic acid [561], naphthalene disulfonic acid [563], gelatin, bakers yeast, casein, glycocoll [564], hydroxyethylcellulose [565], sodium selenite [525, 559], sodium sulfite [560, 561], Rochelle salt, sodium thiosulfate, potassium bromide [561], and the chlorides of As, Bi, Fe, Cr, Sn, Zn, Cd, Pb [482, 563], and alkali metals have been used as brighteners. Additions of small amounts of lead are reported to improve plating properties [566, 567]. Organic acids such as oxalic, lactic, tartaric, malic, and citric or their ammonium or alkali salts also produce some brightening [482].

36.6 Operating Conditions

A copper pyrophosphate plating solution may be operated over a relatively wide range of conditions. The optimum range, similar to that given by Couch and Stareck [568], is shown in Table 11.

36.7 Pyrophosphate/Copper Ratio

The copper pyrophosphate solution is analyzed for both copper and pyrophosphate, and it is therefore convenient to describe the solution in terms of the ratio between

TABLE 11 Optimum Operating Conditions

P_2O_7/Cu ratio	7.0 : 1–8.0 : 1
pH	8.0–8.8
Temperature, °C	50–60
Cathode current density, A dm^{-2}	1–8
Anode and cathode efficiency, %	100
Anode/cathode ratio	1 : 1–2 : 1
Agitation	1–1.5 m^3 / (min) (m^{-2}) of surface

these constituents. For optimum plating, the weight ratio of pyrophosphate to copper should be kept in the range 7 : 1 to 8 : 1. Ratios of 8.5 : 1 or higher promote formation of orthophosphate and thereby decrease the bright plating range. Operation with a ratio of less than 7 : 1 tends to produce a rough-surfaced plate and renders the solution unstable.

36.8 pH

The optimum solution pH is in the range of 8 to 9. As discussed earlier, a pH outside this range results in hydrolysis of pyrophosphate to orthophosphate or the formation of precipitates. Also, if the pH is too high, both anode corrosion and operating current density range decrease. Radin [569] states that a pyrophosphate solution will operate acceptably even at pH values as high as 9.3 provided the copper concentration exceeds 26 g liter^{-1}. When a solution is operated at this high pH, the tendency toward roughness can be eliminated by increasing the P_2O_7/Cu ratio. In practice, a pH of 7 to 9 is easily maintained because at this pH both anode and cathode efficiencies are virtually 100%. Clearly, there is no chemical breakdown of the constituents, and the solution is highly buffered. When adjustments must be made, pyrophosphoric acid is used for lowering pH and potassium hydroxide for raising it.

36.9 Temperature

The solution is usually operated between 50° and 60°C. Temperatures greater than 60°C can lead to rapid formation of orthophosphate.

36.10 Current Density

Cathode current density is a function of the temperature and agitation of the plating solution. However, under standard operating conditions, a current density of 1 to 9 A dm^{-2} is appropriate. The anode current density is fairly critical, and it should be kept between 2 and 4 A dm^{-2}. At too high an anode current density, an insoluble oxide tends to form. The operating current density range may be increased by ultrasonics, by current interruption or reversal, and by increased metal concentration.

36.11 Agitation

Copper pyrophosphate solutions are among those that must be most vigorously agitated. Without sufficient agitation, a brownish deposit is obtained, and the operating current density range is drastically lowered. A solution can be continuously agitated by either one or a combination of three methods: air, mechanical movement of the cathode, or ultrasonics. When air agitation is used, the air should be supplied by a low-pressure blower because compressed air can contain oil.

Gurylev et al. [570] claimed that ultrasonic agitation reduces anode and cathode polarization and permits a 4 to 4.5 times increase in the deposition rate over that obtained with intensive mechanical agitation. The combination of ultrasonics and periodic reverse current increased current efficiency and allowable anode current density. Rutter et al. [571] reported that the use of ultrasonics resulted in improved adhesion and a denser deposit. Safranek and Miller [507, 508] and others who have plated over zinc die castings showed that ultrasonic agitation of copper pyrophosphate solutions displaced air or hydrogen from blind pockets and resulted in more uniform deposits than could be obtained with copper cyanide strikes. Uspenskii and Schluger [572] investigated pH variations in an ultrasonic field and Trofinov and Galushko [573] polarization characteristics. Trofinov [574] concluded that metal distribution from a copper pyrophosphate electrolyte could not be influenced much by an ultrasonic field. Vasudevan et al. [575] reported an improvement in both anodic and cathodic current efficiencies under ultrasonic agitation but no change in hardness of the deposit.

36.12 Equipment

The equipment: a tank, heater, and filter is similar to that used for bright nickel plating. A steel tank coated with rubber, Koroseal (Trademark: B. F. Goodrich Co.), PVC, or a plastisol is suitable, although rubber linings should be leached before use with a dilute solution of potassium hydroxide to remove all alkali soluble matter that can contaminate the plating solution. For heating the solution, either steam heating coils or electric immersion heaters made of stainless steel or Duriron may be used. Continuous filtration helps prevent rough deposits. The filter should be constructed of stainless steel, rubber, or PVC-lined steel or cast iron.

36.13 Anodes

OFHC anodes (Trademark: American Metal Climax, Inc.) are the best available; however, electrolytic sheet and rolled electrolytic copper anodes are also satisfactory. Cast anodes of good purity can also be used, but they should be free of impurities such as lead, nickel, silver, and tin because these tend to promote deposit roughness. Anode bags are not usually used; they are not needed and can restrict circulation of the plating solution around the anode and cause polarization. Anodes can be left in the solution when it is not being used since they do not dissolve in the absence of applied current.

37 MAINTENANCE AND CONTROL

37.1 Analysis

For the analytical determination of copper, pyrophosphate, and orthophosphate, the simplest procedure is that of Konishi [576], where all three constituents are determined by titration with EDTA. Copper can also be determined electrolytically [577] or by titration by the iodine-thiosulfate method [577–579], pyrophosphate by titration with standard alkali [577–579], and orthophosphate gravimetrically [578–579] or colorimetrically [580]. Ammonia is analyzed by distillation and adsorption in standard acid [577, 579] or by cyclic voltammetric stripping (CVS) [581]. Tench and Ogden and their circle of researchers have pioneered in the use of CVS, particularly for controlling additives and contaminants in copper plating solutions for printed wiring board production. Their work with pyrophosphate solutions is covered in references [542–549] and reviewed in detail in [582]. Tam and coworkers [534, 535], also used CVS to study the effects of all constituents on cathodic and anodic reactions.

37.2 Impurities and Purification

Cyanide, oil, lead, decomposition products of the organic additives, residues from photoresists and solvents used in circuit development, epoxy laminate residues, excess orthophosphate, and substances leached from plating tanks are the impurities most often encountered in a copper pyrophosphate plating solution. These impurities cause streaked, dull deposits and also lower the effective current density range. Precautionary methods will prevent these materials from entering the solution. Thorough rinsing, when copper cyanide striking is used, will prevent cyanide contamination; a clean air line for solution agitation will prevent oil contamination; and avoiding the use of lead coils, lining, fittings, or filters will prevent lead contamination.

Cyanide, which can be detected by a copper benzidine test [583], can be removed by treatment with either 1.25 ml liter^{-1} of 30% H_2O_2 or with 0.1 to 0.2 g liter^{-1} potassium permanganate, followed by treatment with activated carbon. The tendency toward streaked, nonuniform pyrophosphate deposits due to residual traces of cyanide left on the surface after copper cyanide striking can be minimized by brief cathodic treatment in a solution of potassium pyrophosphate before it is placed in the plating solution. Oil can be removed by treating the solution with 2.5 to 6 g liter^{-1} of activated carbon at 55°C for 4 to 8 hours. Low current density electrolysis will remove lead.

Copper pyrophosphate solutions are quite tolerant of metallic contamination because many metals besides copper form complexes with the pyrophosphate ion and remain in solution. For example, no harmful effects have been reported from zinc which may be introduced when the solution is used to plate zinc die castings. It has been recommended, however, that a high P_2O_7/Cu ratio (preferably 7.5 : 1) be maintained under this circumstance so as to allow some excess pyrophosphate for complexing with the contaminant. Phosphorus included in the deposit was reported to influence external appearance, microhardness, and microstructure [584]. Iron and lead can alter deposit morphology and concentrations below 200 and 1000 ppm, respectively, are recommended [516]. Chloride ion, which is known to affect anode reactions, can be tolerated at concentrations lower than 0.1 M with vigorous agitation of the solution [515].

Organic contaminants, besides reducing deposit ductility and strength, can also cause loss of throwing power and irregular deposition. The contamination can come from: decomposition products from organic additives, air agitation, leaching of photoresists, and the carbon used for removing contaminants. Rothschild et al. [585] provide an excellent review on carbon treating of pyrophosphate solutions for keeping them free of organic contaminants and recommend the use of Hull cell tests to determine the effectiveness of treatment. In some cases an activated carbon treatment alone is not sufficient purification. To ensure removal of the decomposition products, it is best to carbon treat the solution, H_2O_2 treat, and then carbon treat again. The carbon should be plating grade and sulfur free.

Cyclic voltammetric stripping analysis is an excellent means for detecting a wide range of contaminants in addition to organic additives in pyrophosphate solutions. Excellent correlation has been obtained between the measured contaminant level and plating problems normally attributed to solution contamination such as poor solder adhesion and low deposit ductility [542, 544, 545, 548].

Waste pyrophosphate solutions are difficult to treat because of ammonia in the solutions and also the complexes that are formed [586]. Wastes require low pH hydrolysis to orthophosphate, followed by precipitation of the heavy metals [488].

38 STRUCTURE AND PROPERTIES

The best source for property data is Safranek's treatise on properties of electrodeposits [587]. Another comprehensive reference is that of Lamb et al. [588] who provide data on Young's modulus, fatigue strength, effects of cold rolling and annealing on mechanical properties, tensile and elongation data for temperatures from −78° to 325°C, thermal expansion, texture analysis, and gas content for solutions containing no addition agents.

Copper pyrophosphate deposits show tensile strength variations from 27 to 70 $kg\,mm^{-2}$, yield strengths from 14 to 36 $kg\,mm^{-2}$, hardness from 83 to 250 $kg\,mm^{-2}$, and elongations from 1% to 39%. Several investigators have related these variables to solution content and operating variables with additive concentration being the most important variable [495, 547, 589–595].

Optimum ductility has been obtained by using solutions free of organic additives [588]. In some cases, continuous filtration is also used [589, 596]. High-purity, high-quality deposits have been produced on internal walls of colliders in beam tubes by using multistage purification and pulsed reverse current in solutions with no additives [597, 598].

A comparison of recrystallization temperature for electrodeposited pyrophosphate copper and wrought copper with various degrees of cold work suggests that the electrodeposited copper exhibits behavior expected of 100% cold worked material [592]. Malone [599] reported that pyrophosphate copper electrolytes were capable of producing deposits with excellent mechanical properties. However, he cautioned that the pyrophosphate electrolyte is very difficult to control and maintain in comparison with the acid sulfate solution, and producing deposits consistently within a specific range may prove difficult unless the electroplater has considerable experience with the pyrophosphate solution and the many pecularities associated with it. The influence of pH, temperature, agitation, and current density on throwing power and efficiency can be found in [600].

39 PLATING OF PRINTED WIRING BOARDS

The previous edition of this book [484] devoted a section to this topic in the chapter on copper pyrophosphate. Around the time of publication of that volume, high-throwing power acid copper sulfate solutions were developed [601]. Today these solutions are used instead of pyrophosphate for most printed wiring board applications, except in some military and special applications [490]. It is estimated that greater than 95% of all printed wiring boards today are plated in acid copper sulfate solutions [518], and this is discussed in more detail in the section on acid copper plating. A good review of present practice with pyrophosphate plating of printed wiring boards can be found in [490].

REFERENCES

475. A. Roseleur, *Manipulations Hydroplastiques—Guide Practique du Doreur, de L'Argenteur et de Galvanoplaste*, 1st ed., Roseleur, Paris, 1855.
476. A. Gutensohn, British Patent 4784 (1893).
477. A. Brand, *Z. Anal. Chem.*, **28**, 581 (1888).
478. E. Delval, in A. Roseleur, 6th ed., 1892.
479. L. F. Royer, French Patent 381,475 (1908).
480. F. Regelsberger, *Elektrochem. Z.*, **19**, 181 (1912).
481. J. E. Stareck, U.S. Patent 2,081,121 (1937).
482. J. E. Stareck, U.S. Patent 2,250,556 (1941).
483. T. G. Coyle, *Proc. Am. Electroplat. Soc.*, **28**, 113 (1941).
484. F. A. Lowenheim, *Modern Electroplating*, 3rd ed., Wiley, New York, 1974.
485. F. Passal, *Plating*, **46**, 628 (1959).
485. G. C. Van Tilburg, *Plating Surf. Finish.*, **71**, 78 (Apr. 1984).
486. F. A. Lowenheim, *Electroplating*, McGraw-Hill, New York, 1978.
487. L. M. Weisenberger, "Copper Plating," in *Metals Handbook*, 9th ed., vol. 5, *Surface Cleaning, Finishing and Coating*, American Soc. for Metals, 1982.
488. L. M. Weisenberger and B. J. Durkin, "Copper Plating," in *ASM Handbook*, vol. 5, *Surface Engineering*, ASM International, 1994.
489. A. Sato and R. Barauskas, *Metal Finishing, 64th Guidebook and Directory Issue*, vol. 94, no. 1A, 214 (1996).
490. E. F. Duffek, "Plating" in *Printed Circuits Handbook*, 4th ed., C. F. Coombs Jr., ed., McGraw-Hill, New York, 1996.
491. D. M. Lyde, *Met. Ind. (London)*, **101**, 82 (Aug. 1962).
492. L. Missel and M. E. Shaheen, *Met. Finish.*, **62**, 61 (Sept. 1964).
493. H. F. Maddocks, French Patent 2,020,754 (1969).

494. H. Wolfson and B. Thomson, British Patent 600,873 (1948).
495. H. J. Wiesner and W. P. Frey, *Plating Surf. Finish.*, **66**, 51 (Feb. 1979).
496. Y. T. Chen, J. R. DePew, and H. D. McCabe, *J. Vac. Sci. Technol.*, **11**, 777 (1974).
497. A. I. Levin and V. V. Gulylev, *Draht*, **14** (1), 18 (1963); *Met. Finish. Abstr.*, **5**, 51 (1963).
498. A. A. Voronko, S. P. Pilite, and A. M. Molchadskiy, Russian Patent 159,368 (1962); *Met. Finish. Abstr.*, **6**, 52 (1964).
499. J. T. N. Atkinson, U.S. Patent 2,871,171 (1959).
500. R. D. Potlushnewa, *Galvanotechnik*, **54** (3), 142 (1963); *Met. Finish. Abstr.*, **5**, 91 (1963).
501. Pernix, Enthone, French Patent 1,448,839 (1965); *Met. Finish. Abstr.*, **8**, 249 (1966).
502. Yu. Ya. Lukomskiy, L. N. Zapatrina, and A. N. Aleksandrova, Russian Patent 191,983 (1963); *Met. Finish. Abstr.*, **9**, 204 (1967).
503. B. E. Bunce, *Met. Finish.*, **52**, 70 (Jan. 1954).
504. S. Wernick and R. Pinner, eds., *Surface Treatment of Aluminum*, 3rd ed., Robert Draper, Teddington, Middlesex, England, 1964.
505. H. R. Johnson, W. D. Bonivert, and J. Hachman, "Alternatives for Cyanide Copper Plating," Spring Meeting, Materials Research Society (MRS), Symposium W, Environmentally Conscious Materials Processing (May 1991).
506. C. P. Steffani and J. W. Dini, *Int. J. Environmentally Conscious Design Manuf.*, **3**, no. 2, 43 (1994).
507. W. H. Safranek and H. R. Miller, *Plating*, **55**, 233 (1968).
508. W. H. Safranek, *Plating*, **56**, 11 (1969).
509. J. E. Stareck, *Mon. Rev. Am. Electroplat. Soc.*, **30**, 25 (1943).
510. W. V. Sternberger and E. R. Fahy, *Met. Prog.*, **47**, 278 (1945); **48**, 1311 (1945).
511. C. J. Miller, *Met. Prog.*, **49**, 783 (1946).
512. N. Solowjow, in *Met. Finish. Abstr.*, **55**, 71 (Sept. 1957).
513. H. Fischer and H. Baermann, *Korros. Metallschutz*, **16**, 405 (1940).
514. E. E. Dougherty, *Plating*, **51**, 415 (1964).
515. D. DeFilippo, A. Rossi, and M. A. Spezziga, *J. Appl. Electrochem.*, **16**, 463 (1986).
516. D. DeFilippo, L. Dessi, and A. Rossi, *J. Appl. Electrochem.*, **19**, 37 (1989).
517. E. K. Yung, L. T. Romankiw, and R. C. Alkire, *Electrodeposition Technology, Theory and Practice*, Electrochemical Soc., L. T. Romankiw and D. R. Turner, eds., 75 (1987); *J. Electrochem. Soc.*, **136**, 206 (1989).
518. J. P. Langan, private communication, May 1996.
519. L. Lourch, private communication, May 1996.
520. D. L. Snyder, *Met. Finish.*, **89**, 37 (Apr. 1991).
521. D. J. Combs, *Proc. AESF SUR/FIN 80*, Session B, Milwaukee, WI (June 1980).
522. S. Senderoff, *Met. Finish.*, **48**, 59 (July 1950).
523. I. M. Kolthoff, *Pharm. Weekbl.*, **57**, 474 (1920).
524. B. C. Halder, *Sci. Cult. (India)*, **14**, 340 (1949); *Chem. Abstr.*, **43**, 6103 (1949).
525. *Theory and Practice of Bright Electroplating*, transl. from Russian, U.S. Department of Commerce, TT 65-50000 Clearinghouse for Federal Scientific and Technical Information, Springfield, VA., B. A. Purin and E. A. Ozola, 146–163; V. I. Lainer, 181–185 (1962).
526. R. N. Bell, *Ind. Eng. Chem.*, **39**, 136 (1947).
527. H. Bassett, W. L. Bedwell, and J. B. Hutchinson, *J. Chem. Soc. (London)*, 1412 (1936).
528. J. I. Watters and A. Aaron, *J. Am. Chem. Soc.*, **76**, 611 (1953).

529. J. I. Watters, E. D. Loughran, and S. M. Lambert, *J. Am. Chem. Soc.*, **78**, 4855 (1956).
530. J. R. Van Wazer and C. F. Callis, *Chem. Rev.*, **58**, 1011 (1958).
531. N. T. Kudryavtsev and B. M. Dikova, *Prot. Met. (USSR)*, **5**, 147 (1969).
532. H. Konno and M. Nagayama, *Electrochimica Acta*, **22**, 353 (1977).
533. H. Konno and M. Nagayama, *Electrochimica Acta*, **23**, 1001 (1978).
534. T. M. Tam and R. Taylor, *J. Electrochemical Soc.*, **133**, 1101 (1986).
535. T. M. Tam and G. A. Fung, *J. Electrochem. Soc.*, **130**, 874 (1983).
536. T. A. Dickinson, *Met. Finish. J.*, **1**, 343 (1955).
537. T. H. Irvine, *Plating*, **54**, 1251 (1967).
538. T. Kojima, *J. Met. Finish. Soc. Japan*, **21** (4), 166 (1970); *Met. Abstr.*, **3**, 1521 (Nov. 1970).
539. B. R. Rothschild, *Met. Finish.*, **76**, 49 (Jan. 1978).
540. F. H. Wells and D. M. Lyde, British Patent 939,997 (1963).
541. F. H. Wells and D. M. Lyde, U.S. Patent 3,161,575 (1964).
542. D. Tench and C. Ogden, *J. Electrochem. Soc.*, **125**, 194 (1978).
543. D. Tench and C. Ogden, *J. Electrochem. Soc.*, **125**, 1218 (1978).
544. C. Ogden and D. Tench, *Plating Surf. Finish.*, **66**, 30 (Sept. 1979).
545. C. Ogden and D. Tench, *Plating Surf. Finish.*, **66**, 45 (Dec. 1979).
546. C. Ogden and D. Tench, *J. Electrochem. Soc.*, **128**, 539 (1981).
547. C. Ogden, D. Tench, and J. White, *J. of Applied Electrochemistry*, **12**, 619 (1982).
548. B. Lowry, C. Ogden, D. Tench, and R. Young, *Plating Surf. Finish.*, **70**, 70 (Sept. 1983).
549. M. Jawitz, C. Ogden, D. Tench, and R. Thompson, *Plating Surf. Finish.*, **71**, 58 (Jan. 1984).
550. D. M. Lyde, U.S. Patent 3,784,454 (1974).
551. F. H. Wells and D. M. Lyde, British Patent 940,282 (1963).
552. F. H. Wells and D. M. Lyde, U.S. Patent 3,157,586 (1964).
553. Albright & Wilson Ltd. and Wilmot Breeden, Ltd., French Patent 1,329,175 (1962); *Met. Finish. Abstr.*, **5**, 148 (1963).
554. Kamimura Chobei Co., Ltd., Japanese Patent 20,805/66 (1966); *Met. Finish. Abstr.*, **9**, 144 (1967).
555. Albright & Wilson Ltd., Belgian Patent 714,454 (1968); *Met. Finish. Abstr.*, **11**, 13 (1969).
556. D. M. Lyde, U.S. Patent 3,729,393 (1973).
557. Metal and Thermit Corp., British Patent 774,424 (1957).
558. W. Canning & Company, Ltd., British Patent 1,051,150 (1966).
559. Okuno Seiyaku Kogyo, Ltd., Japanese Patent 24,285/68 (1968); *Met. Finish. Abstr.*, **11**, 14 (1969).
560. T. L. Rama Char, *Electroplating*, **10**, 347 (1957).
561. S. K. Panikkar and T. L. Rama Char, *J. Sci. Ind. Res. (India)*, **19**A (June 1960).
562. B. A. Purin and E. A. Ozola, Russian Patent 351,764 (1961).
563. L. Serota, *Met. Finish.*, **58**, 76 (Apr. 1960).
564. I. Hampel, G. Stumpf, and H. Zeng, *Elektrie*, **17** (8), 269 (1963); *Met. Finish. Abstr.*, **5**, 181 (1963).
565. D. M. Lyde, U.S. Patent 3,674,660 (1972).
566. E. Toledo, *Circ. Manuf.*, **11** (1), 46 (1971).
567. E. Fino et al., U.S. Patent 3,775,268 (1971).

568. F. A. Lowenheim, *Modern Electroplating*, 2nd ed., Wiley, New York, 1963.
569. A. Radin, *Prod. Finish.*, **31**, 62 (July 1967).
570. V. V. Gurylev, A. I. Levin, and M. Nasakina, *J. Appl. Chem., (USSR)*, **37**, 1055 (1964); *Met. Finish.*, **67**, 59 (April 1969).
571. E. G. Rutter et al., Russian Patent 121,002 (1958); *Met. Finish. Abstr.*, **1**, 228 (1959).
572. S. I. Uspenskii and M. A. Schluger, *Sov. Electrochem.*, **2**, 226 (1966); *Met. Finish.*, **68**, 40 (July 1970).
573. A. N. Trofinov and A. P. Galushko, *Sov. Electrochem.*, **1**, 877 (1965).
574. A. N. Trofinov, *Sov. Electrochem.*, **1**, 1029 (1965).
575. R. Vasudevan, R. Devanathan, and K. G. Chidambaram, *Met. Finish.*, **90**, 23 (Oct. 1992).
576. S. Konishi, *Met. Finish.*, **63**, 58 (Mar. 1965).
577. K. E. Langford, *Analysis of Electroplating and Related Solutions*, 4th ed., Robert Draper, Teddington, Middlesex, England (1971).
578. D. Foulke and F. Crane, *Electroplater's Process Control Handbook*, Reinhold, New York, 1953.
579. L. J. Durney, ed., *Electroplating Engineering Handbook*, Van Nostrand Reinhold, New York, 1984.
580. C. H. Fiske and Y. Subborow, *J. Biol. Chem.*, **66**, 375 (1925).
581. T. M. Tam and J. S. Zevely, *J. Electrochem. Soc.*, **131**, 109 (1984).
582. J. W. Dini, *Electrodeposition: The Materials Science of Coatings and Substrates*, Noyes Publications (1993).
583. F. Feigl, *Spot Tests in Inorganic Chemistry*, 5th ed., Elsevier, Amsterdam, 1958, 276.
584. G. N. Sorkin, Yu. G. Lavrent'ev, and R. Yu. Bek, *J. Appl. Chem. (USSR)*, **44**, 679 (1971).
585. B. F. Rothschild, J. E. Semar, and H. K. Omata, *Plating Surf. Finish.*, **64**, 53 (Nov. 1977).
586. F. Altmayer, private communication, April 1996.
587. W. H. Safranek, *The Properties of Electrodeposited Metals and Alloys*, 2nd ed., American Electroplaters and Surface Finishers Society, Orlando, FL, 1986.
588. V. A. Lamb, C. E. Johnson, and D. R. Valentine, *J. Electrochem. Soc.*, **117**, 281C, 341C, 381C (1970).
589. M. H. Smith, "Failure Analysis of Plated Through-Holes in Multilayer Printed Wiring," McDonnell Douglas Astronautics Co., Paper WD 1023, Jan. 1970, presented to IPC, Washington, DC, Apr. 15, 1969.
590. C. Schmitz, *Insul./Circ.*, **16** (11), 35 (1970).
591. P. E. Hinton, "Structure Modifications Produced in Electrodeposited Copper by an Organic Compound in the Electrolyte," M.S. thesis, University of Arizona, Tucson, Arizona (1968).
592. D. E. Sherlin and L. K. Bjelland, *Circuit World*, **4**, 22 (Oct. 1977).
593. W. T. Hollar, *Prod. Finish.*, **31**, 58 (Mar. 1967); 62 (July 1967).
594. NASA Tech Brief 67-10358, "Steel Test Panel Helps Control Additives in Pyrophosphate Copper Plating," Clearinghouse for Federal Scientific and Information, Springfield, VA.
595. C. J. Owen, *Plating*, **57**, 1012 (1970).
596. C. J. Owen, H. Jackson, and E. R. York, *Plating*, **54**, 821 (1967).

597. J. Cl. Puippe and W. Fluhmann, *Supercollider* 4, J. Nonte, ed., Plenum Press, New York, 1992, 381.

598. W. Fluhmann and W. Saxer, *Plating*, **59**, 1140 (1972); U.S. Patent 3,660,251 (1972).

599. G. A. Malone, "Investigation of Electroforming Techniques," NASA CR 134959 (Dec. 1975).

600. J. W. Dini, H. R. Johnson, and J. R. Helms, *Plating*, **54**, 1337 (1967).

601. B. F. Rothschild, *Plating Surf. Finish.*, **66**, 70 (May 1979).

Part E
Copper Composites

Present-day structural materials have two major shortcomings: loss of strength at elevated temperatures and relatively low elastic moduli [602]. One way to solve these problems is in strengthening materials by incorporating high-strength and high-modulus particles, fibers, or continuous filaments in a metal matrix. The major disadvantages of present methods (internal oxidation, powder metallurgy, high pressure bonding, and infiltration) in producing dispersion strengthened and fiber reinforced composites are the high process temperatures and pressures required. Degradation of the fibers due to handling and difficulties encountered in machining the finished composite structure are also concerns.

A less common method of producing these composites is electrodeposition. There is no requirement in electrodeposition for high temperatures that can damage the fibers. The strengthening of electrodeposits can be accomplished by the encapsulation of inert particles, fibers (whiskers), or filaments during plating to produce high-performance coatings [602–604]. Two terms commonly used to refer to use of electrodeposition to produce composite materials are *composite plating* and *electrocomposites*. Electrodeposition of copper and nickel are the two plating processes most commonly used to produce composites; copper will be covered in this section. Reviews on composite plating can be found in [602–610].

Copper composite materials have typically been produced in sulfate solutions, although acid fluoborate and alkaline cyanide solutions have also been used. Particles that have been codeposited with copper include: alumina, barium sulfate, borides, carbides, carborundum, corundum, graphite, molybdenum disulfide, polystyrene, phosphorus, PTFE, quartz, titania, tungsten disulfide, and zirconium oxide. Mechanical or ultrasonic agitation was required to uniformly disperse the particles which ranged from 50 to 200 g liter^{-1} in the solution.

A variety of techniques can be used for codeposition of particles. Namely occlusion plating-particles are allowed to codeposit on the cathode surface while some intermittent vigorous mechanical stirring of the plating solution is applied [609], sediment codeposition-particles are allowed to sediment on a flat cathode as electrolysis proceeds and are only occasionally agitated [611–614], and composite plating particles are deposited on a vertical cathode [609]. Particles are kept in suspension by either mechanical or chemical means. Suspension by mechanical means is accomplished by use of a vibrating perforated bottom plate or by the use of air agitation. Suspension by chemical means typically requires the addition of appropriate surfactants to the plating solution.

The optimum conditions for preparing particle suspensions and codepositing a designated volume fraction of uniformly dispersed inert particles have not been firmly established. Most of the literature data have been derived empirically. Particle diameters have ranged from 0.02 to 100 μm. Quantities dispersed in solution have varied from 50 to 200 g liter^{-1} [604]. Wetting agents, monovalent cations and aliphatic amines absorbed on particle surfaces, promoted encapsulation and improved dispersion [609, 615, 616]. Particle encapsulation was less during operation of a copper sulfate solution, in comparison with the amounts reported for codeposition in cyanide solutions [604, 609, 615–617].

40 ALUMINA

The copper-alumina system has been investigated by many researchers [604, 605, 607, 618–641]. Both alpha alumina (0.3 to 1.0 μm) and rutile titania (0.3 μm) were readily codeposited with copper in an acid solution, but gamma alumina (0.2 μm) and anatase titania were not. Calcining (20 hours at 1125°C) promoted the codeposition of gamma alumina particles in copper [618]. This effectively transformed the surface of gamma particles into an alpha structure, as verified by X-ray studies. Lakshminarayanan et al. [619–621] concluded that alpha and gamma alumina could be codeposited provided that the plating solution was free of chloride. By contrast, Roos et al. [622] reported that both gamma and alpha alumina could be codeposited in an acid sulfate solution with or without the presence of chloride ions in the solution. However, under identical conditions, much smaller quantities of gamma alumina were codeposited as compared with alpha alumina.

With pulsed or periodically reversed current, increasing the current density increased the alumina content of copper [623, 624]. Ultrasonic agitation reduced agglomeration, improved particle distribution, and the strength retention of copper-alumina composites [625].

41 PROPERTIES

Lakshminarayanan et al. [620] reported that codeposition of 1.2%, by volume, of 0.03 μm gamma alumina increased the yield and ultimate tensile strengths of copper from 97 and 197 MPa to 128 and 246 MPa, respectively, while elongation was reduced from 32% to 20%. An increase in the alumina content to 2.4% increased the yield and tensile strengths to 217 and 345, respectively, while the elongation was 15% [620]. A fairly high yield strength of 50 MPa was retained after the copper-alumina composites were annealed at 420°C. Stankovic and Gojo [641] noted that tensile strength increased to about 290 MPa with increasing alumina particle content. Further increases in the particle content did not affect the tensile strength significantly. As regards electrical resistivity, a 2.2% alumina composite exhibited a moderate increase over pure copper (1.86 vs. 1.72 microhm-cm) [620].

Abrasive wear was reduced 83% to 85% by codepositing 5.5% or 7.3% of carborundum particles [642]. Based on friction measurements, copper-graphite composites appeared to be the best self-lubricating coating when compared with composites containing molybdenum disulfide or PTFE particles [643]. Silicon carbide codeposited particles provided a high hardness which was retained up to 900°C. Abrasion and oxidation resistance also improved as a result of inclusion of silicon carbide [605]. A summary on property data can be found in reference [606].

42 MECHANISM

A two-step mechanism was proposed by Guglielmi [644] in which the combined effect of adsorption and electrophoretic attraction is held responsible for the encapsulation of particulate matter in a growing electrodeposited layer. The validity of Guglielmi's model has been verified for different codeposition systems including copper with alumina from acidic sulfate solutions, with and without the addition of thallium. However, Guglielmi's model does not allow prediction of the manner in which process parameters such as size, type, and pretreatment of the particles, composition, temperature, and pH of the plating solution affect the electrolytic codeposition [609].

Foster and Kariapper [627] proposed a mathematical model in 1974 that could describe the effect of hydrodynamics on codeposition but only a limited amount of quantitative work has been done to prove it's validity. Snaith and Groves [645–647] also reported on mechanisms.

In 1987 Celis et al. [639] proposed a model that contains measurable parameters so that the prediction of the amount of codeposited particles for a given system becomes feasible. As described, "This mathematical model is developed on the basis of two fundamental postulates related to the mechanism of codeposition, namely: 1—an adsorbed layer of ionic species is created around particles at the time these particles are added to the plating solution or pretreated in ionic solutions and 2—the reduction of some of these adsorbed ionic species is required for the incorporation of particles in the metallic matrix." [639].

Figure 3 shows that the particle has to proceed through five stages: (1) the adsorption of ionic species upon the particle, (2) the movement of the particle by convection toward the hydrodynamic boundary layer at the cathode, (3)the diffusion of the particle through the diffusion double layer, (4) the adsorption of the particle with its adsorbed ionic cloud at the cathode surface, and (5) the reduction of some adsorbed ionic species by which the particle becomes irreversibly incorporated in the metal matrix. The model has been demonstrated to be valid for the codeposition of alumina particles with copper from acid copper sulfate plating solutions [609, 639].

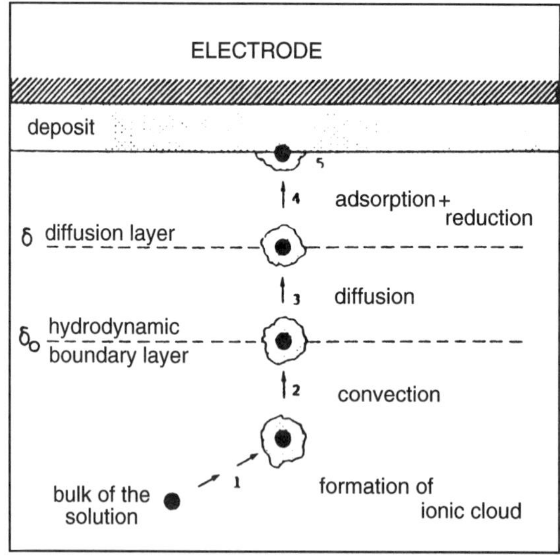

Figure 3 The five stages in the codeposition of a particle, from Celis et al. [639]. Reproduced by permission of the Electrochemical Society, Inc.

43 CONTINUOUS FIBER-REINFORCED COMPOSITES

Acid copper solutions have been used in fabricating fiber-reinforcing metal composites. In these processes, simple salt solutions work satisfactorily, whereas highly complexed solutions such as cyanide do not [648]. Continuous, unidirectional filament reinforced electrocomposites that have been produced include 25 μm tungsten in copper and 100 μm boron in copper [604]. A composite containing 40% volume of tungsten filaments, produced by continuously winding tungsten wire on the cathode during copper deposition exhibited a tensile strength of 1325 MPa [649].

Graphite/copper composite panels have been fabricated by weaving copper-coated graphite fibers into a fabric and then consolidating by hot pressing. This offers a potential technique for low-cost processing of metal matrix composite electronic heat sinks [650]. Carbon fibers are continuously coated with copper in a pyrophosphate-HEDP (hydroxyethylidene diphosphonic acid) electrolyte [651].

REFERENCES

602. V. P. Greco, "Electrocomposites: The Strengthening of Metals with Particles or Fibers by Electrodeposition Techniques and Other Methods," Course Notes, American Electroplaters and Surface Finishers Soc., Orlando, FL (Feb. 1987).
603. V. P. Greco, *Plating Surf. Finish.*, **76**, 62 (July 1989).
604. V. P. Greco, *Plating Surf. Finish.*, **76**, 68 (Oct. 1989).
605. R. Narayan and B. H. Narayana, *Rev. Coatings Corrosion*, **4** (2), 113 (1981).
606. W. H. Safranek, *The Properties of Electrodeposited Metals and Alloys*, 2nd ed., American Electroplaters and Surface Finishers Soc., Orlando, FL (1986).
607. I. Rajagopal, "Composite Coatings," in Surface Modification Technologies, T. S. Sudarshan, ed., Marcel Dekker, New York, 1989.
608. C. E. Johnson and M. Browning, *Proc. AESF SUR/FIN 90*, Session T, Boston, MA (July 1990).
609. J. R. Roos, J. P. Celis, and M. De Bonte, in *Materials Science and Technology: A Comprehensive Treatment*, R. W. Cahn, P. Haasen, and E. J. Kramer, eds., VCH, New York, vol. 15, *Processing of Metals and Alloys*, R. W. Cahn, eds., 481 (1991).
610. L. C. Archibald and P. R. Ebdon, *Met. Finish.*, **90**, 23 (Feb. 1992).
611. M. Viswanathan and K. S. G. Doss, *Met. Finish.*, **70**, 83 (Feb. 1972).
612. M. Ghouse, M. Viswanathan, and E. G. Ramachandran, *Met. Finish.*, **78**, 31 (Mar. 1980).
613. M. Ghouse, M. Viswanathan, and E. G. Ramachandran, *Met. Finish.*, **78**, 55 (Nov. 1980).
614. M. Ghouse and E. G. Ramachandran, *Met. Finish.*, **79**, 85 (June 1981).
615. J. E. Hoffman and C. L. Mantell, *Trans. AIME*, **236**, 1015 (1966).
616. T. W. Tomaszewski, L. C. Tomaszewski, and H. Brown, *Plating*, **56**, 1234 (1969).
617. E. A. Brandes and D. Goldthorpe, *Metallurgica*, **195** (Nov. 1967).
618. E. S. Chen, G. R. Lakshminarayanan, and F. K. Sautter, *Met. Trans.*, **2**, 937 (1971).
619. G. R. Lakshminarayanan, E. S. Chen, and F. K. Sautter, *Plating Surf. Finish.*, **63**, 38 (Apr. 1976).
620. G. R. Lakshminarayanan, E. S. Chen, and F. K. Sautter, *Plating Surf. Finish.*, **63**, 35 (May 1976).

621. G. R. Lakshminarayanan, "Codeposition of Alumina with Copper," Manufacturing Technology Note NTN-78/0746, U.S. Army Materiel Development and Readiness Command, Alexandria, VA (Oct. 1978).

622. J. R. Roos, J. P. Celis, and J. A. Helsen, *Trans. Inst. Met. Finish.*, **55**, 113 (1977).

623. V. V. Grinina and Y. M. Polukarov, "Effect of Periodic Current on the Incorporation of Foreign Particles in Electrolytic Copper Deposits," Institute of Physical Chemistry, Academy of Sciences of the USSR, Moscow, Russia (Nov. 1985).

624. M. M. Davliev, G. P. Petrov, I. A. Abdullin, V. A. Golovin, and P. A. Norden, *Elektronnaya Obrabotka Materialov*, no. 3, 26 (1986).

625. E. S. Chen and F. K. Sautter, *Plating Surf. Finish.*, **63**, 28 (Sept. 1976).

626. J. Hoashi, *J. Met. Finish., Soc. Jpn.*, **15** (7), 258 (1964).

627. J. Foster and A. M. J. Kariapper, *Trans. Inst. Met. Finish.*, **51**, 27 (1973).

628. A. M. J. Kariapper and J. Foster, *Trans. Inst. Met. Finish.*, **52**, 87 (1974).

629. J. M. Sykes and D. J. Alner, *Trans. Inst. Met. Finish.*, **52**, 28 (1974).

630. J. P. Celis and J. R. Roos, *J. Electrochem. Soc.*, **124**, 1508 (1977).

631. J. R. Roos, J. P. Celis, and H. Kelchtermans, *Thin Solid Films*, **54**, 173 (1978).

632. J. P. Celis, H. Kelchtermans, and J. R. Roos, *Trans. Inst. Met. Finish.*, **56**, 41 (1978).

633. C. White and J. Foster, *Trans. Inst. Met. Finish.*, **56**, 92 (1978).

634. J. R. Roos, J. R. Celis, H. Kelchtermans, M. Van Camp, and C. Buelens, *Proc. of Interfinish 80*, Kyoto, Japan 203 (Oct. 1980).

635. C. White and J. Foster, *Trans. Inst. Met. Finish.*, **59**, 8 (1981).

636. C. Buelens, J. P. Celis, and J. R. Roos, *J. Appl. Electrochem.*, **13**, 541 (1983).

637. C. C. Lee and C. C. Wan, *J. Electrochem. Soc.*, **135**, 1930 (1988).

638. H. Hayashi, S. Izumi, and I. Tari, *J. Electrochem. Soc.*, **140**, 362 (1993).

639. J. P. Celis, J. R. Roos, and C. Buelens, *J. Electrochem. Soc.*, **134**, 1402 (1987).

640. M. H. Fawzy, M. M. Ashour, and A. M. Abd El-Halim, *Trans. Inst. Met. Finish.*, **73**, 132 (1995).

641. V. D. Stankovic and M. Gojo, *Surf. Coatings Technol.*, **81**, 225 (1996).

642. R. S. Saifullen and L. N. Akulova, *Prot. Metals*, **8**, 74 (1972).

643. V. Bhalla, C. Ramasamy, N. Singh, and M. Pushpavanam, *Plating Surf. Finish.*, **82**, 58 (Nov. 1995).

644. N. Guglielmi, *J. Electrochem. Soc.*, **119**, 1009 (1972).

645. D. W. Snaith and P. D. Groves, *Trans. Inst. Met. Finish.*, **50**, 95 (1972).

646. D. W. Snaith and P. D. Groves, *Trans. Inst. Met. Finish.*, **55**, 136 (1977).

647. D. W. Snaith and P. D. Groves, *Trans. Inst. Met. Finish.*, **56**, 9 (1978).

648. A. A. Baker, M. B. P. Allery, and S. J. Harris, *J. Materials Sci.*, **4**, 242 (1969).

649. J. C. Withers and E. F. Abrams, *Plating*, **55**, 605 (1968).

650. D. A. Foster, *SAMPE Quarterly*, **21**, 58 (Oct. 1989).

651. Y. Gan, *Plating Surf. Finish.*, **79**, 81 (June 1992).

3 Electrodeposition of Nickel

GEORGE A. DI BARI

INTRODUCTION

Nickel electroplating is a commercially important and versatile surface finishing process. Its commercial importance may be judged from the amount of nickel in the form of metal and salts consumed annually for electroplating, now roughly 100,000 metric tonnes worldwide; its versatility, from its many current applications [1]. The applications of nickel electroplating fall into three main categories: decorative, functional, and electroforming.

In decorative applications, electroplated nickel is most often applied in combination with electrodeposited chromium. The thin layer of chromium was first specified to prevent the nickel from tarnishing. It was originally deposited on top of a relatively thick, single layer of nickel that had been polished and buffed to a mirror-bright finish. Today decorative nickel coatings are mirror-bright *as-deposited* and do not require polishing prior to chromium plating. Multilayered nickel coatings outperform single-layer ones of equal thickness and are widely specified to protect materials exposed to severely corrosive conditions. The corrosion performance of decorative, electroplated nickel plus chromium coatings has been further improved by the development of processes by which the porosity of chromium can be varied and controlled on a microscopic scale (microdiscontinuous chromium). Modern multilayered nickel coatings in combination with microdiscontinuous chromium are capable of protecting steel, zinc, copper, aluminum, and many other materials from corrosion for extended periods of time. The complexity of modern-day nickel plus chromium coatings is more than offset by the greatly improved corrosion resistance that has been achieved without significantly increasing coating thickness and costs.

There are many functional applications where decoration is not the issue. Instead, nickel and nickel alloys with matte or dull finishes are deposited on surfaces to improve corrosion and wear resistance, or modify magnetic and other properties. The properties of nickel electrodeposits produced under different conditions of operation are of particular interest in this connection.

Electroforming is electroplating applied to the fabrication of products of various kinds. Nickel is deposited onto a mandrel and then removed from it to create a part made entirely of nickel. A variation of this is electrofabrication where the deposit is not separated from the substrate and where fabrication may involve electrodeposition through masks rather than the use of traditional mandrels.

Modern Electroplating, Fourth Edition, Edited by Mordechay Schlesinger and Milan Paunovic.
ISBN 0-471-16824-6

The many current applications of nickel electroplating are the result of developments and improvements that have been made almost since the day the process was discovered. This is evident in the following retrospective on the development of nickel electroplating solutions, as well as in subsequent sections that deal with basics, decorative electroplating, functional applications and deposit properties, nickel electroforming, nickel anode materials, quality control, and pollution prevention.

1 RETROSPECTIVE ON NICKEL ELECTROPLATING SOLUTIONS

Bottger developed the first practical formulation for nickel plating, an aqueous solution of nickel and ammonium sulfates in 1843, but earlier references to nickel plating can be found. Bird apparently deposited nickel on a platinum electrode in 1837 from a solution of nickel chloride or sulfate, and Shore patented a nickel nitrate solution in 1840 [2, 3]. The solution developed by Bottger remained in commercial use for seventy years, however, and he is acknowledged to be the originator of nickel plating [4].

Dr. Isaac Adams Jr., a medical doctor educated at Harvard University and at the Ecole de Medicine in Paris, was one of the first to commercialize nickel plating in the United States, and his patented process gave his company a virtual monopoly in commercial nickel plating from 1869 to 1886. His patent covered the use of pure nickel ammonium sulfate. Although Adams's solution was similar to Bottger's, his emphasis on operating the bath at neutral pH was undoubtedly vital for controlling the quality of the nickel deposited, since excessive amounts of ammonia would tend to lower cathode efficiency and embrittle the deposit. Largely as a result of the publicity generated by Adams, nickel plating became known worldwide, and by 1886, the annual consumption of nickel for plating had grown to about 135 metric tons [5, 6].

Remington, an American residing in Boston, attempted to market a nickel ammonium chloride electroplating solution in 1868, but perhaps of greater significance, in view of subsequent developments, were his attempts to use small pieces of electrolytic nickel as an anode material in a platinum anode basket [7]. Weston introduced the use of boric acid and Bancroft was one of the first to realize that chlorides were essential to ensure efficient dissolution of nickel anode materials [8, 9].

Professor Oliver P. Watts at the University of Wisconsin, aware of most of these developments, formulated an electrolyte in 1916 that combined nickel sulfate, nickel chloride, and boric acid and optimized the composition of the nickel electroplating solution [10]. The advantages of his hot, high-speed formula became recognized and eventually led to the elimination of nickel ammonium sulfate and other proprietary solutions. Today the Watts solution is widely applied, and its impact on the development of modern nickel electroplating technology cannot be overstated.

Decorative nickel plating solutions are variations of the original Watts formulation, the main difference being the presence in solution of organic and certain metallic compounds to brighten and level the nickel deposit. Because of his use of organic additives like benzene and naphthalene di- and trisulfonic acids, Schlotter's decision to market a bright nickel plating solution in 1934 is a milestone in the commercial development of decorative nickel plating [11].

The introduction of coumarin-containing, semi-bright nickel-plating solutions by DuRose in 1945 was a major contribution because it subsequently led to the development of double- and triple-layer nickel coatings with greatly improved corrosion resistance [12]. His introduction of semi-bright nickel processes came at a critical time in the history of decorative nickel electroplating, when confidence in the ability of single-layer, bright nickel coatings to prevent corrosion of automotive components was at a relatively low point.

The dominant position of Watts solutions has been challenged from time to time, but the only ones that have been adopted on a substantial scale are nickel sulfamate solutions [13]. The Watts solution is the basis for most decorative nickel plating solutions, although considerable variation in the chloride content of different proprietary decorative processes may be specified by the suppliers of those processes. Sulfamate solutions are rarely, if ever, used for decorative plating. Watts and nickel sulfamate solutions are used for functional plating and for electroforming, but sulfamate is more popular for the latter.

2 BASICS

Nickel electroplating is similar to other electrodeposition processes that employ soluble metal anodes; that is, direct current is made to flow between two electrodes immersed in a conductive, aqueous solution of nickel salts. The flow of direct current causes one of the electrodes (the anode) to dissolve and the other electrode (the cathode) to become covered with nickel. The nickel in solution is present in the form of divalent, positively charged ions (Ni^{++}). When current flows, the positive ions react with two electrons ($2e^-$) and are converted to metallic nickel (Ni^0) at the cathode surface. The reverse occurs at the anode where metallic nickel is dissolved to form divalent, positively charged ions which enter the solution. The nickel ions discharged at the cathode are thus replenished by those formed at the anode.

2.1 Application of Faraday's Laws to Nickel

The amount of nickel deposited at the cathode and the amount dissolved at the anode are directly proportional to the product of the current and time and may be calculated from the following expression:

$$m = 1.095\,(a)\,(I)\,(t) \tag{1}$$

where m is the amount of nickel deposited at the cathode (or dissolved at the anode), in grams, I is the current that flows through the plating tank in amperes, t is the time that the current flows, in hours, and a is the current efficiency ratio (see chapter 1 for the definition of current efficiency). The proportionality constant, (1.095) grams per ampere hour, equals M/nF, where M is the atomic weight of nickel (58.69), n is the number of electrons in the electrochemical reaction (2), and F is Faraday's constant, equal to 26.799 ampere-hours (more commonly given as 96,500 coulombs).

The proportionality constant must be multiplied by the actual electrode efficiency ratio if precise values are required. The anode efficiency for nickel dissolution is almost always 100% under practical electroplating conditions; that is, $a = 1$ when estimating anode weight loss. If the pH of the solution is too high and/or the chloride ion concentration too low, hydroxyl ions may be discharged in preference to the dissolution of nickel, and oxygen will be evolved. Under those unusual conditions the nickel anode becomes passive, and the efficiency of anode dissolution is close to zero.

The cathode efficiency of different nickel plating solutions may vary from 90% to 97% and accordingly, a will vary from 0.90 to 0.97. A small percentage of the current is consumed in the discharge of hydrogen ions from water. That reduces the cathode efficiency for nickel deposition from 100% to approximately 96% in an additive-free nickel electroplating solution. The discharged hydrogen atoms form bubbles of hydrogen gas at the cathode surface. Cathode efficiencies as low as 90% are characteristic of some bright nickel plating solutions that are formulated to give highly leveled, mirrorlike deposits rapidly; that is, at thicknesses below 12 μm [14]. An average cathode efficiency of 95.5% is commonly used to make estimates when precise values are not essential.

Because the anode and cathode efficiencies are not exactly equal, the nickel ion concentration and the pH of the solution will slowly increase as plating proceeds. The rate of increase in nickel ion concentration depends on the difference between cathode and anode efficiencies. Because cathode efficiencies may vary from 90% to 97%, whereas anode efficiency is almost always 100%, the rate of increase in nickel ion concentration depends on the cathode efficiency and the nature of the plating solution, not on the type of soluble nickel anode material that is used. The use of a special insoluble anode in combination with soluble anodes has been developed to help control the rate of increase of the nickel ion concentration [15].

2.2 Average Coating thickness

An expression for calculating nickel thickness, s in micrometers, can be derived by dividing Eq. (1) by the product of the density of nickel, d, ($8.907\,\mathrm{g\,cm^{-3}}$), and the surface area to be electroplated, A, and multiplying by 100 to obtain the thickness in micrometers, as shown below:

$$s = \frac{(m)\,(100)}{(d)\,(A)} = \frac{(109.5)\,(a)\,(I)\,(t)}{(8.097)\,(A)} = \frac{(12.294)\,(a)\,(I)\,(t)}{(A)} \qquad (2)$$

The ratio, $(I)/(A)$, is the current density and thus the above expression shows that the coating thickness depends on the *current density* and time, whereas the amount or mass of nickel deposited, Eq. (1), depends on the *current* and time. Equation (2) is the basis for the electrodeposition data compiled in Table 1, which gives the time *in minutes* required to deposit a nickel coating of specified thickness at different values of current density. The expression above and Table 1 provide a means of estimating the *average* coating thickness.

2.3 Current and Metal Distribution

The *actual* thickness at any point on the surface of a shaped article is dependent on the current density at that point. The current density at any point is determined by how the current is apportioned over the surface of the article being electroplated. In

TABLE 1 Nickel Electrodeposition Data

Deposit	Weight per	Amp Hours	Time (min) to Obtain Deposit at Various Current Densities ($A\,dm^{-2}$)									
Thickness (μm)	Unit Area ($g\,dm^{-2}$)	per Unit ($Ah\,dm^{-2}$)	0.5	1	1.5	2	3	4	5	6	8	10
2	0.18	0.17	20	10	6.8	5.1	3.4	2.6	2.0	1.7	1.3	1
4	0.36	0.34	41	20	14	10	6.8	5.1	4.1	3.4	2.6	2
6	0.53	0.51	61	31	20	15	10	7.7	6.1	5.1	3.8	3.1
8	0.71	0.68	82	41	27	20	13	10	8.2	6.8	5.1	4.1
10	0.89	0.85	100	51	34	26	17	13	10	8.5	6.4	5.1
12	1.1	1.0	120	61	41	31	20	15	12	10	7.7	6.1
14	1.2	1.2	140	71	48	36	24	18	14	12	8.9	7.1
16	1.4	1.4	160	82	54	41	27	20	16	14	10	8.2
18	1.6	1.5	180	92	61	46	31	23	18	15	11	9.2
20	1.8	1.7	200	100	68	51	34	26	20	17	13	10
40	3.6	3.4	410	200	140	100	68	51	41	34	26	20

Note: Based on 95.5% cathode efficiency.

nickel plating, the current distribution is largely determined by geometric factors that is, by the shape of the part, the relative placement of the part with respect to the anode, how the parts are placed on plating racks, and the dimensions of the system. Because almost all except the simplest shapes to be electroplated have prominent surfaces that are nearer to the anode than recessed areas, a uniformly thick nickel coating is difficult to produce. The current density at prominences is greater due to the shorter anode-to-cathode distance and the lower resistance to current flow that implies. Conversely, recessed areas, being further away from the anode, will have a lower current density because of increased resistance to current flow. This inevitably means that prominent areas will have thicker coatings than recessed ones.

Because geometric factors exert the greatest influence on current density at localized areas in the case of nickel plating, current distribution is virtually the same as metal distribution. Thus shields and auxiliary anodes can be used effectively to obtain acceptable thickness uniformity. Shields are made of nonconductive materials, and they may be placed on the anode, on the cathode, or between electrodes to block or control current flow. Auxiliary anodes may be either soluble or insoluble, and are placed closer to the cathode than principal anodes so as to direct the current to a recessed or relatively small area on the cathode. The analysis of geometric effects by computer modeling has received attention [16].

2.4 Throwing Power

In addition to the geometric factors, metal distribution is influenced by cathode polarization, the cathode efficiency-current density relationship, and the electrical conductivity of the solution [17]. The complex relationship between the factors that influence current distribution, and hence metal distribution is called *throwing power*. A solution with a high-throwing power is capable of depositing almost equal thicknesses on both recessed and prominent areas. For example, in copper cyanide solutions, high cathode polarization and the favorable cathode efficiency-current density relationship (cathode efficiency is lower at high than at low current densities) result in a solution with excellent throwing power. As already implied, cathode polarization and current efficiency do not significantly affect the throwing power of acid nickel plating solutions formulated with simple salts. The cathode polarization is low and the cathode efficiency is high and relatively constant above $1\,A\,dm^{-2}$.

Throwing power can be measured to give relative values [18]. Measurements indicate that the throwing power of nickel electroplating solutions can be somewhat improved by lowering the current density, increasing the electrical conductivity of the solution, increasing the distance between anode and cathode, and by raising the pH and the temperature. Table 2 compares the throwing power of various plating solutions; it is based on the work of Watson who applied a Hull cell to make the measurements. The nickel plating solution with the best throwing power contained a high concentration of anhydrous sodium sulfate; its composition is given in Table 3, which also indicates that throwing power decreases as the current density is increased [19].

2.5 Internal Stress

Internal stress refers to forces created within the deposit as a result of the electrocrystallization process and/or the codeposition of impurities such as

TABLE 2 Throwing Power of Various Electroplating Solutions

Solution	Average Current Density		Throwing Power (%) at Primary Current Density Ratios		
	$A\,dm^{-2}$	$A\,ft^{-2}$	5:1	12:1	25:1
Watts nickel	4.3	40	8	7	14
Sulfamate nickel	4.3	40	11	13	19
All-chloride nickel	4.3	40	18	18	27
Na/high sulfate	4.3	40	23	31	40
Mg/high sulfate	4.3	40	16	18	32
Proprietary bright nickel A	4.3	40	1	−12	−6
Proprietary bright nickel B	4.3	40	3	−12	−6
Acid copper	4.3	40	0	−29	−61
Rochelle copper	4.3	40	86	91	93
Conventional chromium	16	150	−42	−48	−100

Source: Watson [19].

Note: The 100% represents a uniform thickness over the cathode surface; −100 indicates the opposite extreme, where recessed areas are thinly plated, and prominent areas are thickly plated.

TABLE 3 Composition and Throwing Power at Various Current Densities of a High-Sulfate Solution

Average Current Density		Throwing Power (%) at Primary Current Density Ratios		
$A\,dm^{-2}$	$A\,ft^{-2}$	5:1	12:1	25:1
0.2	2	63	76	85
1.0	10	38	50	64
4.3	40	23	31	40

Source: Watson [19].

Note: Nickel sulfate—30 g $liter^{-1}$; nickel chloride—38 g $liter^{-1}$; boric acid—25 g $liter^{-1}$; sodium sulfate (anhydrous)—180 g $liter^{-1}$. The 100% represents a uniform thickness over the cathode surface; −100 indicates the opposite extreme where recessed areas are thinly plated, and prominent areas are thickly plated.

hydrogen, sulfur, and other elements [20]. Internal stress is either tensile (contractile) or compressive (expansive) in nature. In tensively stressed deposits, the average distance between nickel atoms in the lattice is greater than the equilibrium value, creating a force that tends to drive the atoms closer together. When a tensively stressed deposit is detached from its substrate, it contracts. In addition, if a thin cathode strip is electroplated on one side only (by painting the back and placing the bare side facing the anode), a deposit stressed in tension will cause the strip to bend or curl towards the anode. In compressively stressed deposits, the atoms are closer together and the force tends to drive them further apart. When detached from the substrate, compressively stressed deposits expand and a thin strip plated on one side only, as described, will bend away from the anode. Dislocation theory provides a logical explanation of the origins of internal stress in electrodeposits [21, 22].

Stress in electrodeposited nickel can vary over a wide range depending on solution composition and operating conditions. In general, nickel electrodeposited from additive-free Watts solutions exhibits a tensile stress that is 125 to 185 MPa for the conditions given in Table 4. Deposits from sulfamate solutions display lower tensile stress within the range of 0 to 55 MPa. Compressively stressed nickel deposits are obtained from solutions that contain sulfur-containing organic additives similar to the carriers that are added to bright nickel plating solutions (discussed below). As far as is known, compressively stressed nickel deposits are almost always associated with the codeposition of sulfur. High-deposit stress can be especially troublesome in electroforming where the adhesion between the electrodeposit and the mandrel is deliberately kept as low as possible to facilitate separation.

2.6 Adhesion

With the exception of electroforming, a high degree of adhesion between the deposit and the substrate is critical in all applications. Under favorable conditions, the crystal lattice of the substrate extends into the deposit (epitaxy). Epitaxial growth rarely occurs in commercial electroplating, however, where adhesion is due to cohesive forces between metal atoms. Atoms of the electrodeposited metal align themselves in opposition to atoms of the substrate and are held to the surface by interatomic forces that result in the formation of metallic, covalent, ionic, polar or other types of bonds. *Perfect* adhesion in electoplating is achieved when the bond strength is greater than the tensile strength of the weaker component; that is, in measuring adhesion by means of a quantitative test, failure does not occur at the interface, but either within the deposit or within the substrate. In commercial electroplating, good adhesion is achieved by means of standard preparation methods that are described in ASTM standards and elsewhere [23, 24].

2.7 Leveling and Microthrowing Power

The ability of an electroplating solution to fill in defects and scratches on the surface *preferentially* is called leveling. The deposit fills defects and tiny scratches, eventually covering them, and as a result the surface becomes smoother as the deposit increases in thickness. The semi-bright and bright nickel electroplating solutions described in the next section have excellent leveling properties. Organic additives in those solutions are adsorbed preferentially at micropeaks; the resulting increase in the local resistance to current flow increases the current density in microgrooves, thereby promoting leveling [25].

Microthrowing power refers to the ability of an electroplating solution to fill tiny crevices with deposits that follow the contour of a defect or scratch without any leveling action whatsoever [26]. Additive-free nickel plating solutions have excellent microthrowing power, but little leveling ability. In the absence of specifically adsorbed organic additives, the current density in micropeaks and microgrooves is

TABLE 4 Nickel Electroplating Solutions

	Electrolyte composition[a] ($g\,liter^{-1}$)		
	Watts Nickel	Nickel Sulfamate	Basic Semi-bright Bath[b]
Nickel Sulfate, $NiSO_4 \cdot 6H_2O$	225–400		300
Nickel sulfamate, $Ni(SO_3NH_2)_2$		300–450	
Nickel chloride, $NiCl_2 \cdot 6H_2O$	30–60	0–30	35
Boric acid, H_3BO_3	30–45	30–45	45
		Operating conditions	
Temperature (°C)	44–66	32–60	54
Agitation	Air or mechanical	Air or mechanical	Air or mechanical
Cathode current density ($A\,dm^{-2}$)	3–11	0.5–30	3–10
Anodes	Nickel	Nickel	Nickel
pH	2–4.5	3.5–5.0	3.5–4.5
		Mechanical properties	
Tensile strength (MPa)	345–485	415–610	—
Elongation (%)	10–30	5–30	8–20
Vickers hardness (100 g load)	130–200	170–230	300–400
Internal stress (MPa)	125–185 (tensile)	0–55 (tensile)	35–150 (tensile)

[a]Anti-pitting agents formulated for nickel plating are added to control pitting.

[b]Organic additives available from plating supply houses are required for semi-bright nickel plating.

[c]Typical properties of *full-bright* nickel deposits are as follows: Elongation percentage—2 to 5; Vickers hardness, 100 gram load—600 to 800; internal stress, MPa—12 to 25 compressive.

uniform because of the low polarization and high cathode efficiency of most nickel electroplating solutions.

3 DECORATIVE ELECTROPLATING

The technology of decorative nickel electroplating has been improved continually over the years due to the development of bright and semi-bright nickel plating solutions, multilayer nickel coatings, and microdiscontinuous chromium. The major effect has been an increase in the corrosion resistance of decorative, electroplated nickel plus chromium coatings. In addition new and improved techniques for plating on plastics, aluminum alloys, and stainless steel have broadened the scope of decorative applications.

3.1 Bright Nickel Solutions

Modern bright nickel electroplating solutions employ combinations of additives carefully formulated to produce bright deposits over a wide range of current density. The deposits have excellent leveling or scratch-filling characteristics, fair ductility, and low internal stress. Modern processes produce bright deposits in areas of low current density, permit use of high average current densities and bath temperatures, are less sensitive to metallic contaminants than some of the solutions first commercialized, permit continuous purification of the plating solution by filtering through activated carbon, and produce breakdown products that can also be removed by activated carbon treatment and are not overly sensitive to anode effects. Bright nickel electroplating solutions have similar compositions as the Watts nickel electrolyte included in Table 4. Because the exact formulations of commercial processes are proprietary, the recommendations of the suppliers of decorative nickel plating processes should be followed.

The additives for bright nickel electroplating fall roughly into three categories: carriers, brighteners, and auxiliary brighteners. Those are the terms preferred by electroplaters, but the terminology is not standardized. To avoid confusion, alternative terms mentioned in the literature are shown in parentheses [27, 28].

Carriers (Brighteners of the First Class, Secondary Brighteners, Control Agents) These are usually aromatic organic compounds. Examples are benzene sulfonic acid; 1,3,6-naphthalene sulfonic acid (sodium salt); *p*-toluene sufonamide; saccharin (*o*-benzoic sulfonimide); thiophen-2-sulfonic acid; benzene sulfinic acid; and allyl sulfonic acid. Carriers are the principal source of the sulfur codeposited with the nickel. Their main function is to refine grain size and provide deposits with increased luster compared with matte or dull deposits from baths without additives. When used by themselves, carriers do not produce mirror-bright deposits. Carriers are used in concentrations of about 1 to 25 g liter^{-1}, the exact concentration depending on the specific compound. Carriers are not consumed rapidly by electrolysis, and consumption is primarily by dragout and by losses during activated carbon treatments. The stress-reducing property of carriers is increased if they contain amido or imido nitrogen. For example, saccharin is a most effective stress reducer and often helps to decrease or eliminate hazes.

Brighteners (Brighteners of the Second Class, Primary Brighteners, Leveling Agents) In combination with carriers and auxiliary brighteners, brighteners produce brilliant deposits having good ductility and leveling characteristics over a wide range of current densities. Some of these compounds include formaldehyde chloral hydrate; *o*-sulfo benzaldehyde; allyl sulfonic acid; coumarin; *o*-hydroxy cinnamic acid; diethyl maleate; 2-butyne-1,4-diol; 2-butyne-1,4-disulfonic acid; ethyl cyanohydrin; *p*-amino azo benzene; thiourea; and allyl thiourea and polyethylene glycols of various kinds. The best-known example in this class may be coumarin (1,2-benzopyrone), the leveling agent introduced by DuRose for producing semi-bright nickel deposits. Because they increase the internal stress and promote brittleness of nickel deposits, brightener concentrations are kept low and carefully controlled. Concentrations of 0.005 to 0.2 g $liter^{-1}$ are generally used. The rates of consumption of these materials may vary within wide limits.

Auxiliary Brighteners Auxiliary brighteners augment the luster attainable with carriers and brighteners, and increase the rate of brightening and leveling. Some examples are sodium allyl sulfonate; zinc, cobalt, cadmium; and 1,4-butyne 2-diol. The concentration of these additives may vary from about 0.1 to 4 g $liter^{-1}$ Sulfobetaines, such as pyridinium propyl sulfonate, are especially active sulfur-containing organic compounds that have a powerful effect on leveling in nickel deposits. However, because they introduce sulfur into the deposit they cannot be used for semi-bright nickel plating [28]. They can be especially effective when added to bright nickel plating solutions in low, carefully controlled concentrations. The inorganic metallic ions—zinc cobalt, cadmium—are not often used anymore as auxiliary brighteners.

3.2 Electrocrystallization

The mechanism of nickel electrodeposition involves surface adsorption of species formed in the cathode film accompanied by inhibition of growth of certain crystal faces. In the absence of organic additives, species like H_2, H_{ads} and $Ni(OH)_2$, that form as a result of the reduction of hydrogen ions, determine most of the micro-structural features of the nickel deposit [29].

The mechanism by which *unsaturated* organic additives modify the electro-crystallization process to yield mirror-bright surfaces also involves adsorption, hydrogenation, and desorption. The organic molecule is adsorbed on the surface via the unsaturated bond, blocking certain sites on the nickel lattice and thus altering the growth rates of different crystal faces. The unsaturated bond reacts with hydrogen in the cathode film and the resulting reduction products are desorbed from the surface and/or incorporated into the deposit. The rates of those processes are influenced by the degree of unsaturation, the size and shape of the organic molecule, the functional groups present, aromatic rings, and other stereochemical factors.

That general mechanism has been confirmed and clarified by studying the effects of individual additives [30]. For example, the active group in a carrier like sodium benzene sulfonate is =C–S. Hydrogenolysis of the =C–S bonds results in the formation of sulfur anions that are adsorbed on the (110) crystallization direction causing the [100] texture to predominate. The sulfur is incorporated into the deposit as the sulfide. Below 25 A dm^{-2} the active groups in saccharin, =C–S and S=O, react with hydrogen to form sulfur anions that are also codeposited and suppress growth in the (110) direction, promoting the formation of the [100] texture and extending the [100] planes in the surface of the deposits, but above that current density, hydrogenation of adsorbed C=C and C≡C bonds also takes place. This promotes reduction of hydrogen ions and increases the pH in the cathode film with the consequent formation of a colloidal suspension of nickel hydroxide. The selective adsorption of colloidal nickel hydroxide inhibits growth in the (100) direction, favoring formation of [211], [211] + [111], and [111] textures. The active groups in sulfur-free brighteners include C=C, C≡C, C≡N, and others, and the mechanism by which these modify crystal growth also involves hydrogenation, increased alkalinity in the cathode film, and precipitation and inhibition of crystal growth by colloidal nickel hydroxide causing [211] and [111] textures to dominate crystallographic features. The two mechanisms by which sulfur- containing and sulfur-free organic addition agents influence crystal growth are thus distinctly different, and complex sulfur-containing compounds like saccharin apparently display features of both. The effect of pulsed reversed current in the presence of organic additives has also been studied [31].

Brightness is attained if the microstructural components of the surface form a plane from which they do not vary by distances greater than the wavelength of light [32]. The criterion for brightness is thus not simply the production of fine-grained deposits, but the creation of *flat* crystals. Exactly how a complex mixture of carriers, brighteners, and auxiliary brighteners act in concert to meet those criteria may need further elucidation, but the synergistic effects first described in the classical work of Edwards are probably involved [33].

3.3 Effects of Codeposited Sulfur

The incorporation of sulfur has several important effects. Sulfur increases the electrochemical reactivity of bright nickel compared to sulfur-free nickel, and that effect is applied in multilayer coatings to improve corrosion performance. Codeposition of sulfur changes the intrinsic internal stress of electrodeposited nickel from tensile (contractile) to compressive (expansive) and the incorporation of sulfide, an anion considerably larger than the nickel atom, must compress the nickel atoms in the crystal lattice. Carriers are thus useful for controlling internal stress. In addition, sulfur increases the hardness and lowers the elongation percentage of electrodeposited nickel (Table 4, note c).

3.4 Semi-bright Nickel

Semi-bright nickel solutions contain nickel sulfate, nickel chloride, boric acid and leveling agents. A typical composition is included in Table 4. Coumarin has, for the most part, been superseded by other organic compounds [34], notably the acetylenics that are now widely employed. Although the ability of coumarin to level nickel deposits is excellent, the acetylenics form reduction products that have little or no effect on the properties of the semi-bright nickel deposit, whereas the melilotic acid formed by the cathodic reduction of coumarin is incorporated in the deposit lowering its ductility and raising its hardness. The effect of melilotic acid must be controlled by frequent treatments with activated carbon. Coumarin-free semi-bright nickel solutions need to be carbon treated less frequently, are easier and more economical to maintain, and produce deposits with consistent properties.

Semi-bright nickel plating processes yield smooth, sulfur-free deposits that are semi-lustrous with columnar structures similar to Watts nickel deposits, in contrast to full-bright nickel deposits that have banded (laminar) structures as a result of the periodicity of sulfur codeposition. The highly leveled surface is easy to polish and buff to a mirror finish. The good ductility of semi-bright nickel deposits is important in those applications where multilayer coatings are stressed in service. Semi-bright nickel deposits are an essential part of multilayer nickel coatings.

3.5 Multilayer Nickel Coatings

Single-layer bright nickel plus chromium coatings proved to be less resistant to corrosion than the polished Watts nickel plus chromium coatings that they replaced. The lower resistance to corrosion was attributed to the activating effect of the sulfur incorporated in bright nickel electrodeposits. The search for coatings with improved corrosion resistance led, first, to the development of semi-bright nickel coatings. Semi-bright nickel deposits, being sulfur-free, are equivalent in corrosion resistance to polished Watts nickel deposits, and for a while (1945–50) were applied as single-layer coatings in the production of automobile bumpers to obtain improved corrosion performance despite the disadvantage of having to polish the coatings after electroplating to achieve full brightness.

The idea of combining layers of bright and semi-bright nickel was conceived later as a means of eliminating the costly polishing operation [35]. The fact that decorative double-layer nickel coatings are much more resistant to corrosion than equivalent thicknesses of single-layer bright nickel was discovered as a result of studies conducted after the double-layer coatings were applied commercially. The introduction of triple-layer coatings occurred later [36]. Today single-layer bright nickel coatings are only specified for mild corrosion service with nickel thickness commonly in the range of 5 to 12 μm. For severe and very severe corrosion service, thicker, multi-layered coatings are specified.

Double-layer nickel coatings have an undercoat of highly leveled, sulfur-free, semi-bright nickel covered with sufficient bright nickel to give a mirror finish without having to polish the coating. The excellent leveling characteristics of the semi-bright nickel layer also helps minimize the requirement for expensive mechanical finishing of the substrate. Because the undercoat of sulfur-free, semi-bright nickel is electrochemically more noble than bright nickel, corrosive attack, when it does occur, is preferentially directed toward the bright nickel and a flat-based pit is formed (Fig. 1). This effectively retards the pitting attack because, before the pit can penetrate the coating further, a considerable portion of the bright nickel layer must be removed. As a result double-layer coatings show a marked improvement in corrosion performance over single-layer coatings.

The electrochemical basis for the improved corrosion performance observed with double-layer coatings was provided by Flint and Melbourne, DuRose, Safranek and others [37–39]. Petrocelli, Hospadaruk, and DiBari [40] constructed Evans-type corrosion diagrams from which the galvanic relationships between nickel and chromium in various test electrolytes were deduced. They showed that galvanically coupling semi-bright nickel to bright nickel caused the semi-bright nickel to become

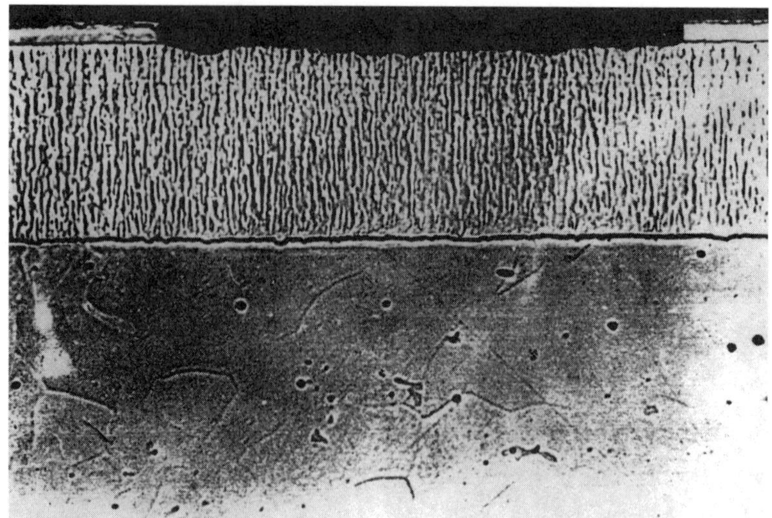

Figure 1 In double-layer nickel coatings, corrosion is initially confined to the top, full-bright nickel layer and penetration through the underlying semi-bright nickel layer is delayed (after 30 months exposure outdoors in an industrial atmoshpere).

cathodically polarized, lowering its rate of corrosion, whereas the bright nickel became anodically polarized, increasing its rate of corrosion. When bright nickel, semi-bright nickel, and chromium were galvanically coupled in the same solution, chromium was cathodic to both the bright and the semi-bright nickel, and as a result coupling to chromium increased the rate of corrosion of both types of nickel.

During electrochemical studies of the behavior of various types of nickel deposits, it became clear that there were basically two types of proprietary bright nickel processes commercially available, one with a significantly higher reactivity than the other [41]. When equal areas of semi-bright nickel were coupled to each type, the highly reactive bright nickel provided greater protection of the semi-bright nickel than the low-reactivity type. The relative merits of using high-reactivity bright nickel in combination with semi-bright nickel was confirmed in outdoor stationary and mobile exposure programs performed with contoured panel [42]. Contoured panels have raised and recessed areas, and the thickness of the coating is greater on the relatively flat portion of the panel than in the recessed areas where the average thickness of the coating is 45% of the nominal value. The fact that recessed areas performed better with the high reactivity bright nickel, not only confirmed the observations made in the laboratory but also showed that the improvement in corrosion performance due to double-layer nickel coatings becomes more important as coating thickness is decreased. Snyder [43] refers to the two types as high- and low-potential bright nickel, and states that the low-potential types may give better performance as single-layer coatings.

In triple-layer coatings, semi-bright and bright nickel layers are separated by a thin, highly active nickel layer containing about 0.15% sulfur, deposited from a special solution [34, 44]. The thickness of the very active nickel layer is of the order of 1.2 to 2.5 μm. The corrosion mechanism is similar to that of double-layer nickel coatings; once a corrosion pit reaches the very active nickel layer, corrosion proceeds laterally, the top layer of bright nickel becomes cathodically polarized, and its rate of corrosion is decreased. Whether or not triple-layer nickel coatings give significantly better corrosion performance compared to double-layer nickel coatings of equal thickness over extended periods of time is still not completely settled, but certainly the appearance in the initial stages of corrosion is improved because much of the corrosion takes place beneath the top layer of bright nickel and is difficult to see. Double- and triple-layer nickel coatings delay penetration of corrosion to the substrate and are effective in improving corrosion performance of nickel plus chromium coatings.

3.6 Microdiscontinuous Chromium

On surfaces that are electroplated with nickel plus conventional chromium, corrosion of the nickel begins at a pore in the chromium coating. Because conventional electrodeposited chromium has relatively few discontinuities, any pore or crack is surrounded by a large cathodic area of chromium that draws current from a relatively small area of nickel, and as a result rate of pit penetration through the nickel can be quite rapid. Corrosion may be under anodic control during the initial stages, but as the exposed nickel anode area increases, the rate of corrosion becomes cathodically controlled.

These observations led some investigators to conclude that a pore-free, electrodeposited chromium coating would lead to significant improvement in the corrosion performance of nickel plus chromium coatings. Crack-free chromium electroplating processes were introduced in the 1960s, but quickly disappeared when it became apparent that crack-free chromium develops porosity in service, similar to the porosity of conventional chromium. (Crack-free chromium has a predominantly hexagonal close-packed structure unlike conventional chromium which is body-centered cubic. The very high wear rate of crack-free chromium was attributed to its different crystal structure [22].)

Saur was one of the first to establish that an inverse relationship exists between the number of pores in the chromium and the degree of substrate corrosion [45]. Based on his evaluation of test panels and parts that had been in service for up to twenty-one years, he concluded that "an increase in chromium discontinuities increases the area of nickel corrosion, the increase in exposed nickel is associated with an increase in basis metal protection, and a minimum chromium imperfection density is beneficial for corrosion protection of the substrate." Saur and his coworkers [46] at the General Motors Research Laboratory also developed the test described in ASTM Standard B 627 [23], *Electrolytic Corrosion Testing (EC Test)*, that applies potentiostatic and galvanostatic polarization in a specified sequence to simulate anodic and cathodic corrosion control and permits a quantitative evaluation of the protectiveness of nickel plus chromium coatings to be made. Use of an interference microscope made it possible to examine specimens subjected to that test; measure pit densities, radii, and depths; and correlate that data to basis metal protection and appearance in nickel chromium coating systems [47]. Saur's work and that of others led to the development of commercial processes for increasing the number of discontinuities in electrodeposited chromium.

Commercial processes for producing microdiscontinuous chromium result either in the creation of micropores that are essentially circular in shape, or in microcracks that form a fine, invisible, continuous network of cracks over the entire surface. One of the most popular processes for producing microporous chromium requires the deposition of a thin layer of bright nickel, prior to chromium plating, from a solution that contains insoluble, extremely fine, nonmetallic particles dispersed throughout the tank. The resulting chromium deposit is completely microporous when applied over the special nickel layer. Methods that involve mild impingement with fine particles after the deposition of chromium are effective in producing microporous chromium [49].

The development of microcracked chromium began with dual chromium which had the requisite crack pattern [50], and the work of Safranek and others [51] who added selenium to conventional chromium plating solutions to induce microcracking; those methods are no longer in commercial use. Still in limited use is a method that requires the deposition of a thin, highly-stressed nickel deposit from an all-chloride nickel plating solution, prior to chromium plating, which gives the desired crack pattern after plating with chromium. One difficulty with the latter process is controlling the crack pattern in low current density areas. The most popular processes for producing microdiscontinuous chromium then are those based on deposition of special nickel strikes *prior* to chromium plating or on particle impingement *after* electroplating with chromium.

The use of microdiscontinuous chromium in combination with decorative multilayer nickel coatings, with or without copper underlayers, has resulted in great improvement in outdoor corrosion performance under severely corrosive conditions. The effect of the microdiscontinuities is to distribute the corrosion current over many tiny cells. The current available at any cell or pit is small, and the rate at which a pit proceeds through the bright nickel layer is correspondingly small. As the defect area increases, the area of exposed nickel (the anode) increases while the area of chromium (the cathode) decreases. Because the corrosion reaction is under cathodic control, reducing the size of the cathode surrounding each pit reduces the rate of nickel corrosion. The mechanism is consistent with the principles of galvanic corrosion. The area of exposed nickel was estimated by assuming that pits are cylindrically shaped, with pit radii and densities described in the literature; at very high chromium defect densities, the exposed nickel area is approximately 3.5% [42]. The rate of pitting of double-layer nickel coatings 40 μm thick electroplated with microdiscontinuous chromium appeared to be approximately 2 μm per year, and steel panels plated with those coatings did not rust in a moderately severe, outdoor marine exposure under static conditions for more than 15 years [52]. Similar results have been obtained in programs conducted by members of ASTM Committee B 8, and others [53–55].

3.7 The STEP Test

The focus on electrochemical aspects of corrosion culminated in the development of the STEP test by Harbulak [56]. The test made it possible to measure potential differences between various layers in a multilayer nickel coating on actual electroplated parts rather than on deposits detached from the substrate. STEP is an acronym for simultaneous thickness and electrode potential measurement. In the test, which is a simple, but brilliant modification of the well-known coulometric method of thickness testing, the electrode potential is monitored continuously as the coating is dissolved anodically. This is made possible by placing a reference electrode in the form of a silver or platinum wire in the coulometric cell used for stripping the coating. (Silver is preferred because the silver-silver chloride electrode has a constant potential whereas the potential exhibited by platinum is sensitive to the exact surface state of the platinum.) By recording the changes in potential with time, the potential differences, as well as the thicknesses, of the individual nickel layers can be measured. Unpublished work on the precision of the STEP test by ASTM Committee B 8 members indicate that potential differences and thicknesses can be measured with a standard deviation of less than 5% on standard reference materials. The STEP test is described in ASTM Standard B 764 [23].

Although there are no universally accepted values for the optimum potential differences between nickel layers, ASTM Standard B 456 [23] provides the following guidelines: (1) The STEP potential difference between the semi-bright and bright nickel layer generally falls within the range of 100 to 200 mV and in all combinations, the semi-bright nickel layer is more noble (more positive) than the bright nickel layer; (2) the STEP potential difference between the high-activity layer and the bright nickel layer in triple-layer coatings has a potential range of 15 to 35 mV, and the high-activity layer is more active (anodic) than the bright nickel layer; and (3) the STEP potential difference between the bright nickel and any thin layer applied just prior to chromium plating for producing microporous or microcracked chromium has a potential range of 0 to 30 mV, and the bright nickel layer is more active. The importance of the latter in minimizing the deterioration in surface appearance that may occur in severe outdoor conditions or during accelerated corrosion testing has been stressed by Tremmel [57].

3.8 Electroplating Plastics, Aluminum, and Stainless Steel

Electroplating on plastics, aluminum, and stainless steel has broadened the scope of decorative nickel plating.

Plastics The discovery that ABS plastics can be chemically etched and rendered electrically conductive made it possible to deposit *decorative* nickel/chromium coatings onto injection molded plastic parts with a level of adhesion not previously attainable [58]. ABS is a terpolymer of acrylonitrile, butadiene and styrene. When parts made of that material are immersed in strong chromic or chromic plus sulfuric acid solutions, the relatively soft, tiny butadiene rubber particles dispersed in the acrylonitrile-styrene matrix are chemically etched. The effectiveness of etching depends on how uniformly the butadiene particles are dispersed throughout the part and other factors related to injection molding parameters. The physical bond achieved by the *keying* of the deposit into microscopic etch pits on the surface no doubt plays an important role in achieving adhesion, but the observation that the hexavalent chromium in the etching solution is reduced to trivalent chromium and must be periodically replenished, led to speculation that the bonding might be chemical in nature [59]. Regardless of the exact mechanism, the adhesion (perhaps, 50 to 120 $g\,mm^{-1}$ peel strength) is sufficient to prevent coatings from blistering at elevated temperatures on parts that are properly processed.

The traditional preparation sequence includes: conditioning, etching, neutralizing, catalyzing, and autocatalytic deposition. Conditioning prior to etching can eliminate certain types of adhesion problems and may be done in a chromic/sulfuric acid solution or in an organic solvent (2,4-pentadione [60] and other ketones are sometimes used). The organic solvent softens and swells the plastic surface making it more receptive to subsequent processing steps. Etchants are either chromic, chromic/sulfuric, or chromic/sulfuric/phosphoric acid types. The purpose of the neutralization step that follows etching is to remove all residual chromic acid from the surface of the parts, and this is accomplished by immersing them in acid or alkaline solutions that may contain complexing or reducing agents. After neutralization, catalyzing is commonly accomplished by immersing in a solution containing colloidal stannous/palladium chloride [61] and an excess of hydrochloric acid. Tiny amounts of palladium are deposited on the surface; the palladium acts as a catalyst for the autocatalytic (electroless) deposition of either copper or nickel. To pass thermal cycle tests, substantial thicknesses of ductile acid copper must be electroplated on top of the autocatalytically deposited metal, and it is expedient to use electroless copper rather than electroless nickel to avoid processing problems, such as immersion deposition of copper and bipolar effects [62]. In addition electroless nickel can sometimes corrode preferentially at the metal-plastic interface causing blistering and lifting of the coating [63]. (That is most likely due to low phosphorus in the electroless nickel. Small amounts of phosphorus activate nickel [64]; large amounts tend to passivate it [65].) Electroless copper followed by electroplating with bright acid copper is now common practice, although electroless nickel is also used successfully. This basic method for preparing plastics for electroplating is discussed in ASTM Standard B 727 and elsewhere [23, 66].

Efforts to simplify the traditional method have been made including gas-etching with sulfur dioxide and the use of permanganates to replace chromates [67]. The activation of plastic surfaces by plasma and ozone etching are other ways to prepare plastics for electroplating [68] .

Decorative nickel plus chromium coatings can be deposited on top of electroplated bright acid copper by established procedures. ASTM Standard B 604 [23] specifies the use of multilayer nickel coatings for the most severe corrosion conditions on the basis of the comprehensive corrosion performance program conducted in the 1970s by members of ASTM and the American Society for Electroplated Plastics (ASEP) that showed that copper underlayers corroded during mobile, static, and accelerated corrosion tests if the nickel deposit was too thin [69]. The coatings specified in the standard are intended to protect the underlying copper layer from corrosion, for extended periods of time. The chromium layer should be either microcracked or microporous for severe and very severe corrosion service.

The original applications for electroplated plastics were established with ABS, but many polymers can be successfully electroplated including polypropylene, polysulfone, polyphenylene oxide, polycarbonate, polyester, nylon, and others [70]. In general, catalysis with palladium is common to most preparation sequences, but different etching solutions and etching times are required to process different plastics. Automotive applications for decorative electroplated plastics appear to be growing [71].

Aluminum Immersion deposition is the most widely used technique for preparing aluminum and its alloys for electroplating. Immersion deposition processes are either alkaline zincate or alkaline stannate solutions that strip the oxide film from the aluminum and deposit a thin film of zinc or tin, respectively, on the surface before the oxide film can re-form. Although immersion deposition films are generally nonadherent, their adhesion on aluminum and its alloys can be excellent; the bond strength can equal or exceed the tensile strength of the aluminum or aluminum alloy [72]. Traditionally, treatment in zincate would be followed by electroplating in a tartrate-type copper cyanide solution, but methods for direct nickel deposition on the zinc surface have been introduced employing nickel glycolate [73] or autocatalytic (electroless) nickel strike [74]. A proprietary tin immersion/bronze strike is reportedly less susceptible to undercutting and blistering in corrosive environments than the zinc immersion/copper cyanide strike [75]. The preparation of aluminum for electroplating is discussed in detail in ASTM Standard B 253 [23].

The performance of decorative, electroplated nickel plus chromium coatings on aluminum alloys was studied in connection with the development of lightweight automobile bumpers in the mid-1970s [76]. The effects on corrosion performance of various preparation methods, alloy type, nickel thickness, and type of chromium coatings were evaluated [77]. In reviewing the results, it was concluded that the use of microdiscontinuous chromium was probably the most important factor contributing to the improved corrosion performance of electroplated aluminum [78]. The experience gained during the development program has led to increased use of plated aluminum on truck bumpers and on styled aluminum wheels for trucks and passenger cars.

Stainless Steel Decorative nickel plus chromium plating of stainless steel is primarily applied in automotive applications, and only to a limited extent. The justification for electroplating stainless steel is the cost and difficulty of achieving a mirror finish on the material by mechanical means and the need to color match bumper inserts and other components placed in proximity to nickel/chromium plated ones. The methods of preparing stainless steel for electroplating described in ASTM Standard B 254 [23] are suitable for decorative plating.

3.9 Standards and Coating Requirements

ASTM Standard B 456 [23] provides information on specific requirements for decorative nickel plus chromium coatings to achieve acceptable performance under different conditions of service. The standard classifies various coating systems according to their resistance to corrosion and recommends specific coatings for different service conditions. The thicknesses that are specified in most cases have been verified by corrosion performance studies conducted by members of the committee.

The requirements for double- and triple-layer nickel coatings included in the standard are summarized in Table 5. The sulfur contents are specified to identify the type of nickel and to control the electrochemical potentials between individual nickel

TABLE 5 Requirements for Double- and Triple-Layer Nickel Coatings

Layer (Type of Nickel Coating)	Specific Elongation (%)	Sulfur content (%, $m\,m^{-1}$)	Thickness percentage of Total Nickel Thickness	
			Double Layer	Triple Layer
Bottom (*s*)	Greater than 8	Less than 0.005	Greater than 60 (but at least 75 for steel)	Greater than 50 (but not more than 70)
Middle (*b*)	—	Greater than 0.15	—	10 max
Top (*b*)	—	Between 0.04 and 0.15	Greater than 10 but less than 40	Equal to or greater than 30

Source: Adapted from ASTM Standard B 456.

Note: *s* designates the semi-bright nickel layer applied prior to bright nickel; *b* designates full-bright nickel that contains the amount of sulfur specified.

layers. Coating classification numbers appropriate for each service condition are given in Table 6 for steel. The classification numbers for aluminum alloys given in Table 7 are adapted from ISO International Standard 1456 [79]. The ones for plastics, Table 8, are from ASTM Standard B 604 [23].

The service condition numbers grade the severity of corrosion environments in service as follows:

- *Service Condition No. SC 5 (extended very severe)*. Service conditions that include likely damage from denting, scratching, and abrasive wear in addition to corrosive environments where *long-time protection* of the substrate is

TABLE 6 Decorative Nickel plus Chromium Coatings on Steel

Service Condition Number	Classification Number	Minimum Nickel Thickness (μm)
SC 5—extended very severe	Fe/Ni35d Cr mc	35
	Fe/Ni35d Cr mp	35
SC 4—very severe	Fe/Ni40d Cr r	40
	Fe/Ni30d Cr mp	30
	Fe/Ni30d Cr mc	30
SC 3—severe	Fe/Ni30d Cr r	30
	Fe/Ni25d Cr mp	25
	Fe/Ni25d Cr mc	25
	Fe/Ni40p Cr r	40
	Fe/Ni30p Cr mp	30
	Fe/Ni30p Cr mc	30
SC 2—moderate service	Fe/Ni20b Cr r	20
	Fe/Ni15b Cr mp	15
	Fe/Ni15b Cr mc	15
SC 1—mild	Fe/Ni10b Cr r	10

Source: Adapted from ASTM Standard B 456.

TABLE 7 Decorative Nickel plus Chromium Coatings on Aluminum

Service Condition Number	Classification Number	Minimum Nickel Thickness (μm)
SC 4—very severe	Al/Ni50d Cr r	50
	Al/Ni35d Cr mp	35
	Al/Ni35d Cr mc	35
SC 3—severe	Al/Ni30d Cr r	30
	Al/Ni25d Cr mp	25
	Al/Ni25d Cr mc	25
	Al/Ni35p Cr r	35
	Al/Ni30p Cr mp	30
	Al/Ni30p Cr mc	30
SC 2—moderate	Al/Ni20b Cr r	20
SC 1—mild	Al/Ni10b Cr r	10

Source: Adapted from ISO Standard 1456.

TABLE 8 Decorative Nickel plus Chromium Coatings on Plastics

Service Condition Number	Classification Number	Minimum Nickel Thickness (μm)
SC 5—extended very severe	Pl/Cu15a Ni30d Cr mc	30
	Pl/Cu15a Ni30d Cr mp	30
SC 4—very severe	Pl/Cu15a Ni30d Cr r	30
	Pl/Cu15a Ni25d Cr mp	25
	Pl/Cu15a Ni25d Cr mc	25
SC 3—severe	Pl/Cu15a Ni25d Cr r	25
	Pl/Cu15a Ni20d Cr mp	20
	Pl/Cu15a Ni20d Cr mc	20
SC 2—moderate service	Pl/Cu15a Ni15b Cr r	15
	Pl/Cu15a Ni10b Cr mp	10
	Pl/Cu15a Ni10b Cr mc	10
SC 1—mild	Pl/Cu15a Ni7b Cr r	7

Source: Adapted from ASTM Standard B 604.

required; for example, conditions encountered by some exterior components of automobiles.

- *Service Condition No. SC 4 (very severe)*. Service conditions that include likely damage from denting, scratching, and abrasive wear in addition to exposure to corrosive environments; for example, conditions encountered by exterior components of automobiles and by boat fittings in salt water service.
- *Service Condition No. SC 3 (severe)*. Exposure that is likely to include frequent or occasional wetting by rain or dew or possibly, strong cleaners and saline solutions; for example, conditions encountered by porch and lawn furniture; bicycle and perambulator parts; hospital furniture and fixtures.
- *Service Condition No. SC 2 (moderate)*. Exposure indoors in places where condensation of moisture may occur; for example, in kitchens and bathrooms.

- *Service Condition No. SC 1 (mild).* Exposure indoors in normally warm, dry atmospheres with coating subject to minimum wear or abrasion.

The classification numbers are a way to specify the coatings that are appropriate for each service condition; for example, the classification number

Fe/Ni30d Cr mp

indicates that the coating is applied to steel (Fe), consists of 30 μm of double-layer nickel (*d*) with a top-layer of microporous chromium (*mp*), 0.3 μm thick. (The thickness of chromium is not included in the classification number unless it differs from 0.3 μm. If different, the number would appear immediately next to the symbol for chromium.)

The type of nickel is designated in the tables by the following symbols: *b* for electrodeposited bright nickel (single-layer); *d* for double- or triple-layer nickel coatings; *p* for dull, satin, or semi-bright nickel that is not polished; *s* for polished dull or semi-bright nickel. The symbol, *a*, is used to designate ductile copper deposited from bright acid-type baths. The type of chromium is designated by the following symbols: *r* for regular or conventional chromium; *mp* for microporous chromium having a minimum of 10,000 pores cm^{-2}; *mc* for microcracked chromium having more than 300 cracks cm^{-1} in any direction over the entire surface.

ASTM Standard B 456 [23] provides test methods for measuring coating requirements. It is a comprehensive guide to controlling the quality of decorative, electroplated nickel plus chromium coatings.

3.10 Decorative Applications and Market Size

Decorative applications may be placed into two categories: automotive and non-automotive. The decline in decorative plating for automotive applications noted in the 1980s has abated as a result of the popularity and increased production of light trucks and recreational vehicles that have retained brightwork, the development of nickel chromium plated steel and aluminum styled wheels, and an increase in the number of automotive applications for plated plastics [1]. Non-automotive decorative applications comprise a large number of consumer products including furniture and building hardware, plumbing fixtures, housewares, hand tools, major appliances, wire goods, bicycles, mopeds, motorcycles, and others. These continue to expand with the growth of the global economy. Approximately 80% of the nickel consumed for electroplating annually is for decorative purposes. The balance, 20%, is consumed for functional electroplating and electroforming.

4 FUNCTIONAL ELECTROPLATING AND DEPOSIT PROPERTIES

Electrodeposited nickel coatings are applied in functional applications to modify or improve corrosion resistance, hardness, wear, magnetic and other properties. Although the appearance of the coating is important and the plated surface should be defect-free, the lustrous mirror like deposits described in the previous section are not required.

Typical formulations for Watts and sulfamate solutions, the two most popular ones for functional applications, have been included in Table 4 along with recommended operating conditions and representative mechanical properties of deposits from each solution. Although the table indicates that the maximum current density for depositing nickel from a Watts solution is 11 $A\,dm^{-2}$, higher deposition rates can be achieved with increased agitation and solution flow-rates. Other nickel electroplating solutions (Table 9) have been applied in functional applications, and deposits from those solutions have useful properties.

4.1 Deposit Properties

The main constituents in Watts solutions affect the properties of electrodeposited nickel. Nickel sulfate improves conductivity and metal distribution, and determines the limiting cathode current density for producing sound nickel deposits. Nickel chloride improves anode corrosion, but also increases conductivity, throwing power, and uniformity of coating thickness distribution. In addition chlorides increase the internal stress of the deposits, and they tend to refine grain size and minimize formation of nodules and trees. A Boric acid is added for buffering purposes and affects the appearance of the deposits. Deposits may be cracked and burnt at low boric acid concentrations. Anionic wetting agents or surfactants that lower the surface tension of the plating solution so that air and hydrogen bubbles do not cling to the parts being plated are almost always added to control pitting and by eliminating porosity, have an indirect effect on corrosion performance.

Operating conditions, such as pH, temperature, current density and chloride content, affect the properties of deposits from Watts solutions [80]. In a Watts solution, hardness, tensile strength, and internal stress increase above pH 5.0 while the elongation percentage decreases (Fig. 2). The hardness increases rapidly at low values of current density (Fig. 3). Increasing the temperature of the plating solution causes hardness and tensile strength to reach minimum values at about 55°C, while the elongation percentage is a maximum at that temperature (Fig. 4). Increasing the chloride ion concentration affects the properties of deposits from Watts solutions; for example, the elongation percentage is at a maximum, and hardness and tensile strength are at minimum values, when the solution contains 25% by weight nickel chloride (Fig. 5). In general, conditions that increase the hardness of a nickel deposit will increase its tensile strength and lower its ductility. Close control of the main constituents and the operating conditions is thus required to produce nickel electrodeposits with consistent and known properties.

The properties of deposits from Watts and sulfamate solutions are affected in different ways by changes in operating conditions, as illustrated qualitatively in Figure 6 [81]. For example, internal stress is not significantly affected by increasing the temperature of a Watts bath, whereas increasing the temperature in a sulfamate solution reduces internal stress significantly. Cathode current density has a relatively small effect on the tensile strength of deposits from a Watts solution, but increasing current density reduces the tensile strength of deposits from a sulfamate solution.

The mechanical properties measured at temperatures from −195° to 870°C of nickel electrodeposited from Watts, sulfamate and all-chloride solutions were

TABLE 9 Other Nickel Plating Solutions and Some Properties of Deposits

Type	Composition[a] (g liter^{-1})	pH	Temperature (°C)	Cathode Current Density (A dm^{-2})	Vickers Hardness, (100 g load)	Tensile Strength (MPa)	Elongation (%)	Internal Stress (MPa)
Fluborate	Nickel fluoboratem, 225–300 Nickel chloride, 0–15 Boric acid, 15–30	2.5–4	38–70	3–30	125–300	380–600	5–30	90–200
Hard nickel	Nickel sulfate, 180 Ammonium chloride, 25 Boric acid, 30	5.6–5.9	43–60	2–10	350–500	990–1100	5–8	300
All-chloride	Nickel chloride, 225–300 Boric acid, 30–35	1–4	50–70	2.5–10	230–260	620–930	4–20	275–340
All-sulfate	Nickel sufate, 225–410 Boric acid, 30–45	1.5–4	38–70	1–10	180–275	410–480	20	120
Sulfate/chloride	Nickel sulfate, 150–225 Nickel chloride, 150–225 Boric acid, 30–45	1,5–2.5	43–52	2.5–15	150–280	480–720	5–25	210–280
High sulfate	Nickel sulfate, 75–110 Soidum sulfate, 75–110 Ammonium chloride, 15–35 Boric acid, 15	5.3–5.8	20–32	0.5–2.5	—	—	—	—

Black nickel (sulfate bath)	Nickel sulfate, 75 Zinc sulfate, 30 Ammonium sulfate, 35 Sodium thiocyanate, 15	5.6	24–32	0.15	—	—	—	—
Black nickel (chloride bath)	Nickel chloride, 75 Zinc chloride, 30 Ammoumium chloride, 30 Sodium thiocyanate, 15	5.0	24–32	0.15–0.6	—	—	—	—
Nickel phosphorus	Nickel sulfate, 170 or 330 Nickel chloride, 35–55 Boric acid, 0 or 4 Phosphoric acid, 50 or 0 Phosphorous acid, 2–40	0.5–3.0	60–95	2–5	—	—	—	—

[a] The formulas of the compounds in the table are as follows: nickel fluoborate, $Ni(BF_4)_2$; nickel sulfate, $NiSO_4 \cdot 6H_2O$; nickel chloride, $NiCl_2 \cdot 6H_2O$; boric acid, H_3BO_3; ammonium chloride, NH_4Cl; ammonium sulfate, $(NH_4)_2SO_4$; sodium sulfate, Na_2SO_4; phosphoric aicd, H_3PO_4; phosphorous acid, H_3PO_3; zinc sulfate, $ZnSO_4 \cdot 7H_2O$; zinc chloride, $ZnCl_2$; sodium thiocyanate, NaSCN.

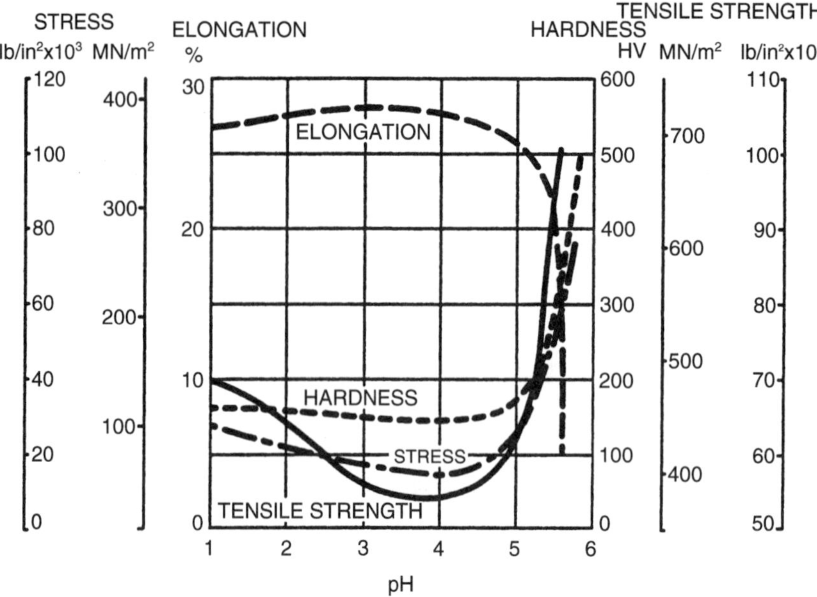

Figure 2 Influence of pH on the internal stress, tensile strength, ductility and hardness of nickel electrodeposited from a Watts solution, at 55°C and 5 A dm^{-2} [80].

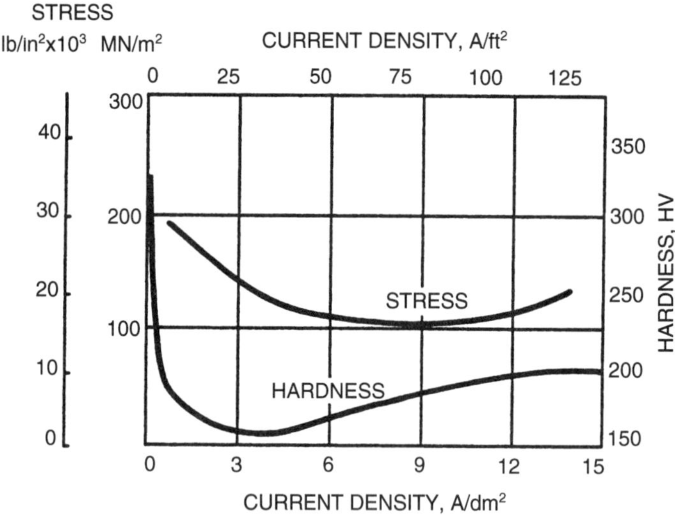

Figure 3 Influence of current density on the internal stress and hardness of nickel electrodeposited from a Watts solution at 55°C and pH 3.0 and [80].

determined and compared to the properties of wrought nickel by Knapp and Sample [82] (Fig. 7). At −195°C, chloride and sulfamate nickel had ultimate tensile strength in excess of 100 kg mm^{-2} compared to a value of 56 kg mm^{-2} for Watts and annealed wrought nickel. The reduction in the elongation percentage above 450°C for the

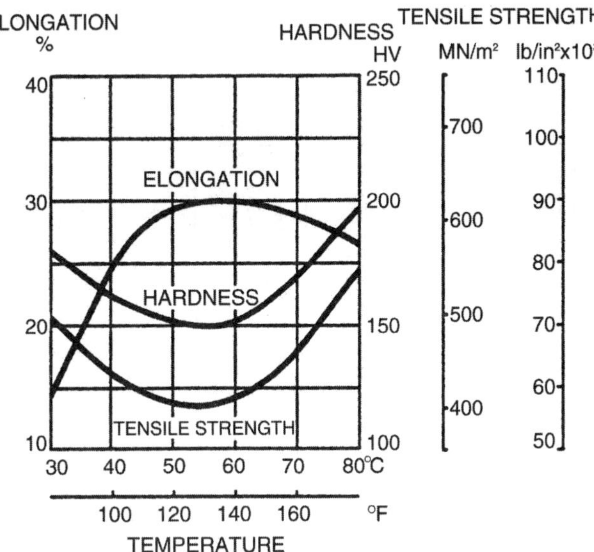

Figure 4 Influence of temperature on the elongation, tensile strength and hardness of nickel electrodeposited from a Watts solution at pH 3.0 and 5 A dm^{-2} [80].

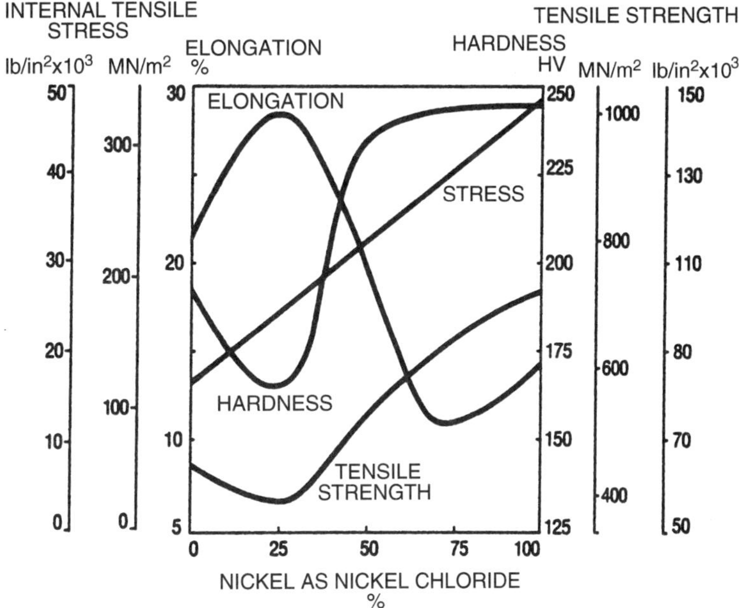

Figure 5 Influence of chloride concentration on the elongation, internal stress, hardness, and tensile strength of nickel electrodeposited from Watts-type solutions at pH 3.0, 55°C and 5 A dm^{-2} [80].

three types of electrodeposited nickel was attributed to the presence of small amounts of sulfur in the electrodeposited nickel. Although wrought nickel has comparable levels of sulfur, it also contains small amounts of manganese that prevent sulfur from migrating to grain boundaries at elevated temperatures where

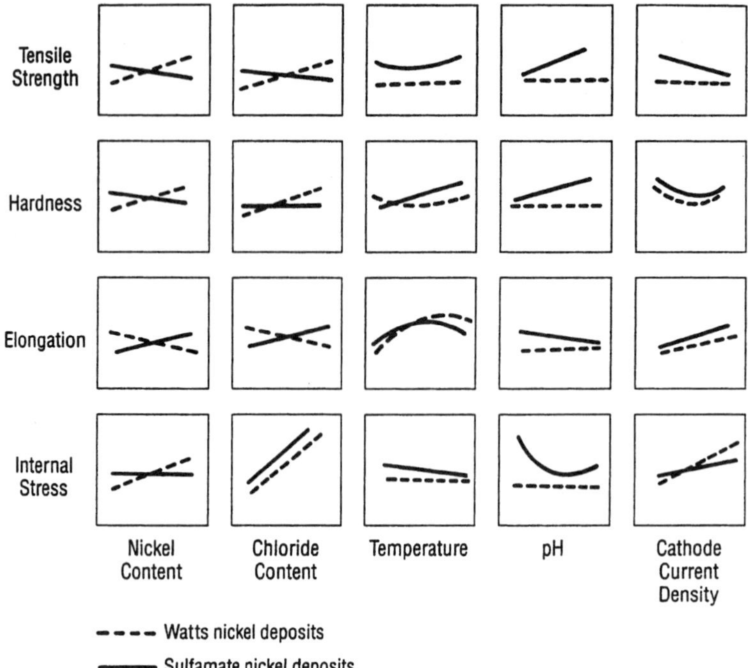

Figure 6 Qualitative effects of operating conditions on the properties of nickel electrodeposited from Watts and sulfamate solutions.

it forms nickel sulfide that reduces high temperature ductility. The codeposition of small amounts of manganese with nickel has since been shown to improve high temperature ductility [83]. Considerable information on the physical and mechanical properties of electrodeposited nickel, nickel alloys, and nickel composite coatings is available [84, 85].

4.2 Coating Requirements in Functional Applications

Corrosion performance in functional applications depends on nickel thickness and other factors, including the condition of the surface prior to plating. The thickness that should be applied depends on the specific application. For example, nickel thickness in optical and electronic applications where nickel is applied as an undercoat prior to electroplating with other metals may be 5 μm. To protect product purity, thickness may be 75 μm thick on processing equipment, such as ferrous containers. For drying cylinders and rolls for paper processing, for condenser and calender rolls for the textile industry, and for externally and internally plated pipe, fittings, and other components for chemical and nuclear plants, nickel coating thicknesses of 125 μm have been specified. Nickel coatings on automotive components, such as hydraulic rams, cylinder liners, and shock absorbers, may be 125 μ thick to provide corrosion and wear resistance. Coatings to prevent fretting corrosion and to enhance wear resistance are generally greater than 125 μm thick and have been used in the automotive and mining industries to coat pistons, cylinder walls, rotary engine

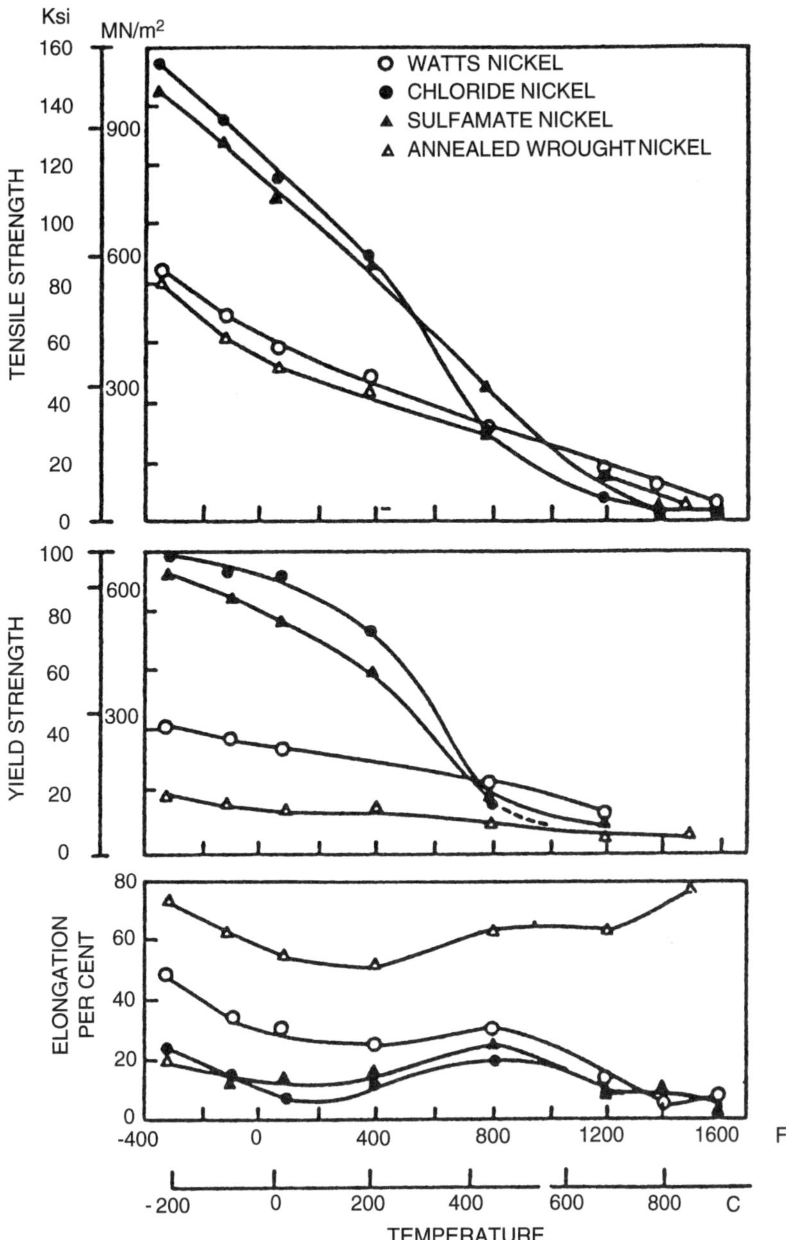

Figure 7 Effect of temperature on the tensile strength, yield strength, and elongation of electrodeposited nickel.

housings, gear shafts, drive shafts, pump rods, and hydraulic pistons. When nickel is applied to salvage worn or mismachined components, thickness is determined by the extent of the repair required. Other requirements for functional nickel coatings are discussed in ASTM Standard B 689 [23]. Nickel coatings used as diffusion barriers beneath precious metal deposits in electronic applications, nickel electroplating of strip and hardware in the production of batteries, and nickel electroplated steel coin blanks are a few examples of functional applications that have grown in importance.

4.3. Other Solutions

Other nickel plating solutions for functional applications are listed in Table 9, along with typical mechanical properties of the deposits. Many of these solutions were developed to meet specific engineering requirements; all are used to a lesser extent than Watts and nickel sulfamate solutions.

Fluoborate The fluoborate solution can be operated over a wide range of nickel concentrations, temperature, and current density, and is relatively simple to control [86]. The fluoborate anion is corrosive, however, and some materials that contact the solution are chemically attacked. The mechanical and physical properties of deposits from a fluoborate bath are similar to those from Watts solutions. The major advantage is that electrodeposition from a nickel fluoborate solution can be performed at high current density.

Hard Nickel Developed especially for functional applications, this solution is applied where controlled hardness, improved abrasion resistance, greater tensile strength and good ductility are required without using sulfur-containing organic addition agents [87]. Close control of pH, temperature, and current density is necessary for this bath to give reproducible results. The internal stress is slightly higher than in deposits from Watts solutions. The disadvantages of the hard nickel bath are its tendency to form nodules on edges, and the low annealing temperature (230°C) of its deposits. Hard nickel deposits are used primarily for buildup or salvage purposes. For optimum results, the ammonium ion concentration should be maintained at 8 g $liter^{-1}$. In applications where the part being plated is not going to be exposed to elevated temperatures in service, it is easier to add organic compounds such as saccharin, *p*-toluene sulfonamide, and *p*-benzene sulfonamide to Watts or sulfamate solutions to achieve hardness without increased internal stress. Since these additives introduce 0.03% sulfur (or more), this approach must not be used for parts to be exposed to high temperatures where sulfur severely embrittles the nickel deposit. Leveling agents (brighteners) are effective in increasing the hardness of electrodeposited nickel without introducing sulfur, but they tend to raise internal stress [88].

All-Chloride The principal advantage of the all-chloride solution is its ability to operate effectively at high cathode current densities. Other advantages include its high conductivity, its slightly better throwing power, and a reduced tendency to form nodular growths on edges [89]. Deposits from this electrolyte are smoother, finer-grained, harder, and stronger than those from Watts solutions, and more highly stressed. Because of the partial solubility of lead chloride, lead cannot be used in contact with the all-chloride solution, and mists from this solution are corrosive to the superstructure, vents, and other plant equipment, if not well protected. The solution has been used for salvaging undersize or worn shafts and gears.

All-Sulfate This solution has been applied for electrodepositing nickel where the principal or auxiliary anodes are insoluble [90]. For example, insoluble auxiliary or conforming anodes may be required to plate the insides of steel pipes and fittings. Oxygen is evolved at insoluble anodes, and as a result the nickel concentration and pH decrease during plating. The pH is controlled and the nickel ion concentration maintained by adding nickel carbonate. Another procedure that has been used in low-pH solutions replenishes the nickel electrolytically by employing a separate replenishment tank with nickel anodes; the current in the replenishment tank is periodically reversed to keep the nickel nodes actively dissolving in the absence of chlorides [90]. The insoluble anodes in all-sulfate solutions may be lead, carbon, graphite, or platinum. If a small anode area is required, solid platinum (in the form of wire) may be used. For large anode areas, platinum-plated or platinum-clad titanium is recommended; iridium-coated titanium is now in commercial use. In some forms, carbon and graphite are too fragile; lead has the disadvantage of forming loose oxide layers, especially if it is immersed in other solutions in the course of a plating cycle. In chloride-free solution, pure nickel is *almost* insoluble and may function as an internal anode if properly bagged.

Sulfate/Chloride The sulfate/chloride solution included in Table 9 has roughly equivalent amounts of nickel sulfate and nickel chloride, and was developed to overcome some of the disadvantages of the all-chloride solution [91]. It has high conductivity and can be operated at high current densities. Although the internal stress of the deposits is higher than in deposits from Watts solutions, the stress is lower than in the all-chloride solution. The other properties are about midway between those for deposits from Watts and all-chloride solutions. Lead may not be used for equipment in contact with this solution because of the high chloride content.

High Sulfate The high-sulfate bath was developed for plating nickel directly on zinc-base die casting [92]. It has been used to plate nickel on aluminum that has been given a zincate or comparable surface preparation treatment. The high-sulfate and low-nickel contents, together with the high pH, provide good throwing power with little attack of the zinc. The deposits are less ductile and more highly stressed than nickel deposited from a Watts bath. In general, the deposition of copper from a cyanide solution directly on zinc-base die castings prior to the deposition of nickel is simpler and more reliable. Nickel glycolate and autocatalytic nickel solutions, as mentioned above, are preferred for depositing nickel directly on aluminum that has been given a zincate treatment.

Black Nickel There are at least two formulations for producing black nickel deposits; these incorporate zinc (Zn) and thiocyanate (CNS-) ions. Table 9 gives the [93] composition and operating conditions for sulfate and chloride black nickel plating baths [93]. The process was developed for decorative reasons: colour matching and blending. The black nickel deposit has little wear or corrosion resistance, and is usually deposited over a layer of nickel deposited from a bright or dull

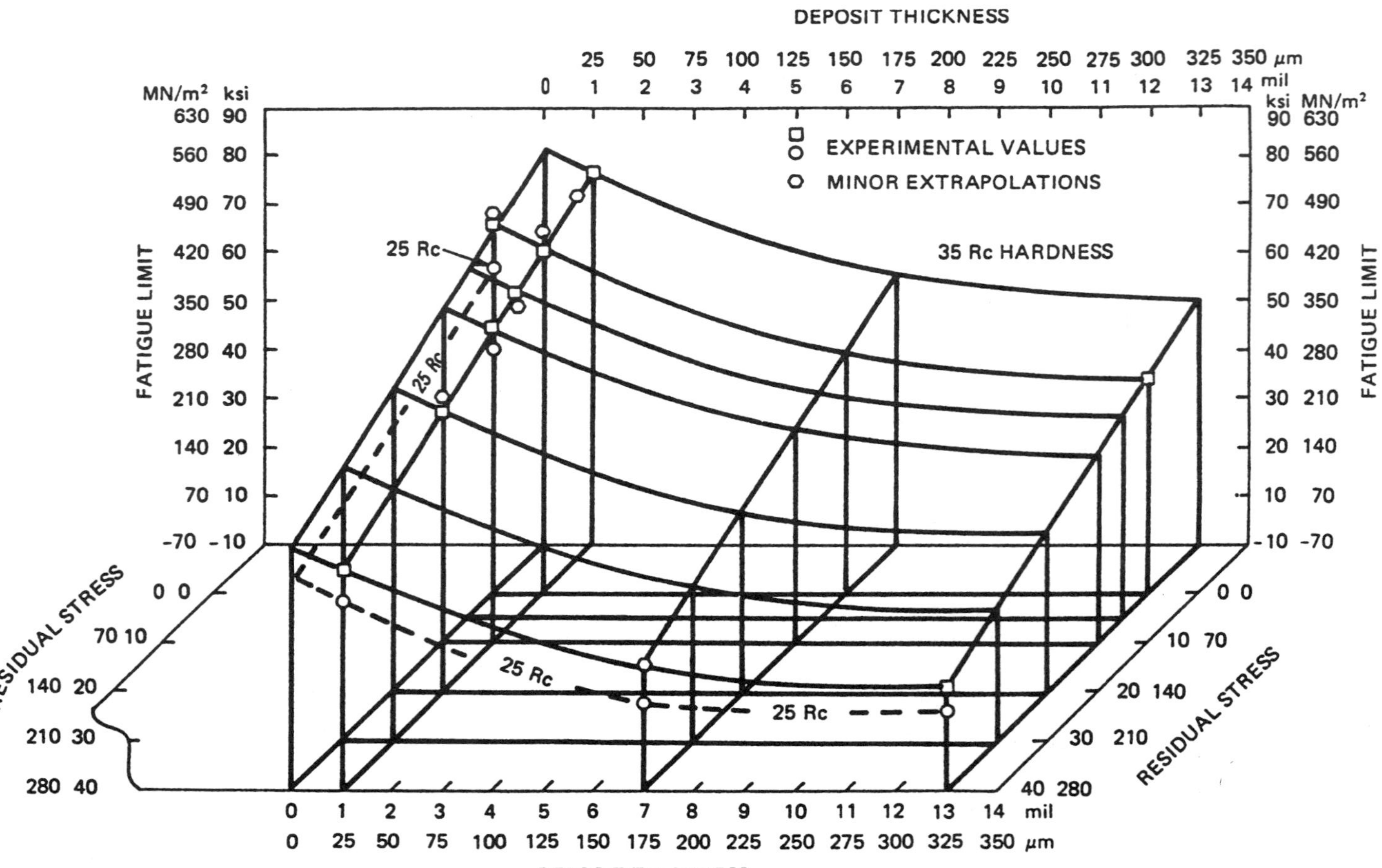

Figure 8 Effect of deposit thickness and internal stress on the 100 million cycle fatigue life of nickel electroplated AISI 4340 steel.

nickel plating solution. Black nickel deposit is still in commercial use, and the deposit is often protected with a clear lacquer coating.

Nickel Phosphorus These solutions result in the electrodeposition of nickel phosphorus alloys that are analogous to those deposited autocatalytically using sodium hypophosphite as the reducing agent [94]. The hardness of the electrolytic deposits can be increased by heat treatment in the same way that autocatalytic nickel deposits can with maximum hardness occurring at 400°C. The phosphorus content of the deposits is controlled by frequent additions of phosphite or phosphorous acid. The electrodeposition of nickel phosphorus alloys is receiving increased attention because deposits with greater than 10% phosphorus are amorphous and therefore have enhanced resistance to corrosion [95].

4.4 Fatigue Strength

Thick nickel deposits applied to steel substrates may cause significant reductions in composite fatigue strength when subjected to cyclical stress loading. The reduction in fatigue strength is influenced by the internal stress and thickness of the coating, and by the hardness and tensile strength of the steel. The fatigue limit of nickel plated steel is reduced almost proportionally to the amount of residual tensile stress in the nickel. Compressively stressed deposits have less of an effect on fatigue, and as a result, steel crankshafts have been nickel electroplated for resistance to wear and corrosion with low-stressed deposits from nickel sulfamate solutions. Specifying the thinnest coating consistent with design requirements is also beneficial for minimizing the effect on fatigue life. In some cases, specifying a steel with increased hardness and tensile strength may be necessary. Shot peening the steel prior to electroplating minimizes the reduction in fatigue strength see ASTM Standard B 851 for details of the shot peening strength. process [23]. The effect of these variables on the fatigue limit of a nickel-plated high-strength steel is shown in Figure 8 [96].

4.5 Hydrogen Embrittlement

Highly stressed, high-strength steels are susceptible to hydrogen embrittlement during normal plating operations. Although nickel plating per se may not be the cause of embrittlement, the preparation of steels for electroplating involves exposures to acids and alkalies that are capable of introducing hydrogen into the steel. In general, parts made from steel with tensile strengths that are equal or greater than 1000 MPa should be heat-treated after electroplating to avoid hydrogen embrittlement; depending on the state of the steel, stress relief prior to electroplating may be required to minimize hydrogen embrittlement susceptibility. Recommended treatments are given in ASTM Standards B 849 and B 850 [23].

5 NICKEL ELECTROFORMING

Electroforming with copper was conceived in 1838 by Jacobi of the Academy of Sciences, St. Peterburg, Russia [97]. Bottger reportedly electroformed nickel in 1843. Today nickel is utilized in the largest number of applications because it is strong, tough, and resistant to corrosion, erosion, and wear and because its mechanical properties can be varied and controlled by changing solution composition and operating conditions, by alloying, and by incorporating particles and fibers within the electrodeposited nickel matrix [98].

5.1 Electroforming and Electrofabrication

Electroforming is electroplating applied to the production of parts. In decorative and functional applications, good adhesion of the coating to its substrate is critical to its function. In electroforming, metal is deliberately electrodeposited non-adherently so that it can be separated from the substrate. The substrate in electroforming is a *mandrel or mold*, and the nickel that is separated from the mandrel becomes the final product. The mandrel is usually a negative or reverse replica of the product to be fabricated, and is often recovered and re-used. Electrofabrication also describes processes that result in the production of components and parts but that do not involve the use of traditional mandrels. In electrofabrication, the substrate often becomes an integral part of the component, and electrodeposition takes place selectively through masks or by means of specially designed cells [99, 100]. The distinction between electroforming and electrofabrication lies in the nature of the mandrel and the role of the substrate.

5.2 Capabilities of Nickel Electroforming

Nickel electroforming processes can reproduce fine surfaces with great accuracy due to the excellent microthrowing power of nickel-plating solutions. In the production of stampers for pressing compact audio and video discs, the accuracy of reproduction is within a fraction of a micrometer. An extension of this property is in the reproduction of complex surface finishes; for example, bright and semi-bright surfaces can be reproduced without the need for machining or polishing individual parts. The combination of electroforming and modern photolithographic techniques for generating patterns makes it possible to produce parts with extreme precision and fineness of detail on a microscopic scale. Parts can be reproduced in quantity with a high degree of dimensional accuracy. The mechanical and physical properties of nickel can be closely controlled by selecting the appropriate solution and the operating conditions. There is virtually no limitation to the size and thickness of the parts that can be electroformed. Shapes can be made that are impossible or too expensive to make any other way.

5.3 Mandrels

The ability to produce an electroformed part depends on the design and fabrication of the mandrel. A key consideration is to facilitate separation of the electroform. Rounding exterior angles, tapering the mandrel where possible, eliminating re-entrant angles, controlling the mandrel's surface finish, and designing a fixture to assist in parting the electroform from the mandrel should be considered at the design stage. Mandrels may be manufactured by casting, machining, electroforming, and conventional pattern-making techniques, and they may be conductors or non-conductors of electricity, and expendable or permanent. The nature of the material will determine how the mandrel is prepared for electroforming; in some cases there is a need to passivate the surface of the mandrel to prevent adhesion. Some of these details are discussed in ASTM Standard B 832 [23].

5.4 Nickel Electroforming Solutions

Nickel sulfamate solutions (Table 4) are popular for electroforming because of the low internal stress of the deposits, the high rates of deposition possible, and the improved throwing power. Watts nickel solutions are utilized in a number of applications, but often with the addition of stress reducing agents.

Nickel sulfamate is similar to nickel sulfate except that one of the hydroxyl groups is replaced by an amido group. The formula of the normal crystallized form of nickel sulfamate may be written, $Ni(SO_3NH_2)_2 \cdot 4H_2O$. Crystallized forms are not readily available in the United States where pre-purified concentrated liquid solutions are preferred. Nickel sulfamate can be prepared by reacting high-purity nickel powder or high-purity nickel carbonate with sulfamic acid under controlled conditions. Because of the high solubility of nickel sulfamate, higher nickel metal concentration is possible than in other nickel electrolytes, permitting higher plating rates. A small amount of nickel chloride is usually added to nickel sulfamate solutions to minimize anode passivity, especially at high current densities. If nickel chloride is not added, sulfur-containing nickel anode materials are required to avoid anodic oxidation of sulfamate anions (discussed below). Prolonged use of sulfamate solutions at temperatures above 70°C or at a pH of less than 3.0 can hydrolyze the nickel sulfamate to the less soluble form of nickel ammonium sulfate. Ammonium and sulfate ions increase the internal tensile stress and hardness of the deposits [101]. As far as is known, there is no *simple* way to remove ammonium ions from sulfamate solutions.

5.5 Anodic Oxidation of Sulfamate Anions and the Ni-Speed Process

A phenomenon that apparently occurs only in sulfamate solutions is anodic oxidation of sulfamate anions to form species that diffuse to the cathode where they can affect the internal stress and the composition of the deposit. This occurs at insoluble anodes operating at high potentials. At an insoluble platinum anode, a stress reducer forms which was identified as an azodisulfonate [102]; it reacts at the cathode, introduces sulfur into the deposit, especially at low current density, and lowers internal stress. The stress can be less than zero (compressive) depending on the amount of current flowing through the small insoluble platinum anode. A small auxiliary platinum anode that drew 1% to 2% of the total current passing through the electroplating cell was effective in controlling stress at compressive levels, but resulted in the codeposition of small amounts of sulfur with the nickel [103]. The codeposited sulfur affected the ductility and other mechanical properties of the deposit, and its tendency to become embrittled at elevated temperatures [104].

The concentrated nickel sulfamate process, *Ni-Speed*, was developed for electroforming at high current densities and low internal stress [105]. Because low to zero-stress conditions can be achieved without organic addition agents, there is no incorporation of sulfur and the deposits do not become embrittled when heated above 200°C. The solution contains about 500 to 650 g $liter^{-1}$ of nickel sulfamate; 5 to 15 g $liter^{-1}$ of nickel chloride hexahydrate, and 30 to 45 g $liter^{-1}$ boric acid. After purification with carbon and permanganate to remove all organic impurities, the solution is conditioned electrolytically with a non activated anode material. When properly conditioned a deposit prepared at 5 A dm^{-2} and at 60°C will be semi-lustrous and have a compressive stress. During operation, the solution is circulated through a separate conditioning tank where it is continuously electrolyzed at low current density; the conditioning tank has 10% to 20% of the capacity of the main tank and is operated with a nonactivated, sulfur-free anode material. The main electroforming tank is operated with an activated sulfur-containing anode material. Controlling the stress at or very close to zero depends on maintaining the solution temperature and current density at specified, *coupled* values; for example, internal stress is approximately zero at 50°C and 8 A dm^{-2}; at 60°C and 18 A dm^{-2}; or at 70°C and 32 A dm^{-2}.

The reaction that occurs at an insoluble platinum anode is apparently different from the reaction that occurs at the anode in the conditioning tank. The potential of a platinum anode in the nickel sulfamate solution is approximately +1 volt versus SCE. The potential of the nonactivated nickel anode material in the conditioning tank will be approximately +0.2 to +0.4 volt versus SCE because of the presence of chlorides, whereas the activated anode in the electroforming tank will dissolve at −0.2 volt versus SCE. It has been postulated that the oxidation reaction that occurs at the high potential is different from the reaction that occurs at intermediate potentials [106]. At an active anode dissolving at −0.2 volt versus SCE, anodic oxidation of sulfamate anions does not occur. The existence of several oxidation species in sulfamate solutions has been reported [107] . The Ni-Speed process has been operated successfully in Europe where it has been used to electroform electrolytic nickel foil at low internal stress continuously on rotating drums at current densities as high as 40 A dm^{-2}. It has not been used to any great extent in North America where it is considered difficult to control and where its reproducibility has been questioned.

Zero-stress can reportedly be maintained in conventional nickel sulfamate solutions by eliminating nickel chloride entirely, using a sulfur-activated nickel anode material, and continuously purifying the solution by means of an auxiliary tank similar to the conditioning tank described above.

5.6 Leveling Agents for Electroforming

The use of a leveling agent, for example, 2-butyne 1,4-diol, can improve metal distribution on the mandrel by suppressing the growth of nodules and by preventing the formation of planes of weakness when electroforming into corners [108]. Leveling agents increase internal stress but the increase may be tolerable when the initial stress is at or close to zero. The breakdown products formed by butyne diol can be removed by continuous filtration of the solution through carbon and this helps control the effect on internal stress.

5.7 Postelectroforming Operations

After electrodeposition is complete, operations, such as machining and finishing, parting and separating from the mandrel, and backing the electroform, are usually required [109].

5.8 Nickel Electroforming Applications

Products made by nickel electroforming include stampers for the production of phonograph records, compact discs, videodiscs, holograms, and information storage discs; textile printing screens; a wide assortment of molds and dies for processing plastics; abrasion strips for helicopter blades; seamless belts for photocopiers and facsimile machines; and porous nickel foam for making battery electrodes [110]. The electrofabrication of thin film heads for magnetic recording, of thin film chip carriers, of micromovable parts, and of cooling channels in rocket thrust chambers involves the deposition of one or more layers of metal onto a substrate that becomes an integral part of the final product [111–113].

6 NICKEL ANODE MATERIALS

Most nickel-plating processes are operated with soluble nickel anode materials. Nickel from the anode is converted into ions which enter the plating solution to replace those discharged at the cathode. In addition the anode distributes current to the parts being electroplated and influences metal distribution. Commercially available nickel anode materials made to strict chemical specifications do not introduce impurities into the electroplating solution and influence quality only to the extent that they affect current and metal distribution.

The simplest way to satisfy anode requirements is to suspend electrolytic nickel strip from hooks placed on an anode bar, so that the nickel, not the hook, is immersed in the plating solution. Although significant quantities of electrolytic nickel strip are still consumed for that purpose, it is the least satisfactory way to satisfy anode requirements. Electrolytic nickel strip dissolves preferentially at its bottom and its sides which results in an ever-changing anode area accompanied by a corresponding increase in anode current density. Strip dissolves nonuniformly becoming fragile and spongy and, as a result, tends to break and disintegrate before it is completely consumed creating scrap that has to be discarded or recycled. The replacement of electrolytic nickel strip usually requires interruption of the plating process, when it yields anode spears or stubs that have to be disposed of. The fact that the nickel strip is available in limited lengths is perhaps its greatest disadvantage. These limitations led to the development of wrought (rolled) products that dominated the market for many years and later to the development of improved primary forms of electrolytic nickel for titanium anode baskets.

6.1 Wrought Nickel Anode Materials

The first anode material developed intentionally to obtain improved performance was wrought depolarized nickel (ca. 1929). That material contains greater than 99% nickel, about 0.5% nickel oxide, and a minute amount of sulfur. In discussing the uniform dissolution and enhanced activity of wrought depolarized nickel, Wesley considered the effect of nickel oxide on sulfur distribution to be the key factor [114]; nickel oxide may prevent or interfere with the formation of nickel sulfide at grain boundaries where it would lead to nonuniform anode dissolution in wrought materials. Wrought depolarized nickel anodes form a brown film and metallic particles during dissolution, and they must be encased in anode bags to prevent those residues from entering the solution and causing roughness at the cathode. Although this anode material dissolves smoothly, it still has some of the disadvantages of electrolytic nickel strip; that is, anode area and current density change during electroplating and operations usually have to be interrupted to replace anodes and anode bags. But because this material is made by hot-rolling, there is no size limitation, and lengths suitable for large-scale electroplating in deep tanks are commercially available.

The wrought carbon anodes introduced in 1938 contain about 0.25% carbon and about 0.25% silicon. During dissolution, silicon is oxidized to silicic acid which, in combination with carbon, forms a thick, highly retentive, black film on the surface which keeps tiny metallic particles in contact with the anode long enough for them to dissolve [115]. Increased acidity of the anolyte due to the silicic acid is believed to promote uniform dissolution of this material. Rolled carbon anodes form little, if any, metallic residue, but they are nevertheless used with anode bags to keep carbon plus silicon residues out of solution. Because the silicon is oxidized, anode efficiency depends on the silicon content of the anode material, which is less than 100%. Work to develop an improved auxiliary anode material for use without anode bags indicated that lowering the carbon to 0.15% and increasing the silicon content to 1% or more was beneficial in eliminating roughness-producing metallic particles [116].

6.2 Anodic Behavior of Primary Nickel

The limitations of primary electrolytic nickel are related to its tendency to become passive when dissolved anodically. In nickel sulfate solutions, the anodic polarization curve for nickel obtained potentiostatically under *quasi-equilibrium* conditions exhibits distinct active, passive and transpassive regions [117]. The departure of the active portion of the curve from linearity as Epp is approached, as illustrated in Figure 9, is the point where an oxide film begins to form on the anode surface. Epp is the critical or principal passivation potential. The Flade potential, Ep, is the point where film formation is complete; that is, where an oxide film completely covers the surface and conducts electricity electronically rather than ionically. In the passive region, the nickel oxide film increases in thickness. In the transpassive region, the current density increases rapidly as a result of the oxidation of hydroxyl ions and the generation of oxygen. Because the value of the current density at Epp is very low, pure electrolytic nickel cannot be used to replenish nickel ions in a nickel sulfate electroplating solution that does not contain chloride ions.

When *E-I* curves are measured *potentiodynamically*, the shapes of the curves are affected by the sweep rate and two peaks corresponding to the formation of NiO and NiO_2 appear at low and high potentials, respectively, as reported by Hart and by Chatfield and Shreir [118, 119]. Whether the higher oxides of nickel are formed under actual electroplating conditions is not known.

When nickel chloride (Fig. 10) is added to nickel sulfate solutions, the active to passive transition region is still observed, but at an applied potential of about +0.2 volt verses SCE, the current on the anode rises rapidly [120]. Hart and coworkers confirmed that the transition from active to passive dissolution occurs even at very high chloride concentrations [121]. The persistence of the active to passive transition

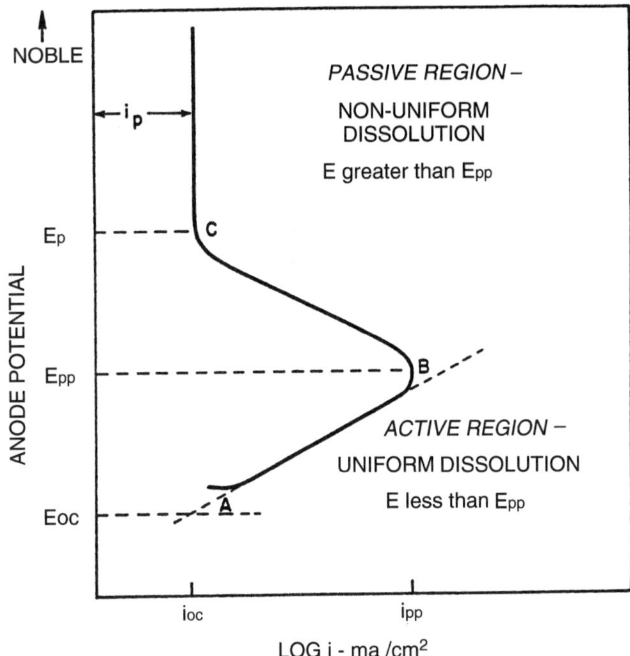

Figure 9 Idealized anodic polarization curve obtained potentiostatically for pure nickel in chloride-free nickel sulfate solutions [120]. E_{oc} is the open-circuit potential; E_{pp}, the principal or critical passivation potential; E_p, the passivation potential; i_{oc} is the corrosion current density; i_{pp} the principal passivation current density; and i_p the current density in the passive region. *AB* is the active region, *BC* the active to passive transition region, and *C* the beginning of the passive region.

region suggests that an oxide film forms in the presence of chlorides, but does not increase in thickness. Under practical plating conditions, pure electrolytic nickel dissolves at a constant potential, +0.2 volt versus SCE, over a wide range of current density when chlorides are present, and the rate of dissolution is virtually independent of the anode potential [120].

The nonuniform dissolution of electrolytic nickel observed in most commercial nickel electroplating solutions, that is, its tendency to dissolve through pits on the surface and to become spongy at advanced stages of dissolution, is due to the oxide film on its surface, and a correlation between the electrochemical and the *physical* dissolution characteristics exists. That is, nickel dissolves nonuniformly when its dissolution potential exceeds the peak potential, Epp, and uniformly below that potential [120].

It is unlikely that the reaction taking place in the active region is simply $Ni \longrightarrow Ni^{++} + 2e^-$. Active dissolution of nickel in a chloride-free nickel sulfate solution involves adsorbed hydroxyl ions and may proceed in three consecutive steps as follows [122].

1. $Ni + OH^- \rightarrow NiOH_{adsorbed} + e^-$
2. $NiOH_{adsorbed} \rightarrow NiOH^+ + e^-$
3. $NiOH^+ \rightarrow Ni^{++} + OH^-$

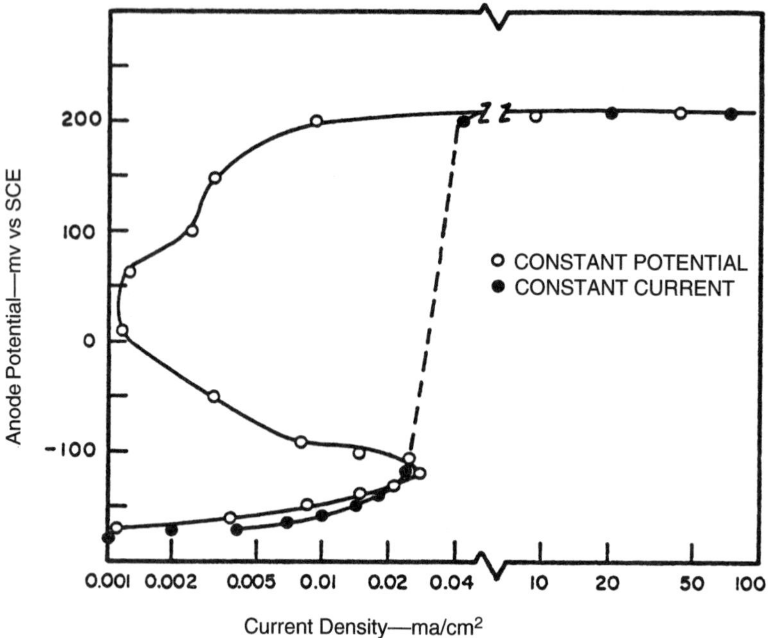

Figure 10 Anodic polarization of pure nickel in a Watts solution containing 14 g liter^{-1} of chloride, at pH 4, 55°C, obtained potentiostatically (constant potential) and galvanostatically (constant current), from [120].

The reactions responsible for the transition from active to passive behavior may be written

4. $NiOH_{adsorbed} + OH^- \rightarrow Ni(OH)_{2\downarrow} + e^-$
5. $Ni(OH)_{2\downarrow} \rightarrow NiO_{\downarrow} + H_2O$

The anodic polarization characteristics of nickel are the result of the competition among these reactions. The rate-determining step changes (from 3 to 2 to 1) as overvoltage is increased. At potentials where reactions 4 and 5 are possible, the percentage of the surface area covered with an oxide film increases, and accordingly, the current density gradually decreases in the transition region. In the presence of chlorides, species like $NiCl_{adsorbed}$ and $NiCl^+$ interfere with the formation of nickel hydroxide and thickening of the oxide film. This *chloride-assisted* dissolution increases anode efficiency to 100%, but dissolution is localized and occurs beneath or through the oxide film.

6.3 Titanium Anode Basket and Primary Forms of Nickel

The introduction of titanium anode baskets [123] transformed nickel anode practices and led to the development of special forms of primary nickel for electroplating.

Titanium Anode Baskets Baskets for nickel electroplating are made of titanium mesh strengthened by solid strips of titanium at tops, bottoms, and edges. The baskets are encased in cloth anode bags, suspended on the anode bar by hooks that are an integral part of the baskets, and loaded with small pieces of nickel. The mesh facilitates the free-flow of plating solution. Baskets that incorporate hoppers at the tops facilitate basket loading and help prevent pieces of nickel from falling into the tank.

Titanium anode baskets were quickly accepted because of their many advantages [124]. The basket anode is large and unchanging, assuring that a uniform anode area gives constant current distribution and consistent thickness for repeat batches of the same work. Anode maintenance can be done without interfering with electroplating, and it involves topping-up the load to keep the baskets filled. Conforming baskets are possible in virtually any size and shape. The anode to cathode distance is constant, contributing to good current distribution. Lowest-cost, primary forms of nickel, can be used to fill the baskets. Baskets can be semi-automatically or automatically filled with nickel, and that practice is growing in progressive electroplating shops. Anode bags last longer with anode baskets, and if the bags are tight-fitting, less metallic residues are formed by electrolytic nickel anode materials.

Titanium can corrode in nickel plating solutions if it is not in contact with nickel metal (titanium is protected by the nickel), or if excessively large, solid areas of titanium are used to fabricate the baskets. Titanium cannot be used in concentrated fluoborate solutions or in solutions containing fluoride ions; small amounts of fluoride in solution activate titanium causing it to corrode [125].

Sulfur-Containing Electrolytic Nickel Work to develop an improved primary form of nickel for use in baskets focused on finding additives that would increase the activity of the metal in electroplating solutions. Sulfur, selenium, tellurium, phosphorus, carbon, and silicon were studied. The most effective additive proved to be sulfur [126]. Sulfur lowers the dissolution potential by at least 0.4 volt compared to pure electrolytic nickel and transforms the mode of dissolution from nonuniform to uniform. Sulfur-containing electrolytic nickel is the most active anode material commercially available as indicated by the data in Figure 11. A black, nickel sulfide film forms on the surface as sulfur-activated nickel anode materials are dissolved. Because nickel sulfide is highly insoluble, sulfur in the anode material does not enter the solution but is retained in the anode bag as part of the residue. The amount of residue is less than 0.1% of the metal dissolved; that is, more than 99.9% of the nickel goes into solution. The lowering of the dissolution potential conserves energy and reduces power costs, as confirmed by tests conducted in the laboratory and in plating shops [127]. Sulfur-containing electrolytic nickel dissolves at 100% efficiency even in the absence of chloride ions, and this makes it possible to control deposit stress by eliminating chlorides from solution.

The mechanism by which sulfur depolarizes or activates electrolytic nickel has not been fully defined. Because the sulfur is present in the anode material as a sulfide and is electronegative, it may lower the effective concentration of hydroxyl ions in the anode film so that the nickel oxide film can only form at high anodic overvoltage *in the absence of chlorides*. Sulfur-containing electrolytic nickel does not appear to form an oxide film when chlorides are present in solution. DiBari and Petrocelli first suggested that adsorbed sulfur-containing anions might be responsible for the

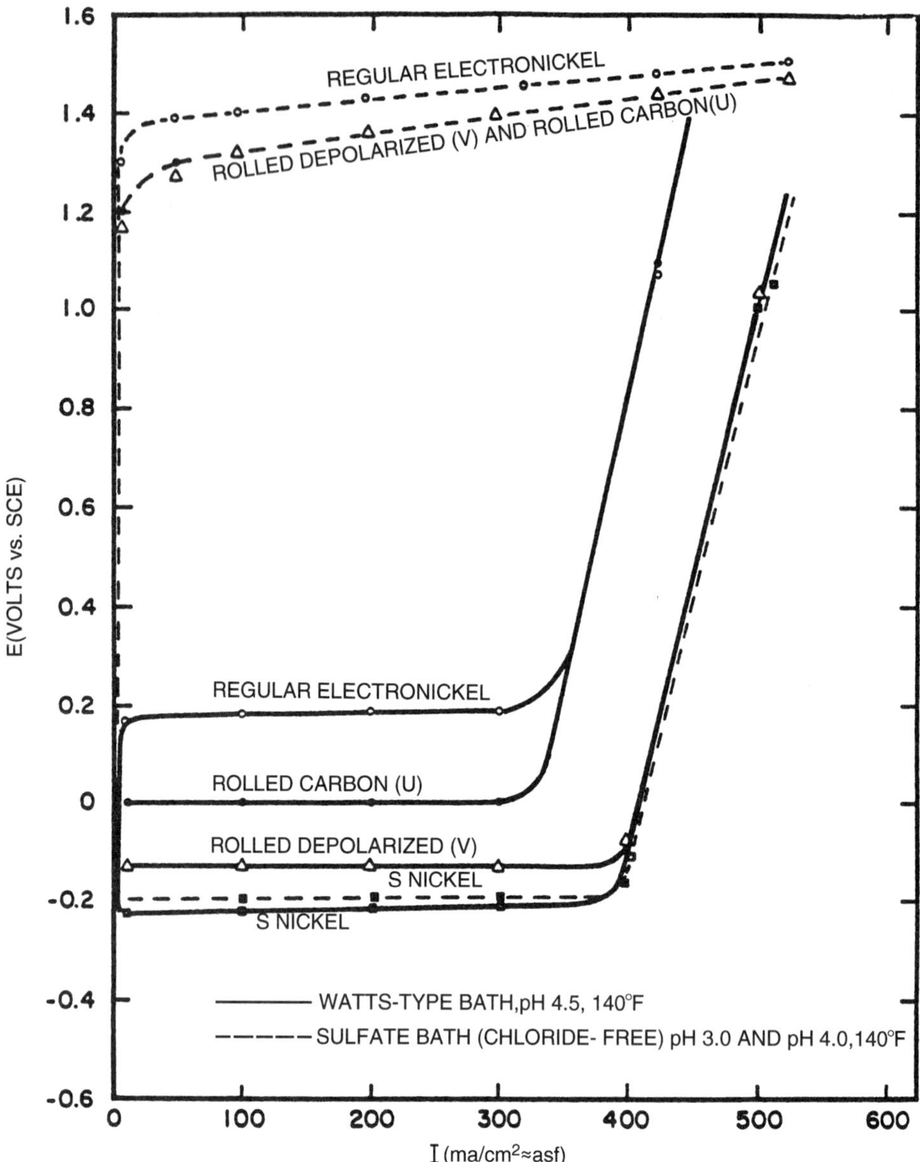

Figure 11 Anodic polarization under galvanostatic conditions of various nickel anode materials in a Watts and a chloride-free nickel sulfate solution.

activation [117]; Morris and Fisher proposed that a tiny amount of the sulfur in the anode material is oxidized to form an anion, perhaps thiosulfate, that prevents oxide film formation by specific adsorption [128].

Electrolytic nickel containing a small amount of sulfur first became available in the form of sheared squares in 1963 [129]. Two forms of sulfur-containing nickel were introduced in 1972: a button-shaped material made by electrowinning and a spherical form made by the decomposition of nickel carbonyl. Electroplaters quickly converted to the round and spherical forms because of their better settling characteristics compared to the electrolytic nickel squares that bridge and hang up at their corners.

Other Primary Forms of Nickel for Baskets Available forms of primary nickel for baskets include nonactivated, high-purity electrolytic nickel in buttonlike shapes [130] and screened nickel pellet made by the decomposition of nickel carbonyl. Electrolytic nickel squares about 25 × 25 mm and, to a less extent, squares that are 100 × 100 mm are used to fill baskets, but that practice is becoming obsolete except in some rapidly developing economies.

7 QUALITY CONTROL

Process quality control involves maintaining the concentrations of the main constituents within specified limits, controlling pH, temperature, and current density and maintaining the purity of nickel electroplating solutions. Product quality control includes eliminating coating defects, properly preparing substrates prior to electroplating, as well as conducting tests to verify that specified product requirements have been met. The basic nickel electroplating facts in Table 10 and the conversion factors in Table 11 are often required for calculations related to quality control.

7.1 Process Control

Controlling the composition of the plating bath is one of the most important factors contributing to the quality of electrodeposited nickel. At the outset, the bath must be prepared to the specified composition, adjusted to the proper pH, and purified before use. Thereafter the composition and pH of the solution must be controlled within specified limits, and contamination by metallic, organic, and gaseous substances must be prevented.

Main Constituents The basic constituents of nickel electroplating solutions that are regularly controlled are the nickel metal content, the chloride concentration, the

TABLE 10 Nickel Electroplating Facts

	Symbol	Ni
	Atomic weight	58.69
	Valency	2
	Specific gravity	8.90
Nickel	Plating rate, at 100% cathode efficiency	1.095 g Am-h^{-1} (0.039 oz am-h^{-1})
Nickel sulfate	$NiSO_4 \cdot 6H_2O$	Contains 22.3% nickel
Nickel chloride	$NiCl_2 \cdot 6H_2O$	Contains 24.7% nickel
Nickel sulfamate	$Ni(NH_2SO_3)_2$	Contains 23.2% nickel
Nickel carbonate	$NiCO_3$	Contains about 46% nickel

TABLE 11 Units and Conversion Factors for Electroplating

Quantity	A Traditional Unit	B SI Unit[a]	To Convert from A to B Multiply by	To Convert from B to A Multiply by
Coating thickness	mil	μm	25.4	0.0394
	inch	mm	25.4	0.0394
Coating mass (weight)	$mg\,in^{-2}$	$g\,m^{-2}$	1.55	0.645
	$mg\,in^{-2}$	$mg\,cm^{-2}$	0.155	6.45
	$oz\,in^{-2}$	$kg\,m^{-2}$	43.9	0.0228
	$oz\,ft^{-2}$	$kg\,m^{-2}$	0.305	3.28
Current density	$A\,ft^{-2}$	$A\,m^{-2}$	10.76	0.0929
	$A\,ft^{-2}$	$A\,dm^{-2}$	0.1076	9.29
	$A\,in^{-2}$	$A\,m^{-2}$	1550	6.45×10^{-4}
	$A\,in^{-2}$	$A\,cm^{-2}$	0.155	6.45
Plating rate	A-hft^{-2}-mil	A-s m^{-2}-mm	1530	6.55×10^{-4}
Volume	gal	m^3	0.00379	264
	gal	liter	3.79	0.264
	fl oz	ml	29.6	0.0338
Mass concentration	oz gal^{-1}	g $liter^{-1}$	7.49	0.134
		$kg\,m^{-3}$	7.49	0.134
Volume concentration	fl oz gal^{-1}	ml $liter^{-1}$	7.81	0.128
		cm^3 $liter^{-1}$	7.81	0.128
Force (internal stress)	1000 psi	MPa ($Mn\,m^{-2}$)	6.89	0.145

boric acid level, and the concentration of all addition agents. Nickel metal concentration is maintained between 60 and 80 g $liter^{-1}$ in most commercial applications. It is desirable to have a minimum of 25 g $liter^{-1}$ nickel chloride in the solution to promote anode corrosion except when sulfur-activated electrolytic nickel anode materials are used. Boric acid is the most commonly used buffering agent for nickel plating baths. Boric acid is effective in stabilizing the pH in the cathode film within the ranges normally required for best plating performance. It is available in a purified form and is inexpensive. Organic addition agents must be controlled within the limits specified by the suppliers of proprietary processes, and they must be replenished due to losses from drag-out, electrolytic consumption, and the effects of carbon filtration (or batch treatment).

Traditional procedures for chemical analysis of nickel, chloride, boric acid, and organic addition agents in nickel electroplating solutions exist that are based on titration, precipitation and other *wet* chemical techniques, but these procedures have been supplanted in many cases by instrumental techniques that are rapid and accurate. Suppliers of decorative nickel electroplating solutions usually provide methods for analyzing and controlling the specific organic additives in their processes. Liquid-chromatography has become popular for controlling organic additives, and electroanalytical methods, like polarography, have been applied to the control of electroplating solutions [131].

Controlling pH, Temperature, Current Density, and Water Quality The pH of the nickel plating solution will rise during normal operation of the bath necessitating regular additions of acid to maintain the pH within the prescribed limits. (A decrease in pH accompanied by a decrease in nickel ion concentration indicates that the process is not functioning properly.) In Watts solutions, sulfuric acid is added for pH adjustment; sulfamic acid is added to control the pH of nickel sulfamate solutions. The pH of nickel plating solutions should be measured frequently, and it is most often done by an electrometric method employing a glass electrode and a saturated calomel reference electrode.

The operating temperature has a significant effect on the properties of the deposits and should be maintained within specified limits (±2°C) of the recommended value. In general, most commercial nickel plating baths are operated between 40° and 60°C.

The nickel plating process should be controlled by estimating the surface area of the parts to be electroplated, and the ampere-hours required to deposit a specified thickness of nickel at a specified current density. The practice of operating the process at a fixed voltage is not recommended. Controlling cathode current density is important for meeting minimum coating thickness requirements, and for producing deposits with consistent and predictable properties.

Since current density determines the rate of deposition, it must be as uniform as possible to achieve uniformly thick nickel deposits. Current distribution is controlled by proper rack design and proper placement of components on racks, by the use of nonconducting shields and baffles, and by the use of auxiliary anodes, when necessary. With care, relatively good thickness distribution can be achieved.

The quality of the water used in making up the bath and in replacing water lost by evaporation is important. Demineralized water should be used, especially if the local tap water has a high calcium content, greater than 200 ppm. Filtering the water before it is added to the plating tank is a useful precaution to eliminate particles that can cause rough deposits.

Controlling Impurities Inorganic, organic and gaseous impurities may be introduced into nickel plating solutions during normal operations. Continuing efforts to eliminate the sources of these impurities from the plating shop can improve the quality of he deposits, as well as productivity and profitability. Inorganic contaminants arise from numerous sources, including nickel salts of technical grade, hard water, carry over from acid dip tanks, airborne dust, bipolar attack of metallic immersion heaters, corrosion of the tank material through cracks in the lining, corrosion of anode bars, dirt from structures above the tank, and from parts that fall into the solution and are not removed.

Table 12 lists maximum limits for inorganic, metallic impurities in nickel plating baths. The degree of contamination by many inorganic materials may be controlled by continuous filtration and low current density electrolysis at 0.2 to 0.5 $A\,dm^{-2}$. This may be accomplished on a batch basis or continuously by installing a compartment and overflow dam at one end of the electroplating tank. Solution from the filter is pumped into the bottom of the compartment, up past corrugated cathode sheets, over the dam, into the electroplating section of the tank, out through a bottom outlet at the far end of the tank, and back to the filter. Solid particles and soluble metallic impurities (e.g., copper, zinc, and lead) are removed simultaneously

TABLE 12 Maximum Concentrations for Contaminants in Nickel Plating Solutions

Contaminant	Maximum Concentration (ppm)
Aluminum	60
Chromium	10
Copper	30
Iron	50
Lead	2
Zinc	20
Calcium	[a]

Note: The limits are different when several contaminants are present at the same time when complexing agents are part of the solution formulation.

[a]Calcium will precipitate at the saturation point, about 0.5 g $liter^{-1}$, but the exact saturation point is dependent on solution pH.

by this procedure. The effects of chromium, copper, iron, zinc, and lead impurities on some properties of nickel deposits were evaluated many years ago, and that information is still useful [132].

Organic contaminants may arise from many sources, including buffing compounds, lubricating oil dropped from overhead equipment, sizing from anode bags, waving lubricants on plastic anode bags, uncured rack coatings or stop-off lacquers, adhesives on certain types of masking tape, decomposition products from wetting agents, organic stabilizers in hydrogen peroxide, paint spray, and new or patched rubber tank linings.

Many organic contaminants can be effectively removed from nickel plating solutions by adsorption on activated carbon on either a batch or a continuous basis. On a batch basis, the solution is transferred to a spare tank, heated to 60° to 71°C, stirred for several hours with a slurry of 6 g $liter^{-1}$ minimum of activated carbon, permitted to settle, and then filtered back into the plating tank. It is usually necessary to do a complete chemical analysis and adjust the composition of the solution after this type of treatment.

For solutions in which organic contamination is a recurring problem, continuous circulation of the solution through a filter, coated at frequent intervals with small amounts of fresh activated carbon, is recommended. When continuous carbon filtration is used, the wetting agent in the solution must be replenished and controlled more carefully to prevent pitting of the nickel deposits. In cases of severe organic contamination, it may be necessary to treat the solution with potassium permanganate prior to treating with carbon [133]. This should be used only as a last resort because of the difficulty of removing manganese dioxide and other precipitated solids; in the case of a proprietary solution, the supplier of the process should first be consulted.

Gaseous contamination of nickel plating solutions usually consists of dissolved air or carbon dioxide. Dissolved air in small amounts may lead to a type of pitting characterized by a teardrop pattern. Dissolved air in the plating solution usually can be traced to entrainment of air in the pumping system when the solution is circulated. If this occurs, circulating pump and valves should be checked and modified, if necessary. Nickel plating solutions can be purged of dissolved air by heating to a temperature at least 6°C higher than the normal operating temperature for several hours. The solution is cooled to the operating temperature before plating is resumed. Dissolved carbon dioxide in a nickel plating solution is usually found after nickel carbonate has been added to raise the pH, and is liberated from warm nickel plating solutions after several hours. If solutions containing carbon dioxide are scheduled for immediate use, they should be purged by a combination of heating and air agitation for approximately 1 h at 6°C or more above the normal plating temperature.

Effects of Impurities on Bright Nickel Plating Metallic and organic contaminants affect bright nickel electrodeposition in the following ways [134]:

- Aluminum and silicon produce hazes, generally in areas of medium-to-high current density, and they may also cause a fine roughness called salt and pepper, or stardust.
- Iron may produce roughness, particularly at high pH.
- Calcium contributes to needlelike roughness as a result of the precipitation of calcium sulfate when calcium in solution exceeds the saturation point of 0.5 g $liter^{-1}$ at 60°C.
- Chromium as chromate causes dark streaks, high current density gassing, and peeling. After reduction to the trivalent form by reaction with organic materials in the solution or at the cathode, chromium may produce hazing and roughness effects similar to those produced by iron, silicon, and aluminum.
- Copper, zinc, cadmium, and lead affect areas of low current density, producing hazes and dark-to-black deposits.
- Organic contaminants may also produce hazes or cloudiness on a bright deposit or result in a degradation of mechanical properties. Haze defects may occur over a broad or narrow current density range.
- Mechanical defects producing hairline cracks, called macrocracking, may be encountered if the coating is sufficiently stressed as a result of solution contamination. These cracks usually appear in areas of heavier plating thickness, but they are not necessarily confined to those areas.

Contamination by zinc, aluminum, and copper is most often caused by the dissolution of zinc-based die castings that have fallen from racks into the plating tank and been permitted to remain there. Inadequate rinsing before nickel plating increases the drag-in of metallics. The presence of cadmium and lead may be attributed to a number of sources, including lead-lined equipment and tanks, impure salts, and drag-in of other plating solutions on poorly rinsed racks. Chromium is almost always carried into the nickel solution on rack tips that have not been chromium stripped, or on poorly maintained racks that have been used in the chromium tank and have trapped chromium plating solution in holes, pockets, and tears in the rack coating.

Purification Techniques and Starting up a New Bath The following procedures can be used singly or in combination to purify nickel electroplating solutions:

- The high-pH treatment consists of adding nickel carbonate to the hot solution until a pH of 5.0 to 5.5 is obtained. This precipitates the hydroxides of metals such as iron, aluminum, and silicon, which in turn frequently absorb other impurities. Addition of hydrogen peroxide oxidizes iron to the ferric state, making it more easily precipitated at high pH, and frequently destroys organic impurities.
- Treatment with activated carbon removes organic impurities.
- Electrolytic purification removes many harmful metallic and organic impurities.

A complete purification procedure for a freshly prepared solution consists of the following steps:

1. Use a separate treatment tank (not the plating tank) to dissolve the nickel sulfate and nickel chloride in hot water at 38° to 49°C to about 80% of desired volume.
2. Add 1 to 2 ml $liter^{-1}$ of 30% hydrogen peroxide; agitate briefly and allow to settle for 1 hour.
3. Add 1.2 to 2.5 g $liter^{-1}$ activated carbon and agitate thoroughly.
4. Heat to 66°C, then add 1.2 to 2.5 g $liter^{-1}$ of nickel carbonate to the solution, with agitation to adjust the pH to 5.2 to 5.5. More nickel carbonate may be required and the mixture should be stirred to assist the dissolution of the carbonate. Allow to settle 8 to 16 hours.
5. Filter into the plating tank.
6. Add and dissolve boric acid; add water to bring bath up to its desired volume.
7. Electrolytically purify by using a large area of nickel-plated corrugated steel sheets as cathodes. The average cathode current density should be 0.5 A dm^{-2} and treatment should continue until 0.5 to 1.5 A-h $liter^{-1}$ have passed through the solution. The solution should be agitated and the temperature held at 49° to 60°C. Prepare deposits at normal current densities at some point to check appearance, stress, and sulfur content; if not acceptable, continue the electrolytic purification until the properties are acceptable.
8. Remove the dummy cathodes and adjust the pH of the solution to the desired value.

7.2 Product Control

The nature of coating defects in parts that have been unsuccessfully electroplated may indicate the source of quality problems. Other aspects of product quality control include preparation of metals prior to electroplating and testing nickel electrodeposits.

Coating Defects Common defects include roughness, pitting, blistering, high stress and low ductility, discoloration, burning at high current density areas, and failure to meet thickness specifications [133].

- Roughness is usually caused by the incorporation of insoluble particles in the deposit. In bright nickel baths, chlorine generated at an auxiliary anode that is close to the cathode can react with organic additives to form insoluble particles. Insoluble particles may arise from incomplete polishing of the basis metal so that silvers of metal protrude from the surface, incomplete cleaning of the surface so that soil particles remain on the surface, detached flakes of deposit from improperly cleaned racks, dust carried into the tank from metal polishing operations and other activities, insoluble salts, and metallic residues from the anode material.
- Roughness from incomplete polishing, cleaning, and inadequate rack maintenance is avoided by good housekeeping, regular inspection and control. Roughness caused by dust can be controlled by isolating surface preparation and metal polshing operations from the plating area, by providing a supply of clean air, and by removing dirt from areas near and above the tanks. Roughness caused by the precipitation of calcium sulfate can be avoided by using demineralized water. Continuous filtration of the plating solution so as to turn over the solution at least once an hour is important for minimizing roughness problems. Anode residues must be retained within anode bags, and care should be taken not to damage the bags or allow the solution level to rise above the tops of the bags.
- Pitting is caused by many factors including adhesion of air or hydrogen bubbles to the parts being plated. Air should be expelled, as already mentioned. Pitting from adherent hydrogen bubbles can result from a solution that is chemically out of balance, as too low a pH, or is inadequately agitated. Incorrect racking of complicated components, too low a concentration of wetting or anti-pitting agents, use of incompatible wetting agents, the presence of organic contaminants, the presence of copper ions and other inorganic impurities, incomplete cleaning of the basis material, incomplete dissolution of organic additives that may form oily globules can all result in pitting. Pitting is therefore avoided by maintaining the composition of the plating solution within specified limits, by controlling the pH and temperature, and by preventing impurities of all kinds from entering the solution.
- Blistering is generally due to poor adhesion as a result of poor or incorrect surface preparation prior to plating (see next section). Blistering may also be related to incomplete removal of grease, dirt, or oxides; formation of metal soaps from polishing compounds; or silica films from cleaning solutions. In the case of zinc-base die castings or aluminum castings, blistering during or immediately after plating may be due to surface porosity and imperfections that trap plating solution under the coating.
- High stress and low ductility usually occur when organic addition agents are out of balance, and also because of the presence of impurities.
- Discoloration in low current density areas is most likely the consequence of metallic contamination of the plating solution. The effects can be evaluated systematically by plating over a reproducible range of current densities on a Hull cell cathode. Hull cells are available from plating supply houses and are shaped so that nickel can be deposited onto a standard panel over a predictable range of current densities. The variation in current density over the face of the panel is achieved by placing the panel at a specified angle to the anode. Bent

panels that are L-shaped and that are plated with the recessed area facing the anode can also be used to assess discoloration at low current density areas, and thay may provide information on roughness problems.

- Burning at high current densities can be caused by applying the full load on the rectifier to the lowest parts on a rack as it is lowered into the tank. This can be controlled by applying a reduced load or ramping the current during immersion of the rack. Burning is sometimes related to the presence of phosphates in solution introduced via contaminated activated carbon. Incorrect levels of organic additives can cause burning.
- Failure to meet thickness specifications is most frequently due to the application of too low a current and/or too short a plating time. This can be avoided by measuring the area of the parts to be plated, then calculating the total current required for a specified current density, and plating for the appropriate time. Another major cause of failure to meet thickness requirements is nonuniform distribution of current leading to insufficient deposit in low current density areas. Poor electrical contacts and stray currents can also cause thin deposits. Anode and cathode bars, hooks and contacts, should be kept clean.

Preparation Prior to Plating Nickel can be deposited adherently on most metals and alloys, plastics, and other materials by following standard methods of preparation and activation, including proper use of intermediate deposits such as cyanide copper, acid copper, and acid (Wood's) nickel chloride strikes. Standard procedures for the preparation of materials prior to electroplating can be found in handbooks and in ASTM standards [23, 24].

Controlling and Testing Deposits Properties The requirements for testing electrodeposited nickel coatings vary depending on the application. In decorative applications the appearance and the thickness of the deposit should be controlled and monitored on a regular basis. The plated surface must be free of defects such as blisters, roughness, pits, cracks, discoloration, stains, and unplated areas. It must also have the required finish: bright, satin, or semi-bright. In the case of decorative multilayered coatings, the sulfur contents of the deposits, the relative thicknesses of individual layers, the ductility of the semi-bright nickel layer, and the differences in electrochemical potentials between individual layers should be controlled. Requirements for corrosion performance and adhesion may also be specified and may require additional testing. In functional and electroforming applications, it may be necessary to monitor hardness, ductility, and internal stress, in addition to thickness and appearance. Some test methods are briefly reviewed below:

- Thickness. The coulometric method described in ISO Standard 2177 [79] and ASTM Standard B 504 [23] can be used to measure the chromium and nickel thicknesses, as well as the thickness of copper undercoats, if present. The coulometric method measures the quantity of electrical energy required to deplate a small, carefully defined area of the component under test. A cell is sealed to the test surface and filled with the appropriate electrolyte; then a cathode is inserted. The component is made the anode, and the circuit is connected to the power supply via an electronic coulometer. By integrating time in seconds with the current passing, the electronic coulometer provides a direct reading in coulombs; modern instruments provide a direct reading of thickness. The completion of the deplating is shown by a marked change in the applied voltage. For routine control of production, it is convenient to monitor nickel thickness nondestructively by means of a magnetic gauge, calibrating the gauge at intervals with standard samples. Instruments for measuring thickness by beta backscatter, X-ray spectrometry, and eddy current techniques are also available. The traditional method of measuring thickness by microscopic examination of a metallographically prepared cross section of the plated part is still employed, but it is time-consuming, expensive, and destructive. ASTM Standard B 659 [23] is a general guide to the measurement of thickness of electrodeposited coatings on different substrates.

- The STEP test. This test was developed to measure potential differences between individual layers of nickel in decorative nickel coatings on production parts. It is similar to the coulometric method just described. By including a reference electrode in the circuit, however, it is possible to measure the potential of the material being dissolved at the same time that the thickness of the individual layers is being measured. With a double-layer nickel coating, a relatively large change in potential occurs when the bright nickel layer has dissolved and the semi-bright nickel layer begins to be attacked. The potential difference is related to overall corrosion resistance, and should be greater than 100 millivolts. Details can be found in ASTM Standard Test Method B-764 [23].
- Corrosion testing. When corrosion performance is specified, the electroplater may be required to perform accelerated corrosion tests on a specified number of production parts. Three accelerated corrosion tests are recognized internationally. They are the Copper-Accelerated Acetic Acid Salt-Spray (CASS), the Corrodkote, and the Acetic Acid Salt Spray tests. The CASS and Corrodkote tests were developed when conventional chromium was the only type of chromium available; when used to evaluate microdiscontinuous chromium coatings, the surface appearance deteriorates more rapidly than in real-world environments. Details of these three tests can be found in ISO Standard 1456 [79] as well as in ASTM Standards B 368 (CASS) and B 380 (Corrodkote) [23]. CASS and other corrosion test requirements are specified in ASTM Standard Specification B 456 [23] for nickel plus chromium coatings applied to steel, zinc alloys, or copper alloys. Similar information for nickel plus chromium coatings on plastics is given in ASTM Standard Specification B 604 [23].
- Thermal cycle testing. Thermal cycle testing is specified for controlling the quality of decorative nickel/chromium electroplated plastics and involves exposure of electroplated parts to low and high temperatures under controlled conditions. Thermal cycle testing may be combined with accelerated corrosion testing. Test procedures are described in ASTM Standard B 604 [23].
- Ductility. The ductility of the semi-bright nickel layer in a multilayered nickel coating is specified to control the properties of the deposit and to check that the solution is in good working condition. The simple test described in ISO Standard 1456 [79] and in ASTM B 489 [23] is based on bending a test strip of the deposit over a mandrel of specified diameter until the two ends of the strip are parallel. A semi-bright nickel deposit that meets the requirements of the

bend test has a elongation percentage greater than 8. Other tests based on hydraulic or mechanical bulge testing are available [135]. The elongation percentage can also be determined by machining a test sample from relatively thick electroformed nickel and subjecting it to the conventional tensile test. Difficulties with conventional tensile testing as applied to electrodeposited coatings and films have been discussed, and improved techniques have been described [136]. Since ductility is affected by the thickness of the coating, ductility should be measured at the actual thickness specified in a specific end-use [80, 104, 136].

- Adhesion. Qualitative tests for adhesion are frequently used in the electroplating shop because quantitative tests are considered costly and time-consuming. The tests described in ASTM B 571 [23] include bend, burnish, chisel, file, and others that are qualitative in nature. The method for determining the peel strength of metal electroplated plastics (ASTM Standard B 533 [23]) is quantitative and provides an average value of the bond strength when a universal tensile test machine is used in making the measurements; a less rigorous variation of the test is often applied for process control. Quantitative results can also be obtained by conical head and ring shear tensile testing, and by ultrasonic, centrifuge, and flyer plate methods, which have been described in the literature [137].
- Internal stress measurements. There are many ways that the internal stress of electrodeposits can be measured, but the one most frequently applied in production is the spiral contractometer method (ASTM Standard B 636 [23]). The method is based on electrodepositing nickel on the outside of a helix formed by winding a strip of metal around a cylinder followed by annealing. During the measurement of stress, one end of the helix is fixed in the contractometer and the free end is attached to an indicating needle that moves as stress develops. Internal stress can be calculated from the degree of needle deflection on the circular dial. Contractometers and pre-coated helices to prevent internal electroplating are commercially available. Modifications of the contractometer method including measuring the needle deflection electronically have been made [138]. Rigid and flexible strip methods are alternative techniques that give reliable results [139]. The strain gage method permits stress to be monitored and controlled throughout the electrodeposition process, and the method has been applied in the production of electroformed optical parts having dimensional accuracies of 0.15 μm [140]. The dilatometer method also allows stress to be monitored continuously [141].
- Microhardness testing. The hardness of electrodeposited coatings can be determined by the methods described in ASTM Standard B 578 [23] and E 384. Measurements are made on the cross section of a deposit of specified thickness using a specified load on the Knoop indenter.

8 POLLUTION PREVENTION

Regulations to prevent pollution and to protect health and safety in the workplace have affected nickel and chromium electroplating technically and commercially.

Technically, a great deal of the research effort over the past 25 years has been devoted to finding substitutes for hazardous materials and making process changes to comply with government regulations. For example, trivalent chromium to replace hexavalent chromium, zinc-nickel alloy deposits to replace cadmium, substitution of permanganates for chromates in the preparation of plastics, elimination of coumarin from nickel electroplating processes, and the development of alkaline cyanide-free copper electrolytes have been driven by pollution prevention considerations to a great extent. Commercially, pollution control has increased electroplating costs and some shops that were only marginally profitable have gone out of business. The number of shops engaged in nickel electroplating in the United States is about 2800 compared to 3500 in 1970 [1].

Although there was great resistance to plating pollution prevention and control in the early 1970s, most electroplaters in the United States and many other parts of the world are complying with existing regulations. In some cases the recycling and recovery of salts and metals, coupled with the conservation of water and energy, have led to economies that partially offset the cost of compliance. Efforts to minimize waste generation in an electroplating plant require good housekeeping and operating practices that tend to improve overall quality.

Strict environmental regulations have been imposed on the electroplating industry to prevent nickel and other metallic ions from entering the environment via plant effluents. Conventional processes for wastewater treatment include precipitation of nickel and other metals as hydroxides (or sulfides). If hexavalent chromium is present, it must first be reduced to the trivalent form before being precipitated. Cyanides are commonly removed by alkaline chlorination using sodium hypochlorite, but ozonation, electrochemical, thermal, and precipitation methods are known. The solid wastes generated in conventional processes for wastewater treatment are then disposed of in landfills. Because disposal of the solid waste is expensive and wasteful, the recovery of metal values by reverse osmosis, ion-exchange, electrowinning, and other methods is becoming important and economically feasible [142]. Nickel producers accept nickel-containing sludges for recycling through smelters or special plants, and companies that collect and recycle electroplating wastes have grown in number.

The adoption and enforcement of strict environmental regulations arises from concern with the possible effects of metal contaminants on human health. Although the general perception is that these health effects are completely understood, the reality is that our knowledge is extremely limited. It is only when metals are present in high concentrations and in very specific forms that they may be toxic.

In a nickel electroplating shop, three types of exposure are possible: nickel and its compounds may be inadvertently ingested, nickel-containing solutions may be allowed to remain on the skin for long periods of time, and nickel and its compounds may be taken into the body by breathing [143]. Nickel and its inorganic compounds are not highly toxic substances and exposure to small amounts does not present serious health risks. Nevertheless, it is advisable to avoid ingesting even small amounts of these substances by taking some simple precautions in the workplace—wearing work gloves, washing one's hands before eating, and by not eating in the workplace. Only one compound, gaseous nickel carbonyl, is known to be acutely toxic. This compound forms under special conditions in the few nickel refineries that produce high-purity nickel using nickel carbonyl as a process intermediate. It is not present or formed in the electroplating shop.

People who have become skin sensitized should avoid contact with nickel and its compounds. To avoid *becoming* sensitized, one must limit contact with nickel and its compounds. In electroplating, this may mean wearing work gloves and washing one's hands immediately after coming in contact with nickel electroplating solutions. The risk of cancer appears to be limited to the inhalation of high concentrations of dusts containing nickel subsulfides and oxides under conditions previously existing in certain nickel refineries. Similar health problems have not been observed in electroplating and other workplaces where nickel is found. Airborne nickel in the workplace should be kept below regulatory or other acceptable exposure limits.

REFERENCES

1. G. A. DiBari, "Nickel Electroplating Applications and Trends," *Plating Surf. Finish.*, **83** (10), 10, 1996; G. A. DiBari and S. A. Watson, "A Review of Recent Trends in Nickel Electroplating Technology in North America and Europe," *NiDI Reprint Series*, no. 14 024, Nickel Development Institute, Toronto, Ontario, Canada, Nov. 1992; G. A. DiBari, "Plenary Paper—Survey of New Applications in Surface Finishing Technology," *Proc. 4th Int. Congress on Surface Technology, Berlin '87*, published by AMK Berlin, Postfach 191740, Messerdam 22, D-1000 Berlin 19.
2. G. Bird, *Phil. Trans.*, **127**, 37 (1837).
3. J. Shore, U.K. Patent 8407 (1840).
4. R. Bottger, "Investigation of Nickel Plating on Metals," *Erdmann's Journal für Praktische Chemie*, **30**, 267 (1843).
5. G. Dubpernell, "The Story of Nickel Plating," *Plating*, **46**, 599 (1959).
6. J. K. Dennis and T. E. Such, *Nickel and Chromium Plating*, 3rd ed., Woodhead Publ., Cambridge, England, 1993.
7. W. H. Remington, U.S. Patent 82,877 (1868).
8. E. Weston, U.S. Patent 211,071 (1878).
9. W. D. Bancroft, *Trans. Am. Electrochem. Soc.*, **9**, 218 (1906).
10. O. P. Watts, *Trans. Am. Electrochem. Soc.*, **29**, 395 (1916).
11. M. Schlotter, U.S. Patent 1,972,693 (1934).
12. A. S. DuRose, U.S. Patent 2,635,076 (1953).
13. L. Cambi and R. Piontelli, *Italian Patent 368,824* (1939).
14. P. C. Crouch and H. V. Hendricksen, *Trans. Institute of Metal Finishing*, **61**, 133 (1983).
15. G. A. DiBari and R. A. Covert, "Nickel Buildup in Plating Baths," *Products Finishing*, Nov 1989, p. 70. See summary of presentation at AESF Sur/Fin '89 by Craig Brown, Exec. VP, Eco-Tech Ltd., Pickering, Ontario, Canada, included in that paper.
16. U. Landau, "Morphology and Thickness Distribution of Electrodeposits," *Proc. Symp. on Electrodeposition Technology, Theory and Practice*, L. T. Romankiw and D. R. Turner, eds., Electrochemical Society, Inc., Pennington, NJ, 1987, p. 589; G. E. Giles, "Electroforming Cell Design Tool Development," *Proc. AESF Electroforming Symposium*, Mar. 27–29, 1996, p. 75.
17. J. Kronsbein, *Plating*, **37**, 851 (1950).
18. H. E. Haring and W. Blum, *Trans. Am. Electrochemical Society*, **44**, 313 (1923).

19. S. A. Watson, *Trans. Institute of Metal Finishing*, **37**, 28 (1950).

20. M. Ya. Popereka, *Internal Stresses in Electrolytically Deposited Metals*, transl. from Russian, Indian National Scientific Documentation Center, New Delhi, National Bureau of Standards and the National Science Foundation, Washington, DC, 1970.

21. R. Weil, "The Origins of Stress in Electrodeposits," *Plating*, **57**, 1231 (1970); **58**, 137 (1971).

22. J. W. Dini, *Electrodeposition — The Materials Science of Coatings and Substrates*, Noyes Publ., Park Ridge, NJ, 1993, ch. 9, p. 279; ch. 11, p. 331.

23. *1996 Annual Book of ASTM Standards*, vol. 02.05, American Society for Testing and Materials, West Conshohocken, PA.

24. "1996 Metal Finishing Guidebook and Directory," *Metal Finishing*, **94** (1a), 105–174 (1996); D. L. Snyder and J. K. Long, "Typical Processing and Operating Sequences," *Electroplating Engineering Handbook*, L. J. Durney, ed., Van Nostrand Reinhold, New York, 1984, p. 174; J. B. Hadju and G. Krulik, "Plastics," ibid, p. 202.

25. S. A. Watson and J. Edwards, *Trans. Institute of Metal Finishing*, **34**, 167 (1957).

26. O. Kardos, *Proc. Am. Electroplaters Soc.*, **43**, 181 (1956).

27. G. A. DiBari, "Nickel Plating," *ASM Handbook-Surface Engineering*, vol. **5**, ASM International, Materials Park, OH, 1994, p. 204.

28. Reference 6, pp. 96–103.

29. R. Weil, "Epitaxial Electrocrystallization under Inhibited Growth Conditions," *Proc. Symp. on Electrocrystallization*, Electrochemical Society, Pennington, NJ, 1981, p. 134; M. Jousellin and R. Wiart, "Anion Dependence of Nickel Electrodeposition in Acidic Electrolytes," ibid, p. 111; K. Raghunathan and R. Weil, *Surf. Technol.*, **10**, 1472, (1973); M. Froment and J. Thevenin, *Metaus, Corrosion, Industrie*, **59** (4), 1 (1975); J. Ambard, M. Froment, and N. Spyrellis, *Surf. Technol.*, **5**, 205 (1977); J. Amblard, I. Epelboin, M. Froment, and G. Maurin, *J. Appl. Electrochem.*, **9**, 233 (1979).

30. J. Macheras, D. Vouros, C. Kollia, and N. Spyrellis, "Nickel Electrocrystallization: Influence of Unsaturated Organic Additives on the Mechanism of Oriented Crystal Growth," *Trans. Institute of Metal Finishing*, **74** (2), 55, (1996).

31. C. Kollia and N. Spyrellis, "Crystal Growth Inhibition in Nickel Electrodeposition under Pulse Reversed Current Conditions," *Trans. Institute of Metal Finishing*, **72** (3), 124 (1994); F. Kotzia, C. Kollia, and N. Spyrellis, "Influence of Butyne-2-diol 1,4 in Nickel Electrocrystallization under Pulse Reversed Current Regime," *Trans. Institute of Metal Finishing*, **71** (1), 34 (1993).

32. R. Weil and R. Paquin, *J. Electrochem. Soc.*, **107**, 87 (1960); H. J. Read and R. Weil, *Plating*, **37**, 1257 (1950).

33. J. Edwards, "Aspects of Addition Agent Behavior," *Trans. Institute of Metal Finishing*, **41**, 169 (1964); ibid, **39**, 33, 45, 52 (1962); ibid, **41**, 140, 147, 157 (1964); ibid, 45, 12 (1967). Several of these papers were coauthored by Margaret J. Levett.

34. D. L. Snyder, "Electroplating in the Nineties," *Asia Pacific Interfinish 90 Proc.*, Nov. 19–22, 1990, Singapore, Australian Institute of Metal Finishing and the Singapore Metal Finishing Society, pp. 21–16.

35. J. F. Vogt and R. J. Herbert, U.S. Patent 2,635,075 (1953); C. N. Isackson, British Patent 684,434 (1952).

36. H. Brown, *Metalloberflache*, **11**, 333 (1962).

37. G. N. Flint and S. H. Melbourne, *Trans. Institute of Metal Finishing*, **39**, 85 (1960).

38. A. H. DuRose, *Proc. AES 47th Annual Conference*, 1960, p. 83; A. H. DuRose and W. J. Pierce, *Metal Finishing*, **57**, 44 (1959).

39. W. H. Safranek, R. W. Hardy, and H. R. Miller, *Proc. AES 48th Annual Conf.*, 1961, p. 156.
40. J. V. Petrocelli, V. Hospadaruk, and G. DiBari, "The Electrochemistry of Copper, Nickel and Chromium in the Corrodkote and CASS Test Electrolytes," *Plating*, **49**, 1 (1962).
41. G. A. DiBari, A. J. Dill, and B. B. Knapp, *Proc. 1st AES Decorative Plating Symp.*, Dearborn, MI, 1973, p. 93.
42. G. A. DiBari, "Corrosion of Decorative Electroplated Nickel Chromium Coatings on Steel, Zinc, Aluminum and Plastics," *Met. Finish.*, **75** (June), 17–20 (1977) and **75** (July), 17–24 (1977).
43. D. L. Snyder, "Quality Decorative Plating," *Prod. Finish.*, **61** (3), 40 (Dec. 1996).
44. B. B. Knapp and H. Brown, *Modern Electroplating*, 3rd ed., F. A. Lowenheim, ed., Wiley, New York, 1974, p. 308.
45. R. L. Saur, "Toward Protective Decorative Chromium Plating," *Plating*, **48**, 1310 (1961); R. L. Saur, "Influence of Pit Density on the Dimensions of Corrosion Pits in Decorative Plating Systems," ibid, **58**, 1075 (1971).
46. R. L. Saur and R. P. Basco, "An Accelerated Electrolytic Corrosion Test and a Corrosion Analysis Procedure for the Nickel-Chromium Plating System. Part I.," *Plating*, **53**, 35 (1966).
47. R. L. Saur, "New Interference Microscope Techniques for Microphotographic Measurements in the Electroplating Laboratory," *Plating*, **52**, 663 (1965).
48. H. Brown and T. W. Tomaszewski, *Proc. International Conf., Surfaces '66 (Basel)*, Forster Verlag A. G., Zurich 1967, p. 88; U.S. Patents 3,152,971 and 3,152,973.
49. T. Malak, D. Snyder, and A. H. DuRose, "Physical Method for Creating Microporosity in Chromium," *Plating*, **59**, 659 (1972); T. G. Kubach, W. H. R. Pritsch, and W. Bolay, U.S. Patent 3,625,039 (1971).
50. W. E. Lovell, E. H. Shotwell, and J. Boyd, *Proc. Am. Electroplaters Soc.*, **47**, 215 (1960); J. H. Lindsay, D. W. Hardesty, and W. E. Lovell, ibid, **48**, 165 (1961); E. J. Seyb, ibid, **47**, 209 (1960).
51. W. H. Safranek, H. R. Miller, and C. L. Faust, *Plating*, **49**, 607 (1962); W. H. Safranek and C. L. Faust, *Trans. Institute of Metal Finishing*, **42**, 41 (1964).
52. G. A. DiBari and F. X. Carlin, "Decorative Nickel Chromium Electrodeposits on Steel—15 Years of Corrosion Performance Data," *Plating*, **72** (5), 1 (1985).
53. D. L. Snyder, "Fifteen Years of Outdoor Corrosion of Trivalent and Hexavalent Chromium Deposits," *Met. Finish.*, **90**, 113 (1992). Reference 34, p. 21–14.
54. Unpublished results of ASTM Corrosion Performance Programs, nos. 8–12.
55. R. J. Clauss and R. W. Klein, *Proc. 7th International Met. Finish. Conf. Interfinish 68*, Deutsche Gesellschaft für Galvanotechnik e. V., Dusseldorf, May 1968, p. 124; E. J. Seyb, *Proc. Am. Electroplat. Soc.*, **50**, 175 (1963); W. H. Safranek and H. R. Miller, *Plating*, **55**, 233 (1968); V. E. Carter, *Trans. Institute of Metal Finishing*, **48**, 16, 19 (1970).
56. E. P. Harbulak, "Simultaneous Thickness and Electrochemical Potential Determination of Individual Layers in Multilayer Nickel Deposits," *Plating Surf. Finish.*, **67** (2), 49 (1980).
57. R. A. Tremmel, "Methods to Improve the Corrosion Performance of Microporous Nickel Deposits," *Plating Surf. Finish.*, **83** (10), 24 (1996).
58. E. B. Saubestre, L. J. Durney, J. Hajdu, and E. Bastenbeck, "The Adhesion of Electrodeposits to Plastic," *Plating*, **52**, 982 (1965); W. P. Innes, J. J. Grunwald, E. D. D'Ottavio, W. H. Toller, and C. Carmichael, "Chromium-Plated ABS and Polypropylene Plastics: Performance in Typical Tests," *Plating*, **56** (1), 51 (1969). See Reference 6, ch. 12, pp. 330–341, and 360–362 for bibliography on electroplating onto plastics.

59. A. Rantell, *Trans. Institute of Metal Finishing*, **47**, 197 (1969); E. B. Saubestre and R. P. Khera, *Plating*, **58**, 464 (1971).
60. The use of 2,4 pentadione is patented.
61. U.S. Patent Numbers 3,011,920; 3,874,772; 3,904,792; 3,672,923; 3,672,938; 3,682,671; 3,960,573; 3,961,109.
62. R. L. Coombes, "Electroless Copper Preplating on ABS Plastics," *Plating*, **57**, 675 (1970).
63. R. R. Wiggle, V. Hospadaruk, and D. R. Fitchmun, "The Mechanism of Adhesion Failure of Plated Plastics in Corrosive Environments," *J. Electrochem. Soc.*, **118** (1), 158 (1971); R. G. Wedel, *Plating*, **62**, 40, 235 (1975).
64. G. A. DiBari and J. V. Petrocelli, *J. Electrochem. Soc.*, **112**, 99 (1965).
65. G. A. DiBari, "Marine Corrosion Performance of Electroless Nickel Coatings on Steel—Final Report of ASTM Program 14," *Proc. EN Conf. '91*, Orlando, FL, Gardner Publ., Cincinnati, OH, 1991.
66. J. L. Adcock, "Electroplating Plastics—An AESF Illustrated Lecture," American Electroplaters and Surface Finishers Society, Orlando, FL, 1978.
67. J. E. McCaskie, "Electroless Plating (Sulfur Dioxide Etching) of Plastic Enclosures for EMI/RFI Shielding," *Proc. AESF Second Electroless Plating Symp.*, American Electroplaters and Surface Finishers Society (Feb. 1984); Anonymous, "Permanganates," *European Surface Treatment* p. 22 (Winter, 1994/95).
68. J. M. Jobbins and P. Sopchak, "Chromic Acid-Free Etching (Ozone Etching)," *Met. Finish.*, **83**, 15 (1985); J. H. Lindsay and V. LaSala, "Vacuum Preplate Process (Plasma Etching) for Plating on Acrlyonitrile-Butadiene-Styrene (ABS)," *Plating Surf. Finish.*, **72**, 54 (1985).
69. "Performance of Decorative Electrodeposited Copper-Nickel Chromium Coatings on Plastics," final report on programs conducted by ASEP and ASTM, on file at ASTM Headquarters, Report Number RR B-8-1003. Also see P. C. Crouch, "The Effect of Nickel Thickness and Copper Undercoats on the Performance of Plated Plastics," *Trans. Institute of Metal Finishing*, **49**, 141 (1971).
70. See Reference 6, pp. 341–344, 360.
71. A. C. Hart, "Decorative Electroplating of Plastics," *Materials World*, May 1996, p. 265.
72. B. B. Knapp, *Met. Finish.*, **47** (12), 42 (1949).
73. L. Missel, *Plating Surf. Finish.*, **64** (7), 32 (1977).
74. D. W. Baudrand, "Electroless Nickel Plating of Aluminum," *Aluminum Finishing Seminar—Technical Papers, Volume II*, Seminar held Mar 30–April 1, 1982, St. Louis, MO; Aluminum Association, Washington, DC, p. 595. (Proprietary solutions are available from plating supply houses.)
75. J. C. Jongkind and E. J. Seyb, "Pretreatment for Plating on Aluminum Using the Stannate Process," *Aluminum Finishing Seminar—Technical Papers, Volume II*, Seminar held Mar 30–April 1, 1982, St. Louis, MO, Aluminum Association, Washington, DC, p. 539.
76. J. C. Jongkind, *Plating Surf. Finish.*, **62**, 1136 (1975); G. A. DiBari, Presentation before AESF Golden West Regional Meeting, San Diego, CA, Mar. 1981.
77. G. A. DiBari, "Plating on Aluminum—Pretreatments and Corrosion Performance," *Plating Surf. Finish.*, **64**, 68 (1977).
78. G. A. DiBari, "Decorative Electroplated Aluminum—Applications and Performance," *Aluminum Finishing Seminar—Technical Papers, Volume II*, Seminar held Mar 30–April 1, 1982, St. Louis, MO, Aluminum Association, Washington, DC, p. 577.
79. Available from International Standards Organization, 1 rue de Varembe, Geneva 20, Switzerland.

80. P. Zentner, Brenner, and Jennings, *Plating*, **39**, 365, 1229 (1952).
81. C. B. Sanborn, "Electroforming Applications—Why They Exist," *Symp. Electrodeposited Metals as Materials for Selected Applications, Metals and Ceramics Information Center*, Battelle Columbus Laboratories, Columbus, OH, MCIC Report/January 1972, p. 65.
82. B. B. Knapp and C. H. Sample, "Physical and Mechanical Properties of Electroformed Nickel at Elevated and Sub-zero Temperatures," *Symp. on Electroforming—Applications, Uses and Properties of Electroformed Metals*, ASTM Special Technical Publication no. 318, 32–43 (1962).
83. J. W. Dini, H. R. Johnson, and L. A. West, "On the High Temperature Ductility Properties of Electrodeposited Nickel," *Plating Surf. Finish.*, **65** (2), 36 (1978); W. R. Wearmouth and K. C. Belt, "Electroforming with Heat-Resistant Sulfur-Hardened Nickel," *Plating Surf. Finish.*, **66** (10), 53 (1979).
84. W. H. Safranek, *The Properties of Electrodeposited Metals and Alloys*, 2nd ed., American Electroplaters and Surface Finishers Society, Orlando, FL, 1986.
85. J. W. Dini, *Electrodeposition — The Materials Science of Coatings and Substrates*, Noyes Publ., Park Ridge, NJ, 1993, ch. 5.
86. W. A. Wesley and E. J. Roehl, *Plating*, **37**, 142 (1950); C. Struyk and A. E. Carlson, *Plating*, **37**, 1242 (1950).
87. W. A. Wesley, U.S. Patent 2,331,751 (1943); W. A. Wesley and E. J. Roehl, *Trans. Electrochem. Soc.*, **82**, 37 (1942).
88. A. J. Dill, "Sulfur-Free Hardening Agents for Electrodeposited Nickel," *Plating Surf. Finish.*, **62**, 770 (1975).
89. W. A. Wesley and J. W. Carey, *Trans. Electrochem. Soc.*, **75**, 209 (1939).
90. W. A. Wesley, D. S. Carr, and E. J. Roehl, "Nickel Plating with Insoluble Anodes," *Plating*, **38**, 1243 (1951).
91. W. L. Pinner and R. B. Kinnaman, *Mon. Rev. of Am. Electroplaters Soc.*, **32**, 227 (1945).
92. M. R. Thompson, *Trans. Am. Electrochem. Soc.*, **47**, 163 (1925).
93. *Sulfate solution*: J. G. Poor, *Met. Finish.*, **4** (11), 694 (1943). *Chloride solution*: W. A. Wesley and B. B. Knapp, U.S. Patent 2,844,530 (1958).
94. A. Brenner, D. E. Couch, and E. K. Williams, "Electrodeposition of Alloys of Phosphorus and Nickel or Cobalt," *Plating*, **37** (1), 36, (1950); **37** (2), 161 (1950).
95. D. S. Lashmore, R. Oberle, and M. P. Dariel, "Electrodeposition of Artificially Layered Materials," *Proc. AESF 3rd Int. Pulse Plating Symposium*, American Electroplaters and Surface Finishers Society, 1986; D. S. Lashmore and J. P. Weinroth, "Pulsed Electrodeposition of Nickel-Phosphorus Metallic Glass Alloys," *Plating Surf. Finish.*, **69** (8), 72 (1982).
96. C. B. Sanborn and F. X. Carlin, "Influence of Nickel Plating on the Fatigue Life of Hardened Steel," *Symp. on Electrodeposited Metals for Selected Applications*, Battelle Columbus Laboratories, Columbus, OH, Nov. 1973.
97. O. I. Pavlova, *Electrodeposition of Metals — An Historical Survey*, U.S. Department of Commerce, Springfield, VA; C. A. Smith, "Early Electroplating, Part 2, Commencement of Industrial Applications (1836–1852)," *Finish. Ind.*, **1** (3), 24 (1978).
98. E. Gnass, "Electroforming with Dispersed Particles," *Eight Ulmer Gesprach-Galvanoformung*, Eugen G. Leuze, Saulgau, 1986, p. 75 (in German); G. Malone, "Electrodeposition of Dispersion Stengthened Alloys," *Symp. on Electrodeposited Metals for Selected Applications*, Battelle Columbus Laboratories, Columbus, OH, Nov. 1973; S. J. Harris, A. A. Baker, A. F. Hall, and R. J. Bache, "Electroforming Filament Winding Process — Method of Producing Metal Matrix Composites," *Trans. Institute of Metal Finishing*, **49** (5), 205 (1971).

99. L. T. Romankiw, "Electroforming of Electronic Devices," *Plating*, **84** (1), 10 (1997); L. T. Romankiw, "Evolution of Plating through Lithographic Mask Technology," *Proc. ECS Symp. on Magnetic Materials, Processes and Devices IV, Applications to Storage and Microelectromechanical Systems (MEMS)*, L. T. Romankiw and D. A. Herman Jr., eds., Electrochemical Society, Pennington, NJ, 1995.

100. E. W. Becker, W. Ehrfeld, P. Hagman, A. Mauer, and D. Muchmayer, "Fabrication of Microstructure with High Aspect Ratios and Great Structural Heights by Synchrotron Radiation Lithography, Galvanoforming and Plastic Molding," *Microelectronic Engineering*, **4**, 34 (1986).

101. J. L. Marti, "Effect of Some Variables upon Internal Stress of Nickel Deposited from Sulfamate Solutions," *Plating*, **53** (1), 61 (1966).

102. A. F. Greene, *Plating*, **55**, 594 (1968); O. J. Klingenmaier, *Plating*, **52**, 1138 (1965).

103. B. B. Knapp, "Notes on Nickel Plating From Sulfamate Solutions," *Plating*, **58**, 1187 (1971).

104. G. A. DiBari, "Evaluation of a Simple, Thin Film Ductility Tester and Review of the Ductility of Nickel Sulfamate Deposits," *Plating*, **79**, 63 (1992).

105. R. J. Kendrick, "High-Speed Nickel Plating from Sulfamate Solutions," *Proc. 6th Int. Conf. on Electrodeposition, Trans. Inst. Met. Finish.*, **41**, 235 (1964).

106. R. J. Kendrick and S. A. Watson, *Proc. Symp. on Sulfamic Acid*, Milan, May 1966, p. 197.

107. Z. Hai-Yan and Z. Liang-Yu, *Proc. AESF Annual Conf.*, Chicago, 1987, Session O.

108. J. M. Notley, "Corner Weakness in Nickel Electroforms," *Trans. Institute of Metal Finishing*, **50** (1), 6 (1972).

109. G. A. DiBari, "Electroforming," *Electroplating Engineering Handbook, 4th ed.*, L. J. Durney, ed., Van Nostrand Reinhold, New York, 1984, p. 474; *Nickel Electroforming*, International Nickel, Saddle Brook, NJ, 1991; booklet available on request.

110. S. A. Watson, "Electroforming today," *Asia Pacific Interfinish '90 Proc.*, Australian Institute of Metal Finishing and Singapore Metal Finishing Society, Singapore, 1990, p. 5–1.

111. L. T. Romankiw and T. A. Palumbo, "Electrodeposition in the Electronics Industry," *Proc. Symp. on Electrodeposition Technology, Theory and Practice*, L. T. Romankiw and D. R. Turner, eds., Electrochemical Society, Pennington, NJ, 1987.

112. I. M. Croll and L. T. Romankiw, "Iron, Cobalt and Nickel Plating for Electronics," ibid, p. 285.

113. S. Harsch, D. Muchmayer, and H. Reinecke, "Electroforming of Movable Microdevices Manufactured by the LIGA-Process," *Proc. Electroforming Session (Toronto)*, American Electroplaters and Surface Finishers Society, Orlando, FL, 1991.

114. W. A. Wesley, "Nickel Atoms, Ions and Electrons," *Trans. Institute of Metal Finishing*, **33**, 452 (1956).

115. B. Wenderott, "The Solubility of Carbonised Nickel Anodes," *Metalloberflache*, **17** (6), 169 (1963).

116. G. A. DiBari, B. B. Knapp, and C. H. Sample, U.S. Patent 3,449,224 (1969).

117. G. A. DiBari and J. V. Petrocelli, "Effect of Composition and Structure on the Electrochemical Reactivity of Nickel," *J. Electrochem. Soc.*, **112** (1), 99 (1965).

118. A. C. Hart, "The Anodic Behavior of Nickel in Electroplating Solutions," *Proc. 9th Int. Metal Finishing Congress*, VOM, Amsterdam, 1976; *Metalloberflache*, **31**, 334 (1977); *Galvanotechnik*, **68** (7), 232 (1977).

119. C. J. Chatfield and L. L. Shreir, "Effect of Sweep Rate on the Active-Passive Transitions of the Ni/H_2SO_4 Systems," *Corrosion Sci.*, **12**, 563 (1972).

120. G. A. DiBari, "Notes on Nickel Anode Materials," *Plating Surf. Finish.*, **66**, 76 (1979); G. A. DiBari, "The Effect of Sulfur, Phosphorus and Silicon Additives on Activity and Type of Corrosion of Nickel Anodes," *Plating*, **53** (12), 1440 (1966).

121. A. C. Hart, W. R. Wearmouth, and A. C. Warner, *Trans. Institute of Metal Finishing*, **54**, 56 (1976); A. C. Hart and S. A. Watson, *Met. Finish. J.*, **19**, 332 (1973).

122. G. Okamoto and N. Sato, *J. Electrochem. Soc.*, **110**, 605 (1963).

123. A. G. Stecker and V. J. Cassidy, *Plating*, **49**, 597 (1962).

124. P. Berger, "Experiences with Primary Nickel and Titanium Baskets," *Electroplating Met. Finish.*, **16** (7), 227, 253 (1963); T. J. Callaghan, "Nickel Plating with Raw Nickel," *Galvanotechnik*, **15** (8), 432 (1963).

125. F. X. Carlin and W. A. Sellers, "The Anodic Behavior of Nickel in Electroplating," *Plating*, **52** (3), 215 (1965).

126. References 117 and 120.

127. G. L. Fisher, "Power Savings Using Sulfur-Activated Nickel Anode Materials," *Plating Surf. Finish.*, **65**, 46 (1978).

128. G. L. Fisher and P. E. Morris, *Trans. Institute of Metal Finishing*, **53**, 145 (1975).

129. G. A. DiBari, B. B. Knapp, F. X. Carlin, and L. S. Renzoni, U.S. Patent 3,437,571 (1969).

130. W. G. Borner, A. J. Dill, and G. L. Fisher, U.S. Patent 4,147,597 (1979).

131. C. Rosenstein and S. Hirsch, "Chemical Analysis of Plating Solutions," *Metal Finishing Guidebook and Directory Issue*, Elsevier Science, New York, 1996, p. 479; J. W. Dini, *Electrodeposition — The Materials Science of Coatings and Substrates*, Noyes Publ., Park Ridge, NJ, 1993, ch. 7, pp. 195–248.

132. "Reports of American Electroplaters' Society Research Project Number 5," *Plating*, **40**, 1391 (1953); **37**, 1157 (1950); **39**, 1343 (1952); **39**, 1033 (1952); **41**, 1307 (1954); G. J. Greenall and C. M. Whittington, "Metallic Impurities in Nickel Plating Solutions," *Plating*, **53**, 217 (1966).

133. *Inco Guide to Nickel Plating*, International Nickel, Saddle Brook, NJ, 1996; booklet available on request.

134. L. Gianelos, "Troubleshooting of Nickel Plating Solutions," *Plating Surf. Finish.*, **64** (8), 32 (1977); **64** (9), 32 (1977); **64** (10), 22 (1977).

135. S. Nakahara, Y. Okinaka, and H. K. Strashil, "Ductility of Plated Films: Its Measurement and Relationship to Microstructure," *Testing of Metallic and Inorganic Coatings, ASTM STP 947*, W. B. Harding and G. A. DiBari, eds., American Society for Testing and Materials, West Conshohocken, PA, 1987, pp. 32–51; R. Rolff, "Significance or Ductility and New Methods of Measuring the Same," ibid, pp. 19–31.

136. I. Kim and R. Weil, "Tension Testing of Very Thin Electrodeposits," ibid, pp. 11–18; T. D. Dudderar and F. B. Koch, "Mechanical Property Measurements on Electrodeposited Metal Foils," *Properties of Electrodeposits, Their Measurement and Significance*, R. Sard, H. Leidheiser Jr., and F. Ogburn, eds., The Electrochemical Society, Pennington, NJ, 1975.

137. J. W. Dini, *Electrodeposition—The Materials Science of Coatings and Substrates*, Noyes Publ., Park Ridge, NJ, 1993, ch. 3, pp. 46–89.

138. R. Weil, "The Measurement of Internal Stress in Electrodeposits," *Properties of Electrodeposits, Their Measurement and Significance*, R. Sard, H. Leidheiser Jr., and F. Ogburn, eds., The Electrochemical Society, Pennington, NJ, 1975, p. 319.

139. B. Stein, "A Practical Guide to Understanding, Measuring and Controlling Stress in Electroformed Metals," *Proc. AESF Electroforming Symp., Mar. 27–29, 1996*, American

Electroplaters and Surface Finishers Society, Orlando, FL, 1996, p. 49; L. Borchert, "Investigation of Methods for the Measurement of Stress in Electrodeposits," *Proc. 50th Annual Conf., American Electroplaters Society*, Orlando, FL, 1963. Also references 21 and 137.

140. R. W. George et al., "Apparatus and Method for Controlling Plating Induced Stress in Electroforming and Electroplating Processes," U.S. Patent 4,648,944 (1987).

141. W. H. Cleghorn, K. S. A. Gnanasekaran, and D. J. Hall, "Measurement of Internal Stress in Electrodeposits by a Dilatometric Method," *Met. Finish. J.*, **18** (4), 92 (1972).

142. G. C. Cushnie, *Pollution Prevention and Control Technology for Plating Operations*, National Center for Manufacturing Sciences, Ann Arbor, MI (1994).

143. *Facts about Nickel and Health*, Inco Limited, Toronto, ON, 1996, pamphlet available on request. *Safe Use of Nickel in the Workplace-Health Guide* (100 pp.); *Safe Use of Nickel in the Workplace-Summary* (12 pp.); and *Safe Use of Nickel in the Workplace-Health Brochure* (7 pp.), available on request from Nickel Development Institute, Toronto, Ontario, Canada.

4 Electrodeposition of Gold

PAUL A. KOHL

INTRODUCTION

While all that glitters may not be gold, it is the most beautiful of all the elements in their pure form. Historically gold was one of the first metals known. Gold has been valuable throughout the ages chiefly because of its physical properties of softness, ductility, corrosion resistance, density, and scarcity. Gold cups and jewelry predating 3500 BC have been found in Iraq. The ancient Egyptians knew how to hammer gold into leaf as thin as 66 nm. The importance of gold as a form of currency was at a high point in the 1900s when most countries were on the gold standard. The bullion price of gold over the past 21 years is shown in Figure 1.

The world demand for gold in 1994 was approximately 89,000,000 troy ounces. Figure 2 shows the breakdown of the gold usage by industry. Jewelry comprises the largest fraction of the usage, however, much of that is not electroplated. For jewelry applications, gold is commonly alloyed with Group IB or IIB metals, particularly copper, silver, platinum, and palladium, principally to improve its strength and wear resistance.

The electrodeposition of gold is a relatively new process; it has been traced to the early work of Brugnatelli in 1805 [1]. The motives behind the use of electroplated gold changed dramatically in the mid-twentieth century when the emerging electronics industry required special-purpose electrical connections. The electronics industry consumed 5,330,000 troy ounces in 1994. Figure 3 shows the growth in the use of gold in the electronics industry by year. The use of electroplated gold in a variety of different functions in the electronics industry has lead to (1) many advances in our fundamental understanding of the electrodeposition process and (2) new electroplating technologies over the past 25 years.

Electrochemically deposited gold has satisfied many of the demands of the electronics industry. Gold has the third best electrical and thermal conductivity of all metals at room temperature. Also it has high ductility and excellent wear resistance, which are important for electrical contacts. The inertness of gold prevents the formation of insulating surface oxides (as compared to metals like aluminum). Group IIB metals (e.g., gold) are not good catalysts for other reactions, thus avoiding certain problems. For example, Group IB metals (particularly platinum and palladium) can catalyze the polymerization of organic molecules forming insulating layers. In addition, gold is an excellent metal for wire bonding integrated circuits

Modern Electroplating, Fourth Edition, Edited by Mordechay Schlesinger and Milan Paunovic.
ISBN 0-471-16824-6

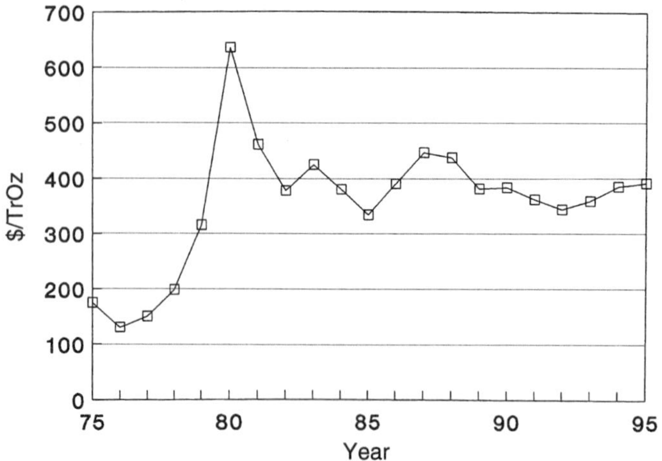

Figure 1 Gold bullion price of gold in U.S. dollars per troy ounce during the period from 1975 to 1995.

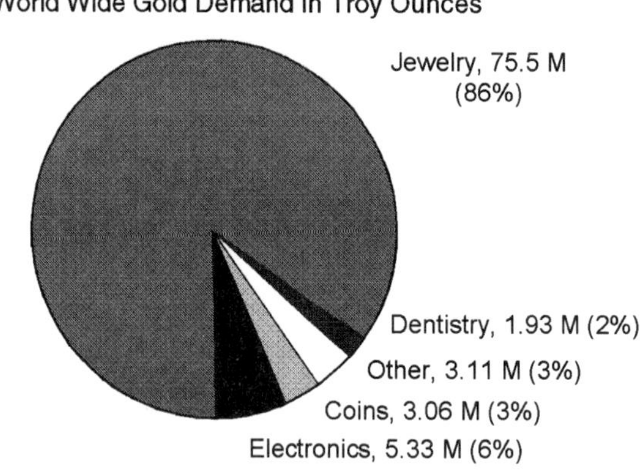

Figure 2 World gold demand in millions of troy ounces and percentage of market share by industry in 1994.

(ICs). Gold wires can be bonded to pure, soft gold pads by thermocompression bonding (300° to 400°C at high pressure to form a weld) or thermosonic bonding (150° to 200°C with ultrasonic energy to form a weld). Aluminum wires can be attached by ultrasonic bonding (ambient temperature with ultrasonic energy). These benefits have justified the high cost of gold in the packaging and interconnection of ICs.

A geographical breakdown of the use of gold in the electronics industry is also shown in Figure 3. Japan is largest consumer of gold for electronic applications, followed by North America, Western Europe, and the Pacific rim (excluding Japan). Although the use of gold for electronic interconnections has maintained steady growth over the past two decades, its use has not kept pace with the more rapid growth in the microelectronics industry, primarily because of gold's high cost. The microelectronics industry has increased speed, performance and packing density

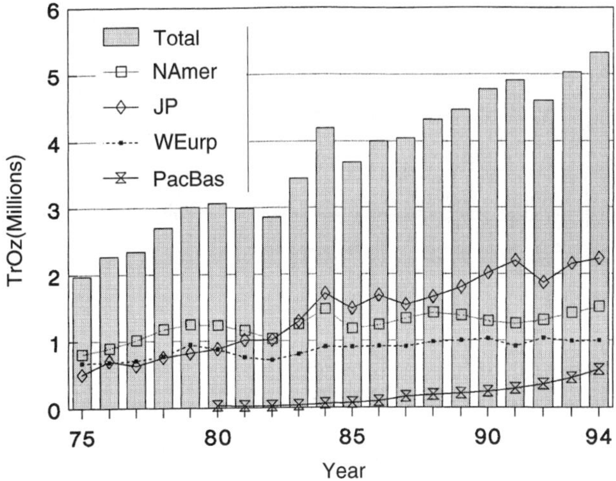

Figure 3 Gold demand in the electronics industry for Japan, North America, Western Europe, and the Pacific Basin (not Japan).

(number of transistors per unit area) while maintaining nearly a constant cost per unit area for ICs. This has been achieved in part by reducing the consumption of expensive materials (e.g., gold), by closely controlling the amount of material needed for specific functions (thickness and area), and finding less expensive replacements for some functions. Palladium and palladium alloys have replaced gold in some electrical contact applications without a significant drop in performance.

The worldwide demand for gold has been met by recycling existing gold supplies and through the mining of metal reserves, as shown in Figure 4. The leading producer of mined gold is the Republic of South Africa followed by the United States and Australia.

Gold exists primarily in the +1 and +3 oxidation states. The standard potential versus the normal hydrogen electrode (NHE) for a variety of gold (I) and (III) complexes is given in Table 1 [2, 3]. The most important ion for electrodeposition is

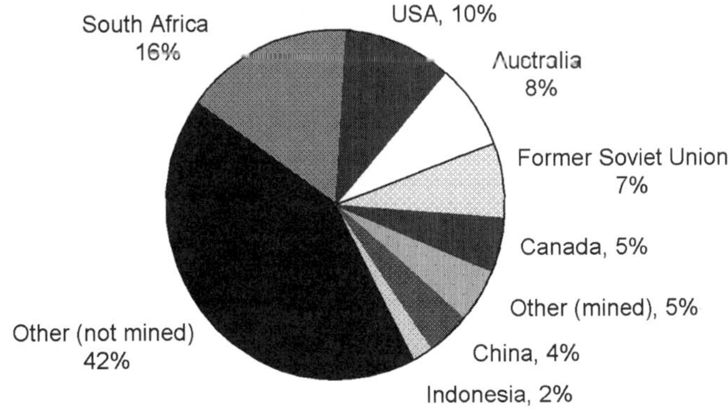

Figure 4 Gold supply in 1995 by percent. The total supply in 1995 was 98 million troy ounces.

$[Au(CN)_2]^-$. The stability of the gold (I) cyanide complex is reflected in the shift of the reduction potential for Au (I) from 1.71 V (aquo complex) to −0.611 V (cyanide complex). The stability constant for $[Au(CN)_2]^-$ is 10^{39} [4]. Two other gold complexes are of interest for electrodeposition: gold sulfite ($K = 10^{10}$) and gold thiosulfate ($K = 10^{28}$) [5]

$$Au^+ + 2CN^- \rightarrow [Au(CN)_2]^- \qquad K = 10^{39} \tag{1}$$

$$Au^+ + 2S_2O_3^{2-} \rightarrow [Au(S_2O_3)_2]^{3-} \qquad K = 10^{28} \tag{2}$$

$$Au^+ + 2SO_3^{2-} \rightarrow [Au(SO_3)_2]^{3-} \qquad K = 10^{10} \tag{3}$$

The potentials for the reduction of Au(III) to Au, and Au(III) to Au(I), as shown in Table 1, are important because Au(III) can also be used as the source of gold in baths in place of Au(I), and Au(III) can be formed at the anode during plating through the oxidation of Au(I).

In the next section, the chemical formulation of gold plating baths will be presented followed by comments on the mechanism of gold deposition and other issues concerning the deposition process.

1 TYPICAL DIRECT CURRENT (DC) PLATING BATHS

Numerous proprietary gold plating baths and additives are used industrially. In this section the composition and operating parameters of representative baths are

TABLE 1 Standard Reduction Potentials for Gold Ions (V vs. NHE)

Half-reactions	Potential
Au(I) ⟶ Au	
$Au^+ + e^- \longrightarrow Au$	1.71–1.85
$AuCl_2^- + e^- \longrightarrow Au + 2Cl^-$	1.15
$AuBr_2^- + e^- \longrightarrow Au + 2Br^-$	0.96
$AuI_2^- + e^- \longrightarrow Au + 2I^-$	0.58
$[Au(SCN)_2]^- + e^- \longrightarrow Au + 2SCN^-$	0.66
$[Au(S_2O_3)_2]^{3-} + e^- \longrightarrow Au + 2S_2O_3^{2-}$	0.15
$[Au(CN)_2]^- + e^- \longrightarrow Au + 2CN^-$	−0.61
Au(III) ⟶ Au	
$Au^{3+} + 3e^- \longrightarrow Au$	1.71–1.85
$AuCl_4^- + 3e^- \longrightarrow Au + 4Cl^-$	1.0
$AuBr_4^- + 3e^- \longrightarrow Au + 4Br^-$	0.85
$[AuI_4]^- + 3e^- \longrightarrow Au + 4I^-$	0.56
$[Au(SCN)_4]^- + 3e^- \longrightarrow Au + 4SCN^-$	0.64
Au(III) ⟶ Au(I)	
$Au^{3+} + 2e^- \longrightarrow Au^+$	1.40
$[AuCl_4]^- + 2e^- \longrightarrow [AuCl_2]^- + 2Cl^-$	0.92
$[AuBr_4]^- + 2e^- \longrightarrow [AuBr_2]^- + 2Br^-$	0.80
$[AuI_4]^- + 2e^- \longrightarrow [AuI_2]^- + 2I^-$	0.55
$[Au(SCN)_4]^- + 3e^- \longrightarrow [Au(SCN)_2]^- + 2SCN^-$	0.62

presented. The cyanide-based baths are divided into three classifications in Table 2: (1) the alkaline gold cyanide bath ($pH > 8.5$), (2) acidic, buffered baths (pH between 1.8 and 6), and (3) the neutral, buffered gold cyanide bath (pH between 6 and 8.5), The fourth group of baths, (4) noncyanide plating baths, are discussed after the three groups of cyanide baths. High-purity potassium gold cyanide, 68% metal (weight percent), can be obtained from commercial vendors in high purity.

Gold readily forms alloys with other metals during the deposition process or by diffusion with the gold substrate onto which it was plated. Gold deposits can be hard or soft, dull or bright, depending on the impurities and deposition conditions.

The potassium form of each salt is preferred over the sodium analogue because the solubility is higher. In cases where solubility is not a concern, sodium salts are sometimes used. Excessively high concentrations of gold in the baths are sometimes avoided due to the cost of the gold lost during drag-out of the salts.

The pH range between 8 and 10 is a critical region because the pK_a of hydrogen cyanide is 9.46. The equilibrium constant for HCN(aq) going to HCN(g) is $10^{1.4}$, making the pK_a for HCN(g) 8.06 [6].

$$HCN(g) \rightarrow H^+ + CN^- \qquad pK_a = 8.06 \tag{4}$$

Thus, at $pH > 10$, the equilibrium in Eq. (4) is shifted to the right, and free cyanide is stable in the bath. At $pH < 8$, the predominate form of cyanide is HCN(g), which evolves as a gaseous product. HCN gas is highly toxic, and its evolution from plating baths is a health concern. Appropriate ventilation is required for all cyanide baths. The presence of significant concentrations of free cyanide in the bath, such as in the case of alkaline baths, is also a health risk because accidental cyanide ingestion or injection can occur.

The presence of free cyanide in the bath is a major consideration in the selection of the anode and metal ion replenishment method. In the presence of free cyanide, gold metal can be used as a consumable anode because it can be electrochemically oxidized forming the gold cyanide complex. In the absence of free cyanide, gold is not oxidized to any appreciable extent, so it acts like an inert electrode. The ramifications of this will be included for each bath.

TABLE 2 Typical Cyanide-Based Plating Baths (concentrations in g liter^{-1})

Bath	Alkaline Cyanide		Buffered Citrate		Buffered Phosphate	
Operation	Rack	Barrel	Rack	Barrel	Rack	Barrel
$KAu(CN)_2$	12	6	20	20	20	20
KCN	20	30				
K_2HPO_4	20	30			40	40
KH_2PO_4					10	10
K_2CO_3	20	30				
K_2H citrate			50	50		
T(°C)	50–60	50–65	60–70	60–70	60–70	60–70
I(mA cm^{-2})	1–5	1–5	1–2	4–6	0.7–2	4–6
pH	11–11.5	11–11.5	4–5.8[a]	4–5.8[a]	6–8	6–8

[a]pH may drift during use. Citric acid or KOH can be used to adjust pH.

1.1 Alkaline Cyanide Baths

Alkaline cyanide baths operate at high pH and with an excess of free cyanide. The potassium gold cyanide concentrations used in the low current density baths shown in Table 2 range from 2 to 12 g $liter^{-1}$, with 6 g $liter^{-1}$ often used. Higher gold concentrations will support higher current densities.

Potassium cyanide is often used as the source of the free cyanide in these baths. Since there is appreciable free cyanide present, the cyanide released during the deposition reaction does not significantly alter the cyanide concentration in the bath. Thus Nernstian shifts in the reduction potential of gold (I) cyanide due to accumulation of free cyanide (even in the vicinity of the electrode surface) do not occur. In lower pH baths, large shifts in the free cyanide concentration at the electrode surface have many ramifications including a shift in the reduction potential during deposition. The excess cyanide permits the use of gold anodes for the replenishment of the metal plated from the bath.

$$\text{Cathode}: \quad [Au(CN)_2]^- + e^- \rightarrow Au + 2CN^- \tag{5}$$

$$\text{Anode}: \quad Au + 2CN^- \rightarrow [Au(CN)_2]^- + e^- \tag{6}$$

$$\text{Full cell}: \quad \text{No change to plating solution} \tag{7}$$

The main changes that occur in the bath are due to oxidation or reduction side-reactions, drag-out of plating salts, and drag-in of water or impurities. The free cyanide promotes the corrosion of the gold anode (forming $[Au(CN)_2]^-$), increases the throwing power, and improves the conductivity. Free cyanide also retards the codeposition of some metals because the stability of their cyanide-complexes shifts their reduction potential to values more negative than those used during deposition. The absorption of atmospheric carbon dioxide build up the concentration of carbonate in the solution as a result of the reaction between cyanide and CO_2. The carbonate can help to stabilize the pH and improve throwing power slightly.

Nonsoluble, dimensionally stable anodes, like platinized titanium or stainless steel, are sometimes used. Oxygen gas and hydrogen ions are the dominant products at the insoluble, dimensionally stable anodes, as shown in Eq. (9):

$$\text{Cathode}: \quad 4[Au(CN)_2]^- + 4e^- \rightarrow 4Au + 8CN^- \tag{8}$$

$$\text{Anode}: \quad 2H_2O \rightarrow 4H^+ + O_2 + 4e^- \tag{9}$$

$$\text{Full cell}: \quad 4[Au(CN)_2]^- + 2H_2O \rightarrow 4Au + 8CN^- + 4H^+ + O_2 \tag{10}$$

The hydrogen ions produced in Eq. (9) and shown in Eq. (10) for the full cell can either be neutralized by hydroxide (lowering the pH of the solution) or can form HCN(g) and evolve from the bath as a gas depending on the operating pH, as shown previously in Eq. (4). Thus the pH and the gold cyanide ion concentration of the bath in Eq. (10) will drop as current is passed, as shown by the two half-reactions. Both these effects must be reversed through the addition of plating salts and KOH. The addition of KOH and $KAu(CN)_2$ as a replenishment for the right-hand side of Eq. (10) will result in the accumulation of KCN (in the form of K^+ and CN^- until the KCN solubility product is reached) in the bath. The buildup of KCN is somewhat mitigated by drag-out. An insoluble anode is convenient because the cathode current distribution is reproducible and consistent over long periods of time. Also the anode does not have to be replaced as it corrodes. Thus, anode maintenance is replaced by bath maintenance (KOH and $KAu(CN)_2$ additions).

The cathodic current efficiency for the deposition of the metal from alkaline baths can be as high as 90% to 100% with adequate gold content in the bath and sufficient agitation [7, 8]. For example, 100% efficiency has been obtained at $10\,mA\,cm^{-2}$ and $12\,g\,liter^{-1}$ $KAu(CN)_2$ under mild agitation, whereas the current efficiency was only about 50% at $4\,g\,liter^{-1}$ $KAu(CN)_2$ (otherwise similar conditions) [7].

The preferred operating temperature of the baths is 50° to 60°F. The codeposition of impurities (e.g., antimony) will increase the hardness and brightness of the deposits [9]. The codeposition of silver, nickel, copper and cobalt has been used to alter the color of the deposit. This effect is used mostly for decorative purposes.

Alkaline formulations are generally not compatible with many polymers used in the microelectronics industry. The high-pH and high-cyanide concentration degrades photoresist and organic laminates. Thus there has been considerable work on developing neutral pH formulations. The alkaline, soft gold deposits are not acceptable for sliding electronic contacts in telephones where wear-resistance is critical. The parts will gall and fail quickly.

1.2 Acid Cyanide Baths

The nonalkaline, cyanide-based baths which contain other metals normally use a citrate buffer and operate around pH = 4. Low-pH buffered baths were originally developed for the jewelry trade but now find wide application throughout the electronics industry for contact surfaces, corrosion protection, bonding surfaces, and special electroforming. The pH of the baths allows the use of photoresist and other polymers. Typical formulations are given in Table 2. Anodes in these baths are usually platinized titanium, gold, or gold-plated platinized titanium. The gold-plated electrodes are found not to dissolve readily at low current densities because of the low free-cyanide concentration. This, however, is not the case at high current densities and low solution volume where appreciable transient concentrations of free cyanide can build up and lead to the oxidation and dissolution of a pure gold anode. Stainless steel and carbon anodes are found to introduce contamination, and their use is discouraged. The use of insoluble anodes requires that plating salts be added to the bath to replenish the gold content (e.g., $KAu(CN)_2$) and maintain pH control. Although K^+ removal and pH control can be accomplished by an ion exchange method, drag-out and salt addition are the usual method of bath maintenance [10].

The codeposition of other metals with gold is easily accomplished in acid cyanide baths, resulting in marked changes in the physical properties (especially hardness) of the deposits. Hard gold is of great importance to electronic components where low contact resistance, pore-free deposits, wear resistance, and chemical inertness are functional requirements. Such applications include the multiple-insertion electrical contacts found on printed wiring board contacts or spring contacts.

Bright, hard deposits can be produced from alkaline and acid cyanide baths containing cobalt, nickel, indium, silver, arsenic, or cadmium. However, the most reliable connector finish is from acid cyanide baths with nickel of cobalt salts as the brightener [11–13]. A typical range of concentrations is given below [14]:

$KAu(CN)_2$	12 to 15 g liter^{-1}
Citric acid	90 to 115 g liter^{-1}
Cobalt	0.07 to 0.1 g liter^{-1}
(added as acetate or sulfate)	
pH (adjust with KOH)	3.6 to 4.7
Temperature	40° to 65°C

To establish the proper pH, KOH is added to the bath. A concentration of 50 g liter^{-1} KOH will produce a bath of about pH = 4.0 [14]. Baths that produce pure gold deposits often operate at close to 100% current efficiency [10]. A yellow color that develops in the bath has been attributed to the formation of the $[Co(CN)_6]^{3-}$ complex [14]. The typical grain size of cobalt-hardened gold is 225 to 275 Å [16].

The high current density used in high-speed baths can produce a temporary buildup of free cyanide in the bath that may lead to the corrosion of a gold anode or even the slow oxidation of a platinum anode. It has also been found that $[Au(CN)_4]^-$ can be formed at the anode especially under high current and high cobalt concentrations at a platinum anode [17]. The buildup of Au(III) results in lower current efficiencies at the cathode for the production of gold metal because it takes three equivalents per mole to reduce $[Au(CN)_4]^-$ to Au whereas it only takes 1 equivalent per mole to reduce $[Au(CN)_2]^-$ to Au.

One of the limiting factors for the current density is the rise in pH in the vicinity of the electrode during the reduction process, either through gold deposition or hydrogen gas production. For example, at pH = 3.5 and 15 g liter^{-1} $KAu(CN)_2$, the current efficiency is 20% to 30% for current densities less than 50 mA cm^{-2} [13]. Hydrogen gas is the main side-product of the cathodic reaction. The reduction of the gold cyanide complex is inhibited by the cobalt ion as seen by the Tafel slopes (see Section 2 on deposition mechanisms), significantly dropping the current efficiently [13, 18]. The cobalt content of the pH = 3.5 bath is a complex function of current density and mass transport [13]. The typical range of the incorporated cobalt is around 0.05% of the gold at high current density (> 50 mA cm^{-2}) and low agitation (Reynolds number (Re) < 3000) to 0.6% of the gold at low current (25 mA cm^{-2}) and high agitation (Re > 13,000).

At higher pH or higher temperature, the cobalt and carbon content of the deposit decrease [19]. An increase in pH above 5.5 virtually eliminates the cobalt from the deposit, which demonstrates the ability of acid-cyanide baths to codeposit metals with gold. Lower current densities (smaller rate of consumption of protons at the cathode) or higher mass transport (higher rate of delivery of protons from the solution) will mitigate the rise in pH in the electrode boundary layer during deposition. Thus the higher the solution agitation, the greater is the allowable current density. The cobalt or nickel ions are known to codeposit in the gold in at least two different forms, the cyano-complex and as a substitutional alloy [20–22]. The increase in hardness is due to the grain refining effect of the codeposited cyano-complex [23].

The porosity of the deposited gold also decreases with higher mass transport and lower current density [13]. The lowest porosity cobalt-hardened gold deposits were obtained at 2 mA cm^{-2} and Reynolds numbers >9000.

Although cobalt salts are the most common hardening agents for acid, cyanide baths, the addition of nickel acetate (ca. 1 g liter^{-1} nickel) to the citrate bath also results in the hardening of the deposit and drop in the current efficiency [12]. The wear properties of high speed gold deposits were studied as a function of nickel or cobalt concentration, the form of the cobalt complex in solution (strongly chelating versus weakly chelating), and current density using commercial baths [24]. The stronger chelating agent was shown to prevent the formation of cyanide complexes. Strongly chelating nickel was found to have the widest operating window. Cobalt concentrations from 1.0 to 2.0 g liter^{-1} and current densities from 100 to 500 mA cm^{-2} were acceptable. As stated above, the weakly chelating cobalt complex required only 0.1 g liter^{-1}. Most deposits showing good wear characteristics were bright; however, all bright deposits were not wear resistant. Although specific ranges for cobalt or nickel were studied, no fundamental basis for the wear properties was established. Generally speaking, empirical relationships for the baths parameters are used to optimize wear properties.

The addition of iron salts to the plating bath will harden the gold deposit, just as cobalt and nickel do. However, iron-hardened gold is much more brittle than cobalt- or nickel-hardened gold, and unacceptable in many applications where ductility is critical, such as in electrical contacts. The accidental contamination of baths with iron is a concern whenever iron or steel parts are used in bath construction, piping, or pretreatment equipment.

Liljestrand et al. studied the effect of nickel concentration and other plating parameters on wear for nickel-hardened gold at 60°C, 16 g liter^{-1} gold concentration, and pH 3.8 to 4.8 [25]. The nickel concentration in the bath was varied and the resulting content in the deposit ranged from 0.2 to 2.5 wt%. In the higher current density range from 10 to 40 mA cm^{-2}, the wear changed from severe to mild and then to brittle upon increasing the nickel content. However, in the current density range from 1 to 16 mA cm^{-2}, the wear changed from severe to "adhesive brittle" and then to mild with increasing nickel content. The terms "severe," "adhesive brittle," and "mild" refer to the physical conditions of the surface. The severe wear at low nickel content was characterized by cold welding between the spring and pin used in the test. During the first wear cycles, the gold was smeared from one part to the other followed by early failure (less than 200 wear cycles). At high nickel content, plated at low current density, brittle wear occurred characterized by cracks reaching down to the nickel underplate. This clearly shows the need for tight control over bath and plating parameters with metal-hardened gold. X-ray diffraction was used to show an increase in the lattice contraction strain with increasing nickel content. Since the solubility of nickel in gold is about 5%, it was concluded that most of the nickel is in a solid solution and that the hardness is due to lattice strain caused by the soluted nickel atoms.

The codeposition of an insulating film from cobalt- or nickel-hardened acid-gold baths has caused significant problems for electrical contacts. High electrical resistivity can be observed upon thermal aging (e.g., 150°C) especially in films plated at high speeds with high cobalt concentration [26]. The subject has received considerable attention since the late 1960s when high contact resistance on telephone spring contacts was correlated with an orange-brown film on gold-plated contacts. Munier was the first to isolate the transparent film on the cobalt-hardened gold and presumed it to be an organic polymer [27]. The indiscriminate use of the word "polymer" to refer to either the electrodeposited inclusion or the extract from the plated metal is not correct. Nevertheless, most of the literature published on the identification and elimination of the material refers to it as an insulating codeposited polymer.

Microchemical determination of the carbon content of the gold deposited from cobalt- or nickel-hardened baths based on citrate buffered gold cyanide provides a clear link between a carbonaceous impurity and the insulating film, as shown in Table 3 [27]. In the absence of cobalt or nickel, or at high pH (i.e., $pH > 10$ using a phosphate buffer and potassium gold cyanide) in the presence of cobalt salts which are not codeposited, no carbon is detected in the deposit.

Carbon-14 labeled complexes were used to quantity the carbon content in the deposits at levels in excess of 1% [28]. Electron microscopy studies showed the deposits to exist in discrete pockets usually 1000 Å and smaller in size with a few as large as 25,000 Å [29]. A study of gold deposited from pure potassium gold cyanide as a function of temperature, of current density, and with anode isolation (in a separate compartment) led to the conclusions that (1) the carbonaceous contamination was minimized at temperatures above 65°C, (2) deposits plated at lower temperature (ca. 25°C) had more carbon and were much harder and brittle, and (3) the carbon was the result of a cathodic process only, and not the re-reduction of an oxidized species released from the anode [30]. However, it led the authors to the inaccurate speculation of the origin of the carbon as potassium cyanide, gold (I) cyanide, potassium cyanoaurate, or hydrogen cyanide. Mossbauer spectroscopy was later used to show that no AuCN, $KAu(CN)_2$, or $KAu(CN)_4$ were in the deposit [31].

A detailed study of the gold appearance, and carbon and cobalt content as a function of pH clearly showed the drop in codeposited cobalt and carbon as the pH was increased to $pH = 6$ [19]. The slow change in the cobalt-containing hard gold solution from reddish-pink, the color of Co^{2+} citrate, to yellow, the color of $[Co(CN)_6]^{3-}$ was noted. This occurred as free cyanide was released from the reduction of $[Au(CN)_2]^-$.

A study by Okinaka et al. [14] clarified many issues. It was shown that the carbon-containing deposits obtained from dissolving cobalt-hardened gold in aqua regia were not an organic material but the cobalt complex: $Co_3{}^{II}[Co^{III}(CN)_6]_2 \cdot xH_2O$. When dissolved in mercury, $K_3Co^{III}(CN)_6$ was found. It was concluded that there is an additional source of both cobalt and carbon in the gold deposit apart from the

TABLE 3 Determination of Carbon in Gold Deposits

Gold Bath		Hardener	Carbon in Gold(%)
Potassium gold cyanide	13 g liter^{-1}	None	0.01–0.04
Potassium hydroxide	15 g liter^{-1}	Cobalt	0.20–0.30
Citric acid	90 g liter^{-1}	Nickel	0.19–0.43
Cobalt or Nickel citrate	0.3 g liter^{-1}		
pH	3.6–4.5		

cobalt cyanide complex. Substitutional metallic cobalt is the additional source of cobalt.

1.3 Neutral Cyanide Baths

Nonalkaline baths exhibit a wider range of physical and chemical properties than possible at higher pH. In the previous section on acid baths, it was related that the codeposition of other metals from the bath provided metallurgical hardening of the deposit. The hardness is the result of a reduction in the grain size of the gold caused by the codeposit increasing the rate of grain nucleation. It was also noted that the amount of codeposited metal (e.g., cobalt or nickel) decreased to nearly zero at $pH > 5.5$. Although the metals used to harden gold are not readily codeposited from neutral pH baths, it has been found that hard gold (as well as soft gold) can be formed in neutral baths by the careful selection of the plating parameters without the use of additives. In the neutral baths the gold salt used is the same as in alkaline and acid baths, $KAu(CN)_2$. The neutral pH and absence of free cyanide make these formulations the preferred ones for use with photoresist and other polymeric materials.

The neutral baths shown in Table 2 produce pure, soft gold at relatively low current densities, typically 2 to 5 mA cm^{-2} and 60° to 70°C. The phosphate salts serve only as the supporting electrolyte and pH control, so that strict control of their concentration is not critical. The phosphate buffer can be used for acid, neutral or basic conditions with pH adjustment carried out by the addition of KOH or H_3PO_4. Control of pH, temperature and gold concentration is critical.

Additive-free hard gold (AFHG) can be produced from neutral baths using phosphate as the buffer, as shown in Table 4 [32]. Although the "low-speed" column is under acid conditions, it is included here for completeness because of its similarities with the high-speed AFHG bath. The advantages of AFHG baths over metal hardened baths are (1) no need to control additive concentration, (2) high current efficiency, (3) greater tolerance for heavy metal impurities due to the insolubility of many metal phosphates, and (4) high ductility and thermally stable contact resistance for electronic components. AFGH plating was first observed in 1974 [33]. It was shown that the grain size of the deposits from the AFGH bath was 250–750 Å, which is slightly larger than those from the cobalt hardened bath (225–275 Å) but much smaller than the grains from soft gold (1 to 2 μm) [16]. AuCN is codeposited in AFHG and accounts for about 50% to 70% of the carbon content in the deposit. The deposited AuCN forms a polymeric structure and hardens the gold by

TABLE 4 Bath Composition and Plating Conditions for Additive-Free, Neutral Gold Plating Bath

	Low Speed	High Speed
$KAu(CN)_2$	40 g $liter^{-1}$	44–59 g $liter^{-1}$
KH_2PO_4	100 g $liter^{-1}$	100 g $liter^{-1}$
pH (KOH of H_3PO_4)	4.3–4.5	6.5–7.5
Temperature	25 ± 2°C	40 ± 2°C
Agitation	Mild	Vigorous
Current density	10–20 mA cm^{-2}	200–300 mA cm^{-2}

promoting the nucleation of grains, or by inhibiting the growth of the existing gold grains, in a role analogous to cobalt hardening of acid gold baths mentioned in the previous section [34]. Thus the conventional bath, Table 2, that produces soft gold at 70°C will produce hard gold at 25°C. The population density of the AuCN in AFHG gold plated from a bath containing $KAu(CN)_2 = 44\,g\,liter^{-1}$, $KH_2PO_4 = 100\,g\,liter^{-1}$, $KOH = 28\,g\,liter^{-1}$, and pH = 7.0, was $10^{17}\,cm^{-3}$ with a volume ratio of 1.3% to 1.9% [34].

Data on the plating efficiency and deposit appearance of an additive-free, neutral bath is shown in Table 5 [32]. At low current density, the hardness was found to be independent of temperature in the range of 10° to 45°C (ca. 160 KHN_{25}), Whereas the hardness decreased to 104 KHN_{25} at 70°C. At higher plating speeds, 86–252 mA cm^{-2} and 30–43°C, the hardness was about 180 KHN_{25}. Wear tests using lubricated sliding contacts showed severe galling for the soft gold plated at 70°C (104 KHN_{25}), whereas the harder deposits formed at lower temperatures (22–45°C) showed excellent wear. The AFHG bath was shown to have superior tensile strength, total elongation (high ductility), and more stable contact resistance than comparable deposits from a cobalt-hardened acid bath. However, a small increase in contact resistance was found when the deposit was heated to > 100°C due to the surface segregation of AuCN [34].

The effect of gold cyanide concentration on current efficiency and the appearance of the finish for the high-speed bath are shown in Table 5 [32, 35]. The conditions were the same as the "high-speed" portion of Table 4 except for the gold cyanide concentration. At $KAu(CN)_2$ concentrations below 23 g $liter^{-1}$, a matte finish was obtained with cathodic current efficiencies of 80%–85%. At $KAu(CN)_2$ concentrations from 23–44 g $liter^{-1}$, a bright finish was obtained and the current efficiency was 90% to 95%.

Certain metallic additives can be used as a grain refiner in the baths, such as thallium or arsenic, to increase the brightness/smoothness range of the deposit. The military specification for bondable gold, MIL-G-45204B, requires that chromium,

TABLE 5 Appearance and Plating Efficiency of High-Speed Additive-Free Hard Gold

Temperature (°C)	Current Density (mA cm^{-2})	Appearance	Plating Efficiency (%)	Hardness
KHN_{25}				
20	121	0	78	
25	198	0	72	
25	155	2	81	
25	138	3	81	
25	121	4	86	
30	155	2	83	193
30	138	4	85	
43	259	2	87	
43	172	5	85	188
61	172	0	91	
61	86	0	82	

Note: Visual appearance defined as 0 = burned appearance and 5 = completely bright.

copper, tin, lead, silver, cadmium, or zinc not be present in the deposit at a concentration greater than 0.1%.

Anodes in these baths are normally platinized titanium, gold, or gold-plated platinized titanium. Stainless steel and carbon anodes are not recommended because of contamination. Gold or gold-plated anodes do not dissolve readily because the concentration of free cyanide is very low. However, at high current density and small solution volumes (or close anode-to-cathode spacings), the transient concentration of free cyanide can become significant and promote the oxidation and dissolution of a gold anode. High current density baths, especially those with high gold cyanide concentrations, are susceptible to the formation and buildup of Au(III) as $[Au(CN)_4]^-$ [17]. The presence of $[Au(CN)_4]^-$ will result in a lower apparent current efficiency at the cathode because it requires three equivalents per mole to deposit Au from Au(III) as opposed to one equivalent for Au(I). This is different from the acid gold plating case where hydrogen gas was the side-product responsible for lowering the plating efficiency. This creates a process control problem. As the Au(III) concentration builds up in the bath, the plating time, or current (number of coulombs cm^{-2}) will have to be increased to achieve the same deposit thickness. Figure 5 shows the current efficiency and percent Au(III) in the bath as a function of weight of gold plated or number of bath turnovers for three different anodes [17]. The cathodic current density was 230–300 mA cm^{-2}, the anodic current density was 150–200 mA cm^{-2}, and the bath was the same as in Table 4 (high-speed bath). The gold anode did not result in a significant build up of Au(III) because it was oxidized resulting in dissolution caused by the high transient concentration of free cyanide. The use of a gold anode is usually an unacceptable solution to the Au(III) problem because of the anode maintenance and changing cathodic current distribution as the anode corrodes and changes shape. The buildup of Au(III) is somewhat mitigated by the dilution effects due to drag-out. Other possible solutions to the Au(III) problem are to use a RuO_2/TiO_2 (often called dimensionally stable anode, DSA), adjust the current density to correct for the lower current efficiency, or chemically treat the Au(III) by reduction (e.g., hydrazine) or removal (activated carbon) [17].

1.4 Noncyanide Baths

The gold (I) cyanide complex is by far the most important for electrodeposition. However, there are several shortcomings to the use of $[Au(CN)_2]^-$ which have stimulated the investigation and commercialization of other gold complexes in plating baths [36]. The stability of the gold cyanide complex causes the reduction potential to occur at very negative potentials resulting in the co-reduction of hydrogen ions, which lowers the plating efficiency and makes the development of electroless plating baths difficult [37]. The release of free cyanide during the reduction of $[Au(CN)_2]^-$ can be incompatible with positive photoresists used in the microelectronics industry [38, 39]. It has also been found that the residual stress of the plated gold can be controlled in noncyanide baths [40, 41]. Low-stress deposits are of particular interest for X-ray lithography masks because they must be thin, flat, and made from high atomic number elements. Last, the health and safety of workers and the environmental impact of the wide scale use of cyanide is a concern.

The operating parameters of commercial gold sulfite baths have been presented with special attention given to the effect of plating temperature and current density

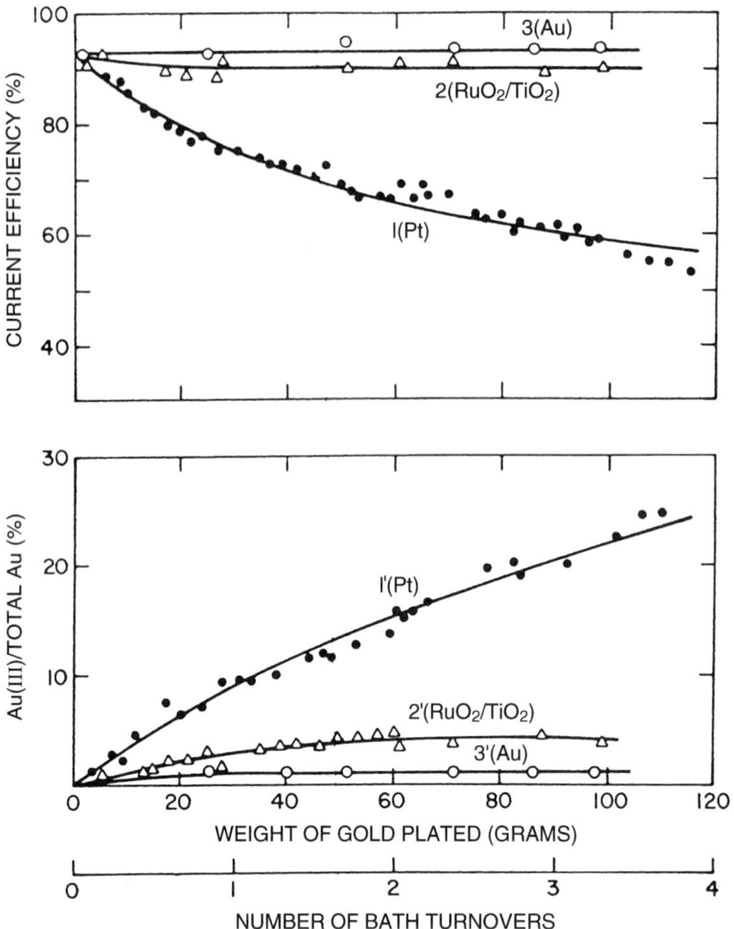

Figure 5 Variation in plating efficiency and Au(III) content with bath age (AFHG bath). (Curves 1, 1′: Pt anode; curves 2, 2′: RuO_2/TiO_2 anode, curves 3, 3′: Au anode.) (From Okinaka and Wolowodiuk [17])

on residual stress in the deposited metal [40, 41]. For example, Figures 6 shows the stress change from compressive to tensile as the temperature or current are raised. Thallium was used as the stress reducing agent at concentrations up to 80 ppm, a value that can be more easily maintained than the low thallium concentration baths [40]. Alkaline sulfite baths can also be used without the use of thallium, which itself is a health hazard [41].

Although gold (I) sulfite has been used in commercial baths, the complex is susceptible to disproportionation forming Au(III) and metallic gold. This spontaneous decomposition of the bath has lead commercial baths to use proprietary stabilizing additives. Gold (I) sulfite ($pK_a = 10^{10}$) is particularly troublesome at pH < 7 where sulfite protonates forming bisulfite [42–45]. Gold (I) thiosulfate has a stability constant ($pK_a = 10^{28}$) which is between that of gold (I) sulfite and gold (I) cyanide ($pK_a = 10^{39}$). The gold (I) thiosulfate complex is stable in weakly acidic solutions because of the low pK_a for thiosulfate:

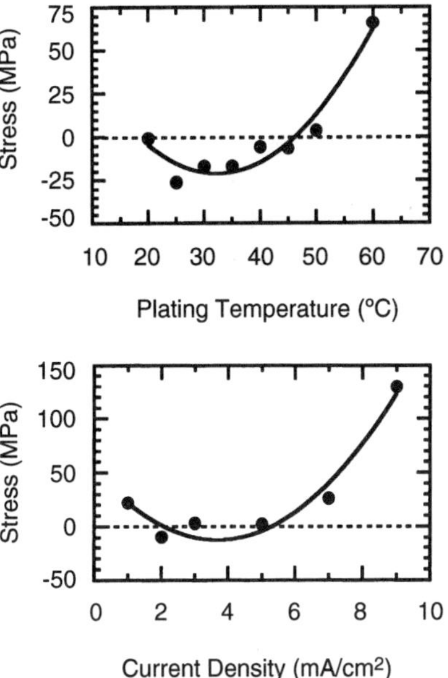

Figure 6 (*Top*) Stress versus plating temperature for samples plated at 3 mA cm^{-2} form a bath containing 75 ppm thallium. (*Bottom*) Stress versus current denisty for samples plated at 50°C from a bath containing 75 ppm thallium. (From Dauksher et al. [40])

$$H_2S_2O_3 \rightarrow H^+ + HS_2O_3^- \qquad pK_{a1} = 0.3 \tag{11}$$

$$HS_2O_3^- \rightarrow H^+ + S_2O_3^{2-} \qquad pK_{a2} = 1.7 \tag{12}$$

The reduction of gold (I) thiosulfate was shown to have an overall reaction producing thiosulfate and gold [46]:

$$[Au(S_2O_3)_2]^{3-} + e^- \rightarrow Au + 2S_2O_3^{2-} \tag{13}$$

The standard heterogeneous rate constant for the reduction was found to be 1.6×10^{-3} cm s^{-1}, a transfer coefficient α, of 0.23, and a diffusion coefficient of 7×10^{-6} cm^2 s^{-1} [37]. The effect of mixed thiosulfate-sulfite electrolytes has been studied and shown to produce soft gold deposits suitable for forming gold "bumps" on integrated circuits and in electronic interconnections [38]. The Vickers hardness of the electrodeposited gold was 80 kg mm^{-2} in the as-deposited state and 50 kg mm^{-2} after annealing. The optimum bath with thallium was composed of

$NaAuCl_4 \cdot 2H_2O$	0.06 M
Na_2SO_3	0.42 M
$Na_2S_2O_3 \cdot 5H_2O$	0.42 M
Na_2HPO_4	0.03 M
Tl^+ (added as Tl_2SO_4)	0.03 M
pH	6

Temperature	60°C
Current density	$5\,\mathrm{mA\,cm^{-2}}$

The deposition of gold from the iodide-thiosulfate bath was also studied [39]. It was suggested that the chemical reaction preceding the electron transfer step is the dissociation of the $[Au(S_2O_3)_2]^{3-}$ to form $[Au(S_2O_3)_2]^-$ and $S_2O_3^{2-}$.

In each of the studies involving a mixed thiosulfate bath (sulfite-thiosulfate and iodide-thiosulfate), it was observed that the mixed salt gold complex is more stable and harder to reduce than either of the single-salt complexes [36–39]. That is, the gold (I) sulfite-thiosulfate complex is reduced at potentials more negative than either the gold (I) sulfite or gold (I) thiosulfate complex.

2 MECHANISM OF DEPOSITION

Electrodeposition from the $[Au(CN)_2]^-$ complex is by far the most common, and well studied. Two distinct mechanisms for gold deposition have been proposed [47–49]. The cathodic deposition of gold under activation control can be described by Butler-Volmer kinetics. At lower currents or less negative potentials the first step is the chemical adsorption of the $[Au(CN)_2]^-$ complex

$$[Au(CN)_2]^- \rightarrow (AuCN)_{ad} + CN^- \tag{14}$$

The electron transfer step is the rate limiting one:

$$(AuCN)_{ad} + e \rightarrow (Au^\circ\,CN)^-_{ad} \tag{15}$$

$$(Au^\circ\,CN)^-_{ad} \Leftrightarrow Au_{lattice} + CN^- \tag{16}$$

The chemical desorption and crystallization is the final step, as shown in Eq. (16) [48, 49]. The equilibria depend on the crystal orientation of the substrate and presence of impurities such as nickel or cobalt that can effect the nucleation rate and hardness of the gold. The Tafel slope at pH = 4.4 for additive-free soft gold and nickel-hardened gold were found to be $-0.35\,\mathrm{V\,decade^{-1}}$ and $-0.47\,\mathrm{V\,decade^{-1}}$, respectively, as determined by chronopotentiometry and chronoamperometry [47]. The Tafel slope found for soft gold by linear sweep voltammetry was slightly smaller, $-0.39\,\mathrm{V\,decade^{-1}}$. The smaller Tafel slope for nickel-hardened gold shows the strong inhibiting influence of nickel ions on the deposition process.

At more negative potentials, a second Tafel slope was obtained indicating a different mechanism for deposition, where the direct electroreduction of the gold cyanide ion takes place. Steeper Tafel slopes (more negative slopes) at higher currents were observed for soft gold ($-0.154\,\mathrm{V\,decade^{-1}}$) and nickel-hardened gold ($-0.186\,\mathrm{V\,decade^{-1}}$), as measured by chronopotentiometry and chronoamperometry. The two Tafel slopes were not observed by linear sweep voltammetry because nucleation begins gradually at less negative potentials and then the rate of crystal growth and nucleation increases, whereas the rates of nucleation and crystal growth are initially high for constant potential or constant current methods.

In weakly acidic oxalate-phosphate baths, two Tafel regions were also found (-0.315 and $-0.110\,V\,decade^{-1}$) for soft gold. Only one Tafel region was reported for hard gold ($-0.200\,V\,decade^{-1}$); however, the nickel concentration was two orders of magnitude lower than that used in [47]. The linear sweep voltammograms show a strange dependence on the mass transport rate of the solution. In addition to the expected supply of reactants to the electrode surface, CN^- or HCN produced at the electrode surface can inhibit the reactions. Higher mass transport rates will reduce the concentration of these inhibitors at the surface.

The structure of electrodeposited soft gold is referred to as columnar because etched cross sections show long crystallites whose length axis coincides with the growth direction of the deposit. The nucleation rate of soft gold is low, and the tendency toward preferred growth on existing grains is high. In contrast, hard gold deposits have a high nucleation rate and submicrometer grain size, as noted earlier. Thus the principal difference between hard and soft gold is the grain size which is derived from the rates of nucleation. That is, the effect of both impurities, current density, and other plating parameters on the mechanical properties of the deposit can be understood in terms of the nucleation and growth rate.

The surface roughness of gold deposits from a commercial gold cyanide solution was examined by scanning tunneling microscopy using smooth (111) gold surfaces [51]. It was found that the surface of the gold was roughened by simple immersion into the plating bath, an effect of the exchange current (the zero applied current) condition. The result of 100 Å of electrodeposited gold was "rolling hills" of gold 200 Å wide and 30 Å high.

3 PULSE PLATING

Pulse plating usually yields finer grain deposits than DC baths because of the high pulse current density and resulting high nucleation rate. Low porosity has been observed with pulse-plated gold due to grain refinement [52]. The growth of gold on crystalline copper is initially dependent on the structure of the substrate under pulse or DC plating. When the deposit reaches 50 to 100 nm in thickness, the growth gradually changes from substrate controlled to transport controlled [53]. In studies on orientated copper substrates, single-crystal growth was observed for [001] and [011] gold grains. The nuclei forming these grains coalesce when they come in contact with each other, which leads to single-crystal grain growth. For larger nuclei, the growing grains obtain equal orientation. These observations for pulse-plated gold are probably due to desorption of impurities during the relaxation time between successive pulses. Coalescence is more likely to occur if the growing surfaces are clean; if not, impurities are pushed in from of the growing surface and prevent adjacent grains from coalescing, resulting in the formation of a grain boundary. Although the nucleation rate is high in pulse plating, the grain refinement is due to the different surface diffusion conditions. During pulse plating, the diffusion paths are blocked due to a high adion concentration resulting in decreased surface mobility, and the structure becomes columnar. In DC plating, the adion mobility is high and a crystalline structure is developed [53]. The ductility, tensile strength, and yield strength were found to be improved in alternating current electrolysis; however, higher contact resistance was also reported [52]. The current distribution for selective pulse plating was quantitatively described as a function of peak current and anode to cathode ratio [54]. The current distribution is generally less uniform at small Wagner numbers. One problem common to all pulse plating strategies is the availability and cost of power supplies. The formation of high current pulses is expensive, and the rarity of users limits the benefits of mass production.

4 SUBSTRATE PREPARATION

Gold plating is often the last step in long series of processing steps, making it immediately suspect when difficulties are encountered. However, in many cases the problems that appear at the gold plating step are due to the cleaning and preparation processes preceding gold deposition. Poor, incomplete, or inappropriate cleaning can result in inadequate gold deposits. While problems do occur in the gold plating process, only a good engineering evaluation of the defective deposits can elucidate the failure mechanism.

4.1 Cleaning of Plating Base

Copper, brass, and silver parts are easily plated once cleaned. The normal practice may require degreasing in an alkaline spray or soak, or by a solvent vapor degreaser. A copper part may require a bright dip (10% sulfuric acid in water) to remove the native oxide. Beryllium copper spring material or castings should be carefully cleaned by a 15 minutes soak in boiling NaCN and NaOH solution (38 g NaCN, 38 g NaOH, and 1 liter H_2O) followed by a water rinse and bright dip and final rinse. Other procedures use a dip in 60% by volume H_2SO_4 at 50°C for 30 seconds followed by a water rinse.

Nickel parts must be activated before gold plating or the plate may peel. Activation can be accomplished by a dip in warm 1:1 HCl in water. It is very important not to allow the nickel parts to dry after activation before gold plating. Nickel alloys like Kovar can be cleaned by soaking for many minutes in hot 18% HCl. The oxide-free surface is prepared for plating by bright dipping for several seconds in 1:3 nitric acid: acetic acid with 15 mL HCl per liter of solution at 70°C. Kovar can be first plated with Woods or Watts nickel before gold plating or to gold strike the nickel alloy directly at high current in a low gold content bath followed by normal gold plating.

Gold plating onto aluminum and aluminum alloys has become more common, particularly for electronics and space applications. The surface is first degreased, such as in trichloroethylene vapors, followed by alkaline cleaning in trisodium orthophosphate (Na_3PO_4 : 12 g liter^{-1}, H_2O : 25 g liter^{-1}, and Na_2CO_3 : 20 g liter^{-1}) at 60°C for 2 to 3 minutes. After water rinsing, the aluminum surface is cleaned in a mixed acid solution (sulfuric acid: 10 mL/L^{-1}, hydrofluoric acid (40%): 12.5 mL/L, and nitric acid: 25 mL/L) for 2 to 3 minutes. A zinc displacement deposition process can precede the electroless nickel underlayer (nickel sulfate: 30 g liter^{-1}, zinc sulfate: 40 g liter^{-1}, sodium hydroxide: 106 g liter^{-1}, potassium cyanide: 10 g liter^{-1}, potassium bitartarate: 40 g liter^{-1}, copper sulfate: 5 g liter^{-1}, and ferric chloride 2 g liter^{-1}) for 1 minute. The first zinc layer is stripped in 50% nitric acid, followed by a second zinc layer by immersion for 30 seconds [55]. Extreme caution should be practiced when handling hydrofluoric acid and potassium cyanide, especially when the cyanide is in the same vicinity as acids.

4.2 Barrier Metals

Service requirements may need an underplate between the base metal and the gold electroplate. Specifications for gold-plated copper parts exposed to temperatures $>60°C$ often require an intervening nickel-plated barrier to prevent copper-gold interdiffusion and subsequent copper bleed through to the surface. The barrier may be 5 to 10 μm of low-stress nickel, like that plated from a sulfamate bath. Gold plating on zinc-based alloys should have at least 8 μm of nickel over 8 μm of copper. Tin or lead should have 8 μm of nickel over 2 μm of copper.

4.3 Gold Strike

The use of a gold strike is common before the deposition of a thick gold plate. The thickness of a gold strike may be 0.02 to 0.25 μm, depending on the thickness and requirements of the full gold thickness. A gold strike bath is usually a dilute version of a normal gold plating bath. A high current density is used to overcome activation control of the plating process. A large fraction of the gold cyanide ions at the electrode surface are reduced to gold at a high nucleation rate. Thus a uniform deposit is formed over the electrode surface compensating for inhomogeneities across it, which may have created problems under normal gold plating conditions. The gold strike surface has increased adhesion, decreased porosity, and less contamination is dragged into the main plating bath. A typical strike bath would have 0.8 g liter^{-1} gold versus 8 g liter^{-1}, and current densities two to ten times higher than the conventional bath.

4.4 High-Speed Strip Plating

There has been extensive use of high-speed plating, and selective plating in the semiconductor industry since around 1970. In particular, reducing cost and materials in the reel-to-reel electroplating of electrical connectors has led to many innovations in high-speed plating. Reel-to-reel plating of connectors includes facilities for cleaning and preparation of parts, strike plating, and transport of parts and chemicals. The sequence of the processes is similar to those in batch plating; however, size, space and time limitations have stimulated a fine-tuning of the processes, which make their review worthwhile. The strip plating process for electrical connectors is composed of the following processes: (1) ultrasonic clean, (2) electropolish, (3) acid dip, (4) nickel plate, (5) gold strike, (6) hard gold plate, (7) hot rinse, and (8) dry [56]. Cleaning is limited to nonfoaming solutions because of the need for rapid recirculation of the liquid from a reservoir tank. Ultrasonic agitation speeds up the cleaning process by dislodging the contaminants from the surface through cavitation. Electropolishing is an anodic process where metal dissolves (cleaning and deburring) in proportion to the anode current density. Copper alloys used for electrical connectors electropolish well in 1:1 phosphoric acid in water. Current densities as high as 4000 mA cm^{-2} can be used to accommodate high-speed processing with small electroplating cells. A major problem in electropolishing is the deposition of copper powder at the cathode, which must be removed periodically to avoid cell shorting. While it is desirable to have the electrodes in close proximity to reduce solution heating and voltage, wider spacing (e.g., 1.5 cm) is necessary to mitigate cell shorting from powdery copper loosely adherent to the cathode. The copper phosphate residue can be removed by a hot water rinse followed by a rinse in 10% sulfuric acid (by volume) at 60°C. The nickel underplate (ca. 0.14 μm) can be

	Gold Strike	High-Speed Bath
$KAu(CN)_2$	4 g $liter^{-1}$	44 g $liter^{-1}$
K_2HPO_4	120 g $liter^{-1}$	—
KH_2PO_4	30 g $liter^{-1}$	100 g $liter^{-1}$
KOH	—	28 g $liter^{-1}$
Temperature	65°C	40°C
pH	7.0	7.0

from a nickel sulfate or nickel sulfamate bath. The additive free hard gold (AFHG) baths used in reel-to-reel gold strike are shown below:

The current density for the gold strike is 30 mA cm^{-2}. The hardness of the final deposit is achieved by producing small grains, which is the result of high current density at low temperature. The hardness of the gold plated from the high-speed bath given above is 180 KHN_{25} for current densities between 150 and 250 mA cm^{-2}. The hardness drops off at lower current densities, to 150 KHN_{25} for 100 mA cm^{-2}, and 125 KHN_{25} at 50 mA cm^{-2} [56]. Others have used even higher current densities, 800 to 2000 mA cm^{-2}, although the conditions are different [57]. A final hot rinse and dry complete the strip plating process. The plating process can be accomplished in relatively short length plating cells (ca. 18 inches long) with insoluble anodes that allow precise placement of the hardware in an attempt to selectively plate the gold only on the functional area of the connectors. Although a majority of the gold can be deposited on the contact area of connectors, it is extremely difficult to prevent some plating on nonfunctional areas immersed in the plating bath. Laser-assisted plating has been shown to assist in the selective plating process. Laser processes have been developed to improve the selectivity of the deposition process. The direct irradiation by a pulsed laser has been shown to cause extremely rapid, surface heating increasing the agitation and deposition rate. It also increases the interdiffusion of the deposited atoms [58, 59]. Lasers have also been used in a patterning technique where the whole component was coated with an insulator, such as glycol phthalate, paraffin wax, or photoresist, and a YAG laser was used to burn off the material in small areas which were then plated [59]. One problem is the precise alignment of the laser to the part either in a moving or a step-and-repeat process.

5 STAINS

Staining of precious metal plated surfaces is a critical concern, especially since chloro-fluoro-carbon (CFC) usage is discouraged. The CFC displacement method did not dry the plated metal surface, it displaced the water from the plated surface. The plating stains are often caused by potassium cyanide or other plating salts precipitating on the surface as the water evaporates. Stains are often seen as irregularly shaped "hollow" blemishes where the stained area is smaller than the original water drop size, and the center of the stain is not very discolored. Droplets form on the wetted areas as the plated parts are removed from the rinse water. The rinsing water simply dilutes the plating salts, leaving a low concentration of salt. As the rinse water dries from the surface, the volume of the original water droplet is reduced through evaporation. The concentration of the salt in the droplet increases as the volume decreases. When the concentration of the salts exceeds their solubility, they precipitate onto the surface. The "hollow" center is formed because the salts continue to precipitate onto the existing deposit. The oxidation of the gold surface is not an issue as it would be with less noble metals. There are many things which can be done to reduce the staining of electroplated gold parts: (1) use high-quality rinse water, (2) countercurrent rinsing techniques are desirable, (3) removing the final rinse water from the surface by CFC displacement, (4) removal of rinse water by hot air knives, centrifugal dryers, or alcohol absorption. Isopropyl alcohol is often preferred, especially over methanol because the rate of evaporation is slower. Rapid evaporation can result in cooling of the surface below the ambient dew point resulting in condensation of water and nucleation of spots.

6 TEST METHODS

The physical properties of significance for the gold deposits are appearance, wear resistance, hardness, brittleness, residual stress, porosity, thickness, density, and deposit uniformity. The chemical properties of importance include the chemical composition of the deposits and solutions, segregation of impurities, and inertness to oxidation. Each of the physical and chemical properties can be tested.

6.1 Physical Properties

Appearance tests are often subjective and reflect only the surface texture and color of the deposit. Color match and uniformity are critical for decorative plating. The appearance is best controlled through strict control of the plating variables. Changes in the deposit are often the result of unintentional codeposition of unwanted metals. Lead from solder baths and iron from plating fixtures are common problems. Burned deposits can be the result of a depleted bath, excessively high current density, or nonuniform current distribution such as would occur from defective or misshaped anodes.

Wear resistance is an important attribute for gold finishes in high wear situations, such as sliding contacts. An abrasive test applicable to jewelry applications has been described [60]. Wear test applicable to electrical parts have been described [24, 25, 60, 61]. The purpose of the tests is to determine how fast a deposit fails under laboratory-imposed mechanical loads, with commercial testers providing a convenient method for comparison. The variables include shape of the parts exposed to wear, cleaning, lubrication, and load.

Deposit hardness can be determined by mechanical deformation of test specimens. Commercial instrumentation and scales of hardness exist [62]. Two of the most common instruments use pyramidal diamond indenters with equal (Vickers) or unequal (Knoop) diagonals. The use of these for gold parts has been discussed [63]. Hardness is usually determined using cross sections of suitable thickness. Indenting normal to the surface often leads to erroneous results because surface roughness can lead to nonuniform loads and thin samples give hardness representative of the base substrate and not the plated gold. Electroplated soft gold has a Knoop hardness of about 60 under 25 g load. Hard gold can have Knoop values near 200.

Brittleness is the result of the presence of some particular phase or compound in the gold, such as can occur by the alloying of gold, either intentional or unintentional. Nickel and cobalt are intentionally added to improve hardness, whereas tin can cause brittle solder joints [64]. When all the gold does not dissolve in the solder, brittle failure can occur. Identification of brittle failures can be performed through metallographic examination. Gold hardened by codepositon of iron at the 0.1% level leads to stressed deposits that are brittle and that fail under mechanical loads milder than acceptable for printed wiring boards and connectors. Iron contamination is a particular concern because of the ease of contamination of gold baths from steel components. The degree of brittleness can be quantified in a flex test of the deposit, such as bending mandrill tests. Other tests include uniaxial tensile tests and diamond scribe tests [32].

Pure gold is electrodeposited in virtually a zero-stress state. The gold can be brought to a low-stress state if it is deposited onto a substrate whose coefficient of thermal expansion is different from gold and the deposition occurs at other than room temperature. During cooling from the deposition temperature, the different rate of contraction of the gold and its substrate produces residual stress. Temperature cycling of plated gold onto a substrate can result in significant stress, depending on the temperature excursion. At sufficiently high temperatures (ca. 150°C), gold can recrystalize reducing the stress at high temperature via grain growth. Upon cooling, the process is not reversible. That is, the coefficient of thermal expansion mismatch will produce a high-stress state in the gold.

There are several common test methods for porosity [65]. Chemical tests are used to determine the area of the exposed base metal by visual examination of the quantity of accumulated corrosion products. The base metal is oxidized by anodic dissolution or by chemical attack. This class of tests are easy to perform. Common forms of the test use polysulfide solutions [66], or atmospheres containing moist sulfur dioxide [67, 68], or nitric acid [65, 69]. The electrooxidation of the gold-plated part using a moist paper or a gel as a support for the electrolyte containing a dye or complexing agent can by used to quantify the porosity of the gold [70]. In the test, the complexed gold forms a colored species that can easily be identified and counted. Specific tests have been developed for copper and nickel.

6.2 Chemical Analysis

Atomic absorption (AA) spectrophotometry is a rapid and easy method for quantifying the chemical composition of gold deposits. The gold plate must first be separated from the base metal by immersing the part in an etchant that oxidizes and dissolves the base metal without oxidizing the gold or leaching out its impurities. The gold is then dissolved and the solution is analyzed by AA using the appropriate standards. Such tests are needed in meeting specifications such as Military Specification MIL-G-45204B for pure, bondable gold where metallic impurities such as chromium, copper, tin, lead, silver, cadmium, or zinc should not be present in the deposit at a concentration greater than 0.1%. Polarography can be used to analyzed solutions for various impurities and well as the gold baths themselves. The concentration of Au(III) has been quantified by polarography [17]. More exotic techniques can be used for specific analysis. For example, Mossbauer spectroscopy of Fe^{57} was used to uncover the form of the cobalt in gold deposits [22]. Radioactive Co^{57} was added to the common Co^{59} additive in the cobalt-hardened bath. The Co^{57} decayed to Fe^{57}, which could be studied by Mossbauer spectroscopy. Another example of a specific analytical method is the analysis of carbon. The carbon content of the deposits has been determined by combustion, converting it to CO_2 which could be measured by coulometric titration [71].

Surface analytical techniques, such as Auger electron spectroscopy and X-ray photoelectron spectroscopy, are often used to determine the elemental analysis of the gold surface as well as thin surface films when combined with sputter etching. Rutherford backscattering (alpha particle backscattering) is a powerful technique for identifying impurities and the distribution of impurities with depth in a rapid and nondestructive test. The thickness of thin films of gold can also be determined [72]. β-ray scattering can also be used for thickness measurements, and the apparatus is quite simple. X-ray fluorescence can also be used for thickness and impurity analysis; however, standards and tight control of analysis geometry are essential for accurate results. Structural aspects of gold deposits, such as grain size, are best determined by transmission electron microscopy. While sample preparation can require considerable effort, the results are irreplaceable [34, 40]. A technique known as through-focus imaging was used to observe organic and gaseous inclusions in gold, which cannot be done by conventional imaging techniques [34]. The availability of scanning probe techniques has provided new tools for examining growth and nucleation of deposits [51]. Last, surface contamination of electrical parts has created a need for evaluating and monitoring the contact resistance of electrical parts. Typical values of about 1 mΩ are obtained for electrical contacts. The contact force and area of the contact need to be controlled. Contact resistance increases of 2 to 50 fold have been observed for aged cobalt-hardened gold [32].

REFERENCES

1. G. Langbein and W. T. Brannt, *Electroplating of Metal*, 4th ed. Henry Carey Baird and Co., Philadelphia, 1902, p. 3.
2. M. Pourbaix, *Atlas of Electrochemical Equilibra in Aqueous Solutions*, Pergamon, New York, 1966.
3. W. M. Latimer, *The Oxidation States of the Elements and Their Properties in Aqueous Solutions*, Prentice-Hall, Englewood Cliffs, NJ, 1985.
4. G. M. Schmid and M. E. Curly-Florino, *in Encyclopedia of Electrochemistry of the Element*, vol. 4, A. J. Bard, ed., Marcel Dekker, New York, 1978.

5. R. Puddephatt, in *Comprehensive Inorganic Chemistry*, vol. 4, J. C. Bailar Jr., H. J. Emeleus, R. Nyholm, and A. F. Trotman-Dickenson, eds., Pergamon Press, Elmsford, NY (1973).
6. G. H. Kelsall, *J. Electrochem. Soc.*, **138**, 108 (1991).
7. W. T. Lee, *Corrosion Technol.*, **39** (April 1963).
8. E. D. Winters, *Plating*, **59**, 213 (1972).
9. S. A. Losi, F. L. Zuntini, and A. R. Meyer, *Electrodeposition Surf. Treatment*, **1**, 3 (1972).
10. L. Silverman, B. Bernauer, and F. Pettinger, *Met. Finish.*, **68**, 48 (Aug. 1970).
11. W. A. Fairweather, *Met. Finish.*, **64**, 15 (May 1986).
12. J. M. Leeds and M. Clarke, *Trans. Inst. Met. Finish.*, **47**, 163 (1969).
13. H. Angerer and N. Ibl, *J. Appl. Electrochem.*, **9**, 219 (1979).
14. Y. Okinaka, F. B. Koch, C. Wolowodiuk, and D. R. Blessington, *J. Electrochem. Soc.*, **125**, 1745 (1978).
15. E. D. Winters, *Plating*, **59**, 213 (1972).
16. Y. Okinaka and S. Nakahara, *J. Electrochem. Soc.*, **123**, 1284 (1976).
17. Y. Okinaka and C. Wolowodiuk, *J. Electrochem. Soc.*, **128**, 288 (1981).
18. A. Knodler, *Metalloberflache*, **28**, 465 (1974).
19. C. J. Raub, A. Knodler, and J. Lendvay, *Plating*, **63**, 35 (1976).
20. Y. Okinaka, F. B. Koch, C. Wolowodiuk, and D. R. Blessington, *J. Electrochem. Soc.*, **125**, 1745 (1978).
21. H. Leideiser Jr., A. Vertes, M. L. Varsanyi, and I. Czako-Nagy, *J. Electrochem. Soc.*, **126**, 391 (1979).
22. R. L. Cohen, F. B. Koch, L. N. Schoenberg, and K. W. West, *J. Electrochem. Soc.*, **126**, 1608 (1979).
23. C. C. Lo, J. A. Angis, and M. R. Pinnel, *J. Appl. Phys.*, **50**, 6887 (1979).
24. K. J. Whitlaw, J. W. Souter, I. S. Wright, and M. C. Nottingham, *IEEE Trans. CHMT*, **8**, 46 (1985).
25. L.-G. Liljestrand, L. Sjohren, L. B. Revay, and B. Asthner, *IEEE Trans. CHMT*, **8**, 123 (1985).
26. J. H. Thomas III and S. P. Sharma, *J. Electrochem. Soc.*, **126**, 445 (1979).
27. G. B. Munier, *Plating*, **56**, 1151 (1969).
28. L. Holt, R. J. Ellis, and J. Stanyer, *Plating*, **60**, 910 (1973).
29. M. Antler, *Plating*, **60**, 468 (1973).
30. H. A. Reinheimer, *J. Electrochem. Soc.*, **121**, 490 (1974).
31. R. L. Cohen, K. W. West, and M. Antler, *J. Electrochem. Soc.*, **124**, 342 (1977).
32. F. B. Koch, Y. Okinaka, C. Wolowodiuk, and D. R. Blessington, *Plating*, **67**, 50 (1980).
33. H. A. Reinheimer, *J. Electrochem. Soc.*, **121**, 490 (1974).
34. S. Nakahara and Y. Okinaka, *J. Electrochem. Soc.*, **128**, 284 (1981).
35. F. B. Koch, Y. Okinaka, C. Wolowodiuk, and D. R. Blessington, *Plating*, **67**, 43 (1980).
36. M. Paunovic and C. Sambucetti, in *Electrochemically Deposited Thin Films*, *Proc.*, vol. 94–31, M. Paunovic, ed., pp. 34–44, Electrochemical Society, Inc., Pennington, NJ (1995).
37. A. M. Sullivan and P. A. Kohl, *J. Electrochem. Soc.*, **144**, 1686 (1997).
38. T. Osaka, A. Kodera, T. Misato, T. Homma, Y. Okinaka, and O. Yoshioka, *J. Electrochem. Soc.*, **144** (1997).
39. X. Wang, N. Issaev, and J. G. Osteryoung, *J. Electrochem. Soc.*, **144** (1997).

40. W. J. Dauksher, D. J. Resnick, W. A. Johnson, and A. W. Yanof, *Microelectronics Engineering*, **23**, 235 (1994).
41. W. Chu, M. L. Schattenburg, and H. I. Smith, *Microelectronics Engineering*, **17**, 223 (1992).
42. H. Homma and Y. Kagaya, *J. Electrochem. Soc.*, **140**, L135 (1993).
43. T. Inoue, S. Ando, H. Okudaira, J. Ushio, A. Tomizawa, H. Takehara, T. Shimazaki, H. Yamamoto, and H. Yokono, *Proc. 45th Electronic Components and Technology Conf.*, Page 1059, IEEE (1995).
44. M. Kato, K. Nikura, S. Hoshino, and I. Ohono, *J. Surf. Finish. Soc. Jpn.*, **42**, 69 (1991).
45. Y. Okinaka and T. Osaka, in *Advances in Electrochemical Science and Engineering*, vol. 3, H. Gerischer and C. Tobias, eds., p. 55, VCH Publishing, New York (1994).
46. A. M. Sullivan and P. A. Kohl, *J. Electrochem. Soc.*, **142**, 2250 (1995).
47. W. Chirzanowski, Y. G. Li, and A. Lasia, *J. Applied Electrochem.*, **26**, 385 (1996).
48. E. T. Eisenmann, *J. Electrochem. Soc.*, **125**, 717 (1978).
49. G. Holmbom and B. E. Jacobson, *J. Electrochem. Soc.*, **135**, 787 (1988).
50. P. Bindra, D. Light, P. Freudenthal, and D. Smith, *J. Electrochem. Soc.*, **136**, 3616 (1989).
51. J. Schneir, V. Elings, and P. K. Hansma, *J. Electrochem. Soc.*, **135**, 2774 (1988).
52. G. Decaraj and S. Gurrviah, *Materials Chemistry and Physics*, **25**, 439 (1990).
53. G. Holmbom and B. E. Jacobson, *J. Electrochem. Soc.*, **135**, 2720 (1988).
54. D. T. Chin, N. R. K. Vilambi, and M. K. Sunkara, *Plating*, **76**, 74 (1989).
55. A. K. Sharma, *Trans. Inst. Met. Finish.*, **67**, 87 (Aug. 1989).
56. D. R. Turner, *Met. Finish.*, **63**, 41 (Nov. 1987).
57. D. W. Endicott and G. J. Casey Jr., *Plating*, **67**, 58 (1980).
58. R. J. von Gutfeld, M. H. Gelchins, and L. T. Romankiw, *J. Electrochem. Soc.*, **130**, 1840 (1983).
59. I. R. Christie, *Trans. Inst. Metal Finishing*, **64**, 47 (1986).
60. R. Duva and D. G. Foulke, *Plating*, **55**, 1056 (1968).
61. A. J. Solomon and M. Antler, *Plating*, **57**, 812 (1970).
62. V. E. Lysaght, *Indentation Hardness Testing*, Reingold, New York, 1948.
63. R. G. Baker and T. R. Palumbo, *Plating*, **58**, 791 (1971).
64. F. G. Foster, *ASTM Spec. Tech. Pub.*, **319**, 3113 (1963).
65. A. A. Khan, *Plating*, **56**, 1374 (1969).
66. F. I. Nobel, B. D. Ostrow, and D. W. Thomson, *Plating*, **52**, 1001 (1965).
67. R. W. Beattie, G. Forshow, and E. N. Loney, *Proc. IEE (London)*, **B-109**, suppl. 21, 109 (1962).
68. M. Clarke and J. M. Leeds, *Trans. Inst. Met. Finish.*, **46**, 81 (1968).
69. F. Knoop, C. G. Peters, and W. E. J. Emerson, *J. Res. Natl. Bur. Stand.*, **23**, RP 1220 (1939).
70. F. V. Bedette and R. V. Chiarinzelli, *Plating*, **53**, 305 (1956).
71. E. N. Wiese, P. W. Gilles, and C. A. Reynolds, *Anal. Chem.*, **25**, 1344 (1953).
72. J. Zahavi, *Met. Finish.*, **83**, 61 (July 1985).

5 Electroless and Electrodeposition of Silver

MORDECHAY SCHLESINGER

INTRODUCTION

Silver metal has been known since ancient times. It is mentioned in Exodus and Genesis. There exist strong indication that man has been able to separate silver from lead as early as 3000 BC. Silver may be found native and in ores. Lead, lead zinc, copper, and other ores are principal sources. Silver may also be recovered during electrolytic refining of copper. Pure silver has the highest electrical and thermal conductivity of all metals. It retains that property in the form of electroplated thin film as well. This makes electroplated silver an important component of many printed circuit systems.

There is indication that silver was plated as far back as the beginning of the nineteenth century. If correct, it must have been done in connection with producing mirrors. The earliest patent for silver plating, however, was not granted until 1840 [1]. That may be viewed as signifying the start of the electroplating industry. That bath is still one of the more used ones to this day, and it is the double-silver cyanide complex [$KAg(CN)_2$] with excess free cyanide. Over the years many other baths have been proposed [2] such as the ones involving nitrate, iodide, thiourea, thiocyanate, sulfamate, and thiosulphate. The usages of silver plating are many. Those include mirrors, deposits for tableware due to its decorative effect as well as resistance to corrosion in its contact with foods. There are a number of industrial uses as well. Those include, but are not limited to, electronic component applications, bearings, hot gas seals, to name but a few. Silver may be plated using either electrodeposition methods or electroless methods. We discuss here both methods, starting with the latter.

1 ELECTROLESS DEPOSITION

It is often necessary to silver coat insulators such as glass or ceramics. While a number of methods may be employed for this purpose (e.g., chemical vapor deposition, ion-sputtering), chemical methods, and in particular the methods of electroless deposition, offer an effective, easy to operate, path. In general, electrochemical

Modern Electroplating, Fourth Edition, Edited by Mordechay Schlesinger and Milan Paunovic.
ISBN 0-471-16824-6

methods may be classified into two types. One is galvanic exchange. Here the potential difference between metals is used to replace one metal layer with another (e.g., in "zincation"). The other is electroless plating where reducing agents are used. Electroless plating is the main focus, in this presentation. As mentioned in a number of other chapters in this book, the electroless methods are effective because of the simplicity of the apparatus used, and they can be done on complex substrates; last, but not least, the methods are suitable for mass production.

1.1 Pretreatment

The substrate pretreatment for electroless silver plating involves, besides cleaning and degreasing, immersion in $SnCl_2$ solution and subsequent rinse in water. If, for instance, the substrate is glass, then the next step is immersion in the silver metallizing bath. Fine silver particles that are positively charged in solution will adhere to the glass surface, and the Sn^{+2} species will reduce the silver ions to metallic silver by being oxidized themselves to the Sn^{+4} state. Since silver is autocatalytic for the electrodeposition of itself, plating will continue as long as the substrate is immersed in the bath. For substrates other than glass, there will be the need to add the step of immersion in $PdCl_2$ before proceeding to the metallizing stage.

1.2 Metallizing Baths Compositions

In practice, the metallizing bath is made by combining two separate baths. One is a silver ion solution, while the other is a reducing agent solution. In what follows we will denote them as A an B respectively.

1. Rochelle salt based [3] (most often used)
   A: Silver nitrate 450 g, saturated ammonia solution 350 ml water 5.5 liter
   B: Rochelle salt 1600 g Epsom salt 110 g water 3.6 liter
2. Glucose method [4]
   A: Silver nitrate 3.5 g, ammonia solution as required, water 60 ml, sodium hydroxide solution 2.5 g / 100 ml
   B: Glucose 45 g, tartaric acid 4 g, water 1 liter, ethyl alcohol 100 ml
3. Hydrazine method [3] (used for spray method, see below)
   A: Silver nitrate 110 g, ammonia solution (as required) 230 ml
   B: Hydrazine sulfate 43 g, ammonia solution 45 ml

In case 3, the A and B are diluted to 4.5 liter and then mixed to the ratio 1 : 1.

1.3 Bath Properties

Electroless silver plating baths are very unstable and so short-lived. It is desirable to stabilize the baths for electroless deposition of nickel and copper. In the case of silver this becomes imperative. Indeed, if the bath can be made more stable, it will be that much more useful. It has been found [5] that a small amount of 3-iodotyrosine or 3,5-di-iodotyrosine added to a bath (referred to as DIT bath) can work as stabilizer.

There seems to be a strong indication that the electroless plating reaction progresses with a combination of a cathodic reaction of a cation, that is, $M^{n+} + \text{ne} \rightarrow M$, and an anodic oxidation of the reducing agent $R \rightarrow \text{O} + \text{ne}$. Here it is assumed that a metal ion is cathodically reduced to metal on the surface that is activated by a catalyst. This state of affairs is valid for silver as well as for other metals on which autocatalytic reaction will occur. A number of examples with representative reducing agents are given above. In all silver oxide precipitation is avoided because the electroless silver plating is carried out in a basic solution. The complexing agent is added to the solution.

Not at all surprising is the fact that bath stability and plating rate are markedly affected by the pH of the bath. Consequently the parameters affecting the bath pH were investigated [5]. It was found, for instance, that the pH changes with the addition of silver nitrate. The change is most noted when the pH values are below 10.5. The complex formation between silver ion and ethylene diamine (en) is most likely as follows:

$$Ag^+ + 2H_2O + 2en \longleftrightarrow [Ag(OH)_2en_2]^- + 2H^+$$

Equilibrium consideration of the reaction indicates that the bath pH changes with concentrations of Ag^+ and en. Even in case of a stabilized bath (using 3,5-di-iodotyrosine), the plating reaction all but stopped after 24 hours and the pH of the bath dropped. To clarify the foregoing, we present in Figure 1 the electroless plating rate of silver as a function of bath pH. From the figure it is evident that a slight drop in pH value may drastically reduce the plating rate.

Other stabilizers have been reported in literature. They include metal ions (Co^{2+}, Ni^{2+}, Co^{2+}) [6] Na-2-3-mercaptopropane sulfonate ($NaBH_4$, N_2H_4, CN^- System bath) [7] and a number more.

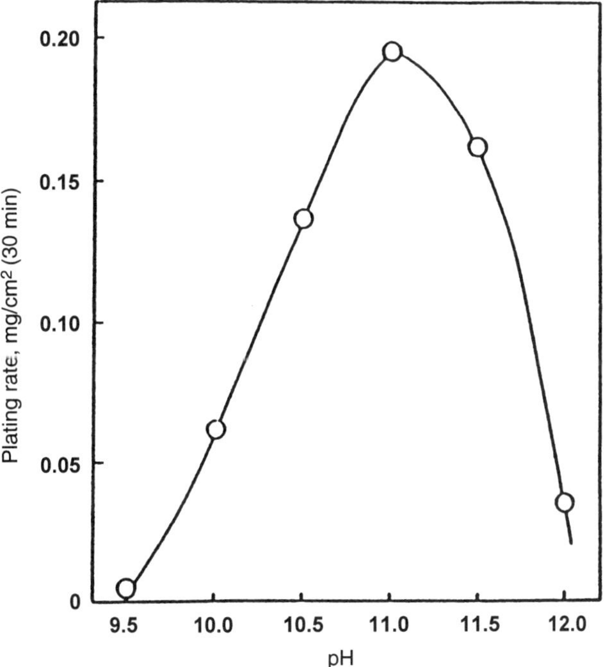

Figure 1 Electroless plating rate of silver as a function of bath pH: Silver nitrate 2.9×10^{-3} M, ethylenediamine 1.8×10^{-2} M Rochelle salt 3.5×10^{-2} M 3,5-diiodotyrosine 4×10^{-5} M, 35°C. Reproduced with permission of AESF.

1.4 Polarization

The specific effects of pretreatment and the mechanism of silver plating can be inferred from the partial anodic and cathodic polarization curves [5]. Both the partial cathodic polarization curve for a solution without Rochelle salt, as the reducing agent and the partial anodic curve for a solution without silver nitrate are given in Figure 2. In these the stabilized, DIT bath was used. The result for the (electrode) surface pretreated with a $SnCl_2$ solution is shown by a solid line, while that for a nontreated one is shown by a dotted line. The difference between the partial cathodic polarization curves in Figure 2 is minimal. This is not so with the difference between the anodic polarization curves of the treated and the untreated electrodes is plainly evident. In the case of the treated electrode, an anodic current appears also in the range of 100 to 600 mV versus SCE. At the same time the rest potential of the untreated electrode becomes 25 mV versus SCE while that of the treated becomes −81 mV versus SCE. The difference in the rest potential may be attributed to the adsorption of Sn^{2+} ions on the electrode. In other words, it is assumed that the

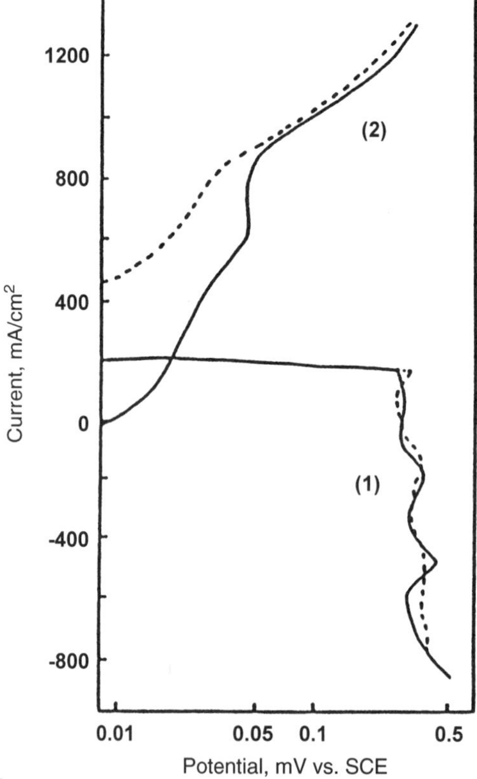

Figure 2 Partial polarization curves for various electrodes. (1) Partial cathodic polarization curve of electroless silver plating solution without Rochelle salt—$AgNO_3$ (0.88 M), ethylenediamine (0.054 M), DIT (4×10^{-5} M), 35°C, pH 10.0; (2) partial anodic polarization curve for electroless silver plating solution without silver nitrate—ethylenediamine (0.054 M), Rochelle salt (0.035 M), DIT (4×10^{-5} M), 35°C, pH 10.0. (*Solid line*) Pt electrode sensitized with $SnCl_2$ solution; (*dotted line*) Pt electrode, not sensitized. Reproduced with permission of AESF.

anodic current due to oxidation of Sn^{2+} flows at potentials of 100 to 600 mV versus SCE. When the anodic and cathodic polarization curves for the treated electrodes are combined a point of intersection appears at 0.02 mA cm^{-2} of current density. When this value of current density is converted to plating rate, it is found to be equivalent to 0.04 mg cm^{-2}/30 min. This value is found to be close to the actual plating rate as determined experimentally. In the case of the untreated electrode, the current density at the intersection is extremely small. Thus, at least in the case of the DIT bath, it may be assumed that the electroless silver plating reaction proceeds only on the nucleation centers established by the $SnCl_2$ surface treatment and does not occur on the untreated surface.

1.5 Deposition in Practice

In the practical silvering operation the silver ion solution and the reducing agent solution are prepared separately. They are combined just before plating. Upon addition of the ammonia solution to the silver nitrate solution, a dark black-brown precipitate begins to form. The ammonia solution should be added until the precipitate all but dissolves. The solution ends up dark brown and clear.

In the spray method a number of different especially designed guns [8] are used. One such gun has two tubes. The two solutions are sprayed separately and mixed on the substrate. There are a number of different final baths [5, 9–15] suggested in the literature, given at the end of this chapter.

Displacement methods [16] can be used in electroless silver plating on metal. For instance, the bath composition for silver plating on copper alloy is made of silver nitrate 7.5 g liter^{-1}, sodium thiosulfate 100 g liter^{-1} and ammonia 75 g liter^{-1}. There are many other formulations [17] given in the literature.

The applications of electroless silver deposits are many but they may be divided into two categories:

1. Decorations and optical uses
2. Electrical conductivity and undercoating prior to other plating

Mirrors [18] are an example of the first type, and undercoating for electroless gold plating or for silver plating on ceramics is an example of the second application type. One other important application is the plating of silver on ceramics for the purposes of SEM (scanning electron microscopy) measurements of surface properties [19]. Still other applications include quantitative analysis for trace amounts of arsenic [20], for the determination of pore diameter of a membrane [21], and for the recovery of mercury.

2 ELECTRODEPOSITION

2.1 Bath Composition

The specific composition of silver (cyanide) plating baths has to fit the required silver plate. For instance, silver plate for decorative purposes is best achieved from baths with a lower silver content than those used for engineering purposes where thicker deposits are needed. Higher silver concentrations make higher current densities possible, and as a consequence more economic plating speeds can be employed.

It is worth noting out here that many present-day silver plating baths are similar to that described in the very first patent from over 150 years ago. Table 1 gives the silver deposition bath compositions that, are in common use to date. Indeed, even a typical bath composition offered, recently [22] in the literature, for silver electroplating is made of

AgCN	0.0185 M
$K_4P_2O_7$	0.3027 M
KCN	0.1228 M

Such a bath may be operated at pH 10.8 with a value of $-100\,\mathrm{mV}$ (vs. Ag) polarization.

A number of attempts have been made to replace AgCN and thus the cyanoargentite ions by different silver electrolytes based on nitrate (see Section 1 on electroless deposition), thiourea, and other compounds. These solutions, however, exhibit instability, and the resulting deposits are of low quality in terms of adhesion, and the like. So cyanide solutions are used at the present time despite their toxicity. The only way to minimize the cyanide concentration is to employ a lower amount as the formulation in [22] suggests. This approach requires sophisticated deposition methods such as pulse plating.

2.2 About Bath Constituents

It is evident from the discussion above that alkaline silver cyanide provides the silver for deposition on the cathode. From a practical point of view low silver content and high free cyanide yields improved throwing power. On the other hand, higher silver concentrations allow the use of higher current densities. Thus for this reason baths low in silver content are used for deposits of thickness between 2 and 5 μm. The alkali cyanide increases the conductivity, and the cathode polarization provides good anode corrosion, not to mention its function in forming the silver complex ion. Also KCN and not NaCN is used because it induces higher solution conductivity, higher

TABLE 1 Typical Bath Composition (in g liter^{-1})

	Bright Silver (Conventional)	High Speed (Thick Deposit)
Silver (metal)	20–45	35–120
Silver cyanide	31–55	45–150
Potassium cyanide (total)	50–80	70–230
Potassium cyanide (free)	35–50	45–160
Potassium carbonate	15–90	15–90
Potassium nitrate	—	40–60
Potassium hydroxide	—	4–30
Current density (A dm^{-2})	0.5–1.5	0.5–10.0
Temperature (°C)	20–28	35–50

Note: Brighteners are added as required.

solubility of the resultant potassium carbonate buildup, and higher limiting current densities. Potassium carbonate too increases the conductivity of the bath and also the anode and cathode polarizations which improves throwing power. The concentration of the carbonate rises with the decomposition of the cyanide. Thus initially the concentration added to the bath is kept to the recommended minimum.

Silver cyanide forms a complex $KAg(CN)_2$. One gram of AgCN requires about 0.5 g KCN in addition to the free cyanide indicated in Table 1. Over the years we have found that the most convenient way to prepare a tank full of silver plating bath is as follows: Dissolve the bath constituents other than the potassium silver cyanide, and dissolve the possible brighteners in a separate container of water equal to about half of the bath volume. Heat the solution to about 70°C the solution can be heated with the presence of activated carbon (1.5 g liter^{-1}). Stir until the constituents dissolve (about 40 to 45 minutes), and then filter it into the operating tank. Fill the tank up to almost to the full volume, and then add the $KAg(CN)_2$ to the tank while stirring. This way no silver will be removed with the carbon. At this stage brighteners can be added and the tank filled up to the right volume. Now, at the risk of repeating what has already been stated above, we offer the following advice: Silver deposition is a well-studied technology in the area of metal deposition. While electron transfer to silver ion should be rapid, ideal silver surfaces may have slow nucleation, as shown by Budevski et al. [23]. A majority of substrates and simple silver solutions, dendritic silver is produced through nucleation and 3D growth of individual metal grains with little or no tendency to merge and compact. Thus, to obtain smooth deposits of silver, plating from cyanide complexes is practiced in most cases.

2.3 Anodes and Anodic Reactions

Cyanide silver plating baths are fairly stable and easy to maintain. Cathode and anode current efficiencies are close to 100%, and the solution can be kept in balance for long periods of operation. For this reason the use of high-quality silver anodes (purity of 99.98% or better) is recommended.

It may be useful at this point to elucidate the process of silver dissolution in cyanide solutions. Cyanide leaching, first introduced in 1887 [24], remains to this day, the common means to recover gold and silver from ores. The reaction was recognized by early workers to be electrochemical in nature [25]. The kinetics of silver cyanidation has been widely studied, very little has been said about the reaction mechanism. Vielstich and Gerischer [26] using the potential step method verified the Butler-Volmer equation (see *Fundamentals* Chapter 6, Eq. 6.45) with transfer coefficient values $\alpha = 0.5$ and 0.44 for $[CN^-] > 0.2$ M and < 0.1 M, respectively. The predominant complex ion of silver was assumed in their study to be $[Ag(CN)_3^{2-}]$, and it did not depend on cyanide concentration. Nachaev and Beck [27], on the other hand, studied the discharging of silver in cyanide solutions measuring the Faraday impedance. They found that through all potential values more negative than -0.45 V silver was deposited through the discharging of $[Ag(CN)_2^-]$ ions independent of the composition of the bath and the conditions of electrolysis.

It was found [28, 29] that a surface-enhanced Raman scattering (SERS) spectrum may be obtained from cyanide adsorbed on a silver electrode. As a result silver has been reaffirmed as an electrode of much interest. Lately a number of studies have concentrated on the method and efforts to correlate SERS with electrochemical methods prevail. For instance, voltammograms for the silver cyanide system with the corresponding SERS spectra of adsorbed silver cyanide were presented by Benner and coworkers [30]. Fleischman and coworkers [31] obtained SERS spectra of $^{12}CN^-$ and $^{13}CN^-$ adsorbed on silver electrodes. These spectra were shown to arise from a complex species whose coordination number did not change with the electrode potential. The species were proposed to be $[Ag(CN)_2^-]$. Blondeau and coworkers [32], using radio chemical means, found that after several dissolution redeposition cycles the quantity of cyanide adsorbed remained constant and the silver-cyanide reaction was reversible. New information from SERS spectra prompted Baltruschat and Vielstich [33] to study the system again, via potentiostatic pulse methods. Using the exact methods of mathematics rather than extrapolating, they obtained better accuracy for exchange current densities. Their results indicated that neither the anodic dissolution of silver in cyanide solution nor its reverse process involved simple charge transfer reactions: that is, (1) the reaction orders are not integers, (2) the apparent transfer coefficient is dependent on concentration or potential, and (3) the rate-determining step is not purely charge-transfer controlled. They also showed that pH has no influence on the process. A more recent work by Wadsworth and coworkers [34] is also of some interest. They point out that up to the early 1990s (1) research had focused on concentrated cyanide solutions (above 0.1 M), (2) the mechanisms require further clarification, (3) the combined effect of intrinsic kinetics and physical processes (e.g., diffusion) are yet to be measured, and (4) the adsorption of cyanide ions on silver surfaces is a steady-state phenomenon. Their work, in which electrochemical techniques were used to examine the kinetics of silver cyanidation in oxygen-free alkaline solutions, drew the following conclusions: The reaction between silver and cyanide may be separated into the two elementary steps: a charge-transfer reaction followed by a chemical reaction. Since the charge-transfer step, which includes the diffusion of cyanide ions from the bulk solution to the silver surface, is rate limiting, the rate of dissolution is calculated using the combined Butler-Volmer and Levich equations.

From the electrochemical measurements the diffusion coefficient of cyanide ions in 0.5 M KNO_3 solutions ($pH = 11$; $[Ag] = 5 \times 10^{-5}$ M and $[CN] = 5 \times 10^{-4} - 1 \times 10^{-2}$ M) is found to be $1.6 \times 10^{-5}\,cm^2\,s^{-1}$ at 24°C and the activation energy 13.7 kJ. The apparent exchange current density in the solutions has been concluded to be having an activation energy of 21.1 kJ mol^{-1}. The transfer coefficient has been found to be close to 0.5, but it changes somewhat with temperature (between 0.56 and 0.43).

2.4 Deposition on Metals

The pretreatment of the metal surfaces that will be silver plated includes the standard steps of cleaning and etching to which is added a silver strike. The explanation for this additional treatment is that most basic metals are less noble than silver (as is even gold in the cyanide system). Consequently they will precipitate silver by immersion, from the regular silver plating baths, and cause the silver deposits to poorly adhere. The strike baths should contain low silver metal and high free cyanide concentrations. In this way the tendency for electrochemical displacement by the base metal is greatly lowered. Besides, the process involves low efficiency hydrogen activation. The procedure for steel surfaces is to use a double strike. The first in a solution containing low silver content with some copper cyanide, and the second in a regular strike solution. The latter may be used as the first strike solution for copper, its alloys, nickel, and nickel silver. Solution composition are given in Tables 2 and 3. The strike can also be used for surfaces made of more than one metal or with soldered parts. The recommended immersion times range from 10 to 25 seconds for decorative silver plating, while for thicker deposits it can range from 15 to 40 seconds. It is also recommended that the object to be plated be made cathodic before immersion in either the strike or the plating bath.

2.5 Additives

In general, small quantities of certain substances, mostly organic, can markedly affect the form or structure of the deposits. Since it can be assumed that adsorption takes place at the cathode surface during electrolysis, one added molecule affects many thousand metal ions. Often the added effects are undesirable. If, however, the result is a bright or level surface or if internal stress is reduced, the added substance is not considered a contaminant. A decrease in the activation overpotential is observed, for instance, for carbon disulfide in the cyanide silver bath. (The effect is the opposite in other metal's bath.) The subject of additives and their nature is discussed in detail in Chapter 10 of *Fundamentals*. For this reason we will only make a few general practical observations and leave it to the reader to consult that text.

The susceptibility of the different metals to the inhibition of electrodeposition by many additives is nearly parallel to their melting point, hardness, and strength, increasing in this order: Pb, Sn, Ag, Cd, Zn, Cu, Fe, Ni. This sequence also fits the order of the increasing tendency of their ions to form complexes. The sequence further fits the order of increasing activation polarization in electrodeposition from their aquated ions.

The additives are generally consumed in the deposition process. They may be decomposed, and the products in parts incorporated into the deposit, or released

TABLE 2 First Strike Bath for Steel

Silver cyanide	1.5–2.5 g liter^{-1}
Copper cyanide	10–15 g liter^{-1}
Potassium cyanide	75–90 g liter^{-1}
Temperature	22°–30°C
Current density	1.5–3.0 A dm^{-2}
Volts	4–6

TABLE 3 Strike for Nickel and Nonferrous Metals, and Second Bath for Steel

Silver cyanide	1.5–5 g liter^{-1}
Potassium cyanide	75–90 g liter^{-1}
Temperature	22°–30°C
Current density	1.5–3.0 A dm^{-2}

back into the electrolyte. As a rule we note that they will affect the internal stresses in the deposits positively or negatively.

Carbon disulfide and thiosulfate have been widely used as additives (brighteners) in silver plating baths. A host of other materials have been used as well. Some are gums, sugars, unsaturated alcohols, sulfonated aliphatic acids, turkey red oil, rhodamine red, and compounds of antimony and bismuth. The majority are sulfur-bearing organic compounds or else reaction products of sulfur and organic compounds. There exist a good number of effective proprietary additives that yield highly desirable deposits in terms of mirror brightness. Such products, when available, obviate the need for buffing the final plate, which not only saves precious metal but also assures a more uniform thickness.

2.6 Physical Properties

One of the underlying concepts of thin film science is that of surface energy (SE). This is the energy required to create a surface. Interestingly metals, as a rule, have a high SE, while oxides have a low SE. The relative values of SE determine whether one material wets another and thus forms a uniform adherent layer as required in the case of electroplated metal. Material with low SE will tend to wet a material of high SE. Conversely, if the deposited material has a higher SE it will tend to form clusters; that is, it will "ball up" on the low SE substrate. Water proofing is a way of manipulating surface energies. Organic materials tend to have low SE; thus a car waxed with an organic substrate will cause water droplets to form on its wet surface. As stated above, SE is defined as the energy spent to create a surface. It is a positive quantity, since energy is added to the system. A liquid balls up to reduce its surface area, and crystals will facet in order to expose their lowest energy surfaces. When one breaks a solid, two new surfaces are created and bonds are broken. It should thus be clear that SE is related to bond energy and to the number of broken bonds creating surfaces. This dynamic in turn is related to the binding energy of the material.

The binding energy is defined as the energy needed to transform one mole of solid or liquid into gas at a low pressure. That is nearly the same as the energy of sublimation (transforming solid to gas) or the energy of evaporation (transforming liquid to gas), except that the gas is measured at the pressure of one atmosphere rather than at low pressure. These energies are related to the interatomic potential energy between atoms.

Using "nearest neighbor only interaction" approximation, we are able to relate heat energy of sublimation, evaporation, melting, and crystallization to surface energies. The binding energy per mole of a substance is expressed as

$$E_b = 1/2 n_c N_A \varepsilon_b$$

where

n_c stands for the coordination number, which is the number of nearest neighbors
N_A stands for Avogadro's number
ε_b stands for the binding energy per single bond

The factor of 1/2 comes as a result of having counted bonds twice

To evaluate surface energy from the point of view of nearest neighbor bonds, we first define N_S to be the number of atoms per unit area, and E_S/A to be the surface energy per unit area. Across any atomic plane each atom has an average $n_c/2$ nearest neighbors, so the number of bonds to be broken per unit area of cleavage along the plane is $1/2n_cN_S$. Now, since we create two surfaces in cleaving, the surface energy per unit area is $n_cN_S\varepsilon_B$. Finally, the ratio of SE per atom to binding energy per atom $1/4n_c\varepsilon$ is given by

$$\frac{E_S/AN_S}{E_B/N_A} = \frac{1/4n_c\varepsilon_b}{1/2n_c\varepsilon_b} = \frac{1}{2}$$

From measured values shown in Table 4, we can calculate the ratio of surface energy per atom to bond (or latent heat) per atom. Indeed, from the table we can infer whether it is easier to deposit silver on copper or whether it is easier to deposit copper on silver.

Silver metal is noted for its extremely high electrical conductivity, 0.0162 ohm-cm^2 m^{-1}. It is somewhat unique in that even in its thin film form it exhibits nearly the same property, 0.017–0.024 ohm-cm^2 m^{-1}, depending on deposition conditions and additives. For instance, sulfur and selenium containing electrolytes yield deposits with conductivities of up to 90% that of bulk silver metal.

Another physical property of importance in case of electrodeposits is micro structural instabilities due to creep (relief) of stress with ensuing time. An optically flat mirror made of silver deposit will deform in a matter of a few months. This is due to room temperature relief of stress in the silver electroplate. Brass mirrors made optically flat and subsequently silver plated show deformation as a result of stress in the silver plate. Removing the silver by machining will restore the original flatness of the brass. Heat treating the silver-plated part at 150°C for one hour seems to eliminate the problem.

2.7 Fractals

In general, electrodeposits made using baths containing no additives are dendritic or treelike. These types of deposits are of no practical value. In recent years these types of deposits contributed to the study of fractals. A fractal is an object with a pattern that under magnification exhibits repetitive levels of structures; a similar structure exists on all levels. In other words, a fractal looks the same whether it is viewed on the scale of a meter, a millimeter, or a micrometer. In the human body, for instance, fractal-like structures make up the blood vessels, nerves, and so on. In nature,

TABLE 4 Relationship between Surface Energy and Latent Heat

Metal	Surface Energy (erg cm^{-2})	Latent Heat Evaporation (Kcal mole^{-1})	Ratio SE/LH	Interatomic Potential (eV atom^{-1})	Cohesive Energy (Kcal mole^{-1})
Copper	1700	73.3	0.22	0.58	80.4
Silver	1200	82	0.15	0.65	68
Gold	1400	60	0.24	0.47	87.96

Figure 3 Electrodeposited clusters (about 0.5 cm in length) photographed 15 minutes after start of growth. Reproduced with permission.

coastlines and tree branching are the obvious examples. In the material sciences, the importance of fractals in that it leads the way to the study and quantitative analysis of the microstructure of metals in fashion hitherto thought impossible. Kahanda and Tomkiewicz [35], for instance, undertook the issue of fractality in regard to the impedance of electrochemically grown silver deposits. To illustrate the fractal nature of some electrodeposits, we present in Figure 3 a photograph of metal clusters obtained after 15 minutes in a bath with no additives.

REFERENCES

1. G. Elkington and H. Elkington, British Patent 8447 (1840).
2. S. R. Natarajan and K. Krishman, *Met. Finish.*, **50** (2), 51 (1971).
3. S. K. Gupata, *Cent. Glass. Ceram. Res. Int. Bull.*, **14**, 18 (1967).
4. T. Ishibashi, Less Me. Finish., **9**, 42 (1968).
5. A. Kubota and N. Koura, *J. Met. Finish. Soc. Jpn.*, **37**, 131 and 694 (1986).
6. T. Hata and T. Hanada, U.S. Patent 3,337,350 (1967).
7. A. Vaskelis and O. Diemontaite, *TSR Mokslu Akad. Darb.*, **B4**, 23 (1975).
8. I. Cross, *Met. Finish*, **48**, 77 (1950).
9. S. Noguchi, Japan Patent 7,593,234 (1975).
10. S. Horst, *Ger. Offen* 2,063,334 (1972).
11. T. Mukaiyama, Japan Patent 7,216,819 (1972).
12. S. Werner, Ger. Offen 2,639,287 (1978).
13. Max Ernes, Ger. Offen 950,230 (1956).

14. D. J. Hood, U.S. Patent 2,120,203.
15. Y. Mori, Japan Patent 7,606,135 (1976).
16. T. Kanabe and H. Ise, *Ind. Met. Sur.*, **7**, 64 (1979).
17. B. A. Bochkarev and E. Z. Napukh, *Elektrokhimiya*, **18**, 6, 843 (1982).
18. G. Maurice, Ger. Offen 2,153,837 (1971).
19. V. J. Pakolenko, *Zanod. Lab.*, **48**, 10, 89 (1982).
20. L. V. Markova, *Z. Anal. Khim.*, **25**, 8, 1620 (1970).
21. K. Wekua and B. Hoffman, *Furbe u. Lack*, **58**, 250 (1952).
22. H. Sanches, E. Chainet, B. Nguyen, P. Ozil, and Y. Meas, *J. Electrochem. Soc.*, **143**, 2799 (1996).
23. E. Budevski, W. Bostanoff, T. Vitanoff, Z. Soinov, A. Kotzen, and R. Kaischev, *Phys. Status Solidi.*, **13**, 577 (1966).
24. J. V. N. Dorr, *Cyanidation and Concentration of Gold and Silver Ores*, McGraw-Hill, New York, 1936, ch. 1.
25. V. Kudryk and H. H. Kellog, *J. Metals*, **6**, 541 (1954).
26. W. Vielstich and H. Gerischer, *Z. Physik Chem.* (new series), **4**, 10 (1955).
27. E. A. Nachaev and R. Yu. Beck, *Electrokhimia*, **2**, 150 (1966).
28. A. Otto, *Surf. Sci.*, **75**, L392 (1978).
29. J. Billman, G. Kovacs, and A. Otto; *ibid.*, **92**, 153 (1980).
30. R. E. Benner, R. Dornhouse, R. K. Cheng and B. L. Laube, *ibid.*, **101**, 341 (1980).
31. M. Fleischman, I. R. Hill, and M. E. Plembe, *J. Electroanal. Chem.*, **136**, 361 (1982).
32. G. Blondeau, J. Zerbino, and N. Jaffrezic-Renault; *ibid.*, **112**, 127 (1980).
33. H. Baltruschat and W. Vielstich, ibid, **154**, 141 (1983).
34. X. Zhu, J. Li, D. M. Bodily, and M. E. Wadsworth, *J. Electrochem Soc.*, **140**, 1927 (1996).
35. G. L. M. K. S. Kahanda and M. Tomkiewicz, *J. Electrochem Soc.*, **137**, 3423 (1990). See references therein regarding fractals in general.

6 Tin and Tin Alloys for Lead-Free Solder

YUN ZHANG and JOSEPH A. ABYS

INTRODUCTION

Soldering is technically defined as the joining of two base materials through the use of a third "filler" metal with a melting temperature well below those of the substrates and typically less than 450°C. Solder alloys are categorized as either soft or hard. Soft solders typically contain tin and lead, although indium, cadmium, and bismuth are also found in soft solders. Hard solders contain metals such as gold, zinc, aluminum, and silicon. A more rigorous classification is based on the melting temperature where soft solders melt below approximately 350°C, while hard solders melt above 350°C.

1 A BRIEF HISTORY OF SOLDERS

The historical use of solders is illustrated by the time line in Figure 1, which extends from approximately 4000 BC, through the Bronze and Iron Ages, and into the present so-called Silicon Age. The earliest archaeological evidence of soldering dates back to 4000 BC, when gold-based, hard solders were developed by artisans in Mesopotamia and soon spread into Egypt (3600 BC), Ur (3400 BC), Greece (2600 BC), and other Mediterranean regions. The Etruscans clearly refined hard soldering techniques during the Bronze Age (ca. 2500 BC). A gold pendant example illustrates their mastery of the step-joining technique in which three different melting temperatures are formed over a distance of 5 mm. There is also evidence of high-quality, gold-based hard solder joints being used by the La Tolita Indians of South America in AD 20.

Tin-lead soft solder compositions and their uses in ancient cultures are obscure, largely because the reduction of the tin component to stannic acid makes solder identification quite difficult. Furthermore written descriptions of soldering technique are very limited, since this craft was performed by artisan-slaves. Such endeavors were deemed unimportant by the upper ruling class and were not well documented. Some well-preserved soft-soldered artifacts were discovered in King Tut's tomb,

Modern Electroplating, Fourth Edition, Edited by Mordechay Schlesinger and Milan Paunovic.
ISBN 0-471-16824-6

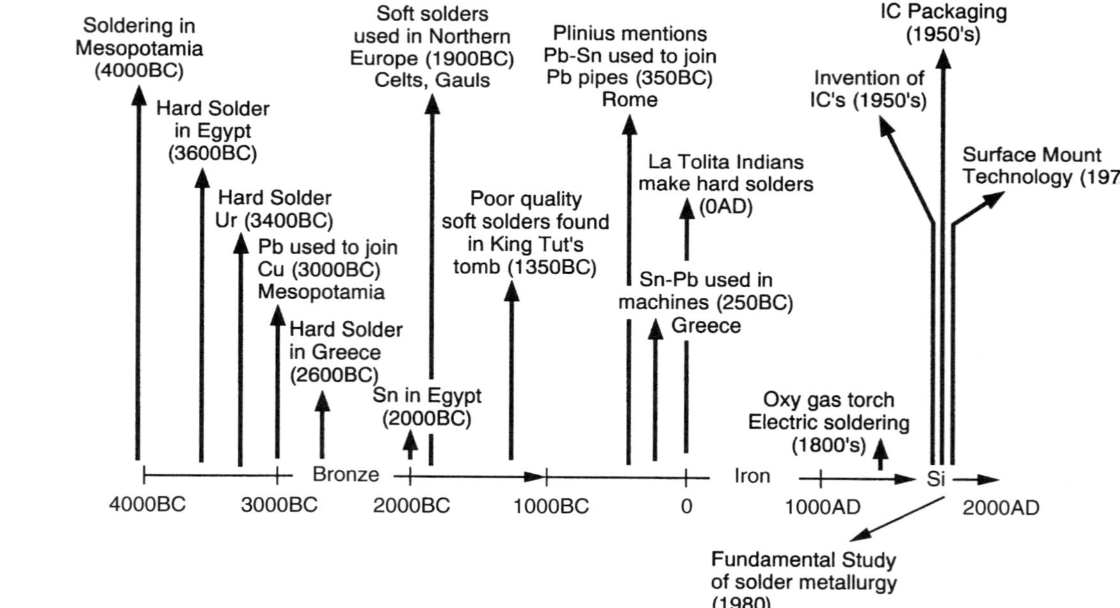

Figure 1 A time line illustrating the use of solders as determined by archaeological evidence. Reprinted with permission from P. T. Vianco et al., *JOM*, pp. 14–19, July 1993.

which dates to 1350 BC. The pieces were of crude construction, suggesting that soft soldering was not as well developed as hard soldering techniques in the Mediterranean cultures. Historians now believe that soft soldering was also developed in Northern and Central Europe by Celts and Gauls in about 1900 BC. The high quality of the Celtic artifacts constructed with tin-lead soft soldering was noted by the first Romans to enter the region. Access to rich tin ores in Northern Europe is cited as the reason for their exceptional soft soldering practices; in contrast, tin was not readily available to the Mediterranean cultures.

In ancient times, tin-lead solders were used primarily for the manufacture and repair of cooking utensils and metal tools. In 350 BC the Greeks sealed their bronze-based water pumps, air pumps, and organs with tin-lead alloys. The Romans used tin-lead solders extensively in the construction of aqueducts. The lead sheets that lined the interior of the water passages were joined together by soft soldering. As is the case today, the higher cost of tin relative to lead provided the Roman engineers with the economic incentives to use the less expensive tin-lead alloys.

Very little changed in soldering from the Iron Age to the Middle Ages. The technique was used primarily by artisans in the fabrication of jewelry and other crafts. Tin-lead solders were used to make joints in objects comprised of copper and brass.

The Industrial Revolution greatly expanded the use of soldering technology. For example, the development of portable heat sources such as the oxy-gas torch and electricity (the soldering iron) facilitated the use of soft soldering in applications such as plumbing (first lead and then copper pipes and tubing), the sealing of food and water containers, and later, the fabrication of larger structures (e.g., automobile radiators). It was in the early twentieth century that soldering entered the electronics industry as a reliable means of connecting copper wires for power and signal transmission. In those early applications, the solder joint served primarily to ensure electrical continuity.

With the advent of silicon-chip technology, the electronics revolution has fostered further miniaturization and increased functionality of devices. Soft soldering has been adapted to micro-chip packaging and higher-level system assembly. However, the reduction of package-size requirements and demand for solder joints to be a mechanical and an electrical connection have challenged the strength as well as the creep and fatigue resistance of tin-lead solders. The advent of surface mount technology clearly illustrated the limitations of tin-lead solder technology. Reliability losses in many electronic systems were identified with failure of the solder joints rather than device malfunctions.

2 DRIVING FORCES OF LEAD-FREE SOLDERS DEVELOPMENT

The selection of materials and processes in modern manufacturing will no longer be driven solely by product performance, reliability, and cost but will be profoundly influenced by environmental and health considerations. Tin-lead solders have been the most prominent material for the interconnection and packaging of electronic components, primarily due to the low cost and convenient material properties. However, in view of increasing environmental and health concerns, the time has come for the materials science community to seriously consider developing new alternatives. In order to successfully replace lead from tin-lead solders, understanding its role in solders is essential.

2.1 Role of Lead in Solders

The role of lead in solders can be understood by examining its effects on the materials properties of its alloys with tin. Melting temperature, electrical and thermal conductivity, coefficient of thermal expansion, fluidity, surface tension, mechanical properties, microstructure, and metallurgical reactions are some of the most important properties of interest. For example, lead lowers the melting temperatures for tin-rich solders as shown in tin-lead phase diagram in Figure 2. When its content increases in tin-lead solders, both electrical and thermal conductivity decrease as shown in Figure 3 [1], whereas density and coefficient of thermal expansion of tin-lead solders increase as shown in Figure 4 [1]. The viscosity of the tin-lead solders is significantly reduced when lead content goes up [2], but the surface tension is lowered relative to pure tin, improving the wetting of solder into gaps and holes. Sometimes a

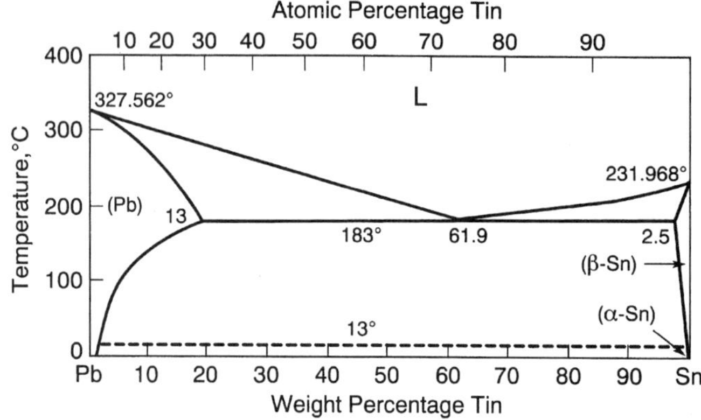

Figure 2 Phase diagram of tin-lead binary solder, from [1]. Sn melting point 231.968°C; Pb melting point 327.562°C.

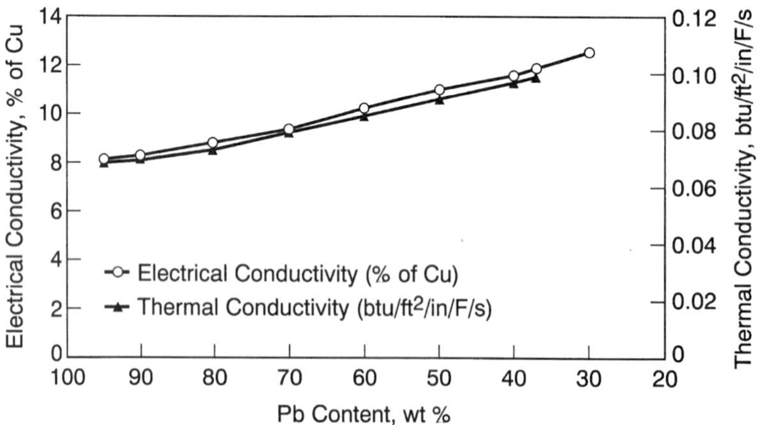

Figure 3 Electrical and thermal conductivities of tin-lead solders, from [1].

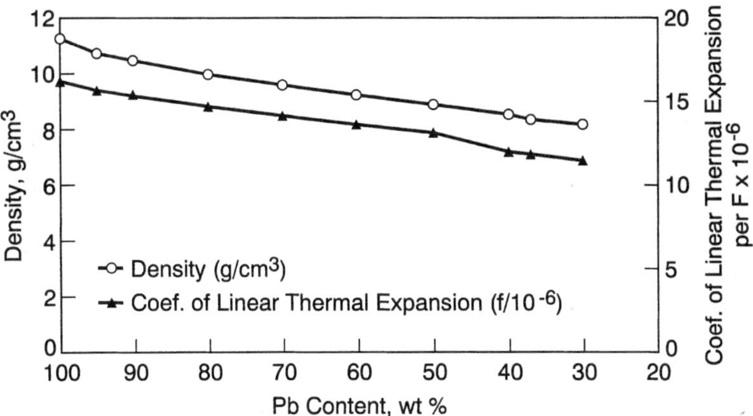

Figure 4 Density and thermal expansion properties of tin-lead solders, from [1].

term called "spreading" is discussed in the context of solder evaluation. Wetting and spreading are related in the sense that both depend on the surface tension of a solder. However, the latter is also associated with the fluidity (or viscosity) of the solder. One of the most important aspects of lead in tin-lead solders is its impact on the mechanical strength, usually represented by tensile and shear strength. The effect of lead on tensile and shear strength in tin-lead solders is illustrated in Figure 5 [3] and Figure 6 [1].

In failure mode analysis of tin-lead solders in electronics, it is often observed that decrease in solder joint strength is closely related to microstructure changes. Upon solidification of eutectic solder 63Sn37Pb, a two-phase microstructure is developed, whose morphology is dependent on the rate of solidification [4]. The microstructure changes at room temperature, and it includes precipitation of tin colonies within lead-rich phase, as well as an overall coarsening of lead-rich particles contained within the tin-rich matrix. A decrease in the strength of the solder is observed with these microstructural changes [5], thereby counteracting improvements

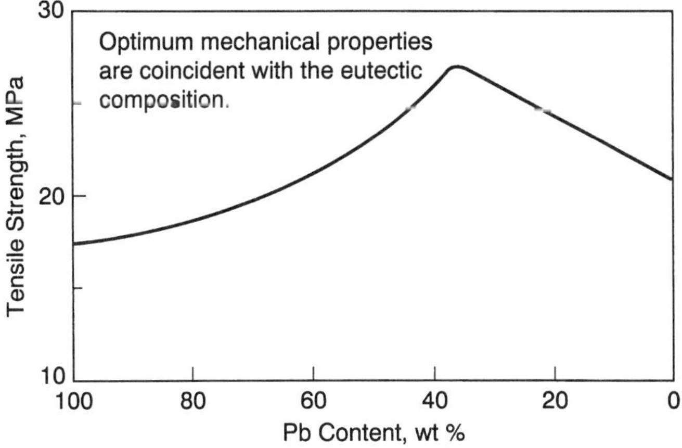

Figure 5 Tensile strength of cast bars of tin-lead alloys, from [3]. Eutectic solder, 63% Sn, 37% Pb.

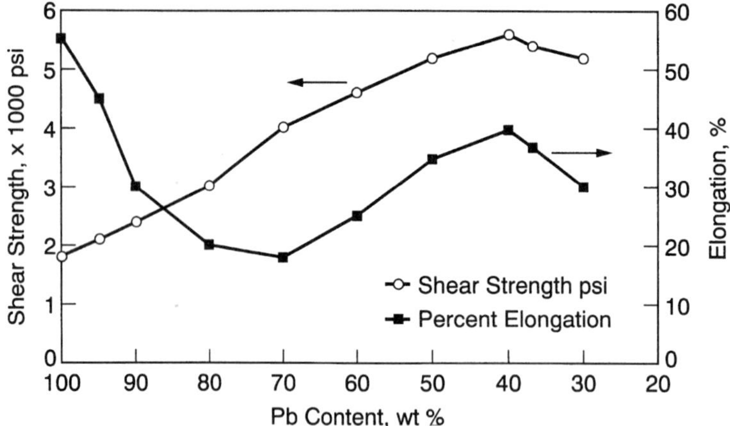

Figure 6 Shear strength of Cu joints soldered with tin-lead solders, from [1]. Eutectic solder, 63% Sn, 37% Pb.

in the mechanical strength realized through the finer morphology generated by faster cooling rates.

Metallurgically, lead is relatively "inert" comparing to tin. For example, when tin-lead solders are deposited on copper, lead itself does not react with copper, but it affects the solid-state growth kinetics of the tin-copper intermetallics. While comparing pure tin and eutectic tin-lead solders, even though the total intermetallic compound thickness ($Cu_6Sn_5 + Cu_3Sn$) is quite similar, the percentage of Cu_3Sn, the bad actor in solderability because it is brittle and nonsolderable, differs significantly [4]. In tin-lead coatings, tin segregates strongly to the surface because of its higher affinity for oxygen, forming tin oxide which is one of the culprits in solderability degradation. Lead oxides, on the other hand, improve the corrosion resistance of tin-lead solders [6].

2.2 Environmental, Health, and Safety Concerns of Lead-Bearing Solders

Lead in the electronic industry today accounts for approximately 0.6% of total consumption. The largest use is in storage batteries, which accounts for 80% of the

TABLE 1 Materials Properties of Tin, Lead, and the Eutectic Solder

Materials Properties	100Sn	63Sn37Pb	100Pb
Melting point (°C)	232	183	328
Electrical conductivity ($ohm^{-1} cm^{-1}$)	9.1	6.9	4.8
Thermal conductivity ($W m^{-1} \cdot K$)	66	50.9	35
Coefficient of thermal expansion ($10^{-6}/°C$)	22	24.1	28.9
Surface tension, ($dyn\,cm^{-1}$)	545	490	439
Shear strength (psi)	4000	3450	2000
Tensile strength (Mpa)	21.5	27.5	17.3
Elongation (%)	< 0.82	40	55

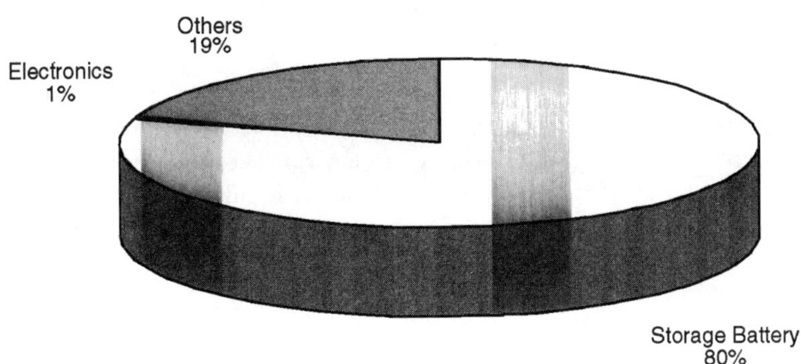

Figure 7 Lead consumption, from [12].

total consumption, as shown in Figure 7. Other major uses include ammunition, paints, cable sheathing, and sheet lead [7].

Lead and its compounds are one of the top 17 chemicals that impose the greatest threat to human health [8, 9, 10], including disorder of nervous and reproductive systems, delayed neurological and physical development, reduced production of hemoglobin, anemia, and hypertension [8]. Lead poisoning is particularly damaging to the neurological development of young children [11].

Lead is heavily regulated in plumbing, paint, and gasoline industries. Currently the electronics industry is exempted from existing regulations. However, on-going concerns regarding environmental pollutants dictate its removal from this industry in the near future.

The public mood and political climate appear to exert increasing pressure on the use of lead regardless of industry and end-use products. As to recent U.S. regulations, the Lead Exposure Reduction Act (S 729) was introduced in April 1993, and passed the Senate floor on May 25, 1994. Furthermore the Lead Tax Act (HR 2479) introduced in June 1993 imposes a $0.45 per pound tax on all lead smelted in the United States and on all imported lead-containing products.

3 CANDIDATES FOR LEAD-FREE SOLDERS

3.1 Materials Properties Considerations

Leaded solders are used primarily as mechanical, electrical and thermal interconnections, and as surface coatings for corrosion protection and solderability. Modern technology demands improved solders that possess enhanced mechanical and metallurgical properties. This provides opportunities to develop materials that (1) have a solid/liquids temperature equivalent or close to that of tin-lead solders to be comparable with existing processing technologies, (2) equal or better mechanical properties in terms of shear and tensile strength, (3) equal or better physical properties in terms of electrical/thermal conductivity, thermal expansion coefficient, and wettability, and (4) improved shelf life and microstructure stability, and (5) equal or better corrosion resistance, ductility, and solderability. See Tables 2 and 3 for a comparison of materials properties of tin-lead and lead-free alloys.

3.2 Rules of Selection

As can be surmised from the discussion above, any alternative to lead-bearing solder must meet certain criteria to be considered as a serious contender. Furthermore any element that will replace lead has to be nontoxic or significantly less toxic, relatively abundant in supply, and not significantly more costly than lead. In addition to the materials properties consideration mentioned above, the lead-free solders should be compatible with existing flux systems. Finally, the reliability of lead-free solder joints has to be tested under various environmental conditions, for example thermal cycling from −55° to 180°C. Models and test methods have to be established to predict the long term reliability. An extensive review of these issues can be found in [12].

As can be seen, after taking into consideration all the criteria, there are only a handful of lead-free solder alternatives that can be considered. For binary alloys, SnBi and SnAg appear to be the only choices. In the right soldering temperature range, there are also some ternary and quaternary lead-free alloy alternatives such as SnBiAg, SnAgIn, SnBiAgCu and SnAgSbBi. To date, they only exist in bulk, metallurgical alloy forms.

TABLE 2 Binary Lead-Free Solder Candidates Compared with 63Sn37Pb

Materials Properties	63Sn37Pb	58Sn42Bi	96.5Sn3.5Ag
Melting point (°C)	183	138	221
Electrical resistivity ($\mu\Omega \cdot$ cm)	14.6	36.4	12.3
Thermal conductivity ($W m^{-1} \cdot K$)	50.9	19	55.3
Coefficient of thermal expansion (10^{-6}/°C)	24.1	14	23
Surface tension ($dyn\,cm^{-1}$)	490	300	460
Wetting, contact angle (time, s)	17°(3.8)	43°(3.3)	36°(2.0)
Shear strength (Mpa)	42	63	55
Ultimate tensile strength (MNm^{-2})	19–56	54–73	20–56
Creep resistance	Poor	Excellent	Good

Sources: References [13–16].

TABLE 3 Ternary and Quaternary Lead-Free Solder Candidates

System	Known Composition	Melting Temperature (°C)
SnBiAg	83.5 Sn/14 Bi/2.5 Ag	142–218
	91.8 Sn/4.8 Bi/3.4 Ag (E)	
SnInAg	77.2 Sn/20 In/2.5 Ag	178
	> 90 Sn/ < 8 In/Ag	195
SnBiAgCu	95.9 Sn/1 Bi/0.1 Ag/3 Cu	206–223
	90 Sn/7.5 Bi/2 Ag/0.5 Cu	207–212
SnAgSbBi	93 Sn/3 Ag/2 Sb/2 Bi (E)	219
	90 Sn/0.5 Ag/4.5 Bi/5 Sb	228–234

As discussed above, the metallurgical and mechanical properties of lead-free solders are of utmost importance. However, if electrodeposition is the method of application, then just as important is an understanding of the chemistry and electrochemistry of lead-free alloys. From an environmental prospective, the electrolyte and other constituents in an electroplating chemistry should be environmentally friendly. From an electroplating perspective, the redox potentials of the metals have to be relatively close. However, the redox potentials of lead-replacing elements usually are significantly different from that of lead, which is close to the tin reduction potential. Hence the corresponding alloys will be difficult to plate. Theoretically speaking, it is possible to electrodeposit ternary and quaternary alloys; practically, it is almost impossible to maintain a plating bath with such complicated systems. All of these will be discussed in detail in Section 6. An understanding of electrodeposition of tin, which will be considered in the next section, is prerequisite for the understanding of electrodeposition of tin alloys.

4 ELECTRODEPOSITION OF TIN

4.1 Physical Properties of Tin

Tin metal is silver white, soft, ductile, nontoxic, with excellent corrosion resistance in air, lubricity, and an ability to form many useful alloys. With its low melting point and high boiling point, tin has a liquid range exceeded by few metals. It readily alloys with many metals and forms several intermetallic compounds of commercial importance. Copper, nickel, silver, gold, and palladium are soluble in liquid tin. Molten tin wets and adheres readily to clean iron, steel, copper, and copper-base alloys.

There are two allotropic forms of tin: white (β) and gray (α). White, or ordinary tin is the familiar form, processing properties that make it useful; it crystallizes in the body-centered tetragonal system. Gray tin has a diamond lattice; it is considerably less dense than β-tin and is nonmetallic in appearance and properties. It is a semiconductor. The allotropic transformation occurs at 13°C and is extremely slow. In fact the existence of gray tin was not even discovered until the mid-nineteenth century, when some tin organ pipes in Moscow were found to have disintegrated during an exceptionally cold winter. The allotropic change is known as "tin pest," probably because it appears to spread from the center of "infection." The transformation is inhibited or prevented by the incorporation into the tin of a few tenths of a percent of antimony, lead, or bismuth.

Tin whiskers were first reported in 1951 [21]. They are single crystals of tin filaments, usually having diameters of 0.5 to 5 μm and lengths up to a few millimeters. They can be straight or kinked, and during their growth, which occurs by extrusion from the tin electrodeposits, their orientation with respect to the substrate may vary. Tin whisker growth rate has been reported in numerous studies in the range of 0.03 to 9 mm per year. This phenomenon has been observed from a few hours to a few years after electrodeposition. They are not easily detached from the substrate and can carry currents as large as 25 mA before they burn out. They can therefore sustain potential drops to a few tenths of a volt per millimeter. There are some remedies for the tin whiskers. Reflowing the deposit by heating it is one method. Others include hot air solder leveling (HASL), hot dipped tin, and alloying tin with a small percentage of lead or other metals [92].

4.2 Chemical Properties of Tin

Tin has the atomic number of 50, and its atomic weight is 118.69. It has the electronic configuration of $4d^{10}5s^25p^2$. Thus there are four electrons available for bonding. Accordingly tin is tetravalent in many of its compounds. As with its homologies germanium and lead, however, the 5s electrons may act as an "inert pair," and tin is also divalent. In this respect tin is intermediate between germanium and lead; with the former, bivalency is uncommon, and lead exhibits tetravalency only in its organic compounds. For tin, the two valence states are almost equally stable and readily interconvertible. Solutions of stannic tin or tin(IV) are readily reduced to stannous tin or tin(II) by many reducing agents, especially metals such as antimony and nickel. Solutions of tin(II) are just as readily oxidized to tin(IV) by common oxidants, including air.

Tin is amphoteric: It reacts with acids and bases, and is relatively resistant to neutral solutions. The overpotential of hydrogen on tin is quite high, about 0.75 V, so that attack by acids and bases is slow unless an oxidizing agent is present to depolarize the evolution of hydrogen. Distilled water has no effect on tin, which has been the preferred medium for preparing and storing it.

In acidic solutions, tin(II) compounds probably exist in the form of Sn^{2+} aquo ion, but Sn^{4+} probably does not exist as such. It is either hydrolyzed or complexed, as in $SnCl_6^{2-}$ and $Sn(OH)_6^{2-}$.

In alkaline media, tin(IV) is the most stable; alkaline stannite, or stannate(II) solutions disproportionate according to

$$2[Sn(OH)_4]^{2-} \rightarrow [Sn(OH)_6]^{2-} + Sn + 2OH^-$$

This reaction is important in plating from alkaline stannate solutions, as will be shown in Section 4.4. All tin compounds tend to hydrolyze in aqueous solution: Alkaline solutions must be stabilized by the presence of excess alkali, acid solutions of excess acid.

4.3 Electrochemistry of Tin and Lead

Thermodynamic Treatment of Tin and Lead One of the most thorough and complete thermodynamic consideration of the electrochemistry of tin and lead can be found in Chapter 8 of the handbook '*Standard Potentials in Aqueous Solution*' [22]. At room temperature, the standard potentials for the deposition of tin and lead are −0.136 V and −0.125 V, respectively. Lead is the more noble metal. The reduction potentials of the two are very close together; therefore the tin-lead alloy is one of the easiest to be codeposited.

The half reactions for Sn^{2+} and Pb^{2+} under standard condition (i.e., $[M^{2+}] = 1$ M, temperature = 298 K, 1 atmosphere pressure) are

$$Sn^{2+} + 2e^- \rightarrow Sn, \quad E^0 = -0.136\,V$$

$$Pb^{2+} + 2e^- \rightarrow Pb, \quad E^0 = -0.125\,V$$

In noncomplexing acidic media, in order to maintain simple Sn^{2+} and Pb^{2+} ions in solution, the pH of the solution needs to be less than 1.

Different oxidation states of tin and lead and their interconversions are displayed in the following potential diagrams. Data are only available in acidic solutions.

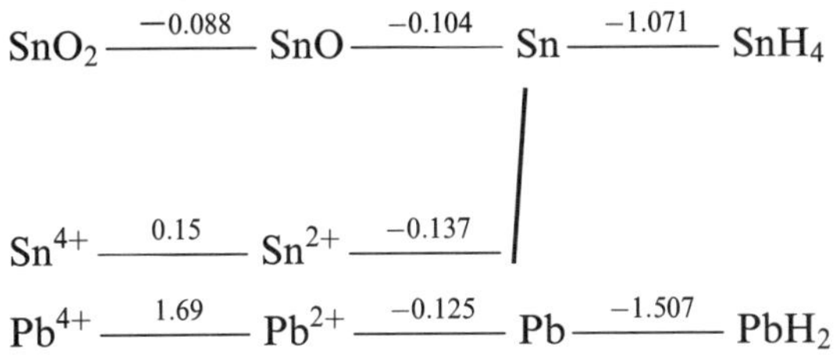

$$SnO_2 \overset{-0.088}{\text{———}} SnO \overset{-0.104}{\text{———}} Sn \overset{-1.071}{\text{———}} SnH_4$$

$$Sn^{4+} \overset{0.15}{\text{———}} Sn^{2+} \overset{-0.137}{\text{———}} (Sn)$$

$$Pb^{4+} \overset{1.69}{\text{———}} Pb^{2+} \overset{-0.125}{\text{———}} Pb \overset{-1.507}{\text{———}} PbH_2$$

Normal electrodeposition conditions deviate significantly from the standard or equilibrium condition. The reduction potential of tin or lead at any given temperature and concentration can be expressed by the following equation:

$$E = E^0 + \left(\frac{RT}{nF}\right) \times \ln a + C$$

where

E is in unit of volts
R is the gas constant, $R = 8.31441\,\mathrm{J\,mol^{-1}\,K^{-1}}$
T is the temperature, in unit of K
n is the number of electrons that are involved in the electrodeposition process
F is the Faraday constant, $F = 96485\,\mathrm{C\,mol^{-1}}$
a is the concentration or activity of the metal ion, in this case, tin or lead
C is a term that includes all the other factors that contribute to the polarization of the electrodeposition process

As is easy to see, changing the temperature or the concentration of tin or lead, or using a complexant that contributes to the C term, will affect the reduction potential. Section 6 on the electrodeposition of tin alloys will discuss this issue in more detail.

Reduction Potentials of Tin and Lead in Different Acidic Media Listed in Table 4 are onset reduction potentials for tin, lead, and tin-lead obtained by cyclic voltammetry in sulfate [23], fluoborate [24, 25], and methane sulfonic acid solutions (MSA) [26]. (Onset potential is the potential where an electrochemical process initiates.) It is important to point out that the potentials reported in the table were not the half-wave or peak potentials, but the onset potentials. In order to obtain direct correlation with an actual plating bath, high metal concentrations were utilized in the electrochemical experiments. This can have a distortion effect on the redox waves caused by a combination of factors such as large solution resistance, IR drop, or complicated chemical and electrochemical reaction kinetics that make it difficult to obtain a clear-cut peak potential or a half-wave potential. Therefore onset potentials are more useful. A discussion of this phenomenon is beyond the scope of this work and can be found in standard electrochemical textbooks [27–30].

TABLE 4 Onset Reduction Potentials (V) of Tin, Lead and Tin-Lead in Various Acidic Media

	Sulfate[a]		Fluoborate[b]		MSA[a]	
Half-reaction	No BA	BA	No TX/Ph	TX/Ph	No Additives	Additives
$Sn^{2+} \rightarrow Sn$	−0.53	−0.55	0.07	—	−0.38	−0.50
$Pb^{2+} \rightarrow Pb$	—	—	0.08	—	−0.40	−0.60
$M^{2+} \rightarrow M$	—	—	0.07	0.0	−0.39	−0.33
M = Sn, Pb						

Solution Compositions: (1) Sulfate/sulfuric solution: 0.01 M Sn^{2+} + 0.56 M H_2SO_4; (2) fluoborate solution: 0.14 M Sn^{2+} + 0.05 M Pb^{2+} + 5 M HBF_4; (3) Methane sulfonic acid solution: 0.21 M Sn^{2+} + 0.058 M Pb^{2+} + 1 M MSA. BA: benzaldehyde; TX/Ph: Triton X-100 and phenolphthalein; Starter: organic additives.

[a]Reference electrode, SCE.

[b]Reference electrode, $Pb/PbSO_4$, −0.65 V versus. SCE.

Nonetheless, the data shown in Table 4 have practical implications and are useful for comparative analysis. For instance, in MSA, the presence of "additives" that contain organic compounds such as quaternary ammonium surfactants and aromatic aldehydes, makes tin and lead deposition more negative when these two metals are deposited separately. However, when tin-lead alloys are deposited, the presence of "additives" makes it more positive. More can be learned if one examines the work of Kohl [24] and Zhang [26]. Kohl used cyclic voltammetry to evaluate the effect of Triton X-100 and phenolphthalein in fluoborate solutions shown in Figure 8. As can be discerned from the figure, both Triton X-100 and phenolphthalein significantly suppress the deposition of tin and lead. The effect becomes more pronounced when both additives are used. Another example is given in Figure 9. In this case the "additives" significantly suppress the electrodeposition of tin and lead.

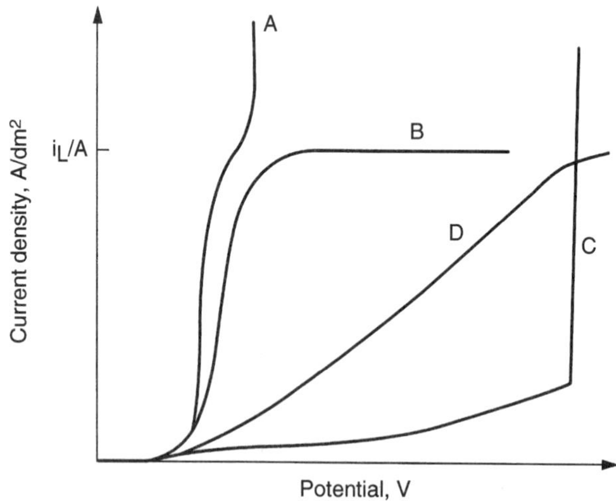

Figure 8 Current-voltage behavior of a Sn electrode in solder baths with (*a*) no additives, (*b*) Triton X-100, (*c*) phenolphthalein, (*d*) Triton X-100 and phenolphthalein. Reprinted with permission from P. A. Kohl, *P&SF*, p. 45, August 1981.

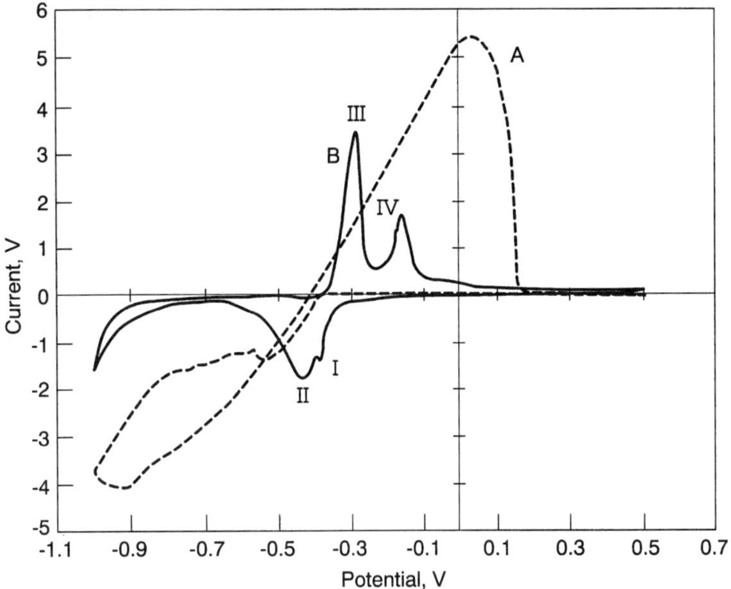

Figure 9 Cyclic voltammograms of a Sn electrode in a solder solution with (*a*) no starter additive, (*b*) starter additive. Wave I: $Sn^{4+} \rightarrow Sn^{2+}$; wave II: $Sn^{2+} \rightarrow Sn$; wave III: $Sn \rightarrow Sn^{2+}$; wave IV: $Sn^{2+} \rightarrow Sn^{4+}$. The current scale for {*a*} is 10 mA/V and for {*b*}, 1 mA/V.

Second, the onset potential for tin-lead was shifted to more positive values, which means that the tin-lead alloy could be plated out at lower current densities comparing to the situation when additives were absent. The redox waves were much better defined and it was correlated with a bright, smooth tin-lead surface. This phenomenon will be discussed more later.

Important Electrode Reactions in Tin Electrodeposition In electrodeposition, metal ions are reduced at the cathode via the following reactions:

$$\text{pH} < 1, \quad Sn^{2+} + 2e^- \rightarrow Sn$$

$$\text{pH} > 13, \quad [Sn(OH)_6]^{2-} + 4e^- \rightarrow Sn + 6OH^-$$

Under ideal conditions these should be the only cathodic reactions, and the current efficiency should be 100%. In practice, other side reactions have to be taken into consideration. For instance, discharge of hydrogen occurs at current densities close to or exceeding the limiting current density of the deposited metal, especially in acidic media. Decomposition of some organic additives is another possible side reaction. Therefore it is very rare that the cathode current efficiency is 100%.

At the anode, one of the following oxidation reactions can occur depending on the type of anodes. If a soluble anode is used, the anodic oxidation reactions are

$$\text{pH} < 1, \quad Sn \rightarrow Sn^{2+} + 2e^-$$

$$\text{pH} > 13, \quad Sn + 4OH^- \rightarrow [Sn(OH)_4]^{2-} + 2e^-$$

The stannite ion so formed is not stable in alkaline solution and most likely will disproportionate into Sn and $Sn(OH)_6^{2-}$ via the following reaction:

$$[2Sn(OH)_4]^{2-} \rightarrow [Sn(OH)_6]^{2-} + Sn + 2OH^-$$

The thus formed tin is loose, spongelike. Undesirable deposit may be obtained if it gets incorporated in. This is why an insoluble anode is usually preferred in alkaline tin plating (*vide infra*).

If an insoluble anode is used, then the anodic oxidation reactions are

$$\text{pH} < 1, \quad 2H_2O \rightarrow O_2 + 4H^+ + 4e^-$$

$$\text{pH} > 13, \quad 4OH^- \rightarrow O_2 + 2H_2O + 4e^-$$

Overall reactions are

$$\text{pH} < 1, \quad 2Sn^{2+} + 2H_2O \rightarrow O_2 + 2Sn + 4H^+$$

$$\text{pH} > 13, \quad [Sn(OH)_6]^{2-} \rightarrow Sn + O_2 + 2OH^- + 2H_2O$$

Disregarding the pH of a solution (i.e., acidic or basic), oxygen will form at anode when an insoluble anode is used. As will be discussed later, in acidic solutions, this will cause excessive formation of Sn(IV). Therefore, whenever possible, a soluble anode should be used. By contrast, in alkaline media, the oxidation of Sn(II) to Sn(IV) is beneficial because Sn(II) is not desirable.

It is equally true that side reactions will occur at the anode. For example, the oxidation of organic compounds which are used as wetting agents, grain refiners, and brighteners is especially susceptible if an insoluble anode is used. Generally speaking, breakdown products from the organic additives are often generated from anodic reactions rather than cathodic reactions.

Once the electrodeposition takes place, the deposition rate is governed by the Faraday law, expressed as

$$Q = I \times t = \frac{nFW}{M}$$

where

Q stands for coulombs
I is the current in amperes
t is the time in seconds
n is the number of electrons involved in the electrodeposition process
F is the Faraday constant, equals 96485 C mol^{-1}
W is the weight of deposit in grams
M is the atomic weight of deposited metal

As can be seen, the same amount of metal is deposited by increasing the current while shortening the plating time, or by reducing the current but increasing the plating time. Of course, in production the first procedure is the better choice. However, this is not always possible to implement, since it is sometimes limited by the plating processes. Examples will be given in Section 4.4 and where specific plating processes are discussed. Furthermore it is shown that in order to deposit a fixed amount of metal, twice as much is required when the number of electrons involved is four (as in alkaline tin plating bath) comparing to two (as in acid tin plating baths). This is why acid tin plating chemistries are more efficient than alkaline tin plating chemistries and are most preferred method to deposit tin or tin-lead.

4.4 Tin Electroplating Solutions and Processes

Considerations of Tin Electrodeposition Processes

Equipment Depending on whether the electroplating process is acidic or alkaline, requirements on the plating equipment are different. In acidic media it matters whether the acid is fluoborate, sulfuric, or methane sulfonic. The fluoborate-based system mandates the use of the most acid resistant materials. Sulfuric acid and MSA-based systems, although corrosive by definition, are much less severe on equipment. Polypropylene or rubber-lined steel can be used to construct plating tanks. Heaters made from porcelain or quartz with polypropylene guard can be used in sulfuric acid and MSA media. For alkaline plating processes, a plain welded steel tank is generally employed, the tank being fitted with the necessary anode and cathode rods and provided with means of solution heating, such as a plain steel steam coil or steel cased electric immersion heaters [53].

Anodes In tin plating, both insoluble and soluble anodes are used. Soluble anodes are usually used in acidic solutions while insoluble anodes find their use in alkaline solutions. Depending on the type of anodes used, the anodic reactions are different (*vide supra*).

At the insoluble anode, oxygen is generated. In acidic solutions, this leads to a pH decrease and a buildup of free acid, which may cause anode passivation as in the case of sulfuric/sulfate chemistry. This occurs because when free sulfuric acid is present in large excess, the conductivity of the solution is reduced significantly. Because of the common ion effect, the solubility of stannous sulfate is also substantially reduced. As a result the stannous sulfate will "salt out" at the anode, causing anode passivation. In alkaline solutions, electrodeposition of tin leads to a pH increase as shown by the overall reaction in Section 4.3 and subsequently causes substantial reduction in cathodic current efficiency, which will be discussed later in this section. Therefore it is important to monitor the acid or the base concentration closely if an insoluble anode is used.

Different types of insoluble anodes are used in electrodeposition of tin. In alkylsulfonic acid media, platinized titanium is usually used. This type of anode should not be used in sulfate-based chemistry, since sulfuric acid will attack the material. In sulfuric acid, tantalum or niobium are the recommended materials.

The purity of tin anodes is the subject of standards in many countries. The most likely impurities are Sb, As, Bi, Cu, Fe, Pb, Al, Cd, and Zn. The tin content should not be less than 99.9%. Impurity levels permitted by typical standards are

Impurity	Sb	As	Bi	Cu	Fe	Pb	Al + Cd + Zn
Max %	0.03	0.03	0.01	0.02	0.01	0.04	0.002

Metal Distribution Metal distribution or throwing power and covering power is very important in some of the applications of tin electroplating. Throwing power is a term associated with the entire spectrum of the current density, and it is a measure of deposit thickness uniformity of a plated part. Covering power reveals whether the deposition will occur at low current densities.

There are some theoretical treatments on the throwing power [31, 32] in addition to empirical treatments such as Haring-Blum [33], Heathley [34], and Field [35] equations. Although it is arguable that the theoretical model given by Tan [31, 32] allows good correlation to the observables, the experimental procedures are tedious and require computer simulation to obtain the final throwing power number. Alternatively, the Haring-Blum cell is more popular for its simplicity. However, it does have some severe limitations as pointed out by Abys and his coworkers [36]. Recently Kohl [25] used an empirical equation and based the ratio of electrode resistance and solution resistance, R_e/R_s, to represent the throwing power of a high-speed electroplating solder solution. The concept is simple and experiments are relatively easy to run. Since the measured properties have direct correlation with solution characteristics such as conductivity, metal and additive concentrations, it is much easier to interpret the data.

Additives The generic term "additives" covers a wide variety of chemicals that affect plating processes and deposits characteristics. The additives can be organic or metallic, ionic or nonionic, and are adsorbed on the plated surface and often incorporated in the deposits. The use of additives in aqueous electroplating solutions is extremely important, owing mainly to the interesting and important effects produced on the growth and structure of deposits. The potential benefits of additives include brightening the deposit, reducing grain size, reducing the tendency to form dendritic growth, increasing the current density range, promoting leveling, changing mechanical and physical properties, reducing stress and reducing pitting. However, it is equally important to point out that when additives are out of control, the opposite of all of the above can happen. The striking effects on electrocrystallization of small concentrations of addition agents, ranging from a few mg liter^{-1} to a few percent but generally in a concentration range of 10^{-4} to 10^{-2} M, point to their adsorption on a high-energy surfaces, producing a poisoning or inhibiting effect on the most active growth sites [37]. The results obtained with additives seem to be out of proportion to their concentration in solution; one added molecule may affect many thousands of metal ions. Their function and mechanism of interaction is not yet clearly understood, and their investigation so far has been mostly empirical. Nonetheless, plating additives are extremely important, and establishing the proper agents most often determines the success or failure of a given plating process [38]. Among the additives that are used in plating chemistries, the ones in nickel plating are the most studied [39, 40]. There are much fewer studies for tin and tin-lead plating [24, 41, 42].

Additives act as grain refiners and brighteners because of their effects on (1) electrode kinetics and (2) the structure of the electrical double layer at the electrode/electrolyte interface. Modern analytical and electroanalytical methods make it possible to understand these phenomena at a molecular level [43–45]. Numerous mechanisms have been suggested to explain behavior of additives such as (1) surface modification by blocking, (2) modification of Helmholtz potential, (3) surface complex formation including adsorption and ion bridging, (4) changes in surface work function and film formation of the electrode, (5) hydrogen adsorption and evolution, and (6) the effect of intermediates. These are discussed in detail in a comprehensive review by Frankin [46]. Additional excellent coverage on mechanisms of additives can be found in a more recent paper by Oniciu and Muresan [47].

Regardless of the mechanism by which additives function, their effects can be found in (1) influencing the mechanical properties such as tensile strength and elongation [48], (2) appearance and surface structure [49], (3) purity [50], (4) hardness [50, 51], and (5) solderability of the deposits [52, 53]. More will be discussed when materials properties of electroplated tin coatings are considered in Section 5.

Lack of control of the constituents of a plating solution is a major problem that leads to reduced reliability and increased costs for plated parts. One reason is the difficulty in performing quantitative analysis on the additives, which are often a mixture of two or more compounds (not to mention the numerous additive breakdown products that accumulate with time) in the ppm and ppb ranges in the presence of high concentrations of electrolytes. Techniques such as wet chemical analysis, UV-visible spectroscopy, and HPLC have been commonly used. Other practical tools such as Hull cell or hydrodynamically controlled Hull cell (HCHC) [54, 55] are also found useful in monitoring and controlling additive concentrations. However, interpretation of Hull cell results are usually not straightforward; hence it is better when coupled with chemical analysis. It is possible to quantitatively analyze the additives and their breakdown products in an acid tin plating chemistry [56]. An example will be given when MSA chemistry is discussed.

Oxidation of Tin(II) to Tin(IV) In acidic solutions, tin(II) and tin(IV) are equally stable, and tin(II) can be easily oxidized to tin(IV) by atmospheric and anodic oxidation. The loss of stannous tin, if not corrected, causes loss in deposition rate and productivity. Furthermore Sn(IV) forms fine colloidal particulates which may be incorporated in the deposits causing surface roughness and degradation of solderability.

Most of the proprietary tin plating processes contain antioxidants to minimize the oxidation of the divalent tin by atmospheric oxygen. Another way to minimize tin(IV) formation is to add a piece of tin sponge in the plating tank according to the equation:

$$\mathrm{Sn(IV)} + \mathrm{Sn} \rightarrow 2\mathrm{Sn(II)}$$

The mechanisms [57] of inhibition for Sn(IV) formation by antioxidants can be summarized by one of the following: (1) The additives form stable complexes with Sn(II), (2) the additives reduce oxygen solubility in the plating solution, thereby reducing the rate of oxidation, (3) the additives "tie up" the soluble oxygen in the solution and thereby lower the oxidation rate [58].

Bath Maintenance and Process Control Any high-quality electrodeposition process demands good quality control. Unfortunately, this is not a reality for most plating shops where only tin and electrolyte concentrations are monitored regularly. It is common knowledge that in many cases process performance deteriorates because the organic additives are out of control or due to the presence of harmful breakdown products. Therefore more attention should be directed toward monitoring and controlling the organics in a plating bath. Efforts have been made at the authors' labs to demonstrate the importance as well as the benefits of analyzing all the bath constituents by conventional analytical methods [56].

Major Acid Electroplating Tin Chemistries The most commonly encountered acidic electroplating tin chemistries are based on one of the following acids: fluoboric acid, sulfuric acid, phenolsulfonic acid (PSA), hydrochloric/hydrofloric acid, and MSA. PSA chemistries are the most widely used based on volume, since they are used in the production of tin-plate in the "Ferrostan" process [59]. Some 70% of the 130 to 140 continuous tin-plate lines around the world use this process. Sulfuric acid based chemistry is mainly found in bright tin plating. In this section a brief description of the advantages and disadvantages of each chemistry will be presented.

Fluoborate Chemistry It is one of the oldest electroplating acid tin chemistry, usually used in high-speed plating. Tin tetrafluobarate, $Sn(BF_4)_2$ is very soluble. Therefore it permits the use of high current densities. Organic additives most commonly used in fluoborate chemistry are peptone, gelatin, β-naphthol, catechol, and hydroquinone, the last two are generally used as antioxidants. With the aforementioned organic additives, smooth, fine-grained deposits are usually obtained. In the last decade or so, there has been some development in proprietary additive formulation for this chemistry. Typical bath compositions from this electrolyte are summarized in Table 5.

Fluoborate ion undergoes some hydrolysis in solution as shown below:

$$BF_4^- + H_2O \rightarrow [HO-BF_3]^- + HF$$

Formation of fluoride ion can be reduced by adding boric acid,

$$H_3BO_3 + 4HF \rightarrow HBF_4 + 3H_2O$$

This is especially important in the case of tin-lead deposition because of the precipitation of PbF_2.

TABLE 5 Typical Bath Composition and Operating Parameters of a Fluoborate Chemistry

Parameters	Rack and Barrel	Reel to Reel
$Sn(BF_4)_2$ (g liter^{-1})	75–113	225–300
Sn (g liter^{-1})	30–45	45–60
HBF_4 (g liter^{-1})	188–263	225–300
H_3BO_3 (g liter^{-1})	22.5–37.5	22.5–37.5
Anode current density (ASF)	20–25	Do not exceed 25
Cathode current density (ASF)	1–80	Up to 300
Temperature (°C)	30–55	30–55
Typical additives	Peptone, β-naphthol, hydroquinone	
Anode	Pure tin, bagged with dynel or polypropylene	
Agitation	Mild, mechanical	
Filtration	Constant filtration using a nonsilicated filter aid is desirable, although not necessary; such treatment keeps the solution clean and affords agitation	

Source: Reference [60].

The advantages of fluoborate chemistry are its ability to operate at high current densities; its high-throwing power and high current efficiencies at both anode and cathode. The major disadvantages of fluoborate are the environmental concerns for the fluoride and the boric ions, hence the high cost of the waste treatments. It is also the most corrosive of all acidic tin electroplating solutions.

Sulfate/Sulfuric Acid Electrolyte The characteristics of this chemistry is its high cathode (100%) and anode current efficiencies under normal operating conditions, namely current densities <30 ASF. Although 100 to 150 ASF have been achieved in some reel to reel applications. It is a low initial cost chemistry, and it is relatively easy to maintain and control because of the following reaction:

$$SnSO_4 + H_2O \rightarrow [Sn(OH)]^+ + HSO_4^-$$

A sulfuric acid concentration that provides pH near zero or less than zero should be maintained. It is important to avoid adding too much sulfuric acid since concentrations exceeding 400 g liter^{-1} decrease the conductivity of the solution drastically [61]. This suppresses the solubility of the tin complex, limiting accessible current density range, and it also leads to anode passivation. A typical tin sulfate bath formulation and operating conditions are found in Table 6.

The most advantageous feature of sulfate/sulfuric acid system is the low initial cost and its relatively high throwing power. Deposits ranging from matte, semi-bright to bright can be obtained from sulfate chemistry by using additives such as phenol- or cresol-sulfonic acid, gelatin, β-naphthol, and resorcinal. Proprietary additives from electroplating chemical suppliers are also available. The major disadvantages of sulfate chemistry are anode passivation at high current densities (CD $>$ 30 ASF), oxidation of Sn(II) to Sn(IV), and corrosivity of the solution to the plating equipment. Furthermore its use is restricted to pure tin plating because of very limited solubility of $PbSO_4$. Unlike fluoride-containing baths, where tin(IV) compounds exist as the hexafluostannate ion, SnF_6^{2-}, no such complex can form in sulfate chemistry. Therefore turbidity, which is caused by fine colloids of SnO_2, is expected.

In bright tin plating, it is essential that chloride ion concentration be kept as low as possible. $SnSO_4$ should not contain chloride higher than 0.1%. 100 ppm Cl^- causes a marked decrease in brightness of the deposits [53].

TABLE 6 Typical Bath Composition and Operating Parameters for a Sulfate/Sulfuric Acid Based Electroplating Chemistry

Parameters	Range
$SnSO_4$ (g liter^{-1})	15–45
Sn (g liter^{-1})	7.5–22.5
Sulfuric acid (g liter^{-1})	135–210
Additives [62]	Alkylphenol, imidazoline, heterocyclic aldehydes
Anodes	Pure tin
Anode current density (ASF)	25 max
Cathode current density (ASF)	1–25
Temperature	Room temperature
Agitation	Mechanical, cathode rod

Source: Reference [53].

Phenolsulfonic Acid Electrolyte/Halogen Electrolyte These two chemistries are mostly used to plate pure tin in the continuous steel strip plating industry. Typical bath formulations and operating conditions of the two chemistries can be found in Tables 7 and 8.

An attractive feature of these two processes is their ability to operate at very high current densities, which is very important in the steel industry. The PSA process uses a vertical cell design with either soluble or insoluble anode, whereas halogen process utilizes a horizontal cell design with soluble anode. The two processes are not interchangeable without major plant reconstruction. The PSA process usually runs at pH less than one, whereas the halogen process runs at pH around 3.4.

Sludge formation is a major problem in halogen chemistry where typically more than 20% of tin goes to sludge. Another problem is related to the waste treatment of the sludge because of ferro-ferric cyanide hazardous waste. Because of the steel strip and the horizontal cell design, the halogen bath is always contaminated with ferrous ions. The ferrous ions react with oxygen to form ferric ions, and they oxidize the stannous to stannic ions thus reducing ferric ions to ferrous ions whereby the cycle repeats. The result is that even at a low concentration of iron, there is a large loss of stannous ions and a loss of plating efficiency and quality of deposit. It is desirable to keep the concentration of iron in the electrolyte in the range of 5 to 7 g liter^{-1}. Usually sodium ferricyanide is used in the halogen bath to precipitate ferric ions. The reaction product, ferro-ferric cyanide solids which are highly toxic, is part of the sludge. Therefore sludge removal for halogen chemistry is quite costly.

TABLE 7 Typical Bath Composition and Operating Conditions of a Phenolsulfonic Acid Based Chemistry

Parameters	Range
Sn (g liter^{-1})	20–35 g liter^{-1}
Phenolsulfonic acid (g liter^{-1})	40–80 g liter^{-1}
Additives [53, 63]	Ethoylated β-naphtholsulphonic acid
Antioxidant[a]	—
Current density (ASF)	200–500
Temperature (°C)	30–40

[a]Phenolsulfonic acid acts as an antioxidant.

TABLE 8 Typical Bath Composition and Operating Parameters of a Halogen Chemistry

Parameters	Concentration or Condition
NaF (g liter^{-1})	30
$NaHF_2$ (g liter^{-1})	31
SnF_2 (g liter^{-1})	19
$SnCl_2 \cdot 5H_2O$ (g liter^{-1})	22
$Na_4Fe(CN)_6 \cdot 10H_2O$ (g liter^{-1})[a]	2–4
Additives [53]	Naphtholsulphonic acid and polyalkylene oxides
Antioxidant [64]	p-$NH_2C_6H_4NHCOMe$
pH	3–4
Current density (ASF)	200–500
Temperature (°C)	55–65

[a]$Na_4Fe(CN)_6 \cdot 10H_2O$: a reagent to precipitate ferric ions.

Compared with the halogen process, sludge formation is much less severe for the PSA process. This is because PSA, which is the main electrolyte, also acts as an antioxidant. A drawback of PSA process is the toxicity of the phenol group which is released during the electrodeposition process.

In the last decade, MSA based chemistry has been considered an alternative for PSA and halogen chemistry because it is environmentally more benign. MSA is twice as conductive as PSA chemistry; hence lower tin concentration can be used, which reduces significantly the loss of tin due to drag-out. Though in theory MSA and PSA chemistries can be interchanged without any major plant reconstruction, the initial *chemical* cost for MSA chemistry is significantly higher than PSA. To convert halogen chemistry to MSA chemistry requires some major plant modification, partly because MSA chemistry ($pH \sim 0$) is more corrosive than the halogen chemistry ($pH \sim 3$). Even though MSA process can be operated with a horizontal plating line, the construction materials need to be polypropylene type; concrete and steel can not be used unless they are coated with acid resistance materials. The chemical cost of a halogen chemistry is again less than that of MSA.

Methane Sulfonic Acid Electrolyte In the 1940s Proell [65] recognized the utility of alkane sulfonic acids for electroplating applications. Alkane sulfonic acids, having between one and five carbon atoms in the alkyl group, form water soluble salts of various metals (mesylates). The alkane sulfonates do not undergo any appreciable hydrolysis, regardless of the process temperatures. Additionally the alkane sulfonates are stable in acidic, neutral or alkaline solutions. Proell indicated that it is possible to plate many metals from alkane sulfonate baths, including cadmium, lead, nickel, silver, and zinc.

Although the usefulness of the alkane sulfonic acid system was known for several decades, it only gained commercial acceptability during the early 1980s and became a desirable electrolyte for tin, lead, and tin-lead plating. Various plating baths and additives have been formulated for MSA electroplating systems and many patents have been granted [66–69]. Obata [66] described tin, lead and tin-lead plating baths that remained stable over a broad pH range (2.0–9.0) and were used over a wide current density range.

MSA chemistry shows clear advantages over fluoborate, sulfate, halogen, and phenolsulfonic acid chemistries. It is less corrosive than fluoborate and sulfate chemistries, Sn(II) to Sn(IV) conversion is less a problem, and it is less costly than fluoborate, PSA, and halogen chemistries for effluent treatment and disposal. It is environmentally more friendly.

An additional attractive feature of MSA chemistry is its ability to access very high current densities. At authors' facility, current densities as high as 1800 ASF has been achieved with a simple additive system. The throwing and covering power of this chemistry are adequate to applications such as PWB and connectors with restricted areas [80]. Insoluble anodes can be used in MSA chemistry.

Table 9 summarizes three typical bath compositions and operating parameters for PWB, high-speed reel-to-reel and wire applications utilizing a satin bright MSA chemistry. Bright deposits are not desirable for PWB and wire applications, instead, matte or semi-bright finishes are preferred because of ductility, solderability, and corrosion resistance requirement. However, for some applications, appearance is the most critical consideration and a bright tin chemistry needs to be considered.

TABLE 9 Typical Bath Compositions and Operating Conditions of a Satin Bright MSA Based Tin Chemistry for Various Applications

Parameters	PWB	Reel to Reel	Wire
Sn (g liter^{-1})	15–25	30–50	50–70
MSA (ml liter^{-1})	225–275	175–225	175–225
Surfactant [56] (ml liter^{-1})	20–30	30–45	30–45
Grain refiner [56] (ml liter^{-1})	10–20	20–30	20–30
Antioxidants (g liter^{-1})[a]	—	—	0.75
Current density (ASF)	< 100	25–250	50–700
Temperature (°C)	30–60	30–60	30–60

[a]The antioxidants are proprietary addition agents.

For a bright tin chemistry, the most important aspects to take into account other than throwing/covering power of the bath are (1) foaming, especially in controlled depth cell applications, (2) Sn(IV) formation, and (3) brighteners control.

Foaming can be generated from (1) high surfactant/brightener concentration or (2) anodic or cathodic gas formation. Foaming can be corrected if lower additive concentrations may be used. When that is not possible, an antifoaming agent is used. Alcohols act as antifoaming agents, but care has to be taken because they can be fire hazards. In addition their effects are short-lived, since high agitation, or anodic and cathodic gas formation accelerate their dispersion from solution. Another well-known class of antifoaming agents is silicon-based surfactants. They are not stable in MSA based chemistry because the operating pH is normally less than one, which is too acidic for this class of materials.

When anodic or cathodic gas formation becomes the predominant source of foam, usually consisting of tiny bubbles instead of large bubbles as generated by mechanical agitation, simply adding antifoaming agents will not solve the problem. Organic additives that selectively suppress the gas formation at the electrodes are required. This usually requires an understanding of the overall plating chemistry, and the effects of the organic foam suppressers on deposit properties.

Another concern in high-speed reel-to-reel application is the Sn(IV) formation, especially in flood cell. Sn(II) to Sn(IV) conversion may be severe due to constant and extensive exposure of the plating solution to air. A common remedy is to use antioxidants. In MSA-based chemistries, hydroquinone and its derivatives have been used extensively for this purpose. Because of the active nature of this type of compounds and its potential detrimental effects on deposit properties, care has to be taken to maintain optimum concentrations.

The difficulties in controling the brightener levels in the plating bath make it almost impossible to control the amount of included organic matters in the deposits. Organic matter causes a number of problems such as inconsistent appearance or color, brittle coating, and poor solderability. It is therefore sensible to use a satin bright or semi-bright tin chemistry when those properties are of concern.

The biggest difference between a satin or semi-bright tin chemistry and a bright chemistry is in the organic compounds used as additives. In a satin or semi-bright tin chemistry only wetting agents and grain refiners are utilized. Of course, depending on the number and chemical nature of the organic compounds used as wetting agents and grain refiners, one may find that some satin or semi-bright chemistries are rather complicated as well.

The appearance of connectors plated from the satin bright chemistry summarized in Table 9 is shown in Figure 10. Ductility and solderability are good and consistent. Figure 11 displays the control of the additives by high performance liquid chromatography (HPLC). Among the two additives, only the grain refiner was

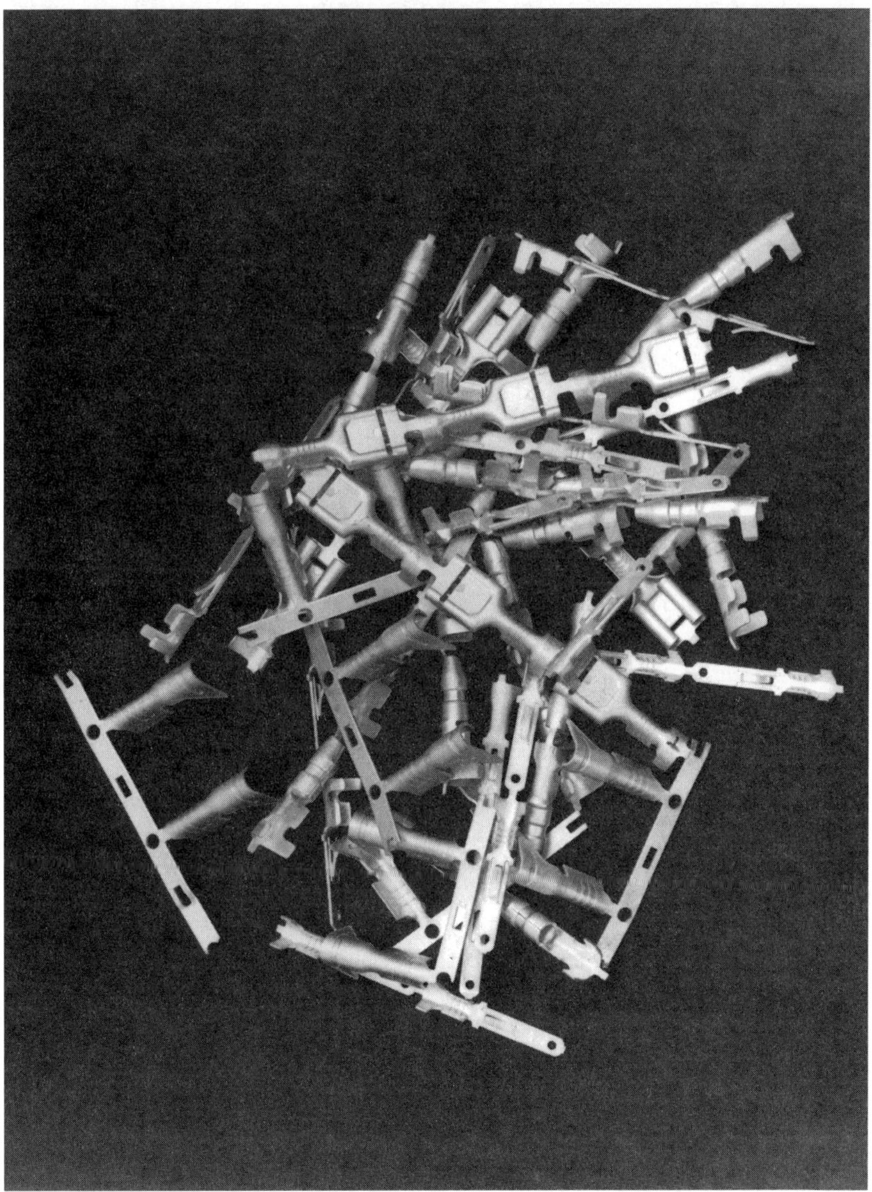

Figure 10 Connectors obtained from a satin bright tin chemistry. Photo courtesy of Electroplating Chemicals and Services, Lucent Technologies Bell Labs.

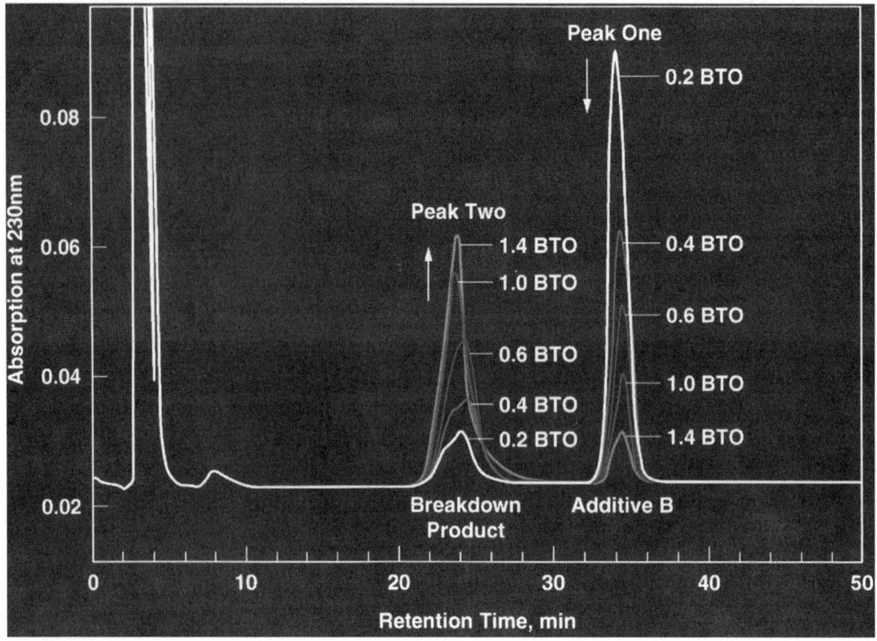

Figure 11 HPLC monitoring of additives and breakdown products in satin bright tin electroplating chemistry as a function of bath turnovers (BTO), from [56].

consumed during the electrodeposition, and there was very little consumption of the surfactant. The slight decrease of surfactant concentration was mainly caused by drag-out. The HPLC data were also useful in determining the concentration of breakdown products [56].

Alkaline Tin Plating Chemistries and Processes There are essentially two types of alkaline tin plating chemistries based on either potassium or sodium stannate [53, 70, 71]. Typical bath formulations and operating parameters can be found in Tables 10 and 11. Of the two types, the potassium stannate formulation is superior. Potassium stannate is more soluble than sodium stannate, and it allows operation at higher current densities with higher current efficiency. The choice between them depends on a balancing of the superior operating properties of the potassium bath against the lower chemical cost of the sodium.

In general, these processes are simple and can operate without additives. According to Haring and Blum [33], the throwing power of alkaline tin plating bath

TABLE 10 Typical Bath Formulations and Operating Conditions of Potassium Stannate Baths

Parameters	Barrel	Rack
$K_2Sn(OH)_6$ (g liter^{-1})	95–110	390–450
KOH (g liter^{-1})	13–19	15–30
Cathode current density (ASF)	3–10	Up to 40
Temperature (°C)	65–90	80–90

Source: From reference [71].

TABLE 11 Typical Bath Formulation and Operating Conditions of a Sodium Stannate Bath

Parameters	Range
$Na_2Sn(OH)_6$ (g liter^{-1})	95–110
NaOH (g liter^{-1})	7.5–11.5
Cathode current densities (ASF)	0.5–3
Temperature (°C)	60–85

Source: From reference [71].

is among the highest. These bath are usually operated at high temperatures, which allows higher cathodic current density and faster anodic dissolution of tin. A minimum temperature of 60°C is maintained.

Nonetheless, there are some concerns in operating alkaline baths. In order to maintain reasonable cathode current efficiency, the free alkaline concentration has to be under tight control for the following reasons: If the alkalinity is too low, hydrolysis of tin(IV) occurs because of the reaction

$$[Sn(OH)_6]^{2-} \leftrightarrow SnO_2 + 2OH^- + 2H_2O$$

On the other hand, if the alkalinity is too high, the reduction potential for $Sn(OH)_6^{2-}$ will be pushed to more negative value according to the equations

$$[Sn(OH)_6]^{2-} + 4e^- \rightarrow Sn + 6OH^-$$

$$E = E^0 - \frac{RT}{4F} \ln \frac{[OH^-]^6}{[Sn(OH)_6]^{2-}}, \quad [OH^-] > [Sn(OH)_6]^{2-}$$

Consequently the codeposition of hydrogen becomes relatively more important, thereby reducing the cathode current efficiency for the electrodeposition of tin.

As we noted early in our discussion of *anodes*, if an insoluble anode is used, it is inevitable that the pH will increase, so the excess OH^- has to be neutralized. Acetic acid is commonly employed for this purpose. Accompanying the increase in alkalinity, the stannate concentration decreases as a result of electrodeposition, so potassium or sodium stannate has to be replenished. The resulting salt buildup decreases stannate solubility and current efficiency, and at some point the bath can not be rejuvenated.

In cases where a soluble anode is used, it has to be "filmed" first. Lack of the proper film will cause tin to dissolve as Sn(II), which is the main cause of unsatisfactory deposits. In order to film tin anode, a higher-than-normal current density—a "surge"—must be applied for a few seconds to a minute; the current is then reduced to its operating value. If this process is carried out properly, a yellowish film will form and will be maintained as long as current flows. The film dissolves fairly quickly on shutdowns and must be re-formed on start-up. If current density is too high for the operating conditions, the film will thicken and turn black; the tin is now covered with an impervious film of oxides and is essentially inert and insoluble. If the current is too low, the film will be lost and Sn(II) will form in the bath. Ninety percent of the problems encountered in operating alkaline bath are attributed to anode reactions [71].

Compared with acid tin electroplating chemistries, alkaline tin chemistries are not as widely used, mostly because the current densities are limited.

5 MATERIALS PROPERTIES AND APPLICATIONS OF ELECTROPLATED TIN

Any discussion on electrodeposition will not be complete if the materials properties of the deposits are not covered. In this section effort is made to correlate materials properties such as appearance, surface morphology, ductility, purity, and hardness of electrodeposited tin to the electroplating chemistries and process parameters.

5.1 Appearance

Appearance is a generic term used here to describe the overall quality of a deposit when examined visually, or with the help of some quantitative instrument such as a glossmeter [72]. It is one of the most important properties of a deposit, and it is critical to many applications. For instance, it is a primary indicator of the deposit quality and the long-term performance of the chemistry. Despite of its importance, very little work has been done to establish a system that will quantitatively evaluate appearance of a surface and/or rank various surface finishes for their brightness. In the electroplating industry, people loosely use terms such as dull, matte, satin bright, semi-bright, and bright finishes. There exist a lot of inconsistencies. This is especially true for electroplating using tin chemistries.

The glossmeter appears to be a simple, quantitative tool to evaluate appearance. It is based on the measurement of the specular component of the reflected light. The measurement values are related to a highly polished, black glass standard. The black glass standard has an assigned specular gloss value of 100 (calibration). The specular component of the reflected light is mostly associated with the brightness of a surface. By measuring the specular reflectance percentage of high-, semi-, or low-gloss surfaces, equivalent to, bright, semi-bright, or matte surfaces, a quantitative relationship can be established between the glossmeter reading and the scale of brightness.

In tin electroplating, brighteners usually have the most profound effect on appearance, though metal and acid concentrations also exert some effects but not with the same magnitude. In terms of plating parameters, for a nonbright chemistry, changing current densities can make a deposit look matte to semi-bright. For example, of the MSA-based chemistries, the one illustrated in Section 4.4 yields brighter deposits with higher current densities until the limiting current is reached. Pulsed current also has significant effect on appearance as well. Principles and applications of pulse plating can be found elsewhere [73]. Although there are extensive publications on the use of pulse plating in electroplating industry, most of the work are done with gold, silver, copper, and nickel chemistries. Little was published on tin and tin-lead chemistries. As will be shown in the next section, pulse current modifies the surface morphology substantially, which is reflected by the change in the appearance of the deposit.

5.2 Surface Morphology and Texture

As discussed by Dini [39], the properties of all materials are determined by their structures. Even minor structural differences often have profound effects on the properties of electrodeposited metals [74]. Therefore it is important to obtain structural information of deposits at a microscopic level. Scanning electron microscopy is commonly used to study surface morphologies. Besides grain size and shape which can be obtained from surface morphological studies, the texture of deposits (defined as preferred distribution of grains having a particular crystallographic orientation with respect to a fixed reference frame) is also important. Texture is usually studied by X-ray diffraction.

The three most often encountered structures of electrodeposited tin are (1) columnar, (2) polygonized, large-grained, and (3) fine-grained. Columnar structures are characteristic of deposits from simple ion acidic solutions containing no or low additives such as tin from sulfate, fluoborate, or MSA solutions, operated at elevated temperature or low current density. An example of this type of structure is shown in Figure 12, which was obtained from a MSA matte tin chemistry.

Well-polygonized, large-grained tin deposits have been reported from a satin bright tin chemistry [56] and a bright tin chemistry [75] under different deposition conditions; details can be found in the respective references. Figure 13*a* shows an example of this structure where the main features are well-polygonized grains in the range of roughly 1 to 8 µm. As can be discerned in Figure 13, the pulse current changes the surface morphology significantly from a columnar type structure to a well-polygonized large grained structure. As demonstrated by Kakeshita, et al. [75], this type of structure is instrumental in preventing whisker growth, and it will be discussed in Section 5.8.

Figure 12 A typical columnar type of structure of electrodeposited tin coatings, from [53].

Figure 13 Electrodeposited tin coatings obtained from a satin bright tin chemistry, obtained (*a*) by pulsed current plating [73], (*b*) by direct current plating.

Fine-grained structures are normally obtained from complex ion solutions such as cyanide copper solution or with the presence of brighteners in the case of tin plating. Fine-grained deposits usually are bright, whose grain sizes fall below 0.15 μm. They are less pure, harder, and more brittle, exhibiting higher electrical resistance due to organic inclusions. The biggest problem with bright deposits is solderability degradation, caused by high organic inclusions, intermetallic compounds formation, and surface oxidation. They should not be used in applications that require reflow.

It is well known that organic inclusions have profound effects on deposit properties such as ductility and hardness. Texture or preferred orientation play an important role as well on these properties. The later is not fully recognized and less studied. Tin deposited from alkaline stannate solutions was reported to have a preferred orientation with (100) crystal planes, whereas a (110) orientation was observed from a acid tin solution [76].

Nonetheless, surface morphology and texture of a deposit are influenced by plating parameters. The effect of metal ion concentration, additives, current density, temperature, agitation, and polarization on surface morphology is illustrated in Figure 14 [77]. The effect of the same plating parameters on texture is more complex and less studied, so it is difficult to show even pictorially how individual plating variables influence texture.

5.3 Purity

Purity of electrodeposited metals is mostly affected by the specific electroplating chemistries and anode materials. The conventional methods for purity analysis of electrodeposited tin coatings are (1) elemental and (2) differential thermal analysis. Organic contents such as carbon, sulfur, and hydrogen are usually reported in wt% (C, S) or in ppm (H) in the elemental analysis. Differential scanning calorimetry has

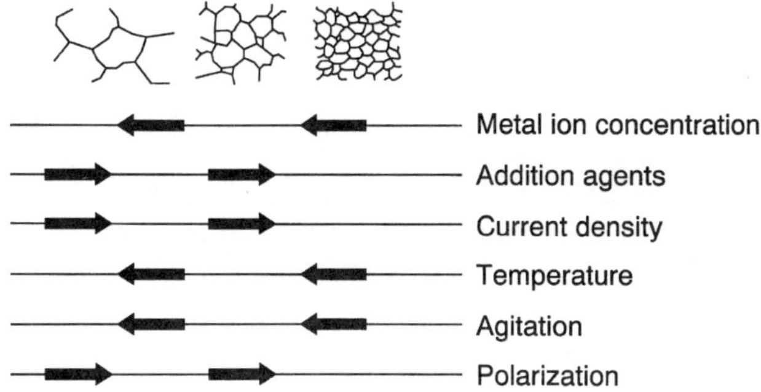

Figure 14 Relation of structure of electrodeposits to operating conditions of solutions, from [77, p. 147].

been employed to measure the purity of a satin bright tin deposit [56] and to determine solder alloy composition and liquidus temperature [78, 79]. The accurate assessment of liquid temperature, at which a metal or an alloy totally melts, is very important in setting up reflow profiles.

Typically organic inclusions from a nonbright tin chemistry are less than those from a bright tin chemistry. It was reported [56] that the carbon content in a satin bright tin deposit was $< 0.004\%$. Carbon contents in bright coatings are usually $> 0.1\%$ and are typically much higher. High organic inclusions are usually associated with poor ductility, higher hardness, and poor solderability and reflow characteristics. On the contrary, the satin bright tin deposit mentioned earlier has excellent ductility ($\%E > 26\%$), excellent solderability, and reflows without dewetting.

Although most of the impurity problems are related to organic inclusions, metallic and inorganic impurities can be just as harmful. Common metallic impurities in electroplating pure tin are copper, iron, zinc, and nickel. In acidic tin plating solutions, copper, zinc, and iron usually come from substrates. Nickel is usually a result of drag-in from the nickel plating tank. These contaminates can have profound effects on the brightness of tin deposits as demonstrated by Zhang and coworkers [80]. Among inorganic species, it is known that ppm level of Cl^- has dramatic effect on appearance of either satin bright [80] or bright tin coatings [53]. It also reduces covering power of a pure tin bath [80].

5.4 Solderability

Solderability is defined as the ability of a base metal surface to be wetted by molten solder under specified conditions of time, temperature, and environment exposure. Four factors must be considered during a soldering operation:

1. Thermal demand. The thermal characteristics of the component must allow heating of the solder joint area to the desired temperature within the time available for the soldering operation.
2. Wettability. The surface must allow wetting by molten solder within the time available for the soldering operation and without subsequent dewetting.

Wettability is generally measured by degree of coverage of a surface (i.e., lack of dewetting), or by some measure of the wetting angle, time or force.

3. Resistance to soldering heat. The temperature of the solder and the associated thermal stresses must not affect the function of the component.
4. Resistance of metalization. This issue revolves around the dissolution of the surface finish, which can lead to the exposure of an underlying nonsolderable surface.

For tin-lead coated components solderability evaluation, conditions 2 and 4 are especially important.

The major solderability tests for components are the dip-and-look test, wetting-balance test, and sequential electrochemical reduction analysis (SERA) [81]. Of the three, the dip-and-look test is still the most popular, but the wetting-balance test is the most quantitative and informative. SERA is a relatively new test method, and it has just started to gain recognition.

The dip-and-look test is the oldest, simplest, cheapest, and most widely used solderability test. The specimen is fluxed and then dipped vertically into a molten solder bath. Both the rate of immersion and withdrawal must be controlled as well as the dwell time and the temperature of the solder. The test is frequently performed manually, but it can be automated. The measured value is the percent coverage of the surface. The evaluation is normally done by the naked eye. The subjectivity of this measurement is the principal drawback.

The wetting-balance test was invented by Duis in 1967, and acceptance and use of this methodology was initially much better in Europe than in the United States or the Pacific Rim. In this method the specimen is suspended from a very sensitive balance over a solder pot. Subsequently the pot is raised at a controlled speed, and the specimen is immersed to a specified depth in the molten solder. After a specified dwell time, the pot is lowered. The output of the test is a curve of the measured weight change of the specimen during the test. Wetting force and speed can be obtained from the wetting curve, which can be correlated to solderability. The principle issues concern repeatability, both for individual machines and from site to site, and pass/fail criteria. Interpretation of the results can be arguable, and for this reason the ANSI states that the wetting balance is a test "without established accept/reject criteria," and leaves it to the suppliers and the users to determine the acceptance criteria.

Solderability is one of the most important properties of tin-lead solders for obvious reasons. The single most important parameter for unacceptable solderability is organic inclusions in the deposits. Although the mechanisms proposed on the effects of organic content on solderability remain a matter of debate, the simple fact is that bright finishes have worse solderability than matte or satin/semi-bright finishes, especially after storage under ambient condition. Therefore, when solderability is of concern, a satin bright or semi-bright finish should be sought.

Another matter to take into account is the use of underlayer as a means to preserve solderability. For instance, for electronics applications, the following thickness specification is not uncommon: 100 μin Sn or SnPb over 60 μin Ni on a copper alloy substrate. The purpose of a nickel underplate is to reduce the rate of Sn/Cu intermetallic compound formation. Nickel forms intermetallic compounds with tin under a much slower rate [82]. The over 100 μin tin or tin-lead thickness specification ensures that the intermetallic compounds and/or substrate will not be exposed to air because of pores generated in the deposit during plating or that the tin in the deposit is not completely depleted because of the formation of intermetallics.

The most commonly encountered solderability failures in the dip-and-look test are nonwetting and dewetting, which are often attributed to the formation of surface oxides, organic inclusions, and intermetallic compounds. Of the three, organic content is largely dependent on the type of chemistry chosen. Surface oxides and intermetallic compound formations are time and temperature dependent. In their recent work [83], Zhang and coworkers demonstrated that the extent of surface oxide formation are comparable for a bright and a semi-bright 90/10 tin-lead deposit. Specifically, the same surface species (SnO_2) and thickness of the oxide were identified after 48 hours steam aging by Auger depth profiles for both finishes. However, the intermetallic compounds layer was three times thicker in the case of the bright 90/10 tin-lead deposit. In both cases there was no evidence of the substrate and intermetallic compounds at the surface. The solderability failure, which was nonwetting, was therefore attributed mostly to the surface oxide formation.

The detrimental effect of organic inclusions was illustrated in a controlled study with a bright 90/10 tin-lead finish and a semi bright 90/10 tin-lead finish [80]. After plating, done under identical conditions, both finishes were subjected to a thermal bake test for 4 minutes, the resulting surfaces were then examined visually and by a SEM. As shown in Figure 15, the bright finish had gross dewetting, whereas the semi-bright finish had excellent wetting characteristics.

In summary, it is not only important to know how to perform and interpret solderability tests, but it is more important to choose the proper plating processes and control plating parameters to ensure good solderability.

5.5 Ductility

Ductility, as evidenced by elongation percentage, is a measure of plastic deformation of a thin coating prior to fracture. Tin and tin-lead coatings are usually known for

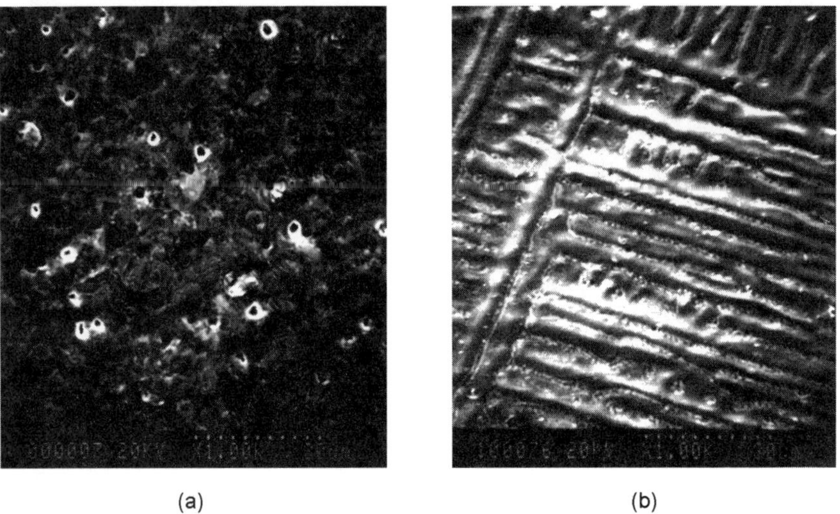

(a) (b)

Figure 15 SEM micrographs of (*a*) a bright 90/10 tin-lead coating after thermal bake, (*b*) a satin bright 90/10 tin-lead coating after thermal bake.

their high ductility or elongation. This property is highly dependent on the specific chemistry and the plating parameters. In general, it is true that bright deposits normally have lower ductility [84].

Other than the plating chemistry, the ductility of a thin coating is also influenced by the substrate and the thickness of the coating. For instance, ductility is usually higher when the deposit is attached to its substrate on which it has been plated. Lower ductilities of the foils tested without their substrate are probably caused by the more severe local plastic deformation in the region where fracture subsequently occurs. Murphy [85] has showed that the elongation percentage of electrodeposited copper increased significantly as the thickness of the deposit increased.

5.6 Hardness

Tin and lead coatings are known to be soft [86]; lead is softer than tin, as shown in Figure 16. As far as tin-lead coatings are concerned, matte finishes usually have lower hardness values ($\sim$8–10 $kg\,mm^{-2}$) than those of bright finishes ($\sim$15–20 $kg\,mm^{-2}$) [53]. Bright deposits often result in cracking when they are bent, exposing the substrate so that both solderability and corrosion resistance suffer. In general, the hardness of an electrodeposited metal is proportional to the amount of organic content in the deposit. However, texture or preferred orientation of the crystal growth sometimes plays a more important role. It has been shown that, by changing the preferred orientation, which can be accomplished by manipulating the electroplating conditions, the hardness of a copper foil changed drastically [87], Figure 17. Therefore it is possible for a deposit containing higher organics to have lower hardness value, depending on the texture. This phenomenon has been observed in authors' labs on bright tin-lead deposits [88].

5.7 Applications of Electroplated Tin Coatings

Tin deposits are widely used in the manufacture of food processing and shipping equipment and containers, pump parts and automotive pistons, copper and steel wire, and in the plating of electrical components and printed wiring boards. Tin is

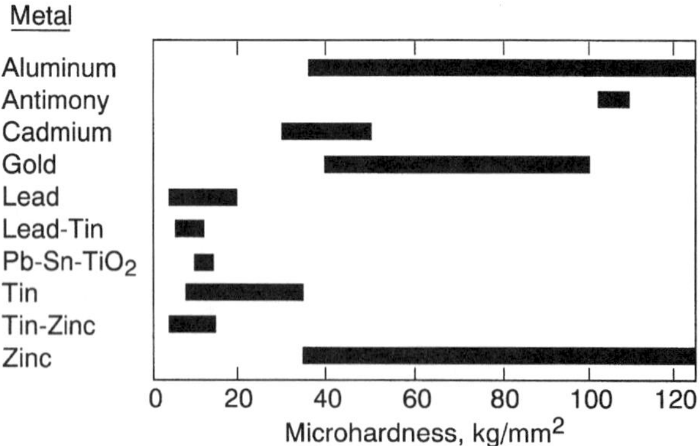

Figure 16 Microhardness ranges for relatively soft electrodeposited metals and alloys, from [86].

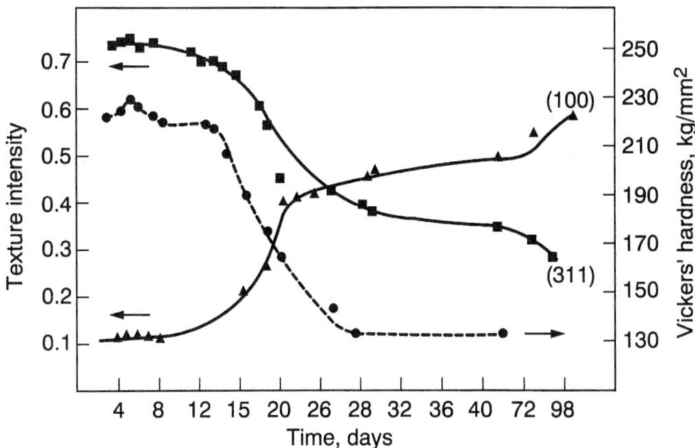

Figure 17 The relationship between the texture and hardness of electrodeposited copper, from [87, p. 167].

appealing to the food processing industry because it is a nontoxic, ductile, and rather corrosion resistant metal. The excellent ductility of the metal allows a tin-coated base metal sheet to be formed into a variety of shapes without damage to itself. When used to protect steel, the deposit should be pore free. If the deposit has pores, scratches, or other discontinuities, severe rusting of the base metal will result when the part is exposed to moisture. Thicker deposits up to 30 μm are generally required in food processing equipment and shipping containers to prevent this phenomenon.

Tin is widely used in electronics industry because of the metal's ability to protect the base metal (typically copper, nickel, and tin-nickel) from oxidation. Tin thus preserves solderability of the base material, a critical requirement when plating for electronic applications.

Although tin is used primarily for corrosion resistance and electronic applications and to a minor extent for decorative purposes, the coating is also used in a variety of engineering applications. Thick deposits (50 to 250 μm) are electroplated on pump parts and piston rings. Because of the excellent lubricity of the metal, it is plated on oil well pipe couplings. The alkaline stannate tin plating process is generally used in this application because of the fear of hydrogen embrittlement with acidic tin plating processes.

It should, however, be pointed out that as plated, tin coatings are usually compressively stressed. Tin whiskers may grow spontaneously which may cause electrical shorting of closely spaced components. The remedies to overcome tin whisker growth will be discussed next.

5.8 Prevention of Tin Whiskers

The root cause of whisker formation is internal stress generated during the electroplating process or induced under external mechanical stress. Various factors in an electroplating process can influence the properties of a tin coating. For instance, the effects of electrolyte, current density, temperature, brighteners [89, 90], substrate [91] and inorganic "impurities" such as Pb [91, 92] and Zn [93], have been reported. Thickness [89] and grain structures [94–97] of the electroplated coatings are also discussed in the context of whisker prevention. Among all the variables that are related to the electroplating processes and deposit properties, two factors seem to play a more important role: the grain structure and organic impurities in the deposits. These two factors are closely associated with the nature of the electroplating chemistry and the plating process.

External mechanical stress has been shown to have a significant effect on the whisker formation from electroplated tin coatings [95–97]. Tin whiskers grow under compressive internal stress that represents a thermodynamically unstable condition. Thus one way to reduce the probability of whisker growth is to electrodeposit tin close to their equilibrium condition. If a tin deposit is fine grained and under stress, it is favorable to recrystalize and form large grains to lower the total free energy. For pure tin, the recrystalization can take place at a temperature as low as −4°C [95], and it induces the formation of tin whisker single crystals which have diameters in the range of a few microns [92]. Alternatively, if the grain size of the tin electrodeposit is in the few micron range, the driving force to recrystalize is eliminated; hence the probability of whisker formation is greatly reduced. Similar conclusions were reported by Kakeshita, et al. [95] and Cunninghem et al. [96]. For instance, Kakeshita has shown that tin deposits having grains in the range of 0.2 to 0.8 μm formed whiskers easily, whereas no whiskers were found for deposits having well polygonized grains in the range of 1 to 8 μm.

Furthermore fine-grained deposits have faster rate of intermetallic formation due to the effect of diffusion. Diffusion can occur by various mechanisms: diffusion through bulk, diffusion through grain boundaries, and diffusion through surface defects such as pores and voids. The dominant mechanism at ambient temperature is through grain boundaries. The finer the grains, the more grain boundaries these are and the more intermetallics formation. Intermetallics increases the internal stress, thus increases the probability of whisker formation.

Another important factor, which warrants some elaboration, is the effect of impurities on whisker growth. Although the effect of organic impurities has been discussed extensively in the literature in the context of bright tin-lead plating and properties such as ductility and solderability, the negative effect of organic inclusions has not been fully recognized in increasing the internal stress thus promoting whisker growth.

Besides internal stress, external mechanical stress, which is usually compressive and inhomogeneous in nature, also induces whisker formation. Under such conditions even 90/10 tin-lead coatings are not totally immune from whisker formation. It was shown that after 17 days, whiskers were found at the point of mechanical stress from a bright 90/10 tin-lead coating [96]. In the same study, it was reported that whiskers found of a bright tin coating caused electrical failure after 11 days. The whisker incubation period for a satin bright tin deposit is much longer than that of a bright tin deposit. It took 180 days to see a whisker on the satin bright tin coating shown in Figure 18.

Even though tin whisker growth is an issue in PWB and surface mount technologies, it is possible to reduce the potential danger of shorting by reflowing after electrodeposition, which of course lowers or eliminates the internal stress. However, it is important to be aware that very few types of deposits can be reflowed

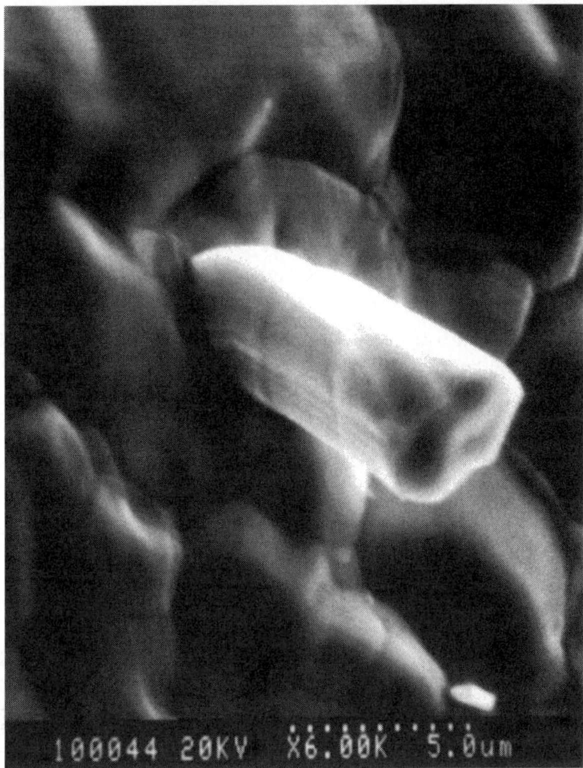

Figure 18 SEM micrograph of a tin whisker observed at the bend area after 180 days, from [56].

without complications. For example, when deposits have high organic content, dewetting usually occurs. Therefore a good policy when dealing with reflow operation is to use a chemistry that produces deposits with low organic inclusions; usually this implies a matte or satin/semi-bright tin chemistry [56].

The "greener" future of the world dictates that lead will be eventually removed from the electronics industry just as what we have witnessed in paint and plumbing industries, although there are speculations indicating reluctance and possible resistance of this change [98]. Although there exist lead-free solders, their uses do not have broad acceptance because of the following facts: First of all, none have all the characteristics of tin-lead solders, even though in some particular aspects they can be better than tin-lead. They are not totally compatible with tin-lead solders [99] because a small amount of lead will significantly weaken the solder joints formed between components and printed circuit boards [98]. Second, none of the lead-free solders have been electroplated successfully because of the difficulties that will be discussed in Section 6. This points researchers to seriously consider electroplated tin as a replacement for tin-lead. In addition, as one can see, pure tin coatings can be applied in other applications in which whiskers are not a concern. Therefore it is reasonable to predict a bright future for electroplated pure tin.

6 CHALLENGES IN ELECTRODEPOSITION OF TIN ALLOYS OTHER THAN TIN-LEAD

The development of lead-free solder plating chemistries has been slow because of the inherent difficulties in plating these alloys. The binary lead-free candidates are tin/bismuth and tin/silver. As described in Section 1.3, 52Sn48Bi and 96.5Sn3.5Ag were chosen as candidates to replace eutectic tin/lead for solder applications mainly for their wetting characteristics and mechanical properties. Tin/bismuth has a melting temperature of 138°C and tin/silver has a melting temperature of 221°C. The former is used mostly in multiple step wave soldering in electronic packaging, while the latter is mostly used for automotive high-temperature applications. Presently they are applied as solder pastes and preforms. To authors' knowledge, electroplated tin/bismuth and tin/silver coatings are not commercially available. The development of these lead-free solder electroplating chemistries requires a consideration of the followings: phase diagrams of tin/bismuth and tin/silver, redox potentials, thermodynamic and kinetic considerations.

6.1 Phase Diagrams of Tin/Bismuth and Tin/Silver

Figures 19 and 20 [100] display the phase diagrams of tin/bismuth and tin/silver, respectively. The lowest melting point for tin/bismuth is at eutectic point of 138°C which requires 48 wt% bismuth. Adding Ag into Sn reduces the melting temperature from 232° to 221°C and requires 3.5 wt% Ag to form a eutectic tin/silver alloy. Since the "wells" are not so deep, it should be relatively easy to obtain eutectic compositions based on phase diagrams.

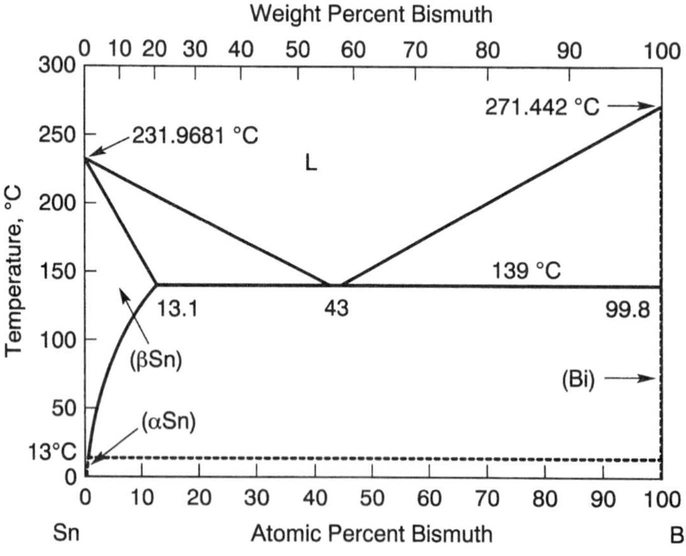

Figure 19 Phase diagram of tin-bismuth, from [98].

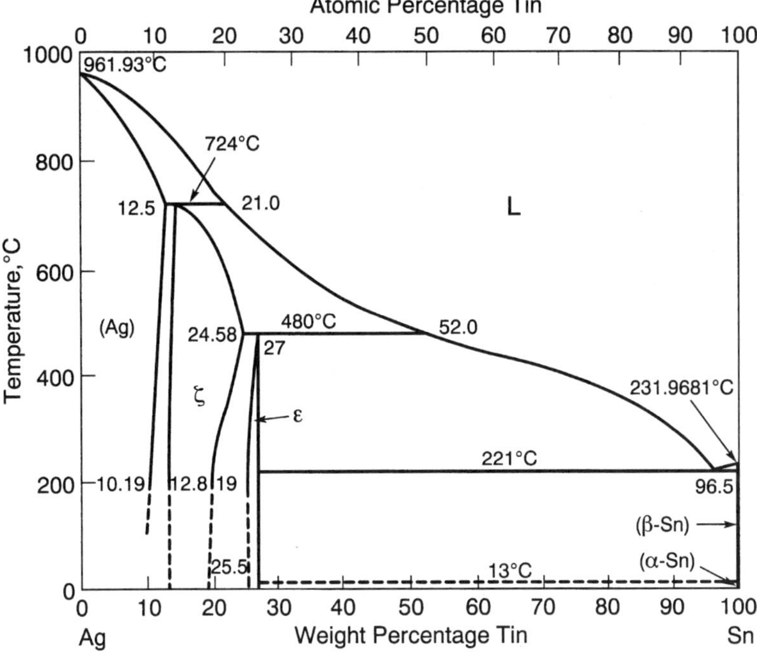

Figure 20 Phase diagram of tin-silver, from [98].

6.2 Redox Potentials

Tin-lead solders are among the easiest to be plated from various electrolytes compared to other tin alloys. The fundamental reason is that the redox potentials of tin and lead in aqueous solutions are very close allowing for efficient codeposition. For instance, in 1 M solution of acidic electrolyte, there is only 12 mV difference in deposition potentials between the two. Table 12 lists the standard potentials of various metals in aqueous acidic solution [22].

6.3 Thermodynamic and Kinetic Considerations

For electrodeposition of an alloy the respective metals should have similar reduction potentials, preferably within 200 mV. When looking at the standard potentials of stannous ion and bismuth ion, one realizes that in noncomplexing aqueous solution, bismuth will preferentially plate out, because the standard deposition potential of Bi^{3+} is much more positive than that of Sn^{2+} (see Table 12). Since the difference in deposition potentials is 454 mV, one may not be able to obtain an alloy at all. However, certain approaches may be taken to narrow the potential difference between the metal components. For instance, one may chose the following to modify the deposition potentials of each metal, (1) adjusting the concentration of one metal component. In other words, making the concentration of the more noble metal diffusion limited so that the less noble metal can plate out before the complete depletion of the more noble metal. This would be difficult for tin/bismuth, but it may be possible for tin/silver. (2) use of complexants, and (3) use of organic additives to poison or block the electrodeposition of the more noble metal.

TABLE 12 Standard Potentials in Aqueous Acidic Solutions

Half-reaction	E°/V
$In^{3+} + 3e^- \rightarrow In$	−0.338
$Sn^{2+} + 2e^- \rightarrow Sn$	−0.137
$Pb^{2+} + 2e^- \rightarrow Pb$	−0.125
$Sn^{4+} + 2e^- \rightarrow Sn^{2+}$	0.150
$SbO^+ + 2H^+ + 3e^- \rightarrow Sb + H_2O$	0.204
$Bi^{3+} + 3e^- \rightarrow Bi$	0.317
$Cu^{2+} + 2e^- \rightarrow Cu$	0.340
$Ag^+ + e^- \rightarrow Ag$	0.799

According to the modified Nernst equation, the potential of a metal ion in solution is given by the sum of its standard potential and a term expressing its activity (a), shown below:

$$E = E^0 + \left(\frac{RT}{nF}\right) \times \ln a + C$$

where C includes all the factors which contribute to the polarization in the electrodeposition process. In the case where two metals are simultaneously deposited, the two equations for the two species can be expressed as follows:

$$E_1 = E_1^0 + \left(\frac{RT}{nF}\right) \times \ln a_1 + C_1$$

$$E_2 = E_2^0 + \left(\frac{RT}{nF}\right) \times \ln a_2 + C_2$$

The difference in potential between the two metals involved can be expressed by

$$\Delta E = E_1 - E_2 = E_1^0 - E_2^0 + \left(\frac{RT}{nF}\right) \times \left(\ln \frac{a_1}{a_2}\right) + (C_1 - C_2)$$

For a successful alloy deposition, ΔE must be minimized. The codeposition condition can be met either when the two values E_1^0 and E_2^0 are similar, such as in the cases for Sn and Pb, or by adjusting the concentration of one species in solution, so altering its activity, or by arranging a suitable difference in the term C, which can be achieved by introducing complexants. The most commonly used method for modifying the deposition potentials is to use complexants, especially when the difference between the standard potentials is large. The activity term is log dependent and therefore difficult to alter potentials significantly. The complexant may react with both metal ions or only react selectively with one or the other. No matter what is the case, the end result should be that ΔE be minimized.

If all of the above fail, one may consider using organic additives to preferentially inhibit the deposition of the more noble metal. Caution, however, has to be taken when this approach is considered. As discussed throughout this chapter, organic additives can have detrimental effects on deposit properties such as ductility and solderability. Furthermore there are no generic recipes for choosing organic additives or "poisons." And in most cases, a trial-and-error approach is used because of the lack of fundamental understanding on the role of individual additives at a molecular level during the electrodeposition process.

Even though thermodynamics determines whether an electrodeposition process can take place, it is the kinetics that determines how it happens and what kind of deposit one will obtain. In order to simplify the following discussion, we now consider the electrodeposition of one metal.

For any electrodeposition to take place, a current has to flow through an electrochemical cell. This may be limited by any of the three factors:

1. Conductance of the bulk solution
2. Transport of reactants to the electrode or products away from the electrode
3. Rate of electron transfer and/or chemical reactions coupled to the electron transfer step

When a net current flows through the cell, the cell is not at equilibrium. The cathodic potential can be expressed by

$$E_{\mathrm{ca}} = E_{\mathrm{e}} + \eta$$

where η is called overpotential and E_{e} is the equilibrium potential (i.e., at which no current flows through the cell, or there is no electrodeposition). η is contributed by all the factors that make the cathode deviate from its equilibrium condition. For instance, it is influenced by the solution and interfacial processes that are involved in the three major parameters outlined above, and also by the external electrical field that is applied in an electrodeposition process. In order to get a physical meaning for the significance of the overpotential, it is helpful to say that small η generally means slow electron transfer reactions, whereas large η yields fast electron transfer reactions.

The conductance or conductivity of the bulk solution is important because it is a measure of the rate of charge transport through solution under the influence of an electric potential gradient. Typically this parameter is controllable, and one can obtain sufficient conductivity of the solution by adding electrolytes that contain ions for electrolytic conductance. Under normal circumstances, step (2) is usually the rate limiting step. The electron transfer rates (step 3) for the reduction of simple metal ions are usually much faster than step 2. In electrodeposition it is generally not desirable to have step (2) be the rate-limiting step. Dendritic deposits will result from the lack of metal ions transported to the electrode interface to support further cathodic reduction. Under such circumstances the incipient fine crystals recrystalize and grow larger to lower the total energy of the system. At the electrode interface, the dimension that has more degree of freedom is the direction perpendicular to the electrode surface. This is normally how dendrites are formed. In order to avoid dendritic growth, it is necessary to make the electron transfer process (step 3) the rate-limiting step. Common practice includes plating at lower current density, that is providing small η or using organic molecules to inhibit or slow down the electron transfer process by blocking the high-energy sites of the electrode surface and thus increasing the activation energy for step 3. The latter approach is used widely in the surface finishing industry to obtain smooth, dendrite-free deposits. With the help of organic additives, sometimes it is possible to electrodeposit a metal close to its limiting current, which is desirable in view of productivity of a plating operation.

Obviously the considerations above are further complicated when binary or ternary alloys are involved. As discussed previously, one must consider the relative reduction potentials of each respective metal and provide a electrolyte/complexant that allows a significant overlap of the reduction potentials, thus ensuring codeposition. Additionally one wants to prevent plating at or close to limiting current of each metal to avoid the formation of dendrites or undesirable alloy composition. Needless to say, the processes must be environmentally friendly and robust enough to be used in a manufacturing operation. No doubt progress on the development of lead-free alloys has been hampered by the issues discussed above.

Next we will discuss specifically the electrodeposition of tin/bismuth and materials properties of tin/bismuth.

6.4 Electrodeposition and Materials Properties of Tin/Bismuth

There are very few studies reported on the electrodeposition of eutectic tin/bismuth alloy [101–103], although codeposition of $< 1\%$ bismuth with tin has been documented for more than two decades [104–107]. The small amount of bismuth in the latter case was aimed at stopping the harmful phase transformation of pure tin from α-phase to β-phase at temperatures below 13°C, sometimes referred to as tin pest. Plating baths of this type are largely sulfuric acid based systems.

The challenges of depositing eutectic tin/bismuth can be attributed to the following facts: (1) The difference between the reduction potentials of the two is too large, and (2) displacement of tin by bismuth according to the electrochemical reaction

$$2Bi^{3+} + 3Sn \rightarrow 2Bi + 3Sn^{2+}$$

The first difficulty can be lessened by using complexant, as explained earlier in Section 6.3. As a matter of fact, both ethylenediaminetetraacetic acid (EDTA) [105] and diethylenetriaminepentaacetic acid (DTPA) [102] have been used to plate eutectic tin/bismuth. Even though Cavallotti [102] and his coworkers obtained encouraging results for the eutectic tin/bismuth coatings by using DTPA as complexant, the plating process was not easy to control. At low current densities, bismuth plated out preferentially. On the other hand, at high current densities, tin plated out preferentially. The alloy composition was also very sensitive to the plating temperature. If a pure tin anode is used, bismuth ion in the solution will replace tin from the anode. If care is not taken, a black slime will form at the anode, causing anode passivation. One remedy to this problem is to use tin anode doped with small percentage of bismuth [108]. At the same time it is important to not leave plated pieces immersed in the solution when deposition is done because the deposit will become gray-black due to the displacement reaction.

Recently Katayama [109] et al. reported another approach to obtain eutectic tin/bismuth; that is, they plated bismuth and tin layers separately with a total thickness of around 10 μm. Then the double-layer deposit was reflowed at 195°C. The reflowed coating had adhesive strength comparable to that of eutectic tin/lead coatings on copper. The thickness ratio of bismuth/tin was a critical parameter in order to obtain crack-free reflowed eutectic tin/bismuth. A similar approach was used before [110], but it was not very successful due to a relatively weaker bond between the sandwich and the copper substrate.

Because of the lack of data on the electrodeposition of tin/bismuth with high bismuth content, very little can be found on the deposit properties. Therefore the discussion on the materials properties are mainly based on the studies done with metallurgically prepared eutectic tin/bismuth.

When bismuth is added into tin, the phase diagram of tin/bismuth indicates that the solubility of bismuth is reduced sharply from 21 wt% at the melting point (138°C) to about 4 wt% at 20°C. Tin/bismuth forms a well-defined eutectic microstructure at all solidification rates, which is very similar to the performance of eutectic tin/lead. Unlike eutectic tin/lead, the tin- and bismuth-maps are complementary. Inside the tin-rich phase small precipitate particles are visible, while the bismuth-rich phase areas are very clean. The metallurgically prepared eutectic tin/bismuth alloy is strong and brittle, and it was related to its lamellar microstructure at room temperature. Because of the poor ductility, it can not be drawn as solid or cored solder wire.

Bismuth lowers surface tension of molten alloys; that is why it has excellent mobility in the molten state. The wetting characteristics of tin/bismuth, while not as good as eutectic tin/lead (see Table 2), is acceptable. On bare copper, tin/bismuth does not wet as well as eutectic tin/lead. On nickel-gold plated substrate, tin/lead wets the best, but tin/bismuth is also acceptable. Au dissolves more slowly into molten tin/bismuth solder than into tin/lead. The wettability of tin/bismuth is significantly improved if the soldering operation is done under nitrogen, presumably because the oxidation of the surface is reduced. Bismuth disrupts the protective tin oxide layer that forms on tin and tin/lead alloys. Thus it is relatively more difficult to preserve solderability of tin/bismuth solder than that of tin/lead. Flux activity is normally low at melting temperature of tin/bismuth.

When eutectic tin/bismuth was used as a solder material for low-temperature wave soldering in packaging such as surface mount applications, it showed stronger shear strength and longer joint life than the tin/lead [110]. It was also found that the intermetallic layer formed between copper and the solder is really formed with tin but not with bismuth. The copper/bismuth binary phase diagram shows that there is no intermetallic phase between bismuth and copper; they are essentially insoluble under 270°C [111]. When used in conjunction with tin/lead solders, it was found that tin/bismuth had very low tolerance for lead contamination. Because it will form a low melting lead/tin/bismuth ternary eutectic, which melts at 95°C, this phase freezes last and concentrates at grain boundaries, thus causing localized low shear strength, grain boundary sliding, and high creep rates [110].

6.5 Electrodeposition and Materials Properties of Tin/Silver

Little has been published in relation to electrodeposition of silver/tin alloys. The aim of such work was to minimize the tendency of silver to tarnish by addition of small amounts of tin. Because of the large difference in their deposition potential (0.94 V; see Table 12), the alloy can not be obtained from simple ion solutions. Since silver is the more noble metal, it is desirable to use complexant to bring its reduction potential closer to that of tin. Silver forms complexes readily with simple anionic species such as CN^-, SCN^-, $S_2O_3^{2-}$, and $S_2O_8^{2-}$. The reduction potentials of these silver complexes are in the range of 0.21 V to −0.31 V versus NHE. Taking into consideration of the chemical nature (i.e., reactivity) of these anionic species in acidic medium, the best are SCN^- and $S_2O_8^{2-}$. Attempt to electrodeposit tin/silver from a MSA based chemistry was not successful due to the reaction between $Sn(MSA)^{2+}$ and the silver complexes of SCN^- and $S_2O_8^{2-}$. Therefore the option now seems to be limited to alkaline medium using CN^-.

Electroplating baths that contain cyanide for silver/tin electrodeposition have been reported [112, 113]. Tin was introduced as tin sulfate ($SnSO_4$) or tin pyrophosphate ($Sn_2P_2O_7$) and silver as potassium silver cyanide. In these cases the pH of the solutions were around 9, and alloys contained from 10% to 16% of tin.

As described earlier, for solder applications, 96.5Sn3.5Ag is one of the candidates that is proposed to replace tin/lead solder. Because of the high tin content, previously reported bath formulations may not be suitable for the eutectic composition. However, it is reasonable to believe that silver cyanide perhaps has to be used as the component for silver. The authors are not aware of any processes for the electrodeposition of eutectic tin/silver except one patent filed in 1996 by Oono et al. [114].

As in the case of tin/bismuth, the materials properties discussed below on tin/silver were obtained from the work done on tin/silver bulk solder.

Even though adding silver into tin does lower the melting point, the eutectic temperature 221°C is not significantly lower than that of tin (232°C). Hence its use will be restricted to high-temperature applications. Unlike bismuth, silver increases the surface tension of tin thus the wettability or spreading of tin/silver is poor comparing with eutectic tin/lead and tin/bismuth. The 96.5sn3.5Ag wets copper rather poorly. On Ni-or Au-plated substrates, while inferior to that of tin/lead, the wetting of tin/silver is acceptable. The interdiffusion rate of gold into tin/silver solder is slower than into eutectic tin/lead.

Eutectic tin/silver has comparable elongation and shear strength comparing with eutectic tin/lead. It has far superior room temperature isothermal fatigue behavior. Its higher melting temperature and superior thermal fatigue property comparing with tin/lead makes it a good candidate for automotive applications.

6.6 Considerations of Electrodeposition of Ternary and Quaternary Tin Alloys

From the discussion above we can surmise that the potential difficulties of plating ternary and quaternary lead-free solders are significantly more challenge than those of binary alloys. Take ternary SnBiAg, for example. The standard potentials are −0.137 V, 0.317 V, and 0.799 V, for tin, bismuth, and silver, respectively. The smallest ΔE is 0.454 V, and the largest ΔE is 0.94 V. According to the discussion in Section 6.3, the difference is too large for codeposition to take place without a complexant or some kinetic maneuvers. If using the complexant is a method of choice, for an alloy composition 83.5Sn14Bi2.5Ag one needs to find a complexant that will bind silver the strongest and tin the weakest so that the reduction potentials can be pushed closer together. Or if one chooses to maneuver the kinetics of the electrodeposition processes, one wants to force silver ion reduction under transport limiting rate while tin is deposited under kinetic control. This type of deposition is sensitive to the concentrations of the respective metal ions, the solution agitation, and the surface features of the parts. Therefore it may not be practical in many applications.

Still it is possible to electrodeposit ternary and quaternary tin alloys listed in Table 3. There have not been any reports or published data on this subject, so the authors are unable to provide any practical examples. However, what is discussed in Sections 6.3 and 6.6 can be used as a general guideline in designing experiments that can obtain the ternary and quaternary alloys of interest.

ACKNOWLEDGMENTS

The authors wish to thank W. I. N. Piccu and many other colleagues whose work has been cited throughout this chapter for helpful discussions.

REFERENCES

1. *Soldering Manual*, 2nd ed., American Welding Society, Miami, FL, 1977.
2. H. J. Fisher and A. Phillips, "Viscosity and Density of Liquid Lead-Tin and Antimony-Cadmium Alloys," *J. Inst. Met.*, **11**, 1060 (1954).
3. H. Inoue, Y. Kurthara, and H. Hachino, "Pb–Sn Solder for Die Bonding of Silicon Chips," *IEEE Trans. Components, Hybrids Manuf. Technol.*, **9**, 190 (1986).
4. P. T. Vianco and D. R. Frearr, "Issues in the Replacement of Lead-Bearing Solders," *JOM*, 14 (1994).
5. R. J. Klein Wassink, "Soldering in Electronics," 2nd ed., Electrochem. Pub., Ayr. Scotland, 1989, pp. 141–147.
6. H. H. Manko, *Solders and Soldering*, 3rd ed., McGraw-Hill, New York, 1992.
7. W. D. Woodbury, *Lead*, U.S. Department of the Interior, Bureau of Mines, Washington, DC, Apr. 1992, p. 22.
8. EPA, Washington, DC.
9. N. Irving Sax, *Dangerous Properties of Industrial Materials*, 6th ed., Van Nostrand Reinhold, New York, 1984.
10. H. H. Manko, "Lead Poison, Solder and Safety in the Workplace," *EP&P*, 92 (1992).
11. S. Waldman, "Lead and Your Kids," *Newsweek*, July 15, 1991, p. 42
12. L. C. Lee, "Lead-Free Soldering," *SMI Workshop*, Sept. 1995, San Jose, CA.
13. J. Glazer, "Metallurgy of Low Temperature Lead-Free Solders for Electronic Assembly," *Int. Materials Rev.*, **40**, 65 (1995).
14. L. E. Felton, C. H. Raeder, and D. B. Knorr, "The Properties of Tin-Bismuth Alloy Solders," *JOM*, 28 (1993).
15. G. Humpston and D. M. Jacobson, *Principles of Soldering and Brazing*, Materials Park, OH, ASM International, 1993.
16. P. T. Vianco, F. M. Hoskings, and D. R. Frear, "Lead-Free Solders for Electronics Applications—Wetting Analysis," *Conf. on Materials Developments in Microelectronics Packaging: Performance and Reliability*, Montreal, Quebec, Canada, 373 (1991).
17. I. Artaki, A. M. Jackson, and P. T. Vianco, "Evaluation of Lead-Free Solder Joints in Electronic Assemblies," *J. Electronic Materials*, **23**, 757 (1994).

18. U. R. Kattner and W. J. Boettinger, "On the Sn–Bi–Ag Ternary Phase Diagram," *J. Electronic Materials*, **23**, 603 (1994).
19. J. A. Slattery and C. E. T. White, "Lead-Free Alloy Containing Tin, Silver, and Indium," Patent EP0568952A1, May 3, 1993; Nov. 10, 1993.
20. S. Tulman, "Tin Base Lead-Free Solder Composition Containing Bismuth, Silver and Antimony," U.S. patent 4,806,309, Jan. 5, 1988.
21. K. G. Campton and A. Mendizza, *Corrosion*, **7**, 327 (1951).
22. *Standard Potentials in Aqueous Solution*, A. J. Bard, R. Parsons, and J. Jordan, eds., Marcel Dekker, New York, 1985.
23. G. S. Tzeng, "Effects of Additive Agents on the Kinetics of Tin Electrodeposition from An Acidic Solution of Tin(II) Sulfate," *P&SF*, 67–71 (1995).
24. P. A. Kohl, "The High Speed Electrodeposition of Sn/Pb Alloys," *J. Electrochem. Soc.*, **129**, 1196 (1982).
25. P. A. Kohl, "High-Speed Solder Plating Baths," *P&SF*, 45 (1981).
26. Y. Zhang, unpublished results.
27. J. O'M. Bockris and A. Reddy, Modern Electrochemistry, Plenum Press, New York, 1970.
28. P. H. Reiger, *Electrochemistry*, 2nd ed., Chapman and Hall, New York, 1993.
29. A. J. Bard and L. R. Faulkner, *Electrochemical Methods*, Wiley, New York, 1980.
30. J. O'M. Bockris and S. U. M. Khan, *Surface Electrochemistry—A Molecular Level Approach*, Plenum Press, New York, 1993.
31. T. C. Tan, "Model for Calculating Metal Distribution and Throwing Power of Plating Baths," *P&SF*, 67 (1987).
32. T. C. Tan, "A Novel Experimental Cell for the Determination of the Throwing Power of an Electroplating System," *J. Electrochem. Soc.*, **44**, 3011 (1987).
33. H. Haring and W. Blum, *Trans. Electrochem. Soc.*, **44**, 313 (1923).
34. A. H. Heathley, *Trans. Electrochem. Soc.*, **44**, 283 (1923).
35. S. Field, *J. Electrodepos. Tech. Soc.*, **7**, 83 (1932).
36. J. J. Maisano, I. Kadija, and J. A. Abys, "A Rotating Cylinder Throwing Power Electrode," *Proc. Annual Conf., American Electroplaters and Surface Finishers Society* (June 1992).
37. M. Schwartz, "Deposition from Aqueous Solutions: An Overview," in *Deposition Technologies for Films and Coatings*, R. F. Bunshah, ed., Noyes Pub., Park Ridge, NJ, 1982, ch. 10.
38. U. Landau, "Plating-New Prospects for an Old Art," in *Electrochemistry in Industry, New Directions*, U. Landau, E. Yeager, and D. Kortan, eds., Plenum Press, New York, 1982.
39. J. W. Dini, "Electrodeposition—The Materials Science of Coatings and Substrates," in *Additives*, Noyes Publ., Park Ridge, NJ, 1993, ch. 7.
40. H. Brown, "Addition Agents, Anions, and Inclusions in Bright Nickel Plating," 9th William Blum Lecture, *Plating*, **55**, 1047 (1968).
41. A. H. DuRose, *Modern Electroplating*, 2nd ed., F. A. Lowerheim, ed., Wiley, New York, 1963, p. 377.
42. A. K. Graham and H. L. Pinkerton, "Properties and Comparative Performance of Electrodeposited Terne Alloys," *Plating*, **54**, 1967.
43. Y. Zhang and M. J. Weaver, "Application of Surface Enhanced Roman Spectroscopy to Organic Electrocatalytic Systems: Decomposition and Electrooxidation of Methanol and Formic Acid on Gold and Platinum-Film Electrodes," *Langmuir*, **9**, 1397 (1993).

44. X. Gao, Y. Zhang, and M. J. Weaver, "Observing Redox-Induced Surface Molecular Transformations by Atomic-Resolution Scanning Tunneling Microscopy: Sulfide Electrooxidation on Au(111) in Aqueous Solutions," *J. Phys. Chem.*, **28**, 4156 (1992).
45. N. Batina, T. Will, and D. M. Kolb, "Study of Initial Stages of Copper Deposition by in-situ Scanning Tunneling Microscopy," *Faraday Discuss.*, **94**, 93 (1992).
46. T. C. Frankin, "Some Mechanisms of Action of Additives in Electrodeposition Processes," *Surf. Coatings Technol.*, **30**, 415 (1987).
47. L. Oniciu and L. Muresan, "Some Fundamental Aspects of Leveling and Brightening in Metal Electrodeposition," *J. Appl. Electrochem.*, **21**, 565 (1991).
48. C. Ogden, D. Trench, and J. White, "Tensile Properties and Morphologies of Pyrophosphate Copper Deposits," *J. Appl. Electrochem.*, **12**, 619 (1982).
49. R. Weil and R. Paquin, "The Relationship between Brightness and Structure in Electrodeposited Nickel," *J. Electrochem. Soc.*, **107**, 87 (1960).
50. Y. Zhang, K. Murski, and J. A. Abys, unpublished results.
51. M. Jordan, *Electrodeposition of Tin and Its Alloys*, Eugen G. Leuze Publ., Saulgau/Wurtt, Germany, 1995, p. 64.
52. D. A. Vermilyea, "Additives and Grain Refinement," *J. Electrochem. Soc.*, **106**, 66 (1959).
53. C. C. Roth and H. Leidheiser Jr., "The Interaction of Organic Compounds with the Surface during the Electrodeposition of Nickel," *J. Electrochem. Soc.*, **100**, 553 (1953).
54. I. Kadija, J. A. Abys, V. Chinchankar, and H. K. Strashil, *P&SF*, **78**, 60 (1991).
55. J. A. Abys and I. Kadija, U.S. patent 5,228,978.
56. Y. Zhang and J. A. Abys, "An Alternative Surface Finish for SnPb Solders: Pure Tin," *Proc. Annual Conference American Electroplaters and Surface Finishers Society* (June 1996).
57. R. Moshohoritou, I. Tsangaraki, and C. Kotira, *P&SF*, **81**, 53–58 (1994).
58. S. Meibuhr and P. R. Carter., *Electrochem. Tech.*, **2**, 267 (1964).
59. M. Carano, "Tin Plating," *P&SF*, **82**, 72, 1995.
60. S. Hirsch, "Tin-Lead, Lead and Tin Plating," in *Metal Finishing, Guidebook and Directory Issue for 1994*, p. 282.
61. From "The Electrodeposition of Tin and Its Alloys," Eugen G. Leuze Publ., 1995, p. 54.
62. DOS1,956,144.
63. W. R. Johnson, *Stahl und Eisen*, **95**, 301 (1975).
64. R. N. Steinbicker, U.S. Patent 4,073,701 (1978).
65. W. A. Procell, U.S. Patent 2,525,942 (1950).
66. K. Obata, et al., U.S. patent 4459185.
67. F. I. Nobel et al., U.S. patent 4565609.
68. F. I. Nobel et al., U.S. patent 4717460.
69. G. Karustis et al., U.S. 3850765.
70. J. C. Jongkind, *Plating*, 901, September 1970.
71. F. A. Lowenheim, "Alkaline Tin Plating," in *Modern Electroplating*, F. A. Lowenheim, ed., McGraw-Hill, New York, 1978, p. 310.
72. G. K. Boeckler, "Measurement of Gloss and Reflection properties of Surfaces," *Met. Finish.*, 28 (May 1995).
73. N. M. Osero, "An Overview of Pulse Plating," *P&SF*, **73**, 20 (1986); Y. Zhang, U.S. Patent 5,750,017 (1998).

74. W. H. Safranek, *The Properties of Electrodeposited Metals and Alloys*, 2nd ed., American Electroplaters and Surface Finishers Society, Orlando, FL, 1986.
75. T. Kakeshita, K. Shimizu, R. Kawanaka, and T. Hasegawa, "Grain Size Effect of Electroplated Tin Coatings on Whisker Growth," *J. Mat. Sci.*, **17**, 2560 (1982).
76. T. Sonosa, H. Nawafune, and S. Mizumoto, "Properties of Bright Tin-Lead Alloy Deposits from Neutral Gluconate Baths," *P&SF*, 66 (1995).
77. J. W. Dini, "Deposit Structure," *P&SF*, **75**, 11 (1988).
78. V. J. Kuck, "Analysis of Tin/lead Solders Using Differential Thermal Analysis," *Proc. of the North American Thermal Analysis Society Conf.*, Sept. 23–36, 1984, p. 306.
79. V. J. Kuck, "Determination of the Liquidus Temperature and Composition of Tin/Lead Solders Using Differential Thermal Analysis," *Thermochimica Acta*, **99**, 233 (1986).
80. Y. Zhang, K. Murski, F. Humiec, and J. A. Abys, "Throwing and Covering Power Evaluations of a Pure Tin Chemistry by Hydrodynamically Controlled Rotating Cylinder Throwing Power Electrode and Hull Cell Studies," Abstract, SUR/FIN'97.
81. MIl-STD 202F, method 208 and J-STD-002 and 003.
82. J. Haimovich, "Intermetallic Compound Growth in Tin and Tin/Lead Plating over Nickel and Its Effects on Solderability," *Welding Res. Suppl.*, 102 (March, 1989).
83. Y. Zhang, K. Murski, G. F. Breck, J. Maisano, and J. A. Abys, "A Unique Electroplating Solder Chemistry," *Proc. of IICIT*, Boston, Sept. 1996, p. 255.
84. R. R. Vandervoort, E. L. Raymond, H. J. Wiesner, and W. P. Frey, "Strengthening of Electrodeposited Lead and Lead Alloys, II-Mechanical Properties," *Plating*, **57**, 362 (1970).
85. T. I. Murphy, "The Structure and Properties of Electrodeposited Copper Foil," *Finish. Highlights*, 71 (1978).
86. W. H. Safranek, *The Properties of Electrodeposited Metals and Alloys*, 2nd ed., 1986.
87. T. F. Battey, N. Nelson, and P. A. Martens, "Etching Electropolished Copper Foil," *PC Fab*, **10**, 60 (1987).
88. K. Murski, Y. Zhang, and J. A. Abys, unpublished results.
89. D. B. Blackwood, W. Brouillette, W. Leyshon, A. Tardone, C. Byrns, F. Poe, and L. Zakraysek, "Whisker Growth from a Bright Acid Tin Electrodeposit," *P&SF*, **64**, 38 (1977).
90. Y. Nakamura, N. Kaneko, and H. Nezu, "Surface Morphology and Crystal Orientation of Electrodeposited Tin from Acid Stannous Sulfate Solutions Containing Various Additives," *J. Appl. Electrochem.*, **24**, 569 (1994).
91. S. C. Britton, *Trans. Ins. Metal Finishing*, **52**, 95 (1974).
92. S. M. Arnold, "Repressing the Growth of Tin Whiskers," *AT&T Bell Labs. Technical Memorandum*, June 1965.
93. R. Kawanaka, K. Fujiwara, S. Nango, and T. Hasegawa, "Influence of Impurities on the Growth of Tin Whiskers," *Jpn. J. Appl. Phys.*, **22**, 917 (1983).
94. M. C. Lin, "Tin Whisker Growth in IC Lead Finish—A Review," *AT&T Bell Labs. Technical Memorandum*, 1984.
95. T. Kakeshita, K. Shimizu, R. Kawanaka, and T. Hasegawa, "Grain Size Effect of Electroplated Tin Coatings on Whisker Growth," *J. Mat. Sci.*, **17**, 2560 (1982).
96. K. M. Cunungham and M. P. Donahue, "Tin Whiskers: Mechanism of Growth and Prevention," *4th Int. SAMPL Electronics Conf.*, June 1990, p. 569.
97. J. Haimovich, "Hot Air Leveled Tin: Solderability and Some Related Properties," *39th Electronic Components Conference*, May 22–24, 1989, p. 107.

98. D. Endicott, "The Challenge of a Lead-Free Finish for Semiconductor Pins," *Met. Finish.*, 50 (1997).

99. W. F. Iverson, *Circuit Assembly*, **1**, 26 (1995).

100. *Binary Alloy Phase Diagrams*, Thaddeus B. Massalski et al., eds., 1987, ASTM, OH, vol. 1, and Max Henson, *Constitution of Binary Alloys*, McGraw-Hill, New York, 1958.

101. N. T. Kudryatsev, "Electrodeposition of Alloys," *IPST* (*Jerusalem*), 87 (1964).

102. P. L. Cavallotti, L. Nobili, G. Zangari, and V. Sirtori, "Tin and Bismuth Substitutes for Electrodeposited Lead," *Proc. of SURFIN'94*, p. 499, 1994.

103. V. V. Kuznetsov and T. N. Kuzovkova, Deposited Doc., 1975, VINITI 3588-75; *Chem. Abstr.* 88: 43103.

104. A. V. Kasyanov, *Tekhnol. Organ. Proizvod.*, 60 (1970).

105. G. I. Medvedev and G. S. Solovev, *Zh. Prikl. Khim.*, **51**, 2734 (1978).

106. J. C. Jongkind, *Plating*, 722 (1968).

107. V. V. Orekhova and L. V. Trubnikova, *Khim. Teknol.*, **25**, 734 (1982).

108. A. V. Kasyanov, *Teknol. Organ. Proizvod.*, 58 (1973).

109. J. Katayama and K. Okuno, "Solderability of Bismuth/Tin Double Layer Deposits," *Met. Finish.*, 12 (1996).

110. T. Murphy, "Tin Bismuth Alloy Plating, a Fusible Low Temperature Etch for High Aspect Ratio PC Boards," *IPC Techn. Rev.*, April 1992.

111. Z. Mei and J. W. Morris Jr., "Characterization of Eutectic Tin/Bismuth Solder Joints," *J. Electronic. Mat.*, **21**, 599 (1992).

112. N. Kubota, T. Horikoshi, and E. Sato, *Met. Finish.*, 53 (1984).

113. J. C. Puippe and W. Fluehmann, *P&SF*, 46 (1983).

114. K. Oono, Y. Muramatsu, D. Yoshiaki, and Y. Yada, JP 08013185 A2, Jan. 16, 1996.

7 Electrodeposition of Chromium

NENAD V. MANDICH and DONALD L. SNYDER

INTRODUCTION

Electroplated chromium deposits rank among the most important plated metals and is used almost exclusively as the final deposit on parts. Without the physical properties offered by electroplated chromium deposits, the service life of most parts would be much shorter due to wear, corrosion, and the like. Parts would have to be replaced or repaired more frequently, or they would have to be made from more expensive materials, thus wasting valuable resources.

The thickness of electroplated chromium deposits falls into two classifications: decorative and functional. Decorative deposits are usually under 0.80 μm in thickness. They offer a pleasing, reflective appearance while also providing corrosion resistance, lubricity, and durability. Decorative chromium deposits are typically plated over nickel but are occasionally plated directly over the substrate of the part.

Functional "hard chrome" deposits have a thickness customarily greater than 0.80 μm and are used for industrial, not decorative, applications. In contrast to decorative deposits, functional chromium is usually plated directly on the substrate and only occasionally over other electrodeposits, such as nickel. Industrial coatings take advantage of the special properties of chromium, including resistance to heat, hardness, wear, corrosion, and erosion, and a low coefficient of friction. Even though it has nothing to do with performance, many users want their functional chromium deposits also to be decorative in appearance. Functional deposits are also used on parts such as cutting tools and strip steel and are even thinner than decorative deposits.

The most common and oldest commercial type of chromium process utilizes hexavalent chromium (Cr^{+6}) in an aqueous solution containing one or more catalysts. The commercial process of hexavalent chromium plating resulted principally from the work in 1923 and 1924 of Fink and Eldridge [1, 2]. Liebreich [3] made similar discoveries more or less simultaneously, but his work was masked by an overemphasis of the importance of the trivalent chromium ion.

Early in the 1970s aqueous trivalent chromium (Cr^{+3}) solutions started to attain commercial success for decorative applications. Even though much work has been done on functional trivalent chromium deposits, there are only very restricted number of applications due to limitations in the physical properties of thick deposits.

Noteworthy improvements in hexavalent chromium plating came with the introduction of double and organic catalyzed systems. Double catalyzed (mixed-catalyst)

Modern Electroplating, Fourth Edition, Edited by Mordechay Schlesinger and Milan Paunovic.
ISBN 0-471-16824-6

systems introduced in 1949 generally contain sulfate and silicofluoride in forms that are either self-regulating [4] or operator regulated. In comparison to the initial commercial processes that were only sulfate catalyzed, double catalyzed processes offer higher plating speeds and help activate the part prior to plating by mildly etching the substrate. Fluoride compounds with limited solubility supply the free fluoride catalyst in self-regulating processes. Undissolved fluoride compounds stayed in the bath until they dissolved to increase the concentration of free fluoride. Operator regulated baths depend upon the proper additions of free fluoride from outside the tank.

Organic catalyzed processes have increased plating speeds and improved deposit physical properties, and they do not etch iron substrates. Proprietary organic additives are added from outside the tank to maintain the proper concentration. This process is well suited for functional applications.

For more details on the history of chromium plating, see [1, 5–7]. Blum and Hogaboom [7] emphasize the effect of the introduction of chromium plating on other electroplating processes.

1 PRINCIPLES

Unlike most other platable metals, chromium cannot be deposited from an aqueous solution containing the metal ions only. Chromium processes must contain one or more acid radicals that act as catalysts (for hex-chromium) or complexers (for tri-chromium) to bring about or aid in the cathodic deposition of chromium metal. The catalysts most commonly used for double (mixed) catalyzed, hex-chromium processes are sulfate and fluoride. The fluoride is generally in the form of a complex such as silicofluoride (SiF_6^{2-}) [8], since simple fluorides are effective in such small quantities that process control becomes difficult. For successful continuous operation, the ratio (by weight) of chromic acid to total catalyst acid radicals must be maintained within definite limits: preferably about 100 to 1 in the case of sulfate [2, 3].

Proprietary organic additives introduced in the mid-1980s are used in conjunction with sulfate for organic catalyzed, high-speed hexavalent-chromium processes. Since fluoride is not used, these processes do not etch steel, which could contribute to a buildup of iron contaminate. Concentrations of hexavalent-chrome, sulfate, and organic acids must be controlled within range to operate. These processes can also operate at higher temperatures than other hexavalent chromium processes permitting the use of higher current densities to obtain faster plating speeds.

The conductivity and density of pure chromic acid solutions are shown in Figure 1. They are based measurements made at NIST [9] (originally named the National Bureau of Standards, NBS). Small amounts of chromium (III) (Cr^{+3}, trivalent chromium) and other cations decrease the conductivity. The maximum conductivity is not achieved until a concentration of 400 to 500 g $liter^{-1}$ chromic acid is reached. Commercial chromium plating processes generally use baths containing 200 to 400 g $liter^{-1}$ chromic acid in order to obtain the best conductivity possible, along with acceptable current efficiency, satisfactory deposits, and stable, easy to maintain, solution composition. In practice, chromic acid concentrations are increased to overcome the effect of contaminates that decrease bath conductivity.

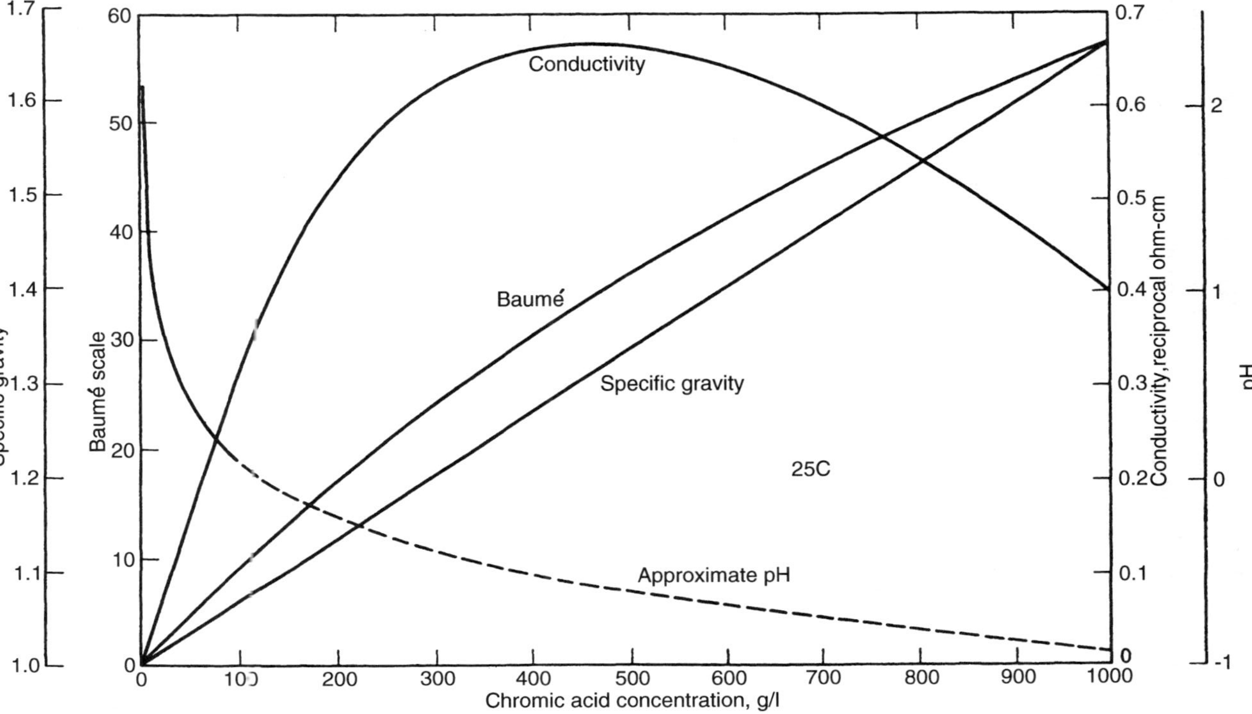

Figure 1 Some physical properties of chromium plate.

Higher chromic acid concentrations increase solution losses due to higher drag-out, resulting in an increased requirement for solution recovery or waste treatment. The specific gravity of the baths provides a rough measure of the concentration of chromic acid, especially if due allowance is made for other salts known to be present.

Sulfate is ordinarily present in all hexavalent chromium plating baths, because even the best commercial grades of chromium trioxide (CrO_3) contain sulfate as an impurity. Chromium trioxide, also called chromium anhydride, is the most common source of hexavalent chromium ions. Even though chemically incorrect, chromium trioxide is usually referred to as chromic acid. The acid is actually formed in aqueous solution.

Sulfuric acid and sodium sulfate are the most common sources of additional sulfate; fluorosilicic acid and silicofluorides [8] are the most common sources of fluoride. References to the amount of catalytic agent or acid radical in a bath usually mean the total quantity of sulfate and fluoride ions, although the method of determining those may vary among the processes and needs to be well understood when using a particular process.

Although the current efficiency in hexavalent chromium plating baths is low (generally between 10% and 25%, depending on the process), a fairly high rate of bright plate is obtained when relatively high current densities are used. Figure 2 show deposits offering the best physical properties. In this figure the semi-dashed lines *A* and *B* circumscribe bright plate areas of low and high concentrations of chromium tri-oxide, respectively. The complete bright plating area is circumscribed by line *X*.

The voltages required are higher than in most other electroplating processes, generally 4 to 12 volts depending on operating conditions. Also the high current and voltage must be applied with very low ripple (the percentage of ac current superimposed on the dc current). Consequently rectifiers used must have a lower ripple and

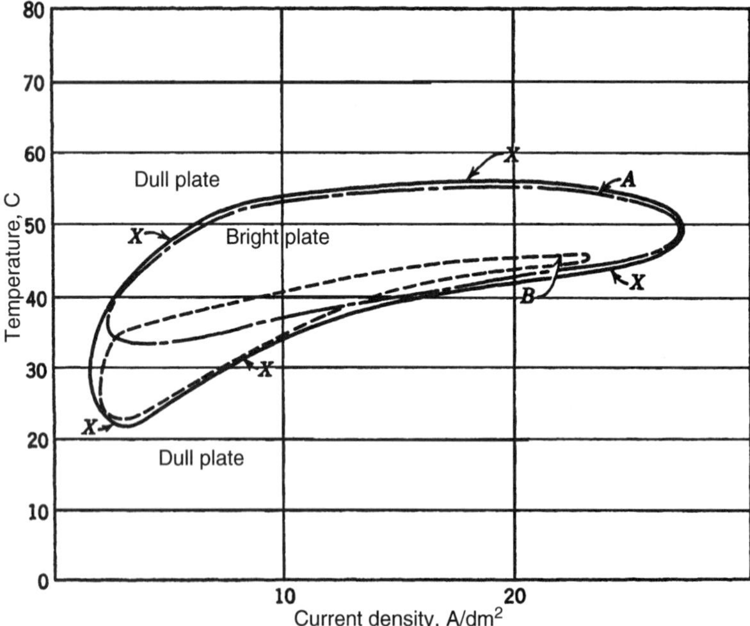

Figure 2 Bright plating range.

higher capacity for chromium plating than is required for most other metal plating, but this disadvantage has not seriously hindered the widespread and increasing use of this process.

The throwing power (coating distribution over the part's current density range) of hexavalent chromium plating is relatively poor compared to most other platable metals such as nickel. Trivalent chromium processes have much greater throwing power and closely approximate that of Watts-based nickel processes. However, usable coverage is obtainable with hexavalent chromium processes even in the plating of irregular shaped articles, if the optimum bath conditions are carefully maintained. Special auxiliary anodes, masks, and shields are sometimes used in order to cover deep hollows or recessed portions, and especially used to obtain more uniform thickness. These techniques are similar to those used with other plating processes but must be designed in accord with established principles of ample size for current-carrying requirements and proper spacing for uniform current distribution.

Proprietary complexers in the form of organic acid radicals stabilize the trivalent chromium ions in an aqueous, pH 2 to 4, solution [10, 11]. Sulfates and chlorides, in varying amounts, are used to increase bath conductivity. Commercial trivalent chromium formulations are much more complex than hexavalent chromium formulations and, at present, are all proprietary. Trivalent processes plate between two and three times faster than hexavalent chromium processes at much lower current densities.

The current altering techniques used in hexavalent chromium processes are not typically necessary with trivalent chromium processes, since both the throwing and covering (ability to plate in low current density) powers are better than those of hexavalent chromium processes. Since they are already present, sulfate and chloride ion concentrations do not have to be tightly controlled in trivalent chromium processes as is required in hexavalent chromium processes. Hexavalent chromium processes, on the other hand, are less sensitive to metallic contamination than trivalent processes, but metallic contaminates are easily removed from trivalent processes by continuos bath circulation through ion-exchange resins. Organic contamination can usually be removed by carbon filtration. Bath operation and maintenance for trivalent chromium processes are much closer to what is required for nickel processes than for hexavalent chromium processes. Very little process information has been published for trivalent chromium processes because it is still kept proprietary, much new, and several different chemistries are commercially available.

2 THEORY OF CHROMIUM ELECTRODEPOSITION

A typical transition element, chromium forms many compounds that are colored and paramagnetic. Chromium has oxidation states as follows: −2, −1, 0, +1, +2, +3, +4, +5, +6; the highest oxidation state, +6, corresponds to the sum of the numbers of 3d and 4s electrons. The lowest, −2, −1, 0, and +1 are formal oxidation states displayed by chromium in compounds such as carbonyls, nitrosyls, and organometallic complexes.

Divalent chromium in the oxidation state +2 was not considered in the past to be of particular interest for electrodeposition mechanisms. It does play a role, however, in the passivation of chromium. Recently it has been recognized that, probably, it plays a role in the deposition and dissolution mechanisms. The outstanding characteristic of the Cr^{+2} ion (sky blue in aqueous solution) is its strength as a reducing agent $Cr^{+3} + e \leftrightarrow Cr^{+2}$ $E_0 = 0.41$ V. Because it is easily oxidized by oxygen, preservation of the solution requires exclusion of air. Even under such conditions the Cr^{+2} ion is oxidized by water with the formation of hydrogen. The rate of oxidation depends on several factors, including the acidity and anions present.

It has been known for some time [12] that pure chromium (usually obtained electrolytically) dissolves in acids to form Cr^{+2} with no (or very little) Cr^{+3}, if the solution is protected from air; impurities apparently catalyze formation of Cr^{+3}. Chromium (+2) solutions may also be obtained [13, 14] by electrolytic reduction of Cr^{3+} chromium (+3).

Chromium (+3) is the most stable and most important oxidation state of the element. The E_0 values [15] show that both oxidation of Cr^{+2} to Cr^{+3} and the reduction of Cr^{+6} to Cr^{+3} are favored in acidic aqueous solutions. The preparation of Cr^{+3} compounds from either state presents few difficulties and does not require special conditions [16].

The chemistry of Cr^{+3} in aqueous solutions is coordination chemistry. It is demonstrated by the formation of kinetically inert outer orbital octahedral complexes. The bonding can be explained by d^2sp^3 hybridization; a great number of complexes have been prepared. The kinetic inertness results from the $3d^3$ electric configuration of Cr^{+3} ion [17]. The type of orbital charge distribution makes liquid displacement and substitution reactions very slow and allows separation, persistence and/or isolation or Cr^{+3} species under thermodynamically unstable conditions.

Chromium (+3) is characterized by a marked tendency to form polynuclear complexes. Literally thousands of Cr^{+3} complexes have been isolated and characterized and, with a few exceptions, are all hexacoordinate. The principal characteristic of these complexes in aqueous solution is their relative kinetic inertness. Ligand displacement reactions of Cr^{+3} complexes have half-times in the range of several hours. It is largely because of this kinetic inertness that so many complex species can be isolated as solids and that they persist for relatively long periods in solution, even under conditions of marked thermodynamic instability.

The hexaaqua ion $[Cr(H_2O)_6]^{3+}$, which is a regular octahedral, occurs in numerous salts, such as the violet hydrate $[Cr\ (H_2O)_6]\ Cl_3$ and in an extensive series of alums, $MCr(SO_4)_2$, $\exists 12H_2O$, where M usually is NH_4^+ or K^+ ion. The aqua ion is acidic (pK = 4), and the hydroxo ion condenses to give dimeric hydroxo bridged species.

On further addition of base, a precipitate is formed that consists of H-bonded layers of $Cr(OH)_3(H_2O)_3$ which readily redissolves in acid. Within one minute, however, this precipitate begins "aging" to an oligomeric or polymeric structure that is much less soluble [18–20].

The Cr^{+3} ion may also polymerize, as a result of hydrolysis and associated reactions, to form bridged complexes with a certain composition, whose existence is indicated by indirect, but substantial, evidence. Complexes of this type range from dimers through polymers of colloidal dimensions to precipitated Cr^{3+} hydroxide. Except under special circumstances, such reactions are inevitable in neutral and basic solutions and highly probable in slightly acid solutions.

What makes the chemistry of Cr^{3+} complexes interesting, and often difficult for researchers, is the large number of steps and mechanisms possible. The processes include aquation, hydrolysis, olation, polymerization, and oxolation and anion penetration.

2.1 Aquation

Chromium salts (chloride, sulfate, nitrate, etc.) are aqua complexes characterized by ions such as $[Cr(H_2O)_6]^{+3}$, $[Cr(H_2O)_5Cl]^{+2}$, and $[Cr(H_2O)_4Cl]^{+2}$. In aqueous solutions the replacement of coordinated groups by water molecules (aquation) is a common reaction: $[CrA_5X]^{+2}+H_2O$ □ $[CrA_5H_2O]^{+}$, where A is a singly coordinated neutral molecule and X is a singly charge coordinated negative ion (e.g., Cl, CN, CNS).

The extent of aquation depends on several factors, including the relative coordinating tendencies of H_2O and X and the concentration of X. Accordingly every aqueous solution of Cr^{+3} is potentially a solution of aqua complexes. The Cr^{+2} ion (whose complexes are labile) catalyzes such reactions, which are usually quite slow otherwise. Electron-transfer reactions between Cr^{+2} and $[Cr(H_2O)_{5X}]^{+}$ proceed predominantly through bridged intermediates $\{[Cr\text{–}X\text{–}Cr]^{+4}\}$. Ligand transfer accompanies electron transfer. In the investigations establishing these conclusions, the reaction conditions have generally been characterized by relatively low $[Cr^{+2}]$ and relatively high $[H^{+}]$. With relatively high $[Cr^{+2}]$ and relatively low $[H^{+}]$, another pathway is available [21] with the rate determining reaction involving a hydroxy bridged complex: $[(H_2O)_4\text{–}X\text{–}CrOHCr_6]^{+3}$.

The role of Cr^{+2} is very important, however, in industrial "hard" chromium applications when plating thick layers of chromium or Cr–Ni and/or Cr–Ni–Fe alloys from trivalent chromium solutions, as an alternative for Cr(+6)-based solutions. Failure to control the transient levels of Cr^{+2} is recognized as the reason for Cr(+3)-based solutions not to sustain heavy deposition with an appreciable deposition rate. The problem is recognized as massive olation, catalyzed by a buildup of Cr^{+2} in the high-pH region in the vicinity of the cathode. Although the bulk of the electrolyte can be about pH 2, the diffusion layer can reach pH 4. At this pH and with Cr^{+2} promoted catalysis, oligomeric species are released into the bulk of the electrolyte, where they can build up and reduce the level of active species and, consequently, the deposition rate [22–25].

2.2 Hydrolysis

The behavior of aqua complexes as acids leads to far-reaching consequences. The acidity of such solutions arises because of the $[Cr\ (H_2O)]^{+3}$ □ $[Cr\ (H_2O)]^{+2}+H^{+}$ reaction. The equilibrium can be displaced to the right by heating and, of course, by the addition of base. The order of magnitude of the first hydrolysis constant is $K=10^{-4}$. As the pH of the Cr^{+3} solution is raised, the equilibrium is shifted, and more of the coordinated water molecules are converted to OH groups, which brings into the picture a new process called *olation*.

2.3 Olation

Olated compounds are complexes in which the metal atoms are linked through bridging with OH groups. Such a group is designated as an *ol* group to distinguish it from the *hydroxo* group (i.e., a coordinated OH linked to only one metal atom). The process of formation of ol compounds from hydroxo compounds is called olation. Olation results from the formation of polynuclear complexes consisting of chains or rings of Cr^{+3} ions connected by bridging OH groups. The first step of this process may be as follows [26]:

$$[Cr(H_2O)_6]^{+3} + [Cr(H_2O)_5OH]^{+2} \rightarrow \left[(H_2O)_5Cr \overset{\overset{H}{\uparrow}}{\underset{/\ \backslash}{O}} Cr(H_2O)_5\right]^{+5} + H_2O \quad (1)$$

$$2[Cr(H_2O)_5OH]^{+2} \rightarrow \left[(H_2O)_4Cr \overset{\overset{H}{\uparrow}}{\underset{/\ \backslash}{O}} \underset{\underset{H}{\downarrow}}{\overset{\backslash\ /}{O}} Cr(H_2O)_4\right]^{+4} + 2H_2O \quad (2)$$

Because the *diol* produced by the reaction shown by Eq. (2) is stabilized by the four-member ring, there is a driving force tending to convert the singly bridged to a doubly bridged complex. This diol is produced by polymerization of $[Cr(H_2O)_5OH]^{+2}$, oxidation of Cr^{+2} by molecular oxygen, warming an equimolar mixture of Cr^{+3} and NaOH and boiling an aqueous solution of $[Cr\ (H_2O)_6]^{+3}$.

The diol and any other polynuclear products containing water molecules (or a group that can be displaced by water molecules) can still act as acids, releasing hydrogen ions and leaving coordinated OH groups.

2.4 Polymerization

Instead of reaching a definite termination, the reaction of Eq. (2) may continue, with the formation of larger and larger molecules, the polymers, as a continued process of olation. This will occur if the product of each successive step contains aqua or hydroxo groups. The ultimate consequence is precipitation of chromium hydroxide, $Cr(OH)_3 \times H_2O$, a tri-dimensional olated complex [27]. Olation reactions are pH and time dependent. At moderate acidity they are quite slow. It can take days for higher oligomers to be formed after addition of the base to aqueous Cr^{3+} solution, but they will subsequently decay, contributing to pH stabilization after a few weeks [28].

The continued process of olation starts with the hydrolysis of salts of such metals as Al or Cr. The acidity of solution of such salts results from conversion of aqua to hydroxo groups: $[Cr(H_2O)_6]^{+3} \leftrightarrow [Cr(H_2O)_6OH]^{+2} + H^+$. The degree of hydrolysis increases as the temperature is raised, and this relationship depends on the nature of the anion, and especially on the pH of the solution. If alkali is added to a warm solution of hydrolyzed chromium salt, but not enough for complete neutralization, polymerization occurs instead of precipitation of the basic salt or hydroxide.

Because of the octahedral configuration of complexes of metals such as chromium, the bonds of a given metal occur in pairs, each of which lies in a plane perpendicular to the planes of the other two pairs. Accordingly such cross-linked polymers are three-dimensional.

The process of olation is favored by an increase in concentration, temperature, and basicity. The process reverses slowly when the solution of olated complexes is diluted or when the solution is cooled (i.e., olation decreases the reactivity of coordinated OH groups).

2.5 Oxolation

Oxolation may accompany or follow olation, particularly if the reaction mixture is heated. This reaction converts bridging OH^- groups to O^- groups. Olation and oxolation account for changes in reactivity of chromium hydroxide as it ages. Freshly precipitated chromium hydroxide usually dissolves quite rapidly in mineral acids, but after standing some hours, it becomes difficult to dissolve. Presumably olation continues in the precipitate; because bridged OH^- groups react more slowly with acids than singly coordinated OH^- groups, the reactivity of the precipitated hydroxide progressively diminishes. If the hydrate is heated, there is a drastic decrease in reactivity as a result of oxolation, a process even more difficult to reverse than olation. While olation and oxolation are both reversible, the long times required for the acidity of solutions, which have been heated and then cooled, to return to the original values, leads to the conclusion that deoxolation is extremely slow. In general, ol groups are more readily depolymerized than oxo compounds, because protons react more rapidly with oxo groups.

2.6 Anion Penetration

It is well known that the addition of neutral salts to a solution of basic sulfate changes the hydrogen ion concentration. Coordinated water molecules, OH^- groups, OH bridges, or other ligands are replaced by anions in the solution. The extent to which anion penetration occurs with ol complexes is determined by the relative coordination tendencies of the entering anions and the groups that they replace, and the length of time that the solutions are allowed to stand [29]. Anions that can enter the coordinated sphere easily and displace OH groups can effectively prevent olation. Penetration by anions into basic chromium complexes decreases in the following order [30]:

Oxalate > glycinate > tartaratecitrate > glucolate > acetate

> monochloracetate > formate > sulfate > chloride > nitrate > perchlorate

Consequently, if a solution of $[Cr(H_2O)]^{+3}$ is required, the only anion that should be weakly coordinated is nitrate or perchlorate, because anions of greater coordinating tendency will displace one or more of the coordinated molecules. In a stock solution of basic chromium sulfate, Serfas et al. [29] found ionic species having molecular weights of 68,000.

2.7 Reaction Rates

In a system containing Cr^{+3} complexes, after a parameter is changed, the corresponding change in composition of the complexes generally occurs only slowly.

Heating a solution (or dispersion) of such complexes promotes olation and oxolation, both of which reverse at a low rate when the system is cooled. Reversal of oxolation is much slower than reversal of olation. If the pH of a solution containing olated complexes is reduced to a value at which normally only monometric Cr^{+3} complexes would exist, it may take a long time for the state of aggregation corresponding to the new pH to be attained.

3 HEXAVALENT CHROMIUM

The mechanisms of the electroreduction of chromic acid are of great interest, not only from a theoretical point of view but also for their application in industry as well. The vast majority of decorative, and almost all hard, chromium plating is carried out using CrO_3 as the electrolyte. The fact that chromium can be deposited from Cr^{+6} solutions but not from simple aqueous solutions of lower valency salts is a disadvantage for the following reasons:

1. Because the electrochemical equivalent of Cr in a CrO_3 solution is 0.3234 g h^{-1} and cathode current efficiency is typically 10% to 20%, the passage of current of one *A*-hour yields only 0.032 to 0.064 g of metal. This is 15 to 30 times less than for nickel, 18 to 36 times less than for copper from acid solution, and 63 to 126 times less than for silver. The only way to offset this is to increase the working current density via increase in mass transport and temperature and/or plating time.
2. The minimum current density at which electrodeposition takes place is two to three orders of magnitude larger than in the case of other metals (Zn, Ni, Sn, Ag, Au, etc.).
3. The electrodeposition of chromium is more sensitive to operating conditions (temperature and current density) than any other deposition process.
4. In contrast to other processes, the cathodic current efficiency varies inversely with temperature but is proportional to current density (which causes low throwing power).
5. Chromium will plate only in the presence of a catalyst (e.g., H_2SO_4), whose concentration influences the plating rate.
6. On the positive side, hexavalent chromium electrolytes are relatively less sensitive to the presence of impurities, and the anode material is lead or lead alloys, which can easily be made to conform to any shape.

Despite its paramount technological importance and with all the advances of modern science and instrumentation, the exact mechanisms of chromium electrodeposition are still open to considerable conjecture. The main difficulty is the necessary formation and presence of a cathodic film on the surface of the metal being plated. The argument whether the reduction of Cr^{6+} ions to chromium is direct or indirect developed during the last decade into a discussion of whether or not the cathodic film is useful (and in what way should it be modified to improve the process, inasmuch as the existence of this film is no longer in question).

Because of the absence of complete understanding of the deposition mechanism, it is important to understand the chemistry of chromium with all its intricacies of condensation, polymerization, number of different valence states, ability to make anion/cation compounds [e.g., $Cr_2(Cr_2O_7)_3$], existence of a number of double-salts (alums), isomers, oxyhydrates, and so on. Virtually all Cr^{+3} compounds contain a Cr–O unit.

3.1 Chromic Acid

The primary Cr–O bonded species is chromium (+6) oxide, CrO_3, which is better known as chromic acid, the commercial and common name. This compound is also known as chromic oxide and chromic acid anhydride. Chromium (+6) forms a large number and considerable variety of oxygen compounds, most of which may be regarded as derived from Cr^{+6} oxide. These include the oxy-halogen complexes and chromyl compounds, chromates, dichromates, trichromates, tetrachromates, and basic chromates. All these Cr^{+3} compounds are quite potent oxidizing agents, although kinetically they cover a wide range.

Chromic trioxide has a molecular weight of 100.01 and forms dark red prismatic crystals belonging to the orthorhombic system, the bipyramidal subclass. The density of the solid is 2.79 g cm^{-3}. It melts with some decomposition at 197°C. CrO_3 is very hygroscopic. Its solubility in water varies from 61.7% at 0°C to 67.5% at 100°C. Oxidation potentials of CrO_3 and chromate solutions are augmented by increasing the acidity of the solution. Chromic acid, H_2CrO_4, or CrO_3, is not known except in solution, where it shows a marked tendency to form polyacids by elimination of water [31].

The change from $H_2Cr_2O_7$ to $H_2Cr_2O_7$ is rapid, but further polymerization takes measurable time. The color of CrO_3 indicates that it is itself highly polymerized, for it is redder than the di- or trichromates and is approached in color by the tetrachromates. De-polymerization of CrO_3 solution in water is very rapid. It also seems to de-polymerize on heating.

3.2 Chromates and Dichromates

Chromates are salts of the hypothetical chromic acid H_2CrO_4. Salts of the hypothetical polybasic chromic acids, $H_2Cr_2O_7$, $H_2Cr_3O_{10}$, $H_2Cr_4O_{13}$, known as dichromates, trichromates, tetrachromates, and so on.

The chromate ion and most of the normal solid chromates are yellow, but upon on acidifying, the solutions change colors through orange to red. The dichromates are red in the solid state and in solution. The higher polychromates are even of deeper red than the dichromate in the solid state. Although the various ions, CrO_4^-, $Cr_2O_7^{2-}$, $Cr_3O_{10}^{2-}$, $Cr_4O_{13}^{2-}$, and so on, exist together in equilibrium in solution, the ions higher than dichromate exist only in the most concentrated solutions. Water is easily added to the higher polychromate ions, causing them to revert to the dichromate. On further dilution, even the dichromate ion adds water, forming the chromates. The $HCrO_4^-$ ion exists in quantity only in dilute solution, according to Udy [32], but more recently Raman spectroscopy proved nonexistence of $HCrO_4^-$ ions in dilute and concentrated solutions [33–36].

3.3 Polychromates

Polychromate ions are of particular interest because of their role in chromium plating from hexavalent solutions. It is recognized and accepted that chromium cannot be electrodeposited from Cr^{+6} solutions without the addition of a catalyst, usually in the form of the sulfate. Because the strength of commercial solutions is customarily 1 to 3 molar, at this concentration, considering the low pH and taking into account the dark red color of the solution, at least the tri- and possibly the tetrachromate ions are present. It should be noted that in the absence of electric current, the pH of the chromium plating solution is subject to considerable variation, depending on the initial concentration of chromic and sulfuric acids. If the amount of CrO_3 is increased from 10 to 300 $g\,liter^{-1}$ (0.1–3 M), the pH changes from 1.4 to 0.08. Martens and Carpeni [37] using radioactive chromium measured the autodiffusion coefficients of isopolychromates at 25°C in aqueous solution as a function of concentration. They found that in the plating operating ranges ($1.5 < CrO_3 < 3.5\,mol\,liter^{-1}$), the predominant species are di- and trichromateions.

The dominant role of trichromates in chromium deposition is advanced by Hoare [38]. According to his model, in the absence of the bisulfate ion (or sulfate, which at low pH dissociates to bisulfate) the trichromate ion will in successive steps (of electron transfer and loss of oxygen and reaction with H_3O^+ ion) decompose to chromous hydroxide and dichromates, which in turn may undergo condensation with other chromates to regenerate trichromates. The process then includes an intermediate step of formation of chromic (+3), then chromous (+2) dichromates, finally discharging at the cathode as black chromium at very low current efficiency. In the presence of sulfates, the next step of the reduction mechanism is the formation of a complex between the Cr^{+2} hydroxide and the bisulfate through hydrogen bonding:

$$Cr(OH)_2 \Leftrightarrow Cr=O + H_2O \tag{3}$$

$$Cr=O + HSO_4^- \Leftrightarrow {}^{+\delta}Cr-OH\ldots O-SO_3^- \tag{4}$$

where the ellipses represent the hydrogen bond and $^{+\delta}$ represents a dipole generated on the chromium end (left side) of the complex. Now, the positively charged complex may be specifically adsorbed on the cathode, two electrons transferred to this endon configuration with formation of metallic Cr and regeneration of HSO_4^-:

$$^{+\delta}Cr-OH\ldots O-SO_3^- \xrightarrow{2e^-,2H^+} Cr^0 + HSO_4^- + H_2O \tag{5}$$

According to this model, the chromic-dichromate complex is necessary to protect the Cr^{-3} from forming stable Cr (3^+) aquocomplexes. As a refinement of this model, the HSO_4^- ion has a dual role—it also "blocks" other chromium atoms in trichromate ions from being reduced (leading to Cr^{+3} aquocomplexes formation). The ideally protected trichromate ion would be

This would leave one end [the right side of (6)] protected, preventing formation of unwanted dichromatic chromate complex, decomposition of which would lead to unwanted $[Cr(H_2O)_6]^{+3}$ formation. This also explains the narrow range (CrO_3:

$HSO_4 = 100 : 1$) of bisulfate concentration in the chromium plating solution. Too little HSO will cause insufficient protection of the Cr at the right end of the trichromate ion (undercatalization); too much will block the left end Cr, which is necessary for reactions (3) to (6) and Cr deposition (overcatalization).

According to Hoare [39–40] for fluoride-catalyzed CrO_3-based plating systems, almost the identical mechanism is proposed in which F^- plays the role of blocking agent and catalyst. Although not complete, this mechanism is the most accepted to date. The incompleteness of his remarkable theory is that it treated the chromium deposition mechanism without reference to the structure and influence of the liquid layer adjacent to the cathode (L-film), which is formed at the beginning of the cathodic process and is continuously forming and reforming in the steady-state condition.

Research originating in Russia is extensive on the L-film formation and reactions that occur in the film. They recognized quite early its decisive importance for the deposition mechanism in general and for current efficiency in particular. On the other hand, it is a well-known fact that halide ions (X) such as Cl^- and F^- have a marked improving effect on the cathode current efficiency of chromium electrodeposition as recently reported [41].

Because the hydration of halide anions is incomplete, they can penetrate the hydrogen layer and be absorbed onto the metal surface. XPS results [41] show that F^- and Cl^- ions, which are stable in the chromic acid bath, may participate in the film formation. The probable activation steps of halides are the absorbed halide first penetrates the hydrogen layer at the chromium surface and then forms a bridged transition surface complex. The electrons on the cathode are transferred to Cr^{+3} through halides, and Cr^{3+} is reduced to metallic chromium. By the formation of the transition complex, the activation energy of the reduction of Cr^{3+} to Cr^0 is decreased. The overpotential of chromium deposition apparently is decreased, which facilitates chromium electrodeposition. The of reaction follows a first order rate equation [42]. In case of a rotating cylinder, the specific reaction rate constant was found to increase with increasing rotation speed up to a limiting value which is reached with further increase in the rotation speed. A study of the reaction mechanisms has shown that at a relatively low rotation speed, the reduction of chromium is partially controlled by diffusion; at higher speeds, the reaction becomes kinetically controlled. Agitation (cylinder rotation) increases the rate of chromium reduction by decreasing the degree of cathode coverage by hydrogen bubbles, consequently increasing the effective cathode area [43–44]. In this sense it seems that nonstationary currents can be of great advantage, since the current interruptions and/or current reversal can promote hydrogen liberation [45]. In addition the use of current pulses interrupts the nucleation and resulting crystal growth. Each pulse enables a fresh re-nucleation with the net effect of refining the structure and size of grains. Grain consolidation appears to interfere with the accumulation of internal stresses and to act as an inhibitor of crack formation, as noted in earlier studies [46].

As reported in a recent paper [35], a X-ray diffraction study was done to identify the predominant species in an industrial CrO_3–H_2O system. Structural analysis showed that dichromate ions may have maximum likelihood, but that linear trichromate ions may also exist in significant concentrations. This study also concluded that formation of a complex shown in Eq. (6) can be hardly assumed because of steric hindrance, and that it is more realistic that one HSO_4 ion reacts with polychromate.

A recent paper [47] studied the existence of various chromium complexes in CrO_3/H_2SO_4 plating solutions for different $X = CrO_3/H_2SO_4$ ratios. They concluded that although five different chromium complexes exist, the reduction to metal proceeds only from the following type of complexes $[HSO_4]_n^- \cdot [Cr_2O_7]_m^{2-}$, where $n = 1$, $m = 1$ and $25 < X < 150$–200. They concluded that those complexes are characterized by a single hydrogen bond between two ions in the complex.

In another recent paper on chromium mechanisms [48], potentiodynamic and impedance measurements are used to further corroborate their mechanism of deposition, based on formation of a cathode film (with solid and liquid phases) consisting of oxide-hydroxide Cr^{+3} compounds. It is felt that an in situ method is needed to study the deposition mechanisms under both transient and steady-state conditions.

Pressure from environmentalists is leading to research regarding the issue of replacing Cr^{6+} solutions by the less toxic Cr^{3+}. At the same time it becomes obvious that mechanisms of deposition from trivalent and hexavalent solutions are rather intertwined and that in both cases chromium coordination chemistry is heavily involved.

Despite the flurry of research on chromium deposition mechanisms in the 1950s, 1960s, and 1970s, the flow of papers on chromium was later reduced to a trickle. The reason is the complexity of the problem and the difficulties involved in the highly colored, highly concentrated solutions of chromium salts, the number of different valence states involved, and general lack of in-depth information regarding chromium coordination chemistry.

What further complicated the matter is that at the onset of the deposition process, one set of reactions occurs—formation of a compact film independent of the anions present with a rather thin profiles, $5\,mg\,m^{-2}$. Some Russian workers use the term "product of partial reduction of $Cr^{6+} \rightarrow Cr^{3+}$," and this film forms in the first branch of the chromium polarization curve, at potentials up to about 700 mV. Once this film is formed (the C-film, short for compact film), another cathodic film (layer) is formed on the surface of the C-film and closer to the bulk of the solution-the L-film (short for liquid film).

Yoshida et al. [49] studied the behavior and composition of a cathode film with the help of radioactive tracers in the form of S^{35} radioactive-labeled sulfuric acid and over-the-counter, high-grade CrO_3 treated with radiation to obtain Cr^{51} as a tracer. A special, rather simple plating cell was constructed with a rapid rinsing station. In essence, a steel cathode was plated for a short time, so that the C- and/or L-films were formed and could be analyzed. Because the L-film is liquid and soluble in either hot plating solution or hot alkali, by dissolution or simple brushing, its formation and influence on the deposition of metallic chromium was studied. By initially forming the C- and L-films with radio-active-labeled Cr or H_2SO_4, and plating in pure (unlabeled) solution, and vice versa, they came to these important conclusions:

1. The cathode film is composed of two layers with different forming properties in terms of thickness and composition. The outer layer referred to as the L-film, and the inner layer, the C-film, differ in that the L-film contains sulfate ions and dissolves easily in the electrolyte and is about 10 times thicker than the C-film.

2. The C-film has a mass of about 5 mg m^{-2}, contains very few sulfate ions, and does not dissolve easily in the electrolyte.
3. The cathode film itself is not reduced to metallic chromium, which is deposited from a separate chromium complex compound that passes through the cathode films (C and L) from the bulk of the solution.
4. In the electrolyte, the L-film vigorously repeats the dissolving and forming cycles, while the C-film remains constant, once formed.
5. The cathode film may be a chromium hydroxy *aquo* complex, or primarily an oxolated version of this compound. Assuming that the cathode film is formed from such chromium complexes, the authors suggest that the L-film is a compound with lower molecular weight, while the C-film is a large complex with a high degree of polymerization.

Kimura and Hayashi [50] also used sulfates labeled with radio- active S^{35} to overcome the difficulties of determining the amount of sulfate in the cathode film. They used standard analytical methods (because of the relatively small content) to study sulfate content in the cathodic film which is formed during potentiostatic polarization of 0.4, 1.5, and 2.5 M CrO_3 baths on Fe, Au, and Pt cathodes. They found that the sulfate content is directly related to the potential in the region of −0.6 to −1.0 V, which in turn is controlling the state of the cathode surface (L-film formations) and the accompanying electrochemical reactions. In the region of −0.2 to −0.8 V, where current is increasing (C-film), sulfate content was negligible for Pt, Au, and Fe cathodes. In the region >-0.8 V, where current starts to decrease and L-film starts to form, sulfate concentration increases sharply. In the potential regions between −1.0 and −1.1 V (beginning of Cr deposit region), the sulfate concentration in the film drops as a result of liberation of sulfates from the complexes. At potentials more than −1.1 V, the sulfate concentration increases slightly again because of inclusion in the cracks and imperfections in metallic chromium deposits. They also found that as sulfate concentration in a 0.4 M CrO_3 bath is increased from 0.002 M (200 : 1) to 0.008 M (50 : 1), sulfate content in the L-film tends to increase. A temperature increase has a similar effect, while an increase in CrO_3 concentration at constant ratio has the opposite effect. At any given CrO_3 concentration, the maximum amount of sulfates in the L-film is predictably in a 100 : 1 ratio of sulfuric acid. The authors also investigated the influence of other anions in addition to H_2SO_4. Specifically, HCl or KBr (0.01 M) added to 1.5 M CrO_3+0.01 M H_2SO_4 solution considerably increased the sulfate content of the L-film, while 0.01 M Na_2SiF_6 addition had the opposite effect, demonstrating the substantial film dissolution effect of Na_2SiF_6. The effects of HCl, KBr, and Na_2SiF_6 on the film were also proportional to increases in their respective concentrations.

Nagayama and Izumitani [51] studied the coordination chemistry of chromium complexes as related to deposition mechanisms. They started with the observations made by Levitan [52] that during galvanostatic ($I=75\,\mathrm{mA\,cm^{-2}}$) chromium deposition from a sulfate-catalyzed bath, a chromic acid dimer is formed together with a polymer of unknown structure, as well as mononuclear $[Cr(H_2O)_6]^{+3}$ the stable aquocomplex. Rather than use the galvanostatic method, where current is kept constant and potential changes, they chose to keep potential fixed at −0.75 V (vs. SCE). Here only $Cr^{6+}\rightarrow Cr^{+3}$ and $2H^+\rightarrow H_2$ reactions are in progress (for the

$Cr^{+6} \rightarrow Cr^{+3}$ reaction to happen, this potential is too positive). During electrolysis (0–60 min) they took samples at different time intervals, and with the use of anion and cation exchange chromatography, they separated the mononuclear, binuclear, and polynuclear Cr^{+3} complexes. They found that the complex formation rate for mononuclear complexes increases linearly with time, while for the other two complexes, the rate increase is more gradual. They concluded that each complex is forming at its own constant rate.

They repeated the experiment at $-1.10\,V$ (Cr^0 formation region) and obtained similar results. The authors concluded that the cathode layer, made of the dense film of various Cr^{+3} complexes, is a necessary condition for the deposition reaction $Cr^{+6} \rightarrow Cr^{+3} \rightarrow Cr^{+2} \rightarrow Cr^0$ to happen. The catalyst (e.g., H_2SO_4) promotes formation and dissolution of binuclear and polynuclear soluble Cr^{+3} complexes, thus maintaining a film of constant thickness where deposition proceeds via intermediate Cr^{+3} (inner orbital) complex rather than through the extremely stable $[Cr(H_2O)_6]^{+3}$ (outer orbital) complex.

Okada [53] holds that SO_4^{-2} ions will penetrate an olated compound, to form a complex, and that from this complex metallic chromium is deposited. According to Okada, reduced solubility of the L-film causes the OH cross-linking level to rise together with the increase in pH.

Yoshida, Tsukahara, and Koyama [54] used ESCA to further elaborate their previous research, in which they noted that there are two layers, the L- and C-films, within the cathode film. They obtained depth profile of these films and demonstrated that the C-film is a highly polymerized complex, with very few anions present, if any. The L-film appears to be mostly in the Cr^{+3} state, but the exact valence could not be established, suggesting the possibility of two- and four-valence states, as well.

They suggested that Cr^{+3} complexes are the main constituents of L-film and that metallic chromium does not deposit from this cathode film but from the Cr^{+6} state. That contributes to the formation of the cathode film and also forms olated complexes, hydroxy aquocomplexes, and polymers of higher molecular weight. These olated complexes will penetrate the cathode film from the bulk of the solution before being reduced to metallic chromium [55].

4 METHODS OF OPERATIONS OF CHROMIUM PLATING SOLUTIONS

4.1 Constituents of Chromium Baths and Their Actions

The chromium plating bath, used for decorative and hard chromium baths, is still mostly of the type originally investigated by Sargent. It is the simplest plating bath to make up, and it consists of two essential ingredients: (1) A water soluble salt of chromium and (2) a small, but critical amount of an anion, which for want of better name is called the catalyst. The catalyst is supplied in the form of sulfuric acid alone or in combination with another acid radical, or anion(s), usually fluoride or fluoroborate or a mixture of them. Relatively recently an organic acid radical in the form of alkene-sulfonic acid, (e.g., methane disulfonic acid, $CH_2(SO_3H)_2$ or one of its alkali metal salts) has been successfully included in the high energy efficiency formulations (HEEF) introduced by Atotec Company (USA) [56].

Because chromium metal will not serve satisfactorily as an anode, owing to its close to 100% anodic dissolution efficiency, insoluble anodes are used generally as a lead alloy. The source for the chromium trioxide, CrO_3 (chromic anhydride), is commonly referred to as chromic acid. It is a deep red to reddish-brown crystal that volatilizes at 110°C. It is highly soluble in water (165 g/100 g at 0°C and 206 g/100 g at 100°C), producing a solution containing a mixture of $H_2Cr_2O_7$ and polychromic acids. Many manufacturers are now aware of the effect of even small amounts of catalyst acid radicals, and they furnish a pure grade of chromic acid especially suited for chromium plating. This chromic acid is made to meet specifications that require it to contain not more than a small fraction of a percent of sulfate that is free from other catalysts such as chloride.

A most popular solution containing 250 g liter^{-1} chromic acid, contains about 50% or 125 g liter^{-1} chromium metal. With complete current utilization, which is never the case, and no losses, 200 g of chromium would be sufficient to cover an about 110 m^2 surface with a deposit 0.156 μ thick.

The conversion of a pure chromic acid solution to a chromium plating bath solution constitutes a sulfate catalyst. With a given set of conditions of bath temperature, current density, and chromic acid concentration, too low amounts of catalyst will result in either no current flow, at first, or no plate or in an iridescent to brown oxide stains. Too high a catalyst content will result in an adverse effect: either partial plating with poor throwing power or, with great excess, no plate at all. The latter effect is due to depolarization action or easy formation of chromium (III) at the cathode. By increasing the current density and temperature to a sufficiently high value, plating can be accomplished with a very low ratio (up to 10 : 1). The essential criterion of bath composition for chromium plating from the congenital chromic acid–sulfate solution is the ratio, by weight, of chromic acid to sulfate. This ratio should be kept within the limits of 50 : 1 to 250 : 1, and preferably at about 100 : 1. A ratio of 90 : 1 is common; ratios of 70 : 1 to 80 : 1 are common in hard chromium baths, especially at higher temperatures.

A typical formula for chromium plating using a sulfate as the catalyst acid radical is presented in Table 1.

Although concentrations of chromic acid from about 50 g liter^{-1} up to saturation (about 900 g liter^{-1}) can be used, most commercial baths are operated between 150 and 400 g liter^{-1}. Still higher concentration gives very low current efficiencies. The important requirement is the proper ratio already mentioned.

Baths containing 200 g liter^{-1} chromic acid have a slightly higher current efficiency than more concentrated solutions. They also have a lower conductivity and therefore require a higher voltage for a given current density. The more dilute baths are also more sensitive to the changes of catalyst acid radicals from drag-in

TABLE 1 Basic Chromium Plating Baths

	Dilute Bath		Standard Bath		Concentrated Bath	
	g liter^{-1}	Molarity	g liter^{-1}	Molarity	g liter^{-1}	Molarity
Chromic acid, CrO_3	100	1.0	250	2.5	400	4.0
Sulfate, SO_4^{2-}	1.0	0.001	2.5	0.026	4	0.042
Ratio CrO_3/SO_4	100		100		100	

and drag-out. Hence they require more frequent and more careful adjustment for maintenance. Usually the more concentrated solutions are favored for decorative applications and the more dilute baths for heavy hard chromium plating.

Silicofluoride has had wide use as a catalyst in chromic acid baths since Fink and McLeese first proposed it in 1932 [8]. Such solutions were difficult to analyze and maintain. Yet those baths have definite advantages compared to sulfate only catalyzed baths. They have inherently higher current efficiency, can be operated at higher deposition rates, and produce somewhat harder and brighter deposits. Fluorides or rare earth's metals give better throwing and covering power. On the other hand, there are some important disadvantages. These metals are sensitive toward changes in composition and toward impurities such as iron and aluminum, and consequently more careful attention to bath purification, frequent analytical control, and housekeeping is required. Also analytical control of simple or complex fluorides are relatively more complicated, and finally, those solutions will attack or etch the base metal at low current density as well as unmasked areas such as blind holes. If masking is less than optimum, which sometimes cannot be avoided, proper attention must be paid to the possible etching effect. The solutions are aggressive toward plating equipment such as tank liners and heating/cooling coils.

4.2 High-Efficiency Chromium Plating Baths

The extra efficiency available from fluoride containing baths still resulted in chromium deposition rates which, in relation to the high current densities employed, were much lower than for most other plating baths. However, in 1986, proprietary plating solutions were introduced that had higher cathodic efficiencies than obtainable from the fluoride containing bath and these baths were established as viable industrial processes. They are based on chromic acid solutions which do not contain any fluorides or other halogens. Their chromic acid content is between 250 and 300 g liter^{-1}. The only other constituent of these solutions which is known is the primary catalyst which is a sulfate ion, within the ratio of 100 : 1, and 1% to 3% of alkene sulfonic acid as secondary catalyst. These proprietary solutions provide extra high cathodic efficiencies of up to 25%. These constituent catalysts have either been patented [56–59] or kept secret. The properties of the deposits have been documented [60–61], together with optimum operating parameters of these plating processes. The solutions are usually operated at temperatures between 55° and 60°C and typical cathodic current densities are 30 to 50 A dm^{-2}. Even at these high current densities, deposit distribution is superior to that obtained from conventional baths, with less edge buildup. The deposits have good hardness (1000–1150 HV), and retain it better than conventional chromium when heated. The chromium plate always microcracked, having 200 to 400 cracks per centimeter.

One of the greatest benefits of these fluoride-free plating solutions is that they do not attack steel on those portions of the cathodes where the current density is too low for chromium to be deposited. This low current density etching is especially detrimental when complex shaped steel objects are hard chromium plated for a long period of time in fluoride containing baths. The fluoride ion dissolves the protective oxide film off those portions of the substrate steel exposed to low current densities and thus the acid solution can then dissolve it with consequent iron buildup. This etching attack has been a limiting factor in the use of those baths for hard chromium deposit. Consequently many decorative and hard chromium platers prefer to use either conventional or non-fluoride high-speed chromium solutions despite their lower cathodic efficiency or the HEEF bath. The ability to plate at higher current efficiencies without this detrimental attack at low current density areas has been the major feature of HEEF bath which has resulted in these processes gaining a significant role in the hard chromium plating field.

In practice, relatively high concentrations of chromic acid are used, e.g. from 250 to 400 g $liter^{-1}$(33 and 53 oz gal^{-1}) of CrO_3. This increase in concentration increases the conductivity up to a maximum but decreases the cathode efficiency. In some cases these two factors, concentration and conductivity, may offset each other at the higher current density obtainable at a given voltage in a more concentrated chromic acid bath and may not yield a greater weight of deposited chromium.

Bright Plating Range The wide use of decorative chromium coatings depends largely on the fact that under appropriate conditions it is possible to obtain bright smooth deposits at a fair range of current densities. The conditions under which bright deposits are obtainable are often defined as the plating range for bright chromium. This dependence of the appearance on the conditions of deposition makes it necessary in chromium plating to hold the temperature nearly constant. For example, if a decorative bath is operated at 45°C (113°F), this temperature should be kept, preferably by automatic control, between 44 and 46°C (111° and 115°F).

It is also desirable to keep the current density as nearly uniform as practicable. On flat sheet cylinder rods, or nearly symmetrical articles, there is no difficulty in obtaining a nearly uniform cathode current density. However, on irregular shapes the ratio of the maximum to the minimum current density is usually at least 2, and it may be 5 or larger. The bright range for chromium deposits seldom covers a current density greater than about 3 to 1. Hence with more irregularly shaped articles, it is not possible to produce bright deposits or, in some cases, any deposit in the areas with low current densities without obtaining burnt deposits on the more exposed areas. In all such cases efforts must be made to make the primary current distribution more nearly uniform by (1) conforming anodes, (2) intermediate or bipolar anodes, (3) thieves to detract current from points or edges, or (4) shields to obstruct current to more exposed areas. Much of the success in chromium plating has resulted from ingenious applications of these methods. The acidity of chromium plating baths is very high; it is not ordinary controlled or measured. The measurements that have been made (with glass electrode) indicate values for acidity off the usual pH scale, and in the range of small negative values of pH [62].

5 MIXED CATALYSTS AND SELF-REGULATING BATHS

If a fluoride, silicofluoride, or fluoroborate anion is added (mixed) into the sulfate catalyzed bath, a *mixed catalyst* bath is obtained. Although higher speed and other beneficial effects are obtained, the difficulty of controlling these baths, due to their reactivity analytical problems, speaks against their wider commercial use. Self-regulating high-speed, or simply SRHS, baths, developed by United Chromium, Inc (now, Atotech, USA), is an attempt to simplify the handling of the catalyst by automatically controlling CrO_3/catalyst ratio by virtue of the solubility characteristics of the chemicals used. The advantages and results to be obtained were described by Stareck, and Passal [4, 63, 64]. The influence of cryolite [65], fluoroborate [66], magnesium [67], calcium [68], and ammonium fluoride [69] is presented in the literature. Little is available in literature regarding the theoretical aspects of the role of the fluorides in electrochemical reactions related to the deposition mechanism [70, 71].

The main advantage of SRHS baths is their higher current efficiency. In addition they are less sensitive to current interruptions, less subject to chemical control, have a wider plating range, yield brighter and slightly harder deposits, and exhibit better ability to activate passive nickel surfaces.

Moderate disturbances of the bath balance, asby drag-in or contamination, are minimized by the nature of the system. If fresh a catalyst is needed, a reservoir is normally present on the bottom of the tank in the form of undissolved salts, which can be readily dissolved, as desired, by heating and stirring.

When a bath gives low-catalyst results such as good coverage but a dull or rough deposit, and it is desired to increase the catalyst concentration, it is only necessary to dilute a little and then stir to re-establish saturation. This yields a slightly higher catalyst concentration, which together with the lower chromic acid concentration gives the desired adjustment.

Similarly, if a bath gives poor coverage and is overcatalyzed, it can be adjusted by increasing the concentration a little and stirring to establish equilibrium. This will lower the catalyst and increase the chromic acid concentration.

Increasing the temperature of a self-regulating bath has two simultaneous effects: (1) The warmer solution has a wider bright range and less tendency to burning, as with ordinary baths and (2) the increased solubility of catalyst gives a pronounced decrease in ratio. In general, these effects are of about the same magnitude, or the latter may be greater. If it is desired to operate at a higher temperature, it may be necessary to maintain a higher concentration so as to balance the increased catalyst content. Conversely, when operating at a lower temperature, some dilution of the bath may be in order.

The first self-regulating bath to be introduced [63] was a simple combination of chromic acid with an excess of low-solubility catalysts, which gave bath systems with a proper catalyst concentrations in the range of about 350 to 400 g liter^{-1} CrO_3. In order to produce bath systems with good catalyst balance in lower concentration ranges as 200 to 250 g liter^{-1} CrO_3, the solubility of the catalyst salts was suppressed by means of the common-ion effect [64]. Thus, if the sulfate was furnished by saturating the bath with strontium sulfate, the solubility was suppressed by adding strontium chromate or strontium carbonate. Similarly, if silicofluoride was furnished by saturating with potassium silicofluoride, the solubility was suppressed by addition of potassium dichromate. Postins and Longhand [72] have briefly discussed the operation of self-regulating solutions containing suppressants.

While formulas are sometimes given in the literature for self-regulating solutions, it should be realized that these are not simple formulations where all the constituents are completely soluble. Generally, the formulas include an excess quantity of catalyst which only partially dissolves.

Romanowski and Brown [73] have patented the use of fluorides and complex fluorides of lanthanum, neodymium, and praseodymium, and their mixtures. These rare earth fluorides, especially the silicofluorides, are only slightly soluble in the chromic acid baths and therefore are self-regulating with respect to the complex fluoride ion. The ratio of CrO_3 to SO_4 generally used with these fluorides is about 160 : 1.

A self-regulating bath system was adapted to the production of crack-free chromium deposits by Dow and Stareck in 1953 [74]. This process involves the use of rather high temperature (65°C) and gives smooth, satiny deposits that can be buffed to a higher luster. They are used where a combination of corrosion and wear resistance is required, as on washing machine shafts [75], high-temperature-resistant coatings [76, 77], gun barrels [78], and gas turbine buckets [79]. They have been employed for the decorative plating of zinc die-cast parts without undercoating [80], and two-tone, brushed finishes may be obtained by buffing part of the surface through a mask.

Seyb et al. [81] have developed a self-regulating bath for thicker decorative coatings for bright crack-free chromium plating up to thickness of 1.3 to 2.5 μm, compared to the maximum of about 0.5 μm which had been used previously. This process was of interest in the search for more corrosion-resistant decorative coatings and had the added advantage of better throwing or covering power than had been obtained previously [82]. Similar procedures were advocated with sulfate catalyzed baths [83]. Similar procedures were found by Safranek and Faust [84] to give substantially improved corrosion resistance to decorative deposits on zinc die castings when the bright crack-free chromium deposits were 2 μm thick. Wiener [85] confirmed the improvement obtained with bright crack-free chromium.

Complex fluoride catalysts have the advantage of giving higher current efficiencies at higher temperature of operation, whereas sulfate-catalyzed baths decrease markedly in efficiency at higher temperature [86–89].

Bilfinger [90] and Hood [91] discussed the operating characteristics of mixed catalyst baths. Bilfinger's curves of the current efficiency due to sulfate, fluoride, and silicofluoride catalysts are given in Figure 3. The best bright plate range is indicated by feather-like markings on each curve.

It should be noted that simple fluoride is a much "stronger" catalyst than sulfate, and that silicofluoride is much "weaker", thus making its effect easier to control because of the larger quantities required [92, 93].

Fluoride is not only a very powerful catalyst; it is also a very reactive one. The hydrofluoric acid, which it generates in the strongly acid chromic acid bath, is unstable as a catalyst. It will etch glass and dissolve silica in any form, thus converting to the weaker silicofluoride. It combines readily with boric acid, which may be dragged in from nickel baths, and the list of other "complexing agents" is long. A method of determining the relative strength of complex fluoride catalysts by means of the rate of solution of aluminum has been patented [94].

An unexpected behavior of solutions catalyzed largely with fluoride or silicofluoride is that the current efficiency increases with increasing chromic acid concentration in the usual commercial range, whereas in sulfate-catalyzed baths the reverse is true. This behavior is portrayed in Figure 4. The main disadvantage of the SRHS solution are its corrosive nature, which shortens the life of the plating equipment (e.g., heating/cooling coils) and its sensitivity to the iron, aluminum, boric acid, and chloride contaminations.

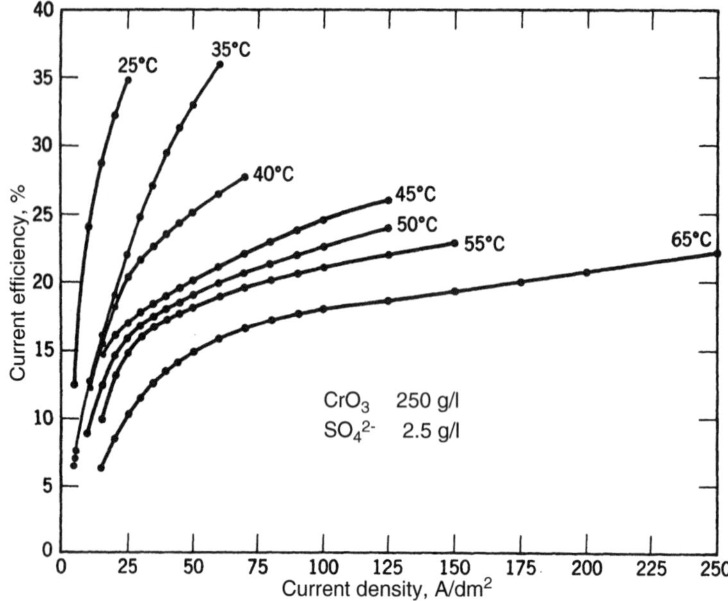

Figure 3 Current efficiency in 250 g liter^{-1} CrO_3 bath with sulfate, fluoride, and silicofluoride catalysts.

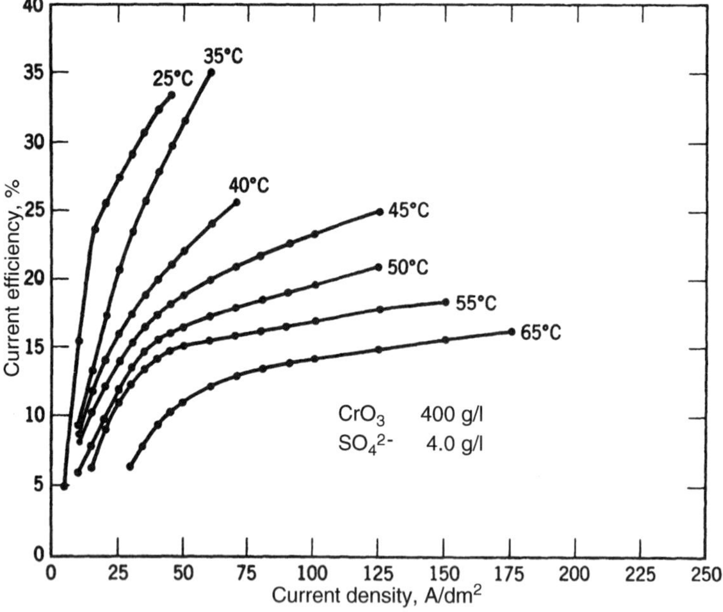

Figure 4 Current efficiency versus CrO_3 concentration.

6 CHROMIC ACID BATHS—OPERATING CONDITIONS

In general, bright plate is obtained by keeping temperature and current density within definite limits, taking into account the chromic acid concentration of the bath and the catalyst ratio. A convenient chart [2] showing the conditions for bright plating, is given in Figure 2. In that figure the semi-dashed line *A* circumscribes the bright plate area for solutions containing about 250 g liter^{-1} CrO_3, and the dashed line *B* circumscribed by the line *X* is typical of the behavior of most chromium plating baths. Thus, to produce bright deposits from a solution containing 250 g liter^{-1} chromic acid and 2.5 g liter^{-1} sulfate at a temperature of 40°C, cathode current densities between 3 and 16 A dm^{-2} must be used; at 45°C the current densities must be 50% higher.

If faster plating is desired and sufficient dc power is available, the temperature is often increased to about 55°C and the current density to about 30 A dm^{-2}. These conditions, when used with the 250 g liter^{-1} solution for building up a heavy plate for industrial purposes, result in a plating speed of almost 25 μm of chromium per hour. Higher plating speeds can be obtained at higher current densities at 80 : 1 ratio, but the deposits are prone to become slightly rough and nodular, which presents no problem if the parts are going to be ground.

Figures 5 to 8 [95] show the current efficiencies and plating speeds of two chromium plating solutions as determined in the laboratory. Actual speeds in productions plating will not conform with these exactly because shop conditions never exactly duplicate laboratory conditions. Other current efficiency data were published by Griffin [96].

The current efficiency increases regularly as the concentration of CrO_3 decreases down to 75 g liter^{-1}. The average increase in efficiency on diluting the solution, with a 100 : 1 ratio, 55°C, and current density 30 A dm^{-2}, was about 0.25% per 10 g liter^{-1} decrease in concentration of chromic acid [97]. These figures are useful in calculating the plating speed variation with small changes in chromic acid concentration.

The acidity of chromium plating baths is very high; it is not ordinary controlled or measured. Those measurements that have been made (with glass electrode) indicate values for acidity off the usual pH scale, and in the range of small negative values of pH [98]. (See Fig. 1.)

7 THROWING POWER

The poor throwing power of chromic acid plating solutions frequently refers to three different but related phenomena: (1) the covering power, (2) throwing power, and (3) bright plating range, previously discussed. By virtue of their effectual interrelation, the terms "covering" and "throwing" power are at times erroneously used synonymously, perhaps because plating baths with poor throwing power, generally exhibit poor covering power and vice versa.

7.1 Covering Power

The covering power (CP) of chromium plating solutions refers to ability to initiate deposition over the entire cathode surface at varying current densities (CD). While other plating baths may permit metal deposition at very low current densities, chromium will not generally deposit below 1 A dm^{-2} (10 Aft^{-2}), depending of course on specific plating conditions. Below the critical density, a manyhued nonmetallic deposit is sometimes formed. It has the appearance of a rainbow—a pleasant name

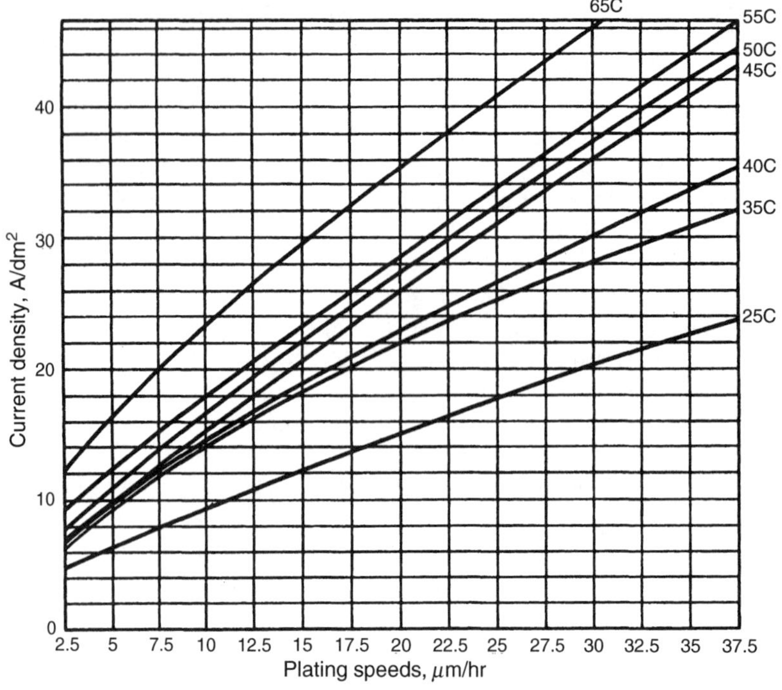

Figure 5 Current efficiency in 250 g liter^{-1} CrO_3 bath.

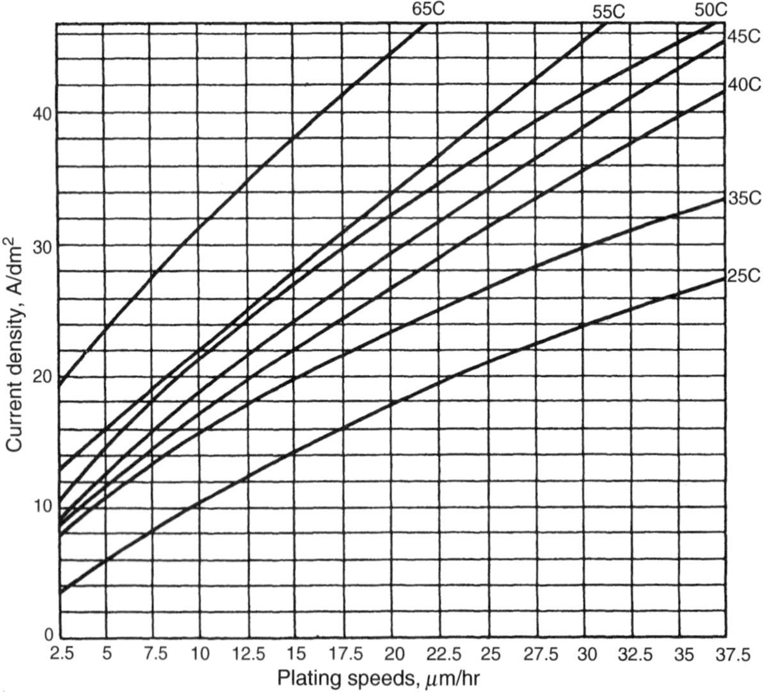

Figure 6 Current efficiency in 400 g liter^{-1} CrO_3 bath.

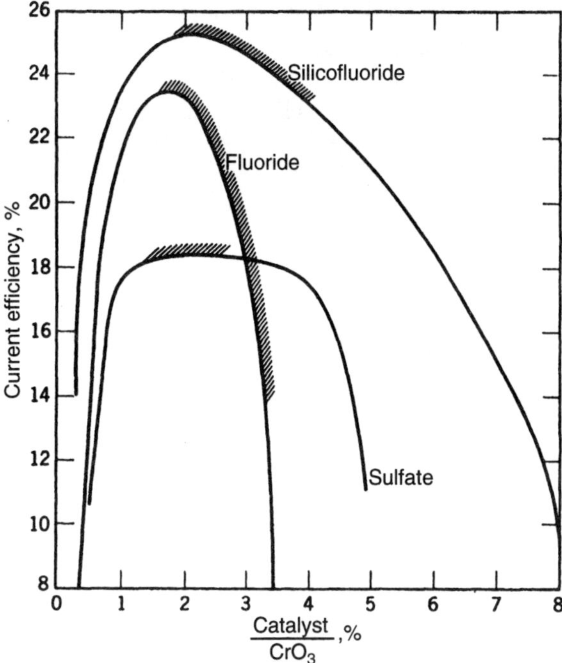

Figure 7 Chromium plating speed in 250 g liter^{-1} 100 : 1 solution (25 μ = 1 mil).

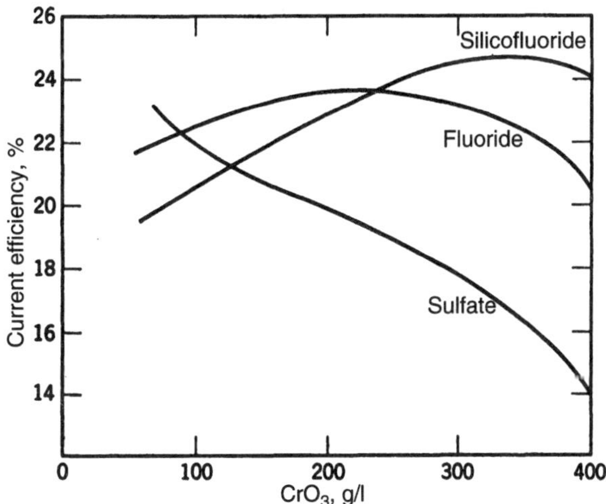

Figure 8 Chromium plating speed in 400 g liter^{-1} 100 : 1 solution (25 μ = 1 mil).

for a troublesome phenomenon. According to L.C. Pan [99], angle cathodes with different angles and various side lengths, slotted cathodes and slit cells are normally used to determine the covering power.

The covering power depends on the electrolysis conditions and on it's nature, any pretreatment, and surface condition of the base metal. CP improves with increasing

CD, and this fact is used particularly for chromium plating. Thus for a short time the plating is carried out at higher than normal CD ("strike" or "covering" current). Then plating is continued at normal CD, as soon as the part has been completely covered with a thin chromium layer. The CP of the chromium bath is less on aluminum and aluminum alloys, for example, than on copper or iron, although it can be improved by the use of an appropriate intermediate layer. On iron, it is better than on copper or nickel.

7.2 Throwing Power

The ability of a plating bath to uniformly deposit metal on the cathode surface is a measure of its TP. The major factors influencing TP in a chromium plating bath are the primary current distribution, polarization, secondary current distribution, and cathode efficiency. The primary current distribution is a function of the geometrical properties of the system, that is, the shape and distance of the anode to the cathode. The electrochemical properties of the system transform the primary current distribution into the secondary current distribution, which is determined by all the factors that influence the polarization during chromium deposition. Polarization is the change in potential on the cathode, which is mainly due to concentration gradients and the rates of electrochemical reactions, and in the case of chromium it is fairly constant. In general, polarization increases with an increase in CD. As the polarization becomes greater at the projected (higher CD) areas, the net result is a more uniform secondary current distribution. Unfortunately, in chromium plating, the cathode efficiency increases with the CD, and its effect on metal distribution can counteract any benefit of polarization. In the operation of a chromium plating bath the TP is generally improved by increasing the bath temperature and plating CD. As opposed to covering power, higher ratio baths have a tendency to improve TP. Increasing the anode to cathode gap is helpful, but this usually requires higher operating voltages; an initial high strike current may be necessary in order to obtain adequate coverage. On irregular cathodes, the CD varies widely, being highest on corners, edges, and areas closest to the anode; it is lowest in recesses, re-entrant angles, concavities, and areas farthest from the anode. It is then evident that cathode efficiencies being highest at the high CD areas result in heavier deposition, whereas the low CD areas have thinner deposits.

Consequently the major variables connected with throwing power in bright chromium plating are the current efficiency and the bright plating range. If a certain set of conditions gives the widest possible bright plate range, and the plating is done at an average cathode current density near the upper limit of current density for this bright plate range, optimum throwing power will be attained.

In the conventional throwing power cell, cyanide copper plating baths with good throwing power generally have a rating of around 20% to 40%, whereas most nickel plating and acid copper plating baths have a rating near zero. The throwing power in chromium plating has been found to vary from around −13% under the best conditions to −100% and even lower [7, 97, 100]. The relative throwing power of chromium plating baths is often estimated by an empirical test as the Hull cell test, described in the next section.

8 METALLIC IMPURITIES

They are two general classes of impurities in a chromium bath: (1) inorganic impurities such as chlorides, excess of sulfates or fluorides, boric acid, and organic matter, and (2) metallic impurities. The former are most common in the form of detrimental impurities such as iron, copper, and zinc. These metals enter the solution from parts accidentally dropped into the bath and not recovered, from the attack of the solution on racks and fixtures, attack on anode or cathode bars, and from corrosion of plating tanks through pinholes in the tank linings. The maximum acceptable concentrations of these metals will depend somewhat on the overall bath composition and the type of work being processed. However, approximate limits are for iron $15\,g\,liter^{-1}$ and for copper $0.2\,g\,liter^{-1}$. In or near these concentrations, copper and iron will restrict bright coverage in low CD areas. It should be noted that harm done by metallic impurities is very much synergetic; while one impurity alone even at high concentrations is not necessarily highly damaging, a combination of impurities even with less total concentration, certainly is.

Other cations that may commonly be present in chromium plating baths include chromium (III), which usually results when baths are operated with too large a cathodic area and too small an anode area, or when organic matter is introduced. The chromium (III) content can be kept down by increasing the area of the lead anodes used, relative to the cathode area, or, where this is not practical, by electrolyzing the solution for a time with relatively large anode area and a small cathode area.

Of all the impurities that can be present, chromium (III) is particularly detrimental, although contrary statements have been made [101]. The tradition has emerged that a small amount of chromium (III) is beneficial when added to a new bath. This is due to making such additions in the early days in the form of chromic sulfate or chromium hydroxide precipitated from chromic sulfate and containing some sulfate, thus affecting the catalyst content of the bath. A number of investigations have failed to indicate any improvement in new baths with the addition of small amounts of chromium (III), and there is no need to electrolyze a properly made up new bath for this purpose.

Buildup of metallic impurities can be corrected only by two procedures: discarding a portion of the bath (only as a last resort) and sending it to the authorized landfill or purification through a properly selected method [102]. The rapidly growing problem of chromium waste disposal has both helped and aggravated the problem of metallic contamination. Extensive use of drag-out tanks or the use of closed-looped systems accumulates and concentrates the impurities in the bath that might otherwise reduce itself in normal drag-out. On the other hand, the incentive for the plater to install chromium recovery and reclamation units has given many plants the facilities for electrolytic, selective membrane, or ion-exchange treatment of contaminated baths.

Periodic checks with a magnet, and if nonferrous parts are processed, periodic pumping out the bath, should be performed on any chromium installation to determine and remove fallen or broken anodes, processed parts, racks, tools, sludge, and the like.

9 MAINTENANCE AND CONTROL

Chromium plating baths are very stable in use, and their composition can be readily maintained by physical analysis or by more accurate chemical or instrumental analysis. If the bath is meant to be operated faultlessly, continuous correction in the control of bath composition is inevitable. Since chromium deposition is sensitive and controlled by such small amounts of catalyst it was recognized even in the early times that maintaining the proper catalyst ratio is of utmost importance.

Analysis of the catalyst concentrations in such small range, while routine to the fully equipped analytical laboratory, are complicated and of debatable accuracy for the average electroplating plant setting. Even if this were not the case, the validity of such tests is often subject to questions, in view of the fact that a bath's operating characteristics depend on the effect of the concentrations of the total catalyst content in the bath. Analysis for sulfate alone may neglect traces of chlorides, fluorides, and so on, which mole for mole have a much more marked effect on the bright electroplating characteristics from a similar bath using high-chloride content tap water.

Analysis for sulfate content can be a complicated procedure. The simplest procedure is to use the centrifuge method. This method is adequate for routine control and, if desired, may be checked occasionally by the classic but more elaborate gravimetric method given in the literature [103–106]. Since gravimetric methods require equipment that may not be available in the average electroplating plant laboratory, the straightforward titrimetric method may be of value [107]. It requires no special equipment and it is relatively fast. The most accurate is ion chromatographic (IC) method [108, 109].

Analytical testing methods for radicals other than sulfate are quite complicated, which is one of the drawbacks of these solutions. Data on these testing methods can be found in the literature or in the process manuals of a particular solution supplier. Organic catalyst determinations, as in the case of HEEF-25, need an accurate, reliable, and rather specialized equipment, which is the IC chromatograph. This relatively new and elegant instrumental technique can simultaneously determine many common ions. The attractive feature of the IC is for process control, where in a matter of a fraction of an hour, CrO_3, sulfates, fluorides, silicofluorides, and chloride anions can be determined in a single experiment [110]. The separation mechanism is based on differential absorbency and affinities toward the material used in the anion separating column. Concentration of each anion is determined with highly selective conductivity detector. Trivalent chromium is determined in cation separation column using ultraviolet detector. For determinations of alkenesulfonic acids, which are constituents of modern high-speed baths, this is the only practical method. These analyses are usually performed by the supplier.

Chemical analysis for the chromic acid content is not an excessively complicated procedure, and typical methods can be found in the literature. Hydrometer readings based on the density of the solution are a common and quite accurate method of chromic acid determination in new and relatively pure baths. A wide difference in hydrometer compared with analysis readings is not inevitable. Unfortunately, this is the case in most baths which, with age, build up in trivalent chromium, metallic, and organic impurities. Thus daily or at least weekly hydrometer checks, coupled with a periodic analytical determination, will form an acceptable and reliable control procedure for a chromium bath if done frequently enough. The standard hydrometer, calibrated to read solution density, or better, to read directly in oz gal^{-1} of chromic acid at the normal bath operating temperatures, is an essential tool and can be placed directly in the bath. The specific gravity is a good indication of the chromic acid content with new baths, but it may show considerable deviation as the bath is used and accumulates metallic impurities. Sulfate is often determined centrifugally, but this method is not always reliable. Excess sulfate is commonly precipitated by the addition of barium carbonate, but it can also be counterbalanced by the addition of chromic acid, if convenient.

In the absence of silicofluoride-containing chromium, plating baths were found difficult to analyze and control. Many methods were proposed, but they were not generally dependable for the total fluorine content. There is considerable confusion and inaccuracy in the literature about fluoride- and silicofluoride-containing chromium baths. Despite the difficulties, present-day silicofluoride containing baths with or without self-regulating features still have more than compensating advantages. The estimation of fluorine can be done in a comparatively simple method with the aid of fluoride sensitive electrode [62].

Probably in no other electroplating solution, except perhaps that of bright nickel, is the value of electroplating tests greater than in the chromium baths. In most electroplating baths, control tests (e.g., the Hull cell) are important additions to chemical analysis. In chromium baths, when properly employed, control tests can supersede and often eliminate some of the routine chemical analysis. The reason for this is the extreme sensitivity of the bath to relatively minute changes in a catalyst content. A standard bath with of 2.5 g $liter^{-1}$ of sulfates will exhibit detectable narrowing or widening of the bright electroplating range when this concentration is altered 10% (0.25 g $liter^{-1}$) or a mere 250 ppm. The interpretation of Hull cell or other electroplating cells panels requires only a moderate amount of experience and average amount of skill. It is an invaluable tool for day-to-day solution control, especially for troubleshooting.

Hull Cell electroplating tests are fast, effective and rather simple [111–114]. On the other hand, the chromium electroplating operations that thrive on excellence should also employ advanced versions of electroplating cell tests like Hanging Hull Cell, Jiggle Cell, and Rotating Cathode Cell [16–20]. The ultimate testing cell accomplice is the Hanging Hull Cell which eliminates the major drawbacks of regular Hull Cell, namely a lack of correlation with actual bath agitation, preplating cycle and possible rectifier ripple, bipolar effects and stray currents. Since the Hanging Hull Cell operates directly in the electroplating solution, a much closer picture of the actual electroplating range present in the bath is obtained from these test panels than from the laboratory type Hull Cells.

Some of the chromic acid is reduced to chromium (III) concentration at a relatively low figure under usual operating conditions, especially if the area of the lead anodes is sufficient [115]. If iron or other nonlead anodes are used for special purposes, they do not reoxidize the chromium (III) to chromic acid as well as do lead anodes, and a higher equilibrium concentration of chromium (III) is reached after the bath has been used for some time. Furthermore these other anodes, unless highly insoluble as lead, introduce contaminating metals such as iron into the solution, and therefore should generally be avoided. A rise of chromium (III) content can be seen by darkening of the color of the solution.

Wetting agents are frequently used to suppress the mist of solution carried into the atmosphere by the hydrogen evolution at the cathode, rather than to prevent pitting as in other plating baths. A great variety of wetting agents have been developed to minimize the fumes evolved during plating; the prospective user of such compounds should satisfy himself about their stability under his/her particular conditions. If they are used, surface tension measurements may become desirable for control, although visual observation of the fume suppression or amount of foaming may be sufficient.

Due to the high oxidizing power of chromic acid, plating baths are seldom filtered, although filtering has been recommended [116, 117]. If some clarification is desired, it can be accomplished by settling overnight and decanting. If desired, a chromium solution may be filtered through a pad of glass or through fiber glass filter cloth. Filtering cloths of Vinylite (Vinyon) and Saran are also available and have substantially complete resistance to chromic acid.

9.1 Anodes

While it possible to use anodes made of solid chromium, there are four serious objections: (1) Chromium anodes are much more expensive than chromium purchased in the form of chromic acid. (2) Chromium metal dissolves with much higher anodic efficiency (85–100%) than the prevailing cathode current efficiency (12–24%) and hence rapidly increase the chromium content in the bath. (3) Unlike lead anodes, there may be reoxidation of Cr(III) to Cr(VI), an unfortunate reaction that proceeds in parallel with the main, chromium deposition reaction. (4) No metal can match ease of lead for forming and joining together ("burning") when making conforming anodes.

While many anode materials such as iron, steel, stainless steel, nickel, and titanium can be used for special purposes, as auxiliary or conforming anodes, they are unsatisfactory for extended use, since they dissolve, and contaminate the solution and also increase the Cr(III) content. Although pure lead can be used for its ease for conforming, platers generally prefer more corrosion-resistive anode material, such as lead alloyed with silver, tin, or antimony. Many anode materials other than lead alloys have been tried, but nothing better has been found [119]. Recently, it was revealed that bismuth doped lead anodes can triple the anodic reoxidation of Cr(III) [120]. Pure iron, such as Armco or electrolytic iron, dissolves less when used as an anode than steel, nickel, stainless steel, or similar alloys. Iron anodes have occasionally been used, particularly in industrial chromium plating, in special instances where greater strength and rigidity than are obtainable with lead are desired. Their continued use, however, leads to the accumulation of iron and chromium (III) in the bath. Small platinum wire anodes can be used for special purposes, such as plating the insides of very small openings, such as those of wire-drawing dies. Antimonial lead anodes are preferable to chemical lead due to the greater corrosion resistance and strength, but they do not eliminate the formation of copious amount of lead chromate sludge. Lead-tin alloys have higher corrosion resistance but less rigidity than antimonial lead anodes, and these are widely use. A good compromise is achieved by using Pb–Sb–Sn ternary alloy. The best are silver containing lead and lead alloys [121], but they are more expensive. Those anodes not only have the advantage of increased durability, these also give quick or immediate startup after down time, without special reactivation.

Lead and lead alloys serve two key functions in the chromic acid plating bath: (1) They provide effective current distribution and (2) reoxidize Cr(III) to Cr(VI). Lead peroxide film, which forms on these anodes during use, causes continuous reoxidization of the chromium (III), forming chromic acid and thereby keeping its concentration at a low, acceptable value [122].

Lead and lead-alloy anodes of varying shapes and cross sections have been proposed and used from time to time. They have to be thick enough to conduct the high currents required. Anodes that are too thin will overheat in use and will corrode and warp excessively. This difficulty can be avoided by the use of solid round copper core in the center of lead anodes. The copper core aids in rigidity and securing good current distribution because considerable current can come from the back as well as the front.

Auxiliary conforming anodes are sometimes used in hard chromium plating or through the complete cycles of decorative plating. Improved coverage is obtained on difficult shapes, and more uniform plate distribution is achieved on large surfaces where a minimum plate thickness is required, as for the production of microcracked chromium. Pure nickel anodes are perhaps the best for this service, and cast nickel is sometimes used to produce a number of anodes of a special shape. Platinized titanium anodes are also used for this service, but they have the disadvantage of a limited life and insufficient indication of when they are becoming inoperative, except for increasing rejects. Nickel dissolves slowly in use and can be replaced when visibly worn away.

Lead anodes used in chromium plating cannot have too heavy or irregular a coating of lead dioxide on them, or the current distribution may be affected. It is beneficial to clean the anodes regularly, especially those used in heavy hard chromium plating that conform closely to the article being plated. The cleaning is done by acid dips and scratch brushing, but the process is difficult and time-consuming; frequently not all the semi-insulating coating is removed. Hyner [123] developed the method of electrolytic reduction of the coating to metallic lead by cathodic treatment in an alkaline pyrophosphate solution. Lead anodes coated with lead dioxide tend to become somewhat passive after standing idle for some time. Some platers would electrolyze the bath with full tank voltage from several minutes to an hour for reactivation in order to reestablish the original conductivity. However, other platers found it unnecessary to do this. This passivation tendency, investigated by Hardesty [124], is presumably due to insulating effects of insoluble lead compounds such as lead chromate and lead fluoride. The only way found to avoid this effect, aside from recleaning as described, is to remove the anodes from the tank promptly after use and to permit the solution to dry on them rather than rinsing it off. This gives good results but is not always practical. Platinized titanium anodes with thermally deposited iridium dioxide are proposed [125]. Titanium anodes coated with PtO_2 are recently recommended for chromium plating [126]. Practical experience with platinum plated titanium anodes in chromium plating solution is described [127].

9.2 Materials of Construction

Most tanks for chromium plating are made from steel and lined with some kind of acid-resistant material. In the past, chromium plating tanks were made of lead or antimonial lead lined steel. Acid-proof brick linings have also proved very satisfactory for chromium plating tanks made of steel [128], although they are seldomly used any more. The type of lining now used, which gives satisfactory service, consists of flexible synthetic resin sheets (plasticized polyvinyl chloride, PVC) cemented to the steel tank and welded at seams and corners. This type of lining saves space compared to a brick lining but is generally not recommended for temperatures above 60°C. Special insulating materials of the vinyl type (Corroseal) have been developed [129] to withstand the action of hot chromium plating solutions, and thus have good mechanical properties. These insulators are used in sheet, rod, tube, tape, and other solid forms in the construction of composite racks and in liquid form for coating ordinary racks or for stop-offs. The use of insulated racks results in a saving of power and chemicals and gives much better plating. Microcrystalline, high-melting paraffin wax, and still higher melting chlorinated naphthalene wax compounds are also used for stopping off in industrial chromium plating.

Tanks can be heated and cooled by lead alloy coils submerged in the solution. Titanium, columbium, tantalum, Teflon, or Teflon-coated stainless steel coils and heat exchangers have come to be used extensively for their extremely long life and efficient operation. More durable than titanium, tantalum is probably the most suitable material. Its high initial cost can be justified by a long period of trouble-free operations. Titanium coils are not the best choice for fluoride based solutions.

Owing to the relatively high current densities used in chromium plating, it is necessary for all compounds of the circuit to be of sufficient size to carry the amperage required without overheating or excessive voltage drop. The plating tanks have to be of such a size that the parts being are positioned 10 to 25 cm (4 to 10 in) from the sides and the bottom of the tank as well from the surface. When designing the tank and the rectifiers capacity , consideration should be given to an ideal current loading of 1 to 1.5 amps liter^{-1} of the tank volume. This will save energy required for heating and cooling in improperly designed tanks.

9.3 Safety and Health Considerations

The chromium metal and trivalent chromium compounds are nontoxic in comparison with the much more hazardous sixvalent compounds. The chromic acid is sharply irritating and corrosive to the mucous membranes of the nose and throat. This spray therefore requires removal or suppression to protect the workers and equipment, and adequate exhaust facilities must be provided for the purpose. Carcinogenic factors are also suspected. Skin contact can cause ulcers and dermatitis; different persons may react differently to dermatitis effects. Chromium plating solutions emit mist as they are used. The mist contains, in addition to hydrogen and oxygen gas, basically the same ingredients as the plating solution and therefore presents a health hazard to the workers and the community. The mist must be therefore be captured and removed from the air and the tank to protect the workers and equipment, and adequate exhaust facilities must be provided for the purpose. The U.S. Environmental Protection Agency (EPA) regulates the amount of chromium that may remain in the air discharged from a hard chromium plating to either 0.015 mg dscm^{-1} for large and new facilities and 0.03 for small facilities.

Many measures have been proposed to replace or supplement the necessary exhaust hoods, and to prevent some of the chromic acid and heat losses these entail.

The proposed use of a layer of floating plastic beads [130, 131] or of stable wetting agents may in some cases offer a partial solution to the problem, but they may not make it possible to dispense with an adequate exhaust system. The development of completely stable perfluorinated sulfonate wetting agents [132, 133] made an important difference. They are effective in reducing the emission between 93% to 98% depending on the operating conditions and the type of foaming/wetting agent used. These wetting agents would not induce, but may accentuate, basis metal pitting in thick, hard chromium plating deposits [117, 134–137]. Their use is quite widespread in decorative baths and permit economies in chromic acid and heat losses, among other advantages. Floating balls pose of problem of traveling from tank to tank and also can become stuck in the crevices of the parts.

Studies of industrial dermatitis arising in workers exposed to chromates or chromic acid have been published [138, 139]. The remedies suggested include avoidance of contact with the irritating chemicals, cleanliness, thorough washing, use of protective and healing salves and ointments, visits to a physician when necessary.

The disposal of waste waters containing chromic acid is a problem of increasing importance [140, 141]. Chromic acid-based plating solution also presents a fire hazard when in contact organic matter such as paper.

9.4 Bulk Chromium Plating

Chromium plating barrels of both the batch and continuous types have been described [142–145] and are operating successfully in a number of plants. The barrel plating time, for decorative purposes, is about 5 to 10 min, but heavy hard chromium plates can also be reduced in the barrel by using longer plating times [146]. Deposition from trivalent baths have been recently described [147]. Round or cylindrical articles, which roll easily and are not too light, lend themselves best for barrel plating. The lower limit is about 3 g, although the most important aspect is the ratio between area and weight. The preferred solution is sulfate free chromic acid baths containing silicofluorides as a catalyst. This solution has the lowest threshold current density requirement (as low as $0.6\ \mathrm{A\,dm^{-2}}$), has relatively the best activating properties, and can withstand short current interruptions, which occur repeatedly when plating in a barrel. Small parts such as screws, nuts, bolts, and rivets can be chromium plated in wire mesh baskets or by stringing them on wires. Stringing on wires or racking is convenient for articles of moderate size, perhaps 25 mm long or longer. For basket plating, horizontal copper wire mesh trays are generally used with a rim about 13 mm high, soldered to a frame for suspension from the cathode rod. The small parts should be spread in a thin layer on a tray so that they do not cover each other. The whole may be plated, at as high a current density as possible without burning, for 5 to 10 min. Generally the entire basket is shaken or jarred a little a few times during plating, to cause the parts to shift position and avoid contact marks. Flat art, which fit closely on top of one another, do not lend themselves readily to basket plating.

9.5 Preparation of Basis Metals

In order to ensure the satisfactory adhesion of chromium deposits, the parts must be almost perfectly clean and free of any grease. If parts are transferred from nickel or other baths to the chromium bath without unmerited delay, only an acid dip and a water rinse may be enough. On the other hand, if the nickel is buffed or the parts are handled or stored for a time, further treatment may be necessary before chromium plating.

The cleaning of work to be chromium plated for bright or decorative finish (as distinguished from work for thick deposits or for industrial applications) may be divided into tree general classifications—solution cleaning, dry cleaning, and vapor degreasing. Typical solution cleaning procedures are detailed in the chapter on preparation for plating. Dry cleaning consists of wiping the work on a buff wheel or by hand with pumice powder, without dipping it in solutions of any type.

Where solution cleaning is feasible, it generally gives better results than dry cleaning. It helps to remove any oxide or tarnish on a nickel surface, whether visible or invisible, and results in "activating" the nickel or making it easier for chromium plate. Nickel surfaces are considered "passive" if they are oxidized and difficult to cover with bright chromium plate. Cathodic alkaline cleaning is quite effective in removing this condition if it is not too severe. Acid dipping is even more effective. Typical acid immersion procedures for maximum nickel activation are

1. Chemical activation in 30% to 50% (volume) HCl for 30 to 60 s.
2. Chemical activation in 5% to 20% (volume) H_2SO_4 for about 2 to 5 min.
3. Cathodic activation in 5% (volume) H_2SO_4 at 4 to 6 V for about 15 s.

Where wet cleaning is not feasible, the plater must sometimes resort to dry cleaning. Success of this procedure depends on the fact that the chromium plating solution itself serves to some extent as both cleaner and acid dip. The vigorous evolution of gas during plating, together with the strong cleansing action of the hot chromic acid, tends to remove light soil films. If the dirt, grease, and oxide are excessive, the cleansing action of the plating solution is overtaxed, with the result that the chromium plate is defective.

The importance of a satisfactory wet cleaning procedure for nickel surfaces has been confirmed by Tucker and Flint [148]. They reviewed some of the previous work in the field. Cathodic electrolytic cleaning is also helpful, and special solutions and procedures are sometimes used [149]. Mandich discussed recently the practical and theoretical aspects of nickel and chromium activation as well as chromium reverse etching [150]. Anodic cleaning in the usual alkaline cleaners must be scrupulously avoided, since it tends to oxidize nickel surfaces and make them impossible to chromium plate.

Plating over stainless steel also requires wet cleaning and activation. The surface should be freshly buffed, and not permitted to stand from one day to another. Often it can be placed after a short dwell time, particularly if a silicofluoride type solution is used under conditions of higher than normal catalyst concentration or low ratio and high temperature. Heavy chromium deposits used in industrial or hard chromium plating usually require extraordinary good adhesion to the basis metal because the plated articles are often subject to severe stress in service. A high degree of adhesion of chromium to steel is the normal result of plating in a hot chromic acid bath, but adhesion tendencies can best be treated using electrolytic cleaning or etching of the steel surface before chromium plating. A satisfactory etch is obtained by treatment of steel parts as anode at 6 V for about 1 min in chromic acid solution or in the plating bath. Anodic etching in sulfuric acid (sp. 1.53 g) at about 25°C for about 1 min gives the highest adhesion. Similar results are obtained by electropolishing [151]. Additional details for the preparation of steel for heavy chromium plating are given in a *Recommended Practice of ASTM* [152], Greenwood [153], Morisset [154], Guffie [155], Peger [156], and Mandich [150]. Levy [157] and Dini [158] give procedures for plating on some special alloys.

Zmihorski [159] has investigated the adhesion of heavy chromium deposits on steel by means of a shear test; he found that an adhesion of about 40 to 45 kg mm^{-2} is obtained by the usual hard chromium plating procedures. He also found that etching in sulfuric acid gives somewhat better adhesion than etching in chromic acid, that a low current density gives somewhat better adhesion than high current densities, that thin deposits are better than thick, that deposits from pure solutions containing no iron and chromium (III) are better than those from contaminated solutions, that silicofluoride solutions are better than sulfate solutions, and that heat treatment appears to have no effect on adhesion.

The adhesion of thick chromium deposits is difficult to measure because it is commonly greater than the tensile strength of the coating, which will fail before it can be pulled off the base metal. Williams and Hammond [160] made direct measurements on the range of 16 to 32 kg mm^{-2}. Beams [161] used a centrifugal force method. Chessin and Poor [162] first use an indentation method for adhesion and later [163–164] developed a "push-out" test in which a 6 mm diameter hole was pushed out of the basis metal from underneath the coating, and the nature of the fracture around the hole was examined. Dini and Johnson [165] described a "flyer plate" test for measuring the adhesion under dynamic conditions. Methods for measuring adhesion have been critically reviewed by Davies and Whitaker [166] and Ploog [167].

High-carbon cast irons and steels may be difficult to chromium plate directly if acid pickled before plating. Pickling apparently develops a low-overvoltage surface which makes it easy to deposit hydrogen and difficult to plate chromium. It is therefore recommended that acid pickling be avoided in such cases, and that sandblasting or other methods of cleaning be used.

Zinc and zinc base die-castings are commonly chromium plated for decorative purposes after previous copper and nickel plating. If the castings are satisfactorily nickel plated, the chromium plating is the same as for any other nickel-plated basis metals. If the nickel plate, directly applied on a zinc base article, does not completely cover it or is too thin, it will be difficult or impossible to deposit chromium at or near the bare or thin points. A remedy for such a difficulty is to plate a substantial thickness of copper under the nickel. There is a certain amount of chromium plating directly on zinc die-castings, generally for wear purposes.

9.6 Hard Chromium Plating

Dubpernell's book treates the behavior of the sulfate and fluoride type catalysts [168]. Morriset [6, 169], and Weiner and Walmsley [170] have published some general books on the subject. A practical plant manual has been published by Greenwood [171, 172], Guffie [155], and Peger [156]. Dennis and Such [174] covered both nickel and chromium plating in their book. Racking for hard chromium plating is well elaborated by Logozzo [175] and Peger [156]. Mandich treated practical problems in hard chromium plating in a series of papers [176].

The success of chromium plate in industrial applications may be attributed to its unique combination of properties not possessed by any other single, material available commercially. The most important of these are hardness, adhesion, corrosion resistance, nongalling and nonwetting qualities, low coefficient of friction and high melting point. These properties make hard chromium invaluable for industrial and engineering purposes. The hardness alone, although approaches that of diamond, would not be sufficient to secure widespread use, because a number of other hard materials or hardening processes are available. It is the combination of very high degree of hardness with extremely good corrosion resistance (equal or even superior under most conditions to that gold or platinum), and very low coefficient of friction or unique surface qualities, which has given remarkable results in many applications of chromium plate. To these should also be added the relative ease of application and control, which ensures maintenance of fixed standards of quality and durability, together with moderate cost. There is also the ease of stripping and replanting for repeated salvage in cases where the plate wears beyond suitable limits.

The benefit of the harness of chromium deposits is not efficiently obtained unless the coating is deposited on a sufficiently hard basis metal and to satisfactory thickness. Generally, hardened steel is used for the basis metal. Even a relatively heavy deposit of chromium may be crushed or indented if applied over a soft basis metal such as copper. The best possible adhesion is also important in many uses where the surface may be subjected to severe stress or shock, and any chipping of the deposit would be injurious.

The low coefficient of friction and desirable surface properties of chromium are realized for the most part only on relatively smooth surfaces, although the advantages of certain types of interrupted surfaces are also described in the section on porous chromium. Frequently chromium deposits are ground or lapped to size. The deposits are easily ground but are sensitive to the heat generated and usually need to be ground with very light cuts [177–178].

Sometimes a bright deposit is applied to a smooth surface and used without further mechanical treatment. By means of careful operation it is possible to plate to size within very close limits. Worn machine parts are salvaged by chromium plating oversize and grinding back to size.

Some of the outstanding applications of industrial chromium plate include gages, tools, and machine parts generally, both new and worn parts, which are plated or replated for, salvage purposes. Taps, reamers, drills, saws, milling cutters, burnishing tools, and so on, have all been successfully plated. Molds for plastics and rubber are plated to reduce wear and sticking and to improve appearance. Drawing dies and mandrels, coinage dies, rolls for cold-rolling metals to high luster, calendar rolls for various materials and printing and engraving dies are other examples of common uses. Oswald [179] reported an increased life of over four times for rolls used for the cold-rolling of steel, and eight to ten times for printing cylinders. Wilson [180] mentions sometimes getting 3,000,000 copies from rotogravure cylinders without apparent wear.

Gun barrels are frequently plated for the maintenance of accuracy over a long period of use [181–186]. It is reported that the life of machine gun barrels is increased 30 times by chromium plating [187]. Oil drilling rods, pump shafts, and the cylinders of internal combustion engines have been plated with good results. The list of special uses could be greatly extended [128].

In each application the most desirable thickness of chromium and hardness of the basis metal have to be determined empirically, with the aid of previous experience. If high corrosion resistance is desired in addition to wear resistance, as with rotary dryers for corrosive chemicals and paper mill machinery, relatively thick deposits are required. Sometimes substantial undercoats of nickel or copper are used in such applications.

Hard chromium plate has been found useful on basis metals of widely varying hardness, although the basis metal should generally be as hard as possible. Thus, on one end of the scale, good results have been obtained by chromium plating cutting tools tipped with tungsten carbide [188]. On the other hand, zinc alloy dies for auto stamping have been chromium plated for longer life [189, 190], and the hard chromium plating of aluminum has been developed considerable [191–193], especially in connection with small internal-combustion engine cylinders [191, 194, 195].

Additional information on hard chromium plating is given in other sections of this chapter.

9.7 ETCH Prevention in Hard Chromium Plating

The biggest disadvantage that self-regulating baths have and common to all baths containing fluorides or complex fluorides is a tendency to etch the areas where cathodic current density is low, or areas not covered with chromium. Because of this risk, fluoride containing baths are often discouraged. This tendency to etching is especially noted on steel; conversely, the ordinary baths with sulfate catalyst have a strong tendency to etch copper, bronze and brass. This etching tendency is especially marked in overcatalyzed baths and can be overcome to a considerable extent by keeping the bath in proper balance, or operating at as low a catalyst concentration as possible. In addition this tendency results in the increase of the iron content of the bath, which when it reaches a certain level, may slow down the plating rate, produce roughness, and if not checked make bath unusable. Stareck and Dow [196] overcame etching by prefilming the surface to be plated as cathode in a plain chromic acid solution. This procedure is effective only in higher ratio baths.

Etching and pitting of the basis metal is sometimes encountered in hard chromium plating even in ordinary baths with only sulfate catalyst if stray current is permitted to leave nonplating areas such as the exterior of cast iron diesel cylinder liners, they will be plated on the inside only. This can be prevented by greater care in maintaining full insulation on racks and fixtures, or by insulating the surface to be protected [150, 197, 198].

Bedi [199] found that immersion deposits of noble metals such as platinum and palladium on steel would prevent the etching that occurs in low current density recesses, particularly in baths containing fluoride catalysts. Such immersion deposits of noble metals create a low overvoltage condition on the steel surface, which favors hydrogen evolution and tends to prevent other reactions. They may even prevent chromium deposition at higher current densities, and Bedi and Dubpernell [200] demonstrated the possible use for stopping-off purposes.

9.8 Black Chromium Deposits

Black chromium plating now has reached the stage where it is a completely practical plating operation that can be conducted in any existing plating plant. This process received proper amount of attention in last few decades. Black chromium selective surfaces have held the promise of being the most suitable coating for a wide range of low to medium temperature application as solar selective coatings because of their excellent optical properties and apparent high degree of stability under diverse operating conditions. Mcdonald [201] reported that black chrome is not susceptible to degradation in humid atmospheres and possesses excellent selectivity. Mattox [202] reported excellent selectivity and thermal stability in air and in vacuum up to 350°C, and these authors have reported similar characteristics for black chrome.

The ideal selective absorber will have a high absorbency to incident solar radiation with wavelengths below 3000 nm, and a low emittance beyond 3000 nm; this means that it will absorb solar radiation but simultaneously will emit little long-wave thermal radiation. Black chromium comes close to this ideal and will retain more heat energy than other types of black coatings.

The coating has good thermal stability. At temperatures below 480°C (900°F) there is no effect by the coating; at temperatures up to 590°C (1100°F) there is a slight graying of the deposit but the color reverts to black on cooling. The use of black chromium is not recommended on components subject to temperatures in excess of 700°C (1300°F).

Black chromium deposits have a high degree of microporosity, and this produces a corrosion resistance that is better than standard bright chromium. The same porosity gives it the ability to absorb and retain oil and paint films, which make it useful for the machine tool and electronic industries and these properties are retained even after such operations as stamping, forming drawing, and welding. An early "black" was produced by using a high current density in a cold bath principally chromic and acetic acids [203–205].

Modifications of this process for "black" chromium plating have proposed [206–211]. Current interruption [212] and nitrate, borate and fluorosilicate [213] sulfate-free acid with a fluoride or complex fluoride catalyst [214] have been used. Mechanisms and structure of black chromium were studied [215–217]. There is some controversy among authors in the area of surface structure and composition, with the only real agreement being that black chromium is in homogeneous deposit of chromium metal and its compounds. The coating is generally believed to consist of particular chromium metal and chromium oxides and compounds. Baths based on trivalent chromium [218] and tetrachromates have been proposed [219, 220].

9.9 Postplating Treatments

Postplating treatments are not commonly used on chromium plate, since the great passivity and tarnish resistance of the metal usually make them unnecessary for most commercial parts. The normal "air passivity" of the metal, however, can be increased considerably by treatment with an oxidizing agent such as nitric acid, where this is practical. Flasch [221] treated thick chromium deposits with hot nitric acid, chromic acid, or permanganate solutions for many hours. He found that the passive chromium would no longer dissolve in hydrochloric acid, and was nobler than platinum. Except for nitric acid, such treatments tend to discolor the chromium surface.

Willson [222] found that electrodeposited chromium foils 250 µm or more thick (produced by stripping the brass basis metal from them in concentrated nitric acid and then baking overnight at 165°C) resisted corrosion for two hours in concentrated hydrochloric acid. It should be noted, however, that this is a metastable condition, and any slight disturbance or handling may result in a sudden violent attack of the chromium by the hydrochloric acid with the evolution of hydrogen gas.

Although treatment in a sodium hydroxide-sodium nitrite solution has been recommended for improving the corrosion resistance of thick chromium deposits on steel, other workers have disputed the efficacy of this procedure. Electrolytic polishing of the steel before chromium plating was effective in giving hard chromium deposits greater corrosion resistance [223].

Giesker and Britton [224] improved the corrosion resistance of chromium-plated steel surfaces by treating them cathodically in 50 g liter^{-1} sodium dichromate solution of pH 4.5 at about 95°C and 0.3 to 0.5 A dm^{-2} for one to two minutes. Similar results were obtained by simple immersion in the same solution for about two hours [225].

Safranek and coworkers [226] improved the corrosion resistance of chromium-plated parts with supplementary surface films applied by cathodic treatment for about one minute in a solution of 50 g liter^{-1} sodium dichromate and 1g liter^{-1} chromic sulfate at 85° to 95°C and pH 2.0 to 2.5, using 0.32 to 0.64 A dm^{-2}. The extent and permanence of the improved corrosion resistance are not certain. There are indications that the effect tends to be lost after a year or two of outdoor exposure. Occasionally chromium plate is used as an undercoat [227, 228]; for example, chromium is about the only undercoat that readily bonds to molybdenum [229]. In such cases, or whenever it is desired to deposit another metal over chromium, special precautions are necessary, owing to chromium's passive surface condition. This can be readily overcome by dipping in strong hydrochloric or other strong acid until the chromium starts to etch and hydrogen is briskly evolved. Quick rinsing and plating will then result in an adherent electrodeposit of almost any metal.

This etching procedure has the disadvantage of removing more chromium than is permissible, or results in greater dulling and roughening of the surface than is desired. In such cases the chromium can be activated without etching by cathodic cleaning in alkali and followed by a short dip in mild acid such as dilute sulfuric at about 25°C, or else by cathodic treatment in weak acid. One may Follow with a strongly acid nickel strike [227, 228], and no difficulty should be experienced in obtaining a good bond to the chromium. A strongly acid copper chloride strike was used to obtain adherent copper coatings on chromium many years ago [230].

If the current is interrupted even briefly during hexavalent chromium plating, the subsequent layer of chromium may peel from the first due to the intervening passivity. The subsequent layer of chromium is also likely to be dull. This difficulty of plating one layer of chromium on another can be decreased by using a high catalyst or low ratio composition of the chromium bath, silicofluoride along with the sulfate catalyst, and high bath temperatures. If a current interruption occurs, chromium can be plated on the first layer of chromium by permitting it to come up to the temperature of the bath while submerged without current in the bath. The current can then be applied slowly in increments over a period of a minute or two, starting with less than the plating current [231]. The hot, strong chromic acid solution has considerable activating action that only needs to be assisted by gentle gassing at just below the current density for the beginning of chromium deposition.

Current interruptions during trivalent chromium electroplating will not passivate the chromium deposit. Removing the part from the plating solution, rinsing, inspecting, and then returning it to the tank and continuing with the plating process can be done. Thus activating the surface before plating is not typically required. There will also be no loss of appearance. Trivalent chromium processes typically plate over passive deposits much easier than hexavalent chromium processes. The activation steps developed for plating hexavalent chromium deposits might not be necessary when trivalent chromium processes are used.

Occasionally organic coatings need to be applied over chromium and once again obtaining good adhesion is of major concern. In general, organic coatings baked at a high temperature will adhere to chromium, whereas air-dried coatings will not. A detailed investigation of the adhesion of organic coatings to chromium plate was published by Safranek and Miller [232]. Practical experience is reported by a number of workers [234]. One requirement appears to be to apply the organic coating to the fresh clean chromium surface before it can become contaminated with grease, dirt, or other foreign matter. Waiting even a few hours after chromium plating to apply organic coatings increases the degree of difficulty of application. Additional experience with the application of lacquer to chromium-plated surfaces can be found in the section on tin-free steel (TFS).

9.10 Stripping

Every chromium plater is eventually faced with necessity of stripping chromium plate. Chemical, electrochemical and mechanical methods are used to remove chromium plate.

In view of its passivity and usefulness, chromium is remarkable for ease of stripping. The metal dissolves readily as an anode in almost any aqueous solution to form chromic acid (Cr^{+6}) with close to 100% efficiency. It is only necessary to select a solution in which the basis metal suffers little attack from anodic action, or in which the action is not too injurious for the purpose at hand. The most common strip is a dilute alkaline solution, such as an alkaline cleaner used with reverse (anodic) current. The cleaner becomes contaminated with chromate, and it is therefore best to use separate solutions for cleaning and stripping. A solution containing 70–100 gr/l of sodium carbonate or 50–100 gr/l of caustic soda can be used. It is important that the solution is free of chloride to prevent attack on the basis metal. For chromium thicknesses above 0.010 inch (250 μ) is more economical to grind the chromium deposit, rather than to use any stripping method. Soft grinding wheel with the lot of coolant is required.

Chromium deposits from trivalent chromium processes containing metallic contamination might not strip completely in alkaline solutions. Activating the deposit in hydrochloric acid prior to the anodic alkaline strip is effective if dissolving it completely in hydrochloric acid is not desirable.

Simple immersion in hydrochloric acid from 10% (vol) to the concentrated acid of about 1.18 specific gravity, at 25°C, will dissolve a thin decorative chromium coating with hydrogen gas evolution in a few seconds, and will generally leave the underlying nickel in condition for immediate replating. Care must be taken to rinse thoroughly to avoid contaminating the hexavalent chromium solution with chloride. Stareck [235] developed an alkaline pyrophosphate strip for chromium and other metals.

This has the merit of leaving the steel, cast iron, or other basis metal clean and ready for further finishing operations. An alkaline strip tends to oxidize or passivate the underlying nickel and to make it difficult or impossible to replate with chromium. If, however, the strip is used cold at low concentration, low current density, and for the shortest possible time, replating is sometimes possible.

Thick chromium deposits are sometimes stripped in hydrochloric acid, but the action tends to slow down or stop where the deposit is thickest and to be too corrosive to the basis metal. Reverse current in a plain chromic acid solution of about 100 to 400 g liter^{-1} is a better and safer stripping method for heavy chromium deposits on steel.

Another useful thickness test for thin decorative coatings is the anodic solution method [236].

10 TESTS OF DEPOSITS

10.1 Porosity and Cracking

Dubpernell [168] reported that acid copper plating over chromium up to about 25 µm thick could detect pores and cracks. Under the conditions of the test, acid copper does not deposit on chromium, owing to the passivity of the surface, and the copper deposits only in pores or cracks of sufficient magnitude to permit the solution to penetrate them to the basis metal. This makes the pores and cracks visible to the naked eye and easy to study at low magnifications.

Baker and Pinner [237] used this test to study the porosity and cracking of decorative chromium coatings. An important precaution is to apply it only to articles completely covered with chromium, or to insulate all areas not chromium plated. Otherwise, all the copper will deposit on the areas not covered by chromium, and none will plate on the pores or cracks in the chromium plate. Another way of treating this difficulty is to increase the voltage during the acid copper plating step until some copper is deposited on the chromium-plated areas. However, too much voltage will plate copper directly on the chromium, thus hiding the pores and cracks. Masking off the unplated areas is preferred over increasing the voltage.

Another important precaution is to rinse the chromium plate in hot water to establish the equilibrium amount of cracking for the particular plate being tested. If this is not done, the deposit may appear relatively crack free and yet develop cracks with time, and the test will fail to give reproducible results. A certain amount of low-temperature heat treatment is necessary to stabilize the crack structure. Chessin and Seyb [238] used a 2 minutes immersion in boiling water. Some inconsistencies in the literature may possibly be due to the use of a hot-water rinse by some workers, and the drying of specimens after a cold-water rinse under laboratory conditions by others.

Additional studies of the porosity and cracking of chromium plate were reported [5]. The cracks in thicker deposits were made evident by Gebauer [239] by anodic etching in 10% caustic soda solution followed by microscopic examination. Anodic treatment in chromic acid solution has been used by Dubpernell as a test for cracks in heavy chromium deposits. Almost any brief treatment is sufficient, a typical one being 60 A dm^{-2} for 15 seconds. These tests are of much shorter duration than the anodic treatment described below for the production of porous chromium deposits. Covering the surface with the ink from a solvent-based, dark felt pen, wiping off the excess surface ink, and examining the ink retained in the cracks under low magnification (100–150×) is a very fast method usable on some deposits.

Wyllie [240] used the copper plating test to show the cracks in chromium deposits after tensile testing. Cohen [241] studied the film of chromium compounds found in the crack network of heavy chromium deposits, particularly after heat treatment of the metal.

10.2 Corrosion Resistance

The need in recent years for more durable decorative chromium deposits has led to something of a revolution in the specification and testing of such deposits. This in turn has led to several new developments to help meet the new requirements with thicker bright copper, nickel, and chromium coatings.

Until about 1960 decorative chromium deposits were generally confined to thicknesses of 0.5 μm or less since macrocracking would begin at this thickness. Under normal service conditions, macrocracks in thick deposits would contribute to a reduction in corrosion resistance [237]. Since the development of bright macrocrack-free chromium plating [242, 243], it has been demonstrated that if the undercoating of nickel or copper and nickel is of adequate thickness (generally more than 25 μm), a thicker decorative chromium plate adds substantially to the corrosion resistance. This is evidenced by accelerated corrosion and outdoor exposure tests. This development was foreshadowed as early as 1934 [244, 246], but it remained for the development of better accelerated tests and intensive testing to demonstrate the usefulness of thicker chromium coatings.

An electrolytic corrosion (EC) test was developed by Saur and Basco [247]. This test is highly accelerated and permits the examination of decorative chromium-plated parts in a few minutes as compared with many hours in other tests such as CASS. Even though this test is fast, it has not received wide acceptance.

Seyb and Rowan [248] showed the improvement that could be obtained by means of thicker decorative chromium deposits, whether cracked or crack free. This demonstration was followed by the recommendation of duplex chromium coatings [249–252] for greater corrosion protection of both steel and zinc die castings. Because of the color change and the difficulty of plating decorative duplex chromium, this approach lost importance for extended corrosion resistance. The use of decorative microporous chromium over two or more layers of nickel is now preferred.

10.3 Microcracked and Microporous Deposits

A substantial improvement in the corrosion resistance, afforded by thicker decorative chromium deposits, was made with the introduction of "duplex" or "dual" chromium plating procedures in 1959 [249]. By first plating with "crack-free" chromium with good throwing or covering power, and then with a second layer of highly cracked or microcracked chromium, a convenient means was found for applying heavier decorative coatings with good corrosion resistance. Seyb [250] has explained the corrosion resistance of duplex chromium coatings as depending on many small electrochemical cells with a relatively low rate of corrosion of the nickel undercoat, and decreased tendency to basis metal corrosion. Lovell and coworkers [251], who detailed the same explanation, also report their experience with dual chromium systems and emphasize the need for a substantial copper layer under the nickel and chromium for best results.

The duplex chromium plate provides a convenient means of increasing the thickness and obtaining a microcracked plate with existing equipment, as well as combining the properties of different types of coatings. The most commonly used duplex system is a first plate of bright "crack-free" chromium with good coverage and adequate thickness in recesses, followed with an ordinary cracked or microcracked chromium plate. The total thickness of the two deposits generally ranges from 0.75 to 2.5 μm. Specifications usually call for a minimum of 0.8 μm [252, 253].

Even with the improvements in duplex chromium plating, this system was difficult to consistently produce. "Crack-free" chromium deposits quickly develop cracks in service thus reducing their corrosion resistance. It was also found in practice that it was difficult to plate two layers of chromium and that the color of the deposits changed because of the two layers of chromium. This approach has almost been completely replaced by 0.25 μm of micro-discontinuous chromium thick over two or more layers of nickel [254–257]. Another development is the use of an undercoat of chromium [228, 258, 259].

The need for a minimum thickness, about 0.75 μm, to obtain microcracking brings with it a need for good plate distribution into recesses in order to avoid unduly prolonged plating times. This need was lessened by Chessin and Seyb [238] who produced low current density microcracking by programming the current density during plating.

Microcracking of chromium in conventional thickness (0.25–0.75 μm) was obtained by several workers using a thin substrate of highly stressed nickel over the bright nickel layer before chromium plating [260, 261]. Du Rose et al. [260] also used hot-water treatment and the addition of selenium compounds to the chromium bath to increase microcracking in thinner chromium deposits.

The use of thin microporous chromium plate to improve corrosion resistance originated through the work of Brown and Tomaszewski [262]. They produced decorative, satin-finished nickel-chromium coatings by means of particle inclusion in the bright nickel undercoat. Chromium did not plate where the inclusions were on the nickel surface. Odekerken [263] and Chessin [264] have described other methods for producing microporous and thin microcracked chromium deposits. Improvement in the corrosion resistance of microporous chromium plate can also be effected by a low impact impingement of fine hard particles onto the brittle chromium-plated surface [265, 266].

Du Rose [267], Ogburn and Schlissel [268], and Willson [269] reported that chromium may be anodic or cathodic to the nickel, depending on exposure conditions. Willson further suggested that residual stresses in the chromium-nickel system may be a factor in determining the type of corrosion that occurs.

The large number of papers that have appeared dealing with the general subject of improved corrosion resistance indicates the intense interest in these developments. The copper undercoat, the nature and thickness of the nickel, and the nature and thickness of the chromium as well as the galvanic effects between the dissimilar metals have all been discussed, and it appears that all play a part in the final result

[261]. These improved decorative chromium deposits are being used more widely on a commercial scale and are included in recent specifications [220, 221]. Beacom [271] points out that the same kinds of systems that are used on die castings and steel are also desirable for plating over plastics.

The importance of the potential difference between adjacent nickel layers has been demonstrated in actual service. Micro-discontinuous chromium is now required by most decorative exterior automotive specifications, with microporous chromium being preferred. Harbulack developed an easy to use STEP test that simultaneously measures nickel thickness and adjacent electrochemical potential on completed parts [272]. An ASTM Standard details the operating conditions of the test [273].

10.4 Structure of Deposits

Chromium deposits are somewhat unique among the commercially used electroplated metals due to the deposit structure's influence on its performance. Many variations in structure and physical properties can be obtained by proper adjustment of the plating conditions and postplating treatments.

Chromium deposits 0.5 µm or less thick are normally crack free but porous, whereas thicker deposits are generally cracked after rinsing in hot water or after time in service. Up to a certain thickness, which varies depending on the plating conditions, the surface of the deposit is usually smooth and bright appearing to the unaided eye. When viewed under the microscope, however, the surface is revealed as having numerous domelike projections. Figure 9 shows a typical hard chromium

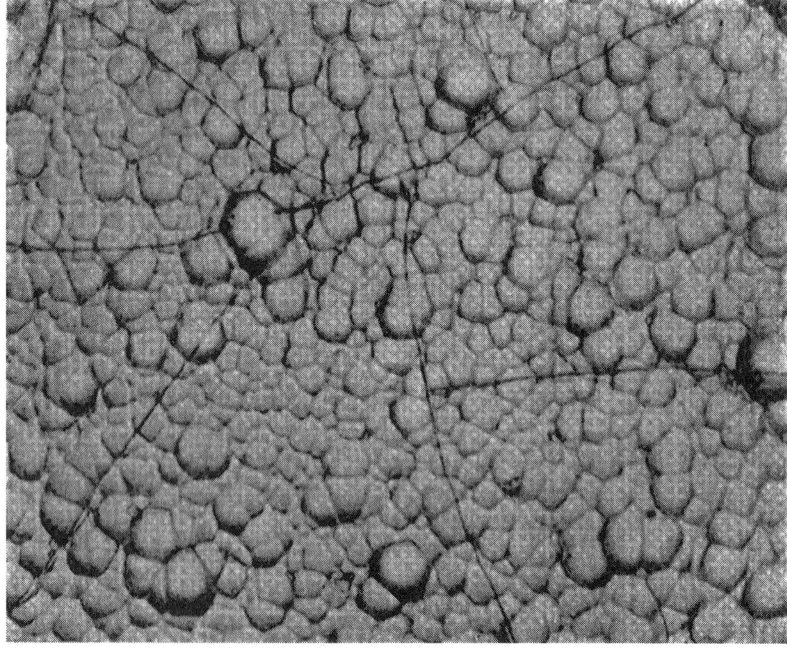

Figure 9 Semi-bright hard chromium plate, 1000×.

surface. These projections become more prominent as the thickness increases, up to the point that they become visible to the unaided eye.

Figure 10 shows ordinary bright chromium deposit built up over the usually decorative thickness. The crack pattern is clearly shown, along with vestiges of "plated-over" cracks. All conventional bright plates more than 0.5 μm thick are cracked in this fashion. Figure 11 shows a crack-free, slightly "milky" type of plate such as is produced by the Mahlstedt process [274], plating at high temperature and low current density. Such deposits crack readily when heat treated, however.

Although other causes have been suggested, the basic cause of the cracks in deposits was believed by Snavely [275] to be related to the formation of unstable chromium hydrides of variable composition during the plating operation. Hydrides can be electrodeposited in either the hexagonal crystal form (formula Cr_2H to CrH) or the fcc crystal form (from CrH to CrH_2). The hexagonal hydride is most likely the one formed under normal plating conditions. It decomposes to bcc chromium and free hydrogen, even at 25°C. Some of the hydrogen escapes during the deposition process; the remainder is included in the deposit.

Data on the structure of the chromium hydrides were reported by Snavely and Vaughn [276]. The normal structure of hexavalent chromium electrodeposits is bcc and is typically formed through deposition from warm solutions. Trivalent chromium electrodeposits are microporous under 0.8 μm and microcracked over this thickness. The structure is amorphous as plated but becomes bcc if heated. The amount of heating required is time and temperature dependent, but one hour at 200°C is sufficient.

The decomposition of either hexagonal or fcc chromium hydrides to bcc chromium may involve a volume shrinkage of over 15%. Because the deposit is

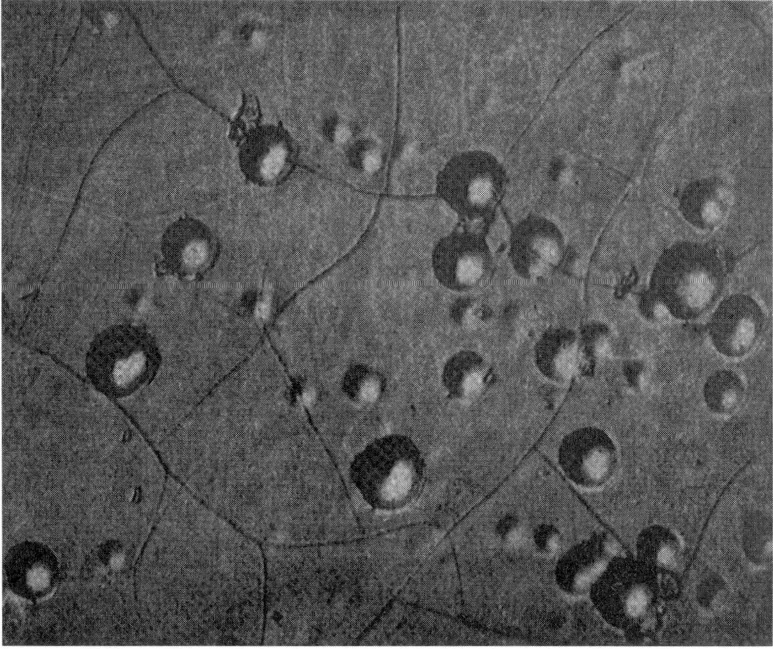

Figure 10 Thick bright chromium plate, 1000×.

Figure 11 Crack-free "milky" low current density chromium plate, 1000×.

restrained in the plane of the basis metal, surface cracks form normal to that surface. The chemical constituents making up the cathode film are drawn into these cracks and may be bridged over by newly deposited chromium. These crack-filling constituents are the inclusion films studied by Cohen [241].

In addition to the oxide film inclusions in the cracks, which are readily detectable by metallographic examination of as-deposited chromium, experimental evidence indicates that there may be additional quantities of oxide or oxide-forming compounds finely dispersed through the plate. This dispersed oxide appears to be agglomerated on heating and then may be detected by the microscope [86, 277, 278].

Wood [279] measured the grain size of various chromium deposits by X-ray diffraction and arrived at a figure of 14×10^{-6} mm for bright plate made at 50°C. As a comparison, the smallest grains producible by cold-working metals are about 10^{-3} mm in average diameter. The extremely fine grains in chromium plate have been explained as resulting from the hydride decomposition [275].

Hardesty [280] reviewed cases of epitaxial chromium deposition in crystalline form. By working at over 80°C in a solution containing 250 g liter^{-1} CrO_3 and 5 g liter^{-1} fluoride ions with 500 to 700 mA cm^{-2}, he demonstrated the continued growth of the crystals of an electropolished nickel basis metal in the cross section of a thick chromium deposit.

Structural changes taking place on beating chromium plate are those normal for a highly stressed, fine-grained metal [277, 281]. Recrystallization takes place on prolonged heating in the temperature range 300° to 500°C or on shorter exposure at higher temperatures. The new grains are elongated normal to the basis metal. On prolonged exposure at temperatures of about 1100°C, large equiaxed grains are formed. During this heating the inclusion films remain in their original positions, although they are agglomerated into lines of spheroidal particles.

Brittain and Smith [282] examined the grain size of annealed chromium plate under the electron microscope and concluded that it increased to about 10^{-4} mm at 450°C in one hour, and to about 2×10^{-3} mm in the same time at 1000°C.

Bright hexavalent chromium plate is strongly grain oriented, with the (111) plane parallel to the basis metal [283]. This preferred orientation persists even after recrystallization [281]. The initial layer of plate is unoriented, and the preferred orientation is achieved as plating continues. Chromium plates on polished brass or electropolished steel attain the preferred orientation rapidly, whereas plates on a machine-ground basis metal receive a relatively thick layer of unoriented plate before the preferred orientation appears.

The laminations or striations in chromium deposits parallel to the basis metal [281] were found [284] to have an important connection with the crack structure; many cracks start and stop in the striations. Cracking occurred periodically, corresponding to the striations in deposits [285].

X-ray and other work on the structure of hexavalent chromium deposits were reviewed in 1967 [286]. Chromium generally has a bcc structure (alpha-chromium). Hexagonal chromium (beta-chromium) was shown [266] to be an unstable chromium hydride rather than an allotropic modification. These hydrides decompose into the normal bcc, chromium at 25°C in three to eight weeks, but decomposition is complete in one hour at 150°C. Snavely [275] gives the best conditions for plating the hexagonal close-packed (hcp) hydride in dull gray form as 600 g $liter^{-1}$ chromic acid, 3 g $liter^{-1}$ sulfate, and 10 g $liter^{-1}$ sugar to reduce some chromic acid to chromium (III). The temperature was maintained below 12°C and current densities were 4.6 to 9.3 A dm^{-2}. These were essentially the conditions reported earlier by Wood [287].

"Cold" chromium plate, deposited from solutions of widely varying composition at 25°C or less, has a distinctive, velvet-smooth, mouse-gray appearance, and is obtained at current efficiencies up to about 35% to 40%. Such cold chromium plate is generally crack free, and has an entirely different aspect from ordinary bright plate from warm solutions. Nevertheless, both cold chromium and bright plate were said to be entirely of the normal bcc structure [275]. Cold chromium plates significantly slower than regular chromium processes. One would think that cold chromium plate would be formed by the decomposition of the hexagonal hydride, and the results of a number of workers [288, 289] have confirmed chromium plate to consist primarily of it.

Shluger and Kazakov [290] reported a thick cathode film on deposits of the hexagonal type made at 20°C, but a thin film on cubic deposits made at higher temperatures. Okada and Ishida [291] found mixed crystals of hexagonal and cubic chromium in black chromium deposits. Matsunaga [292] obtained hexagonal chromium from Wood's solution at lower temperatures and higher current densities, and cubic chromium under other conditions. He observed the oxide film on all of his deposits on X-ray examination, unless they were first etched with hydrochloric acid.

Cold chromium plate has the disadvantage of poor adhesion, in addition to being relatively soft and expensive to buff to a high luster. The poor adhesion can be partially overcome by special care in the preparation of the basis metal, as by electropolishing [293].

Cold chromium has been used to a limited extent commercially, particularly in Europe and North America. A common method is the Bornhauser process [294, 295], often called the tetrachromate or D-chromium bath (direct chromium plating).

Snavely [275, 296] also found some fcc chromium hydride in his hexagonal hydride deposits. By using a bath containing 1020 g $liter^{-1}$ chromic acid, 300 : 1 ratio of chromic acid to sulfate, and 20 g $liter^{-1}$ sugar, at 0° to 4°C and 12.4 A dm^{-2}, Snavely produced a black plate was obtained which was the fcc hydride. This deposit was stable for about a week before X-ray tests showed appreciable decomposition. Dubpernell [119] studied such black deposits in 1942 but found the current efficiency to be extremely low, and thick deposits were difficult to produce by prolonged plating. The hardness was about 600 Brinell.

The dull deposits produced with interrupted current from single-phase rectification were partly hexagonal [297]. X-Ray results [298] showed that dull deposits from interrupted current and from periodic reverse current were cubic, but that the current from a half-wave rectifier gave a mixture of cubic and hexagonal chromium.

Metallographic investigation of deposits of different structures showed that the hexagonal or beta modification could be identified in the unetched cross section by means of polarized light [291]. Some deposits were found to be partly hexagonal and partly cubic. It was necessary to increase the chromium (III) content of Wood's [279] bath by adding 32.4 g $liter^{-1}$ sugar instead of 10 g $liter^{-1}$ in order to get completely hexagonal chromium [300]. The hydrogen overvoltage was considerably lower on hexagonal than on cubic chromium, although the same source later reported the opposite [301, 302]. Stress measurements [289] indicated that cubic chromium absorbs hydrogen and converts to hexagonal at least partly when treated as cathode in N Na_2SO_4.

11 PHYSICAL PROPERTIES OF CHROMIUM PLATE

A number of convenient summaries of the physical and chemical properties of hexavalent chromium and of chromium deposits are available [286, 303–305].

11.1 Hardness and Wear Resistance

The hardness and wear resistance of chromium deposits are generally quite good in service, regardless of moderate variations in the conditions of plating. This is true as long as a satisfactory thickness is applied for the intended use. There also has to be good coordination between the hardness of the basis metal and the pressures encountered [306]. Frequently service is at elevated temperatures, and some of the as-plated hardness is lost. On the other hand, relatively soft deposits produced from hot solutions have often given especially good service, possibly owing to freedom from cracks or to greater strength and cohesion of this type of plate. Trivalent chromium deposits become harder when heated. Chrome carbides are thought to form by a reaction between the chromium and the codeposited carbon.

Hardness measurements are difficult to make and not very rewarding, since so many other factors are involved in service life. Unfortunately, hardness is not a simple elementary property, it depends on other properties in a little-known manner

[307]. For these reasons the hardness measurement is not often used in chromium plate specifications.

Generally, the hardness of bright hexavalent chromium deposits is given as about 900 to 1000 kg mm^{-2}. Approximately the same values are obtained whether on the Brinell, Knoop, Vickers, or diamond pyramid scale [6]. Organic catalyzed hexavalent chromium deposits can reach 1100 Knoop or Vickers with a 50 to 100 g load. As-plated trivalent chromium deposits fall into the same hardness range. After heat treatment the hardness can reach 1500 Knoop or Vickers (50 to 100 g loads).

The microhardness number depends to some extent on the test load used, and the deposit must be thick enough to withstand it or be conducted on a thick enough cross section. On a soft basis metal, a thin chromium deposit will show only the hardness of the basis metal; it would have to be extremely thick to indicate its full hardness. Chromium deposits 25 λm or more thick appeared satisfactory on a hardened steel base using 50 to 1000 g liter^{-1}loads; it is recommended that the plate thickness should be at least 14 times the depth of penetration of the indenter [308]. Low loads often result in considerable increase in the hardness number, and greater accuracy. Tests are sometimes made on the cross section of the deposit if it is thick enough and well supported [281]. If the deposit cracks around the indention, a smaller load should be used.

Brenner et al. [86] reported Knoop hardnesses for as-deposited chromium ranging from 300 to 1000, the softest deposits being from boiling solutions. Hosdowich [309] converted scratch hardnesses to the Brinell scale and reported figures from 640 to 1165 Brinell. The softest deposits were dull plates obtained at 15°C, and the hardest were burnt nodules. The hardness of the deposits was closely correlated with the appearance. All the bright deposits had a hardness of about 1000 Brinell.

Eilender and Arend and coworkers [310, 311] also made an extensive investigation. Wahl and Gebauer [312] determined Vickers hardnesses of 390 to 1280 with a 50 g load on deposits from 250 g liter^{-1} solutions at 20 to 80°C and 10 to 200 A dm^{2}.

Brenner et al. [86] found in some tests with fluoride-catalyzed baths at 85°C that the deposits were harder than corresponding ones from sulfate-catalyzed baths. Wahl and Gebauer [312] found slightly higher hardnesses for deposits from solutions containing silicofluoride catalyst. In general, deposits from solutions containing fluoride or complex fluoride catalysts seem to be about 100 to 200 units harder than corresponding deposits from solutions containing only sulfate catalyst. There also seems to be definite evidence of better retention of hardness by such deposits when exposed to heat.

The wear resistance is perhaps even more difficult to measure than the hardness. Two types of measurement have been used: abrasive wear with relatively light load similar to grinding [309, 312], and rubbing or frictional wear under higher pressures against a harder material such as a tungsten carbide wheel. There is considerable evidence that the hardest deposits do not necessarily give the greatest wear resistance. Thus Hosdowich [309] found the bright or slightly frosty deposits around 1000 Brinell to have the greatest wear resistance, and the burnt deposits up to about 1165 Brinell had somewhat less, perhaps because of brittleness. Likewise deposits of medium hardness around 750 to 800 Vickers were found to have the best frictional wear resistance, whether obtained as deposited or by moderate heat treatment of still harder deposits [310, 311].

Radioactive chromium plate is used to follow the wear of chromium-plated piston rings and to check the chromium transferred to the cylinder walls [313] or the lubricating oil [314, 315]. See the Section 12.4 on porous chromium plating.

11.2 Coefficient of Friction

The low coefficient of friction of chromium plate against other metals is an important factor in its use on shafting, piston rings, internal-combustion engine cylinders, and similar applications. Table 2 illustrates the superiority of chromium in terms of this property.

Bright chromium plate against cast iron gave a lower coefficient of friction than mat or burned plates [318]. The coefficient of friction of chromium plate against steel or cast iron increases rapidly with temperature [319]. The increase may be avoided by final polishing after heat treating.

11.3 Coefficient of Expansion

Hindent [320] found an average coefficient of expansion of 6.8×10^{-6} °C^{-1} for annealed electrolytic hexavalent chromium in the temperature range of 20° to 100°C. The coefficient of expansion at any temperature t between −75° and +650°C was shown to be expressed by the formula $a_t = (5.88 + 0.01584t - 0.00001163t^2)10^{-6}$. Within the temperature range −100° to 700°C, the length of a chromium plate specimen is expressed by the formula $L_t = L_{0a}[l + (5.88t + 0.00774t^2 - 0.00000388t^3)10^{-6}]$. Hindent also noted a linear shrinkage of approximately 1.1% during heating the deposits to 500°C for the first time. During subsequent heating and cooling cycles, the expansion and contraction were normal. Snavely [275] ascribed the initial shrinkage to relief of internal stress and closing up of voids between crystallites. These voids are a result of the decomposition of chromium hydrides.

TABLE 2 Coefficient of Friction for Various Metal Combinations

Metal	Static Coefficient	Slinding Coefficient
Reference [316]		
Chromium-plated steel on chromium-plated steel	0.14	0.12
Chromium-plated steel on babbitt	0.15	0.13
Chromium-plated steel on steel	0.17	0.16
Steel on babbitt	0.25	0.20
Babbitt on babbitt	0.54	0.19
Steel on steel	0.30	0.20
Reference [317]		
Bright chromium plate on cast iron		0.06
Bright chromium plate on bronze		0.05
Bright chromium plate on babbitt		0.08
Hardened steel on cast iron		0.22
Hardened steel on bronze		0.11
Hardened steel on babbitt		0.19

Because of this unusual behavior there is little hope that a basis metal can be found that will match the expansion and contraction characteristics of the chromium plate from chromic acid baths during heating cycles.

11.4 Melting Point

Sully and Brandies [286] tabulate measurements ranging from 1560° to 1920°C. Udy [303] selects the high value of 1930 ± 10°C based on results in [321] and [322]. Currently the American Society for Metals [304] has adopted the value of 1875°C, after work at the National Bureau of Standards [305]. Sully and Brandies [286] discuss much of the work in this area and state that it is not possible to come to any final conclusion, but that a value close to 1878 ± 22°C seems to be the most likely.

11.5 Density

The density of chromium plate varies according to the amount of inclusions in the plate, the number and size of cracks, and the magnitude of internal strain. Brenner and coworkers [86] reported a systematic study of the density of chromium deposited under various conditions. Values from 6.90 to 7.21 $g\,cm^{-3}$ were obtained. The oxide content decreased as the density increased. After annealing at 1200°C the density of the deposits increased to within the range 7.09 to 7.22 gcm^{-3}. The density of pure chromium is 7.20 $g\,cm^{-3}$, as calculated from its lattice parameter. Therefore the reported values over 7.20 can be considered anomalies related to the precision of the measurements. Hinder [320] reported a density of 6.93 $g\,cm^{-3}$ for as-deposited chromium, and this value is considered representative of most commercially deposited chromium.

Knoedler [288] reported a density of 6.143 for cold chromium or hexagonal hydride as deposited at 12° to 15°C, compared to 7.017 for as-deposited cubic chromium and 7.148 after annealing two hours at 900°C.

11.6 Reflecting Power

Coblentz and Stair [323] studied the reflectivity of hexavalent chromium plate over a light range from ultraviolet to infrared. For the visible range of light, 4000 to 7000 Å in wavelength, they obtained reflectivity values between 62% and 72%. For ultraviolet light, the reflectivity ranged from 55% to 70%, and for infrared from 62% at 7000 Å to 88% at 40,000 Å. These high reflectivity values are usually retained over prolonged periods of exposure of chromium plate because of its corrosion and tarnish resistance. The reflectivity may be seriously reduced when the plate is exposed to highly corrosive atmospheres.

11.7 Electrical Resistivity

Electrical resistively, like density, is a measure of the continuity, purity, and general soundness of a metal. The number, distribution, and size of the inclusion-filled cracks in chromium are related to the plating conditions. Therefore the electrical resistively varies according to these conditions. Brenner and coworkers [86] reported electrical resistively values for a wide range of deposition conditions. They showed that a resistively of about 50 to 60 microhm-cm at 28°C may be expected for conventional chromium plate, with much lower values down to 14 microhm-cm for deposits from hot solutions. After annealing at 1200°C, the oxide film inclusions are spheroidized, and the cracks in which they originated are no longer continuous. As a result the resistively of annealed electrolytic chromium approaches a common value of 13 microhm-cm at 28°C, regardless of conditions of deposition [119].

11.8 Internal Stress

According to the theory of chromium hydride formation and decomposition during chromium plating [275], the cracks in the plate result from internal stresses that exceed the cohesive strength of the metal. Cracking relieves these stresses to the point where they are no longer of sufficient magnitude to extend the cracks. Most thick plates are cracked, and contain residual internal stress. Thin plates may contain even higher stress because they are restrained from cracking by the basis metal, and transfer their stress to it.

Brenner et al. [86] reported stress values as high as $56\,\mathrm{kg\,mm^{-2}}$ for very thin chromium deposits which were not cracked. Conventional plating practices produced thicker cracked plates having internal stress of about $12\,\mathrm{kg\,mm^{-2}}$. Plates from a dilute bath at 85°C were crack free but contained stresses of $45\,\mathrm{kg\,mm^{-2}}$.

Stareck et al. [324] investigated deposits up to 100 μm thick, and found that the stress in highly cracked deposits might become negative or compressive with increasing thickness. Compressive stress as high as $-12\,\mathrm{kg\,mm^{-2}}$ was found. This was explained as due to a wedge effect of chromium plated into previously formed cracks. Williams and Hammond [325] confirmed the presence of moderate compressive stress in some thick chromium deposits in the as-plated condition.

Nishihara et al. [289] showed that the stress was lower in the hexagonal hydride but that cathodic treatment of ordinary bcc deposits in N Na_2SO_4 at 40°C and $1\,\mathrm{A\,dm^{-2}}$ decreased the stress in less than one hour to less than that of the cold chromium (hexagonal hydride). Additional work on stress in chromium deposits has been reviewed [326, 327].

11.9 Effect on Fatigue Strength of Basis Metal

The stress in various hexavalent chromium deposits was correlated with the crack structure and thickness [324], and it was found that because of a wedge effect of chromium deposited in previously formed cracks the stress frequently decreased with thickness, and heavy deposits even developed compressive stress. This was in turn correlated with the effect of the deposits on the fatigue strength of steel [328]. Two types of deposits that had a minimum effect on the reduction of fatigue strength were found: highly cracked deposits with low stress as plated, and deposits from high concentration baths that could cause stress damage as plated but could be heat treated with good results.

Chromium plate generally reduces the fatigue strength of steel markedly [329–331]. Stareck et al. [324, 328, 330] found that the decrease in strength was due to the stress in the deposits that weakened the basis metal. They found several ways to overcome or minimize the effect, such as by producing deposits of low or compressive stress or by heating to high temperatures to eliminate the stresses as far as possible. These results were confirmed and extended by Williams and Hammond [332].

Shot peening [333], roller burnishing [334], and grit blasting [335] of the basis metal have also been shown to be helpful. In this way there is introduced into the surface before plating a compressive stress which tends to counterbalance any tensile stresses in the deposit.

Continuing investigations of these matters are quite extensive. German work has been reviewed [336–339]. Effects of surface finishes in gun barrel manufacture were investigated by Greco and Pennell [340]. The detrimental effect of chromium plate on other basis metals and methods of overcoming it have also been investigated for aluminum [341–344] and titanium [345].

11.10 Ductility

No ductility was found in chromium deposits from aqueous solutions by Wyllie [240] or Brenner et al. [86], although the latter found tensile strengths of 6 to 56 $kg\,mm^{-2}$. Deposits from fused salt baths have, however, been found to be ductile [304].

Klopp [346] reviews progress in improving the ductility and strength of chromium and chromium base alloys. Brandes and Whittaker [347] reported a tensile strength of 20 $kg\,mm^{-2}$ and an elongation of 17% on electrolytic chromium at room temperature, after annealing in hydrogen at 1600°C.

12 CHEMICAL PROPERTIES

12.1 Oxidation and Tarnish Resistance

Chromium plate normally has a very thin oxide film on its surface. This film is so stable, tenacious, refractory, and self-healing that it protects the metal underneath from further oxidation. The plate remains bright at temperatures up to 260°C. On prolonged heating of chromium plate to temperatures of about 315°C in air, the oxide film thickens and darkens. At higher temperatures, temper colors are produced, and a black or green-black oxide layer is finally formed. At temperatures around 1000°C, an oxide layer forms on the surface, and an extremely hard chromium nitride layer forms between the oxide and the chemically unaffected portion of the plate [281]. Pure trivalent chromium deposits undergo similar color changes due to heating. Deposits containing metallic codeposits such as iron and copper darken faster and to a greater degree.

The thin oxide film on chromium plate forms quickly when plating is completed, or is present during plating, so tarnishing of the plate is not likely to be encountered. The chromium oxide is a satisfactory protection against sulfides, which cause serious tarnishing of silver, copper, or nickel.

12.2 Chemical Resistance

The chemical resistance of chromium plate is not so great as might be supposed from its performance in the atmosphere. Chromium is readily attacked by mineral acids and by reducing solutions in general. It is resistant to nitric acid, which heals the protective oxide film, and nitric acid may be used to dissolve other metals such as copper or nickel away from chromium plate. Smith and Dubpernell [348] found it possible to improve the acid resistance by anodizing.

Christov and Pangarov [300] found their cold chromium plate (beta- or hexagonal chromium) to be more passive and corrosion resistant than bright alpha cubic chromium plate deposited at 45°C. Thus a pH of about 1 was required to dissolve hexagonal chromium, while cubic chromium dissolved at pH 2; cubic chromium stopped dissolving when the pH was increased to about 2.6, while hexagonal chromium stopped dissolving around pH 1.7 to 1.8.

The chemical resistance of chromium plate may be used to the best advantage only if the underlying metal is completely covered [349]. For that reason hard or industrial chromium plates for corrosive service should be at least 25 to 50 μm thick to ensure that the cracks are not continuous to the basis metal. Otherwise, a crack-free type of deposit should be used. Many trivalent chromium deposits tend to have more substrate exposed than hexavalent chromium deposits. Thin (under 0.8 μm) trivalent chromium deposits are microporous. Over 0.80 μm the deposits are microcracked. Over about 2 μm the deposits tend to develop macrocracks, many of which extend to the substrate.

In general, hexavalent chromium plate may be used in the same types of corrosion-resistant service as the high-chromium stainless steels at ordinary temperatures, depending on the physical properties required of the basis metal. Even though trivalent chromium deposits contain the same type of oxide film, these deposits may not have the same chemical resistance. Codeposited metals with chromium from trivalent chromium can have a significant influence on chemical resistance.

Resistance to corrosion and wear in high-purity water at high temperatures in atomic reactors is a field where hexavalent chromium plate appears to have some utility, but many special application problems are involved [340]. Suss [351] found contradictory results on chromium-plated stainless steel, probably because of electrochemical effects. The general behavior of chromium exposed to corrosion in aqueous media at 25°C has been comprehensively outlined on a thermodynamic basis by means of potential pH diagrams [352, 353].

A considerable number of specific corrosion tests on chromium and chromium plates were made [8, 245, 329]. These tests included a wide variety of acid and salt solutions, and organic compounds and acids as well, at 12° and 58°C.

12.3 Chromium-Plated Strip Steel–Tin-Free Steel (TFS)

Chromium-plated strip steel for the production of "tin cans" originated commercially in Japan in 1962, as described by Uchida and coworkers of the Fuji Iron and Steel Company [354–356]. The process originally consisted in chromium plating for 10 s to get a thickness of 0.05 Å, followed by a chemical dip treatment in 1% chromic acid or 2% to 3% sodium bichromate to improve the corrosion resistance, and finally lacquering with a high-temperature baking lacquer.

Some of the further developments in the United States as well as in Japan have been reviewed [357–362]. Commercial production began in the United States in 1967. The resultant tin-free steel (TFS) has been used for the manufacture of beverage cans, a large proportion of which is now made from this material. The tin-free steel lacks the easy solderability of tin plate for high-speed can production, and has necessitated the use of processes for cementing or welding the seams of cans. The resulting cans are more readily recyclable because they contain no tin.

In newer procedures the chromium thickness has been cut down to 0.005–0.0075 Å (35–54 $mg\,m^{-2}$) [363] but is sufficient to prevent filiform or underfilm rusting in the corrosion tests used [359] when combined with the oxide coating produced in a final cathodic post treatment solution compatible with the plating bath [359]. The latter oxide is about five times the thickness of the natural oxide film present on most chromium plate, which is about 3.75 $mg\,m^{-2}$ [357, 359, 365] in terms of weight of chromium in the oxide. Not much more than 16 $mg\,m^{-2}$ of chromium as oxide can be applied without getting a colored coating instead of a transparent one.

In plating the steel strip moving at 300–550 $m\,min^{-1}$, Seyb and coworkers [366] found it necessary to use fluoride or complex fluoride containing electrolytes, and they formulated the first practical baths for this application, which have become the standard in commercial use. Current efficiencies of about 25% are obtained in practice. Typically the moving strip passes between four pairs of 38 cm long sections of anode placed vertically, or six pairs counting the somewhat longer cathodic posttreatment. This gives a plating time of about 0.05 s in front of each pair of anodes, or a total plating time of about 0.30 s to produce the finished product, which is then rinsed, dried, oiled, and lacquered.

The reception of this product has been good, and its use has grown rapidly, resulting in the saving of substantial quantities of tin for other purposes. Tin-free steel is less expensive than comparable tin plate [362].

12.4 Porous Chromium Plate

This name has been given to modified chromium deposits with oil-retaining properties, used on internal-combustion engine cylinders and piston rings. Such deposits were used especially on aircraft and diesel engine to make the engines last longer. Today most parts plated with hard chromium benefit from the cracks.

Three main types of "porous" chromium plate have come into common use. The first is the "mechanical" type produced by grit blasting the basis metal, chromium plating, and finally finishing to size by grinding, honing, or polishing [367]. The second and third types of plate are those with "pitted" and "channel" porosity. Both of these are obtained by treating the chromium deposit in an etching solution. The type of porosity obtained depends on careful control and regulation of the conditions of deposition. Numerous publications and patents describe the production of all these types of porous chromium plate and the results obtained with them [368–376].

Another variation is to etch pits into the surface of the deposit through a plastic mask [371] or a photoresist [372]. Etching by ion bombardment through a screen [372] and with alternating current [379] is proposed. Raymond [380] plates at extremely high current densities from 232 to 1160 $A\,dm^{-2}$ at 30° to 55°C to produce a porous deposit directly. Still another variation is to impregnate the porosity with Teflon to provide the possibility of dry lubrication [381], but this does not seem to have become important commercially. Patents were issued to Forestek [382] and Letendre [383]. Zubrisky [384] patented the final grit blasting of polished or honed porous chromium surfaces to provide better breaking in qualities.

The "pit" type of porous chromium may be produced, for example, by plating under ordinary hard chromium plating conditions. Baths containing 250 $g\,liter^{-1}$

chromic acid and 2.5 g liter^{-1} sulfate, at 50°C are used. Plating took place at 46 to 54 A dm^{-2} for a minimum of two to three hours to get a deposit at least 100 Å thick. They were then treated as an anode or cathode in a suitable etching solution, or by simple immersion in acid. A typical anodic treatment [365] is about 150 A-min dm^{-2}, but this may be prolonged or repeated if it is desired to remove more metal or to obtain deeper porosity. After the deposit has been heavily attacked, numerous cracks are found to be eaten away, and a surface crust of undermined metal remains. When this crust is ground, honed, or polished away to the extent of 25–50 Å, numerous pits remain in the chromium plate.

Good conditions for producing the "channel" type of porous chromium plate are 60°C and a ratio of chromic acid to sulfate of 115 : 1. The usual current densities of 46–62 A dm^{-2} are employed for a deposit thickness of at least 100 or 125 μm. After treatment in the etching solution, the deposit does not have a loose surface crust but only a network of fissures so that grinding, polishing, or honing off about 25 μm leaves channels, with dense chromium " plateaus" or "lands" between. This type of porous chromium has been largely used for aircraft engine cylinders, whereas the pit type has been more extensively employed in diesel engines and on piston rings.

Several specifications have been issued on porous chromium plating [385]. It is a well-established procedure conducted on a large scale in a limited number of plants. The merits of the product are confirmed by carefully controlled tests. Thus Kishi et al. [386] with radiotracer techniques determined that the wear on porous chromium and ordinary hard chromium was about equal under low loads. The ordinary hard chromium showed a rapidly increasing weight loss, the porous chromium showed very little wear.

13 TRIVALENT CHROMIUM BATHS

13.1 Trivalent Chromium Processes

Historically trivalent chromium baths have always been the first and favorite approach to chromium plating because of the increased safety and health properties [387]. It was not until the mid-1970s [388] that a commercial process was available and increased productivity became more important than health and safety in choosing this process.

The United States Bureau of Mines developed a process for electrowinning of pure chromium metal from the ore or from ferrochromium, using a mixed bivalent and trivalent chromium sulfate solution [389–391], but the process has not been adapted successfully for plating purposes. A two-compartment cell has been used at the Union Carbide Corporation, Marietta, Ohio plant. Continuous operation on a large scale seems necessary for good results [392]. In another review of the operation [392a], Bacon mentions that low efficiencies always prevail during the startup of a cell.

Success was very limited in efforts to use such baths for plating [393–403]. A chromium ammonium chloride solution was used for brush plating [399, 400]. Better results in brush plating were obtained with Gregory's salt, ammonium chromium oxalate [406–409].

More recent efforts were made to commercialize the use of chromium chloride solutions in, or containing, organic solvents. Bharucha and Ward [410] of the British

Non-Ferrous Metals Research Association obtained several patents and published several articles [411, 412]. Diamond Shamrock Chemical Company has offered a similar process [413]. The deposits are darker than those from chromic acid solutions and some chlorine gas is generally evolved at the anode. Levy and Momyer of Lockheed [414] reported tests with chromium thiocyanate and chromium format dissolved in organic solvents. Brown and Tomaszewski also give a brief report of their work with trivalent baths [262].

In recent years, there has been many references on trivalent chromium research [415–419]. Snyder [420] reviewed the deposit's physical properties of one commercial formulation. Corrosion, a concern with decorative deposits, have also been extensively studied. The same author [421, 422] published corrosion data showing that except for thin nickel applications, trivalent chromium deposits have equal or better corrosion resistance than hexavalent chromium deposits. The inability of trivalent ions to "chromate" metal, as is possible with hexavalent chromium ions, makes trivalent chromium less corrosion resistance over thin nickel or bear steel. Carter and coworker [423] also reviewed trivalent chromium corrosion studies using a different formulation to produce the deposits. One major advantage of some trivalent chromium formulations is that metallic impurities can be easily removed by directly passing the plating solution through an ion-exchange [424].

13.2 Chromium Alloy Plating

One major advantage of trivalent chromium chemistry versus hexavalent chromium chemistry is the ability to easily produce chromium alloys. Chrome-iron [425, 426] and chrome-nickel-iron [427, 428] are the most typical alloys produced.

Chromium alloy plating can be considered a subheading under trivalent baths; there is almost no alloy plating possible from hexavalent solutions. There has been a great deal of work, and some reviews are available [429–431], but nothing of commercial importance seems to have been developed. Now that trivalent chromium processes are commercial, it is expected that there will be more research of chromium alloys. It is very easy to alloy metals such as nickel, iron, copper, and zinc from a trivalent chromium electrolyte.

One exception to the rule of no alloy plating from chromic acid baths is the work of Vagramyan and his collaborators [430]. Alloys of up to 37% selenium, 15% manganese, 2% molybdenum, and 1% rhenium were obtained with cold chromium deposits at about 20°C. These alloys are, on the whole, no longer obtained as the temperature is raised, so presumably they are alloys with the dull hexagonal hydride produced at low temperatures, and not with the bcc deposits which form ordinary bright plate.

Snavely and coworkers [432] reported the physical data for chromium alloys containing iron, molybdenum, nickel, phosphorus, or tungsten from trivalent chromium baths. Alloys containing 6% to 10% iron were harder after heating to 800°C than hexavalent chromium from conventional baths. However, they were softer than conventional deposits prior to heating. Alloys containing 6% iron retained their hardness of 600–700 kg mm^{-2} after heating to 600°C [433]. Alloys with 15% iron exhibited a hardness of 1000–1025 kg/mm^{-2} prior to heating [434]. Chromium alloys with up to 60% iron exhibited stress of 36,000–38,000 psi [435], even though they had many cracks. Coefficients of expansion for chromium alloys containing 6% to 18% iron were slightly higher than those for conventional chromium [436].

Much work has been done, mainly in France, to investigate claims of improved wear resistance of chromium-molybdenum alloys produced from chromic acid solutions [437, 438]. It appears that bright deposits generally contain less than 1% molybdenum, and this could possibly result from solution contamination of the deposit instead of alloying. Hard chromium plating baths are frequently deficient in catalyst due to too strict adherence to the 100 : 1 ratio of chromic acid to sulfate. The improved wear resistance of deposits from solutions containing molybdenum compounds might in fact be due to the catalytic imbalance.

Chromic acid plating baths containing molybdenum salts have been employed for obtaining deposits requiring high abrasion resistance and wear [439–422]. Abrasion resistance was improved 200% to 300% over conventional deposits with the inclusion of molybdenum. A deposit containing 3% molybdenum had a reported hardness from 1000–1300 kg cm^{-2}. The hardness increased as the concentration of molybdate increased to 100 g $liter^{-1}$ [434]. X-ray diffraction studies showed the presence of molybdenum trioxide indicating a dispersion [439].

Chromium-ammonium sulfate solutions containing sodium hypophosphite produced deposits with increasing hardness as the phosphorous content increased up to 15% [444]. Hassion and coworkers studied formulations containing magnesium oxide, zirconia, and thoria and found that dispersions were produced [445]. The hardness of the deposits increased in every case. Additions of sodium tungstate or magnesium oxide [446] increased hardness. The same was reported for the addition of titanium oxide [447].

14 OTHER SPECIAL TYPES OF CHROMIUM PLATE

A "frosty" or satin-finish plate in between cold chromium and bright plate was found desirable for press plates [448]. Such smooth-bubbly or natural rounded nodular plate has been found useful for handling textile materials. Trist [449] used a special cold chromium plate produced in refrigerated electrolytes for printing plates.

Carveth [450] obtained a black color on chromium deposits by immersion in molten cyanide. The carburizing of chromium deposits for greater hardness has also been frequently attempted [451–453]. Although the plate is first softened by the heat, it does appear possible to obtain very hard chromium carbide coatings.

Bohlman [454] was successful in spot plating with a jet of ordinary chromic acid solution and achieved extremely high local rates of deposition. Chessin and Walker [455] developed a bath with organic additions to give a uniformly iridescent chromium plate, and Chessin and Gempel [456] obtained similar results with additions of molybdenum compounds.

REFERENCES

1. G. Dubpernell, *Plating*, **47**, 35 (1960).
2. C. G. Fink, U.S. Patents 1,581,188 (1926); 1,802,463 (1931).
3. Elektro-Chrom-G.m.b.H. (assignee of E. Leibreich), German Patent 448,526 (1927); British Patent 237,288 (1925); French Patent 601,059 (1926); Swiss Patent 118,632 (1927).

4. J. Stareck, F. Passal, and H. Mahlstedt, *Proc. Am. Electroplat. Soc.*, **37**, 31 (1950).
5. G. Dubpernell, in F. A. Lowenheim, ed., *Modern Electroplating*, 2nd ed., Wiley, New York, 1963, pp. 80–140.
6. P. Morisset, J. W. Oswald, C. R. Draper, and R. Pinner, *Chromium Plating*, Robert Draper, Teddington, Middlesex, England, (1954).
7. W. Blum and G. B. Hogaboom, *Principles of Electroplating and Electroforming*, 3rd ed., McGraw-Hill, New York, 1949, p. 344.
8. C. G. Fink and H. D. McLeese (to United Chromium, Inc.), U.S. Patent 1,844,751 (1932).
9. H. R. Moore and W. Blum, *Natl. Bur. Stand. Res. Papers*, **198**, 255 (1930).
10. U.S. Patents: 3,954,574 (1976); 4,038,160 (1977); 4,053,374; 4,093,521; and 4,054,494 (1980).
11. D. L. Snyder, *AESF Chromium Colloquium*, (1987).
12. J. Koppel, Z. *Anorg. u. Aligem. Chem.*, **45**, 359 (1905).
13. F. Hein and S. Herzog, in *Handbook of Preparative Inorganic Chemistry*, 2nd ed., vol. 2, Academic Press, New York, 1965; pp. 1361–1370.
14. *Inorganic Synthesis*, vol. 10, McGraw-Hill, New York, 1967, p. 26.
15. F. A. Cotton and G. Wilkinson, *Advanced Inorganic Chemistry*, 5th ed., Wiley, New York, 1985, pp. 679–697.
16. *Kirk-Othmer Encyclopedia of Chemical Technology*, 4th ed., vol. 6, Wiley, New York, p. 267.
17. F. Basal and R. G. Pearson, *Mechanisms of Inorganic Reactions*, Wiley, New York, 1967, pp. 141–145.
18. L. Spicia, H. Stoeckli-Evans, H. Marty, and R. Giovanoli, *Inorg. Chem.*, **26**, 474 (1987).
19. L. Spicia and W. Marty, ibid., **25**, 266 (1986).
20. D. Rai et al., ibid., **26**, 474 (1987).
21. D. E. Pennington and A. Haim, ibid., **5**, 1887 (1966).
22. M. R. El-Sharif, S. Ma, and C. U. Chisholm, *Trans. Inst. Met. Finish.*, **73** (1), 19 (1995).
23. M. R. El-Sharif, A. Watson, and C. U. Chisholm, *Trans. Inst. Met. Finish.*, **66**, 34 (1988).
24. A. Smith, A. Watson, and D. Waughan, ibid., **71** (3), 106 (1993).
25. M. R. El-Sharif, S. Ma, C. U. Chisholm, and A. Watson, *Proc. AESF SUR/FIN'93.*
26. J. C. Blair, *Comprehensive Inorganic Chemistry*, vol. 3, Pergamon Press, New York, 1970.
27. J. E. Earley and R. D. Cannon, *Aqueous Chemistry of Chromium (III) in Transition Metal Chemistry*, vol. 1, Marcel Decker, New York, 1966, p. 64.
28. H. Stunzi, L. Spicia, F. P. Rotzinger, and W. Marty, *Inorg. Chem.*, **28**, 66 (1989).
29. E. Serfas, E. Theis, and T. Thovensen, *J. Amer. Leather Chem. Assoc.*, **43**, 132 (1948).
30. E. Serfas, G. Wilson, and E. Theis, ibid., **44**, 647 (1949).
31. J. W. Mellor, *A Comprehensive Treatise on Inorganic and Theoretical Chemistry*, vol. 11, Longmans, London, 1931.
32. M. J. Udy, *Chromium*, vol. 1, Reinhold, New York, 1956.
33. G. Michel and R. Cahay, *J. Raman Spectroscopy*, **17**, 76 (1986).
34. G. Michel and R. Machiroux, *J. Raman Spectroscopy*, **14**, 22 (1983).
35. T. Radnai and C. Dorgai, *Electrochim. Acta*, **37** (7), 1239 (1992).
36. N. V. Mandich and N. V. Vyazovikina, *Extend. Abstracts, 49th Meeting of Int. Soc. Electrochem.*, Kitakyushu, Japan (1998).

37. A. Martens and G. Carpeni, *J. Chim. Phys.*, **60**, 534 (1963).
38. J. P. Hoare, *Plating Surf. Finish.*, **76**, 46 (1989); *J. Electrochem. Soc.*, **126** (2), 190 (1979).
39. J. P. Hoare, *Plating Surf. Finish.*, **76** (9), 46 (1989); *J. Electrochem. Soc.*, **126** (2), 190 (1979).
40. J. P. Hoare and M. A. La Boda, *J. Electrochem. Soc.*, **132** (4), 798 (1985).
41. J.-L. Fang, N.-J. Wu and Z.-W. Wang, *J. Appl. Electrochem.*, **23** (5), 495 (1993).
42. A. Radwan, A. El-Kiar, H. Farag, and G. Sedahmed, *J. Appl. Electrochem.*, **22** (12), 1161 (1922).
43. D. Landolt, R. Acosta, R. H. Mullar, and C. W. Tobias, *J. Electrochem. Soc.*, **117**, 839 (1970).
44. H. Voght, *Electrochim. Acta*, **3**, 633 (1987).
45. N. V. Mandich, Ph.D. thesis, Aston University, Birmingham, Great Britain (1996).
46. T. C. Saiddington and G. R. Hoey, *J. Electrochem. Soc.*, **120**, 1475 (1973).
47. V. Guro, M. Schluger, O. Khodzhaev, and Sh. Ganiev, *Electrokhimiya*, **30** (2), 251 (1994).
48. Z. A. Soloveva, Yu. V. Kondashov, and S. V. Vashenko, *Elektokhimiya*, **30** (2), 228 (1994).
49. K. Yoshida, A. Suzuki, K. Doi, and K. Arai, *Kinzoku Hyomen Gijutsu*, **30**, 338 (1979).
50. H. Kimura and T. Hayashi, *Denki Kagaku*, **37**, 5 (1969).
51. M. Nagayama and T. Izumitani, *Kinzoku Hyomen Gijutsu*, **21**, 505 (1970).
52. J. Levitan, *J. Electrochem. Soc.*, **111** (3), 286 (1964).
53. H. Okada, *Kinzoku Hyomen Gijutsu*, **11**, 623 (1960); *Kinzoku Kagaku*, **4**, 1 (1967).
54. K. Yoshida, Y. Tsukahara, and K. Koyama, *Kinzoku Hyomen Gijutsu*, **30** (9), 457 (1979).
55. N. V. Mandich, *Plating Surf. Finish.* (a) **84** (6), 97 (1997); (b) **84** (5), 108 (1997).
56. H. Chessin and K. Newby, U.S. Patent 4,588.481 (1986).
57. N. Martyak, U.S. Patent 4,810.336 (1989).
58. W. Korbach and W. McMullen, U.S. Patent 4,828.650 (1989).
59. W. Korbach, U.S. Patent 4,790.674 (1989).
60. A. R. Jones and A. Neidgrer, *Proc. 2nd AESF Chromium Colloquium*, Miami (1990).
61. K. Newby, *Proc. AESF Sur/Fin* 1999.
62. M. S. Frant, *Plating Surf. Finish.*, **54**, 102 (1967).
63. F. Passal (to United Chromium, Inc.), U.S. Patent 2,640,021 (1953).
64. J. E. Stareck (to United Chromium, Inc.), U.S. Patent 2,640,022 (1953).
65. R. Chellapa and N. V. Pathasaradhy, *Met. Finish.*, **75** (2), 93 (1981).
66. S. R. Natarjan, ibid., **79** (5), 93 (1981).
67. R. Krishnan, ibid., **84** (11), 31 (1981).
68. S. Sriveraraghavan, ibid., **94** (2), 68 (1946).
69. R. M. Krishnan, and N. V. Pathasaradhy, ibid., **69** (9), 59 (1971).
70. J. P. Hoare and M. A. La Boda, *J. Electrochem. Soc.*, **132** (4), 798 (1985).
71. Z. A. Soloveva and A. E. Lapshina, *Soviet Electrochemistry*, **1** (8), 840 (1965).
72. C. C. Postins and J. E. Longland, *Prod. Finish.*, **22**, 51 (1969).
73. E. A. Romanowski and H. Brown, U.S. Patent, 3,334,033 (1967).
74. R. Dow and J. E. Stareck, *Proc. Am. Electroplat. Soc.*, **40**, 53 (1953); *Plating*, **40**, 987 (1953); J. E. Stareck and R. Dow, U.S. Patents, 2,686,756 (1954); 2,787,588; 2,787,589 (1957).

75. Maytag Company, *Steel*, **136**, 122 (1955); *Iron Age*, **175**, 140 (1955).
76. W. H. Safranek and G. R. Schaer, *Proc. Am. Electroplat. Soc.*, **43**, 105 (1956).
77. S. Zirinsky and D. S. Carr, *Proc. Am. Electroplat. Soc.*, **45**, 97 (1958).
78. V. A. Lamb and J. P. Young, "Experimental Plating of Gun Bores to Retard Erosion," PB Rep. 151,405, Natl. Bur. Stand. Tech. Note 46, Office of Technical Services, Washington, DC (1960).
79. W. M. Spurgeon and O. Isaacs, *Proc. Am. Electroplat. Soc.*, **45**, 145 (1958).
80. T. P. McFarlane, *Iron Age*, **176**, 103 (1955).
81. E. J. Seyb, A. A. Johnson, and A. C. Tulumello, *Proc. Am. Electroplat. Soc.*, **44**, 29 (1957); see also J. E. Stareck, E. J. Seyb, A. A. Johnson, and W. H. Rowan, U.S. Patents 2,916,424 (1959); 2,952,590 (1960).
82. H. Mahlstedt, *Automot. Ind.*, **118** (10), 48 (1958).
83. H. Brown, M. Weinberg, and R. J. Clauss, *Plating*, **45**, 144 (1958).
84. W. H. Safranek and C. L. Faust, *Plating*, **45**, 1027 (1958).
85. D. E. Weimer, *Met. Finish. J.*, **5**, 89 (1959).
86. A. Brenner, P. Burkhead, and C. Jennings, *Proc. Am. Electroplat. Soc.*, **34**, 32 (1947); Res. Natl. Bur. Stand., **40**, 31 (1948); Res. Paper 1854.
87. N. E. Ryan, F. Henderson, S. T. M. Johnstone, and H. L. Wain, *Nature*, **180**, 1406 (1957); N. E. Ryanand, E. J. Lumley, *J. Electrochem. Soc.*, **106**, 388 (1959); N. E. Ryan, ibid., **107**, 397 (1960).
88. P. C. Good, D. H. Yee, and F. E. Block, "High Purity Chromium by Electrolysis," U.S. Bur. Mines Rep. Invest. 5589, (1960).
89. M. J. Ferrante, P. C. Good, F. E. Block, and D. H. Yee, *J. Met.*, **12**, 861 (1960).
90. R. Bilfinger, *Das Hartverchromungs— Verfahren*, Herm. Beyer Verlag, Leipzig, 1939, p. 84; 2nd ed. (1942), pp. 39, 134–135; republished by Edwards Bros., Ann Arbor, MI, (1946).
91. T. A. Hood, *Plating Notes (Aust.)*, **4**, 31 (1952); *Met. Finish.*, **50**, 103 (1952).
92. E. J. Seyb Jr., R. E. Woehrle, and J. G. Neitzel, U.S. Patent 3,498,892 (1970).
93. E. M. Baker and P. J. Merkus, *Trans. Electrochem. Soc.*, **61**, 327 (1932).
94. A. A. Johnson and P. G. Kenedi, U.S. Patent 3,393,980 (1968).
95. T. H. Webersinn and J. M. Hosdowich, "Research Report No. 128," United Chromium, Inc. (1932).
96. J. L. Griffin, *Plating*, **53**, 196 (1966).
97. W. L. Grube and F. L. Clifton, *Mon. Rev. Am. Electroplat. Soc.*, **34**, 140, 471 (1947).
98. W. H. Hartford, *Ind. Eng. Chem. (Anal. Ed.)*, **14**, 174 (1942); *Ind. Eng. Chem.*, **41**, 1993 (1949).
99. L. C. Pan, *Trans. Electrochem. Soc.*, **58**, 423 (1930); *Met. Ind. (NY)*, **28**, 271 (1963).
100. H. L. Farber and W. Blum, *Natl. Bur. Stand. Res. Papers*, **131**, 27 (1930).
101. E. A. Ollard and E. B. Smith, *Sheet Met. Ind.*, **23**, 1129 (1946).
102. N. V. Mandich, J. R. Selman, C.-C. Lee, *Plating Surf. Finish.*, **84** (12), 82 (1997).
103. K. E. Langford, *Analysis of Electroplating and Related Solutions*, Electrodeposition and Metal Finishing, Teddington, U.K., 1951.
104. D. G. Foulke and F. E. Crane, *Electroplaters Process Control Handbook*, Van Nostrand, New York (1963).
105. A. F. Bogenschutz and U. George, *Analysis and Testing*, Finishing Publications Ltd, Teddington, U.K. (1985).
106. *Metal Finishing Guidebook*, vol. 95, no. 1A, Elsevier, New York (1998).

107. W. W. White and M. C. Henry, *Plating Surf. Finish.*, **59** (5), 429 (1972).
108. M. Traficante, "Analytical Methods For Hexavalent Chromium Electroplating Solutions," *Proc. AESF Chromium Colloquium*, Orlando, Fl (Jan. 1994).
109. S. S. Heberling, D. Campbell, and S. Carson, "Analysis of Chromium Electroplating Solutions and Wastewaters by Ion Chromatography," *Proc. 2nd Chromium Colloqium*, AESF, Miami, Fl (Feb. 1990).
110. Technical Note 24, "Determination of Chromium" (1987), Dionex Corporation, Sunnyvale, CA.
111. R. O. Hull, U.S. Patent, 2,149.344 (1935).
112. W. Nohse, *The Hull Cell*, Robert Draper, Teddington, U.K. (1966).
113. R. O. Hull and J. B. Winters, *Proc. AES 38th Annual Conf.* (1951), 133.
114. A. K. Graham and H. L. Pinkerton, *Proc. AES 50th Annual Conf.* (1963), 13.
115. R. Seegmiller and V. A. Lamb, *Proc. Am. Electroplat. Soc.*, **35**, 125 (1948).
116. R. F. Ledford and L. O. Gilbert, *Proc. Am. Electroplat. Soc.*, **42**, 33 (1955).
117. V. Massuet Grau, *Met. Finish. J.*, **4**, 467 (1958).
118. N. V. Mandich, *Proc. AESF Tech. Conf. SUR/FIN' 96.*
119. G. Dubpernell, *Trans. Electrochem. Soc.*, **80**, 589 (1941).
120. K. Mondal, N. V. Mandich, and S. Lalvani, *J. Appl. Electrochem.*, submitted for publ.
121. J. B. Niles, U.S. Patent 2,398,110 (1946); W. Cibulskis, M. Shaeat, and H. Mahlstedt, U.S. Patent 2,840,523 (1958).
122. F. I. Danilov and A. B. Velichenko, *Electrochim. Acta*, **38**, 437 (1993).
123. J. Hyner, U.S. Patent 2,456,281 (1948).
124. D. W. Hardesty, *Plating*, **56**, 705 (1969).
125. F. Hine, K. Takayasu, and N. Koyagani, *J. Electrochem. Soc.*, **133** (2), 346 (1986).
126. M. Veda, *Electrochim. Acta*, **40**, 817 (1995).
127. R. M. Krishan, et al., *Met. Finish.*, **93** (9), 46 (1995).
128. J. M. Hosdowich, in Reference 32, vol. 2, pp. 65–92; see also K. G. LeFevre, *Proc. Am. Electroplat. Soc.*, **43**, 34 (1956).
129. C. F. Corfe, *Electroplating*, **13** (5), 48 (1960).
130. J. E. Molos, *Ind. Med.*, **16**, 404 (1947); *Proc. Am. Electroplat. Soc.*, **34**, 270 (1947).
131. A. C. Stern, L. P. Benjamin, and H. Goldberg, *J. Electrochem. Soc.*, **93**, 67 (1948); *Plating*, **35**, 565, 958 (1948).
132. H. Brown, U.S. Patent 2,750,334 (1956); see also U.S. Patents 2,750,335; 2,750,336; 2,750,337 (1956); 2,846,380; 2,857,295 (1958); 2,913,377 (1959).
133. G. Hama, W. Frederick, D. Millage, and H. Brown, *Am. Ind. Hyg. Assoc. Q.*, **15**, 211 (1954).
134. D. E. Weimer, *Met. Finish.*, **5**, 89 (1959).
135. P. J. Ramsden, *Electroplat. Met. Finish.*, **10**, 152 (1957).
136. D. Millage and W. Hague, *Proc. Am. Electroplat. Soc.*, **45**, 118 (1958).
137. A. L. Jones, "Sources of Nodules and Pits in Hard Chromium Plating," AESF Hard Chromium Workshop (1992).
138. L. Schwartz, *U.S. Public Health Service Bull.*, **229** (1936); **249** (1939).
139. A. R. Wilkerson, *J. Am. Leather Chem. Assoc.*, **39**, 90 (1944).
140. C. L. Faust, in Reference 124, pp. 108–126.
141. "Recommended Practice for Chromium Plating on Steel for Engineering Use," ASTM designation B177-68 (1968).

142. G. Dubpernell and S. M. Martin U.S. Patent 2,624,728 (1953).
143. R. G. Bikales, U.S. Patent 2,868,709 (1959).
144. L. F. Howard, U.S. Patent 2,898,293 (1959).
145. Fuji Plant Industrial Co., U.S. Patent, 3,425.926 (1969), U.K. Patent 1,062.360 (1967).
146. J. Hyner, *Met. Finish. Guidebook*, **40**, 263 (1972).
147. G. E. Shahin, *Plating Surf. Finish.*, **79** (8), 19 (1992).
148. W. M. Tucker and R. L. Flint, *Trans. Electrochem. Soc.*, **88**, 335 (1945).
149. E. M. Relitz, Canadian Patent 378,303 (1938).
150. N. V. Mandich, *Plating Surf. Finish.*, **85** (12), 91 (1998).
151. C. L. Faust, in Reference 124, pp. 108–126.
152. "Recommended Practice for Chromium Plating on Steel for Engineering Use," ASTM Designation B177-68 (1968).
153. J. D. Greenwood, *Hard Chromium Electroplating*, 3rd ed., Robert Draper Ltd., Teddington, U.K. (1984).
154. R. Morisset, *Chromium Electroplating*, Robert Draper Ltd., Teddington, U.K. (1954).
155. R. K. Guffie, *The Handbook of Hard Chromium Electroplating*, Gardener Publ., Cincinnati, OH (1986).
156. C. H. Peger, *Chrome Electroplating Simplified*, 4th ed., Hard Chrome Plating Consultants, Cleveland, OH, 1981.
157. C. Levy, *Proc. Am. Electroplat. Soc.*, **43**, 219 (1956).
158. J. W. Dini and H. R. Johnson, *Plat. Surf. Finish.*, **68** (10), 64 (1981).
159. E. Zmihorski, *J. Electrodepositors' Tech. Soc.*, **23**, 203 (1948).
160. C. Williams and R. A. F. Hammond, *Trans. Inst. Met. Finish.*, **31**, 124 (1954).
161. J. W. Beams, *Proc. Amer. Electroplat. Soc.*, **43**, 211 (1956).
162. H. Chessin and J. G. Poor, *Plating*, **43**, 913 (1956).
163. H. Chessin and J. G. Poor, *Plating*, **46**, 1037 (1959).
164. J. G. Poor, H. Chessin, and C. L. Alderuccio, *Plating*, **47**, 811 (1960).
165. J. W. Dini and H. R. Johnson, *Met. Finish.*, **54** (4), 48 (1977).
166. D. Davies and F. A. Whittaker, *Met. and Mater.*, **1** (2), (1967).
167. H. Ploog, *Galvanotechnik*, **61**, 155 (1970).
168. G. Dubpernell, *Electrodeposition of Chromium from Chromic Acid Solution*, Pergamon Press, New York (1977).
169. P. Morisset, *Chromage, Dur et Decoratif* (Chromium Plating-Hard and Decorative), Centre d'Information du Chrome Dur, Paris, 1961.
170. R. Weiner and A. Walmsley, *Chromium Electroplating*, Finishing Publ., Teddington, U.K. 1980.
171. J. D. Greenwood, *Hard Chromium Plating*, 2nd ed., Robert Draper, Teddington, Middlesex, England, 1971.
172. J. D. Greenwood, *Heavy Deposition*, Robert Draper, Teddington, Middlesex, England, 1970.
173. C. H. Peger, *Hard Chromium Fixtures*, Hard Chromium Consultants, Cleveland, OH, 1982.
174. J. K. Dennis and A. Such, *Nickel and Chromium Electroplating*, 3rd ed., Woodhead Publ., Cambridge, U.K. 1993.
175. A. W. Logozzo, *Plating*, **56**, 1019 (1969).
176. N. V. Mandich, *Met. Finish.*, **97** (6) 100; **97** (7), 42; **97** (8), 42 (1999); **97** (9), 72, 97 (10), 30 (1999).

177. H. J. Wills, *Amer. Mach.*, **81**, 247 (1937).
178. F. J. Benn, *Grits and Grinds*, **37** (5), 8 (1946).
179. J. W. Oswald, *Trans. Inst. Met. Finish.*, **48**, 169 (1870).
180. H. T. Wilson, *Plating*, **58**, 345 (1971).
181. W. Blum, *33rd Conv. Am. Electroplat. Soc.*, 1946, p. 16; *Trans. Electrochem. Soc.*, **90**, 85 (1946); *Metal. Finish.*, **45**, 66 (1947).
182. J. L. Vaughan and I. A. Usher, *Can. Met.*, **7**, 20, 53 (1944).
183. A. L. Fry, *Met. Finish.*, **44**, 467, 478 (1946).
184. V. A. Lamb and J. P. Young, *Proc. Am. Electroplat. Soc.*, **43**, 260 (1956); "PB 151", 405, *Natl. Bur. Stand. Tech. Note No. 46* (1960); *Ordnance*, **45**, 725 (1961).
185. R. A. F. Hammond, *Trans. Inst. Met. Finish.*, **34**, 83 (1957).
186. R. J. Girard and E. F. Koetsch Jr., *Proc. Am. Electroplat. Soc.*, **47**, 199 (1960).
187. F. O'Neill, *Plating*, **58**, 19 (1971).
188. A. W. Ehlers, *Tool and Die J.*, **98**, 120 (1945).
189. S. Menton, *Steel*, **120**, 124 (1947).
190. W. G. Patton, *Iron Age*, **161**, 78 (1948).
191. P. Csokan, *Metalloberflache*, **13**, 81, 113 (1959).
192. S. Wernick and R. Pinner, *The Surface Treatment and Finishing of Aluminum and Its Alloys*, 2nd ed., Robert Draper, Teddington, Middlesex, England (1959).
193. E. Herrmann, *Galvanotechnik*, **51**, 336 (1960).
194. *Automob. Eng.*, **35**, 508 (1945); **38**, 236 (1948).
195. A. W. Mall, *Mach. Des.*, **20**, 136 (1948); *Iron Age*, **164** (22), 86 (1949).
196. J. E. Stareck and R. Dow, U.S. Patent 2,812,297 (1957).
197. V. E. Guersney, *Plating Surf. Finish.*, **63** (2), 38 (1976).
198. V. E. Guersney, *Plating Surf. Finish.*, **63** (3), 44 (1976).
199. R. D. Bedi, *Plating*, **55**, 238, 365 (1968).
200. R. D. Bedi and G. Dubpernell, *Plating*, **55**, 246 (1968).
201. G. E. Mcdonald, *Solar Energy*, **17**, 119 (1975).
202. D. M. Mattox, *Plating Surf. Finish.*, **63**, 55 (1976).
203. Siemens and Halske A. G., German Patents 607,420 (1929); 611,200 (1931); 612,255 (1933); French Patent 754,360 (1933); British Patent 408,097 (1934).
204. A. Ungelunk, J. Fischer, and H. Endrass, U.S. Patent 1,975,239 (1934).
205. E. A. Ollard, *J. Electrodepositors Tech. Soc.*, **12**, 33 (1936).
206. L. O. Gilbert and C. C. Buhman, U.S. Patent 2,623,847 (1952).
207. M. F. Quaely, *Plating*, **40**, 982 (1953); *Proc. Am. Electroplat. Soc.*, **40**, 48 (1953); U.S. Patents 2,739,108-9 (1956); 2,772,227 (1956); 2,824,829 (1958).
208. G. E. Oleson and R. M. Woods (to Corilliurn Corp.), U. S. Patent 3,414,492 (1968); R. M. Woods and D. R. Moul, U.S. Patents: 3,418,221 (1968); 3,442,777 (1969); 3,454,474 (1969).
209. K. S. Willson, U.S. Patent 3,620,935 (1971).
210. J. P. Branciaroli and P. G. Stutzman, *Plating*, **56**, 37 (1969).
211. J. P. Branciaroli, *Trans. Inst. Met. Finish.*, **48**, 172 (1970).
212. J. H. Hunter and L. K. Kosowsky, U.S. Patent 2,826,538 (1958).
213. V. Grips, S. Rajagopalan, and I. Rajagopal, U.S. Patent 5,019.223 (1991).
214. A. K. Graham, *Proc. Am. Electroplat. Soc.*, **46**, 61 (1959).

215. B. N. Popov, R. E. White, D. Slavkov, and Z. Koneska, *J. Electrochem. Soc.*, **139** (1), 91 (1992).

216. M. Driver, *Solar Energy Mat.*, **4**, 179 (1981).

217. F. J. Monteio, F. Oliveira, R. Reis, and O. Pavia, *Plating Surf. Finish.*, **79** (1), 46 (1992).

218. M. Selvam, *Met. Finish.*, **80** (5), 107 (1982).

219. K. J. Kathro, ibid., **76** (10), 57 (1978).

220. L. Sivaswamy, S. Gowri, and B. S. Shenoi, ibid., **78**, 48 (1974).

221. H. Flasch, German Patent 722,364 (1942).

222. K. S. Wilson, *Plating*, **54**, 49 (1967).

223. W. Eilender, H. Arend, and F. Sadrizil, *Metalloberflache*, **3**, 32 (1949).

224. W. C. Giesker and R. K. Britton, U.S. Patent 2,746,915 (1956).

225. W. C. Giesker and R. K. Britton, U.S. Patent 2,788,292 (1957).

226. W. H. Safranek, H. R. Miller, and C. L. Faust, *Plating*, **50**, 507 (1963).

227. B. B. Knapp, *Trans. Inst. Met. Finish.*, **35**, 139 (1958).

228. H. Brown and M. Weinberg, *Proc. Am. Electroplat. Soc.*, **46**, 128 (1959); see also M. Weinberg and H. Brown U.S. Patent 2,871,550 (1959).

229. A. Korbelak, *Plating*, **40**, 1126 (1953); *Proc. Am. Electroplat. Soc.*, **40**, 90 (1953).

230. D. Gray and B. K. Northrop, U.S. Patent 1,892,051 (1932).

231. C. G. Fink and C. H. Eldridge, U.S. Patent 1,942,356 (1934).

232. W. H. Safranek and H. R. Miller, "Adherence of Paint on Chromium-Plated Zinc Die Castings, Reprint No. 668C, Society of Automotive Engineers," New York (Mar. 1963); *Prod. Finish.*, **27**, 38 (1963).

233. B. K. Dent, *Plating*, **50**, 1100 (1963).

234. W. S. Garwood, *Plating*, **53**, 1323 (1966); **54**, 403 (1967).

235. J. E. Stareck, Canadian Patent 434,632 (1946).

236. C. F. Waite, *Proc. Am. Electroplat. Soc.*, **40**, 113 (1953); *Plating*, **40**, 1245 (1953).

237. E. M. Baker and W. L. Pinner, *J. Soc. Aut. Eng.*, **22**, 331 (1928); Discussion, **23**, 50 (1928).

238. H. Chessin and E. J. Seyb, *Plating*, **55**, 821 (1968).

239. K. Gebauer, *Korros. Metalischutz*, **16**, 297, table 1 (1940); *Oberflhchentech.*, **18**, 2, 11, 19, 31 (1941); *Metallwirtschaft*, **21**, 468 (1942).

240. M. R. J. Wyllie, *Trans. Electrochem. Soc.*, **92**, 519 (1947).

241. J. B. Cohen, *Trans. Electrochem. Soc.*, **86**, 441 (1944).

242. E. J. Seyb, A. A. Johnson, and A. C. Tulumello, *Proc. Am. Electroplat. Soc.*, **44**, 29 (1957); see also J. E. Stareck, E. J. Seyb, A. A. Johnson, and W. H. Rowan, U.S. Patents 2,916,424 (1959); 2,952,590 (1960).

243. H. Brown, M. Weinberg, and R. J. Clauss, *Plating*, **45**, 144 (1958).

244. W. Blum, P. W. C. Strausser, and A. Brenner, *J. Res. Nat. Bur. Stand.*, **13**, 331 (1934), Res. Paper 712.

245. W. A. Wesley, in *Corrosion Handbook*, H. H. Uhlig, ed., Wiley, New York, (1948), p. 817.

246. J. Chadwick, *Electropl. Met. Finish.*, **6**, 451 (1953).

247. R. L. Saur and R. P. Basco, *Plating*, **53**, 35, 320, 981, 1124 (1966).

248. E. J. Seyb and W. H. Rowan, *Plating*, **46**, 144 (1959).

249. E. J. Seyb, *Prod. Finish.*, **23**, 64 (1959); *Plating*, **46**, 852 (1959); *Metal. Prog.*, **76**, 113 (1959); see also D. E. Weimer, *Electropl. Met. Finish.*, **12**, 340 (1959).

250. E. J. Seyb, *Proc. Am. Electroplat. Soc.*, **47**, 209 (1960).

251. W. E. Lovell, E. H. Shotwell, and J. Boyd, *Proc. Am. Electroplat. Soc.*, **47**, 215 (1960).
252. *ASTM Designation B456-71, Book of Standards*, p. 7, 718 (1971).
253. *British Standard* 1224, (1970), *Specification for Electroplated Coatings of Nickel and Chromium*, British Standards Institution, London, 1970.
254. M. M. Beckwith, *Plating*, **47**, 402 (1960); see also M. N. Beckwith, Canadian Patent 558,991 (1958).
255. M. R. Caldwell, *Met. Prog.*, **75**, 78 (1959); *Precision Met. Molding*, **17**, 47 (1959); *Inco Nickel Topics*, **12** (7), 6 (1959).
256. D. M. Bigge, *Plating*, **47**, 1263 (1960).
257. H. Brown, *Electroplating*, **15**, 398 (1962); H. Brown, U.S. Patent 3,090,733 (1963).
258. H. Brown and D. Millage, *Trans. Inst. Met. Finish.*, **37**, 21 (1960).
259. D. Millage, E. Romanowski, and H. Brown, "Paper No. 24", First National Die Casting Exposition and Congress, Detroit, Mich., Nov. 8–11, (1960).
260. A. H. DuRose, K. S. Willson, and G. C. Tejada, U.S. Patent 3,563,864 (1971).
261. G. D. DeCastelet, U.S. Patent 3,620,936 (1971).
262. H. Brown and T. W. Tomaszewski, *Proc. Int. Conf. Surfaces 66* (Basel), Nov. 1966, Forster-Verlag AG, Zurich, (1967), p. 88; U.S. Patents 3,152,971-3 (1964); see also T. W. Tomaszewski, R. J. Clauss, and H. Brown, *Proc. Am. Electroplat. Soc.*, **50**, 169, 201 (1963).
263. J. M. Odekerken, *Electropl. Met. Finish.*, **17**, 2 (1964).
264. H. Chessin, U.S. Patents 3,574,068; 3,595,762 (1971).
265. T. G. Kubach, W. H. R. Pritsch, and W. Bolay, U.S. Patent 3,625,039 (1971).
266. T. Malak, D. Snyder, and A. H. DuRose, *Plating*, **59**, 659 (1972).
267. A. H. DuRose, *Proc. Am. Electropl. Soc.*, **48**, 83 (1960).
268. F. Ogburn and M. Schlissel, *Plating*, **54**, 54 (1967).
269. K. S. Willson, *Plating*, **59**, 226 (1972).
270. W. Blum, *Plating*, **48**, 613 (1961).
271. S. E. Beacom, *Prod. Finish.*, **33**, 130 (June 1969).
272. E. P. Harbulak, *Plating Surf. Finish.*, **67**, 49 (1980); U.S. Patent 4,310,389.
273. "ASTM B 764," *Amer. Soc. for Metals*, 100 Bar Harbor Drive, West Conshohocken, PA 19428.
274. H. Mahlstedt, U.S. Patent 1,967,716 (1934).
275. C. A. Snavely, *Trans. Electrochem. Soc.*, **92**, 537 (1947).
276. C. A. Snavely and D. A. Vaughan, *J. Am. Chem. Soc.*, **71**, 313 (1949).
277. J. J. Dale, *Proc. 3rd Int. Electrodeposition Conf. Electrodepositors' Tech. Soc.*, London (1947), p. 185.
278. J. J. Dale, A. Brenner, C. A. Snavely, and C. L. Faust, Discussion: *J. Electrochem. Soc.*, **97**, 466 (1950).
279. W. A. Wood, *Trans. Faraday Soc.*, **31**, 1248 (1935).
280. D. W. Hardesty, *J. Electrochem. Soc.*, **116**, 1194 (1969).
281. C. A. Snavely and C. L. Faust, *J. Electrochem. Soc.*, **97**, 99 (1950).
282. C. P. Brittain and G. C. Smith, *Trans. Inst. Met. Finish.*, **31**, 146 (1954).
283. W. Hume-Rothery and M. J. R. Wyllie, *Proc. Roy. Soc. (London)*, **A181**, 331 (1943); **AIS2**, 415 (1943).
284. H. Fry, *Trans. Inst. Met. Finish.*, **32**, 107 (1955).
285. C. P. Brittain and G. C. Smith, *Trans. Inst. Met. Finish.*, **33**, 289 (1956).

286. A. H. Sully and E. A. Brandes, "Chromium" 2nd ed., Plenum Press, New York (1967).

287. W. A. Wood, *Phil. Mag.*, **24**, 511, 772 (1937).

288. A. Knoedler, *Metalloberflache*, **17**, 161, 331 (1963).

289. K. Nishihara, M. Kurachi, Y. Sakamoto, and M. Nishihara Jr., "Interfinish, Deutsche Gesellschaft fur Galvanotechnik," Dusseldorf (1968), pp. 118–123.

290. M. A. Shluger and V. A. Kazakov, *Zh. Prikl. Khim.*, **33**, 644 (1960); *Chem. Abstr.*, **54**, 18126 (1960).

291. H. Okada and T. Ishida, *Nature*, **187**, 496 (1960); *Met. Finish. Abstr.*, **2**, 168 (1960).

292. M. Matsunaga, *Sci. Pap. Inst. Phys. Chem. Res. (Tokyo)*, **54**, 177 (1960); *Chem. Abstr.*, **55**, 5188 (1961).

293. A. J. Steiger, *Met. Finish.*, **56**, 56 (1958).

294. O. Bornhauser, U.S. Patent 1,985,308 (1934).

295. F. Taylor, *Electropl. Met. Finish.*, **5**, 109 (1952).

296. C. A. Snavely, U.S. Patent 2,635,993 (1953).

297. M. E. Beckman and F. Maass-Graefe, *Metalloberflache*, **5**, 161A (1951).

298. V. G. Kakovkina and A. N. Sysoev, *Trudy Kharkovsk. Politekh. Inst., Ser. Khim. Tekhnol.*, **18** (5), 107 (1958); *Chem. Abstr.*, **54**, 24001 (1960).

299. L. Koch and G. Hein, *Metalloberflache*, **7**, 145A (1953).

300. S. G. Christov and N. A. Pangarov, *Z. Elektrochem.*, **61**, 113 (1957); see also *Izv. Khim. Inst. Bulgar. Akad. Nauk*, **7**, 237, 257 (1960); *Chem. Abstr.*, **55**, 2309 (1961).

301. S. G. Christov, I. P. Nenov, and R. G. Raichev, *Proc. 3rd Int. Congr. on Metallic Corrosion*, vol. 1, Moscow, 1969, pp. 389–395.

302. I. P. Nenov, S. G. Christov, R. Raicev, and Z. Georgiev, *Electrochim. Acta*, **12**, 1537 (1967).

303. M. C. Udy, in *Chromium*, vol. 2, M. J. Udy, ed., Reinhold, New York (1956), p. 101.

304. American Society for Metals, *Ductile Chromium*, Cleveland, OH (1957).

305. F. E. Bacon, *ASM Metals Handbook*, vol. 1, American Society for Metals, Metals Park, OH (1961), p. 1200.

306. T. G. Coyle, *Proc. ASTM*, **43**, 556 (1943).

307. A. Kutzelnigg, *Testing Metallic Coatings*, Robert Draper, Teddington, Middlesex, England (1963).

308. C. G. Peters and F. Knoop, *Met. Alloys*, **12**, 292 (1940).

309. J. M. Hosdowich, *Proc. Am. Electroplat. Soc.*, **36**, 103 (1949).

310. W. Eilender, H. Arend, and E. Schmidtmann, *Metalloberflache*, **2**, 49 (1948).

311. W. Eilender, H. Arend, and E. Schmidtmann, *Metalloberflache*, **3**, 57 (1949).

312. H. Wahl and K. Gebauer, *Metalloberflache*, **2**, 25, (1948).

313. J. T. Burwell and S. F. Murray, *Nucleonics*, **6** (1), 34 (1950).

314. R. G. Abowd Jr., and C. E. Alsterberg, Paper presented at the *Symposium on Applications of Radioactivity at Traceflabs, Inc.*, Boston, MA (1956).

315. R. G. Abowd Jr., Preprint No. 57LC-3, *Meeting Am. Soc. Lubrication Engineers*, Toronto, Canada, Oct. 7, 1957; also Preprint No. 171, "Nuclear Eng. and Sci. Conf.," Chicago, IL., March 17–21, 1958; and Paper No. 8U, "S.A.E. Annual Meeting," Detroit, MI, Jan. 12–16, 1959.

316. Worthington Pump and Machinery Corp., "Industrial Chromium Plating," Bull. S-2001-A (1931); F. E. Allen, *Worthington News* (Worthington Pump and Machinery Corp. house organ), **2** (6), 2 (Oct. 1930); **2** (7), 1 (Nov. 1930); **2** (8), 1 (Dec. 1930); **2** (9), 1 (Jan. 1931); **2** (10), 2 (Feb. 1931); **2** (11), 1 (Mar. 1931).

317. V. I. Arkharov, A. M. Tagrubskii, and S. A. Nemnonov, *Vestn. Metalloprom.*, **20**, 13 (1940).

318. M. Kontorovich and V. L. Arkharov, *Vestn. Metalloprom.*, **20**, 10 (1940).

319. R. Graham, K. R. Williams, and R. W. Wilson, *Engineering*, **167**, 241 (1940); 265 (1940).

320. P. Hindent, *J. Res. Nat. Bur. Stand.*, **26**, 81 (1941).

321. D. S. Bloom and N. J. Grant, *Trans. Am. Inst. Met. Eng.*, **188**, 4 (1950); **191**, 1009 (1951).

322. R. M. Parke and F. P. Bens, *Symposium on Materials for Gas Turbines*, ASTM Spec. Publ., 1946, p. 80.

323. W. W. Coblentz and R. Stair, *J. Res. Natl. Bur. Stand.*, **2**, 343 (1929).

324. J. E. Stareck, E. J. Seyb, and A. C. Tulumello, *Plating*, **41**, 1171 (1954); *Proc. Am. Electroplat. Soc.*, **41**, 209 (1954).

325. C. Williams and R. A. F. Hammond, *Trans. Inst. Met. Finish.*, **32**, 85 (1955).

326. R. Walker, *Internal Stress in Electrodeposited Metallic Coatings*, published by Metal Finishing Journal, London, (1968).

327. M. Ya. Popereka, *Internal Stresses in Electrolytically Deposited Metals*, TT68-50634, Clearinghouse for Federal Scientific and Technical Information, Springfield, VA (1970).

328. J. E. Stareck, E. J. Seyb, and A. C. Tulumello, *Plating*, **42**, 1935 (1955); *Proc. Am. Electroplat. Soc.*, **42**, 129 (1955).

329. M. Kolodney, "Chromium Plating," Information Release No. 4, War Metallurgy Comm., Nat. Acad. Sci., Res. Council, (1943); American Documentation Institute, Washington, DC, *Document* No. 1806.

330. J. E. Stareck et al., U.S. Patents 2,800,436-7-8; 2,800,443 (1957).

331. R. A. F. Hammond and C. Williams, *Metal. Rev.*, **5** (18), 165 (1960).

332. C. Williams and R. A. F. Hammond, *Trans. Inst. Met. Finish.*, **32**, 85 (1955); **34**, 317 (1957); *Proc. Am. Electroplat. Soc.*, **46**, 195 (1959).

333. B. Cohen, *Proc. Am. Electroplat. Soc.*, **45**, 33 (1958).

334. G. P. Kotlyarevskii, *Metalloved. i Obrab. Metal.*, No. **7**, 52 (1958).

335. R. Brunetaud, R. Chevalier, and P. Lessop, *Chrome Dur 1959–1960*, Centre d'Information du Chrome Dur, Paris, 1959, pp. 3–14.

336. H. Wiegand and U. H. Furstenberg, *Hartverchromung-Eigenschaften and Auswirkungen auf den Grundwerkstoff* (Hard Chromium Plate-Properties and Effects on the Basis Metal). Maschinenbau-Verlag GmbH, Frankfurt, 1968.

337. H. Speckhardt, *Metalloberflache*, **23**, 272 (1969).

338. A. Buch, *Metalloberflache*, **24**, 124 (1970).

339. H. Wiegand and U. H. Furstenberg, *Metalloberflache*, **25**, 123 (1971).

340. V. P. Greco and B. L. Pennell, *Plating*, **58**, 35 (1971).

341. H. Wiegand, *Metallwirtschaft*, **20**, 165 (1941).

342. A. Beerwald, *Laftfahit-Forsch.*, **18**, 368 (1941); Translations: *Eng. Digest*, **3**, 252 (1942); *Sheet Met. Ind.*, **16**, 1889 (1942); **17**, 68 (1943).

343. E. Raub, *Metallforschung*, **2**, 121 (1947).

344. C. J. Morgan and F. E. Brine, *Trans. Inst. Met. Finish.*, **47**, 77 (1969).

345. C. J. Morgan and W. A. Marshall, *Trans. Inst. Met. Finish.*, **46**, 144 (1968).

346. W. D. Klopp, *J. Met.*, **21** (11), 23 (1969).

347. E. A. Brandes and J. A. Whittaker, *Engineer*, **220**, 929 (1965).

348. P. J. Smith and G. Dubpernell, Canadian Patent 791,698 (1968).
349. H. Silman, *Met. Ind. (London)*, **78** (17), 327 (1951).
350. D. J. De Paul, *Corrosion*, **13**, 91 (1957).
351. H. Suss, *Corrosion*, **16**, 105 (1960).
352. M. Pourbaix, *Atlas of Electrochemical Equilibria*, Pergamon Press, New York (1966).
353. E. Deltombe, N. de Zoubov, and M. Pourbaix, "Electrochemical Behavior of Chromium-Potential-pH Equilibrium Diagram of the System Chromium-Water at 25°C" (in French), Tech. Rep. No. 41, Centre Beige d'Etude de la Corrosion (1956).
354. H. Uchida and A. Horiguchi, *Met. Prog.*, **83**, 113 (1963).
355. H. Uchida, O. Yanabu, A. Horiguchi, and H. Sato, *Trans. Inst. Met. Finish.*, **40**, 212 (1963); **42**, 138 (1964).
356. H. Uchida and O. Yanabu, U.S. Patent 3,113,845 (1963).
357. E. J. Smith, *Iron Steel Eng.*, **44**, 125 (1967).
358. G. W. Ward and S. E. Rauch Jr., *Blast Furnace and Steel Plant*, **56**, 229 (1968).
359. G. G. Kamm, A. R. Willey, and N. J. Linde, *J. Electrochem. Soc.*, **116**, 1299 (1969).
360. J. E. Allen, *Iron Steel Eng.*, **47**, 88 (1970).
361. M. W. Dippold, *Iron Steel Eng.*, **48**, 62 (1971).
362. H. Uchida, *J. Met. Finish. Soc. Jpn.*, **21**, 341 (1970).
363. E. J. Smith, L. C. Beale Jr., and L. W. Austin, U.S. Patent 3,526,486 (1970).
364. C. E. Roberts and G. W. Ward, U.S. Patent 3,574,069 (1971).
365. O. Yanabu, T. Saito, K. Doi, T. Enari, K. Miyakawa, and H. Kawasaki, *Interfinish 1968*, Conference Report, Deutsche Ges. fur Galvanotechnik, pp. 312–320.
366. E. J. Seyb Jr., R. E. Woebrle, and J. G. Neitzel, U.S. Patent 3,498,892 (1970).
367. R. E. Cleveland, U.S. Patent 2,114,072 (1938); E. R. Granger and R. E. Cleveland, U.S. Patent 2,248,530 (1941).
368. H. van der Horst, *Proc. Am. Electroplat. Soc.*, **31**, 56 (1943).
369. T. G. Coyle, *Proc. Am. Electroplat. Soc.*, **32**, 20 (1944).
370. R. Pyles, *Proc. Am. Electroplat. Soc.*, **32**, 136 (1944).
371. H. van der Horst, U.S. Patents 2,314,604 (1943); 2,412,698 (1946); British Patent 518,694 (1940); Dutch Patent 57,383 (1946).
372. T. H. Webersinn and J. Hyne. Forestek, U.S. Patent 3,279,936 (1966).
373. T. C. Jarrett and R. D. Guerke (to Koppers Co., Inc.), U.S. Patent 2,433,457 (1947).
374. F. Passal, U.S. Patent 2,450,296 (1948).
375. W. S. Bohlman and D. S. Bruce, U.S. Patent 2,453,404 (1948).
376. United Chromium Inc., British Patents 600,818; 600,850; 601,065 (1948).
377. S. C. Wilsdon, U.S. Patent 2,620,296 (1952); British Patent 561,788 (1943).
378. H. M. Bendler, C. E. Bleil, T. W. Hertzog, M. A. LaBoda, and M. Ben, U.S. Patent 2,968,555 (1961).
379. J. A. Andrisek and T. R. Gill, U.S. Patent 2,947,674 (1960); Canadian Patent 593,405 (1960); British Patent 835,821 (1960).
380. L. W. Raymond, U.S. Patent 2,830,015 (1958).
381. Van der Horst Corp. of America, *Steel*, **152**, 52 (1963).
382. C. W. Forestek, U.S. Patent, 3,279.936 (1966).
383. C. O. Letendre, U.S. Patent, 3,341,348 (1967).
384. W. J. Zubrisky, U.S. Patent 3,063,763 (1962), U.S. Patent 2,430,750 (1947).

385. "Army Air Force Specif. 20031A" (1944); "Bureau of Ships, Navy Dept.," ("MIL-P-20218C," Dec. 29, 1959); *Society of Automotive Engineers Specif. AMS 2407A* (1951).
386. S. Kishi, Ti Furusawa, H. Yamaguchi, and H. Sugano, *Bull. Doc. Cent. Inform. Chrome Dur*, pp. 1–10 (1971).
387. A. K. Sidney and H. Salem, *J. Appl. Toxicol.*, **13** (3), 1992.
388. J. C. Crowther and S. Renton, *Electroplat. Met. Finish.*, **28** (5), 6 (1975).
389. R. S. Dean, *U.S. Bur. Mines Rep. Invest. 3547* (1941); *Rep. Invest. 3600* (1941).
390. R. R. Lloyd et al., *Trans. Electrochem. Soc.*, **89**, 443 (1946); *Eng. Min. J.*, **148** (7), 95 (1947); *J. Electrochem. Soc.*, **94**, 122 (1948); **97**, 227 (1950); U.S. Patents 2,507,475-6 (1950).
391. J. B. Rosenbaum, R. R. Lloyd, and C. C. Merrill, *U.S. Bur. Mines Rep. Invest. 5322* (1957).
392. F. E. Bacon, "Chromium Electrowining," in *Encyclopedia of Electrochemistry*, C. A. Hampel, ed., Reinhold, New York, 1964, pp. 198–201.
392a. F. E. Bacon, "Chromium and Chromium Alloys," in *Kirk-Othmer Encyclopedia of Chemical Technology*, 2nd ed., vol. 5, Wiley-Interscience, New York, 1964, p. 451.
393. C. A. Snavely, C. L. Faust, and J. E. Bride, U.S. Patent 2,693,444 (1954).
394. W. H. Safranek, U.S. Patent 2,822,326 (1958).
395. G. R. Schaer, U.S. Patent 2,927,066 (1960).
396. L. D. McGraw, J. A. Gurklis, C. L. Faust, and J. E. Bride, *J. Electrochem. Soc.*, **106**, 302 (1959).
397. T. Yoshida and R. Yoshida, *J. Chem. Soc. Japan, Ind., Chem. Sect.*, **58**, 89 (1955); *Chem. Abstr.*, **49**, 14540 (1955); U.S. Patents 2,704,273 (1955); 2,766,196 (1956).
398. M. R. Zell, *Met. Finish.*, **55**, (1) 57 (1957); U.S. Patent 2,801,214 (1957).
399. H. Nitto, U.S. Patent 2,938,842 (1960).
400. J. A. Bodrov and N. T. Kudryavtsev, *Plating*, **46**, 157 (1959) (Abstract only).
401. K. Firoyu, *Bull. Inst. Politekh. Bucuresti*, **20** (1), 53 (1958) (in Russian); *Bull. Cent. Inform. Chrome Dur*, pp. 1–13 (Dec. 1960) (in French); *Chem. Abstr.*, **54**, 5290 (1960).
402. W. Machu and M. F. M. El-Ghandour, *Werkst. Korros.*, **10**, 556 (1959); **11**, 274, 420, 481 (1960).
403. G. R. Sherwood, M. R. Holmes, and F. Bergishagen, *J. Electrochem. Soc.*, **106**, 204C (1959) (Abstract only).
404. Warner Electric Company, *Business Week, No. 757*, **81** (1944).
405. L. W. Skala (to M. M. Warner and H. Blech), U.S. Patent 2,470,378 (1949).
406. J. G. Icxi, U.S. Patent 2,748,069 (1956); French Patent 1,007,691 (1952); German Patent 933,720 (1955); British Patent 697,225 (1948); Canadian Patent 493,784 (1953).
407. M. Rubinstein, *Mater. Methods*, **40** (12), 98 (1954); *Proc. Am. Electroplat. Soc.*, **43**, 246 (1956).
408. H. D. Hughes, *Trans. Inst. Met. Finish.*, **33**, 424 (1956).
409. W. Machu, *Galvanotechnik*, **49**, 467 (1958).
410. N. R. Bharucha and J. J. B. Ward, British Patents 1,144,913 (1969); 1,213,556 (1970).
411. N. R. Bharucha and J. J. B. Ward, *Prod. Finish.*, **33**, 64 (1969).
412. J. J. B. Ward, I. R. A. Christie, and V. E. Carter, *Trans. Inst. Met. Finish.*, **49**, 97, 148, 172 (1971).
413. J. E. Bride, *Plating*, **59**, 1027 (1972); U.S. Patents 3,706,636-43 (1972).
414. D. J. Levy and W. R. Momyer, *Plating*, **57**, 1125 (1970); *J. Electrochem. Soc.*, **118**, 1563 (1971).

415. C. Barnes, J. Ward, and J. R. House, *Trans. Inst. Met. Finish.*, **55** (2), 73 (1977).
416. L. Gianelos, *Plating Surf. Finish.*, **66** (5) 56 (1979).
417. J. E. Bride, *Plating Met. Finish.*, **59** (11), 1027 (1972).
418. A. K. Hsieh, T. H. Ee, and K. N. Chen, *Met. Finish.*, **91** (4) (1993).
419. A. Watson, A. M. H. Anderson, M. R. el-Sharif, and C. U. Chisholm, *IMF Ann. Tech. Conf.*, York, UK, 1990, 26.
420. D. L. Snyder, *Proceed. AESF 2nd Chromium Colloquium*, Miami, FL (1990).
421. D. L., Snyder, *Plating Surf. Finish.*, **66** (6), 60 (1979).
422. D. L. Snyder, *Prod. Finish.*, **53** (3), 56 (1988); **54**(8), 61 (1989); **55**(12), 42 (1990).
423. V. E. Carter and I. R. A. Christie, *Trans. IMF*, **51** (1973), 41.
424. V. Opaskar and D. Crawford, *Met. Finish.*, 49 (1991).
425. G. Shahin, *Plating Surf. Finish.*, **79**, 19 (1992).
426. G. Shahin, U.S. Patent 5,294,326 (?).
427. H. Tadao and A. Ishihama, *Plating Surf. Finish.*, **66** (9), 36 (1979).
428. A. Ishihama and T. Hayashi, *Proc. of Interfinish 80 (?)*.
429. A. T. Vagramyan and N. T. Kudryavtsev, eds., *Theory and Practice of Chromium Electroplating*, TT65-.50001, Clearinghouse for Federal Scientific and Technical Information, Springfield, VA, 1965.
430. A. T. Vagramyan, in *Electrodeposition of Alloys*, V. A. Averkin, ed., OTS64-11015, Office of Technical Services, U.S. Department of Commerce, Washington, DC (1964), pp. 182–189.
431. A. Brenner, *Electrodeposition of Alloys*, vol. 2, Academic Press, New York, 1963, pp. 110–136.
432. C. A. Snavely, C. L. Faust, and J. E. Bride, U.S. Patent 2,693,444 (1954).
433. C. A. Snavely and G. R. Shear, U.S. Patent 2,693.444 (1954).
434. L. D. McGraw, J. A. Gurklis, C. L. Faust, and J. E. Bride, *J. Electrochem. Soc.*, **106** (4) 301 (1951).
435. P. L. Elsie, S. Govri, and B. Shenoi, *Met. Finish.*, **68** (11), 52 (1970).
436. W. H. Safranek and G. R. Schaer, *Proc. Am. Electroplaters' Soc.*, **34**, 32–73 (1947).
437. C. Fridmann and J. Royon, *Bull. Doc. Cent. Inform. Chrome Dur*, pp. 1–46 (June 1970).
438. A. Kaichinger, *Bull. Doc. Cent. Inform. Chrome Dur*, pp. 25–38 (July–Aug. 1970).
439. K. Aotani and K. Nishimoto, *Kinzoku Hyomen Gijutsu (Japan)*, **21** (7), 356 (1970).
440. P. A. Jacquet and P. Lepetit, *Bull. Doc., Centre Inform. Chrome Dur*, pp. 1–26 (Sept.–4 Oct. 1965).
441. P. A. Jacquet, J. J. Galbrun, and A. Popoff, *Chrome Dur*, **46** (1959–60).
442. N. V. Korovin, *Electropl. Metal Finish.*, **17** (6), 188 (1964).
443. B. A. Shenoi and K. S. Indira, *Met. Finish.*, **63** (5), 56; (6), 94 (1965).
444. V. V. Bondar and I. I. Potapov, *Zashchita Metallov*, **5** (3), 346 (1969).
445. F. X. Hassion and J. Szanto, *Springfield Armory, U.S. Dept. Comm., AD-621 920* (1965), p. 43.
446. Yu. N. Petrov, L. I. Dekhtyar, and A. E. Beznosov, "Effect on Microhardness of Alloying of Electrodeposited Chromium with Tungsten and Magnesium," *Izv. Akad. Nauk. Mold. SSR*, **6**, 74–78 (1967).
447. V. P. Greco and W. Baldauf, "Electrodeposition of Ni-Al_2O_3, Ni-TiO_2 and Cr-TiO_2 Dispersion-Hardened Alloys," *Plating*, **55** (3), 250–257 (1968).
448. W. H. Mason, U.S. Patent 1,844,921 (1932).

449. A. R. Trist, British Patents 475,902 (1937); 491,530 (1938); U.S. Patent 2,203,849 (1940).
450. H. R. Carveth, U.S. Patent 1,937,629 (1933).
451. C. A. Marlies and G. E. White, U.S. Patent 2,048,276 (1936).
452. V. I. Arkharov and S. A. Nemnonov, *Bull. Acad. Sci. URSS Classe Sci. Tech. (9110)*, 32 (1943); also V. I. Arkharov and V. N. Konev, *Vestn. Mashinostr.*, **35** (11), 55 (1955).
453. A. F. Gerds and M. W. Mallett, *Trans. Am. Soc. Met.*, **52**, 1027 (1960).
454. W. S. Bohlman (to Chromium Corp. of Am.), U.S. Patent 1,809,826 (1931).
455. H. Chessin and P. D. Walker (to M&T Chemicals, Inc.), German Patent Publ., DOS 2,025,751 (1970).
456. H. Chessin and R. F. Gempel (to M&T Chemicals, Inc.), German Patent Publ. DOS 2,005,254 (1970).

8 Electrodeposition of Lead and Lead Alloys

MANFRED JORDAN

INTRODUCTION

The standard potential for lead in aqueous solutions is −0.12 V. This metal has a high hydrogen overpotential, which means that it is easily deposited electrolytically from strongly acidic solutions with a cathodic efficiency approaching 100%. The values for the hydrogen overpotential are dependant on the surface and structure of the electrode and are given in the literature as varying between 0.84 V for 99.8% pure Pb [1] to about 1.2 V for electrolytically deposited coatings [2].

The electrochemical equivalent for the reaction:

$$Pb^{2+} + 2e \rightarrow Pb \tag{1}$$

is 3.865 g Ah^{-1}. The lead deposition can be used for coulometric determinations because of the high cathodic current efficiency.

Electrolytic deposition of lead has been investigated using several different processes. Important prerequisites for such a process is a high solubility of the lead salts to be used and stability of the anion against hydrolysis in the electrolyte. Table 1 lists some data on the solubility of the most relevant lead salts.

In the past the two most commonly used lead electrolytes were the fluoroborate- and fluorosilicate-based processes. In more recent years (since the 1980s) electrolytes based on methanesulphonic acid are becoming the prefered systems, particularly for electrodeposition of tin-lead alloys.

Electrolytes based on nitric acid are not recommended for the electrodeposition of metals because the nitrate anion undergoes cathodic reduction, which leads to the occurrence of unfavorable side reactions. Nitrate contamination can also be very detrimental to other lead electrolytes, leading to a reduction of the cathodic efficiency and poor coverage in the low current density areas [3].

The sections that follow give a brief description of some processes that are used for the electrodeposition of lead.

Modern Electroplating, Fourth Edition, Edited by Mordechay Schlesinger and Milan Paunovic.
ISBN 0-471-16824-6

TABLE 1 Solubility of Lead-Salts

Lead-Salt	Solubility (g in 100 ml solution)
Lead nitrate, $Pb(NO_3)_2$	36
Lead perchlorate, $Pb(ClO_4)_2$	226
Lead fluoroborate, $Pb(BF_4)_2$	No data available in literature; commercial
Leadfluorosilicate, $PbSiF_6$	68.9
Leadacetate, $Pb(CH_3COO)_2$	30.7
Leadchloride, $PbCl_2$	0.97
Leadsulphate, $PbSO_4$	0.004
Leadmethanesulphonate, $Pb(CH_3SO_3)_2$	54
Leadamidosulphonate, $Pb(NH_2SO_3)_2$	70.9

1 ELECTROLYTE TYPES

1.1 Perchlorate Electrolytes

Perchloric acid electrolytes are very suitable for electrodeposition of lead because of the high solubility of lead in such electrolytes and the high electrical conductivity of the solution. However, such electrolytes are not widely used in practice, partly because of the high price of the acid and also the acid's hazardous nature which can lead to the risk of explosion. There are varying descriptions of this hazardous nature in the literature [4, 5]. More recent reports in literature [6] describe fatal accidents where perchloric acid has penetrated through cracks in stone or concrete floors and reacted with wood in the underlying construction to form cellulose-perchlorates, which are very shock-sensitive. Since many plating shops frequently have catwalks supported by wooden frameworks, such hazards would be virtually unavoidable at some point in time. For these reasons the use of perchloric based electrolytes is not generally recommended.

1.2 Amidosulphonate Electrolytes

There are several articles in the literature which refer to electrodeposition of lead from amidosulphonic acid based electrolytes [7–11]. The use of such electrolytes for electrolytic refining of raw lead, which is contaminated with high levels of antimony and bismuth is also described [12, 13]. One problem that can occur in this particular application of such electrolytes is the tendency for the sulphamate anion to be hydrolyzed into ammonia and sulphate. This reaction causes the precipitation of leadsulphate and thus a high formation of sludge and a loss of lead in the electroraffination [12].

The rate of hydrolysis of amidosulphonic acid in aqueous solutions increases rapidly once the temperature exceeds 60°C and the pH of the solution is less than 1 [14].

1.3 Hexafluorosilicate Electrolytes

The hexafluorosilicate electrolyte was developed for electrolytic refining of lead by a process known as the Betts Process [15, 16]. This process is suitable for the purification of raw lead that is contaminated with high levels of As, Sb, Cu, Ag, and Au. The hexafluorosilicate electrolyte is considered to be more stable (and hence more economical) than the sulphamate electrolyte [13]. These electrolytes are not however used for electrodeposition of lead. Typical parameters of such an electrolyte for electrolytic refining of lead are given in Table 2 [17].

In the recent literature no papers were published dealing with the use of hexafluorosilicate electrolytes for electroplating of lead. As general information in Table 3, the electrolyte and process parameters are given as they were published by Wiesner [31].

1.4 Tetrafluoroborate Electrolytes

The tetrafluoroborate electrolyte is the most common process for the electrolytic deposition of lead. Lead tetraflouroborate is a stable lead salt that is very soluble in

TABLE 2 Parameters of a Lead Electrorefining Plant

Electrolyte composition	
H_2SiF_6	70–120 g liter^{-1}
Pb	60–100 g liter^{-1}
Additives	Aloes extract
	Lignin sulphonate
Current density	1.3–2.2 A dm^{-2}
Voltage	0.3–0.5 V
Temperature	35–42°C
Anode composition (wt%)	
Sb	1.0–1.2%
As	0.3–0.6%
Cu	0.04–0.07%
Bi	0.1–0.2%
Ag	0.1–0.6%
Anode sludge (wt%)	
Sb	40–50%
As	20–30%
Pb	10–20%
Bi	5–15%
Ag	5–15%
Electrorefining cells	800, each 2700 liters
production capacity	120,000 to/a

TABLE 3 Compositions of Hexafluorosilicate Electrolytes

	Electrolyte 1	Electrolyte 2
Lead	75 g liter^{-1}	180 g liter^{-1}
Total fluorosilicate	150 g liter^{-1}	140 g liter^{-1}
Glue	0.2 g liter^{-1}	5.4
Temperature	35–40°C	
Cathodic current density	0.5–8 A dm^{-2}	
Anodic current density	0.5–3 A dm^{-2}	
Voltage at 1 A dm^{-2}	0.1–0.2 V	
Cathodic current efficiency	100%	
Anodic current efficiency	100%	

water. Tetrafluoroboric acid is formed in the reaction between hydrofluoric acid and boric acid according to the reaction

$$4\,HF + H_3BO_3 \rightarrow HBF_4 + 3\,H_2O \tag{2}$$

It is possible that in the reverse reaction the tetrafluoroborate anion can be hydrolyzed into free fluoride and boric acid, which could lead to the possibility of the formation and precipitation of lead fluoride PbF_2. The hydrolysis of tetrafluoroboric acid occurs through the following sequence of reactions:

$$HBF_4 + H_2O \rightarrow HBF_3(OH) + HF \tag{3}$$

$$HBF_3(OH) + H_2O \rightarrow HBF_2(OH)_2 + HF \tag{4}$$

$$HBF_2(OH)_2 + H_2O \rightarrow HBF(OH)_3 + HF \tag{5}$$

$$HBF(OH)_3 + H_2O \rightarrow [HB(OH)_4] + HF \tag{6}$$

The tendency for this hydrolysis to occur can be reduced by the addition of boric acid. For this reason lead or lead alloy electrolytes based on tetrafluoroboric acid usually contain 20 to 30 g liter^{-1} boric acid, which shifts the equilibrium in the reaction in the equations (3) to (6) above toward the left-hand side of each reaction.

1.5 Methanesulphonate Electrolytes

Electrolytes based on short chained alkylsulphonic acids (in particular methanesulphonic acid), Table 4, are gaining increasing popularity since about 1980. The use of methanesulphonic acid for deposition of lead was first mentioned in a patent by Proell in 1950 [17], but there was not practical application of the process at that time.

Fluoride-free processes were initially developed for environmental reasons because they offer less hazardous alternatives to the fluoroborate systems and produce waste streams that are more easily dealt with. Since their introduction to the marketplace, these processes have proved to be more stable and easier to operate and in some applications technically superior, which explains the rapid growth in their popularity.

TABLE 4 Composition of a Methanesulphonate Lead Electrolyte

	Deposition of Thin Layers (< 50 μm)	Deposition of Thick Layers (up to 200 μm)
Methanesulphonic acid	100 g liter^{-1}	100 g liter^{-1}
Lead	50 g liter^{-1}	200 g liter^{-1}
Organic additives	Proprietary	Proprietary
Electrolyte temperature	20–30°C	20–30°C
Cathodic current density	< 1.5 A dm^{-2}	3–6 A dm^{-2}
Deposition rate	0.85 μm min^{-1} at 1.5 A dm^{-2}	1.7 μm min^{-1} at 3 A dm^{-2}

1.6 Other Electrolytes

Several other electrolytes have been investigated for deposition of lead. A detailed discussion on all would go beyond the scope of this chapter. The only electrolytes that have reached any significant importance for industrial applications are those already mentioned. A detailed survey of all possible lead electrolytes is given in Gmelin's *Handbook of Inorganic Chemistry* [33].

Alkaline Electrolytes The two most usual alkaline electrolytes used for the deposition of lead are the pyrophosphate and plumbite processes.

Pyrophosphate Electrolytes As an example for the deposition from alkaline solutions, Table 5 gives the parameters for pyrophosphate process. The pyrophosphate electrolytes are renowned for having particularly good throwing power.

Plumbitelectrolytes Electrodeposition of lead from alkaline plumbite(II)-electrolytes is possible. But opposite to the tin deposition, where the alkaline stannate processes still have some technical importance, the alkaline plumbite processes do not find technical applications. In the alkaline solution the lead ion is present in a complexed form. The electrodeposition has therefore a higher overpotential compared to the acidic solution, where the lead ion is uncomplexed. For this reason alkaline solutions can be operated without inhibitors and show normally a high-throwing power. Additionally alkaline solutions are less corrosive against the plants and equipment. The main disadvantages are process engineering problems. The

TABLE 5 Lead Pyrophosphate Electrolyte

Lead (g liter^{-1})	Pyrophosphate (g liter^{-1})	Cation	Additive	Temperature (°C)	Current density (A dm^{-2})	References
20.7	79.2	Na	Glue	50–60	≤ 5	[34]
20.3	78	Na		60	≤ 4.2	[35]
20.3	78	K	Glue	60	≤ 3.5	[35]
41.4	87	K	Phenols	20–40	0.5–2	[36]
20.7	65	K	Dicyandiamide/formaldehyde/hexamethylentetramine-reaction product	60	0.5–5	[37]

TABLE 6 Composition of an Alkaline Plumbite-Electrolyte

Pb	40 g liter^{-1}
NaOH	200 g liter^{-1}
Na-acetate	23.4 g liter^{-1}
Na-tartrate	26.2 g liter^{-1}
Temperature	60–90°C
Cathodic current density	1–2 A dm^{-2}
Additive	6 g liter^{-1} colophony

cathodic current densities, that can be reached from alkaline solutions are even at high temperatures up to 90°C low and do not normally pass $2\,A\,dm^{-2}$. A smooth dissolution of the anodes is only possible at low anodic current densities. Above 1 or $2\,A\,dm^{-2}$ there is a steep increase of the potential, and lead dioxide, PbO_2, is formed on the anode surface [38]. One example for an alkaline lead electrolyte is given in Table 6 [39].

2 GENERAL INFORMATIONS ON THE ELECTRODEPOSITION OF LEAD

The softness and dull finish of lead deposits limits their use commercially. They are not suitable for decorative purposes. As a coating material for corrosion protection, lead can only be used with some limitations. The standard potential of lead is more noble than steel, and the base material therefore cannot be protected anodically. An effective corrosion protection is only given, when the coating can be deposited in a pore-free form. The main use of lead as a coating material is for applications where lead coming into contact with those aggressive media, forms insoluble, tightly adhering layers. The preferred field of application of lead coatings is therefore the corrosion protection against sulphuric acid containing media. The surface of the lead coating is covered by a tightly adhering layer of lead sulphate, which protects the underlying metal against any further attack by the sulphuric acid. For such applications coatings with a thickness up to 200 μm are normally used. This has to be considered by making up the electrolytes. Some details on the makeup and operating parameters of the more important lead electrolytes are given in the sections that follow.

2.1 Additives

Lead is one of the group of metals that can be deposited with a low overpotential from aqueous solutions. In additive-free acidic electrolytes, lead will be deposited in a dendritic form, as seen in Figure 1. Large isolated lead crystals are formed especially around the edges of the cathode. Deposition of dense pore-free coatings is therefore possible only when the electrolyte contains special inhibitors in a sufficient quantity. Table 7 lists some organic chemicals that can be used as grain refiners in lead electrolytes. This table lists compounds published in technical literature or in patents. It can be assumed that there are several other chemicals not mentioned in publications that are used in proprietary additives. In the patent literature, in particular, that relating to methanesulphonic acid electrolytes, there are many additives listed for use for the deposition of lead, tin, and tin-lead deposition. In most of the patent applications the effect of the additives is described with reference to the tin-lead deposition.

Extensive experimentation was carried out by Graham and Pinkerton [19, 20] to select suitable additives for the lead deposition. Their findings are included in Table 7.

2.2 Anodes

Insoluble anodes cannot be used in lead plating electrolytes as lead dioxide, PbO_2, will form on the surface of the anodes. The purity of the soluble lead anodes used

Figure 1 Lead deposition from an acidic, additive free electrolyte. The electrodeposition occurs under this condition in a typical dendritic form. (SEM-picture by courtesy of Heinz Brennenstuhl, Esslingen)

determines the extent to which a film forms on the surface of the anodes. These layers are caused by the metallic impurities in the anodes, which are more noble than lead and are therefore not dissolved anodically. The typical contaminants found in lead anodes are bismuth, copper, antimony, silver, arsenic, and tin. If the level of these contaminants in the lead anodes is not too high, anode passivation is not problematic in the more common lead electrolytes (fluoroborate, fluorosilicate, and methanesulphonate based electrolytes).

Lead can also be dissolved by chemical dissolution in acidic electrolytes, which results in an equivalent current efficiency of greater than 100%. However, because of the high hydrogen overpotential, such dissolution occurs only when oxygen is available. The lead anodes are therefore immersed completely in the electrolyte to prevent dissolution. If any part of the anode is not covered by the electrolyte, chemical dissolution of the lead will occur at the interface between the electrolyte and air.

2.3 Temperature

Increasing the electrolyte temperature will increase the limiting current density and thus allow the electrolyte to be operated with a higher current density [40]. The electrolyte temperature for the electrodeposition of lead is usually between 20° to 40°C. In most common lead electrolytes an increase in the electrolyte temperature leads to a coarser grain structure [41] to a greater or less extent, depending on the grain-refiner used in the electrolyte.

TABLE 7 Additives for the Lead Deposition in Acidic Electrolytes

Compound	Electrolyte Type	Concentration ($g\,liter^{-1}$)	References
Quinone	HBF_4	10	[19]
Quinhydrone	HBF_4	10	[19]
1,4-naphthoquinone	HBF_4	<1	[19]
Hydroquinone	HBF_4	10	[19]
1,2-naphthoquinone	HBF_4	10	[19]
Eugenole	HBF_4	1	[19]
Ligninsulphonic acid	HBF_4	1	[19]
6-amino-1-naphthol-3-sulphonic acid	HBF_4	2	[19]
1,4-naphtholsulphonic acid	HBF_4	1	[19]
Coumarin + ligninsulphonic acid	HBF_4	2+1	[19]
Glue	HBF_4	1	[21]
Resorcinol + peptone	HBF_4	Each 0.23	[22]
Dibenzenesulphoneamide	HBF_4	0.5–5	[23]
Gelatine	HBF_4; MSA	0,2	[24, 27]
Ethoxilated alkylphenols	HBF_4	0.1–10	[25, 26]
Hydroxylated cellulose	MSA	2	[27]
Polyethyleneglycol	MSA	0.25	[27]
Peptone	HBF_4	1–10	[28]
Cetytrimethylammoniumbromide	$H_2N\text{-}SO_3H$	2–15	[29]
Phenol, cresol	H_2SiF_6	No exact data given	[30]

2.4 Electrolyte Agitation

Electrolyte agitation influences the maximum cathodic current density at which smooth dendrite free deposition is possible. Graham and Pinkerton [42] investigated the influence of cathode rotation on the limiting current density. During their experiments they noted that at constant rotation speed of the cathode the limiting current density can also be effected by the type of inhibitor used.

Agitation of the electrolyte is also important to keep the lead concentration in the electrolyte constant. Without agitation, the lead which is dissolved from the anode can build up at the bottom of the plating cell because of the higher density of this solution.

2.5 Troubleshooting

The electrodeposition of lead is mainly used where corrosion-resistant coatings are required. In such applications the coatings have to be dense and porefree. The most frequent failures that occur in lead plating are defects in the deposits such as nodules, "growths" or treeing. The main cause for such failures and suitable countermeasures are described in the following paragraphs.

Treeing The formation of treeing is normally seen on edges of the cathode. It can happen when there is a lack of additives in the electrolyte. A typical example of treeing is shown in Figure 1 for lead deposited from an additive-free electrolyte. Such defects can be corrected by additions of additive according to the instructions given for the electrolyte system.

Nodules or "Growths" During the deposition of thick lead coatings (up to 200 μm) formation of nodules or "growths" can occur, as seen in Figure 2 of a lead deposit from a methanesulphonate electrolyte. In this electrolyte the additive system consisted of a mixture of nonionic surfactants and hydroquinone as grain refiner.

This failure does not generally occur with a freshly made-up solution, and when it does occur, it can in most cases be rectified by a purification of the electrolyte with activated carbon. The cause for the problem is therefore probably the contamination of the electrolyte by breakdown products of the organic additives.

In practice, such failures had been seen together with a rapid decrease of the lead concentration in the electrolyte. The drop of the lead concentration from 200 to 135 g $liter^{-1}$ was parallel to a dark discoloration of the electrolyte. The anodic current efficiency was reduced, which caused the drop of the lead content in the bath. By the passivation of the anode, lead dioxide was formed on the anode surface, which caused a partial oxidation of the organic additives.

This failure mechanism can be confirmed by using insoluble anodes. On the anode surface a layer of lead dioxide forms that leads either to anodic oxidation of the organic additives on the anode surface during operation of the electrolyte,

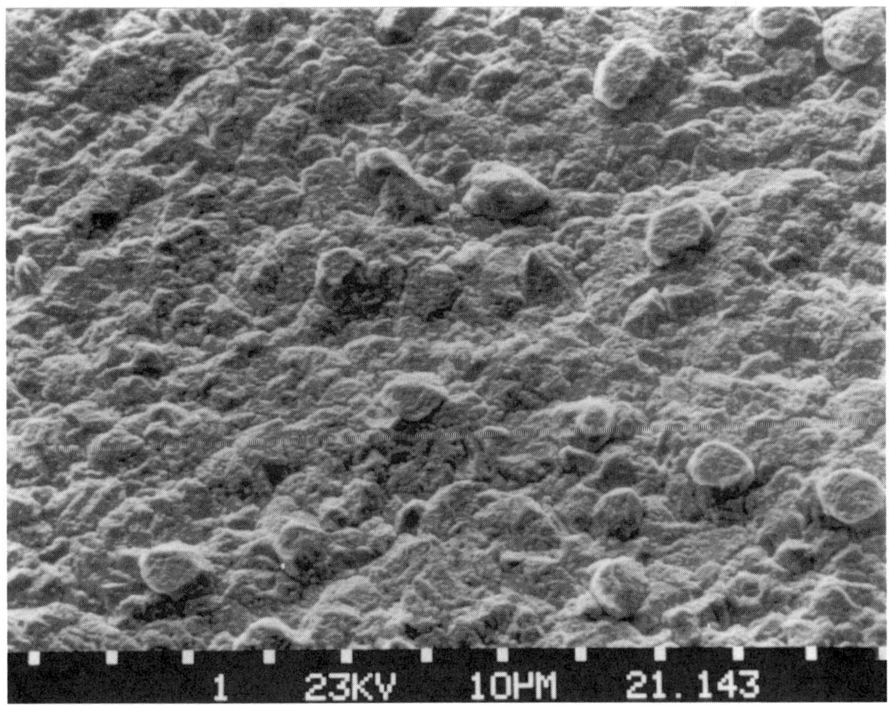

Figure 2 Lead deposition from a methanesulphonate-based electrolyte with hydroquinone as additive. The formation of dendrites and nodules can be seen. Especially when plating thick layers (< 200 μm) from such systems, the formation of nodules can be severe.

or during idle periods by chemical oxidation of the organic additives by the lead dioxide [43].

In contrast to the problem of treeing, this formation of dendrites and "growths" can normally not be corrected by more additives, since it is not caused by a lack of additives but by a contamination with breakdown products in the electrolyte. The only remedy for such problems is purification of the electrolyte by a carbon treatment.

Streaking or Step Plating Dini and Helms [44] and Wiesner and coworkers [45] referred to defects in lead plating in the form of step plating that sometimes followed the flow pattern of the solution across the plated surface. In both papers fluoroborate electrolytes were investigated in which resorcinol and peptone or hydroquinone were used as inhibitors. The experiments carried out confirmed that the problem of streaking or step plating is caused by oxidation of the resorcinol or hydroquinone. Dini observed that the failure only occurred after the electrolyte had been operated for several ampere-hours and could be corrected by a carbon treatment. The use of organic additives, which can either be modified by anodic oxidation or by the contact with the oxygen from the air, is therefore not recommended for the electrodeposition of lead. Wiesner [45] recommends a mixture of ligninsulphonic acid and coumarin.

Figure 3 Lead deposition from a methanesulphonate-based electrolyte. Opposite to the sample shown in picture 2, the additive system didn't contain hydroquinone or similar chemicals, which can be oxidised during bath ageing and thus cause the formation of a nodular deposition.

Figure 3 shows the morphology of the lead surface deposited from an electrolyte based on methanesulphonic acid. The additive system is a proprietary mixture, and it does not contain compounds like resorcinol or hydroquinone [46]. In comparison to Figure 2, which shows the surface of a lead layer plated from a MSA-based electrolyte containing hydroquinone, Figure 3 shows very little sign of dendritic deposition.

Figure 4 shows the surface of a lead layer that was deposited from a commercial fluoroborate electrolyte. Although no nodules or dendrites are seen, the comparison with Figure 3 clearly shows that a finer-grained deposit can be obtained from MSA-electrolytes with suitable additive systems.

Pores (Voids) Lead is electropositive relative to iron which means that lead cannot offer sacrificial anodic corrosion protection to steel. Therefore lead coatings used to protect steel from corrosion must be completely pore (void) free.

The number of pores in an electrodeposited layer of lead depends on the additive system and the thickness of the layer. The formation of pores can also be caused by contamination in the electrolyte, which will reduce the throwing power of the electrolyte. Such failures can be caused by contamination of nitrates or halide ions. Because nitrate is reduced at the cathode, an increase in the level of nitrate contamination leads to an reduction of the cathodic current efficiency of the lead deposition, which can eventually lead to the situation where there is no lead deposition in the low current density at all [3, 47].

Figure 4 Lead deposition from a fluoroborate-based electrolyte. The big-shaped crystal show that under this plating condition an additive system with insufficient grain-refining was used.

Contamination with chloride ions can lead to a severe reduction of coverage and widespread formation of pores. In a sulphamate electrolyte contamination with 300 to 500 mg $liter^{-1}$ chloride causes dendritic plating [48].

In one particular application, where cable shoes manufactured from steel were being plated with lead using an MSA-based electrolyte, the deposit appeared very porous, which led to a corrosion in form of red rust after only several hours [43]. The pores were caused by a chloride contamination in the electrolyte in the order of 500 mg $liter^{-1}$ which was caused by a drag-over of a hydrochloric acid pickling solution from the pretreatment of the steel parts. Such a problem would not occur if the same pretreatment sequence were used in conjunction with a tetrafluoroborate electrolyte. This is because additions of chloride to a tetrafluoroborate-based lead electrolyte leads to a precipitation of the white, crystalline compound PbClF. In the fluoride-free MSA-electrolyte such a precipitation of chloride cannot occur, so there is no "self-curing" mechanism for chloride contamination. Precipitation of chloride from an MSA-based electrolyte is possible by the addition of sodium fluoride to the electrolyte. Figure 5 shows the reduction of chloride in a MSA-based lead electrolyte in relation to the addition of sodium fluoride.

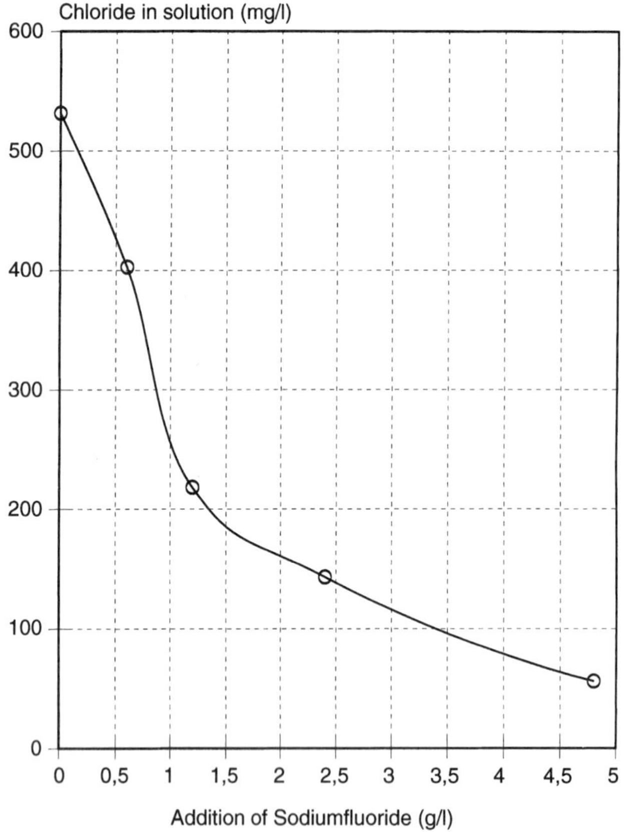

Figure 5 Removal of chloride ions from a methanesulphonate-based leadelectrolyte by addition of sodium fluoride. Chloride is precipitated as PbClF.

Discoloration Lead deposits are generally not used for decorative applications. The color and appearance of the coatings are therefore of little importance as long as the deposit meets the specified technical requirements.

Copper contamination can, however, lead to an unacceptable discoloration of the deposit. Copper that is present in the electrolyte will be codeposited with the lead during electrolysis, but this codeposition will not cause discoloration of the deposit. Discoloration of the lead deposit to a brown or black color occurs due to deposition of copper onto the lead surface by a displacement reaction that can happen in the electrolyte if the current is left switched off while the parts are still immersed, or in the rinses if they are heavily contaminated with copper.

If the plating parameters and rinsing conditions are properly controlled, lead electrolytes will seldom show copper contamination in excess of 50 mg liter^{-1}, even when copper parts are plated.

3 PROPERTIES OF ELECTRODEPOSITED LEAD COATINGS

3.1 Hardness

The data for the hardness of electrodeposited lead layers given in the literature differ considerably. Because of the creep behavior of lead, it is not easy to compare all the results. They depend on the load and the measuring time. Typical values for the Knoop hardness with a load of 25 g are in the range of 4 [49]. For a load of 5 g, the hardness is given as 7 to 8 [44].

3.2 Elongation

The elongation of electrodeposited lead coatings with a thickness of 750 μm is 50% to 52%. The elongation was calculated by measuring the change of length of the sample in a section of 1 inch around the fracture in the measurement of the tensile strength [44].

3.3 Tensile Strength

The tensile strength of electrodeposited lead layers is about 14 to 16 kg mm^{-2} and is thus comparable with that of a sample produced from the molten metal [44].

3.4 Density

The density of conventionally electrodeposited lead layers is not different to that from a molten sample, a value of 11.34 g cm^{-3} is given. Under high-speed conditions deposited layers showed a slightly lower density of 11.22 to 11.34 g cm^{-3} [50].

3.5 Electrical Resistivity

The electrical resistivity of electrodeposited lead layers is 20 to 22.9 μΩ · cm, and thus it is slightly higher than that of a sample prepared from molten lead (20.6 μΩ · cm) [50].

3.6 Corrosion Resistance

The corrosion resistance of electrodeposited lead coatings was investigated by Graham and Pinkerton [49] in atmospheric exposure tests in severe industrial, rural and maritime climates. Coatings with a thickness of 25 and 50 μm on steel panels did not show base material corrosion after three years.

To have a better comparison of the behavior of lead in the different test climates, the thickness was reduced to 6.4 μm. Even with such a thin deposit less than 20% base material corrosion was observed after 8 months in the maritime, 16 months in the industrial, and 24 months in the rural climate. The influence of the additive system on the corrosion resistance of the lead deposit was shown by using glue instead of hydroquinone as the inhibitor. The corrosion resistance of the same thickness of deposit in this case was reduced to 4, 9, and 16 months, respectively.

Deposition of thin layers of tin or iron prior to lead plating reduced the corrosion resistance compared to those samples with the lead coating directly plated on the steel. But the deposition of a thin copper layer (1.3 μm) prior to lead plating improved the corrosion resistance considerably.

4 DISPERSION PLATING

The codeposition of suspended solid particles in electrodeposited lead coatings was investigated by Pini and Weber [51], Vandervoort and coworkers [52], and Wiesner and coworkers [45]. Fluoroborate electrolytes were used for lead deposition with hydroquinone [51] or ligninsulphonic acid and coumarin [45] as the inhibitors.

Pini and Weber investigated the dispersion plating of boron carbide or titanium carbide. The particle size of the suspended solids was 3.3 to 18 μm and 20 to 50 μm for boron carbide and 20 to 60 μm for titanium carbide. With 150 g liter^{-1} each suspended material a rate of codeposition of about 20% was reached. The codeposition of the carbides doubled the hardness of the deposit. The corrosion resistance was slightly improved by the codeposition of the small boron carbide particles, but the codeposition of the bigger particles caused a significant reduction of the corrosion resistance.

Wiesner and coworkers investigated the codeposition of PbO, PbO_2, Pb_3O_4, Al_2O_3, TiO_2, and $BaSO_4$ particles. The lead oxides and the alumina agglomerated in the electrolyte. The codeposition of $BaSO_4$ and TiO_2 was possible at rates of codeposition of 0.49% for TiO_2 and 2.7% for $BaSO_4$. Codeposition of TiO_2, especially when codeposition of about 0.1% tin was also included in the deposit, improved the mechanical properties of the deposited layer considerably.

5 LEAD ALLOYS

The most important electroplated lead alloy is the tin-lead alloy. This will therefore be treated in a separate section. All the other lead alloys described in the paragraphs that follow have limited applications.

5.1 Lead-Antimony

Deposition of lead-antimony alloys was investigated from fluoroborate electrolytes with pepton and resorcinol as additives [53]. This electrolyte differed from the known fluoroborate electrolytes for the deposition of lead or lead-alloys in that the electrolyte for the deposition of lead-antimony alloys contained free fluoroboric acid and also free hydrofluoric acid. Codeposition of antimony at a concentration of between 0 and 15% increases the hardness from 8 to 21.5 and increases the tensile strength from 15 to 65 $kg\,mm^{-2}$, but the ductility severely reduced. The electrolyte composition for a lead-antimony deposition is shown in Table 8.

A tetraflouroboric acid based electrolyte with diethanolamine and tartaric acid [96] was used to investigate the improvement that could be achieved in the corrosion resistance of lead deposits against sulphuric acid by codeposition of antimony. It was established that an antimony content of 10% gave the maximum of corrosion resistance.

5.2 Lead-Bismuth

Brenner [55] makes only one reference to deposition of lead-bismuth deposition. There is only one other reference in recent literature [56] to such an electrolyte. Codeposition of bismuth was investigated to improve the corrosion resistance, and it was confirmed that the optimum alloy was found to contain 17% bismuth. The electrolyte used was the HBF_4-electrolyte with additions of Trilon B and glue.

5.3 Lead-Thallium

Thallium was investigated as a possible coating for metal bearings [57]. Attention then turned to the possibility of using lead-thallium for such applications using a sulphamate electrolyte with pepton and gelatine as additives. With a ratio of 3 : 1 lead : thallium in the electrolyte at a current density of 0.5 to 1 $A\,dm^{-2}$, deposits with 2% to 35% Tl could be achieved.

However, because the alloy composition in the deposit has a high dependency on the applied current density, such an electrolyte will only find a use in applications, where the geometry of the anode to cathode can be precisely controlled to ensure a uniform distribution of the current density. This requirement can normally be met in bearings.

TABLE 8 Composition of a Lead-Antimony Electrolyte

Pb	62 g $liter^{-1}$
Sb	11.5 g $liter^{-1}$
HBF_4	74 g $liter^{-1}$
Boric acid	20.6 g $liter^{-1}$
Pepton	0.25 g $liter^{-1}$
Resorcinol	0.25 g $liter^{-1}$

5.4 Lead-Cadmium

Investigations into the deposition of lead-cadmium alloys were made with the aim of a possible application for bearings. Lead-indium alloys are a good bearing alloy because of the high fatigue strength and corrosion resistance against oxidation products of the lubricating oils, but there is still demand for alternatives because of the high price of indium.

This led to the development of lead-cadmium alloys with a cadmium content up to 30%. Because the standard potentials of Pb (−0.126 V) and Cd (−0.403 V) are relatively close, codeposition from uncomplexed solutions should be possible [58–60]. Investigations were made into solutions based on fluorosilicates [58], fluoroborates [59], sulphamates, and a citrate-acetate electrolyte [60]. The electrochemical measurements were made using polarographic techniques and the alloy composition was measured using atomic absorption spectroscopy after dissolving the deposits in HNO_3/HCl-mixtures.

No further evaluations on the suitability of those layers as bearing metals were described in the cited papers. In the fluorosilicate electrolyte with an electrolyte composition of 220 g $liter^{-1}$ Cd, 50 g $liter^{-1}$ Pb and 6 g $liter^{-1}$ resorcinol alloys with 10% to 20% Cd were plated at a current density of 2 to 5 A dm^{-2}. It was noted that the alloy composition is very sensitive to variations in the plating parameters.

In the fluoroborate system an electrolyte composition with 220 g $liter^{-1}$ Cd and 50 to at most 75 g $liter^{-1}$ Pb deposited alloys with a maximum of 30% Cd. Resorcinol, pepton, and polyethylenglycol were tested as possible additives in this system. The temperature was kept at 30° to 60°C, and the deposition was made under potential-controlled conditions at −600 mV against SCE (corresponding to a cathodic current density of about 10 A dm^{-2}). The properties of the electrolyte changed within several days, probably because of the formation of oxidation products of the resorcinol.

The experiments using the sulphamate electrolyte, in which resorcinol at a concentration of 6 g $liter^{-1}$ was also used, gave essentially the same results. Deposition of alloys with up to 20% Cd is possible, but the alloy composition is very dependant on the plating parameters. No codeposition of cadmium was possible from the electrolyte based on citrates and acetates.

5.5 Lead-Silver

The deposition of lead-silver alloys was investigated as a possible option for coating metal bearings. Small additions of silver (1–2%) increases the stability of lead anodes in their use in sulphate-based electrolytes.

Deposition of the lead-silver alloy is made from electrolytes containing cyanides. Brenner [55] has compiled a list of lead-silver electrolytes that have been investigated. The cyanideion is a selective complexing agent for silver, so an increase in the cyanide concentration will lead to a stronger inhibition of the silver deposition. Lead is complexed in such electrolytes by tartrate.

The electrolyte formulations discussed in the literature yield deposits of approximately 10% lead. The lead content in the alloy increases with increasing current density, especially when the electrolyte contains a high concentration of free cyanide. Codeposition of lead in silver causes a considerable increase of the hardness from about 70 to 100 kg mm^{-2} to values of about 200 kg mm^{-2}. The maximum

TABLE 9 Composition of a Lead-Silver Electrolyte

Ag (as $KAg(CN)_2$)	8–16 g liter^{-1}
Pb (as $Pb(NO_3)_2$)	4–16 g liter^{-1}
KSCN	120 g liter^{-1}
K-Na-tartrate ($KNaC_4H_4O_6 \cdot 4H_2O$)	240 g liter^{-1}
NaOH	To adjust pH 9

hardness is achieved by codeposition of about 2% Pb. Table 9 shows the composition of an electrolyte for the deposition of a lead-silver alloy [61].

While silver-lead alloys with a lead content up to 10% were investigated mainly for their possible application as a coating for metal bearings, recent publications refer to lead-silver alloys with up to 10% silver for the production of lead dioxide. Such coatings can be deposited from cyanide-based electrolytes that use glycerol as a complexing agent for lead. In comparison to the electrolytes using tartrate as the complexing agent that were mentioned earlier, the glycerol-based electrolytes exhibit higher cathodic current efficiencies [62]. The typical composition for such electrolytes is about 4 mol liter^{-1} potassium hydroxide, 0.31 to 0.36 ml liter^{-1} lead, 0.031 to 0.063 mol liter^{-1} silver, 0.62 to 0.73 mol liter^{-1} free cyanide and about 60 ml liter^{-1} glycerol. A modified electrolyte formulation for the deposition of alloys with about 2.5% Ag is claimed by the same authors in a Japanese patent application [63].

Lead anodes with a silver content of about 1% are used in the electrolytic production of zinc. To save silver it is suggested that a lead-silver alloy with about 0.45 to 1% Ag and a thickness of 100 to 200 μm as the active anode surface on a lead anode will suffice [64].

Deposition of lead-silver alloys from cyanide-free electrolytes was also investigated. Electrolytes using potassium hexacyanoferrate (II), $K_4Fe(CN)_6$, and polyethylenpolyamines as complexing agents were tested [65, 66].

5.6 Zinc-Lead Alloys

Deposition of zinc-lead alloys with codeposition of about 10% lead was investigated with the aim of improving corrosion resistance in comparison to pure zinc coatings [67]. The authors investigated the deposition from alkaline zincate, plumbite, pyrophosphate electrolytes and acidic perchlorate, fluoroborate and acetate electrolytes.

Deposits from the alkaline electrolytes tended to be porous and spongy when operated at usually working temperatures (30–80°C). The deposits from the perchlorate electrolyte were dark and rough, and codeposition of zinc only occurred in traces. The optimum parameters for the acetate electrolyte are quoted in the literature as pH 3 to 3.5 and electrolyte temperature 30°C. With a ratio of 0.5 M Zn and 3.10^{-4} M Pb in the electrolyte, codeposition of 16% Pb can be achieved. Lead is present in the acetate solution in an uncomplexed form, and it is preferentially deposited as the more noble element.

The optimum parameters for the fluoroborate electrolyte are a pH of 3 to 3.5 and temperatures from 30° to 40°C. With similar molar ratios of Zn and Pb as given above, alloy codeposition of 1.9% Pb can be achieved with this electrolyte. The best results are achieved from a pyrophosphate electrolyte at a temperature of 30° to

60°C. Lead is also the more noble element in this electrolyte. To achieve the desired alloy composition of about 10% Pb, the ratio of Zn : Pb has to be about 100.

The corrosion resistance of the coatings with the different Pb-contents were tested using salt-spray tests and atmospheric exposure tests. Deposits with 9.1% Pb were inferior in all tests to pure zinc coatings. An improvement of the corrosion resistance was found for alloys with a lead content between 2.5% to 4.75%. The corrosion resistance was tested on samples without any posttreatment.

Deposition of zinc-lead deposits with up to 5.6% Pb from weakly acidic electrolyte systems (chloride, fluoroborate, or fluorosilicate electrolytes) is referred to in the literature with the use of ethoxilated or mixed ethoxilated/propoxilated alkylsubstituted phenols as additives [68]. There is one reference in the literature [68] to the use of zinc-lead alloys for coating bearings that claims an improved wear resistance when this coating is applied.

5.7 Lead-Copper

The deposition of copper-lead alloys had been investigated also as a possible coating to improve wear resistance on metal bearings. Electrodeposition of this alloy also offers the possibility to produce lead-copper alloys that cannot be produced by melting the individual metals because of limited solubility. A comprehensive list of lead-copper electrolytes has been compiled by Brenner [55].

Because of the large difference in the standard potentials between copper and lead, complexing agents are necessary in the electrolyte to bring the standard potentials closer to one other. Cyanide is a suitable complexant for such electrolytes, which reacts selectively with copper to form the complex-anion $[Cu(CN)_3]^{2-}$. Lead can be complexed by hydroxycarbonic acids or pyrophosphate. Alloys with up to 60% lead can be deposited from such systems, its composition being very dependant on the operating parameters of the electrolyte. Because such systems have not found any commercial applications, they will not be discussed in any detail in this text. For further information, refer to the literature by Brenner [55].

In the recent literature further investigation into the deposition on lead-copper alloys is described. All papers are concentrated essentially on investigation of the deposition of copper-lead alloys from cyanide-free systems. The deposition of a copper-lead alloy with a lead content up to 6% is possible in an electrolyte based on sulphamate/citrate/tartrate at pH 9 to 12 [69]. The use of ethylendiamine as a complexing agent for copper was first mentioned by Brockman and Mote [70], or perhaps Roszkowski and coworkers [71], and it has been further investigated in recent years [72–74]. Deposits with a lead content up to 40% can be deposited from such electrolytes. Deposits with a lead content up to 20% are red to yellow in color, and those with a higher lead content have a dark, matt appearance [74].

Acidic electrolytes which contain the metals in an uncomplexed form were also investigated for the deposition of copper-lead alloys (Brenner [55] also makes reference to these electrolytes). However, deposition of lead-copper alloys from acidic solutions is difficult because of the large difference in the standard potentials of the two metals. In analytical chemistry this difference is used for the quantitative separation of copper and lead. In solutions containing free nitric acid, the copper can be deposited at the cathode, while lead is deposited as lead dioxide at the anode. During the production of lead dioxide anodes, copper is very often added to the solution to prevent the dendritic deposition of lead at the cathode [75]. By additions of cetyltrimethyl ammonium bromide, Udupa [75] was successful in plating a copper-lead alloy, as long as the copper to lead ratio in the electrolyte was not too high.

In the recent literature only a few papers have been published [76–78]. There were no technically important application's found in these systems.

Electrodeposited copper-lead layers with lead contents up to 6% have a microhardness of 250 to 350 $kg\,mm^{-2}$. They are thus much harder than the corresponding alloys produced by melting the two metals together, which show a hardness of about 40 $kg\,mm^{-2}$. References have been made in the literature with regard to the superconductivity of supersaturated alloys of copper and lead [75, 76].

5.8 Lead-Indium

Lead-indium alloys have some importance as a material for coating bearings. The inclusion of up to 10% indium into a lead deposit will improve the corrosion resistance. The usual way to produce a lead-indium alloy is to plate separate layers of lead and indium and then form the alloy by heating the sample to 150°C. Layers produced by this diffusion mechanism show a gradient of the indium concentration from the outside to the inner side.

For this reason electrolyte formulations for the electrodeposition of homogeneous alloys were examined. The standard potentials for lead and indium differ by only about 0.2 V. Acidic electrolytes based on tetrafluoroboric acid, sulphamic acid, and perchloric acid as well as solutions containing complexing agents, weakly acidic, and alkaline electrolytes were all investigated. Brenner describes a number of electrolyte formulations [55]. There are some further references to such electrolytes given in more recent literature, including those electrolytes that use complexing agents such as tartrates [81], diethylene-triamine pentaacetate/gluconate [82], diethylenetriamine-pentaacetate [83], and polyethylenepolyamine [84].

Eastham compared the properties of lead-indium and the ternary lead-tin-copper alloys as wear resistant coatings for bearings [85]. He found that the lead-indium layer has a higher fatigue strength but a lower wear resistance compared with the ternary alloy.

There is still some production of lead-indium alloys for steel-backed silver-lead or lead-bronze bearings. In such applications the lead-indium layer is still produced by plating the two individual layers and diffusing [86].

5.9 Lead-Cobalt, Lead-Nickel

A U.S. patent [87] refers to the deposition of a cobalt-nickel-lead alloy onto aluminium. The function of this deposit is to enable the aluminium parts to be welded with each other or with other metals. The electrolyte used for such deposition is a citrate-based electrolyte at pH 10.5.

Investigations into the magnetic properties and the structure of cobalt-lead alloys with 20% to 70% Co were described by Surzhko and Korolinko [88]. A pyrophosphate electrolyte with tartrate and citrate as complexing agents operated at pH 9.7 to 10, and 60°C was used to deposit the alloy.

The influence of thiocyanate ions on the deposition of nickel-lead or cobalt-lead alloys was investigated by Franklin and coworkers [89]. The thiocyanate-ion

TABLE 10 Electrolyte Composition and Working Parameters for the Lead Dioxide Deposition

Lead nitrate, $Pb(NO_3)_2$	350 g liter^{-1}
Copper Nitrate, $Cu(NO_3)_2{\cdot}3H_2O$	20 g liter^{-1}
Substrate for the anode	Titanium
Cathode	Copper
Electrolyte temperature	60°C
Current density	4–8 A dm^{-2}

catalyzes the deposition of the transition metals with a corresponding higher rate of codeposition in the alloy.

6 ELECTRODEPOSITION OF LEAD DIOXIDE

Electrodeposition of lead dioxide is of commercial importance because it is the method used for the production of lead dioxide anodes. PbO_2 has a high electrical conductivity ($0.5{\cdot}10^4$ ohm^{-1}) and a high overpotential for the production of oxygen. Such anodes are therefore used in many technical applications (production of zinc by electrolysis from electrolytes containing sulphates, persulphate production, etc.).

Lead can be used as anode material. During anodic polarisation the lead surface passivates and forms lead dioxide. The lead for the reaction comes from the underlying layers of lead. If lead anodes, which contain 1% silver, are used as the starting material, the lead dioxide anodes subsequently produced will be extremely stable.

For other applications, deposition of lead dioxide can be made on suitable substrates, especially titanium [90] or graphite [91]. The valve function of titanium during anodic polarization is prevented either by deposition of thin layers of platinum or palladium or by addition of fluoride [90] to the electrolyte. The electrolyte is normally a solution of lead nitrate and copper nitrate. The addition of copper is made to prevent unwanted cathodic deposition of lead. Table 10 shows a typical electrolyte composition and the working parameters for the deposition of lead dioxide [92].

The combination of the cathode and anode reaction leads to a liberation of nitric acid. In order to keep the metals in the solution and the pH constant, the addition of lead oxide, PbO, and copper metal is recommended [92]. Investigation into the formation of stresses in electrolytically deposited lead dioxide layers were reported by Gnanasekaran and coworkers [93]. The existence of internal stresses can lead to cracks in the anodes and thus can have an adverse effect on the stability and the shelf-life of the anodes. The operating parameters of the electrolyte must therefore be carefully controlled.

7 DEPOSITION FROM NONAQUEOUS SOLUTIONS

Investigations into the deposition of lead alloys from nonaqueous solutions were made for those metals that cannot be plated from aqueous solutions because of their negative standard potential.

7.1 Depositions from Aprotic Solvents

Deposition of a wide range of lead-aluminium alloys from organic solvents is described in many papers. Alloys with a lead content of 32.6% to 99.3% have been plated from an electrolyte based on ethylbenzene/toluene with $AlBr_3$ (3 mol liter^{-1}), tetraethyl ammonium bromide (0.5 mol liter^{-1}) and lead in form of lead tartrate. The molar ratio between aluminium and lead has to be greater than 100 [94].

In an earlier publication the same authors investigated the deposition of aluminium-lead alloys from benzene, toluene, or ethylbenzene electrolytes with $AlBr_3$, $PbBr_2$, and tetraethyl ammonium bromide or KBr as conducting salts. Lead deposition is favored (90–95%). To achieve alloys with a higher aluminium-content, the lead concentration has to be limited to maximum 0.005 mol liter^{-1}, when the Al-concentration is 2 to 3 mol liter^{-1}. It is also possible to complex the lead by addition of tartrate [95].

The anodic dissolution of aluminium-lead alloys in electrolytes based on $AlBr_3$, $PbBr_2$, and KBr were investigated using cyclic voltammetry [96]. Investigation into the cathodic deposition by cyclic voltammetry shows that the potential difference between lead and aluminium in ethylbenzene electrolytes is only about 300 mV, compared to 1.5 V in aqueous solutions [97].

It is not possible to plate aluminium-lead alloys from electrolytes based on $AlCl_3$ and $LiAlH_4$ in tetrahydrofurane or diethylether, which are the electrolytes used for deposition of aluminium [95]. Capuano [98] gives a review on the electrolytic deposition of Al-alloys with specific reference to the use of plating technologies such as pulse plating and barrel plating.

7.2 Depositions from Molten Salts

Deposition of lead-calcium alloys from molten salts based on mixtures of $CaCl_2$–CaF_2–$PbCl_2$ have been described in the literature. Deposition of calcium is favored from such mixtures, since the concentration of lead in the mixture is limited to 3% to 5% by weight [99].

Production of a lead-calcium alloy by using molten lead as the cathode and depositing calcium on this from a molten salt mixture of $CaCl_2$–CaF_2–CaO at 750°C is referenced in a Japanese patent [100]. A similar reaction for the deposition of a lead-zinc alloy that involves the deposition of zinc from a mixture of LiCl–KCl–$ZnCl_2$ at 400°C has also been described in the literature [101].

Ternary alloys of lead such as K–Na–Pb, Mg–Na–Pb, or K Mg–Pb can be deposited from molten salt mixtures of Na_2CO_3, KCl and NaCl [102], NaCl/$MgCl_2$, or KCl/$MgCl_2$ using liquid lead as a cathode. The current efficiency of such deposition of sodium or potassium can be reduced by sulphate ions, probably by a reaction of the alkalimetals with sulphate to form alkalioxides and sulphides in accordance with Eq. (7):

$$8\,Na_{(liq.)} + Na_2SO_4 \rightarrow 4\,Na_2O + Na_2S \tag{7}$$

8 UNDERPOTENTIAL DEPOSITION

The study of the literature under the keywords "electrodeposition" and "electroplating" of lead shows a lot of references to underpotential deposition (UPD) of lead. Underpotential deposition is a deposition of a metal ion on the surface of a metal that is electropositive in relation to that metal. The deposition occurs at potentials that are considerably more anodic than the theoretical thermodynamic value. The energy for this process is given by a high energy of adsorption of the metal ion onto the surface of the base metal. Investigation into the UPD of lead have been made on silver or gold substrates. Besides theoretical investigations to study the phenomenon itself, experiments were carried out in an attempt to modify the catalytic properties of gold electrodes by the adsorption of lead layers for electrochemical reductions [104, 105].

The UPD of lead onto stainless steel was examined [106, 107]. It was discovered that the evolution of hydrogen on the cathode and thus the adsorption in the steel (known as hydrogen embrittlement) could be considerably reduced.

Underpotential deposition is a process which is different from the objectives of the lead deposition in the conventional meaning of electroplating. A detailed discussion of this process is beyond the scope of this text.

9 APPLICATIONS OF ELECTRODEPOSITED LEAD

The main applications of electrodeposited lead are a result of corrosion resistant properties against such corrosive chemicals like sulphuric acid, chromic acid, or phosphorus acid. In the atmosphere several corrosion products, mainly carbonates and alkaline carbonates, form protective layers that prevent the underlying metal against further attack.

A huge field of application of electroplated lead is in the production of lead accumulators. All the parts that come into contact with sulphuric acid are plated with 200 μm lead. The base material is most usually copper, but there is also reference made to the use of aluminium because of the weight reduction of the components [108].

Some new applications for lead deposition have been developed in the field of electronics and component manufacture in recent years. In PCB (printed circuit board)-manufacture layers of metal are deposited to act as etch resists. In the past bright tin deposits, which remained on the surface after copper etching, or matt tin-lead coatings, which were reflowed, were used as metal resists. The so-called copper-only-technique (SMOBC technique: solder mask over bare copper) is the standard process today. This means that the metal resists are removed after copper etching. The same electrolyte types that had been used for the deposition of the metal resists that remained on the boards still can be (and are) used for the deposition of metal resist for SMOBC techniques.

To reduce the costs of this production step, the use of pure lead layers as metal resist was proposed [109–112]. At the time when the proposals were made the application of lead compared to pure tin had a clear cost advantage. However, in the last ten years the tin price dropped considerably, and the cost advantage of lead was therefore mitigated.

From the technical point of view, the application of lead as a metal resist in comparison to tin or tin-lead alloys has the advantage that no intermetallic layers can be formed. The intermetallic layer Cu_6Sn_5 will cause problems in the subsequent hot-air-leveling process. This layer must therefore be completely removed during the stripping process. This means that stripping processes used will show slight attack on the copper surface. Since lead does not form an intermetallic layer with copper, the chosen stripper solutions must not attack the copper.

A suitable electrolyte for deposition of lead for printed circuit board manufacture is one based on methanesulphonic acid containing 25 to 50 g liter^{-1} lead. Throwing power comparable to that of conventional tin-lead electrolytes can be achieved with the use of suitable organic additives [113].

Such pure lead deposits on printed circuits are less corrosion resistant compared to pure tin or tin-lead deposits. The lead deposit can corrode when the boards are stored between the lead deposition and the subsequent manufacturing steps. If the boards are not carefully dried after the lead plating (e.g., if water residues still remain in small through holes in the board), lead corrosion can occur. Residues of the alkaline solution used to strip the dry film or dry film residues soaked with alkaline solution on the sides of the tracks can cause a corrosion of the lead deposit when the boards are stored between dry-film stripping and copper etching. This is particularly noticeable when lead deposition occurs over the dry film (mushroom effect). After copper etching, one will see a slightly etched structure of the track ("mouse-bits"). A low hardness lead surface will scratch easily, and this means that it is just as likely to develop etching voids and breaks in the track.

These disadvantages combined with the reduced cost of tin makes lead a less than attractive option for use as a metal resist in PCB manufacture. Deposition of thin undercoatings of pure lead under a tin-lead layer on reflowed PCB's improves their solderability. This process is described in Chapter 9 on tin-lead alloys.

Lead is a superconducting metal with a jump temperature of 7.2 K. The use of electrodeposited lead layers for applications in the field of superconductivity is described in the literature [114–116].

REFERENCES

1. J. O. M. Bockris, *Trans. Faraday Soc.*, **43**, 417 (1947).
2. G. Milazzo, *Electrochemistry*, Amsterdam, 1963, p. 230.
3. N. Marinkov and J. Zagorov, *Mashinostroene (Sofia)*, **17**, 113 (1968); CA 69:32425.
4. E. R. Thews, *Chem. Tech.*, **18**, 41 (1945).
5. T. Richards, *Met. Finish.*, **51**, 59, 62 (1953).
6. W. Hasenpusch, *Galvanotechnik*, **86**, 64 (1995).
7. F. Mathers and F. Forney, *Trans. Electrochem. Soc.*, **76**, 371 (1939).
8. E. Schweikher, *Am. Electroplat. Soc.*, 90 (1942).
9. R. Piontelli, *J. Electrochem. Soc.*, **94**, 106 (1948).
10. K. S. Indira and H. V. K. Udupa, *Met. Finish.*, 94 (1971).
11. L. Domnikov, *Met. Finish.*, 67 (1968).
12. G. Tremolada and L. Abduini, *Proc. Symp. on Sulfamic Acid and Its Electrometallurgical Applications*, Milan, 1966, p. 353.
13. E. R. Freni, *Proc. Symp. on Sulfamic Acid and Its Electrometallurgical Applications*, Milan, 1966, p. 367.
14. C. H. Huang, *Plating Surf. Finish.*, 64 (1964).

15. A. Betts, U.S. Patent 679,824 (1901).
16. A. Betts, U.S. Patent 713,277 (1902).
17. D. L. Thomas, C. J. Krauss, and R. C. Kerby, AIME, TMS Paper Selection A81-8, 1981.
18. W. A. Proell, U.S. Patent 2,525,942 (1950).
19. A. K. Graham and H. L. Pinkerton, *Proc. Am. Electroplat. Soc.*, **50**, 139 (1963).
20. A. K. Graham and H. L. Pinkerton, *Proc. Am. Electroplat. Soc.*, **50**, 135 (1963).
21. H. J. Wiesner, W. P. Frey, R. R. Vandervoort, and E. L. Raymond, *Plating*, 358 (1970).
22. J. W. Dini and J. R. Helms, *Met. Finish.*, **67**, 53 (1969).
23. Harshaw Chemical Co., U.S. Patent 2, 773,819.
24. A. F. Orlov and N. B. Pleteneva, *Tsvetn. Met.*, **23**, 24 (1950).
25. I. Rajagopal and K. S. Rajam, *Met. Finish.*, 51 (1978).
26. F. Elser and E. Raub, *Metalloberfläche*, **31**, 171 (1977).
27. M. Goodenough and K. J. Whitlaw, *Trans. Inst. Met. Finish.*, **67**, 44 (1989).
28. T. M. Tam, *J. Electrochem. Soc.*, **133**, 9 (1986).
29. K. S. Indira and H. V. K. Udupa, *Met. Finish.*, 94 (1971).
30. T. Richards, *Met. Finish.*, **51**, 59 (1953).
31. H. J. Wiesner, Modern Electroplating.
32. M. Jordan, *Galvano-Trommel* (company journal, Dr.-Ing. Max Schlötter), **36**, 12 (1984).
33. *Gmelin's Handbook of Inorganic Chemistry*, **47**, B2.
34. T. L. Rama Char, *Electroplat. Met. Finish.*, **10**, 347 (1957).
35. V. Sree, *J. Sci. Ind. Res. (India)*, **A18**, 478 (1959).
36. M. A. Loshkarev and Dýachenko, *J. Appl. Chem. USSR*, **37**, 79 (1964).
37. N. V. Gudovich and O. K. Kudra, *Ukr. Khim. Zh.*, **27**, 121 (1961).
38. N. P. Fedotév, B. P. Artamonov, and N. I. Rasmerova, *J. Electrodepositors' Tech. Soc.*, **13**, Paper No. 11, 6 (1937).
39. F. C. Mathers, *Metal Finishing Guidebook-Directory*, 27th ed., Westwood, NJ, 1959, 353.
40. A. K. Graham and H. L. Pinkerton, *Plating*, **49**, 1071 (1962).
41. T. Richards, *Met. Finish.*, **51**, 59 (1953).
42. A. K. Graham and H. L. Pinkerton, *Trans. Inst. Met. Finish.*, **40**, 249 (1963).
43. Unpublished investigations, Dr.-Ing. Max Schlötter, Geislingen.
44. J. W. Dini and J. R. Helms, *Met. Finish*, 53 (1969).
45. H. J. Wiesner, W. P. Frey, R. R. Vandervoort, and E. L. Raymond, *Plating*, 358 (1970).
46. Lead Bath MSN-10, Dr.-Ing. Max Schlötter.
47. F. A. Lowenheim, *Electroplating*, McGraw-Hill, New York, 1978, p. 325.
48. N. B. Pleteneva and T. V. Globa, *Tsvetn. Met.*, **27**, 53 (1954).
49. A. K. Graham and H. L. Pinkerton, *Plating*, 367 (1967).
50. W. H. Safranek and C. H. Layer, *Trans. Inst. Met. Finish.*, **53**, 121 (1975).
51. G. C. Pini and J. Weber, *Oberfläche-Surf.*, **19**, 54 (1978); 9th World Congress on Metal Finishing Abstracts, Amsterdam, 1976.
52. R. R. Vandervoort, E. L. Raymond, H. J. Wiesner, and W. P. Frey, *Plating*, 362 (1970).
53. J. W. Dini and J. R. Helms, *J. Electrochem. Soc.*, **117**, 269 (1970).
54. V. V. Bogoslovskii, K. M. Tyutina, and L. T. Kudryavtsev, *Zash. Met.*, **11**, 499 (1975).
55. A. Brenner, *Electrodeposition of Alloys*, Academic Press, New York, 1963.
56. K. M. Tyutina and G. G. Svirshchevskaya, *Tr. Mosk. Khim.-Tekhnol. Inst.*, **81**, 130 (1974).

57. C. P. S. Johal, D. R. Gabe, and D. R. Eastham, *Surf. Technol.*, **35**, 181 (1988).
58. D. Eyre, D. R. Gabe, and D. R. Eastham, *Trans. Inst. Met. Finish.*, **62**, 113 (1984).
59. D. Eyre, D. R. Gabe, and D. R. Eastham, *Plating Surf. Finish.*, 74 (1985).
60. D. Eyre, D. R. Gabe, and D. R. Eastham, *Trans. Inst. Met. Finish.*, **63**, 22 (1985).
61. I. Krastev, M. E. Baumgärtner, and C. J. Raub, *Galvanotechnik*, **86**, 731 (1995).
62. M. Ueba and A. Watanabe, *Kagaku Gijutsu Kenkyusho Hokoku*, **86**, 177 (1991); CA 116:64568.
63. M. Ueba and A. Watanabe, JP 61, 12890; CA 105:10515.
64. G. Haberstroh, EP 194321; CA 105:234518.
65. A. M. Kolomoets and F. I. Kukoz, USSR 386032; CA 80:221641.
66. F. I. Kukoz, A. M. Kolomoets, and V. P. Kolomoets, *Elektrokhim. Osazhdenic. Primen. Pokrytii. Dragotsennymi Redk. Met.*, 33 (1972); CA 80:115373.
67. J. P. G. Farr and S. V. Kulkarni, *Electrodeposition Surf. Treat.*, **3**, 307 (1975).
68. N. Doi, K. Obata, and T. Sonoda, JP 60 152693; CA 104:138265n.
69. R. Piontelli, P. Cavalotti, and L. Giuliani, *Proc. Symp. on Sulfamic Acid and Its Electrometallurgical Applications*, Milan, 1966, p. 329.
70. C. J. Brockman and J. H. Mote, *Trans. Electrochem. Soc.*, **73**, 371 (1938).
71. E. S. Roszkowski, H. R. Hanley, W. T. Schrenk, and C. Y. Clayton, *Trans. Electrochem. Soc.*, **80**, 235 (1941).
72. J. Gala, A. Budniok, and P. Rehlich, *Rudy Met. Niezelaz.*, **24**, 525 (1979).
73. J. Gala and R. Kopiec, *Metall.*, **35**, 882 (1981).
74. J. Gala, A. Budniok, and P. Rehlich, PL 115265; CA 99:79041q.
75. H. V. K. Udupa, K. C. Narasimham, and P. S. Gomathi, *Plating Surf. Finish.*, 1150 (1975).
76. Y. M. Polukarov and V. V. Grinina, *Primen. Elektrokhim. Pokrytii. Splavami Kompoz. Mater. Prom-sti. Mater. Semin.*, 1981, 50; CA 99:29938s.
77. V. M. San'kov, A. N. Latyshov, and V. A. Evgrafov, USSR 503941; CA 84:128070z.
78. V. V. Kuznetsov, V. P. Grigorév, O. A. Osipov, and S. P. Shpan'ko, USSR 490869 CA 84:51524r.
79. Ch. J. Raub and E. Raub, *Z. Physik*, **186**, 310 (1965).
80. E. M. Savitskii, J. V. Jefimov, Ch. J. Raub, H. R. Khan, T. M. Frolova, and G. T. Omarova, *J. Less Common Metals* (1982).
81. Y. P. Perelygin, *Sashch. Met.*, **26**, 683 (1990).
82. W. J. Waterman, U.K. Patent Appl. 2039298; CA 94:92622w.
83. W. J. Waterman and B. J. Woolford, Ger. Offen. 2415653; CA 84.51510h.
84. A. V. Ryabchenkov, A. A. Gerasimenko, and M. P. Krivoruchko, USSR 305205; CA 75:104543u.
85. D. R. Eastham, *J. Eng. Gas Turbines Power*, **115**, 706 (1993).
86. Indium Corporation of Europe, private communication.
87. K. F. Dockus, U.S. Patent 4,028,200; CA 87:45974h.
88. O. A. Surzhko and V. A. Korolenko, *Tr. Novocherk. Politekh. Inst.*, **322**, 52 (1976).
89. T. C. Franklin, J. Chappel, R. Fierro, A. I. Aktan, and R. Wickham, *Surf. Coat. Technol.*, **34**, 515 (1988).
90. GB 1189183 (BASF, 1966).
91. F. D. Gibson, U.S. Patent 2,945,791 (1958).

92. P. Krishnaswamy, E. G. Usha Bhai, and Y. Koteswara Rao, *Plating Surf. Finish.*, 70 (1985).

93. K. S. A. Gnanasekaram, K. C. Narasimham, and H. V. K. Udupa, *Electrochimica Acta*, **15**, 1615 (1970).

94. M. Galova, L. Lux, and K. Zupcanova, *Zb. Ved. Pr. Vys. Sk. Tech. Kosiciach*, 199 (1988).

95. M. Galova, K. Zupcanova, and L. Lux, *Hutn. Listy*, **39**, 340 (1984).

96. M. Galova and L. Lux, *Chem. Pap.*, **42**, 457 (1988).

97. M. Galova, L. Lux, and K. Zupcanova, *Chem. Pap.*, **42**, 281 (1988).

98. G. A. Capuano, *J. Electrochem. Soc.*, **138**, 484 (1991).

99. J. P. Millet, H. Pham, G. Pourcelly, and M. Rolin, *Ext. Abstr., Meet.—Int. Soc. Electrochem.*, **30**, 228 (1979).

100. J. Tsuruki and Y. Yoshida, JP 79,162619; CA 92:219622n.

101. M. Kamaludeen, K. Balakrishnan, G. Singh, and N. S. Rawat, *Bull. Electrochem.*, **4**, 491 (1988).

102. S. A. Zaretskii, V. Busse-Macukas, F. I. L'vovich, and A. G. Morachevskii, *Ref. Zh., Khim.* 1971, Abstr. No. 14L331; CA 77:55619d.

103. F. I. L'vovich, V. Busse-Macukas, A. G. Morachevskii, and V. I. Markin, *Zh. Prikl. Khim.*, **47**, 1654 (1974); CA 81:130168u.

104. G. Kokkinidis, A. Papoutsis, and G. Papanastasiou, *J. Electroanal. Chem.*, **359**, 253 (1993).

105. A. Papoutsis and G. Kokkinidis, *J. Electroanal. Chem.*, **371**, 231 (1994).

106. B. N. Popov, G. Zheng, and R. E. White, *Proc. AESF Annu. Tech. Conf.*, **80**, 809 (1993).

107. G. Zheng, B. N. Popov, and R. E. White, *J. Electrochem. Soc.*, **140**, 3153 (1993).

108. R. L. Seth, *Electroplat. Met. Finish.*, 5 (1972).

109. W. Tastl, *Galvano-Trommel* (company journal, Dr.-Ing. Max Schlötter), **37**, 9 (1985).

110. Anon., *Galvanotechnik*, **75**, 680 (1984).

111. Anon., *Leiterplatten*, **4–5**, 12 (1984).

112. E. Michel, *Oberfläche-Surf.*, **26**, 365 (1985).

113. G. Strube, *Galvano-Trommel* (company journal, Dr.-Ing. Max Schlötter), **37**, 5 (1985).

114. E. Lüders, *Abstracts of 3. Ulmer Gespräch*, German Society for Surface Technology, 1989, 79.

115. R. A. Alvaret, D. Birx, D. Byrne, M. Mindonca, and R. M. Johnson, *IEEE Trans. Magnetics*, **17**, 935 (1981).

116. J. R. Delayen, G. J. Dick, and J. E. Mercereau, *IEEE Trans. Magnetics*, **17**, 939 (1981).

9 Electrodeposition of Tin-Lead Alloys

MANFRED JORDAN

INTRODUCTION

The deposition of tin-lead alloys has the most technical importance among all tin or lead alloys. Processes for the electrolytical deposition of tin-lead alloys are known since 1921 and were first developed for the coating of naval torpedos [1, 2]. A second field of application evaluated was the deposition of tin-lead alloys for bearings [3, 4, 5]. Intensive research in the field of tin-lead plating has been stimulated by the different requirements in the production of electronic components such as PCB (printed circuit board), semiconductors, connectors, and the like. A further impulse has come since about 1980 with the development of fluoride free tin-lead electrolytes.

The standard potentials for tin and lead are very close together (Pb : −0.126 V; Sn : −0.136 V). Alloy deposition is therefore possible in all alloy compositions. Both metals have a high hydrogen overvoltage, so deposition of tin-lead alloys is possible from strong acid solutions without complexing agents with high current efficiencies. The current efficiency can be reduced by adding a range of different types of organic additives to such electrolytes as brightener systems.

1 ELECTROLYTE SYSTEMS

Electrolyte systems based on tetrafluoroboric or methanesulphonic acids are mainly used for the deposition of tin-lead alloys. For special applications weakly acidic systems with special complexing agents are also on the market. Electrolytes can also be divided into those for the deposition of matt tin-lead coatings and for the deposition of bright deposits. A further divison can then be made into processes for conventional plating such as rack and barrel plating and those for high-speed applications such as wire-plating, reel-to-reel processes for connectors, and continuous processes for IC-leadframe finishing.

Modern Electroplating, Fourth Edition, Edited by Mordechay Schlesinger and Milan Paunovic.
ISBN 0-471-16824-6

2 ALLOY COMPOSITIONS

Although tin-lead alloys can form over the entire compositional range, they tend to fall into three classes in terms of technical applications:

- High lead alloys
- Eutectic or near eutectic alloys
- Alloys with 5% to 15% Pb

In the first category are alloys with around 93% Pb and 7% Sn that are used as coatings for fuel tanks and on various bearings. For bearing applications there is also usually a codeposition of about 2% copper. The deposition is in a form of a matt coating. In this category fluoroborate electrolytes are still the most frequently used systems.

Eutectic alloys are those that possess the lowest melting point, which makes them suitable for solderable coatings. The eutectic alloy contains 63% tin. A considerable field of application has been the production of PCB using the metal resist technique. When the electrodeposited tin-lead layer is used as a reflowed coating, which remains on the board, an alloy composition very close to the eutectic composition is important so that the boards are reflowed with the lowest possible temperature. The main function of the reflow-process is to eliminate the overhangs of tin-lead, which are caused by underetching during the copper-etching. By reflowing the electroplated tin-lead layer this is completely converted into a liquid form. The overhangs are thus removed. In the liquid state the tin-lead is covering the side-walls of the tracks and provides thus an additional corrosion protection.

The reflow-process can be made by immersing of the boards into a heatbath of synthetic or natural oils above the melting point of the tin-lead alloy or by heating the boards by IR-radiation.

The process causes a high thermal stress to the boards. Because of this and also other disadvantages this process has been substituted by other techniques. A detailed description is given in [11].

The importance of the reflow process has diminished in recent years. When tin-lead is used as a metal resist, the alloy composition is no longer so important. For reflow processes matt-depositing electrolytes are used because, in general, bright coatings cannot be reflowed as the codeposited organic additives will cause a dewetting. Eutectic bright tin-lead coatings are used in some applications for electronic components. The lower melting point of the coating can be beneficial for reflow soldering processes.

Alloys in the third category, with 5% to 15% Pb, are mainly used for coating electrical or electronic assemblies. In the past many of these components had been plated with pure tin. Because of the risk of formation of whiskers, the codeposition of some percent of lead became almost a standard for such applications. Electrolytes to deposit low lead-containing alloys are now the most widely used tin lead electrolytes. Because of growing environmental and health concerns, in the future the use of lead will be reduced to a minimum, and the application of low lead-containing alloys will further increase. Both matt and bright plating systems are available for the deposition of low lead-containing alloys. The decision on which deposit to use depends on technical requirements, specification limitations, and the properties of the coatings.

3 BATH COMPOSITIONS

Because their deposition potentials are so close to one another, the ratio of metal ions in the electrolyte can, to a first approximation, be taken as the same as the ratio of the two metals in the electrodeposited alloy. The proportion of either metal in the electrolyte is thus expressed as

$$\%\mathrm{Sn} = \frac{\mathrm{Sn}(\mathrm{g\,liter}^{-1})}{[\mathrm{Pb}+\mathrm{Sn}](\mathrm{g\,liter}^{-1})} * 100 \tag{1}$$

The slightly more positive electrodeposition potential of lead favors deposition of this metal, especially at low current densities. The deposition potential difference between the two metals can be increased somewhat by the use of certain organic additives. Under certain circumstances the order of deposition potentials can even be reversed, making tin the more noble metal. By means of such manipulations, Pb/Sn-alloys of the same composition can be deposited from baths with quite different Pb and Sn ionic ratios, and a change in the type of organic additive used can be reflected in the composition of the electrodeposited alloy [6].

The total metal ion concentration range from 10 g liter^{-1} for barrel plating systems up to 100 g liter^{-1} for high-rate deposition processes. Electrolytes for deposition of thick deposits of 93% Pb, 7% Sn may contain up to 150 g liter^{-1} total metals. These levels may appear high compared with, say nickel or copper plating solutions. However, the atomic weights of tin (118.7) and lead (207) are very high in comparison to copper (63.5) and nickel (58.7) and comparisions on molar basis would be more valid.

3.1 Tetrafluoroboric Acid Electrolytes

Until about 1980 tetrafluoroboric acid baths were the most widely used for tin-lead alloy deposition. The basic components, tetrafluoroboric acid, tin and lead fluoborate salts, are not known in their pure form and are commercially available as their aqueous solutions. These are usually 50 wt% for the free acid, 27 wt% for the lead salt, and 20 wt% for the tin fluoroborate. Typical specifications for these are shown in Table 1 [7].

Tetrafluoroboric acid for use in tin-lead plating can be treated with a lead salt to remove any sulphate present by precipitation as the lead sulphate. This may give rise to a slight excess of lead ions in the acid. Although this is not a problem when operating a tin-lead plating bath, it may be a problem if the acid is to be used for deposition of other metals.

As a rule, fluoroborate tin-lead plating bath are operated with addition of 10 to 25 g liter^{-1} boric acid (H_3BO_3). The fluoroboric acid has the tendency to hydrolyse, releasing free HF as shown below:

$$\mathrm{HBF_4 + H_2O \rightarrow HBF_3OH + HF} \tag{2}$$

Boric acid reacts with so formed free HF to regenerate the fluoroboric acid:

$$\mathrm{4HF + H_3BO_3 \rightarrow HBF_4 + 3H_2O} \tag{3}$$

TABLE 1 Typical Compositions of Fluoroborates

	Colorless Clear Liquids			
Appearance	HBF_4, Technical Grade	HBF_4, Special Grade	$Pb(BF_4)_2$	$Sn(BF_4)_2$
Density, D_{20}	1.40	1.40	1.78	1.65
Composition (min)	49.5%	49.5%	27.2% (as Pb) 470 g liter^{-1} Pb	20.3% (as Sn) 335 g liter^{-1} Sn
Boric acid	max 2.5%	max 2.5%	max 2.5%	max 2.5%
Fe	$\leq$0.01%	$\leq$0.005%	$\leq$0.005%	$\leq$0.001%
Cu		$\leq$0.001%	$\leq$0.002%	$\leq$0.001%
Ni		$\leq$0.001%	$\leq$0.002%	$\leq$0.001%
Zn		$\leq$0.001%	$\leq$0.002%	$\leq$0.001%
Chloride		$\leq$0.005%	$\leq$0.005%	$\leq$0.005%

This reaction prevents the precipitation of the sparingly soluble lead fluoride. However, fluoride ions can also serve to retain tin in solution in the form of its Sn(IV) hexafluorostannate ion

$$Sn^{4+} + 6F^- \rightarrow [SnF_6]^{2-} \quad (4)$$

Thus fluoroborate electrolytes containing boric acid close to its solubility limit are more prone to form sludges based on Sn(IV) species than solutions with a lower boric acid concentration, which then split off fluoride ions to complex four-valent tin.

In the third edition of *Modern Electroplating* Wiesner [8] provides solution compositions for various tin-lead alloys. Compositions are given here for high lead-containing alloys with a maximum tin content of 60%. Additives recommended include gelatine or glue, resorcinol or peptone.

Table 2 lists typical bath formulations for deposition of various Sn-Pb alloy compositions including high tin-containing alloys that have been published in the recent literature. The two different versions for tin-lead electrolytes with high throwing power show the influence of the additives on the throwing power. With conventional peptone a very high acid concentration is necessary to achieve good results. The second bath uses a proprietary additive system based on nonionic wetting agents, and this electrolyte can be used with much lower acid concentration.

3.2 Alkylsulphonic acid electrolytes

Electrolyte systems based on alkylsulphonic acids for the deposition of tin-lead alloys have been developed about 1980. In principle, a number of alkylsulphonic acids are suitable for tin-lead electrolytes, such as methanesulphonic acid, alkylsulphonic acids with more than one carbon atom in the chain (ethane-, propane-, butanesulphonic acid), β-hydroxysubstituted derivatives like hydroxyethane-, 2-hydroxypropanesulphonic acid, or alkyldisulphonic acids such as methanedisulphonic acid. Of all these possible derivatives, methanesulphonic acid is by far the most commonly used. Tin- and lead methanesulphonates are highly soluble in water. For subsequent effluent treatment, the alkyl side-chain should be as short as possible, to keep the COD (chemical oxygen demand) value of the solution to a minimum. Obviously methanesulphonic acid best matches this criterion. Methanedisulphonic acid would give even lower COD-values, but this type of acid is not commercially available in sufficient quantities. Alkylsulphonic acids are strong acids, comparable with sulphuric acid and they provide electrolyte solutions with very high conductivity. Conductivity of methanesulphonic acid is plotted in Figure 1 [11]. As with sulphuric acid, conductivity increases sharply with concentration, it reaches a maximum and then falls more slowly, which is reflecting a supression of dissociation at the high concentrations. From Figure 1 it can be seen that a concentration of up to 400 g liter^{-1}methanesulphonic acid could be used in electrolyte applications, but for economic reasons one will seldom go higher than 200 g liter^{-1} (calculated as 100% MSA).

Likewise, as for fluoroborate-based electrolytes, the ratio of the metal ions in the methanesulphonate solutions can, as a first approximation, be taken as the same as the ratio of the two metals in the electrodeposited alloy. Typical electrolyte formulations for methanesulphonate electrolytes are given in Table 3.

TABLE 2 Solution Compositions for Tin-Lead Fluoroborate Electrolytes

Alloy Composition	HBF_4 (g liter^{-1})	Sn (g liter^{-1})	Pb (g liter^{-1})	Temperature (°C)	Current Density (A dm^{-2})	References
93/7 Pb/Sn rack and barrel	15–30	10–20	195–239	21–38	2–7	[9]
60/40 Sn/Pb rack and barrel	100–150	53–60	23–30	21–29	2.5–3.5	[9]
60/40 Sn/Pb high throw	350–500	12–20	8–14	21–38	1.5–2.5	[9]
60/40 Sn/Pb high throw	50–100	20–30	9–14	20–60	0.5–2	[10]
90/10 Sn/Pb rack and barrel	150–200	70–80	8–12	21–38	0.1–8	[9]

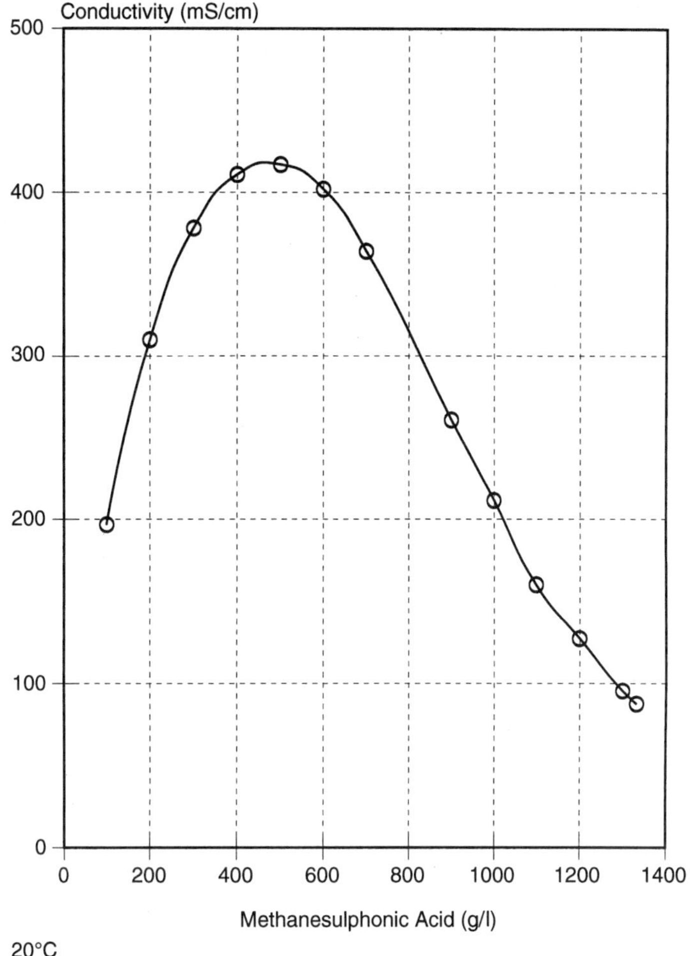

Figure 1 Electrical conductivity of methanesulphonic acid solutions.

3.3 Tin-Lead Deposition from Moderate pH Solutions

Apart from strongly acidic solutions there are some applications of electrolytes for tin-lead deposition, which are operated in a moderate pH range, mainly between 4 to 9. At these pH values tin is only soluble in the presence of complexants. Commonly used complexing agents are the different types of hydroxycarbonic acids, like gluconic, citric, tartaric acid or diphosphates.

Compared with acid type baths, these near-neutral electrolytes show no significant advantages. Because they contain complexing agents, the waste water treatment is more critical. The electrical conductivity of these solutions is also lower than that of acid types, which increases the energy required for the deposition. In general, because of the lower conductivity, such systems can only be operated at lower current densities compared with acidic electrolytes.

In certain cases, the neutral solutions can have advantages, especially when plating electronic components incorporating ceramics or special types of glass. These

TABLE 3 Solution Compositions of Tin-Lead Methanesulphonate Electrolytes

Alloy Composition	MSA (g liter^{-1})	Sn (g liter^{-1})	Pb (g liter^{-1})	Temperature (°C)	Current Density (A dm^{-2})	References
90/10 SnPb rack and barrel	200–250	17–25	2–4	20–50	0.1–25	[9]
90/10 Sn/Pb high speed	100–200	35–70	3–7	20–60	5–25	[9]
60/40 Sn/Pb rack and barrel	200–250	12–20	6–10	21–29	0.1–25	[9]
60/40 Sn/Pb high speed	100–200	35–70	12–25	20–60	5–25	[10]
93/7 Pb/Sn rack and barrel	30–50	4–5	56–72	21–29	0.1–4	[9]

TABLE 4 Parameters of Tin-Lead Electrolytes Operating at Moderate pH Values

Complexing Agent	pH Range	Temperature (°C)	Current Density ($A\,dm^{-2}$)	References
Citric acid; ammoniumtartrate	6			[14]
Aliphatic or aromatic sulphocarboxylic acids	3–8.5	20–30	0.5–3	[15]
Gluconate, citrate, malate, tartrate, malonate	1.5–5.5	25	0.5–1	[16]
Gluconate				[17]

can be sensitive to acid or alkali chemical attack which may result in acidic or alkaline media. Table 4 gives some bath compositions with their operating data for the tin-lead deposition from moderate pH electrolytes.

4 ADDITIVES

In general, the same organic additives can be used for fluoroborate- or methane-sulphonate electrolytes, but it can be seen in many cases that the same type of additive is more effective in the methanesulphonate systems and requires a lower dosage.

For the deposition of matt-finish surfaces from fluoroborate baths, the earliest addition agents to be described were combinations of bone glue and resorcinol [18], bone glue, gelatine and resorcinol [19], or peptone as a single additive [20] (peptone remains widely used to this day). These additives are not used in methanesulphonate electrolytes. Bulwith [21] reported in 1988 about defects associated with incorporation of peptone. Peptone is a protein and, as such, not very stable in acid solutions, so the peptide linkages can become split in this medium. The breakdown products build up in solution, and periodic treatment with activated carbon is necessary. Since some of these breakdown products are quite malodorous, conditions for operating staff can become unpleasant. Hobson [22] has examined this topic in detail. Possible alternatives to peptone are the nonionic surfactants, used in conjunction with Janus green or phenolphthalein [23].

In recent patents, mainly those dealing with modern methanesulphonate systems, additive combinations consisting of different types of nonionic and cationic surface active components together with different types of grain refiners are mentioned. The nonionic surface active compounds are mostly ethoxilated derivatives of linear or branched aliphatic alcohols, alkylsubstituted phenols or ethoxilated naphthols.

For deposition of fully bright deposits, special brighteners are necessary. These systems also normally contain different types of nonionic wetting agents. In these systems they have a double function: They inhibit the formation of nodules, and they also act as a solubilizer for the brighteners, which in most cases are not soluble in water. Typical brighteners for tin-lead depositions are aromatic aldehydes or aromatic ketones, which are insoluble in water if they do not have a functional group, like a sulpho group, in their molecule, which provides a water solubility. The water insoluble brighteners are emulsified by the nonionic wetting agents.

TABLE 5 Troubleshooting for Lead-Tin Electrolytes

Defect	Possible Cause	Remedy
Dendritic treelike growth	Deficiency of organic additives, esp. surfactants	Addition of additives
	Limiting current density exceeded by	
Rough dark deposits in high current density region ("burning")	Too high cathodic current density	Reduce current density
	Too low metal content	Increase metal content
	Bath temperature too low	Increase temperature
	Insufficient agitation	Increase work movement or bath circulation
Rough deposits over the entire current density range	Additive concentration too low	Increase additive concentration stepwise
	Suspendend matter in the bath (e.g., anode slimes or lead sulphate)	Check filtration, anode bags locate source of sulphate contamination
	Organic impurities	Activated carbon treatment necessary
	Insufficient coverage power due to	
No deposition at low current density region	Chloride impurities	Locate source of chloride ions; precipitation of chlorides is difficult; bath dilution may be the only answer
	High bath temperature	Cool the bath, perhaps reduce bath loading
	High metal content	Dilute electrolyte, perhaps use insoluble anodes
	Low concentration of additives	Increase additive concentration stepwise
	Contamination by nitric acid	Remove nitric acid containing pretreatment; nitrate can partly be removed by dummying the bath
Poor metal distribution	High metal concentration	Dilute electrolyte or use insoluble anodes were possible
	Low acid concentration	Add acid
	Low additive concentration	Increase additive concentration

Gassing at the cathode (in some electrolytes this can be a normal reaction due to additive system)	Exceeding limiting current density, see second entry	See second entry
	Excess brightener concentration	Add surface active compounds; dummy bath where possible
Gassing at the anode	*Anode passivation (in lead-tin electrolytes unusual)*	
	Anode bag clogged	Check anode bags
	Copper contamination in the anode	Check anode quality; prevent copper contamination in the bath, which can lead to chemial deposition of copper on the lead-tin anodes
	Anode current density too high	Monitor anode surface area
Uneven, patchy plating; blistering of deposit	Insufficient pretreatment of the base material	Check degreasing and activating solutions
Step plating especially at regions of high electrolyte flow	Overdosing of brightener	Check brightener concentration, reduce brightener by dummy plating; add surfactants
Brown stains on bright lead-tin coatings in low current density region	Copper contamination	Remove by plating out at low current density; locate source of copper contamination
Flaking of deposits (normally only with bright deposits)	Overdosing of brightener	Add surfactants

All modern electrolytes are based on proprietary additive systems. It is therefore not possible to give recommendations for the maintenance. This can be made only as per instruction of the supply houses. Some general information on trouble-shooting is given in Table 5.

Besides these grain-refiners and brighteners tin-lead electrolytes contain in most cases antioxidants to prevent the formation of stannic tin in the electrolyte. Although this formation of stannic tin is normally not harmful for tin-lead deposition, it is an unwanted reaction because it causes a loss of tin from the bath as the Sn(IV) precipitates in form of stannic oxide; therefore the formation of stannic tin must be prevented in the plating reaction.

In acidic tin electrolytes based on phenolsulphonic acids prevention of the oxidation of tin is well known [24]. The actual antioxidants are phenolic compounds, which are present as by-products in technical grades of phenolsulphonic acid. It is therefore obvious that phenolic compounds themselves are also used as antioxidants [25]. Besides their function as antioxidants, such compounds are also active as grain refiners; they are also often used in pure lead electrolytes.

Chi Pong Ho [26] describes a new system of prevention of the oxidation of Sn(II) to Sn(IV). It depends on the use of reducing agents based on vanadium pentoxide in divalent tin sulphate-sulphuric acid solutions to limit the sludge formation. This reaction mechanism was also investigated in MSA-based solutions to prevent the formation of tin-sludge especially in the presence of high contaminations of iron [27].

5 ANODES

5.1 Soluble Anodes

Normal practice is to use anodes with the same composition as that which is desired for the deposit. These anodes dissolve anodically in fluoroborate- and methanesulphonate electrolytes even at high current densities in an uniform manner without passivation. In the electrolyte, a modest rate of oxidation of tin (II) must always be assumed. This loss of tin in its depositable form can, to some extent, be countered by using anodes with a slightly higher tin content than what actually is required in the deposit. However, increase in lead concentrations in solution should be avoided as much as possible, and it is often necessary to add tin salts to maintain the correct balance. Depending on bath formulation used, around 2 to 10 g $liter^{-1}$of tin (in salt form) should be added for each excess g $liter^{-1}$ Pb. While such measures maintain the correct Pb : Sn ratio in solution, the total metal content will increase, and metal distribution (throwing power) may suffer.

It is possible to operate continously using pure tin anodes, with additions of lead salts to the solution as required. However, since the anodic dissolution of tin is 100% efficient while its deposition efficiency at the cathode is lower (since part of the current is used to deposit lead ions), a progressive buildup of tin in solution may occur. This method of operation is only realistic for alloys with up to 5% Pb because drag-out losses and tin (II) oxidation serve to keep the tin in solution more or less constant.

Under normal operating conditions a black slime will form on the anodes. This is caused by the metallic impurities in the anode material. Most commonly found

TABLE 6 Tolerable Metallic Contaminations in Anodes

Metallic Contaminant	Maximum Permissible Contamination (%)
Antimony	0.03
Arsenic	0.03
Bismuth	0.01
Copper	0.02
Iron	0.01
Aluminium + cadmium + zinc	0.002

impurities such as Sb, Bi, Cu, and As are not dissolved anodically, as they are more noble than Pb or Sn. Above a certain impurity concentration, especially for antimony, a dark spongelike coating forms on the anode surface. Should this impurity pass into the electrolyte, the result will be rough deposit on the cathode. Use of anode bags made of an acid-resistant material is thus strongly recommended. Table 6 lists the tolerable metallic contaminations for tin-lead anodes recommended for use in tin-lead electrolytes.

Anodes are used in forms of rectangular or oval shapes. The anode hooks must be made of acid-resistant material. For fluoroborate systems normally polymer-coated hooks are used. One of several advantages of methanesulphonate electrolytes is the stability of titanium in those systems, which can therefore be used as material for anode hooks. It is also possible to use titanium baskets filled with Pb/Sn-balls or chunks. This form is normally used in high-speed installations because it allows an easy maintenance and anode control. The stability of titanium as an anode hook or basket is very sensitive to traces of fluorides, which attack the passive layer of titanium oxides. Any contamination has therefore to be avoided carefully. When, for example, a MSA-based electrolyte is installed in a line that previously contained a fluoroborate, it is very difficult to remove all traces of fluoride from the tank and equipment.

5.2 Insoluble Anodes

The use of insoluble anodes is not very common in tin-lead plating. In this case in the anodic reaction oxygen is formed by the reaction

$$2H_2O - 4e \rightarrow O_2 + 4H^+ \tag{5}$$

The first prerequisite to enable insoluble anodes to be used is a very effective antioxidant to prevent the formation of stannic tin by the anodically produced oxygen, the second is a stability of the additives themselves against anodic oxidation.

When using insoluble anodes the free acid concentration increases in the electrolyte, Eq. (5). To compensate for this increase, Sn^{2+} is either replenished by stannous oxide or in an electrolytic cell with separate anode- and cathode compartments. In both reactions the hydrogen ions, produced in the anodic reaction (5), are consumed by the following reactions:

- Replenishing Sn^{2+} by SnO,

$$SnO + 2H^+ \rightarrow Sn^{2+} + H_2O \tag{6}$$

- Electrolytical replenishing,

$$\text{Anodic reaction:} \quad Sn - 2e \rightarrow Sn^{2+} \tag{7}$$

$$\text{Cathodic reaction:} \quad 2H^{+} + 2e \rightarrow H_2 \tag{8}$$

Insoluble anodes are used for tin-lead plating in special plants under jet-plating conditions, or to balance the metal concentration in an electrolyte, when it increases in the bath due to a reduced cathodic current efficiency. In this case both soluble and insoluble anodes are used in the same systems. Because the formation of oxygen on the insoluble anodes occurs at higher voltage than the dissolution of Sn and Pb, insoluble and soluble anodes cannot be installed in one circuit, separate rectifiers have to be used. Standard material for insoluble anodes is platinum-coated titanium or niobium.

6 MAINTENANCE AND CONTROL

6.1 Electrolyte Composition

Metal Concentrations The metal concentration in a tin-lead electrolyte has to be controlled in terms of the total metal content and the ratio between Sn and Pb. While the total metal content has influence on the limiting cathodic current density and the throwing power of the electrolyte, the ratio will of course influence the alloy composition of the coating. In bright systems a high metal content will cause problems with dullness in low current density area.

If the electrolyte is working with a high cathodic current efficiency, there will be a balance between the anodically dissolved and cathodically deposited metals. In general, there will be a steady loss of metals in the bath due to drag-out. This has to be replenished accordingly by the metal salts in form of their liquid concentrates. The ratio between tin and lead in the bath is kept by using the appropriate alloy-composition in the anodes. Since in the electrolyte there will always be some loss of tin by formation of tin-oxides, a slightly higher Sn content in the anodes will compensate this loss (e.g., anode composition 70/30 Sn/Pb when the desired alloy composition is 65/35 Sn/Pb). Corrections in the Sn/Pb ratio in the electrolyte can also be made by using a part of the total anodes in form of either 100% Sn or 100% Pb anodes. These measures do not correct the deposited alloy composition immediately; this only happens when the required Sn/Pb ratio in the electrolyte is achieved.

When the organic additive system causes a reduced cathodic current efficiency, the metal content in the bath will increase correspondingly. To keep the total metal content in the specified range, either dilutions have to be made or the surplus of metal is worked out by using insoluble anodes.

Acid Concentration The acid concentration in the tin-lead electrolyte is important for the throwing power and especially in high speed electrolytes for the conductivity of the solution. Similar to the metal concentration there will be a loss of free acid by drag-out, when anode- and cathode efficiency are both 100%. If the cathodic efficiency is lower, there will be a loss of free acid by the cathodic side reaction according to

$$2H^+ + 2e \rightarrow H_2 \tag{9}$$

6.2 Electrolyte Temperature

Tin-lead electrolytes for rack- and barrel plating are used normally at temperatures between 15° to 30°C. The low temperatures are especially recommended for bright electrolytes. With the lower temperature consumption of additives can be reduced considerably. This improves the solderability of the deposited coatings. Such electrolytes need efficient cooling systems. Stainless steel with thin polyethylene coatings is suitable as cooling coils. Bare stainless steel can be used, as long as the electrolyte is not contaminated by chloride ions and the temperature does not go above 50°C. Fluoropolymer heat exchangers can be used, but since the thermal conductivity of this material is low, specially designed coils have to be used for effective cooling.

Matt-depositing electrolytes are less sensitive to higher temperature; such systems are often operated at temperatures between 40° and 60°C in high-speed plating [12]. The higher temperature increases the limiting current density. But, since the temperature can influence the alloy composition, the maximum usuable temperature also depends on the additive system.

6.3 Agitation

Electrolyte agitation is a very important parameter in tin-lead plating. As in any electrolyte, increasing agitation improves the maximum permissible cathodic current density. Intensive flow rates are therefore essential for all high-speed plating processes. Table 7 gives some data on the thickness of the diffusion layer and thus the limiting current density in relation to the electrolyte agitation [55].

TABLE 7 Limiting Current Densities and Diffusion Layer Thicknesses

Hydrodynamic Regime	Limiting Current Density ($A\,dm^{-2}$)	Diffusion Layer Thickness (μm)
Natural convection to vertical electrode	1.44	200
Natural convection to horizontal electrode	3.65	80
Electrolyte flow along the plane of the electrode ($v = 25\,cm\,s^{-1}$; laminar)	3	100
Force flow cell ($25\,m\,s^{-1}$), turbulent flow	365	0.8
Rotating cylinder (180 rpm, peripheral velocity, $94\,cm\,s^{-1}$)	8.1	36
Gas-evolving electrode		
$13\,cm^3$ gas $cm^{-2}\cdot min$	72	4
$1\,cm^3$ gas $cm^{-2}\cdot min$	19.4	15
Gas feed through frit		
65 μm pore size, 0.1 liter gas min^{-1}	2.76	100
65 μm pore size, 17 liter gas min^{-1}	13.2	20
Ultrasound ($7\,watt\,cm^{-2}$)	50	6

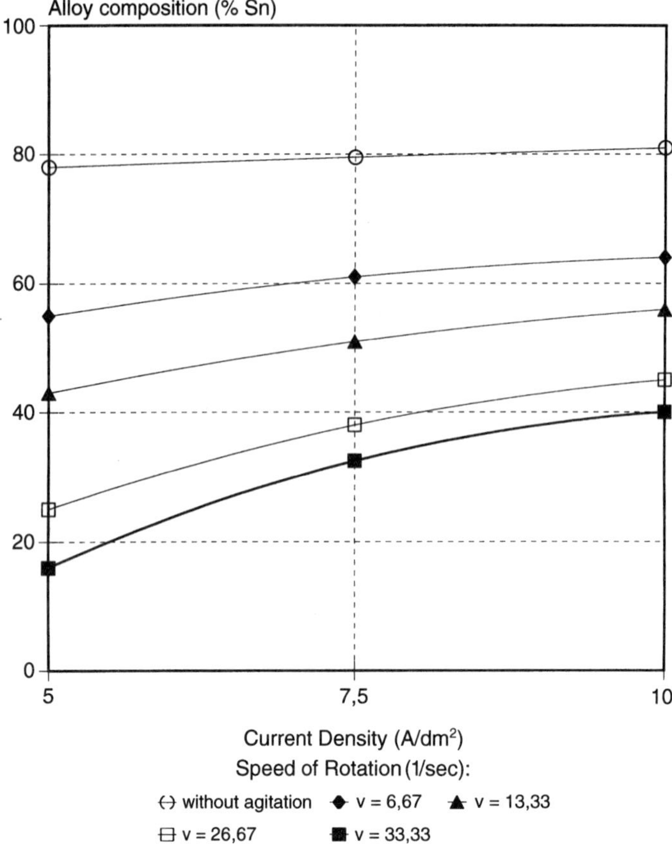

Figure 2 Effect of rotation rate using a rotating disc electrode on composition of Pb-Sn alloy from a fluoroborate bath (after data from Madry [56]).

The flow rate can influence the alloy composition. How sensitive this can be depends also on the type of additive system and the temperature. Some data on the alloy composition in relation to the current density at different rotation speeds are shown in Figure 2.

6.4 Contaminants

The most common sources of contaminants in tin-lead plating solutions are as follows:

- Impurities in bath chemicals
- Drag-in of pretreatment solutions
- Contaminants from the work to be plated
- Breakdown products of organic additives
- Impurities from anode reactions

Supply houses aim to provide the basic bath chemicals with a guaranteed purity level more than sufficient for use in plating systems. When using commodity chemicals, the buyer must make sure that their purity is high enough, either by having analyses done or by obtaining specifications from the supplier. Impurities may be present in a dissolved form or as insoluble particles. Thus it may be useful to filter a solution before using it in the plating bath.

Dragged-in impurities can come from cleaner or activating solutions or from other plating baths, in cases where an intermediate layer is deposited. Since chloride ions can lead to problems in most tin-lead electrolytes, acid pickling bath based on HCl should not be used. Chloride ions are harmful to the throwing power of such baths, and in the case of matt-plating solutions, they promote dendritic growth.

Drag-in of sulphates can also cause problems resulting in roughness due to inclusions of insoluble lead sulphate particles. If tin-lead coatings with such inclusions are reflow melted, a de-wetting can take place. Thorough filtration is the only means of avoiding such problems. Eliminating chloride ions is much more difficult since the solubility of lead chloride is too high to allow precipitation as a means for their removal. Though silver ions can be used to form the insoluble AgCl, the cost of such treatment is usually prohibitive.

Metallic Contaminants Nickel and copper are the most commonly found contaminants, since this metals are widely used to form interlayers in the substrate. Drag-in from activating solutions has also been mentioned and these can introduce zinc or iron into the bath. Especially in the case of barrel plating, bipolarity effects at the commencement of each tin-lead plating operation can result in anodic dissolution of the substrate metal and thus provides a further source of contaminant.

Because of its more positive deposition potential, copper will be codeposited in any tin-lead plating bath. In the case of matt deposits, up to 100 mg liter^{-1} can be tolerated without ill effects. With bright deposit baths, copper can lead to darkening or staining especially at low current densities. In the case of matt tin-lead deposits, copper can act to give a finer grain structure. In the case of peptone-containing baths, copper contamination can lead to increased levels of organic inclusions that are harmful if the deposit is then reflowed [21]. In the molten state, copper reacts with tin to form an intermetallic phase, which then crystallizes out during cooling, giving a rough appearance. Typical incorporation rates of copper as a function of Cu^{2+} concentration in solution are plotted in Figure 3 for a fluoroborate bath with peptone addition and for a methanesulphonate bath with synthetic additives.

Nickel contamination stems from drag-in from nickel-plating baths that are specified immediately prior to the tin-lead plating stage to form an interlayer. Since nickel is not deposited from strongly acid type bath, the presence of nickel ions is not usually a problem here. However, most nickel plating bath contain chloride ions to improve anode solubility. In the case of the widely used Watts nickel bath, this is present at 10 to 15 g liter^{-1} chloride. Problems associated with chloride have been previously discussed, and drag-in from nickel plating baths is a source of problems mainly only as far as chloride ions are involved.

Iron can enter the system by being dragged in or by dissolution of the substrate metal. A common problem is small parts falling off the racks to the bottom of the tank, where they slowly dissolve. Iron, nickel, and zinc as possible contaminations in

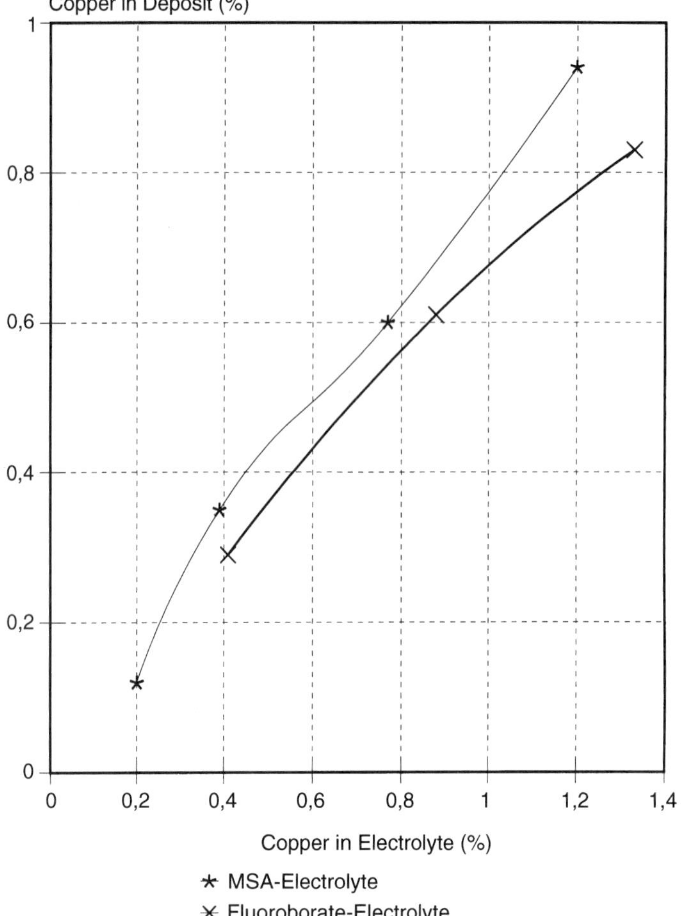

Figure 3 Codeposition of copper from tin-lead alloy plating baths based on methanesulphonic acid or tetrafluoroboric acid.

lead plating solutions are not codeposited, even when their concentration is as high as 10 g liter^{-1}. Indirect problems can be caused by incorrect analysis results, when the presence of lead is established in the solution by a complexometric titration. During the analysis any iron contaminants will also be complexed, giving incorrect results. Because of the large difference in between the atomic weights lead and the contaminants, a contamination with 1 g liter^{-1} Ni will simulate 3.5 g liter^{-1} Pb.

High contaminations of iron can cause an increased rate of formation of stannic tin. Iron will enter the solution in its divalent form, which is easily oxidized to Fe(III) by air. This Fe(III) reacts with Sn(II) to form stannic tin according to

$$2\text{Fe(III)} + \text{Sn(II)} \rightarrow 2\text{Fe(II)} + \text{Sn(IV)} \tag{10}$$

It follows that iron as a contaminant should be held as low as possible.

Tin (IV) Concentration Depending on the stabilizer system used, a given concentration of tin (IV) will be found in the solution. In case of fluoroborate bath, Sn(IV) is soluble up to around 20 g liter^{-1}. In the case of methanesulphonic acid baths, Sn(IV) is partially soluble, and if the cathode efficiency of the tin-lead deposition is less than 100%, some of this will be reduced to Sn(II). Sn(IV) will precipitate from fluoroborate bath when insufficient free fluoride ion is present to complex it. In methanesulphonate bath, Sn(IV) exists in a form of metastannic acid. Splitting off water and ageing, in general, converts this to tin dioxide, SnO_2. Both of these mentioned species can exist in a colloidal form which has a very high specific surface area. They can adsorb the organic additives used in solution and so cause problems because their effective concentration in solution is thereby decreased [28]. If necessary, the stannic tin can be precipitated by means of special flocculating agents as demonstrated in Figure 4. They are proprietary chemicals, available by the supply houses.

Decomposition Products of Organic Additives The properties of electrodeposited tin-lead alloys can be affected as a result of changes in the original composition of the organic additives added to the solution. These changes can arise from chemical reactions within solution components or as a result of electrochemical reactions of additives at anode or cathode. For example, peptone, widely used as an additive in

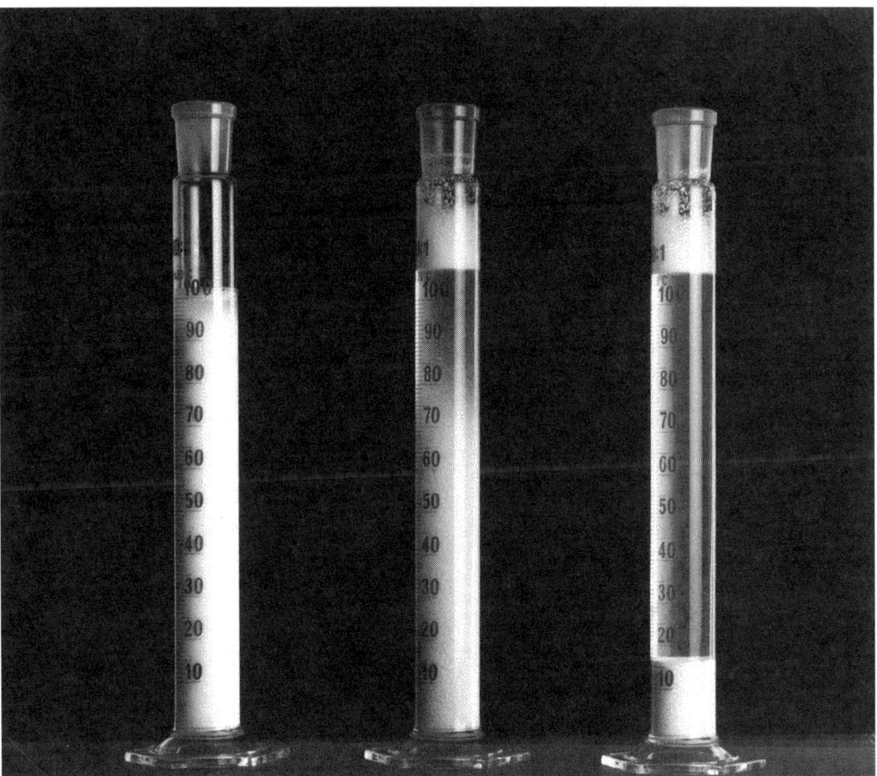

Figure 4 Precipitation of Sn(IV) compounds from an acid tin-lead plating solution. Left: original sample; centre: commencement of precipitation after addition of precipitating agent; right: after completion of precipitation.

fluoroborate baths, can be hydrolytically split. Peptone breakdown products may be incorporated into a deposit to a greater extent than peptone. This can result in a de-wetting on the surface during reflowing operations. To prevent the deposition of such breakdown products from peptone type baths, a regular activated carbon treatment on a three-month cycle is recommended [29].

The additives in recently developed electrolytes are chemically more stable and, in the case of matt deposit baths, activated carbon treatment is only necessary if organic impurities are dragged in from external sources.

In bright deposits a variety of by-products can form depending on the additives used, particularly in the solutions that use formaldehyde. As concentrations of these extraneous organic species increase in solution, the extent to which they become incorporated into the deposit will increase, so solderability may suffer.

Impurities Resulting from Anodic Reactions Tin-lead anodes over the entire compositional range dissolve readily in fluoroborate or methanesulphonate baths. On the anode surface a black slime will form, which is caused by the metallic impurities in the tin-lead anode, which are more electropositive (Sb, Bi, Cu, As, Ag). These metals exist in the anode as intermetallic compounds with tin and in consequence, their volume of their slimes is higher than one would predict on the basis of their volumes as pure metals. Should this slime contamine the electrolyte, it will cause rough deposits to form. Antimony is specially prone to form such sludges so its concentration in the anode should be as low as possible.

Where soluble anodes are used, anodic oxidation of organic additives in solution is not usually significant. However, where insoluble anodes are used, a range of side reactions is possible either by direct anodic oxidation or by chemical reaction with the dissolved oxygen formed as oxygen gas evolves at the anode.

7 ELECTROCHEMICAL DEPOSITION EQUIVALENT FOR TIN-LEAD ALLOYS

The electrochemical deposition equivalent for any tin-lead alloy may be calculated using the following formula [30]:

$$\text{Tin-lead deposition equivalent} = \frac{3.861 * 2.214}{f_1 * 3.861 + f_2 * 2.214}$$

where

$3.861 =$ deposition equivalent for lead (g Ah^{-1})
$2.214 =$ deposition equivalent for tin (g Ah^{-1})
$f_1 =$ weight fraction of tin in alloy
$f_2 =$ weight fraction of lead in alloy
$f_1 + f_2 = 1$

The data for the electrochemical deposition equivalents for the entire range of the tin-lead system are given in Table 8 together with the densities and the deposition rates.

TABLE 8 Data on Electrochemical Deposition Equivalent, Density, and Deposition Rate for Tin-Lead Alloys

Alloy composition (% Pb)	Density ($g\,cm^{-3}$)	Electrochemical Deposition Equivalent ($g\,Ah^{-1}$)	Deposition Rate ($\mu m\,min^{-1}$)
0	7.28	2.214	0.507
5	7.41	2.262	0.509
10	7.55	2.312	0.51
15	7.69	2.365	0.512
20	7.84	2.421	0.515
25	8.00	2.478	0.516
30	8.16	2.539	0.518
35	8.33	2.6	0.520
40	8.50	2.669	0.523
45	8.68	2.74	0.526
50	8.87	2.814	0.529
55	9.07	2.893	0.532
60	9.28	2.975	0.534
65	9.50	3.063	0.537
70	9.72	3.156	0.541
75	9.96	3.256	0.545
80	10.21	3.361	0.549
85	10.47	3.473	0.553
90	10.75	3.594	0.557
95	11.05	3.723	0.562
100	11.36	3.861	0.566

8 DENSITY OF TIN-LEAD ALLOYS

8.1 Determination of Density of Tin-Lead Alloys

For the density of a deposited alloy to be determined, a foil of the alloy is formed by deposition on a highly polished and passivated stainless steel cathode. The foil is stripped from the cathode, washed, dried, and weighed. It is then introduced into a pyknometer filled with water thermostatically controlled at 20°C. The density of the alloy is then given by the expression

$$d_{\text{Alloy}} = \frac{m}{(M+m) - M'} * d'$$

where

m = weight of foil
M = weight of water-filled pyknometer
M' = weight of water-filled pyknometer + foil
d' = density of water at 20°C

The density of a tin-lead alloy can be calculated as

$$\frac{1}{d_{\text{Alloy}}} = \frac{f_1}{d_1} + \frac{f_2}{d_2}$$

where

d_{Alloy} = density of alloy
d_1 = density of component 1
d_2 = density of component 2
f_1, f_2 = weight fractions of components 1 and 2, respectively
$f_1 + f_2 = 1$

This expression may be used for the calculation of any tin-lead alloy, because a study of the determination of the density of samples produced by mixing the molten metals showed only a low volume shrinkage [31].

9 DEPOSITION RATE

The theoretical deposition rate for tin-lead alloys can be calculated from the mass of metal plated per unit surface, the plating time and the calculated density. The data are shown in Table 8. The values range from 0.505 $\mu m\ min^{-1}$ for 100% Sn to 0.57 $\mu m\ min^{-1}$ for 100% Pb at 1 $A\ dm^{-2}$ assuming 100% cathodic efficiency. It can be seen that for practical purposes, one can consider deposition rate of about 0.5 $\mu m\ min^{-1}$ low-lead containing tin-lead alloys.

10 PROPERTIES OF ELECTROPLATED TIN-LEAD FILMS

10.1 Corrosion Resistance

Long-Term Corrosion Tests Long-term corrosion resistance of tin-lead alloys in atmospheric exposure tests was studied by Jostan et al. [32]. The aim of these investigations was to test the behavior of tin-lead coatings in electronic component applications. The corrosion resistance was therefore mainly judged by its effects on contact resistance. The coatings perform well in aggressive and moderately severe industrial environments, with electrical contact resistance increasing slowly. In a marine environment, tin-lead surfaces are rapidly attacked, the contact resistance reaches values in the k-ohm range after relatively short exposure.

Short-Term Corrosion Tests Short-term corrosion tests were conducted by the same authors using environmental chambers with artificial atmospheres. As with the atmospheric tests, the most damaging conditions are those where chlorine is present, such as HCl.

Oxide Formation The formation of oxides on tin-lead surfaces affects their solderability. On initial contact with air a freshly electroplated tin-lead layer forms

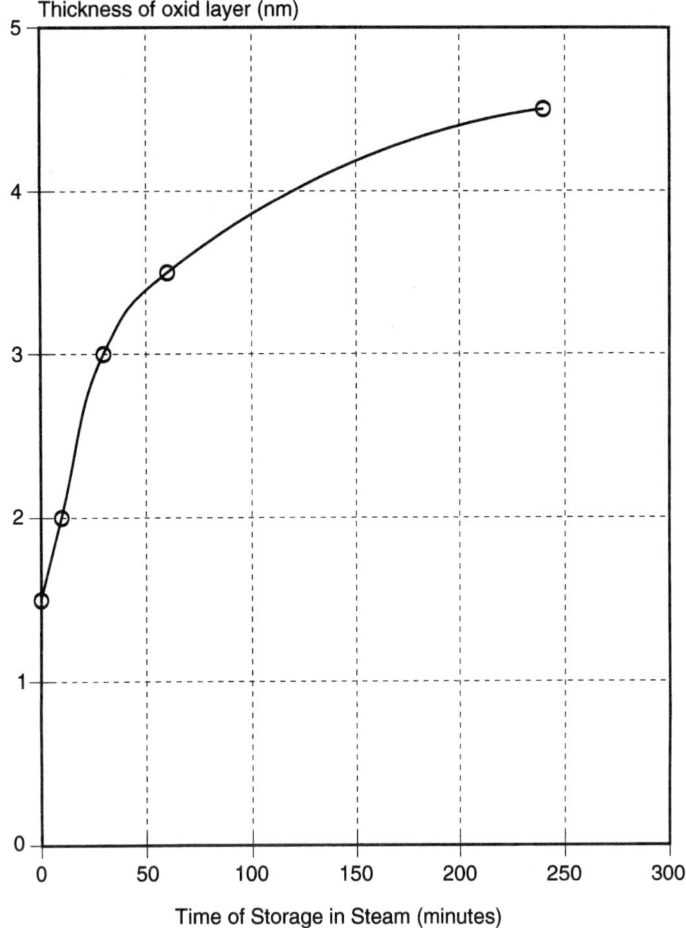

Figure 5 Oxide formation on lead-tin alloy surfaces during exposure over boiling water.

an oxide layer of about 1.5 nm [33]. These oxide layers are very stable and further growth is extremely slow. After one year, it will typically be around 3 nm.

If the metal is stored at higher temperatures, say 200°C, an oxide layer around 30 nm thick will form in 24 hours. The oxidation rate under these conditions is twice as fast as at 100°C. In humid atmospheres, the growth rate is accelerated. Figure 5 shows the oxide formation as a function of time over boiling water [33].

The oxides on a tin-lead surface, which are formed during heating in normal atmosphere, consist mainly of tin oxides, as it is shown by the Auger-spectrum in Figure 6 of a 90/10 Sn/Pb-alloy heated to 175°C for 24 hours.

11 SOLDERABILITY

As mentioned above, there is a close correlation between the formation of oxides and the solderability of tin-lead coatings. It is therefore now standard practice to test

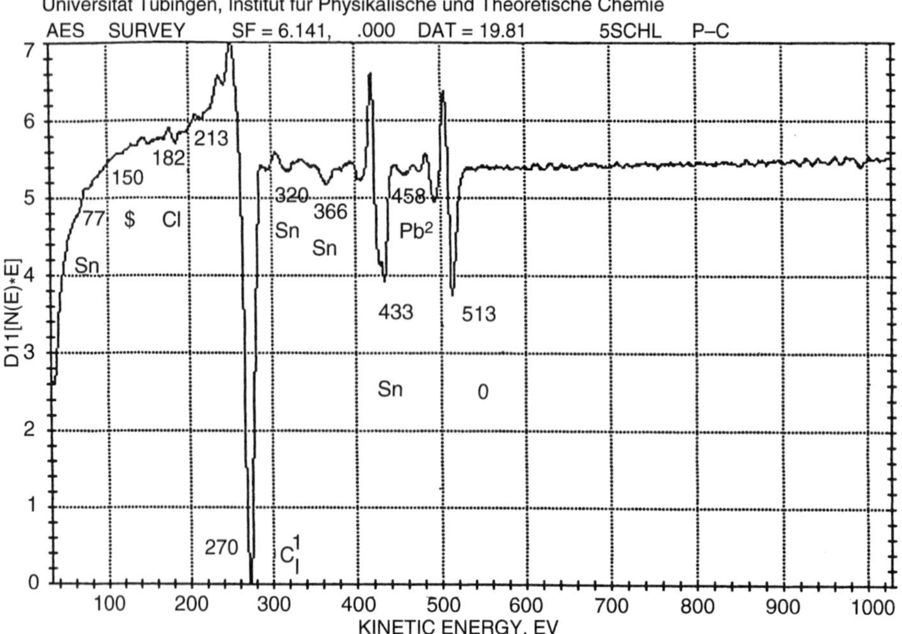

Figure 6 Auger-spectrum of a bright lead-tin surface after 24 hrs heat-treatment at 175°C The enrichment of carbon compounds at the surface can be seen.

solderability after accelerated ageing tests, which simulate some reactions which can occur on the surface of tin-lead coatings during storage. Ageing tests also show the influence on solderability of the formation of intermetallic layers.
The standard ageing tests are as follows:

- Dry heat ageing at 155°C or 175°C for 4, 8, 16 hours
- Steam ageing up to 8 hours
- Storage in moist heat (40°C, 93% R.H.) for 4, 10, 21, or 56 days
- Storage in a distilled water cyclic test sequence (24 hour cycle; from 25°C to 55°C in 3 hours, 9 hours at 55°C, then cooling to 25°C at 95% R.H.)

11.1 Carbon Content of Tin-Lead Deposits

Because all acid tin-lead electrolytes use organic additives as grain refiners or brighteners, there is always some codeposition of this organics with the tin-lead coating. These organics can cause a deterioriation of the solderability, the amount of codeposition permitted is therefore limited to 0.05%, calculated as carbon [34]. Matt tin-lead deposits have usually much lower carbon contents, typical values are 0.005% [35, 12]. Bright coatings show generally higher codepositions, which are sometimes higher than the 0.05% value. Levels as high as 0.35% have been reported in the literature [36]. In bright electrolytes there is in addition often a problem that in the solution a buildup of breakdown products of the organic brighteners can occur, which leads to a further increase of codeposition of organics. Bright tin-lead coatings are therefore not allowed for applications required to comply to MIL-STD-81728A. This specification calls for the use of matt or reflowed coatings.

11.2 Solderability Test

Solderability is tested either by the dip-and-look test (MIL-STD-202; DIN 32506 part 3) or by the wetting-balance method (MIL-STD 883; DIN 32506, part 4). The standard test conditions are

Solder-bath	60/40 Sn/Pb
Temperature	235°C
Flux	Nonactivated; R-type (pure colophony in isopropanol)
Immersion time	5 seconds

The criterion for acceptance is a wetting of not less than 95% of the immersed surface.

12 ELECTRICAL CONTACT RESISTANCE

Changes in electrical contact resistance of electrodeposited coatings used to form protective coatings in electronics were studied by Jostan et al. [32]. Exposed to H_2S or SO_2 containing atmospheres, tin-lead surfaces show good performance with only slight increase in contact resistance. In chloride-containing atmospheres, tin-lead behaves better than tin, possibly because formation of lead chloride layers inhibits further attack. Tin-lead surfaces perform very well when exposed to NO_2 or NH_3 containing atmospheres. Figure 7 shows the change in electrical contact resistance after exposure to various test gas mixtures.

Storage in air at 120°C changes the electrical contact resistance very little. For bright tin-lead coatings the authors observed an increase of the electrical contact resistance during storage at 120°C; it reached a maximum in the first 10 hours, after which the values fall slowly away and then they increase again. The behavior is shown in Figure 8.

A possible explanation given by the authors is that by grain-boundary diffusion, the incorporated organics migrate to the surface with a resulting increase of electrical contact resistance. This is followed by their evaporation or oxidative destruction which is reflected in the fall in resistance values. Finally, progressive oxide growth causes resistance to increase again. The Auger spectrum in Figure 6 shows how carbon-containing compounds build up at the surface.

13 HARDNESS

Hardness values for electrodeposited tin-lead alloys are low. In the case of matt deposits, the value is largely independent of alloy composition, typically being 8 to 10 Vickers hardness with 15 g load. According to Raub and Blum [28], addition of

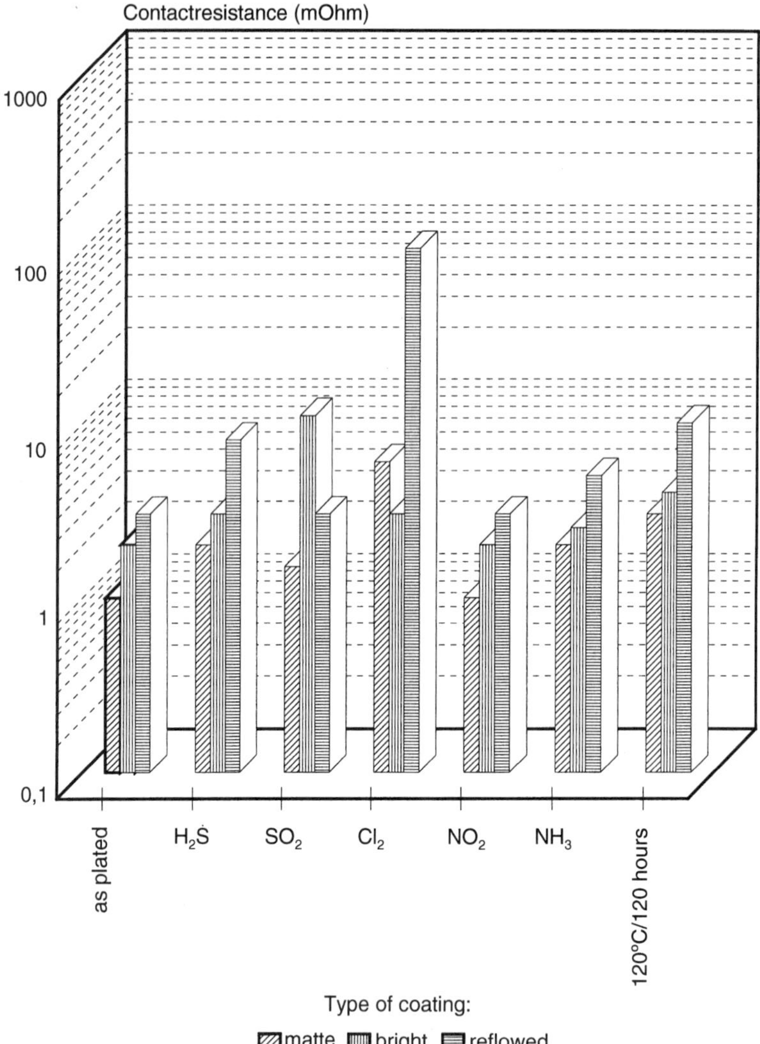

Figure 7 Change in electrical contact resistance of lead-tin alloy surfaces after exposure to various gas mixtures (Data after Jostan and coworkers [32]).

tin results in an increased hardness compared with pure lead with a maximum reached at about 10% Sn.

Bright tin-lead coatings show significantly higher hardness values. They depend strongly on alloy composition and the brightener used. Up to 10% lead, Vickers hardness values HV_{15} are between 15 and 20. By using increased concentrations of additives hardness can be raised to 25 HV. With increasing lead content hardness decreases markedly though not linearly. In several systems a sudden decrease in hardness was found at around 15% to 20% Pb alloy values. At this point hardness values drop from 15 to 5 to 6 (HV_{15}).

The hardness of tin-lead alloys can be increased by codeposition of small amounts of copper, antimony, or bismuth. The use of the first two is well-known in the

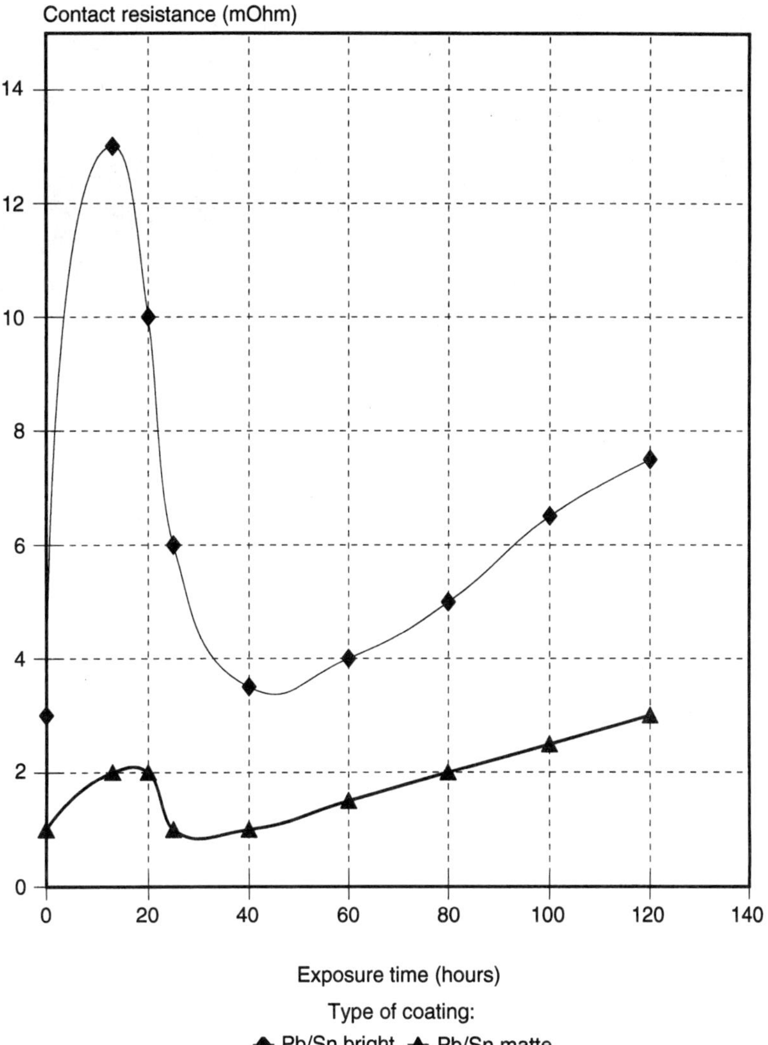

Figure 8 Change in electrical contact resistance of a lead-tin alloy surface after storing at elevated temperature (Data after Jostan and coworkers [32]).

bearing industry, whereas the role of bismuth is mainly considered as a side effect when this metal is used to improve brightness in bright deposit electrolytes [37].

The move from matt to bright alloy deposits is accompanied by a doubling in hardness. At the same time, the crystal structure is seen to change. Matt Pb/Sn deposits exhibit a predominantly nonorientated growth. In the case of bright Pb/Sn with hardness values around 20 HV_{15}, a definite field-orientated columnar crystal growth can be seen.

Since the harder deposits are also more brittle, these changes in crystal structure can lead to problems if the plated component are to be subjected to any kind of mechanical deformation. Cracking can take place so much that the base metal can be exposed, and oxidation within these cracks results in poor solderability. In

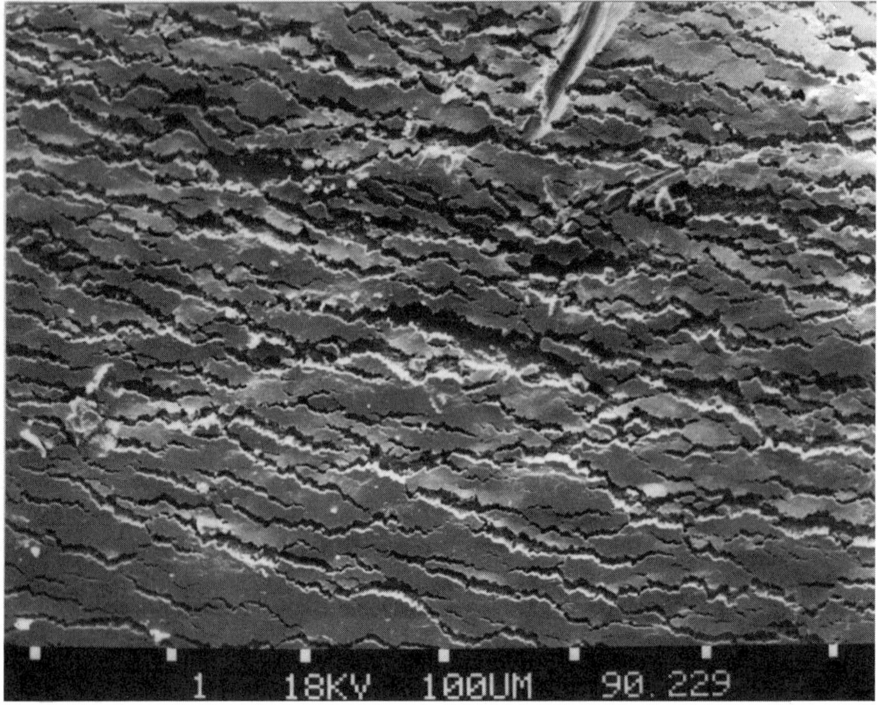

Figure 9 Formation of cracks in a bright tin-lead coating during a bending process of leads of electronic components. By this crack formation the base material surface can be exposed, the solderability can be detoriated, when these cracks are formed in the area, where the solder joint has to be formed; for example at the tip of a "J"-formed lead of a SMT-device. Bright tin-lead coatings are therefore not recommended, when a mechanical bending process follows after plating.

surface-mounted electronic components the cracking would occur close to the soldered connections [38]. For these applications the use of matt tin-lead deposits is preferred. Figure 9 shows cracking in a bright tin-lead deposit. Note that the cracking extends right down to the substrate.

14 APPLICATIONS OF TIN-LEAD COATINGS

14.1 Terne Plate

Steel sheet coated with tin-lead of composition 93% Pb, 7% Sn is known as Terne plate. The main consumer of this product is the automotive industry where it is used for fuel tanks, filter casings, sump pans and radiator components. Global production is around 600,000 tonnes [39], and up to 1973, hot-dipping was the only method of production. This was when an electroplated equivalent was developed. In 1983 a second plant started production [40]. As with tin-plate, the newer method offered better control of coating thickness. It also enabled the production of single-side coated sheet.

The electrolytes developed were based on fluoroborate and used hydroquinone as an inhibitor, according to suggestions by Graham and Pinkerton [41].

14.2 Bearing Alloys

Alloys of about the same composition are used for running surfaces on bearing shells. The shell itself is of steel, and the bearing alloy is cast onto it. Various alloy compositions are used, with aluminium bronzes among the most favored. The shell is mechanically finished and a nickel interlayer laid down. The bearing alloy is Pb/Sn or Pb/Sn/Cu and is typically 25 μm thick. One particular composition is 88% Pb, 10% Sn, and 2% Cu and is specially effective as an anti-seize coating [42].

14.3 Anodes for Chromium Plating Bath

Tin-lead alloys with 7% Sn are used as insoluble anodes for chromium plating from chromic acid solutions. In most cases, the anodes are made by melting the solid metals in their correct proportions. However, in certain cases such as auxiliary anodes for plating internal surfaces, there are advantages in the use of shaped anodes that are plated with tin-lead [43]. Using a total metal concentration in solution of 150 to 200 g/liter^{-1} and using suitable inhibitors, pore-free deposits of up to several millimeters can be formed.

14.4 Eutectic Alloys

Tin-lead coatings with the eutectic composition (63% Sn, 37% Pb) are used in the production of printed circuits and as a solderable coating for electronic components. Flip-chip technology, a new application, requires the deposition of Sn/Pb-bumps of either 95Pb/5Sn or 65Sn/35Pb composition on the Si-wafer for the connection of unpackaged chips directly to PCBs [44].

In the PCB production, eutectic tin-lead coatings have been used as metallic etch-resists; after copper etching the tin-lead coating was reflowed. While tin-lead is still used widely today as etch-resist, the quantity of boards with reflowed coatings is now very small. In the newer technologies the tin-lead coating is only used as a protective coating in the copper-etching process. After this step it is removed by special tin-lead strippers. The PCB is then coated with a solder-resist, which leaves only those parts uncovered where in the assembling process solder connections have to be formed. To preserve the solderability of those areas, the PCB is selectively coated with tin-lead by dipping into molten solder. By this hot-air-leveling (HAL) process only the unprotected copper can react with the molten solder. The solder-resist, which is an organic resin, is not wetted by the molten solder. Consequently there is almost no real need for tin-lead deposits for metal-resist applications, and it has been substituted by pure tin deposits in many cases [45]. But for special applications the specifications still demand the use of reflowed tin-lead on PCBs.

A problem with coatings that are normally produced from reflow tin-lead processes is a very uneven thickness distribution after reflow. On the corner or knee of the plated through-holes the thickness of tin-lead is less than 1 μm. This causes solderability problems [46], especially if also required is a solder-mask that needs a high temperature curing, for example, 2 hours at 160°C. The low tin-lead coating on the corner will be converted to intermetallic Cu/Sn layer that is no longer solderable with nonactivated or mildly activated fluxes. To overcome this problem, a special sequence for reflowing has been developed. It consists of a preheating of the boards at 185°C for 45 seconds and the actual reflow then takes place in a second bath at 205°C with a short dip time of about 5 seconds [47].

Another approach to guarantee a minimum thickness of 1 μm after reflow as specified by the ESA-specification PSS-50 QRM-17E, is a two-stage plating of a pure lead layer of about 2 μm followed by a 8 to 10 μm thick tin-lead layer (70% Sn) (SLOTOLON-Process [48, 49]). This combination of deposits is reflowed using any of the normal processes. The result is a Pb/Sn-coating which, at the edge of the through-hole, is at least 2 μm thick. During reflowing, an alloying of the tin into the lead layer takes place and a copper-tin diffusion layer forms, as in conventional reflowing. The improved coating thickness around the hole edge is presumably due to increased viscosity due to the lead layer, which prevents the molten layer from flowing down from the edge. An increase in temperature will again result in a retreat of the coating from the hole edge.

14.5 Tin-Lead Deposits for Electronic Components

A major application for electroplated tin-lead coatings is in the production of electronic components. Tin-lead coatings are a prerequisite for reliable soldering when making inter-connection on PCBs.

In terms of the type of connections, components can be grouped under four headings:

1. Wire-lead components with axial connections
2. Components with radial connectors for insertion mounting (DIL, dual-in-line-, or DIP, dual in-line package-components)
3. Components with radial connections suitable for surface mounting
4. Leadless connections

Within the first group are the traditional components that feature connecting leads for through hole insertion. These leads or wires are normally made from preplated wire, although in some cases plating is carried out on the completed components. In the second group there are the DIP or DIL type components. These have two rows of pins that are inserted in plated through-holes on the PCB. In the third group are mainly packaging types similiar to those in the second group except that the connector pins are mechanically formed for surface mount application into a "gull-wing" if they are turned out or in a J-format beneath the package. Leadless components are either square-shaped or cylindrical components such as chip-resistors and MELF (metal electrode face) components. Their mating surfaces are metallized and serve as solder faces for solder connections.

All components require a coating if solderability is to be assured. The type of coating depends on various factors, but the following coating types are mainly used:

- Matt tin-lead (approx. 5–20% Pb)
- Matt tin-lead (approx. 35% Pb)

- Bright tin-lead (approx. 5–20% Pb)
- Bright tin-lead (approx. 35% Pb)

Various factors govern the choice from this range, and consideration must be given to the mechanical processing of components, solderability, and long-term storage. In the past pure tin-coatings have been used for such applications. While these are still being used to a considerable extent, tin-lead deposits have been introduced primarily to avoid whisker formation and the most common electrolytes give matt deposits with approximately 15% Pb. Bright tin-lead coatings suffer from the problem of cracking during bending and are not recommended therefore for applications where any trim-and-form steps follow after plating. Bright solutions are best use only where no mechanical deformation of the components is involved. A big application here is in the field of the finishing of leadless components.

Matt tin-lead deposits have a rougher surface than their bright counterparts. The typical morphologies of a matt and a bright electroplated tin-lead surface are shown in Figures 10 and 11. The rougher surface of matt coatings, together with their lower hardness, results in a degree of metal loss in the trim-and-form process. In addition the relative softness of tin-lead coatings results in the formation of solder peeling and scratches in the trim-and-form process. This problem cannot be overcome by modifications to the deposit characteristics alone but only in combination adjustments to the bending process. Figure 12 (on page 419) shows typical scratches

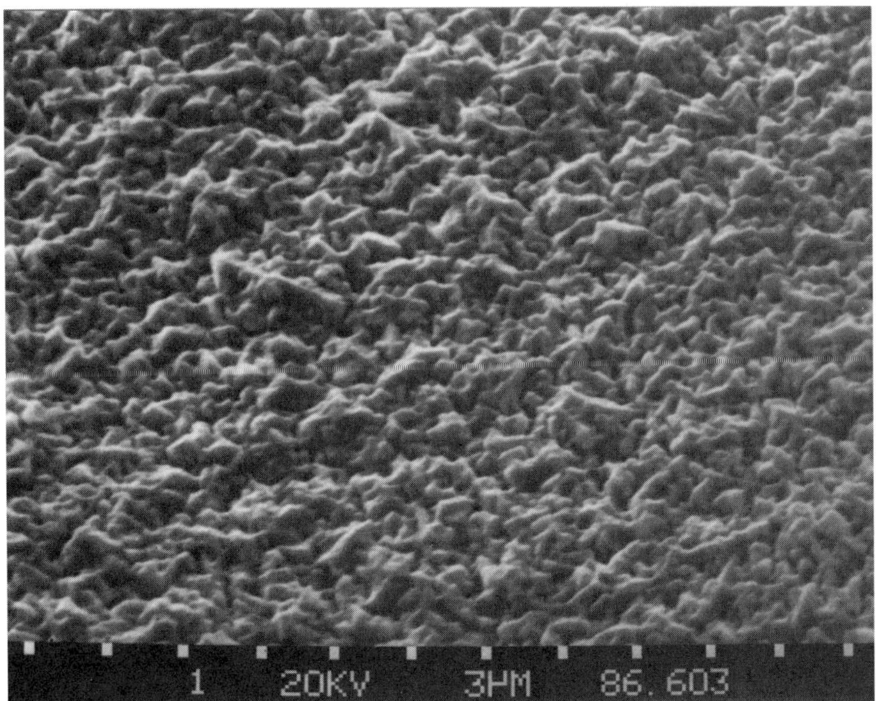

Figure 10 Morphology of a matt lead-tin alloy electrodeposit from an alkylsulphonic electrolyte.

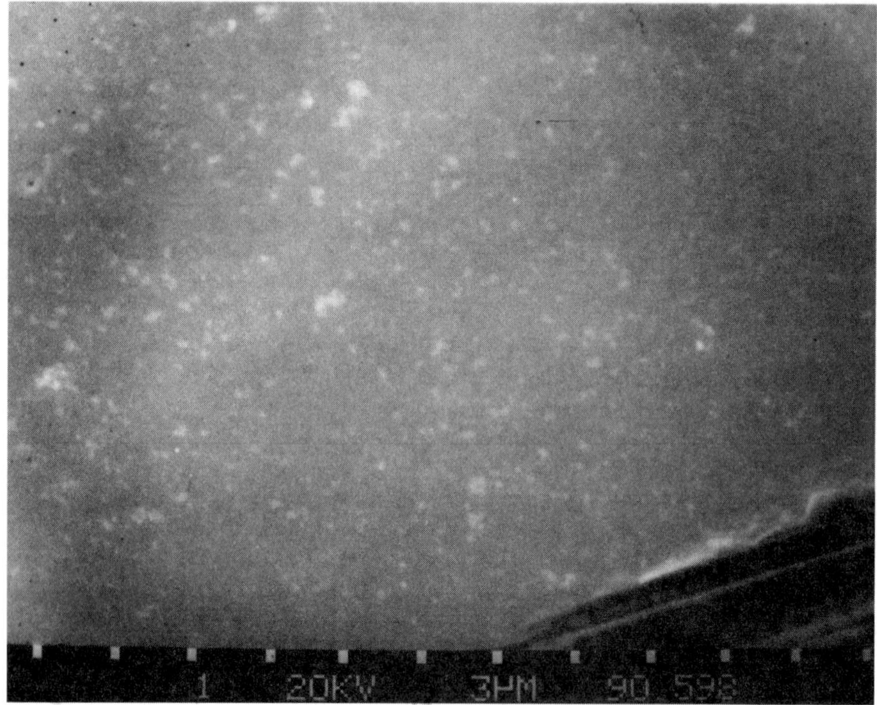

Figure 11 Morphology of a bright lead-tin alloy electrodeposit from an alkylsulphonic electrolyte.

in the surface of the tin-lead layer after trim and form. This abrasion of tin lead metal causes problems by contaminating of the trim and form tools. This foreign material can be pressed into areas during trim and form. Solderability can also be affected when cracking occurs, during the bending process, in the area where the pin comes into contact with the solder on the solder pad, for example, the shoulder of a J-formed lead.

The formation of cracks after bending of a bright tin-lead coating is shown in Figure 9. It can easily be seen that in the cracks the base material is exposed, and therefore oxidation of the base material can occur here. Reliable solderability cannot be asured in such cases. The formation of cracks in the bending of bright tin-lead coatings is caused by their rather brittle, field-orientated structure. Figure 13 (on page 420) shows the structure of such an layer after cooling in liquid nitrogen and bending. Unlike this field-orientated structure, matt tin-lead coatings are deposited in an unorientated dispersion type (UD) structure. They do not form cracks under normal trim-and-form conditions. Here again, the coating was made brittle by cooling in liquid nitrogen to demonstrated the orientation of crystallisation during electrodeposition, as shown in Figure 14 (on page 421).

Coating Thickness For electronic components terminations, the thickness of electrodeposited tin-lead coatings should not be less than 5 μm and the required value depends on the subsequent processing of the component. The upper limit is

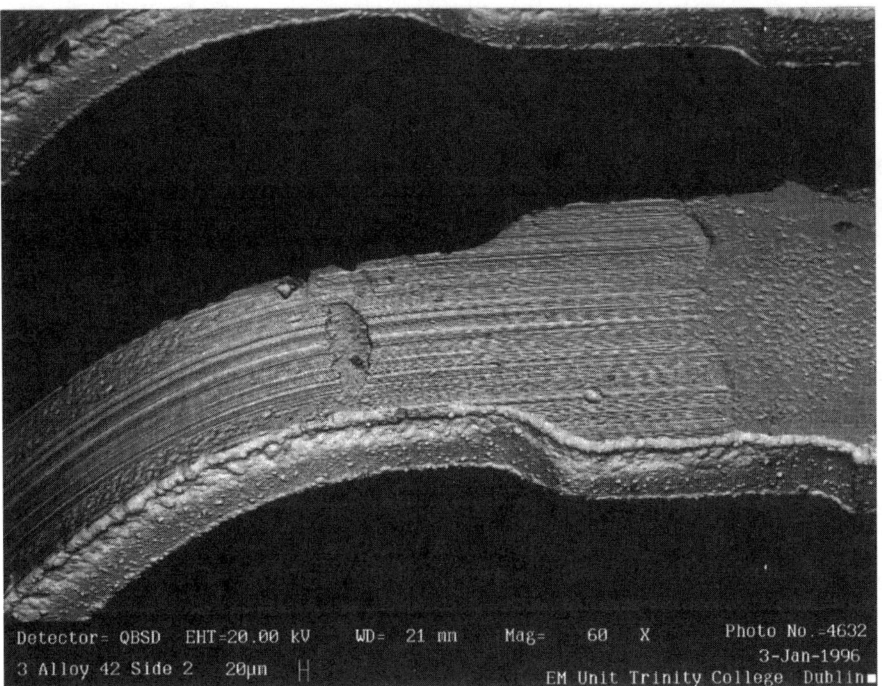

Figure 12 Scratches in an electroplated lead-tin deposit formed during the trim-and-form process of leads. The solderability is normally not detoriated by this scratches, but the material removed from the surfaces can cause problems by contaminations of the bending tools. Lead-tin particles can thus be transferred to following components and can cause short-circuits between the leads.

about 20 μm. Too high a thickness can enhance problems in the trim-and-form process due to increased solder shedding. In their deposited state, and without any posttreatment coatings as thin as 2 μm are normally pore free and have sufficient solderability under normal storage conditions for up to three months. But since most coatings undergo processes involving exposure to heat after plating, such as applying identifaction markings and burn-in tests for high reliabilty components, such thin coatings will normally not be used.

14.6 Interlayers

Interlayers have to be plated on brass. If this is not done, there will be a rapid diffusion of zinc from the brass into the tin-lead surface, causing severe detoriation of its solderability. Interlayers of copper or nickel can be used. The minimum thickness is 2 μm for nickel and 3–5 μm for copper [50]. Copper is plated using either cyanide or acid electrolytes, nickel using additive free system such as sulphamate nickel or Watts-nickel. Interlayers of copper are sometimes also recommended for special copper alloys used as base materials when they are difficult to activate or contain alloying elements, which can interfere with the subsequent tin-lead plating (types of bronze with different tin-content, berylliumcopper alloys, copper alloys containing traces of zirconium, silicon, nickel, etc.).

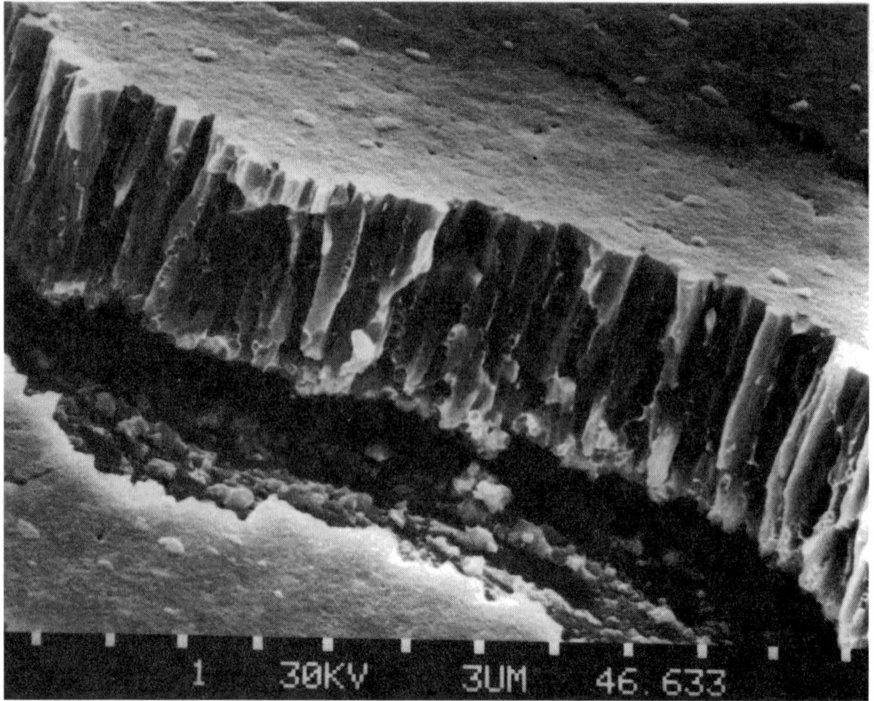

Figure 13 Fracture surface of a bright electroplated lead-tin deposit. The layer shows a typical field-orientated growth (fibrous structure). This type of structure is very sensitive to form cracks during bending as shown in Figure 9.

14.7 Tin-Lead Coatings for Plug Connectors

Tin-lead alloys under atmospheric conditions form thin oxide layers that lead to increased contact resistance. However, since these oxides are very soft, the oxide layer is readily fractured and metal-to-metal contact is restored. Tin-lead coatings are only suitable for plug connectors when the insertion force is high enough to break these oxide layers and to make a reliable contact. Because of its low hardness values and the need for a somewhat higher insertion force than with other contact materials, a degree of wear does take place. Therefore tin-lead coated contacts are only suitable for applications where a modest number of total insertions and withdrawals are foreseen. The tribological aspects of tin coatings were reported by Hamman and Schedin with CuSn4 (bronze) as a substrate [51]. Use of a suitable lubricant can reduce the wear behavior and the insertion force required [52]. Use of composite coatings in which graphite particles are dispersed in a tin matrix also give lubricant action [53].

Tin-lead coated plug connectors can also suffer from wear corrosion. Where a contact is capable of oscillation or other movement in operation, the oxides formed are repeatedly broken up. The pure metal exposed on the surface then become oxidized. This process continues until an oxide layer has formed that is so thick that it cannot be broken by the contact force of the mating contact. The result is a contact resistance so high that failure occurs. Studies by Ireland et al. [54] suggest that in the case of relative movement of 75 μm at a 50 Hz frequency, failure will occur after some 50,000 cycle, and that under current flow conditions, this may reduce to 30,000

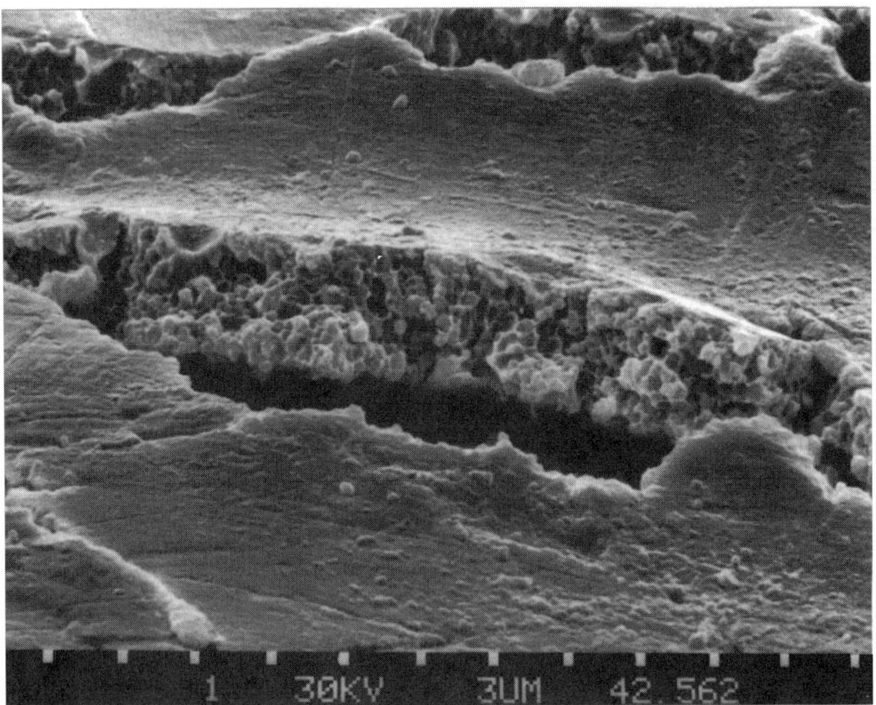

Figure 14 Fracture surface of a matt electroplated lead-tin deposit. The layer shows a growth in the form of an unorientated structure (UD-type, unorientated dispersion type). Such coatings are very ductile and don't form cracks during bending. For the preparation of this fracture the sample was cooled in liquid nitrogen.

cycles with temperature increases also occuring. As a criterion for failure, an increase in contact resistance from the original value of 0.5 mOhms to 200 mOhms has been established.

REFERENCES

1. J. F. Groff, U.S. Patent 1,364,051 (1920).
2. C. H. Chandler, U.S. Patent 1,373,488 (1921).
3. R. Schaefer and J. B. Mohler, U.S. Patent 2,461,350 (1949).
4. D. L. Cox, *Plating*, **51**, 976 (1964).
5. R. Putman and E. Roser, *Plating*, **42**, 1133 (1955).
6. C. Rosenstein, *Met. Finish.*, 17 (1990).
7. H. Benninghoff, *Oberfläche*, 8 (1970).
8. F. A. Lowenheim, ed., *Modern Electroplating*, 3rd ed., Wiley, New York, 1974.
9. S. Hirsch and C. Rosenstein, *Metal. Finish. Guidebook and Directory*, **94**, 289 (1996).
10. G. Strube, *Galvanotechnik*, **73**, 1330 (1982).
11. M. Jordan, *The Electrodeposition of Tin and its Alloys*, Eugen G. Leuze Verlag, Saulgau, Germany, 1995, p. 103.
12. R. Schetty, *Electron. Packaging Prod.*, 114 (February 1994).
13. M. Jordan, *Galvanotechnik*, **84**, 797 (1993).

14. JP 89.152295.
15. EP 0192273.
16. U.S. Patent 4,691,670.
17. T. Sonoda, H. Nawafune, and S. Mizumoto, *Plating Surf. Finish.*, **82**, 66 (1995).
18. A. H. DuRose and D. M. Hutchinson, *Plating*, 470 (1953).
19. J. B. Mohler, *Met. Finish.*, 45 (1971).
20. U.S. Patent 3,554,878
21. R. A. Bulwith, *PC Fab.*, 46 (1988).
22. W. T. Hobson, *Plat. Electron. Ind. Symp.*, 161 (1973).
23. P. A. Kohl, *Oberfläche-Surf.*, **23**, 190 (1982).
24. J. A. McCarthy, *Plating*, 805 (1960).
25. Lea-Ronal, U.S. Patent 4,717,460 (1987).
26. C.-P. Ho, *Materials Protect.*, **24**, 20 (1991).
27. Lea-Ronal, U.S. Patent 5,378,347 (1993).
28. E. Raub and W. Blum, *Metalloberfläche* (part A), **9**, 54 (1955).
29. G. Cavanaugh and J. P. Langan, *Proc. AES 2nd Electronics Symp.* (February 1969).
30. A. G. Gray, ed., *Modern Electroplating*, Wiley, New York, 1953, p. 92.
31. J. Goebel, *Z. Metallkunde*, **14**, 357, 388, 425, 499 (1922).
32. J. L. Jostan, W. Mussinger, and A. F. Bogenschütz, *Korrosionsschutz in der Elektronik*, Eugen G. Leuze Verlag, Saulgau, Germany, 1986.
33. U. R. Evans, *The Corrosion and Oxidation of Metals*, Arnold, London, 1960, Suppl. 1968.
34. MIL-STD-38510 H.
35. M. Jordan, *Galvanotechnik*, **78**, 480 (1987).
36. B. F. Rothschild, *Met. Finish.*, 35 (1980).
37. Lea-Ronal, U.S. Patent 4,701,244.
38. J. Schütt and R. Schmid, *Galvanotechnik*, **81**, 255 (1990).
39. B. Meuthen, *Metall.*, **36**, 958 (1982).
40. Company brochure, Messr. Rasselstein, 1990.
41. A. Graham and H. L. Pinkerton, *Plating*, **52**, 309 (1965).
42. D. R. Eastham and C. S. Crooks, *Trans. Inst. Met. Finish.*, **60**, 9 (1982).
43. R. Subramaman and S. Ramachandran, *Met. Finish.*, 53 (1966).
44. H. Richter, K. Rueß, A. Gemmler, and W. Leonhard, *83rd AESF Annual Technical Conf. SUR/FIN'96, Proc.*, 247.
45. M. Jordan, *Galvanotechnik*, **79**, 3053 (1988).
46. R. W. Woodgate, *PC Fabrication*, 56 (1991).
47. B. D. Dunn, *Trans. Inst. Met. Finish.*, **58**, 26 (1980).
48. DE 3440668, Dr.-Ing. Max Schlötter.
49. G. Strube, *Galvanotechnik*, **76**, 417 (1985).
50. DIN 50965; ASTM B 545-85.
51. T. Hamman and E. Schedin, *Proc. 6th Nordic Symp. on Tribology, NORDTRIB '94*, Uppsala, 1994.
52. S. Karpel, *Tin and Its Use*, no. 146, 1 (1985).
53. EP 195995, Siemens AG.
54. T. P. Ireland, N. A. Stennett, and D. S. Campbels, *Trans. Inst. Met. Finish.*, **67**, 127 (1989).
55. N. Ibl, *Oberfläche-Surface*, **14**, 367 (1973).
56. K. Madry, *Oberfläche-Surface*, **27** (10), 9 (1986).

10 Electrodeposition of Zinc and Zinc Alloys

RENÉ WINAND

INTRODUCTION

Zinc has a standard reversible potential (−0.76 V/SHE) that is more negative than that of iron (Fe/Fe^{+2}−0.44 V/SHE). For this reason zinc is used for sacrificial cathodic protection of steel against corrosion.

According to Fischer [1], in metal electrodeposition, zinc can be considered as an intermediate metal, giving rise to medium values of overpotentials due to secondary inhibition resulting from adsorbed hydrolized species.

In acid sulfate solutions without an organic additive, BR metallographic cross-sectional structures are obtained (BR: basis reproduction; coherent deposits with coarse crystals whose diameter increase with the deposits thickness; see Fig. 1). In chloride electrolytes, even coarser grains are observed, because of the more activating character of chloride ions against sulfate ions [2] and despite the complex formation in solution [3, p. 464]. In cyanide baths, on the other hand, much finer grains are obtained, and the deposits pertain to the FT (field-oriented texture: coherent deposits, with elongated crystals perpendicular to the substrate, with almost constant diameter throughout the deposit thickness) or UD types (UD: unoriented dispersion type; coherent deposit, with small crystals showing three dimensional nucleation to occur all the time during electrodeposition).

Figure 2 shows a deposit obtained in a cyanide solution without an organic additive: the structure is initially FT, but it becomes progressively bad. Going from a cyanide bath to noncyanide alkaline baths should result in still worst deposits because electrolyte complexation decreases [3] and exchange current density increases [4, pp. 528–543].

Accordingly, when rather thick deposits are needed and/or if a bright surface is required, organic additives are always present at high concentration (1 g $liter^{-1}$ or more). This always occurs in barrel-and-rack plating of small pieces of equipment, because the local current densities will vary over a wide range. For continuous plating on steel sheets or wires, constant current density and rather good hydrodynamic control may be achieved over the whole cathodic surface, and organic additives are not so much in use.

Modern Electroplating, Fourth Edition, Edited by Mordechay Schlesinger and Milan Paunovic.
ISBN 0-471-16824-6

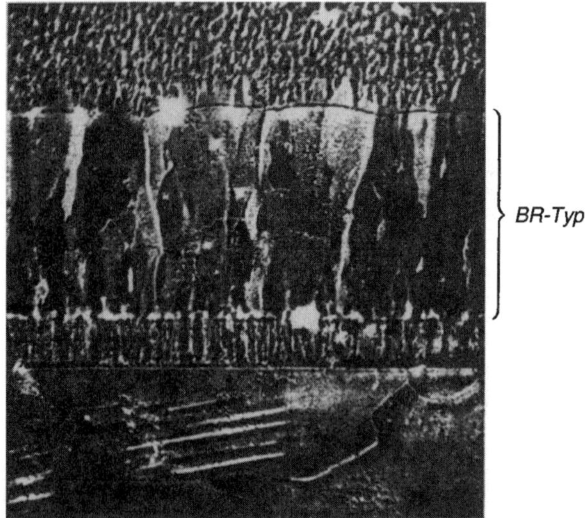

Figure 1 Zinc electrodeposited from a sulfate electrolyte (×400). After (1), p. 619 – BR deposit.

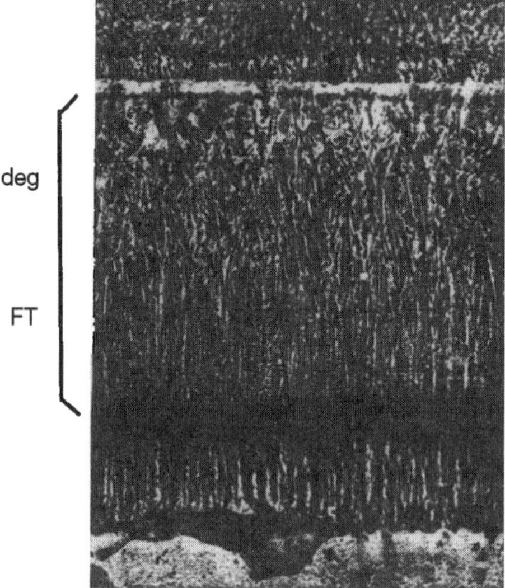

Figure 2 Zinc deposited from a cyanide electrolyte without organic additive (×400). After (1), p. 476 – Initial FT deposit progressively degenerated.

Whatever the pH, zinc is always more negative than hydrogen [5]. In electrolytes where zinc is complexed, its standard reversible potential is still more negative; for instance, for cyanide baths, the reversible potential for reaction

$$[Zn(CN)_4]^{2-} + 2e \leftrightarrow Zn + 4CN^-$$

in alkaline solution is -1.26 V/SHE [6]. Consequently hydrogen evolution occurs at the cathode as a competitor to zinc deposition, resulting in a decrease of current efficiency and eventually in some atomic hydrogen diffusion into the substrate.

Finally, in industrial processes electrolytic plating is an alternative to hot dip galvanizing. Both processes having advantages and disadvantages. Continual technological improvements make it difficult to conclude in favor of one process against the other. However, hot dip includes always some surface alloying by diffusion, and the deposit thickness is less easily controlled than in electroplating.

In this chapter, barrel-and-rack plating will be discussed separately from continuous plating on sheets and wires. The reason is that the electrolytes and required properties of the deposits are different. Of course some overlap is inevitable.

Since the last edition of *Modern Electroplating* [7], the main trends in zinc and zinc alloys plating are as follows:

- Replacement of cyanide baths in barrel-and-rack plating by noncyanide alkaline baths or by acid baths
- Large development of high current density plating of zinc and zinc alloys on continuous steel sheets or wires, with a very active competition against hot dip
- Enforcement of environmental constraints

It should be emphasized that approximately half of the world's consumption of zinc is used to coat steel by various processes. This is a huge business, representing each year about 4 millions tons of zinc and, in the case of sheets, more than 40 millions tons of steel. Electroplating of steel sheets covers between 25% and 30% of zinc production [8, 9].

1 BARREL AND RACK PLATING [7, 10–13]

Zinc is often used for coating iron and steel parts when protection from either atmospheric or indoor corrosion is the primary objective. Without subsequent treatment, electroplated zinc becomes dull gray after exposure to air, so that bright zinc is always given a chromate conversion coating or a coating of a clear lacquer (or both) if a decorative finish is required.

Commercial zinc plating is accomplished in cyanide baths, alkaline noncyanide baths, and acid chloride baths. In the 1970s most commercial zinc plating was done in conventional cyanide baths, but the strong international effort to lower pollution emissions has led to the development of other processes. Today alkaline noncyanide and acid chloride baths comprise most of the production.

1.1 Cyanide Zinc Baths

Bright cyanide zinc baths can be divided into four broad classifications based on their cyanide content:

- Regular cyanide zinc baths
- Midcyanide baths
- Low cyanide baths
- Microcyanide baths

Table 1 summarizes the various compositions of cyanide zinc baths [10].

TABLE 1 Composition of Cyanide Zinc Baths (g liter^{-1})

Constituent	Regular	Mid	Low	Micro
$Zn(CN)_2$	60	30	10	(a)
NaCN	40	20	8	1
NaOH	80	75	65	75
Na_2CO_3	15	15	15	—
Na_xS_y	2	2	—	—
Brightener	1–4	1–4	1–4	1–5

(a): Zinc anodes are dissolved to the desired concentration of zinc metal in solution, that is, 7.5 g liter^{-1}.

Normally temperature ranges from 20° to 40°C. In barrel plating, the average current density is 0.6 A dm^{-2} with 12 to 25 V. In rack plating, the current density is between 2 and 5 A dm^{-2} with 3 to 6 V.

As can be seen in the table, most cyanide baths are prepared from zinc cyanide, sodium cyanide, and sodium hydroxide, with sodium polysulfide acting as a bath purifier and thus precipitating heavy metals such as lead and cadmium that enter the bath as soluble anode impurities. In looking at the chemistry of these baths, it should be emphasized that zinc cyanide is practically insoluble in water. Nevertheless, when added to sodium cyanide, the zinc cyanide dissolves to produce a complex salt, sodium zinc cyanide, as follows:

$$Zn(CN)_2 + 2NaCN \rightarrow Na_2Zn(CN)_4$$

On the other hand, when used with sodium hydroxide, zinc cyanide yields sodium zincate and sodium cyanide:

$$Zn(CN)_2 + 4NaOH \rightarrow Na_2Zn(OH)_4 + 2NaCN$$

Clearly, an equilibrium exists in solution between the cyanide zinc complex and the zincate ion according to

$$Na_2Zn(CN)_4 + 4NaOH \leftrightarrow Na_2Zn(OH)_4 + 4NaCN$$

or

$$Zn(CN)_4^{2-} + 4OH^- \leftrightarrow Zn(OH)_4^{2-} + 4CN^-$$

Thus the addition of sodium hydroxide reduces the amount of cyanide zinc complex and increases the zincate ion concentration. Accordingly the cyanide baths may be considered mixtures of two different solutions: pure cyanide and pure zincate. The dissolution constant of $Zn(CN)_4{}^{2-}$ is very low (in the range of 10^{-9} to 10^{-20}; see [1, p. 174], and the concentration of free Zn^{++} ions in solution is very low. In zincate solutions at high pH, if some dissociation of $Zn(OH)_3^-$ and $Zn(OH)_2$ is observed, the concentration of free Zn^{++} ions in solution is also very low as shown in Figure 3 [3, p. 464].

An investigation of the cathode reaction mechanism at a mercury electrode has shown that in cyanide and in alkaline noncyanide baths, the reacting specie at the

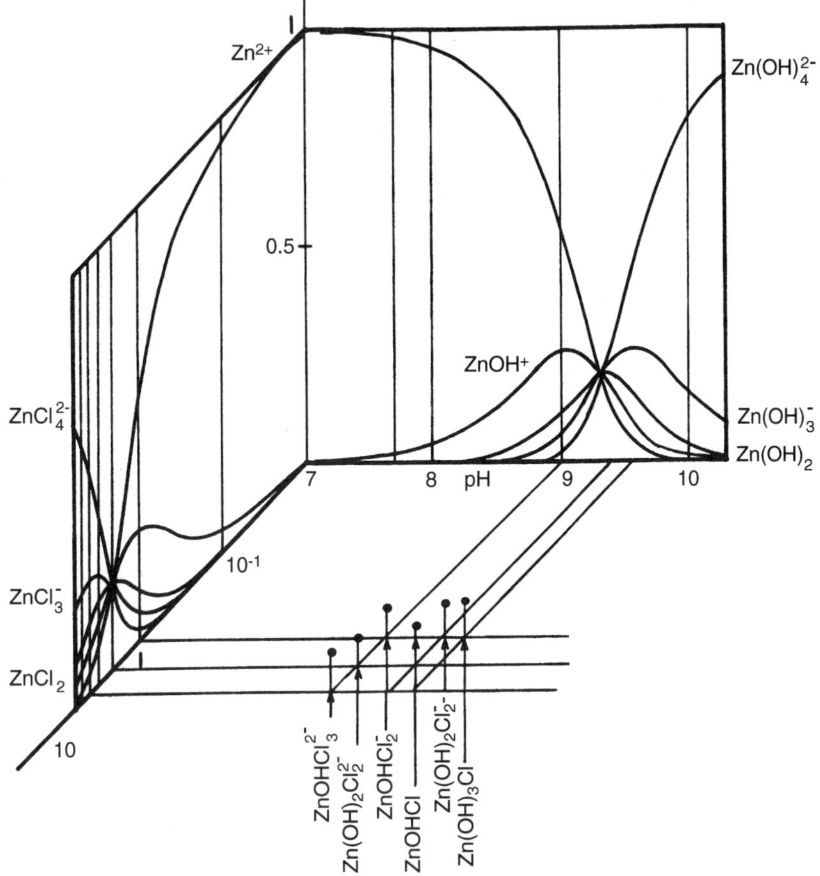

Figure 3 Relative concentrations of basic zinc and zinc chloride species in aqueous solution as a function of pH (3).

surface of the electrode remains the same. Therefore the reaction mechanism was suggested to be as follows [4, pp. 536, 539]:

For cyanide baths,

$$Zn(CN)_4^{2-} + 2OH^- \leftrightarrow Zn(OH)_2 + 4CN^-$$

$$Zn(OH)_2 + 2e \rightarrow Zn + 2OH^-$$

For noncyanide alkaline baths,

$$[Zn(OH)_4]^{2-} \leftrightarrow Zn(OH)_2 + 2OH^-$$

$$Zn(OH)_2 + 2e \rightarrow Zn + 2OH^-$$

A more detailed examination of the cathode mechanism revealed that in the reaction ZnOH is adsorbed [3, pp. 477, 480]. Therefore the second reaction in each above-mentioned formula is replaced by

$$Zn(OH)_2 + e \rightarrow [Zn(OH)_2]^-$$

$$[Zn(OH)_2]^- \leftrightarrow ZnOH + OH^-$$

$$Zn(OH) + e \leftrightarrow Zn + OH^-$$

In this last row of reactions at low current density (or overpotential) the chemical reaction is the rate determining step, whereas at high current density (or overpotential), the last electrochemical reaction is rate determining.

From a practical point of view, passing from regular to low-cyanide zinc baths is a continuous process, without drastic changes in bath behavior. In every case the current efficiency drops from about 90% at $1\,A\,dm^{-2}$ to less than 60% at $6\,A\,dm^{-2}$. This achieves very good throwing and covering power. Microcyanide baths represent essentially a retrogression from alkaline noncyanide baths; cyanide acts as an additive that can be replaced by organic ones.

Zinc plating bath brighteners are almost exclusively proprietary mixtures of organic additives, usually combinations of polyproxyamine reaction products, polyvinyl alcohols, aromatic aldehydes, and quaternary nicotinates. Quaternary pyridinium compounds are also used as brighteners [14]. All additives are formulated at high concentrations in order to achieve brightness at both low and high current density areas and to maintain stability at elevated temperatures.

1.2 Alkaline Noncyanide Zinc Baths

These baths are a logical development in the effort to produce a nontoxic cyanide-free zinc electrolyte. Initially these baths were considered as able only to give spongy or powdery deposits, which were of interest mainly for electrochemical power sources. The same type of electrolyte was used to investigate the dendritic electrocrystallization of zinc [15] with exchange current densities in the range of $10\,A\,dm^{-2}$. From a more practical point of view, it was shown in a flow-through channel cell that various types of deposits (flat, bulbous, or dendritic) could be obtained as a function of current density and Reynolds number [16]. Further research on the subject concentrated on the deposit morphology [15–23], with the aim to improve the control of surfaces with powdery deposits, and also on the quality of dense deposits made, first, by pulsated current or by superimposing ac on dc and, second, with organic additives or on the electrode mechanism [24–32]. To summarize, because of the high value of the exchange current density (see [30, p. 2657]; values in the literature are between 2 and $31\,A\,dm^{-2}$), surface diffusion should not be rate-determining; diffusion in solution is more likely to play this role [18]. A very important role is played by good hydrodynamic control, and it is also important to concentrate on brighteners that are necessary in industrial practice. A drawback is that some contamination of the zinc deposits is possible, and these may be up to 0.3% and 0.5% carbon originating from the additives. This is 10 times the amount of carbon found in deposits using the cyanide system.

Table 2 summarizes the composition characteristics of alkaline noncyanide zinc baths. They pertain to two different ranges of concentrations, called low chemistry (LC) and high chemistry (HC) [11]. Operating temperatures are between 15° and

TABLE 2 Composition Characteristics of Alkaline Noncyanide Zinc Baths ($g\,liter^{-1}$)

Constituent	LC	HC
Zn	6–9	13.5–22.5
NaOH	75–105	120–150
Proprietary additives	(as specified, 1–3%)	

45°C (optimum 25–30°C; see [3, pp. 477–480]. For barrel plating, cathode current density ranges from 2 to $4\,A\,dm^{-2}$ with cell voltage 12 to 18 V. For rack plating, cathode current density is around $0.6\,A\,dm^{-2}$ with 3 to 6 V.

In comparison with cyanide systems, noncyanide alkaline zinc baths have a narrower range of optimum operating zinc concentration. For instance, if the zinc metal content is around $7.5\,g\,liter^{-1}$ (LC) at $3\,A\,dm^{-2}$, a bright deposit is obtained at 80% current efficiency. However, at $5.3\,g\,liter^{-1}$ zinc, at the same current density, the current efficiency is already lower than 60%. Still, raising the zinc content above $17\,g\,liter^{-1}$ will result in dull grey deposits. Also increasing sodium hydroxide concentration increases current efficiency but causes metal buildup on sharp cornered edges [10, p. 246]. Accordingly, for the adopted two types of composition—low chemistry and high chemistry—the zinc and sodium hydroxide concentrations are increased at the same time.

Of course additives are essential [14, pp. 33–35]. In the first commercial alkaline noncyanide zinc baths, it was necessary to replace the complexing effect of cyanide ions by other complexing agents like EDTA, gluconate, tartrate, and triethanolamine. However, this created new effluent problems (e.g., EDTA can prevent precipitation of zinc following pH adjustement and can also delay precipitation of other metals in mixed effluents). This approach has fallen out of use, and the alkaline noncyanide zinc baths may now be considered zincate baths. The deposition of zinc from such an electrolyte follows a four-step mechanism [36]:

$$[Zn(OH)_4]^{2-} \leftrightarrow [Zn(OH)_3]^- + OH^-$$

$$[Zn(OH)_3]^- + e \rightarrow [Zn(OH)_2]^- + OH^-$$

$$[Zn(OH)_2]^- \leftrightarrow Zn(OH) + OH^-$$

$$ZnOH + e \rightarrow Zn + OH^-$$

The second step is the rate-determining step, and it reduces the rate of the reaction to give zinc time to form a dense deposit instead of powdery nonadherent deposits.

Organic additives fall in two categories: carriers and brighteners. The most commonly used carrier is polyvinyl alcohol (PVA). Owing to the polarity of the carbon-oxygen bond, it is possible for this material to be present in significant amounts in the cathode film, forming a weak physical barrier that hinders zinc deposition. More efficient carriers are polyaliphatic amines, aliphatic polyamines containing tertiary or quaternary nitrogen atoms, and heterocyclic amines (imidazole, pyrazole with epichlorohydrin, etc.). When tertiary and particularly quaternary nitrogen atoms are introduced into the polymer chain, extreme brilliance is obtained, even at high current densities. Less effort has been put toward studying brighteners, and hence aromatic aldehydes are still in use.

At the start of the 1980s, zincate plating had not proved to be as reliable as plating from cyanide baths. Four main problems remained to be solved:

- Intermittent deposit blistering (due to initial considerable hydrogen evolution and adsorption into the steel base). This was solved by efficient pretreatment and activation of the substrate, higher solution temperature, higher ratio NaOH/Zn, and avoiding additives with excessive cationic character leading to high carbon content in the deposits that are highly stressed.
- Burning or pitting. This was solved by solution agitation and higher solution temperature.
- Zinc anode passivation [37, 38] due to saturation of the surface by $Zn(OH)_4^{2-}$ and $Zn(OH)_3^-$ forming a solid film progressively converted to $ZnO_{1-x}(OH)_{2x}$, OH^- doped ZnO, and finally Zn-doped ZnO. This can be solved by solution agitation and by chosing not too high a value for the anode current density.
- Chromating problems. This was solved by a very efficient rinse of the plated parts to eliminate any residual zincate plating solution.

Today zincate plating is a highly successful process. In 1970, only 4% of zinc electroplating was achieved by this method, and in 1990, 30%. During the two decades cyanide plating decreased from 97% to 20%; the large percentage is now covered by acid plating, mainly from chloride solutions [39]. Other comments in this direction can be found in [40–43].

1.3 Acid Chloride Zinc Baths

This type of bath radically changed the technology of zinc plating since the early 1970s and constitutes about 50% of all zinc baths for rack and barrel plating in most developed nations.

Despite zinc complexation in solution shown in Figure 3 [44], it was recognized rather early that the exchange current density was high in concentrated solutions: at $2\,g\,liter^{-1}$ Zn^{II}, $J_0 = 0.7\, a_{Zn^{II}}$ in $A\,cm^{-2}$, resulting in $J_0 = 70\,A\,dm^{-2}$ if $a_{Zn^{II}}$ is unity (in [45] values ranging from 5 to 10 $A\,dm^{-2}$ were found at pH = 2 to 3.5, after converting from 10^{-2} to 1 $mole\,liter^{-1}$ Zn^{II}). Concerning the cathode reaction mechanism, it was considered [46] that a fraction of the electrode is covered by adsorbed atomic hydrogen, partially consumed during hydrogen evolution:

$$H^+ + e \rightarrow H_{ads}$$

$$H^+ + H_{ads} + e \rightarrow H_2$$

For zinc electrocrystallization, Zn^I_{ads} adions, weakly bounded to the metal and able to diffuse along the electrode surface were considered, undergoing the following reactions:

$$Zn^{II} + Zn^I_{ads} + e \leftrightarrow 2Zn^I_{ads}$$

$$Zn^{II} + e \rightarrow Zn^I_{ads}$$

$$Zn^{I}_{ads} + H_{ads} \rightarrow Zn + H^{+}$$
$$Zn^{I}_{ads} + e \rightarrow Zn$$

The last two reactions lead to the growth of the deposit. Other growth sites were also considered, but this is beyond the scope of this chapter.

Because of the high values of the exchange current density, noncoherent deposits (dendritic or powdery) are expected if organic additives and/or close hydrodynamic control are not in use [47–50]. In fact this type of electrolyte was used to study the growth of zinc dendrites [51]. Even for electrowinning, the need of organic additives was recognized [52–55], also at high current density (up to $31\,A\,dm^{-2}$; see [56]). However, while for electrowinning rather low concentrations of organics are enough (less than $0.1\ g\,liter^{-1}$), such is not the case in plating if bright deposits are required.

Table 3 gives the composition and operating characteristics of acid chloride zinc plating baths [10, p. 251; 11, pp. 330–335]. These are the electrolytes used at temperatures ranging from 15° to 55°C; the pH values are between 5 and 6. In barrel plating, at current densities between 0.3 and $1\,A\,dm^{-2}$, the cell voltage is between 4 and 12 V, while in rack plating, at current densities between 2 and $5\,A\,dm^{-2}$, the cell voltage lies between 1 and 5 V.

Although organic additives are proprietary, brighteners are again subdivided into two categories [57–61]: carriers (polyalcohol, polyamine, fatty alcohols, polyglycolether, and quaternary nitrogen bonds) and primary brighteners (aliphatic, aromatic, and heterocyclic carbonyl bonds). Carriers, acting as wetting agents, are often necessary to solubilize the primary brighteners.

Bright chloride acid zinc baths have a number of intrinsic advantages:

- Waste disposal is minimized (neutralization at pH 8.5 to 9).
- Current efficiencies are high, even at high current densities.
- In the presence of organic additives, leveling ability is obtained, resulting in superb out-of-bath brightness.
- Cast iron, malleable iron and carbonitrided parts, almost impossible to plate in cyanide and zincate baths, are readily plated in acid zinc baths.
- Electrical conductivity is much higher, resulting in substantial energy savings.
- Hydrogen embrittlement is reduced.

TABLE 3 Composition and Operating Characteristics of Acid Chloride Zinc Plating Baths ($g\,liter^{-1}$)

	All	Low Ammonium		No Ammonium
Constituent	Ammonium Chloride	Potassium Chloride	Sodium Chloride	Potassium Chloride
Zn	15–30	15–30	15–30	22–38
NH_4Cl	120–180	30–45	30–45	—
KCl	—	120–150	—	185–225
NaCl	—	—	120	—
H_3BO_3	—	—	—	22–38
Carrier brightener	4 vol%	4 vol%	4 vol%	4 vol%
Primary brightener	0.25%	0.25%	0.25%	0.25%

There are however a few drawbacks:

- Acid chloride electrolyte is very corrosive.
- Eventual bleed out of entrapped solution in the deposit will ruin the part so that special care should be taken when plating complex assemblies.

As can be seen in Table 3, acid chloride baths are principally of two types: with ammonium chloride or with potassium chloride. Ammonium-based baths were developed first. They can be operated at high current density, but ammonium ions act as complexing agents for nickel and copper in waste streams and can require expensive chlorination for disposal. This is the main reason for the development of the potassium chloride baths.

Other acid baths were developed: sulfate [10, p. 253; 58, 59, 62], fluoborate [10, p. 253], sulfamate, acetate [63], perchlorate [64], and even very low melting point molten salts [65–67]. They are not much in use in barrel-and-rack plating, except for the sulfate bath for specific applications (e.g., complex assemblies).

1.4 Some Further Considerations Concerning Zinc Barrel-and-Rack Plating

The main characteristics of the various processes used for zinc barrel-and-rack plating can be compared: cyanide, zincate, and acid chloride baths. Figure 4 [42, 68]

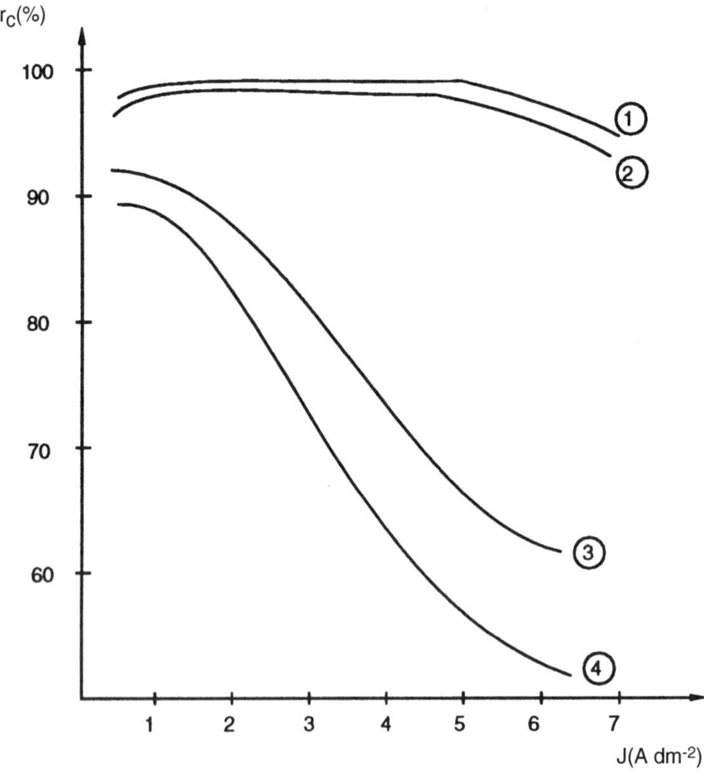

Figure 4 Current efficiency in function of current density for various electrolytes used in zinc barrel and rack plating. 1: acid chloride; 2: acid sulfate; 3: cyanide; 4: noncyanide (zincate).

TABLE 4 Properties of Bright Zinc Deposits

Parameter or			Bright Deposit from NonCyanide	
Property	Dimension	Acid Chloride	Alkaline	Cyanide
Crystal structure	—	hex	hex	hex
Deposit structure	—	UD	FT-UD	FT-UD
Crystal diameter	nm	70 ± 30	90 ± 40	130 ± 50
Microinternal stress	$N\,mm^{-2}$	80	120	150
Macrointernal stress	$N\,mm^{-2}$	−40	+80	−40
Micro Vickers hardness	—	100 ± 30	130 ± 30	140 ± 40
Ductility	%	0.5–7.5	1–5.5	0.2–3.2

Source: Paatsch [42, p. 2635].

shows that acid electrolytes (including the sulfate high acid one) have high current efficiencies (above 95%) over a wide range of current densities, while cyanide electrolytes, and especially the zincate electrolytes, show a sharp decrease in current efficiency at increasing current density. The throwing power, and especially for bright deposit, can be estimated by penetration inside a tube [68, 42]. Table 4 [42] gives some properties of the bright zinc deposits.Further readings on the subject are in [69, 70]. Internal stresses were studied in [71, 72].

Pulsated or periodically reversed currents were assumed to decrease the grain size and increase brightness, eventually without additives [73–77]. Interestingly, the multilayers of zinc are obtained both by pulse plating [77] and by self-oscillations of current at constant cathode potential [78]. Ultrasonic agitation was considered to smooth the surface [79–80]. The influence of impurities and additives in solution was studied especially in acid sulfate solutions, where electroplating conditions are not very far from electrowinning [81–101]. From a more technical point of view, special care must be taken in plating tank construction, especially for acid chloride baths. Tanks may not be constructed of unlined steel in this case, whereas steel is suitable for alkaline noncyanide and for cyanide baths [11, p. 328].

1.5 Zinc Alloy Plating

In barrel-and-rack plating, pure zinc remains the principal coating material for steel [12; 10, p. 285, and 39]. However, during the last two decades alloys were introduced to meet demands for higher quality finishes and longer lasting finishes. The push came mainly from the automotive industry and from aerospace field in its demand for fastener and electrical components. Today, there is talk to replace cadmium because of its toxic nature [102].

Zinc alloy plating will be discussed further in the continuous plating section. In this section, only the characteristics and properties specific to barrel-and-rack plating will be considered.

Zinc-nickel alloys can be plated from acid or alkaline noncyanide solutions. The acid bath yields a higher nickel content than an alkaline bath (10–14% vs. 6–9%). Corrosion protection of steel increases with increasing nickel content up to approximately 15%. Beyond that level, the deposit becomes more noble than the substrate, losing its sacrificial properties. Also above 10% the deposit becomes more passive and accordingly less receptive to form a good chromate conversion coating. Finally, the deposit from the acid bath has a poor thickness distribution, and significant alloy variation is observed from low to high current density areas. On the other hand, the plating rate is lower with the alkaline bath. Finally, the columnar structure of the alkaline bath is preferred to the laminar structure of the acid bath, especially when the part must be mechanically formed after plating. Table 5 summarizes bath parameters for both types.

Zinc-cobalt alloys are plated from conventional low ammonium or ammonium-free acid chloride baths, with the addition of a small amount of cobalt. The alloy contains generally up to 1% cobalt. The current efficiency is high and so is the plating rate. There is reduced hydrogen embrittlement compared with alkaline systems, but the thickness distribution varies with the current density. An alkaline bath is also available. Zinc-cobalt alloys accept blue, bright yellow iridescent, and nonsilver black chromate conversion coatings. Table 6 summarizes bath parameters for both types of baths.

Zinc-iron alloys are mainly used for their ability to produce a nonsilver black chromate. At iron concentrations between 0.3% and 0.9%, they proved to have a very good resistance to the new methanol fuel blend [39]. They are mostly deposited from alkaline baths. Table 7 gives the bath parameters.

Tin-zinc alloys [12] are deposited from acid, alkaline, and neutral baths. In general, alloys range from 70% to 75% tin, and are accordingly more on the tin side. They have excellent resistance to SO_2 atmospheres, frictional properties, and ductility allowing forming after plating. However, they are very soft, and so are susceptible to mechanical damage. They have low electrical contact resistance and do not undergo bimetallic corrosion. As a result they are used for steel fasteners for aluminum alloy panels.

TABLE 5 Bath Parameters for Zinc-Nickel Barrel and Rack Plating (g liter^{-1})

	Acid Baths		
Parameter	Rack	Barrel	Alkaline Bath
Zinc chloride	130	120	—
Zinc metal	—	—	8
Nickel chloride	130	110	—
Nickel metal	—	—	1.6
Potassium chloride	230	—	—
Sodium hydroxide	—	—	130
pH	5–6	5–6	
Temperature (°C)	24–30	35–40	23–26
Cathode current Density (A dm^{-2})	0.1–4	0.5–3	2–10
Anodes	Zinc and nickel separately, eventually with separate rectifiers		Mix: nickel-plated steel 70%–zinc 30%

Source: Budman and Sizelove [12].

TABLE 6 Bath Parameters for Zinc-Cobalt Barrel-and-Rack Plating ($g\,liter^{-1}$)

	Acid Baths		
Parameter	Rack	Barrel	Alkaline Bath
Zinc metal	30	30	6–9
Potassium chloride	180	225	—
Ammonium chloride	45	—	—
Caustic soda	—	—	75–105
Cobalt metal	1.9–3.8	1.9–3.8	0.03–0.05
Boric acid	15–25	15–25	—
pH	5–6	5–6	
Temperature (°C)	21–38	21–38	21–32
Cathode current Density ($A\,dm^{-2}$)	0.1–5.0	1–50	2–4
Anodes	Pure zinc	Pure zinc	Steel (passivated)

Source: Budman and Sizelove [12].

TABLE 7 Alkaline Bath Parameters for Zinc-Iron Alloy Barrel-and-Rack Plating ($g\,liter^{-1}$)

Parameter	Value
Zinc metal	20–25
Iron metal	0.25–0.50
Caustic soda	120–140
Temperature (°C)	18–23
Cathodic current Density ($A\,dm^{-2}$)	1.5–3.0
Anodes	Steel (passivated)

Source: Budman and Sizelove [12].

Brass (zinc-copper alloy) [10, p. 285] can be plated from pure zinc to pure copper. It is used in many applications:

- For bright decorative finishes on wire goods, picture frames, cosmetic items, builders hardware as a thin layer on bright nickel or on other bright plates.
- For antique or other dark finishes, in heavier thicknesses on items that are then burnished, brushed, or buffed.
- For engineering applications as a good drawing lubricant on steel sheet and to provide good adherence of rubber to steel (e.g., steel tire cord wire).

Commercial brass plating solutions are cyanide based. Noncyanide solutions suffer usually from a lack of stability, resulting in difficulties with alloy control on wide ranges of current density. It should be remembered that the reversible standard potentials of the two metals differ by more than one volt. Some sodium carbonate may be added to the cyanide bath to achieve some buffering action and thus keep the color of the alloy constant.

The critical factor is the ratio of cyanide to zinc. The copper content is usually between 15 and 30 g liter^{-1} and the zinc content between 4 and 10 g liter^{-1}. Current density ranges from near 0 to 16 A dm^{-2}, at temperatures between 30° and 80°C (according to the bath composition) at pH 10. Proprietary additives are available, but if plated on a bright base metal, brass deposits are fully bright up to thickness of 2.5 micrometers. Alkaline tartrates [103, 104] and glucoheptonates [105, 106] have been evaluated to replace the cyanides.

2 CONTINUOUS PLATING

Considerable interest has developed over the last 20 years in high current density continuous plating of zinc on steel sheets. Current densities ranging from 90 to 150 A dm^{-2} are currently used, and values up to 300 A dm^{-2} were successfully used at large pilot scale for 5 to 15 μm thick deposits [107, 108, 47–49], while 450 A dm^{-2} was reported for lower thicknesses, around 1 μm [47]. These processes, used now at large industrial scale, are mainly for protecting car bodies against corrosion, and they require a close control of current density and hydrodynamic conditions at the surface of the continuous plate. For this reason equipment design and process control become essential.

Electrocrystallization studies have shown that in simple acid baths with no organic additive, the zinc deposit metallographic structure remains the same up to very high current densities, provided that the ratio of the cathode current density to the diffusion limiting current density is kept constant. In Figure 5 the fields of stability of the main zinc deposit structures are a function of this ratio, and if the inhibition intensity stays at the same position while the current density and agitation are increased simultaneously, the point characterizing the electrodeposition conditions stays at the same position [109, 110]. Further studies on hydrodynamics related to high current density plating can be found in the literature [111–117].

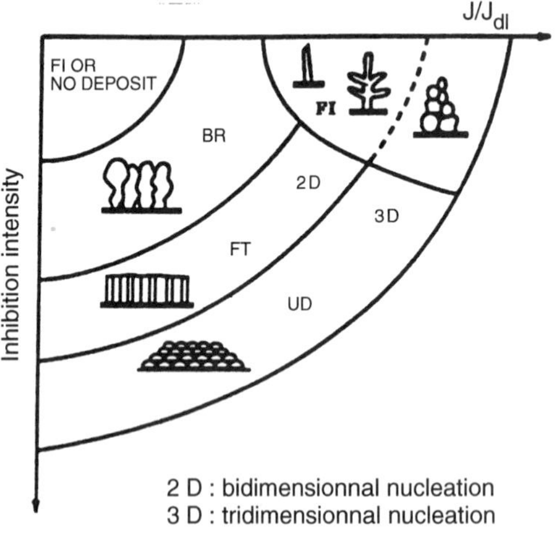

FI : field oriented crystals type (whiskers, dendrites or loose crystalline powder)

BR : basis reproduction type (coherent deposits; grain size and surface roughness increase with deposit thickness)

FT : field oriented texture type (coherent deposits; small grain size, almost constant throughout the deposit)

UD : unoriented dispersed type (coherent deposits; small grain size; new crystals generated throughout deposit thickness)

Figure 5 Fields of stability of the main deposit metallographic structures (109–110).

Many electrolytic cells were designed, respecting more or less the above-mentioned condition on the whole surface of the cathode. Some of them are shown in Figures 6 to 17. There are four main types of cells:

- Those where the steel sheet is pressed against a rotating conducting cylinder feeding the current directly inside the electrolytic cell (Radial Cell [112], Carosel Cell [112], and Kawasaki KC Cell [118]: Figs. 6, 7, and 8).
- Those where the steel sheet moves vertically between anodes, the current feeder rolls are outside the electrolytic cell (Sumitomo Vertical Cell [119],

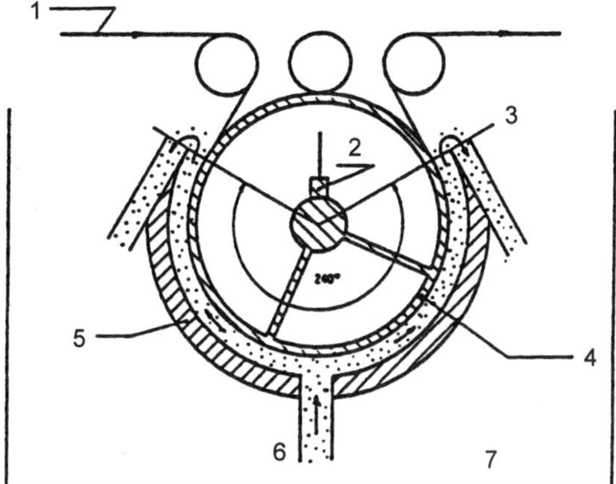

Figure 6 RADIAL CELL (US STEEL) (112): the electrolyte is injected in the middle and flows in directions co-or countercurrent to the steel sheet. 1: Steel sheet; 2: electrical brushes; 3: electrolyte outlet; 4: cathode conducting cylinder; 5: anodes; 6: electrolyte inlet; 7: electrolyte tank.

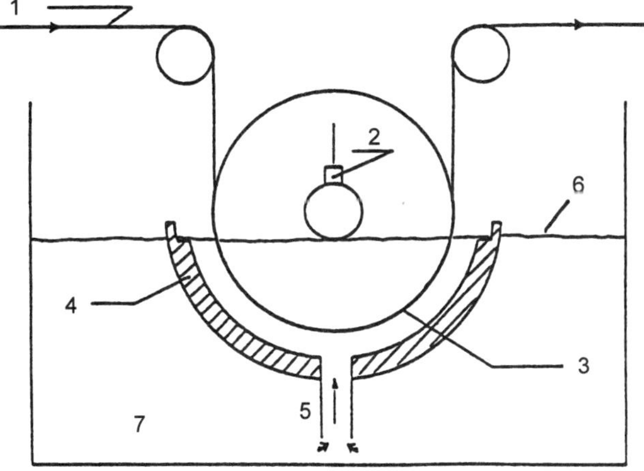

Figure 7 CAROSEL CELL (US STEEL) (112): the electrolyte is also injected in the middle and flows in directions co-or countercurrent to the steel sheet but overflows directly in the electrolyte tank. 1: steel sheet; 2: electrical brushes; 3: cathode conducting cylinder; 4: anodes; 5: electrolyte inlet; 6: electrolyte level; 7: electrolyte tank.

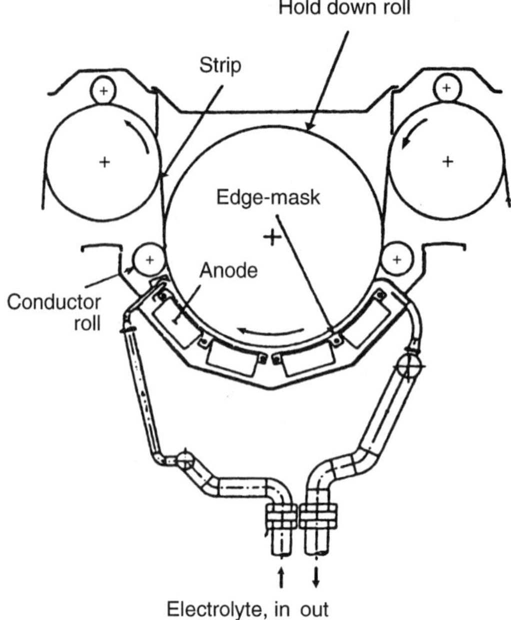

Figure 8 KAWASAKI KC-CELL (118): the electrolyte is injected on one side in counter-current to the steel sheet.

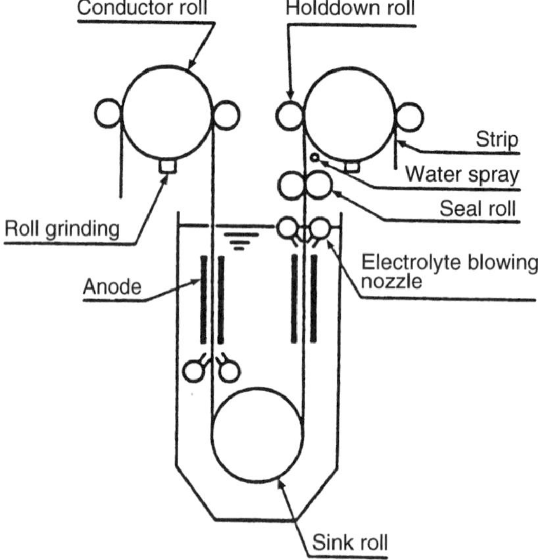

Figure 9 SUMITOMO VERTICAL CELL (119).

Andritz-Ruthner Gravitel Cell [120], and Rasselstein Vertical Hydrojet Cell [121]: Figs. 9, 10, and 11).

- Those where the steel sheet moves horizontally between the anodes, the current feeder rolls being outside the electrolytic cell (Nippon Steel Jet Cell [112],

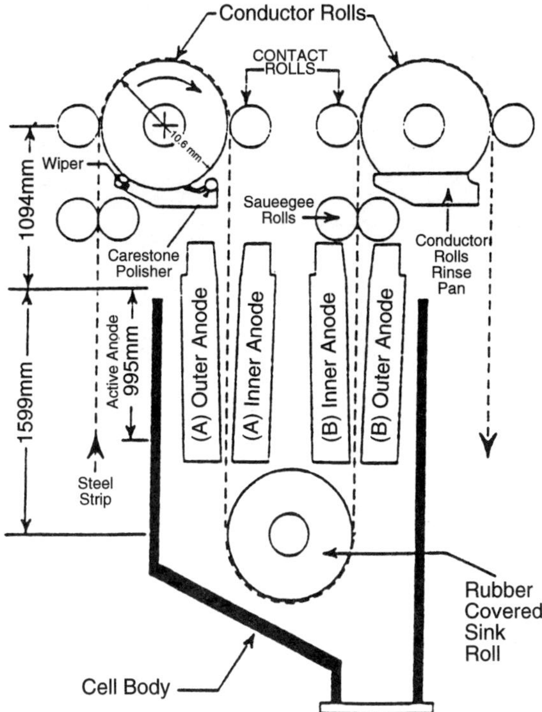

Figure 10 ANDRITZ-RUTHNER GRAVITEL CELL (120) with hollow anode boxes through which the electrolyte is fed by overflow on the upper side of the channels between the steel sheet and the catalytic anodes. The electrolyte flows by gravity in these channels.

Nippon Steel Liquid Cushion Cell [112, 122, 123], Sumitomo horizontal cell and its modified version [124], Cockerill-Sambre Flash Cell [47]: Figs. 12 to 16).

- Those where the steel sheet moves horizontally, pressed against a cathode feeder facing the anode directly in the electrolytic cell (Cockerill-Sambre Belt Cell [47]: Fig. 17).

Most electrolytic cells are able to deposit zinc only on one side of the steel sheet. This has proved to be an advantage for spot welding [125–129].

Electrolytes are acid, chloride, or sulfate, usually without organic additive. Solutions can be either highly acidic (pH lower than one: eventually 130 g liter^{-1} H_2SO_4) or slightly acidic (pH around 3.5). In the latter case, salts of aluminum, sodium, or ammonium are added to increase conductivity. Chloride electrolytes have higher electrical conductivity, but they have drawback, unlike sulfate baths, that they cannot operate with the insoluble anodes because chlorine evolves. Owing to high electrolyte agitation (Reynolds numbers between 15,000 and 60,000 are currently achieved), diffusion-limiting current densities are very high (from 210 to more than 900 A dm^{-2}). At the same time, due to the high hydrogen overpotential on pure zinc, cathodic current efficiencies higher than 95% are obtained, as shown in Figure 18 [107, 110], despite of rather high temperatures (50°C). Figure 19 [108, 47] shows very

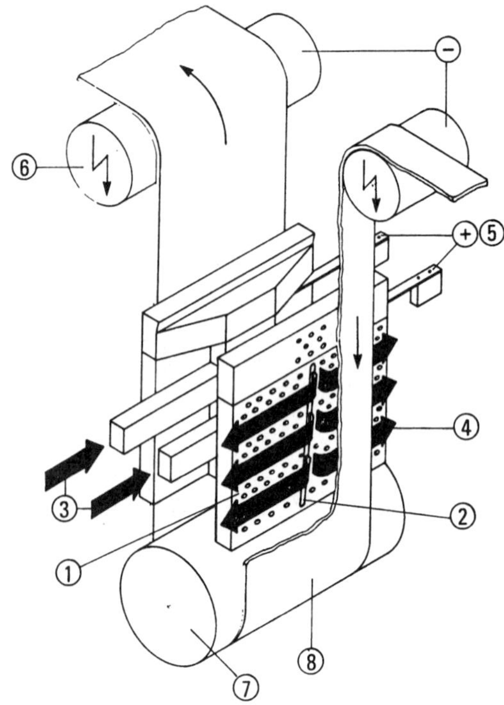

Figure 11 RASSELSTEIN VHC CELL (121). 1: titanium anode basket, containing eventually granulated zinc and nickel if operated in the soluble anode mode; if not, the basket is provided with a perforated insoluble anode; 2: nozzle in anode basket through which the electrolyte is injected between the steel sheet and the anode; 3: electrolyte inlet; 4: electrolyte in anode/cathode gap; 5: anode current supply; 6: cathode conductor rolls; 7: deflector roll; 8: steel sheet.

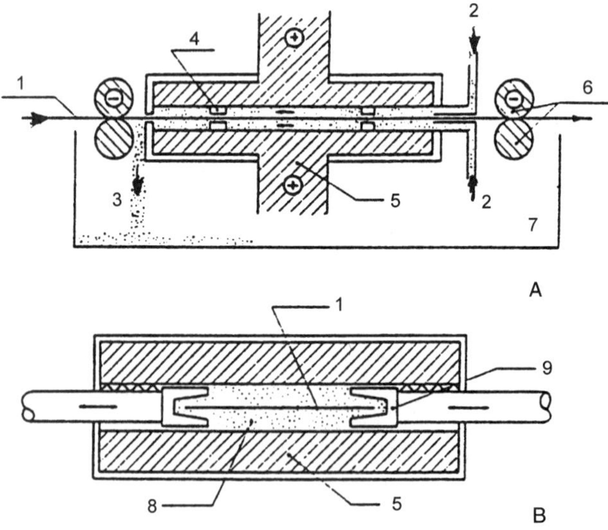

Figure 12 NIPPON STEEL JET CELL (112). A: lateral view; B: transverse section. 1: steel sheet; 2: electrolyte inlet. 3: electrolyte outlet; 4: insulators; 5: anodes; 6: cathode conducting rolls; 7: electrolyte tank; 8: electrolyte; 9: side masks.

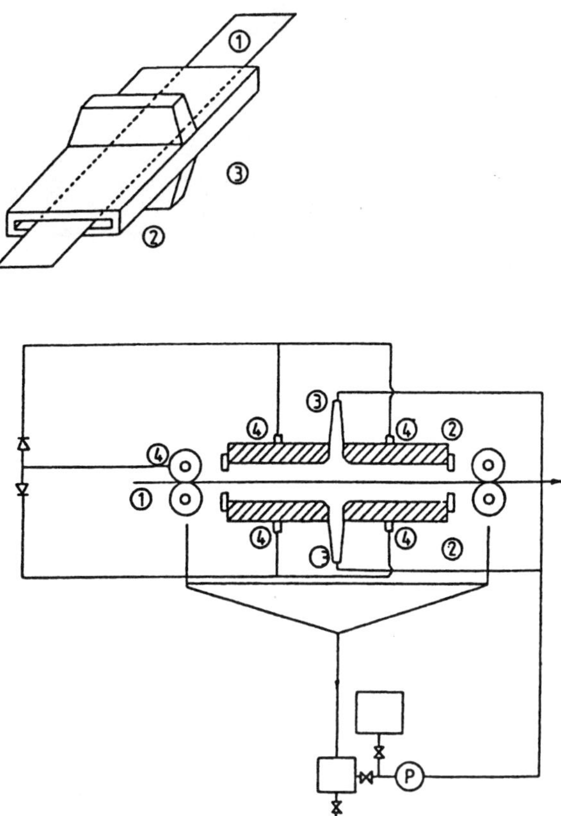

Figure 13 NIPPON STEEL LIQUID CUSHION CELL (112, 122, 123). 1: steel sheet; 2: anodes; 3: electrolyte injectors (the electrolyte flows in co-or countercurrent to the steel sheet); 4: cathode and anode feeders.

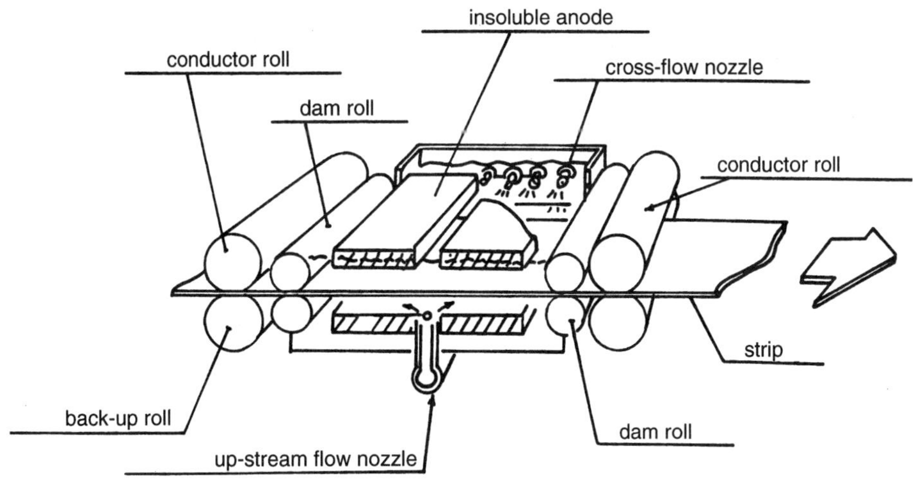

Figure 14 SUMITOMO horizontal cell (124). Initial design with central up-stream flow nozzle and lateral cross-flow nozzles.

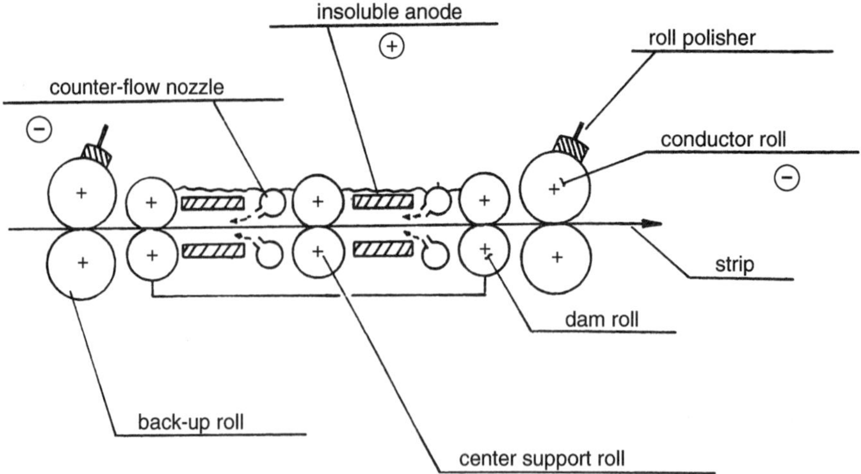

Figure 15 SUMITOMO horizontal cell (124). Improved design with central support roll to avoid oscillations of the steel sheet and counterflow nozzle for injection of the electrolyte.

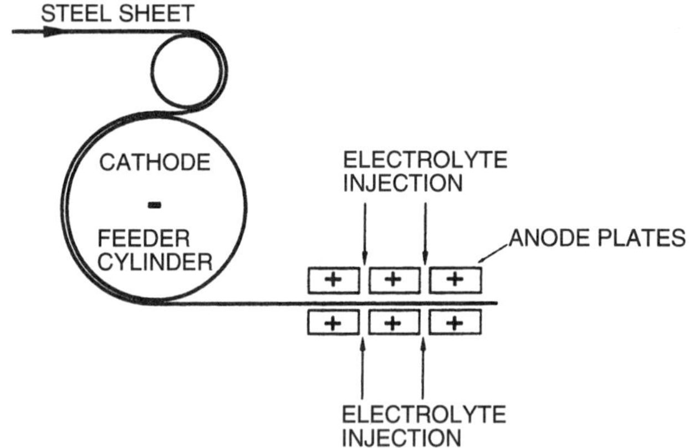

Figure 16 COCKERILL-SAMBRE FLASH CELL (47) designed for very high current density fully covering thin films.

clearly how a surface structure of 10 μm thick zinc deposits is improved at high current density compared with low current density plating in the same electrolyte, without the organic additive. It should be emphasized that bright deposits are not required. A good phosphate conversion posttreatment is always applied for good paint adhesion.

It is important to remember that electroplated steel sheets are rarely used in that form. In most applications (automobile, construction, household, electricity, etc.) they undergo surface conversion (phosphatation, chromatation) and are painted. To evaluate the characteristics of an electroplating process, the properties of the electroplated sheets must be checked according to the manufacturing system creating the final product and according to the market requirements for this final product [130]. For instance, Figure 20 shows a picture of a generalized multilayer coating

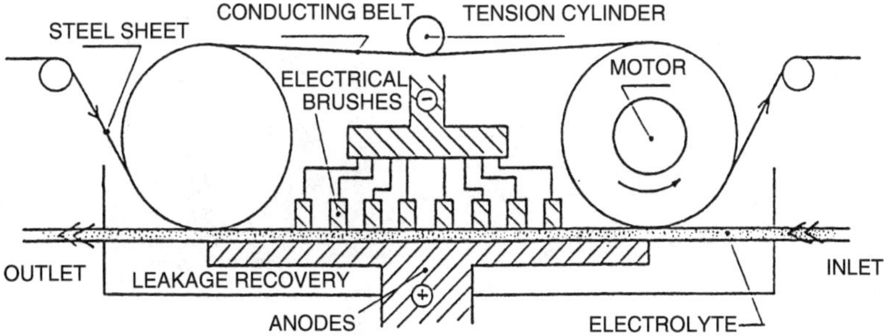

Figure 17 COCKERILL-SAMBRE BELT CELL (47) reproducing a rectangular flat flow-through channel cell with perfect control of the ratio of the current density to the diffusion limiting current density.

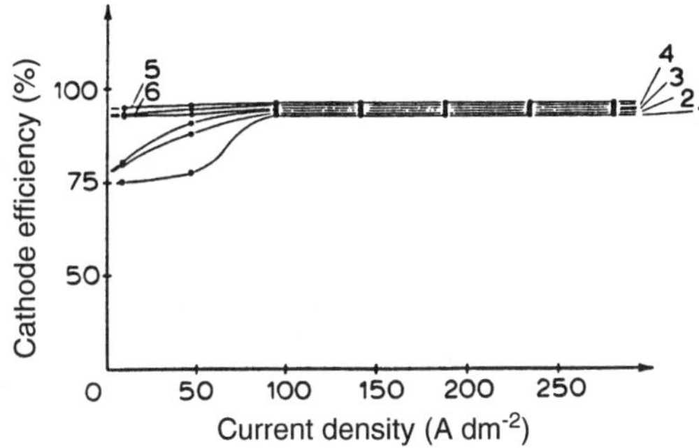

Figure 18 Cathode current efficiency as a function of current density for zinc plating for 80 g liter^{-1} Zn^{++}; 135 g liter^{-1} H_2SO_4, channel cell all at 50°C and electrolyte speed at 4 m s^{-1}. Theoretical deposit thickness: (1) 5 micrometers; (2) 10 micrometers; (3) 25 micrometers; (4) 50 micrometers; (5) 100 micrometers; (6) 200 micrometers.

system for automotive body panels. The steel sheets are stamped or deep drawed after electroplating prior to surface conversion, and the surface morphology of zinc influences the forming accuracy. The zinc deposit surface consists of a collection of discrete hexagonal-shaped grains (Fig. 19), and their size and orientation influence the friction coefficient with the tooling.

The grain size usually decreases with increasing current density, all other factors being fixed. The crystal orientation in sulfate electrolytes is strongly influenced by pH [131] and by electrolyte flow rate [115, 132]. The surface micro-roughness is influenced also by zinc ion supply (agitation) and current density [133]. Studies were made of the die-sheet metal friction for zinc deposits whose grains are oriented with basal planes parallel or approaching normal to the substrate surface, using a light mill oil lubricant and higher friction coefficient (0.19) was measured with the basal planes parallel to the substrate (0.13 in the other case). The more efficient lubricant reduced the difference. Other authors observed similar results and noted the higher degree of surface galling associated with the parallel basal orientation [134, 135]. The characterization of the surface of the electroplated sheet is therefore very important [136]. It is beyond the scope of this chapter to discuss in detail all the other factors that must be considered in terms of deposit integrity after forming and problems

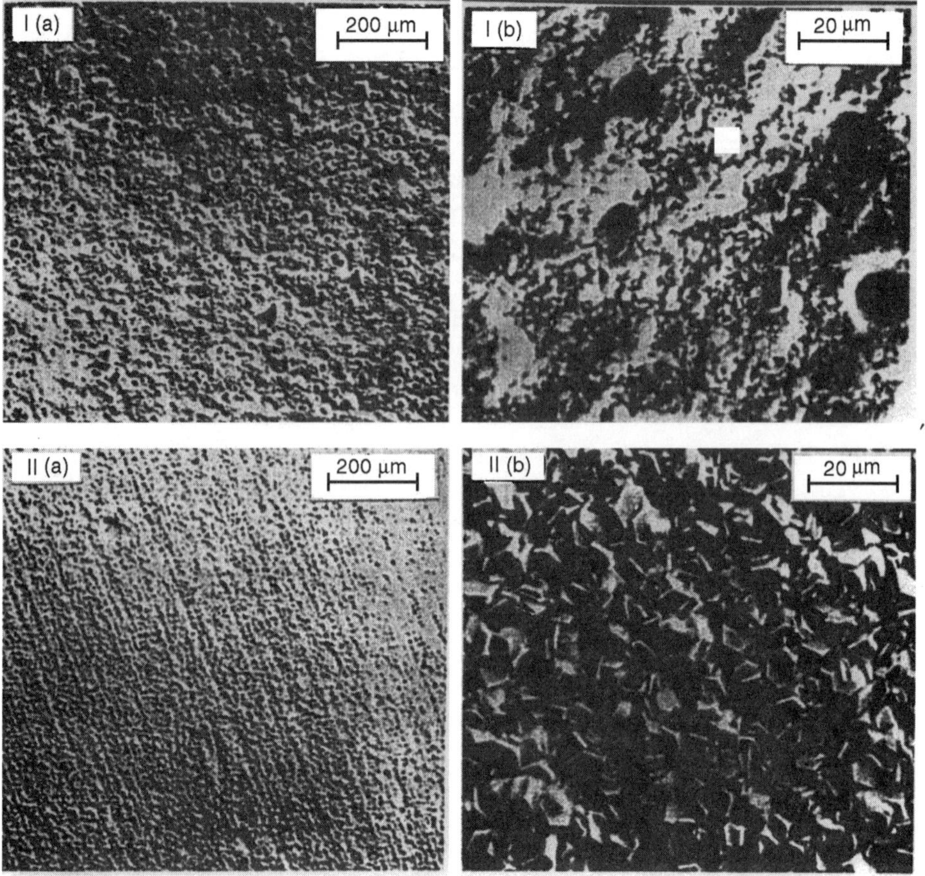

Figure 19 Surface of 10 micrometers thick zinc deposit on steel. Pure sulfate solution: 80 g liter^{-1} Zn^{++}, 135 g liter^{-1} H_2SO_4, 50°C. Electrolyte flow rate: 3 m s^{-1}. I: 10 A dm^{-2} (a) × 100; (b) × 1000. II: 200 A dm^{-2} (a) × 100; (b) × 1000.

connected to spot welding (especially electrode life), phosphatation (phosphophyllite $ZnFe_2(PO_4)_2 \cdot 4H_2O$ is preferred to hopeite $Zn_3(PO_4)_2 \cdot 4H_2O$, and commercial attack solutions now contain the nickel, iron, and/or manganese ions to regulate crystal nucleation and growth; the results are independent of the zinc crystal size and orientation), and painting (priming by electrophoresis generates crating if a threshold voltage of about 250 V is exceeded on zinc coated substrates instead of 400–450 V for cold-rolled steel; interaction of the curing of the base and top coats can occur in the phosphate layer and/or the electrodeposit). Corrosion studies are also beyond the scope of this chapter.

A number of alloys were electroplated [137], and research in the field is still very active, with the aim to improve the general behavior of the plate in connection with customers requirements, mainly deformability, weldability, paintability, and corrosion resistance after surface conversion and painting.

Zinc-nickel alloys were at first investigated at low current densities (up to 5 A dm^{-2}). A complete study of the system in sulfate solutions was performed

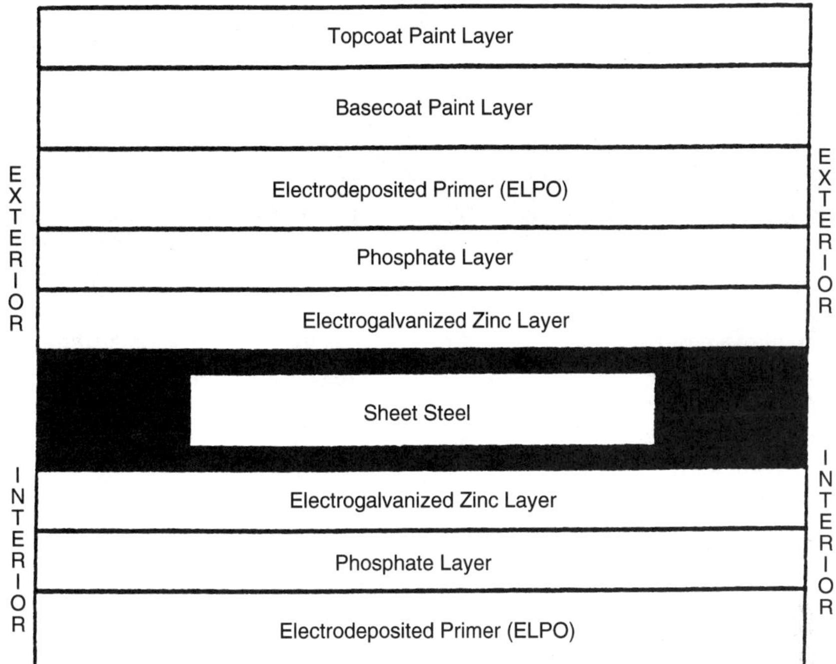

Figure 20 Generalized multilayer coating system for automotive body panels (130).

(138–144). It was shown that in a large field of composition (relative zinc concentration between 0.1 and 0.8) the cathode deposits have a complex structure corresponding to so called γ-phase. A flat and rather bright surface was obtained that had a silvery white appearance. This γ-phase behaves like a monometallic electrode, has a higher hydrogen overpotential than pure zinc, is electronegative versus iron, and has compressive internal stress. This alloy was considered as a possible substitute to cadmium [145, 146, 102]. More recently the same alloy was prepared at high current densities from various electrolytes: sulfates at medium pH [147, 148, 121, 149, 150] (Ni^{2+}: 50–70 g liter^{-1}; Zn^{2+}: 15–30 g liter^{-1}; Na^{+}: 20–30 g liter^{-1}; pH 1–3), at low pH [151, 152] (Ni^{2+}: 30–40 g liter^{-1}; Zn^{2+}: 40–50 g liter^{-1}; H_2SO_4: 35–50 g liter^{-1}), or chloride [153–156]. Again, a zinc-nickel alloy pertaining to the γ-phase at about 13% nickel was found to show the best corrosion resistance among the nickel-zinc alloys and was said to be four times better than pure zinc. Moreover coating adhesion and weldability were excellent [157]. However, current efficiencies were lower than for zinc deposition due to hydrogen evolution, 75% to 90%. Phosphatability and paintability were somewhat poorer. It was easy to obtain almost constant composition of the alloy over a wide range of electrodeposition conditions.

Though more difficult to plate than zinc-nickel, zinc-iron alloys were also studied, mainly with the idea of improving phosphatability and paintability, the two weaknesses of zinc-nickel alloys. Again, various electrolytes are possible: sulfates at moderate pH [150, 158–164] ($FeSO_4 \cdot 7H_2O$: 300 g liter^{-1}; $ZnSO_4 \cdot 7H_2O$: 200 g liter^{-1}; Na_2SO_4: 30 g liter^{-1}; $CH_3COONa \cdot 3H_2O$: 12 g liter^{-1}; pH 1–3; 50 A dm^{-2}; 50°C; flow rate of electrolyte: 0.7–3 m s^{-1}), at low pH [151, 152, 165–167, 109, 110]

(Zn^{2+}: 15–35 g liter^{-1}; Fe^{2+}: 40–60 g liter^{-1}; H_2SO_4: 35–60 g liter^{-1}; Fe^{3+}: <4 g liter^{-1}), or chlorides [168, 169]. Without going into the details [170–172, 159, 164], Figures 21 and 22 show that if excellent workability and corrosion resistance are achieved in the range of iron content between 15% and 25%, good paintability is obtained when the iron content is higher than 50% [173, 174]. Accordingly dual-layer iron-zinc alloys were considered: a first layer at about 18% iron and a top layer at more than 50% iron; and eventually a first layer of nickel-zinc alloy and a top layer of the iron rich zinc alloy [175, 176, 137]. Close control of the hydrodynamic conditions was found to be essential in order to achieve alloy composition homogeneity on the steel strip [177]. Figure 23 shows the difficulties associated with zinc-iron alloys electroplating. Current efficiency is very sensitive to current density and electrolyte composition at constant electrolyte flow rate. Figures 24 and 25 illustrate metallographic structures and Figure 26 their field of stability. Their surface microscopic structure is given in Figure 27 [110].

Zinc-cobalt alloys were also investigated, at first at low current density (lower than 7.5 A dm^{-2}) and in two ranges of cobalt concentrations in the alloy (higher than 1.5% [178], and lower than 0.9% [179]. Alloys at low cobalt concentration, around 0.3%, show already interesting properties, and they have been considered for high current density plating in a ternary zinc-nickel-cobalt alloy [180]. On further research performed in the field, see [181–192].

More fundamentally all the zinc alloys with a metal from the iron group (nickel, iron, cobalt) are obtained under so-called anomalous codeposition [193], that is, with preferential deposition of the less noble zinc. Recent investigations on the mechanism of the reaction [194–197, 165, 149, 198, 167, 199–208] show that this phenomenon is linked to a local rise in pH at the surface of the cathode, due to hydrogen evolution, resulting in some zinc hydroxide adsorbed specie through which the discharge of the iron metal is slowed down. Moreover, these alloys show large discrepancies of crystal structure with thermal equilibrium ones as shown in Figure 28 [196] and Figure 29 [110].

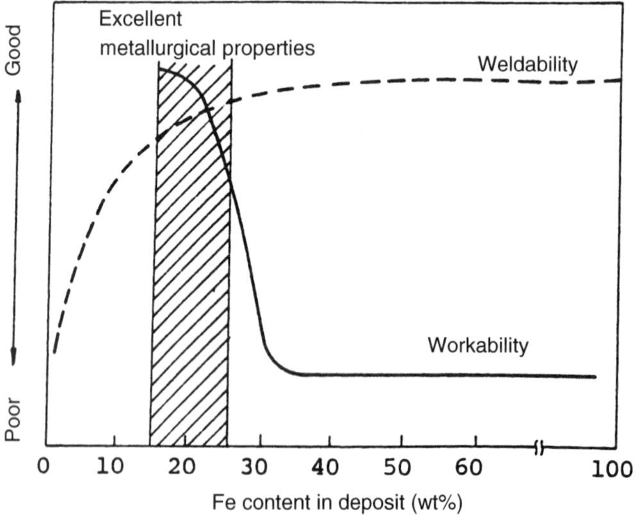

Figure 21 Metallurgical properties of zinc-iron alloys versus their iron content (174).

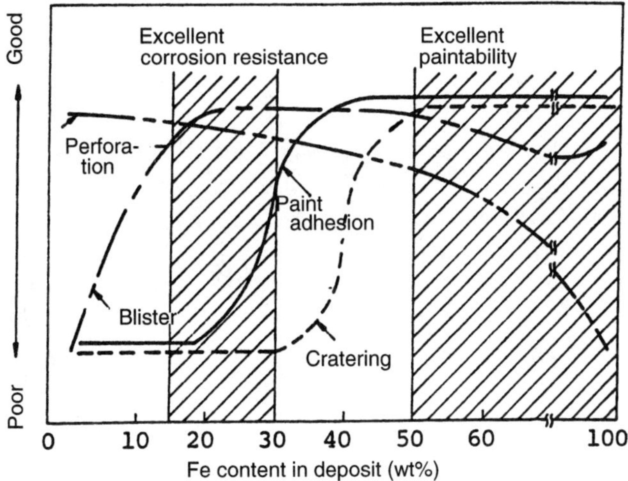

Figure 22 Performance after painting of zinc-iron alloys versus their iron content (174).

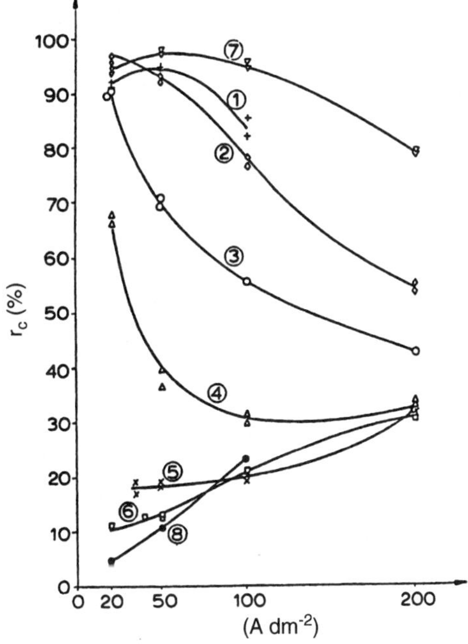

Figure 23 Cathode current efficiency as a function of current density for zinc-iron alloy plating. Channel cell. Deposit thickness: 5–10 micrometers. Electrolyte velocity: 2 m/s^{-1} for 0.2 M H_2SO_4. T: 50°C (ref. 110).

Zinc-manganese alloys have recently received attention for their extremely high corrosion resistance without the aid of painting, for the formation of dense corrosion product of γ-phase Mn_2O_3, for their excellent corrosion resistance after painting with thin alloy coating layers, and for their excellent paintability properties allowing

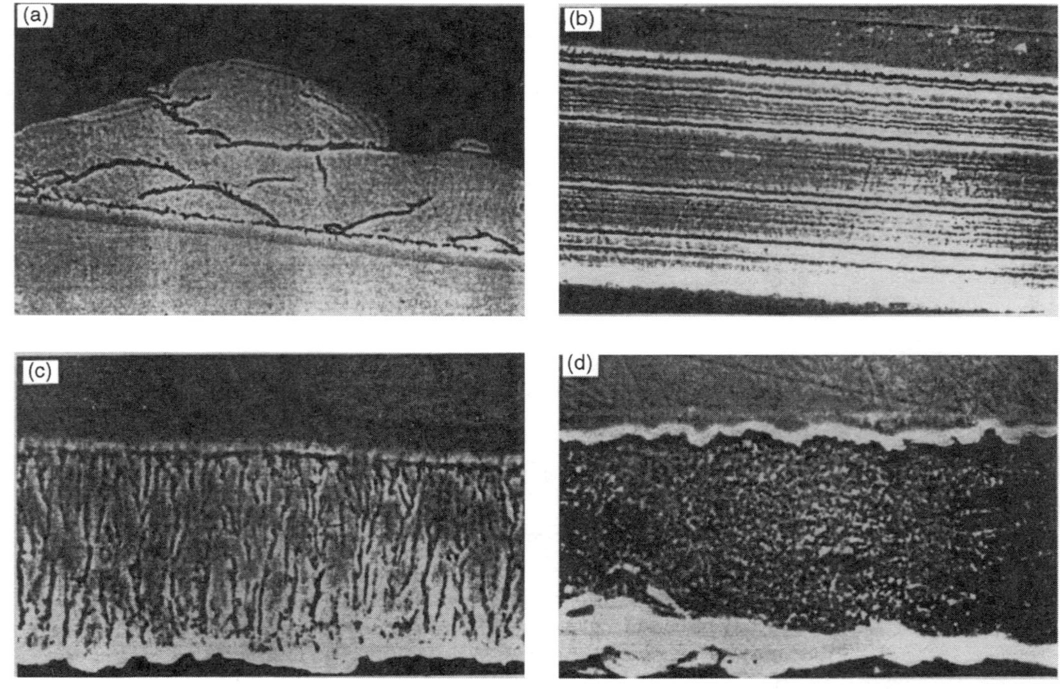

Figure 24 Various metallographic structures observed in the system Fe-Zn: N, RL, FT and pseudo U (later shown as BR) structures are observed: (a) Nodular (N); (b) Rythmic Lamellar (RL); (c) Field oriented texture (FT); (d) unoriented dispersed (UD) (ref. 110).

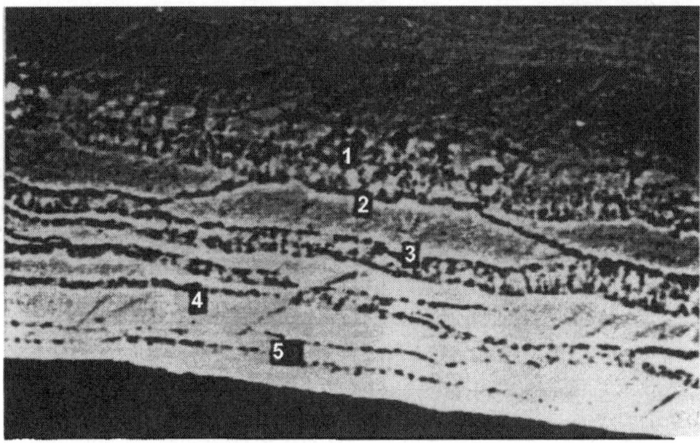

Multiphase deposit (Fe_d=35%)

Figure 25 Detail of a metallographic structure obtained for a multiphase Fe-Zn deposit at 35% Fe (ref. 110).

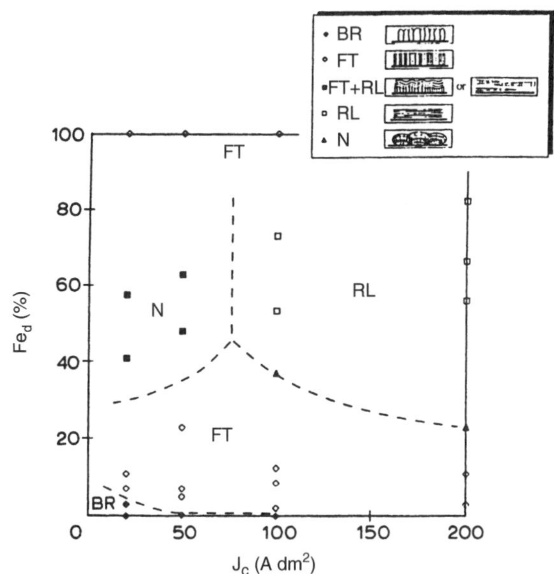

Figure 26 Cross section metallographic structures of zinc-iron alloy deposits as a function of current density and composition. Channel cell: 0.2 M H_2SO_4 with Fe^{++} and Zn^{++} as sulfates at 50°C. Electrolyte speed: $2\,m\,s^{-1}$ (ref. 110).

them to meet the requirements for both interior and exterior surfaces of auto body panels [209]. Electroplating needs complexation because zinc and manganese have reversible potentials different by more than 0.4 V and the system Zn–Mn does not show anomalous deposition. Citrate is usually used as the complexant in solutions of sulfates [210–215]. The difficulties arising from low cathode efficiency and poor bath stability seems to explain why no commercial applications have appeared so far.

In the field of multilayer electrogalvanized steel sheet, zinc-chromium oxide (so-called Zn-Cr-CrO_x or zincrox) is worth mention [216–218]. This coating is made of a

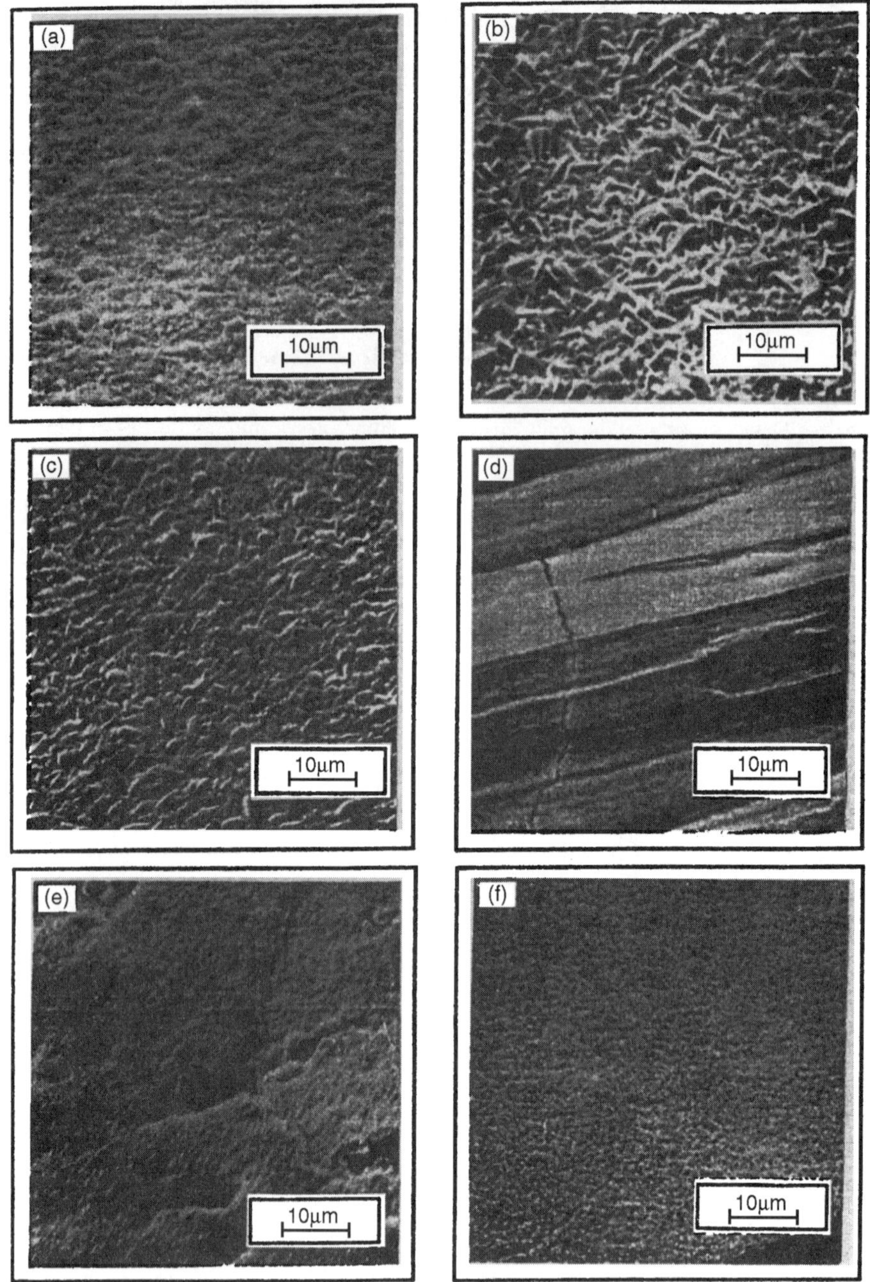

Figure 27 Surface microscope examination of various Zn-Fe deposits. Channel cell: 0.2 M H_2SO_4 with Fe^{++} and Zn^{++} as sulfates at 50°C. Electrolyte velocity: $2\,m\,s^{-1}$. Current density: $100\,A\,dm^{-2}$. (a) $Fe_d = 1\%$; (b) $Fe_d = 10\%$; (c) $Fe_d = 13\%$; (d) $Fe_d = 36\%$; (e) $Fe_d = 54\%$ and (f) $Fe_d = 100\%$.

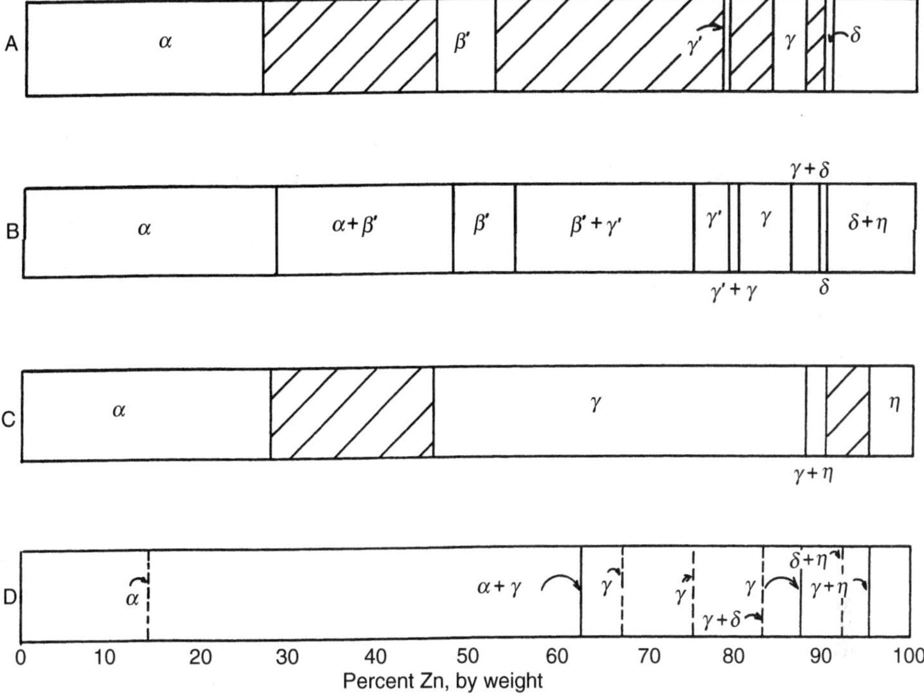

Figure 28 Phase versus composition diagrams for pyrometallurgically (A, B) and electrochemically (C, D) prepared Zn-Ni alloys (no phase determination reported in shaded regions) (196).

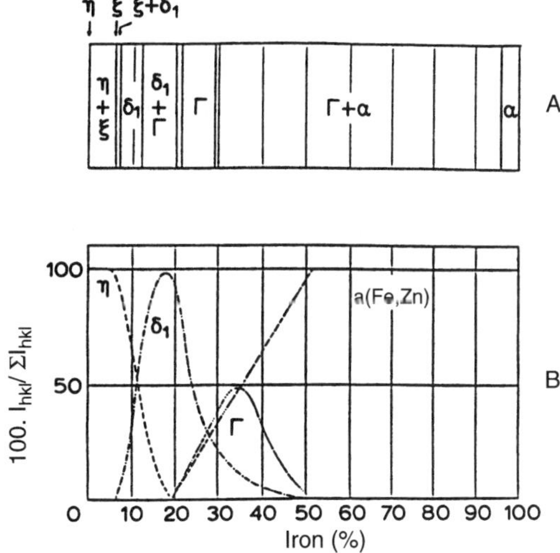

Figure 29 Comparison of electrodeposited phases (B) with thermal equilibrium (A) for iron-zinc alloys (110).

first layer of sacrificial zinc (99% of the coating weight); a second layer of a metal capable of protecting zinc from too early sacrificial corrosion, from acid atmosphere, and from contact with the welding electrodes, and consisting of chromium (0.9% of the coating weight); and a final layer of a very effective paint bonder, consisting of chromium oxide (0.1% of the total coating weight). All layers are obtained by continuous electrolytic process. It should be added that the zinc thickness is in the usual range of about 3 to 11 μm (18–80 $g\,m^{-2}$), the chromium layer (0.5 $g\,m^{-2}$) is only 70 nm thick, while the chromium oxide is in the range of 5 to 10 nm thick as in stainless steel.

So-called composite layers are also mentioned in recent literature. The objective is to improve the mechanical properties of the electroplated zinc or zinc alloy to facilitate forming and also to obtain better temporary protection before phosphatation. This can be achieved by two different methods:

- Include in the zinc or zinc alloy deposits tiny particles of oxides, chromates, carbides [219–227] or even polystyrene [228].
- (Or more easily) Place a thin layer of some organic composite on top of the zinc or zinc alloy deposit [137, 229–232].

Zinc-nickel and zinc-cobalt alloys have been electrodeposited on steel wires to replace the current brass coating [233, 234]. Zinc-manganese coating has been used as reported [235].

Hydrogen embrittlement of the steel substrates, especially for high-strength low-alloyed steel, is a matter of concern. However, it was shown that the substrates are mainly contaminated by atomic hydrogen during the period when full coverage by the first layers of zinc (or zinc alloy) is not achieved. Zinc and zinc alloys constitute a good diffusion barrier to hydrogen. If full coverage is readily achieved, this is an advantage. It can be a drawback, however, if the substrate is charged with atomic hydrogen. As the hydrogen is removed by heat treatment, blistering occurs [236–251]. Acid baths and high current density plating aid in decreasing the hydrogen embrittlement risk.

Insoluble lead-based anodes have been replaced by catalytic dimensionally stable anodes for oxygen evolution. IrO_2-based catalytic layers appear to be favored on titanium substrates [252–256]. They might be challenged by Magnéli phase titanium oxides [257].

Finally, some effort has been devoted to zinc plating on aluminum substrate. The objective is to allow phosphatation of car bodies made partly of steel and partly of aluminum. Surface conditioning is a critical step [258, 259].

REFERENCES

1. H. Fischer, *Elektrolytische Abscheidung und Elektrokristallisation von Metallen*, Springer Verlag, Berlin, 1954, pp. 618–623.
2. A. T. Vagramyan and M. A. Zhamagortsyants, "Electrodeposition of Metals and Inhibiting Adsorption," translation from Russian, U.S. Dept. of Commerce, NTIS, Springfield, VA, Contract NSF C-466, 1974, pp. 40–54.

3. A. Despic, in *Comprehensive Treatise of Electrochemistry*, vol. 7, B. E. Conway, J. O'M. Bockris, E. Yeager, S. U. M. Khan, and R. E. White, eds., Plenum Press, New York, 1983, Ch. 7.
4. K. J. Vetter, *Elektrochemische Kinetik*, Springer Verlag, Berlin, 1961.
5. M. Pourbaix, *Atlas d'équilibres électrochimiques*, Gauthier-Villars, Paris, 1963, p. 409.
6. R. Parsons, *Handbook of Electrochemical Constants*, Butterworths Scientific, London, 1959.
7. F. A. Lowenheim, ed., *Modern Electroplating*, 3rd ed., J. Wiley, New York, 1974.
8. T. Asamura, *Proc. 4th Int. Conf. on Zinc and Zinc Alloy Coated Steel Sheet* (Galvatech, 98), 1998, Chiba, Japan, The Iron and Steel Institute of Japan, pp. 14–21.
9. F. E. Goodwin, ibid., pp. 31–39.
10. H. Geduld, in *Metals Handbook*, 9th ed., vol. 5, American Society for Metals, 1982, pp. 244–255.
11. Cl. Biddulph and M. Marzano, in the *99 Guide and Directory Issue of Metal Finishing*, New York, 1999, pp. 328–337.
12. E. Budman and R. R. Sizelove, ibid., pp. 338–345.
13. R. Winand, in *Electrodeposition Technology, Theory and Practice*, vol. 17, L. T. Romankiw and D. Turner, eds., The Electrochemical Sociey, Permington, NJ, 1987, pp. 207–232.
14. B. S. James and W. R. McWhinnie, *Trans. Inst. Met. Finish*, **58**, 72 (1980).
15. J. Diggle, A. Despic, and J. O'M. Bockris, *J. Electrochem. Soc.*, **116**, 1503 (1969).
16. R. Naybour, *J. Electrochem. Soc.*, **116**, 520 (1969).
17. I. Justnijanovic and A. Despic, *Electrochim. Acta*, **18**, 709 (1973).
18. D. Drazic, S. Jordanov, and Z. Nagy, *Croatica Chem. Acta*, **45**, 199 (1973).
19. M. Froment and G. Maurin, *Electrodep. Surf. Treat.*, **3**, 245 (1975).
20. J. Jovicevic, D. Drazic, and A. Despic, *Electrochim. Acta*, **22**, 589 (1977).
21. R. Walker and N. S. Holt, *Plating Surf. Finish*, **68**, 44 (1981).
22. D. T. Chin, R. Sethi, and J. McBreen, *J. Electrochem. Soc.*, **129**, 2677 (1982).
23. K. Popov and N. Krstajic, *J. Appl. Electrochem.*, **13**, 775 (1983).
24. I. Epelboin, M. Ksouri, and R. Wiart, *J. Electrochem. Soc.*, **122**, 1206 (1975).
25. A. Despic, D. Jovanovic, and T. Rakic, *Electrochim. Acta*, **21**, 63 (1976).
26. J. Jovicevic, D. Drazic, and A. Despic, *Electrochim. Acta*, **22**, 589 (1977).
27. J. Bressan and R. Wiart, *J. Appl. Electrochem.*, **9**, 43 (1979).
28. J. Bressan and R. Wiart, *J. Appl. Electrochem.*, **9**, 615 (1979).
29. C. Cachet, U. Stroder, and R. Wiart, *J. Appl. Electrochem.*, **11**, 613 (1981).
30. M. Abyaneh, J. Hendrikx, W. Visscher, and E. Barendrecht, *J. Electrochem. Soc.*, **129**, 2654 (1982).
31. C. Cachet, U. Stroder, and R. Wiart, *Electrochim. Acta*, **27**, 903 (1982).
32. J. Hendrikx, A. van der Putten, W. Visscher, and E. Barendrecht, *Electrochim. Acta*, **29**, 81 (1984).
33. J. Darken, *Trans. Inst. Met. Finish.*, **57**, 145 (1979).
34. L. Oniciu and L. Muresan, *J. Appl. Electrochem.*, **21**, 565 (1991).
35. F. Galvani and I. A. Carlos, *Met. Finish.*, **95**, 70 (1970).
36. J. O'M. Bockris, Z. Nagy, and A. Damjanovic, *J. Electrochem. Soc.*, **119**, 285 (1972).
37. P. L. Cabot, M. Cortes, F. Centellas, and E. Perez, *J. Appl. Electrochem.*, **23**, 371 (1993).

38. M. Cai and S. M. Park, *J. Electrochem. Soc.*, **143**, 2125 (1996).
39. Th. Natorski, *Met. Finish.*, **90**, 15 (1992).
40. W. Mandel and V. van der Putten, *J. Oberflächentech.*, **33**, 48 (1993).
41. G. D. Wilcox and P. J. Mitchell, *Trans. Inst. Met. Finish.*, **65**, 76 (1987).
42. W. Paatsch, *Galvanotechnik*, **83**, 2633 (1992).
43. S. J. Wake and T. Pearson, *Proc. AESF Ann. Tech. Conf.*, **83**, 157 (1996).
44. R. Winand, *Hydrometall.*, **27**, 285 (1991).
45. H. Sierra Alcazar and J. Harrison, *Electrochim. Acta*, **22**, 627 (1977).
46. J. Bressan and R. Wiart, *J. Appl. Electrochem.*, **9**, 43 (1979).
47. R. Winand, *Hydrometall.*, **29**, 567 (1992).
48. R. Winand, *Electrochim. Acta*, **39**, 1091 (1994).
49. R. Winand, *Electrochim. Acta*, **43**, 2925 (1998).
50. R. Winand, *Parker Int. Symp. on Surf. Technologies*, Nihon Parkerizing, Tokyo, 1998, pp. 85–92.
51. Y. Oren and U. Landau, *Electrochim. Acta*, **27**, 739 (1982).
52. D. J. Mackinnon, J. Brannen, and V. Lakshmanan, *J. Appl. Electrochem.*, **9**, 603 (1979).
53. D. J. Mackinnon, J. Brannen, and V. Lakshmanan, *J. Appl. Electrochem.*, **10**, 321 (1980).
54. D. J. Mackinnon, J. Brannen, and R. Morrison, *J. Appl. Electrochem.*, **12**, 39 (1982).
55. D. Mackinnon and J. Brannen, *J. Appl. Electrochem.*, **12**, 21 (1982).
56. B. Thomas and D. Fray, *J. Appl. Electrochem.*, **11**, 677 (1981).
57. G. K. Schwartz, *Galvanotechnik*, **68**, 874 (1977).
58. M. Pushpavanam and B. A. Shenoi, *Met. Finish.*, **75**, 29 (1977).
59. M. Harsanyi, J. Harsanyi, E. Grunwald, and C. Varhelyi, *Galvanotechnik*, **72**, 123 (1981).
60. E. Jeannier, *Oberflächen Werkst*, **34**, 14 (1993).
61. Th. C. Franklin, T. Williams, T. S. N. Sankara Narayanan, R. Guhl, and G. Hair, *J. Electrochem. Soc.*, **144**, 3064 (1997).
62. S. Morisaki, T. Mori, and S. Tajima, *Plating Surf. Finish.*, **68**, 55 (1981).
63. S. S. Abd el Rehim, S. M. Abd el Wahaab, E. E. Fouad, and H. H. Hassan, *J. Appl. Electrochem.*, **24**, 350 (1994).
64. S. Sreeveeraraghavan, R. M. Krishnan, S. R. Natarajan, and B. Balakrishnan, *Trans. SAEST*, **31**, 41 (1996).
65. W. R. Pitner and Ch. L. Hussey, *J. Electrochem. Soc.*, **144**, 3095 (1997).
66. Y. F. Lin and I. W. Sun, *J. Electrochem. Soc.*, **146**, 1054 (1999).
67. Y. F. Lin and I. W. Sun, *Electrochim. Acta*, **44**, 2771 (1999).
68. W. Paatsch, *Galvanotechnik*, **68**, 392 (1977).
69. H. Benninghoff, *Metalloberflache*, **34**, 445 (1980).
70. G. Strube, *Galvanotechnik*, **77**, 1318 (1986).
71. S. A. Armyanov and G. S. Sotirova, *Surf. Tech.*, **79**, 311 (1979).
72. S. A. Armyanov and G. S. Sotirova, ibid., **79**, 325 (1979).
73. A. R. Despic and K. I. Popov, *J. Appl. Electrochem.*, **1**, 275 (1971).
74. H. D. Hedrich, W. Gunther, and Ch. J. Raub, *Metalloberflache*, **34**, 462 (1980).
75. W. Paatch, *Metalloberflache*, **41**, 39 (1987).
76. Y. Plin and J. R. Selman, *J. Electrochem. Soc.*, **138**, 3525 (1991).
77. V. A. Zabludovsky, N. V. Fedotova, and E. F. Shtapenko, *Trans. IMF*, **74**, 106 (1996).

78. H. Yan, J. Downes, P. J. Boden, and S. J. Harris, *J. Electrochem. Soc.*, **143**, 1577 (1996).
79. R. Walker and N. S. Holt, *Plating Surf. Finish.*, **67**, 92 (1980).
80. R. Walker and N. S. Holt, *Plating Surf. Finish.*, **68**, 44 (1981).
81. D. J. Mackinnon and J. M. Brannen, *J. Appl. Electrochem.*, **7**, 451 (1977).
82. D. J. Mackinnon, J. M. Brannen, and R. C. Kerby, *J. Appl. Electrochem.*, **9**, 55 (1979).
83. D. J. Mackinnon, J. M. Brannen, and R. C. Kerby, *J. Appl. Electrochem.*, **9**, 71 (1979).
84. M. Saloma, J. Avila-Mendoza, G. Alonso-Viveros, and A. Ramirez-Delgado, *Hydrometall.*, **20**, 121 (1988).
85. V. Srinivasan, J. S. Cuzmar, and Th. O'Keefe, *Metall. Trans. B*, **21B**, 81 (1990).
86. S. Rashkov, M. Petrova, and C. Bozhkov, *J. Appl. Electrochem.*, **20**, 11 (1990).
87. C. Bozhkov, M. Petrova, and S. Rashkov, *J. Appl. Electrochem.*, **20**, 17 (1990).
88. R. Wiart, C. Cachet, C. Bozhkov, and S. Rashkov, *J. Appl. Electrochem.*, **20**, 381 (1990).
89. C. Bozhkov, I. Ivanov, and S. Rashkov, *J. Appl. Electrochem.*, **20**, 447 (1990).
90. N. Masse and D. Piron, *J. Appl. Electrochem.*, **20**, 630 (1990).
91. D. J. Mackinnon, R. M. Morrison, J. E. Mouland, and P. E. Warren, *J. Appl. Electrochem.*, **20**, 728 (1990).
92. D. J. Mackinnon, R. M. Morrison, J. E. Mouland, and P. E. Warren, *J. Appl. Electrochem.*, **20**, 955 (1990).
93. C. Cachet and R. Wiart, *J. Appl. Electrochem.*, **20**, 1009 (1990).
94. C. Bozhkov, M. Petrova, and S. Rashkov, *J. Appl. Electrochem.*, **22**, 73 (1992).
95. C. Cachet and R. Wiart, *J. Electrochem. Soc.*, **141**, 131 (1994).
96. J. C. Lin and S. L. Tsai, *J. Appl. Electrochem.*, **24**, 1044 (1994).
97. R. Ichino, C. Cachet, and R. Wiart, *J. Appl. Electrochem.*, **25**, 556 (1995).
98. B. C. Tripathy, S. C. Das, G. T. Hefter, and P. Singh, *J. Appl. Electrochem.*, **27**, 673 (1997).
99. S. C. Das, P. Singh, and G. T. Hefter, *J. Appl. Electrochem.*, **27**, 738 (1997).
100. B. C. Tripathy, S. C. Das, G. T. Hefter, and P. Singh, *J. Appl. Electrochem.*, **28**, 915 (1998).
101. T. Dobrev, C. Cachet, and R. Wiart, *J. Appl. Electrochem.*, **28**, 1195 (1998).
102. M. D. Thomson, *Trans. IMF* (*Bull*), **74** (3), 3 (1996).
103. S. S. Abd el Rehim and M. E. el Ayashy, *J. Appl. Electrochem.*, **8**, 33 (1978).
104. S. S. Abd el Rehim, *J. Appl. Electrochem.*, **8**, 569 (1978).
105. Y. Fujiwara and H. Enomoto, *Surf. Coatings Technol.*, **35**, 101 (1988).
106. Y. Fujiwara and H. Enomoto, *Surf. Coatings Technol.*, **35**, 113 (1988).
107. A. Weymeersch, R. Winand, and L. Renard, *Plating Surf. Finish.*, **68**, 56 (1981).
108. A. Weymeersch, R. Winand, and L. Renard, *Plating Surf. Finish.*, **68**, 118 (1981).
109. R. Winand, *Proc. Int. Conf. on Zinc and Zinc Alloy Coated Steel Sheet* (Galvatech'89), Iron and Steel Inst. of Japan, Tokyo, 1989, pp. 27–36.
110. R. Winand, *J. Appl. Electrochem.*, **21**, 377 (1991).
111. B. Meuthen and D. Wolfhard, *Metalloberflache*, **36**, 70, (1982).
112. R. Winand, *Oberfläche-Surface*, **25** (11), (1984).
113. M. Kawabe, M. Sagiyama, A. Tonouchi, Y. Okubo, T. Adaniya, and T. Hara, *Trans. ISIJ*, **25**, B-260 (1985).
114. R. Winand, *Interfinish*, **88**, Paris (Oct. 1988), vol. 1, pp. 189–202.
115. J. M. Ting, D. Duffy, R. Y. Lin, F. H. Guzzetta, and T. R. Roberts, *Proc. Int. Conf. on Zinc and Zinc Alloys Coated Steel Sheet* (Galvatech'89), ISIJ, Tokyo, 1989, pp. 179–185.

116. H. M. Wang, S. F. Chen, T. J. O'Keefe, M. Degrez, and R. Winand, *J. Appl. Electrochem.*, **19**, 174 (1989).
117. L. Van Hee, J. C. Monnier, R. Winand, and M. Stanislas, *J. Appl. Electrochem.*, **24**, 303 (1994).
118. A. Komoda, A. Kibata, S. Murakami, H. Kimura, K. Mochizuki, and R. Muko, *Proc. Int. Conf. on Zinc and Zinc Alloy Coated Steel Sheets* (Galvatech'89), ISIJ, Tokyo, 1989, pp. 170–177.
119. H. Onishi, T. Shiohara, T. Hattori, and T. Motoyama, *AESF Continuous Steel Strip Plating Symp.*, Dearborn, May 1993, pp. 23–29.
120. M. S. Blaser and E. T. Nowak, ibid., pp. 31–39.
121. K. H. Killian, K. Taffner, F. Weber, and H. U. Weigel, *AESF, 5th Continuous Steel Strip Plating Symp.*, Dearborn, May 1987, p. T1-22.
122. K. Sakai, H. Nitto, M. Kamada, and R. Yoshihara, *AESF 4th Continuous Steel Strip Plating Symp.*, Chicago, 1984, p. J1-25.
123. K. Sakai, H. Nitto, R. Yoshihara, and M. Kitayama, *Iron Steel Eng.*, **62** (8), 49 (1985).
124. T. Nonaka, H. Oishi, H. Nagasaki, S. Shoda, A. Shibuya, and T. Hattori, *AESF 5th Continuous Steel Strip Plating Symp.*, Dearborn, May 1987, p. D1-20.
125. A. Matthews, Eurozinc 1979, Report on seminar Zinc in Europe, London, Apr. 1979, pp. 39–41.
126. E. Gorl and B. Meuthen, *Stahl und Eisen*, **103** (17), 790 (1983).
127. G. J. Harvey and P. N. Richards, *Met. Forum*, **6** (4), 234 (1983).
128. S. Dinda and R. J. Traficante, *Automotive Eng.*, **92** (12), 38 (1984).
129. B. Meuthen and J. H. Meyer zu Bexten, *Iron Steel Eng.*, **62** (6), 26 (1985).
130. J. H. Lindsay, *Proc. 78th AESF Ann. Tech. Conf.*, pp. 25–43 (1991).
131. J. H. Lindsay, R. F. Paluch, H. D. Nine, U. R. Miller, and T. J. O'Keefe, *Plating Surf. Finish.*, **76** (3), 62 (1989).
132. S. Alota, N. Azzeri, R. Bruno, M. Memmi, and S. Ramundo, *Metall. Ital.*, **79** (3), 217–224 (1987).
133. T. R. Roberts, F. H. Guzzetta, and R. Y. Lin, *Proc. AESF 5th Continuous Strip Plating Symp.*, Dearborn, May 1987, pp. J1–22.
134. S. J. Shaffer, A. M. Philip, and J. W. Morris Jr., *Proc. Int. Conf. on Zinc and Zinc Alloy Coated Steel Sheet* (Galvatech'89), ISIJ, Tokyo, 1989, pp. 338–344.
135. S. F. Shaffer, W. E. Nojima, P. N. Skarpelos, and J. W. Morris Jr., *Zinc-Based Steel Coating Systems: Metallurgy and Performance*, G. Krauss and D. K. Matlock, Eds., TMS, Warrendale, PA, 1990, pp. 251–262.
136. V. Leroy, *Proc. Int. Conf. on Zinc and Zinc Alloy Coated Steel Sheet* (Galvatech'89), ISIJ, Tokyo, 1989, pp. 399–409.
137. R. Winand, *Surf. Coatings Technol.*, **37**, 65 (1989).
138. M. Kurachi and K. Fujiwara, *Denki Kagaku*, **38**, 600 (1970).
139. Y. Imai and M. Kurachi, ibid., **45**, 728 (1977).
140. Y. Imai, Sh. Ohsumi, and M. Kurachi, ibid., **46**, 264 (1978).
141. Y. Imai, T. Watanabe, and M. Kurachi, ibid., **46**, 202 (1978).
142. Y. Imai and M. Kurachi, ibid., **47**, 89 (1979).
143. M. Kurachi and K. Fujiwara, *Trans. J. I. M.*, 11, 311 (1970).
144. M. Kurachi, K. Fujiwara, and T. Tanaka, *Proc. 8th Congress of the Int. Union for Electrodep. and Surf. Finish*, pp. 152–157 (1973).

145. J. Dini and H. Johnson, *Met. Finish.*, **77** (8), 31 (1979).
146. J. Dini and H. Johnson, ibid., **77** (9), 53 (1979).
147. R. Noumi, H. Nagasaki, Y. Foboh, and A. Shibuya, SAE Technical Paper Series 820332, Detroit, Feb. 1982, pp. 22–26.
148. S. G. Fountoulakis, R. N. Steinbicker, and T. W. Fisher, *AES 4th Continuous Strip Plating Symp.*, Chicago, May 1984, pp. F1–21.
149. H. Fukushima, T. Akiyama, K. Higashi, R. Kammel, and M. Karimkani, *Metallwissenschaft Technik*, **42** (3), 242 (1988).
150. T. Akiyama, H. Fukushima, K. Higashi, M. Karimkhani, and R. Kammel, *Proc. Int. Conf. on Zinc and Zinc Alloy Coated Steel Sheet* (Galvatech'89), ISIJ, Tokyo, 1989, pp. 45–50.
151. A. Weymeersch, L. Renard, J. J. Conreur, R. Winand, M. Jordan, and C. Pellet, *Proc. 4th AES Continuous Strip Plating Symp.*, Chicago, May 1984, pp. B1–20.
152. Ibid., *Plating Surf. Finish.*, **73**, 68–73 (1986).
153. J. Beers, Bethlehem Steel Research Department Report, Apr. 15, 1983.
154. A. Matsuda, T. Yoshihara, K. Miyachi, A. Komoda, and S. Kikuchi, *Trans. ISIJ*, **24**, B267 (1984).
155. A. Komoda, A. Matsuda, T. Yoshihara, and K. Kimura, *Proc. 4th AES Continuous Strip Plating Symp.*, Chicago, May 1984, pp. C1–29.
156. J. Keller, G. Colin, P. Seurin, and E. Millon, *Proc. Int. Conf. on Zinc and Zinc Alloy Coated Steel Sheet* (Galvatech'89), Tokyo, 1989, pp. 161–177.
157. A. Shibuya, T. Kurimoto, Y. Hoboh, and N. Usiki, *Trans. ISIJ*, **23**, 923 (1983).
158. T. Watanabe, M. Ohmura, T. Honma, and T. Adaniya, *SAE*, **90** (3), 30 (1982).
159. T. Hara, T. Adaniya, M. Sagiyama, T. Honma, A. Tonouchi, T. Watanabe, and H. Ohmura, *Trans. ISIJ*, **23**, 954 (1983).
160. T. Tsuda, K. Asano, and A. Shibuya, *Trans. ISIJ*, **26** (1), 53 (1986).
161. K. Kondo, S. Hinotani, and Y. Ohmori, *J. Appl. Electrochem.*, **18**, 154 (1988).
162. K. Kondo, *ISIJ Int.*, **30**, 464–468 (1990).
163. A. van Cauter, J. Dilewijns, B. C. de Cooman, and A. de Boeck, *Proc. Int. Conf. on Zinc and Zinc Alloy Coated Steel Sheet* (Galvatech'95), Chicago, 1995, pp. 393–398.
164. K. de Wit, J. Dilewijns, A. de Boeck, and B. C. de Cooman, ibid., pp. 399–406.
165. A. Rodriguez Fajardo, R. Winand, A. Weymeersch, and L. Renard, *Proc. 5th AES Continuous Strip Plating Symp.*, Dearborn, May 1987, pp. S1–28.
166. A. A. Rodriguez Fajardo, M. Degrez, and R. Winand, *Oberfläche-Surf.*, **30** (8), 20 (1989).
167. M. Degrez, A. A Rodriguez Fajardo, and R. Winand, *Oberfläche Surf.*, **30** (9), 14 (1989).
168. A. Matsuda, T. Yoshihara, K. Miyachi, A. Komoda, K. Kyono, and K. Yamato, *Trans. ISIJ*, **24**, B266 (1984).
169. T. Irie, K. Kyono, H. Kimura, T. Honjo, K. Yamata, T. Yoshihara, and A. Matsuda, *Proc. 4th AES Continuous Strip Plating Symp.*, Chicago, May 1984, pp. G1–24.
170. T. Irie, K. Kyono, H. Kimura, T. Honjo, and K. Yamato, Research Rep. Kawasaki Steel Co., p. 31 (Mar. 1984).
171. T. Hada, T. Kanamaru, M. Nakayama, K. Arai, T. Fujiwara, Y. Suemitsu, M. Sato, and Y. Ogawa, Nippon Steel Tech. Rep. no. 25, pp. 11–18 (Apr. 1985).
172. T. Adaniya, M. Sagiyama, and T. Honma, Nippon Kokan Tech. Rep. Overseas, no. 43, pp. 33–39 (1985).
173. T. Adaniya, T. Hara, M. Sagiyama, T. Honma, and T. Watanabe, *Plating Surf. Finish.*, **72**, 52–56 (1985).

174. T. Adaniya, T. Hara, M. Sagiyama, T. Honma, and T. Watanabe, *Proc. 4th AES Continuous Strip Plating Symp.*, Chicago, May 1984, pp. H1–15.
175. N. Miura, T. Saito, T. Kanamura, Y. Shindo, and Y. Kitazawa, *Trans. ISIJ*, **23**, 913 (1983).
176. S. Nomura, H. Sakai, H. Nishimoto, T. Uegaki, M. Sakaguchi, M. Iwai, and I. Kobuko, *Trans. ISIJ*, **23**, 930 (1983).
177. T. Tsuda, K. Asano, and A. Shibuya, *Trans. ISIJ*, **26**, 53 (1986).
178. R. Srivastava, O. Gupta, and Ja. La Luddin, *Metalloberfläche*, **34**, 76 (1980).
179. W. Verberne, *Trans. Inst. Met. Finish.*, **64**, 30 (1986).
180. T. Saito, R. Wake, J. Oka, and M. Kitayama, Nippon Steel Technical Report no. 25, pp. 1–10 (Apr. 1985).
181. W. Siegert, *Metalloberfläche*, **41** (6), 259 (1987).
182. Ch. Karwas and T. Hepel, *J. Electrochem. Soc.*, **136**, 1672 (1989).
183. Th. E. Sharples, *Proc. 78th AESF Ann. Tech. Conf.*, 73–86 (1991).
184. T. Zhen-Mi, Z. Jing-Shuang, L. Wennliang, Y. Zhe-Long, and A. Maozhong, *Trans. IMF*, **73** (2), 48, 1995.
185. R. Fratesi, G. Roventi, G. Giulani, and C. R. Tomachuk, *J. Appl. Electrochem.*, **27**, 1088 (1997).
186. I. Kirilova, I. Ivanov, and S. Rashkov, *J. Appl. Electrochem.*, **27**, 1380 (1997).
187. A. Stankeviciute, K. Leinartas, G. Bikulcius, D. Virbalyte, A. Sudavicius, and E. Juzeliunas, *J. Appl. Electrochem.*, **28**, 89 (1998).
188. E. O. S. Carpenter and J. P. G. Farr, *Trans. IMF*, **76** (4), 135 (1998).
189. I. Kirilova, I. Ivanov, and S. Rashkov, *J. Appl. Electrochem.*, **28**, 637 (1998).
190. I. Kirilova, I. Ivanov, and S. Rashkov, *J. Appl. Electrochem.*, **28**, 1359 (1998).
191. V. Narasimhamurthy and B. S. Sheshadri, *Trans. IMF*, **77** (1), 29 (1999).
192. S. S. Abd el Rehim, M. A. M. Ibrahim, S. M. Abd el Wahaab, and M. Dankeria, *Trans. IMF*, **77** (1), 31 (1999).
193. A. Brenner, *Electrodeposition of Alloys*, vol. 2, Academic Press, New York, 1963, pp. 194–228.
194. K. Higashi, H. Fukushima, T. Urakawa, T. Adaniya, and K. Matsudo, *J. Electrochem. Soc.*, **128**, 2081 (1981).
195. H. Fukushima, T. Adaniya, and K. Higashi, *J. Jpn. Met. Finish. Soc.*, **34** (9), 446 (1983).
196. D. E. Hall, *Plating Surf. Finish.*, **70** (11), 59 (1983).
197. S. Swathirajan, *J. Electroanal. Chem. Interfacial Electrochem.*, **221** (1–2), 211 (1987).
198. Ch. Karwas and T. Hepel, *J. Electrochem. Soc.*, **135**, 839 (1988).
199. H. Fukushima, *Proc. Int. Conf. on Zinc and Zinc Alloy Coated Steel Sheet* (Galvatech'89), ISIJ, Tokyo, 1989, pp. 19–26.
200. H. Fukushima, T. Akiyama, K. Higashi, R. Kammel, and M. Karimkhani, *Metall.*, **44** (8), 754 (1990).
201. K. Kondo, *ISIJ Int.*, **30** (6), 464 (1990).
202. E. Chassaing and R. Wiart, *Electrochim. Acta*, **37**, 545 (1992).
203. H. M. Wang and T. J. O'Keefe, *J. Appl. Electrochem.*, **24**, 900 (1994).
204. A. M. Alfantazi, J. Page, and U. Erb, *J. Appl. Electrochem.*, **26**, 1225 (1996).
205. V. G. Roev and N. V. Gudin, *Trans. IMF*, **74** (5), 153 (1996).
206. F. J. Fabri Miranda, O. E. Barcia, O. R. Mattos, and R. Wiart, *J. Electrochem. Soc.*, **144**, 3441 (1997).

207. Ibid., *J. Electrochem. Soc.*, **144**, 3449 (1997).

208. F. Elkhatabi, M. Benbella, M. Sarret, and C. Muller, *Electrochim. Acta*, **44**, 1645 (1999).

209. T. Hara, M. Sagiyama, and T. Urakawa, Nippon Kokan Tech. Rep. Overseas, 48, 29 (Feb. 1987).

210. M. Sagiyama, T. Urakawa, T. Adaniya, T. Hara, and Y. Fukuda, *Proc. 5th AES Continuous Strip Plating Symp.*, Dearborn, May 1987, pp. V1–16.

211. M. Sagiyama, T. Urakawa, T. Adaniya, T. Hara, and Y. Fukuda, *Plating Surf. Finish.*, **74** (11), 77 (1987).

212. T. Urakawa, Y. Sugimoto, M. Sagiyama, and T. Watanabe, *Proc. Int. Conf. on Zinc and Zinc Alloys Coated Steel Sheet* (Galvatech'89), ISIJ, Tokyo, 1989, pp. 51–58.

213. H. Yoshikawa, H. Iwata, T. Urukawa, and M. Sagiyama, ibid., pp. 59–65.

214. G. D. Wilcox and B. Petersen, *Trans. IMF*, **74** (4), 115 (1996).

215. B. Bozzini, F. Pavan, G. Bollini, and P. L. Cavalotti, *Trans. IMF*, **75** (5), 175 (1997).

216. A. Catanzano and M. Palladino, TEKSID Technical note no. 2 (1983) and United Nations Economic Commission for Europe, Steel Seminar 8/R35, Apr. 1982.

217. A. Catanzano, S. Borsari, V. Matullo, G. Santunione, and N. Zagni, *Proc. 4th AES Continuous Strip Plating Symp.*, Chicago, May 1984, pp. I1–21.

218. V. Ferrari, L. Pacelli, E. Severini, and M. Campioni, *Proc. 2nd Int. Conf. on Zinc and Zinc Alloy Coated Steel Sheet* (Galvatech'92), CRM, Amsterdam, 1992, pp. 511–517.

219. K. Mori, H. Ishii, T. Miyawaki, and Y. Matsushima, *Proc. Int. Conf. on Zinc and Zinc Alloy Coated Steel Sheet* (Galvatech'89), ISIJ, Tokyo, 1989, pp. 66–72.

220. S. Umino, C. Kato, T. Komori, A. Yasuda, and K. Yamato, ibid., pp. 73–79.

221. J. P. Celis, J. R. Roos, C. Buelens, and J. Fransaer, *Proc. 2nd Int. Conf. on Zinc and Zinc Alloy Coated Steel Sheet* (Galvatech'92), CRM, Amsterdam, 1992, pp. 506–510.

222. M. Yoshida and T. Izaki, *Proc. AESF Continuous Steel Strip Plating Symp.*, Dearborn, May 1993, pp. 111–122.

223. V. John, *Proc. 3rd Int. Conf. on Zinc and Zinc Alloy Coated Steel Sheet* (Galvatech'95), ISS, Chicago, Sept. 1995, pp. 249–262.

224. D. Aslanidis, J. Fransaer, and J. P. Celis, *J. Electrochem. Soc.*, **144**, 2352 (1997).

225. A. Sachian, M. Blidariu, E. Roman, and C. Raducanu, *Trans. IMF*, **75** (6), 213 (1997).

226. D. Aslanidis, J. Fransaer, J. P. Celis, U. Meers, and L. Renard, *Proc. 4th Int. Conf. on Zinc and Zinc Alloys Coated Steel Sheet* (Galvatech'98), ISIJ, Chiba, 1998, pp. 515–520.

227. H. Nakakoji and K. Mochizuki, ibid., pp. 570–574.

228. A. Hovestad, R. J. C. H. Heesen and L. J. J. Janssen, *J. Appl. Electrochem.*, **29**, 331 (1999).

229. F. M. Androsch, K. Kosters, and W. Schiefermuller, *Proc. AESF Continuous Steel Strip Plating Symp.*, Dearborn, May 1993, pp. 69–79.

230. T. Ichida, *Proc. 4th Int. Conf. on Zinc and Zinc Alloy Coated Steel Sheet* (Galvatech'98), ISIJ, Chiba, 1998, pp. 272–279.

231. L. Conde Moragues, ibid., pp. 280–283.

232. T. Kubota, K. Sasaki, Y. Sugimoto, M. Yamashita, and M. Sagiyama, ibid., pp. 284–289.

233. J. Giridhar and W. J. van Ooij, *Surf. Coatings Technol.*, **52**, 17 (1992).

234. H. Yan, J. Downes, P. J. Boden, and S. J. Harris, *Trans. IMF*, **77** (2), 71 (1999).

235. B. Bozzini, F. Pavan, and P. L. Cavallotti, *Trans. IMF*, **76** (5), 171 (1998).

236. J. A. Zehnder, J. Hadju, and J. Nagy, *Plating Surf. Finish.*, **62** (9), 862 (1975).

237. H. Zeilmaker, L. C. van den Boogaard, and E. Barendrecht, *Metalloberfläche*, **31** (8), 342 (1977).

238. W. Paatsch, *Galvanotechnik*, **70** (8), 706–710 (1979).

239. W. Paatsch, *Metalloberfläche*, **34** (4), 174 (1980).

240. K. N. Srinivasan, M. Selvam, S. Venkata Krishna Iyer, *J. Appl. Electrochem.*, **23**, 358 (1993).

241. J. H. Payer, S. L. Amey, and G. M. Michal, *Proc. AESF Continuous Steel Strip Plating Symp.*, Dearborn, May 1993, pp. 259–264.

242. M. J. Carr and M. J. Robinson, *Trans. IMF*, **73** (2), 58 (1995).

243. L. Mirkova, C. Tsvetkova, I. Krastev, M. Monev, and S. Rashkov, *Trans. IMF*, **73** (2), 44 (1995).

244. D. H. Coleman, G. Zheng, B. N. Popov, and R. E. White, *J. Electrochem. Soc.*, **143** (6), 1871 (1996).

245. M. Pushpavanam and K. Balakrishnan, *Trans. IMF*, **74** (1), 33 (1996).

246. W. Paatsch, *Plating Surf. Finish.*, **83** (9), 70 (1996).

247. W. Paatsch, W. Kautek, and M. Sahre, *Trans. IMF*, **75** (6), 216 (1997).

248. M. Monev, L. Mirkova, I. Krastev, H. Tsvetkova, S. Rashkov, and W. Richtering, *J. Appl. Electrochem.*, **28**, 1107 (1998).

249. D. H. Coleman, B. N. Popov, and R. E. White, *J. Appl. Electrochem.*, **28**, 889 (1998).

250. M. Ramasubramanian, B. N. Popov, and R. E. White, *J. Electrochem. Soc.*, **145**, 1907 (1998).

251. F. Yuse and T. Nakayama, *Proc. 4th Int. Conf. on Zinc and Zinc Alloy Coated Steel Sheet* (Galvatech'98), ISIJ, Chiba, 1998, pp. 553–557.

252. K. L. Hardee, L. K. Mitchell, and E. J. Rudd, *Plating Surf. Finish.*, **76** (4), 68 (1989).

253. R. H. Newnham, *J. Appl. Electrochem.*, **22**, 116 (1992).

254. K. L. Hardee, L. M. Ernes, C. W. Brown, and T. E. Moore, *Proc. AESF Continuous Steel Strip Plating Symp.*, Dearborn, May 1993, pp. 239–252.

255. S. Kulandaisamy, J. Prabhakar Rethinaraj, S. C. Chockalingam, S. Visvanathan, K. V. Venkateswaran, P. Ramachandran, and V. Nandakumar, *J. Appl. Electrochem.*, **27**, 579 (1997).

256. R. Otagawa, K. Soda, S. Yamauchi, M. Morimitsu, and M. Matsunaga, *Proc. 4th Int. Conf. on Zinc and Zinc Alloy Coated Steel Sheet* (Galvatech'98), ISIJ, Chiba, 1998, pp. 768–771.

257. J. R. Smith, F. C. Walsh, and R. Clarke, *J. Appl. Electrochem.*, **28**, 1021 (1998).

258. M. Turner and J. K. Dennis, *Trans. IMF*, **64**, 94 (1986).

259. T. Xue, W. C. Cooper, R. Pascual, and S. Saimoto, *J. Appl. Electrochem.*, **21**, 231 (1991).

11 Electrodeposition of Iron and Iron Alloys

MASANOBU IZAKI

INTRODUCTION

Iron plating is used principally for numerous applications due to the desirable physical properties of iron and its low cost. The early literature on iron deposition, which was concerned with both its commercial applications and its electrorefining, has been comprehensively surveyed in [l]. Production of electrotypes in Russia [2], and of driving bands for shells in Germany during World War I, electroforming of iron sheets and tubes [3], and building up of worn parts [4, 5] were among the early applications of electrodeposited iron. Good magnetic properties that may be obtained with electrolytic iron led to its use by Western Electric Company [6] in the cores of Pupin induction coils prior to the development of nickel-iron alloys with even better magnetic properties.

An application is a process reported in 1930 [7] in which intaglio plates for printing government currency and bonds were made by depositing a nickel face backed by a heavy deposit of electrolytic iron from a hot chloride bath.

Iron plating was used during World War II to make electrotypes and to coat stereotypes in order to conserve nickel and copper [8, 9]. During the same period the United States Rubber Company electroformed iron molds for rubber, glass, and plastics [10]. Soldering tips are plated with iron commercially, and undoubtedly there are many other small-scale applications. Electrodeposition of iron as a means of producing iron powder for powder metallurgy is an application that has continued [11].

Interest in iron plating persists despite the fact that many of the applications were short-lived. This is attested by several publications on, for example, production of iron strips [12, 13], electrowinning [14, 15], electroforming of phonograph record stampers [16, 17], rolls and molds [18], and the use of iron for restoring worn parts [19]. Continuous iron strip produced by iron plating has been used for composite roofing materials, low-wattage resistance heating elements, and several magnetic products such as laminated cores for coils, magnetic display boards, and magnetic shielding [20–22]. Hard iron plating was employed on an aluminum piston to enhance wear resistance, to eliminate iron or steel sleeves, and to allow die-casting method [23]. Over five-million pistons were plated with hard iron deposit in General

Modern Electroplating, Fourth Edition, Edited by Mordechay Schlesinger and Milan Paunovic.
ISBN 0-471-16824-6

Motors Corporation [24]. A 1973 patent mentioned the effectiveness of intermediate thin iron deposit with low stress to obtain thick and highly stressed iron deposit on a soft material without peeling or spalling of the iron deposit [25].

There are several reasons for this persistent interest in iron plating. Iron is cheap and abundant. It can be deposited as a hard and brittle metal which, by heat treatment, can be rendered soft and malleable, or as a soft and ductile metal to which surface hardness can be imparted by curburizing, cyaniding, and nitriding. The fatigue strength of surfaces prepared by case hardening electrodeposited iron has been reported equivalent to the best of commercial rolling-element bearing materials [26]. Electrodeposited iron can be welded readily, other metals can be easily plated on it, and in the soft state it has superior drawing properties [27]. Electrodeposited iron is relatively resistant to corrosion, as would be expected from its high purity; the contrary opinion that has been quoted in the literature [28] is probably due to failure to rinse deposits completely free to electrolyte traces. The throwing power of iron baths is comparable to that of nickel baths.

The productions of several iron alloy plating have been of the important development of the last decade in metal finishing industry. Iron-nickel alloy plating has been widely used for decorative coating [29], surface coating of continuous casting mold in steel making industry [30], and for magnetic devices [31–37]. Iron-zinc alloy plating has been used for the corrosion protection of iron-based materials [38] in rack or barrel operations [39] and for continuous plating operation in steel industry [40].

An important problem with iron and iron alloy plating that has limited its usage to specialized or to high-volume applications is that despite usually lower costs for anodes or solutions, the expenditures for capital equipment and maintenance may be higher for iron plating than for other more commonly used plating baths. Special high-temperature or corrosion-resistant equipment may be required to heat, agitate, filter, or ventilate the iron plating bath. Also, unless used regularly, the solution will oxidize gradually. The time and effort required to restore the electrolyte to an operable condition may overweigh the economics of depositing a lower cost metal.

1 PRINCIPLES

Practically all iron is plated from acidic solutions of iron (II) (ferrous) salts. The presence of iron in the iron (III) (ferric) state in these baths, in an appreciable concentration, is undesirable because it lowers the cathode efficiency for depositing the metal and it may cause deposits to be brittle, stressed, and pitted. In practice, it is not difficult to maintain the concentration of iron (III) ion at harmless level.

Until recently practically all iron was plated from baths containing iron (II) sulfate, iron (II) chloride, or mixture of the two. Iron (II) fluoborate and iron (II) sulfamate baths have been used to some extent. Research on plating from solutions of other iron salts has not been neglected. In 1887 Watt [41] investigated deposition from solutions of a large number of iron salts, and concluded incorrectly that only sulfate baths were practical. Baths based on the iron salts of several fluorine-substituted phosphoric acids and fatty acids were investigated, but they did not yield satisfactory deposits [42]. Selected baths of other types are described in a later section of this chapter.

The most frequent addition to both the sulfate and chloride bath is conducting salts, such as a corresponding alkali and alkaline earth salts. Many other salts, such as those of various organic acids, have been recommended. Oxycarboxylic and dicarboxylic acids such as citric, malic, malonic, tartaric acid, and ascorbic acid have been used to prevent the formation of ferric hydroxide precipitation [43]. These acids may give codeposition of an impurity element that affect the properties of an iron deposit and may be justified in special circumstances. The simple bath combinations described next should fill most requirements, and they have been used generally for the production of an iron strip.

2 THE FERROUS SULFATE BATH

The iron (II) (ferrous) sulfate bath produces deposits that are smooth and normally light gray in color. There is little tendency toward pitting, and thick deposits can be produced. Disadvantages of the bath are that it yields brittle deposits, the deposition rate is slow, and the current density at which burning occurs is about one half that of a hot chloride bath. An advantage is that it can be operated at room temperature (about 25°C), in contrasts with the ferrous chloride–calcium chloride bath.

The most common sulfate bath is the one containing the double salt, ferrous ammonium sulfate. It has been used principally for building up undersized machine parts [4, 5] and also to apply a hard facing to stereotype [8].

Iron (II) sulfate may also be used alone, or other salts such as sodium, magnesium, or aluminum sulfate may be added [1, 44]. Addition of a small amount of ammonium fluoborate to the high-pH sulfate bath is claimed to make the slimy iron (III) hydroxide precipitate that accumulates in it more easily filterable [45].

The different added salts probably have minor specific effects on the properties of the deposits, but no data are available. The presence of ammonium ion appears to reduce the rate of air oxidation of iron (II) ion and the internal stress.

It will be noted in Table 1 that the two ranges of pH are shown for the sulfate bath at 25°C. The two distinct ranges result from the fact that iron (III) hydroxide precipitates at a pH of about 3.5, whereas iron (II) hydroxide precipitates at a pH of about 6. In the low pH range, even in a well-reduced bath, some iron (III) ion is present because of air oxidation; therefore operation at a pH too close to 3.5 results in dark-colored, excessively stressed deposits, probably caused by inclusion of basic iron (III) salts in the deposits. Operation at a pH below the recommended minimum for the low pH range results in lower cathode efficiency and increased deposit stress.

In the high pH range a very low concentration of iron (III) ion is automatically maintained (equal to the solubility of iron (III) hydroxide), since the pH is above that at which iron (III) hydroxide precipitates. The bath in this pH range therefore tends to be "sludgy." The sludge does not cause significant roughness of deposits up to a few tens of micrometers in thickness, but if heavy deposits are to be produced, filtration may be required. The high-pH sulfate bath has better covering power and yields deposits that are less stressed than those from the low-pH bath. Deposits with minimum stress are obtained at pH value in the range 4.0 to 5.0. The internal stress of the deposit generally increases with the increase in current density.

At the higher operating temperatures, sludging due to air oxidation is rapid at high pH. Therefore only the lower pH range is recommended for operation at

TABLE 1 Composition and Operating Conditions for Iron Plating Baths

Solution	Composition	Operating Conditions
Sulfate	$FeSO_4 \cdot (NH_4)_2SO_4 \cdot 6H_2O$:250–300 g liter^{-1}	Low pH 2.8–3.4 or High pH 4.0–5.5 2 A dm^{-2} 25°C
	$FeSO_4 \cdot 7H_2O$:250 g liter^{-1} $(NH_4)_2SO_4$:120 g liter^{-1}	pH 2.1–2.4, 4–10 A dm^{-2} 60°C
(for production of strip)	$FeSO_4 \cdot 7H_2O$: 600 g liter^{-1}	pH 1.4, 6.7 A dm^{-2} 47°C
Chloride (Fischer-Langbein)	$FeCl_2 \cdot 4H_2O$: 300 g liter^{-1} $CaCl_2$: 335 g liter^{-1}	pH 0.8–1.5, 6.5 A dm^{-2} 90°C
	$FeCl_2 \cdot 4H_2O$: 300–450 g liter^{-1} $CaCl_2$: 150–190 g liter^{-1}	pH 0.2–1.8, 2–9 A dm^{-2} 88–99°C
(for electrotype)	$FeCl_2 \cdot 4H_2O$: 240 g liter^{-1} KCl: 180 g liter^{-1}	pH 5–5.5, 2–5 A dm^{-2} 25–40°C
(for production of strip)	Ferrous chloride 120–150 g liter^{-1} as Fe^{2+}	pH 0.5–4.7, 33–40 A dm^{-2} 98–106°C
Sulfate-chloride (for electrotype)	$FeSO_4 \cdot 7H_2O$: 250 g liter^{-1} $FeCl_2 \cdot 4H_2O$: 42 g liter^{-1} NH_4Cl: 20 g liter^{-1}	pH 3.5–5.5 5–10 A dm^{-2} 40–43°C
(for production of strip)	$FeSO_4 \cdot 7H_2O$: 500 g liter^{-1} NaCl: 50 g liter^{-1}	pH 2.5, 3–27.5 A dm^{-2} 80°C
Sulfamate	Iron (II) sulfamate: 250 g liter^{-1} Ammonium sulfamate: 30 g liter^{-1}	pH 3.2–15 A dm^{-2} 50–70°C
Fluoborate	$Fe(BF_4)_2$: 226 g liter^{-1} NaCl: 10 g liter^{-1}	pH 2–3, 2–10 A dm^{-2} 55–60°C

elevated temperatures. The advantage of an elevated temperature is the higher permissible current density. The disadvantages of operating at temperature higher than about 60°C probably outweigh any gain in permissible current density. Deposits from the sulfate bath do not become significantly ductile even if the bath is operated at the boiling point.

3 THE FERROUS CHLORIDE BATH

The important and distinctive characteristics of the chloride baths is that they permit high current density or fast deposition rates and yield ductile deposits when operated at temperatures higher than about 85°C. The most commonly used bath is a solution of iron (II) and calcium chlorides, which has been referred to as the Fischer-Langbein solution [46]. This bath yields dark-colored, hard, highly stressed deposits at 25°C. At increasingly higher temperatures, the deposits gradually become lighter colored.

The composition of the Fischer-Langbein solution as modified by Thomas and Blum [7] and that modified in industrial use [47] are shown in Table 1. The lower concentrations introduced by Thomas and Blum permit the use of moderate current densities and do not lead to crystallization of the salts when the bath is cooled. The presence of calcium chloride in the hot chloride bath to increases the bath conductivity and cathode efficiency, and does not appear to have significant effects on the structure and physical properties of deposits [48]. The importance of the hygroscopic character of calcium chloride in reducing evaporation of water from this type of bath, and in raising boiling point, seems overstressed in the literature [46]. Elimination of calcium chloride from the bath permits the use of higher ferrous chloride concentrations with resultant higher operable current densities. For rapid deposition at current density as high as 40 A dm^{-2}, a solution of ferrous chloride alone has been used, at concentration from 120 to 150 g $liter^{-1}$ as Fe^{2+} [22]. For lower stressed deposits, concentration as high as 725 g $liter^{-1}$ $FeCl_2 \cdot 4H_2O$ have been used [49].

Numerous modifications of the hot chloride bath have been described, in which calcium chloride has been replaced by other alkaline earth chlorides [1, 50]. The presence of a low concentration of manganese (II) chloride has been claimed to result in deposits of finer grain size [51], and a bath containing a large amount of this salt has been recommended for plating machine parts [52]. The presence of $AlCl_3$, $BeCl_2$, or $CrCl_2$ in low concentration has been reported to render deposit softer and more ductile [53]. The presence of 20 to 100 g $liter^{-1}$ $AlCl_3$ has been claimed to produce better stability [54]. The maintenance of a small concentration of iron (III), up to about 0.5 g $liter^{-1}$, has been stated to result in better throwing power [55], and in the elimination of pitting [16]. The presence of iron (III), however, causes deposits to be harder, less ductile, and more highly stressed than those from well-reduced baths. From the bath with concentration ratio of iron (III) to iron (II) below 0.09, a deposit with good quality is produced [21]. The internal stress increases with an increase in the current density or with a decrease in the bath temperature. A deposit with excellent ductility is obtained at higher temperatures, lower pHs, and high current densities.

3.1 The Fluoborate Bath

Iron plating from fluoborate solution is not new, but since a variety of fluoborate salts became commercially available, interest in their use in plating was stimulated, and several iron (II) fluoborate baths have been described [56–60]. The composition shown in Table 1 has been recommended for plating stereotypes [56] and is probably satisfactory for general-purpose baths. A bath prepared from the commercially available concentrated solution will contain a small excess of fluoboric and boric acids, which are present in the concentrate. Boric acid appears to have the desirable buffering action. In the iron fluoborate bath described by Kudryavtsev and Mel'nikova [40], the recommended concentration of boric acid is 18 g $liter^{-1}$. Fluoboric acid may be used to adjust the pH.

The fluoborate bath had good stability, high conductivity, and high tolerance to metallic impurities. The disadvantage of this bath is that it is more expensive than chloride or sulfate bath. The fluoborate bath yields deposits similar in brittleness and in most other properties to those from the sulfate bath.

3.2 Other Baths

Solution Used for Iron Electrotypes The two baths listed in Table 1 were developed for making iron electrotypes [8, 9]. Although the ferrous chloride-potassium chloride bath appears to be similar to the Fischer-Langbein type, it differs from the latter in performance in yielding light-gray, stress-free deposits at 25°C. The deposits are, however, brittle. This difference in performance appears to be caused specifically by the potassium ion. The sulfate-chloride bath yields brittle deposits that are under moderate stress. It has somewhat better "covering power," which is improved by the presence of ammonium ion, as is also the case with nickel electrotyping solutions.

Both iron plating solutions operate satisfactorily at higher temperatures than are shown in the table. The upper limits of temperature were imposed because of the low softening points of wax or plastic electrotype molding media.

Solution Used for Production of Iron Strip The three baths of simple sulfate [61], chloride [22, 62], and sulfate-chloride bath [63] listed in Table 1 have been developed for production of continuous iron strip. These baths contained ferrous salts at higher concentration than those for rack operation, and generally operated at higher current density, higher bath temperature, and high flow rate of the solutions by using special electrolytic cell which will be mentioned latter in this chapter. Chloride bath is favorable for this purpose because of the allowance of higher current density as high as $100\,A\,dm^{-2}$ at which satisfactory deposit was obtained. A mixed fluoborate-sulfamate iron (II) electrolyte has been used by Levy and Hutton for the deposition of high strength iron strips [64].

Other Special-Purpose and Less Common Baths Many iron plating baths based on unusual salts, or special formulations for specific purposes, have been described.

Good performance was claimed by Piontelli [65], Barrett [66], Ueno [67], and Lowrie [47] for iron (II) sulfamate baths. Current densities up to $10\,A\,dm^{-2}$ and cathode efficiency up to 96% were described by Misra and Rama Char [68] in a bath containing iron (II) sulfamate and ammonium sulfamate. The effectiveness of saccharin addition to decrease internal stress was mentioned in sulfamate-ammonium fluoride acidic bath [67]. Iron sulfamate in the concentrated solution is available commercially.

An iron bath of a mixed chloride-sulfate type, containing also boric acid and sodium formate, has been recommended for depositing a starting plate on active basis metals such as aluminum, beryllium, or uranium [69], and for applying an intermediate layer of iron between steel and electrodeposited antimony deposit [70].

Two types of alkaline iron plating baths have been described. In one, iron (III) is complexed with triethanolamine and ethylendiaminetetraacetic acid [71, 72]. In the other, the main constituent of the bath is iron (III) pyrophosphate [73]. Only thin deposits have been reported from these baths. Both suffer from the disadvantage that iron does not dissolve in them anodically. Other baths based on complexes of iron (III), such as oxalate or citrate, have been described [74]; this type of bath, however, is useful primarily for electroanalysis rather than for electroplating.

Iron powder obtained by electrodeposition is used commercially. Conditions for depositing iron in powder form have been defined by several authors [75–78]. The bath described in [76] is unusual in that it consists of an aqueous solution of iron (III) oxide and sodium hydroxide, in which the iron is probably present as sodium ferrate. Procedures for producing iron powders by crushing brittle deposits were reviewed by Shafer and Harr [11], who report that sintered parts made from electrolytic iron powder have higher density and better physical properties than those made from nonelectrolytic powders.

4 IRON ALLOYS

The electrodeposition of binary and ternary iron alloys have been reviewed in several articles [79–82]. Twenty-two elements including carbon have been electrodeposited with iron. Electrochemical investigation of anomalous deposition behavior in iron alloy systems have been discussed in many papers [83–86]. Bath composition and operating conditions for production of iron alloy deposits are represented in Table 2.

The greatest interest among iron binary alloys in recent years, as indicated by the number of publications, has been in iron-nickel alloy plating. Iron-nickel alloy deposit is favorable to decorative and protective deposit as undercoats of thin chromium deposit, because operating costs could be reduced somewhat compared with that for nickel plating [87, 88]. The alloy deposits are used for appliance hardware, tubular furniture, plumbing goods, jewelry, and wire products. Most of baths used for decorative nickel-iron alloy plating contain nickel sulfate, nickel chloride, ferrous sulfate, boric acid, brightener and stabilizer of hydroxycarboxylic acids type, such as citric acid, to prevent the oxidation of ferrous ion. Saccharin or specific aromatic sulfonamides are used as primary brightener and give codeposition of sulfur in the deposits. Heterocyclic amine, nitrils, and secondary acetylnic alcohol have been used as secondary brightener. These brighteners tend to increase the stress and hardness. Nickel content ranging from 10% to 45% have no detrimental effects on brittleness, leveling characteristic, and successive chromium plating. Bath temperature, ferric ion concentration, and buffering ability of added stabilizer have effects on the leveling and cathode efficiency. The alloy deposits have been used as a replacement for bright nickel for mild or moderate service conditions, since corrosion protection of the alloy deposits were equivalent to those from comparable nickel/chromium deposits [88, 89].

The iron (4–10%)-nickel alloy deposit has been used for inner surface coating of continuous casting mold to prolong mold life, to improve product quality, and to reduce the running cost. The same holds for other coatings of chromium plating, thick and tapered nickel coating, and multiple-layer coating of nickel/nickel-phosphorous/chromium plating [90]. The alloy deposit has twice the hardness and half the elongation of nickel plating, and has thermal expansion coefficient that is the same as that of copper mold base. Kanayama and co workers [91] found iron-nickel alloy deposit lead to prolong mold life about 1.2 to 1.7 times that for a nickel plating or multiple-layer coating.

The iron(20%)-nickel alloy (Pearmalloy) deposit has been of considerable interest for magnetic devices such as thin film recording head, magnetic yokes in printer heads, bubble memories, shielding, logic devices, and magnetic cores in micro-solenoids because of its magnetic properties [92, 93]. When employed for discrete element, it is frequently deposited as a full film and the magnetic pattern is then defined with an etch resist [94]. In using an electroforming technique of iron-nickel alloy deposit, Lochel and Maciossek [95] developed a micromachining technique for the production of microcantilever, three-dimensional microcoils with an integrated magnetic core, a magnetically driven microvalve, and a microturbine. Bath compositions and operating conditions for the production of magnetic iron-nickel alloy plating were surveyed in a literature [96] and are represented in Table 2. The deposition of an iron (39%)-nickel alloy with low expansion coefficient (Invar) was produced from an sulfamate bath [97] and has been discussed for use of wave-guides [98]. The active area of research in the iron-nickel alloy deposit involves the improvement of magnetic properties by codeposition of a third elements such as phosphorous [99], arsenic [100], sulfur [101], molybdenum [102], copper [103], cobalt [104], chromium [105], and indium [106]. The another active area is in developing technique to control the composition within tighter desirable range. Most of baths

TABLE 2 Bath Composition and Operating Conditions of Iron Alloy Plating

Alloy	Purpose	Bath Composition	Operating Conditions	Characteristics
Iron-nickel	Decorative [88]	$NiSO_4 \cdot 6H_2O$: 150 g liter^{-1} $NiCl_2 \cdot 6H_2O$: 90 g liter^{-1} $FiSO_4 \cdot 7H_2O$: 20 g liter^{-1} H_3BO_3: 45 g liter^{-1} stabilizer and brightener	60°C pH 3.5 4 A dm^{-2}	Fe–50%Ni Hardness: 508 Knoop
	Magnetic [37]	Nickel sulfate: 16 g liter^{-1} Nickel chloride: 40 g liter^{-1} Ferrous sulfate: 1 g liter^{-1} Boric acid: 25 g liter^{-1} Sodium saccharin: 0.375–3 g liter^{-1} Sodium lauryl sulfate: 0.2 g liter^{-1}	26°C pH 2.5 1.5 A dm^{-2}	Fe–80% Ni (705 nm thick) Coercive force: 0.25 Oe
	Invar [97]	Nickel sulfamate: 0.75 mol liter^{-1} Ferrous chloride: 0.25 mol liter^{-1} Boric acid: 0.5 mol liter^{-1} Sodium dodecyl sulfate: 0.5 g liter^{-1} Ascorbic acid: 1 g liter^{-1} Saccharin: 2 g liter^{-1}	50°C pH 2.5 10 A dm^{-2}	Fe–39% Ni: thermal expansion coefficient of 8.1×10^{-6} °C^{-1} $4–5 \times 10^{-6}$ °C^{-1} (after annealing at 680°C)
Iron-zinc	Corrosion protection [40]	$FeSO_4 \cdot 7H_2O + ZnSO_4 \cdot 7H_2O$: 500 g liter^{-1} Na_2SO_4: 30 g liter^{-1} $CH_3COONa \cdot 3H_2O$: 20 g liter^{-1} $C_6H_8O_7$: 5 g liter^{-1}	50°C pH 2.5 3 A dm^{-2}	
Iron-cobalt	Antifriction [115]	$FeCl_2$: 227 g liter^{-1} $CoCl_2 \cdot 6H_2O$: 8 g liter^{-1} H_3BO_3: 10 g liter^{-1} NH_4Cl: 75 g liter^{-1} NaCl: 75 g liter^{-1} NH_4BF_4: 10 g liter^{-1}	50°C pH 1 5–40 A dm^{-2}	Fe–6% Co: Hardness: 640 Vickers

Iron-chromium-nickel	Corrosion protection [128]	$K_2Cr_2(SO_4)_4 \cdot 25H_2O$: 326 g liter^{-1} Nickel sulfate: 84 g liter^{-1} Ferrous sulfate: 56 g liter^{-1} Glycine: 150 g liter^{-1}	20°C pH 2–2.2 8.5 A dm^{-2}	9.8–11% Cr 7.3–6% Ni
Iron-phosphorous	Corrosion protection [111]	Fe $Cl_2 \cdot 4H_2O$: 240 g liter^{-1} KCl: 180 g liter^{-1} Hypophosphorous acid	40°C pH 2 40 A dm^{-2}	Fe–0.4% P
	Magnetic [112]	$FeSO_4 \cdot 7H_2O$: 1.0 mol liter^{-1} $(NH_4)_2SO_4$: 0.1 mol liter^{-1} $C_6H_7O_6$: 0.1 mol liter^{-1} $NaH_2PO_2H_2O$: 0.01 mol liter^{-1}	25°C pH 2.5 1 A dm^{-2}	Fe–6% P Coercive force: 2 Oe
Iron-boron	Magnetic [113]	$FeSO_4 \cdot 7H_2O$: 0.07 mol liter^{-1} KBH_4: 0.3 mol liter^{-1} $KNaC_4H_4O_6 \cdot 4H_2O$: 0.6 mol liter^{-1} $(NH_4)_2SO_4$: 0.08 mol liter^{-1} NaOH: 0.4 mol liter^{-1}	30°C pH 12.5 1 A dm^{-2}	Fe–8% B Coercive force: 3.6 Oe Saturation magneto striction: 27×10^{-6}
Iron-carbon	Antifriction [120]	$FeSO_4 \cdot 7H_2O$: 40 g liter^{-1} L-ascorbic acid: 3.0 g liter^{-1} Citric acid: 1.2 g liter^{-1}	50°C pH 2.5 3 A dm^{-2}	Fe–1% C Hardness: 800 Vickers
Iron-carbon-boron	Antifriction [122]	$FeCl_2 \cdot 4H_2O$: 1.0 mol liter^{-1} Malic acid: 5×10^{-3} mol liter^{-1} Boric acid: 40 g liter^{-1} Dimethylamineborane: 2.5 g liter^{-1}	50°C pH 3.5 5 A dm^{-2}	Fe–0.2% C–0.11% B Hardness: 800 Vickers

used for the purpose contained saccharin, which resulted in decrease in the grain size and in reduction of coercive force [96]. Pearmalloy deposit with coercive force below 1 Oe [37], which is lower than that of iron deposit is obtained. Effects of current waveform such as a pulsating current, a reversing current, or a sinusoidal alternating current superimposed on a direct current, cathode efficiency, composition, or phase structure of the alloy deposits were mentioned in literature [107–110].

Iron-zinc alloy deposit is one of the important development for the corrosion protection of iron-based materials [38]. Iron(0.2–0.5%)-zinc alloy plating are conducted by rack and barrel operations [39]. Single-layer deposit of iron(15–25%)-zinc alloy deposit has been used commercially in Japan for the inside surface of fenders, doors, trunk lids, and other unexposed automotive body parts, because of its good weldability, formability, and corrosion resistance after painting [40]. Iron-rich iron(> 40%)-zinc alloy deposits have been used for upper layer on zinc-rich iron(20%)-zinc or zinc-nickel alloy deposit with good corrosion resistance, since the deposits have good paint adhesion and compatibility with cationic electropainting [40].

Iron–0.4% phosphorous alloy plating has been used as upper layer on zinc-iron or zinc-nickel alloy deposits to enhance the paintability and phosphatability of steel strips for automotive body parts in Japan for several years [111]. Amorphous iron-phosphorous(> 6%) alloy plating is under industrial evaluation for soft magnetic materials, because of its low coercive force and high magnetic flux density [112]. Bath formulation to obtain an amorphous iron–8% boron alloy was published at 1998 [113]. The iron-boron alloy plating is under industrial evaluation for a strain sensor and microcantilever, because of the very large magnetomechanical coupling factor, large magnetostriction, and high permeability [114].

A 1983 patent mentioned the electrodeposition of low stress, hard iron-6% cobalt alloy deposit produced from a chloride bath with high cathode efficiency of 96% [115].

Attempts to produce steel by codepositing carbon in iron deposit, followed by subsequent heat treating were reported even in the early literature on iron plating. A reported bath for this purpose consists of a hot ferrous chloride solution containing a relatively high concentration of glycerol, sugar, or starch [116]. Deposits are described that contain 0.7% carbon and have a hardness of 477 Brinell after heat treatment. Bath formulation to obtain hard iron–1% carbon alloy deposit at high current efficiency of above 70% was published at 1989 [117]. The bath was a ferrous sulfate solution containing small amounts of carboxylic acids as carbon source and L-ascorbic acid to prevent the oxidation of ferrous ion in the solution. The deposits with carbon content of 0.13%, 0.56%, 0.6%, and 1% are obtained by oxalic, succinic, malic, and citric acid, respectively [118]. The cathode efficiency ranging from 70% to 90% were obtained, depending on the added carboxylic acids. The iron-carbon(> 1%) alloy deposit showed black and bright appearance, and showed high hardness of above 800 Vickers [119–121]. The iron-carbon alloy deposits containing small amounts of phosphorus or boron showed the enhanced ductility as well as high hardness [122]. The iron-carbon alloy deposits have been under industrial evaluations as hard coating or an alternative of thermal surface hardening such as carburizing or nitriding steels, because of its comparable hardness and less thermal distortion [123].

Active area involves attempts to electrodeposit iron-nickel-chromium alloys; electrolytes for producing stainless steel deposits for decorative or corrosion resistant coating [124–128] are described. The alloy deposits containing 2.9% to 29% chromium and 8% to 54% nickel were obtained, depending on the type of bath and operating conditions. In general, the deposits are highly stressed and are limited to a thickness of less than 30 μm. Cathode efficiency is generally low (18–35%) [128]. Cathode efficiency rapidly dropped, and deposit quality deteriorated with the passage of the deposition time. Matsuoka, Kammel, and Landau [129] found that an iron–18% chrominum–8% nickel alloy deposit with a high current efficiency of 60% showed ferro-magnetic character and showed poor corrosion resistance, that was different from thermally prepared stainless steel. An amorphous iron-chromium-nickel-phosphorous-carbon alloy deposit with good corrosion resistance was described in a literature [130]. Most of the baths used for iron-chromium-nickel alloy plating contained glycine, which acted not only as a pH buffer but also as a complexing agent. Glycine also prevents the codeposition of basic salts.

5 PREPARATION, MAINTENANCE, AND CONTROL

Commercial iron (II) salts, and even reagent grade salts, are likely to contain a significant amount of iron (III). Therefore it is usually necessary to reduce the iron (III) ion before using a freshly prepared bath. This is done by adding degreased iron turnings or steel wool to the bath, together with sufficient acid to lower pH to about 0.5 (sulfuric or hydrochloric acid for the sulfate or chloride bath). The reduction treatment requires 24 to 48 hrs, during which further addition of acid may be required to maintain a low pH. Alternatively, the iron (III) may be reduced by dummying the bath at a low pH. Completion of a reduction treatment is indicated by the color of the solution, which should be a clear green, free from any yellow tint.

An operating bath, if used steadily, remains fully reduced as a result of cathodic reduction of iron (III). If a bath is idle, maintenance of a small excess of acid helps to prevent oxidation. Since excess acid is rapidly depleted by reaction with anodes, the anodes should be removed from a bath that will be idle more than a day or so. Cubes of gum rubber floating on the surface of the bath reduce oxidation, conserve heat, and limit evolution of spray [131]. Expanded polyethylene chips or hollow balls of polypropylene serve the same purpose and are especially useful for hot plating baths. For several applications such as production of iron or iron alloy plated steel strips and electroplating of magnetic iron-nickel alloy (Pearmalloy) deposit, the anion exchange membrane [132] or regeneration tank [133] has been employed to keep ferric ion concentration below the desirable level.

5.1 Impurities

Iron baths are similar to nickel baths in that small amounts of metallic and organic impurities can cause deposits to be brittle, stressed, or pitted. Therefore a freshly prepared bath requires purification, which is best performed after the mentioned reduction treatment. Organic impurities are removed by adsorption on activated carbon, and the carbon is removed by filtration. If the pH of the solution during treatment with carbon is adjusted between 5.0 and 5.5, contaminant metals that precipitate as hydroxide in this pH range are also removed. The pH may be adjusted by adding sodium, potassium, or ammonium hydroxide, or by suspending iron (II) carbonate in the bath. High-pH treatment is not recommended for the more concentrated chloride and fluoborate baths in that rapid oxidation of iron (II) occurs with subsequent precipitation of iron (III) hydroxide. Metallic impurities such as copper, lead, and nickel are removed by dummying the bath at an average cathode current density of $0.5\,A\,dm^{-2}$ with a corrugated cathode. Details of these procedures are similar to those described for purifying nickel baths, with the notable exception that hydrogen peroxide, or other oxidizing agents, often added to nickel baths during treatment with activated carbon, must not be added to an iron (II) solution.

Not much is known concerning effects of specific impurities. It was found that more than $0.2\,g\,liter^{-1}$ zinc in the iron (II) chloride–potassium chloride electrotyping bath causes no harm in the hot chloride bath [51], and that the iron-zinc alloy can be produced from sulfate baths [134, 135]. Low tolerance limits have been reported for copper, lead, arsenic, tin, and molybdenum in the hot chloride bath [51]. Copper, at a concentration of $0.2\,g\,liter^{-1}$ in the hot chloride bath, has been reported to cause spongy roughness in areas where the current density is high, and to decrease ductility [17]. Experience with the hot chloride bath indicates that copper or lead concentrations above $0.1\,g\,liter^{-1}$ and nickel or cobalt concentrations above $0.2\,g\,liter^{-1}$ may result in roughness and poor throwing power in low current density areas ($<2.5\,A\,dm^{-2}$). In a sulfamate-ammonium fluoride bath, zinc ($>0.1\,g\,liter^{-1}$), stannous ($0.1\,g\,liter^{-1}$), and manganese ($>0.5\,g\,liter^{-1}$) increased the internal stress and chromium, and copper ($>1\,g\,liter^{-1}$) gave black deposit with rough surface [67]. In general, continuously worked baths tend to remain free of impurities in harmful amounts, but occasionally they may require treatment for reduction or purification, as already described.

5.2 pH

As indicated in the discussion of the various baths, maintaining control of pH is essential. The pH can be measured most conveniently with a glass electrode. When operated at pH below 3.5, the acid content of the bath is slowly depleted because the anode efficiency is higher than the cathode efficiency: 100% and 80% to 99%, respectively. It is therefore necessary to add the appropriate acid to maintain the pH.

In the high-pH range, both anode and cathode efficiencies are close to 100%. Air oxidation of iron (II), however, followed by precipitation of iron (III) hydroxide, results in a decrease in pH; it may be maintained by adding the appropriate hydroxide (KOH or NH_4OH). Iron (II) carbonate suspended in the bath has been used to control the pH in this range.

5.3 Surfactants

Although a highly purified and well-worked iron bath does not usually yield pitted deposits, pitting is sometimes encountered. Use of a wetting agent tends to increase deposit stress, but it may be helpful if used judiciously. Sodium lauryl sulfate has been reported by several authors to be a suitable wetting agent and seems to be compatible with all types of iron and iron-nickel plating solutions. The addition of

1 g liter^{-1} of a condensate of sodium naphthalene sulfonate and formaldehyde to the iron (II) chloride bath is reported to eliminate pitting regardless of iron (III) content [23]. Reference has already been made to the control of pitting by maintaining a small concentration of iron (III) in the hot chloride bath [17]. Stirring of the solution or mechanical agitation of the cathode may also reduce pitting. A "bumping" type of motion is most effective. Air agitation should not be used because it results in excessive oxidation of iron (II).

5.3 Analytical Technique

The approximate concentration of iron is determined by measurement of the specific gravity. Only an occasional analysis for iron is necessary. Standard analytical procedures are used to determine iron (II), chloride, and sulfate [136]. The quantitative analysis of iron may be performed by titrimetry or spectrophotometry [137].

6 EQUIPMENT

All the iron plating solutions described, except the alkaline baths, require acid-proof material for tanks and auxiliary equipment. Equipment similar to that used for acid pickling is satisfactory. For most relating, rubber-coated steel tanks are adequate; they may be further lined with acid-proof brick for heavy service. Steel tanks with vinyl resin or fiber reinforced plastics (FRP) linings are satisfactory at temperatures up to 60°C. Glass and glass-lined equipment is resistant to corrosion in sulfate or chloride electrolytes, but it is not rugged enough for commercial use unless precautions are taken to protect the cell surface from thermal and mechanical shock. Rubber-lined pumps and pumps constructed of impregnated carbon, Teflon, or titanium parts have given satisfactory service with the hot chloride bath. Vinyl plastisols are satisfactory as rack coatings and stop-offs, even in the hot chloride bath.

When the heat developed from the passage of the plating current through the bath is sufficient, or nearly so, to maintain the proper working temperature, it has been found practical to inject live steam directly into the solution as the simplest means of initial heating. If there is too much condensate, heat exchangers of tantalum, titanium, zirconium, or Teflon may be used. Quartz immersion heaters can be used in all solutions, and carbon immersion heaters can be used in fluoborate or fluoride solutions. If the nature of the work permits, mechanical agitation of the bath makes possible the use of higher current densities and facilitates the formation of a more even deposit.

For iron plating on large volume articles as molds and rolls, deep tank with the depth above 4 m is used to settle the ferric hydroxide precipitation on the bottom. A 1977 patent mentioned the process for exchanging a surface layer of a moving steel strip with iron deposit in an iron plating solution [138]. A special plating cell such as a paddle cell and rotating cathode cell have been employed for production of magnetic devices such as thin film Pearmalloy (iron-nickel alloy) heads to maximize thickness uniformity. Several types of continuous plating cell [139] have been employed for production of iron strip and iron alloy plated steel strips in industrial fields. Subramanyan and King [22] used a rotating-strippable-drum system for the production of an iron strip and allowed a current density as high as 40 A dm^{-2}. The thickness of the deposit was determined by the rotating rate of drums and deposition rate. In the drum system, a bare or chromium plated stainless steel drum is used as the cathode and iron and steel scraps are used as anode. Iron alloy plating on a steel strip has been carried out at high flow rate ranging from 0.1 to 0.4 $m\,s^{-1}$ and at a narrow strip–electrode distance ranging from 9 to 50 mm [140, 141] by using insoluble anode. When an insoluble anode is used, the anion exchange membrane is effective in preventing oxidation of the ferrous ion in the bath, if it is used for separating the plating cell into two compartments of catholyte and anolyte. A regeneration reactor [133] filled with iron scraps has been used for rapid reduction of the ferric ion by circulating the used anolyte solution.

7 ANODES

Iron anodes of high purity, such as ingot iron, wrought iron, or Swedish iron, are preferred, but anodes of steel or cast iron have been used. A high-purity iron anode is necessary to obtain ductile deposits, since a hot chloride bath is easily contaminated by impurities contained in the anode materials. All of these dissolve with high efficiency but produce some insoluble sludge residue that may cause rough deposits. It is therefore usually desirable to bag the anodes. Glass cloth, although fragile, can be used, and it has been reported to be quite satisfactory when coated with phenolic resin [51]. Porous stoneware diaphragms have been used, but they are rather cumbersome and have a relatively high electrical resistance. The synthetic fabrics Orlon and Dynel are satisfactory bag materials. These fabrics are more durable when used as a cover on a frame, rather than as loose bags. Polypropylene cloth has excellent chemical resistance and durability in all iron plating baths and is commonly used in the hot chloride electrolyte. The necessity for bagging may be obviated by continuous filtration. The passivity of iron anodes has not been reported, even in high-pH sulfate baths. For the production of iron strip and iron alloy plated steel strips, insoluble materials such as platinum-plated titanium sheets or graphite sheets have been used as anodes.

8 CHARACTERISTICS OF DEPOSITS

The data available in 1935 on the physical properties of electrolytic iron, including mechanical, magnetic, and electrical properties, were summarized by Cleaves and Thompson [1]. The data on mechanical and physical properties of iron and iron alloy deposits have been comprehensively surveyed in [142]. A representative selection of data on physical and mechanical properties is given in Table 3. In general, an iron deposit is an aggregate of columnar grains with a body-centered cubic lattice that corresponds to that in equilibrium state at ambient temperature. In general, hardness, tensile strength, and internal stress decreases, and the elongation increases with the increase in bath temperature or decrease in current density. The hardness ranging from 120 to 350 Vickers is obtained from simple chloride or sulfate bath. A high hardness of 450 Vickers is obtained at a low bath temperature, and a high current density in a hot chloride bath. The effects of current density and boric acid addition on the oxygen content were reported for a chloride bath [143]. Codeposited oxygen in the form of

TABLE 3 Properties of Iron Deposits

	Bath Composition	pH	Temperature (°C)	Current Density (A dm^{-2})	Tensile Strength (Kg mm^{-2})	Elongation (%)	Hardness	Coercive Force
Iron	Sulfate bath	3.4	20	0.5			263	
	$Fe(NH_4)_2(SO_4)_2$	3.4	20	2.0			354	
	[148]	4.4	19	0.5			182	
		4.4	41	2.0			240	
							(Brinell)	
	Chloride bath, $FeCl_2$	3.5–4.7	100–106	400	50–56	5–15		9.5–11.0 (Oe)
	(120–150 g $liter^{-1}$ as Fe^{2+}) [22]							
	Chloride bath	0–3	86	20	45	2.1		19.5
	$FeCl_2{\cdot}4H_2O$: 400 g $liter^{-1}$		96	10	42	3.33		18
	$CaCl_2$: 80 g $liter^{-1}$		96	20	43	4.67		17.5
	Wetting agent: 2 ml $liter^{-1}$		96	30	43	5.54		225
	[21]		108	20	36	2.67		17.5
								(Atm^{-1})
	Chloride bath	0.2–0.4	70	15			450	
	$FeCl_2{\cdot}4H_2O$: 465 g $liter^{-1}$			5			300	
	H_3BO_3: 38 g $liter^{-1}$			2.5			170	
	[24]		95	15			150	
				5			150	
				2.5			120	
							(Vickers)	
	Chloride bath	0	60	2.0			280	
	$FeCl_2{\cdot}4H_2O$: 1.57 mol dm^{-3}	1		2.0			250	
	$CaCl_2$: 2.04 mol dm^{-3}	2		2.0			200	
	[146]						(Vickers)	
	Sulfate-chloride bath	2.5	80	8	46	4.3		
	$FeSO_4{\cdot}7H_2O$: 500 g $liter^{-1}$			15	56	5.3		
	NaCl: 50 g $liter^{-1}$ [63]			27	64	2.7		

TABLE 3 ***(Continued.)***

	Bath Composition	pH	Temperature (°C)	Current Density ($A\ dm^{-2}$)	Tensile Strength ($Kg\ mm^{-2}$)	Elongation (%)	Hardness	Coercive Force
	Deposits from either sulfate or chloride baths, annealed at 900°C [1]					40	70–90 (Brinell)	
Iron-carbon	$FeSO_4 \cdot 7H_2O$: 40 g $liter^{-1}$ L-ascorbic acid: 3.0 g $liter^{-1}$	2.5	50	3				
	Citric acid: 1.2 g $liter^{-1}$ [117]						800	
	Or							
	Malonic acid: 0.56 g $liter^{-1}$ [118]						480 (Vickers)	

iron oxide was observed to deteriorate the hardness and ductility of an iron-carbon alloy deposit [120]. After small amounts of hypophosphorous acid or dimethylamineborane were added to the iron-carbon alloy plating bath, the oxygen content was reduced to a harmless level, and this enhanced ductility [122]. Sulfur and carbon contents in iron-nickel alloy deposits were quantitatively determined for sulfate/chloride bath containing saccharin [144]. The data in Table 3 show that annealing of electrolytic iron at 900°C results in properties approaching those of thermally prepared iron ingot. Heating the deposits at 200° to 300°C has no significant effect on their properties. A high hardness of 800 Vickers was obtained for iron-carbon(>1%) alloy deposits that had a martensitic phase with the body-centered tetragonal lattice. With annealing at 350°C, the iron-carbon alloy deposit gained the high hardness of 1200 Vickers [145]. An iron-nickel alloy deposit has the phase structure of an α-iron with a body-centered cubic lattice, or a γ-nickel structure with a face-centered cubic lattice, or a mixture of the two, depending on the alloy's composition [97, 147].

REFERENCES

1. H. E. Cleaves and J. G. Thompson, *The Metal-Iron*, Alloys of Iron Research Monograph Series, McGraw-Hill, New York, 1935.
2. M. Klein, *Chem. News*, **18**, 133 (1868).
3. D. Belcher, *Trans. Am. Electrochem. Soc.*, **45**, 455 (1924).
4. W. A. MacFadyen, *Trans. Faraday Soc.*, **15** (3), 98 (1920).
5. D. R. Kellog, *Min. Metall.*, **3**, 61 (1922).
6. G. F. McMahon, *Chem. Metall. Eng.*, **26**, 639 (1922).
7. C. T. Thomas and W. Blum, *Trans. Am. Electrochem. Soc.*, **57**, 59 (1930).
8. V. A. Lamb and W. Blum, *Proc. Am. Electroplat. Soc.*, **39**, 106 (1942).
9. R. M. Shaffert and B. W. Gonser, *Trans. Electrochem. Soc.*, **84**, 319 (1943).
10. A. W. Bull, J. W. Bishop, M. H. Orbaugh, and E. H. Wallace, *Met. Ind. (NY)*, **37**, 461 (1939).
11. W. M. Shafer and C. R. Harr, *J. Electrochem. Soc.*, **105**, 413 (1958).
12. Montau Union, Handelsgesellshaft m.b.H., German Patent 1,018,286 (1957).
13. H. Silman, *Met. Finish.*, (Westwood), **62** (12), 36 (1969); *Met. Finish. Abstr.*, **12** (2), 65 (1970).
14. E. H. Konrad and W. E. C. Eustis (to Sulfide Ore Process Co.), U.S. Patent 2,587,630 (1952).
15. P. D. Vasudeva Rao, H. V. K. Udupa, and B. B. Dey, *Symp. Electrodeposition Met. Finish.*, Bangalore, India Sect., Electrochem. Soc., Indian Inst. Sci., **77** (1960).
16. L. K. Kosowsky (to Columbia Records. Inc.), U.S. Patent 2,758,961 (1956).
17. A. M. Max and G. R. Van Houten, *Proc. Am. Electroplat. Soc.*, **136**, 43 (1956).
18. Technical Catalogue "Purealone," Fuso Company, Ltd., Hyogo, Japan.
19. M. M. Lekhikoinen, *Dokl. Akad. Nauk Tadzhik SSR*, **20**, 86 (1957).
20. J. A. James (to British Thompson-Houston Co., Ltd.), U.S. Patent 2,792,340 (1957).
21. S. F. Harty, J. A. McGeough, and R. M. Tulloch, *Surf. Tech.*, **12**, 39 (1981).
22. P. K. Subramanyan and W. M. King, *Plating Surf. Finish.*, **69** (2), 48 (1982).
23. O. J. Kllingenmaier (to General Motors Corp.), U.S. Patent 3,404,074 (1968).

24. O. J. Kllingenmaier, *Plating*, **61**, 741 (1974).
25. O. J. Kllingenmaier (to General Motors Corporation), U.S. Patent 3,753,664 (1973).
26. R. K. Kepple, E. R. Mantel, R. L. Mattson, and O. J. Kllingenmaier, Paper No.70-Lub.S-20, ASME Spring Symp. (1970).
27. G. P. Fuller, *Trans. Am. Electrochem. Soc.*, **50**, 371 (1926).
28. W. E. Hughes, *Chem. Metall. Eng.*, **29**, 536 (1923).
29. Technical Catalogue "NIRON", Udylite Co., Japan.
30. Technical Catalogue "Tough Alloy-I," Nomura Plating Co. Ltd., Osaka, Japan.
31. I. W. Wolf and W. P. McConell, *Proc. Am. Electroplat. Soc.*, **43**, 215 (1956).
32. R. S. Smith, L. E. Godycki, and J. C. Lloyd, *J. Electrochem. Soc.*, 108, 996 (1961).
33. W. O. Freitag and J. J. Mathias, *Electroplat. Met. Finish.*, **17** (2), 42 (1964).
34. A. F. Bogenshutz and J. L. Josten, *Galvanotechnik* (*Saulgau*), **61** (4), 279 (5), 362 (1970); *Met. Finish. Abst.*, **12** (4), 183 (1970).
35. H. V. Vekatasetti, *J. Electrochem. Soc.*, **117**, 403 (1970).
36. S. H. Liao and S. E. Anderson, *J. Electrochem. Soc.*, **140**, 208 (1993).
37. P. Duke, T. Montelbono, and L. Missel, *Plating Surf. Finish.*, **69** (9), 61 (1982).
38. R. Sard, *Plating Surf. Finish.*, **74** (2), 30 (1987).
39. G. W. Loar, K. R. Romer, and T. J. Aoe, *Plating Surf. Finish.*, **78** (3), 74 (1991).
40. T. Adaniya, T. Hara, M. Sagiyama, T. Homma, and T. Watanabe, *Plating Surf. Finish.*, **72** (8), 52 (1985).
41. A. Watt, *Electrician*, **20**, 6 (1887–88).
42. H. Conner and V. A. Lamb, *Plating*, **48**, 388 (1961).
43. A. Ohwada, Japan Kokai Tokkyo Koho, JPN 78,103,935 (1978).
44. N. T. Kudryavtsev and L. A. Yakovleva, *Tr. Mosk. Khim-Tekhol. Inst.*, No. 22, 135 (1956); *Chem. Abstr.*, **51**, 16142a (1957).
45. R. E. Harr (to Western Electric Co., Inc.) U.S. Patent 2,710,832 (1955).
46. F. Fischer, *Z. Electrochem.*, **15**, 595 (1909), German Patent 212,994 (1908), U.S. Patent 992,951 (1911).
47. C. F. Lowrie, *Metal Finishing Guide-Book-Directory*, p. 247 (1996).
48. C. Kasper, *J. Res. Natl. Bur. Stand.*, **18**, 535 (1937).
49. J. D. Thomas, O. J. Kllingenmaier, and D. W. Hardesty, *Trans. Inst. Met. Finish.*, **47**, 209 (1969).
50. M. P. Melkov, *Vestn. Mashinostr.*, **33** (3), 50 (1953); *Chem. Abstr.*, **47**, 10375 (1953).
51. W. B. Stoddard Jr., *Trans. Electrochem. Soc.*, **84**, 305 (1943).
52. V. P. Revyakin and P. D. Myaniskov, *Vestn. Mashostr.*, **37** (4), 64 (1957); *Chem. Abstr.*, **51**, 11887b (1957).
53. Fr. Muller, E. Heuer, and O. Witnes, *Z. Electrochem.*, **47**, 135 (1941).
54. M. P. Melkov and B. V. Namakonov (Moskow Motering Inst.), Russian Patent 238, 981 (1968); *Met. Finish. Abstr.*, **11** (4), 159 (1969).
55. E. H. Wallace and R. K. Iler, U.S. Patent 2,306,917 (1943).
56. S. Okada, Japanese Patent 329 (1950).
57. K. L. Kossler and R. R. Sloan, *Electrotypers' and Stereotypers'Mag.*, **38** (7), 7 (1952).
58. J. Poor (to Van der Horst Corp. of Am.), U.S. Patent 2,745,800 (1956).
59. N. T. Kudryavtsev and M. M. Mel'nikov, *Nauchn. Dokl. Vyssh. Shk. Khim. i Khim. Tecknol.*, no. 1, 173 (1958); *Chem. Abstr.*, **53**, 913c (1959).

60. M. E. Farmer, D. C. West, and C. D. Darlington, *Plating*, **56**, 699 (1969).
61. Y. Kondo and K. Koike, *J. Surf. Finish. Soc., Japan*, **25**, 62 (1974).
62. S. Matsubara, N. Nakamura, K. Takagi, T. Omi, and K. Miyanami, *J. Iron Steel Inst. Japan*, **76**, 2167 (1990).
63. K. V. Gow, S. P. Iyer, H. H. Wu, K. M. Castelliz, and G. J. Hutton, *Surf. Tech.*, **8**, 333 (1979).
64. E. M. Levy and G. J. Hutton, *Plating*, **55**, 138 (1968).
65. R. Piontelli, *Korros. Metallschutz*, **19**, 110 (1943).
66. R. C. Barret, *Proc. Am. Electroplat. Soc.*, **47**, 170 (1960).
67. H. Ueno, H. Hayashi, S. Takagi, S. Miwa, and T. Hayashi, *J. Surf. Finish. Soc. Japan*, **23**, pp. 72, 186, 252, 316 (1972).
68. S. S. Misra and T. L. Rama Char, *J. Sci. Indai Res.*, **20D** (1), 43 (1961).
69. J. G. Beach, *Plating*, **43**, 616 (1956).
70. G. R. Schaer, U.S. Patent 2,918,415 (1959).
71. E. F. Foley Jr., H. B. Linford, and W. R. Mayer, *Plating*, **40**, 887 (1953).
72. A. Nakagawa, *J. Met. Finish. Soc. Japan*, **11** (5), 85 (1960); *Met. Finish. Abstr.*, **2** (4), 130 (1960).
73. V. Sree and T. L. Rama Char, *Bull. India Sect. Electrochem. Soc.*, **9**, 59 (1960).
74. M. Deutsch, J. R. Downing, L. G. Elliott, J. W. Irvine Jr., and A. Roberts, *Phys. Rev.*, **62**, 3 (1942).
75. *Met. Bull.*, 34 (Sept. 14, 1973).
76. G. Wranglen, *J. Electrochem. Soc.*, **97**, 353 (1950).
77. A. I. Levin and S. A. Pushkareva, *J. Appl. Chem. (USSR)*, **29**, 1323 (1956) (in English); *Chem. Abstr.*, 51, 8551e; 13613h (1957).
78. J. A. M. Le Duc, R. E. Loftfield, and L. E. Vaaler, *J. Electrochem. Soc.*, **106**, 659 (1959).
79. S. S. Misra and T. L. Rama Char, *Bull. India Sect. Electrochem. Soc.*, **11**, 64 (1962).
80. M. Sarojamma and T. L. Rama Char, *Electroplat. Met. Finish.*, **18** (2), 51 (1965).
81. R. Sevakuma and T. L. Rama Char, *Electroplat. Met. Finish.*, **24** (1), 14 (1971).
82. A. Krohn and C. W. Bohn, *Plating*, **58** (3), 237 (1971).
83. J. O'M. Bockris, D. Drazic, and A. R. Despic, *Electrochim. Acta*, **4**, 325 (1961).
84. H. Dahms and I. M. Croll, *J. Electrochem. Soc.*, **112**, 771 (1965).
85. J. Horkans, *J. Electrochem. Soc.*, **128**, 45 (1981).
86. K. Y. Sasaki and J. B. Talbot, *J. Electrochem. Soc.*, **142**, 775 (1995).
87. R. J. Clauss, R. A. Tremmel, and R. W. Klein, *Trans. Inst. Met. Finish.*, **53**, 22 (1975).
88. R. A. Tremmel, *Plating Surf. Finish.*, **68** (1), 22 (1981).
89. M. Pushpavanam, S. Jayakrishnan, S. R. Natarajan, and R. Subramanian, *Met. Finish.*, **84** (1), 32 (1986).
90. T. Nakashima, M. Kawamoto, M. Kouzuma, K. Sekine, and S. Kurihara, Technical Report of Sumitomo Metal Industries, Ltd., 43, No. 2, 77 (1991).
91. H. Kanayama, A. Ichihara, Y. Watanabe, G. Hattori, and K. Suzuki, *Kawasaki Sttel Giho*, **14** (4), 12 (1982).
92. R. L. White, *Plating Surf. Finish.*, **75** (4), 70 (1988).
93. K. Komaki, *J. Electrochem. Soc.*, **140**, 529 (1993).
94. S. Herrera, *Trans. Inst. Met. Finish.*, **73** (1), 12 (1995).
95. B. Lochel and A. Maciossek, *J. Electrochem. Soc.*, **143**, 3343 (1996).

96. S. N. Srimathi, S. M. Mayanna, and B. S. Sheshaeri, *Surf. Tech.*, **16**, 277 (1982).
97. D. L. Grimmett, M. Schwartz, and K. Nobe, *J. Electrochem. Soc.*, **140**, 973 (1993).
98. A. F. Bogenshutz, J. L. Jostan, and W. Hemmrich, *Oberflache (Berlin)*, no. 12, 506 (1970); *Met. Finish. Abstr.*, **13** (2), 84 (1971).
99. K. Mukasa, M. Sato, and M. Maeda, *J. Electrochem. Soc.*, **117**, 22 (1970).
100. W. O. Freitag, G. V. DiGuilio, and J. S. Mathias, *J. Electrochem. Soc.*, **113**, 64 (1970).
101. F. E. Luborsky, *IEEE Trans. Magn.*, **4**, 19 (1968).
102. W. O. Freitag and J. S. Mathias, *J. Electrochem. Soc.*, **112**, 64 (1965).
103. J. P. Leekstin, *IEEE Trans. Magn.*, **1**, 246 (1965); IBM Corp, British Patent 1,055,452 (1967).
104. E. Toledo and R. Mo, *Plating*, **57**, 43 (1970).
105. T. M. Harris, G. M. Whitney, and I. M. Croll, *J. Electrochem. Soc.*, **142**, 1031 (1995).
106. H. V. Venkatasetty, *J. Electrochem. Soc.*, **120**, 618 (1973).
107. M. D. Maksimovic, *Surf. Tech.*, **31**, 325 (1987); **35**, 21 (1988).
108. D. L. Grimmett, M. Schwartz, and K. Nobe, *J. Electrochem. Soc.*, **137**, 3414 (1990).
109. D. L. Grimmett, M. Schwartz, and K. Nobe, *Plating Surf. Finish.*, **75** (6), 94 (1988).
110. N. Phan, M. Schwartz, and K. Nobe, *Plating Surf. Finish.*, **75** (8), 46 (1988).
111. T. Honjo, K. Kyono, K. Yamato, T. Ichida, and T. Irie, *J. Iron Steel Inst. Japan*, **72**, 976 (1986).
112. T. Osaka, M. Takai, A. Nakamura, and F. Aso, *Denki Kagaku*, **62**, 453 (1994).
113. N. Fujita, M. Inoue, K.-I. Arai, P. B. Lim, and T. Fujii, *J. Appl. Phys.*, **83**, 7294 (1998).
114. N. Fujita, M. Inoue, K.-I. Arai, M. Izaki, and T. Fujii, *J. Appl. Phys.*, **85**, 4503 (1999).
115. O. J. Kllingenmaier (to General Motors Corporation), U.S. Patent 4,388,379 (1983).
116. Yu. Petrov, *Avtomob.*, **29** (2), 31 (1951); *Dokl. Akad. Nauk Tadzh. SSR*, no. 20, 67 (1957); *Chem. Abstr.*, **53**, 8879d, 8880b (1959).
117. M. Izaki, H. Enomoto, and T. Omi, *J. Surf. Finish. Soc. Japan*, **40**, 1304 (1989).
118. Y. Fujiwara, T. Nagayama, A. Nakae, M. Izaki, H. Enomoto, and E. Yamauchi, *J. Electrochem. Soc.*, **143**, 2584 (1996).
119. M. Izaki, H. Enomoto, and T. Omi, *J. Japan Inst. Metals*, **56**, 191 (1992).
120. M. Izaki and T. Omi, *Met. and Mater. Trans. A*, **27A**, 483 (1996).
121. Y. Fujiwara, M. Izaki, H. Enomoto, T. Nagayama, E. Yamauchi, and A. Nakae, *J. Appl. Electrochem.*, **28**, 855 (1998).
122. M. Izaki, Y. Fujiwara, H. Enomoto, A. Nakae, and N. Miyamoto, *Kokai Tokkyo Koho*, JPN 9-202991 (1997).
123. Technical Catalogue "Carboplus®," Fuso Chemical Company, Ltd., Osaka, Japan.
124. A. L. Rotinyan, L. A. Aytner, and N. P. Fedot'ev, *Zh. Prikl. Khim.*, **41** (10), 2201 (1968); *Met. Finish. Abstr.*, **11** (1), 8 (1969).
125. I. M. Katser, O. A. Pterova, and A. I. Vitkin, *Zh. Prikl. Khim.*, **41** (10), 2300 (1968); *Met. Finish. Abstr.*, **11** (1), 8 (1969).
126. W. H. Cleghorn, S. Gowrie, P. L. Elsie, and B. A. Shenoi, *Met. Finish.*, **67** (8), 65 (1969).
127. L. Domnikov, *Met. Finish.* (*Westwood*), **68** (2), 57 (1970); **68** (4), 56 (1970); **68** (5), 56 (1970); *Met. Finish. Abstr.*, **12** (3), 119 (1970); **12** (4), 181 (1970).
128. C. U. Chisholm and R. J. G. Carnegie, *Plating*, **59** (1), 29 (1972).
129. M. Matsuoka, R. Kammel, and U. Landau, *Plating Surf. Finish.*, **74** (10), 56 (1987).
130. J.-C. Kang and S. B. Lalvani, *J. Appl. Electrochem.*, **25**, 376 (1995).

131. C. T. Thomas, *Mon. Rev. Am. Electroplat. Soc.*, **30**, 720 (1943).

132. N. Suzuki, A. Shibuya, T. Tsuda, T. Izuo, and Y. Terada, *J. Iron Steel Inst. Japan*, **72**, 932 (1986).

133. Y. Miwa, S. Matsubara, K. Takagi, T. Omi, and K. Miyanami, *Powder Technol.*, **64**, 168 (1991).

134. S. Jepson, S. Meecham, and F. W. Salt, *Trans. Inst. Met. Finish.*, **32**, 160 (1955).

135. W. H. Safranek (to Rockwell Spring and Axle Co.), U.S. Patents 2,809,156 (1957); 2,832,729 (1958).

136. *Guide Book Directory 1987*, *Metal Finish.*, 562 (1987).

137. J. Bassett, R. C. Denney, G. H. Jeffery, and J. Mendham, *Vogel's Textbook of Quantitative Inorganic Analysis*, 4th ed., Longman, London, England, 1986.

138. O. J. Kllingenmaier (to General Motors Corporation), U.S. Patent 4,050,996 (1977).

139. K. Sakai and R. Yoshihara, *J. Iron Steel Japan*, **72**, 940 (1986).

140. K. Matsuzuka, *J. Iron Steel Japan*, **72**, 891 (1986).

141. A. Weymeersh, L. Renard, J. J. Conreur, R. Winand, M. Jorda, and C. Pettet, *Plating Surf. Finish.*, **73** (7), 68 (1986).

142. W. H. Safranek, "The Properties of Electrodeposited Metals and Alloys" p. 195, American Electroplaters and Surface Finishers Society, 1986.

143. S. Gadad, and T. M. Harris, *J. Electrochem. Soc.*, **145**, 3699 (1998).

144. M. Kounchenova, G. Raichevski, and S. T. Vitkova, *Surf. Tech.*, **31**, 137 (1987).

145. M. Izaki, H. Enomoto, A. Nakae, S. Terada, E. Yamauchi, and T. Omi, *J. Surf. Finish. Soc. Japan*, **45**, 1302 (1994).

146. S. Yoshimura, S. Yoshihara, T. Shirakashi, and E. Sato, *Trans. Inst. Met. Finish.*, **73** (1), 31 (1995).

147. K. Nakamura, M. Umetani, and T. Hayashi, *Surf. Tech.*, **25**, 111 (1985).

148. D. J. Macnaughton, *J. Iron Steel Inst. (London)*, **109**, 409 (1924).

12 Electrodeposition of Palladium and Palladium Alloys

JOSEPH A. ABYS and CONOR A. DULLAGHAN

INTRODUCTION

Palladium metal was discovered and named by the English medical doctor, W. H. Wollaston, in 1803 while he was conducting experiments on the isolation and purification of platinum. Dr. Wollaston named the metal in honor of the newly discovered asteroid Pallas. However, the name Pallas has its roots in Greek mythology and relates to the "gift by Zeus" of the wooden statue of Pallas Athena that stood at the gates of Troy to safeguard the city. It was believed that as long as the Trojans kept the statue, the citadel would remain "safe." Mythology tells us that during the Trojan Wars, Odysseus and Diomedes stole the statue and carried it to the Greek camp, after which Troy was conquered.

In Roman legend, Pallas Athena was obtained by Aenasus during the sacking of Troy and taken to Italy where it was installed in the Temple of Vesta in Rome. The Romans venerated the statue and believed it would ensure the existence of the empire. Thus the word palladium became synonymous with "safeguard" and more specifically the "safeguard of liberty" [1–2].

An excellent review of the discovery and history of palladium was written by Ian E. Cottington [3] of Johnson Matthey.

1 GEOLOGICAL OCCURRENCE [4–6]

Unlike other metals, the platinum group metals (PGMs) exist in the earth's crust mostly as metallic alloys especially of nickel and iron, and only a few mineral compounds are known. The natural abundance of the six metals in the earth's crust is very low with palladium being the most abundant [5]; see Table 1. Interestingly enough, PGMs have been found in high concentrations in meteorites rich in iron and nickel as opposed to the "stony" or rock type meteors. This has led geochemists to the supposition that on earth, a similar metallurgy took place and that PGMs concentrated during the earth's formation mainly in the iron-nickel core. This accounts for their relatively low abundance in the lithosphere.

Modern Electroplating, Fourth Edition, Edited by Mordechay Schlesinger and Milan Paunovic.
ISBN 0-471-16824-6

TABLE 1 Concentration of Platinum Group Metals - Earth's Crust

Metal	Concentration in Crust (ppm)
Palladium	0.01
Platinum	0.005
Rhodium	0.001
Iridium	0.001
Ruthenium	0.001
Osmium	0.001

2 SUPPLY, DEMAND AND USES OF PALLADIUM

The economically significant sources of PGMs are South Africa, North America, and Russia. The mines from these countries are primary deposits usually associated with copper and nickel mining. Table 2 provides the estimated concentration of PGMs for each geographic region [7]. The Russian sources of palladium are the most concentrated, and indeed Russia is the largest supplier of palladium today. Figure 1 is a 10 year history of the palladium supply [7] which reached 6.34 million troy ounces* in 1995. As can be seen, palladium shipments increased almost 50% over the last 2 years. Virtually all the additional demand was supplied from Russian government reserves. The need for "hard" currency was the principal motive, although the Russians might be concerned that higher prices might trigger increased substitution of palladium in the electronics sector, which in the long term would reduce the demand for this metal.

Figure 2 shows the 10 year history of demand by geographic region; clearly seen is the sharp increase since 1992. For example, demand for palladium [7] increased 25.2% to reach a total of 6.1 million troy ounces in 1995. Japan remains the largest consumer for electronics and dental applications. Meanwhile the increases in North America and Europe have been spurred by autocatalysts. The supply-demand figures are shown in Figure 3, indicating that cumulatively from 1986 to 1995 there exists an approximate eight hundred thousand troy ounce surplus. However, this surplus has been fueled by sales of Russian reserves and not mining output, and this trend could easily be reversed.

TABLE 2 Concentration of Platinum Group Metals - Geographic Distribution (wt%)

Metal	Canada	Russia	South Africa
Platinum	43.4	30	64.02
Palladium	42.9	60	25.61
Iridium	2.2	2	0.64
Rhodium	3.0	2	3.20
Ruthenium	8.5	6	6.40
Approximate ratio, Pt : Pd	1 : 1	1 : 2	2.5 : 1

*One troy ounce = 31.1 grams.

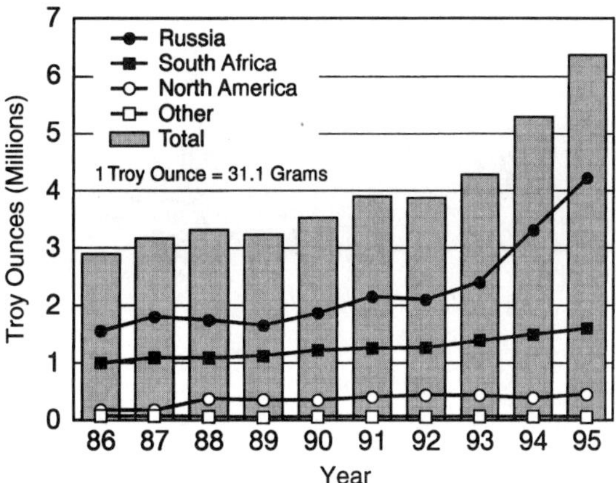

Figure 1 World palladium supply by geographic region from 1986 to 1996, in troy ounces.

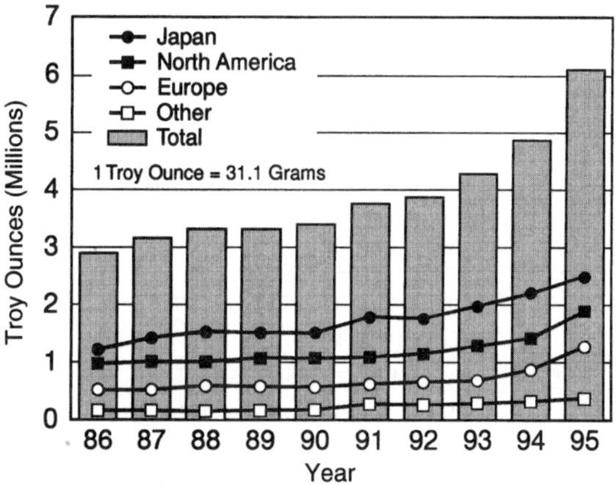

Figure 2 World palladium demand by geographic region from 1986 to 1995, in troy ounces.

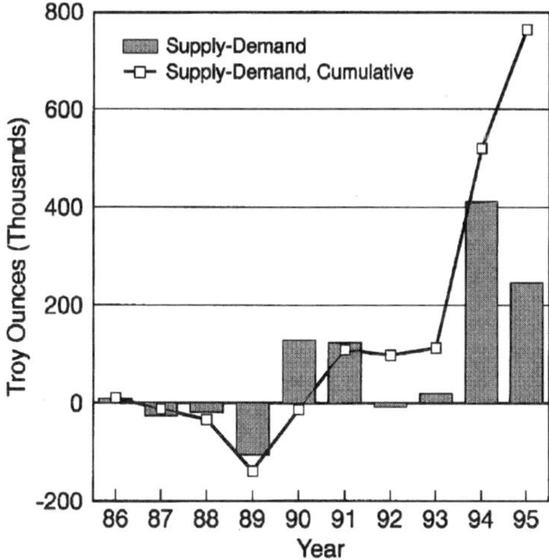

Figure 3 World palladium supply-demand from 1986 to 1995, in troy ounces.

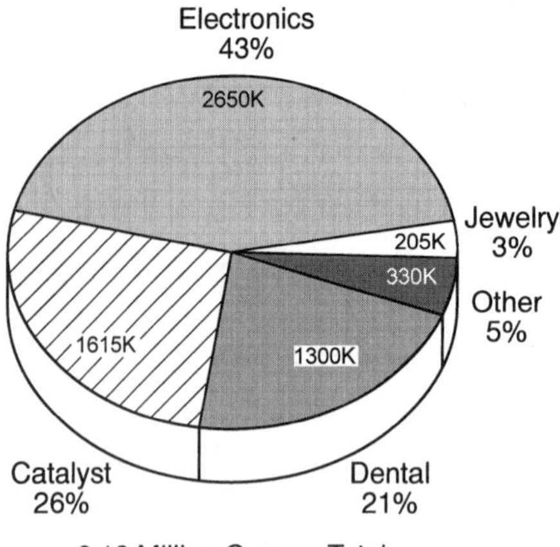

Figure 4 World palladium demand by industry for 1995, in troy ounces.

The consumption of palladium by individual applications [7] can be seen in Figure 4. Electronics is the largest consumer accounting for approximately 43% of total demand. Palladium components are used in most types of electronic devices from basic consumer products to complex military hardware. Although each of these components contains relatively small quantities of metal, they are produced in the billions accounting for the substantial demand of about 2.65 million troy ounces in 1995. More than 75% of this metal was consumed as "pastes" in the manufacture of multilayered ceramic capacitors; smaller amounts were used in hybrid integrated circuits and for plating operations.

The next largest application is in catalysts, and this accounts for about 26% of total palladium demand. Autocatalyst utilize approximately 75% for this application, and demand is rising rapidly among the European and North American automotive manufacturers. Dental, jewelry (nonplated), and other miscellaneous applications, including plated palladium, account for the remaining 31% of consumption.

3 A BRIEF HISTORY OF ELECTROPLATED PALLADIUM

It is believed that the earliest example of plated palladium are in the Percy collection at the Science Museum of London. It is a thin sheet of copper coated on both sides with palladium prepared by a Mr. T. H. Henry circa 1855 [35]. Henry used the method of Smee who recorded in his first edition of his *Electro-metallurgy* the use of a palladium nitrate electrolyte. In the second edition of the same work, he recommended an ammoniacal solution of the "ammonia-muriate of palladium." Interestingly enough, today the most utilized processes for palladium deposition remain those based on ammoniacal electrolytes. An excellent historical review including numerous recipes for plating palladium was written by Atkinson and Raper in 1933 [35].

Electroplated palladium represents about 5% to 8% of the worldwide consumption of this metal. Figure 5 is a chronological overview of its major

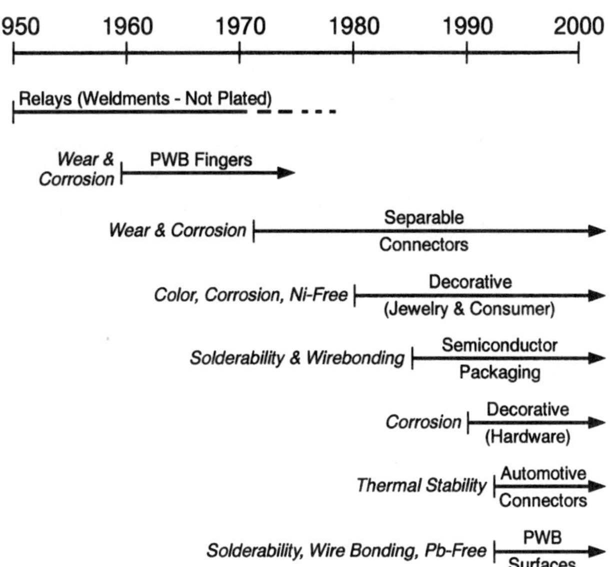

Figure 5 History of palladium electroplating by application from about 1960 to 1997.

applications and indicates that it was not until the mid-1970s that the plating of palladium became of technological significance. Since that time electroplated palladium has found numerous uses in interconnection products especially for telecommunication, computer and automotive connectors, semiconductor packaging, and most recently in the printed circuit industry. Furthermore, since the emergence of legislation in Europe that bans the use of nickel due to the phenomena of nickel dermatitis, palladium has found wide acceptance in the decorative industry.

The use of thick electroplated gold films for electronic applications was the standard in that industry until the mid-1970s. In the past the relatively low cost of gold and the availability and reliability of existing gold electrodeposition technologies precluded the use of any alternate material. However, the deregulation of gold in the early 1970s coupled with the political and economic events of that era brought about an astronomical increase in its price in the late 1970s and early 1980s, Figure 6. Since then gold has traded in the range of $320 to $440 per troy ounce, which is approximately a factor of 10 higher than prior to deregulation. Thus the major impetus for substituting gold in electronic components was economic. This spurred significant research in the reduction of gold usage [19, 20] and in the search for a suitable, more economic alternative [8–34].

Palladium, one of the PGMs, is viewed as a suitable alternative. The lower price of palladium, Figure 7, coupled with its lower density, Table 3, implies considerable cost savings on utilizing it to replace gold on electronic components. Today not only the economic but the technological advantages of substituting palladium or palladium alloys for gold are generally recognized [8–27]. The material properties (e.g., hardness, ductility, and thermal stability) of palladium are in many instances superior to hard gold. For example, the higher hardness of electroplated palladium is

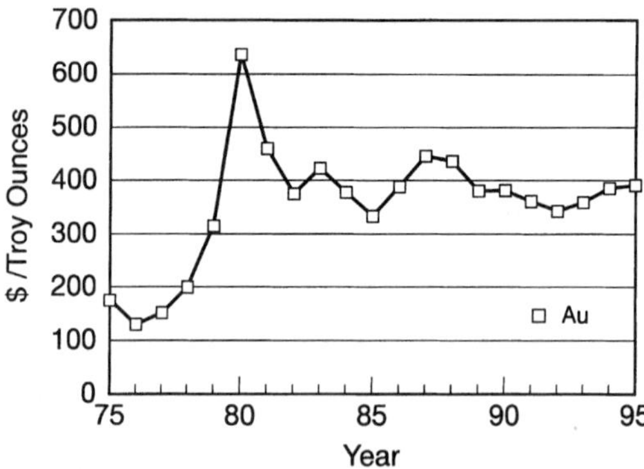

Figure 6 Gold metal price in dollars per troy ounce from 1975 to 1995.

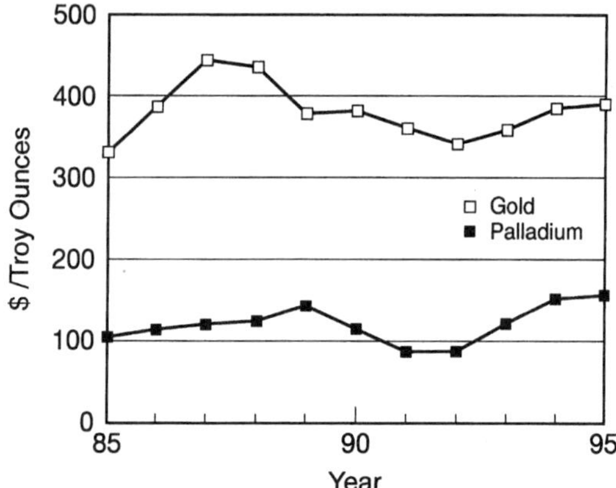

Figure 7 Gold and palladium metal price in dollars per troy ounce from 1985 to 1995.

beneficial for wear resistance, which can be further enhanced by a thin coating of electroplated gold that acts as a solid lubricant. The lower porosity of electroplated palladium alloys (e.g., PdNi) enhances the corrosion resistance of plated articles.

Thus electroplated palladium has found numerous applications during the last decade. However, historically there has not been an appreciable demand for electroplated palladium. This was due to the relative difficulty of plating palladium with satisfactory material properties, and a lack of impetus to develop this technology due to the preponderance of gold.

As mentioned previously, the electrodeposition of palladium from ammoniacal solutions was first reported by T. H. Henry circa 1855. Up until the mid-1980s, Pd electrodeposition technology evolved slowly. Several critical reviews [35–40] on this subject were published in the 1960s and 1970s. Although an extensive amount of literature was available, it was not until the early to mid-1980s that palladium processes suitable for large-scale manufacturing became available. This was especially true for high-speed ($> 50\,\mathrm{mA\,cm^{-2}}$) plating operations, which remain a challenge even today.

TABLE 3 Physical Properties of the Platinum Group Metals

Properties	Palladium	Platinum	Gold
Atomic structure	Kr-$4d^{10}5s^{0}$	Xe-$4f^{14}5d^{9}6s^{1}$	Xe-$4f^{14}5d^{10}6s^{1}$
Atomic number	46	78	79
Atomic weight	106.4	195.09	196.967
Crystal structure	fcc	fcc	fcc
Atomic radius (Å)	1.37	1.39	1.46
Covalent radius (Å)	1.28	1.30	1.34
Ionic radius (Å)	0.896 (+2)	0.96 (+2)	1.37 (+1)
Density (g/cc)	12.16	21.4	19.3
Hardness, DPN:			
Annealed	37	37	
Electrodeposited (KHN_{50})	250–400	400–500	140–200[a]
Ultimate tensile strength (annealed) $lb\,in^{-2}$	25,000	18,000	
Young's modulus, $lb\,in^{-2} \times 10^6$	16	25	
Coefficient of linear thermal expansion 20°C, μ-in/°C	11.8	8.9	14.2
Thermal conductivity (0–100°C) joules cm^{-1}, $cm^2\,°C\,s^{-1}$	0.76	0.73	
Electrical resistivity, μ ohm-cm	10.8 (20°C)	9.83 (0°)	2.19 (0°)
Reflectivity (average over visible spectrum), %	62	67	
Electronegativity	2.2	2.2	2.4
First ionization potential (k-cal $g\text{-}mol^{-1}$)	192	207	213
Common oxidation states	+2, +4	+2, +4	+1, +3
Melting point (°C)	1552	1769	1063
Boiling point (°C)	3980	4530	2970

[a]Ni/Co hardened gold.

4 PHYSICAL AND CHEMICAL PROPERTIES OF PALLADIUM

4.1 Physical Properties of Palladium, Platinum, and Gold [41–43]

Table 3 describes the more significant physical properties of palladium and compares them to platinum, the most widely recognized PGM, and gold, the "target metal" that palladium has been substituting. These properties are descriptive only insofar as possible to obtain the purest available specimens. It is well understood that the material properties of PGMs such as hardness, ductility, and electrical conductance are greatly affected by the presence of impurities. Furthermore, since the PGMs have a marked tendency to absorb gases such as hydrogen, it is understandable that obtaining accurate quantitative data is problematic. Nonetheless, Table 3 provides needed information that is utilized for the practical application of PGMs. For additional information on the properties of other PGMs, see [41–49].

Palladium is a silvery white, malleable, and ductile element with a face-centered cube (fcc) crystal structure. Its symbol is Pd, and its electronic structure is (Kr)$4d^{10}5s^0$; its element number is 46 with an atomic weight of 106.4 $g\,mol^{-1}$, a density of 12.16 $g\,cm^{-3}$, and a melting point of 1555°C. Metallurgically, palladium is known to form alloys with other metals of group VIII and IB that display distinct advantages over the pure metals [45]. In general, the alloying elements tend to increase the resistivity, hardness, and tensile strength of palladium. Copper, nickel, gold, iridium, rhodium, and ruthenium have been used to manufacture palladium alloys for many practical applications. For example, an alloy of silver (60/40 wt% Pd/Ag) is commonly used in electric contacts.

4.2 The Chemical Properties of Palladium

General Reactivity PGMs are relatively inert, with respect to chemical attack by oxygen or many acids, and this is one of the properties that makes them of practical value. The direct oxidation [43] of PGMs is summarized in Table 4. With the exception of osmium, which forms a considerably volatile oxide at room temperature, and to a lesser extent palladium, which forms an oxide when heated to about 350°C, the other PGMs require temperatures of >700°C to form oxides. This provides some understanding of their chemical inertness. However, it must be emphasized that the chemical reactivity of the PGMs is greatly affected by the state of the subdivision of the metal or the particle size (i.e., surface area). Thus, palladium sponge is more readily attacked than the compact metal. Also, if it is alloyed with another metal, namely lead or silver, it is more reactive. Palladium, finely dispersed on a supporting media such as silica gel, is still more reactive and displays remarkable catalytic properties.

The "inertness" of the PGMs including palladium stems from their strong atomic bonds in the solid state. The increase in reactivity exhibited by samples with a high surface area, compared to the bulk, is attributed to an increase in the number of atoms with the higher energy associated with surface sites, or "dangling bonds." This becomes especially significant in electroplated palladium where the grain size can be on the order of 50 to 250 Å, and thus the reactivity at the grain boundaries can be significantly higher than the bulk metal by the argument presented above.

Palladium, of all the PGMs, has the highest capacity to absorb hydrogen [47]—as much as 900 times its own volume. The uptake of hydrogen corresponds roughly to the composition Pd_2H, but modern studies appear to have largely ruled out the formation of such discrete substance, Figure 8. Instead, it is inferred that below 300°C there are two phases, each consisting of a solid solution, whereas above this critical temperature there is only a single solution phase. In each phase, hydrogen atoms are held interstitially in such a way as to involve actual chemical bonding, as deduced from changes in electric conductance and magnetic susceptibility. To a smaller degree, platinum and rhodium exhibit a similar absorption of gaseous hydrogen.

The ability of palladium to absorb hydrogen poses a significant problem for the electrodeposition of this metal from aqueous solution. This issue, that is, 'the hydrogen embrittlement problem,' will be discussed in some detail in Section 5.2.

TABLE 4 Reaction of Platinum Group Metals with Oxygen

Metal	Extent of Oxide Formation (25°C)	Oxide Formed	Formation Temperature (°C)	Decomposition Temperature (°C)
Platinum	Negligible	Platinum (IV) oxide, PtO_2	< 1000	—
Palladium	Superficial	Palladium (II) oxide, PdO	> 350	> 870
Rhodium	Superficial	Rhodium (III) oxide, Rh_2O_3	~ 700	1100
Iridium	Superficial	Iridium (IV) oxide, IrO_2	~ 700	1140
Osmium	Considerable	Osmium (VIII) oxide, OsO_4	200	—
Ruthenium	Superficial	Ruthenium (IV) oxide, RuO_2	700	—

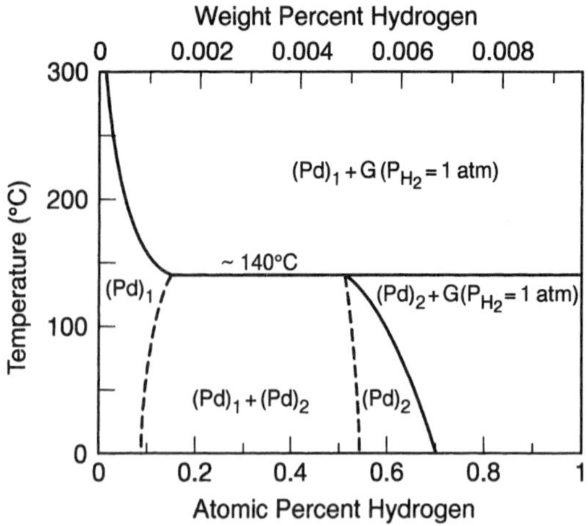

Figure 8 Binary alloy phase diagram of the palladium-hydrogen system.

TABLE 5 Attack of Platinum Group Metals by Mineral Acids

Metal	Form	Nature of Attack
Palladium	Compact	Attacked by hot concentrated nitric acid & boiling sulfuric acid; dissolved by aqua regia
	Sponge	Dissolved by all the above acids
Platinum	Compact or sponge	Not attacked by single mineral acids; dissolved by aqua regia
Rhodium	Compact	Attacked by boiling sulfuric acid or hydrobromic acid; not dissolved by aqua regia
Iridium	Compact	Practically unattacked by hot mineral acids or aqua regia
	Sponge	Dissolved in Carius tube by hot hydrochloric acid plus an oxidizing agent (nitric acid or sodium chlorate)
Ruthenium	Any	Virtually unattacked by hot mineral acids or aqua regia
Osmium	Any	Virtually unattacked by hot mineral acids or aqua regia

Table 5 summarizes the action of acids on the platinum-group metals [43]. Pd is more reactive than the other metals. It is dissolved by nitric acid, giving $Pd^{IV}(NO_3)_2(OH)_2$; in bulk form the attack is slow, but it is accelerated by oxygen and oxides of nitrogen. As a sponge, Pd also dissolves in HCl in the presence of chlorine or oxygen. The action of aqua regia on palladium yields chloropalladic acid, H_2PdCl_6; however, on evaporation of this solution, the polymeric dichloride, $(PdCl_2)_x$, is the compound recovered.

Thus $PdCl_2$ is significant because it can be the starting material for the synthesis of most electrolytes used to electrodeposit palladium.

Oxidation States and Coordination Chemistry The principal oxidation state of palladium is +2, although less common there exists a rich chemistry for Pd^{+4}. There is also some important chemistry in the monovalent state, where metal-metal bonds are involved, and in the zero-valent state for certain tertiary phosphine, CO or other π-acid organometallic compounds [48–50].

The coordination chemistry of platinum and palladium has attracted considerable attention, largely because these metals have been a source of compounds of great intrinsic value. Use as catalysts and more recently as anticancer agents has spurred interest. Furthermore the square-planar geometry of the bivalent state made possible the study of cis and trans isomers [51–52]. The coordination chemistry of palladium (II) is of great interest, since electrodeposition from aqueous solution is basically the chemistry of the bivalent state. Therefore the rest of this discussion will be restricted to the Pd(II) state.

Palladium +2 is the most common oxidation state and possesses a d^8 electronic structure [49]. Almost all of these complexes have a coordination number of 4 and form stable 16 electron complexes. The tendency of transition metals to form complexes in which the metal has an "effective number" corresponding to the next higher inert gas has long been recognized [53]. If one restricts oneself to the group VIII elements, then essentially most stable, well-characterized compounds have either 16 or 18 valence electrons [54, 55]. Furthermore the structure of tetracoordinated palladium is square planar rather than tetrahedral, since in this case the ligand field stabilization energy is relatively more important than valence shell electron pair repulsion, which would dictate a tetrahedral structure. A detailed discourse on the chemistry and structure of palladium compounds can be found in [43–48].

5 THE ELECTROCHEMISTRY OF PALLADIUM

5.1 General Electrochemistry — Thermodynamics

One of the most complete thermodynamic considerations of the electrochemistry of palladium can be found in Chapter 12 of the handbook *Standard Potentials in Aqueous Solutions* [56]. Under standard conditions in solutions of zero ionic strength, the standard potentials for the electroreduction of Pd^{+2} has been reported by various investigators to occur between 0.915 and 0.979 V.

The half-cell potential of the Pd^{+2}/Pd couple in $HClO_4$ solutions at zero ionic strength, was reported by Izatt [57] as follows:

$$Pd^{+2}(aq) + 2e^- \rightarrow Pd(s) \quad E^0 = 0.915 \pm 0.005\,V \tag{1}$$

However, Templeton [58, 59] reported a value of $E^0 = 0.987\,V$ from a $HClO_4$ solution at ionic strength $I = 4\,mol\,liter^{-1}$ which, when adjusted to zero ionic strength [59], yields $E^0 = 0.945\,V$. Other measurements by various investigators [60] have yielded $E^0 = 0.978 \pm 0.0005\,V$ while polarographic [61] measurements confirmed the value of Izatt.

Pourboix et al. [62] used thermodynamic data to calculate the formal potential of the formation of palladium from palladium hydroxide.

$$Pd(OH)_2(s) + 2H^+ + 2e^- \rightarrow Pd(s) + 2H_2O \quad E^0 = 0.897\,V \tag{2}$$

Table 6 provides available thermodynamic data for various aqueous complexes of palladium (II) along with either the measured or calculated E^0. For example, the electroreduction of palladium from $PdCl_4^{-2}$ in HCl(aq) media was reported by several authors [58, 63] to be

$$PdCl_4^{-2}(aq) + 2e^- \rightarrow Pd(s) + 4\,Cl^-(aq) \quad E^0 = 0.62\,V \tag{3}$$

and $E^0 = 0.60$ V for 1 M KCl. These values were confirmed [64] for 1 M HCl, $E^0 = 0.59$ V, and by the study of the equilibrium constants for the stepwise dissociation [65]

$$Pd^{+2}(aq) + 4\,Cl^-(aq) \rightarrow PdCl_4^{-2}(aq) \quad \log\,\beta_4 = 12.2 \tag{4}$$

For the amine complex, which is very important for the electrodeposition of palladium, the following potential has been calculated:

$$Pd(NH_3)_4^{+2}(aq) + 2e^- \rightarrow Pd(s) + 4NH_3 \quad E^0 = 0.0\,V \tag{5}$$

on the basis of the stability constants ($\log\,\beta_4 = 30.5$) measured for the equilibrium in ammonia solutions at $I = 1$ mol liter^{-1}.

For palladium in the IV state, one of the potentials for the half-reaction is

$$PdO_2(s) + 2H^+(aq) + 2e^- \rightarrow PdO(s) + H_2O \quad E^0 = 1.263\,V \tag{6}$$

This was calculated by Pourboix [62] from thermodynamic data. Furthermore, for Pd^{+4} in the important ethylenediamine systems, the following potential was obtained from *emf* determinations [66] at 25°C and 10^{-5} to 10^{-6} mol liter^{-1} nitrate solutions:

$$Pd(en)_2Cl_2(aq) + 2e^- \rightarrow Pd(en)_2^{+2}(aq) + 2Cl^-(aq) \quad E^0 = 1.15\,V \tag{7}$$

TABLE 6 Thermodynamic and Electrochemical Data on Palladium (II)

Formula (Aqueous)	ΔH° kJ^{-1} mol^{-1}	ΔG° kJ^{-1} mol^{-1}	S° J^{-1} mol^{-1}K^{-1}	$\log\,\beta_x$	E^0-V
Pd^{+2}	149.0	176.5	– 184	< 3	0.915
$Pd(SO_4)$				3.16	0.82
$PdCl_4^{-2}$ (1 M HCl)	– 550.2	– 417.1	16.7	12.2	0.62
$PdBr_4^{-2}$	– 384.9	– 318.0	247	15	0.49
PdI_4^{-2}		– 159.0		25	0.18
$Pd(NO_2)_4^{+2}$		– 68			0.34
$Pd(OH)_2$					0.07
$Pd(en)_2^{+2}$				26.9	
$Pd(NH_3)_4^{+2}$		– 75		30.5	0.0
$Pd(CNS)_4^{-2}$		410.5			
$Pd(CN)_4^{-2}$		628		51.6	– 1.52

For palladium in the VI state, from data on the anodic formation of PdO_3 in alkaline solutions, Pourboix [62] found:

$$PdO_3(s) + 2H^+(aq) + 2e^- \rightarrow PdO_2(s) + H_2O \quad E^0 = 2.030\,V \tag{8}$$

Thus the overall potential diagram for the electroreduction of palladium is

$$Pd^{+6} \xrightarrow{2.03} Pd^{+4} \xrightarrow{1.263} Pd^{+2} \xrightarrow{0.915} Pd^0$$

However, normal electrodeposition conditions deviate significantly from the standard, or equilibrium condition. The reduction potential at any given temperature and concentration can be expressed by

$$E = E^0 + \left(\frac{RT}{nF}\right) \ln a + C \tag{9}$$

In addition to the standard or formal potential, the reduction potential for palladium or any other species (e.g., H_3O^+) can be affected by the temperature, the concentration of the species at the electrode surface, and any other factors that contribute to the C term such as complexing ligands for metals, and organic additives that may affect the overpotential for the metal deposition.

Finally, the deposition of palladium from acqueous solutions is usually accompanied by the codeposition of hydrogen and the formation of palladium hydrides, PdH_x. As will be seen in Section 5.2, the successful electrodeposition of palladium and its alloys is highly dependent on the avoidance of this phenomenon. In this regard we will attempt to understand it by treating the electrode surface as the scene of more than one electrochemical reaction (see Section 5.3).

5.2 The So-called Hydrogen Embrittlement Problem

Hydrogen embrittlement is a generic term used to describe a variety of fracture phenomena having a common relationship to the presence of hydrogen in a metal or an alloy as a solute element [67]. The ramifications of hydrogen embrittled metals and alloys can be very serious and, in some cases, such as aircraft parts or nuclear reactor components, even life threatening.

It appears that most materials can become embrittled if hydrogen is introduced under "pressure" by means such as electrodeposition [69]. It has been said that "one of the most spectacular and memorable sounds associated with zinc plating is standing in a quiet storage room next to drums of freshly zinc-plated steel springs and listening to the metallic shriek as the springs slowly destroy themselves in order to release occluded hydrogen" [71]. Likewise, the author has plated palladium with such large amounts of codeposited hydrogen and found that after electrodeposition, the plated coupons violently twisted and released enough heat from the exothermic release of hydrogen to burn his fingers. The effects of this phenomenon are discussed in [67–71].

The phase diagram of Pd and hydrogen is seen in Figure 8. This system has been well studied [47], and recently there has been renewed interest triggered by the speculative phenomena of "cold-fusion" [72–74]. Actually some classical fundamental work on this system was carried out by Lacher [75] and Gillepsie [76–77] from 1935 to 1939. They determined that palladium hydride exists in an fcc α-phase up to an atomic concentration of [H]/[Pd] = 0.03. Flanagan and Simons [78] found that the lattice of the α-PdH_x increases slightly from that of pure palladium, which is 3.889 to 3.893 Å due to hydrogen absorption. However, the pure β-phase with a minimum atomic ratio of [H]/[Pd] = 0.57 prevails at higher hydrogen content. This phase, which is also fcc, has a larger lattice constant of 4.04 Å. Furthermore this β-phase is dependent on the temperature of the system and on the hydrogen pressure. For example, Frumkin and Aladjalowa [79] state that an atomic ratio of [H]/[Pd] = 0.68 is obtained from one atmosphere of hydrogen pressure; Hoor and Arrowsmith state that this ratio equals 0.61 between 20° and 30°C. In the concentration range of ratios from 0.03 to 0.57 [H]/[Pd], the two phases are at heterogeneous equilibrium.

The hydrogen embrittlement phenomenon stems from the thermodynamic instability of the β-PdH_x phase. Immediately after deposition, β-PdH_x spontaneously decomposes, evolving H_2 gas, and it is transformed to the thermodynamically stable α-PdH_x. This transformation is accompanied by corresponding lattice contractions, Figure 9; it changes in the mechanical properties of the Pd films. During the contraction the deposits become highly stressed, and when the tensile components of this stress exceed the yield point, the stress is relieved by cracking. The deleterious effects of such a phenomenon are seen in Figure 10.

The electrodeposition and material properties of palladium are thus greatly affected by the chemistry and deposition parameters (e.g., temperature and pH), since they determine the amount of codeposited hydrogen. Some recent investigations on the electrochemistry and mechanisms of hydrogen inclusion during palladium electrodeposition can be found in [72–74] and [81–87]. This phenomenon and the electrochemical parameters that affect it must be understood if one is to successfully electroplate palladium.

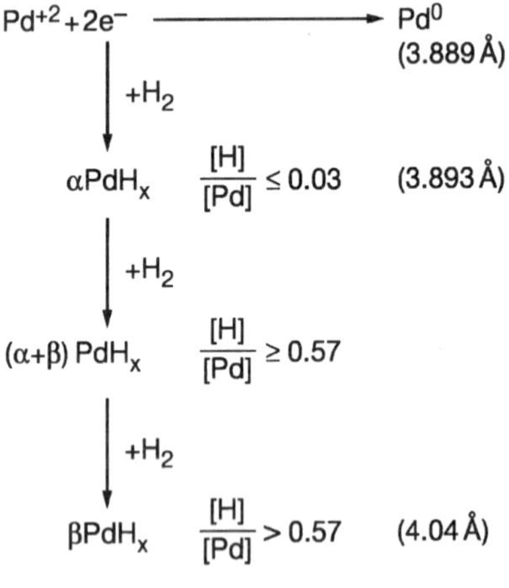

Figure 9 Schematic of palladium-hydrogen embrittlement phenomenon.

Figure 10 Scanning electron micrograph of hydrogen embrittled palladium film.

5.3 Electrochemistry of Simultaneous Reactions — Electrokinetics

The "hydrogen embrittlement" of electrodeposited palladium is a matter of great concern when one attempts to deposit films of technological use. Obviously these films must be crack-free and of low stress. Thus it is important to eliminate or inhibit the codeposition of hydrogen with palladium. Alternatively, the electrodeposition of palladium alloys such as PdNi, PdCo, and PdAg where the alloying metals are in the range of 10% to 40% are of significant technological interest. Thus there exists the situation where in the later case a high degree of codeposition for alloy formation is desired, and in the former case where codeposition is undesirable. Identical electrochemical considerations can be utilized to obtain an understanding of both phenomena in which the interface is the scene of more than one electrochemical reaction.

The treatment of simultaneous reactions at an electrode interface [88] is based on an important fact: Though the current densities of $i_1, i_2, \ldots, i_n$ may be different for each reaction, they are produced by the same absolute Galvani potential difference, ΔE, at the interface. In the case of two simultaneous reactions, the following equation will determine the relative current densities for each electroreduction:

$$\frac{i_1}{i_2} = \left[\frac{i_{0,1}}{i_{0,2}} e^{\left(\frac{F}{RT}\right)(\vec{a}_1 \Delta E_{e,1} - \vec{a}_2 \Delta E_{e,2})}\right] e^{\left(\frac{F}{RT}\right)(\vec{a}_1 - \vec{a}_2)\Delta E} \tag{10}$$

Thus the value of the current-density ratio, and the degree of "codeposition," depends not only on the exchange-current densities ($i_{0,i}$), the transfer coefficients (α_i), and the equilibrium potentials ($\Delta E_{e,1}$, $\Delta E_{e,2}$) but also, in an exponential way, on the potential difference across the interface (ΔE). Thus, even if the deposition of a second species (i_2) is inhibited by lower exchange current density (i.e., $i_{0,2} < i_{0,1}$) and more negative equilibrium potentials (i.e., $\Delta E_{e,2} < \Delta E_{e,1}$), it can have a higher deposition-current density if its transfer coefficient $\vec{\alpha}_2$ is greater than that of the other, $\vec{\alpha}_1$.

Another situation in which a species with a more negative equilibrium potential can be deposited to a greater extent occurs when the deposition of the other entity is limited by mass transport. Under such conditions

$$i_1 = i_L = \frac{DnFC_1^0}{\delta} \tag{11}$$

where D is the Diffusion Coefficient, n is the number of electrons, F is Faraday's Constant, δ is the diffusion layer thickness, and C_1^0 is concentration in the bulk. If i_2 is not transport limited, then i_2 can, most likely, be greater than i_1.

The Electrokinetics of Hydrogen Evolution The arguments in Section 5.3 are equally valid for a situation where metal deposition occurs along with any other cathodic reduction such as the formation of hydrogen [88]. From a technological point of view, in most cases one seeks to electrodeposit a metal at high current efficiencies (i.e., limit hydrogen formation). The reasons are obvious, since hydrogen embrittlement, as discussed in Section 5.2, can be problematic. Furthermore a high current efficiency implies that the desired reaction occurs at a faster rate, and less time is utilized to deposit the desired amount of metal at a given total current density. Technologically speaking, this is of great importance since it "speeds up" the process and increases manufacturing throughput.

In order to achieve this objective, it becomes important to understand the current density ratio i_H/i_M or the percentage of current efficiency $[i_M/(i_M + i_H)] \cdot 100$ that is attributed to metal deposition. The intent is to utilize the current only for metal deposition so that cathodic current efficiency = 100%.

The basic factors determining current efficiency are revealed by considering i_1 as the hydrogen evolution current and i_2 as the metal deposition current in the Eqs. (10) and (11). Thus the same principles apply: That is, the current going toward hydrogen evolution depends on the ratio of the exchange-current densities, the equilibrium potentials, and the relative transfer coefficients of hydrogen evolution, and metal deposition reactions. In general, the deposition of metal is favored by a more positive equilibrium potential with respect to hydrogen, a higher exchange-current density, and a lower value of the transfer coefficient. Even, if these parameters are unfavorable for the deposition of a metal, it may still be possible to obtain high current efficiencies by insuring that i_H is transport limited. Utilizing Eq. (11) for i_L, it becomes clear that one needs to limit $C^0_{H^+}$ which can be accomplished by manipulating the pH. Conversely, if one exceeds the i_L for metal deposition (there is always a limit), hydrogen evolution will replace it. In aqueous solutions, hydrogen will always be evolved if the total deposition current is made sufficiently high; the hydrogen may come not only from H_3O^+ but, if limited at high pH, from water itself.

The effect of hydrogen codeposition on the crystallography of metal deposition are rather complicated and beyond the scope of this chapter. However, several ramifications of hydrogen evolution must be noted. As discussed in Section 5.2, hydrogen may codeposit with a metal and change its mechanical properties. It may also adsorb more fully on certain crystal planes and block them so that the metal grows preferentially on others. Finally removal of H_3O^+ from the diffusion layer may significantly increase the pH. If the solution at the interface becomes sufficiently alkaline, it may cause the solubility product of a hydroxide of the metal ion present to be exceeded, which could cause inclusion of such a species in the deposit. This can be problematic for mechanical properties such as stress and ductility. Alternatively, it could cause the formation of relatively thick (10^{+2}–10^{+3} Å) films on plated parts, thus "passivating" the cathode surface.

In summary, hydrogen codeposition usually leads to many undesirable effects in metal deposition. Therefore selective control over the rate of hydrogen evolution is vital, hence the need to understand the reaction. Electrode kinetics, as discussed above, is important in determining the i_H/i_M and the factors that affect it; however, the "hydrogen overpotential" is also an important phenomenon that determines the parameters under which a metal can be deposited.

Hydrogen Overpotential An operating electrolytic cell that has net current flowing through it is not at equilibrium [88]. The potential will differ from equilibrium:

$$\eta = E_i - E_q \tag{12}$$

where E_i is the potential when current is flowing, E_q is the equilibrium potential, and η is the overpotential or overvoltage. This overpotential may be caused by several factors. For example, concentration polarization at the electrode surface caused by the depletion of reactants at an electrode surface can be described by the Nerst equation:

$$\eta_{conc.} = \frac{RT}{nF} \ell_n \frac{a_c}{a_o} \tag{13}$$

this is referred to as concentration overpotential, where a_c is the activity of the depositing ion at the electrode surface and a_o that of the bulk solution. In the case of hydrogen evolution, a concentration overpotential occurs when surface hydrogen concentration changes (i.e., depleted) with surface pH changes. Alternatively, an overpotential may be required to overcome various kinetic barriers to a reaction; this is referred to as activation overpotential [88]. This is the minimum energy or potential barrier that must be overcome for the reaction to proceed, and for the hydrogen evolution reaction it is referred to as the "hydrogen overpotential."

If it were not for hydrogen overpotential, many metals could not be deposited from aqueous solutions. For example, at pH of 4, which is a common pH for nickel plating, the E^0 for hydrogen evolution is about -0.236 V; in alkaline solutions with pH = 10, it is -0.59 V. These potentials are significantly more positive than the reduction potentials of many metals plated from these solutions. Activation overpotential is extremely high in the evolution of hydrogen, and it is this that enables metal deposition or inhibits hydrogen evolution.

The hydrogen overpotential is dependent on the nature (i.e., electronic structure) and structure (morphology) of the electrode surface. For example, at cathodes of tin, zinc, or lead, it can be more than IV negative than its equilibrium potential, and this is why these metals can be deposited from very acidic solutions, pH < 1, where the concentrations of H_3O^+ are very high. Alternatively, it is extremely low for metals such as platinum or palladium. On platinum black it is negligible, which is why it is used in setting up the hydrogen electrode. Therefore it is a challenge to plate palladium from aqueous solution without the interference of hydrogen, and one cannot rely on the "hydrogen overpotential" to avoid codeposition.

An additional factor that can aid in suppressing hydrogen formation is the use of additives or "poisons" that adsorb on an electrode surface but are essentially electrochemically inactive. Such impurities adsorbed on the cathode surface may increase hydrogen overvoltage. Furthermore such impurities often include addition agents that "poison" selected sites and are useful for leveling or brightening the deposit. A detailed discussion of the mechanisms of hydrogen formation including the use of "poisons" is beyond the scope of this chapter and can be found in [88].

Minimization of Hydrogen Codeposition So for the discussion of this section has demonstrated, in a theoretical sense, what is required to either minimize or maximize the current ratio, i_1/i_2, of two simultaneous reactions occurring at an electrode interface. It now becomes important for us to understand what factors, chemical, physical, and/or electrochemical, can be utilized to achieve the end objective, which in the case of palladium plating is to minimize i_1 and maximize i_2. It is stated by the earlier arguments in this section that the following parameters must be considered:

1. The exchange-current densities, i_0
2. The transfer coefficients, α
3. Equilibrium potentials, ΔE_e
4. Potential difference, ΔE, across the interface
5. $i_{L,1}$ versus $i_{L,2}$, limiting current
6. Hydrogen overvoltage modifiers

- Metallics (e.g., Ni for PdNi alloy)
- Organic (surfactants, brighteners, etc.)

The fact of the matter is that the exchange-current densities, the transfer-coefficients, and the equilibrium potentials are, for the most part, determined by the nature of species 1 and 2, in this case palladium and hydrogen, and the nature of the electrode material which in this case will be palladium. These are constants at a given temperature and cannot be manipulated. In this situation only the relative concentration of species 1 and 2 at the electrode surface and the temperature can be altered significantly. However, what can be modified by changing the chemical system and the plating conditions (e.g., solution agitation) are the relative reduction potentials of each species, the concentration of each at the electrode surface, and potentially the overpotential of hydrogen and/or metal deposition at a "modified" palladium surface through the use of metallic or organic additives.

As stated in Section 5.1, the reduction potential of an element at any given temperature and concentration is described by the sum of its standard potential and a term expressing its activity in solution as shown below:

$$E_1 = E_1^0 + \frac{RT}{nF} \cdot \ln(a_1) + C_1 \tag{14}$$

$$E_2 = E_2^0 + \frac{RT}{nF} \cdot \ln(a_2) + C_2 \tag{15}$$

The difference in potential between the two elements involved can be expressed as

$$\Delta E = E_1 - E_2 \tag{16}$$

For successful alloy deposition, ΔE must be minimized. The condition can be met either when the two values E_1^0 and E_2^0 are similar, or by adjusting the concentration of one species in solution, so altering its activity, or by arranging a suitable difference in the C term, which can be achieved by introducing complexants. The latter is the most common manipulation, especially when the difference between the standard potentials is large.

Alternatively, to avoid or inhibit hydrogen codeposition one must maximize ΔE. That is, the reduction potential for palladium must be as positive as possible while the hydrogen reduction potential must be as negative as possible. In this regard we utilize cyclic voltammetry* to examine the reduction characteristics (peak potentials, E_p) of various chemical systems to demonstrate how to achieve the above objective.

For a totally irreversible system, the expression for the peak potential, E_p is the following [90]:

$$E_p = E^{0'} - \frac{RT}{\alpha nF}\left[0.78 + \ln\left(\frac{{D_0}^{1/2}}{k^0}\right) + \ln\left(\frac{\alpha nFv}{RT}\right)^{1/2}\right] \tag{17}$$

where $E^{0'}$ is the formal potential (see Eq. 27), α is the transfer coefficient, n is the number of electrons, F is Faraday's constant, D_0 is the diffusion coefficient, k^0 is the standard heterogeneous rate constant, and v is the linear potential scan rate.

The objective is to obtain maximum separation of $E_p(\mathrm{Pd})$ and $E_p(\mathrm{H_2})$ as shown below:

$$\Delta E_p = E_p(\mathrm{Pd}) - E_p(\mathrm{H_2}) \tag{18}$$

Examination of the expression for E_p tells us that this will depend on the formal potential of each species and to some extent and as discussed before, the temperature, the transfer coefficients, the diffusion coefficients and the standard rate constants for each reaction. The reduction potential, E, can be described through the Nerst equations below:

$$E(\mathrm{Pd}) = E_{\mathrm{Pd}}^{0'} + \frac{0.0591}{2}\log[\mathrm{Pd}^{+2}]_0 \tag{19}$$

$$E(\mathrm{H_2}) = E_{\mathrm{H_2}}^{0'} + 0.0591\log[\mathrm{H}^+]_0 \tag{20}$$

in which $E^{0'}$ are formal potentials and $[\mathrm{Pd}^{+2}]_0$, $[\mathrm{H}^+]_0$ the concentrations at the electrode surface.

Formal potentials, of course, are dependent on the standard potential (see Eq. 27) and the relative concentration of each species at the electrode surface, which in turn is dependent on the equilibrium constants for specific palladium chemistries (see Table 6) and on the pH for the hydronium ion concentration. In Section 5.4 we will

*For a discussion on cyclic voltammetry, see Reiger and Bard [89, 90].

see how to manipulate these factors to maximize ΔE_p. However, prior to this discussion, we must examine the equilibrium constants of various palladium complexes and other factors that are important in these considerations.

Theoretical and Practical Considerations for Plating Palladium The standard potential, E^0, of the Pd^{+2}/Pd^0 as described in Section 5.1 ranges from $+0.915$ to $+0.979$ V. This highly positive value, as would be expected for a noble metal, implies facile reduction of Pd(II) to Pd(0) from an aqueous solution of its simple salts, namely noncomplexed ion. For example, for the $[Pd(SO_4)]^{2-}$ system in Table 6, log $\beta_4 = 3.16$, which indicates that this aquo complex is of low stability. As a matter of fact, tetra-aquo palladium (II), present in acidic media, can be reduced by less noble metals (e.g., Cu) as a result of charge exchange or "displacement" plating. In this case, even though the H_3O^+ concentration is high, it can be plated with 100% current efficiency and thus avoid hydrogen codeposition. However, the palladium films produced as a product of electrodeposition coupled with displacement plating are powdery and nonadherent [37]. Furthermore electrodeposition from solution of $pH < 1$ is problematic in a practical sense.

On the other hand, electrodeposition from aqueous solutions of highly stable complexes such as $[Pd(CN)_4]^{-2}$ is impractical. The stability constant, β_4, of tetracyanopalladate is

$$\beta_4 = \frac{[(Pd(CN)_4)^{-2}]}{[Pd^{+2}][CN^-]^4} = 10^{51.6} \tag{21}$$

which indicates that this complex is much too stable, and that its reduction potential is more negative than the onset of hydrogen evolution. This prevents polarization to a potential where metal reduction would occur and hydrogen is preferentially discharged. It has been reported [35] that only traces of palladium can be deposited from this complex even if the electrolyte is strongly alkaline and a metal with a high hydrogen overvoltage is employed as the cathode. Smee in 1919 made use of this fact to separate gold from palladium by electrolysis of a solution containing the complex cyanides of both metals [34]; he also stated that palladium was not deposited from such a solution until the cyanide species had been decomposed. The relationship between β_x and E^0 can be seen in Table 6.

Complexes with nitrogen-containing ligands, such as ammonia, are well suited for Pd deposition; they are stable enough to prevent immersion plating and yet amenable to cathodic reduction. For example, palladium tetra-ammine complex has the following stability constant:

$$\beta_4 = \frac{[Pd(NH_3)_4{}^{+2}]}{[Pd^{+2}][NH_3]^4} = 10^{30.5} \tag{22}$$

which is more than 20 orders of magnitude lower than the cyano complex. Likewise, mono amines such as CH_3–NH_2 and multidentate ligands such as diamines (e.g., 1,3 diaminopropane), can be used to electrodeposit palladium. Thus palladium complexes within an intermediate range of stability are very useful in practical electroplating processes.

In the following section we will discuss electrodeposition chemistries for alkaline, neutral, and acid electrolytes. We will begin with the examination of a highly alkaline (pH $\sim$ 12), additive "free" chemistry that utilizes 1,3 diaminopropane as the ligand of choice. This analysis will demonstrate how to manipulate the chemical and physical properties of the process in order to avoid hydrogen embrittlement and deposit "smooth," "crack-free" palladium films.

6 THE ELECTRODEPOSITION OF PALLADIUM

Palladium has been plated from a wide variety of electrolytes, which are far too numerous to include in this chapter. This work will summarize what the author believes to be the most important processes, from a practical sense and also a historical perspective. In an attempt to organize this information, the processes will be broadly classified as alkaline (pH 8–13), neutral (pH 5–8), and acidic (pH $<$ 1 to 5).

6.1 Alkaline Electrolytes (pH 9–13)

In the presence of ammonia or amines, palladium ions exhibit a very strong tendency to form stable complexes. As shown in Table 6, the log β_4 and log β_2 for ammonia and ethylenediamine are 30.5 and 26.9, respectively. While certainly not as stable as the cyano complex, they are nonetheless more stable than the aquo or halide complexes and are not susceptible to displacement plating.

The amine complexes are readily formed in palladium chloride solutions by the general equation:

$$\mathrm{Pd}X_2 + x\mathrm{NH}_2 - R \rightarrow [\mathrm{Pd}(\mathrm{NH}_2R_x)]^{+2} + 2X^- \tag{23}$$

where

X = halides (Cl^-, Br^-, I^-); NO_3^-; SO_4^{-2}
R = H; CH_3; $C_yH_{2y} - CH_3$; $C_yH_{2y} - NH_2$
x = 2, 4 (depending on whether its a diamine or a monoamine)
y = 1, 2, ...

Table 7 is a summary of various alkaline processes that have been found in the literature (see the references).

In this pH range ($>$9), only the diamine processes can be considered of practical use. While the ammonia systems certainly can plate palladium with respectable properties; additives and wetting agents are essential if one requires smooth, bright to semi-bright deposits. In the absence of additives, the deposits are nodular, of dull appearance and susceptible to "fingerprinting" or easily stained [37, 40]. Furthermore at pH $>$ 8, the presence of high concentrations of ammonia, especially at elevated temperatures, becomes problematic due to its excessive outgassing from solution, which in many cases is being subjected to aggressive movement or agitation. This makes it difficult to control pH, requires extensive replenishment with ammonium hydroxide, and poses an obvious environmental challenge. Thus the only

TABLE 7 Palladium Electroplating Processes - Alkaline Electrolytes (pH 9–13)

Parameter	Bath Type						
	A[71,37,98,35]	B[92]	C[8,99]	D[81,40,96]	E[93,94]	F[97]	G[98]
Source of Pd	$PdCl_2$	$PdCl_2$	$PdCl_2$	$Pd(NH_3)_2(NO_2)_2$	$Pd(NH_3)_2Br_2$	$Pd(NH_3)_2(NO_2)_2$	$Pd(OOCCH_2NH_2)_2$
Pd Metal ($g\,liter^{-1}$)	1–10	28	1–40	4–25	25–35	4–18	1–50
Ligand	NH_3	en[a]	pn[a]	NH_3	NH_3	NH_3	$OOCCH_2NH_2$
Amino Acetic Acid						20	1–150
Ammonium Bromide					45		
Ammonium Chloride	60–10						
Ammonium Hydroxide		25–75		Adjust pH	Adjust pH		
Ammonium Nitrate				90–100			
EDTA							5
Hydrochloric Acid			Adjust pH				
Potassium Hydroxide			Adjust pH			Adjust pH	Adjust pH
Potassium Phosphate			50–100				
Sodium Sulfate		140					
Sodium Nitrate				9–12			
Tetrapotassium Pyrophosphate						5–300	
Additives	See Ref.	None	None	See Ref.	See Ref.	See Ref.	See Ref.
pH	9–16	11–12	10–13	9–10	9–9.5	8.5–11	7–12
Temperature (°C)	25–10	25	40–70	25–55	25–55	35–45	60
Current Density (mA/cm^2)	1–25	20	1–500	1–300	2.5–100	2–50	10

[a]en = 1,2 ethylenediamine, pn = 1,3 diaminopropane.

real contenders in this pH range are the "ammonia-free" systems. The remainder of this section will be restricted to the diamine processes (C, Table 7), which is an excellent vehicle to examine the "hydrogen embrittlement" issues discussed above.

1,3 diaminopropane Electrolyte [8, 99] The ligand of choice in the formulation of bath C is 1,3-diaminopropane, (pn), whose formula is $NH_2C_3H_6NH_2$. It is a bidentate nitrogen containing ligand that forms a 16 electron, square planar coordination complex, $[Pd(pn)_2]^{+2}$.

The choice of 1,3 diaminopropane for this specific process was based on the analysis of numerous possibilities. Other ligands considered were additional aliphatic diamines (e.g., 1,4 diaminobutane), aliphatic triamines (e.g., diethylenetriamine), substituted diamine and triamine-species (e.g., 2-hydroxy-propanediamine), numerous substituted and unsubstituted monoamines (e.g., triethanolamine), pyridines, and numerous substituted and unsubstituted aromatic amines (e.g., diphenylamine). 1,3 diaminopropane was chosen based on an exhaustive analysis that took into account the requirements of performance, cost, and environmental considerations.

Counter Ion Effects Equation (23) above describes the synthesis of an amine palladium complex where the counter ion can be Cl^-, Br^-, NO_3^-, and SO_4^{-2}. Cyclic voltammograms of the 1,3 diaminopropane electrolytes are shown in Figure 11. The Pd^{+2} reduction peak potential (−0.91 V versus SCE) is essentially unaffected by the change in anion, and it is concluded that the anion has little or no effect on the electrochemical reduction of palladium. This phenomenon was recently studied in an ammonia system by LePenven et al. [100] for Br^-, Cl^-, and NO_2^- electrolyte.

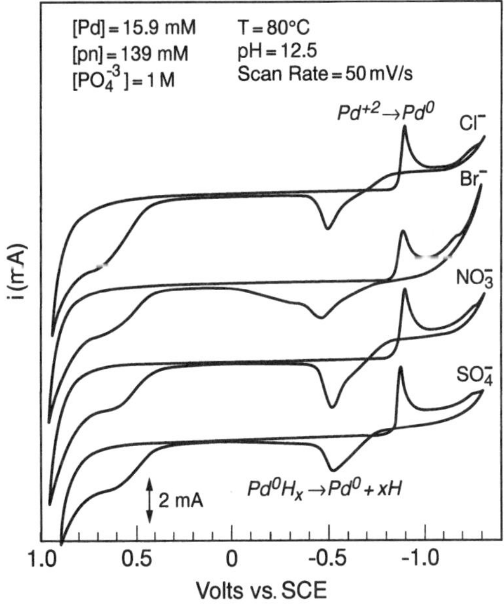

Figure 11 Cyclic voltammograms: Palladium-diaminopropane with chloride, bromide, nitrate, and sulfate counter ions.

However, there are other factors that must be considered in choosing the counter ion. For example, NO_3^- is not electrochemically stable. This is also true for the nitrite systems where ammonium nitrate is subsequently formed by electrolysis but is itself unstable [101]. The SO_4^{-2} ion is electrochemically stable, but due to the difficulty of preparing $PdSO_4$, the cost becomes prohibitive. The Br^- is easily oxidized at an insoluble anode, and one would like to avoid the formation of Br_2.

The most reasonable choice from a technical and cost perspective is chloride. It is "relatively" electrochemically stable at a Pt anode, very soluble, and easily transported across anion permeable membranes [36].* Furthermore the starting chemical, $PdCl_2$, is readily available and the least expensive of the palladium salts. The remainder of this discussion will deal solely with $Pd(pn)_2Cl_2$.

pH and Temperature Effects Figure 12 shows a cyclic voltammogram of the $Pd(pn)_2Cl_2$ at a gold rotating disk electrode (RDE 0.49 cm^2; $\omega = 0$) at 25°C. At pH = 11.5, the Pd reduction peak potential (−1.04 V versus SCE) and the H^+ reduction peak potential (−1.10 V versus SCE) are separated by only (~60 mV). Formation of PdH_x is evident by both the reduction peak (labeled PdH_x) and the hydrogen oxidation peak at −0.32 V observed on the reverse waves.

As the pH is varied from 11.5 to 12.5, the hydrogen reduction peak is shifted to more negative potentials. This shift (~0.059 V/pH) decreases the overlap with Pd reduction peak and reduces the formation of PdH_x as evident by the relative decrease in the height of the PdH_x peak and the corresponding decrease in the hydrogen oxidation peak. Thus ΔE_p (see Eq. 18) is increased, and an overall reduction in hydrogen incorporation is accomplished by controlling the $C^O_{H^+}$ as discussed in Section 5.3.

A further increase in the value of ΔE_p is achieved by thermal destabilization of the Pd chelate complex which shifts Pd reduction to more positive potentials and away from the hydrogen reduction as illustrated in Figure 13. Temperature variation from 25° to 80°C shifts the peak potential, E_p, from −1.04 to −0.91 V or approximately

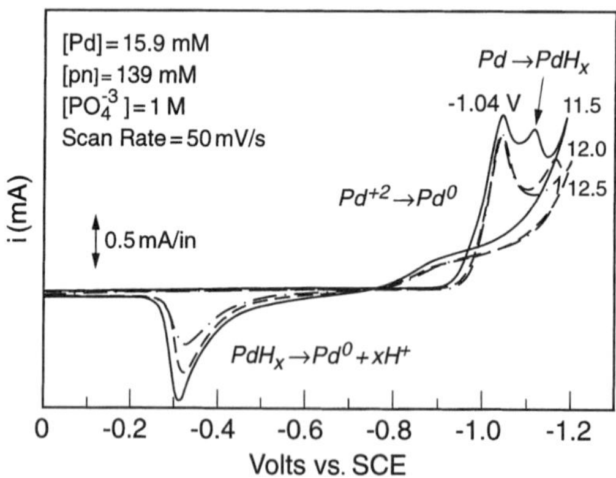

Figure 12 Cyclic voltammograms: Palladium-diaminopropane system; variation of pH 11.5 to 12.5.

*This property becomes important when removal of excess anion is necessary.

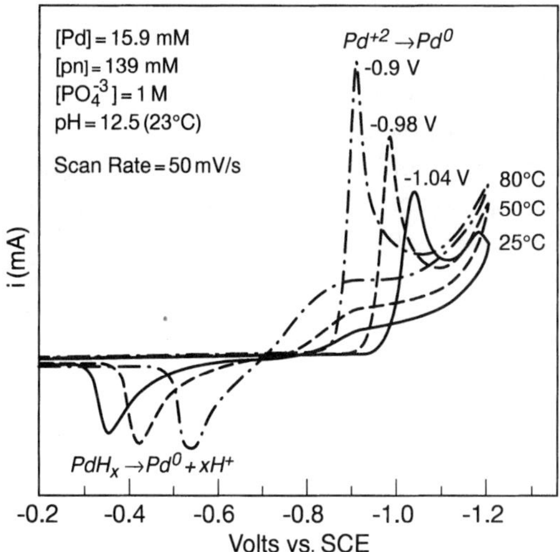

Figure 13 Cyclic voltammograms: Palladium-diaminopropane system; variation of temperatures, 25° to 80°C.

130 mV. A corresponding increase in the peak current, i_p, is realized due to an increase in the diffusivity of ions.

The shift in E_p with temperature is mostly due to the thermodynamic destabilization of the palladium chelate complex which changes $E^{0'}$, and to a lesser extent changes in RT and k^0 which affect $E^{0'}$ and E_p, as predicted by Eq. (17).

The pH dependence at elevated temperature is dramatically demonstrated in Figure 14. Voltammograms taken at 80°C and various pH show a decrease in the

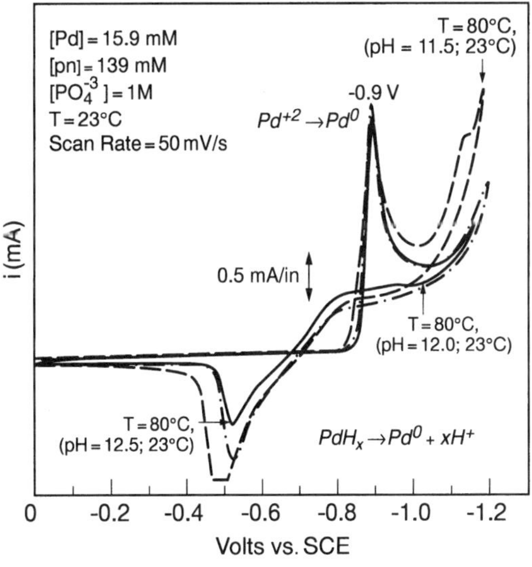

Figure 14 Cyclic voltammograms: Palladium-diaminopropane system; variation of pH 11.5 to 12.5 at 80°C.

formation of PdH_x and the corresponding hydrogen oxidation peak with decreasing $[H^+]$.

Therefore, through the use of $[Pd(pn)_2]^{+2}$ as the complex electrolyte, the exclusion of additives that cause strong kinetic inhibition of the Pd reduction, and through manipulation of temperature and pH, one can separate the Pd^{+2} reduction peak from the hydrogen reduction peak and obtain a ΔE_p of approximately 330 mV. Figure 15 demonstrates that one can reach and exceed the limiting current for reduction of Pd^{+2} and avoid hydrogen incorporation as evident by the absence of a hydrogen oxidation peak.

Ligand Concentration Effects and Replenishment Schemes The overall reaction for Pd electrodeposition using a nonconsumable anode is

$$\mathrm{Pd\,(pn)_2\,Cl_2 + H_2O \longrightarrow Pd^0(s) + 2H^+ + 2Cl^- + 2\,pn + \tfrac{1}{2}O_2} \tag{24}$$

This indicates that for every Pd^{+2} reduced 2pn molecules and $2Cl^-$ ions remain in solution, and $2H^+$ are generated, lowering the pH of the solution.

Figure 16 shows voltammograms of $[Pd(pn)_2]^{+2}$ as the pn concentration is varied from stoichiometry to $100\times$ excess. The Pd^{+2} reduction peak shifts to more negative potentials and thus closer to the hydrogen regime. This can be predicted through use of the Nerst relationship (see Eq. 19) and stability constants.

The formation of the Pd diamine complex is represented by the following equilibrium:

$$\mathrm{Pd^{+2} + 2\,pn \Leftrightarrow [Pd(pn)_2]^{+2}} \tag{25}$$

for which the stability constant is

$$\beta_2 = \frac{[\mathrm{Pd(pn)_2}]^{+2}}{[\mathrm{Pd^{+2}}][\mathrm{pn}]^2} \tag{26}$$

Combining Eq. (19) and (26) yields the following at $T = 25°C$:

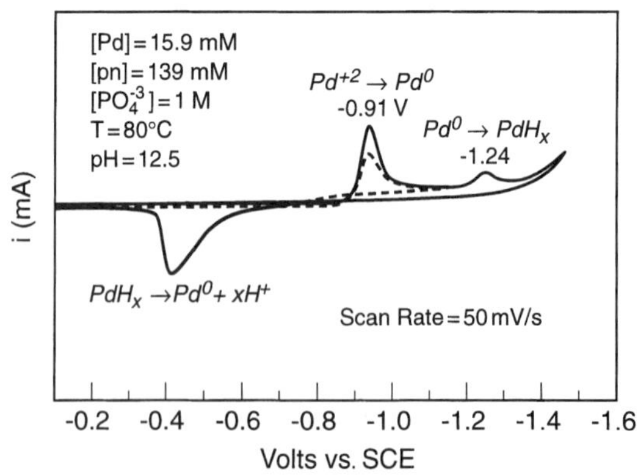

Figure 15 Cyclic voltammograms: Palladium-diaminopropane system.

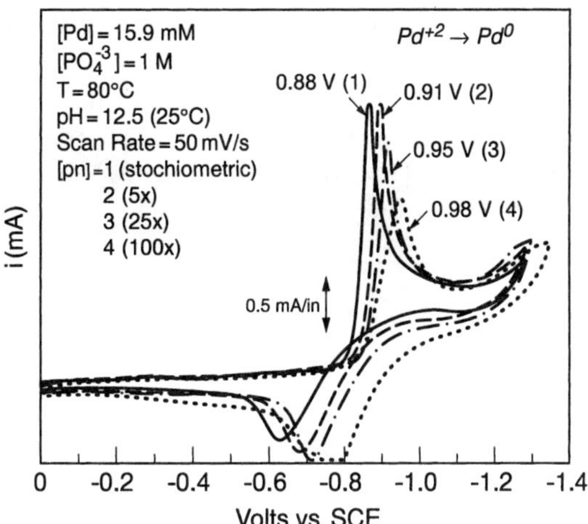

Figure 16 Cyclic voltammograms: Palladium-diaminopropane system; ligand concentration effects.

$$E^{0'} = E^0 - \frac{0.0591}{2}\log\beta_2 - \frac{0.0591}{2}\log[\mathrm{pn}]^2 + \frac{0.0591}{2}\log[\mathrm{Pd(pn)_2}^{+2}] \quad (27)$$

This equation predicts a negative shift in $E^{0'}$ as the pn concentration increases. This of course shifts E_p in a similar direction (see Eq. 17) as is experimentally observed in Figure 16.

In order to avoid the accumulation of ligand, $PdCl_2$ is used as the starting and replenishment chemical. The formation of the palladium complex is shown in the equation below:

$$\mathrm{PdCl_2} + y(\mathrm{pn}) \longrightarrow \mathrm{Pd(pn)_2Cl_2} + (y-2)\,\mathrm{pn} \quad (28)$$

The electrochemical reactions are

$$\mathrm{Pd(pn)_2Cl_2} + (y-2)(\mathrm{pn}) + 2e^- \longrightarrow \mathrm{Pd^0} + 2\mathrm{Cl^-} + y(\mathrm{pn}) \quad (29)$$

$$\mathrm{H_2O} \longrightarrow 2e^- + 2\mathrm{H^+} + \tfrac{1}{2}\,\mathrm{O_2} \quad (30)$$

where the overall reaction is

$$\mathrm{Pd(pn)_2Cl} + (y-2)(\mathrm{pn}) + \mathrm{H_2O} \longrightarrow \mathrm{Pd^0} + y(\mathrm{pn}) + 2\mathrm{HCl} + \tfrac{1}{2}\,\mathrm{O_2} \quad (31)$$

$$\downarrow \mathrm{PdCl_2}$$

$$\mathrm{Pd(pn)_2Cl_2} + y - 2(\mathrm{pn}) + 2\mathrm{HCl}$$

For this chemical system, it is optimum to maintain the [pn] in excess of stochiometry (y varies from 3 to 5 excess) which is useful in the dissolution of $PdCl_2$. However, the pn concentration remains relatively constant, and the problem of the Pd^{+2} reduction shifting closer to the hydrogen is eliminated. As can be seen, the Cl^- buildup and pH control remain potential problem areas.

The pH is controlled through use of a buffer system and periodic additions of KOH. While large amounts of Cl^- (KCl) do not seem to adversely affect the process, eventually salts may precipitate necessitating filtration of the system. In this regard a soluble form of $PdO \cdot xH_2O$ as the starting material and replenishment salt solves both problems and is preferred [102–104]. Equation (32) describes the system if a soluble form of $PdO \cdot xH_2O$ could be utilized:

$$\begin{array}{ccc} PdO \cdot xH_2O + y(pn) & \xrightarrow[\Delta]{H_2O} & Pd(pn)_2(OH)_2 + (y-2)(pn) \\ \uparrow & & \downarrow + 2e^- \\ & Pd^0(s) + y(pn) + \tfrac{1}{2}O_2 + 2(OH)^- + 2H^+ & \end{array} \quad (32)$$

Buffer System and Supporting Electrolyte The buffer of choice is K_2HPO_4/K_3PO_4, since the pH range desired is from 10.5 to 12.5. It is fortuitous that the phosphate system is electrochemically and thermally stable and does not significantly affect the morphology of the deposit. In this case the buffer salts also act as the supporting electrolyte.

Limiting Currents, i_L The concentration and mobility of metal ions, and the degree of solution agitation dictate the limiting current. Figure 17 shows current-voltage profiles of the Pd reduction process, and demonstrates how i_L varies with rotation speed (ω) at a RDE. The Levich equation is

$$i_L = nFAC_T D^{2/3} \omega^{1/2} \nu^{-1/6} \quad (33)$$

where i_L is the limiting current, n is the number of electrons, F is Faraday's constant, A is the area of the electrode, C_T is the total bulk concentration of Pd^{2+}, D is the diffusion coefficient, ω is the rotation speed, and ν is the kinematic viscosity. The equation implies that a process is mass transfer controlled when a linear relationship exists between i_L and $\omega^{1/2}$ where ω is the only changing variable. Figure 17*b* shows

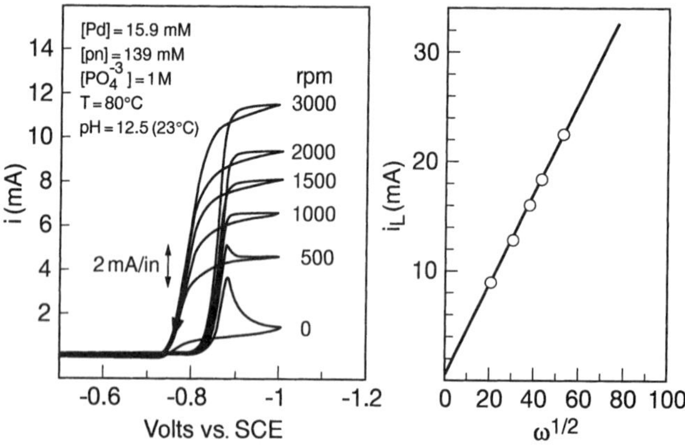

Figure 17 (*a*) Cyclic Voltammograms: Palladium-diaminopropane system; i_L versus rotation speed, ω. (*b*) Plot of i_L (mA) versus $\omega^{1/2}$.

that this reduction process is mass-transfer controlled. Furthermore the relatively "flat" limiting current plateau implies that all metal ions reaching the surface are reduced, that no other reaction is taking place, and that the surface of the deposit is macroscopically smooth.

Quantitative Hydrogen Analysis Figure 18 demonstrates qualitatively how the pn system behaves regarding hydrogen incorporation. The limiting current is reached with little or no incorporation of H_2 as is evident by the smooth plateaus and the absence of a hydrogen oxidation peak until the termination potential of the cathodic scan reaches approximately -1.1 V. An operating window (ΔE_w), which is described as the difference in potential between E_{i_L}(Pd) and E_{ONSET} (H_2)

$$\Delta E_w = E_{i_L}(\mathrm{Pd}^{2+}) - E_{ONSET}(\mathrm{H}_2) \tag{34}$$

of approximately 220 mV (compare to $\Delta E_p \sim 330$ mV) is realized. Thus, to a first approximation, ΔE_p and ΔE_w yield information regarding the probability of a system to avoid hydrogen incorporation. Maximization of ΔE_p and ΔE_w is desired.

However, a quantitative, albeit more time-consuming, analysis is obtained via the anodic stripping of hydrogen described elsewhere [8, 86, 102]. Figure 19 is a plot of $\%H_2$ versus i/i_L* at pH = 12.5. The increase in temperature (25–70°C) reduces the

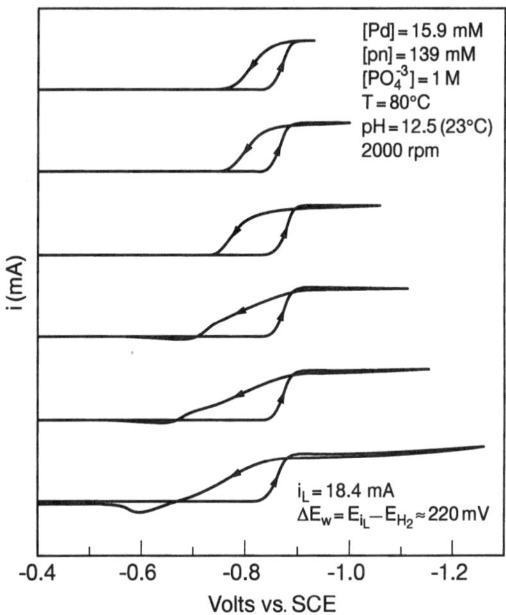

Figure 18 Cyclic voltammograms: Palladium-diaminopropane system; determination of ΔE_w.

*The parameter i/i_L is more informative than merely i(mA) for it takes into account the operating parameters of ω, [Pd], and T. It is believed that morphology and materials properties are best correlated with i/i_L and not simply i(mA).

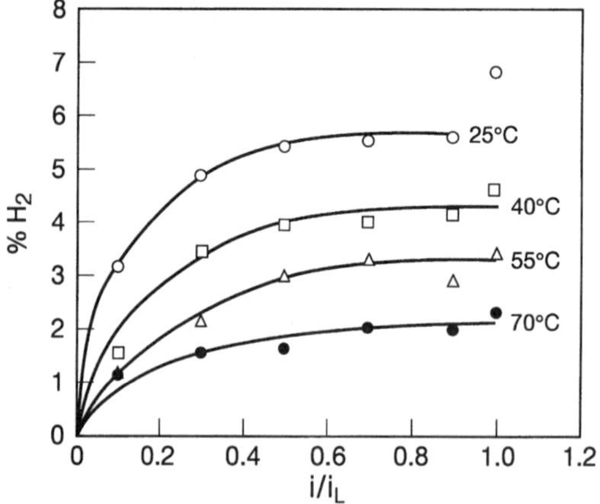

Figure 19 Percentage of hydrogen versus i/i_L at various temperatures.

amount of hydrogen incorporated in the Pd films substantially. It must be kept in mind that the [Pd] is 15.9 mM and that operating at higher Pd concentrations and greater solution agitation allows much higher limiting currents making high rates of deposition possible with reduced hydrogen content (see arguments in Section 5.3). Note also that the Pd films deposited under these conditions were bright to semi-bright, adherent, and not cracked up to thicknesses exceeding 2.5 μm. Figures 20 and 21 are scanning electron micrographs of palladium deposited at high speeds. As can be seen, the deposits are smooth, uniform and crack-free at current densities of 500 mA cm^{-2}.

Materials Properties [8, 102] Deposits plated from the diaminopropane system have been found to exhibit excellent properties for technological applications. The deposits are smooth, and the brightness can be controlled by the bath temperature, pH, and operating current density. In general, the brightness, hardness, and internal stress are increased by operating at lower temperatures (25°C) and lower pH (10.5) and higher current densities; (i.e., $i/i_L > 0.6$).

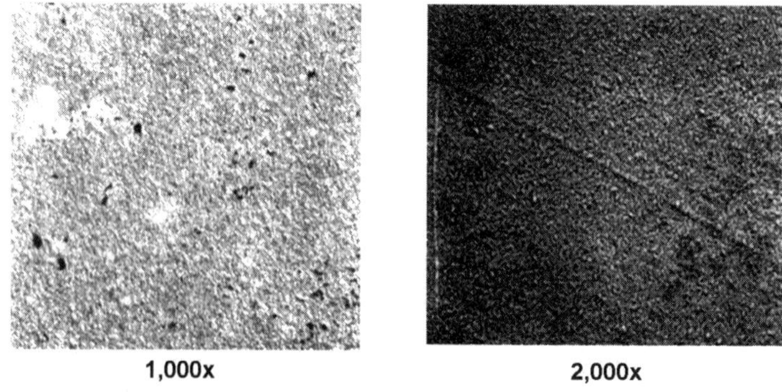

Figure 20 Scanning electron micrograph: Palladium plated at 500 mA cm^{-2} and 2.5 μm in thickness.

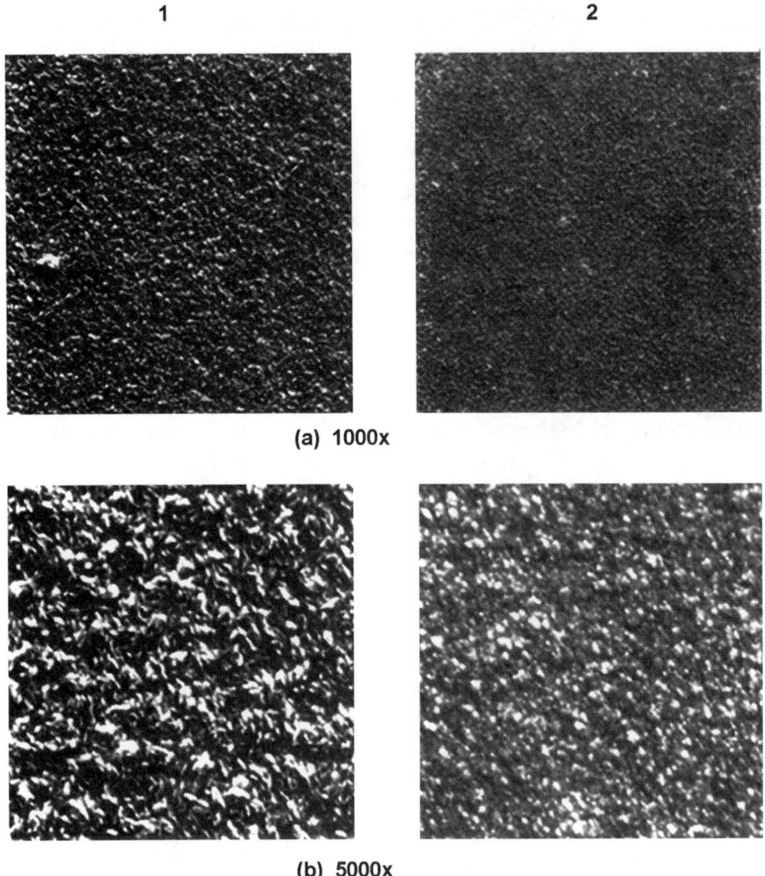

Figure 21 Scanning electron micrograph: Palladium plated at $(i/i_L) = 0.1, 0.3$, and 2.5 μm in thickness.

The deposit hardness ranges from 250 to 350 KHN_{50} (Knoop hardness), where the values increase as the i/i_L increases. Ductility, measured by the method of Nakahara et al. [105] is in the range of 3% to 7% elongation for films of 1.25 μm thickness. In general, the films are harder and more ductile than hard gold.

Transmission electron microscopy indicates that the grain growth is not epitaxial and that grain size varies with the operation current ratio, i/i_L. Preliminary results show that for i/i_L of 0.1 the grain size ranges from 150 to 800 Å, while for i/i_L of 0.9 the grain size ranges from 30 to 50 Å. Thus, as the operating i/i_L increases, one obtains smaller-grain, smoother deposits. Comparably hard gold grain size ranges from 150 to 350 Å.

Mass spectrographic evolved gas analysis performed on various Pd films electroplated at various i/i_L indicate that the amount of occluded impurities is independent of the operating current ratio. A total weight loss ranging from about 0.8% to 1.7% occurred when the films were treated at 260° to 270°C; all other gases evolved above 1000°C. Overall the Pd films are substantially cleaner than electrodeposited gold films [113]. Environmental and standard wear tests [27] and fretting wear tests [29] have demonstrated that this process produces deposits that can be utilized for connector and other applications.

6.2 Acid Electrolytes (pH < 1 to 5)

Table 8 provides the recipe for a number of acid electrolyte processes.

Acid Chlorides (pH < 1) The simplest palladium bath to prepare is an acid chloride system (bath A, Table 7) prepared according to the following equation:

$$Pd^{+2} + 4Cl^- + 2H^+ \Leftrightarrow H_2PdCl_4 \tag{35}$$

In these systems [37] of high acid and chloride content, the equilibrium constant is

$$\beta_4 = \frac{[H_2PdCl_4]}{[Pd^{+2}][Cl^-]^4[H^+]^2} = 10^{12.2} \tag{36}$$

This relatively weak palladium complex implies that immersion plating on a non-noble metal would occur quite readily.

A cyclic voltammogram of $PdCl_2$ in 1 M HCl electrolyte, is seen in Figure 22 [114]. The $E_p(Pd^{2+})$ is about 0.1 V relative to SCE, and the PdH_x peak potential can be seen at about 0.27 V before bulk hydrogen begins to evolve. On the return scan, the hydrogen is anodically stripped [8, 86, 102] from the palladium hydride, E_p (oxidation) + 0.03 V. Finally, palladium metal is dissolved at a potential of about 0.57 V.

Palladium deposition from acid electrolytes is typically 97% to 100% efficient. Compact, dense films that are crack-free even at thicknesses of 10 μm have been reported (see the reference in Table 8). The deposits generally have a satin finish extending to full bright and gray. Depending on the deposition conditions, they have a microhardness of 250 to 350 HV (Vickers hardness) [37].

According to a study by Raub [37], in the current density of 1 to 100 mA cm^{-2} at temperatures of 10° to 50°C, palladium is deposited at +0.1 to +0.5 V relative to NHE. There is a temperature dependence of the cathodic overpotential that can be easily observed on the basis of total current potential characteristics; see Figure 23.

If the Pd is deposited in an electrolyte with lower chloride content, then the cathodic overvoltage for its reduction is reduced; see Figure 24. This occurs because the Pd-free ion concentration remains higher, since the lower chloride and hydrogen concentration reduce the stability of the palladium (II) chloro-complex; see Eqs. (35) and (36) and arguments in Section 5.3.

For this system hydrogen codeposition has been reported to be low at potentials +0.5 to +0.1 V. Hydrogen discharge complies with Nerst's law at an equilibrium potential, Eq. (20). Thus, even in strong chloride solutions with 10 normal hydrogen ion concentration, the H_2 evolution starts only at an equilibrium of +0.059 V.

At current densities between 1 and 5 mA cm^{-2}, the hydrogen content is below 3 NmlH_2/gPd, which makes the $[H]/[Pd] < 0.03$. This result implies that films deposited for this system are in the homogeneous range of the α-phase.

TABLE 8 Palladium Electroplating Processes - Acid Electrolytes pH (< 1 to 5)

	Bath Type				
Parameter	H[109,37,39,35]	I[35]	J[39,110]	K[35]	L[111]
Source of Pd	$PdCl_2$	H_2PdCl_4	$Pd(NO_3)_2$	$Na_2Pd(NO_3)_4$	$PdCl_2$
Pd Metal (g liter^{-1})	5–50	5	2–15	5–10	0.1–3.0
Pd Sulfite (g liter^{-1})			0.2–2		
Ligand	H_2O	Cl^-	H_2O	NO_2	Organic Diamines
Ammonium Chloride (g liter^{-1})	20–50				
Boric Acid (g liter^{-1})	As Needed	10–30			
Hydrochloric Acid (g liter^{-1})		As Needed			
Oxalic Acid (g liter^{-1})	As Needed				
Phosphoric Acid (g liter^{-1})	As Needed				
Sodium Chloride (g liter^{-1})		40		10–40	10–60
Sodium Nitrite (g liter^{-1})		14			
Sulfuric Acid (g liter^{-1})			98		
Additives				See Ref.	See Ref.
pH	0–0.5	4.5–6	0	4.9–8	3–4
Temperature (°C)	25–60	50	20–35	40–50	25–45
Current Density (mA/cm^2)	10	4–10	5–80	~1	5–50

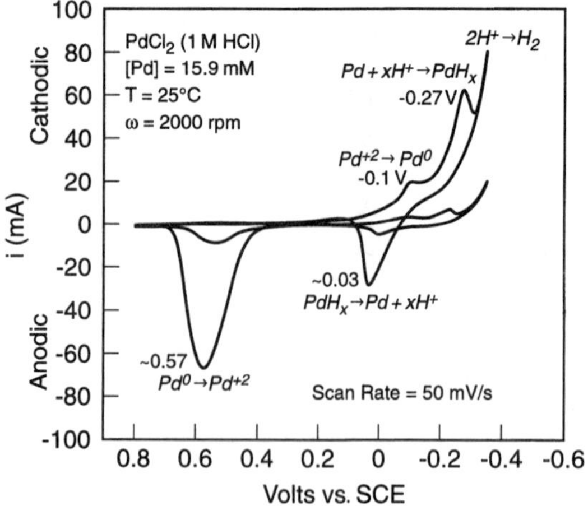

Figure 22 Cyclic voltammograms: $PdCl_2$ in HCl. 0 and 2000 rpm.

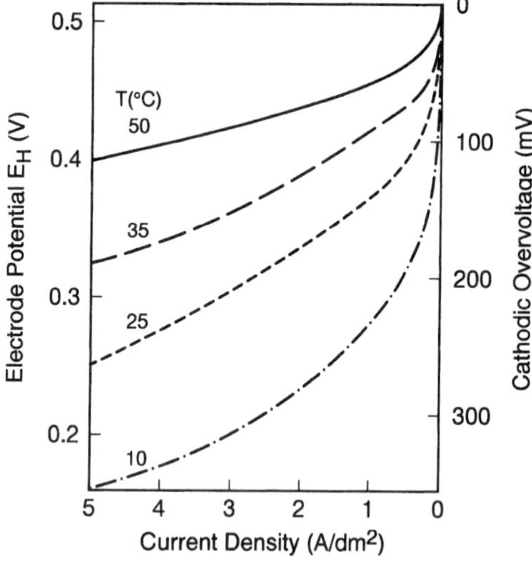

Figure 23 Current-voltage curves: 0.1 M $PdCl_2$ in 5 M HCl, $di/dt = 0.2\,\mathrm{A\,dm^{-2}\cdot s}$ at various temperatures.

X-ray analysis of Pd films deposited from 0.1 M $PdCl_2$ solution with 5 N HCl at current densities of 1 to 3 mA cm^{-2} show a slight expansion of the α-Pd lattice to 3.895 Å compared to 3.890 Å for pure Pd. These films are fine crystalline and irregularly oriented with no preferential orientation. They exhibit very low internal stress which decreases with increasing film thickness, Figure 25.

It has been stated by Raub [37] that acid palladium chloride electrolytes are well suited for depositing thick palladium films of low internal stress at about 100% current efficiency. The disadvantages of these systems are that non-noble base metals cannot be directly coated and that the electrolytes are very aggressive in their attack of metals used for substrates and in plating equipment. An excellent analysis of this system, including the measurement of exchange current densities and the

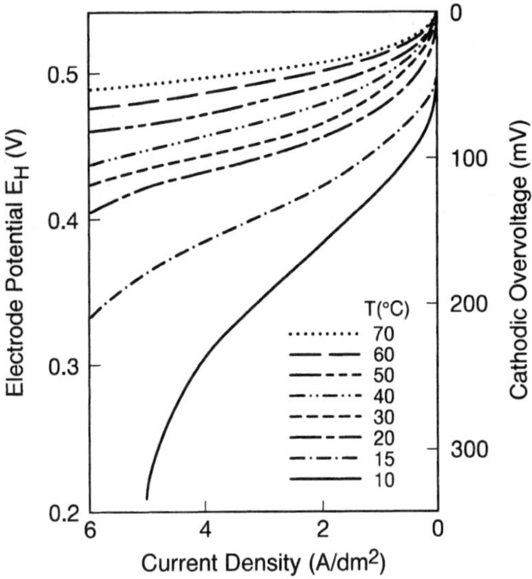

Figure 24 Current-voltage curves: 0.1 M $PdCl_2$ in 0.5 M HCl at pH = 5.0 for $di/dt = 0.2\,A\,dm^{-2} \cdot s$.

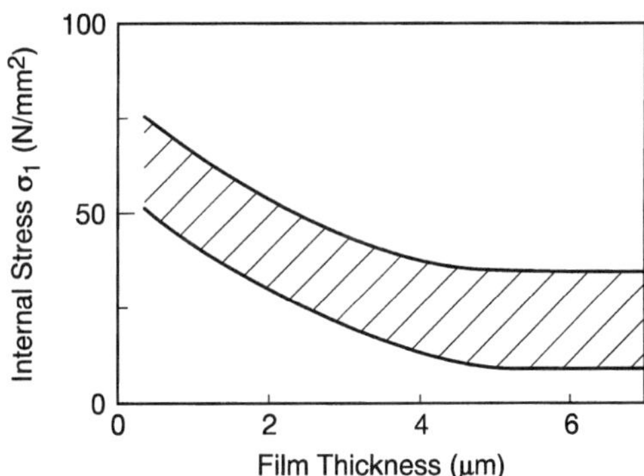

Figure 25 Internal stress versus film thickness: 0.1 M $PdCl_2$ in 5 M HCl at 1 and 3 A dm^{-2}.

determination of the activation energy for the charge transfer reaction of Pd^{+2}/Pd is presented by Raub [37].

Acid Sulfates (pH < 1) Sulfuric acid electrolytes (bath J, Table 8) were investigated by Raub [39] and Simon [110]. It was reported that all of the drawbacks of the Cl^- systems are eliminated. A solution of palladium nitrate in sulfuric acid produces black, spongy, loosely adhering palladium films. Alternatively, smooth, lustrous palladium deposits can be obtained only after adding 5% to 20% palladium sulfite. The formation of a palladium-sulfite complex ($[Pd(SO_3)_2]^{-2}$) decreases the Pd-free ion concentration and shifts its reduction potential to a more negative point. The increased cathodic overvoltage for metal deposition has a positive effect on the film properties.

Smooth, lustrous films are obtained which contain relatively little hydrogen and are of low stress [39]. Palladium deposited from this system have a hardness of about 350 HV. Sulfite is reduced to sulfide during deposition of the palladium and adsorbed on the cathode. Palladium sulfide inhibits palladium deposition and is incorporated during crystallization. The sulfur content increases with electrolyte sulfite content and cathodic current density. The properties of these films are largely determined by the sulfur content. These films are in general fine grain, hard, and with a pronounced ⟨100⟩ orientation.

The disadvantages of this process are the necessity of a noble metal strike (e.g., gold) to prevent attack on non-noble metal surfaces, and the necessity for continuous filtration to reduce the tendency for the reduction of palladium ions to metal according to the following equation [39]:

$$Pd^{+2} + SO_3{}^{-2} + H_2O \longrightarrow Pd + H_2SO_4 \qquad (37)$$

Acid Strikes (pH 3 to 4) An acidic palladium "strike" process was developed by Straschil et al. [111, 112] (bath L, Table 8). This process uses organic diamines to stabilize a palladium complex in the 3 to 4 pH range. Unlike the highly acidic baths with pH $\leq$ 1, this chemistry does not undergo "displacement" plating or attack on non-noble metal substrates. However, it promotes the removal of surface oxides especially on nickel which makes it ideal as a "strike" process. This process also reduces the porosity of subsequent noble metal layers by increasing the nucleation and "coverage" of substrate surfaces. It exhibits good chemical and electrochemical stability along with stable pH due to the high buffer capacity. The preparation and operation of this bath are described by Straschil [112]. The limitations of this process is the inability to plate greater thicknesses than 0.2 μm without the formation of cracks due to hydrogen embrittlement [112]. However, for strike purposes only thin layers are required.

6.3 Neutral Electrolytes (pH 5 to 9)

Numerous chemical systems have been investigated in this pH range, including oxalates, sulfamates, and the like. While in many cases these systems possessed the proper current-potential profiles to deposit palladium and avoid hydrogen codeposition, and while the initial deposits from a "non-aged" electrolyte could provide films of excellent hardness, ductility, and surface appearance, these systems did not stand up during extensive aging.

Hydrogen embrittlement is not the only issue that must be addressed when developing a plating process for commercial use. In these cases the chemical and electrochemical stability of the ligand system was the main concern. Furthermore simple replenishing systems were difficult to devise due to slow attack of these ligands on salts such as $PdCl_2$. The decomposition of the ligands and the buildup of ligands in solution lead to an aggravation of the hydrogen problem and microcracking. However, palladium electrolytes based on ammonia ligand have proved to be viable, and they are used almost exclusively in the pH range of 5 to 9.

Hedrich and Raub [37, 39, 40] studied nonorganic additive ammonia systems extensively and found that they are "well suited to deposit palladium films for technological applications," baths N and O, Table 9. The reduction takes place at

TABLE 9 Palladium Electroplating Processes Neutral Electrolytes (pH 5–9)

Parameter[a]	Bath Type M[36]	N[36,40]	O[36,40]	P[91]	Q[106]	R[107]	S[108]
Pd Source	$Pd(NH_3)_4(NO_3)_2$	$Pd(NH_3)_2(NO_3)_3$	$Pd(NH_3)_4Cl_2$	$Pd(NH_3)_4Cl_2$	$Pd(NH_3)_4Cl_2$	$Pd(NH_3)_2Cl_2$	$Pd(NH_3)_4Cl_2$
Pd Metal ($g\,liter^{-1}$)	10	4	20	10–20	20–30	10	5–25
Ligand	NH_3	NH_3	NH_3	NH_3	NH_3	NH_3	NH_3
EDTA						20	
Ammonium Carbonate ($g\,liter^{-1}$)			10				
Ammonium Chloride ($g\,liter^{-1}$)			10	60–90	30–60		90
Ammonium Hydroxide ($g\,liter^{-1}$)	Adjust pH	Adjust pH	Adjust pH	Adjust pH			Adjust pH
Ammonium Nitrate ($g\,liter^{-1}$)	90–100						
Ammonium Phosphate ($g\,liter^{-1}$)							90
Ammonium Sulfamate ($g\,liter^{-1}$)					30–40		
Ammonium Sulfate ($g\,liter^{-1}$)			20				
EDTA ($g\,liter^{-1}$)						20	
Hydrochloric acid ($g\,liter^{-1}$)							to pH
Potassium Phosphate ($g\,liter^{-1}$)						90	
Sodium Nitrate ($g\,liter^{-1}$)		10–11					
Sodium Sulfite ($g\,liter^{-1}$)					1–1000 ppm	28	
Sulfuric Acid ($g\,liter^{-1}$)							
Additives	–	–	See Ref.	See Ref.	See Ref.	See Ref.	See Ref.
pH	9.1	7.5–10	7.5–9	8.0–9.5	~9	8.5–9	7–8
Temperature (°C)	60–80	40–55	20–40	25–50	25–40	35	30–50
Current Density mA/cm^2	5–20	1–40	5–15	1–25	15–25	50	1–>250

[a] Diaphragm process.

essentially less noble potentials than in acid systems, Figure 26. For example, the Pd potential E_H in 0.1 M $PdCl_2$ solution is +0.55 V, whereas the palladium tetrammine complex of the same concentration is +0.22 V. In palladium diamino dinitro-electrolyte, P-Salt, Pd reduction takes place in approximately the same potential range as the ammonia system.

Tetrammine Electrolytes. "Nonadditive" Systems Amine complexes are formed in palladium chloride solutions in the presence of ammonium ions at pH values above 4.5, and they are completely transformed into the square planar complex at pH = 7.5. When these palladium tetrammine solutions are acidified with hydrochloric acid, the complex is transformed step-by-step into palladium diammino complex, which is stable in the acid range. Starting from pH 6, yellow $Pd(NH_3)_2Cl_2$ is precipitated in cold 0.1 M solution, the precipitation process is complete at pH 4.

The following composition was chosen as standard electrolyte for electrodeposition experiments:

$$0.1\,\text{M PdCl}_2 + 1.0\,\text{M NH}_4\text{Cl} + x\text{M NH}_3{\cdot}\text{H}_2\text{O} \qquad (38)$$

The pH value was set between 8 and 10 with NH_4OH.

Pd is deposited from the tetrammine complex at current densities from 1 to 5 mA cm^{-2} at overpotential of 300 and 600 mV, Figure 27. The current-potential characteristics shown in Figure 27 exhibit a strong temperature dependence, since the stability of $[Pd(NH_3)_4]^{2+}$ decreases with increasing temperature. The current efficiency is below 100%, so the cathode always represents a double electrode. Current efficiency measurements in agitated electrolytes indicate that hydrogen evolution is strongly dependent on the current density and electrolyte temperature, Figure 28. The current yield increases with increasing cathodic current density and at

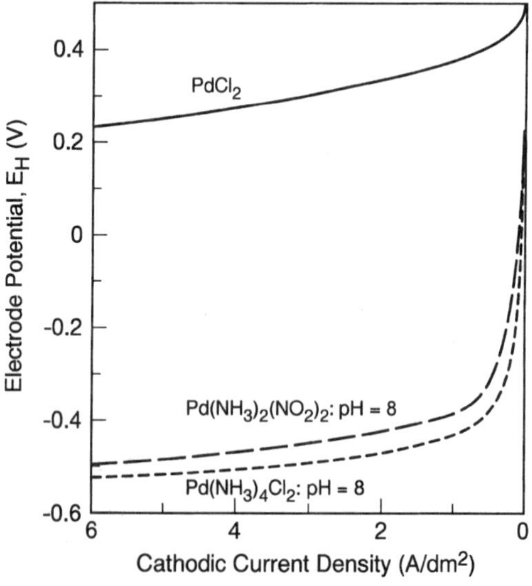

Figure 26 Current-voltage curves: Acid Pd versus ammoniacal Pd chemistries, $di/dt =$ 0.2 A dm$^{-2}\cdot$s.

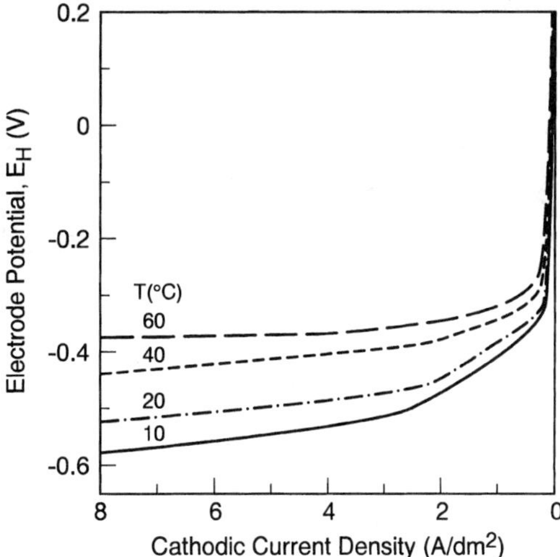

Figure 27 Current-voltage curves: $Pd(NH_3)_4Cl_2$ at various temperatures, $di/dt = 0.2\,A\,dm^{-2} \cdot s$.

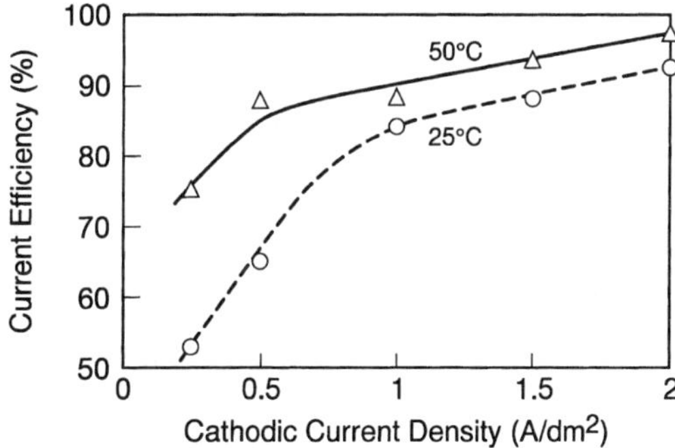

Figure 28 Current efficiency versus current density: $Pd(NH_3)_4Cl_2$ at 25° and 50°C.

higher electrolyte temperatures [37]. The palladium tetrammine complex stability increases with increasing pH values, so the cathodic overpotential required for Pd deposition increases distinctly, Figure 29.

The hardness of these Pd deposits varies according to the deposition conditions (usually measured ~25 μm thick films). Pd films deposited at electrolyte temperatures between 15° and 25°C and at current densities from 10 to 20 mA cm^{-2} have hardness values between 270 and 320 HV directly after deposition. Pd deposits obtained in the same current density range from the same electrolyte, at 60°C exhibit hardnesses between 150 and 180 HV.

Stress measurements indicate that the Pd deposits exhibit tensile stresses that decrease with increasing electrolyte deposition temperature, Figure 30, and decrease with increasing film thickness. Hydrogen codeposition is associated with this phenomenon, which results in considerable stress in the lattice as revealed by X-ray analysis [37].

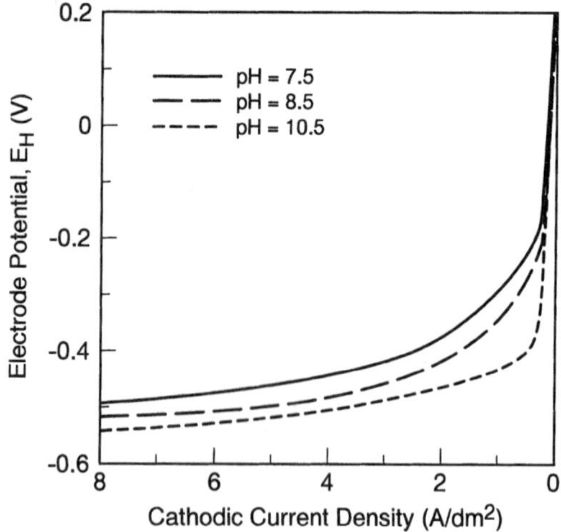

Figure 29 Current-voltage curves: $Pd(NH_3)_4Cl_2$ at various pH, $di/dt = 0.2\,\mathrm{A\,dm^{-2}\cdot s}$.

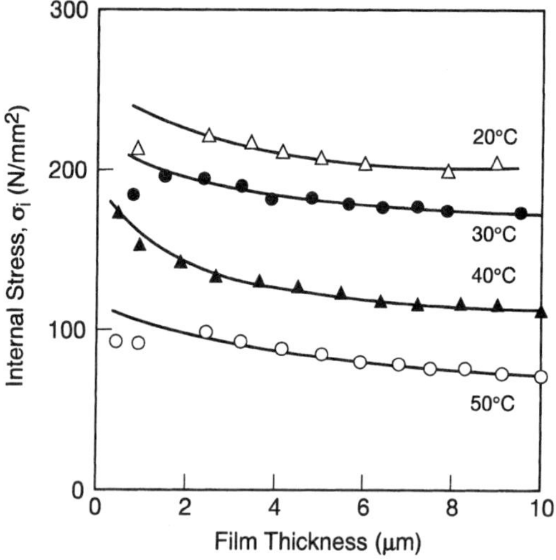

Figure 30 Internal stress versus film thickness: $Pd(NH_3)_4Cl_2$; pH = 9; at $2\,\mathrm{A\,dm^{-2}}$.

Quantitative hydrogen analysis via anodic stripping indicates that the hydrogen content is strongly affected by current density and deposition temperature, Figure 31. At low temperatures, $<40°C$, the hydrogen content increases rapidly with current densities; whereas at temperatures greater than 40°C, the increase is much lower.

Raub [37] found that Pd films with a hydrogen content below 10 Nml H_2/g Pd or atomic H/Pd ratios below 0.1 have a metallic brilliance, and are suited for decorative and technological applications. For higher H/Pd ratios, the Pd deposits become dull and dark. For an atomic H/Pd ≥ 0.8 ratio they are black and powdery, that is, burned deposits.

Palladium deposits from tetramine electrolytes without additives are in general dull and gray in appearance. Viewed under a scanning electron microscope, the

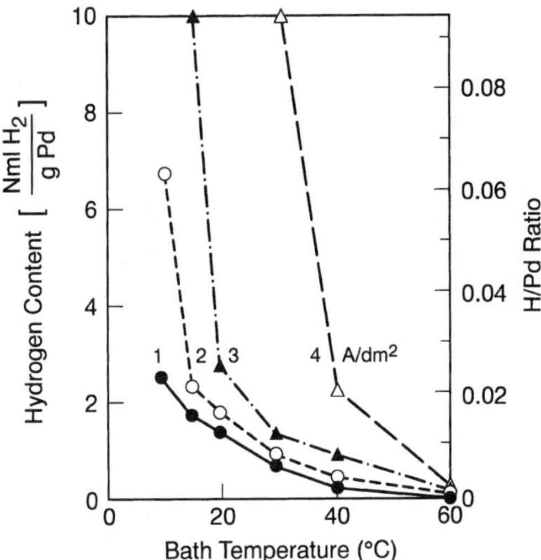

Figure 31 Hydrogen content versus bath temperature at various current densities $Pd(NH_3)_4Cl_2$, 1 min after termination of electrolysis.

deposit surface is "budded" and nodular [40]. These deposits are very sensitive to "finger prints" and other contaminants since "oils" readily adhere to microrough surfaces.

In terms of processing technology, electrolytes high in ammonia content are difficult to operate and control at high pH (> 7.5) and temperature (> 35°C). Ammonia vaporization losses are significant and require that the pH be monitored and controlled. In the presence of halides, significant attack on metals such as copper, brass, and stainless steel (in plating equipment) occurs. Furthermore the resulting metallic impurities degrade the appearance and mechanical properties, and they decrease the partial current density and covering power for Pd deposition.

Hedrich and Raub [115] also studied the effects of "additives" in these systems to improve the surface appearance and reduce "roughness" of the deposits. They studied the effects of citrates, benzoates and salicylate, nicotinic acid, and nicotinamide. In general, these additives all displace the reduction potential to more negative values closer to the hydrogen regime and increase the hydrogen content of the films.

Aliphatic citrates produce an irregular crystallization, whereas benzoates and salicylates produce a ⟨110⟩ preferred orientation; however, they also increase the hydrogen content and tensile stress in the plated films, resulting in microcracked deposits. The electrodeposition of smooth, full bright, adherent, crack-free palladium films from tetrammine systems requires the use of additives. In the next section we will discuss some of these systems.

Tetrammine Chloride Electrolytes with Additives There exist numerous patents on the use of additives in palladium-ammonia electrolytes whose function is to deposit dense, full bright palladium films of low porosity without microcracks at thicknesses in excess of 25 μm. For example, Deuber [107] patented a process (bath Q, Table 9) using an alkylene diamine phosphonate derivative of the following formula:

$$[(MO)_2 - \mathrm{OP} - R]_2 - \mathrm{N} - R - \mathrm{N}[\mathrm{R}' - \mathrm{OP}(\mathrm{O}M)_2]_2 \qquad (39)$$

where R is an alkylene group of 2 to 6 carbon atoms, each R' is an alkylene group of 1 to 4 carbon atoms, and each M is a nondeleterious cationic moiety that does not interfere with the plating process, such as hydrogen, alkali metals, and ammonium. This phosphonic compound is believed to act as a buffering compound, a conducting salt, and a complexing agent for contaminant metal ions that it is not desired to plate out. The concentration range of the phosphonic compound is from 5 g liter^{-1} up to 100 g liter^{-1}. Dequest 2041 is ethylene diamine tetramethylphosphonic acid which contains approximately 10% water; it is one of the preferred phosphonates.

The pH of the electroplating solution should be maintained at a value of from 4.5 to 12 in order to avoid stability problems. Values from 4.5 to 7.0 are preferred for strike plating. For ordinary electroplating, a pH value from 7 to 10 is preferred. The bath parameters are as follows:

COMPONENT	CONCENTRATION (g LITER^{-1})
Ethylene diamine tetramethylphosphonic acid	45
$Pd(NH_3)_2Cl_2$	1–10 (metal)
Sulfamic acid	40
Ammonium chloride	50–150
Ammonium monohydrogen phosphate	25–75
Ammonium hydroxide	pH 6.5–10

The bath may be employed to plate palladium or its alloys. Deuber [107] also makes use of class I and class II organic brighteners with a palladium-amine system at pH value of 4.5 to 12.

The palladium is supplied as any electrodepositable complex (e.g., $[Pd(NH_3)_4]^{+2}$). The stability is improved if a palladous complex is employed, such as the urea or an amine complex, with the following counter ions: chloride, bromide, nitrite, and sulfite. The palladium content normally is in the range of 0.1 to 50 g liter^{-1}. For a strike plate a concentration of 1 to 5 g liter^{-1} is preferred, and for ordinary plating a concentration of 5 to 15 g liter^{-1} is preferred.

The class I brighteners are generally unsaturated sulfonic compounds where the unsaturation is in the α or β position with respect to the sulfonic group. Such compounds have the formula

$$A - \mathrm{SO_3} - B \qquad (40)$$

where A is an aryl or alkylene group, substituted or unsubstituted, and B may be OH, OR, OM, $NH_{2'} >$ NH, H, R with M being an alkali metal, or else ammonium, and R being an alkyl group of not more than six carbon atoms.

The class II organic brighteners are generally unsaturated or carbonyl organic compounds. Examples are compounds containing $>$C=O; $>$C=C$<$; –C≡N; $>$C=N–; –C≡C–; –N=N–. Concentration of the individual brighteners may range from 0.001 to 25 g liter^{-1}. Some compounds may fall within the description of both class I and class II. Such compounds may be employed alone, but particularly improved results are obtained when a second different compound of either class is also employed.

The temperature of the bath should be maintained between room temperature and approximately 50°C. Current densities of from 0.1 to 50 mA cm^{-2} are suitable. For rack plating, a current density from 5 to 15 mA cm^{-2}, may be employed. For barrel plating, the preferred range is from 0.5 to 3 mA cm^{-2}.

An example of such a bath is as follows:

COMPONENT	CONCENTRATION (g LITER^{-1})
EDTA	20
$(NH_4)_2HPO_4$	80
Na_2SO_3	28
Pd as $Pd(NH_3)_2Cl_2$	10
Class I brightener	0.005–2
Class II brightener	0.005–2

The bath pH is maintained between 8.5 and 9 and plating performed at a temperature of 40°C at a current density of 5 mA cm^{-2}.

When neither brightener was present, only hazy deposits were obtained. The class I brightener employed was methylenebisnaphthalene sodium sulfonate, and the class II was benzaldehyde-*o*-sodium sulfonate. The class I brightener was added first and was ineffective at concentrations all the way up to 2 g liter^{-1}.

With the class II brightener, improved deposits of semibright quality were obtained at 0.02 g liter^{-1} and bright deposits were obtained from 0.37 to 2.0 g liter^{-1}. Similar results were obtained when 2-butene-1,4-diol was employed as the class II brightener. Unlike the work of Hedrich and Raub which provides extensive electrochemical and material properties, no such information has been found in regard to these systems. Furthermore, in neither Hedrich and Raub's or Deuber's work is there any information on the aging of such systems. Abys et al. [12, 108] developed a process to electroplate full bright pure palladium to thicknesses $> 25\ \mu m$ with low porosity and high hardness.

This chemistry employs palladium tetrammonium chloride in a buffered electrolyte system containing a combination of two proprietary additives. Specific bath composition ranges for low-speed and high-speed operations are shown below:

CHEMICAL CONSTITUENT	LOW SPEED	HIGH SPEED
Pd (as metal)	2–10 g liter^{-1}	15–40 g liter^{-1}
NH_4Cl	50–100 g liter^{-1}	50–200 g liter^{-1}
K_2HPO_4	80–100 g liter^{-1}	80–100 g liter^{-1}
Surfactant	2–25 ml liter^{-1}	2–50 ml liter^{-1}
Brightener	0.1–20 ml liter^{-1}	0.2–20 ml liter^{-1}
Specific gravity	1.05–1.15	1.10–1.25

The surfactant is used to improve wettability and to reduce the effect of organic contaminants. It also enhances the efficiency of the brightener, which imparts luster and leveling to the deposit. Potassium phosphate and ammonium chloride are used as buffering agents and as supporting electrolytes, respectively.

This chemistry was developed to operate under generally mild conditions of low temperature (~35°C) and a neutral to slightly alkaline pH. Operating parameters for the low- and high-speed plating baths are shown below:

OPERATING PARAMETER	LOW SPEED	
	Typical	*Range*
Temperature	35°C	30–40°C
pH	7.5	7.0–8.0
Current density	$5\,mA\,cm^{-2}$	$2–10\,mA\,cm^{-2}$
Agitation	$5\,cm\,s^{-1}$	$5–50\,cm\,s^{-1}$
Cathode efficiency	87%	75–98%
Plating rate	$0.12\,\mu m\,min^{-1}$	$0.04–0.25\,\mu m\,min^{-1}$
Anode/cathode area	2:1	1:1–5:1

OPERATING PARAMETER	HIGH SPEED	
	Typical	*Range*
Temperature	35°C	30–40°C
pH	7.5	7.0–8.0
Current density	$100\,mA\,cm^{-2}$	$40–\geq 500\,mA\,cm^{-2}$
Agitation	$100\,cm\,s^{-1}$	$50–\geq 200\,cm\,s^{-1}$
Cathode efficiency	87%	75–98%
Plating rate	$2.2\,\mu m\,min^{-1}$	$0.75–12.4\,\mu m\,min^{-1}$
Anode/cathode area	2:1	1:1–5:1

The current density can be varied from $<5\,mA\,cm^{-2}$ to $>500\,mA\,cm^{-2}$ with the appropriate selection of metal content, additive balance, and agitation conditions. Thus this chemistry can be used in barrel, rack, and reel-to-reel applications. A general description of the process performance and maintenance procedures can be found in Kadija et al. [12].

An important aspect of any electrodeposition technology is to determine the physical characteristics of the plated films. Kadija et al. [12] reported the materials properties of the Pd films deposited from this process.

Grain Size Transmission electron microscopy studies show that the grain size depends on the presence or absence of additives, Figure 32. While in the additive-free solution, the grain size ranges from 500 to 5000 Å, it decreases to between 50 and 200 Å in the presence of additives. This change in crystallinity contributes to the brightness and hardness of the deposit.

Surface Appearance Deposit brightness varies with plating conditions and is often a measure of the surface microroughness. Reflectivity measurements were made using a bifurcated fiber system with a green LED source ($\lambda_{max} \sim 560$ nm, λ-range ~ 535–625 nm). Both specular and diffuse reflectivity were measured. The measurements were normalized to the reflectivity of a highly polished bulk Pd metal standard.

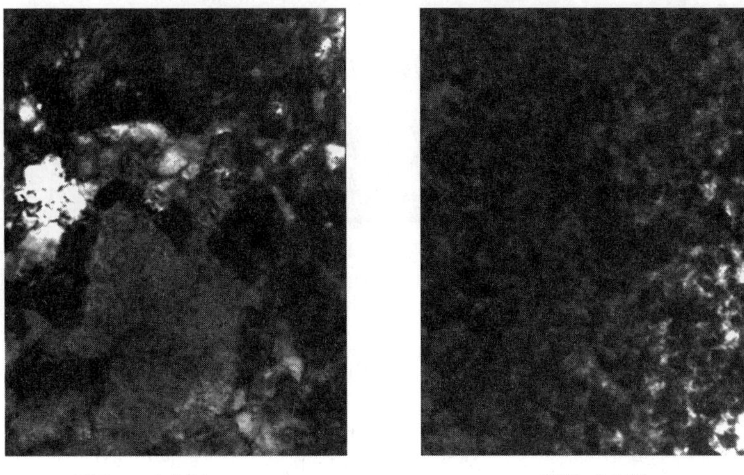

Figure 32 Transmission electron micrographs at 100,000× : $Pd(NH_3)_4Cl_2$ electrolyte, effect of additive.

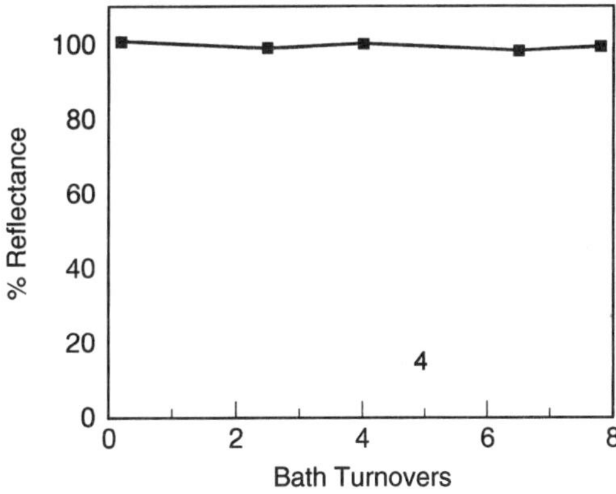

Figure 33 Percentage of reflectance versus bath turnovers: $Pd(NH_3)_4Cl_2$ electrolyte.

Figure 33 shows the reflectivity of Pd films as a function of bath aging. As can be seen, the reflectance is maintained at about the 98% level, indicating that a specular surface is achieved and that no degradation of the surface appearance is evident with aging.

Leveling Effect Figure 34 exhibits the leveling effect of this plating process. The scanning electron micrograph (SEM) of the 5 μm thick deposit clearly shows the rolling marks of the copper substrate (0 μm), which disappears as the thickness increases to 25 μm.

Density The density of the plated deposit measured by pycnometry is approximately 11.75 g cm^{-3}, that is, 98% of the theoretical value. Values for low- and high-speed deposited films are shown below:

BATH TYPE	DENSITY OF DEPOSITS	
	Plated Density *($g\,cm^{-3}$)*	*Theoretical Density* *($g\,cm^{-3}$)*
High speed ($\sim 250\,mA\,cm^{-2}$)	11.79	12.02
Low speed ($\sim 5\,mA\,cm^{-2}$)	11.71	12.02

Deposit Purity Thermogravimetric analysis was performed on Pd films plated at $100\,mA\,cm^{-2}$. The films were plated at 0, 1.07, and 3.5 bath turnovers. The films were heated from 30° to 950°C at the rate of $20°C\,min^{-1}$ under an atmosphere of nitrogen.

As can be seen from the data below, essentially no weight loss was observed:

Pd SAMPLES (BATH TURNOVER)	THERMO GRAVIMETRIC ANALYSIS (WT%)	WEIGHT LOSS (%)
0	99.79	0.21
1.07	99.68	0.32
3.5	99.89	0.11

These data imply that the purity of the deposit is high (>99%) and that the contaminant inclusion is low (<1%) and remains low throughout the lifetime of the bath.

Thermal Stability The thermal stability was measured by differential scanning calorimetry from 25° to 450°C at a scan rate of $10°C\,min^{-1}$. No thermal reaction was detected in palladium deposits. These findings may imply good bulk thermal stability of the deposit up to 450°C.

Solderability Solderability was determined by the IPC method TM-650-2.414.1 which entails immersion in molten solder. Pd deposits were exposed to 90% humidity at 90°C for 8 and 36 hours. The results shown below indicate that both 0.075 and 0.15 μm thick palladium provide excellent solderability:

Palladium Solderability-Test Solder Coverage (IPC-TM-650)

SAMPLE PREPARATION	%COVERAGE		
Cu SUBSTRATE	*As Is*	*8 Hours*	*36 Hours*
0.75 μm Ni	0	0	0
0.75 μm Ni/0.075 μm Pd	99	99	98
0.75 μm Ni/0.15 μm Pd	99	99	99
1.5 μm Ni/0.15 μm Pd	99	99	97

Ductility Ductility was measured by the ASTM-B-489-85 bend method. This method involves bending samples plated on a thin Cu substrate through a series of bend angles over mandrils of various diameters. An inspection for cracks is conducted at a magnification of 10. The angle of bend was 180°, and the diameter of the mandrils varied from 3.3 to 0.5 mm. A more rigorous inspection was carried out

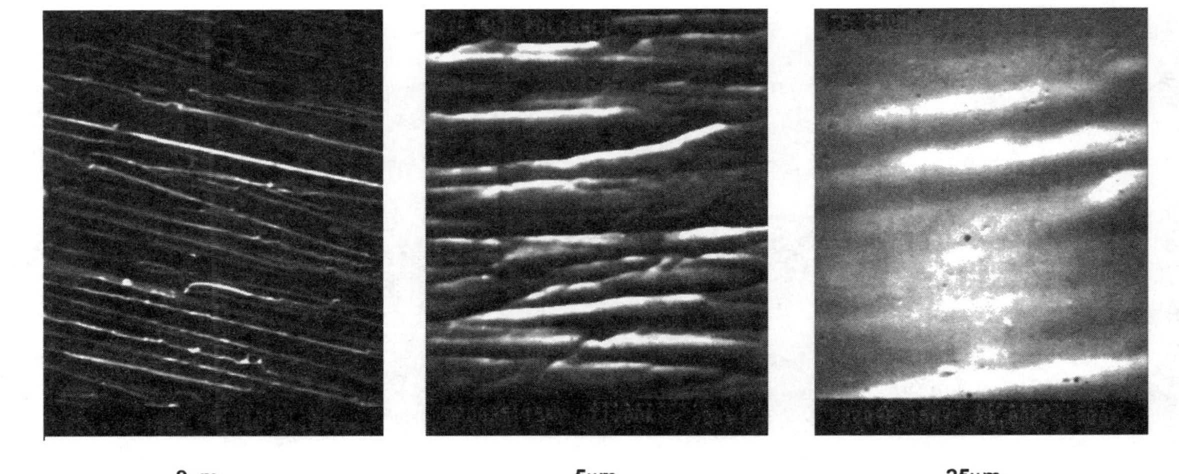

Figure 34 Scanning electron micrographs at 1000 ×: Levelling effect at 0, 5, and 25 μm film thickness; $Pd(NH_3)_4Cl_2$ electrolyte.

at 1000 magnification. Figure 35 shows the SEM photos of the 1.25 mm diameter bend of a 5 μm thick deposit.

Five micrometer thick deposits from a low-speed bath pass the test at 0.5 mm diameter, while the high-speed samples pass at 1.25 mm diameter. In both instances, the deposit elongation was greater than 9%. This ductility is comparable to that of soft gold and far superior to the ductility of cobalt-hardened gold deposits.

Hardness The Knoop hardness was measured by a Tukon hardness tester using a diamond indenter at 50 g load. Figure 36 shows the effect of the brightener concentration on the hardness for a low-speed process. For example, starting from

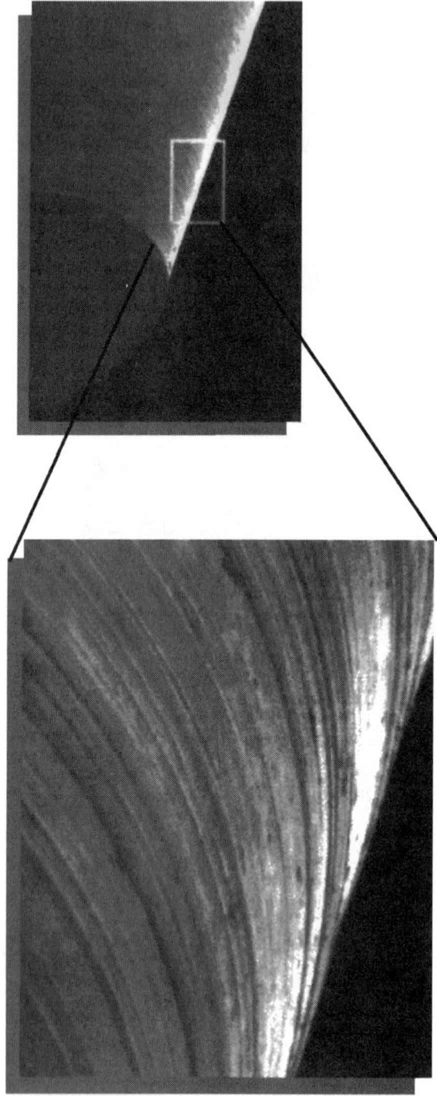

Figure 35 Scanning electron micrographs of bent 5 μm Pd deposit: $Pd(NH_3)_4Cl_2$ electrolyte.

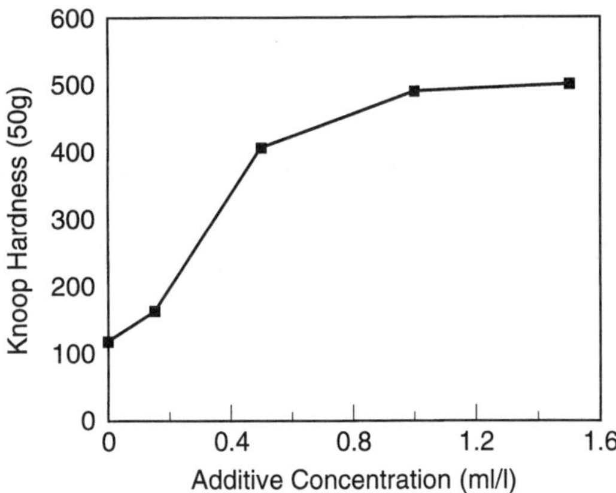

Figure 36 Hardness versus additive concentration: $Pd(NH_3)_4Cl_2$ electrolyte.

the range of 100 to 150 KHN_{50} for the additive-free bath, the hardness increases to the level of 400 to 500 KHN_{50} at 1 ml $liter^{-1}$ of brightener and remains unchanged at higher additive concentrations.

Overall, this process is reported to produce adherent, mechanically stable, specular Pd films at plating rates ranging from $< 5\,mA\,cm^{-2}$ to $> 500\,mA\,cm^{-2}$. The deposits are smooth and crack free.

7 PALLADIUM ALLOYS

As discussed in Section 2.1, palladium is known to form alloys with other metals of group VIII and IB that display distinct advantages over the pure metals [45]. The alloys tend to exhibit higher hardness, tensile strength, and ductility. They can also be plated with a smaller grain size and brighter than palladium even in the absence of additives. Furthermore, and of great importance, is the ability of the alloying element, in some cases, to increase the hydrogen overpotential and thus decrease the amount of codeposited hydrogen.

The arguments for the technological use of palladium could thus be even more valid for palladium alloys. Accordingly alloys such as PdNi (80/20 wt%) have been extensively studied [8–33]. This alloy has been reported to exhibit a brighter appearance, a harder and more ductile deposit, much lower susceptibility to hydrogen embrittlement, lower porosity, and better wear characteristics than either pure palladium or gold. Overall, the general material properties of palladium-nickel suggest that it should be an excellent contact material, and indeed it is in significant commercial use for connector applications worldwide. Likewise metallurgically prepared PdAg (60/40 wt%), exhibits excellent material properties but has not been successfully plated in commercial applications [116–118]. The reasons of course lie in the fundamental differences among the chemistry of palladium, nickel, and silver.

The development of electrodeposition processes for plating palladium alloys requires considerations of the following: phase diagrams, coordination chemistry in the aqueous phase, redox potentials, thermodynamics, and kinetics. For the electrodeposition of an alloy, as discussed in Section 5.3, the ΔE (Eq. 16) must be minimized in order to successfully control the alloy composition. In general, the respective metals should have similar reduction potentials, preferably less than 200 mV apart. As previously discussed, this is highly dependent on the specific metal and the chemical system employed.

Table 10 exhibits standard reduction potentials for various metals from aqueous media. As can be seen, in most cases the relative standard potential for palladium and the alloying metals are significantly different. This requires manipulation of the complexing ligand and of the additive systems in order to codeposit the alloy. In this section we will describe the chemical systems that have been reported for palladium alloys. We will pay specific attention to palladium-nickel because of its technological importance and briefly describe other systems that have been reported.

7.1 PdNi (80/20 wt%)

While acidic palladium-nickel electroplating baths have been proposed [119], most processes appear to be ammoniacal solutions in the pH range from 7 to 9. Recommended metal sources for palladium are its complexes with ammonia or organic amines, for nickel, the chloride, sulfate, hydroxide, and sulfamate. A palladium-nickel plating bath listed by R. J. Morrissey [91] contains 6 g liter^{-1} $Pd(NH_3)_2(NO_2)_2$, 3 g liter^{-1} nickel sulfamate, 90 g liter^{-1} ammonium sulfamate and ammonium hydroxide to pH of 8 and 9. This bath operates at 20° to 40°C at current densities between 5 and 10 mA cm^{-2} and it suitable for plating an alloy of about 25% Ni. Other chemistries use plating temperatures of up to 60°C in the pH range of 7 to 9, metal concentrations of up to 25 g liter^{-1} Pd and 30 g liter^{-1} Ni, and up to four different proprietary additives.

Such processes exhibit a number of disadvantages in their operation. Perhaps the most troublesome (and at the same time most frequent) complaint is the lack of pH stability that results in undesirable fluctuations in the alloy composition. Another concern are the environmental and occupational hazards connected with ammonia vapors during plating. This is particularly true in most high-speed operations where the pH, temperature, and solution agitation tend to be high, thus favoring the release of ammonia. Additionally the loss of deposit brightness with bath aging can shorten the useful life of the bath. Last, but not least, highly stressed and cracked deposits are also frequently encountered, especially in high-speed operations.

TABLE 10 Standard Potentials in Aqueous Media

Half Reaction	Acid	Neutral/Basic
$Pd^{+2} + 2e^- \longrightarrow Pd(s)$	0.915	0
$Ag^+ + 1e^- \longrightarrow Ag(s)$	0.799	0.342
$Cu^{+2} + 2e^- \longrightarrow Cu(s)$	0.340	−0.100
$Bi^{+3} + 3e^- \longrightarrow Bi(s)$	0.317	0.072
$Sn^{+2} + 2e^- \longrightarrow Sn(s)$	−0.137	−0.631
$Ni^{+2} + 2e^- \longrightarrow Ni(s)$	−0.257	−0.49
$Co^{+2} + 2e^- \longrightarrow Co(s)$	−0.277	−0.733
$Fe^{+2} + 2e^- \longrightarrow Fe(s)$	0.44	−0.8

Chemistries and Material Properties of PdNi Figure 37 is a phase diagram of PdNi; it demonstrates that it forms a solid-solution across all compositions. The lowest melting point of 1237°C occurs at a palladium concentration of 45.4% by weight. This implies that one should be able to plate PdNi of any composition. However, the 80/20 PdNi alloy is preferred, although in some cases a 90/10 alloy is utilized commercially.

In general, one would suspect that the more nickel is incorporated into the alloy, the less noble the deposit and the more susceptible it is to surface oxidation and corrosion. However, the author is not aware of any exhaustive study that compares material properties of PdNi to the alloy composition. This area of study is open to new investigations.

Table 11 exhibits four PdNi bath compositions, and in general, they utilize a palladium diamine or tetramine chloride salt and a nickel hexamine sulfate salt. The

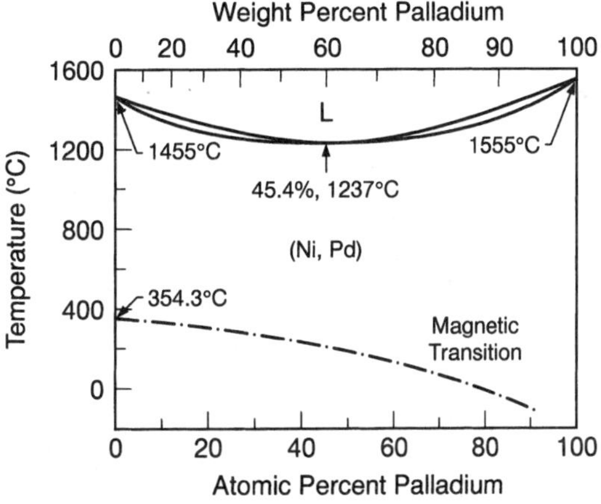

Figure 37 Binary alloy phase diagram of palladium-nickel system.

TABLE 11 PdNi Alloy Plating

	Bath Type			
Parameter	T[120]	U[121–123]	V[124]	W[125]
Pd Source	$Pd(NH_3)_2Cl_2$	$Pd(NH_3)_4Cl_2$	$Pd(NH_3)_4Cl_2$	Pd-Sulfamate
Pd Metal (g liter^{-1})	25	20	20	10.6
Ni Source	$Ni(NH_3)_4SO_4$	$Ni(NH_3)_6SO_4$	$Ni(NH_3)_4SO_4$	Ni-Sulfamate
$(NH_4)_2SO_4$	50	10	9	5.9
NH_4Cl		50	10	
NH_4OH(ml liter^{-1})	100	To pH	To pH	
Ligand (Other)				en 80 g liter^{-1}
Additives				
3-pyridine sulfonic acid (g liter^{-1})	5	See Ref.	See Ref.	
pH	7.74	8.5	8	4.5–6.0
Temperature (°C)	60	30	25	20–60
Current Density mA/cm^2	3–250	10	10	10–30

supporting electrolytes are either ammonium sulfate or ammonium chloride, and ammonium hydroxide is used to adjust pH which is maintained in the range of 6 to 8.5. The temperature is in the range of 25° to 60°C and the maximum current density [120] reported is 250 mA cm^{-2}. Organic surfactants, brighteners, levelers, and stress reducers are commonly added to the bath composition, but in most situations they are proprietary and are either patented or kept as a trade secret.

The only nonammonia system that was extensively studied was reported by Walz [125]; it is based on ethylene diamine. It was reported that bright palladium-nickel deposits up to 40% nickel could be obtained over a wide current density range, Figure 38. The advantages of this system are that it does not suffer from the evaporation of ammonia and subsequently has a stable pH. There is less attack on substrate metals such as copper and brass, and thus less chance of contaminating the plating solution. However, in general, this system has been reported to suffer from higher tensile stress.

Boguslavsky [126] recently reported the development of a proprietary PdNi process that seems to overcome many deficiencies. The deposits exhibit excellent alloy stability (±2 wt%) over a wide current density range (~ 50–800 mA cm^{-2}), Figure 39, are full bright in appearance, of high hardness (~ 550 KHN_{50}), and good

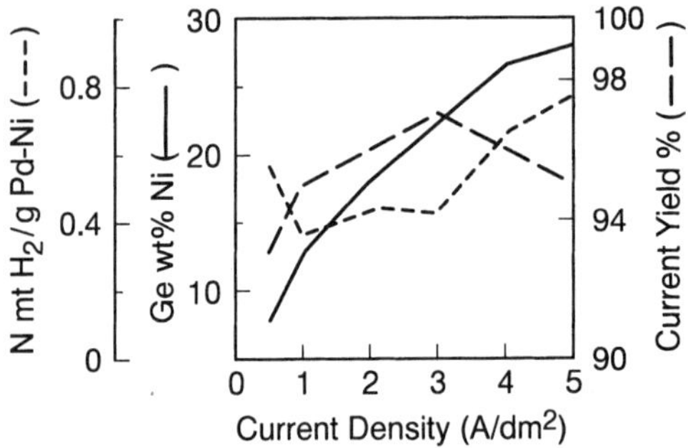

Figure 38 H_2 content versus current density at various wt% Ni content: $Pd(en)_2Cl_2 + Ni$ sulfamate system at 50°C, pH = 4.5.

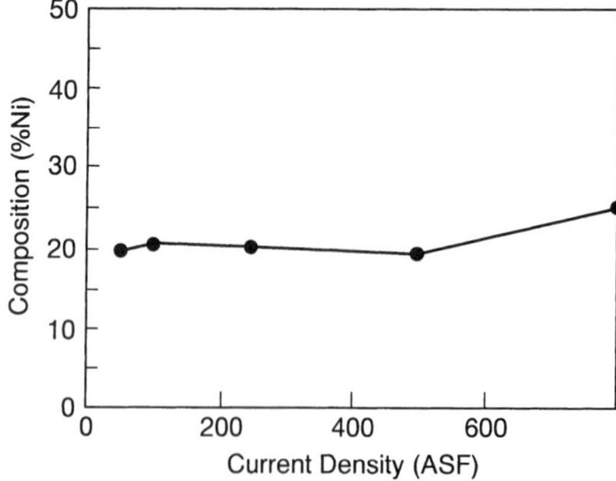

Figure 39 Percentage by weight of nickel versus current density: Pd–Ni system from ammoniacal electrolyte.

ductility. This process is reported to be less corrosive than previous processes and environmentally benign.

7.2 Palladium-Cobalt

Palladium-cobalt, like palladium-nickel, forms a solid solution over the entire composition range and exhibits similar materials properties. Unlike nickel, cobalt has a +2 and +3 aqueous chemistry, and the standard potentials for cobalt reduction are significantly more negative than palladium, Table 10. However, the stability of cobalt complexes with various ammonia electrolytes at relatively neutral pHs indicate that its reduction potential could be more noble than palladium in the same electrolyte. This becomes problematic when attempting to plate a consistent alloy composition over a wide range of operating parameters.

Kume [127] reports on the electrochemical behavior and the effect of plating parameters on the alloy composition of PdCo from an ammoniacal electrolyte. The basic bath composition is as follows:

PARAMETER	CONCENTRATION
$Pd(NH_3)_2Cl_2$	0.1 M
$CoSO_4 \cdot 7H_2O$	0.03 M
$(NH_4)_2SO_4$	0.4 M
NH_4OH	pH 7–10

Kume states that the standard potentials for Pd and Co are, respectively, + 0.987 V and − 0.250 V, the difference being 1.23 V. However, in ammoniacal plating baths, the potential difference is much smaller, for example, 0.27 V at current densities of 10 mA cm^{-2}. Figure 40 shows polarization curves for the deposition of Pd, Co, and Pd–Co alloys. Figure 41 is a graph of alloy composition as a function of current

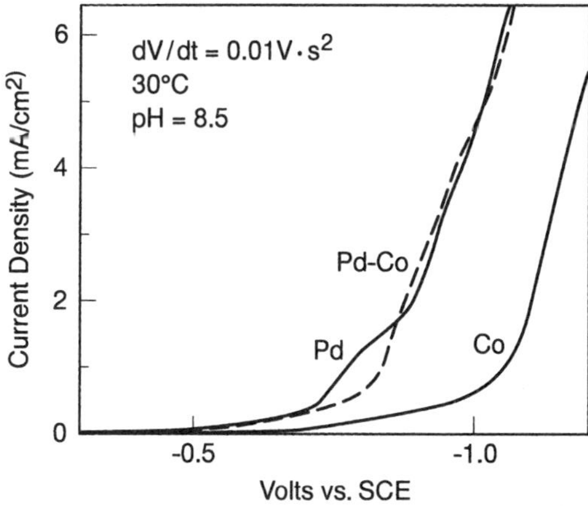

Figure 40 Current-potential curves: Palladium, cobalt, and palladium-cobalt from ammoniacal electrolyte.

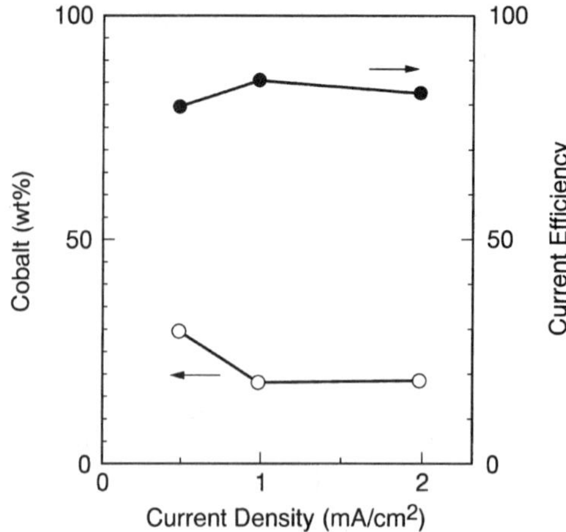

Figure 41 Percentage by weight of cobalt versus current density: Pd–Co system from ammoniacal electrolyte.

density. It shows that the cobalt varies between 20% and 30% by weight between 5 and 20 mA cm^{-2} and clearly demonstrates that a higher cobalt content is deposited at lower current densities. This seems in conflict with Kume's assertion that the cobalt deposition is either at the same or more negative potential than Pd. When one considers that the palladium concentration is three times greater than cobalt, once again the conclusion must be that the electrochemistry needs to be re-examined. Cavallotti plated Pd–Co from electrolytes based on cobalt amino-citrates and palladium amino-nitrite complexes [147].

7.3 Palladium-Iron

Pd–Fe alloys prepared metallurgically are known to be hard magnetic materials [1]. These alloys can also be used as CO sensors in catalytic combustion chambers [129].

The galvanic deposition of palladium-iron alloys is reported by Juzkis et al. [130] from an ammoniacal electrolyte containing a complexing agent for Fe(III). The electrolyte composition and operating temperature is as follows:

PARAMETER	CONCENTRATION/OPERATING CONDITIONS
$PdCl_2$	0.05–0.2 M (as Pd)
$Fe_2(SO_4)_3$	0.1–0.3 M (as Fe)
Sulfosalicylic acid	0.2–0.6 M
NH_4OH	0.3 M
Temperature	25–50°C
Current density	5–300 mA cm^{-2}

The palladium-iron alloys deposited with the above electrolytes produce a light gray satin finish. Figure 42 shows the dependency of palladium content on the current density as the iron content is kept constant. Thus a 0.1 M Pd and 0.15 M Fe bath composition allows the Pd content to be about $80 \pm 5\%$. However, in utilizing

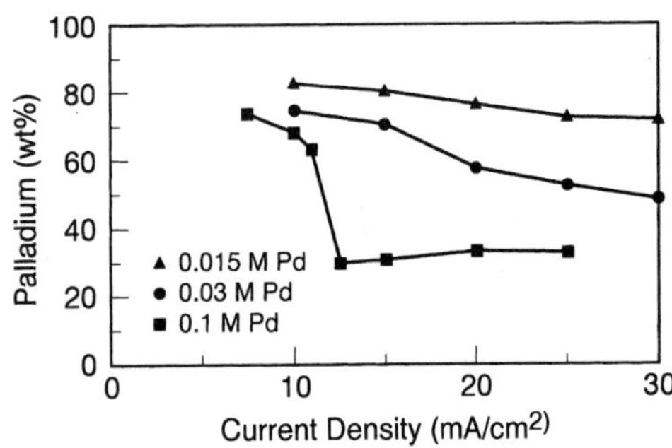

Figure 42 Percentage by weight of Palladium versus current density at various [Pd]: Pd–Fe system from ammoniacal electrolyte.

direct current plating at 50°C, the current efficiency becomes about 75%, and a significant amount of hydrogen is codeposited. The H/Pd + Fe ratio is about 0.15. This relatively high hydrogen content causes the deposits to be brittle and stress cracked, which cannot be avoided by manipulating the operating parameters. However, with pulsed electrolysis the deposits contain less hydrogen, they are brighter in appearance and can be plated to 10 μm crack free.

7.4 Palladium-Gold

A slightly alkaline gold plating solution, free of cyanide and phosphates, containing a sodium gold sulfite complex and a palladosamine chloride complex is used to plate a palladium-gold alloy [131]. A typical electroplating formulation is as follows:

PARAMETER	CONCENTRATION/OPERATING CONDITIONS
$Pd(NH_3)_2Cl_2$	1.8 g liter^{-1}
$NaAu(SO_3)_2$	10 g liter^{-1}
$(NH_4)_2SO_3$	15 g liter^{-1}
$CuSO_4$	0.095 g liter^{-1}
$Na_2As_2O_3$	0.03 g liter^{-1}
pH	8.2
Temperature	58°C

At 5 mA cm^{-2} a plating rate of 100 mg per ampere minute can be achieved which yields a fully bright, gold flesh toned deposit.

7.5 Palladium-Indium

Interest in palladium-indium goes back to the late 1970s and early 1980s [132–137]. According to these studies, Pd–In alloy crystallizes at 1285°C in a simple cubic structure of the CsCl type with a lattice parameter of 3.22 Å. This alloy belongs to compounds of the colored β-brass group and, depending on the indium content has a yellow, golden or rose-lilac color. An article by Reshetnikova [138] describes the acquisition of various properties including coefficients of friction, resistivity, contact resistance, and the like. A summary of their results is shown in Table 12.

TABLE 12 Palladium-Indium Materials Properties

Coating type	Indium content (mass %)	Resistivity (μΩ·cm) at 20°	Micro-hardness (kg/mm²)	Constant resistance (Ω) at 10 mA current under load (g)		Mass loss in wear-resistance rest (mg)	Coefficient of friction versus steel	Internal stress (log/cm²)	Spreading area of POS-61 lead-tin solder (mm²)
				50	100				
Solid solution	3–6	59	290	0.0035	0.0023	20.9	0.18–0.19	3200	56–79
Intermetallic	45–55	88	450	0.0240	0.0740	2.7	0.12–0.13	8700	5–6
Palladium	0	20	260	0.0020	0.0015	28.3	0.2	3600	44–77

Another study by Dzhandubaeva [139] provides the following recipes:

PARAMETER	CONCENTRATION/OPERATING CONDITIONS
Palladium	20 g $liter^{-1}$
Indium	20 g $liter^{-1}$
Trilon β	60 g $liter^{-1}$
NH_4Cl	25 g $liter^{-1}$
pH	$\sim$8.5
Temperature	20°C

It was reported that film thickness of 10 μm could be deposited. A study of the deposits from this electrolyte using direct and pulsed current is the object of this chapter.

7.6 Palladium-Silver

The properties of Pd–Ag alloys are known to be suitable for contact materials [140–144]. In particular, savings in material costs and comparable corrosion behavior to either gold or palladium have led a number of researchers to investigate palladium-silver.

The Pd-Ag alloy system is a homogeneous solid solution phase over its entire composition range. The mechanical and electrical properties therefore will not be susceptible to precipitation of a second phase as in a multiphase alloy, and erratic, abrupt, or irreproducible changes. Small changes in alloy composition have only a minor effects on such properties since, in a homogeneous phase, properties, in general, vary gradually with composition. Cohen et al. [142–145] chose to utilize a concentrated lithium chloride bath, which contains simple ions and no additives. The electrolyte consists of the following:

PARAMETER	CONCENTRATION/OPERATING CONDITIONS
$PdCl_2$	3 g $liter^{-1}$
AgCl	12 g $liter^{-1}$
LiCl	500 g $liter^{-1}$
HCl	20 ml $liter^{-1}$
Current density	0.5–20 mA cm^{-2}

From this electrolyte it was possible to plate dense, uniform, single-phase alloys with Ag content of 40% to 70% at about 100% current efficiency. The Ag–Pd alloy properties appear to be favorable for contact applications possessing a microhardness of about 200 KHN_{25}, very low compressive internal stress $< 1\ kg\ mm^{-2}$ and similar wear, friction, and corrosion properties to hard gold. However, this electrolyte composition is extremely corrosive and undergoes displacement plating with non-noble metal substrates.

Nobel [145–146] claimed to have developed a proprietary process feasible for electroplating this alloy over a wide range of current densities. Deposits from this proprietary alloy are said to be "pore-free" and comparable to hard-gold, possess excellent wear and fretting resistance, contact resistance, and corrosion resistance which is better than palladium-nickel alloy.

Kume [148] reports plating a Pd–Ag alloy from ammoniacal alkaline solutions. Silver deposition takes place under mass-transport controlled conditions. His conclusion is that Pd–Ag is a useful substitute for hard gold in contact applications. However, Kume does not discuss the possibility of the formation of silver-fulminates (from ammonia-type systems), which could potentially be explosive. The author is not aware of plated PdAg being used in any commercial application, and plating palladium-silver with the desired composition remains a challenge today.

7.7 Other Palladium-Alloys

Numerous palladium binary and ternary alloys have been reported in the literature or in patents; however, it is beyond the scope of this chapter to discuss them all. The most notable are a palladium-rhenium alloy [149] plated from ammoniacal solutions with Trilon as an additive. The Re content is about $\sim$24% and current efficiency 63% to 90%. Palladium-tin alloys plated from pyrophosphate baths containing diethylenetriamine pentaacetic acid [150]. Palladium-antimony alloy deposition was reported in a patent [153] where a palladium salt was dissolved in Trilon β at 60° to 70°C and pH = 4. In each case the intent was to improve the materials properties of palladium for technological applications such as contact material.

Palladium-arsenic alloy processes were patented by Abys [151] in 1991 and Wang [152] in 1994. This alloy, usually 20 wt% arsenic, is extremely ductile and can be electroformed into various shapes such as bellows which can be “flexed” without cracking.

8 RECENT ECONOMICS; TECHNICAL ADVANCES AND EMERGING APPLICATIONS

8.1 Current Economic Concerns

Early in 1998, the bullion prices of gold and palladium converged and consequently diverged as depicted in Figure 43. The simple economic principles of supply and demand can rationalize this inversion of metal prices. With respect to demand, there has been a strong trend towards increased use of palladium in automotive catalysts, e.g., in 1996 auto catalyst demand was approximately 2.4 million troy ounces, in 1998 this figure almost doubled and was valued at 4.5 million troy ounces [154]. Notwithstanding the increased demand from the automotive industry, there has been an increase in demand from the electronics sector [154], as traditionally gold plated components switch to palladium and its alloys for technological reasons. Relatively speaking, gold prices have remained steady, trading in a narrow range of $270 to $350 per troy ounce.

Supply has also played a key role in the recent price hike for palladium bullion. As stated earlier, see Section 1.2, almost two thirds of the world’s supply of palladium originates in Russia. Recent socioeconomic issues have limited, and in some instances, halted palladium shipments. An entirely predictable increase in metal price transpired as a result of the announcements that shipments were suspended. There has however never been a problem is delivering palladium containing products to the consumer.

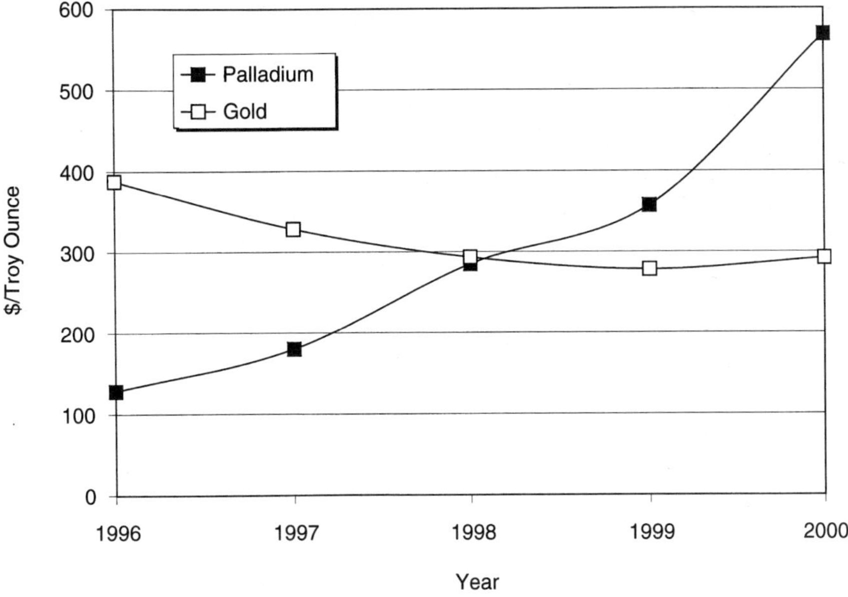

Figure 43 Gold and palladium metal price in dollars per troy ounce from 1996–2000* (03/10/2000 – second london fix)*.

The outlook for a sharp downward correction in the price of palladium is good. Factoring the average palladium price by year (1996–1999), the projected value for 2000 is $435 per troy ounce. There are abundant amounts of palladium resources in the world. Current US Geological Survey reports [155] indicate that there is a total of 100 million kilograms of PGMs in reserves (metal in the ground at current mine sites) and in the reserve base (metal in the ground, but having no current mining operation). Based on the aforementioned numbers and current demand levels, there is enough palladium available for the next 150 years.

There are also intense automotive industrial research programs in progress, looking to identify alternative metals and alloys capable of competing with catalytic activities of palladium based systems. One GM purchasing director is quoted as saying that alternative systems that were not economically viable when Pd traded at $200 per troy ounce are becoming "viable very quickly" [156].

8.2 Chemistry and Material Properties of PdCo

Since the commercialization of a readily controllable and stable plating chemistry capable of depositing an 80/20 alloy of palladium/cobalt in ca. 1996, numerous papers concerning its deposit material properties and new applications have appeared in the literature [157–163]. An attempt will be made here to summarize and identify the key features of PdCo. Throughout this discussion, comparison to plated hard gold deposits will be made.

With respect to bath chemistry, Table 13 provides a snap shot of the key components and their respective concentrations. The chemical manipulation, which was central to the production of a plating system capable of producing a good

TABLE 13 80/20 PdCo Alloy Plating Chemistry

Component	Typical
Pd as metal (g liter^{-1})	40
Co as metal (g liter^{-1})	8
Conducting salt (g liter^{-1})	40
Additives (ml liter^{-1})	20
Ligand (ml liter^{-1})	150

quality and stable alloy, involved the choice of an appropriate ligand. This produced a system where ΔE, as discussed in Section 3.3, was sufficiently close. The palladium cobalt alloy composition is also highly dependent upon the following; metal concentrations, pH, current density, temperature and solution agitation. For a detailed discussion on how the alloy composition varies as a function of any of the aforementioned variables, the reader is pointed to reference [163].

The goal of any electroplating chemistry development work is to produce a system capable of producing deposits with certain physical properties and manage to do it on a consistent basis with a minimum of bath maintenance. Similar to the discourse on the material properties for PdNi alloy deposits, Section 5.1.1, illustrated examples of how palladium cobalt deposits outperform plated hard gold will be given. Again, for more details on experimental conditions, testing methods, etc., the reader should consult the appropriate references.

A summary of important material properties for both gold-flashed palladium cobalt (GFPdCo) and hard gold is given in Table 14 [160]. Especially noteworthy elements contained within Table 14 include:

Density The value for plated hard gold is 17.3 g cm^{-3} as compared to the value obtained, via a pycnometric method, for an 80/20 palladium cobalt deposit of 10.8 g cm^{-3}. Thus, for an equivalent surface area, an immediate cost saving of 40% can be realized. Figure 44 is useful in illustrating the significant contributions both alloying and density can have on the cost of plated parts. Here it can be seen that with gold at approx. $300 per ounce, Pd prices close to $600 have to be exceeded; then and only then, will all metal price based cost savings be eroded. Figure 44 assumes the density values listed above, an 80/20 alloy of palladium and equivalent thickness. The following hardness and porosity summaries will support further cost

TABLE 14 Comparative Materials Properties

Material Properties	Hard Au	GFPdCo
Density (g cm^{-3})	17.3	10.8
Thermal Stability (°C)	150	395
Hardness (KHN_{25})	140–200	590–640
Ductility (%E at 2.5 μm)	<3	3–9
Porosity Index		
Connector Pins (1.5 μm/2.5 μm Ni)	0.70	0.02
Wear properties (load: 100 g)		
Cycles to failure (× 1000)	20	> 80
Coefficient of Friction at 10 K cycles	0.60	0.43

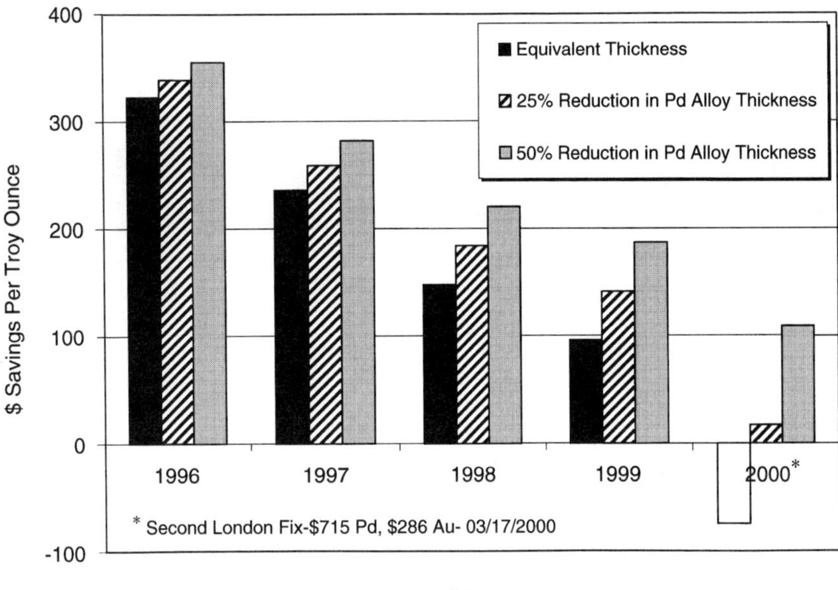

Figure 44 Cost savings per ounce – 80/20 Pd alloy versus gold from 1996 to 2000* (03/10/2000 – second london fix), for an equivalent thickness, a 25% reduction in Pd alloy thickness and a 50% reduction in Pd alloy thickness.

savings through a reduction in deposit thickness due to the superior material properties of palladium alloys when compared to hard gold.

Hardness A Knoop hardness value in the range of 590 to 640 for PdCo can be obtained using current plating chemistry [160]. An average hardness value for "hard" gold is on the order of 150 Knoop. The significance of this higher hardness number may manifest itself in increased wear resistance. To test this hypothesis, a sliding wear test (the conditions and experimental details are given elsewhere [160]) was conducted. The results obtained are summarized in Figure 45. A gold flashed palladium cobalt deposit exhibits higher wear resistance when compared to hard gold (or even gold flashed palladium nickel). The precious metal thickness was the same for all finishes tested.

Porosity One of the most common failure mechanisms, where for electronic connectors this typically implies a change in contact resistance greater than 10 milliohms, is pore corrosion [164]. Thus, a porosity test – ASTM B799-88 is invariably conducted during any part qualification program. The test involves exposure to SO_2, followed by ammonium sulfide for fixed periods of time. A porosity index scale, described elsewhere by Kudrak [160, 165], can be used for comparative purposes. Using this porosity index scale, gold flashed palladium cobalt is found to possess the lowest number. Thus it can be stated unequivocally that GFPdCo is superior to hard gold with respect to intrinsic porosity.

In summary, palladium cobalt is a fine-grain, single-phase alloy. It possesses properties such as high hardness, low porosity and high thermal stability. All of these

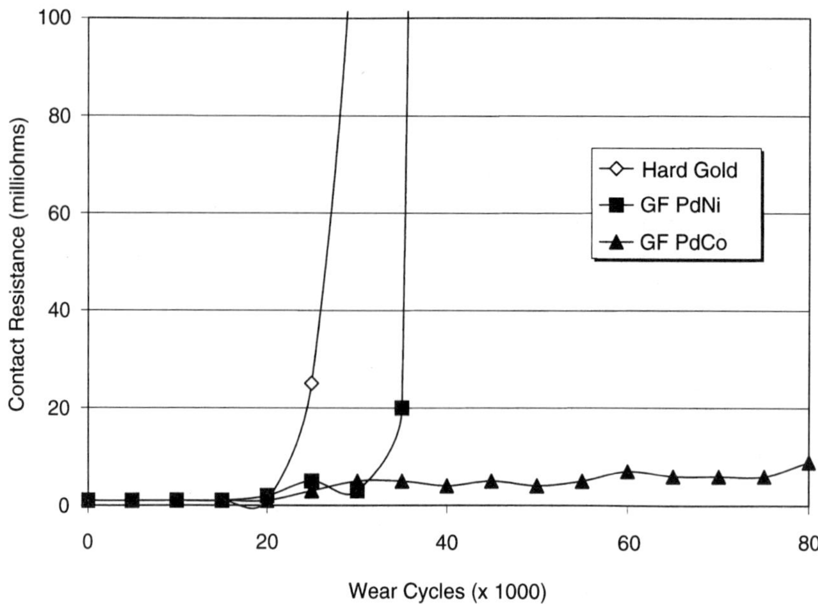

Figure 45 Sliding wear–contact resistance (100 g load).

attributes clearly support the assertion that palladium cobalt is a new alloy, capable of outperforming hard gold and in some instances, pure palladium and palladium nickel for certain applications.

8.3 The use of Palladium for PWB Applications

Once again, a very detailed discussion of the uses of palladium, with particular reference to printed wiring board (PWB)-type applications is beyond the scope of this chapter. As technology advances within the PWB industry, so too have the demands on manufacturing processes that serve this sector. External demands have also been placed on PWB manufacturing operations, e.g., national and international legislation will ban the use of lead in all electronics by 2004 [166]. Certain Original Equipment Manufacturers or OEMs have self-imposed lead-reduction/elimination dates which are even more aggressive, i.e., some wish to eliminate lead from their products by 2001. Lead is currently one of the principal components of soldering systems.

In order for a new surface finish to be viable, the following items must be taken into consideration:

- Uniform thickness distribution,
- Solderability,
- Wirebondability,
- Use of lead-free materials,
- Environmental acceptability,
- Favorable economics.

The idea of using palladium in printed wiring board assembly is not new and was proposed as early as 1984 by Wilkinson and O'Hara [167]. It was however dismissed just a quickly, as the technology for plating pure palladium at the time was not as advanced as it is today, i.e., deposits suffered from micro-cracking and low ductility. The current situation, especially considering how far palladium plating chemistry has evolved, is however very different and numerous examples of production proven processes exist. Thin palladium and gold flashed palladium deposits, i.e., < 0.5 microns in thickness, can be used to good effect as an etch resist and final finish for PWBs. Literature exists to support this claim [168–170].

The benefits to board manufacturers are; 1) elimination of lead, 2) improved processing times (reduction in cycle times), 3) higher yields and 4) ionically cleaner substrate. The benefits of using Pd as a final finish are; 1) good solderability, 2) good wire and die bondability, 3) excellent diffusion/migration barrier properties, 4) thermally stable surfaces, 5) component co-planarity requirements are easily satisfied, 6) compatible with active devices, and 7) compatible with lead free solders [170].

8.4 Palladium Preplated Leadframes (PPFs)

Propositions for the use of palladium in preplated leadframes were advanced in the mid-1980s [171–174]. The technology utilizes high speed nickel and palladium plating over the entire leadframe surface; as opposed to selective silver plating for die attachment and wirebonding, and solder plating of the external leads for solderability. Schematically the two very different technologies are illustrated in Figure 46.

There are four distinct areas where Pd PPF technology offers significant advantages [168, 175];

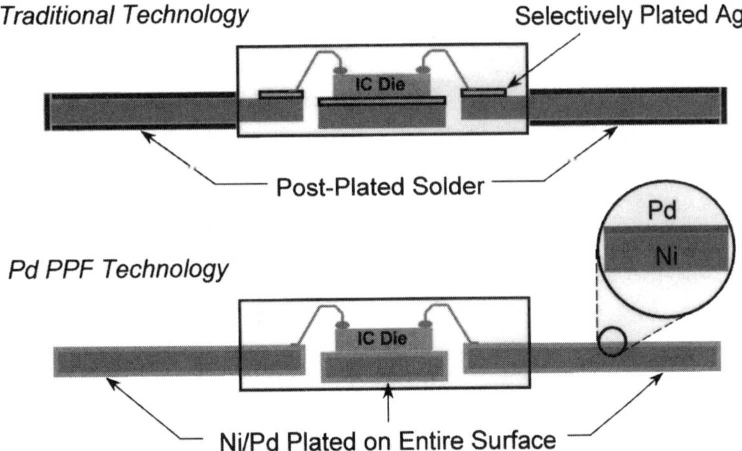

Figure 46 Schematic cross-sections of leadframes plated with silver/solder or nickel/palladium, after die attachment, wirebonding and encapsulation and before trimming and forming.

1) Plating process improvements – Whereby the need for costly selective plating is removed, yields and throughput are increased and all wet processes are completed prior to package assembly,
2) Quality improvements – Insofar as co-planarity is improved and can be maintained, the elimination of solder bridging and no possibility for silver migration,
3) Environmental improvements – Both cyanide and lead are eliminated from the plating process and,
4) Overall packaging cost reduction – The cost reductions stem from the simplification of the plating process, the elimination of post-assembly processing, reductions in waste treatment and extension of the leadframe shelf-life.

TABLE 15 New Acid Strike Plating Chemistry

Component	Typical
Pd as metal (g liter^{-1})	5
Acid Electrolyte (ml liter^{-1})	150
Additive (ml liter^{-1})	5

Figure 47 Deposits of varying thickness from new acid strike.

Note: The use of a conformable nickel underlayer is necessary for high reliability applications. Arguments supporting this assertion are already published [172].

8.5 A Novel, Strongly Acidic Palladium Strike

Building upon Section 4.2.3, the authors would like to make note of a newly developed acid strike system [176]. The new pure palladium strike combines the "activating" power of a very low pH (< 1) with the advantage of room temperature operation to deposit bright Pd layers onto various substrates with surprisingly high cathode efficiencies. The system leads to improved adhesion on easily passivated surfaces, e.g., nickel and real porosity reductions of subsequent palladium and palladium alloy deposits.

The chemical make-up of this new plating bath is summarized in Table 15. As is, the system is simple and bath maintenance is thus minimal. The chemistry is robust and can withstand relatively high levels of foreign metal contamination, e.g., copper and nickel. The components are also claimed [176] to be compatible with chloride- and sulfate-based palladium plating systems and do not exert deleterious effects on the performance of the main electrodeposition operation. One other important advantage over prior art is that relatively thick deposits can be plated, as can be seen from Figure 47. There is also a marked leveling effect.

9 CONCLUSION

Although the electrodeposition of palladium was first reported by Henry around 1855, it took well over one hundred years for this technology to mature. Knowledge of the electrochemistry of palladium and the effects of hydrogen codeposition were essential in devising chemical systems that could plate palladium under a variety of conditions in a manufacturing environment. Furthermore extensive reliability data on the use of palladium as a contact finish was instrumental in overcoming the "gold" standard in electronic components.

Palladium and it alloys have thus emerged as competitive substitutes for gold in electronic applications, and for rhodium and nickel in the decorative industry. The electrodeposition of palladium alloys, including nickel, remain problematic. While many chemistries have been reported, the challenge is to electroplate alloys with controllable composition and material properties under a variety of plating conditions.

ACKNOWLEDGMENTS

The authors wish to thank Y. Zhang, I. Kadija, and Y. Okinaka for their many helpful discussions and useful recommendations. A special note of gratitude to M. Lake for the long hours spent in the preparation of this chapter.

REFERENCES

1. *Encyclopedia Americana*, International edition.
2. *Compton's Interactive Encyclopedia*, 1993; 94 Compton's News Media, Inc.

3. I. E. Cottington, *Platinum Met. Rev.*, **35** (3), 141–151 (1991).
4. J. H. Crockett, "Platinum," in *Handbook of Geochemistry*, vol. 2, H. H. Wedepohl, ed., Springer-Veiloy, New York, 1969, sections B–G, K, M, O.
5. T. L. Wright and M. Fleischer, "Geochemistry of the Platinum Metals," *Geological Survey Bulletin*, 1214–A, U.S. G. P. Office, Washington, DC, 1965, p. 24.
6. B. Mason, *Principles of Geochemistry*, 2nd ed., Wiley, New York, 1951, p. 310.
7. *Platinum*, Johnson Matthey, 1996.
8. J. A. Abys, "A Unique Palladium Electrochemistry—Characteristics and Film Properties," Second American Electroplaters Society Symp. on Economic Use of and Substitution for Precious Metals in the Electronics Industry, Sept., 1980, Danvers, MA.
9. H. K. Straschil, I. Kadija, J. J. Maisano, and J. A. Abys, *PCIF, Circuit World*, **17** (2), 9–15, (1991).
10. E. J. Kudrak, J. A. Abys, I. Kadija, and J. J. Maisano, *Proc. Annual Conf. American Electroplaters and Surface Finishers Society*, July 1990, Boston, MA.
11. H. K. Straschil, I. Kadija, J. J. Maisano, and J. A. Abys, *Proc. Annual Conf. American Electroplaters and Surface Finishers Society*, July 1990, Boston, MA.
12. I. Kadija, J. A. Abys, V. Chinchankar, and H. K. Straschil, *Proc. Annual Conf. American Electroplaters and Surface Finishers Society*, July 1990, Boston, MA.
13. J. A. Abys, H. K. Straschil, I. Kadija, E. J. Kudrak, and J. Blee, *Met. Finish.*, **89** 43 (1991).
14. J. Stevenson and L. Mayer, *IICIT Annual Symp.*, 1989, Philadelphia, PA.
15. J. A. Abys, J. J. Maisano, C. Wolowodiuk, and H. K. Straschil, *Proc. Connectors '89 Conf.*, Mar. 1989, Coventry, England.
16. M. Antler, *Platinum Met. Rev.*, **31** (1), 13–19 (1987).
17. W. A. Fairweather, *Trans. Inst. Met. Finish.*, **64** (3), 1 (1986).
18. A. Graham and S. Updegraff, "Properties of Palladium-Nickel Alloy and Pure Palladium for Connector Applications," *Proc. NEPCON Conf.*, 1984, Anaheim, CA.
19. American Electroplaters Society Symp. on Economic Use of and Substitution for Precious Metals in the Electronics Industry, 1982, Danvers, MA.
20. American Electroplaters Society Symp. on Substitution for Gold, 1980, Milwaukee, WI.
21. D. Gravitz, *Electronic Packaging and Production*, Feb. 1982.
22. I. H. Brockman et al., *Proc. Holm Conf.*, 1988, pp. 73–83.
23. H. Grossman et al., *Proc. Holm Conf.*, 1984, pp. 551–560.
24. D. Ruhlicke et al., *Proc. Holm Conf.*, 1984, pp. 83–88.
25. S. W. Updegraff, *Proc ICEC*, 1986, pp. 399–403.
26. A. H. Graham, *Proc. Holm Conf.*, 1984, pp. 61–67.
27. G. J. Russ, *IEEE Trans. Component, Hybrid, Manuf. Tech.*, Vol-CHMT 6, no. 4, Dec. 1983, pp. 389–395.
28. E. J. Kudrak, J. A. Abys, I. Kadija, and J. J. Maisano, *Proc. Annual Conf. American Electroplaters and Surface Finishers Society*, July 1990, Boston, MA.
29. M. Antler, *Platinum Met. Rev.*, **3** (1), 13–19 (1987).
30. T. Sato et al., *Proc. 26th Annual Holm Conf. on Electrical Contacts*, 1980, pp. 41–47.
31. A. H. Graham, *Proc. 30th Annual Holm Conf. on Electrical Contacts*, 1980, pp. 61–67 (1980).
32. T. Sato, *Plating Surf. Finish.*, **74**, 55–59 (1987).
33. M. Antler, ***ASLE Trans.***, **26** (3), 376–380 (1983).

34. W. O. Freitag, *Proc. of Holm Seminar on Electrical Contacts*, Chicago, pp. 17–23 (1975).
35. R. H. Atkinson and A. R. Raper, *J. Electrodepositors Tech. Soc.*, **8** (10), 1 (1933).
36. F. H. Reid, *Plating*, **52**, 531 (1965).
37. H. D. Hedrich, and C. J. Raub, *Metalloberflache*, **311** (11), 512–520 (1977).
38. R. Weil, *Plating*, **60**, 1231 (1970); **61**, 50, 127 (1971).
39. H. D. Hedrich and C. J. Raub, *Galvanotechnik*, **70** (10), 934–939 (1979).
40. H. D. Hedrich and C. J. Raub, *Surface Technology*, **8**, 347–362 (1979).
41. F. H. Reid, *Metall. Rev.*, **8** (30), 167–211 (1963).
42. Periodic Table of the Elements, Sargent-Welsh Scientific Company, 7300 Linder Avenue, Shokie, Illinois 60076.
43. *Platinum Group Metals*, Printing and Publishing Office, National Academy of Science, 2101 Constitution Avenue, N.W., Washington, DC 20418.
44. A. R. Powell, "The Platinum Metals in the Periodic System: A Comparative Study of the Transition Metals," *Platinum Met. Rev.*, **4**, 144–149 (1960).
45. E. M. Wise, *Palladium Recovery, Projection and Use*, Academy Press, New York, 1968, p. 187.
46. W. Hume-Rothery, "The Platinum Metals and Their Alloys: A Review of Their Electronic Structure and Constitution," *Platinum Met. Rev.*, **10**, 94–100 (1966).
47. F. A. Lewis, *The Palladium-Hydrogen System*, Academic Press, New York, 1967, p. 178.
48. F. A. Cotton and G. Wilkinson, *Advanced Inorganic Chemistry*, 4th ed., Wiley, New York, 1980.
49. F. R. Hartley, *The Chemistry of Platinum and Palladium*, Wiley, New York, 1973, p. 544.
50. S. Otsuka, S. Y. Tatsuno, and K. Ataka, "Univalent Palladium Complexes," *J. Am. Chem. Soc.*, **93**, 6705–6706 (1971).
51. R. A. Reinhardt and W. W. Monk, *Inorg. Chem.*, **9**, 2026–2030 (1970).
52. F. Bosolo and R. G. Pearson, *Proc. Inorg. Chem.*, **4**, 381–453 (1962).
53. N. V. Sidgwick, *The Electronic Theory of Valency*, O.V.P., London, 1929, p. 163.
54. F. Basolo and R. G. Pearson, *Mechanisms of Inorganic Reactions*, Wiley, New York, 1967, p. 526.
55. G. E. Coates, M. L. H. Green, and K. Wade, *Organometallic Compounds*, vol. 2, *The Transition Elements*, Methuen, London, 1968, p. 6.
56. A. F. Bard, R. Parsons, and J. Jordan, eds., *Standard Potentials in Aqueous Solutions*, Marcel Dekker, New York, 1988.
57. R. M. Izatt, R. Eatough, and J. J. Christensen, *J. Chem. Soc.*, **A**, 1301 (1967).
58. D. H. Templeton, G. W. Watt, and C. S. Garner, *J. Am. Chem. Soc.*, **65**, 1608 (1943).
59. R. N. Goldberg and L. H. Hepler, *Chem. Rev.*, **68**, 229 (1968).
60. R. M. Izatt et al., *J. Chem. Soc.*, **A**, 2514 (1970).
61. E. Jackson and D. A. Pantony, *J. Appl. Electrochem.*, **1**, 283 (1971).
62. M. Pourboix, ed., *Atlas d'equilibres Electrochimiques @ 25°C*, Gothier-Villars, Paris, 1963, p. 359.
63. A. B. Fasman, G. G. Kutyyukov, and D. V. Sokalski, *Zh. Neorg. Khim.*, **10**, 1338 (1965).
64. V. I. Kravstov and M. I. Zelenski, *Elektroklimiya*, **2**, 1138 (1966).
65. T. Ryhl, *Acta. Chem. Scand.*, **26**, 2961 (1972).
66. A. V. Babaeva and E. Y. Khananova, *Zh. Neorg. Khim.*, **10**, 2579 (1965).
67. H. K. Birnbaum, "Hydrogen Embrittlement," *Encyclopedia of Material Science and Engineering*, M. B. Bever, ed., Pergamon Press, New York, 1986.

68. J. W. Dini, *Electrodeposits, The Material Science of Coatings and Substrates*, Noyes Publication 367, Noyes Publ., Park Ridge, NJ, 1993.
69. S. Nakahara and Y. Okinaka, "Microstructure and Ductility of Electroless Copper Deposits," *Acta Metall.*, **31**, 713 (1983).
70. L. J. Durney, "Hydrogen Embrittlement: Baking Prevents Breaking," *Prod. Finish.*, **49**, 90 (1995).
71. H. Geduld, *Zinc Plating*, ASM International, 203 (1988).
72. M. Fleischmann, S. Pons, and M. Hawkins, *J. Electroanal. Chem.*, **261**, 301 (1989).
73. S. E. Jones et al., *Nature*, **338**, 737 (1989).
74. J. M. Rosamilia, J. A. Abys, and B. Miller, *Electrochimica Acta*, **36** (7), 1203–1208 (1991).
75. J. R. Lacher, *Proc. Roy. Soc.*, **A161**, 525 (1937).
76. L. J. Gillespie and L. S. Galstoun, *J. Am. Chem. Soc.*, **58**, 2565 (1935).
77. L. J. Gillespie and W. R. Downs, *J. Am. Chem. Soc.*, 2496 (1939).
78. J. W. Simons and T. B. Flanagan, *J. Phys. Chem.*, **69**, 3581, 3773 (1965).
79. A. N. Frumkin and N. Aladjalowa, *Acta Physiochem., USSR*, **19**, 1 (1944).
80. T. P. Hoan and D. F. Arrowsmith, *Electroplating Met. Finish.*, **10**, 5, 1 (1957).
81. M. Baldouf and D. M. Kolb, *Electrochim Acta*, **38** (15), 2145–2153 (1994).
82. V. D. Jovic et al., *J. Serb. Chem. Soc.*, **57** (12), 951–962 (1993).
83. M. Enyo and P. C. Biswas, *J. Electroanal. Chem.*, **335** (1–2), 309–319 (1993).
84. L. J. Vracar and M. Stojanovic, *J. Serb. Chem. Soc.*, **61** (7), 567–575 (1996).
85. B. I. Podlovchenko, E. A. Kolyadko, and S. Lu, *J. Electroanal. Chem.*, **399** (1–2), 21–27 (1995).
86. I.-Y. Wei and J. Brewer, *Proc Annual Technical Conf. on American Electroplaters and Surface Finishers Society*, pp. 321–332 (1995).
87. A. V. Smolin, Y. M. Maksimov, and B. F. Podlovchenko, *Russ. J. Electrochem*, **31** (6), 520–525 (1995).
88. J. O. M. Bockris and A. K. N. Reddy, *Modern Electrochemistry*, 10.2.17, Plenum Press, New York, 1970, pp. 1223, 1227, 1053.
89. P. H. Rieger, *Electrochemistry 2nd ed.*, Chapman and Hall, London, 1984, pp. 183–194.
90. A. J. Bard and L. R. Faulkner, *Electrochemical Methods*, Wiley, 1980.
91. R. J. Morrisey, *Metal Finishing Guidebook-Directory 97*, p. 288.
92. USSR Patent 519,497 (June 30, 1976).
93. F. A. Lowenheim, *Electroplating*, McGraw-Hill, New York, 1978, p. 299.
94. J. M. Stevens, *Trans. Inst. Met. Fin.*, **46**, 26 (1968).
95. H. L. Grube, *Metalloberflache*, **B5** (4), 61 (1953).
96. W. Keitel and H. Zschiegner, *Trans. Electrochem. Soc.*, **59**, 273 (1931).
97. T. F. Davis, U.S. Patent 4,092,225 (May 30, 1978).
98. H. J. Schuster and K. D. Heppner, U.S. Patent 4,144,141 (Mar. 3, 1979).
99. J. A. Abys, U.S. Patent 4,486,274 (Dec. 4, 1984).
100. R. LePenven, W. Levanson, and W. Pletcher, *J. Appl. Electrochem.*, **22** (51), 421–424 (1992).
101. L. A. Heathcote, *Platinum Met. Rev.*, **9** (3), 80–82 (1965).
102. J. A. Abys, *The Electrodeposition of Pure Palladium with a Palladium Hydroxide Replenishment System*, American Electroplaters Society Symposium AESF Week, 1991.
103. J. A. Abys and Y. Okinaka, U.S. Patent 4,468,296 (Aug. 28, 1984).

104. J. A. Abys and Y. Okinaka, U.S. Patent 5,135,622 (Aug. 4, 1992).
105. S. Nakhara et al., *J. Testing Eval.*, **5**, 178 (1977).
106. J. J. Caricchio and E. R. York, U.S. Patent 4,076,599 (Feb. 28, 1978).
107. J. M. Deuber, U.S. Patent 4,098,656 (July 4, 1978).
108. J. A. Abys, V. Chinchankar, V. T. Eckert, I. Kadija, E. J. Kudrak, J. J. Maisano, and H. K. Straschil; U.S. Patent 4,911,799 (Mar. 27, 1990).
109. U.S. Patent 2,457,021 (1948).
110. F. Simon and W. Zilske, *Plating Surf. Finish.*, **69**, 86 (1982).
111. J. A. Abys and H. K. Straschil, U.S. Patent 5,178,745 (Jan. 12, 1993).
112. H. K. Straschil, J. A. Abys, E. J. Kudrak, J. J. Maisano, and S. Nakahara, "New Palladium Strike Improves Adhesion and Porosity," *Met. Finish.*, **90**, 42–47 (1992).
113. P. K. Gallagher, *Thermochimica Acta*, **41**, 323–327 (1980).
114. J. A. Abys, Unpublished data.
115. H. P. Hedrich and Ch. J. Raub, *Metalloberflache*, **33** (8), 308–315 (1979).
116. F. I. Nobel et al., *Plating Surf. Finish.*, **73** (6), 88–93 (1986).
117. U. Cohen and R. Sard, U.S. Patent 4,269,671.
118. U. Cohen et al., *J. Electrochem. Soc.*, **130**, 1987 (1983).
119. D. Walz and Ch. J. Raub, *Metalloberflache*, **40**, 162–166, 199–203 (1986).
120. T. Hirano and M. Tsukemato, U.S. Patent 5,342,504 (Aug. 30, 1994).
121. K. S. Berge, U.K. Patent Application 2,115,440A (Sept. 22, 1982).
122. K. S. Berge, U.S. Patent 4,416,740 (Nov. 22, 1983).
123. K. S. Berge, U.S. Patent 4,430,172 (Feb. 7, 1984).
124. P. Wilkinson, U.K. Patent Application 2,115,440A (Sept. 7, 1983).
125. D. Walz and Ch. J. Raub, *Metalloberflache*, **40** (6), 239–241 (1986).
126. I. Boguslavsky, J. A. Abys, H. K. Straschil, H. Tsuruta, J. J. Maisano, and V. T. Eckert; *Proc. Annual Conference American Electroplaters and Surface Finishers Society*, June 1997, Detroit, MI.
127. M. Kume, *76th Technical Conf. of the Metal Finishing Society of Japan*, Paper 30C-7, 1987.
128. G. T. Rado and H. Suhl, *Magnetism*, vol. 5, Academic Press, New York, 1973.
129. *Platinum Met. Rev.*, **37** (2), 120 (1993).
130. P. Juzkis, M. U. Kittel, and Ch. J. Raub, Private communications. Research Institute for Precious Metals, 7070 Schwabisch Gmund.
131. P. Stevens, U.S. Patent 4,048,023 (Sept. 13, 1977).
132. S. N. Vinogradow and P. Yu, *Perelygin, Zashch. Met.*, **16** (4), 507 (1980).
133. A. A. Tikhonov, P. M. Vyacheslavov, G. K. Burkat, and V. P. Bursin, *Modern Methods of Deposition of Electrolytic and Chemical Coatings* (in Russian), Izd. Mosk. Doma Nauchn.-Tekhn. Propagandy Im. F. E. Dzerzhinskogo (1979), p. 79.
134. E. M. Savitskii et al., *Metallic Single Crystals* (in Russian), Izd. Nauka, Moscow (1976), p. 127.
135. E. M. Savitskii et al., *Alloys of Noble Metals* (in Russian), Izd. Nauka, Moscow (1977), p. 224.
136. L. M. Nomerovannaya et al., *Dokl. Akad. Nauk SSSR*, **246** (3), 585 (1979).
137. A. M. Meretskii et al., *Metally*, **5**, 193 (1976).
138. N. F. Reshetnikova and K. S. Pedan, *Zh. Prikladoni Khimii*, **55** (9), 1966–1999 (1982).

139. F. M. Dzhandubaeva, P. M. Vyacheslavov, and G. K. Burkat, *Zashchita Metallov*, **18** (3), 427–430 (1982).
140. R. Long and K. F. Bradform, *Proc. 8th International Conf. on Electrical Contact Phenomena*, Tokyo, Japan (1970).
141. W. A. Crossland and E. Knight, *Proc. Seminar on Electrical Contacts*, p. 247, Chicago (1973).
142. U. Cohen and R. Sard, U.S. Patent 4,269,671.
143. U. Cohen, F. Kohn, and R. Sard, *J. Electrochem. Soc.*, **130**, 1987 (1983).
144. U. Cohen, K. R. Walton, and R. Sard, *J. Electrochem. Soc.*, **131** (11), 2489 (1988).
145. F. I. Nobel, *IEEE Trans. Components, Hybrids Manuf. Technol.*, **8** (1), (1985).
146. F. I. Nobel et al., *Plating Surf. Finish.*, **73**, 88 (1986).
147. P. L. Cavallotti et al., *Proc. Electrochem. Soc.*, **6**, 261–276 (1994).
148. M. Kume and U. Tadeo, *Hyomen Gijutsu*, **46** (17), 663–666 (1995).
149. S. N. Vinogradov and S. V. Plokohov, *Gal'venotekh Obrab. Poverkhn*, **1** (1–2), 32–34 (1992).
150. M. Noro et al., *Kagaku to Kogoyo*, **69** (3), 89–94 (1995).
151. J. A. Abys and H. K. Straschil, U.S. Patent 5,024,733 (June 18, 1991).
152. L. Wang, *Cailiao Baohu*, **27** (5), 24–25 (1994).
153. Soviet Union Patent 535,378 (Nov. 26, 1976).
154. Platinum 1998, Published by Johnson Matthey.
155. U.S. Geological Survey, Mineral Commodity Summaries, Jan. 1999.
156. M. Lovatt, *Bridge News*, Feb. 16, 2000.
157. C. Fan et al., *Proc. Annual Conference American Electroplaters and Surface Finishers Society*, June 2000, Chicago, IL.
158. E. J. Kudrak et al., *Proc. 32nd IICIT Connector and Interconnection Technology Symposium*, Anaheim, CA, Sept. 1999.
159. J. A. Abys et al., *Trans. of the Institute of Metal Finishing*, July 1999.
160. J. A. Abys et al., *Connector Specifier*, Feb. 1999, p. 12.
161. J. Abys, *Fleck Materials Conf.*, Indian Wells, CA, Sept. 1998.
162. I. Boguslavsky et al., *Proc. Annual Conference American Electroplaters and Surface Finishers Society*, Session L, June 1998, Minneapolis, MN.
163. J. A. Abys et al., *Plating and Surface Finishing*, **86**, 108, Jan. 1999.
164. M. Clarke, Chap. 8, "*Properties of Electrodeposits*", R. Sard, H. Leidheiser, and F. Ogburn Eds, The Electrochemical Society, Inc., Princeton, NJ (1975), pp. 122.
165. Western Electric Manufacturing Standard 17000, Sect. 1275 (1982), based on ASTM B-799.
166. G. C. Munie et al., *Journal of SMT*, **15**, Jan. 2000.
167. P. Wilkinson and B.O'Hara, *IPC PC World Convention*, May 1984.
168. L. J. Mayer et al., *Proc. IPC – Fall Session*, Sept. 1986.
169. I. Kadija et al., *Proc. Annual Conference American Electroplaters and Surface Finishers Society*, June 1993, LOCATION, ??, and references therein.
170. I. Kadija et al., *Plating and Surface Finishing*, **82** (2), 56, Feb. 1995.
171. T. Mc Guiggan and E. E. Benedetto, *Circuits Assembly*, Nov. 1997.
172. C. Fan et al., *Proc. Symposium on Environmental Aspects of Electrochemical Technology: Applications in Electronics*, M. Datta, J. M. Fenton, and E. W. Brooman, Eds., PV 96-21,

p. 188, The Electrochemical Society Proceedings Series, Pennington, NJ, 1997, and references therein.

173. European Patent Application 87305080.1 by Texas Instruments in 1987.

174. European Patent Application 89302939.7 by Texas Instruments in 1989.

175. D. C. Abbott et al., *IEEE Transactions on Components, Hybrids and Manufacturing Technology*, **14** (3), 567, Sept. 1991.

176. C. A. Dullaghan et al., *Proc. Annual Conference American Electroplaters and Surface Finishers Society*, Session I, June 2000, Chicago, IL.

13 Nickel Alloys, Cobalt, and Cobalt Alloys

GEORGE A. DI BARI

INTRODUCTION

Over the past 25 years, new alloy electroplating processes, and applications have been developed and commercialized, often involving the codeposition of nickel either as the major or minor constituent of a binary alloy coating. Despite the fact that nickel and nickel alloy electroplating solutions can often be transformed into cobalt ones by substituting the equivalent cobalt salt for the nickel, electrodeposited cobalt and cobalt alloys are not as commercially important as their nickel counterparts, probably because most end-uses can be satisfied by nickel or nickel alloys at considerably less cost. A cobalt or cobalt alloy electrodeposit must thus confer a unique property to justify its use. For example, the applications of electroplated cobalt and cobalt alloys in the fabrication of recording discs and other devices for computers take advantage of the metal's magnetic properties. Electroplated zinc-cobalt alloys applied to autobody steel coil as a base for paint is justified by the significantly improved corrosion performance conferred by the small amount of codeposited cobalt.

Electrodeposited nickel alloys are reviewed in the first part of this chapter; cobalt and cobalt alloys, in the second. The emphasis throughout is on binary alloy electrodeposition; ternary alloys are treated perfunctorily because they have not been applied to any great extent.

1 NICKEL ALLOYS

Table 1 lists the electrodeposited binary alloys of nickel mentioned in Brenner's [1] treatise on alloy electrodeposition, briefly summarizes the results, and provides some references to work done prior to 1960 [2–20]. The deposition of nickel alloys is discussed by Dennis and Such who provide references through 1991 [21]. The book by Safranek includes a review of electrodeposited nickel alloys and summarizes their physical and mechanical properties [22].

Part 1, *Nickel Alloys*, was written especially for this edition. Part 2, *Cobalt and Cobalt Alloys*, is a revision of the chapter by F. R. Morrall and W. H. Safranek that appeared in the third edition.

Modern Electroplating, Fourth Edition, Edited by Mordechay Schlesinger and Milan Paunovic.
ISBN 0-471-16824-6

TABLE 1 Summary of Some of the Early Work on Binary Electroplated Nickel Alloys [1]

Alloying Element	Comments and References
Aluminum	Codeposition of 4–5% from aqueous solution reported but never confirmed [4].
Arsenic	Codeposition of 1–20% [5]. No known applications.
Cadmium	Small amounts added as brightener in early bright nickel plating solutions. Large amounts of cadmium can be codeposited, but deposit properties are poor [6]. Jet engine parts electroplated with alternate layers of nickel and cadmium are heated to produce diffused nickel cadmium alloy coatings [7].
Chromium	Addition of nickel salts to hexavalent chromium solutions results in the codeposition of tiny amounts of nickel. Alternatively, the addition of chromic acid or chromium sulfate salts to concentrated nickel (or cobalt) solutions results in little or no chromium being deposited [1]. Trivalent chromium solutions are more promising.
Cobalt	Alloys over the entire range can be prepared from simple salt solutions.
Copper	Codeposition occurs from cyanide, citrate, pyrophosphate and thiosulfate solutions. The incentive of the early work was to match the marine corrosion resistance of Monel metal. Recent developments suggest renewed interest.
Gallium	Largely unsuccessful. Traces of gallium codeposit in the presence of citrate at pH 3.5. Ammoniacal baths operated at very high current density yielded poor deposits containing gallium [8].
Germanium	10–60% germanium codeposited with nickel from an ammoniacal oxalate solution containing ammonium sulfate [9]. Work was done to recover germanium from dilute aqueous solutions, so coating properties were of no concern.
Gold	Nickel in small amounts is codeposited with gold from cyanide solutions to produce hardened, white alloys.
Iron	The first electroplated binary nickel alloy to be electroplated because of presence of iron in first nickel anodes. Iron codeposits with nickel from simple salt solutions. Decorative nickel iron alloy plating processes were commercialized in the 1970s and nondecorative processes with about 20% iron are exploited for their magnetic properties.
Lead	Up to 16% lead codeposited from nickel solutions containing citrate ion. Deposit properties deteriorate with increasing lead content [10].
Lithium	Lithium detected spectrochemically but too small to detect chemically [11].
Manganese	Small amounts, less than 0.5%, codeposit from simple salt solutions; an ammoniacal solution permits codeposition of higher amounts [12]. The early work aimed to develop a nickel-rich coating that corroded sacrificially vis-à-vis steel. Recent interest stems from the fact that manganese acts as a getter, preventing sulfur embrittlement of nickel at elevated temperature.
Molybdenum	Significant amounts can be codeposited with nickel from acid and alkaline tartrate solutions, from ammoniacal solutions containing citrates and tartrates, and from inorganic alkaline and pyrophosphate solutions [1]. Interest in these alloys stems from the need for coatings that retain their strength at elevated temperatures.
Palladium	Palladium alloys containing up to 35% nickel were electrodeposited from a simple solution of unspecified composition [13]. Alloys

(*Continued*)

TABLE 1 (***Continued***)

Alloying Element	Comments and References
	containing 10–50% nickel have been produced. Nickel-palladium alloy electrodeposits have become important substitutes for gold in electronics applications.
Phosphorus	The electrodeposition of alloys of nickel and phosphorus has been studied [14], the properties of the deposits being similar to those prepared by autocatalytic deposition. Renewed interest arises from the amorphous nature of deposits containing high levels of phosphorus and the need for nonmagnetic materials for electroforming.
Platinum	Patents issued in 1908 claim deposition of platinum-nickel coatings for high-temperature applications [15].
Rhenium	Deposits with 6–78% rhenium can be prepared from Watts-type solutions with and without citric acid by addition of potassium perrhenate [16].
Rhodium	Patents on the deposition of nickel alloys with at least 25% rhodium exist [17].
Selenium	Small amounts of selenium can be codeposited with nickel from Watts nickel solutions.
Silver	Small amounts of nickel can apparently be codeposited with silver from cyanide and noncyanide solutions; early work was aimed at increasing hardness and improving the tarnish resistance of silver [18, 19].
Sulfur	Bright nickel deposits containing up to 0.15% sulfur may be considered alloys. Alloys containing greater amounts of sulfur can be electrodeposited from solutions containing sodium thiosulfate [20].
Tellurium	Small amounts of tellurium can be codeposited with nickel from Watts nickel solutions.
Thallium	Mentioned in the early literature [21].
Tin	The fluoride-chloride process developed by the Tin Research Institute (England) is used commercially and yields a deposit that is an intermetallic compound with 35% nickel, 65% tin; this unique coating is hard, brittle, tarnish resistant, and has excellent corrosion resistance. Sometimes used as a substitute for decorative chromium.
Tungsten	Can be codeposited with nickel from solutions similar to those formulated for the electrodeposition of nickel molybdenum alloys [1].
Zinc	Alloys with 10–20% nickel have increased resistance to corrosion and became commercially important in the early 1980s. The earliest reference to electrodeposited zinc-nickel alloy dates to 1905 [22].

The interest in nickel alloy electroplating is reflected in the patent, as well as the technical, literature [23]. Fifty percent of the nickel electroplating patents issued between 1980 and 1991 deal with alloys, notably electrodeposition of nickel-zinc, nickel-iron, nickel-cobalt, and amorphous nickel-phosphorus alloys. Palladium-nickel alloy electroplating processes are a relatively new development, and the coatings are now being used in the electronics industry as substitutes for gold and for pure palladium [24]. A review of alloy plating developments in Japan reveals keen interest in nickel alloy electroplating including the electrodeposition of nickel with boron, molybdenum, sulfur, phosphorus, aluminum, antimony, or cobalt plus boron [25]. Multilayer coatings produced by cyclic modulation of the deposition potential during electroplating are a *new class* of material, sometimes displaying extraordinary properties as a result of the artificially layered alloy structure [26].

The electroplated binary nickel alloys that are commercially significant or in which there is renewed interest are discussed first. These include nickel alloyed with cobalt, copper, gold, iron, manganese, molybdenum, palladium, phosphorus, sulfur, tin, tungsten, or zinc. Those that have not found wide industrial use or that are relatively new are described under other nickel alloys.

1.1 Nickel-Cobalt

The electrodeposition of nickel-cobalt alloys occurs readily from simple acid solutions (solutions that do not contain complexing or chelating agents) made from sulfates, chlorides, or sulfamates. The addition of cobalt sulfate to a typical Watts bath, for example, will yield satisfactory alloy electrodeposits. Binary alloys over the complete composition range can be prepared. The cobalt ion concentration can be controlled by apportioning the current between separate nickel and cobalt anodes using two power sources or variable resistors. Malone described a technique for controlling the cobalt content with one power supply by switching the current back and forth between separate nickel and cobalt anode arrays [27]. By varying the time that each anode is dissolved in proportion to the desired composition, he was able to vary and control the cobalt content in solution and in the deposit over a broad range.

The first commercially significant application of nickel-cobalt alloy plating was in the production of decorative coatings. Although the process developed by Weisburg and Stoddard [28] is now of historical interest only, it is probably the only commercial bright nickel plating process that produces a mirrorlike deposit that is completely sulfur free. In the past, cobalt has been substituted for some of the nickel in decorative nickel electroplating solutions when nickel was in short supply, but nothing was gained economically or technically thereby, except for the conservation of nickel.

Endicott and Knapp studied the electrodeposition of nickel-cobalt alloys from conventional sulfamate solutions in considerable detail, and showed that the nickel : cobalt ratio was the dominant factor determining the composition of the deposit [29]. By maintaining the metal ion concentration in solution constant and varying the ratio of nickel to cobalt within the limits given in Table 2, deposits containing 5% to 98% cobalt were prepared. Increasing the current density decreased the cobalt content of the deposit at nickel:cobalt ratios of 5 to 33, but deposit composition was less sensitive to current density at low ratios (0.11–2.33) and at high (100).

Belt, Crossley, and Kendrick studied the influence of cobalt in a concentrated nickel sulfamate solution, Table 2, and established the relationship between current density and alloy deposit composition shown in Figure 1 [30]. The relationship between the macrohardness of the deposits electroplated at $5.4\,\mathrm{A\,dm^{-2}}$ and the cobalt in the deposit is shown by the dotted line in Figure 2. The maximum hardness occurs at about 35% cobalt. The solid line in Figure 2 gives the corresponding concentration of cobalt in solution. Figure 3 shows the effect of current density on the internal stress of deposits as influenced by various cobalt concentrations in solution. Internal stress increases as the current density, and cobalt ion concentration are increased. The variation of hardness with heat-treatment temperature is

TABLE 2 Representative Sulfamate Solutions for Electrodepositing Nickel-Cobalt Alloys

	Composition (g liter^{-1})		
Constituent	Conventional Sulfamate	Concentrated Sulfamate	80% Co–20% Ni (Barrett)
Metal (Ni plus Co) each added as sulfamates	78.8		
Nickel : cobalt ratio	0.11–100		1:1
Nickel sulfamate, $Ni(SO_3NH_2)_2 \cdot 4H_2O$		500–600	225
Cobalt sulfamate, $Co(SO_3NH_2)_2$*		0.2–7.5	225
Nickel chloride, $NiCl_2 \cdot 6H_2O$		5–15	
Nickel bromide, $NiBr_2$	14		
Boric acid	30–45	30–45	30
Magnesium chloride			15
Wetting agent (sodium lauryl sulfate)	To 30 dynes cm^{-1}		To 30 dynes cm^{-1}
		Operating conditions	
Temperature (°C)	40–60°C	60–70	25°C
Agitation	Air or mechanical	Air or mechanical	Mechanical
Cathode current density (A dm^{-2})	2–5	Up to 90	1.5–3
pH	4	3.5–4.5	2–4

shown in Figure 4 for pure nickel and nickel alloys with 20% to 35% cobalt. Although the alloy deposits begin to soften when heated above 200°C, the hardness of the alloy deposits is always greater than that of similarly heated pure electrodeposited nickel. The improvement in high temperature properties that occurs at 35% cobalt has been applied in electroforming aerospace components.

The electrodeposition of a third element with nickel and cobalt, for example, tungsten, molybdenum, or manganese would no doubt further improve high temperature properties. Electrodeposited nickel/cobalt/iron alloys should display martensitic aging and hardening reactions when treated with heat. Cobalt-nickel alloys with 20% nickel, Table 2, prepared from sulfamate solutions developed by Barrett [31], have useful magnetic properties.

1.2 Nickel-Copper and Composition-Modulated Alloy Coatings

The early incentive for studying the electrodeposition of nickel-copper alloys was to develop a coating with increased resistance to marine corrosion. Although there has been considerable work in the past, no practical process for the electrodeposition of nickel-copper alloys exists [32]. A solution consisting of nickel and copper sulfate, and sodium citrate operated at pH 3 to 4 produced excellent deposits that were copper colored at very low current density and nickel-colored at high [33].

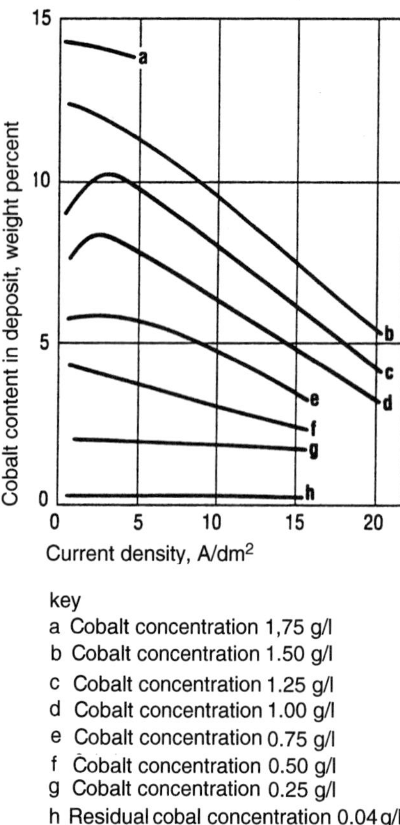

Figure 1 Relationship between current density and alloy deposit composition for alloy deposits produced from a concentrated nickel sulfamate solution with various cobalt concentrations in solution. From [30].

A renewed interest in nickel-copper alloy electrodeposition stems from studies of composition-modulated alloy electrodeposition. Compositionally modulated alloy coatings can be prepared by two methods. One method involves exposing the cathode to two different electrolytes either by rapidly draining and replacing the solution in one cell, or by transferring the cathode back and forth between two cells. This is difficult to do successfully because of the possibility of cross-contamination, poor adhesion between lamellae, and other practical problems. In the second method, the alloying elements are present in the same solution and are sufficiently different electrochemically that cyclic pulsing of the current or of the potential results in the deposition of one element at low current density and the other at high. The nickel-copper system is one of the first to be studied in this connection. Copper, being more noble, deposits preferentially at low current density; whereas nickel deposits at high current density. Compositionally modulated nickel-copper alloys have greater tensile strength than pure nickel, are likely to display exceptional ductility, and may have enhanced resistance to corrosion [34].

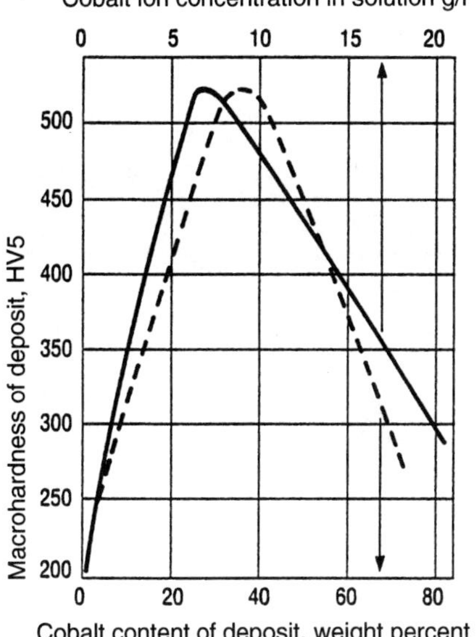

Figure 2 Relationship between macrohardness of deposits electroplated at 5.5 A dm^{-2} from a concentrated nickel sulfamate solution, and the relationship between the cobalt in the deposit and in the solution. From [30].

1.3 Gold-Nickel

Nickel is a minor but effective constituent of the electrodeposited gold coatings specified in the electronics industry. Nickel-hardened electroplated gold is used for coating contacts and connectors where as little as 0.1% to 0.2% nickel increases the hardness and improves the appearance of the deposits [35]. The most important solutions are the acid cyanide types that deposit relatively thick, hard, bright coatings. These solutions contain citrates or other complexing agents to control the amount of nickel that codeposits. The nickel-hardened golds are often electrodeposited over electroplated nickel coatings that improve the performance of electronic components by reducing porosity of the gold coating and by preventing undesirable elements from diffusing from the substrate to the surface. The alkaline cyanide baths used for depositing relatively thin, decorative gold coatings contain nickel to control the color of the coating. The nickel content is varied from 5% to 15% to produce pale yellow to white gold coatings [36].

1.4 Nickel-Iron

Brenner suggests that the early work on the electrodeposition of nickel-iron alloys involved efforts to eliminate various defects in nickel deposits—high internal stress, exfoliation, staining, and microcracking. These defects were attributed to the presence of iron that most likely came from the impure nickel anodes and salts available in the late 1800s. Today electrodeposited nickel-iron alloy processes are used for functional and decorative purposes.

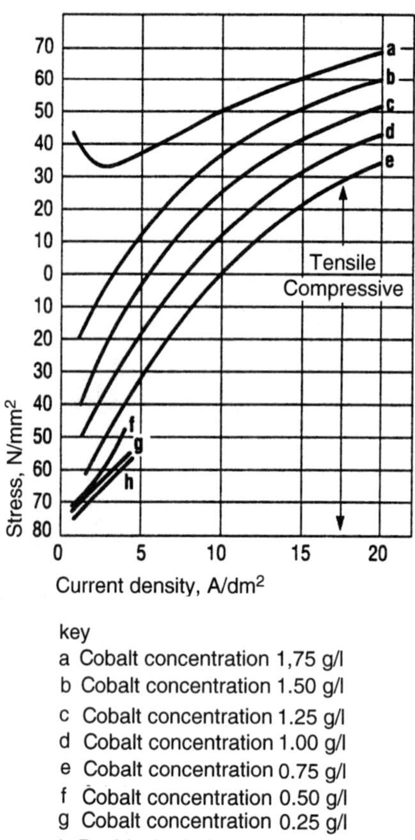

Figure 3 Relationship between the current density and deposit stress for various cobalt concentrations in the concentrated sulfamate solution. From [30].

In functional applications the electrolytes may be simple acid solutions containing sulfates, chlorides or sulfamates, along with boric acid, saccharin and wetting agents. Table 3 gives representative solution compositions for electrodepositing *soft* magnetic coatings with about 19.5% iron. The solutions are operated at pH 2 to 3 which in the absence of complexing agents prevents precipitation of ferric hydroxide. Solutions made from nickel and ferrous sulfates (or from the corresponding chlorides) operate at low current densities, $2\,A\,dm^{-2}$, whereas the sulfamate solution can be operated at a current density of $11\,amps\,dm^{-2}$. The nickel-iron alloys produced by means of these processes have low coercivity (less than 25 oersteds), high permeability, very rapid response, and allow very short switching times of the order of a few nanoseconds. They have zero magnetostriction and are applied in high-speed memory storage devices in computers and other electronic equipment, for example, in the electrofabrication of magnetic recording heads [37]. The importance of third elements, for example, phosphorus, sulfur, and others, and the need to employ thermal treatments to stabilize and optimize magnetic properties have been stressed by Romankiw and Thompson [38].

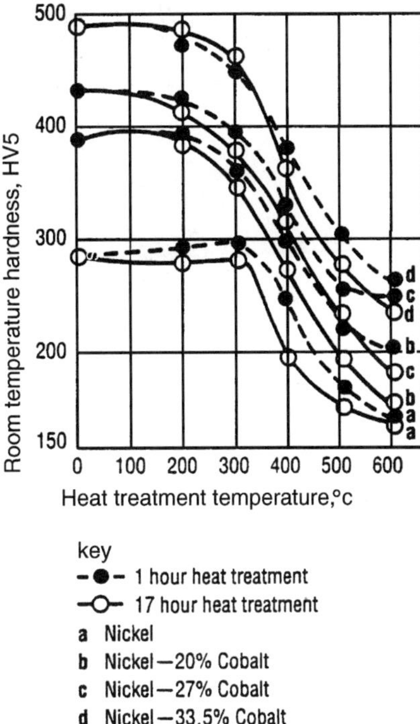

Figure 4 Variation of room-temperature hardness with heat-treatment temperature for electrodeposited pure nickel and nickel-cobalt alloys containing 20%, 27%, and 33.5% cobalt. From [30].

The first process for electrodepositing decorative nickel-iron alloy coatings of good quality is the one described by DuRose and Pine [39]. The addition of an organic sulfonate to control stress, and operation at low pH to control the incorporation of basic matter made it possible to produce sound deposits. Deposits containing 30% to 40% iron were hard, tough, and ductile. Deposits containing 10% to 30% iron prevented the rusting of steel better than an equivalent thickness of nickel.

The proprietary, decorative nickel-iron alloy electroplating processes introduced in the mid-to late 1970s contain carboxylic complexants or reducing agents like ascorbic acid to prevent precipitation of iron and to control the composition of the deposits. In most other respects, these processes are similar to conventional bright nickel electroplating ones that contain organic addition agents to control internal stress, brightness, and leveling of the deposits. The corrosion performance of chromium-electroplated, decorative nickel, and nickel-iron alloy electrodeposits with 15% to 33% iron was studied in marine and industrial environments and by means of accelerated tests [40]. Because of the rapid staining that occurs in marine and industrial environments, nickel-iron alloy electrodeposits are not suitable for decorative applications in moderate and severe corrosion service but may be suitable for mild corrosion service. The cost savings associated with substituting iron for nickel proved controversial. Although nickel was conserved, the added cost of addition agents and of process and pollution control offset some of

TABLE 3 Typical Solutions for Electrodepositing *Soft* Magnetic Coatings (80% Ni–20% Fe)

	Composition (g liter^{-1})		
Constituent	Sulfate	Chloride	Sulfamate
Nickel sulfate, $NiSO_4 \cdot 7H_2O$	213		
Nickel chloride, $NiCl_2 \cdot 6H_2O$		200	
Nickel sulfamate, $Ni(SO_3NH_2)_2$			200
Ferrous sulfate, $FeSO_4 \cdot 7H_2O$	3.5		
Ferrous chloride, $FeCl_2 \cdot 4H_2O$		4	
Ferrous sulfamate, $Fe(SO_3NH_2)_2$			5
Boric acid	25	25	25
Saccharin	1	1	1
Sodium lauryl sulfate	0.4	0.4	0.4
		Operating conditions	
pH	3	3	2
Temperature (°C)	30	30	85
Current density (A dm^{-2})	2	2	11

Source: Jacobs [2].

the savings. Today these processes are used to a limited extent to electroplate components for service in mildly corrosive environments, for example, tubular furniture.

1.5 Nickel-Manganese

Manganese is electrochemically more active than nickel and early efforts to codeposit manganese were aimed at developing a nickel-rich alloy deposit that would corrode sacrificially with respect to steel. It is difficult to deposit large amounts of manganese with nickel from acid solutions, but small amounts of manganese, less than 0.5%, codeposit with nickel when manganese sulfate is added to acid nickel sulfate or sulfamate solutions. Ammoniacal solutions yield deposits with much higher levels of manganese, but the deposits may be highly stressed and rough [41]. A practical process for producing a sacrificial nickel-manganese alloy coating for steel was never developed.

The codeposition of manganese has more recently been evaluated to improve the high-temperature ductility of electrodeposited nickel. The loss in ductility on heating electrodeposited nickel was attributed to the effect of tiny amounts of sulfur that migrate to grain boundaries to form nickel sulfide that embrittles the nickel. Because manganese is added to molten nickel during the production of wrought nickel and is a well-known sulfur *getter*, codeposition of manganese should improve the high-temperature ductility of electrodeposited nickel. Dini, Wearmouth, Malone and others [42–44] confirmed that and applied the idea in producing electroformed articles to be exposed to elevated temperatures in service. The effectiveness of the manganese is related to the amount codeposited and to the temperature to which the nickel is exposed.

The cracking of electrodeposited nickel coatings on steel adjacent to the heat affected zone during arc welding was also attributed to sulfur embrittlement at high temperature. Experiments done in the 1960s with electrodeposited nickel containing about 0.4% manganese showed that manganese did not prevent cracking during the welding of thick nickel coatings on steel [45]. Either the cracking accompanying arc welding was not due to sulfur embrittlement or arc welding was too severe for the manganese to be effective. Subsequent development of alternate welding techniques eliminated the problem.

1.6 Nickel-Molybdenum and Nickel-Tungsten

Some typical solutions for electrodepositing nickel with either molybdenum or tungsten [1] are given in Table 4. Ternary alloys (nickel-cobalt-molybdenum and nickel-cobalt tungsten) are also possible. The interest in these alloys stems from their increased hardness and improved high-temperature oxidation resistance. The alloys can be deposited from relatively dilute, ammoniacal nickel sulfate solutions operated at pH 6 to 10 containing either sodium tungstate or sodium molybdate along with sodium citrate or tartrate as complexing agents. The properties of deposits containing 25% to 40% tungsten have been summarized by Safranek [3]. Nickel-tungsten and cobalt-tungsten alloys have recently been considered as possible substitutes for decorative chromium. Recent work on the electrodeposition of nickel-molybdenum alloys established conditions for depositing composition-modulated multilayer alloy deposits and concluded that deposits containing greater than 42% molybdenum are theoretically possible [46].

1.7 Palladium-Nickel

The substitution of palladium (or gold-electroplated palladium) for pure gold as a contact material in electronics was a response to the rapid escalation of the price of gold that began in the early seventies and peaked in 1980. The development of palladium-nickel alloy electroplating processes occurred shortly thereafter. Palladium-nickel alloys are not only less costly than pure palladium, but the addition of nickel increases the hardness, lowers the tendency toward hydrogen embrittlement, and improves the ductility and the brightness of the deposits [47]. It has also been reported that palladium-nickel alloy deposits have better wear resistance than pure gold or palladium, and are intrinsically less porous. Many of the electrolytes for electrodepositing nickel-palladium are alkaline ammoniacal ones [48, 49], but acidic solutions are available and the one described by Straschil and coworkers appears capable of depositing alloys with 10% to 50% nickel [50].

1.8 Nickel-Phosphorus

The electrodeposition of nickel-phosphorus alloys was described shortly after the discovery of autocatalytic (electroless) nickel processes for producing

TABLE 4 Nickel-Molybdenum and Nickel-Tungsten Alloy Electroplating Formulations

	Composition (g liter^{-1})			
Constituent	Ammoniacal Ni–Mo (18% Mo)	Acid Ni–Mo (3–15% Mo)	Alkaline Ni–W (10% W)	Alkaline Ni–W (35% W)
Nickel sulfate, $NiSO_4 \cdot 7H_2O$	86	140	120	20
Sodium molybdate, $Na_2MoO_4 \cdot 2H_2O$	48	13		
Sodium citrate, $Na_3C_6H_5O_7$	105			
Citric acid,				66
Sodium tungstate, $Na_2WO_4 \cdot 2H_2O$			90	50
Sodium potassium tartrate (Rochelle salt), $KNaC_4H_4O_6 \cdot 4H_2O$		100	400	
Ammonium chloride, NH_4Cl			50	
		Operating conditions		
pH	10.5 adjusted with ammonia	0.2–1.4 adjusted with sulfuric acid	9 adjusted with ammonia	8 adjusted with ammonia
Temperature (°C)	25	90	95	70
Current density (A dm^{-2})	10	3–10	2	7–15
Cathode efficiency (%)	70	15–45	93	45

nickel-phosphorus alloys with 2% to 12% phosphorus [51]. Electrolytic processes are capable of depositing nickel-phosphorus alloys with higher phosphorus levels and at higher rates than those attainable by autocatalytic deposition. Electrodeposited nickel-phosphorus coatings respond to heat treatment in much the same way as do autocatalytic nickel-phosphorus alloys, but unlike the latter, the electrodeposited alloys show increasing hardness with increasing phosphorus content. Thermally treated alloys are resistant to abrasion and have excellent wear resistance. Alloy deposits containing greater than 10% phosphorus have better corrosion resistance than those with 3% phosphorus.

Deposits containing 20% to 30% phosphorus are amorphous and that characteristic has attracted attention because of the improved corrosion performance that amorphous materials generally exhibit. Composition modulated nickel/nickel-phosphorus alloy coatings have been mentioned in the literature and may have interesting magnetic properties [52]. Electrodeposited nickel-phosphorus alloys have been applied in the electrofabrication of bellows [53]. The nonmagnetic nature of alloys containing greater than 10% to 11% phosphorus is a useful property in certain electroforming applications.

1.9 Nickel-Sulfur

Electroplated bright nickel deposits prepared from solutions formulated with organic additives usually contain 0.05% to 0.15% sulfur. The codeposited sulfur plays an important role in attaining the mirror-bright appearance of decorative nickel coatings. The codeposited sulfur increases the hardness and strength, and reduces the ductility and internal stress of the coatings. In addition to the effects on the physical and mechanical properties, sulfur increases the electrochemical reactivity of the coatings. Attempts to develop a sulfur-free bright nickel electroplating process have been made over the years because a nickel deposit without sulfur should have increased resistance to corrosion, but as far as is known, a nickel process that produces a fully bright, highly leveled electrodeposit that is completely devoid of sulfur is not available commercially. Although alloys containing 20% or more sulfur can be produced by electrodeposition from thiosulfate solution [54], the resulting nonmetallic sulfide deposits apparently have no known uses.

1.10 Tin-Nickel

The chloride-fluoride process for electrodepositing tin-nickel alloys produces a coating that is an intermetallic compound containing 35% nickel–65% tin. Although electrolytes that have been investigated include cyanide, pyrophosphate, fluoborate and acetate, the traditional solution given in Table 5 containing nickel chloride, stannous fluoride, and ammonium bifluoride is the one most frequently used. The alloy deposit is semi-lustrous, hard, and has a rose-pink color; it is resistant to corrosion and tarnishing. The excellent resistance to corrosion has been attributed to the passive, polystannate film that forms on outdoor exposure [55]. It has been used in engineering applications including automotive braking systems, heavy duty switch gears, and mixing valves [56]. The chloride-fluoride solution has excellent throwing power, and as a result tin-nickel alloy deposits have been used in electronics applications for electroplating through-hole printed circuit boards. Tin-nickel alloy

TABLE 5 Typical Tin-Nickel Alloy Electroplating Solution

	Composition ($g\,liter^{-1}$)	
Constituent	Nominal	Range
Stannous chloride, anhydrous $SnCl_2$	50	42–55
Nickel chloride, $NiCl_2 \cdot 6H_2O$	300	250–340
Ammonium bifluoride, NH_4HF_2	55	50–65
Ammonium hydroxide, NH_4OH	Adjust pH to 2.5	
	Operating conditions	
Temperature (°C)	65	65–70
pH	2.5	2–2.5
Current density	2	1–3

Note: Solution makeup: (1) Add two-third of the required water; (2) heat to about 65°C; (3) add and completely dissolve the nickel chloride; (4) add and completely dissolve the ammonium bifluoride; (5) add and dissolve the stannous chloride; (6) adjust the pH with either ammonium hydroxide or hydrofluoric acid; and (7) low current density electrolysis (*dummying*) to condition the solution.

deposits have also been applied as etch resists and as diffusion barriers between copper and tin/lead coatings on printed circuit boards [57]. Proprietary processes for electrodepositing tin-nickel alloys based on the original chloride-fluoride formulation exist. One that is operated at pH 4.2 to 4.8 contains organic additives that reportedly help produce bright, ductile coatings [56]. Electrodeposited tin-nickel alloys have been used as substitutes for decorative chromium in the rack electroplating of tubular furniture parts, but the alloy process is especially effective in barrel electroplating where its superior throwing power compared to decorative, electroplated chromium is an obvious advantage.

1.11 Zinc-Nickel

Electrodeposited zinc-nickel coatings with 8% to 20% nickel have better corrosion resistance than pure electrodeposited zinc. The addition of nickel makes zinc less active, and as a result the potential difference between steel and a zinc-nickel alloy coating is smaller than that between steel and pure zinc. The alloy coating is *less sacrificial* and galvanic corrosion thus proceeds at a lower rate compared to pure zinc. The effect on the rate of galvanic corrosion is proportional to the nickel content of the alloy coating, suggesting that corrosion can be minimized by optimizing the nickel content. The optimum nickel content is probably around 20% because the coating reportedly becomes cathodic to steel when the nickel content is 24% [58]. Zinc-nickel alloy electroplating processes thus provide an opportunity to match the electrochemical characteristics of the coating to those of the substrate by varying the nickel content of the alloy coating. Nickel-rich alloys can undoubtedly be prepared but are not available commercially.

Interest in zinc-nickel alloy electroplating is not new. A proprietary process for electrodepositing zinc-nickel coatings on steel strip has produced material for making auto-rocker panels since the 1960s. The literature between 1960 and 1983 is reviewed in a chapter by Hall [59], and that from 1983 to 1987 by Watson [60].

There are basically two kinds of zinc-nickel alloy plating processes available, acid and alkaline types [61]. The acid type consists of a mixture of zinc sulfate and nickel sulfate (or zinc chloride and nickel chloride), for example, 0.5 mol $liter^{-1}$ of each. The use of buffering and complexing agents like sodium acetate, sodium citrate, boric acid, or succinic acid is useful, and addition of sodium acetate, a complexing agent, may be especially beneficial [62]. The solution is operated at pH 4.5, 40°C, and current densities of 0.4 to 50 A dm^{-2}. Metallic nickel and zinc anodes can be connected to separate rectifiers or to variable resistors to control independently the current density on each anode. Nickel will deposit on soluble zinc anodes, but this does not appear to interfere with process control as long as the anodes are bagged. It is also possible to control these processes by using soluble zinc anodes and replacing the nickel by periodic additions of nickel carbonate. Operation with insoluble anodes is also practiced, but leads to evolution of chlorine and oxygen gases, and increases the voltage and hence the power costs.

The alkaline solution contains sodium hydroxide, zinc sulfate (or chloride), and is operated with soluble zinc and insoluble steel anodes to control zinc buildup in solution [61]. The solution contains 6 to 12 g $liter^{-1}$ of zinc and 0.7 to 1.5 g $liter^{-1}$ of nickel. Nickel is controlled by means of a liquid additive that contains a brightening agent, as well as a complexant to prevent precipitation of nickel hydroxide. Alkaline processes were introduced after the acid types and may offer advantages in rack and barrel plating of individual parts and components. Other solutions for electrodepositing zinc-nickel alloys are described by Hall [59] and by Dennis and Such [2].

The commercial processes developed in the early 1980s were introduced as part of the effort to eliminate auto-body rusting of new automobiles for at least 10 years. Automobile manufacturers evaluated electroplated zinc, zinc-nickel, zinc-iron, and zinc-cobalt coatings on steel for the fabrication of auto body panels, and the results of those studies confirmed that zinc-nickel and zinc-cobalt are superior to pure zinc in their corrosion performance. Nickel-iron also improves performance, but the benefits do not appear to be as long-lasting. Japanese automakers favored the use of zinc-nickel coatings and began to produce zinc-nickel coatings on steel coil continuously on a large scale. The preferred coating in the United States is pure electroplated zinc, but the zinc-cobalt alloy is also being used. Because of the significantly improved corrosion resistance of zinc-nickel alloys, coating thickness is usually less than that specified for pure zinc. The thinner coating improves the appearance after painting by eliminating *orange-peel* and other surface defects that sometimes arise during fabrication of automobile body panels by stamping.

Although the major application for zinc-nickel coatings is in the continuous electroplating of steel coil for auto-body panels, processes for rack-and-barrel plating have been introduced and are becoming popular for processing individual components and parts. Conversion coatings can be applied to zinc-nickel alloy plated parts to further improve corrosion resistance. Electroplated zinc-nickel alloy coatings have long been mentioned as possible replacements for cadmium coatings. It has been estimated that 2700 to 3200 metric tonnes of nickel are consumed annually in the production of zinc-nickel alloy electroplated-steel, and demand for the alloy coating is expected to increase [23].

1.12 Other Nickel Alloys

The electroplated alloys described in this section have been used to a limited extent, but they are of interest nonetheless. The heat treatment of electrodeposited composites as a means of producing homogeneous alloy coatings is, also, described.

Nickel-Chromium and Iron-Chromium-Nickel The desire to develop electroplated nickel-chromium and iron-nickel-chromium alloy coatings is based on the expectation that the electrodeposited coatings will display the same physical and mechanical properties as their metallurgical counterparts, as well as the excellent corrosion resistance of thermally prepared alloys like 80% nickel–20% chromium and the nickel stainless steels. The early work suggests that significant quantities of nickel and iron are unlikely to be codeposited with chromium from chromic acid-based solutions. Better results are obtained with trivalent chromium solutions. For example, nickel-chromium and nickel-chromium-iron alloy deposits with compositions comparable to thermally prepared commercial alloys have been prepared from trivalent chromium-dimethylformamide electrolytes; the importance of controlling the concentrations of divalent and hexavalent chromium present in those solutions to sustain deposition has been stressed and some interesting substrate effects on coating composition have been observed by el-Sharif, Watson, and Chisholm [63]. Lashmore, Weishauss, and Pratt described a pulse-plating process for electrodepositing smooth, dense nickel alloys containing up to 20% chromium from aqueous trivalent solutions containing formic acid, sodium citrate, sodium bromide and ammonium chloride [64]. Microstructural studies suggest that the electrodeposited alloys are multiphase in nature, having structures different from thermally prepared alloys of the same composition [63]. Progress has thus been made in controlling the composition of electrodeposited nickel-chromium and iron-chromium nickel alloys, but controlling the physical and mechanical properties, including the phases present, may require additional work. These alloys have not achieved commercial application to any significant extent, but a proprietary process for depositing thin, decorative nickel-chromium alloy coatings has been announced [65]. Brooman recently reviewed this subject in detail and provides a summary of solution compositions and plating parameters for nickel-chromium and other alloys [66].

Nickel-Germanium A germanium alloy plating solution prepared by dissolving 0.5 g $liter^{-1}$ of germanium oxide and about 18 g $liter^{-1}$ nickel hydroxide in an ammoniacal solution of ammonium chloride produced deposits that contained about 5% germanium. The nickel-germanium deposits had microhardness values of 450 to 800 kg mm^{-2} depending on cathode current density [67].

Nickel-Indium These alloys can be electroplated from sulfate/citrate/chloride solutions [68]. Deposits containing between 5% and 10% indium exhibited a layered structure. The coefficient of friction of the coatings decreased with increasing indium content, resulting in low contact resistance, prompting evaluation of nickel-indium coatings as substitutes for electrodeposited gold in some electronic applications. Nickel-indium alloy coatings have not been extensively used in the electronics industry, because the properties change with time due to formation of oxide films.

Nickel-Ruthenium The codeposition of ruthenium and nickel has been reported [69]. Deposits containing 15% to 20% ruthenium exhibited hardnesses of 800 to 1000 kg mm^{-2}.

Nickel-Silver Alloy electrodeposits containing 5% to 40% silver were prepared from pyrophosphate solutions. They displayed increased hardness and electrical resistivity [70].

Nickel-Thallium The electrodeposition of nickel-thallium alloys has been reported by Russian investigators [71].

Heat-Treated Electroplated Composites Composite coatings that contain solid particles dispersed throughout an electrodeposited matrix can be heat-treated to form homogeneous alloy coatings. For example, alloys of nickel with 22% chromium were prepared by codepositing chromium carbide particles with nickel followed by heat treatment at 1000°C for 24 hours in hydrogen to remove most of the carbon [72]. Hardness after heat treatment was 223 Vickers compared to 55 Vickers for similarly heat-treated pure nickel. By codepositing chromium carbide particles in an electrodeposited matrix of either nickel-cobalt or nickel-iron, ternary alloys of chromium-nickel-cobalt or chromium-nickel-iron were produced by heat treatment in hydrogen in a similar way. Although almost complete decarburization occurred with the chromium-nickel-cobalt alloy, the alloy of chromium-nickel-iron retained about 0.8% carbon and as result had a hardness of 332 Vickers. This approach is one way to produce alloy coatings that are difficult or impossible to make by electrodeposition from aqueous solutions; for example, nickel-aluminum coatings were made by codepositing alumina particles with nickel [73]. Nickel-aluminum alloys have also been produced by heat treatment of multilayer coatings consisting of electrodeposited nickel and aluminum electrodeposited from fused salts [74].

2 COBALT AND COBALT ALLOYS

Approximately 26,000 tonnes of electrolytic cobalt are produced annually, most of which is consumed in the production of nickel-cobalt-iron superalloys, hard materials (carbides, diamond tooling), magnets, tire adhesives, catalysts, colorants, chemicals, and batteries. The amount consumed in electroplating is not known and is probably small, perhaps 1% to 3% of the total annual production. New applications, however, such as cobalt-electroplated steel for batteries and electroplated cobalt-zinc alloys as a base for paint on auto-body steel promise to increase markets for cobalt for electroplating purposes.

The magnetic properties of electrodeposited cobalt and cobalt alloys are of great interest in electronic applications, especially in the computer industry where they are used in fabricating permanent data storage devices. Electroplated cobalt has been used in the manufacture of magnetic recording disks, although sputtered cobalt appears to be preferred. Because continued growth of the computer industry is a virtual certainty, increased demand for electroplated cobalt-base alloys is expected.

Electrodeposited cobalt-molybdenum, cobalt-tungsten, and cobalt-nickel alloys have been evaluated for electroformed aerospace components because of their high-temperature oxidation resistance along with composites consisting of alumina, thoria, or chromium carbide particles dispersed in an electroplated cobalt matrix. These and other applications are discussed in this survey of cobalt and cobalt alloy electrodeposition.

2.1 Cobalt

Cobalt can be electrodeposited from all-sulfate, sulfate-chloride, all-chloride, sulfamate, ammonium sulfate, and fluoborate solutions. Typical solution formulations and operating conditions are given in Table 6. The addition of chlorides or other halides to those solutions to ensure efficient dissolution of cobalt anode materials is not necessary because, unlike pure nickel, cobalt dissolves at 100% anode efficiency in the absence of halides. High-purity electrolytic cobalt produced by electrowining is now available from nickel and cobalt producers in the form of rounds and other shapes suitable for use in titanium anode baskets. Cobalt with high levels of iron is not recommended for electroplating because tiny amounts of iron lower cathode efficiency and may affect the properties of the deposits adversely. Cathode efficiency is slightly higher than that of a comparable nickel solution.

The effect of pH and temperature on the mechanical properties of cobalt electrodeposited from a sulfate-chloride solution is illustrated by the data in Table 7 [75]. The cobalt solution is similar to a Watts solution in which the nickel is replaced by the corresponding cobalt salt, and the data thus make it possible to compare the properties of cobalt with those of nickel electrodeposited under comparable

TABLE 6 Some Typical Cobalt Electroplating Solutions

	Composition (g liter^{-1})				
Constituent	Sulfate/ Chloride	Chloride	Sulfamate	Ammonium Sulfate	Fluoborate
Cobalt sulfate, $CoSO_4 \cdot H_2O$	330–565				
Cobalt chloride, $CoCl_2 \cdot 6H_2O$	0–45	90–105			
Cobalt sulfamate, $Co(SO_3NH_2)_2$			450		
Cobalt ammonium sulfate, $Co(NH_2)_2(SO_4)_2 \cdot 6H_2O$				175–200	
Cobalt fluoborate, $Co(BF_4)_2$[a]					115–160
Boric acid	30–45	60		25–30	15
Formamide, $HCONH_2$			30		
Sodium or potassium chloride					
Total cobalt metal	70–130	22–25	105	26–30	30–40
	Operating conditions				
pH	3–5	2.5–3.5	3–5	5–5.2	3.5
Temperature (°C)	35–40	50–55	20–50	25	50
Current density (amps dm^{-2})	2–5	3–4	1–5	1–3	5–10

[a]Available as an aqueous solution only.

TABLE 7 Mechanical Properties of Electrodeposited Cobalt

Constituent			Composition (g $liter^{-1}$)		
Cobalt sulfate, $CoSO_4 \cdot H_2O$			330		
Cobalt chloride, $CoCl_2 \cdot 6H_2O$			45		
Boric acid			30		
			Operating conditions		
pH	1.5	2.0	2.0	4.0	4.0
Temperature (°C)	60	60	70	70	60
Current density (amps dm^{-2})	4	4	4	4	4
			Mechanical properties		
Tensile strength (MPa)	1187	657	598	530	775
Significant Strain[a]	0.019	0.017	0.017	0.006	0.010

[a]At fracture, measured by hydraulic bulge testing. A strain value of 0.022 is approximately equal to 1% elongation.

conditions. Maximum tensile strength occurred at pH 0.9 to 1. Increasing the pH from 1.5 to 4 lowered the tensile strength of the electrodeposited cobalt. Increasing the pH above 4.0 increased the tensile strength, but only to a small extent. Increasing the temperature from 60° to 70°C decreased tensile strength. The significant strain as measured by hydraulic bulge testing was low (the elongation percentage being less than 1%). The relationship between the mechanical properties and operating conditions is different from that observed with nickel. Increasing the temperature from 60° to 70°C increases the tensile strength of electrodeposited nickel, increasing the pH from 1 to 4 reduces tensile strength, and further increases in pH increase the tensile strength of nickel considerably. Electrodeposited cobalt is thus stronger and less ductile than nickel electrodeposited from a comparable additive-free Watts bath (see Table 4, Chapter 3, *Electrodeposition of Nickel*).

Cobalt from additive-free baths is intrinsically stressed in tension and the values can be quite high under certain conditions of electrodeposition. The internal stress of deposits from a cobalt sulfamate solution was 140 MPa, that from a chloride solution with saccharin, 95 MPa [76]. The hardness of cobalt decreases rapidly when it is heated at temperatures above 400°C [77].

Although it is possible to deposit mirror-bright coatings by introducing organic additives into cobalt plating solutions, electroplated cobalt coatings have not been used to any extent in decorative applications. Decorative cobalt coatings are not as corrosion resistant as their nickel counterparts, and the black corrosion products that form on outdoor exposures are unattractive. The higher cost of cobalt metal compared to nickel makes it unlikely that decorative applications will be developed in the future.

Functional applications are important, and hence knowledge of the properties of cobalt as affected by electroplating parameters is desirable. The effects of plating variables on the physical and mechanical properties of electrodeposited cobalt have not been studied as thoroughly and systematically as those of electrodeposited nickel and a work on cobalt as comprehensive as the one done many years ago for nickel by Jennings, Brenner, and Zentner [78] does not exist, as far as is known. The exception is study of the magnetic properties of electroplated cobalt and its alloys as affected by the crystal lattice structure, electroplating parameters, and other factors [79].

Cobalt metal normally has a hexagonal close-packed lattice at temperatures below 421°C, but mixtures of hexagonal closed-packed and face-centered cubic lattices have been reported in the case of the electrodeposited metal [80]. Face-centered cubic cobalt forms from solutions that have pH values less than 2.4, and it is believed to involve formation and codeposition of cobalt hydride [81]. The cobalt hydride is unstable and will decompose into hexagonal close-packed cobalt on outgassing of the hydrogen. At high-solution pH (above 3 or 4), the hexagonal close-packed structure is stable. High-solution temperature also favors the formation of the hexagonal close-packed lattice [82]. The transition of the hexagonal-close-packed lattice to the face-centered cubic lattice occurred at about 425°C when electroplated cobalt was heated [83].

The presence of the face-centered cubic crystal structure in electrodeposited cobalt affects magnetic properties [84]. For example, *hard* magnetic coatings have a high coercivity (greater than 250 oersteds). Cobalt electroplated from a solution containing 560 g liter^{-1} of cobalt sulfate heptahydrate, 10 g liter^{-1} of boric acid, and 65 g liter^{-1} of sodium sulfate decahydrate, and a pH of 4.8 and temperature of 40°C, yielded deposits with coercivities of 1000 to 1200 oersteds. Cobalt deposited from a solution containing 350 g liter^{-1} of cobalt chloride hexahydrate and 20 g liter^{-1} of boric acid, at a pH of 4 and a temperature of 90°C, yielded deposits with coercivities of 600 oersteds. These deposits probably retained a small amount of cubic crystals because the theoretical coercivity of alpha (hcp) cobalt is 5000 oersteds. Because of the high coercivity, low remanence, and slow response times, *hard* magnetic cobalt electrodeposition has been applied in the fabrication of permanent data storage devices [85]. Cobalt can reportedly be electrodeposited from all-sulfate solutions as needle-shaped crystalline particles with the long axis of the close-packed hexagonal crystal perpendicular to the surface [86]. Because the long axis is more readily magnetized, perpendicular recording becomes possible. *Hard* magnetic electrodeposits, however, are more likely to be cobalt-based binary or ternary alloys prepared electrolytically (or autocatalytically).

2.2 Cobalt Alloys

Table 8 provides comments and references on the binary cobalt alloy electroplating processes described by Brenner [1] and is a brief guide to work done prior to 1960. Safranek [3] discusses the physical and mechanical properties of electroplated cobalt and cobalt alloys, and his book is a useful source of information on work done up to about 1983. Comparison of Tables 1 and 8 suggests that more research has been done on nickel alloys than on cobalt alloys. Morral has correlated the properties of cobalt and cobalt alloy deposits with the solutions from which they were prepared [93].

The binary cobalt alloys that are commercially important include those with gold, molybdenum, nickel, phosphorus, tungsten and zinc. The *hard* magnetic ternary alloys are important in computer applications.

TABLE 8 Summary of Some of the Early Work on Electrodeposited Binary Cobalt Alloys [1]

Alloying Element	Comments and References
Germanium	Deposits with 10–50% germanium prepared from solutions containing ammonium chloride, ammonium oxalate, sodium metabisulfate. The metal contents in solution and deposit properties were not reported [9].
Gold	Small amounts of cobalt harden gold deposits. Cobalt does not codeposit with gold from alkaline cyanide solutions [87] but is a constituent of modern, proprietary acid and acid cyanide solutions.
Iron	Deposits with 20–80% iron were prepared from simple salt solutions [88]. The iron percentage in the coating is close to that in solution. Co–Ni–Fe is also known. Of interest because of magnetic properties.
Magnesium	Reports of codeposition of magnesium and cobalt in 1904 by Siemens and others were refuted.
Molybdenum	Cobalt-molybdenum alloys can be prepared from ammoniacal, alkaline pyrophosphate, carbonate and hydroxyacetic acid baths [1]. Continuing interest is related to their strength at elevated temperatures.
Nickel	Alloys over the entire range of composition can be prepared from simple salt solutions.
Palladium	Palladium alloys containing 10% cobalt were reported by Atkinson [89].
Phosphorus	Brenner, Couch, and Williams [14] prepared deposits with up to 15% phosphorus. Magnetic properties are important and of continuing interest.
Rhenium	Deposits with 80% rhenium were prepared from cobalt sulfate and potassium perrhenate solutions [16] containing citric and boric acids at pH 7–8 and 5 $A\,dm^{-2}$.
Rhodium	Patents dating from 1904 claim deposition of alloys containing 25–95% rhodium with cobalt [17].
Tin	Tin-cobalt alloy plating was studied by Sree and Rama Char who showed that deposits containing 1–70% cobalt, balance tin, could be prepared from pyrophosphate solutions by varying the pH and the cobalt content in solution [90]. Quality of the deposits was reportedly good.
Titanium	Cobalt with 10–30% titanium was reportedly deposited from a fluoride bath at high current densities. Cobalt was added as the sulfate; titanium as freshly precipitated hydroxide [91].
Tungsten	Cobalt-tungsten alloy deposits are readily prepared as sound, smooth and thick coatings with 30–60% tungsten from ammoniacal solutions containing citrates and tartrates; deposits from acid solutions are less attractive [1].
Zinc	Kochergin and Pobedimshii deposited alloys from ammoniacal and simple sulfate solutions [92]. Zinc codeposited less readily from ammoniacal than from acid sulfate solutions. Deposits over the full-range of compositions were apparently produced. Zinc alloy electrodeposits with small amounts of cobalt are being used to coat auto-body sheet steel by continuous-coil electroplating.

Gold-Cobalt Electrodeposited gold coatings for engineering use contain greater than 99 mass % gold, may contain small amounts of cobalt that functions as a hardening agent and can be produced from acid cyanide solutions [94]. The coating may be relatively thick, 1 to 40 μm. There are many electrical- and electronic-related applications in which cobalt-hardened gold coatings are specified, the most important being printed and etched circuits, semiconductors, contacts, and connectors.

Cobalt-Nickel and Cobalt-Nickel-Iron The solution for electrodepositing cobalt alloys with 50% nickel included in Table 9 is basically a Watts bath in which cobalt has replaced some of the nickel. It would be a relatively simple matter to modify that solution to obtain deposits with less than 50% nickel by raising the cobalt ion concentration in solution to greater than $10\,\mathrm{g\,liter^{-1}}$. Cobalt-nickel and cobalt-nickel-iron electrodeposited alloys have useful magnetic properties for information storage devices in the computer industry. The importance of minor amounts of nonmagnetic elements like molybdenum, chromium, tungsten, zinc, phosphorus, antimony, arsencic, and bismuth in controlling the magnetic properties has been mentioned [95].

Cobalt-Molybdenum The solution included in Table 9 is capable of producing electrodeposits that contain about 40% molybdenum, balance cobalt, and is typical of the ammoniacal solutions containing sodium citrate that have been studied extensively [96]. The deposits have increased hardness and improved resistance to high-temperature oxidation compared to pure cobalt. The extent to which these are being used industrially is not clear, but a proprietary process may be available.

Cobalt-Phosphorus Deposits containing up to 15% phosphorus were described by Brenner and coworkers [97]. The solution in Table 9 allows deposition of cobalt alloys with about 5% phosphorus. The magnetic properties of cobalt-phosphorus alloys have been applied in the computer industry. Alloys with up to 7% phosphorus have high coercivities ($>$ 800 oersteds) [98]. The coercivities can be increased by thermal treatments. Cobalt-phosphorus alloy deposits, like their nickel-phosphorus counterparts, have exceptional hardness and resistance to wear, especially when heat treated.

Cobalt-Tungsten Electroplated alloys of cobalt and tungsten have been extensively studied because of their high-temperature strength and resistance to oxidation. Alloy deposits containing 30% tungsten can be prepared from the solution given in Table 9. Deposits with 28% tungsten had a retained strength of 500 MPa when heated at 700°C [99].

Zinc-Cobalt Zinc-cobalt alloy coatings with less than 1% cobalt are electrodeposited on auto-body steel coil by continuous plating processes as a base for paint [100]. They are more corrosion resistant than pure electrodeposited zinc, and as a result thinner coatings can be specified than with pure zinc. This alloy coating has many of the advantages of zinc-nickel coatings and is being used industrially.

Cobalt Composites Composites consisting of chromium carbide particles dispersed within an electroplated cobalt matrix have been described by Kedward [101] and have been applied in aircraft parts because of their increased wear resistance at elevated temperatures. Composites consisting of alumina particles within a cobalt matrix can also be prepared [102].

TABLE 9 Some Typical Cobalt Alloy Electroplating Solutions

Constituent	Composition ($g\ liter^{-1}$) Cobalt-Nickel (50% Ni)	Cobalt-Molybdenum (40% Mo)	Cobalt-Phosphorus (5% P)	Cobalt-Tungsten (30% W)
Cobalt sulfate, $CoSO_4 \cdot 7H_2O$	30	85		
Nickel sulfate, $NiSO_4 \cdot 7H_2O$	300			
Nickel chloride, $NiCl_2 \cdot 7H_2O$	50			
Boric acid, H_3BO_3	30			
Sodium molybdate, $Na_2MoO_4 \cdot 2H_2O$		48		
Sodium citrate, $Na_3C_6H_5O_7$		100		
Cobalt chloride, $CoCl_2 \cdot 6H_2O$			180	100
Phosphoric acid, H_3PO_4			50	
Phosphorous acid, H_3PO_3			15	
Sodium tungstate, $Na_2WO_4 \cdot 2H_2O$				45
Sodium potassium tartrate (Rochelle salt), $KNaC_4H_4O_6 \cdot 4H_2O$				400
Ammonium chloride, NH_4Cl				50
Operating conditions				
pH	3.7–4 adjusted with H_2SO_4	10.5 adjusted with ammonia	0.05–2 adjusted with cobalt carbonate	8.5–8.7 adjusted with ammonia
Temperature (°C)	66	25	75–95	82–90
Current density ($A\,dm^{-2}$)	3.8–4	10	5–40	2–5
Anodes	Cobalt and nickel	Cobalt	Cobalt	Cobalt with 30% tungsten
Cathode efficiency (%)	99.5	About 50	About 80	About 90

2.3 Closing Observations on Nickel and Cobalt Alloy Electroplating

The available techniques—electrodeposition from aqueous solutions, production of artificially layered alloy coatings by cyclic modulation of deposition potentials, heat treatment of electroplated multlilayers, and composites—suggest that just about any nickel or cobalt alloy can be produced by electrochemical technology. If one includes techniques not covered in this chapter, such as electrodeposition from fused salts and organic electrolytes, and autocatalytic deposition of nickel polyalloys, that observation is hard to refute.

There is also little doubt that nickel alloy electroplating has grown in importance and is more widely practiced than cobalt alloy electroplating. Alloy plating, in general, is being performed on a scale unknown 25 years ago. The most notable commercial successes are electroplated zinc-nickel and nickel-palladium alloy coatings for engineering applications.

The *hard* magnetic properties of cobalt-base alloys containing small amounts of nickel, iron, and other elements, and the *soft* magnetic properties of nickel-iron alloy electrodeposits have been of utmost importance in the computer industry. Indeed, without electrochemical technology the computer industry would not have developed as well, nor as fast, as it has.

Successful commercialization of alloy plating processes has often been influenced by nontechnical factors. The growth of nickel-palladium alloy plating was influenced by rapid increases in the price of gold. The electrodeposition of nickel-zinc alloys was influenced by environmental concerns, for example, the need to find substitutes for cadmium, and by the competitive need to improve the quality of automobiles made in the United States. The continuing interest in tin-nickel alloy electroplating is partly a response to environmental and health concerns, but it also takes advantage of that process' excellent throwing power. The growth in the use of nickel-iron and cobalt-based electroplated alloys was spurred by the special needs of the computer industry.

Alloy electroplating is likely to become more important in the future, the most promising area at the moment being research and development on composition-modulated alloy coatings which display unusual magnetic properties and may find application as interference coatings for neutron scattering, for magnetic recording devices and for magnetoresistive detectors and sensors of various kinds [103].

REFERENCES

1. A. Brenner, *Electrodeposition of Alloys-Principles and Practice*, vols 1 and 2, Academic Press, New York, 1963. See Chapter 26, vol. 2, p. 110–136, for nickel-chromium alloy electrodeposition. See ch. 33, vol. 2, p. 347–412, for the codeposition of tungsten with nickel, cobalt, and iron; and ch. 34, vol. 2, p. 413–456, for the codeposition of molybdenum with nickel, cobalt, and iron.
2. C. B. Jacobs, *J. Am. Chem. Soc.*, **27**, 972 (1905).
3. Yu. M. Polukarov and K. M. Gorbunova, *Zhur. Fiz. Khim*, **30**, 871 (1956).
4. E. Raub and M. Wittum, *Korrosion und Metallschutz*, **15**, 127 (1939).
5. L. D. McGraw, J. P. Spenard, and C. L. Faust, *Proc. Amer. Electroplater's Soc.*, 209 (1956).

6. R. W. Moeller and W. A. Snell, *Proc. Amer. Electroplater's Soc.*, 189 (1955); D. Singh and U. B. Singh, *Ind. J. Technol.*, **14** (4), 189 (1976).
7. D. E. Couch and I. Krivickas, Unpublished work. National Bureau of Standards, 1958.
8. P. R. Subbaraman and J. Gupta, *J. Sci. Ind. Research (India)*, **15B** (6), 306 (1956).
9. E. Raub and M. Wittum, *Korrosion und Metallschutz*, **13**, 261 (1937).
10. A. Siemens, *Z. Anorg. Chem.*, **41**, 249 (1904).
11. D. N. Gritsan and N. S. Tsvetkov, *Zhur. Priklad. Khim.*, **22**, 600 (1949).
12. Yu. D. Kondrashev, I. P. Tverdovskii, and Zh. L. Vert, *Doklady Akad. Nauk SSSR.*, **78**, 729 (1951).
13. A. Brenner, D. E. Couch, and E. K. Williams, *J. Res. Natl. Bur. Standards*, **44**, 109 (1922).
14. M. Baum, German Patents 201,664 and 201,666 (1908).
15. L. E. Netherton and M. L. Holt, *J. Electrochem. Soc.*, **98**, 106 (1951).
16. P. T. Smith and J. A. Smith, U.S. Patent 2,461,933 (1949).
17. F. C. Mathers and A. D. Johnson, *Trans. Electrochem. Soc.*, **74**, 229 (1938).
18. K. Masaki, *Bull. Chem. Soc. Japan*, **7**, 158 (1932).
19. D. C. Gernes, G. A. Lorenz, and G. H. Montillon, *Trans. Electrochem. Soc.*, **77**, 177 (1940).
20. E. Raub and F. Sautter, *Metalloberflache*, **13**, 129 (1959).
21. J. K. Dennis and T. E. Such, *Nickel and Chromium Plating*, 3rd ed., Woodhead Publ., Cambridge, England, 1993, ch. 13. Published in North America by ASM International, Materials Park, OH 44073.
22. W. H. Safranek, *The Properties of Electrodeposited Metals and Alloys*, 2nd ed., American Electroplaters' and Surface Finishers' Society, Orlando, FL, 1986, ch. 6, 14, and 15.
23. G. A. DiBari and S. A. Watson, "A Review of Recent Trends in Nickel Electroplating Technology in North America and Europe," *NiDI Reprint Series No. 14 024*, Nickel Development Institute, Toronto, Ontario, Canada.
24. M. Antler, *Proc. Connectors '91 Symp.*, G. D. Wilcox and D. H. Ross, eds., Institute of Metal Finishing, Birmingham, England, 1991.
25. T. Hayashi, "Trends in Alloy Plating in Japan," *Int. Technical Conf. Proc.*, American Electroplaters' and Surface Finishers' Society, Orlando, FL, June 1993.
26. J. P. Celis, J. R. Roos, B. Blanpain, and M. Gilles, "Pulse Electrodeposition of Compositionally Modulated Multilayers," *Proc. 12th World Conf. on Surface Finishing*, Paris, France, 435 (1988).
27. G. A. Malone, *Proc. AESF Electroforming Symp.*, American Electroplaters' and Surface Finishers' Society, Orlando, FL (1996).
28. L. Weisburg and W. B. Stoddard, U.S. Patent 2,026,718 (1936); L. Weisburg, *Trans. Electrochem. Soc.*, **73**, 435 (1938); **77**, 223 (1940).
29. D. W. Endicott and J. R. Knapp, "Electrodeposition of Nickel-Cobalt Alloy; Operating Variables and Physical Properties of the Deposits," *52nd Annual Proc. American Electroplaters Society*, American Electroplaters and Surface Finishing Society, Orlando, FL (1965); *Plating*, **53**, 43 (1966).
30. K. C. Belt, J. A. Crossley, and R. J. Kendrick, "Properties of Electrodeposits from a Concentrated Nickel Sulfamate Solution," *Proc. 7th Int. Metal Finishing Conf.*, Hanover, 1968.
31. R. C. Barrett, "Plating of Nickel, Cobalt, Cadmium and Iron from Sulfamate Solutions," *47th Annual Proc. American Electroplaters Society*, American Electroplaters' and Surface Finishers' Society, Orlando, FL (1960).

32. E. Chassain, K. Vu Quang, and R. Wiart, *J. Appl. Electrochem. Soc.*, **17**, 1267 (1987); M. Cherkaoui, E. Chassain, and K. Vu Quang, *Proc. Interfinish '88 Conf., Paris*, October 1988.
33. G. A. DiBari, Unpublished research, Inco Bayonne Research Laboratory, Bayonne, NJ, on an electrolyte developed by B. H. Priscott, Inco European Research and Development Centre, Birmingham, England, ca. 1962.
34. J. Yahalom, D. F. Tessier, R. S. Timsit, A. M. Rosenfeld, D. F. Mitchell, and P. T. Robinson, "Structure of Composition Modulated Cu/Ni Thin Films Prepared by Electrodeposition," *J. Materials Res.*, **4**, 755 (1989); D. Tench and J. White, "Enhanced Tensile Strength of Nickel-Copper Multilayer Composites," *Metall. Trans.*, **15A**, 2039 (1984); A. W. Ruff and Z. X. Wang, "Sliding Wear Studies of Ni–Cu Composition Modulated Coatings on Steel," *Wear*, **131**, 259 (1989).
35. A. M. Weisberg, "Gold Plating," *Metal Finishing Guidebook and Directory*, Elsevier Science, New York, 1997, p. 243; S. J. Helmsley, *Plating Surf. Finish.*, **77** (11), 27 (1990).
36. Y. N. Sadana and Z. Zhang, *Plating Surf. Finish.*, **77** (11), 27 (1990).
37. L. T. Romankiw and T. A. Palumbo, "Electrodeposition in the Electronic Industry," *Electrodeposition Technology and Practice*, L. T. Romankiw and D. R. Turner, eds., Electrochemical Society, Pennington, NJ, 1987, p. 13.
38. I. M. Croll and T. Romankiw, "Iron, Cobalt and Nickel Plating for Magnetic Applications," ibid., p. 285; L. T. Romankiw and D. A. Thompson, "Magnetic Properties of Plated Films," *Properties of Electrodeposits, Their Measurement and Significance*, R. Sard, H. Leidheiser Jr., and F. Ogburn, eds., Electrochemical Society, Pennington, NJ, 1975, p. 389.
39. A. H. DuRose and P. R. Pine, *Steel*, **114** (24), 124 (1944).
40. G. A. DiBari, G. Hawks, and E. A. Baker, "Corrosion Performance of Decorative Electrodeposited Nickel and Nickel-Iron Alloy Coatings," *Atmospheric Corrosion of Metals, ASTM STP 767*, S. W. Dean Jr. and E. C. Rhea, eds., American Society for Testing and Materials 1982, p. 186.
41. D. N. Gristan and N. S. Tsvetkov, "Conditions for the Electrolytic Deposition of Manganese-Nickel Alloys," *Zhur. Priklad. Khim.*, **22**, 600 (1949); D. N. Gristan, E. I. Vail, and V. P. Moskovets, "Electrolytic Production of a Cobalt-Manganese Alloy," *Chem. Abstr.*, **51**, 1752i (1957); R. I. Agladze and M. Ya. Gdzekishvili, "Results on Experiments on the Electrolytic Production of Manganese Alloys," *Soobh. Akad. Nauk Gruzinskoi SSR*, **5** (10), 976 (1944). These papers are reviewed in volume 2, p. 138–150, of Brenner's books on alloy electrodeposition.
42. J. W. Dini, H. R. Johnson, and L. A. West, "On the High-Temperature Ductility Properties of Electrodeposited Sulfamate Nickel," *Plating Surf. Finish.*, **65** (2), 36 (1978); J. W. Dini, H. R. Johnson, and H. J. Saxton, "Influence of Sulfur Content on the Impact Strength of Electroformed Nickel," *Electrodep. Surf. Treat.*, **2** (3), 165 (1974).
43. W. R. Wearmouth and K. C. Belt, "Electroforming with Heat-Resistant Sulfur-Hardened Nickel," *Plating Surf. Finish.*, **66**, 53 (1979); W. R. Wearmouth, "Nickel Alloy Electrodeposits for Non-decorative Applications," *Trans. Inst. Met. Finish.*, **60** (2), 68 (1982).
44. G. A. Malone, "New Developments in Electroformed Nickel-based Structural Alloys," *Plating Surf. Finish.*, **74** (1), 50 (1987); "Multi-stage Electroforming of High Energy Chemical and Eximer Laser Components," ibid., **75** (7), 58 (1988); N. Atanassov and H. W. Schils, "Deposition of Nickel-Based Alloys with Addition of Manganese and Sulfur," ibid., **83** (7), 49 (1996). Reference [27] contains some mechanical property data on an electrodeposited nickel-manganese alloy (0.2% Mn).
45. G. A. DiBari, Unpublished research at the Inco Bayonne Research Laboratory, Bayonne, NJ (1962).

46. E. J. Podlaha and D. Landolt, "Induced Codeposition-1: An Experimental Investigation of Ni-Mo Alloys; 2. A Mathematical Model Describing the Electrodeposition of Ni-Mo Alloys," *J. Electrochem. Soc.*, **143** (3), 885–899 (1996).
47. J. A. Abys et al., *Met. Finish.*, **87** (7), 43 (1991); U.S. Patent 4,911,798 and 4,911,799 (1990).
48. M. Pushpavanam, S. R. Natarajan, and K. Balakrishnan, "Electrodeposited Palladium-Nickel Alloy for Electronic Applications," *B. Electrochem.*, **5** (6), 430 (1989).
49. R. J. Morrissey, "Palladium and Palladium-Nickel Alloy Plating," *Metal Finishing Guidebook - Directory*, Elsevier Science, New York, 1997, p. 288.
50. H. K. Straschil, I. Kadija, J. Maisano, and J. A. Abys, "Electroplating of Thick and Ductile Palladium-Nickel Alloys," *Circuit World*, **17** (2), 9 (1991); I. Boguslavsky, J. A. Abys, E. J. Kudrak, M. A. Williams, and T. C. Ong, "Pd–Ni Plated Lids for Frame-Lid Assemblies," *Plating Surf. Finish.*, **83** (2), 72 (1996); J. A. Abys, J. J. Maisano, I. V. Kadija, E. J. Kudrak, and S. Nakahara, "Annealing Behavior of Palladium-Nickel Alloy Electrodeposits," ibid., **83** (8), 43 (1996).
51. A. Brenner, E. Couch, and E. K. Williams, "Electrodeposition of Alloys of Phosphorus with Nickel and Cobalt," *J. Res. Natl. Bur. Standards*, **44**, 109 (1950). Also see D. A. Luke, *Trans. Inst. Met. Finish.*, **64** (3), 99 (1986).
52. D. S. Lashmore, R. Oberle, and M. P. Dariel, "Electodeposition of Artificially Layered Materials," *Proc. AESF 3rd Int. Pulse Plating Symp.*, American Electroplaters' and Surface Finishers' Society, Orlando, FL, 1986.
53. Xiao Chang Geng, *Nickel*, **5** (4), 3 (1990); published by the Nickel Development Institute, Toronto, Ontario, Canada.
54. W. T. Young and H. Kersten, "A Study of Electrodeposits Containing Nickel and Sulfur," *Trans. Electrochem. Soc.*, **71**, 225 (1938).
55. F. A. Lowenheim, "The Corrosion Resistance of Tin Alloy Electrodeposits," *44th Annual Technical Proc. AESF*, American Electroplaters' and Surface Finishers' Society, Orlando, FL, 1957.
56. S. K. Jalota, "Tin-Nickel Alloy Plating," *Met. Finish.*, **84** (2), 81 (1986).
57. R. J. Baker, *Nickel and Tin-Nickel Plating for Electronic Applications—An AESF Electronics Lecture*, American Electroplaters' and Surface Finishers' Society, Orlando, FL, 1984.
58. K. R. Baldwin, M. J. Robinson, and C. J. E. Smith, "The Corrosion Resistance of Electrodeposited Zinc-Nickel Alloy Coatings," *Corrosion Sci.*, **35** (8), 1267 (1993).
59. D. E. Hall, "Electrodeposited Zinc-Nickel Alloy Coatings—A Review," *Plating Surf. Finish.*, **70** (11), 59 (1983).
60. S. A. Watson, "Zinc-Nickel Alloy Electroplated Steel Coil for Use by the Automotive Industry," *NiDI Review Series No. 13 001*, Nickel Development Institute, Toronto, Ontario, Canada, 1988; "Electrodeposited Zinc-Nickel Alloy Coatings," *Proc. Asia Pacific Interfinish '90 Conf.*, Singapore, 1990.
61. N. Zaki, "Zinc-Nickel Alloy Plating," *Met. Finish.*, **76** (6), 57 (1989).
62. K. Higashi et al., "A Fundamental Study of Corrosion-Resistant Zinc-Nickel Alloy Electroplating," *NiDI Technical Series No. 10 036*, Nickel Development Institute, Toronto, Ontario, Canada, 1990.
63. M. R. el-Sharif, A. Watson, and C. U. Chisholm, "The Sustained Deposition of Thick Coatings of Chromium/Nickel and Chromium/Nickel/Iron Alloys and Their Properties," *Trans. IMF.*, **66**, 34 (1988); "The Role of Chromium II and VI in the Electrodeposition of Chromium Nickel Alloys from Trivalent Chromium-Amide Electrolytes," *Trans. IMF*, **64**, 149 (1986).

64. D. S. Lashmore, I. Weishauss, and K. Pratt, "Electrodeposition of Nickel-Chromium Alloys," (AESF Research Project 58), *Proc. AESF Annual Conf., Session P*, American Electroplaters' and Surface Finishers' Society, Orlando, FL, 1985.
65. M. Law, "Electrodeposition of a Ternary Alloy of Chromium," *Finish*, **8** (7), 30 (1984).
66. E. W. Brooman, "Chromium Alloy Plating," *ASM Handbook*, vol. 5, *Surface Engineering*, ASM International, Materials Park, OH, 1994, p. 270.
67. F. K. Andryushchenko and N. N. Gavyrina, "Electrolytic Deposition of Nickel-Germanium Alloys," *Protective and Metallic and Oxide Coatings—Metal Corrosion and Electrochemistry*, N. P. Fedot'ev, ed., Israel Program for Scientific Translations, Jerusalem, 1968, p. 159.
68. G. J. Sammuels, *Plating Surf. Finish.*, **74** (3), 66 (1987); J. R. Roos, J. P. Cellis, and E. Vacouille, ibid., **77** (2), 60 (1990); A. Davidson and Y. N. Sadana, *Plating*, **58** (11), 1007 (1971).
69. L. I. Kadaner, R. B. Avakyan, and Z. N. Kiryukhina, "Electrochemical Deposition of Ruthenium-Cobalt, Ruthenium-Nickel, Ruthenium-Cobalt-Nickel and Platinum-Nickel Alloys," *Elektrolit. Osazhdenie Splavov*, **2**, 43 (1968).
70. G. T. Burkat, N. P. Fedot'ev, and P. M. Vyacheslavov, "Electrodeposition of Nickel-Silver Alloys," *J. Appl. Chem. (USSR)*, **41** (2), 405 (1968).
71. G. R. Pobedimmskii and A. I. Zhikharev, "Texture of Electrodeposited Nickel-Thallium and Nickel-Cobalt-Thallium Alloys," *Tr. Kazansk. Khim.-Tekhnol. Inst.*, **36**, 278 (1967).
72. W. R. Wearmouth, "Nickel Alloy Electrodeposits for Non-decorative Applications," *Trans. IMF*, **60** (2), 68 (1982); A. C. Hart and W. R. Wearmouth, U.K. Patent 1,476,099 (June 1977).
73. G. P. Cammarota et al., "Alloys of Nickel-Iron-Alumina," *Galvano Tec.*, **30** (6), 117 (1979).
74. D. E. Couch and J. H. Connor, "Nickel-Aluminum Alloy Coatings Produced by Electrodeposition and Diffusion," *J. Electrochem. Soc.*, **107**, 272 (1960).
75. J. H. Lindsay and H. J. Read, "Some Properties of Electrodeposited Cobalt," *Plating*, **57**, 497 (1970).
76. D. W. Endicott and J. R. Knapp, *Plating*, **53**, 43 (1960); R. D. Fisher, *J. Electrochem. Soc.*, **109** (6), 479 (1962); R. C. Barrett, *Tech. Proc. Am. Electroplaters Soc.*, **47**, 170 (1960).
77. W. Chubb, *J. Met., AIME Trans.*, **203**, 189 (1955).
78. V. Zentner, A. Brenner, and C. W. Jennings, "Physical Properties of Electrodeposited Metals, Part 1—Nickel," *Plating*, **39** (8), 865 (1952).
79. V. V. Bondar et al., "Electrodeposition of Magnetically Hard Alloys—1. Electrodeposition of Nickel-Phosphorus Alloys; 2. Electrodeposition of Cobalt-Nickel-Phosphorus and Cobalt-Manganese-Phosphorus," *Inform. Sistemy, Moscow*, pp. 117–127 (1964); I. Tsu and M. C. Fritsch, "Electrodeposition of Magnetic Cobalt-Nickel Alloys," U.S. Patent 3,111,463 (1963). See Romankiw and coworkers, References [38], and [3, pp. 83–88].
80. S. Nakahara and E. C. Felder, "The Influence of pH on Microstructure of Electrodeposited Cobalt," *J. Electrochem. Soc.*, **129**, 45 (1982).
81. I. M. Croll and B. A. May, "Effect of Buffering Agents on Structure and Magnetic Properties of Plated Cobalt Films," *Electrodeposition Technology, Theory and Practice*, Electrochemical Society, Pennington, NJ, 1987.
82. L. Reiner, "Magnetic Investigations of Electrolytically Deposited Thin Cobalt Films," *Zeitscrift für Naturforschung*, **12a**, 1014 (1957); I. Bursuc, A. Ojog, and V. Tutovan, "Contributions to the Study of Magnetism of Electrodeposited Cobalt," *Physica Status Solidi*, **4**, 161 (1964).

83. W. Chubb, "Contribution of Crystal Structure to the Hardness of Metal," *J. Met., AIME Trans.*, **203**, 189 (1955).

84. N. T. Gofman and T. N. Nikolaishvili, "Electrolytic Production of Cobalt Deposits with Given Characteristics," *Izvestiya Vysshikh Uchebnykh Zavedeity SSSR, Khimiya I Khimicheskaya Tekhnologiya*, **8** (1), 104 (1965).

85. See Reference [37].

86. Reference [37, p. 24].

87. N. P. Fedot'ev, N. M. Ostroumova, and P. M. Vyacheslavov, *Zhur. Priklad. Khim.*, **27** (1), 43 (1954).

88. S. Glasstone and J. C. Speakman, *J. Electrodepositors' Tech. Soc.*, **8**, 11 (1932).

89. R. H. Atkinson, U.S. Patent 1,981,715 (1934).

90. V. Sree and T. L. Rama Char, *Bull. India Sect. Electrochem. Soc.*, **9**, 13 (1960).

91. S. M. Kochergin and R. G. Pobedimskii, *Zhur. Priklad. Khim.*, **33**, 238 (1960).

92. S. M. Kochergin and R. G. Pobedimskii, ibid., **31**, 1432 (1958).

93. F. R. Morral, *Plating*, **54**, 693 (1967); *Met. Finish.*, **62** (6), (1964), 82; ibid., **62** (7), 59 (1964).

94. A. M. Weisberg, "Gold Plating," *ASM Handbook, Volume 5, Surface Engineering*, ASM International, Materials Park, OH, 1994, p. 247.

95. L. T. Romankiw and D. A. Thompson, "Magnetic Properties of Plated Films," *Properties of Electrodeposits—Their Measurement and Significance*, Electrochemical Society, Pennington, NJ, 1975, p. 399.

96. D. W. Ernst, Ph.D. thesis, University of Wisconsin, 1956, University Microfilms Publ. No. 18,082, Ann Arbor, MI; D. W. Ernst, R. F. Amlie, and M. L. Holt, *J. Electrochem. Soc.*, **102**, 461 (1955).

97. A. Brenner et al., *Plating*, **37**, 36, 161 (1950).

98. I. Tsu and R. J. Cappell, "The Influence of Plating-Bath Variables on the Composition and Coercivity of Cobalt-Phosphorus Electrodeposits," *Plating*, **52**, 1123 (1965); J. S. Sallo and J. M. Carr, "Studies of High-Coercivity Cobalt-Phosphorus Electrodeposits," *J. Appl. Phys.*, suppl., **33** (3), 1316 (1965).

99. A. Brenner, P. Burkhart, and E. Seegmiller, "Electrodeposition of Tungsten Alloys Containing Nickel and Cobalt," Res. Paper 1834, *J. Res., Natl. Bur. Standards*, **39** (4), 351 (1947).

100. W. M. J. C. Verberne, "Zinc-Cobalt Alloy Electrodeposition," *Trans. IMF*, **64**, 30 (1986).

101. E. C. Kedward and K. W. Wright, "The Wear Control of Aircraft Parts Using a Composite Electroplate," *Plating Surf. Finish.*, **65**, 38 (1978).

102. G. R. Lakshminarayanan, E. S. Chen, and F. K. Sautter, "The Effect of Oxide Dissolution on Electrodepositing Dispersion-Hardened Cobalt and Ni–Al_2O_3 Alloys," *J. Electrochem. Soc.*, **122**, 1589 (1975).

103. M. Alper, W. Schwarzacher, and S. J. Lane, "The Effect of pH on the Giant Magnetoresistance of Electrodeposited Superlattices," *J. Electrochem. Soc.*, **144**, 2346 (1997).

14 Electrodeposition of Semiconductors

T. E. SCHLESINGER

INTRODUCTION

The electrodeposition of semiconductors has been investigated in detail and demonstrated by a large number of workers over many years. This effort is primarily motivated by the fact that electrodeposition is a relatively simple and inexpensive deposition technology that may be scaled up easily. In general, the films deposited by this method do not possess the crystalline perfection or low levels of electrically active impurities of single crystal epitaxial films deposited by techniques such as molecular beam epitaxy or chemical vapor deposition. Nonetheless, in applications where large areas of semiconductors are required, such as photovoltaic power generation or corrosion protection as opposed to integrated circuit fabrication, the low cost and comparatively low material demands make this deposition technology attractive in terms of ultimate commercialization. In addition for application in solar cells, electrodeposition allows one to easily alter both bandgap and lattice constant by composition modulation through the control of growth parameters such as applied potential, pH, and temperature of the bath. Thus it is, at least in principle, possible to easily grow large areas of tandem cells designed for the most efficient conversion of the solar spectrum.

Typical arrangements for electrolytic cells, as well as the fundamental electrochemistry and thermodynamics associated with the deposition process are discussed in detail in Paunovic and Schlesinger [1]. Since a large number of semiconductors have been electrodeposited with varying degrees of success this chapter will focus on the historical development of what has been accomplished in the area of the deposition of semiconductor layers. An attempt is made to indicate the state of the art in each instance. Short summaries as well as the particulars of some of the methods used and an extensive list of references to the literature is provided.

1 SILICON (Si)

The electrodeposition of silicon must employ a nonaqeous medium because of its very large reduction potential and the high reactivity of most of its compounds to water. A great deal of effort has been reported on the electrodeposition of silicon. Many workers have employed either silica or fluorosilicates as the solutes [2, 3, 4–5,

Modern Electroplating, Fourth Edition, Edited by Mordechay Schlesinger and Milan Paunovic.
ISBN 0-471-16824-6

6, 7] and obtained either amorphous silicon or metal silicides. Cohen and Huggins [8] were the first to prepare both mono and polycrystalline silicon by electrolysis of the alkaline fluorosilicates in a LiF–KF eutectic. A review of the deposition of continuous films of silicon from LiF–KF–NaF–K_2SiF_6 melts is provided by Elwell et al. [9]. Deposition of silicon also includes attempts to deposit silicon electrochemically from silicates at temperatures above its melting point [10] and Monnier [11]. Some effort has also been focused on the deposition of silicon powders, for example, the work of Andriiko and coworkers [12], since there are applications where a source of pure Si powder is required. Agrawal and Austin [13] deposited amorphous silicon from nonaqueous baths using $SiHCl_3$ as the silicon source in a bath of 1.0 M $SiHCl_3$ in propylene carbonate containing 0.1 M tetrabutyl ammonium chloride as the supporting electrolyte. Deposits were made potentiostatically at −2.5 V versus a platinum reference electrode with a vitreous carbon counter-electrode at temperatures of 35° to 145°C and on a variety of materials as substrates including Pt, Ti, silicon, and others. With increasing solute concentration and increasing temperature thicker deposits were obtained. Postdeposition anneals at 470°C drove off bonded hydrogen in these films. These amorphous silicon layers were developed as a possible inexpensive material for solar cell applications. Tekeda and coworkers [14] deposited amorphous silicon on a nickel substrate by the electrolysis of a solution of tetraethylorthosilicate in acetic acid. Lee and Kroger [15] deposited amorphous silicon containing fluorine and carbon (and doped with boron or phosphorous) cathodically from solutions of K_2SiF_6 in acetone with HF. Boen and Bouteillon [16] deposited silicon using alkali fluorosilicates as the solute and succeeded in depositing layers of up to 1 mm thickness on silver electrodes and on graphite using a pulsed current technique. These films were polycrystalline with preferred orientations ⟨111⟩, ⟨311⟩, and ⟨220⟩.

Gobet and Tannenberger [17] investigated the electroplating of silicon from solutions of $SiHCl_3$, $SiCl_4$, $SiBr_4$, $Si(CH_2CH_3)_4$, $Si(OCH_2CH_3)_4$, $Si(OOCCH_3)_4$, and $Si[N(CH_3)_2]_4$ in tetrahydrofuran and $LiClO_4$ (0.3M), TBAP (0.3 M), or TBAB (0.2 M) as supporting electrolyte. Depositions were carried out at room temperature in a three-electrode cell. Platinum was used as the counter electrode while the cathode was a polished disk of platinum, gold, or other material. The reference electrode was Ag/$AgClO_4$. Though the Si–O, Si–C, and Si–N bonds are apparently too stable to be reduced, Si films deposited from $SiHCl_3$ and $SiCl_4$ were smooth up to a thickness of 0.25 μm and 0.12 μm, respectively. These films were amorphous as determined by X-ray diffraction, and their thicknesses were probably not uniform. Bulk concentrations of Si ~ 82, O ~ 8, C ~ 8, and Cl ~ 1.5 atomic percent were measured by sputter Auger spectroscopy and a Si : H ratio of 1 : 0.9 determined by proton scattering.

The electrolysis of alkali metal fluorosilicates has also been investigated as a method to produce pure silicon suitable for photvoltaic applications [18]. The following reduction mechanism of K_2SiF_6 in a LiF–KF eutectic mixture at 750°C has been suggested by these authors:

$$\begin{aligned} &\mathrm{Si(IV)} + 2e \longrightarrow \mathrm{Si(II)} + 2e \longrightarrow \mathrm{Si} \\ &\mathrm{S(IV)} + \mathrm{Si} \overset{k_b}{\longleftrightarrow} 2\mathrm{S(II)} \end{aligned} \tag{1}$$

Metal silicide formation is avoided by using a potential step as small negative as possible with pulsed currents producing the best structure of the deposit as well as the best fit between the electrolysis conditions and the proposed mechanism. Growth rates of several microns per hour have been achieved with a total thickness of 3 μm demonstrated. Neutron activation shows these films to contain impurities such as Fe, Cr, Mb, and alkali metals below the ppm range.

Frolenko and coworkers [19] investigated the structure of silicon films deposited with the use of mixed chloride-fluoride melts. This work showed that both polycrystalline and single crystal silicon layers can be obtained, with the substrate and temperature being key factors in determining the structure of the deposit. For this melt and process a (110) texture is typical.

Silicon is the one semiconductor material that can be grown in very high quality and in large quantities at relatively low cost by a huge industrial infrastructure. It therefore remains a significant challenge for electrodeposition to compete in terms of supplying material for most applications.

2 GALLIUM ARSENIDE (GaAs)

As with other semiconductor materials electrodeposition of GaAs offers the potential advantages of the ease of control of film thickness, morphology, and composition by the control of voltage and current density. The inexpensive equipment that can be used, the use of ambient temperature and pressure, and an aqueous solution reduces operating costs while at the same time eliminates the need for the use of volatile and dangerous materials such as AsH_3 and $AsCl_3$ as is often used in other growth techniques.

The electrode reactions and the corresponding potentials are [20]:

$$\begin{aligned}
&\mathrm{Ga}^{3+} + 3e \longleftrightarrow \mathrm{Ga(s)} \\
&E = E^0_{\mathrm{Ga}} + \frac{RT}{3F}\ \ln\left(\frac{a_{\mathrm{Ga}^{3+}}}{a_{\mathrm{Ga}}}\right) = -0.529 + 0.0197\ \log(a_{\mathrm{Ga}^{3+}}) \\
&\mathrm{HAsO_3} + 3\mathrm{H}^+ + 3e \longleftrightarrow \mathrm{As(s)} + 2\mathrm{H_2O} \qquad (-1 > \mathrm{pH} > 8) \\
&E = E^0_{\mathrm{As}} + \frac{RT}{3F}\ \ln\left(\frac{a_{\mathrm{HAsO_3}}}{a_{\mathrm{As}}}\right) + \frac{RT}{F}\ \ln(C_{\mathrm{H^+}}) \\
&\quad = 0.248 + 0.0197\ \log(a_{\mathrm{HAsO_3}}) - 0.0591\mathrm{pH}
\end{aligned} \tag{2}$$

where E is the electrode potential with respect to the normal hydrogen electrode, T is the temperature, R is the universal gas constant, F is Faraday's constant, E^0 is the standard reduction potential, and $a_{\mathrm{Ga}^{3+}}$, $a_{\mathrm{HAsO_3}}$ are the activities of the ions in the solution while a_{Ga}, a_{As} are the activities of the atoms in the electrodeposits.

DeMattei and coworkers [21] deposited GaAs from a molten salt containing 67.4% B_2O_3, 20.3% NaF, 4.2% Ga_2O_3, and 8.1% $NaAsO_2$ by weight at 720° to 760°C. Deposition at 300°C from $GaCl_3$–KCl melt has also been reported [22], and it was shown that under potentiostatic conditions only As was formed while GaAs was formed in this work under galvanostatic conditions. Room temperature deposition of GaAs using $AlCl_3$: 1-butylpyridinium chloride or $AlCl_3$: 1-methyl-3-ethylimidazolium chloride has also been reported [23]. A room temperature melt of $GaCl_3$ and 1-methyl-3-ethylimidazolium cholride and an $AsCl_3$ solute has also been used [24–25]. Electrodeposition from aqueous solutions of $Ga_2(SO_4)_3$, As_2O_3, and H_2SO_4 has been reported [26] and [27] electrodeposited gallium on copper foil from sodium galate solution and arsenic from arsenite solution followed by an anneal at 250° to 530°C.

GaAs has been deposited from both alkaline and acid aqueous electrolytes by Yang et al. [28]. In this last work, a typical three electrode setup was used using a titanium rod as the working electrode, a platinum counterelectrode, and one of three reference electrodes; Ag/AgCl in saturated KCl (0.197 V versus NHE), Hg/HgO in 0.1 M NaOH (0.157 versus NHE), or saturated calomel (0.241 V versus NHE). The acidic electrolyte employed was 1 M As_2O_3 solution in concentrated HCl in deionized water with reagent grade amoniumhydroxide or KOH solution used to adjust pH to less than 3. The alkaline electrolyte employed was As_2O_3 in Ga_2O_3 solution with the pH adjusted by the addition of KOH solution to greater than 12. For films deposited in alkaline solutions the composition depended greatly on the potential with more negative potentials resulting in more Ga rich deposits. For potentials between −1.65 and −1.73 V (versus Hg/HgO) the deposit was about 64% Ga. Room temperature resulted in constant composition of the deposit with higher temperatures resulting in nonuniform deposits. The substrate also had a major effect on the deposition with Ti being the preferred substrate for its low reactivity with Ga and its high hydrogen overpotential. The as-deposited film was amorphous and was therefore annealed in nitrogen atmosphere at 300°C for 12 hours. X-ray diffraction analysis did reveal peaks associated with GaAs after anneal. In acidic solutions pH 2.5 to 3.0 the Ga content was found to be about 65% for potentials between −1.5 and −1.6 V (versus Ag/AgCl) and 50% Ga could be obtained at lower pH = 0.7 with higher current density (350 m Acm^{-2}) or very negative potentials. An anneal at 260°C for 8 hours resulted in films that did display GaAs peaks in X-ray diffraction spectra. Auger electron spectroscopy indicated that the films were about 3200 Å thick.

Villegas and Stickney [29] have investigated the use of Electrochemical Atomic Layer Epitaxy for the deposition of GaAs. As in the case of the deposition of CdTe, this technique relies on the use of successive underpotential depositions of atomic layers of the different elements. They obtained low coverage ordered layers of As by underpotential deposition from $HAsO_2$ and by oxidative underpotential depositions from AsH_3 solutions on (100) and (110) surfaces of Au. They also achieved Ga reductive underpotential deposition on the low index surfaces of Au as well as ordered structures on Au (100) and (110) surfaces by the successive underpotential deposition of atomic layers of As and Ga. This technique, which is described in more detail in the Section 7 on CdTe is most promising for producing high-quality films.

3 GALLIUM PHOSPHIDE (GaP)

Gallium phosphide has been electrodeposited at 800° to 900°C in a melt containing sodium metaphosphate, sodium fluoride, and gallium oxide [30].

4 INDIUM COMPOUNDS (InP, InAs, InSb)

The first discussion of the deposition of indium antimony alloys was made by Smart [31] and was subsequently sudied by a number of workers. Sadana and Singh [32] provide a good discussion of the deposition of this alloy system up to 1985 along with a number of key references. Okubo and Landau [33] electrodeposited InSb from and aqueous solution of $InCl_3$, $SbCl_3$, citric acid ($H_3C_6H_5O_7$) and sodium citrate ($Na_3C_6H_5O_7$). Ortega and Herrero [34] investigated the deposition of InX (X = P, As, Sb) by electrodeposition. In their work they showed that both InSb and InAs can be grown by direct electroplating, that as-deposited InSb films are crystalline but that InAs layers have poor crystallinity even after anneal at 400°C, and that InP layers may be produced by phosphorization of electroplated In layer in a phosphorous atmosphere. These latter films were shown, by X-ray diffraction, to be in the sphalerite phase and of good crystallinity. They employed, for InSb deposition, a bath of 0.3 M citric acid, 0.033 M $InCl_3$, and 0.047 M $SbCl_3$ at cathode potentials more positive than −0.6 V (versus saturated calomel electrode, SCE) for best results. For In As they used 0.04 M $AsCl_3$ and 0.3 M citric acid.

The electrode reactions and the corresponding potentials for In and Sb given by Pourbaix [20] are:

$$In^{3+} + 3e \longleftrightarrow In(s)$$

$$E = E^{o}_{In} + \frac{RT}{3F}\ \ln\left(\frac{a_{In^{3+}}}{a_{In}}\right) = -0.342 + 0.0197\ \log(a_{In^{3+}}) \quad (3)$$

$$HSbO_2 + 3H^+ + 3e \longleftrightarrow Sb(s) + 2H_2O$$

$$E = E^{o}_{Sb} + \frac{RT}{3F}\ \ln\left(\frac{a_{HSbO_2}}{a_{Sb}}\right) + \frac{RT}{F}\ln(C_{H^+}) = 0.23 + 0.0197\ \log(a_{HSbO_2}) - 0.0591\text{pH}$$

5 INDIUM SULPHIDE (In_2S_3)

In_2S_3 has been produced by the chalcogenization of electroplated metallic indium films on titanium substrates in flowing H_2S [35]. The metallic indium was plated from baths that contained 0.05 M $In_2(SO_4)_3$ and 0.4 M citric acid under potentiostatic conditions at −1.4 V/SCE with an indium anode. Films were deposited between 3 and 5 μm in thickness. Chalcogenization was accomplished in flowing H_2S at 130°C for 3 hours, and then the films were heated in the same atmosphere for about 1 hour at 350° to 400°C. X-ray diffraction spectra of these films showed that good quality β–In_2S_3 was produced.

6 LEAD SULFIDE (PbS)

PbS is an attractive material for a wide variety of applications including infrared detectors and Pb ion sensors. PbS was deposited electrochemically by Scharifker and coworkers [36] and more recently on Ti by Takahashi and coworkers [37]. These latter workers used an aqueous solution of $Na_2S_2O_3$ (1 mM), $Pb(NO_3)_2$ and HNO_3 (pH 2.93) and observed a cathodic current to flow at potentials more negative than −0.2 V (versus Ag/AgCl) with the reduction of $Na_2S_2O_3$ occurring in the presence of Pb^{2+}. They conclude that the formation of the PbS is due mainly to the following:

$$\begin{aligned} &S_2O_3^{2-} + 2H^+ \longrightarrow (S) + SO_2 + H_2O \\ &Pb^{2+} + (S) + 2e \longrightarrow PbS \end{aligned} \quad (4)$$

Sharon and coworkers [38] deposited PbS in a similar manner to Takahashi [37] and obtained similar results.

7 CADMIUM TELLURIDE (CdTe)

Cadmium telluride thin films are currently employed in a broad range of optoelectronic devices including solar cells, infrared detectors, and for X-ray and gamma-ray detectors [39, 40, 41]. This material is one of the more extensively studied semiconductors in terms of deposition by electrochemical means. Cadmium telluride can be deposited by pulse, potentiostatic, and galvanostatic electrodeposition techniques. The CdTe deposition details are described by Panicker and coworkers [42], Fulop and coworkers [43], Gerritsen [44], Sella and coworkers [45]. Verbrugge and Tobbias [46] provide a good overview of the deposition details associated with this material as well as an extensive list of references. Most CdTe electrodeposition makes use of an acqueous cadmium sulfate, tellurium dioxide, sulfuric acid electrolyte, and either cathodic of anodic methods may be employed. Cathodic methods make use of the codeposition of higher valent metal and chalcogenide ions, while anodic methods are based on the corrosion of the metal in a chalcogenide environment. A good review of the deposition of a number of II-VI semiconductor alloys is provided by Rajeshwar [47].

The basis for all research on cathodic electrodeposition of CdTe from aqueous solution is based on the work of Panicker and coworkers [42] in which they showed that Cd and Te can be reduced simultaneously from aqueous solution (1M $CdSO_4$, 0.01–1 mM TeO_2, pH 2.5–3.0 adjusted with H_2SO_4) to grow CdTe according to the overall reaction:

$$\mathrm{Cd^{2+} + HTeO_2^+ + 3H^+ + 6}e \longrightarrow \mathrm{CdTe + 2H_2O} \tag{5}$$

with the electrode reactions and the corresponding potentials [20]:

$$\begin{aligned}
&\mathrm{Cd^{2+}} + 2e \longleftrightarrow \mathrm{Cd(s)}\\
&E = E^o_{\mathrm{Cd}} + \frac{RT}{2F}\ln\left(\frac{a_{\mathrm{Cd^{2+}}}}{a_{\mathrm{Cd}}}\right) = -0.403 + 0.0295\ \log(a_{\mathrm{Cd^{2+}}})\\
&\mathrm{Te^{2-}} - 2e \longleftrightarrow \mathrm{Te(s)} \qquad (\mathrm{pH} > 11)\\
&E = E^o_{\mathrm{Te}} + \frac{RT}{2F}\ln\left(\frac{a_{\mathrm{Te^{2-}}}}{a_{\mathrm{Te}}}\right) = -0.94 + 0.0295\ \log(a_{\mathrm{Te^{2-}}})\\
&\mathrm{Te^{4+}} + 4e \longleftrightarrow \mathrm{Te(s)} \qquad (\mathrm{pH} < -0.5)\\
&E = E^o_{\mathrm{Te}} + \frac{RT}{4F}\ln\left(\frac{a_{\mathrm{Te^{4+}}}}{a_{\mathrm{Te}}}\right) = 0.568 + 0.0148\ \log(a_{\mathrm{Te^{4+}}})\\
&\mathrm{HTeO_2^+ + 3H^+ + 4e} \longleftrightarrow \mathrm{Te(s) + 2H_2O} \qquad (0 < \mathrm{pH} < 5)\\
&E = E^o_{\mathrm{Te}} + \frac{RT}{4F}\ln\left(\frac{a_{\mathrm{HTeO_2^+}}}{a_{\mathrm{Te}}}\right) + \frac{3RT}{4F}\ln(\mathrm{C_{H^+}})\\
&\quad = 0.551 + 0.0148\ \log(a_{\mathrm{HTeO_2^+}}) - 0.443\mathrm{pH}
\end{aligned} \tag{6}$$

CdTe deposition occurs at a more positive potential than for Cd alone because of the negative free energy of formation of CdTe of $-92\ \mathrm{kJmol^{-1}}$.

It is the choice of Cd and Te concentrations and the deposition potential which permits the growth of near stoichiometric CdTe. At the same time the substrate plays a crucial role in determining the quality of the grown film in terms of adhesion, crystallinity, and the like. CdTe has been electrodeposited on Ti and Nesatron glass substrates and characterized using X-ray diffraction, scanning electron microscopy, energy dispersive X-ray analysis, and optical transmission measurements [48]. These workers used a slightly modified bath compared to that used by Panicker and coworkers [42] where they speciated the TeO_2 with NaOH (pH 9–10) and then brought the solution into the acidic range. This approach allows for higher TeO_2 concentrations in the bath for growth of p-type films. Stoichiometric CdTe with strong preferred orientation has been grown on tin oxide coated glass [49], CdS-covered tin oxide coated glass [50–51], GaAs [52], Si [53], and other semiconductors [54]. Indium and copper-doped CdTe have been prepared by potentiostatic co-electrodeposition [42, 54].

Energy dispersive analysis of X-rays (EDAX) and X-ray diffraction (XRD) measurements indicate that CdTe films deposited by pulse techniques have reduced porosity and improved conductivity [55]. The best quality films were produced when the deposition took place at frequencies of 40 to 100 Hz. In these films a more intense $\langle 111 \rangle$ X-ray diffraction peak is observed in films produced by galvanostatic or potentiostatic methods, implying more preferentially oriented films; however, in all films Te and Cd reflections are observed. In addition EDAX indicates that these films are also more stoichiometric [56]. Guo and Deng [57] deposited CdTe on nickel substrates and those films deposited at a potential more positive than that for Cd electrodeposition were p-type while those deposited at a more negative potential were n-type. The use of a nonaqeous bath [58, 59] of anhydrous ethylene glycol 1 M $CdCl_2$, 0.01 M $TeCl_4$, and 0.33 M KI at 160°C for CdTe deposition has been explored. The aim was to eliminate hydrogen discharge, which can cause pinholes, and to allow for use of Te sources with reasonably high solubilities. KI was used to improve adhesion and to hinder ion complexation. Sugimoto and Peter [60] investigated the effects of illumination on the electrodepostion process of CdTe. These workers used a standard three electrode cell with a platinum secondary electrode, saturated calomel reference electrode, a silicon substrate, from a solution containing $2\,\mathrm{mM\,dm^{-3}}$ TeO_2, $0.5\,\mathrm{mol\,dm^{-3}}$ $CdSO_4$, $0.5\,\mathrm{mol\,dm^{-3}}$ H_2SO_4, and $0.5\,\mathrm{mol\,dm^{-3}}$ NH_4F (pH 0.5), at a potential of -0.65 V and with HeNe laser illumination to excite photocurrents during the CdTe deposition. These workers observed a dramatic effect in the growth morphology due to illumination with the nonilluminated areas being featureless, while the illuminated areas were much rougher with features associated with elemental Te observed by energy dispersive X-ray spectroscopy. These workers conclude that although the illumination intensities they used were quite high, some attention should still be paid to illumination conditions during film deposition.

Kampmann and coworkers [51] suggest the following mechanism for the electrodeposition of CdTe where surface sites s are involved:

$$\begin{aligned} HTeO_2^+ + s + 3H^+ + 4e &\longrightarrow Te,s + 2H_2O \\ Te,s + Cd^{2+} + 2e &\longrightarrow CdTe + s \end{aligned} \tag{7}$$

and point out that if the second reaction above is not fast enough the Te on the surface site can be incorporated as a Te interstitial and that for applied potentials more negative than the cadmium deposition potential (−0.4 V vs SHE) elemental Cd is deposited.

CdTe has also been deposited on Si using the method of continuous cycling of the potential [61]. In this work a solution of 0.1 M $CdSO_4$, 2×10^{-4} M TeO_2, and 0.5 M K_2SO_4 was adjusted to a pH of 1.5 ± 0.2 by the addition of H_2SO_4. A conventional three-electrode setup was employed with a saturated (KCl) calomel reference electrode and a platinum guauze as a secondary electrode. The deposition was accomplished by setting the potential at −0.2 V (to minimize the formation of a Te layer) and cycling between this value and −0.66 V. By carrying out the deposition in the dark, the formation of the Te layer is virtually eliminated. These films were annealed in an inert atmosphere at 400°C for 15 minutes but some evidence of crystallinity was found in the as-grown CdTe layer. Each cycle contributes about 50 to 60 Å to the thickness. Meulenkamp and Peter [62] provide a good discussion of the kinetics of the electrodeposition of CdTe. In their work they used solutions of TeO_2, $CdSO_4$, and Na_2SO_4 at 85°C or at room temperature. A standard three-electrode configuration was used with a Pt or Cd counterelectrode. The reference electrode was Cd wire immersed in 1.0 mol dm^{-3} $CdSO_4$ or Ag/AgCl. They conclude that at metallic substrates at which the current is diffusion limited, poor quality material is obtained. At substrates at which the current is or can be kinetically limited (including glassy carbon and silicon), poor quality material is also obtained. At other semiconductor substrates at which the current is kinetically limited, good quality material is obtained.

CdTe has been deposited on indium tin oxide coated glass plates from an acidic aqueous bath of sulphuric acid containing 0.2 M $CdSO_4$ and 1 mM $HTeO_2^+$ [63]. In this work the counter electrode, reference electrode, and substrate were a platinum sheet, saturated calomel electrode, and ITO-coated glass, respectively. A potential between −0.35 to −0.64 V was found to be best for the deposition of CdTe. Films deposited at an electrolyte bath temperature of 18°C were not crystalline, but higher deposition temperatures (80°C) resulted in X-ray diffraction spectra showing peaks corresponding to cubic CdTe. Higher current densities resulted in films with a rough surface and which could be easily removed mechanically from the substrate while films deposited at a lower current density were smooth and adhered well to the substrate.

The goal of achieving epitaxial films with long range order is extremely important for many devices to be fabricated using these semiconductor layers. Recently a new technique has been demonstrated for the deposition of CdTe via Electrochemical Atomic Layer Epitaxy (ECALE) [64, 65, 66]. This technique relies on the alternate deposition of the constitutent elements by electrochemical reduction or oxidation in the liquid phase to form the semiconductor compound. Surface-limited electrochemical reactions, also referred to as underpotential depositions, are exploited to produce single monolayers of the desired material in sequence. These workers have used this technique to deposit CdTe, CdSe, and CdS. The apparatus and details of this system are described in Gregory and Stickney [64], Gregory et al. [67], and Colletti et al. [66]. For the deposition of CdTe, 0.5 M Na_2SO_4 was used, while perclorate (0.5 M $NaClO_4$) was used for both CdS and CdSe depositions in water. 0.5 mM $HTeO_2^+$ solution was prepared with TeO_2 and 10 mM borate buffer (pH 9.2).

A 5.0 mM $HSeO^{-3}$ solution was prepared with SeO_2 and 5 mM acetic acid (pH 4.5), and an S^{-2} solution was made with 1 mM $Na_2S \cdot 9H_2O$ (pH 10.5). Substrates were silicon coated with Au (with a thin intervening layer of Ti or Cr). This technique also allows for the decomposition of the deposition process into a series of individual steps, with each step being independently optimized. Huang and coworkers [67] showed that thick films of CdTe (beyond 10 monoloyers) can be deposited with this technique however they stated that film morphology suffers above 100 cycles. This technology relies on a thorough understanding of the fundamental mechanisms for the deposition of the constituent elements in atomic monolayers, and this was investigated by this group [68, 69, 70], useful experimental data is provided in these references. The technology also been investigated in detail for Te by Hayden and Nandhakumar [71] who observed the structure of well-ordered Te layers by scanning tunneling microscopy through the deposition cycle and for Cd by Niece and Gewirth [72]. Smooth films of CdTe semiconductors have been deposited with predominantly (111) orientation, though these films appear to contain more particulates or crystallites as film thickness is increased. More recently Villegas and Napolitano [73] demonstrated a computer-controlled continuous flow systems for electrochemical atomic layer epitaxy for CdTe. Deposition rates of 30 nm h^{-1} have been demonstrated with good uniformity and composition and with CdTe ordered at the 10 to 100 nm range with preferred (111) orientation. The development of ECALE may be particularly important for the realization of films of the quality required for more demanding applications.

8 CADMIUM SELENIDE (CdSe)

CdSe in both single crystal or thin film form has been investigated for its application in solar cells, photoconductors, thin film transistors, and gamma-ray detectors. The growth of this material via electrodeposition would benefit the production of low-cost large-area devices such as solar cells. There is a fair amount of experience in the electrodeposition of CdSe films [74, 75]. These films have been deposited on Ti and Ni from an aqueous solution of HCl, $CdCl_2$, and SeO_2 [76]. CdSe thin films have been electrodeposited for some time from selenosulfate solutions [77], and a Cd/Se ratio close to 1 was demonstrated [78]. CdSe films electrodeposited on titanium substrates were characterized using a number of techniques including photoluminescence, Raman scattering, and X-ray diffraction [79].

While a number of competing reactions are discussed in the literature the deposition of CdSe proceeds via the net reaction [80]:

$$H_2SeO_3 + Cd^{2+} + 6e + 4H^+ \longrightarrow CdSe + 3H_2O \tag{8}$$

This competes with reactions producing excess Se such as that suggested by [81–82]:

$$\begin{aligned} H_2SeO_3 + 6H^+ + 6e &\longrightarrow H_2Se + 3H_2O \\ 2H_2Se + H_2SeO_3 &\longrightarrow 3Se + 3H_2O \end{aligned} \tag{9}$$

Skyllas-Kazacos and Miller [82] suggested a process to avoid the reaction of H_2SeO_3 and H_2Se by the reduction of $SeSO_3^{2-}$. Using 5 mM Na_2SeSO_3 solutions prepared by introducing 0.1 M Na_2SO_3 in 1 M NH_3/NH_4^+ buffer until the Se dissolved. At a pH 10 one has

$$Se + HSO_3^- \longleftrightarrow SeSO_3^- + H^+ \qquad (10)$$

The $SeSO_3^{-2}$ and SO_3^{-2} ions do not oxidize Se^{-2}, and thus Se contamination of the CdSe is not expected.

To avoid the deposition of excess Se, a sequential monolayer electrodeposition process has been employed [83]. In this work sequential layers of CdSe are deposited by sweeping the potential of Ti or Ni substrates continuously between −0.4 and −0.8 V (versus SCE). A conventional three electrode setup was used using a Pt counter electrode, Ti or Ni working electrode and saturated calomel reference electrode. A typical plating solution contained 0.1 M $CdCl_2$, 0.5 M HCl, and 0.3 mM SeO_2 or 0.3 M $CdSO_4$, 0.25 M H_2SO_4 and 0.3 mM SeO_2. At a scan rate of 10 V s^{-1} a submonolayer of CdSe (and far less Se) is deposited in the time it takes to reach the deposition potential for bulk Cd. During the anodic half of the scan excess Cd is stripped back off leaving behind a small amount of CdSe. This process is then repeated. Films produced in this manner were shiny in appearance and adhered well to the Ni but were easily removed from the Ti substrates (though on unpolished Ti the films adhered better). The films were stoichiometric as determined by electron microprobe analysis. These films were annealed at 650°C, after which the X-ray diffraction spectrum contained sharp lines consistent with the haxagonal wurzite structure of CdSe as had been previously observed in conventionally (potentiostatic or galvanostatic) electrodeposited CdSe [79, 84]. Gutierrez and Ortega [85] studied the corrosion properties of electrodeposited CdSe and found that the selenium layer obtained by photoetching on polycrystalline n-CdSe prevents photocorrosion.

Murali and coworkers [86] and Subramanian and coworkers [87] used a pulsed deposition technique to obtain CdSe at room temperature on SiO_2 and Ti substrates using a bath of 0.5 M $CdSO_4$ and 0.1 M SeO_2. Wyands and Cocivera [88] electrodeposited doped CdSe films and showed that it could be doped via diffusion of Se, Cd, or In during the postdeposition anneal. The Se, Cd, or In was prepared as a layer on the indium tin oxide substrate prior to deposition. The CdSe was deposited from an aqueous solution containing 0.0211 M $CdCl_2$, 0.03 M tri-sodium nitrilotriacetic acid, 0.0368 M Na_2SeSO_3, and 0.54 M Na_2SO_3. Typical conditions used by these workers included a deposition for 40 minutes (1 μm film thickness) at a potential of −1.0 V (versus a saturated calomel electrode) followed by an anneal at 500°C in a nitrogen atmosphere. In these studies carrier concentrations ranged from 10^{19} to $10^9\,cm^{-3}$ with mobilites from 1 to 50 $cm^2\ Vs^{-1}$.

Golan et al. [89] showed that it is possible to obtain epitaxial films of CdSe onto gold films from a nonaqueous electrolyte. Epitaxial CdSe films have been electrodeposited onto (111) InP substrates with the CdSe films being nearly stoichiometric [90]. In this latter work the CdSe was cathodically deposited in a three electrode cell from an acidic solution containing 0.2 M $CdSO_4$ to which various amounts of selenious acid (0.125–5 mM [Se]) were added. The pH was adjusted to 2.2, and the temperature maintained at 85°C. The CdSe layer was 30 to 250 nm thick and the ideal conditions for epitaxial growth ([Se] = 0.5 mM and deposition potential equal to the beginning of the diffusion plateau region of the IV curve −0.95 V versus SSE) were determined from a systematic reflection high energy electron diffraction

(RHEED) study. Pole figures showed six peaks rather than three corresponding to the (220) reflections indicating that the films were twinned and their full width at half maximum (FWHM) was 4°. Transmission electron microscopy (TEM) images of the CdSe/InP interface revealed the presence of planar defects, but that the interface was still quite good and that good epitaxy was achieved in spite of the large (3.6%) lattice mismatch between CdSe and InP.

9 ZINC SELENIDE (ZnSe) AND ZINC TELLURIDE (ZnTe)

Zinc selenide has been studied because of its potential for use in a number of semiconductor devices and as a window material for solar cells since it has a band gap of 2.7 eV. Pramanik and Biswas [91] investigated the deposition of ZnSe including Tl doped ZnSe in a chemical bath. Singh and Rai [92–94] deposited polycrystalline ZnSe films from solutions containing selenium dioxide and zinc sulphate and using limiting currents. Mishra and Rajeshwar [80] investigated the deposition of CdX and ZnX (X = Se or Te), and some of their results have been discussed in this chapter in the context of the deposition of some of these other materials. In the case of ZnTe and ZnSe, they employed a standard three electrode cell with a platinum counter electrode and a Ag/AgCl/1.0 M KCl reference electrode. For ZnSe, the supporting electrolyte was 0.5 M H_2SO_4 with 2 mM H_2SeO_3 and 10 mM $ZnSO_4$. A cathodic wave observed at −0.8 V was assigned by these workers to

$$Zn^{2+} + H_2SeO_3 + 4H^+ + 6e \longrightarrow ZnSe + 3H_2O \tag{11}$$

For ZnTe the conditions employed were the same but the electrolyte contained 1 mM $HTeO_2$ and 1 mM $ZnSO_4$. The primary contribution of this work was to elucidate the unifying features of the CdSe, CdTe, ZnSe, and ZnTe deposition mechanisms. Natrajan and coworkers [95] deposited ZnSe from an aqueous bath using a potentiostatic method and concluded that to get a Zn/Se ratio close to 1 the bath should contain a higher concentration of Zn^{2+} and a low concentration of H_2SeO_3. They employed a bath containing < 0.2 M $ZnSO_4$ and 0.1 to 3 mM H_2SeO_3 at a pH of 2 to 2.5 and at a temperature of 65°C. Stainless steel was the preferred substrate. A potential between −0.6 to −1.0 V was used, and under these conditions the X-ray diffraction spectra reveal peaks associated exclusively with cubic ZnSe. A number of unwanted side reactions can compete with that of Eq. (11). At sufficiently negative potentials H_2Se formed via Eq. (12) below can react with zinc ions leading to the precipitation fo ZnSe, but H_2Se may also react with selenous acid to form elemental Se:

$$\begin{aligned} &Se + 2e + 2H^+ \longrightarrow H_2Se(aq) \\ &Zn^{2+} + H_2Se \longrightarrow ZnSe + 2H^+ \\ &2H_2Se + H_2SeO_3 \longrightarrow 3Se + 3H_2O \end{aligned} \tag{12}$$

This latter reaction can lead to additional Se precipitates in the film and reduction of the Zn/Se ratio.

10 ZINC CADMIUM SELENIDE ($Zn_{1-x}Cd_xSe$)

$Zn_{1-x}Cd_xSe$ (ZCS) is a potential alternative to CdS as a window material for solar cells. Cubic ZCS has been deposited from an acidic solution containing H_2SeO_3 as Se precursor and it was observed in that work that excess Cd and Se incorporation into the film was unavoidable even with very low concentrations of their precursors in the bath [95]. An acidic plating bath was used in this latter work consisting of 0.2 M $ZnSO_4$ and 1.2 mM H_2SeO_3 with various $CdSO_4$ concentrations and the pH adjusted using 0.1 M sulphuric acid. The bath temperature was maintained at 65° to 70°C. Here the reaction proceeds according to

$$\begin{aligned} &H_2SeO_3 + 4H^+ + 4e \longrightarrow Se + 3H_2O \\ &(1-x)Zn^{2+} + xCd^{2+} + Se + 2e \longrightarrow Zn_{1-x}Cd_xSe \end{aligned} \tag{13}$$

ZCS has also been deposited from alkaline solutions by the electrodeposition of CdSe from aqueous solutions containing seleno sulfite ions as a selenium precursor [96]. Natarajan et al. [95] have also demonstrated the electrodeposition of ZCS from basic aqueous selenosulfite solutions on stainless steel and titanium substrates. In this process the reduction of Cd selenocomplex leads to CdSe nucleation followed by Se^{2-} formation from the reduction of $SeSO_3^{2-}$, and its subsequent reaction with metal ions results in the metal selenide formation. These authors used a cyclic photovoltametry process (cyclic voltametry under chopped illumination) and concluded that CdSe semiconductor formation begins at a potential of −1.1 V SCE via

$$Cd^{2+} + xSeSO_3^{2-} \longleftrightarrow Cd(SeSO_3)_x^{2-2x}(aq) \tag{14}$$

The Cd-selenosulfite complex is then reduced either directly to CdSe or to Cd:

$$\begin{aligned} &Cd(SeSO_3)_x^{2-2x} + 2e \longrightarrow CdSe + (x-1)SeSO_3^{2-} + SO_3^{2-} \\ &Cd(SeSO_3)_x^{2-2x} + 2e \longrightarrow Cd + x\,SeSO_3^{2-} \end{aligned} \tag{15}$$

The Cd thus formed may react to form CdSe:

$$Cd + SeSO_3^{2-} \longrightarrow CdSe + SO_3^{2-} \tag{16}$$

At this potential the cubic phase was dominant, while the hexagonal phase increased with an increase in the cathodic deposition potential and was predominant over the cubic phase at −1.4 V. The as deposited films were amorphous and became polycrystalline after 30 minutes, sintering at 600°C.

A second reduction peak at −1.25 V was attributed to the formation of Se^{2-} via

$$SeSO_3^{2-} + 2e \longrightarrow Se^{2-} + SO_3^{2-} \tag{17}$$

This will also lead to the formation of CdSe via

$$Cd^{2+} + Se^{2-} \longrightarrow CdSe \tag{18}$$

Zn^{2+} also forms a complex with $SeSO_3^{2-}$ and undergoes a reduction at $-1.35\,V$ leading to the formation of Zn with a very thin film of ZnSe via

$$\begin{aligned} &Zn(SeSO_3^{2-})_x^{2-2x} + 2e \longrightarrow Zn + SeSO_3^{2-} \\ &Zn + SeSO_3^{2-} \longrightarrow ZnSe + SO_3^{2-} \end{aligned} \tag{19}$$

Zn, however, reacts very weakly with $SeSO_3^{2-}$, and therefore the Zn is not converted efficiently to ZnSe. In the case of ZCS the initial reaction is the reduction of the Cd-seleno complex with the subsequent formation of Se^{2-} leading to the formation of ZCS via

$$Se^{2-} + xCd^{2+} + (1-x)Zn^{2+} \longrightarrow Zn_{(1-x)}Cd_xSe \tag{20}$$

CdS_xSe_{1-x} and $Zn_xCd_{1-x}S$ have also been electrodeposited [97]. These films were deposited onto indium tin oxide (ITO) glass substrates in dimethylsulfoxide electrolyte solution. Edamuta and Muto [97] used for CdS, $CdCl_2$ (0.055 M) and saturated sulphur (0.19 M); for CdS_xSe_{1-x}, they used $CdCl_2$ (0.055 M) saturated selenium (0.01 M) with varying concentration of sulphur between 0.05 M and 0.19 M; and for $Zn_xCd_{1-x}S$, saturated sulphur (0.019 M), $CdCl_2$, and $ZnCl_2$ where the concentration of $ZnCl_2 + CdCl_2$ remained 0.055 M and the [Zn]/[Cd] ratio in the solution varied between around 10/1 and 1/10. The deposition was carried out at 120°C, and the resulting films had an hexagonal structure with the c-axis predominantly perpendicular to the substrate. These workers were also able to determine that the films had a direct bandgap and that their dark resistivities were higher than that of pure CdS and increased with increasing Zn or Se concentration.

11 CADMIUM ZINC TELLURIDE($CD_{1-x}Zn_xTe$)

A two-stage process has been employed for the deposition of $Cd_{1-x}Zn_xTe$ in a manner similar to that employed for CdTe, $CuInSe_2$, and ZnTe. In this process thin layers of Cd, Zn, and Te are sequentially deposited and then anealled in a furnace under a controlled atmosphere [98]. Tellurium was carried out in this work in an acidic solution containing $HTeO_2^+$ (TeO_2 dissolved into a 1 M H_2SO_4 solution) and follows:

$$HTeO_2^+ + 3H^+ + 4e \longrightarrow Te + 2H_2O \tag{21}$$

Cadmium was electrodeposited using acidic $CdSO_4$ electrolytes, but the authors of this work do not give details of the Zn baths employed. Anneals at 550°C for 1 hour of films 0.6 to 1.0 μm thick resulted in $Cd_{1-x}Zn_xTe$ alloys as seen by X-ray diffraction spectra.

12 CADMIUM SULPHIDE (CdS)

CdS is widely used in solar cells in a CdS/Cu_2S heterojunction cell. A concise review of CdS solar cells, their history, and some of the methods used for the deposition of this material and the fabrication of junctions for this application is given by Hill [99]. If deposited on indium tin oxide it also has potential for use as a window material and for light meters. CdS has been formed from the annodization of the metal in alkaline sulfide solution [100] and in buffered sulphide solution [101–102]. CdS has been deposited cathodically from a solution containing cadmium and thiosulfate ions [103, 104]. Lowering the pH by the addition of dilute H_2SO_4 was shown to cause the decomposition of the $S_2O_3^{2-}$ to form colloidal sulphur and improve the deposition by elimination of the anodic stripping peak in the cyclic voltammogram compared to that observed with thiosulfate at its natural pH of 6.7. The overall electrode reaction is

$$Cd^{2+} + S_2O_3^{2-} + 2e \longleftrightarrow CdS + SO_3^{2-} \tag{22}$$

The CdS films have been deposited on Pt, Mo, and Al with the best results reported on Al. The electrode material has a significant influence on the formation of the CdS films in terms of both thickness and stoichiometry, although all films tend to contain sulpher rich regions as well as regions of cadmium rich CdS. Pulsed electrodeposition has also been used to deposit CdS from solutions of cadmium chloride and sodium thiosulfate at 90°C [105] on glass/indium tin oxide substrates. Good results were obtained and the details of this method is reported in this reference. While pulsed electrodeposition has been demonstrated by these workers it does not appear from the literature that is the preferred electrodeposition method. Edamura and Muto [97] deposited Cu-doped CdS onto indium tin oxide galvanostatically at a current density of $1.4\,mA\,cm^{-2}$ using dimethylsulphoxide solutions at 120°C containing $CdCl_2$ (0.055 M), elemental sulphur (0.19 M), and $CuCl_2$ (up to 1 mM). These workers obtained films of hexagonal crystallites with preferred c-axis orientation and with decreasing grain size as Cu content increased. Increasing the Cu content also caused the films to become porous and to adhere poorly to the substrate. Dennison [106] used both potentiostatic and galvanostatic techniques to deposit CdS and to study the reasons for the degradation in performance of the CdS electrodeposition process and to determine the mechanism of CdS formation. They suggested that along with Eq. (22) above that a second overall reaction should be considered, namely

$$2Cd^{2+} + S_2O_3^{2-} + 6H^+ + 8e \longrightarrow 2CdS + 3H_2O \tag{23}$$

Das and Morris [107] prepared CdS/CdTe solar cells by periodic pulse electrodeposition. For CdS deposition, they employed a bath of 0.2 M Cd^{2+} (from $CdCl_2$) and 0.01 M $S_2O_3^{2-}$ from sodium thiosulphate at a pH of 2 by adding HCl. For CdTe deposition, they employed a bath of 2.5 M Cd^{2+} from cadmium sulphate, $HTeO^+{}_2$ introduced from a spectroscopic grade tellurium rod in a solution of pH = 2 using H_2SO_4. The deposited layers were annealed at 400°C and X-ray diffraction spectra showed good CdS and CdTe peaks. The final solar cell efficiency was 10% indicating good quality layers and junctions. Demir and Shannon [108–110] investigated the deposition of CdS on Au via underpotential deposition and found that hexagonal CdS is preferred on both the (111) and (100) low index planes of Au. Shirai and coworkers [111] showed that Raman scattering could be used to characterize the quality of electrodeposited CdS films, and Boone and Shannon [112] employed resonance Raman and photoluminescence to characterize ultrathin films of CdS deposited via electrochemical atomic layer epitaxy. They conclude based on their work and on scanning probe microscopy that layer by layer growth is the growth mechanism in operation in electrochemical atomic layer epitaxy (ECALE), and they suggest that this growth technique may be particularly useful for the growth of samples needed for a variety of materials investigations. For films deposited using a bath similar to that of Fatas and coworkers [103], they observed an LO phonon Raman peak whose full width at half maximum decreased from 18.2 cm^{-1} for as deposited films to 12.7 cm^{-1} for films annealed at 500°C, though this is still larger than 6.5 cm^{-1} expected for single crystal CdS. It was suggested that Raman scattering might be useful as an in situ method for monitoring film quality during deposition.

CdS has also been deposited on ITO-coated glass using the potentiostatic method [113]. In this work an aqueous solution of 0.2 M $CdCl_2 \cdot 2H_2O$ and 0.01 M $Na_2S_2O_3$ was adjusted to a pH in the range of 2 to 3, using HCl and a bath temperature of 90°C. For pH greater than 3, the solution produced precipitates of cadmium hydroxide, and for pH less than 2, CdS precipitates were produced. CdS was deposited via a codeposition process obeying the overall reaction of Eq. (23). Between −0.5 and −0.6 V versus SCE the resulting films were yellow in color, indicating stoichiometric CdS. Films were annealed at 350°C in a hydrogen atmosphere for 30 minutes. For more negative potentials the films became Cd rich and then essentially metallic. Bath temperature was also found to affect growth significantly, with increasing bath temperature resulting in higher growth rates and smoothness and adhesion of the films enhanced. X-ray diffraction analysis of these films showed reflections associated with the hexagonal wurzite α-CdS phase. CdS has also been deposited from a nonaqueous bath using 0.5 M $Na_2S_2O_3$, 0.05 M $CdSO_4$, 0.1 M EDTA (ethylene-diaminetetra-acetic acid tetrasodium salt) in ethylene glycol on stainless steel, titanium and ITO-coated glass [114]. Good quality CdS was obtained with a graphite anode at 90°C and a pH of 8. Polycrystalline CdS was obtained as determined by X-ray diffraction, and no postdeposition anneal was employed by these workers.

Goto and coworkers [114] used a bath of 2 mM $CdSO_4$ and 100 mM $Na_2S_2O_3$ at a pH of 2.5 adjusted with H_2SO_4 to deposit CdS films. They studied the effect of the annealing atmosphere (O_2, air, or N_2) using photoluminescence spectroscopy. Annealing in nitrogen resulted in an increasing concentration of native defects as the anneal temperature was raised, while annealing in oxygen resulted in little native defect formation with the oxygen expected to fill any sulphur vacancies formed, since it is isoelectronic to S. Annealing in air was similar to annealing in oxygen; however, the photoluminescence linewidths were broader indicating poorer quality material.

Other materials have also been electrodeposited, though information on these is not as extensive as some of the other materials covered so far. Among these is SnS [115–116].

13 COPPER SULFIDE (Cu_2S)

The formation of junctions suitable for use as solar cells can be accomplished by replacing Cd ions of the CdS by two cuprous ions. This usually involves the substitution

$$CdS + 2CuCl_2^- \longrightarrow Cu_2S + CdCl_3^- + Cl^- \tag{24}$$

by dipping cadmium sulphide into a solution of cuprous chloride. A variant of the above suggested by Nakayama [117] is a cathodic reduction of a cupric ions in solution:

$$2Cu^{2+} + 2e + CdS \longrightarrow Cu_2S + Cd^{2+} \tag{25}$$

But this reaction is disturbed by the deposition of Cu and the formation of cupric sulphide according to

$$Cu^{2+} + CdS \longrightarrow Cd^{2+} + CuS \tag{26}$$

Vedel and coworkers [118–119] improved on this electrolytic method by using 0.1 M sodium acetate, 1 mM cupric perchlorate and 0.7 mM triethylenetetramine. They minimized Eq. (26) above by use of the triethylenetetramine masking agent with reacts with Cu(II) and with Cd(II). Al-Dhafiri and co workers [120] showed that the phase of the Cu_xS layers formed on CdS by dipping in a solution containing Cu ions can be controlled by applying a potential to the CdS substrate and that the most efficient photovoltaic cells made using this technique were produced when the CdS was biased in a narrow range between 0.01 and 0.02 V. It has been suggested that the applied bias affects the exchange of the Cd^{2+} and Cu^+ ions involved in the formation of the Cu_xS by causing the formation of a Cd-rich layer at the Cu_xS/CdS interface and preventing the diffusion of Cu^+ ions deeper into the CdS [120, 121, 122]. In their earlier work Al-Dhafiri and coworkers [121] suggested that both substrate and postdeposition anneal play an important role in determining the stability of the CdS–Cu_2S photovoltaic devices as well.

14 INDIUM SELENIDE (In_2Se_3)

In_2Se_3 forms a pseudobinary system with Cu_2Se with the formula $(Cu_2Se)_{1-x}(In_2Se_3)_x$, which for $x=0.5$ becomes $CuInSe_2$. In_2Se_3 has been electrodeposited by alternate deposition of selenium and indium from separate baths onto titanium substrates and thermal annealing [123]. The titanium was pretreated with concentrated HNO_3 : HCl (1 : 1) and NaOH 10% by weight in water. The selenium was deposited potentiostatically from an aqueous bath of 1 mM SeO_2, 0.25 M citric acid, and 0.15 M of sodium citrate at a pH of 4 with a plating potential of −0.8 V/SCE at 80°C. In was electrodeposited from 1 mM $InCl_3$, 0.8% of ethanolamine, and 0.4% of NH_3 with a pH of 2 by the addition of HCl at −1.1 vs. SCE. After deposition of In/Se in the ratio 2/3 anneals were carried out under Ar at 150° to 600°C. Films annealed at 500°C displayed X-ray diffraction spectra with the best β-hexagonal crystalline structure. Anneals at 600°C resulted in δ-In_2Se_3.

15 COPPER INDIUM DI-SELENIDE ($CuInSe_2$)

Copper Indium di-Selenide electrodeposition techniques can be divided into two categories: those in which the metals are delivered separately from the Se and those where the Se is incorporated with the metals during material deposition. An example of the first method includes the work of Dongcheng and coworkers [124] in which the authors electroplated Cu and In onto Mo or Ti-coated glass substrates followed by high-temperature anneal in H_2S at 400°C. Electrodeposition within the second category has been demonstrated by a number of workers [92, 125–126, 127, 128, 129]. Bhattacharya [125] used an SeO_2 based bath containing $InCl_3$ and $CuCl_2$ complexed by triethanolamine and ammonia. The deposits obtained were amorphous and an anneal was required to improve the crystallinity. Noufi and coworkers [130] and Hodes and coworkers [131] attempted plating Cu–In–Se from aqueous electrolytes containing SeO_2, $InCl_3$, and $CuCl_2$ without a complexing agent, but this resulted in Cu–Se deposits containing no In. Lokhande [132] employed a pulse plating technique to deposit stoichiometric $CuInSe_2$ films from an acidic bath. Ueno and coworkers [133] reported the deposition of $CuInSe_2$ from a sulphate containing bath and no complexing agent, but the films were microdendritic and required a postdeposition anneal to improve their solar conversion efficiency. Pottier and Maurin [134] pointed out that one of the difficulties associated with the direct deposition of the ternary compound is that the conditions which are favorable for the deposition of one metal differs from those necessary for the other constituent elements. For example, a difficulty arises from the very different equilibrium potentials. The electrochemical deposition reactions of the three metals are

$$
\begin{aligned}
&Cu^{2+} + 2e \longleftrightarrow Cu(s) \\
&E = E^o_{Cu} + \frac{RT}{2F} \ln\left(\frac{a_{Cu^{2+}}}{a_{Cu}}\right) = -0.2998 + 0.0295 \log(a_{Cu^{2+}}) \\
&In^{3+} + 3e \longleftrightarrow In(s) \\
&E = E^o_{In} + \frac{RT}{3F} \ln\left(\frac{a_{In^{3+}}}{a_{In}}\right) = -0.9800 + 0.0197 \log(a_{In^{3+}}) \\
&HSeO_2^+ + 4H^+ + 4e + OH^- \longleftrightarrow H_2SeO_3 + 4H^+ + 4e \longleftrightarrow Se(s) + 3H_2O \\
&E = E^o_{Se} + \frac{RT}{4F} \ln\left(\frac{a_{HSeO_2^+}}{a_{Se}}\right) + \frac{3RT}{4F} \ln(C_{H^+}) = 0.10 + 0.0148 \log(a_{HSeO_2^+}) - 0.043\text{pH}
\end{aligned}
\tag{27}
$$

where E is the electrode potential with respect to a saturated sulphate electrode (Hg/$HgSO_4$,K_2SO_4 (sat)), and $a_{Cu^{2+}}, a_{In^{3+}}, a_{HSeO_2^+}$ are the activities of the ions in the solution, while a_{Cu}, a_{In}, a_{Se} are the activities of the atoms in the electrodeposits. Pottier and Maurin [134] found optimal conditions for the deposition of $CuInSe_2$ by finding the potential range corresponding to a weak or negative slope of the vs. V curve using an acidic (pH 1.5 to 4.5) aqueous solution of CuSO4 (5–10 mM), $In_2(SO_4)_3$ (10–20 mM), SeO_2 (10 mM), and K_2SO_4 (60–80 mM) containing 0 to 80 mM sodium citrate. They suggested that the formation of smooth layers is correlated with a slow surface process. In a somwhat similar manner Sahu et al. [35] deposited $CuInSe_2$ using a solution of $CuCl_2$ (0.37 mM), $InCl_3$ (5.27 mM), and SeO_2 (0.9 mM) in distilled water. This solution was chosen so as to obtain a concentration of the ions and a pH that makes the potentials of the three elements coincident. They obtained $CuInSe_2$ films, though due to the differing nobilities of the ions the composition was not uniform. Annealing at 350°C did improve the stoichiometry and crystallinity of the films. These authors suggested that the compound formation proceeds according to

$$\begin{aligned} &CuCl \longrightarrow Cu^+ + Cl^- \\ &InCl_3 \longrightarrow In^{3+} + 3Cl^- \\ &SeO_2 + H_2O \longrightarrow HSeO_2^+ + OH^- \end{aligned} \tag{28}$$

In the initial stages of the deposition, the reaction is

$$\begin{aligned} &Cu^+ + HSeO_2^+ + 3H^+ + 5e \longrightarrow CuSe + 2H_2O \\ &\text{or} \\ &Cu^+ + 2HSeO_2^+ + 6H^+ + 9e \longrightarrow CuSe_2 + 4H_2O \end{aligned} \tag{29}$$

Later when In begins to deposit, a possible reaction is

$$\begin{aligned} &2In^{3+} + 3HSeO_2^+ + 9H^+ + 18e \longrightarrow In_2Se_3 + 6H_2O \\ &\text{or} \\ &In^{3+} + HSeO_2^+ + 3H^+ + 7e \longrightarrow InSe + 2H_2O \end{aligned} \tag{30}$$

When the three constituents deposit together, a possible reaction is

$$Cu^+ + In^{3+} + 2HSeO_2^+ + 6H^+ + 12e \longrightarrow CuInSe_2 + 4H_2O \tag{31}$$

Mishra and Rajeshwar [136] proposed a mechanism in which $CuInSe_2$ is formed by the reduction of nonstoichiometric cuprous selenide into copper and H_2Se, which then reacts with In(III) in the solution and some non-reduced Cu_2Se to form $CuInSe_2$. Khare and coworkers [137] employed a similar electrolyte $CuCl_2$ (3.7 mM), $InCl_3$ (22 mM), and SeO_2 (3.6 mM) in water at a pH of 1.5. At a potential of −0.8 V (versus SCE) these authors deposited films of 1.5 μm thickness and with grain sizes up to 400 Å after anneal of 400°C in a nitrogen/oxygen atmosphere for 20 minutes. Pern and coworkers [138] conducted a similar study in which they investigated the effects of annealing conditions on the structure of $CuInSe_2$. They showed that annealing and chemical treatments can determine whether the films are copper-rich chalcopyrites or indium-rich spalerites and that the presence of oxygen during anneal is detrimental to the films in that this results in the formation of indium oxides, loss of selenium, and a large increase in the film conductivity. Guillen and coworkers [139] deposited $CuInSe_2$ using a bath containing $CuSO_4 \cdot 5H_2O$, $In_2(SO_4)_3$, and SeO_2 in acqueous citrate solution at a pH of 1.7 (adjusted using H_2SO_4 and NaOH) on titanium substrates at −0.6 V (versus SCE). The $CuInSe_2$ was p-type, and the chalcopyrite phase was observed by X-ray diffraction with the best crystallinity obtained after 400°C anneal for 15 minutes. Gomez et al. [140] reported similar results. Kumar and coworkers [141] investigated the use of a rapid thermal annealing technique to prepare $CuInSe_2$ films from stacked Cu–In/Se layers. In this work the Cu–In film was prepared in a nonaqueous bath and the Se was deposited by thermal evaporation with the rapid thermal anneal taking place in the vacuum deposition system at temperatures of 300° to 500°C with the best results in terms of crystallinity and stoichiometry obtained at 400°C. Thouin and coworkers [142] studied the deposition of copper indium diselenide in the presence of excess In(III), and they showed that the presence of the In(III) in the solution favors the reduction of Se(IV) to Se(II) via the formation of $CuInSe_2$. Edamura and Muto [143] employed a pulse plating technique to deposit $CuInSe_2$ on indium tin oxide substrates. They used an aqueous solution of 2 mM SeO_2, 2 mM $CuSO_4$, and 20 mM $InCl_3$ at a pH of 1.5 at room temperature using an ITO glass substrate as the working electrode, platinum counterelectrode and saturated calomel reference electrode. When annealed above 300°C in air for 1 hour, these films displayed a chalcopyrite structure n-type conductivity with a direct bandgap of 0.99 eV. These films were smooth and uniform in appearance and had a compact columnar structure when viewed in cross section.

More recently Fernandez and coworkers [129] demonstrated co-electrodeposition of the three elements and a layer-by-layer electrodeposition process of the precursor materials In_2Se_3 and Cu_2Se using both the potensiostatic technique and a conventional three-electrode cell on Mo substrates. For codeposition they used different chemical baths with bath temperature maintained at 22°C and an applied potential −0.5 V versus SCE. For In, a plating bath of 50 mM $InCl_3$ with pH 1.5 adjusted with 10% HCl and an applied voltage of −1.0 V versus SCE were used. In the layer-by-layer case the precursor materials were deposited using two different chemical baths: 25 mM $InCl_3$ with 25 mM H_2SeO_3 and the second bath 25 mM of $Cu(SO_4)_2$ with 25 mM of H_2SeO_3. These workers found that the Cu/In ratio can be adjusted in the codeposition case by controlling the $InCl_3$ concentration. In the co-electrodeposited case they observed that a 500 Å Cu layer on the Mo substrate reacts with the In–Se layer to form an indium-rich layer, upon which another Cu–Se layer can be electrodeposited. After an anneal in Ar at 500°C for 10 minutes, the major phase observed via X-ray diffraction is the tetragonal $CuInSe_2$. If the In–Se layer on the Cu/Mo substrate is annealed at 450°C in Ar and then the Cu–Se layer is electrodeposited and annealed again in Ar at 500°C for 10 minutes, stoichiometric $CuInSe_2$ is obtained. The layer-by-layer method results in device quality films more easily than coelectrodeposited films. However, the morphology of the co-electrodeposited films appears better with more dense grains of larger size. A good discussion of both the multistep and single-step processes for the deposition of $CuInSe_2$ is given by Vedel [144].

16 MERCURY CADMIUM TELLURIDE ($Hg_{1-x}Cd_xTe$)

Mercury cadmium telluride (MCT) is well established as an important material for infrared detectors and imagers, and in solar cells where multilayer tandem cells are fabricated for optimal performance. Basol and coworkers [145] deposited MCT in a CdS/MCT thin film solar cell.

MCT has also been deposited onto indium tin oxide coated glass [146]. A bath was maintained at 95°C in a three-compartment cell with a Cd metal counterelectrode and a Ag/AgCl reference electrode. A constant potential of −0.99 V was applied to maintain a current density between 30 to 50 $\mu A\,cm^{-2}$. The electrolyte contained 98.9 mM tri-n-butylphosphine telluride, 0.7 mM HgI_2 and 7.1 mM $Cd(ClO_4)_2$ in propylene carbonate. Ethylenediamine (7.8 mM) was used to complex the Hg^{2+} and Cd^{2+} and to prevent the precipitation of tellurium induced by these ions. As-deposited films were quite smooth and were annealed for 1 hour at 200°C in an argon atmosphere or for 20 minutes at 300°C in air, with the latter resulting in larger grain sizes: 220 Å versus 75 Å. Temperatures above these resulted in significant Hg loss from the films. In this study also the Hg to Te ratio in the deposited films depended linearly on the ratio of Hg^{2+} and Cd^{2+} in the electrolyte. The bath temperature and potential affected the composition with the [Hg] varying from 0.14 to 0.07 when the temperature was varied from 80° to 94°C (with a potential of −1.08 V) or the voltage was varied from −1.08 to −0.98 V (with the temperature at 94°C). A direct optical bandgap was observed.

MCT has also been electrodeposited using a standard three electrode cell [147]. A saturated calomel electrode was used as a reference electrode with a large platinum sheet as the counter electrode and the working electrode was titanium or SnO_2 glass. Films were grown at a constant potential in an aqueous solution of 0.5 M $CdSO_4$, 1 mM TeO_2, and concentrations of $HgCl_2$ from 0.05 to 0.2 mM. The pH was adjusted to 1.6 with H_2SO_4, and the temperature was maintained at 90°C for titanium substrates and at 70°C for SnO_2. Current densities observed were between 2 and 4 mA cm^{-2} for the titanium substrate and 0.2 to 0.4 mA cm^{-2} for the SnO_2. Adhesion was observed to be better for the titanium. Films of 1 μm thickness were produced and the mercury content controlled by either fixing the potential (−0.65 V) but using different values of $[Hg^{2+}]$ or fixing the $[Hg^{2+}]$ and varying the potential with more negative potentials resulting in more mercury in the films. Metallic Cd appears in all films but is decreased when mercury in solution is increased. X-ray diffraction spectra do show peaks associated with MCT, and these authors do not report any postdeposition anneal process.

17 CONCLUSIONS

There has clearly been a great deal of progress in the understanding of how one may deposit a wide variety of semiconductors by electrochemical means. Detailed understanding of the chemistry of the reactions that may be expected, the effects of pH, current density, potential, temperature, and annealing conditions have been investigated by many workers. Of particular note is the improved ability to control the deposition process on an atomic scale through the use of techniques such as ECALE and the demonstration of epitaxial or nearly epitaxial growth in a limited number of cases. The driving force behind this work continues to be its relative simplicity, low cost, and scalability. Improved material quality is still required for these semiconductor layers to be widely used in all the applications for which they are intended.

REFERENCES

1. M. Paunovic and M. Schlesinger, "Fundamentals of Electrochemical Deposition," Wiley, New York, 1998.
2. R. Monnier and D. Barakat, "Contribution a l'étude du comportement de la silice dans les bains de cryo lithe fondue," *Helv. Chem. Acta*, **40**, 2041 (1957).
3. K. Grojtheim, K. Matiasovsky, P. Fellner, and A. Silny, "Electrolytic Deposition of Silicon and Silicon Alloys, Part I: Physiochemical Properties of the Na_3AlF_6–Al_2O_3–SiO_2 Mixtures," *Canad. Met. Quart.*, **10**, 79 (1971).

4. Yu. K. Delimarskii, R. V. Chernov, and I. G. Kovzum, "Phase Diagrams of the System, Potassium Fluorosilicate Potassium and Sodium Halides," *Ukrain. Khim. Zhur.*, **33**, 675 (1967).
5. Yu. K. Delimarskii, N. N. Storchak, and R. V. Chernov, "Electrodeposition of Silicon on Solid Electrodes," *Electtokhimiya*, **9**, 1443 (1973).
6. R. V. Chernov, A. P. Nizov, and Yu. K. Delimarskii, "Cathodic Processes during Isolation of Titanium Disilicide by Electrolysis of Fused Salts," *Ukrain. Khim. Zhur.*, **37**, 422 (1971).
7. A. J. Gay and J. Quarkernaat, "A Study of Electroless Siliconizing of Nickel," *J. Less Common Met.*, **40**, 21 (1975).
8. U. Cohen and R. A. Huggins, "Silicon Epitaxial Growth by Electrodeposition from Molten Fluorides," *J. Electrochem. Soc.*, **123**, 381 (1976).
9. D. Elwell and G. M. Rao, "Electrolytic Production of Silicon," *J. Appl. Electrochem.*, **18**, 15 (1988).
10. R. C. DeMattei, D. Elwell, and R. S. Feigelson, "Electrodeposition of Silicon at Temperatures above Its Melting Point," *J. Electrochem. Soc.*, **128**, 1712 (1981).
11. R. Monnier, "L'obtention et le raffinage du silicium par voie electrochimique," *Chimia*, **37**, 109 (1983).
12. A. A. Andriiko, E. V. Panov, O. I. Boiko, B. V. Yakovlev, and O. Ya. Borovik, "Dependence of the K_2SiF_6 Content in the Cathodic Deposit on the Melt Composition during Electrodeposition of Powder-like Silicon from the KCl–KF–K_2SiF_6 Melt Containing Silicon Dioxide," *Russian J. of Electrochem.*, **33**, 1343 (1997).
13. A. K. Agrawal and A. E. Austin, "Electrodeposition of Silicon from Solutions of Silicon Halides in Aprotic Solvents," *J. Electrochem. Soc.*, **128**, 2292 (1981).
14. Y. Takeda, R. Kanno, O. Yamamoto, T. R. Rama Mohan, C.-H. Lee, and F. A. Kroger, "Cathodic Deposition of Amorphous Silicon from Tetraethylorthosilicate in Organic Solvents," *J. Electrochem. Soc.*, **128**, 1221 (1981).
15. C. H. Lee and F. A. Kroger, "Cathodic Deposition of Amorphous Alloys of Silicon, Carbon, and Fluorine," *J. Electrochem. Soc.*, **129**, 936 (1982).
16. R. Boen and J. Bouteillon, "The Electrodeposition of Silicon in Fluoride Melts," *J. Appl. Electrochem.*, **13**, 277 (1983).
17. J. Gobet and H. Tannenberger, "Electrodeposition of Silicon from a Nonaqueous Solvent," *J. Electrochem. Soc.*, **135**, 109 (1988).
18. J. DeLepinay, J. Bouteillon, S. Traore, D. Renaud, and M. J. Barbier, "Electroplating Silicon and Titanium in Molten Fluoride Media," *J. Appl. Electrochem.*, **17**, 294 (1987).
19. D. B. Frolenko, Z. S. Martem'yanova, Z. I. Valeev, and A. N. Baraboshkin, "Structure of Silicon Deposits Obtained by Electrolyzing Mixed Choloride-Fluoride Melts," *Soviet Electrochem.*, **28**, 1427 (1993).
20. M. Pourbaix, *Atlas d' equilibres electrochimiques*, Gauthiers-Villars, Paris, 1963.
21. R. C. DeMattei, D. Elwell, and R. S. Feigelson, "The Synthesis of GaAs by Molten Salt Electrolysis," *J. Crystal Growth*, **43**, 643 (1978).
22. I. G. Dioum, J. Vedel, and B. Tremillon, "Properties of Arsenic in Molten Potassium Tetrachlorogallate at 300 C: Formation of Gallium Arsenide," *J. Electroanal. Chem.*, **139**, 329 (1982).
23. S. P. Wicelinski and R. J. Gale, Abstract 1495, p. 2040, *Electrochemical Society Extended Abstracts*, vol. 2, Honolulu, HI, Oct. 18–23, 1987.
24. M. K. Carpenter and M. W. Verbrugge, Abstract 257, "Electrodeposition of Gallium Arsenide from a Room-Temperature Chlorogallate Melt," p. 367, *Electrochemical Society Extended Abstracts*, vol. 2, Hollywood, FL, Oct. 15–20, 1989.

25. M. K. Carpenter and M. W. Verbrugge, "Electrochemical Codeposition of Gallium and Arsenic from a Room Temperature Chlorogallate Melt," *J. Electrochem. Soc.*, **137**, 123 (1990).

26. L. Astier, R. Michelis-Quiriconi, and M.-L. Astier, Fr. Demande FR 2,529,713 (Jan. 6, 1984). Appl. 82/11,797 (July 2, 1982).

27. K. R. Murali, M. Jayachandran, and N. Rangarajan, "Review of Techniques on Growth of GaAs and Related Compounds," *Bull. Electrochem.*, **3**, 261 (1987).

28. M.-C. Yang, U. Landau, and J. C. Angus, "Electrodeposition of GaAs from Aqueous Electrolytes," *J. Electrochem. Soc.*, **139**, 3480 (1992).

29. I. Villegas and J. L. Stickney, "Preliminary Studies of GaAs Deposition on Au(100), (110), and (111) Surfaces by Electrochemical Atomic Layer Epitaxy," *J. Electrochem. Soc.*, **139**, 686 (1992).

30. J. J. Cuomo and R. J. Gambino, "The Synthesis and Epitaxial Growth of GaP by Fused Salt Electrolysis," *J. Electrochem. Soc.*, **115**, 755 (1968).

31. C. F. Smart, U.S. Patent 2,755,537 (1956) and German Patent 929,103 (1955).

32. Y. N. Sadana and J. P. Singh, "Electrodeposition and X-ray Structure of Antimony-Indium Alloys," *Plating Surf. Finish.*, **64** (1985).

33. T. Okubo and U. Landau, "Electrodeposition of Indium Antimonide Semiconductor from Aqueous Electrolyte," *Electrochem. Soc. Abs.*, vol. 2, Chicago, October 9–14, 1988.

34. J. Ortega and J. Herrero, "Preparation of InX ($X = \text{P, As, Sb}$) Thin Films by Electrochemical Methods," *J. Electrochem. Soc.*, **136**, 3388 (1989).

35. J. Herrero and J. Ortega, "n-Type In_2S_3 Thin Films Prepared by Gas Chalcogenization of Metallic Electroplated Indium: Photoelectrochemical Characterization," *Solar Energy Mat.*, **17**, 357 (1988).

36. B. Scharifker, Z. Ferreira, and J. Mozota, "Electrodeposition of Lead Sulphide," *Electrochim. Acta*, **30**, 677 (1985).

37. M. Takahashi, Y. Ohshima, K. Nagata, and S. Furuta, "Electrodeposition of PbS Films From Acidic Solution," *J. Electroanal. Chem.*, **359**, 281 (1993).

38. M. Sharon, K. S. Ramaiah, M. Kumar, M. Neumann-Spallart, and C. Levy-Clement, "Electrodeposition of Lead Sulphide in Acidic Medium," *J. Electroanal. Chem.*, **436**, 49 (1997).

39. S. A. Ringel, A. W. Smith, M. H. MacDougal, and A. Rohatgi, "The Effects of $CdCl_2$ on the Electronic Properties of Molecular-Beam Epitaxially Grown CdTe/CdS Heterojunction Solar Cells," *J. Appl. Phys.*, **70**, 881 (1991).

40. C. K. Ard, "*Properties of Narrow Gap Cd-based Compounds*," P. Capper, ed., Inspec, Park Ridge, NJ, 1994, p. 598, and references therein.

41. C. L. Johnson, E. E. Eissler, S. E. Cameron, Y. Kong, S. Fan, and K. G. Lynn, "Crystallographic and Metallurgical Characterization of Radiation Detector Grade Cadmium Telluride Materials," in *Semiconductors for Room Temperature Radiation Detector Applications*, R. B. James, T. E. Schlesinger, P. Siffert, and L. Franks, eds., *Mater. Res. Soc.*, 477 (1993).

42. M. P. R. Panicker, M. Knaster, and F. A. Kroger, "Cathodic Deposition of CdTe from Aqueous Electrolyte," *J. Electrochem. Soc.*, **52**, 566 (1978).

43. G. Fulop, M. Doty, P. Meyers, and C. H. Liu, "High-Efficiency Electrodeposited Cadmium Telluride Solar Cells," *Appl. Phys. Lett.*, **40**, 327 (1982).

44. H. J. Gerritsen, "Electrochemical Deposition of Photosensitive CdTe and ZnTe on Tellurium," *J. Electrochem. Soc.*, **131**, 136 (1984).

45. C. Sella, P. Boncorps, and J. Vedel, "The Electrodeposition Mechanism of CdTe from Acidic Aqueous Solutions," *J. Electrochem. Soc.*, **133**, 2043 (1986).

46. M. W. Verbrugg and C. W. Tobbias, "The Periodic Electrochemical Codeposition of Cadmium Telluride," *AlChE J.*, **33**, 628 (1987).

47. K. Rajeshwar, "Electrosynthesized Thin Films of Group II–VI Compound Semiconductors, Alloys, and Superstructures," *Adv. Mater.*, **4**, 23 (1992), and references therein.

48. R. N. Bhattacharya and K. Rajeshwar, "Electrodeposition of CdTe Thin Films," *J. Electrochem. Soc.*, **131**, 2032 (1984).

49. P. Cowache, D. Lincot, and J. Vedel, "Cathodic Codeposition of Cadmium Telluride on Conducting Glass," *J. Electrochem. Soc.*, **136**, 1646 (1989).

50. A. Kampmann, P. Cowache, B. Mokili, D. Lincot, and J. Vedel, "Characterization of $\langle 111 \rangle$ Cadmium Telluride Electrodeposited on Cadmium Sulphide," *J. Crystal Growth*, **146**, 256 (1995a).

51. A. Kampmann, P. Cowache, J. Vedel, and D. Lincot, "Investigation of the Influence of the Electrodeposition Potential on the Optical, Photoelectrochemical and Structural Properties of As-deposited CdTe," *J. Electroanal. Chem.*, **387**, 53 (1995b).

52. P. Sircar, "Growth of CdTe on GaAs by Electrodeposition from an Aqueous Electrolyte," *Appl. Phys. Lett.*, **53**, 1184 (1988).

53. J. M. Fisher, L. E. A. Berlouis, L. J. M. Sawers, S. M. MacDonald, S. Affrossman, D. J. Diskett, and M. G. Astles, "Growth and Characterization of Electrodeposited Films of Cadmium Telluride on Silicon," *J. Crystal Growth*, **138**, 86 (1994).

54. J. A. Von Windheim and M. Cocivera, "Variation of Resistivity of Copper-Doped Cadmium Telluride Prepared by Electrodeposition," *J. Phys. D.*, **23**, 581 (1990).

55. S. M. Babu, R. Dehanasekaran, and P. Ramasamy, "Electrodeposition of CdTe by Potentiostatic and Pulse Technique," *Thin Solid Films*, **202**, 67 (1991).

56. P. D. Paulson, V. Dutta, and C. Sing, "A Comparative Study of Compositional and Structural Properties of CdTe Thin Films Prepared by Different Electroplating Techniques," *Proc. First World Conf. on Photovoltaic Energy Conversion*, **1**, 331 (1994).

57. Y. Guo and X. Deng, "Electrodeposition of CdTe Thin Films and Their Photoelectrochemical Behavior," *Solar Energy Mater. Solar Cells*, **29**, 115 (1993).

58. R. B. Gore, R. K. Pandey, and S. K. Kulkarni, "Investigation of Deposition Parameters for the Nonaqueous Electroplating of CdTe Films and Application in Electrochemical Photovoltaic Cells," *Sol. Energy Mater.*, **18**, 159 (1989).

59. R. K. Pandey, S. Maffi, and L. P. Bicelli, "Study of CdTe Electrodeposition from Nonaqueous Bath," *Mater. Chem. Phys.*, **35**, 15 (1993).

60. Y. Sugimoto and L. M. Peter, "Photoeffects during Cathodic Electrodeposition of CdTe," *J. Electroanal. Chem.*, **386**, 183 (1995).

61. F. Jackson, L. E. A. Berlouis, and P. Rocabois, "Layer-by-layer Electrodeposition of Cadmium Telluride onto Silicon," *J. Crystal Growth*, **159**, 200 (1996).

62. E. A. Meulenkamp and L. M. Peter, "Mechanistic Aspects of the Electrodeposition of Stoichiometric CdTe on Semiconductor Substrates," *J. Chem. Soc. Faraday, Trans.*, **92**, 4077 (1996).

63. M. Rami, E. Benamar, M. Fahoume, F. Chraibi, and A. Ennaoui, "Formation of CdTe by Electrodeposition: Thermodynamic Aspect," *Ann. Chim. Sci. Mat.*, **23**, 365 (1998).

64. B. W. Gregory and J. L. Stickney, "Electrochemical Atomic Layer Epitaxy (ECALE)," *J. Electroanal. Chem.*, **300**, 543 (1991).

65. B. W. Gregory, D. W. Suggs, and J. L. Stickney, "Conditions for the Deposition of CdTe by Electrochemical Atomic Layer Epitaxy," *J. Electrochem. Soc.*, **138**, 1279 (1991).

66. L. P. Colletti, B. H. Flowers Jr., and J. L. Stickney, "Formation of Thin Films of CdTe, CdSe, and CdS by Electrochemical Atomic Layer Epitaxy," *J. Electrochem. Soc.*, **145**, 1442 (1998).
67. B. M. Huang, L. P. Colletti, B. W. Gregory, J. L. Anderson, and J. L. Stickney, "Preliminary Studies of the Use of an Automated Flow-Cell Electrodeposition System for the Formation of CdTe Thin Films by Electrochemical Atomic Layer Epitaxy," *J. Electrochem. Soc.*, **142**, 3007 (1995).
68. B. W. Gregory, M. L. Norton, and J. L. Stickney, "Thin-Layer Electrochemical Studies of the Underpotential Deposition of Cadmium and Tellurium on Polycrystalline Au, Pt, and Cu Electrodes," *J. Electroanal. Chem.*, **293**, 85 (1990).
69. L. P. Colletti, D. Teklay, and J. L. Stickney, "Thin-layer Electrochemical Studies of the Oxidative Underpotential Deposition of Sulphur and Its Application to the Electrochemical Atomic Layer Epitaxy Deposition of CdS," *J. Electroanal. Chem.*, **369**, 145 (1994).
70. L. P. Colletti, S. Thomas, E. M. Wilmer, and J. L. Stickney, "Thin Layer Electrochemical Studies of ZnS, ZnSe, and ZnTe Formation by Electrochemical Atomic Layer Epitaxy (ECALE)," *Mat. Res. Soc. Symp. Proc.*, **451**, 235 (1996).
71. B. E. Hayden and I. S. Nandhakumar, "In-situ STM Study of Te UPD Layers on Low Index Planes of Gold," *J. Phys. Chem. B.*, **101**, 7751 (1997).
72. B. K. Niece and A. A. Gewirth, "Potential-Step Chronocoulometric and Quartz Crystal Microbalance Investigation of Coadsorbed Cadmium and Sulphate on Au (111) Electrodes," *Langmuir*, **13**, 6302 (1997).
73. I. Villegas and P. Napolitano, "Development of a Continuous-Flow System for the Growth of Compound Semiconductor Thin Films via Electrochemical Atomic Layer Epitaxy," *J. Electrochem. Soc.*, to be Published.
74. Y. Avigal, D. Cahen, J. Manassen, and G. Hodes, "Solar Energy Conversion and Storage by a Photoelectrochemical Storage Cell," *Proc. 1977 Photovoltaic Solar Energy Conf.*, Reidel, Boston, 1302 (1977).
75. S. Chandra and R. K. Pandey, "Photoelectrochemical Cell for Solar Energy Conversion Using Electrocodeposited CdSe Films," *Phys. Stat. Sol.*, **59**, 787 (1980).
76. D. J. Miller and D. Haneman, "Preparation of Stable Efficient CdSe Films for Solar PEC Cells," *Solar Energy Materials*, **4**, 223 (1981).
77. J. P. Szabo and M. Cocivera, "Effect of Annealing Atmosphere on the Properties of Thin-Film CdSe," *J. Appl. Phys.*, **61**, 4820 (1987).
78. J. P. Szabo and M. Cocivera, "Composition and Performance of Thin Film CdSe Electrodeposited from Selenosulfite Solution," *J. Electrochem. Soc.*, **133**, 1247 (1986).
79. F. Cerdiera, I. Torriani, P. Motisuke, V. Lemos, and F. Decker, "Optical and Structural Properties of Polycrystalline CdSe Deposited on Titanium Substrates," *Appl. Phys. A*, **46**, 107 (1988).
80. K. K. Mishra and K. J. Rajeshwar, "A Re-examination of the Mechanisms of Electrodeposition of CdX and ZnX (X = Se, Te) Semiconductors by the Cyclic Photovoltammetric Technique," *J. Electroanal. Chem.*, **273**, 169 (1989).
81. M. Skyllas-Kazacos and B. Miller, "Studies in Selenious Acid Reduction and CdSe Film Deposition," *J. Electrochem. Soc.*, **127**, 869 (1980a).
82. M. Skyllas-Kazacos and B. Miller, "Electrodeposition of CdSe Films from Selenosulfite Solution," *J. Electrochem. Soc.*, **127**, 2378 (1980b).
83. A. M. Kressin, V. V. Doan, J. D. Klein, and M. J. Sailor, "Synthesis of Stoichiometric Cadmium Selenide Films via Sequential Monolayer Electrodeposition," *Chem. Mater.*, **3**, 1015 (1991).

84. M. Abramovich, M. J. P. Brasil, F. Decker, J. R. Moro, P. Motisuke, N. Muller-St., and P. Salvador, "Crystal Structure, Luminescence, and Photoelectrochemistry of Thin Electroplated Cd-chalcogenide Layers," *J. Solid State Chem.*, **59**, 1 (1985).

85. M. T. Gutierrez and J. Ortega, "Photoelectrochemical Study of Electrodeposited Polycrystalline CdSe in Ferro-Ferricyanide System," *J. Electrochem. Soc.*, **136**, 2316 (1989).

86. K. R. Murali, V. Suramanian, N. Rangarajan, A. S. Lakshmanan, and S. K. Rangarajan, "Preparation and Characterization of Pulse Plated CdSe Films," *SPIE Proc.*, **1523**, 121 (1992).

87. V. Subramanian, K. R. Murali, N. Rangarajan, and A. S. Lakshmanan, "Succesively Pulse Plated Cadmium Selenide Films and Their Photoelectrochemical Behaviour," *Bull. Mater. Sci.*, **17**, 10489 (1994).

88. H. Wyands and M. Cocivera, "Hall Effect and Resistivity Characterization of Doped Electrodeposited CdSe," *J. Electrochem. Soc.*, **139**, 2052 (1992).

89. Y. Golan, L. Margulis, G. Hodes, I. Rubinstein, and J. L. Hutchinson, "Electrodeposited Quantum Dots; High Resolution Electron Microscopy of Epitaxial CdSe Nanocrystals on (111) Gold," *Surf. Sci.*, **L633**, 311 (1994).

90. H. Cachet, R. Cortes, M. Froment, and G. Maurin, "Epitaxial Electrodeposition of Cadmium Selenide Thin Films on Indium Phosphide Single Crystals," *J. Solid State Electrochem.*, **1**, 100 (1997).

91. P. Pramamik and S. Biswas, "Deposition of Zinc Selenide Thin Films by Solution Growth Technique," *J. Electrochem. Soc.*, **133**, 350 (1986).

92. K. Singh and J. P. Rai, "Electrosynthesis of Polycrystalline Photoelectroactive p-zinc Selenide," *J. Mater. Sci. Lett.*, **4**, 1401 (1985).

93. K. Singh and J. P. Rai, "Electrosynthesis and Photoelectroactivity of Polycrystalline p-Zinc Selenide," *Phys. Stat. Sol. (a)*, **99**, 257 (1987).

94. K. Singh and J. P. Rai, "Electrosynthesis and Photoelectroactivity of Thallium Doped Polycrystalline Zinc Selenide," *B. Electrochem.*, **5**, 876 (1989).

95. C. Natarajan, H. Matsumoto, and G. Nogami, "Mechanism of Electrodeposition of MSe (M = Cd and Zn) Films from Selenosulfite Solution," *Bull. Electrochem.*, **13**, 123 (1997); C. Natarajan, G. Nogami, M. Sharon, "Electrodeposition of $Zn_{1-x}Cd_xSe$ ($x = 0$–1) Thin Films," *Thin Solid Films*, **261**, 44 (1995); C. Natarajan, G. Nogami, M. Sharon, "Electrodeposition and Photovoltaic Properties of Zinc Cadmium Selenide Thin Films," *Bulletin of Electrochemistry*, **12**, 136 (1996).

96. A. Darkowski and A. Grabowski, "Electrodeposition of Cd–Zn–Se Thin Films from Selenosulphite Solutions," *Solar Energy Mater.*, **23**, 75 (1991).

97. T. Edamura and J. Muto, "Preparation and Properties of Electrodeposited Ternary CdS_xSe_{1-x} and $Zn_xCd_{1-x}S$ Films," *Thin Solid Films*, **226**, 135 (1993).

98. B. M. Basol, V. K. Kapur, and M. L. Ferris, "Low Cost Technique for Preparing $Cd_{1-x}Zn_xTe$ Films for Solar Cells," *J. Appl. Phys.*, **66**, 1816 (1989).

99. R. Hill, "Cadmium-Sulphide/Copper-Sulphide Thin-Film Solar Cells: Review of Methods of Producing the CdS and Cu_2S Layers," *Solid-State Electron. Dev.*, **2**, S49 (1978).

100. B. Miller, S. Menezes, and A. Heller, "Anodic Formation of Semiconductive Sulfide Films at Cadmium and Bismuth Rotating Ring-Disk Electrode Studies," *J. Electroanal. Chem.*, **94**, 85 (1978).

101. L. M. Peter, "The Electrocrystallisation of Cadmium Sulphide Films on Cadmium," *Electrochimica Acta*, **23**, 165 (1978a).

102. L. M. Peter, "The Photoelectrochemical Properties of Anodic Cadmium Sulphide Films," *Electrochimica Acta*, **23**, 1073 (1978b).

103. E. Fatas, R. Duo, P. Herrasti, F. Arjona, and E. Garcia-Camarero, "Electrochemical Deposition of CdS Thin Films on Mo and Al Substrates," *J. Electrochem. Soc.*, **131**, 2243 (1984).

104. G. P. Power, D. R. Peggs, and A. J. Parker, "The Cathodic Formation of Photoactive Cadmium Sulfide Films from Thiosulfate Solutions," *Electrochimica Acta*, **26**, 681 (1981).

105. G. C. Morris and R. Vanderveen, "Cadmium Sulphide Films Prepared by Pulsed Electrodeposition," *Solar Energy Mat.*, **27**, 305 (1992).

106. S. Dennison, "Studies of the Cathodic Electrodeposition of CdS from Aqueous Solution," *Electrochimica Acta*, **38**, 2395 (1993).

107. S. K. Das and G. C. Morris, "Preparation and Properties of CdS/CdTe Thin Film Solar Cell Produced by Periodic Pulse Electrodeposition Technique," *Solar Energy Mat. Solar Cells*, **30**, 107 (1993).

108. U. Demir and C. Shannon, "A Scanning Tunneling Microscopy Study of Electrochemically Grown Cadmium Sulphide Monolayers on Au (111)," *Langmuir*, **10**, 2794 (1994).

109. U. Demir and C. Shannon, "Reconstruction of Cadmium Sulfide Monolayers on Au (100)," *Langmuir*, **12**, 594 (1996a).

110. U. Demir and C. Shannon, "Electrochemistry of Cd at ($\sqrt{3} \times \sqrt{3}$) R30o–S/Au(111): Kinetics of Structural Changes in CdS Monolayers," *Langmuir*, **12**, 6091 (1996).

111. K. Shirai, Y. Moriguchi, M. Ichimura, A. Usami, and M. Saji, "Relationship between Raman Spectra and Crystallinity of CdS Films Grown by Cathodic Electrodeposition," *Jpn. J. Appl. Phys.*, **35**, 2057 (1996).

112. B. E. Boone and C. Shannon, "Optical Properties of Ultrathin Electrodeposited CdS Films Probed by Resonance Raman Spectroscopy and Photoluminescence," *J. Phys. Chem.*, **100**, 9480 (1996).

113. G. Sasikala, R. Dhanasekaran, and C. Subramanian, "Electrodeposition and Optical Characterisation of CdS Thin Films on ITO-Coated Glass," *Thin Solid Films*, **302**, 71 (1997).

114. F. Goto, K. Shirai, and M. Ichimura, "Defect Reduction in Electrochemically Deposited CdS Thin Films by Annealing in O_2," *Solar Energy Mat. Solar Cells*, **50**, 147 (1998).

115. Z. Zainal, M. Z. Hussein, A. Kassim, and A. Ghasali, "Electrodeposited SnS Thin-Films from Aqueous-Solution," *J. Mater. Sci. Lett.*, **16**, 1446 (1997).

116. Z. Zainal, M. Z. Hussein, and A. Ghasali, "Cathodic Electrodeposition of SnS Thin-Films from Aqueous-Solution," *Solar Energy Mater. Solar Cells*, **40**, 347 (1996).

117. N. Nakayama, "Ceramic CdS Solar Cell," *Jpn. J. Appl. Phys.*, **8**, 450 (1969).

118. J. Vedel, P. Cowache, and D. Lincot, "Electrochemical Preparation and Conditioning of Cu_2S for Cu_2S-CdS Solar Cells," *Proc. 4th European Communities Photovolatic Solar Energy Conf.*, Stresa, Italy, May 10–14, 1982.

119. J. Vedel, P. Cowache, and M. Soubeyrand, "Electroplating of Cu_xS on CdS," *Solar Energy Mat.*, **10**, 25 (1984).

120. A. M. Al-Dhafiri, G. J. Russell, and J. Woods, "Electrochemical Control of the Cu_xS phase in Cu_xS–CdS Photovoltaic Cells," *Semicond. Sci. Technol.*, **6**, 983 (1991).

121. A. M. Al-Dhafiri, P. C. Pande, G. J. Russell, and J. Woods, "Electroplated Cu_xS–CdS Photovoltaic Cells," *J. Crystal. Growth*, **86**, 900 (1988).

122. A. C. Rastogi and S. Salkalachen, "Improvements in Stoichiometry and Stability of p-Cu_xS Thin-Film Solar Cells," *J. Appl. Phys.*, **58**, 4442 (1985).

123. J. Herrero and J. Ortega, "Electrochemical Synthesis of Photoactive In_2Se_3 Thin Films," *Solar Energy Mat.*, **16**, 477 (1987).

124. W. Dongcheng, L. Tao, L. Hua, and L. Zuming, "Electrochemical Depositing $CuInSe_2$ Thin Films," *Proc. 1989 Congress of Int. Solar Energy Society*, Kobe City, Japan, Sept. 4–8, 1989, **1**, 232 (1990).

125. R. N. Bhattacharya, "Solution Growth and Electrodeposited $CuInSe_2$ Thin Films," *J. Electrochem. Soc.*, **130**, 2040 (1983).

126. R. N. Bhattacharya and K. Rajeshwar, "Electrodeposition of $CuInX$ (X = Se, Te) Thin Films," *Solar Cells*, **16**, 237 (1986).

127. V. K. Kapur, B. M. Basol, and E. S. Tseng, "Low Cost Methods for the Production of Semiconductor Films for $CuInSe_2$/CdS Solar Cells," *Solar Cells*, **21**, 65 (1987).

128. C. Guillen and J. Herrero, "Optical Properties of Electrochemically Deposited $CuInSe_2$ Thin Films," *Solar Energy Mat.*, **25**, 31 (1991).

129. A. M. Fernandez, P. J. Sebastain, A. M. Hermann, R. N. Bhattacharya, R. N. Noufi, and M. A. Contreras, "Electrodeposition of $CuInSe_2$ Thin Films for Photovoltaic Application," *World Renewable Energy Congress*, A. A. M. Sayigh, ed., vol. 8, 396 (1996).

130. R. Noufi, R. Powell, R. Axton, Y. Chen, and T. Datta, *1984 Annual Report Solid State Photovoltaic Research Branch*, SERI/PR-212-2601 DE85008812, Apr. 1985.

131. G. Hodes and D. Cahen, "Electrodeposition of $CuInSe_2$ and $CuInS_2$ Films," *Solar Cells*, **16**, 245 (1986).

132. C. D. Lokhande, "Pulse Plated Electrodeposition of $CuInSe_2$ Films," *J. Electrochem. Soc.*, **134**, 1727 (1987).

133. Y. Ueno, H. Kawai, T. Sugiura, and H. Minoura, "Electrodeposition of $CuInSe_2$ Films from Sulphate Bath," *Thin Solid Films*, **157**, 159 (1988).

134. D. Pottier and G. Maurin, "Preparation of Polycrystalline Thin Films of $CuInSe_2$ by Electrodeposition," *J. Appl. Electrochem.*, **19**, 361 (1989).

135. S. N. Sahu, R. D. L. Kristensen, and D. Haneman, "Electrodeposition of $CuInSe_2$ Thin Films from Aqueous Solutions," *Solar Energy Mat.*, **18**, 385 (1989).

136. K. K. Mishra and K. Rajeshwar, "A Voltammetric Study of the Electrodeposition Chemistry in the Cu + In + Se System," *J. Electroanal. Chem.*, **271**, 279 (1989b).

137. N. Khare, G. Razzini, and L. P. Bicelli, "Electrodeposition and Heat Treatment of $CuInSe_2$ Films," *Thin Solid Films*, **186**, 113 (1990).

138. F. J. Pern, R. Noufi, A. Mason, and A. Franz, "Characterizations of Electrodeposited $CuInSe_2$ Thin Films: Structure, Deposition, and Formation Mechanism," *Thin Solid Films*, **202**, 299 (1991).

139. C. Guillen, E. Galiano, and J. Herrero, "Cathodic Electrodeposition of $CuInSe_2$ Thin Films," *Thin Solid Films*, **195**, 137 (1991).

140. H. Gomez, R. Schrebler, L. Basaez, and E. A. Dalchiele, "Electrochemical Growth and Characterization of Polycrystalline $CuInSe_2$ Thin Films," *J. Phys. Condens. Matter*, **5**, A349 (1993).

141. S. R. Kumar, R. B. Gore, and R. K. Pandey, "Properties of $CuInSe_2$ Films Prepared by the Rapid Thermal Annealing Technique," *Thin Solid Films*, **223**, 109 (1993).

142. L. Thouin, S. Massaccesi, S. Sanchez, and J. Vedel, "Formation of Copper Indium Selenide by Electrodeposition," *J. Electroanal. Chem.*, **374**, 81 (1994).

143. T. Edamura and J. Muto, "Preparation and Characterization of Pulse-plating Electrodeposited $CuInSe_2$ Thin Films," *J. Mater. Sci: Mater. Electron.*, **5**, 275 (1994).

144. J. Vedel, "An Electrochemical Route for the Preparation of Chalcopyrite Semiconductors," *Inst. Phys. Conf. Set. No. 152: Section B: Thin Film Growth and Characterization.* Presented at 11th Int. Conf. on Ternary and Multilayer Compounds, Salford (Sept. 8–12, 1998), p.261.

145. B. M. Basol, O. M. Stafsudd, and A. Bindal, "Thin Films of Mercury Cadmium Telluride for Solar Cell Applications," *Solar Cells*, **15**, 279 (1985).

146. C. L. Colyer and M. Cocivera, "Thin-Film Cadmium Mercury Telluride Prepared by Nonaqueous Electrodeposition," *J. Electrochem. Soc.*, **139**, 406 (1992).

147. J. Ramiro and A. Garcia Camarero, "Influence of Deposition Potential and Electrolyte Composition on the Structural and Photoelectrochemical Properties of Electrodeposited Mercury Cadmium Telluride," *J. Mat. Sci.*, **31**, 2047 (1996).

15 Deposition on Nonconductors

MORDECHAY SCHLESINGER

INTRODUCTION

At about the same time that electroplating of silver was first being practiced (circa 1840), plating on nonconductors was developed for the purpose of electroforming and for making copper engraving plates. Nearly a century ago the art entered into its artistic phase, producing metallic artistic elements on glass, wood, and the like. The one surviving remnant of this phase is the gold, or copper-plated baby shoe. Earlier in this century the art was put to use in the service of more practical applications. Those include graphic arts, toys, buttons, records, and alike.

In most of these applications a number of stages in the plating process were required. Typically the first of these was a roughening process. That was done mostly mechanically and sometimes chemically. As an example of the former we mention sandblasting, while as an example of the latter, etching of glass with hydrofluoric acid comes to mind. The second stage is that of sealing, if required, such as in the case of wooden objects. The third and most important stage is the application of the conductive layer which then will make eventual electroplating possible. Here a number of methods were available to the plater:

- Bronzing. Metallic powder mixed in varnish is applied with a brush. A silver immersion dip coating, to improve conductivity, follows.
- Graphiting. Graphite powder with or without a lacquer binder is applied with a brush.
- Metallic painting. Fused metallic paint, silver base, is dispersed in a flux, applied, and subsequently fired to above 500°C.
- Metallizing. Metallic coating is directly produced by eletrochemical (other than electroless) methods such as a $SnCl_2$ immersion followed by immersion in a silvering bath (consisting of silver nitrate and ammonium hydroxide). That bath is not autocatalytic.

The next stage was electroplating. Since the initial conductivity is, as a rule, marginal, a copper strike solution was used to deposit the first 8 to 12 μm, and a standard electroplating solution for the rest. (A typical strike solution was made of $75\,g\,liter^{-1}$ $CuSO_4 \cdot H_2O$ as well as $2\,ml\,liter^{-1}$ H_2SO_4, resulting in a solution

Modern Electroplating, Fourth Edition, Edited by Mordechay Schlesinger and Milan Paunovic.
ISBN 0-471-16824-6

pH value of 2–2.5.) The final stage was that of polishing the finished product for appearance. Care had to be exercised not to overheat and thus damage the coating during polishing.

1 RECENT DEVELOPMENTS

In the corresponding chapter by E. B. Saubestre in the previous edition, 1960 is stated as the watershed year. The reason is the number of important developments that took place then and changed the practice of plating on nonconductors from an art to a practical science, along the following lines:

1. Plateable plastics. It become evident to resin (particularly ABS) producers that the composition of the copolymer and the molding conditions could be adapted to facilitate the subsequent exposure to electrolytes used in the preparation of surfaces for electroplating.
2. Chemical conditioners. It was found that variants of older etchants for plastics could be used on plastics (e.g., plateable ABS) properly molded to create enough mechanical and bonding sites to obviate the need for mechanical roughening processes. In addition there became less need or no need at all for a thick copper layer for encapsulation.
3. Electroless plating. It is customary to state that electroless (autocatalytic) plating was invented by Abner Brener in the 1940s, but it was not until about 1960 that improved copper baths showed up on the market (proprietary mostly). Their hallmark was their relative stability against decomposition, and they could be replenished a good number of times. That development made semiautomatic plating on nonconductors a matter of course. In the following few years electroless nickel plating baths followed. Their availability reduced considerably the need for rack plating. It permitted eventually the use of fully automated plating methods.
4. Bright acid copper electroplating. Decorative bright acid electroplating baths for the deposition of copper, possessing brilliance and good leveling properties became available. This all but obviated the need for encapsulation and with it the required coating thicknesses of earlier times.

These developments led to the introduction of other advanced techniques for plating on nonconductors. As a result we have the present state of affairs where there exist a large number of ways to plate nonconductors by a variety of metals and upon a number of nonconductor types.

2 METALIZATION

2.1 Polyimide Metalization

Over the past 20 years or so, significant amount of work has been reported on the electrochemistry of polymers in general. Polyimides being a class of same, in particular, are used in the electronics industry as thin dielectric coating and as thick film substrate for flexible circuitry.

In this context, for instance, macromolecules with known electrochemistry are obtained from organic or metal centered charge transfer. In the electrochemical charge transfer reaction between commercial Kapton^R film and highly reducing cluster anions of the main group metals, the anions are some times referred to as Zintl anions. This type of redox reaction leads to the formation of thin films of many of the main group metals on the polymer film surface. The basic electrochemistry of polyimides has now been studied sufficiently so that different techniques for the preparation of metal films on polyimides based upon the electrochemistry of the substrate is now available [1, 2] via nonaqueous or aqueous solutions. By way of illustration we delineate the procedure for metalizing a Kapton^R film. It is done through the following steps: First, electrochemical charge is injected onto/into the film. Second, the film is oxidized in, say, copper oxalate solution. Third, a thick copper film is formed by immersion in an electroless copper solution. More specifically now. A piece of Kapton^R film may be immersed in a solution of $[(CH_3)_4N]_2$ V (EDTA) for about 2 minutes, and after a rinse the (now greenish) radical anion film is immersed in a solution of cupric oxalate for about 3 to 4 minutes. This step results in a thin coating of copper metal. After a subsequent rinse the film goes into an electroless solution for the buildup of a film of the desired thickness.

2.2 Metalization of Plastics

As stated above, polyimides are a class of polymers. They, as well as many types of plastics, satisfy just about all the requirements for electronics applications. Plastics, in general, possess very good thermal chemical and mechanical properties. This is partly due to the inert nature of the surfaces of plastics/polyimides. For this reason it is somewhat difficult to obtain adhesive film bonds to the surfaces of these class of materials. If, as is the case, the bulk characteristics of the materials are to be conserved, then the answer to the problem of adhesion is surface modification. The modifications may consist in the application of any one or more of ion beams, particle beams, sputtering, plasmas, photo grafting (photolithography), chemical methods, and gamma ray irradiation. In general, however, the science and engineering of surface-modified polymers could use some higher understanding. In particular, there remain questions of the optimum thickness and extent of modifications, extent of cross linking, and structure and orientation of functional groups on the surface. Modern methods of surface characterization (e.g., X-ray photoelectron spectroscopy, XPS) have provided a great deal of information but much more must be resolved. We will consider the different methods that presently exist for the effective metallic coating of plastic surfaces.

Photolithography A good illustration is the practical procedure involving photolitographically controlled deposition of copper on a Mylar surface. First, the Mylar sheet is rinsed in acetone and distilled water. The clean sheet is next placed onto a watch glass carrier filled with H_2PtCl_6 alcohol solution. A laser beam or filtered mercury light source is used to illuminate the Mylar sheet for about 20 minutes (by a focused beam). The possible photochemistry of $[PtCl_6]^{2-}$ in alcohols is discussed in [3]. The reaction, it is stated, gives rise to a spatially well defined, albeit discontinuous, deposit of platinum metal upon the illuminated areas of the substrate. Finally, the sheet is placed into an electroless copper plating bath for a few minutes. The latter contains formaldehyde and formaldehydrate at 65°C and pH of 11.4.

Ion Beam Techniques Ion beam techniques traditionally were useful in semiconductor processes. They are now used in many different applications. Chief among them is the metalization of insulators. The metalization can be achieved in three ways: The first is ion implantation. Using the appropriate vacuum setup, an ion beam is directed to strike the material. The ions penetrate to a depth that depends on the beam's energy and other physical parameters. The ions then interact with the target's polymer molecules, resulting in an expulsion of volatile molecular fractions, on the one hand, and a carbon-rich surface layer, on the other. The surface layer will show enhanced (though not metallic) conductivity. The enhancement enables a subsequent electrodeposition of a metallic layer. The second method is ion beam mixing. By this means, a degree of metallic conductivity can be achieved. The thin metallic layer is deposited by evaporation. The layer is bombarded by fast ions that have to be made able to penetrate the metal/polymer interface. There they cause collisions and/or electronic excitations which cause a broadening of the interface as well as metal/polymer chemical interactions. This promotes adhesion of the metal film. The third method is ion beam assisted deposition. Here the metal deposition and ion bombardment are done simultaneously. At a given stage of film thickness the ions cannot reach the metal/polymer interface. The ion beam now affects the growth of the metallic thin film relieving possible stress, for instance.

The different ion beam methods have their advantages as well as disadvantages. The combination of almost any metal with any surface is made possible, giving rise to good adhesion. The process is also suitable for chemical deposition. As to the disadvantages, relatively complicated vacuum techniques are involved and the "line of sight" process excludes substrates of complex geometric features.

Electrochemical-Autocatalytic Methods Autocatalytic deposition (AD) places metal directly on objects immersed in a process solution, with no external electric current required. The metallic coating is deposited by a controlled electrochemical reduction that is catalyzed by the metal or alloy being deposited. An electrochemical reduction agent in solution provides the electrons, and the process takes place only on catalytic surfaces rather than throughout the entire solution. AD is also referred to as "electroless plating". It possesses several characteristics not shared by other techniques. The process is an integral and necessary step in plating on nonconductors such as plastics and printed circuit boards. Some electroless deposits, it should be noted, have unusual or even unique magnetic properties. AD yields deposits of nickel, cobalt, palladium, platinum, copper, gold, silver, and alloys containing these metals plus phosphorus or boron. This commercially important technology has many desirable attributes for flexible circuit manufacture:

- AD coats objects uniformly.
- With surface catalyzation (activation), AD coates nonconductors.
- With ultraviolet patterning of a catalysed surface, AD reproduces patterns of micron dimensions with high resolution.

For more details, the reader should consult [4–6], as well as the relevant chapters in this volume. The preparatory steps for electroless deposition are discussed in some details in Section 4 below.

Plasma-Induced Deposition In this technique a thin film of copper formate or copper chloride is coated on a substrate (polymeric dielectric substrates including polystyrene, Kapton, and similar materials are treated using these methods [7, 8]). Next this precursor layer is reduced to metallic copper in a hydrogen RF plasma. The reduction process is carried out at a low substrate temperature. Temperature sensitive plastics can be metalized via this methodology. The subsequent annealing is another matter. In general, when large areas must be coated by a thin metallic layer, this method is suitable. Once deposited, a subsequent electroless deposited additional metal layer is always an option.

Sputtering Plastic specimens need proper pretreatment before the coating process so as to avoid surface contaminants which may cause undesired characteristics such as poor adhesion. Coatings are deposited on the high-power sputtering principle. To deposit TiN (which is an important diffusion barrier material), a Ti target is sputtered in a nitrogen reactive gas and argon inert gas atmosphere. The dc mode is the method of choice. Coating thickness is proportional to the coating time. Typically 200 nm is achieved in a matter of 5 minutes, while 10 times the thickness will require about 1 hour. For more details on this method, the reader is advised to consult articles in [9].

3 DEPOSIT CHARACTERIZATION—ADHESION

Once a metal coating has been applied to a surface, it is critical to determine the adhesion properties of the deposit. Of the many and varied properties of thin films, adhesion to the substrate stands out as the most important and the most difficult one to measure in a quantitative fashion. Further, while in most applications good film adhesion between film and substrate is what is required, in some cases weak adhesion is desired so that the film can be removed from the substrate. These two different types of requirements almost never are present in one and the same application.

From a microscopic point of view, adhesion is a measure of the force between the constituent atoms-molecules of the thin film and those of the substrate at the interface. From a macroscopic and practical points of view adhesion is simply a measure of the energy required for the removal of a thin film from the substrate. These two properties do often differ for a number of reasons. Stress or strain, which develop during film growth, can reduce the required energy for film removal. Structural imperfections in either the film or the substrate can reduce adhesion as well. Some types of growth methods result in porous films. The pores allow gas and/or water vapor to reach the interface and decrease the adhesion for practical purposes. These and many other complications prevent accurate measures of adhesion to be made, so the many different tests in practice are simply variants of two rather simple visual tests, the tape and the scratch tests. The first involves, in its fundamental form, the application of an adhesive tape to the film. If upon peeling it off no film material has been removed, it is said that the film has passed the test. The second involves, in its fundamental form, the rubbing of an eraser across the film. The number of strokes needed to expose the substrate is a measure of the film's degree of adhesion. As stated, a number of refined variations of these tests are widely practiced. In case of the peel test, the use of a tensile tester to measure the force necessary to remove a film off a substrate is rather common place. In case of the scratch test, a common refinement is the use of a stylus loaded with a known weight, which then is pulled across the film. The weight necessary in order to scratch through the film is a measure of the degree of film adhesion. The size and past usage of the stylus affect the test results, making it difficult to compare results obtained in two different plating shops. Also, in the tape test, the thickness and elasticity of tape, film, and substrate all influence the results, as do the angle of removal and the rate of pulling the tape. Both tests are in common use despite these kinds of limitations. A number of other methods, however, have been proposed. One, suitable for flexible substrates, involves stretching the substrate in a tensile tester with the onset of film cracks being observed. Another is a laser test. In this, pulses of different energy densities are aimed at different sites on the sample. At the threshold energy of damage visible cracks appear, and the corresponding energy density is a quantitative measure of the film's adhesion.

Table 1 gives a summary of laser-type measurement results by Lindberg et al. [10]. Interestingly they felt it necessary to compare their numbers to those obtained using the tape test method as benchmark. It should be noted that the threshold energy densities for the development of cracks, correlate very well with the tape test results.

4 SENSITIZATION—CATALYSIS (PRIOR TO ELECTROLESS DEPOSITION)

The expression "electroless deposition" is not quite accurate. While there are no external electrodes, there is electric charge transfer involved. In place of an anode there is metal provided by the metal salt in solution; replenishment can be had either by adding salt or by an external loop with an anode that has a higher efficiency than the cathode. In place of a cathode there is a substrate, while the electrons are provided by a reducing agent in solution. The electroless deposition process is autocatalytic, and it is first initiated by active metals upon immersion in an electroless bath.

Since practically every substrate requires a different approach, depositing active metal on the surface of a nonconductor is an art. Such deposition requires one or more of the following steps: cleaning, surface modification (e.g., etching), sensitization, and subsequent catalyzing or catalyzing and activating. Rinsing between steps is usually a must. Figure 1 provides the schematics of the electroless deposition process.

TABLE 1 Silver Evaporated onto PET from Different Manufacturers (at $\sim 43\,\mathrm{nm\,s^{-1}}$ to a thickness of $\sim 650\,\mathrm{nm}$)

Source of PET	Threshold Laser Energy Density ($\mathrm{mJ\,cm^{-1}}$)	Tape Test Amount Removed
Dopont 700D	13	90%
Kodak Estar	43	40%
Hostaphan	47	60%
Melinex	54	10%

Note: PET = ployethyleneterephthalate.

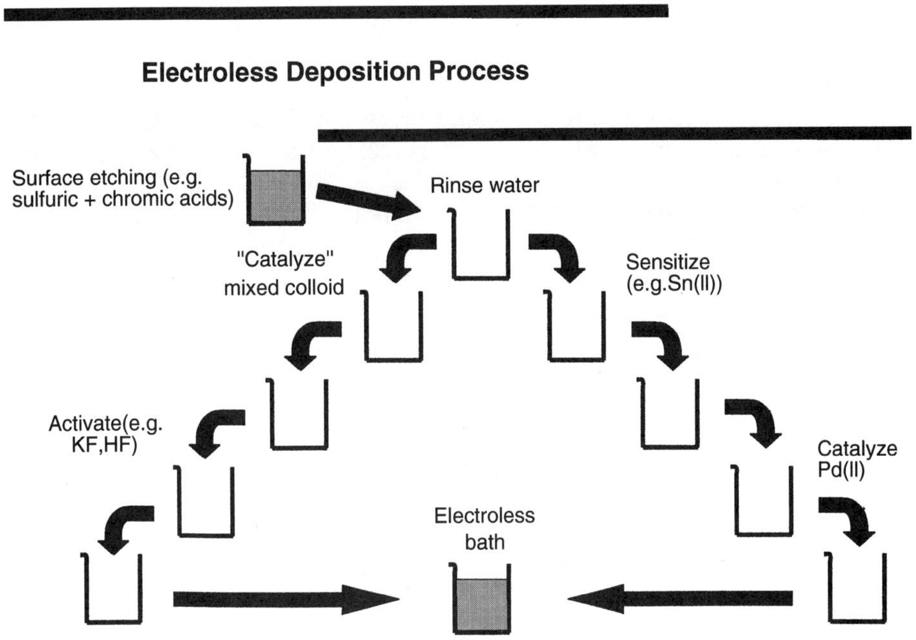

Figure 1 Schematic representation of the electroless deposition process.

Sensitization and catalysis here mean the absorbing of agents from a solution of Sn^{+2} and/or Sn^{+4}. Other sensitizing agents are $AgNO_3$, $AuCl_3$, and metallic Na (in a naphthalene solutions). A simplified model of sensitization and catalysis process is that the 'sensitizing' ion reduces the "active" metal from the catalyst solution, which most often is $PdCl_2$ (again, Au, Pt, Rh, Os and Ag solutions can also be used). Thus

$$Sn^{+2} + Pd^{+2} \longrightarrow Sn^{+4} + Pd^{0}$$

If, as sometimes is the case, a given metal can be reduced by the sensitizing ion, then the substrate is immersed in the electroless bath right after sensitizing and rinsing. Such can be the case when electrolessly depositing Cu or Ag using Sn^{+2}-based sensitizer.

The other path, that of catalysing and activating consists of the use of a mixed colloidal catalyst. The colloid consists of reduced Pd, stabilized by Sn^{+2} and Sn^{+4} ions. The activation step is the removal of the layer formed by the stabilizing agent using HCl, NaOH, or similar. In some instances the activation step may be omitted, but the plating solution becomes contaminated. In practical applications the mixed colloidal technology seems to be preferred. The reasons for this are as follows:

- The technology is more reproducible than the separate Sn–Pd technology.
- An object after removing from the mixed colloidal solution is covered by a Pd layer which is clearly visible. Not so in case of the Sn–Pd process.
- Mixed colloidal solutions are stable and do not exhibit ageing effects as do the Sn solutions.

- Mixed colloidal solutions are rather resistant to impurities.
- Surface preparation is far less critical than in the case of Sn based solutions.

Besides having to be clean, surfaces must be hydrophilic in order to be efficiently wetted by sensitizing solutions. Some workers feel that a rough surface is required, since bonding takes place through mechanical entrapment, and there are many instances of efficient sensitization on very smooth surfaces such as glass slides. Clearly, the surface functional groups are an important factor if not the dominant factor in promoting adhesion on many surfaces.

Despite the above-mentioned advantages of the mixed colloidal technology over the Sn–Pd process, the initial nucleation sites density achieved by the latter method is up to an order of magnitude greater than that obtained by using the former. This means that a continuous film can be obtained at smaller thickness [12]. In practical terms this translates into the following advantages:

- Less metal is needed if the only purpose is to render the surface conductive.
- Better adhesion due to higher density of fastening sites.
- Potential in the development of products where very thin coatings are required.

No discussion regarding sensitization, and the like, can be considered complete without reference to ultraviolet (UV) assisted patterning or selective electroless metal deposition. Schlesinger and co workers [13] showed that it is possible to obtain patterned or selective deposition of, say, electroless copper by UV irradiation of a substrate through a quartz mask at different stages prior to immersion in an electroless bath. Deposition may be made to take place on the nonirradiated areas

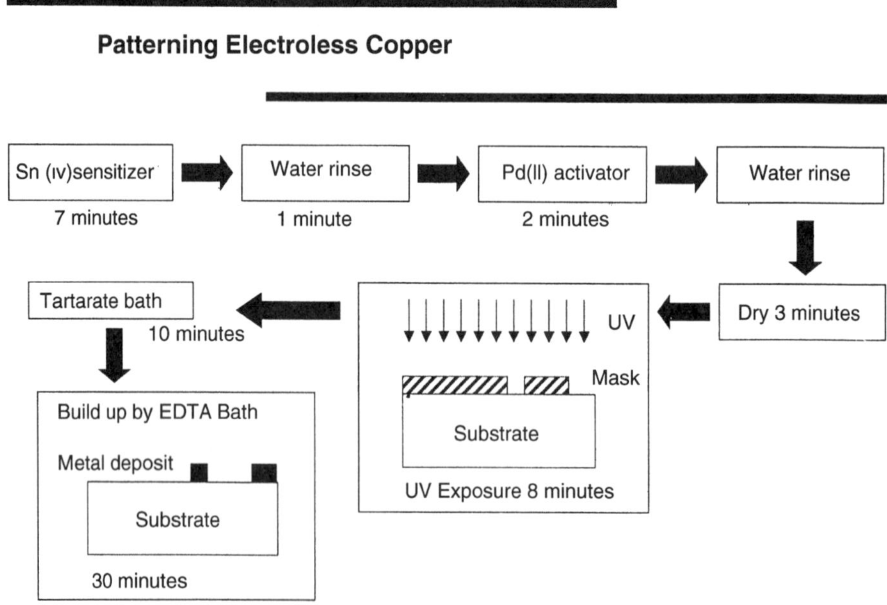

Figure 2 Processing steps involved in (negative) patterning of electroless copper.

only (positive mask image) or on the irradiated areas only (negative mask image). Figure 2 presents the processing steps for the negative-type patterning of electroless copper deposit using a two-stage metalizing procedure.

4 CONCLUSION

The metalization of nonconducting substrates is essential in a large number of industrial applications such as in the fabrication of printed circuits for microelectronics or, in the coating of panels for many other uses. There is now mature technology and many practical methods of electroless plating. These, however, are not always acceptable due to a number of factors, such as slow deposition rates and adhesion quality. There are therefore a number of alternative methods based on direct electroplating on pretreated nonconducting substrates. Specifically, the surface is seeded with a small amount of conductive material such as palladium, carbon, or a thin layer of conductive polymer. Conventional electroplating follows the initial seeding stage with the cathodic lead connected to a metallized region on the substrate.

In practice, it is often observed [11], as depicted in Figure 3, that a thin electrodeposited layer builds up at the edge of the polarized metallic region, and it rapidly propagates along the seeded surface rather than producing buildup on the contact strip, as desired. For a treatment of the issue from a theoretical point of view, the reader is referred to [1].

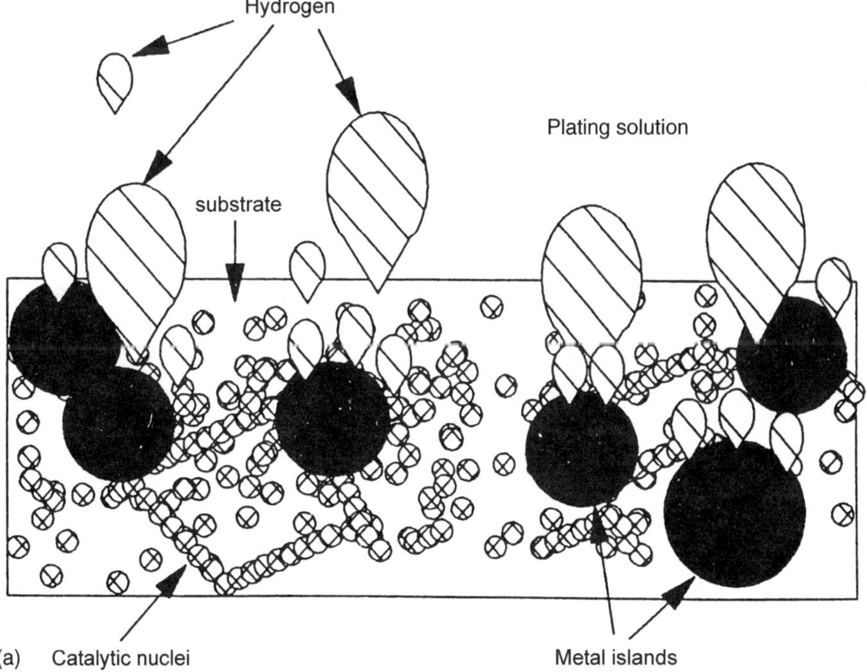

Figure 3 Schematic representation of the deposit propogation along nonconductive substrates: (*a*) Substrate pre-seeded by conductive sites; (*b*) conventional electroplating with nonseeded substrate (1 inch=1000 Å).

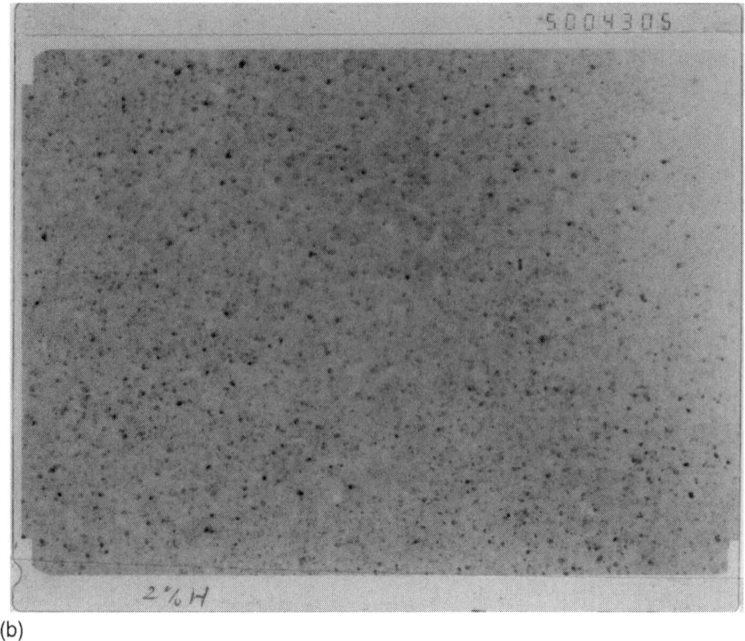

(b)

Figure 3 *(Conitnued).*

REFERENCES

1. S. Mazur and S. Reich, *J. Phys. Chem.*, **90**, 1365 (1986).
2. L. J. Krause and J. A. Rider, U.S. Patents 4,775,556 (1988) and 4,710,403 (1987).
3. M. Zhou, R. Cameron, and F. J. Schwab, in *Metalized Plastics*, vol. 2, K. L. Mittal, ed., Plenum Press, New York, 1991, p. 57.
4. I. Kiflawi and M. Schlesinger, *J. Electrochem. Soc.*, **130**, 872 (1983); and references therein.
5. B. K. W. Baylis, C.-C. Huang, and M. Schlesinger, ibid., **126**, 394 (1979).
6. B. K. W. Baylis, A. Busutil, N. E. Hedgecock, and M. Schlesinger, ibid., **124**, 346 (1977); and references therein.
7. M. Ritz, S. V. Babu, V. Srinivasan, and R. C. Patel, in *Photon, Beam and Plasma Stimulated Processes at Surfaces*, V. M. Donelly, I. Herman, and M. Hirose, eds., vol. 75, Materials Research Society, Orlando, FL 1987, p. 433.
8. A. Gupta and R. Jagannathan, *Appl. Phys. Lett.*, **51**, 2254 (1989).
9. *Deposition Technologies for Films and Coatings*, R. F. Bunsha, ed., Noyes Publ., Park Ridge, NJ, 1982.
10. W.-C. Lee, V. W. Lindberg, P. H. Wojciechowski, and F. J. Duarte, in *Metalized Plastics*, vol. 2, K. L. Mittal, ed., Plenum Press, New York, 1991.
11. D. Weng and U. Landau, *J. Electrochem. Soc.*, **142**, 2598 (1995).
12. J. P. Marton and M. Schlesinger, ibid., **115**, 16 (1968).
13. M. Schlesinger et al., ibid., **119**, 1013 (1972); **126**, 394 (1979); **136**, 872 (1983).

16 Electroplating of Organic Films: Conductive Polymers

TETSUYA OSAKA, SHINICHI KOMABA,
and TOSHIYUKI MOMMA

1 ELECTROPOLYMERIZATION: ELECTROCHEMICAL SYNTHESIS OF CONDUCTING POLYMERS

Electron conductive polymers appeared in 1971. Shirakawa and coworkers [1] synthesized conducting polyacetylene and found that it had considerably a high conductivity compared to organic compounds, 10^3 S cm [1–3]. Since then various conducting polymers and their synthesis methods were studied by many researchers [4]. Their conductivities are shown in Figure 1 in comparison to those of metals, inorganic, and organic compounds. Generally speaking, because conducting polymers posess electrons which are delocalized in π-conjugated system along the whole polymer chain, its electron conductivity is much higher than that of other polymers which have no conjugated system, as shown in Figure 1.

Furthermore, in 1979, Diaz and coworkers synthesized conductive polypyrrole film by electropolymerization [5, 6]. Figure 2 represents a schematic drawing of the electropolymerization. It is similar to electroplating of metal films. Its polymerization reactions occurres electrochemically; that is electrons pull through the electrode/electrolyte interface by the application of a potential to an electrode as shown in Figure 2. A film of conducting polymer is thus obtained in the case of the insoluble polymer which is generated on the electrode.

Essentially all electrochemical reactions may be classified into an oxidative or a reductive reaction regarding electron transfer. Similarly we can classify electrochemical polymerizations. In almost all cases, electropolymerized polymers were synthesized by electrooxidation. It seems that the films obtained by this method contain branching or crosslinking defects [7–9]. On the other hand, electroreductive polymerization performed by using a zero-valent nickel complex is sensitive to the position of halogen atom in the monomer, and it affords polymers with well-defined linkage between the monomer units [10–13].

To date many kinds of organic devices with conducting polymer films have been investigated, and in these formation processes an electropolymerization is used as one of the most interesting methods of preparing functional polymer thin films.

Modern Electroplating, Fourth Edition, Edited by Mordechay Schlesinger and Milan Paunovic.
ISBN 0-471-16824-6

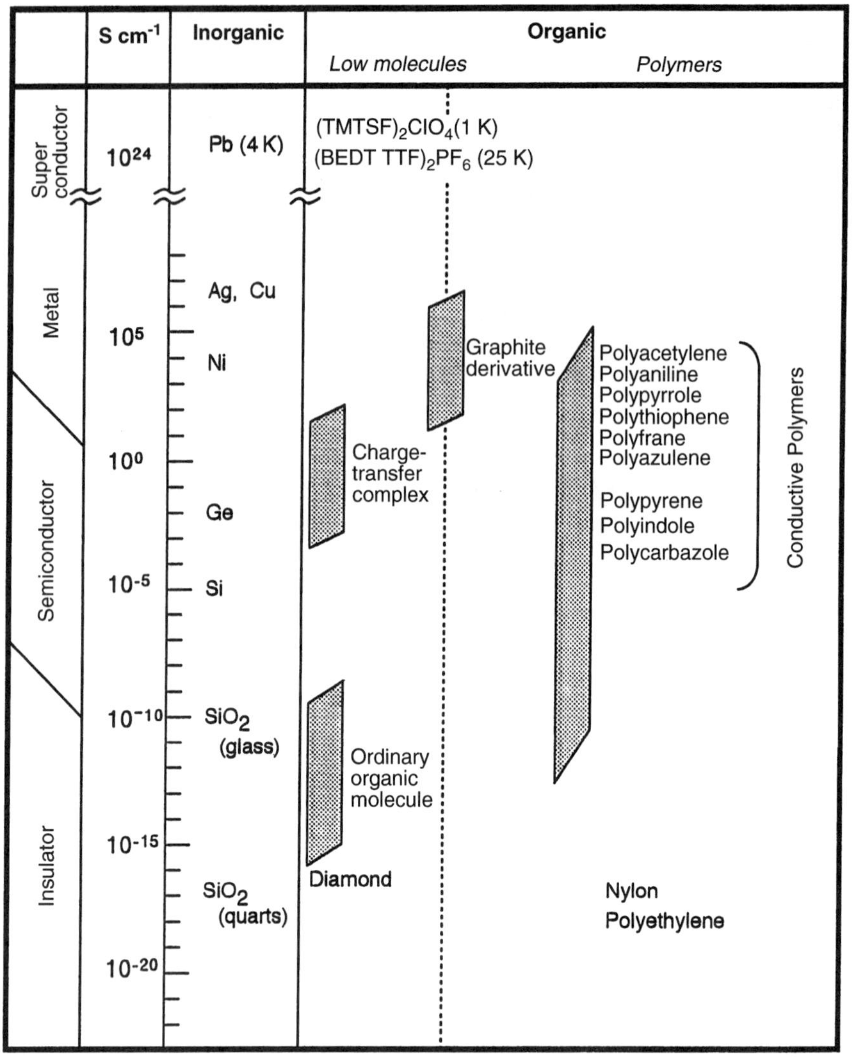

	S cm^{-1}	Inorganic	Organic: Low molecules	Organic: Polymers
Super conductor	10^{24}	Pb (4 K)	$(TMTSF)_2ClO_4$(1 K) (BEDT $TTF)_2PF_6$ (25 K)	
Metal		Ag, Cu		
	10^5	Ni	Graphite derivative	Polyacetylene Polyaniline Polypyrrole Polythiophene Polyfrane Polyazulene Polypyrene Polyindole Polycarbazole (Conductive Polymers)
Semiconductor	10^0		Charge-transfer complex	
		Ge		
	10^{-5}	Si		
Insulator	10^{-10}	SiO_2 (glass)	Ordinary organic molecule	
	10^{-15}	SiO_2 (quarts)	Diamond	Nylon Polyethylene
	10^{-20}			

Figure 1 Conductivity of various organic compounds in comparison with typical inorganic materials.

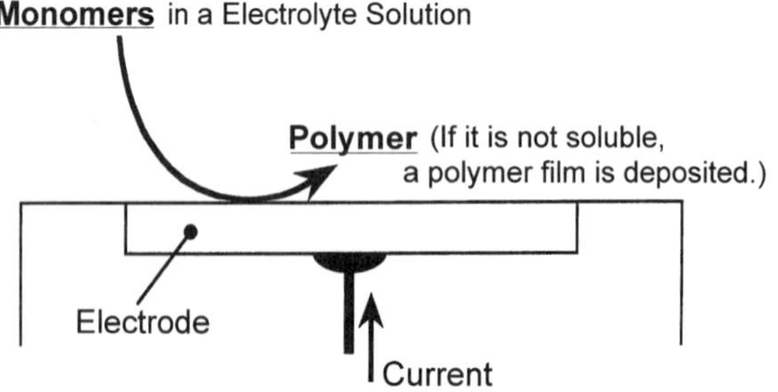

Figure 2 Schematic of electropolymerization.

TABLE 1 Functionalities and Applications of Electropolymerized Polymers

Functionality	Application
Electroactivity	Secondary battery Corrosion prevention Ionic sensing
Semiconducting	Solar battery Optical device Electroluminescence device
Conductivity	Conductor sheet Electrification prevention Diode Transistor Chemical sensor Physical sensor
Electrochromism	Display device Smart window
Ion exchange ability	Ion exchange membrane Chemical sensor
Electrochemical catalysis	Catalyzing electrode

Electropolymerized films deposited directly on various electrodes (e.g., metals, semiconductors) are claimed to be applicable to various devices as shown in Table 1.

Recently we found that electro-inactive nonconducting polymer films are also formed by electropolymerization [14]. When the electropolymerization solution contains nucleophilic reagents, such as, NaOH, $NaHCO_3$, or Na_2CO_3, with monomer reagents, the polymerization first occurs, and then the film propagation stops automatically with the result of changing the nonconductive state by nucleophilic attack reaction. Since the film electrode shows response to the pH value, it can be applied to micro pH sensors [15] and also to biosensors with containing pH change enzymatic reaction [16–18].

In this chapter we discuss the design of functionality of electropolymerized films and its applications to polymer batteries, electric devices, and chemical sensors mainly on the basis of our results. For the other topics on electropolymerization, reader should reffer to the reviews by the others [19].

2 POLYMER BATTERIES

2.1 Introduction

Electropolymerization is a useful method for fabrication of conducting polymer films for application as rechargeable polymer battery [20]. This is so due to several characteristics of electropolymerized polymer films, such as electroactivity, conductivity, and film morphology that may be controlled by film formation conditions [21]. Among these polymers, polyaniline, polypyrrole, and polyazulene all show high electroactivity with good reversibility and chemical stability. Applying the polymer in a battery as the cathode, the battery itself is expected to exhibit both high energy density and mechanical flexibility [22–24].

In this section we describe two typical methods for enhancement of battery characteristics of conducting polymer: the rocking-chair type cathode and the enhancement of the ionic conductvity in the polymer.

2.2 Polymer Cathode for Li Battery

Among many electropolymerized polymers, polyaniline (PAn) and polypyrrole (PPy) can be utilized as one of the promising candidates for cathode materials in future rechargeable lithium cell systems [25]. The candidates for electropolymerized polymer cathodes for Li batteries are p-type polymers whose oxidation and reduction requires anion incorporation and expulsion to compensate for the charge of the polymer film. A lithium battery constructed with a p-type cathode causes the electrolyte concentration to change during the charge and discharge of the cell. Since a decrease in concentration of the electrolyte during charge-up causes the cell resistance to increase, the cell requires a large source of electrolyte solution. However, a cell using a lithium ion-compensated cathode, such as n-type polymers, manganese, cobalt, and nickel oxides, does not cause the concentration of the electrolyte solution to change during cell operations, and so; it can be operated with smaller amounts of electrolyte solution. This property enables the energy density of the battery to be increased.

Many researchers have paid attention to PAn as a cathode material in lithium batteries, because of its high energy density, chemical stability, and high reversibility of the doping-undoping process. Recently disulfide compound with PAn was found to exhibit high energy density cathode materials [26, 27]. In our work we focused on PPy as cathode for PPy/Li metal batteries due to their high cycle life [21]. We found that a battery of Li/gel electrolyte/PPy showed longer cycle life with more than 4000 cycles [28].

Some researchers have examined composite films of conducting polymers and other materials formed by electropolymerization [29–33]. Among those, especially, introducing a polyanion into the p-type conducting polymer, cations compensate for the film charge. The charge of the film is compensated by cations during the oxidation and reduction of the polymer due to the immobilized anion groups incorporated into the polymer chain [33]. By using a composite p-type polymer and a polyanion, Shimidzu and others proposed and developed a rocking-chair type rechargeable battery system, though the system used either an aqueous electrolyte or nonaqueous electrolyte that was not applicable to the lithium battery system [34, 35]. Shimizu and coworkers suggested using an electroactive composite film of polypyrrole and polystyrenesulfonate (PPy/PSS) in a propylene carbonate (PC) electrolyte, which is typical for Li battery, in combination with a carbon fiber electrode with large surface area [36]. Here, we describe the electrochemical properties of a PPy/PSS composite cathode on flat electrode surface for a rocking-chair type rechargeable Li battery system [37, 38].

The PPy/PSS composite film is formed by electropolymerization of pyrrole on Pt with the presence of PSS as supporting electrolyte in an aqueous solution. However, it shows no electroactivity in typical electrolyte solution for Li battery, such as the $LiClO_4$–PC electrolyte solution, without further treatment. When cycling the potential on the PPy/PSS film in an $LiClO_4$-dimethylsulfoxide (DMSO) electrolyte

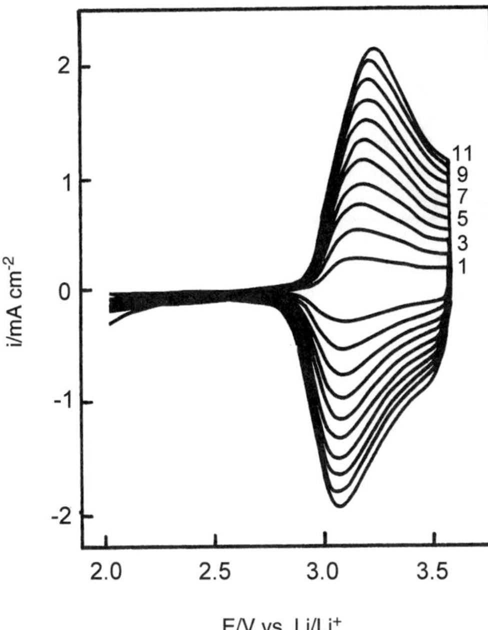

Figure 3 Continuous cyclic voltammograms of PPy/PSS film obtained in a 1.0 mol dm^{-3} $LiClO_4$–DMSO electrolyte solution at 20 mV s^{-1}. Current gradually increased with potential cycling and reached a steady voltammogram. Cycle numbers are shown in the figure.

solution, the faradic current due to the reversible redox of the PPy molecule gradually increases, as shown in Figure 3.

By investigating the optimum preparation condition for a rechargeable battery system, we found that the PPy/PSS composite film obtained at 0.625 V versus Ag/AgCl with 0.5 C cm^{-2} formation charge exhibited the highest redox charge of 51.8 mC cm^{-2} and a good reproducibility in combination with a $LiClO_4$–DMSO electrolyte [38].

The composite film becomes electroactive in organic electrolyte solution when activated, even using PC solvent. PC solvent has an advantage of showing low reactivity between the electrode materials and the electrolyte solution. We examined the activity of two types of films in an $LiClO_4$–PC electrolyte solution. The first film's potential was cycled in the $LiClO_4$–DMSO electrolyte solution until it reached a stable voltammogram as shown in Figure 3. The second film was treated by immersion into an $LiClO_4$-dimethylformamide (DMF) electrolyte solution while keeping its potential at 2.0 V versus Li/Li^+ [37]. While both types of films exhibit electroactivity in the $LiClO_4$–PC electrolyte solution, the treatment of $LiClO_4$–DMF electrolyte enables the activity of the composite film to increase with film formation charge above 0.5 C cm^2. In the $LiClO_4$–PC solution, the composite film shows no loss of activity during continuous potential cycling, in contrast to the voltammograms for a $LiClO_4$–DMSO electrolyte. The maximum redox capacity of the composite film is attained by activating the film in the $LiClO_4$–DMF electrolyte solution, not in the $LiClO_4$–DMSO electrolyte solution.

A battery with a lithium anode was constructed incorporating a $LiClO_4$–PC electrolyte and PPy/PSS composite film, and the cathode performance of the PPy/PSS

TABLE 2 Unit Volume Energy Density of Cathode Material in the Lithium Battery Using the PPy/PSS Cathode and a Rougher PPy/ClO_4 Cathode

Film Formation Change ($C\,cm^{-2}$)	Film	Specific Energy Density (Wh $liter^{-1}$)
8	PPy/PSS	220
	PPy/ClO_4	142
0.5	PPy/PSS	230
	PPy/ClO_4	51.2

Source: Shimizu, Yamanaka, and Kohno [36].

film was evaluated. The cell was charged with the desired charge at a constant current density, and discharged at the same current density. The cell exhibited the potential to be recharged with an output voltage of 3.9 to 2.8 V. The energy density of this cell is shown in Table 2 as the value per volume of cathode. The energy density per unit volume of a PPy/PSS film is superior to that of a PPy/ClO_4 film which is a typical p-type polymer material for cathode. It was confirmed that the Li^+ ions compensate mainly for the charge of the film generated in the PPy/PSS cathode during operation. The difference between the cell performances using PPy/PSS film and PPy/$LiClO_4$ film is small, while the cathode morphology is different. That is, the PPy/ClO_4 film is porous and thicker than the dense morphology of the PPy/PSS film.

From the results of the charge-discharge tests using PPy/PSS or PPy/ClO_4 cathodes, the n-type PPy/PSS composite film has high potential as a cathode material for a rocking-chair type rechargeable battery with high energy density.

2.3 Polymer Cathode in a Solid Polymer Electrolyte

The introduction of solid polymer electrolytes into electrochemical devices is expected to make these devices flexible, thin, and lightweight. Liquid-free devices also have the advantage of long-term reliability. Therefore some researchers have made efforts to improve the interface between PPy and the solid polymer electrolyte [39]. The electrochemical properties of PPy films using aqueous, organic liquid electrolyte, and recently using solid polymer electrolytes have been investigated [39–42]. It has been found that the performance of PPy is influenced by the kinetics of the doping/undoping process of ions in the PPy films. For example, ionic movement has been improved by making PPy/anion composite films for changing the moving ion through the film [39, 43].

We described that a rough and porous PPy film showed much better redox properties using poly(ethylene oxide) (PEO)–$LiClO_4$ [41]. Mastragostino and coworkers have reported that poly-*N*-(oxyalkyl)pyrrole showed much better redox properties than PPy using a solid polymer electrolyte [44]. Here, we would like to mention that the introduction of Nafion molecule, which supports ionic conduction, into the PPy electrode for enhancement of ionic diffusion in the PPy [45, 46].

PPy/Nafion composite film may be formed by electropolymerization of pyrrole at 0.55 V versus Ag/AgCl in 50 ml aqueous solution containing 0.2 mol dm^{-3} pyrrole and 5 ml Nafion 117 solution (Du Pont) as the supporting electrolyte onto a platinum-palladium (8:2) metal sputtered coin-type cathode case. After a PPy/

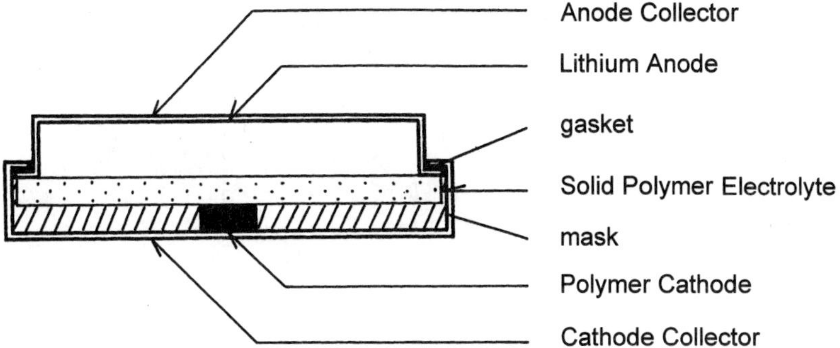

Figure 4 Schematic of a coin-type electrochemical cell for Li/(PEO-$LiClO_4$)/(PPy/Nafion).

Nafion preparation, it is introduced into a coin-type cell, which consisted of the PPy film as the cathode, PEO_8–$LiClO_4$ (PEOn-$LiClO_4$; n is the ratio of ethylene oxide units of PEO to Li ions; that is, where $n = [EO]/[Li]$) as the electrolyte, and Li metal as the anode. The construction of the cell is shown in Figure 4.

Nafion molecule was introduced into a PPy matrix, and the redox performance of the PPy/Nafion electrode was investigated in a PEO_8–$LiClO_4$ solid electrolyte system. A rougher interface between polymer cathode and polymer electrolyte is usually needed for an all-solid battery; however, the PPy/Nafion cathode works well regardless of the flat surface of the PPy/Nafion film. When compared to a PPy film doped with ClO_4^- anions with a similar morphology to PPy/Nafion Figure 5, the PPy/Nafion film demonstrates much better redox performance in the solid polymer electrolyte as is shown in Figure 6. The results of ac impedance spectroscopy and potential-step chronoamperometry confirmed that the improvement in the redox reaction of the PPy/Nafion film is due to the enhancement of the ion diffusion rate in the film. The diffusion coefficients of the PPy/Nafion composite and the PPy/ClO_4 films were 2.78×10^{-11} and $1.63 \times 10^{-14}\,cm^2\,s^{-1}$, respectively. Judging from the change of the electrolyte conductivity by ac impedance measurements, the ClO_4^-

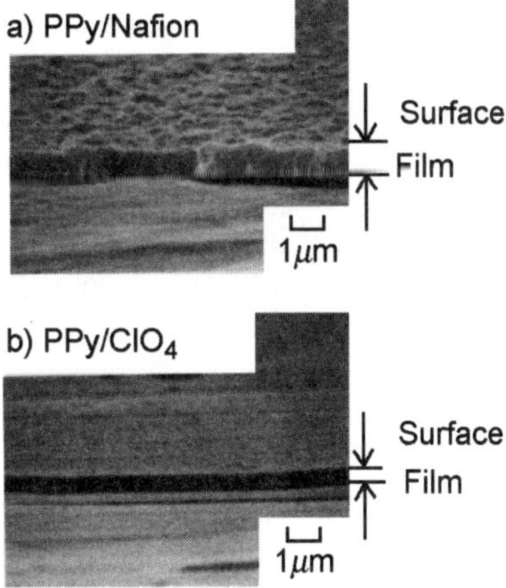

Figure 5 Cross-sectional SEM images of (*a*) the PPy/Nafion film and PPy/ClO4 film prepared by electropolymerization. The charge for the film formation is 100 mC cm^{-2} in both films.

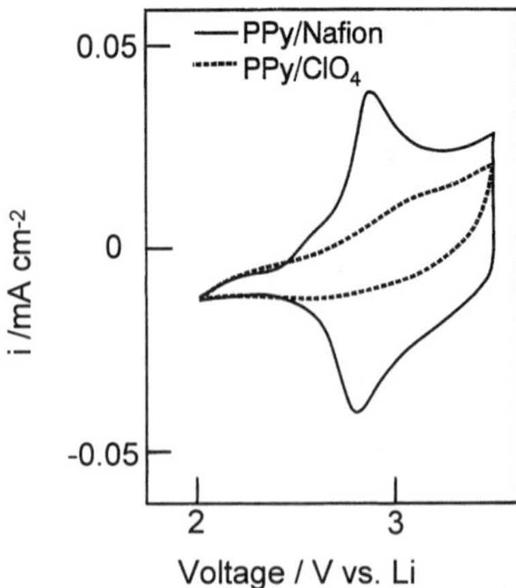

Figure 6 Cyclic voltammograms of PPy/Nafion (solid line) and PPy/ClO_4 (dashed line) at 2 mV s^{-1} in PEO_8-$LiClO_4$ at 80.

anion moves mainly through the PPy/Nafion film for charge compensation, that is, p-type polymer. Additionally the PPy/PSS composite film shows poor redox performance in PEO_8–$LiClO_4$ because of large impedance for ion diffusion process in the film, though the film gives good redox performance in liquid electrolyte, as mentioned above. It is concluded that the improvement in the diffusion rate by incorporating Nafion into the PPy film is due to the structure of the Nafion molecule that promotes ionic movement.

Thus the improvement affects favorably the battery performance of the Li/PEO_8–$LiClO_4$/(PPy/Nafion), and the battery with the PPy/Nafion cathode film showed a coulombic efficiency exceeding 95% up to 0.1 mA cm^{-2} discharge current.

3 ELECTRIC DEVICES

3.1 Introduction

Conjugated polymers have such interesting electric properties as semiconductors because of their delocalized π-electrons as mentioned in the first section [47]. These materials can be regarded as organic semiconductors with energy gaps related to the UV-visible region. Among the various methods of preparation of these polymer films, electropolymerization has great advantages of good processability and ease for forming its thin film and controlling its electrical and optical properties. Some attempts have been made in the past to utilize electropolymerized polymers as high functionality films in the fields of electric devices.

The key function of the electropolymerized polymers is their electroactivity, which is produced by the formation of a polaron, bipolaron, and soliton as a result of the process of doping counter ions into the polymer matrix [48]. The phenomenon of formation of polaron and bipolaron levels in the band structure is revealed in the optical absorption spectra [48]. Hence they can be applied to electrochromic devices [49, 50]. Moreover very low conducting PPy films (about 10^{-12} S cm) formed by electrodeposition, are applied to MIM-type switching devices for large-scale flat panel LCD displays [51, 52].

In 1991 it was reported that π-conjugated polymers could be applied using the fabricated light-emitting diode (LED) (i.e., electroluminescence (EL) devices) as an active layer [53]. In recent years attention has been paid to this possibility. For producing active layers of EL devices, methods of vacuum deposition, spin-casting, and dip-coating have been used [54–61]. Recently we developed an application of electropolymerized poly (3-substituted thiophen) films to the light emission layer and the hole-transporting layer. We discuss this recent application of electropolymerized polymer for polymer LED below.

3.2 Electroluminescence Device

Light Emission Layer [62, 63] In the past several years, EL devices with conjugated polymers were reported as mentioned above. Since semiconducting luminescent polymers must be prepared as uniform thin films for an EL device, they were almost formed mechanically as spin-cast or dip-coat with another polymerization process. "Electropolymerization" is one of the most common methods for forming and synthesizing conjugated polymer films. Its optical characteristics can be varied easily by electrolysis conditions. There is currently considerable interest in the use of such polymer films in electric devices. Furthermore it is expected that realization of polymer EL device using electrochemical process remains an important advantage in simplifying device fabrication. We were successful in the application of electropolymerized polymers to the light-emitting and hole-transporting layers [62–64]. In this section we present results obtained by applying electropolymerized poly(3-substituted thiophen) to EL devices.

Polymerization and film deposition are carried out galvanostatically with an indium-tin oxide (ITO) coated glass as a working electrode. An insulating nitrile butadiene rubber (NBR) film is formed by a dip-coating method using a 2-butanone solution containing 2 wt% NBR. After the polymerization the composite film is undoped by applying a negative potential. The Indium electrode, the rectifying contact, is deposited by vacuum evaporation onto the ITO modified with polymer.

By using various monomers as 3-alkylthiophen (alkyl chain length: $n = 1, 4, 6, 7, 8, 9, 12$) and 3-phenylthiophen, which are soluble in AN, polymer films are deposited on bare ITO glasses by electropolymerization. Poly(3-phenylthiophen) (PPhT), poly(3-hexylthiophen) (PAT6), and poly(3-octylthiophen) (PAT8) films are spread uniformly. The molecular structures of three polymers are shown in Figure 7. Although these polymers were applied to a sandwich-type device, a short current between ITO and indium leaked due to the roughness of the polymer films. We observed no light emission from the device without pre-coating of NBR.

Using the ITO glass coated with NBR film, the poly(3-substituted thiophen)/NBR composite film was prepared by the technique as previously described [65]. By application of this composite to a device, the leakage was prevented due to significant improvement in uniformity of the deposited film. Figure 8 shows the current-bias and bias-luminance characteristics of the device in the case of the PPhT/NBR composite film as an emission layer. The device indicates a rectification characteristic, and forward bias was obtained when the ITO was the positive and the In was grounded. As positive bias increases, light emission turns on at about 10 V, as shown in Figure 8. The orange-red emission becomes visible under ordinary

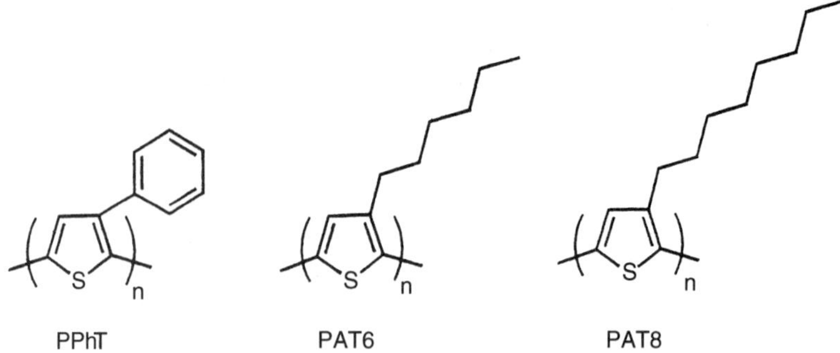

Figure 7 Molecular structures of poly (thiophen) derivatives; PPhT, PAT6, and PAT8.

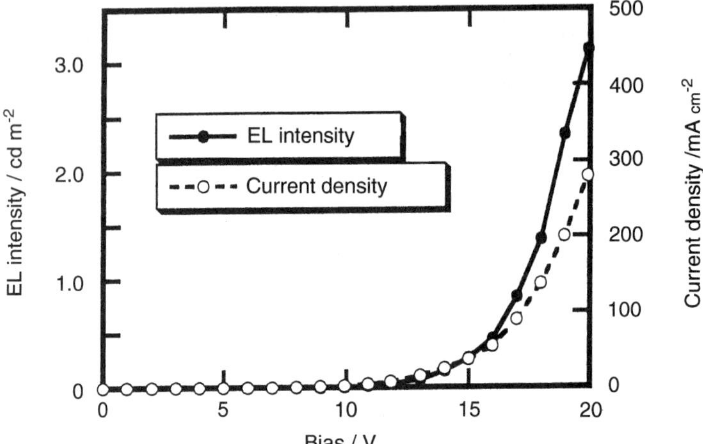

Figure 8 Dependence of current and EL intensity on bias voltage for the device using PPhT/NBR composite film.

dim light. Other composite films for the PAT6/NBR and PAT8/NBR showed similar characteristics. Since the insulating NBR film is nonfluorescent, these poly(thiophen) derivatives are confirmed to act as a light emitter.

The EL spectra of the devices utilizing various poly(thiophen derivative)/NBR composite films are shown in Figure 9. There is little difference in EL emission peaks. The emission peak of the device with the PPhT/NBR composite film appears around 650 nm, while those of the PAT6/NBR and PAT8/NBR composite films appear around 610 and 590 nm, respectively. The peak shift suggests that the EL spectra depend on the substituent in the 3-position of the thiophen. Yoshino et al. reported polymer LED with PAT ($n = 12$) spin-casted films polymerized by oxidizing reagent [66]. In our system, however, shorter emission wavelength is observed due to the difference of the degree of polymerization and/or the conformation of polymer chain. It seems that there is a possibility for enhancement of the EL characteristics by control of the electrolysis condition and selection of emitting materials in the electropolymerization process.

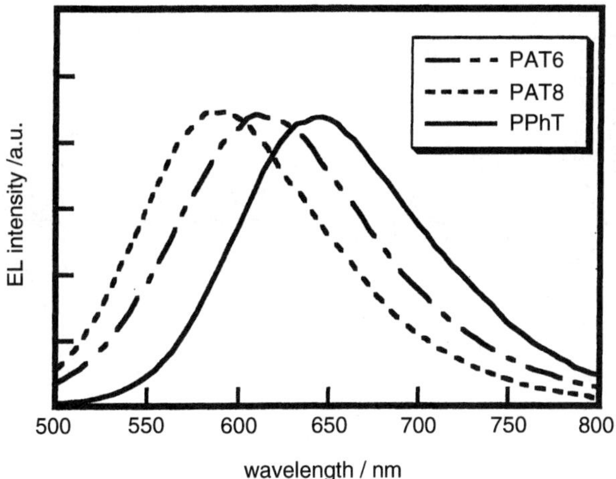

Figure 9 EL spectra of composite films of various poly(thiophen) derivatives with NBR.

Hole Transporting Layer [63, 64] For enhancement of the organic electroluminescence (EL) device, a double-layer construction of a hole-transporting layer was reported. In particular, recently EL devices using hole transporting organic layer were reported to exhibit enhancements of emission efficiency and durability [67, 68]. Yamamoto et al. [69] applied a vacuum-deposited polythiophen film, and Saito et al. [70] attempted an electropolymerized poly(3-methyl-thiophen) film as a hole-transporting layer. Here we show the possibility of EL device using this electropolymerized poly(thiophen derivative) film as hole-transporting layer.

PAT8 film is electrochemically deposited as mentioned above. The EL emission layer, a poly (N-vinylcarbazole) (PVCz) film dispersed with 2,5-bis(1-naphthyl)-1,3,4-oxadiazole (BND) and 3-(2′-benzothiazolyl)-7-diethyl-aminocoumarin (coumarin 6), is formed by dip-coating method [64, 71]. An aluminum electrode for rectifying contact is deposited by vacuum evaporation onto the emitting layer. The film thicknesses of the dip-coated emission layer and the electropolymerized PAT8 hole-transporting layer are to be 50 to about 60 nm and about 20 nm, respectively. Schematic drawings of the device configurations are shown in Figure 10.

Figure 11 shows the dependence of the EL intensity (L) on the bias (V) of the devices of the PVCz layer with or without the PAT8 layer. For the device without the PAT8, the light emission turns on at 13 V, and the maximum intensity is about 10 cd m^{-2}. While the device with the PAT8 layer exhibits more intense emission starting at 8 V and higher intensity of about 100 cd m^{-2}. The turn-on bias is reduced to 5 V, and the EL intensity is increased ten times by adding a PAT8 layer. Furthermore, prior to the formation of the PVCz layer, the electropolymerized PAT8 film was dried under vacuum at 150°C. This drying process of the PAT8 improves extremely the EL characteristics as is seen in Figure 11, (a) and (a)*. The light emission starts at 8 V which is lower than that for the device shown in curves Figure 11, curve (b), and the intensity reaches 700 cd m^{-2}, which is higher by about a factor of seventy. Since the PAT8 film did not affect the emission color and the EL spectrum is identical the PL of coumarin 6, the PAT8 layer acts as a transparent hole-transporting layer. In summary, the combined PVCz/PAT8 double-layer shows

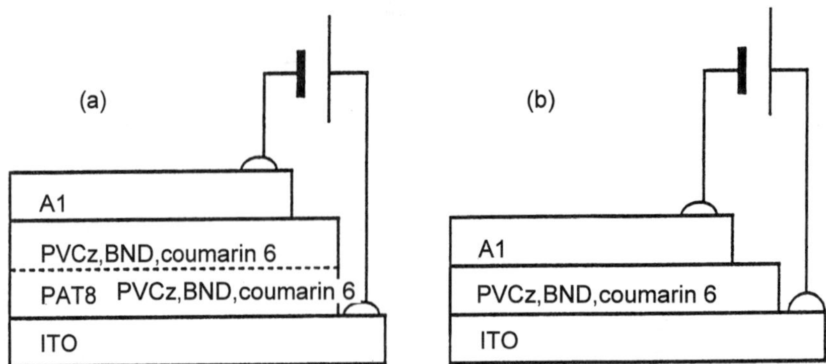

Figure 10 Device configurations of (*a*) double layers of electropolymerized layer and dip-coated layer and (*b*) single layer of dip-coated layer between indium cathode and ITO anode.

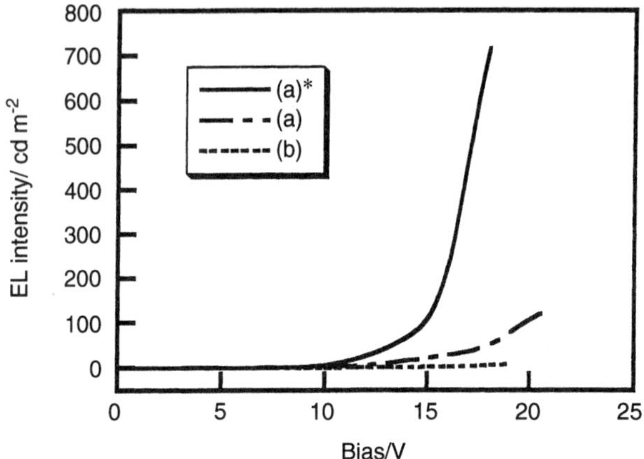

Figure 11 EL characteristics of the devices using (*a*) PAT8 and dip-coated emitting layers, (*a*)* dried PAT8 and dip-coated emitting layers, and (*b*) dip-coated film only as active layer. The thicknesses of PAT8 and PVCz layers are 20 nm (1 min deposition) and 60 nm, respectively.

an intense brightness and a low turn-on bias, where the PAT8 layer acts as a highly effective hole-transporting layer.

4 CHEMICAL SENSORS

4.1 Introduction

The merits of electropolymerization are that new properties can be easily obtained by use of various supporting electrolytes or monomers and the film thickness can be easily controlled by regulating the amount of charge passed. Also the resulting polymer covers the electrode substrate of any shape and size. In addition some functional molecules can be incorporated into the polymer matrix, and such molecules coexist during the electropolymerization. The electropolymerization procedure with such merits is suitable for making microchemical sensors. Electropolymerized films are often utilized for the fabrication of functional electrodes having molecular selectivity [72], such as ion sensors [73–79], bionsensors [79–88], and gas sensors [89–92].

In this section we introduce our recent attempts for cation sensor with the PPy/PSS composite and for urea biosensor with electroinactive PPy film electrode.

4.2 Cation-Sensitive Electrode

PPy film doped with polyanion can be electrodeposited by the coexistence of polyanion in the polymerizing solution. As mentioned above, the composite film of the PPy/PSS shows the high electroactivity with doping/undoping process of cation even in the lithium secondary battery systems. Previously we reported selective potential responses of the PPy/PSS electrode to cationic activities in the test solutions [93]. Here we describe our attempt of the highly selective and stable response to K^+ using a double layer of the PPy/PSS electrode covered with a plasticized polyvinyl chloride (PVC) membrane containing valinomycin.

The PPy/PSS composite films is deposited on Pt disk (diameter of 1 mm) electrochemically. The PVC film is deposited onto a bare Pt or PPy/PSS electrode surface by casting from a cocktail solution containing valinomycin, which is known as an ionphore for potassium ion [94]. The responses of the electrodes to ionic activities of chloride salts were measured at room temperature assembling following cells: Ag/AgCl, KCl(sat.)|0.1 mol dm^{-3} NH_4NO_3|test solution|polymer film on Pt surface.

The selective response to each monovalent cation was investigated for the purpose of construction of double-layer film electrode. By casting, we formed a plasticized PVC layer containing valinomycin as a K^+ ion carrier onto the two electrode. Next we prepared two types of all solid state potassium ion selective electrodes: a single layer of Pt/PVC, and a double layer of Pt/(PPy/PSS)/PVC. As shown in Figure 12, the double-layer electrode of Pt/(PPy/PSS)/PVC responds to K^+ ion even containing Na^+ ion in solution selectively better than the single-layer electrode of the Pt/PVC as a coated-wire type electrode (CWE). The Pt/(PPy/PSS)/PVC exhibits a more stable response and a lower potential drift than the Pt/PVC [95, 96]. Figure 13 shows the mechanism of yielding potential for the CWE and Pt/(PPy/PSS)/PVC electrodes. In the CWE the interface of the PVC and the Pt surface is electrochemically unstable and completely insulated electrically [97] as shown in Figure 13*a*. In case of Figure 13*b*, however, the Pt surface makes contacts electrochemically and electrically with the electroactive and conductive PPy/PSS film. It is expected that the potential of the internal electrode shows a stable value due to the high redox capacity. The stable response is due to the formation of a concentration cell with both sides of the PVC membrane because of the cation exchange ability of the PPy/PSS [95]. The Pt/(PPy/PSS) electrode therefore it acts as both an internal electrolyte solution and internal reference electrode, which is the same as an internal Ag/AgCl reference electrode with KCl solution in the conventional ion-selective electrodes. Furthermore we proved and analyzed the long-term stability of the all-solid-state potassium-selective electrode with the PPy/PSS [96].

There appears to be a strong presence of monovalent cation sensors based on the PPy/PSS composite film. In particular, the double-layer system using n-type active

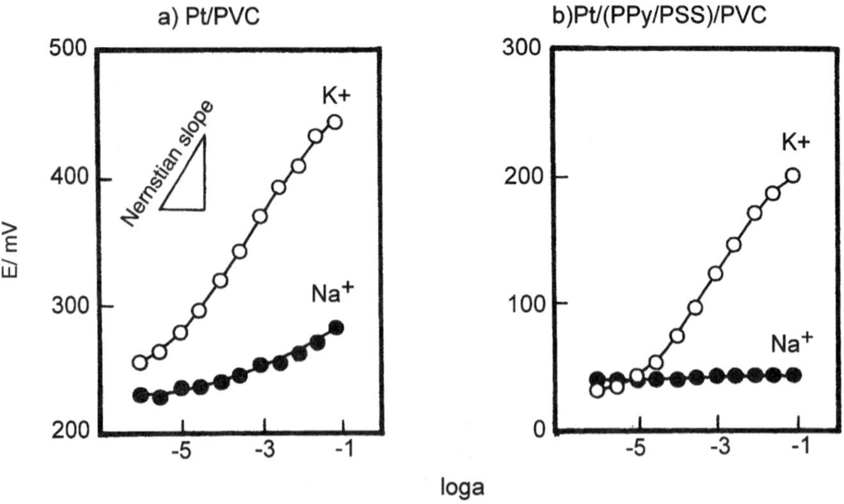

Figure 12 Calibration curves of potential responses to potassium and sodium ions of (*a*) Pt/PVC and (*b*) Pt/(PPy/PSS)/PVC electrodes.

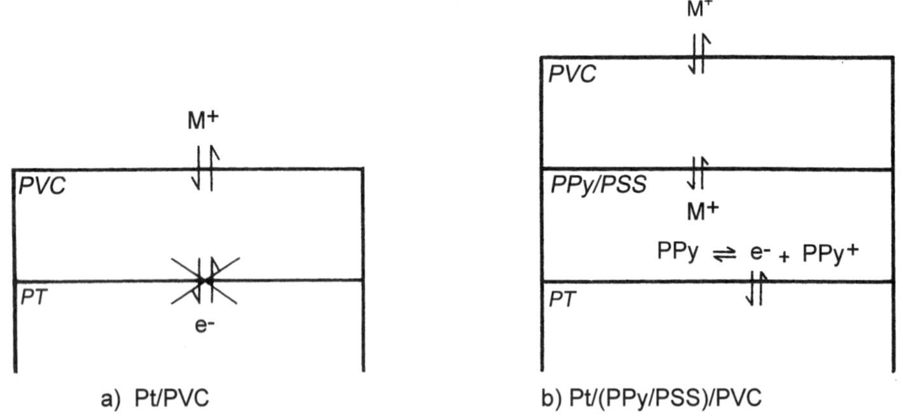

Figure 13 Potential response mechanism for (*a*) Pt/PVC and (*b*) Pt/(PPy/PSS)/PVC electrodes.

and conductive polymer is expected to allow the miniaturization of other important ion-selective electrodes.

4.3 Potentiometric Biosensor for Urea

In the past several years, the idea of incorporating enzymes, such as glucose oxidase, into conducting polymers by electropolymerization was reported [79, 98]. These biosensros were developed with electropolymerized conducting polymers as an amperometric response mechanism to the electrochemical reaction of oxygen and hydrogen peroxide [98]. However, there was an disadvantage with respect to its lifetime: The conductivity decreased by degradation of the polymers [99]. So we focused on the application of pH-sensitive nonconducting PPy to potentiometric biosensor for urea.

Urea is often monitored in blood to obtain information on kidney disease. The monitoring procedure for urea usually involves a catalytic reaction by enzyme. In order to make this monitoring procedure easier, many biosensors for urea, urease (Urs)-immobilized electrodes, have been studied [100–102]. Many urea sensors possess a structure in which an enzyme layer is formed directly on the active surface of a potentiometric transducer, such as an ion-selective membrane electrode [103, 104], a gas sensor [105], or an ion sensitive field effect transistor (ISFET) [106]. Recently immobilization of Urs into a conducting PPy by electropolymerization was reported; however its sensitivity was not sufficient because the Nernstian slope was relatively low [107–109].

When pyrrole is electropolymerized in a solution containing a nucleophilic reagent as a supporting electrolyte, an electroinactive and lowconducting PPy coated electrode is obtained [14, 51, 99]. A Pt electrode coated with this PPy film exhibits a good Nernstian response to the pH value. By use of this electrode, we attempted to fabricate an amperometric micro-pH sensor [15]. Also the film can be applied to potentiometric biosensor for urea, since the pH changes according to the urea hydrolysis catalyzed by the Urs proceeds as follows:

$$NH_2CONH_2 + 2H_2O + H^+ 2NH_4^+ + HCO_3^-$$

Figure 14 shows schematic drawing of potential response mechanism to urea. At this point we proposed urea sensor with three kinds of enzyme loading methods. Their different characteristics were due to variations in the loading amount of enzyme, as are schematically shown in Figure 15.

In the phase 1 of Figure 15, the Urs is entrapped in the insulating PPy film as both Urs and nucleophilic electrolyte are added to the polymerizing solution; Urs-immobilized, electroinactive PPy coated electrode is fabricated [16]. This electrode becomes a urea sensor whose potential response shows a slope of 31.8 mV decade^{-1}. Since the Urs is physically entrapped in the PPy film during its electrodeposition, this type of sensor can be fabricated by one electrochemical step. so the PPy film acts as both pH detector and entrapment matrix. This process, however, seems to have some disadvantages for mobilization of the enzyme. Most of enzyme in the electrolyte solution that electrodeposites the PPy will be wasted, and the enzyme is an expensive

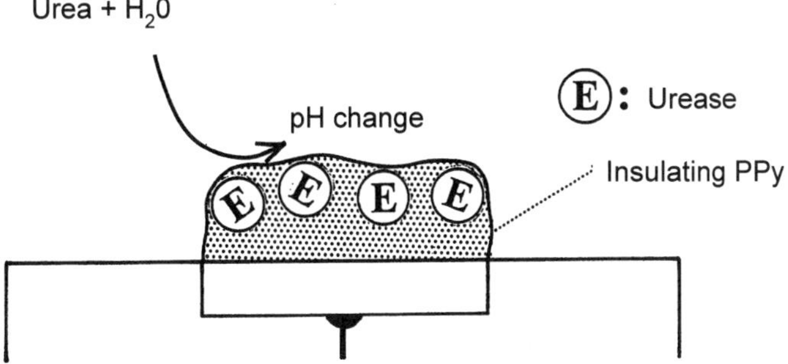

Figure 14 Schematic of the urea potential response.

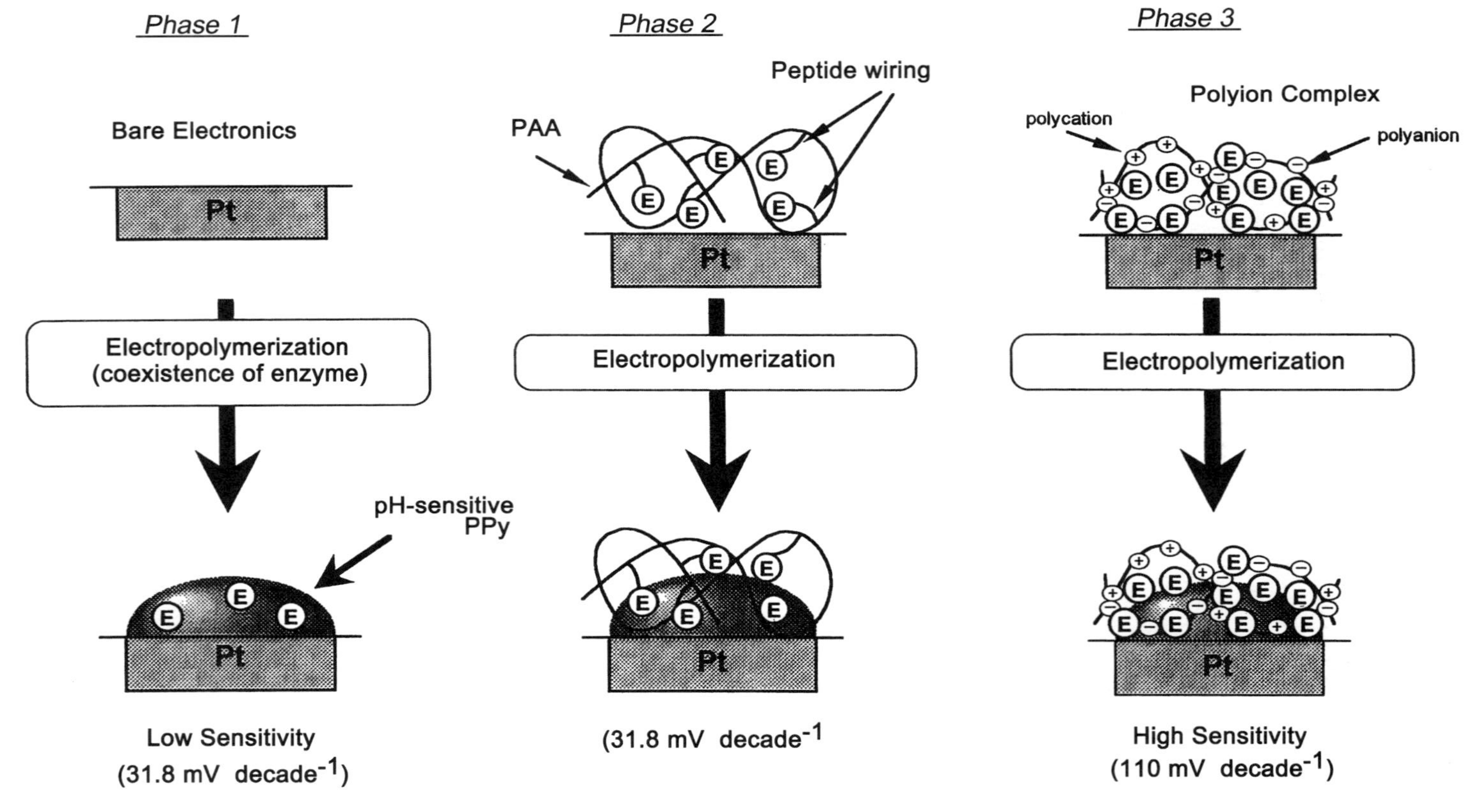

Figure 15 Schematic of the three types of urea biosensor.

chemical. Thus we attempted to improve the design and developed new device that utilizes a composite film consisting of the electroinactive PPy and Urs modified with poly acrylic acid (PAA) by peptide bonding, as shown in the phase 2 of Figure 15 [17].

Now the Urs was first chemically wired with PAA by the condensation reaction of water soluble carbodiimide. Subsequently the electropolymerization of pyrrole was carried out. This type of sensor demonstrated a potential response to urea whose slope was 53 mV $decade^{-1}$; this was higher than that of phase 1 described above. The higher sensitivity was attributed to both the increase in the loading amount of Urs and the covalent peptide bonding of the residual amino group of Urs and carboxyl groups of PAA. However, it is also possible that some of the Urs combined with PAA and became scattered and lost in the aqueous solution of pyrrole prior to electropolymerization, since watersoluble poly acrylic acid was employed in the system. The sensitivity of this type was further improved by using waterinsoluble material for the immobilization.

We fabricated a urea sensor of even highersensitivity based on a composite film consisting of an electroinactive PPy and a poly-ion complex, as shown in the phase 3 of Figure 15. Mizutani et al. reported various amperometric biosensors in which enzyme molecules were incorporated into a waterinsoluble poly-ion complex formed with a poly-cation and a poly-anion [110, 111]. We employed such a poly-ion complex as the material for immobilizing the Urs [18]. After coating the poly-ion complex, electropolymerization of pyrrole was carried out, and a composite film of the PPy and poly-ion complex was obtained. This type of sensor demonstrates the fast potential response to urea by the same detection mechanism as mentioned above. By using this system, the highest urea response with the slope of 110 mV $decade^{-1}$ could be obtained. Such a high sensitivity is attributed to the effective immobilization of large amount of the Urs by electropolymerization on the electrode pre-coated with a poly-ion complex.

Other biosensor could be fabricated with similar architecture. A creatinine sensor was examined with the bi-enzymatic reactions using two enzymes [112]. Also a combination of the biosensor and the flow injection analysis method was attempted to enhance the detection performance [113, 114].

REFERENCES

1. H. Shirakawa, T. Ito, and S. Ikeda, *Polym. J.*, **2**, 231 (1971).
2. T. Ito, H. Shirakawa, and S. Ikeda, *J. Pol. Sci., Polym. Chem. Ed.*, **12**, 11 (1974).
3. T. Ito, H. Shirakawa, and S. Ikeda, *J. Pol. Sci., Polym. Chem. Ed.*, **13**, 1943 (1975).
4. T. A. Skotheim, ed., *Handbook of Conducting Polymers*, vols. 1 and 2, Marcel Dekker, New York, 1986.
5. A. F. Diaz and K. K. Kanazawa, *J. Chem. Soc., Chem. Commun.*, 635 (1979).
6. K. K. Kanazawa, A. F. Diaz, R. H. Geiss, W. D. Gill, J. F. Kwak, J. A. Logan, J. F. Rabolt, and G. B. Street, *J. Chem. Soc., Chem. Commun.*, 854 (1979).
7. G. Tourillon and F. Garnier, *J. Electroanal. Chem.*, **135**, 173 (1982).
8. T. Yamamoto, Z. Zhou, I. Ando, and M. Kikuchi, Makromol. *Chem., Rapid Commun.*, **14**, 833 (1993).

9. L. M. Goldenberg and P. C. Lacaze, *Synth. Met.*, **58**, 271 (1993).
10. G. Schiavon, G. Zotti, and G. Bontempelli, *J. Electroanal. Chem.*, **194**, 327 (1985).
11. G. Zotti, G. Schiavon, N. Comisso, A. Berlin, and G. Pagani, *Synth. Met.*, **36**, 337 (1990).
12. G. Zotti and G. Schiavon, *J. Electroanal. Chem.*, **163**, 385 (1984).
13. G. Schiavon, G. Zotti, and G. Bontempelli, *J. Electroanal. Chem.*, **161**, 323 (1984).
14. T. Osaka, T. Momma, and H. Kanagawa, *Chem. Lett.*, 649 (1993).
15. T. Osaka, T. Fukuda, H. Kanagawa, T. Momma, and S. Yamauchi, *Sensors Actuators B*, **13–14**, 205 (1993).
16. S. Komaba, M. Seyama, T. Momma, and T. Osaka, *Electrochim. Acta*, **43**, 383 (1997).
17. S. Komaba, M. Seyama, K. Tanabe, and T. Osaka, *Denki Kagaku*, **64**, 1228 (1996).
18. T. Osaka, S. Komaba, M. Seyama, and K. Tanabe, *Sensors and Actuators B*, **35–36**, 1 (1996).
19. T. A. Skotheim, *Handbook of Conducting Polymers*, Marcel Dekker, New York, 1986.
20. R. J. Waltman, A. F. Diaz, and J. Bargon., *J. Electrochem. Soc.*, **131**, 740 (1984).
21. T. Osaka, K. Naoi, and S. Ogano, *J. Electrochem. Soc.*, **135**, 1071 (1988).
22. T. Osaka, K. Naoi, S. Ogano, and S. Nakamura, *J. Electrochem. Soc.*, **134**, 2096 (1987).
23. F. Croce, S. Panero, P. Prosperi, and B. Scrosati, *Solid State Ionics*, **28–31**, 895 (1988).
24. T. Osaka, T. Nakajima, K. Shiota, and B. B. Owens, in *Rechargeable Lithium Batteries/ 1989*, S. Subbarao, V. R. Koch, B. B. Owens, and W. H. Smyrl, ed., PV 90-5, p. 170, Proceedings Series, Electrochemical Society, Pennington, NJ, 1990.
25. T. Osaka and K. Ueyama, *Kagaku-Kogyo*, (Chemical Industry), 31 (Mar. 1989).
26. T. Sotomura, H. Uemachi, K. Takeyama, K. Naoi, and N. Oyama, *Electrochim. Acta*, **37**, 1851 (1993).
27. N. Oyama, T. Tatsuma, T. Sato, and T. Sotomura, *Nature*, **373**, 598 (1995).
28. T. Osaka, T. Momma, H. Ito, and B. Scrosati, *J. Power Sources*, in press.
29. M. D. Paoli, R. J. Waltman, A. F. Diaz, and J. Bargon, *J. Polym. Sci. Polym. Chem. Ed.*, **23**, 1067 (1985).
30. B. Kaye and A. E. Underhill, *Synth. Met.*, **28**, C97 (1989).
31. H. Yoneyama, Y. Li, and S. Kuwabata, *J. Electrochem. Soc.*, **139**, 28 (1992).
32. A. Boyle, E. Genies, and M. Fouletier, *J. Electroanal. Chem.*, **279**, 179 (1988).
33. L. L. Miller and Q. X. Zhou, *Macromolecules*, **20**, 1594 (1987).
34. T. Shimidzu, A. Ohtani, and K. Honda, *J. Electroanal. Chem.*, **251**, 323 (1988).
35. C. M. Elliott, A. B. Kopelove, W. J. Alberty, and Z. Chen, *J. Phys. Chem.*, **95**, 1743 (1991).
36. A. Shimizu, K. Yamanaka, and M. Kohno, *Bull. Chem. Soc. Jpn.*, **61**, 4401 (1988).
37. T. Osaka, T. Momma, and K. Nishimura, *Chem. Lett.*, **1992**, 1787.
38. T. Momma, K. Nishimura, T. Osaka, N. Kondo, and S. Nakamura, *J. Electrochem. Soc.*, **141**, 2326 (1994).
39. X. Ren and P. G. Pickup, *J. Phys. Chem.*, **97**, 5356 (1993).
40. K. West, M. A. Careem, and S. Skaarup, *Solid State Ionics*, **60**, 153 (1993).
41. T. Osaka, T. Momma, K. Nishimura, S. Kakuda, and T. Ishii, *J. Electrochem. Soc.*, **141**, 1994 (1994).
42. K. Naoi and T. Osaka, *J. Electrochem. Soc.*, **134**, 2479 (1987).
43. K. Naoi, M. Asada, Y. Hayashi, and Y. Inoue, in *Proc. Symp. on New Sealed Rechargeable Batteries and Supercapacitors*, B. M. Barnett, E. Dowgiallo, G. Halpert,

Y. Matsuda, and Z. Takehara, ed., PV 93-23, p. 86, Proceedings Series, Electrochemical Society, Pennington, NJ, 1993.

44. C. Arbizzani, A. M. Marinangeli, M. Mastragostino, L. Meneghello, T. Hamaide, and A. Guyot, *J. Power Sources*, **43–44**, 453 (1993).
45. T. Momma, S. Kakuda, H. Yarimizu, and T. Osaka, *J. Electrochem. Soc.*, **142**, 1766 (1995).
46. H. Yoneyama, T. Hirai, S. Kuwabata, and O. Ikeda, *Chem. Lett.*, **1986**, 1243.
47. C. K. Chiang, S. C. Gau, C. R. Fincher Jr., Y. W. Park, A. G. MacDiarmid, and A. J. Heeger, **33**, 18 (1978).
48. J. L. Bredas and G. B. Street, *Acc. Chem. Res.*, **18**, 309 (1985).
49. A. Kitani, J. Yano, and K. Sasaki, *J. Electrochem. Soc.*, **209**, 227 (1986).
50. H. T. Chiu, J. S. Lin, and J. N. Shiau, *J. Appl. Electrochem.*, **22**, 522 (1992).
51. T. Osaka, T. Fukuda, K. Ouchi, and T. Nakajima, *Denki Kagaku*, **12**, 1019 (1991), and T. Nakajima, F. Matsushima, T. Fukuda, and T. Osaka, *Denki Kagaku*, **12**, 1074 (1991).
52. T. Osaka, T. Fukuda, K. Ouchi, and T. Momma, *Thin Solid Films*, **215**, 200 (1992).
53. J. H. Burroughes, D. D. C. Bradley, A. R. Brown, R. N. Marks, K. Mackay, R. H. Friend, P. L. Burns, and A. B. Holmes, *Nature*, **347**, 539 (1990).
54. C. W. Tang and S. A. VanSlyke, *Appl. Phys. Lett.*, **51**, 913 (1987).
55. G. Grem, G. Leditzky, B. Ullrich, and G. Leising, *Adv. Mater.*, **4**, 36 (1992).
56. D. Brown and A. J. Heeger, *Appl. Phys. Lett.*, **58**, 1982 (1991).
57. M. Berggren, G. Gustafsson, O. Inganšs, M. R. Andersson, O. Wennerstršm, and T. Hjertberg, *Adv. Mater.*, **6**, 488 (1994).
58. M. Uchida, Y. Ohmori, T. Noguchi, T. Ohnishi, and K. Yoshino, *Jpn. J. Appl. Phys.*, **32**, L921 (1993).
59. G. Grem and G. Leising, *Synth. Met.*, **55–57**, 4105 (1993).
60. Y. Ohmori, M. Uchida, K. Muro, and K. Yoshino, *Solid State Commun.*, **80**, 605 (1991).
61. D. Brown, G. Gustafsson, D. McBranch, and A. J. Heeger, *J. Appl. Phys.*, **72**, 564 (1992).
62. S. Komaba, K. Fujihana, N. Kaneko, and T. Osaka, *Chem. Lett.*, **1995**, 923.
63. T. Osaka, S. Komaba, K. Fujihana, T. Momma, and N. Kaneko, *J. Electrochem. Soc.*, submitted (1997).
64. T. Osaka, S. Komaba, K. Fujihana, N. Okamoto, and N. Kaneko, *Chem. Lett.*, **1995**, 1023.
65. K. Naoi and T. Osaka, *J. Electrochem. Soc.*, **134**, 2479 (1987).
66. Y. Ohmori, M. Uchida, K. Muro, and K. Yoshino, *Jpn. J. Appl. Phys.*, **30**, L1938 (1991).
67. Y. Shirota, Y. Kuwabara, H. Inada, T. Wakimoto, H. Nakada, Y. Yonemoto, S. Kawami, and K. Imai, *Appl. Phys. Lett.*, **65**, 807 (1994).
68. G. Gustafsson, Y. Cao, G. M. Treacy, F. Klavetter, N. Colaneri, and A. J. Heeger, *Nature*, **357**, 477 (1992).
69. T. Yamamoto, T. Inoue, and T. Kanbara, *Jpn. J. Appl. Phys.*, **33**, L250 (1994).
70. S. Hayashi, H. Etoh, and S. Saito, *Jpn. J. Appl. Phys.*, **25**, L773 (1986).
71. Y. Mori, H. Endo, and Y. Hayashi, *Ouyoubutsuri*, **61**, 1044 (1991) (in Japanese).
72. S. Komaba, M. Seyama, and T. Osaka, *Chemical Sensors*, **12**, 97 (1996) (in Japanese).
73. W. R. Heineman, H. J. Wieck, and A. M. Yachnych, *Anal. Chem.*, **52**, 345 (1980).
74. S. Dong, Z. Sun, and Z. Lu, *Analyst*, **113**, 1525 (1988).
75. Q. Pei and R. Qian, *Synth. Met.*, **45**, 35 (1991).
76. S. Daunert, S. Wallace, A. Floride, and L. Bachas, *Anal. Chem.*, **63**, 1676 (1991).

77. T. Okada, H. Hayashi, K. Hiratani, H. Sugihara, and N. Koshizaki, *Analyst*, **116**, 923 (1991).
78. T. Okada, K. Hiratani, H. Sugihara, and N. Koshizaki, *Anal. Chim. Acta*, **272**, 89 (1992).
79. T. Okada, K. Hiratani, H. Sugihara, and N. Koshizaki, *Sensors Actuators B*, **13–14**, 563 (1993).
80. N. C. Foulds and C. R. Lowe, *J. Chem. Soc., Faraday Trans. 1*, **82**, 1259 (1986).
81. A. Begum, H. Tsushima, T. Suzawa, H. Shinohara, Y. Ikariyama, and M. Aizawa, *Sensors Actuators B*, **13–14**, 576 (1993).
82. P. N. Bartlett and P. R. Birkin, *Synth. Met.*, **61**, 15 (1993).
83. P. N. Bartlett and J. M. Cooper, *J. Electroanal. Chem.*, **371**, 1 (1993).
84. G. F. Khan, E. Kobatake, Y. Ikariyama, and M. Aizawa, *Anal. Chim. Acta*, **281**, 527 (1993).
85. F. Pslmisano, D. Centonze, and P. G. Zambonin, *Biosens. Bioelectron.*, **9**, 471 (1994).
86. B. F. Y. Yon Hin and C. R. Lowe, *J. Electroanal. Chem.*, **374**, 167 (1994).
87. M. Aizawa, T. Haruyama, G. F. Kahn, E. Kobatake, and Y. Ikariyama, *Bionsens. Bioelectron.*, **9**, 601 (1994).
88. W. Schhmann, *Biosens. Bioelectron.*, **10**, 181 (1995).
89. J. J. Miasik, A. Hopper, and B. C. Tofield, *J. Chem. Soc., Faraday Trans. 1*, **82**, 1117 (1986).
90. T. Hanawa, S. Kuwabata, and H. Yoneyama, *J. Chem. Soc., Faraday Trans. 1*, **84**, 1587 (1988).
91. T. Hanawa and H. Yoneyama, *Bull. Chem. Soc. Jpn.*, **62**, 1710 (1989).
92. H. Nagase, K. Wakabayashi, and T. Imanaka, *Sensors Actuators B*, **13–14**, 596 (1993).
93. T. Momma, S. Komaba, T. Osaka, S. Nakamura, and Y. Takemura, *Bull. Chem. Soc. Jpn.*, **68**, 1297 (1995).
94. W. Simon and D. Stefanac, *Chimia*, **20**, 436 (1966).
95. T. Momma, S. Komaba, M. Yamamoto, T. Osaka, and S. Yamauchi, *Sensors Actuators B*, **24–25**, 724 (1995).
96. T. Momma, M. Yamamoto, S. Komaba, and T. Osaka, *J. Electroanal. Chem.*, **407**, 91 (1996).
97. B. P. Nikolskii and E. A. Materova, *Ion Select. Electrode Rev.*, **7**, 3 (1985).
98. W. Schuhmann, *Mikrochim. Acta*, **121**, 1 (1995).
99. T. Osaka, T. Momma, S. Komaba, H. Kanagawa, and S. Nakamura, *J. Electroanal. Chem.*, **372**, 201 (1994).
100. M. Mascini and G. Guilbault, *Anal. Chem.*, **49**, 795 (1977).
101. M. Pryzgyt and H. Sugier, *Anal. Chim. Acta*, **237**, 399 (1990).
102. A. M. Nyamsi Hendji, N. Jaffrezic-Renault, C. Martelet, A. A. Shul'ga, S. V. Dzydevich, A. P. Soldatkin, and A. V. El'skaya, *Sensors Actuators B*, **21**, 123 (1994).
103. R. Koncki, P. Leszcynski, A. Hulanicki, and S. Glab, *Anal. Chim. Acta*, **257**, 67 (1992).
104. D. Martorell, E. Martinez-Fabregas, J. Bartoli, S. Alegret, and C. Tran-Minh, *Sensors Actuators B*, **15–16**, 448 (1993).
105. Y. J. Wang, C. H. Chen, G. H. Hsiue, and B. C. Yu, *Biotechnol. Bioeng.*, **40**, 44 (1992).
106. S. Alegret, J. Bartroli, C. Jimenez, E. Martinez-Fabregas, D. Martorell, and F. Valdes-Perezgasga, *Sensors Actuators B*, **15–16**, 453 (1993).
107. S. B. Adeloju, S. J. Shaw, and G. G. Wallace, *Anal. Chim. Acta*, **281**, 611 (1993).
108. S. B. Adeloju, S. J. Shaw, and G. G. Wallace, *Anal. Chim. Acta*, **281**, 621 (1993).

109. E. C. Hernandez, A. Witkowski, S. Daunert, and L. G. Bachas, *Mikrochim. Acta*, **121**, 63 (1995).

110. F. Mizutani, S. Yabuki, and Y. Hirata, *Denki Kagaku*, **63**, 1100 (1995).

111. F. Mizutani, S. Yabuki, and Y. Hirata, *Anal. Chim. Acta*, **314**, 233 (1995).

112. T. Osaka, S. Komaba, and A. Amano, *J. Electrochem. Soc.*, **145**, 406 (1998).

113. S. Komaba, J. Arakawa, M. Seyama, T. Osaka, I. Satho, and S. Nakamura, *Talanta*, **46**, 19 (1998).

114. T. Osaka, S. Komaba, Y. Fujino, T. Matsuda, and I. Satoh, *J. Electrochem. Soc.*, **146**, 615 (1999).

17 Electroless Deposition of Copper

MILAN PAUNOVIC

INTRODUCTION

In part A of Chapter 1, we discussed the overall reaction and the mixed potential theory of electroless metal deposition in general. In this chapter we discuss the specific case of copper: the overall reaction, fundamental, and technological aspects of its electroless deposition.

The overall reaction for electroless copper deposition, with formaldehyde (HCHO) as the reducing agent, is

$$Cu^{2+} + 2HCHO + 4OH^{-} \longrightarrow Cu + 2HCOO^{-} + 2H_2O + H_2 \quad (1)$$

where $HCOO^{-}$ (formic acid) is the oxidation product of the reducing agent. The fundamental aspects of this reaction are presented in the following five sections: (1) electrochemical model, (2) anodic partial reaction, (3) cathodic partial reaction, (4) kinetics of deposition, and (5) modeling. Some special cases of electroless copper deposition solutions are given in an Appendix to this chapter.

1 ELECTROCHEMICAL MODEL

1.1 Mixed-Potential Theory

The mixed potential theory was developed by Wagner and Traud [1] for the purpose of interpreting metal corrosion processes. Paunovic [2] and Saito [3] applied the theory to the interpretation of electroless deposition of copper.

According to the mixed-potential theory, the overall reaction, given by Eq. (1), can be decomposed into a simple reduction reaction, the cathodic partial reaction

$$Cu^{2+}_{solution} + 2e \xrightarrow{\text{catalytic surface}} Cu_{lattice} \quad (2)$$

and one oxidation reaction, the anodic partial reaction

$$HCHO + 2OH^{-} \longrightarrow HCOO^{-} + H_2O + \tfrac{1}{2}H_2O + e \quad (3)$$

Modern Electroplating, Fourth Edition, Edited by Mordechay Schlesinger and Milan Paunovic.
ISBN 0-471-16824-6

Thus the overall reaction Eq. (l) is the result of the combination of two different partial reactions, Eqs. (2) and (3). These two partial reactions, however, occur at one and the same electrode, namely the metal-solution interphase. Each of these reactions strives to establish its own equilibrium potential, E_{eq}. The result of this process is the creation of a steady state with the compromised potential called the *steady-state mixed potential*, E_{mp}.

1.2 Evans Diagram

According to the mixed-potential theory the overall reaction of the electroless copper deposition can be described electrochemically in terms of two current-potential (*i-E*) curves, as shown in Figure 1. Figure 1 was constructed in the following way: First, the current-potential curve of the reduction of cupric ions in the solution containing H_2O, 0.1 M $CuSO_4$, 0.175 M EDTA (ethylenediaminetetraacetic acid) and NaOH to pH 12.50 (no CH_2O present) at 24°C (± 0.5) is determined using a galvanostatic technique. At this electrode only one reaction occurs, the reduction of Cu^{2+}. An electrode with only one electrode process is called a *single electrode*. The result is shown as $i(Cu^{2+}) = f(E)$ in Figure 1. The current-potential curve was recorded starting from the equilibrium potential, $E_{eq}(Cu/Cu^{2+}) = -0.47$ V versus SCE. Second, the current-potential curve for the oxidation of formaldehyde at the single electrode was determined using the galvanostatic technique. The solution in this case contained H_2O, 0.05 M CH_2O, 0.075 M EDTA (excess of EDTA used in the solution

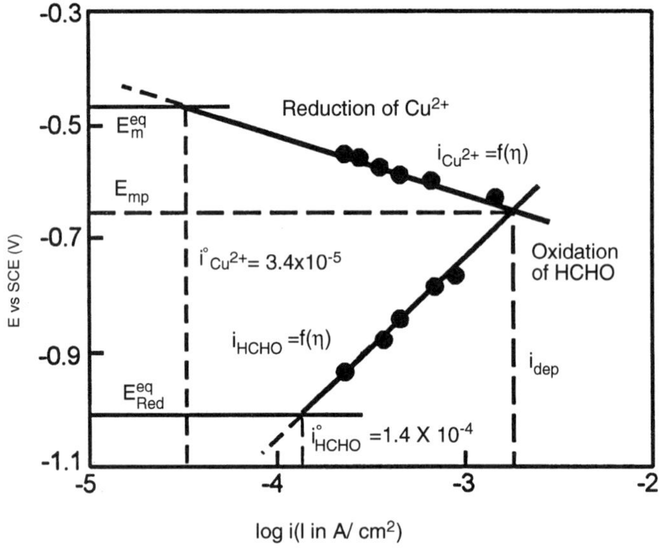

Figure 1 Current-potential curves for reduction of Cu^{2+} ions and for oxidation of reducing agent Red, formaldehyde, combined into one graph (Evans diagram). Solution for the Tafel line for the reduction of Cu^{2+} ions: 0.1 M $CuSO_4$, 0.175 M EDTA, pH 12.50, E_{eq} (Cu/Cu^{2+}) = −0.47 V versus SCE; for the oxidation of formaldehyde: 0.05 M HCHO and 0.075 M EDTA, pH 12.50, E_{eq} (HCHO) = −1.0 V versus SCE; temperature 25°C (±0.50°C). (From Paunovic [2] with permission from the American Electroplaters and Surface Finishers Society.)

for the single cathodic reaction), and NaOH to pH 12.50 (no $CuSO_4$ was present in this solution). The temperature was 24°C (±0.5). The result is shown as $i(CH_2O) = f(E)$ in Figure 1. The current-potential curve was recorded starting from the equilibrium potential, $E_{eq}(CH_2O) = -1.0$ V versus SCE. It is seen from Figure 1 that these two polarization curves, $i(Cu^{2+}) = f(E)$ and $i(CH_2O) = f(E)$, intersect. The coordinates of intersection are (1) abscisa, $i = 1.9 \times 10^{-3}\,A\,cm^{-2}$ and (2) ordinate, $E = -0.65$ V versus SCE. According to the mixed-potential theory, current density $i = 1.9 \times 10^{-3}\,A\,cm^{-2}$ is the rate of the electroless deposition of copper expressed in terms of $A\,cm^{-2}$. The potential $E = -0.65$ V versus SCE is the mixed potential (E_{mp}) of the electroless copper system under study. The rate of deposition expressed in milligrams per hour per square centimeter ($mg\,h^{-1}\,cm^{-2}$) is calculated on the basis of Faraday's law using the equation

$$w = i \times 1.18 \text{ mg h}^{-1}\text{cm}^{-2}$$

where i is given in milliamperes per square centimeter ($mA\,cm^{-2}$). For $i = 1.9 \times 10^{-3}$ $A\,cm^{-2}$ it is $2.2\,mg\,h^{-1}\,cm^{-1}$.

The experimentally determined rate of electroless Cu deposition under the conditions above, using the weight gain method, is $1.8\,mg\,h^{-1}\,cm^{-2}$. This rate is obtained when the time of deposition is counted from the instant of immersion of the copper plate (substrate) into the solution. If the time of deposition is counted from the instant the mixed potential is reached (about 4 minutes after immersion of the Cu substrate), the deposition rate is $1.9\,mg\,h^{-1}\,cm^{-2}$. The experimentally determined mixed potential E_{mp}, for the same conditions, is -0.65 V versus SCE.

Examination of Figure 1 and the results of direct experimental measurements shows that there is a relatively good agreement between the direct experimental and the theoretical values (Evans diagram). Thus we can conclude that the mixed-potential theory is essentially verified for this case of electroless copper deposition. These conclusions are confirmed by Donahue [4], Molenaar et al. [5], and El-Raghy and Abo-Salama [6]. The significance of this conclusion is that on the basis of the mixed-potential theory one can use the kinetic parameters for the partial anodic and cathodic reactions to deduce a variety of predictions and characteristics of the overall reaction of electroless copper deposition. For example, the effect of additives on the overall reaction can be resolved into separate effects on the partial reactions and use these results to select the best conditions for electroless deposition.

1.3 Interaction between Partial Reaction

The original mixed-potential theory assumes that the two partial reactions are independent of each other [1, 2]. In some cases this is a valid assumption. However, it was shown later that the partial reactions are not always independent of each other [7, 8]. For example, Schoenberg [9] has shown that the methylene glycol anion (the formaldehyde in an alkaline solution), the reducing agent in electroless copper deposition, enters the first coordination sphere of the copper tartrate complex and thus influences the rate of the cathodic partial reaction. Ohno and Haruyama [10] showed the presence of interference in partial reactions in terms of current-potential curves.

1.4 Presence of Interfering Reactions

In the presence of interfering (or side) reactions, partial reactions i_a and/or i_c may be composed of two or more components. One example is the electroless deposition of copper from solutions containing oxygen [11, 12]. In this case the interfering reaction is the reduction of the oxygen, and the cathodic partial current density i_c is the sum of two components

$$i_c = i_c(Cu^{2+}) + i_c(O_2)$$

where $i_c(Cu^{2+})$ is the cathodic partial current density for reduction of copper ions Cu^{2+} and $i_c(O_2)$ is that for reduction of the oxygen.

2 ANODIC PARTIAL REACTION

2.1 Overall Reaction

Most electroless copper solutions employ formaldehyde as the reducing agent. The overall reaction of the electrochemical oxidation of formaldehyde at the Cu electrode in an alkaline solution proceeds according to Eq. (3).

2.2 Mechanism

The overall anodic partial reaction, Eq. (3), proceeds in at least two elementary steps:

1. Formation of electroactive species
2. Charge transfer from electroactive species to the catalytic surface (electron injection)

Formation of electroactive species proceeds in three steps:

1. Hydrolysis of H_2CO,

$$H_2CO + H_2O = H_2C(OH)_2 \qquad \text{(methyleneglycol)} \tag{4}$$

2. Dissociation of methylene glycol,

$$H_2C(OH)_2 + OH^- = H_2C(OH)O^- + H_2O \tag{5}$$

3. Dissociative adsorption of the intermediate $H_2C(OH)O^-$ involving breaking of C–H bond,

$$H_2C(OH)O^- = [HC(OH)O^-]_{ads} + H_{ads} \tag{6}$$

where the subscript ads denotes adsorption of species. $[HC(OH)O^-]_{ads}$ is electroactive species.

Charge transfer, the electrochemical oxidation (desorption) of electroactive species, proceeds according to equation

$$[HC(OH)O^-]_{ads} + OH^- = HCOO^- + H_2O + e \quad (7)$$

where $HCOO^-$ (formic acid) is the oxidation product.

The adsorbed hydrogen, H_{ads}, may be desorbed in the chemical reaction

$$H_{ads} = \tfrac{1}{2}H_2 \quad (8)$$

or in the electrochemical reaction

$$H_{ads} = H^+ + e \quad (9)$$

In electroless deposition of copper, when the reducing agent is formaldehyde and the substrate is Cu, H_{ads} desorbs in the chemical reaction, Eq. (8). If the substrate is Pd or Pt, hydrogen desorbes according to the electrochemical reaction, Eq. (9).

2.3 Cannizzaro Reaction

One important side reaction in electroless copper deposition is disproportionation of formaldehyde (Cannizzaro reaction)

$$2HCHO + OH^- = HCOO^- + CH_3OH \quad (10)$$

In this reaction between two molecules of formaldehyde, one molecule is oxidized into formic acid and the other is reduced into methanol. The rate of this reaction increases with increasing pH and temperature [13].

2.4 Kinetics

The major factors determining the rate of the anodic partial reaction are pH and additives. Since OH^- ions are reactants in the charge-transfer step, Eq. (7), the effect of pH is direct and significant [14, 15].

The reduction potential (the rest potential) of formaldehyde increases linearly with pH according to Nernst's equation

$$E = E^0_{csp} - 0.118\ \text{pH} \quad (11)$$

where E^0_{csp} combines the standard electrode potential E^0 and the concentration term in Nernst's equation [14]. The rest potential of a copper single electrode (absence of copper ions) in the solution of formaldehyde, as a function of pH is shown in Figure 2. The average slope $\partial E/\partial$ (pH) of experimental functions in Figure 2 is -0.096 V decade^{-1}. A detailed discussion of the pH effect on the partial anodic reaction is given by Duffy et al. [14].

3 CATHODIC PARTIAL REACTION

3.1 Kinetic Scheme

Examination of the pH dependence of the reduction potential and the rate of oxidation of formaldehyde shows that the pH of the electroless copper solution

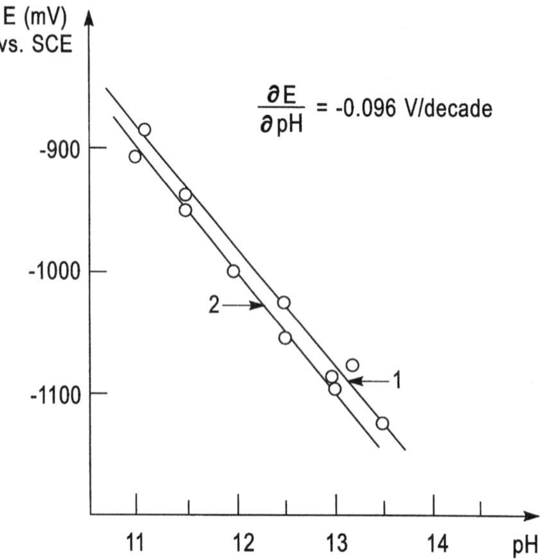

Figure 2 The rest potential of a copper electrode in 0.13 M formaldehyde and 1.0 M KCl as a function of pH. Curve 1: the absence of ligand; curve 2: the presence of 0.05 M EDTP (ethylenedinitrilo-tetra-2-propanol).

should be above 11.0 in order to have practical rates of copper deposition [14]. This pH restriction imposes the use of complexed copper ions in the electroless solution in order to prevent precipitation of copper(II)hydroxide, $Cu(OH)_2 \cdot$ EDTA, EDTP (ethylenedinitrilo-tetra-2-propanol), and tartaric acid are the most commonly used ligands for copper ions [16].

Thus the electroactive species in the partial cathodic reaction may be complexed or noncomplexed copper ions. In the first case the kinetic scheme of the cathodic partial reaction is one of the simple charge transfer

$$Cu^{2+} + e \xrightarrow{RDS} Cu^{+} \tag{12}$$

$$Cu^{+} + e \longrightarrow Cu \tag{13}$$

where RDS stands for rate-determining step (slow step). In the second case the kinetic scheme is of the charge transfer preceded by the dissociation of the complex [16]. The mechanism of the second case involves a sequence of at least two basic elementary steps:

1. Formation of the electroactive species
2. Charge transfer from the catalytic surface to the electroactive species

Electroactive species Cu^{2+} are formed in the first step by dissociation of the complex $[CuL_x]^{2+xp}$:

$$[CuL_x]^{2+xp} \rightarrow Cu^{2+} + xL^{p} \tag{14}$$

where p is the charge state of the ligand L, and $(2 + xp)$ is the charge of the complexed copper ion. The charge transfer

$$Cu^{2+} + 2e \rightarrow Cu_{lattice} \tag{15}$$

proceeds in steps, usually with the first charge transfer (one electron transfer), Eq. (12), serving as the rate-determining step [17].

Thus, from the kinetic aspects, the cathodic partial reaction is an electrochemical reaction, Eq. (15), that is preceded by a chemical reaction, Eq. (14).

Paunovic [16] has shown that in the electroless deposition of copper from the Cu(II)EDTA complex the reduction of the complex is preceded by dissociation of the same. A correlation has been established between the rate of dissociation of the complex and the rate of copper deposition [16].

3.2 Kinetics

The major factors determining the rate of the partial cathodic reaction are the concentration of the copper ions and the ligands, pH of the solution, and the type and the concentration of additives. These factors determine the kinetics of the partial cathodic reaction in a general way, as given by the fundamental electrochemical kinetic equations discussed in Chapter 1.

3.3 pH Effect

The rest potential of the copper electrode in an alkaline solution of cupric ions complexed with EDTP shows a linear pH dependence with a slope $\partial E/\partial(\text{pH}) = -0.066\ \text{V decade}^{-1}$ [14]

$$E = E^0 - 0.066\ \text{pH} \tag{16}$$

This slope is in conformity with the reaction

$$Cu + H_2L' = CuL' + 2H^+ + 2e \tag{17}$$

where $L = \text{EDTP}$, $L' = \text{EDTP} - 2H^+$ [14, 18, 19]. The experimentally observed slope of $-0.066\ \text{V decade}^{-1}$ [14] is in good agreement with the theoretical slope of $-0.059\ \text{V decade}^{-1}$ [17].

In contrast to the anodic partial reaction, the rate of the cathodic partial reaction does not depend significantly on pH, since OH^- ion is not a reactant in the cathodic reaction. Moreover the large concentration of OH^- ions in the metal/solution interphase (the double layer) can hinder the process of reduction of complexed copper ions (CuL), especially if CuL is negatively charged, such as when L is EDTA or tartrate [3].

3.4 Effect of Additives

Schoenberg [9, 20] as well as Paunovic and Arndt [21] have shown that additives may have two opposing effects: acceleration and inhibition. For example, guanine and adenine show accelerating effect on the cathodic reduction of Cu^{2+} ions in the electroless copper solution. The same additives show an increase in the rate of the electroless copper deposition. The accelerating and the inhibiting effects of dipyridyls were examined by Duda [22] as well as by Oita et al. [23]. In another example, the addition of NaCN to the electroless copper solution results in the inhibing effect for reduction of Cu^{2+} ions in an electroless solution. This inhibition increases with an increasing amount of NaCN in solution [24, 25].

4 KINETICS OF ELECTROLESS Cu DEPOSITION

Steady-state electroless copper deposition at mixed potential E_{mp} is preceded by a non-steady-state period, called the *induction period.*

4.1 Induction Period

The induction period is defined as the time necessary to reach the mixed potential E_{mp} at which the steady-state metal deposition start to occur. It is determined in a simple experiment in which a piece of metal is immersed in a solution for electroless deposition of a metal and the potential of the metal recorded from the time of immersion (or the time of addition of the reducing agent), that is, time zero, until the steady-state mixed potential is established. A typical recorded curve for the electroless deposition of copper on copper substrate is shown in Figure 3. The curve has been recorded for the system in an argon atmosphere. For a system in air atmosphere and in the presence of additives in the solution, the duration of the induction period can be considerably longer [16].

The problem of the induction period for the overall process can be resolved into problems of the open-circuit potentials (OCP) of the oxidation and reduction partial reactions, that is, the individual induction period for each process. Paunovic [16] found that the OCP for the Cu/Cn^{2+} system is reached instantaneously. A typical curve representing the change of the open-circuit potential (OCP) with time, for the

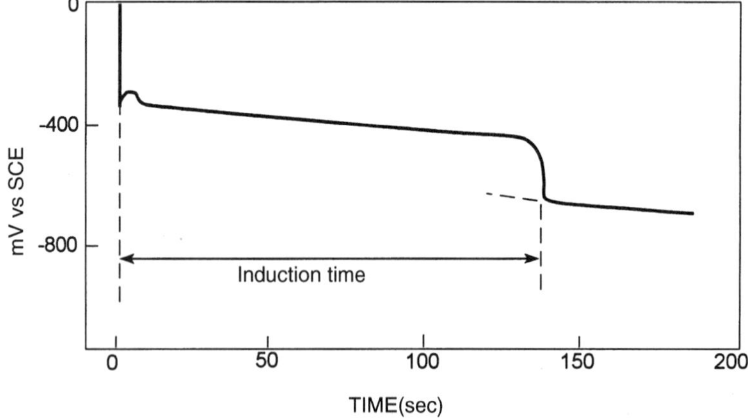

Figure 3 Induction period for the solution 0.3 M EDTA, 0.05 M $CuSO_4$, pH 12.50, 2.5 g liter^{-1} paraformaldehyde, Cu electrode, 2.2 cm^2, 25°C, SCE reference electrode, argon atmosphere. (From Paunovic [16], with permission from the Electrochemical Society.)

reducing agent, is presented in Figure 4. By comparing these OCP values, we can conclude that the setting of the OCP of the reducing agent, CH_2O, is the rate-determining partial reaction in the setting of the steady-state mixed potential in this example of electroless copper deposition.

The major factors that determine the time required to reach the rest potential of the reducing agent are the type and the concentration of the ligand present, and the pH of the solution [16].

4.2 Steady-State Kinetics

There are three electrochemical methods for the determination of the steady-state rate of the electroless deposition of copper at mixed potential. Paunovic and Vitkavage [26, 27] used polarization data in the vicinity of the mixed potential to determine the rate of deposition [26, 27]. Ohno used ac polarization measurements [28]. The third electrochemical method is the use of Evans diagram, as described in Section 1. Ricco and Martin used an acoustic wave device for *in situ* determination and monitoring of the rate of deposition [29]. Various empirical rate equations were determined for electroless deposition of copper [4, 6].

4.3 Effect of pH on the Rate of Deposition

Electroless copper deposition is affected by the pH in two distinct ways. First, OH^- ions are reactants in the overall reaction, Eq. (1), and the partial anodic reaction,

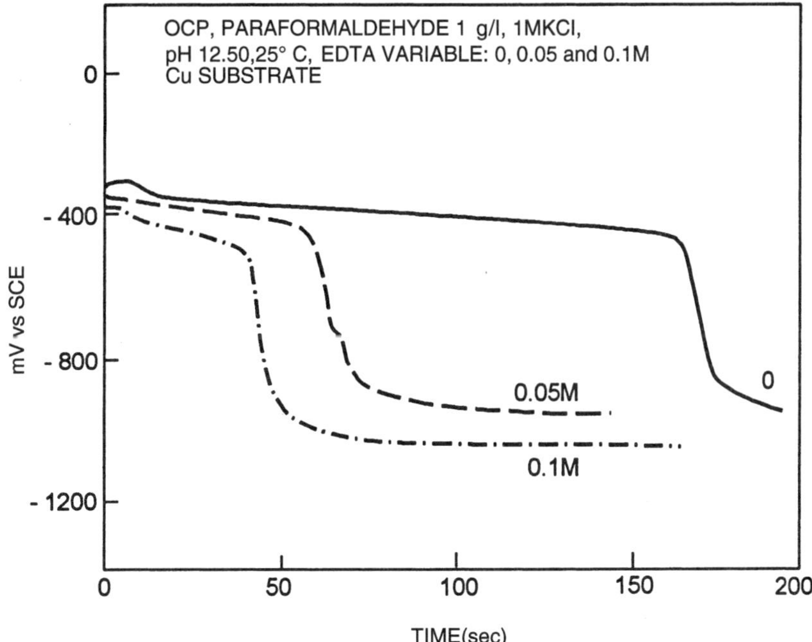

Figure 4 Open-circuit potential for the solution 1 g liter^{-1} paraformaldehyde, pH 12.50, Cu electrode, SCE reference electrode, EDTA variable. (From Paunovic [16], with permission from the Electrochemical Society.)

Eq. (7), and thus influence these reactions in a direct way (primary pH effects). Second, pH affects various phenomena associated with the structure and composition of the metal-solution interphase [14]. Those phenomena include (1) adsorption, (2) the structure of the double layer, (3) the structure of the copper species in the solution, and (4) the ionic strength of the solution. All these phenomena modulate the rate of electroless copper deposition in an indirect way (secondary pH effects).

The primary pH effect is expressed in terms of the reaction order with respect to OH^- ions and graphically as rate against pH. Plots of the experimentally observed plating rates against pH show an initial increase, a maximum value, and then a decrease of the rate with increasing pH. An example of the rate of electroless copper deposition as a function of pH is shown in Figure 5. It can be seen from Figure 5 that the maximum rate of deposition, in this specific case (EDTP solution), is obtained at a pH value of 12.5. The initial increase of the rate is due to the primary pH effect. The maximum and decrease of the rate at high pH values were interpreted in terms of secondary pH effects. The maximum rate of deposition for the tartrate solution is at pH 12.8 [9]. Two secondary effects were suggested so far: (1) change of the relative concentration of the methylene glycol and the hydroxide ions with pH due to the dissociation of methylene glycol [20], and (2) variation of the transfer coefficient for the oxidation of formaldehyde with pH [15].

Interpretation of the pH effect in terms of mathematical models was given by Paunovic [15]. From these it was concluded that the maximum and the falling off of the rate at high pH values are caused by the pH dependence of the kinetic parameters α_{Red} (the transfer coefficient for the oxidation of the reducing agent; Chapter 1, Part A) and i^0_{Red} (the exchange current density for the oxidation of the reducing agent; Chapter 1, Part A). Dissociation of methylene glycol, as proposed earlier, is an important factor in the electroless deposition of copper, but it is not sufficient to explain the pH effect.

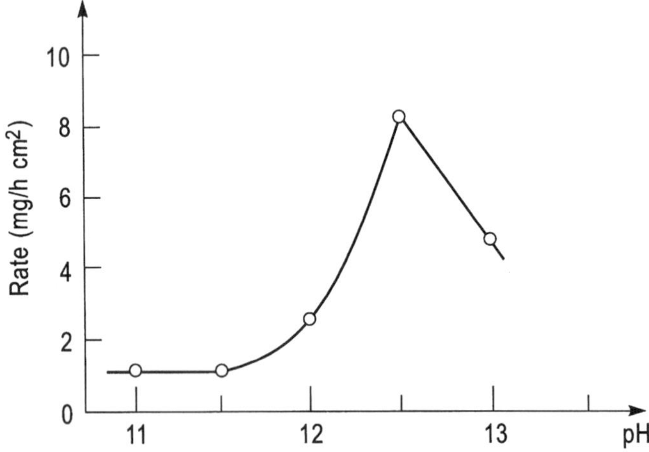

Figure 5 Rate of electroless copper deposition as a function of pH. The electroless copper solution contained 0.05 M $CuSO_4$, 0.15 M EDTP, 0.07 M para formaldehyde, and NaOH to give a desired pH. Oxygen was removed by bubbling argon through the solution. (From Duffy et al. [14], with permission from the Electrochemical Society.)

4.4 Catalysis Phenomena

Catalysis in electroless deposition of copper was studied by Haruyama and Ohno [30] and by Wiese and Weil [31]. Haruyama and Ohno have shown that the catalytic activity of metals for the oxidation of the reducing agent in electroless deposition is mostly determined by the rate constants of the two reaction steps, i.e., the oxidative adsorption and desorption of an anion radical (see Chapter lA, Section 3). Wiese and Weil have shown that copper deposition from EDTA containing solutions is catalyzed by chemisorbed methane-diolate anion.

5 GROWTH MECHANISM

Mechanistically, electroless copper deposition proceeds in two steps: (1) the thin-film stage (up to 3 μm) and (2) the bulk stage.

5.1 Thin-Film Stage

The mechanism of the thin-film formation is characterized by three simultaneous crystal-building processes [32–34]: nucleation (formation), growth, and coalescence of three-dimensional crystallites (TDC).

In the initial stages of electroless copper deposition on *a copper single crystal substrate, (100) plane*, the average density of TDC increases with time of deposition; in this stage the nucleation is the predominant process [32, 34]. Later the average density of TDC reaches a maximum and then decreases with time. In the stage of decreasing density of TDC, the coalescence is the predominant crystal-building process [34]. A continuous electroless film is formed by lateral growth and coalescence of TDC. The process of coalescence deserves special attention, since many physical properties of deposit depend on the type of coalescence. There are two types of coalescence of TDC. Coalescence without the proper filling of the space between TDC results in incorporation of impurities or additives, generation of stress, voids, and dislocations (Fig. 6*b*). Coalescence with filling the space between TDC, favorably joined crystallites (Fig. 6*a*), results in copper of better quality than in the first type of coalescence [35, 36]. The process (type) of coalescence depends to a great extent on the type and concentration of additives in the solution [34]. The initial stages of electroless copper deposition on *Pd activated nonmetallic (nonconducting) substrates* were described by Sard [33] and Rantell [37].

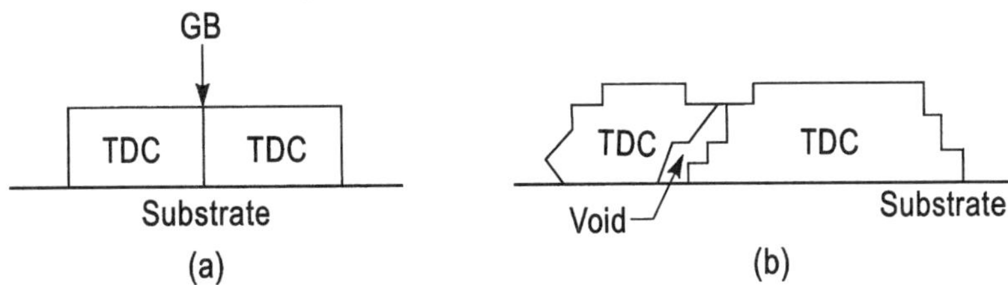

Figure 6 Two types of coalescence of the three-dimensional crystallites (TDC). (a) favorably joined TDC: copper of good quality; (b) improperly joined crystallites: results in incorporation of impurities or additives, generation of stress, voids, and dislocations.

5.2 Bulk Stage

After the formation of the continuous thin film, the deposition of a thick (1–15 μm) film proceeds, in most cases, through the following processes [27, 38–41]: (1) preferential growth of favorably oriented grains, (2) restriction (inhibition) of vertical growth of nonfavorably oriented grains, (3) lateral joining of preferentially growing grains, (4) cessation of growth of initial grains, and (5) nucleation and growth of a new layer of grains.

In the process of vertical and lateral growth, a preferentially growing grain (TDC) increases its width and subsequently joins laterally with other preferentially growing grains. Eventually the width of these grains becomes constant, and during further vertical growth, they develop a columnar shape, Figure 7 [38–41]. Then, at a certain stage, columnar grains no longer grow vertically. This cessation of growth of individual grains is followed by the nucleation and growth of a new layer of grains. Cessation of growth perpendicular to the substrate is influenced by the overpotential and degree of inhibition. This is one of the fundamental relationships in the correlation between (1) structure and variables in the plating solution and (2) structure and electrochemical kinetic parameters of processes composing electroless copper deposition.

The different growth rates on different single crystal substrate orientations were observed experimentally. This experimental observation indicates that certain crystallographic surfaces are more favorable for growth than others [40, 42]. A shift in the preferred direction of the growth was observed depending on the solution composition and concentration of additives [40].

6 STRUCTURE

We discuss two different structures: thin film (up to 1 μm) and thick film (1–25 μm) structure. We also discuss microporosity in electroless copper films.

6.1 Thin Film Structure

Nakahara and Okinaka [38], and Paunovic and Zeblisky [39] have shown that thin films of electroless Cu (up to 1 μm) are characterized by small, nearly equiaxial grains. The average grain diameter appears to be about 0.2 μm.

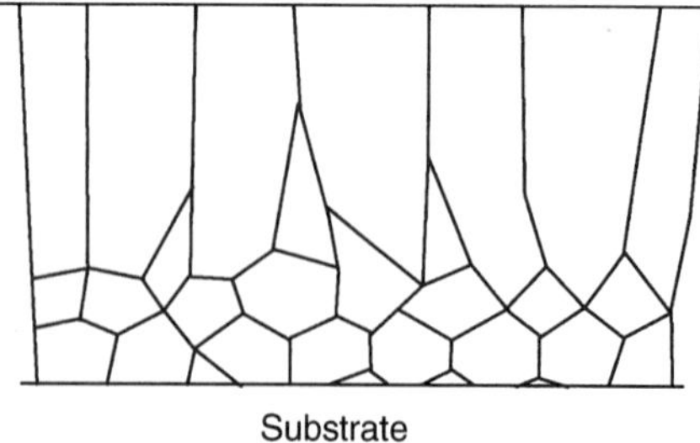

Figure 7 Schematic cross section (perpendicular to the substrate) of the columnar deposit.

6.2 Thick Film Structure

As mentioned in Section 5, a thick film (1–25 μm) of electroless copper has a columnar structure. Paunovic and Zeblisky [39] have shown that electroless copper deposited from EDTA solution containing NaCN, a wetting agent, formaldehyde, and NaOH exhibits a pH value between 10.8 and 12.5, and that it has a columnar structure with an average grain diameter, in a plane parallel to the substrate, between 0.3 and 0.7 μm and an average grain size (height), and in a plane perpendicular to the substrate, between 6 and 7 μm.

6.3 Microporosity

Nakahara [35, 36], using transmission electron microscopy (TEM), has shown that both crystalline and noncystalline films prepared by evaporation, sputtering, electrodeposition, and electroless deposition contain a large number of microscopic voids (pores). The presence of vacancies (voids) in thin films implies that the films contain locally unfilled regions inside the lattice. Studies of the early stages of film formation have shown that most microvoids are generated at the boundaries between three-dimensional faceted crystallites (TDC; Section 5) during their coalescence. The mechanism by which these voids are formed is called "coalescence-induced void formation." The proposed mechanism is based on the assumption that there is a geometrical misfit large enough to be left uncovered during the coalescence of three-dimensional crystallites (e.g., Fig. 6*b*). Voids inside grains could be generated during growth of multiple-atomic steps.

Voids are important lattice defects that influence the physical properties of a film, as is shown in the next section. In one example the number of voids per unit volume was 10^{15} to $10^{16}\,cm^{-3}$ and the average void size was 25 Å [45].

6.4 Hydrogen Incorporation

According to Eq. (1) the deposition of one mole of Cu is accompanied by the evolution of one equivalent mole of H_2. This results in the incorporation of H_2 gas bubbles into the deposit. As shown in Section 2, Eqs. (6) and (8), hydrogen atoms in H_2 originate from the splitting of the C–H bond in the formaldehyde molecule during dissociative adsorption. Nakahara and Okinaka [38, 43 45] studied extensively the incorporation of hydrogen into copper deposit and the effect of hydrogen bubbles on deposit properties. The content of hydrogen in electroless copper can be as high as 930 ppm [43]. Nakahara [45] found, using transmission electron microscopy, that small (20–300 Å) gas bubbles are incorporated uniformly throughout the copper films (25–30 μm), whereas large (~2000 Å) bubbles are trapped at the grain boundaries. Nakahara and Okinaka determined that the population distribution is broad [38]. The size distribution is shown in Figure 8. Thus the density of electroless copper is lower than that of bulk copper due to presence of incorporated hydrogen. Grunwald et al. [46] determined that density of electroless Cu films is in the range from 8.56 to 8.76 $g\,cm^{-3}$. The density of bulk copper is 8.9331 ($\pm$0.0037) $g\,cm^{-3}$.

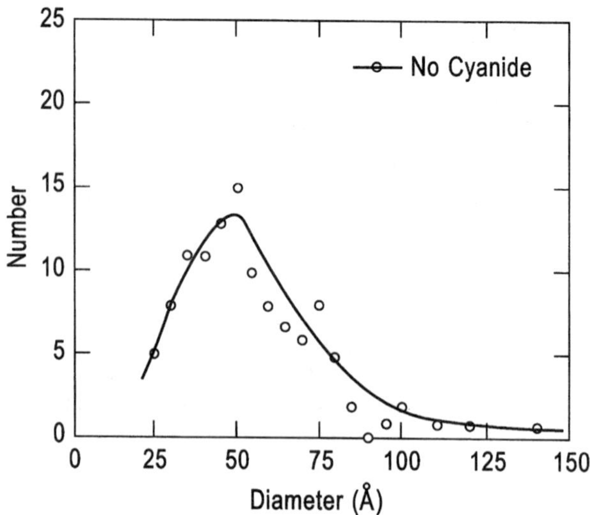

Figure 8 Population distribution of hydrogen gas bubbles as a function of bubble size.

7 PROPERTIES

7.1 Film Ductility

Okinaka and Nakahara [47] showed that the formation of small voids and small gas bubbles containing hydrogen are major factors determining the ductility of electroless copper. Nakahara and Okinaka [38] showed that brittle films contain a large number of small as well as large gas bubbles. They also showed that ductility promoters, such as cyanide ions, and higher temperatures of deposition facilitate desorption of hydrogen gas generated in the reaction given by Eq. (1). Some ductility-promoting additives, for example, 2,2′-dipyridyl and $K_2Ni(CN)_4$, inhibit both the inclusion of hydrogen and the formation of voids [44]. Table 1 shows an example of the difference in properties between Cu deposited from a solution in the absence of ductility promoters (solution A) and a deposit from the solution containing NaCN as a ductility promotor (solution B). It may be seen from Table 1 that the difference in ductility of as-deposited Cu and in gas bubble density ($N\,cm^{-3}$) is significant. The table also shows that there is ductility recovery during room temperature storage. In this example the ductility of the brittle copper recovered to a

TABLE 1 Properties of Electroless Cu Deposit: Plexiglas Substrate

	Solution A (Absence of NaCN)	Solution B (Presence of NaCN)
Ductility (elongation %)		
As–deposited Cu	1.2	3.6
After 6 months	4.8	4.8
Grain size (μm)	0.5–1.0	0.5–1.0
Gas bubble density[a]	9×10^{15}	9×10^{14}

[a] $N\,cm^{-3}$, N = number of gas bubbles.

value comparable to that of the ductile copper. The hydrogen content of the brittle films, obtained from solution A, is in the range from 100 to 200 ppm.

7.2 Ductility Recovery during Room Temperature Storage

The ductility (percent elongation) of electroless copper generally increases during low temperature (100–200°C) annealing. Two mechanisms were proposed to interpret this ductility recovery process.

According to the first mechanism, proposed by Nakahara et al. [48], the ductility improvement observed is attributed to the outdiffusion of hydrogen from the copper lattice. During electroless copper deposition hydrogen can be codeposited in atomic (H) as well as molecular (H_2) form. Most of the hydrogen codeposited in electroless copper is molecular. At room temperature, or at low temperature (100–200°C), annealing the molecular hydrogen diffuses out of copper, interstitially, via a dissociative reaction

$$H_2 \text{ (in the gas-filled void in copper)} \rightarrow 2H \text{ (in copper lattice)} \qquad (18)$$

The annealing removes all the diffusible hydrogen, leaving in copper only residual (non-diffussible) hydrogen. Nakahara et al. [48] distinguish four types of hydrogen incorporated in electroless copper deposit. Details may be found in the original literature.

According to the second mechanism, proposed by Honma and Mizushima [49], ductility improvement is due to structural changes involving recrystallization and grain growth in electroless copper deposit. They point out that the low temperature recrystallization and grain growth are commonly observed in copper films prepared by other growth techniques such as vapor deposition [50], sputtering [51], and electrodeposition [52, 53]. The amount of ductility recovered in electroless copper deposition as a result of low temperature annealing, either by the outdiffusion for the recrystallization mechanism, is determined also by impurity content [44, 54].

7.3 Crack-Free Electroless Copper

The printed circuit (PC) industry requires electroless copper with properties that allow the copper elements comprising the PC boards (PCB) to maintain their integrity during processing and use. One critical step in processing PCB with plated through-holes is the mounting or exchanging components by soldering. In this process the plated copper is subjected to thermal stress. During soldering, the plated copper in the through-holes usually expands less than the substrate. The difference (mismatch) in the thermal expansion depends on the type of substrate and temperature. In the case of the epoxy-glass substrate the difference at the soldering temperature (260°C) is large and the electroless copper in this case must be of high quality in order to maintain its integrity (continuity) during soldering [55].

Depending on the properties of the plating, the copper in the holes either cracks or resists the stress imposed without cracking during soldering. Paunovic and Zeblisjy [39] have shown that when EDTA based electroless copper solutions containing NaCN are used the elongation (ductility) of the 25 μm thick, crack-free copper, ranges from 3% to 11%; the tensile strength of this copper is from 200 to

600 Mpa (30,000 to 87,000 psi). These wide ranges may be subdivided into two smaller ones: class 1 with a high tensile strength and class 2 with a high elongation. A grain diameter in the plain parallel to the substrate is 0.1 to 1 μm, and the grain size in the plane perpendicular to the substrate is 4 to 10 μm, for the case studied of crack-free electroless copper deposit.

7.4 Electrical Resistivity

Patterson et al. determined that an electroless Cu layer of thickness 5000 Å, deposited on titanium nitride, has a resistivity of 2.0 to 2.7 μΩ cm, depending on the solution used [56]. Lopatin et al. [57] reported that the electrical resistivity decreases with the increase of the deposition solution temperature (Fig. 9). Dubin et al. reported that the resistivity decreases down to 1.8 to 1.9 μΩ cm after annealing at 200°C for 2 hours in a H_2 ambient [58]. Electrical resistivity of the bulk copper is 1.7 μΩ cm.

7.5 Electromigration Resistance

The free-electron theory of metals assumes that the valence electrons (the conduction electrons) are virtually free to move everywhere in the metal. In an electric field the electrons drift toward the positive direction of the field, producing an electric current in the metal. The high electronic conductivity of metals is explained in terms of the ease with which the free-electrons move [61]. According to modern quantum electronic theory, electrical resistivity of a metal results from the scattering of electrons by the lattice [61–64]. The scattering does not cause large displacement of the ions in the metal lattice when the current density is low. However, at a high current density (above $10^4\,A\,cm^{-2}$) the transport of electrons (current) can displace metal ions in crystal lattice and cause the transport of mass (positive ions) in the same direction as

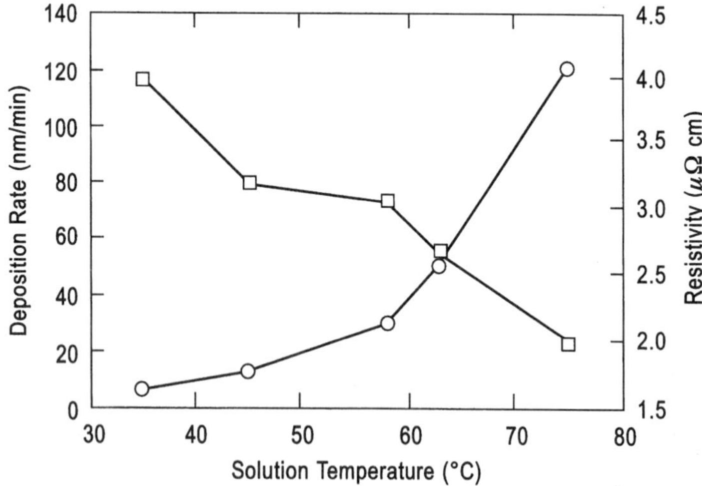

Figure 9 Electroless Cu deposition rate (O) and resistivity (□) versus solution temperature.

the electrons (Fig. 10). This mass transport is called *electromigration.* It occurs in interconnecting conductors (metallic fine lines) in integrated circuits where the current density is very high [65, 66]. For example, when a 1.0 μm wide Al (or Cu), line of 0.2 μm thickness is subjected to a current I of 1 mA, the current density i is $5 \times 10^5 \, A \, cm^{-2}$ (line cross-sectional area A in this case is $0.2 \times 10^{-4} \times 1 \times 10^{-4} = 0.2 \times 10^{-8} \, cm^2$; current density $i = (I/A) = 1 \times 10^{-3} \, A / 0.2 \times 10^{-8} \, cm^2 = 5 \times 10^5 \, A \, cm^{-2}$). Thus in microelectronic devices the transport of electrons (current) can cause the transport of metal ions (mass) in a metal lattice.

At high current densities ($i > 10^4 \, A \, cm^{-2}$) sufficient electron momentum is transferred to metal ions in the metal lattice to physically displace them toward the anode; hence a net mass transport occurs, as shown in Figure 10. This mass transport, electromigration, results in defects formation in conductors in microelectronics. Conductor lines undergo morphological changes due to electromigration where mass depletion (voids) occurs at the cathode and extrusion (hillocks) occurs at the anode.

Aluminum based alloys (Al–Cu, Al–Si) are most widely used as interconnection materials in integrated circuits (IC). One of the problems of the Al alloys is their poor resistance to electromigration (EM) induced failures. One way to express resistance to electromigration is in terms of time-to-failure. *Time-to-failure* is defined as the point at which a 50% increase of the resistance due to the electromigration stressing has occurred. The dc and pulse-dc lifetime of electroless Cu is found to be about two orders of magnitude longer than that of Al–2% Si at 275°C, and about four orders of magnitude longer than that of Al–2% Si when extrapolated to room temperature [67–68].

The another way of expressing resistance to electromigration is in terms of the activation energy for electrotransport. The activation energy is about 0.81 eV for electroless Cu, which is much larger than that typical 0.4 to 0.5 eV for Al alloys [67]. Thus Cu lines in IC are expected to have a larger lifetime than the Al–Si or Al–Cu alloy. For this reason, and because of higher conductivity of Cu, electroless Cu and electrodeposited Cu are considered for application as conductors in IC fabrication [69–72].

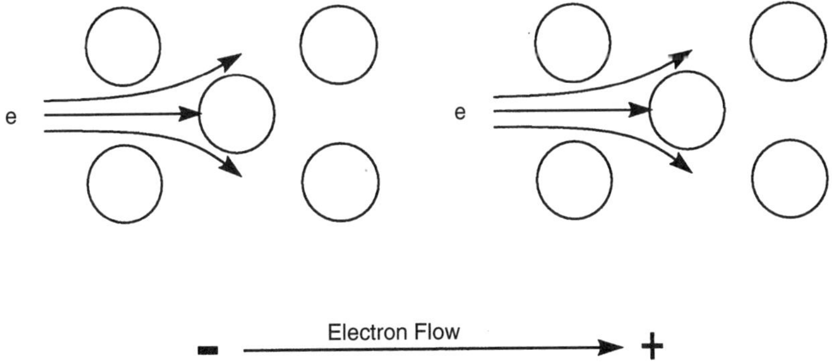

Figure 10 Atomic model of electromigration involving electron momentum transfer to metal ions in the metal lattice during a high current density flow ($i > 10^4 \, A \, cm^{-2}$).

APPENDIX

TABLE A1 Electroless Copper Deposition Solution and Plating Conditions for Electroless Cu deposition from an alkali-free solution

$CuSO_4 \cdot 5H_2O$	0.05–0.1 M
$N(C_2H_5)_4OH$	0.5–1.0 M
EDTA	0.1 M
CH_2O	0.01 M
$N(CH_3)_4CN$	0.01 M
GAF RE–610	0.5–2%
Temperature	45–55°C
pH (adjusted with $N(C_2H_5)_4OH$, tetraethylammonium hydroxide)	11.9–12.3

Source: Shacham-Diamand [73, p. 136].

TABLE A2 Solution Composition and Plating Condition for Electroless Cu Deposition Using Glyoxylic Acid as Reducing Agent

$CuSO_4 \cdot 5H_2O$	0.03 M
EDTA	0.24 M
CHOCOOH (glyoxylic acid)	0.20 M
2,2′-bipyridine	10 ppm
Temperature	60°C
pH (adjusted with NaOH)	12.5
Air agitation, continuous	

Source: Honma and Kobayashi [74] and Burke, Bruton, and Collins [75].

TABLE A3 Solution Composition and Plating Conditions for Electroless Cu Deposition Using Hypophosphite as Reducing Agent

$CuSO_4$	0.024 M
$NiSO_4$	0.002 M
H_3BO_3	0.5 M
NaH_2PO_2	0.27 M
$Na_3C_6H_5O_7$ (sodium citrate)	0.052 M
EDTA	0.026 M
pH	9.2

Sources: Hung and Chen [76] and Saubestre [77].

TABLE A4 Solution Composition and Plating Conditions for Electroless Cu Deposition using *Tartrate as Complexing Agent and Formaldehyde Reducing Agent*

	Solution A	Solution B
$CuSO_4 \cdot 5H_2O$, g liter^{-1}	5	13
$KNaC_4H_4O_6 \cdot 4H_2O$, g liter^{-1} (sodium potassium tartrate)	25	66
NaOH, g liter^{-1}	7	19.3

TABLE A4 *(Continued)*

	Solution A	Solution B
MBT, g liter^{-1} (Mercaptobenzothiazole[a])	–	0.013
HCOOH (37%), (Formaldehyde) mL liter^{-1}	10	38[b]
Temperature, °C	20	25

Sources: Solution A: Goldie [78]; Solution B: Pearlstein [79].
[a]Added as solution of 10 g liter^{-1} MBT in 0.2 M NaOH.
[b]Formaldehyde solution with 12.5% methanol as preservative.

REFERENCES

1. C. Wagner and W. Traud, *Z. Electrochem.*, **44**, 391 (1938).
2. M. Paunovic, *Plating*, **55**, 1161 (1968).
3. M. Saito, *J. Met. Finish. Soc. Jpn.*, **17**, 14 (1966).
4. F. M. Donahue, *Oberflache-Surf.*, **13** (12), 301 (1972).
5. A. Molenaar, M. F. E. Holdrinet, and L. K. H. van Beek, *Plating*, **61**, 238 (1974).
6. S. M. El-Raghy and A. A. Abo-Salama, *J. Electrochem. Soc.*, **126**, 171 (1979).
7. P. Bindra and J. Tweedie, *J. Electrochem. Soc.*, **130**, 1112 (1983).
8. A. Vashkialis and I. Iachiayskene, *Electrochemistry* (Academy of Sciences USSR), **17**, 1816 (1981).
9. L. N. Schoenberg, *J. Electrochem. Soc.*, **118**, 1571 (1971).
10. I. Ohno and S. Haruyama, *Surf. Technol.*, **13**, 1 (1981).
11. T. Hayashi, *Met. Finish*, **85** (6), 85 (1985).
12. J. W. Jacobs and J. M. G. Rikken, in *Electroless Deposition of Metals and Alloys*, M. Paunovic and I. Ohno, eds., *Proceedings*, vol. 12, Electrochemical Society, Pennington, NJ, 1988, p. 75.
13. J. W. Walker, *Formaldehyde*, Reinhold, New York, 1964.
14. J. Duffy, L. Pearson, and M. Paunovic, *J. Electrochem. Soc.*, **130**, 876 (1983).
15. M. Paunovic, *J. Electrochem. Soc.*, **125**, 173 (1978).
16. M. Paunovic, *J. Electrochem. Soc.*, **124**, 349 (1977).
17. E. Mattson and J. O'M. Bockris, *Trans. Faradday Soc.*, **55**, 1586 (1959).
18. J. L. Hall, R. F. Jones, C. E. Delchamp, and C. W. McWillams, *J. Am. Chem. Soc.*, **79**, 3361 (1957).
19. D. A. Keyworth, *Talanta*, **2**, 383 (1959).
20. N. Schoenberg, *J. Electrochem. Soc.*, **119**, 1491 (1972).
21. M. Paunovic and R. Arndt, *J. Electrochem. Soc.*, **130**, 794 (1983).
22. L. L. Duda, *Plating Surf. Finish.*, **85** (7), 60 (1998).
23. M. Oita, Matsuoka, and C. Iwakura, *Electrochim. Acta*, **42**, 1435 (1997).
24. D. Vitkavage and M. Paunovic, *Plating Surf. Finish.*, **70** (4), 48 (1983).
25. M. Paunovic, *J. Electrochem. Soc.*, **132**, 1155 (1985).
26. M. Paunovic and D. Vitkavage, *J. Electrochem. Soc.*, **126**, 2282 (1979).

27. M. Paunovic, in *Electrodeposition Technology, Theory and Practice*, L. T. Romankiw and D. R. Turner, eds., *Proceedings*, vol. 17, Electrochemical Society, Pennington, NJ, 1987, p. 349.
28. I. Ohno, in *Electroless Deposition of Metals and Alloys*, M. Paunovic and I. Ohno, eds., *Proceedings*, vol. 12, Electrochemical Society, Pennington, NJ, 1988, p. 129.
29. A. J. Ricco and S. J. Martin, ibid., p. 142.
30. S. Haruyama and I. Ohno, ibid., p. 20.
31. H. Wiese and K. G. Weil, ibid., p. 53.
32. M. Paunovic and C. H. Ting, ibid., p. 170.
33. R. Sard, *J. Electrochem. Soc.*, **117**, 864 (1970).
34. M. Paunovic and C. Stack, in: *Electrocrystallization*, R. Weil and R. G. Baradas, eds., *Proceedings*, vol. 6, Electrochemical Society, Pennington, NJ, 1981, p. 205.
35. S. Nakahara, *Thin Solid Films*, **45**, 421 (1977).
36. S. Nakahara, ibid., **64**, 149 (1979).
37. A. Rantell, *Trans. Inst. Met. Finish.*, **48**, 191 (1970).
38. S. Nakahara and Y. Okinaka, *Acta Metall.*, **31**, 713 (1983).
39. M. Paunovic and R. Zeblisky, *Plating Surf. Finish.*, **72** (2), 52 (1985).
40. J. Kim, S. H. Wess, D. Y. Jung, and R. W. Johnson, *IBM J. Res. Develop.*, **8**, 697 (1984).
41. H. J. Choi and R. Weil, *Plating Surf. Finish.*, **68** (5), 110 (1981).
42. A. Danjanovic, M. Paunovic, T. H. V. Setty, and J. O'M. Bockris, *Acta Metallurgica*, **13**, 1092 (1965).
43. J. E. Graebner and Y. Okinaka, *J. Appl. Phys.*, **60** (1), 36 (1986).
44. Y. Okinaka and H. K. Straschil, *J. Electrochem. Soc.*, **133**, 2608 (1986).
45. S. Nakahara, *Acta Metall.*, **36** (7), 1669 (1988).
46. J. J. Grunwald, L. Slominski, and A. Landau, *Plating*, **60**, 1022 (1973).
47. Y. Okinaka and S. Nakahara, *J. Electrochem. Soc.*, **123**, 475 (1976).
48. S. Nakahara, C. Y. Mak, and Y. Okinaka, *J. Electrochem. Soc.*, **138**, 1421 (1991).
49. H. Honma and S. Mizushima, *J. Met. Finish. Soc. Jpn.*, **34**, 290 (1983).
50. M. Yoshida, S. Nakahara, and H. Suto, *J. Jpn. Inst. Met.*, **39**, 414 (1975).
51. J. W. Patten, E. G. McClanahan, and J. W. Johnston, *J. Appl. Phys.*, **42**, 4371 (1971).
52. D. S. Stoychev, I. V. Tomov, and I. B. Vitanova, *J. Appl. Electrochem.*, **15**, 879 (1985).
53. I. V. Tomov, D. S. Stoychev, and I. B. Vitanova, *J. Appl. Electrochem.*, **15**, 887 (1985).
54. S. Nakahara, Y. Okinaka, and H. K. Straschil, *J. Electrochem. Soc.*, **136**, 1120 (1989).
55. M. Paunovic, *Plating Surf. Finish.*, **70** (2), 62 (1983).
56. J. C. Patterson, M. O'Reilly, G. M. Crean, and J. Barrett, *Microelectron. Eng.*, **33**, 65 (1997).
57. S. Lopatin, Y. Shacham-Diamand, V. M. Dubin, P. K. Vasudev, B. Zhao, and J. Pellerin, in *Electrochemically Deposited Thin Films*, M. Paunovic and D. A. Scherson, eds., *Proceedings*, vol. 19, Electrochemical Society, Pennington, NJ, 1997, p. 271.
58. V. M. Dubin, Y. Shacham-Diamand, B. Zhao, P. K. Vasudev, and C. H. Ting, *J. Electrochem. Soc.*, **144**, 898 (1997).
59. H.-K. Kang, J. S. H. Cho, I. Asano, and S. S. Wong, *VMIV Conf.*, June 9–10, 1992.
60. J. Tao, N. W. Cheung, C. Hu, H.-K. Kang, and S. S. Wong, *IEEE Electron Device Lett.*, **13**, 433 (1992).
61. M. Paunovic and M. Schlesinger, *Fundamentals of Electrochemical Deposition*, Wiley, New York, 1998, ch. 3.

62. J. Bardeen, *J. Appl. Phys.*, **11**, 88 (1940).
63. P. L. Rossiter, *The Electrical Resistivity of Metals and Alloys*, Cambridge University Press, Cambridge, 1987.
64. M. Paunovic, L. A. Clevenger, J. Gupta, C. Cabral Jr., and M. E. Harper, *J. Electrochem. Soc.*, **140**, 2690 (1993).
65. K.-N. Tu, J. W. Mayer, and L. C. Feldman, *Electronic Thin Film Science*, Macmillan, New York, 1992.
66. M. Ohring, *The Materials Science of Thin Films*, Academic Press, New York, 1992.
67. H.-K. Kang, J. S. H. Cho, I. Asano, and S. S. Wong, *Proc. VLSI Multilevel Interconnection Conf. (VMIC)*, June 9–10, 1992, p. 337.
68. J. Tao, N. W. Cheung, C. Hu, H. K. Kang, and S. S. Wong, *IEEE Electron Device Lett.*, **13** (8), 433 (1992).
69. P. Pai, W. G. Oldham, C. H. Ting, and M. Paunovic, Abstract 481, Electrochemical Society Fall Meeting, Oct. 18–23, 1987.
70. Y. Shacham-Diamand, in *Electrochemically Deposited Thin Films II*, M. Paunovic, ed., *Proceedings*, vol. 31, Electrochemical Society, Pennington, NJ, 1995, p. 293.
71. P. C. Andricacos, *Interface*, **8**, 32 (1999).
72. P. C. Andricacos, C. Uzoh, J. O. Dukovic, J. Horkans, and H. Deligianni, *IBM J. Res. Develop.*, **42**, 567 (1998).
73. Y. Shacham-Diamand, B.-X. Sun, V. Yip, and R. Bielski, in *Electrochemically Deposited Thin Films II*, M. Paunovic, ed., *Proc.*, vol. 31, Electrochemical Society, Pennington, NJ, 1995, p. 136.
74. H. Honma and T. Kobayashi, in *Electrochemically Deposited Thin Films*, M. Paunovic, I. Ohno, and Y. Miyoshi, eds., *Proc.*, vol. 26, Electrochemical Society, Pennyngton, NJ, 1993, p. 344.
75. L. D. Burke, G. M. Bruton, and J. A. Collins, *Electrochim. Acta*, **44**, 1467 (1998).
76. A. Hung and K.-M. Chen, *J. Electrochem. Soc.*, **136**, 72 (1987).
77. E. B. Saubestre, *Proc. Am. Electropl. Soc.*, **46**, 264 (1959).
78. W. Goldie, *Plating*, **51**, 1069 (1964).
79. F. Pearlstein, U.S. Patent 3,222,195 (1965).

18 Electroless Deposition of Nickel

MORDECHAY SCHLESINGER

INTRODUCTION

Electroless (autocatalytic) plating involves the presence of a chemical reducing agent in solution to reduce metallic ions to the metal state. The name *electroless* is somewhat misleading, however. There are no external electrodes present, but there is electric current (charge transfer) involved. Instead of an anode, the metal is supplied by the metal salt; replenishment is achieved by either adding salt or an external loop with an anode, of the corresponding metal, that has higher efficiency than the cathode. There is therefore, instead of a cathode to reduce the metal, a substrate serving as the cathode, while the electrons are provided by a reducing agent. The process takes place only on catalytic surfaces rather than throughout the solution (if the process is not properly controlled, the reduction can take place throughout the solution, possibly on particles of dust or of catalytic metals, with undesirable results).

Brenner and Riddell invented [1] electroless Ni plating in 1946, rather accidentally when they observed that the additive NaH_2PO_2 caused apparent cathode efficiencies of more than 100% in a nickel electroplating bath. This led them to the correct conclusion that some chemical reduction was involved. Further research resulted in the development of the original process that the inventors named *electrodeless* plating. The name soon lost the *de*, and later the name *autocatalytic* was formally adopted, although *electroless* is still widely used. The words are synonymous.

Autocatalytic plating is defined as the deposition of a metallic coating by a controlled chemical reduction that is catalyzed by the metal or alloy being deposited. Such plating has been used to yield deposits of Ni, Co, Pd, Cu, Au, and Ag as well as some alloys containing these metals plus P or B. Electroless Cr deposition has also been claimed. Chemical reducing agents have included NaH_2PO_2 (the one originally used by the inventors for Ni and Cu deposition and still the most important and widely investigated), formaldehyde—especially for Cu—hydrazine, borohydrides, amine boranes, and some of their derivatives.

Electroless plating possesses several characteristics not shared by other techniques and that accounts for it's ever growing popularity. Its throwing power is essentially perfect, at least on any surface to which the solution has access, with no excessive buildup on edges and projections. Deposits may be less porous than electroplates, and hence have better corrosion resistance. Power supplies, electrical contacts, and

Modern Electroplating, Fourth Edition, Edited by Mordechay Schlesinger and Milan Paunovic.
ISBN 0-471-16824-6

the other apparatus necessary for electroplating are not required. The process is usually an integral and necessary step in plating on nonconductors such as plastics (see Chapter 15 of this column). The process is particularly important in the printed circuit industry. Some electroless deposits have unusual, or even unique, magnetic properties.

Experience shows that each substrate requires its own specific techniques, depositing active metal onto the surface of a non(semi)conductor is still somewhat of an art. The surface preparation (i.e., cleaning process) requires very careful selection and application. It must be stressed that cleaning may affect the porosity of the metal deposit. Residues from cleaners and deoxidizers may create inactive spots that will not initiate electroless deposition. This may result in the necessity to have a thicker deposit before continuity is achieved. In extreme cases continuity is never reached.

In general, deposition requires one or more of the following steps (see Fig. 1): (1) cleaning, (2) surface modification, (3) sensitization, (4) catalyzing or (3′) catalyzing, and (4) activation (acceleration). Rinsing is required between the steps. We refer to the steps 3 and 4 (shown in Fig. 1) as *sensitization and catalyzing*. By these terms we mean absorbing ions or molecules from a solution such as acidic Sn(II) and or Sn(IV). Other much less frequently used sensitizing agents are, for example, [2] $AgNO_3$, $AuCl_3$, metallic Na (in naphtalene solution). An oversimplified model assumes that the sensitizing ion can reduce the active metal from the catalyst solution, usually $PdCl_2$ (Au, Pt, Rh, Os, and Ag solutions have also been used), for example,

$$Pd^{2+} + Sn^{2+} \rightarrow Sn^{4+} + Pd^0$$

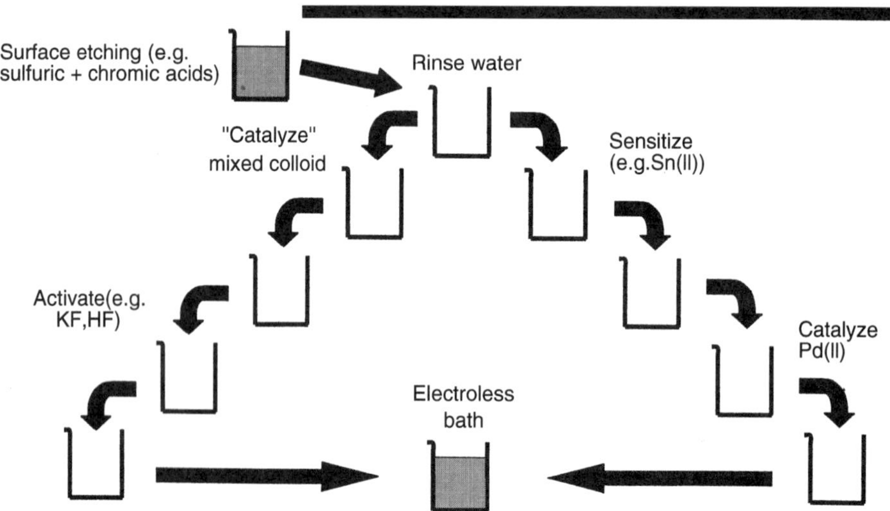

Figure 1 Schematic representation of the electroless deposition process.

If the metal to be deposited electrolessly can be reduced by the sensitizing ion, then it is not necessary to reduce the active metal first. Instead, the substrate is immersed in the electroless bath immediately after sensitizing and rinsing (e.g., electroless copper or silver when using Sn(II) based sensitization).

The alternate method of *catalyzing and activating* makes use of a mixed colloidal catalyst bath. The colloid is reduced Pd stabilized by Sn(II) and Sn(IV) ions. The activation (acceleration) step is the removal of the layer formed by the stabilizing agent with KF, HF, or other chemicals such as HCl or NaOH. In some cases the activation (acceleration) step can be omitted, but then the plating solution may get contaminated. There are also a good number of proprietary and patented processes reported.

1 NUCLEATION

In the process of electroless thin metal film deposition, the density of nucleation sites on the catalyzed substrate determines the adhesion properties of the end product films. Film thickness at which continuity is reached, and to some extent packing rearrangement after heat treatment, are also affected by this parameter. Now the growth initiation upon the catalytic sites is a probabilistic process by nature; hence metal island density and island size distribution are time-dependent functions during the immersion period in the metallizing bath.

In order to evaluate the sensitizer and the subsequent catalyst deposition process, the density of sites capable of nucleation should be determined prior to the metallizing step. If one is to study the final product's physical properties, then the density of nucleation sites at the time when the film reaches continuity is of interest.

Immediately following the catalytic process, before immersion in the metallizing bath, the amount of material present on the surface makes it difficult to obtain reliable data. In addition, in that stage the ultra-thin supporting substrate (formvar) and the deposit are virtually nonconductors; they quickly become charged by the electron microscope's electron beam. As a result they usually disintegrate before an electromicrograph can be taken.

The direct determination of nucleation site density at the time when film continuity is reached is also all but impossible. The reason is clear: By definition, by the time continuity is reached there are no islands to speak of.

In light of these problems the following procedure has to be applied [2]: Using a variety of sensitizing solutions (listed in Table 1) and the catalytic solution (listed in Table 2), a metal (nickel) is deposited electrolessly on the sensitized catalyzed surface by immersion in the metallizing bath for fixed amounts of time. The sensitizers chosen represent a wide range of adsorption properties. The island's size and the density value of island's distribution are then determined. In Table 3 we report the number of the average metal (nickel) island diameter for two (10 and 60 s) immersion times in the metallizing bath. Note that the listing order of sensitizers in Tables 1 and 3 are identical, that is, in order of increasing material adsorption values. A close inspection of Table 3 shows that for a given sensitizer, as it changes its adsorption properties with time (sensitizers 1, 5, 7, 8, 9), the tendency is for the higher nucleation site density to correspond to the smaller average island size. This is particularly evident for the 60 s immersion time. Sensitizers corresponding to higher levels of

TABLE 1 Sensitizers

Sensitizer	0.1 $SnCl_2 \cdot 2H_2O + 0.1$ M HCl plus	Aging (Hours)
1	—	0
2	10 mg liter^{-1} hydroquinone	3
3	10 mg liter^{-1} hydroquinone	48
4	0.05% triton X-100 (weight)	0
5	—	48
6	10 mg liter^{-1} thiourea	72
7	—	288
8	—	96
9	—	144
10	10 mg liter^{-1} thiourea	24
11	0.05% triton X-100 (weight)	240

TABLE 2 Catalyst Solution

$PdCl_2$	0.1 g liter^{-1}
HCl (12 M)	0.1 ml liter^{-1}
H_2O	Balance to one liter
	Temperature = 23°C

TABLE 3 Average Island Diameters

Sensitizer	Time in Plating Bath [=]s	Diameter Average [=]nm	Standard Deviation [=]nm
1	60	140.6	—
2	10	6.4	1.1
2	60	80.6	7.6
3	10	4.9	1.4
3	60	39.7	14.6
4	10	17.7	2.3
4	60	95.3	17.8
5	10	6.3	< 1
5	60	29.7	4.3
6	10	2.1	1.2
6	60	17.2	4.5
7	10	1.5	—
7	60	8.1	2.6
8	10	1.7	—
8	60	11.2	2.8
9	10	< 1	—
9	60	4.5	3.9
10	60	7.6	5.3
11	10	< 1	—
11	60	4.9	1.1

adsorption yield not only smaller averaged-sized islands but also less broadening of the size distribution with the same time spent in the metallizing bath.

At this point we need to explain that in recent years mixed colloidal technology has been favored by most practitioners. The main reason given is that only a single step is necessary for implementation. This is not quite the case. The procedure in fact, often requires two steps, since the activator (accelerator) is made of a separate bath. It is, however, more reproducible than the separate Sn–Pd technology, and that is a good enough reason for it being more popular. In Chapter 15 we compare a number of observations on the two technologies.

To complete the discussion we present in Table 4 a typical combined Sn–Pd activating bath composition. In Table 5 a practical suggested scheme is given, which naturally can be modified as needed. Despite the number of disadvantages in using the separate Sn–Pd process, sometimes the initial nucleation density achieved is up to an order of magnitude greater than what can be expected using the mixed colloidal technology. Thus a continuous film is obtained at a smaller thickness. This translates into the following:

- An economical advantage, since less metal is needed when the only purpose is to render the surface conductive
- Better adhesion due to higher density of fastening sites
- Potential for the development of special products where very thin coatings are needed

Before closing this section it is worth mentioning, in passing, that in electroless nickel deposition on silicon a presensitization step is required. That involves immersion in dilute HF solution. Perhaps more practical is the addition of 450 ml liter^{-1} HF (48%) and 1 ml liter^{-1} HCl to 1 g liter^{-1} $PdCl_2$. Thus an activator solution for

TABLE 4 A Typical Combined Sn/Pd Catalyzing Bath

$SnCl_2$	2 g liter^{-1}
$PdCl_2$	0.2 g liter^{-1}
HCl	10 ml liter^{-1}

TABLE 5a A Suggested Practical Scheme: Sensitizer Solution

Stock	
Stannous chloride, $SnCl_2 \cdot H_2O$	10 g
Hydrochloric acid, HCl	10 ml
Working solution	
0.5 ml of stock solution in 160 ml H_2O	

TABLE 5b A Suggested Practical Scheme: Activator Solution

Stock	
Palladium chloride, $PdCl_2$	10 g
Hydrochloric acid, HCl	10 ml
Working solution	
One ml of stock solution in 200 ml of H_2O	

TABLE 5c A Suggested Practical Scheme: Deposition Process

Solution	Time
Sensitizer	1–2 min
Rinse	30 s
Activator	1–2 min
Rinse	30 s
Metallizing	Temperature-pH dependent

the deposition of electroless nickel on silicon obviates the need for an extra step in the predeposition stage.

2 METALLIZING PRINCIPLES

Chapter 8 in *Fundamentals* deals with the subject of electroless metal deposition. In what follows we will attempt not to duplicate the material presented there. We may begin our discussion here by stating that interest in electroless thin metal deposition, in general, and that in electroless deposition of nickel, in particular, has been steadily growing since its invention. This very special process involves a continuous build-up of metal coating on a substrate by the mere immersion in a suitable aqueous solution. A chemical reducing agent in solution supplies the electrons for converting metal ions to the metal form

$$M^{2+} + 2e \text{ (supplied by reducing agent)} \xrightarrow[\text{Surface}]{\text{Catalytic}} M^0$$

The important point is that the catalytic surface is the only place where this simplified reaction occurs. Once the deposition starts on a surface, the deposited metal must also be catalytic for the deposition to continue. The progress of electroless deposition is linear in time. This means, as has been demonstrated by Marton and Schlesinger [3], that the deposition of, say, Ni–P on a sensitized-activated surface starts at specific activation sites on the surface and continues on these points only. As deposition progresses, islands are formed around these nucleation sites. The islands grow in size until they merge and a continuous film results. Those early observations helped establish the fact that the surface of a Sn–Pd treated dielectric substrate is not completely activated. The activation produces small catalytic sites dispersed on the surface serving as the nuclei for electroless nickel deposition. It was also established [3] that the film mass thickness growth rate in electroless deposition is linear in time once film continuity has been reached. The mass thickness of a deposit is the deposited film's volume per unit area. This thickness factor refers to the total material deposited as a continuous film of even thickness, and it is compared to electrodeposition under conditions of constant current density.

As stated above, Chapter 8 in *Fundamentals* deals with the subject of electroless deposition. Consequently we offer here a short discussion of the electrochemical path leading to the reduction of, nickel.

According to Van Den Meerakker [4] electroless deposition processes may be viewed and understood by, what is referred to as, a universal electrochemical mechanism regardless of the nature of the many possible reducing agents, *R*. Each process can be viewed as made up of a series of elementary anodic and cathodic reactions. The first anodic stage is the dehydrogenation of the reductant. Thus the following four anodic stages in case of alkaline media are recognized:

1. Dehydrogenation, $RH \rightarrow R + H$
2. Oxidation, $R + OH^- \rightarrow ROH + e$
3. Recombination, $H + H \rightarrow H_2$
4. Oxidation, $H + OH^- \rightarrow H_2O + e$

The following two cathodic stages in case of alkaline media are recognized:

5. Metal deposition, $M^{+n} + ne \rightarrow M^0$
6. Hydrogen evolution, $2H_2O + 2e \rightarrow H_2O + 2OH^-$

In acid media stages 4 and 6 are to be written as follows:

4′. Oxidation, $H \rightarrow H^+ + e$
6′. Hydrogen evolution, $2H^+ + 2e \rightarrow H_2$

Thus an electroless deposition reaction can be considered to be the combined result of two independent electrode reactions:

- Cathodic partial reactions such as stage 5 above
- Anodic partial reaction such as stage 2 or 4′ above

Mixed potential theory, advocated by Paunovic [5] in 1968, holds that electroless deposition processes can be predicted from the polarization curves of the partial anodic and cathodic processes. In *Fundamentals* the verifications of the mixed potential theory for a wide number of cases of electroless systems is delineated. Historically deMinjer [6] was the first to test the validity of the mixed potential theory in electroless nickel deposited using hypophosphite as reducing agent.

3 ELECTROLESS NICKEL PLATING BATHS

3.1 Basic Baths

The literature abounds in the number of possible bath formulations for the electroless deposition of nickel using many different reducing agents and under widely different plating conditions. For this reason the reader is advised to consult [3, 7, 8] and the many references therein. We present here some representative formulations. Table 6 gives a summary of possible bath formulations for the electroless deposition of nickel where no additives are suggested (except for baths 3 and 5 where the often used additive Pb^{+2} is included) and with sodium hypophosphite as reducing agent. It is evident from the table that a wide range of baths with pH values from about 4 to about 11 are possible. It is expected that many of the end

TABLE 6 Bath Compositions for Electroless Nickel Deposition Using Hypophosphite Reducing Agent

	Acid Baths				Alkaline Baths			
Bath Constituents (g liter^{-1})	1	2	3	4	5	6	7	8
Nickel chloride, $NiCl_2 \cdot 6H_2O$	30	30	—	21	26	30	20	—
Nickel sulfate, $NiSO_4 \cdot 6H_2O$	—	—	25	—	—	—	—	25
Sodium hypophosphate, $NaH_2PO_2 \cdot H_2O$	10	10	23	24	24	10	20	25
Hydroxyacetic acid, $HOCH_2COOH$	35	—	—	—	—	—	—	—
Sodium citrate, $Na_3C_6H_5O_7 \cdot H_2O$	—	12.6	—	—	—	84	10	—
Sodium acetate, $NaC_2H_3O_2$	—	5	9	—	—	—	—	—
Succinic acid, $C_4H_6O_4$	—	—	—	7	—	—	—	—
Sodium fluoride, NaF	—	—	—	5	—	—	—	—
Lactic acid, $C_3H_6O_3$	—	—	—	—	27	—	—	—
Propionic acid, $C_3H_6O_2$	—	—	—	—	2.2	—	—	—
Ammonium chloride, NH_4Cl	—	—	—	—	—	50	35	—
Sodium pyrophosphate, $Na_4P_2O_7$	—	—	—	—	—	—	—	50
Lead ion, Pb^{2+}	—	—	0.001	—	0.002	—	—	—
pH	4–6	4–6	4–8	6	4–6	8–10	9–10	10–11
Temperature (°C)	100	100	85	100	100	95	85	70

Source: DeMinjer [7].

product deposit's properties will depend on this important parameter's value. Indeed, at low bath (acidic) pH values the resulting film will have an increased phosphorus content, up to 25 atomic % at pH 4. At pH values in the alkaline range less than 1% phosphorus content is rather common place. Electrolessly deposited films do develop stress. Films produced in acidic baths tend to show tensile stress, while those from alkaline bath "shift" their stress in the compressive direction. Another pH dependent property is the fact that films grown, using acidic baths, exhibit good adhesion characteristics to steel, and this is likely why those film types are more common in industry. In addition it is observed that raising a bath's pH value has a marked effect on the deposition rate. Alkaline bath tend to have a higher deposition rate with the concomitant result of decreased stability and as a further result, possible "plate-out". It should be clear that for reliable and repeatable results, it is essential that a constant solution pH be maintained during the entire deposition process.

Another important parameter is that of bath temperature. The deposition rate (everything else being equal) increases exponentially with increased temperature. Indeed, all practical baths will require operating temperatures of 60°C and above. However, bath temperature should rarely, if ever, be above that of 90°C, since solution "plate-out" or bath decomposition then becomes a real possibility. From the foregoing it is not difficult to see why alkaline plating solutions are of relatively little use, say, in the automotive industry. Their use seems to be specific to applications where plating is to be done at relatively low temperatures (no higher than 70°C) such as on polymeric substrates and other applications where a low phosphorus content is a requirement.

In addition to hypophosphite which is useful as a reducing agent for the electroless deposition of nickel, containing phosphorus (Ni–P), there are three others:

1. Sodium borohydride, $NaBH_4$—in the bath pH range of 12 to 14
2. Dimethylamine borane, (DMAB) $(CH_3)_2NHBH_3$—in the pH range of 6 to 10
3. Hydrazine, $N_2H_4 \cdot H_2O$—in the pH range of 8 to 11

The first and second, when acting as reducing agents in a nickel metallizing bath, yield electroless nickel films containing boron (Ni–B). The third yields nitrogen-containing (Ni–N) films.

With respect to type 1 above, in Table 7 we list three different compositions of sodium borohydride electroless nickel plating solutions. The properties such as hardness of the Ni–B films, in particular, bath 1 in the table which contains thallium, have become of considerable interest due to their use in a number of applications. The baths themselves have, however, some important disadvantages. For example, there is the need to maintain pH values above 12 in order to suppress the possibility of nickel boride precipitation. As a consequence only substrates that can withstand the high alkalinity can benefit from these solutions. The inclusion of thallium in bath 1 as a stabilizer (see below) enables the deposition process to take place at a lower practical temperature with a relatively high deposition rate. Note here the dramatic effect that stabilizers can have on the properties of the metallizing baths. Stabilizers and their effects are discussed in the next section. As a final point, environmental considerations seem dictate the type and ultimate compositions of plating baths used in practical industrial settings. For instance, often the use of thallium results in thallium being codeposited in the Ni–B films (up to 6% for some baths).

With respect to type 2 above, it should be noted that replacing hypophosphite as a reducing agent in a plating bath by amine boranes also results in a "workable" solution for the plating of electroless nickel. The formulations of both acid and alkaline plating baths are available in the literature.

TABLE 7 Composition of Sodium Borohydride Electroless Nickel Plating Solutions

Component	Bath 1	Bath 2	Bath 3
Ni^{2+}	2.5	5	6
EDTA disodium salt	35	—	—
Sodium potassium tartarate	—	65	—
Ammonium hydroxide (28%)	—	—	120 ml $liter^{-1}$
Sodium hydroxide	40	40	—
Sodium borohydride	0.5	0.75	0.4
$TlNO_2$ (mg $liter^{-1}$)	50	—	—
$Pb(NO_3)_2$ (mg $liter^{-1}$)	—	10	—
2-MBT (mg $liter^{-1}$)	—	—	20
Temperature (°C)	95	92	60
pH	14	13	12

Source: Mallory and Hajdu [8].
Note: Concentrations in g $liter^{-1}$.

With respect to type 3 above we observe that electroless deposition can be likewise achieved using hydrazine-based baths [7, 9, 10]. The baths consist of a nickel salt complexing agent such as tartrate, malonate, or EDTA and the reducing agent hydrazine. The pH is adjusted to the level desired with sodium hydroxide. The electroless nickel deposits produced from hydrazine solutions contain up to 99% nickel. With oxygen and nitrogen present to the tune of tenths of a percent. One good example is a bath containing 4.8 g $liter^{-1}$ nickel chloride hexahydrate, 32 g $liter^{-1}$ hydrazine, and 4.6 g $liter^{-1}$ sodium tartrate dihydrate at pH 10 and deposition temperature of 95°C. The magnetic properties of such films compare well with those of electrodeposited counterparts.

3.2 Bath Additives

Over the decades of practicing electroless nickel deposition a number of workers have developed special additives to the plating bath that make the bath more practical in various way. Among these are bath stability, deposition rate, product film grain structure, and surface smoothness and brightness.

One family of additives is referred to as complexing agents. These are usually organic acids or their salts, with some notable exceptions. One is the ammonium ion that is added for pH control. Another, which is used in alkaline baths, is the pyrophosphate ion. In Table 8 we list a number of practical complexing agents. These additives play three roles:

1. Help maintain stable pH level
2. Help prevent precipitation of nickel salts such as phosphites
3. Reduce the concentration of free nickel ions

A few words on the concept of complexing are in order here. When nickel ions are in aqueous solution, they are bound to a well-defined number of water molecules. That is the coordination number. In the case of Ni^{2+} ions, the number is 4 or 6. In a situation where water molecules which are coordinated to the hydrated (free, simple) nickel ion are replaced by another ion or molecule, we speak of a nickel complex, and the combining ion or molecular group is the complexing agent. Clearly, the chemical properties of the nickel ions are expected to change as a result of complexation. More specifically, the rate of nickel deposition is proportional to the rate at which the nickel complex dissociates to form free nickel ion. Thus the plating rate is inversely proportional to the complexing ion's stability constant (see Table 8).

TABLE 8 Complexing Agents Commonly Used in Electroless Nickel Baths

Anion	Acid	Stability, pK = $-\log K$
Acetate	CH_3COOH	1.5
Succinate	$HOOCCH_2CH_2COOH$	2.2
Aminoacetate	NH_2CH_2COOH	6.1
Malonate	$HOOCHCH_2COOH$	4.2
Pyrophosphate	$H_2O_3POPO_3H_2$	5.3
Malate	$HOOCCH_2(OH)COOH$	3.4
Citrate	$HOOCCH_2(OH)C(COOH)_2$	6.9

Yet another family of additives are the stabilizers. As every practitioner of electroless plating knows, plating baths can be used for long periods of time without stabilizers. A bath can, however, decompose suddenly without signs of warning. Actually an increase in the volume of the evolving hydrogen gas is followed by the precipitation of fine black particles before the solution caves in completely. The particles that precipitate are nickel phosphide or nickel boride, depending on the reducing agent used. Fortunately, a number of compounds, referred to as stabilizers, exist which can render an electroless deposition bath, stabile or at the least retard precipitation. Besides providing stability, some of them accelerate the rate of plating. In general, there are four "classes" or types of stabilizers in common use:

1. Compounds of group IV elements (Se, Te, etc.),
2. Unsaturated organic acids (Maleic, etc.),
3. Heavy metal cations (Sn^{+2}, Pb^{+2}, etc.)
4. Oxygen-containing compounds (AsO_2^-, MoO_4^{-2}, etc.)

The stabilizer's concentration in a bath is a crucial parameter. For instance, stabilizers of types 1 and 4 are effective at concentrations as low as 0.1 ppm. At concentrations of about 3 ppm, plating is stopped altogether. On the other hand, certain stabilizers of type 1 such as thiourea in the proper concentration range can enhance deposition rates to a considerable degree. Stabilizers of types 3 and 2 are used in concentration ranges of 10^{-3} to 10^{-6} and 10^{-1} to 10^{-3} molar concentration ranges, respectively. In the last analysis, the stabilizers alter the activity of the catalytic substrate. Hence their marked influence at even low concentrations.

In addition to the above, air agitation of the metallizing bath will result in improved bath stability. Specifically, it was observed [11] that when pure oxygen is bubbled through a nonstabilized solution the mixed potential shifted from about $-620\,\text{mV}$ to $-550\,\text{mV}$ (vs. SCE). That is to say, the oxygen-agitated solution is markedly more stable than its quiescent counterpart. A number of methods are available to the practitioner in order to determine solution stability or stabilizer efficiency. One of the best is as follows: Add a couple mL of a 100 ppm palladium chloride solution to a sample of the warm plating solution. The bath is considered stable if no visible black precipitate is in evidence within the first 60 seconds.

Still another family of additives is known as buffers. Buffers are substances or combination of substances that are capable of neutralizing both acid and base alike without changing the pH of the solution by much. A measure of the buffer's efficiency is the amount of, say, acid required to change the solution's pH. Clearly, the greater that amount, the better is the buffer. Titration methods may be used to determine that amount and so the buffer's quality. In Table 9 we list a few acids used as buffer for electroless nickel solutions. The maintenance of the pH value in a deposition bath is as crucial as the deposition rate, and the composition of the end product film is strongly dependent on it. Figure 2 [12] depicts the effect of pH value on the deposition rate of an electroless nickel bath which employs sodium hypophosphite as the reducing agent. The effect of bath pH on the phosphorus content of the end product film is shown in Figure 3. In practice it has been observed that three moles of H^+ are produced for every mole of Ni^{+2} deposited. This means that without any type of buffering, as deposition progresses, the pH value in a bath is

TABLE 9 Acids Commonly Used as Buffer in Electroless Plating Baths

Acid
Acetic
Propionic
Glutaric
Succinic
Adipic

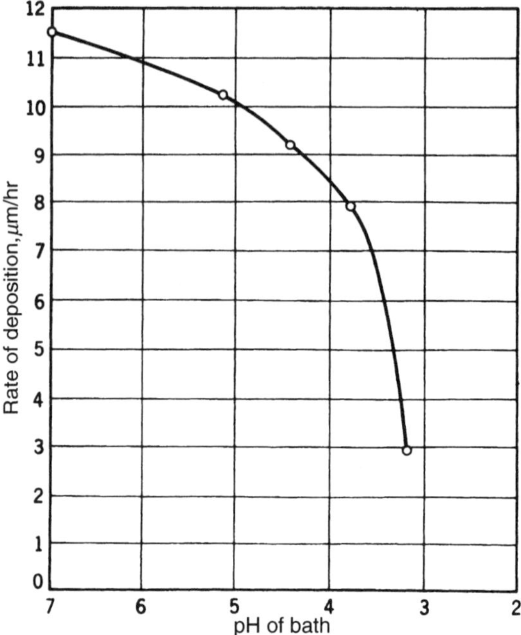

FIGURE 2 Effect of pH on electroless nickel deposition rate. Bath: $NiCl_2 \cdot 6H_2O$, 30 g/l; $NaH_2PO_2 \cdot H_2O$, 10 g/l; sodium glycolate, 10 g/l [12].

expected to become lower. This is reflected in a gradual lowering of the deposition rate as well (see Fig. 2). In addition the film composition will change (see Fig. 3). Whether, in practice, these changes are harmful or undesired depends on the type of application for which the film is required. For instance, if corrosion resistance (low-porosity), or a nonmagnetic state in the freshly as-deposited state is the desired property, then lowering the pH value is an asset rather than a hindrance. If, on the other hand, the phosphorous content must remain stable, then a change in the pH value is to be avoided. As is the case with all rules, there are exceptions. Electroless nickel plating, using DMAB as reducing agent, is accompanied by the evolution of hydrogen ion, hydrogen gas, and the creation of boric acid and dimethylamine. Boric acid and its salts, together with the amine, serve as buffering agents, decreasing the effect of the evolution of hydrogen ions. In some plating baths the pH values were observed to increase slightly. As an aside, but an important practical observation, it ought to be

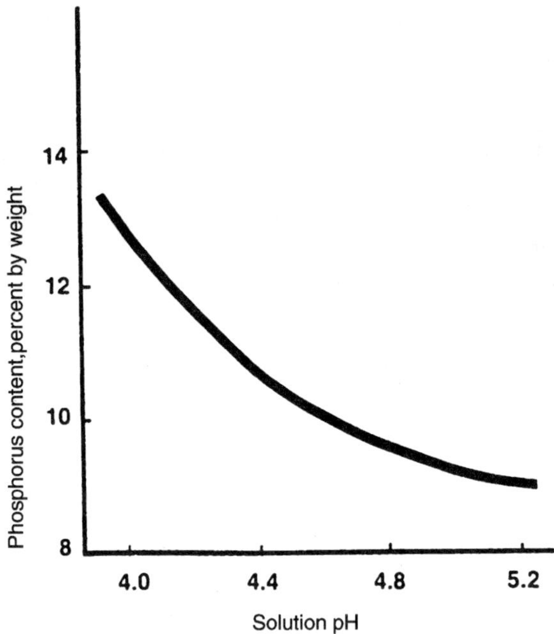

FIGURE 3 Effect of pH on phosphorus content. Reproduced with permission by AESF.

noted that electroless nickel baths using DMAB have a very long operating life. This may be due to the plating reaction's by-product's being soluble in the bath.

4 FILM PROPERTIES

4.1 Structure

Electroless deposits, in general, and electroless nickel, in particular, have a wide number of industrial applications. This is due to their unique properties of corrosion and wear resistance. These and other important properties are a direct result of the structure and chemical makeup of the end product thin films. The properties in turn depend on the deposition bath makeup, and deposition parameters such as temperature and agitation. Yet another important practical advantage offered by electroless deposition is their uniform deposition even on objects of arbitrary shape. For these reasons a discussion of film properties is included here together with a list of relevant references.

Electroless nickel deposits come in two groups, depending on the specific reducing agent used in the deposition process. The first are the nickel–phosphorus alloys, Ni–P; the others are the nickel–boron alloys, Ni–B.

As-deposited electroless nickel is a metastable supersaturated alloy [13]. The structure of electroless nickel grown in acidic bath, using hypophosphate as the reducing agent, is said to be amorphous or liquidlike [3]. Heat treatment to about 330°C was found ([3, 13]; see also *Fundamentals*, ch. 16) to result in semi-crystalline, face-centered cubic (fcc) nickel interspersed with intermetallics such as Ni_3P and Ni_3B. Those intermetallics cannot form during plating, so in the as-deposited state, phosphorus is trapped between nickel atoms in a random fashion. The amount of phosphorus in a Ni–P film is bath-pH dependent, as discussed above (see Fig. 3). In general, it may be stated that the higher the bath-pH value, the lower is the phosphorus content of the film and the higher is the degree of crystallinity of the nickel. That is to say, the lower the phosphorous content, the higher, on the average, are the sizes of the component nickel crystallites making up the film. Therefore phosphorous is thought to act as an inhibitor of crystal formation. This may be understood in simple terms as follows: when the phosphorus atoms are trapped between the nickel atoms, their presence reduces the possibility of contact among nickel atoms such as might form extended nickel crystallites. During deposition and the concomitant hydrogen evolution, the pH value close to the growing film will become higher while subsequent stirring action lowers back that value. This periodic change results in the phosphorus content varying as a function of film layer thickness. This behavior had already been observed by a number of workers in the 1950s [14]. Another phosphorous content dependent property is that of material density. As Figure 4 shows, the material density of films tends to their bulk value at the zero concentration value of phosphorous [15].

4.2 Hardness

Hardness is defined as the reluctance of a material to permanently deform, or indent. This is an easy quantity to measure, and for that reason hardness has been frequently

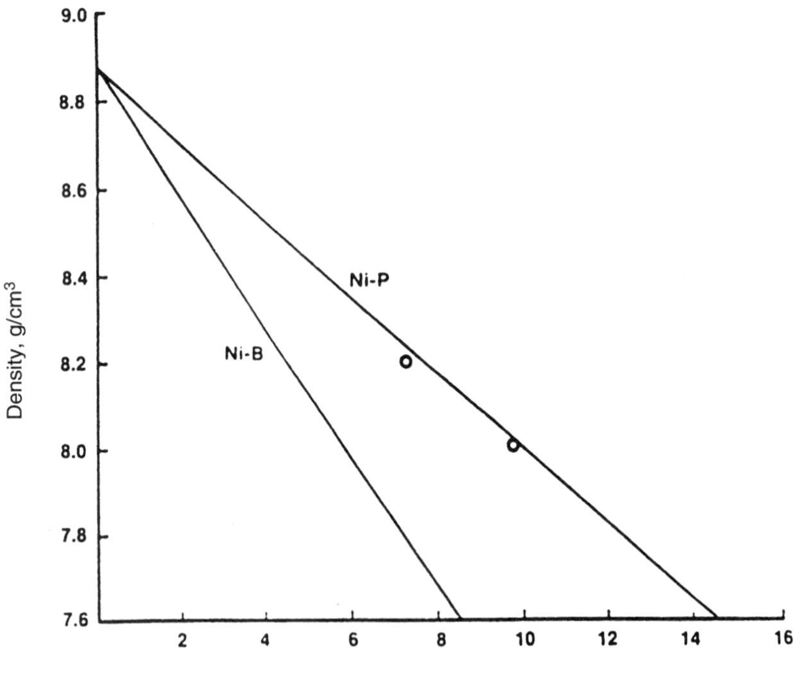

FIGURE 4 Effects of alloy composition on density for Ni–P and Ni–B deposits (29). Reproduced with permission by AESF.

the subject of analysis. Hardness has no direct connection to material strength, but it is a reliable indicator of the degree of abrasive-wear resistance. The hardness of as-deposited Ni–P films is close to 550 kg mm^{-2}, as determined by the Knoop or Vickers indenter using a 100 g load. With increasing phosphorous content a gradual decrease in hardness is observed. As plated, Ni–B films have higher hardness—about 700 kg mm^{-2}. In contrast to Ni–P films, the hardness of Ni–B films is not dependent on composition. Hardness, in general, is affected by heat treatment. The exact functional dependency, however, of hardness versus heat treatment temperature and duration is rather complex, so it will not be discussed here in detail. In general, as-deposited Ni–P films can have their hardness raised from about 550 to nearly 1000 kg mm^{-2} by heating them to 300°C for about 10 minutes. If kept heated longer, the hardness will fall back to the lower value. Similarly at higher temperatures the film loses hardness if kept heated for even shorter periods. In other words, hardness values peak and then decline with increasing heat treatment, and this is likely due to the formation of intermetallics.

The addition of a third element affects hardness as well. Thus it has been noted by many that the addition of molybdenum will markedly increase the hardness of Ni–B films.

4.3 Corrosion and Wear Resistance

It has been observed that electroless nickel generally provides a coating that has lower porosity and more uniform thickness than the equivalent electroplated nickel alloy, such as Ni(P) or Ni(B). This makes the film an effective corrosion-protecting agent. When it comes to actual application of the electroless film as an anticorrosion layer, the substrate pretreatment and the actual plating assure good adhesion and continuity of the coating.

The effectivity of an anticorrosion layer is measured by a neutral 5% salt spray as well as outdoor exposure. Thus it has been observed that Ni–P layers with about 9% by weight phosphorus content provide longer corrosion protection than an electro-plated nickel alloy.

The difference in electrochemical potential between electroless nickel and the substrate is thought to determine what happens in the presence of voids or pores in the coating. That difference is actually dependent on the nature of the corrosive agent. For instance, in an electrolytic cell, the electroless nickel will constitute the cathode against aluminum or steel in most environments. Now, if an area of the substrate (which is the anode) is exposed due to pores in the coat, then the current density i.e. corrosion rate will be high there. More specifically a 300 mV difference will maintain the corrosion of an aluminum or steel substrate if the electroless coating does not provide complete coverage. In cases where the electroless nickel is the anode, the coating is said to corrode sacrificially. The anode area is large and consequently the current density i.e. corrosion rates are rather low. The inclusion of zinc in proper quantities in electroless nickel may tip the balance such that it may become an effective corrosion-protecting agent even for steel.

By way of summary, in engineering applications electroless nickel is used to render protection for metal surfaces exposed to corrosion and/or wear. The factors that set the conditions for the corrosion resistance of electroless nickel may be listed as follows:

1. Surface properties and finish
2. Surface pretreatment
3. Deposit thickness
4. Deposit properties
5. Plating posttreatment
6. Nature of the corrosive medium

Heat treatment to about 350°C, in a view to rendering the coating extra hardness, lowers the corrosion resistance of electroless nickel. This is perhaps due to microcracking. On the other hand, heat treatment up to 650°C improves corrosion resistance, since there is improved bonding to, say, steel. Finally, so many different parameters affect the properties of electroless nickel coating that it is difficult to accurately predict, a priori, their corrosion resistance. Field performance is likewise difficult to base on laboratory observed test results, thus adding to the complexity of the issue at hand.

As indicated above, electroless nickel is used not only for corrosion protection but also for wear protection. If properly applied, electroless nickel is considered to enhance the useful life of pumps, valves, shaft, connector pins, and rotor blades, to name just a few example applications.

Wear is the gradual mechanical deterioration of surfaces in contact. Generally, two types of wear are recognized: adhesive and abrasive. The first is the welding of the two surfaces. The other is the result of the lateral movement shearing the welds. If that does not occur at the original weld, material from one surface can still adhere to another. The resulting weight loss is known as abrasive wear.

Wear, in general, is related to the hardness of the surfaces in contact. Lubrication too reduces friction and wear by reducing the intimate contact between surfaces. When hard particles such as diamond [16] are included in electroless nickel, they form the principal areas of contact with another surface. The result is reduced adhesive wear. If, however, they are pulled out of the matrix, abrasive wear will result, from the mutual surface motion.

Again, it should be clear that wear is a complex process. Among the many factors that affect wear and make it difficult to predict and control are finish and surface hardness, the contact area and its shape, the type of motion and its duration and speed, the temperature and environment, and the type of lubrication. Laboratory tests provide only clues; the actual working conditions under service are the only reliable indicators of how a surface will wear. The following is a summary of the general conditions for wear applications of electroless nickel:

1. Coating should be heat treated to increase hardness.
2. Hardness of coat should be greater on rotating parts than the mating surface.
3. Phosphorous content has to be greater than 10%.
4. Contacting surfaces should be smooth and lubricated.
5. Electroless nickel is not fit to be used under high shear and load conditions.

4.4 ELECTRIC AND MAGNETIC PROPERTIES

As-deposited electroless nickel films from acidic baths possess an amorphous structure, as noted above. The structural changes that occur as a result of heat treatment can be illustrated by observing the changes of film resistivity as heat treatment proceeds. This is seen in Figure 5. In the figure the resistivity changes, in time as heat treatment proceeds are given for acid bath (pH = 5.4) grown films of different thicknesses. From the figure it is evident that as the films change from a solid solution of phosphorous in amorphous nickel to a mixture of polycrystalline nickel metal and the intermetallics Ni_3P, the resistivity decreases. Thermal conductivity usually is in direct (proportional) to electric conductivity, so Figure 5 can be viewed as representing that parameter's characteristics too. Heat treatment changes the nature of the amorphous films from nonmagnetic to weakly ferromagnetic due to the nickel crystallization. The magnetic coercivities of low phosphorous to microcrystalline films increases as well after heat treatment. The reason may be that the paramagnetic intermetallics Ni_3P impedes the movement of domain walls. In

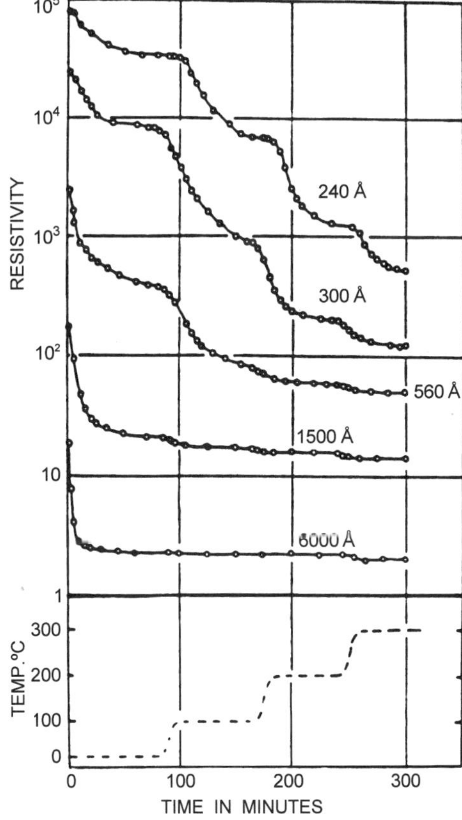

FIGURE 5 Upper curves: the resistivity (in ohm/square) changes in time of Ni–P films during heating. Film thicknesses as indicated. Bottom (dashed) curve: heat-treatment temperature as a function of time.

TABLE 10 Some Properties of Electroless Nickel Films

Solid Solution (%)	Density ($g\,cm^{-3}$)	Resistivity ($\mu\Omega$-cm)
1–3 P	8.6	30
1–2 B		6–12
5–7 P	8.3	50–70
4–5 B	8.5	
8–9 P	8.1	70–100
5 B	8.3	100
>10 P	8	<120
7 B	7.8	200

Table 10 we list some physical properties of electroless nickel films grown in baths with different phosphorous contents.

REFERENCES

1. A. Brenner and G. Riddell, *J. Res. Nat. Bur. Std.*, **37**, 31 (1946).
2. J. Kisel, Ph.D. thesis, University of Windsor, 1988, and M. Schlesinger and J. Kisel, *J. Electrochem. Soc.*, **136**, 1658 (1989).
3. J. P. Marton and M. Schlesinger, ibid., **115**, 16 (1968).
4. J. E. A. M. Van dem Meerakker, *J. Appl. Electrochem.*, **11**, 395 (1981).
5. M. Paunovic, *Plating*, **55**, 1161 (1968).
6. C. H. de Minjer, *Electrodeposition Surf. Treat.*, **3**, 261 (1975).
7. F. Pearlstein, in *Modern Electroplating*, 3rd ed., F. A. Lowenheim, ed., Wiley, New York, 1974.
8. G. O. Mallory and Juan B. Hajdu, eds., *Electroless Plating-Fundamentals and Applications*, American Electroplaters and Surface Finishers Society, chs. 1–11, and references therein.
9. D. J. Levy, *Electrochem. Technol.*, **1**, 38 (1963).
10. J. W. Dini and P. R. Coronado, *Plating*, **54**, 385 (1967).
11. G. Gabrielly and F. Raulin, *J. Appl. Electrochem.*, **1**, 167 (1971).
12. K. M. Gorbunova and A. A. Nikiforawa, *J. Phys. Chem. USSR*, **28**, 883 (1954).
13. M. Schlesinger and J. P. Marton, *J. Phys. Chem. Solids*, **29**, 188 (1968); *J. Appl. Phys.*, **40**, 507 (1969); also S. L. Chow, N. E. Hedgecock, M. Schlesinger, and J. Rezek, *J. Electrochem. Soc.*, **119**, 1614 (1972).
14. A. W. Goldenstein, W. Rostoker, F. Schlossberger, and G. Gutzeit, *J. Electrochem. Soc.*, **104**, 104 (1957); see also F. Ogburn and C. E. Johnson, *Plating*, **60**, 1043 (1973).
15. T. Schmidt et al., *Nucl. Instr. and Methods*, **199**, 359 (1982); see also reference [17].
16. K. Parker, *Plating*, **61**, 834 (1974).
17. M. Schlesinger and J. P. Marton, *J. Appl. Phys.*, **40**, 507 (1969).

19 Electroless Deposition of Cobalt Alloy Films

TETSUYA OSAKA, TAKAYUKI HOMMA
and TOKIHIKO YOKOSHIMA

INTRODUCTION

Electrolessly deposited films are being widely used for functional applications, featuring their mass-producibility and ability to form films on nonflat substrates [1–7]. For these applications it is essential to control the film's microstructure in order to optimize its functional properties. In this chapter we focus on magnetic properties of electroless cobalt alloy films for magnetic recording devices; for other properties the reader should refer to the review by Safranek [8].

1 ELECTROLESS COBALT ALLOY FILMS FOR LONGITUDINAL RECORDING MEDIA

Since their introduction in 1957 [9], magnetic disk drives have been the center of digital data storage devices, featuring large capacity, high access rate, high reliability, and low cost. This recording medium was first realized using a bulk, "coated" material consisting of γ-Fe_2O_3 particles dispersed in an organic binder. Shortly later, thin metal films were developed for recording media in order to achieve even higher performance.

The application of electroless cobalt film in magnetic recording media was first proposed by Fisher et al. in 1962 [10]; this was followed by a number of studies [11–20]. At the same time, electrodeposited cobalt alloy films were also studied extensively [21–27]. Electroless deposited films were once tried for magnetic drum media and analog recording disks in 1970s, since electroless deposition had the potential of use in the mass production of uniform thin films. The electroless deposited film was not actually put into practice for high density hard disk systems until 1981 [28]; in these systems electroless CoNiP media and sputtered γ-Fe_2O_3 media were utilized.

Successively various electroless deposited media were realized, and a precision technology for their mass production was established in practical applications. The

Modern Electroplating, Fourth Edition, Edited by Mordechay Schlesinger and Milan Paunovic.
ISBN 0-471-16824-6

elemental technologies for realizing high reliability production of electroless cobalt alloy media are bath control, substrate pretreatment, overcoating and lubrication, and process control, for example [29–37]. Table 1 gives the typical bath compositions and operating conditions of CoNiP film for longitudinal recording media. Their magnetic properties are listed in Table 2.

Following the development of thin film media by electrolles deposition, research on the surface activity effect enhanced the magnetic properties [30, 38]. However, in order to achieve higher areal recording density, the longitudinal media had to possess higher coercivity, Hc, and less thickness, δ. Such a medium could be attempted by isolating each crystallite [39–41]. The isolated condition proved effective also for noise reduction [42, 43]. In electroless deposited media, it was found that the codeposition of Zn to Co alloy system could effectively isolate the crystallites and obtain a fine-particulated structure, resulting in high Hc and low noise [44, 45]. Figure 1 shows representative MH loops, SEM, and TEM images of the CoNiP and

TABLE 1 Bath Composition and Operating Conditions for Electroless CoNiP and CoNiZnP Longitudinal and Perpendicular Recording Media

Chemicals (mol·dm^{-3})	CoNiP Longitudinal	CoNiZnP Longitudinal	CoNiP Perpendicular	CoNiReP Perpendicular
$NaH_2PO_2 \cdot H_2O$	0.20	0.20	0.20	0.20
$(NH_4)SO_4$	0.10	0.10	0.50	0.50
$CH_2\ (COONa)_2 \cdot H_2O$	0.30	0.30	0.75	0.75
$C_2H_2\ (OH)_2COONa)_2 \cdot 2H_2O$	—	—	0.20	0.20
$C_2H_3OH(COONa)_2 \cdot 0.5H_2O$	0.40	0.40	0.375	—
$CHOH(COOH)_2$	—	—	—	0.05
$(CH_2)_2\ (COONa)_2$	0.50	0.50	—	—
$CoSO_4 \cdot 7H_2O$	0.06	0.06	0.06	0.06
$NiSO_4 \cdot 6H_2O$	0.04	0.04	0.16	0.08
$ZnSO_4 \cdot 7H_2O$	—	0.07	—	—
NH_4ReO_4	—	—	—	0.003
Bath temperature (°C)	80	80	80	80
pH	9.2	9.2	9.5	8.7

TABLE 2 Magnetic Properties of Electroless CoNiP and CoNiZnP Longitudinal and Perpendicular Recording Media

Composition (at%)	CoNiP Longitudinal	CoNiZnP Longitudinal	CoNiP Perpendicular	CoNiReP Perpendicular
Co	75	68	40	32
Ni	19	24	54	55
Re	—	—	—	6
Zn	—	2	—	—
P	6	6	6	7
Ms (emu cm^{-3})	800	750	750	250
$Hc_{(II)}$ (Oe)	590	1100	700	600
$Hc_{(\perp)}$ (Oe)	—	—	1500	1200
K_u ($\times 10^5$ erg cm^{-3})	—	—	-4.8	$+2.9$
$K_{\perp}$ ($\times 10^5$ erg cm^{-3})	—	—	$+30.5$	$+6.8$

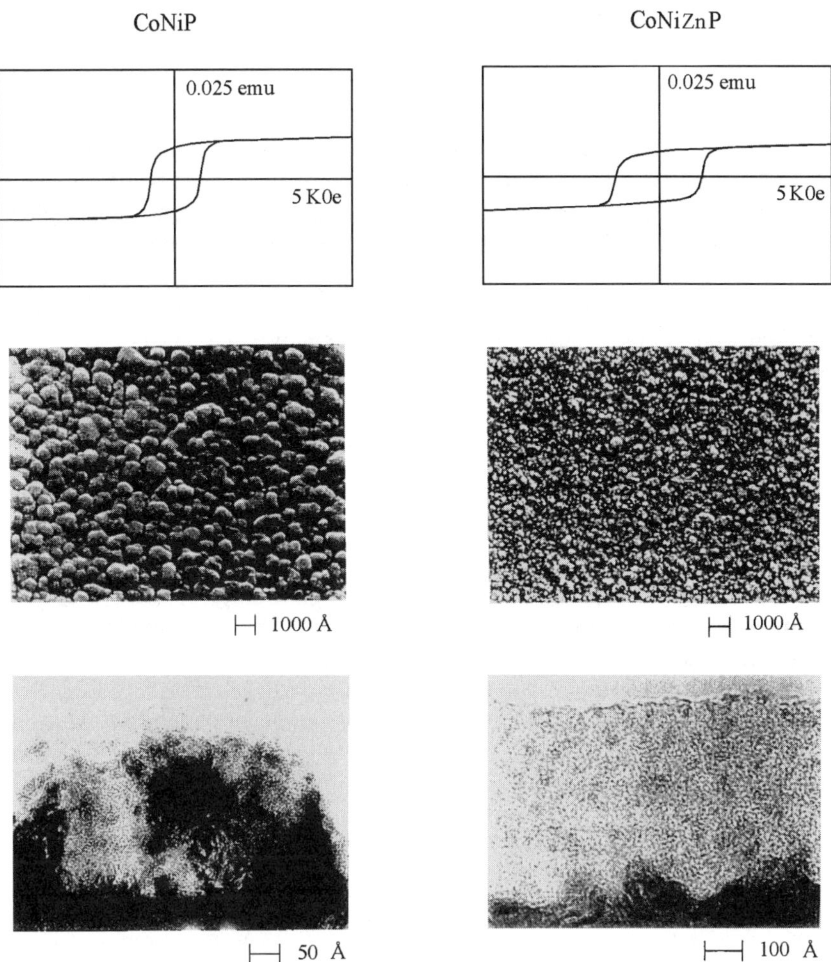

Figure 1 Representative MH loops, SEM top view images, and TEM cross-sectional images for electroless CoNiP and CoNiZnP films for longitudinal magnetic recording media.

CoNiZnP longitudinal recording media. Their magnetic properties are given in Table 2. Recent developments in higher recording density have led to the top coat of carbon sputtered film or SiO_2 spin coated film becoming thinner than 10 nm, and likewise the magnetic film. As a result there have emerged corrosion problems for electroless CoNiP and CoNiZnP thin films. In the late 1980s the plated disk process was superseded by the sputtered disk process, becoming, for example, a CoCrTa or CoCrPtTa medium with a sputtered carbon top coat [46].

2 ELECTROLESS CO ALLOY FILMS FOR PERPENDICULAR MAGNETIC RECORDING MEDIA

The longitudinal media have been constantly improved and their areal recording density has kept increasing; for example, the density of hard disk drives has been increasing at the rate of 100 times decade^{-1} [47–49]. Already achieved is a density higher than the order of Gbits in^{-2} (1.44 Gbits in^{-2} in the mass production level [50] and 5 Gbits in^{-2} for the laboratory level [51, 52]), and a road map for 10 to 20 Gbits in^{-2} has been proposed [53–55]. Recently a 20 Gbits in^{-2} level was announced with a laboratory level at intermag'99 by IBM [56]. Furthermore a post–40 Gbits in^{-2} level has been proved possible by the NEC Corp., using an MR head with higher Bs [57]. Still it is required that longitudinal recording media in which the recorded magnetization is formed along the longitudinal direction to the medium surface possess higher Hc and relatively lower saturation magnetization, Ms, for smaller bit sizes (i.e., the higher areal density to be achieved). Furthermore recently the problem of thermal asperity of such an extremely small bit or "magnet" has arisen, and this indicates the present limit of the longitudinal recording system [58, 59].

In this regard a perpendicular magnetic recording (PMR) system that utilizes a perpendicular mode of magnetization has been proposed to be essentially suitable for high density digital recording [60]. Unlike longitudinal recording, the PMR process is free from the effect of a demagnetizing field that weakens the reproduced voltage in higher density regions. Thus the system is intrinsic to the next generation of ultra-high-density recording, and indicates the possibility of achieving an ultra-high density up to 1000 Gbits in^{-2} [61]. To realize the perpendicular recording system, the medium must have perpendicular magnetic anisotropy; that is, it must be easily magnetized along the direction perpendicular to the medium plane. Sputter-deposited CoCr films have been found to be a good candidate for such media [62, 63]; they can achieve perpendicular anisotropy by the magneto-crystalline anisotropy of their c-axis, which has a perpendicularly oriented hcp structure and an anisotropy shape caused by their fine columnar structure and compositional separation at the nm scale level [64–67]. These unique properties enable the isolation of ferromagnetic crystallites.

The research on electroless deposited media for PMR was first proposed with CoWP films [68, 69], which consist of c-axis perpendicularly oriented hcp crystallites. The properties were continually improved, and various alloy films such as CoMnP, CoNiMnP [70, 71], CoNiReMnP [72–78], and CoNiReP [79–83] were developed. The history of electroless PMR media is reviewed in [2–4, 84, 85]. The typical bath composition and operating conditions for CoNiP and CoNiReP perpendicular magnetic recording media are included in Table 1. Their magnetic properties are also provided in Table 2.

Further improvements have simplified the bath and alloy systems, and CoNiP ternary alloy films with perpendicular magnetic anisotropy were obtained [86–89]. The next section will focus on the correlation between the microstructure and magnetic properties of the CoNiP PMR media.

2.1 Microstructure of Electroless CoNiP PMR Media

Figure 2 shows representative RHEED pattern and TEM bright field image for the electroless CoNiP perpendicular magnetic recording media. As seen in the RHEED pattern, they consist of a c-axis perpendicularly oriented hcp structure, which indicates that their perpendicular anisotropy has originated from the

a)

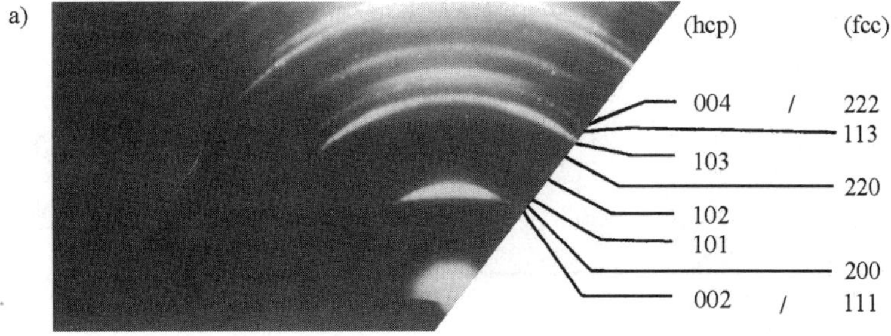

b)

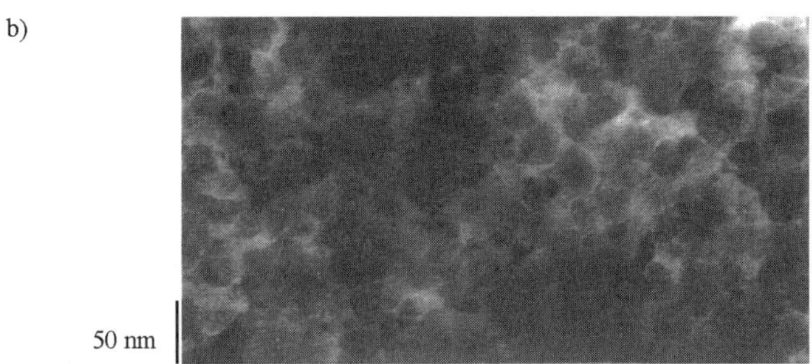

Figure 2 Representative RHEED pattern and TEM bright field images for electroless CoNiP perpendicular magnetic recording media.

magneto-crystalline anisotropy created by such a crystal structure. They also consist of some 30 nm diameter grains with clear boundaries, and they have the columnar structure that causes the micro-shape anisotropy to enhance the perpendicular anisotropy [89].

In order to study their microstructure in detail, we investigated their compositionally separated condition, which is one of the key microstructures for achieving superior characteristics in CoCr sputtered films. In analyzing this condition, we utilized a selective chemical-etching method by which the Co-rich, ferromagnetic component segregated in the grain can be dissolved, leaving the nonmagnetic component [90–92].

Figure 3 shows representative TEM bright field images of the electroless CoNiP film before and after the etching treatment [93]. The as-deposited film consists of some 30 nm diameter grains with clear boundaries, whereas numerous pits several nanometers in size appear in the chemically etched film image. These pits seem to be dispersed hexasymmetrically in each grain, and an asteroid pattern is formed that consists of a core 10 nm in diameter and arms about 20 nm, as shown in Figure 3*c*. Our magnetic property analysis of the as-deposited and chemically etched films indicated that the pits are traces of a Co-rich, ferromagnetic component dissolved by the etching treatment and that the asteroid patterns with hexasymmetry are the nonmagnetic component [91]. On the other hand, an etching pattern could not be

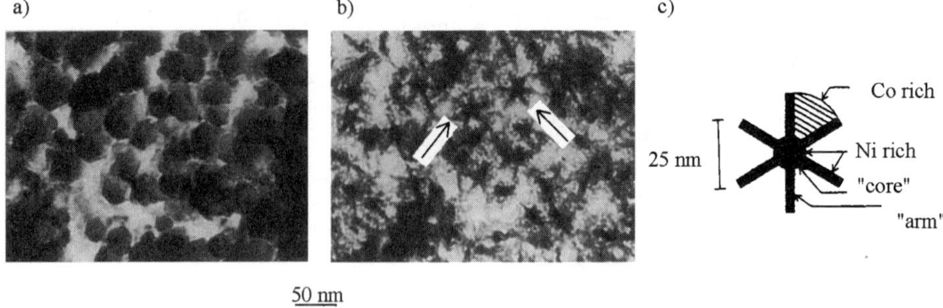

Figure 3 Representative TEM bright field images for the CoNiP films in as-deposited (*a*), and chemically etched (*b*) conditions, and schematic model for the phase-separated microstructure in a grain (*c*).

observed in films with lower Hc values [93, 94], suggesting that a compositionally separated condition is essential for attaining high Hc.

2.2 Spin-Echo ^{59}Co NMR Study for Segregated Condition of CoNiP Media

To investigate such a characteristic structure in more detail, the spin-echo 59-Co NMR method was adopted. This method has proved to be a powerful tool for elucidating compositionally separated condition of the CoCr media [95–97].

Figure 4 gives the representative NMR spectra of the CoNiP films with high and low Hc values and of a $Co_{35}Ni_{65}$ bulk alloy specimen [98]. Previous 59-Co NMR studies on homogeneous CoNi bulk alloy specimen indicated that the main peak resonance frequency of Co decreases linearly from 225 MHz with an increase in Ni content [99]. As can be seen in Figure 4, although the films have almost the same average CoNi compositions as that of the bulk alloy, they exhibit broad spectra,

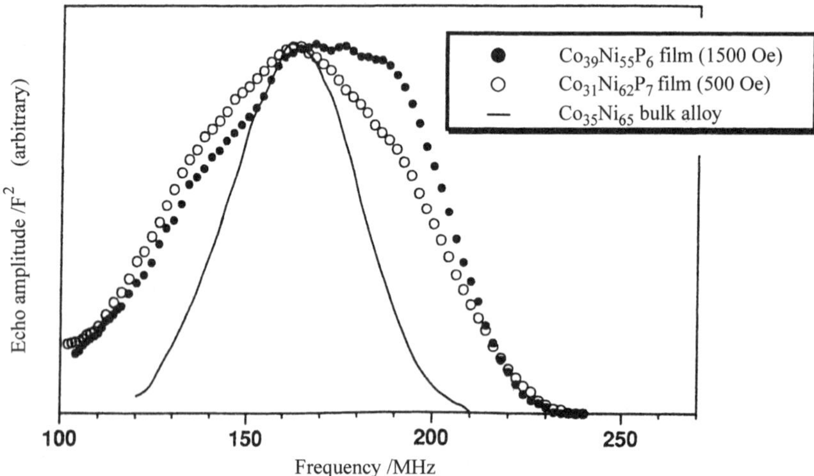

Figure 4 Representative NMR spectra for the CoNiP films with high and low Hc values, and of $Co_{35}Ni_{65}$ bulk alloy.

indicating that they are in an inhomogeneous state. Moreover the spectrum for the film with higher Hc represents the more general condition with a clearer "shoulder" in the higher frequency region around 185 MHz, which nearly corresponds to that of the $Co_{60}Ni_{40}$ alloy and is a richer composition than the average one. Thus it was confirmed that there are Co-rich components in the films and that the film with higher Hc possesses greater compositional inhomogeneity. These results correspond to the chemical-etching results described above.

3 ELECTROLESS CO ALLOY SOFT MAGNETIC THIN FILMS FOR MAGNETIC HEAD APPLICATIONS

Thin film magnetic heads (TFHs) have come into wide use. Their ability to narrow the formation of small gaps has resulted in tracks that achieve higher areal recording density. Electrodeposited permalloy ($Ni_{80}Fe_{20}$) films are exclusively used to form the magnetic core of the TFH. New soft magnetic films with a higher saturation magnetic flux density, Bs, and a high resistivity, ρ, that can achieve higher resolution are in demand future systems which are expected to have extremely high recording densities. Various processes such as sputtering and electrodeposition have been proposed to produce such films. Now, since the electroless deposition process has been widely used for preparing magnetic thin films, as described above, should be noted that this process has the advantage for depositing the films onto a fine selected area of a nonconductive surface. With this in mind, we attempted to develop such soft magnetic films with high Bs and ρ for the core of TFH by using two alternative types of electrochemical process: electroless deposition and electrodeposition.

Table 3 gives representative bath compositions and operating conditions for electroless CoB, CoFeB, and NiFeB soft magnetic films. The electroless deposited $Co_{96}B_4$ (at%) films, which we previously developed, possess a high Bs of 1.4 T and a low saturation magnetostriction, λs. We considered that their permeability, μ, could be improved by annealing in the magnetic field after deposition [100, 101]. Therefore, to further improve the Bs of the CoB films, we attempted codeposition of Fe, by

TABLE 3 Representative Bath Composition and Operating Condition for Electroless CoB, CoFeB, and NiFeB Soft Magnetic Films

	Concentration ($mol\,dm^{-3}$)		
Chemicals	CoB	CoFeB	NiFeB
DMAB	0.025	0.025	0.03
Na-citrate	—	0.05	0.10
Na-tartrate	0.55	0.20	0.10
$(NH_4)_2SO_4$	0.30	0.20	0.30
H_3PO_3	—	0.06	—
$CoSO_4 \cdot 7H_2O$	0.10	0.10	—
$FeSO_4 \cdot 7H_2O$	—	0.01	0.05
$NiSO_4 \cdot 6H_2O$	—	—	0.05
pH adjusted with NaOH		9.0	
Bath temperature		70°C	
Substrate		Cu (50 nm)/Ti (5 nm)/glass	

which we did obtain the CoFeB films (at the composition of $Co_{89}Fe_9B_2$) with high Bs of 1.6 T [102, 103]. Figure 5 shows representative TEM bright field image and THEED pattern of the CoFeB films. As can be seen in the image, the film consists of fine grains of fcc–Co about 10 nm in diameter.

We further attempted to develop the permalloy films, which are usually prepared by electrodeposition by using electroless deposition process [101] based on the electroless NiB system. We found that the electroless $Ni_{70}Fe_{27}B_3$ film exhibits almost the same magnetic properties as the electrodeposited one, except for its rather high λs. The NiFeB films consist of fine-microcrystallites that are almost like those of the CoFeB films described above.

The magnetic and electric properties of the electroless deposited $Co_{96}B_4$, $Co_{89}Fe_9B_2$, and $Ni_{70}Fe_{27}B_3$ films are summarized in Table 4. We have additionally applied the CoFeB films to the next generation core of TFH and the write head core of a merged MR(magneto-resistive) head, featuring high Bs and optimizing the deposition parameters. More recently a micropatterned fabrication process was achieved with CoFeB for the head core application [105].

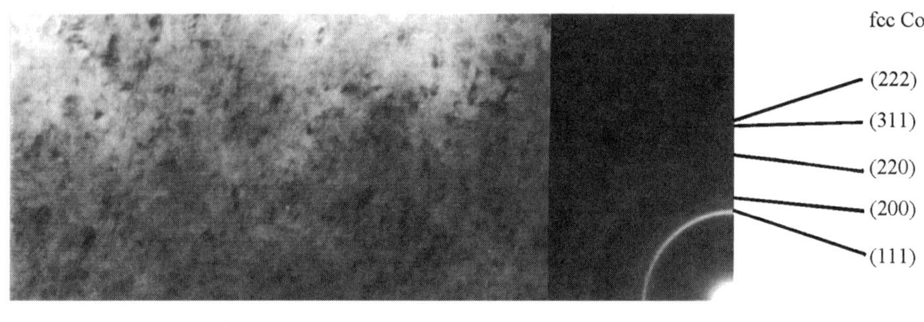

Figure 5 Representative TEM bright field image and corresponding THEED pattern of electroless deposited CoFeB film.

TABLE 4 Magnetic and electric properties of electroless $Co_{96}B_4$, $Co_{89}Fe_9B_2$, and $Ni_{70}Fe_{27}B_3$ films

	Electroless-Plated Films		
Properties	$Co_{96}B_4$ (at%)	$Co_{89}Fe_9B_2$	$Ni_{70}Fe_{27}B_3$
Hc (Oe)	0.7	0.7	0.5
Bs (T)	1.5	1.6–1.7	1.0
μ (at 1 MHz)	1000	650	1000
μ (at 1 MHz) After magnetic annealing	3000	1600	1000
Thermal stability (°C)	300	400	400
ρ (μΩ cm)	—	30	40
λs (× 10^{-6})	− 1.0	+ 5.4	+ 5.0
Structure	hcp Co	fcc Co	fcc Ni

Note: Substrate: Cu (50 nm)/Ti (5 nm)/glass; film thickness: 1.0 μm.

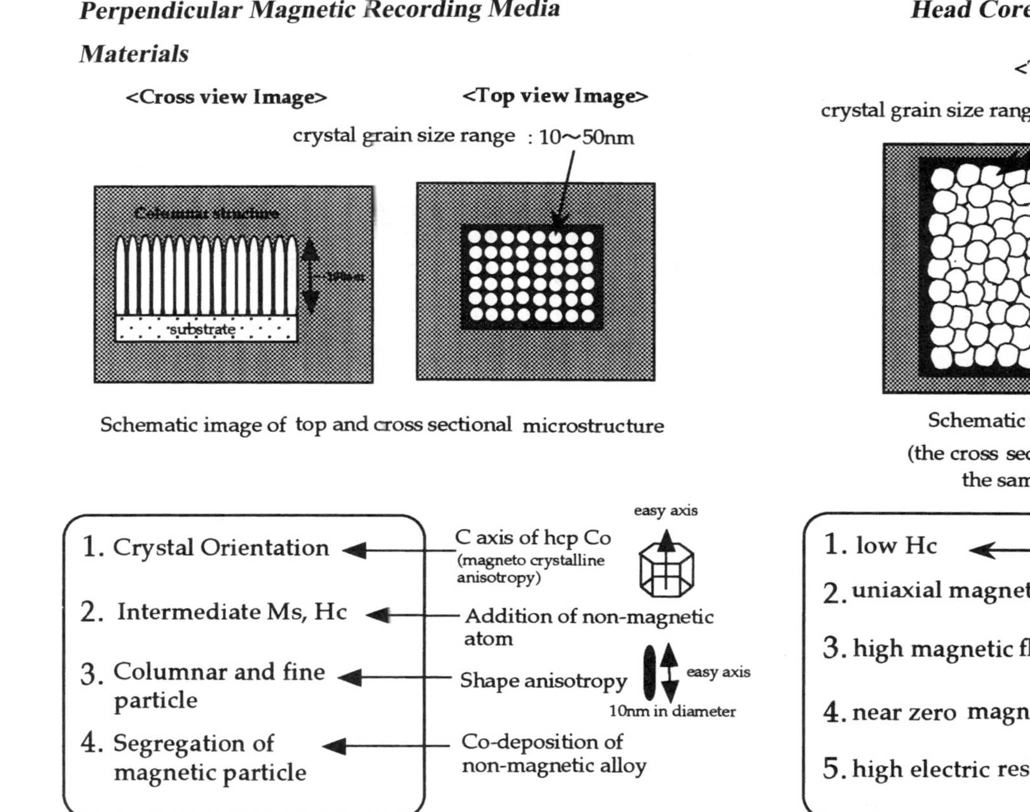

Figure 6 Schematic of perpendicular magnetic recording media and soft magnetic films for head core material.

As a result electrodeposition of the soft magnetic films has become extensively studied, and especially Fe-based alloy films such as $Fe_{73}P_{27}$, which has a high Bs of 1.4 T [106], and $Fe_{84}P_{15}Co_1$, which has a high Bs of 1.6 T with a high ρ of 70 μΩ cm [107]. Also the use of Co-based CoNiFe alloy films [108, 109], which have a high Bs of 1.8 T, has been proposed. To date we have succeeded in obtaining the highest Bs = 2.1 T CoNiFe by controlling a very small amount of the S inclusion in the film [110–112]. From this materials research, the NEC Corp. has succeeded in producing the new generation MR head for future 100 Gbits in^{-2} recording [57].

4 SUMMARY

We reviewed some of recent developments in electroless cobalt films for high-function applications based on our experimental results. The control and design of the film microstructure, such as grain size, crystal orientation, and compositional separation, are important considerations in fabricating films with higher functional properties. Figure 6 schematically presents the conditions for producing perpendicular magnetic recording media and soft magnetic films.

REFERENCES

1. G. O. Mallory and J. B. Hajdu, eds., *Electroless Plating: Fundamentals and Applications*, American Electroplaters and Surface Finishers Society, 1990.
2. V. Brusic, J. Horkans, and D. J. Barclay, in *Advances in Electrochemical Science and Engineering*, vol. 1, H. Gerischer and C. W. Tobias, eds., VCH, 1990, p. 249.
3. T. Osaka and T. Homma, *Trends Electrochem.*, **1**, 13 (1992).
4. T. Osaka and T. Homma, in *New Trends and Approaches in Electrochemical Technology*, N. Masuko, T. Osaka, and Y. Fukunaka, eds., Kodansha-VCH, Tokyo, Japan, 1993, p. 13.
5. Y. Okinaka and T. Osaka, in *Advances in Electrochemical Science and Engineering*, vol. 3, H. Gerisher and C. W. Tobias, eds., VCH, 1994, p. 55.
6. J. E. Williams Jr. and C. Davison, *J. Electrochem. Soc.*, **137**, 3260 (1990).
7. T. Osaka and T. Homma, *Interface*, **4**, 42 (1995).
8. W. H. Safranek, in *Electroless Plating: Fundamentals and Applications*, G. O. Mallory and J. B. Hajdu, eds., American Electroplaters and Surface Finishers Society, Orlando, FL, 1990, p. 463.
9. T. Noyes and W. E. Dickinson, *IBM J. Res. Develop.*, **1**, 72 (1957).
10. R. D. Fisher and W. H. Chilton, *J. Electrochem. Soc.*, **109**, 485 (1962).
11. L. D. Ransom and V. Zentner, *J. Electrochem. Soc.*, **111**, 1423 (1964).
12. D. E. Speriotis, J. R. Morrison, and J. S. Judge, *IEEE Trans. Magn.*, **1**, 348 (1965).
13. J. S. Judge, J. R. Morrison, D. E. Speriotis, and G. Bate, *J. Electrochem. Soc.*, **112**, 681 (1965).
14. D. E. Speliotis, J. S. Judge, and J. R. Morrison, *J. Appl. Phys.*, **37**, 1158 (1966).
15. J. S. Judge, J. R. Morrison, and D. E. Speriotis, *J. Electrochem. Soc.*, **113**, 547 (1966).
16. A. S. Frieze, R. Sard, and R. Weil, *J. Electrochem. Soc.*, **115**, 586 (1968).
17. V. Morton and R. D. Fisher, *J. Electrochem. Soc.*, **116**, 188 (1969).

18. S. L. Chow, N. E. Hedgecock, and M. Schlesinger, *J. Electrochem. Soc.*, **119**, 1614 (1972).
19. J. R. DePew, *J. Electrochem. Soc.*, **120**, 1187 (1973).
20. F. Pearlstein and R. F. Weightman, *J. Electrochem. Soc.*, **121**, 1023 (1974).
21. I. Tsu, *Plating*, **48**, 1207 (1961).
22. J. S. Sallo and J. M. Carr, *J. Electrochem. Soc.*, **109**, 1040 (1962).
23. G. V. Elmore and P. Bakos, *J. Electrochem. Soc.*, **111**, 1244 (1964).
24. G. Bate and D. E. Speliotis, *J. Appl. Phys.*, **34**, 1073 (1964).
25. S. Iwasaki, K. Yasuda, and A. Senda, *IECEJ Tech. Rep.*, **17** (1965).
26. R. Sard, C. D. Schwartz, and R. Weil, *J. Electrochem. Soc.*, **113**, 424 (1966).
27. S. Iwasaki, K. Ouchi, and K. Yasuda, *IECEJ Tech. Rep.*, **18** (1966).
28. S. Hattori, A. Tago, Y. Ishii, A. Terada, O. Ishii, and S. Ohta, *Elec. Comm. Lab. Tech. J.*, **31**, 277 (1982).
29. M. Nagao, Y. Suganuma, H. Tanaka, M. Yanagisawa, and F. Goto, *IEEE Trans. Magn.*, **15**, 1543 (1979).
30. F. Goto, Y. Suganuma, and T. Osaka, *J. Met. Finish. Soc. Jpn.*, **33**, 414 (1982).
31. Y. Hirayama, *J. Met. Finish. Soc. Jpn.*, **38**, 378 (1987).
32. M. Yanagisawa, *Tribology and Mechanics of Magnetic Storage Systems, 2*, ASLE Special Publ., **19**, 16 (1985).
33. Y. Suganuma, H. Tanaka, M. Yanagisawa, F. Goto, and S. Hatano, *IEEE Trans. Magn.*, **18**, 1215 (1982).
34. H. Tanaka, F. Goto, N. Shiota, and M. Yanagisawa, *J. Appl. Phys.*, **53**, 2576 (1982).
35. G. Bate, *J. Appl. Phys.*, **37**, 1164 (1966).
36. G. Bate and J. K. Alstad, *IEEE Trans. Magn.*, **5**, 821 (1969).
37. M. Nagao, *J. Surf. Finish. Soc. Jpn.*, **42**, 273 (1991).
38. T. Osaka, H. Nagasaka, and F. Goto, *J. Electrochem. Soc.*, **128**, 1686 (1981).
39. J. Hokkyo, IECEJ Tech. Rep., MR65-22 (1965).
40. D. E. Speriotis and J. R. Morrison, *IBM J. Res. Develop.*, **9**, 233 (1966).
41. H. Tanaka, N. Shiota, F. Goto, M. Yanagisawa, and Y. Suganuma, *Trans. IECEJ*, J67-C, 82 (1984).
42. D. D. Dressler and J. H. Judy, *IEEE Trans. Magn.*, **10**, 674 (1974).
43. N. R. Belk, P. K. George, and G. S. Mowry, *IEEE Trans. Magn.*, **21**, 1350 (1985).
44. H. Nagasaka, T. Michimori, and H. Takeno, *J. Surf. Finish. Soc. Jpn.*, **42**, 58 (1991).
45. F. Goto and T. Osaka, *Electrochem. Soc. Proc.*, **93** (20), 410 (1993).
46. T. Yogi, C. Tsang, T. A. Nguyen, K. Ju, G. L. Gorman, and G. Castillo, *IEEE Trans. Magn.*, **26**, 2271 (1990).
47. E. Grochowski and D. A. Tompson, *IEEE Trans. Magn.*, **30**, 3797 (1994).
48. T. Howell, D. McCown, T. Diola, Y. Tang, K. Hense, and R. Gee, *IEEE Trans. Magn.*, **26**, 2298 (1990).
49. Y. Miura, *J. Magn. Soc. Jpn.*, **16** (S1), 19 (1992).
50. Travelstar VP 2.5 inch 1.2- and 1.6- GB Hard Disk Drive (DDLA21215/21620), IBM Corp. (shipping year 1996).
51. H. Mutoh, H. Kanai, I. Okamoto, Y. Ohtsuka, T. Sugawara, J. Koshikawa, J. Toda, Y. Uematsu, M. Shinohara, and Y. Mizoshita, *IEEE Trans. Magn.*, **32**, 3914 (1996).
52. I. Okamoto, C. Okuyama, K. Sato, and M. Shinohara, *INTERMAG Digest*, **AB-3** (1996).

53. E. Murdock, R. Simmons, and R. Davidson, *IEEE Trans. Magn.*, **28**, 3078 (1992).
54. M. A. Russak, J. M. Kimmal, and B. B. Lal, *Electrochem. Soc. Proc.*, **18**, 143 (1995).
55. Y. Miura, *J. Magn. Soc. Jpn.*, **20**, 879 (1996).
56. A. Moser and D. Weller, *INTERMAG Digest*, **AA-05** (1999).
57. K. Ohashi, N. Morita, T. Tsuda, and Y. Nonaka, *INTERMAG Digest*, **FB-03** (1999).
58. P.-L. Lu and S. H. Charap, *IEEE Trans. Magn.*, **30**, 4230 (1994).
59. P.-L. Lu and S. H. Charap, *IEEE Trans. Magn.*, **31**, 2767 (1995).
60. S. Iwasaki and Y. Nakamura, *IEEE Trans. Magn.*, **13**, 1272 (1977).
61. Y. Nakamura, *J. Magn. Soc. Jpn.*, **18** (S1), 161 (1994).
62. S. Iwasaki and K. Ouchi, *IEEE Trans. Magn.*, **14**, 849 (1978).
63. K. Ouchi, *J. Magn. Soc. Jpn.*, **13** (S1), 611 (1989).
64. Y. Maeda and M. Asahi, *J. Appl. Phys.*, **61**, 1972 (1987).
65. Y. Maeda and M. Takahashi, *IEEE Trans. Magn.*, **24**, 3012 (1988).
66. Y. Maeda and M. Takahashi, *J. Magn. Soc. Jpn.*, **13** (S1), 673 (1989).
67. Y. Maeda, K. Takei, and D. R. Rodgers, *Oyobutsuri* (Appl. Phys. Jpn.), **63**, 285 (1994).
68. T. Osaka and N. Kasai, *J. Met. Finish. Soc. Jpn.*, **32**, 309 (1981).
69. T. Osaka, F. Goto, N. Kasai, and Y. Suganuma, *Denki Kagaku*, **49**, 792 (1981).
70. T. Osaka, F. Goto, N. Kasai, I. Koiwa, and Y. Suganuma, *J. Electrochem. Soc.*, **130**, 568 (1983).
71. T. Osaka, N. Kasai, I. Koiwa, and F. Goto, *J. Electrochem. Soc.*, **130**, 790 (1983).
72. T. Osaka, I. Koiwa, Y. Okabe, H. Matsubara, A. Wada, and F. Goto, *Denki Kagaku*, **52**, 197 (1984).
73. F. Goto, T. Osaka, I. Koiwa, Y. Okabe, H. Matsubara, A. Wada, and N. Shiota, *IEEE Trans. Magn.*, **20**, 803 (1984).
74. T. Osaka, I. Koiwa, Y. Okabe, H. Matsubara, A. Wada, F. Goto, and N. Shiota, *Bull. Chem. Soc. Jpn.*, **58**, 414 (1985).
75. I. Koiwa, Y. Okabe, H. Matsubara, and T. Osaka, *J. Met. Finish. Soc. Jpn.*, **36**, 204 (1985).
76. T. Osaka, I. Koiwa, Y. Okabe, and K. Yamanishi, *Jpn. J. Appl. Phys.*, **26**, 1674 (1987).
77. O. Takano, H. Matsuda, H. Izumitani, and K. Ito, *J. Met. Finish. Soc. Jpn.*, **35**, 440 (1984).
78. H. Matsuda and O. Takano, *J. Met. Finish. Soc. Jpn.*, **37**, 753 (1986).
79. I. Koiwa, Y. Okabe, H. Matsubara, T. Osaka, and F. Goto, *J. Magn. Soc. Jpn.*, **9**, 83 (1985).
80. I. Koiwa, M. Toda, and T. Osaka, *J. Electrochem. Soc.*, **133**, 597 (1986).
81. I. Koiwa, H. Matsubara, T. Osaka, Y. Yamazaki, and T. Namikawa, *J. Electrochem. Soc.*, **133**, 685 (1986).
82. T. Osaka, I. Koiwa, M. Toda, T. Sakuma, Y. Yamazaki, T. Namikawa, and F. Goto, *IEEE Trans. Magn.*, **22**, 1149 (1986).
83. T. Osaka, H. Matsubara, T. Sakuma, T. Homma, S. Yokoyama, Y. Yamazaki, and T. Namikawa, *J. Magn. Soc. Jpn.*, **12**, 77 (1988).
84. T. Osaka and I. Koiwa, *J. Met. Finish. Soc. Jpn.*, **38**, 362 (1987).
85. T. Osaka and T. Homma, *J. Surf. Finish. Soc. Jpn.*, **42**, 283 (1991).
86. T. Osaka, T. Homma, K. Inoue, and K. Saga, *Denki Kagaku*, **58**, 661 (1990).

87. T. Osaka, T. Homma, K. Inoue, Y. Yamazaki, and T. Namikawa, *J. Magn. Soc. Jpn.*, **13** (S1), 779 (1989).
88. T. Homma, K. Inoue, H. Asai, K. Ohrui, T. Osaka, Y. Yamazaki, and T. Namikawa, *J. Magn. Soc. Jpn.*, **15** (2), 113 (1991).
89. T. Homma, K. Inoue, H. Asai, K. Ohrui, and T. Osaka, *IEEE Trans. Magn.*, **27**, 4909 (1991).
90. Y. Maeda and M. Asahi, *IEEE Trans. Magn.*, **23**, 2061 (1987).
91. D. J. Rogers, Y. Maeda, and K. Takei, *Script. Met.*, **33**, 1553 (1995).
92. M. Takahashi and Y. Maeda, *Jpn. J. Appl. Phys.*, **29**, 1705 (1990).
93. T. Homma, T. Osaka, Y. Yamazaki, and T. Namikawa, *Script. Met.*, **33**, 1569 (1995).
94. T. Homma, K. Ohrui, Y. Iizuka, T. Osaka, Y. Yamazaki, and T. Namikawa, *J. Magn. Soc. Jpn.*, **17**, 105 (1993).
95. K. Yoshida, H. Kakibayashi, and H. Yasuoka, *J. Appl. Phys.*, **68**, 705 (1990).
96. Y. Maeda and K. Takei, *IEEE Trans. Magn.*, **27**, 4721 (1991).
97. Y. Maeda and K. Takei, *J. Magn. Soc. Jpn.*, **16**, 87 (1992).
98. T. Homma, H. Asai, T. Osaka, K. Takei, and Y. Maeda, *Chem. Lett.*, 1783 (1992).
99. S. Kobayashi, K. Asayama, and J. Itoh, *J. Phys. Soc. Jpn.*, **21**, 65 (1966).
100. T. Osaka, T. Homma, N. Masubuchi, K. Satio, M. Yoshino, Y. Yamazaki, and T. Namikawa, *J. Magn. Soc. Jpn.*, **14**, 309 (1990).
101. T. Osaka, M. Takai, and K. Kageyama, *J. Surf. Finish. Soc. Jpn.*, **45**, 213 (1994).
102. T. Osaka, T. Homma, K. Kageyama, and Y. Matsunae, *Denki Kagaku*, **62**, 987 (1994).
103. T. Osaka, T. Homma, K. Kageyama, Y. Matsunae, and O. Shinoura, *J. Magn. Soc. Jpn.*, **18** (S1), 183 (1994).
104. M. Takai, K. Kageyama, S. Takefusa, A. Nakamura, and T. Osaka, *IEICE Trans. Electron.*, **E78-C**, 1530 (1995).
105. T. Yokoshima, D. Kaneko, T. Osaka, S. Takefusa, and M. Oshiki, *J. Magn. Soc. Jpn.*, **23**, 1397 (1999).
106. T. Osaka, M. Takai, A. Nakamura, F. Asa, K. Ohashi, and H. Tachibana, *J. Mag. Soc. Jpn.*, **18** (S1), 187 (1994).
107. M. Takai, A. Nakamura, R. Shigemoto, F. Asa, and T. Osaka, *Electrochem. Soc.*, **18**, 552 (1995).
108. O. Shinoura, S. Kamijo, and K. Narumiya, *J. Mag. Soc. Jpn.*, **18**, 277 (1994).
109. A. Nakamura, M. Takai, K. Hayashi, and T. Osaka, *J. Surf. Finish. Soc. Jpn.*, **47**, 934 (1996).
110. T. Osaka, M. Takai, K. Hayashi, K. Ohashi, M. Saito, and K. Yamada, *Nature*, **392**, 796 (1998).
111. T. Osaka, M. Takai, K. Hayashi, Y. Sogawa, K. Ohashi, Y. Yasue, M. Saito, and K. Yamada, *IEEE Trans. Magn.*, **34**, 1432 (1998).
112. K. Ohashi, Y. Yasue, M. Saito, K. Yamada, T. Osaka, M. Takai, and K. Hayashi, *IEEE Trans. Magn.*, **34**, 1462 (1998).

20 Electroless Deposition of Palladium and Platinum

IZUMI OHNO

BACKGROUND

Much effort has been given to the development of electroless palladium plating along side with rapid technological growth in the electronics industry. In particular, the communications industry requires electroless palladium plating because it can yield uniform deposits on surfaces of small components with intricate designs and can provide a stable contact surface for conductors that operate at elevated temperatures. Palladium coating is useful as a substitute for gold coating and also as a diffusion barrier layer between gold coating and copper or silver substrate in the electronics industry. That is due to its good corrosion resistance, low contact resistance, and lower price.

It was recognized by Brenner [1] that electroless deposition of palladium is possible but also that the progression toward homogeneous deposition of palladium results in a high degree of solution instability. Fortunately the instability of electroless palladium baths can, however, be controlled by a proper selection of additives.

1 ELECTROLESS PALLADIUM PLATING

1.1 Hydrazine-Based Baths

Rhoda developed several electroless palladium plating baths using hydrazine as reducing agent [2–4] (see Table 1). The palladium tetrammine complex used in the baths is prepared by the careful addition of ammonium hydroxide to a palladium chloride solution. The solutions is then heated to dissolve possible precipitates. Ethylenediaminetetraacetic acid added to the bath was found to be an effective stabilizer against spontaneous decomposition of the solution. Bath 1 in Table 1 is recommended for barrel plating [2]. The hardness of the deposits ranges from 150 to 350 Knoop (25 g load), with softer deposits formed at higher deposition rates. The deposits consist of a minimum of 99.4% palladium, and they exhibit a density of $11.96\,g\,cm^{-3}$.

Modern Electroplating, Fourth Edition, Edited by Mordechay Schlesinger and Milan Paunovic.
ISBN 0-471-16824-6

TABLE 1 Composition of Baths for Electroless Palladium Deposition

	Hydrazine Type			Hypophosphite Type	
Bath Consituents (g liter^{-1})	1	2	3	4	5
Palladium (as ammine complex), $[Pd(NH_3)_4]^{2+}$	5.4	7.5	—	—	—
Palladium chloride, $PdCl_2$	—	—	0.023 M	2	0.023 M
Hydrochloric acid, HCl (38%)	—	—	2–8 ml liter^{-1}	4 ml liter^{-1}	2–8 ml liter^{-1}
Ammonium hydroxide, NH_4OH (28% NH_3)	350	280	300 ml liter^{-1}	160 ml liter^{-1}	160 ml liter^{-1}
Ethylenediamine tetraacetic acid disodium salt, Na_2EDTA	33.6	8	0.1 M	—	—
Ammonium chloride, NH_4Cl	—	—	—	27	0.486 M
Hydrazine, $N_2H_4 \cdot H_2O$	0.3	—	0.005 M	—	—
Hydrazine (1 M solution: ml h^{-1})	—	8	—	—	—
Sodium hypophosphite, $NaH_2PO_2 \cdot 2H_2O$	—	—	—	10	0.094 M
Temperature (°C)	80	35	60	55	40
Deposition rate (μm h^{-1})	2.5	0.9	2.5	2.5	0.9
References	[2]	[2, 3]	[5]	[9]	[5]

A disadvantage of the hydrazine-based baths is that the deposition rate decreases drastically after several hours of use, before any significant consumption of bath constituents has taken place. Several modified electroless palladium solutions have been presented [5–7] to prevent this.

The fundamental aspect of electroless palladium deposition using baths 3 and 5 in Table 1 has been studied by Paunovic and Ting [5] who employed electrochemical methods. They reported that the exchange current density for oxidation of hydrazine (7.8×10^{-5} A cm^{-2}) is higher than that for hypophosphite (2.6×10^{-5} A cm^{-2}). Shu and coworkers [8] have studied the autocatalytic effects during electroless palladium deposition in an alkaline bath containing the palladium ammine complex, ethylenediamine tetraacetic acid, and hydrazine using electrochemical polarization measurements. Tetravalent palladium species in the bath can be stabilized through interaction with a coexisting tin oxide on a palladium-tin activated substrate.

1.2 Hypophosphite-Based Baths

In most of the practical electroless palladium plating baths, palladium-ammine complex is used, together with ammonium chloride and hypophosphite as stabilizer and reducing agent, respectively [5, 9–12]. A representative makeup of this type of bath is shown as bath 4 of Table 1. Specifically a stock solution of 20 g liter^{-1} $PdCl_2$–40 ml liter^{-1} HCl (38%) is added to an appropriate quantity of ammonium hydroxide solution while stirring. The resulting solution is allowed to stay at room temperature for at least 20 hours and is filtered before use. Then a hypophosphite solution is added to the solution; following agitation, the resulting solution is brought up to a final volume. Approximately 3 g of sodium hypophosphite monohydrate is consumed in producing 1 g of electroless palladium deposit [13]. Sergienko developed a stable electroless palladium plating bath containing ethylenediamine, ethylenediamine tetraacetic acid, and hypophosphite [14].

1.3 Other Reducing Agent Baths

Stremsdoerfer et al. [15] proposed a new type of electroless bath for depositing palladium and gold-palladium alloys on a bare n-GaAs surface.

The bath, which contains hydroxylamine hydrochloride as a reducing agent and a low concentration of metal species, exhibits a high deposition rate ($0.5\,\mu m\,h^{-1}$) in acidic media at 20°C. The plating bath is composed of two solutions, as shown below. Hydrofluoride and potassium chloride in solutions *A* and *B*, respectively, act as stabilizers. The autocatalytic bath for depositing palladium on bare n-GaAs as a substrate is prepared by adding a small quantity of solution *A* to 40 ml of solution *B*.

Solution *A* for 900 ml,
- 0.3 g $PdCl_2$ dissolved in 9 ml HCl
- 5 ml D.I. water
- 864 ml glacial acetic acid
- 22 ml HF(40%)

Solution *B* for 100 ml,
- $NH_2OH \cdot HCl$, 0.416 (0.06 M)
- KHF_2, 0.937 (0.12 M)
- KCl, 14.912 g (2.0 M)
- Citric acid ($C_6H_8O_7 \cdot H_2O$), 2.101 g (0.1 M)
- D.I. water

The deposits obtained from this bath are well structured and adherent. Further Stremsdoerfer et al. have reported that the palladium deposited from the bath forms a good metal-semiconductor junction (Pd/n–GaAs junction) by subsequent annealing up to 550°C [16].

Nawafune et al. [17] have reported plating conditions on electroless pure palladium plating from palladium-ethylenediamine complex baths using formic acid as a reducing agent. The electroless pure palladium was deposited using the following basic bath composition: 0.01 M palladium chloride, 0.08 M ethylenediamine, and 0.2 M formic acid, pH 6 at 60°C.

Further they have reported electroless palladium plating from ethylenediamine complex baths using phosphite, and trimethylamine borane as a reducing agent [18]. And in case of using phosphite as a reducing agent, the deposits are obtained with minor codeposition of phosphorus.

2 ELECTROLESS PALLADIUM ALLOYS

Electroless palladium-based ternary alloys have been produced by adding sulfate salts of alloying elements in to the electroless palladium bath [9]. Electroless palladium deposits using hypophosphite-based baths (bath 4 in Table 1) contain about 1.5% phosphorus [9]. It is also possible to produce ternary palladium alloys by addition of different metal salts to the bath (bath 4 in Table 1). For example, a palladium-nickel-phosphorus deposit of hardness over 600 Knoop was produced from a hypophosphite bath containing 29.6 g $liter^{-1}$ nickel sulfate hexahydrate. The alloy deposits contained up to 6% nickel. In replacing nickel sulfate hexahydrate by same amount of cobalt sulfate in the bath, less than 10% cobalt was codeposited with palladium. When 36 g $liter^{-1}$ of zinc sulfate were added to the electroless palladium bath (bath 4 in Table 1), about 36% of zinc alloy was produced in deposits in the palladium bath [9].

Electroless palladium-silver [6] and palladium-gold [15, 19] alloys have also been reported as well.

3 PLATING ON VARIOUS SUBSTRATES

Since palladium is a noble metal and has a catalytic activity for anodic oxidation of many reducing agents used in electroless plating [20], electroless palladium is spontaneously deposited at an appreciable rate on many less noble metals and alloys, such as copper, brass, steel, and electroless nickel. Electroless palladium deposits of good quality have been produced on aluminum, chromium, cobalt, gold, iron, molybdenum, nickel, palladium, platinum, rhodium, ruthenium, silver, steel, tin, and tungsten [9]. Electroless palladium has also been applied to various special substrates such as single-crystal (100) GaAs [15, 16], sputtered aluminum [14, 21, 22], and zirconium thin film [7]. Many alloys, such as stainless steel [8], kovar, and brass [23] can be used as substrates for electroless palladium. Initiation of electroless palladium deposition on copper alloys is accelerated if they are activated first by immersing in a solution of 0.1 g $liter^{-1}$ $PdCl_2$–0.5 ml $liter^{-1}$ HCl (38%) at 25°C. Nonconductors, such as glass or plastics, activated by immersing in stannous chloride and then palladium chloride solutions, are also readily plated with electroless palladium [24]. Schlesinger and coworkers [25] reported in detail the effect of UV irradiation on the electroless deposition rate of palladium-tin activated substrates. Laser-irradiation on deposition rate of electroless palladium plating on nickel substrate has also been reported [26].

4 ELECTROLESS PLATINUM PLATING

Electroless platinum plating has been described in the technical and patent literature [27–30]. Rhoda reported on electroless platinum plating bath containing complex alkaline tetravalent platinum hydroxide solutions with hydrazine as a reducing agent as follows:

Disodiumhexahydroxy platinate, $Na_2Pt(OH)_6$, 10 g $liter^{-1}$
Sodium hydroxide, NaOH, 5 g $liter^{-1}$
Ethylene diamine, $C_2H_8N_2$, 10 g $liter^{-1}$
Hydrazine monohydrate, $N_2H_4 \cdot H_2O$, 1 g $liter^{-1}$

Temperature, 35°C
Deposition rate, 12.7 μm h^{-1}

This process involves the addition of hydrazine to the bath either continuously or in portions as the hydrate solution or as the hydrazine salts dissolve in water and initiate platinum deposition. Other baths presented utilize dinitrodiamine platinate, $Pt(NO_2)_2(NH_3)_2$, or potassium tetranitroplatinate, $K_2Pt(NO_2)_4$, as the metal source.

Baths have also been developed that provide electroless plating for platinum alloys with rhodium, iridium [27], and palladium [30]. One example that contains mixed reducing agents is given as follows [30]:

Potassium tetranitroplatinate, $K_2Pt(NO_2)_4$, 0.05 g
Potassium tetranitropalladium, $K_2Pd(NO_2)_4$, 0.05 g
Ammonium hydroxide, $NH_4OH(28\%NH_3)$, 5 mg
Water, H_2O, 50 ml
Hydroxylamine hydrochlorate, $NH_2OH \cdot HCl$ (50%), 0.1 g
Hydrazine monohydrate, $N_2H_4 \cdot H_2O$ (30%), 1.5 ml
Water, H_2O, total 40 ml
pH, 11.7,
Temperature, 60°C

The bath makeup consists in dissolving the two metal salts in the ammonium hydroxide solution while heating, and then adding the other solution containing hydroxylamine and hydrazine. The authors describe the bath as stable. Its activation in a palladium chloride solution was followed by sodium cyanide, and then the platinum-palladium alloy was deposited on copper substrate in 2 μm for 2 hours.

5 SUMMARY

Electroless palladium is becoming increasingly utilized not only as an alternative to electroless gold but also as a plating of choice based on its own merits. In regard to electroless platinum plating, it seems there are various combinations of platinum complexes and reducing agents. For the practical application of electroless platinum plating, however, further research is required with improvements of current baths.

REFERENCES

1. A. Brenner, *Met. Finish.*, **52** (11), 68; **52** (12) 61 (1954).
2. R. N. Rhoda, *Trans. Inst. Met. Finish.*, **36**, 82 (1959).
3. R. N. Rhoda, *J. Electrochem. Soc.*, **108**, 707 (1961).
4. R. N. Rhoda, *Plating*, **50**, 307 (1963).
5. M. Paunovic and C. H. Ting, *Proc. Symp. on Electroless Deposition of Metals and Alloys*, Vol. 12, M. Paunovic and I. Ohno, eds., Electrochemical Society, Pennington, NJ, 1988, p. 170.

6. J. Shu, B. P. A. Grandjean, E. Ghali, and S. Kaliaguine, *J. Membr. Sci.*, **77**, 181 (1993).
7. C. Hsu and R. Buxbaum, *J. Electrochem. Soc.*, **132**, 2419 (1985).
8. J. Shu, B. P. A. Grandjean, E. Ghali, and S. Kaliaguine, *J. Electrochem. Soc.*, **140**, 3175 (1993).
9. F. Pearlstein and F. Weightman, *Plating*, **56**, 1158 (1969).
10. J. P. Abys and N. J. Brindgewater, U.S. Patent, 4,424,241 (Sept. 27, 1982).
11. M. Paunovic and C. Ting, *Proc. Autumn Meeting of the Electrochemical Society*, Abst. 507, Honolulu, Oct. 1987.
12. S. S. Djokic, *Plating Surf. Finish.*, **86**, 104 (1999).
13. F. Pearlstein in *Modern Electroplating*, F. A. Lowenheim, ed., Wiley, New York, 1974, p. 739.
14. A. Sergienko, U.S. Patent 3,418,143 (Dec. 24, 1968).
15. G. Stremsdoerfer, H. Perrot, J. R. Martin, and P. Clechet, *J. Electrochem. Soc.*, **135**, 2881 (1988).
16. G. Stremsdoerfer, D. Nguyen, N. J. Renault, J. R. Martin, and P. Clechet, *J. Electrochem. Soc.*, **140**, 519 (1993).
17. H. Nawafune, S. Nakao, S. Mizumoto, E. Uchida, and T. Okada, *J. Surf. Finish. Soc. Jpn.*, **48**, 474 (1997).
18. H. Nawafune, S. Mizumoto, M. Haga, and E. Uchida, *Trans. Inst. of Metal Fin.*, **74**, 21 (1996).
19. D. Lamouche, P. Clechet, J. R. Martin, E. Haroutiounian, and J. P. Sandino, *Solid-State Electron.*, **29**, 625 (1985).
20. I. Ohno, O. Wakabayashi, and S. Haruyama, *J. Electrochem. Soc.*, **132**, 2323 (1985).
21. C. H. Ting, *Proc. Symp. on Electroless Deposition of Metals and Alloys*, vol. 12, M. Paunovic and I. Ohno, eds., Electrochemical Society, Pennington, NJ, 1988, p. 223.
22. C. H. Ting and M. Paunovic, *Proc. Symp. on Electroless Deposition of Metals and Alloys*, vol. 12, M. Paunovic and I. Ohno, Electrochemical Society, Pennington, NJ, 1988, p. 252.
23. W. Hawk, *Met. Finish.*, **84**, (3), 11 (1986).
24. F. Pearlstein, *Met. Finish.*, **53** (8), 59 (1955).
25. B. K. W. Baylis, C. C. Huang, and M. Schlesinger, *J. Electrochem. Soc.*, **126**, 394 (1979).
26. Y. Sato, M. Yanase, A. Yoshida, T. Nishiyama, K. Kobayakawa, and K. Tennichi, *J. Surf. Finish. Soc. Jpn.*, **46**, 375 (1995).
27. R. N. Rhoda and R. F. Vines, U.S. Patent 3,486,928 (1969).
28. C. W. Walter and F. H. Leaman, U.S. Patent 3,562,911 (1971).
29. F. H. Leaman, *Plating*, **59**, 440 (1972).
30. E. Torigai, K. Takenaka, and Y. Kawami, Japanese Patent 84-80764 (1984).

21 Electroless Deposition of Gold

YUTAKA OKINAKA and MASARU KATO

INTRODUCTION

The first autocatalytic, electroless gold plating bath was developed in 1970 at Bell Laboratories [1] to plate thick, pure soft gold on semiconductors and circuit boards without employing an external source of electric current. The bath contained potassium cyanoaurate (I), $KAu(CN)_2$, as the source of gold, potassium borohydride, KBH_4, or dimethylamine borane, $(CH_3)_2NH \cdot BH_3$, as the reducing agent, potassium cyanide and potassium hydroxide. Typical bath compositions are given in Table 1. The original baths did not contain stabilizing additives, and they were sensitive to minute amounts of impurities, which tended to lead to spontaneous decomposition. Nevertheless, those baths were employed successfully with sufficient stability under carefully controlled conditions [2–5], yielding highly pure, soft gold useful for the purpose of bonding semiconductor devices [6]. Other successful applications documented in the open literature include (1) metallization of GaAs microwave field-effect transistors [7], (2) formation of ohmic contacts consisting of Pd/Sn/Au films on n-GaAs [8], (3) metallization of polyvinylidene fluoride (PVDF) films used for making piezoelectric devices [9], (4) deposition of a conducting layer on the interior surface of wave guide tubes made of an aluminum alloy [10], and (5) deposition of a gold layer on tungsten-metallized ceramics for packaging semiconductor devices [11].

More recently the accelerating trend of miniaturization and the need for greater reliability of electronic components have resulted in the development of new technologies for mounting devices on circuit boards, which in turn created new requirements for compatibility of the plating process with substrate materials and also for properties of gold films to be deposited on semiconductors and circuit boards. Quite generally, the method of electroplating has inherent advantages over that of electroless plating in terms of the ease of maintaining the bath and controlling the film thickness and properties. On the other hand, electroless plating has, in principle, a distinct advantage over electroplating in its capability of minimizing the number of processing steps when gold is needed in areas which are electrically isolated from each other. Under these circumstances, efforts have been made to improve the original borohydride and DMAB baths, especially in their stability and plating rate, and such improved baths are now commercially available. However, those baths have certain properties that are inherently incompatible with the requirements of the

Modern Electroplating, Fourth Edition, Edited by Mordechay Schlesinger and Milan Paunovic.
ISBN 0-471-16824-6

TABLE 1 Examples of Original Borohydride and DMAB Baths Containing No Additives

	Borohydride Bath	DMAB Bath
$KAu(CN)_2$	0.02 M	0.02 M
KCN	0.2	0.02
KOH	0.2	0.8
KBH_4	0.4	—
DMAB	—	0.4
Temperature	75°C	85°C
Plating rate	$0.7\,\mu m\,h^{-1}$	$0.4\,\mu m\,h^{-1}$

Source: Okinaka [1].

current technology. Thus baths with entirely different compositions have been developed in recent years to overcome those problems.

In this chapter the characteristics and problems of the conventional borohydride and DMAB baths will be summarized first, followed by a brief description of the improvements made to those baths. Then, the current status of development of new noncyanide baths will be reviewed.

1 CHARACTERISTICS OF BOROHYDRIDE AND DMAB BATHS

The reactions involved in the borohydride process were studied in detail by means of electrochemical techniques [4]. In the original study it was shown that the species that reduces $[Au(CN)_2]^-$ to Au is BH_3OH^-, which is an intermediate produced during the hydrolysis of borohydride ion, BH_4^-, rather than the latter ion itself:

$$BH_4^- + H_2O \longrightarrow BH_3OH^- + H_2 \tag{1}$$

$$BH_3OH^- + H_2O \longrightarrow BO_2^- + 3H_2 \tag{2}$$

$$BH_3OH^- + 3OH^- \longrightarrow BO_2^- + \left(\frac{3}{2}\right)H_2 + 2H_2O + 3e^- \tag{3}$$

$$[Au(CN)_2]^- + e^- \longrightarrow Au + 2CN^- \tag{4}$$

The overall reaction (5) is obtained by combining the two partial reactions (3) and (4):

$$BH_3OH^- + 3[Au(CN)_2]^- + 3OH^- \longrightarrow BO_2^- + \left(\frac{3}{2}\right)H_2 + 2H_2O + 3Au + 6CN^- \tag{5}$$

The anodic partial reaction (3) was written here under the assumption that this reaction, which was originally derived from a polarographic study at the dropping mercury electrode [12], is also operative on gold. Later Efimov et al. [13, 14] reported that another hydrolysis intermediate, $BH_2(OH)_2^-$, is also involved in the electroless plating process, and that all hydrogen gas produced during the plating results from the hydrolysis reactions and the plating reaction per se does not produce hydrogen.

Thus these authors proposed the following reactions instead of reaction (5) to describe the overall plating process:

$$6[Au(CN)_2]^- + BH_3OH^- + 6OH^- \longrightarrow 6Au + BO_2^- + 12CN^- + 5H_2O \qquad (6)$$

$$4[Au(CN)_2]^- + BH_2(OH)_2^- + 4OH^- \longrightarrow 4Au + BO_2^- + 8CN^- + 4H_2O \qquad (7)$$

Irrespective of the mechanisms of the actual plating reaction, a polarographic analysis of the kinetics of the hydrolysis reactions (1) and (2) has shown that the rate of reaction (2) is almost 500 times faster than that of reaction (1) under the conditions of electroless gold plating. Thus the utilization efficiency of borohydride as the reducing agent is very low, and most of the borohydride is lost by hydrolysis. The utilization efficiency obtained under typical plating conditions amounts to only 2% to 3%.

The hydrolysis reactions above are known to be acid-catalyzed, and therefore they proceed more rapidly at lower pHs. Accordingly, the electroless deposition rate of gold increases with decreasing KOH concentration. However, to prevent spontaneous decomposition, the KOH concentration of the bath must be kept greater than 0.1 M. The dependence of the plating rate on KOH concentration is illustrated in Figure 1.

If DMAB is used in place of borohydride, an opposite dependence is found; that is, the plating rate increases with KOH concentration (Fig. 2). The difference in the effect of KOH concentration on the plating rate between the two baths has been explained by assuming that the actual reducing species in the DMAB bath is also BH_3OH^-, and that this species is produced by the direct reaction of DMAB and OH^- ions:

$$(CH_3)_2NH \cdot BH_3 + OH^- \longrightarrow (CH_3)_2NH + BH_3OH^- \qquad (8)$$

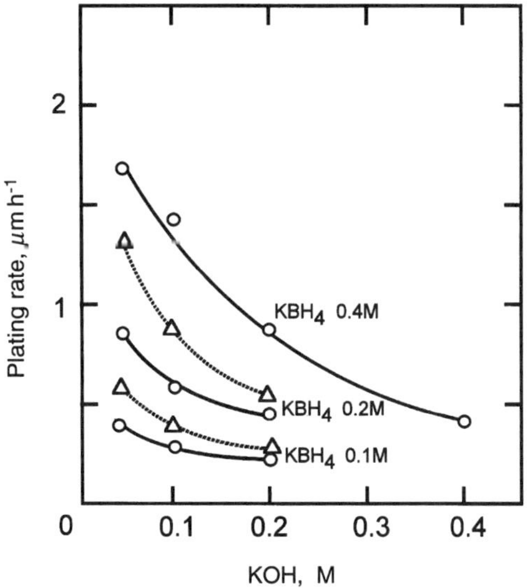

Figure 1 Effect of KOH concentration on plating rate of borohydride bath: 0.02 M $KAu(CN)_2$, 0.2 M KCN (solid line), 0.1 M KCN (broken line), 80°C. (Okinaka [1])

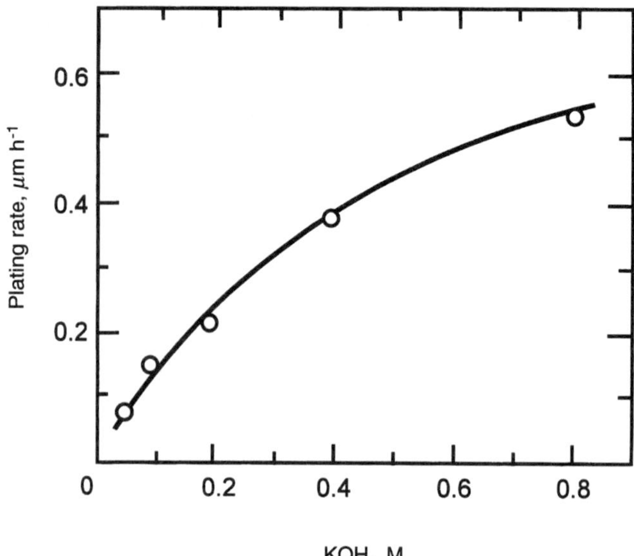

Figure 2 Effect of KOH concentration on plating rate of DMAB bath: 0.02 MKAu$(CN)_2$, 0.02 M KCN, 0.4 M DMAB, 75°C. (Okinaka [1])

Another unique feature of the borohydride system is found in the dependence of plating rate on the concentration of $[Au(CN)_2]^-$; that is, the plot of plating rate against gold concentration exhibits a maximum at approximately 0.003 M. This unusual dependence is attributed to competitive adsorption of BH_3OH^- and a gold-containing species for the same site on the gold surface. Namely the gold species is apparently more strongly adsorbed on gold than the reducing species, BH_3OH^-, thereby acting as a catalytic poison for the oxidation of BH_3OH^-. The adsorbed gold species is actually believed to be AuCN, which has been proposed as an intermediate formed during the reduction of $[Au(CN)_2]^-$ [15, 16]:

$$[Au(CN)_2]^- \longrightarrow AuCN_{ads} + CN^- \tag{9}$$

$$AuCN_{ads} + e^- \longrightarrow Au + CN^- \tag{10}$$

The adsorbed AuCN has recently been positively identified on a single crystal of Au(111), and the structure of the adsorbed layer investigated using low-energy electron diffraction (LEED), Auger electron spectroscopy (AES), and in situ electrochemical scanning tunneling microscopy (ESTM) [17].

The electroless gold deposition from the borohydride and DMAB systems takes place on noble metals such as Pd, Rh, Ag, and Au itself, as well as on base metals such as Cu, Ni, Co, Fe, and their alloys. On noble metals the reaction is catalytic from the very beginning. On the other hand, the gold deposition on the base metals is initiated by galvanic displacement, which results in accumulation of ions of those base metals in the bath. No adverse effect occurs with copper, whereas the introduction of ions of Ni, Co, and Fe into the solution is highly detrimental. Further details of such impurity effects will be described in the subsequent section.

2 PRACTICAL PROBLEMS ASSOCIATED WITH ORIGINAL BOROHYDRIDE AND DMAB BATHS

From the standpoint of practical applications, these basic baths have the following problems:

1. The plating rate is low.
2. The bath stability is insuffiient.
3. The baths are sensitive to contamination with minute amounts of nickel ions.
4. A high pH is required.
5. The baths contain free cyanide.

Much effort has been expended by many investigators to improve on items 1 through 3. On the other hand, items 4 and 5 are inherent to the baths, and they cannot be changed. Both the high alkalinity and the use of cyanide make the baths incompatible with positive photoresists commonly used to delineate patterns on semiconductor devices and circuit boards.

3 IMPROVED BOROHYDRIDE AND DMAB BATHS

3.1 Methods of Increasing the Plating Rate

To obtain a plating rate as large as possible simply by adjusting basic bath compositions and operating conditions without resorting to the special methods described below, the following approaches are useful: (1) increase the agitation speed of the bath; (2) raise the bath temperature; (3) lower the free cyanide concentration; (4) decrease the free alkali hydroxide concentration for the borohydride bath, or increase it for the DMAB bath; (5) increase the reducing agent concentration. However, an attempt to obtain a plating rate greater than $2\,\mu m\,h^{-1}$ by these simple approaches will be futile, because it will lead to spontaneous bath decomposition. The following methods are known to be useful for formulating high-speed baths.

1. Adding a Depolarizer It is well known that the addition of a small amount of Pb^{2+} or Tl^{+} depolarizes the gold electrode in the electrolytic soft gold plating bath [18]; namely these ions shift the gold deposition potential in the positive direction. These ions are known to undergo the so-called underpotential deposition (UPD) on gold, in which the strongly adsorbed metal atoms form on gold at potentials considerably less negative than corresponding equilibrium potentials. The depolarization effect of these UPD ions has been explained by assuming that the adsorbed metal atoms formed by the UPD mechanism chemically react with $Au(CN)_2^-$ by galvanic displacement, which regenerates the original metal ions. Thus the UPD ions act as a catalyst for the reduction of the gold species:

$$M^{n+} + ne^- \longrightarrow M^0_{UPD} \tag{11}$$

$$M^0_{UPD} + n[Au(CN)_2]^- \longrightarrow M^{n+} + nAu + 2nCN^- \tag{12}$$

with the overall reaction being

$$[Au(CN)_2]^- + e^- \longrightarrow Au + 2CN^- \quad (13)$$

When ions of such UPD metals are added to the borohydride or DMAB bath, the reduction potential of $[Au(CN)_2]^-$ shifts in the positive direction, and therefore the plating rate increases.

Matsuoka et al. [19] added $PbCl_2$ or TlCl to a borohydride bath containing $KAu^{III}(CN)_4$ instead of $KAu^{I}(CN)_2$, and they obtained a plating rate 8 to 9 times faster than that obtained in the absence of the additives. Since the $[Au(CN)_4]^-$ ion in this bath is reduced by the borohydride to $[Au(CN)_2]^-$ ion, the plating reaction is basically the same as that of the conventional borohydride bath. However, because the reduction of $[Au(CN)_4]^-$ to $[Au(CN)_2]^-$ produces free CN^- ions, there is no need to add CN^- at the time of bath makeup, which is claimed to be a practical advantage of this system. The composition of the bath proposed by the above authors is given in Table 2. The depolarizing effect of Pb^{2+} on the reduction of $[Au(CN)_2]^-$ is demonstrated in Fig. 3, and the remarkable accelerating effects of Pb^{2+} and Tl^+ are shown in Fig. 4.

2. Adding a Stabilizer Some attempts have been made in the past to increase the plating rate by adding an organic compound which poisons the catalytic activity of gold, while raising the plating temperature or increasing the concentration of the reducing agent to compensate for the loss of plating rate caused by the poisoning effect. For example, a bath stabilized in this manner was operated at a temperature as high as 90°C to obtain 12 to 23 $\mu m\,h^{-1}$ [20]. However, it should be realized that at such a high temperature, the hydrolysis reaction of the reducing agent is greatly accelerated to make the control of bath composition difficult. Thus, for all practical purposes this approach cannot be recommended.

On the other hand, Homma [21] successfully increased the plating rate to 5 $\mu m\,h^{-1}$ without causing bath instability by adding a trace of *p*-dimethylaminobenzylidene rhodanine as a stabilizer while doubling the borohydride concentration. Ott [22] also found that compounds such as ethylene glycol monoethylether, diethyleneglycol monoethylether, and polyethyleneimine also serve as effective stabilizers for the same purpose.

3. Using Cyanoaurate(III) instead of Cyanoaurate(I) Besides the cyanoaurate(III) bath of Matsuoka et al. mentioned above, El-Shazly and Baker [23] also developed a bath containing $KAu^{III}(CN)_4$ as the source of gold and achieved a plating rate ranging from 2 to 8 $\mu m\,h^{-1}$ without any additives. A distinct advantage of using the

TABLE 2 Borohydride Bath with Cyanoaurate (III) Instead of Cyanoaurate (I)

$KAu(CN)_4$	3 g $liter^{-1}$
KOH	11.2
KBH_4	3
$PbCl_2$	0.5 mg $liter^{-1}$
Temperature	70°C

Source: Matsuoka et al. [19].

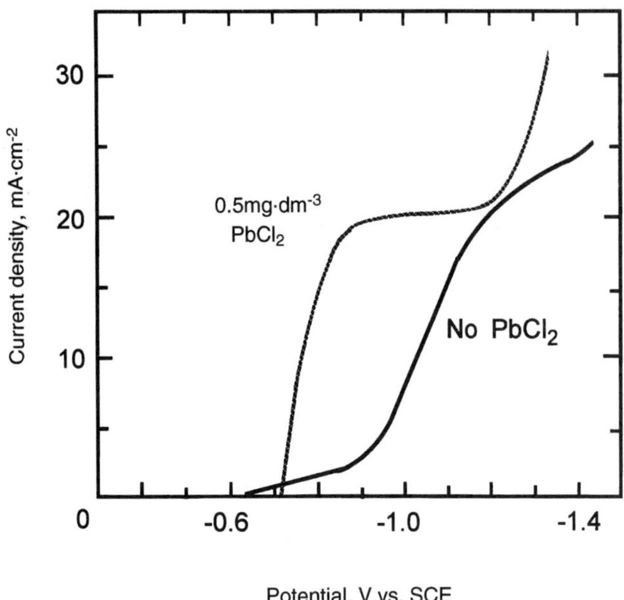

Figure 3 Effect of $PbCl_2$ addition on polarization curve for the reduction of $Au(CN)_2^-$. (Matsuoka et al. [19])

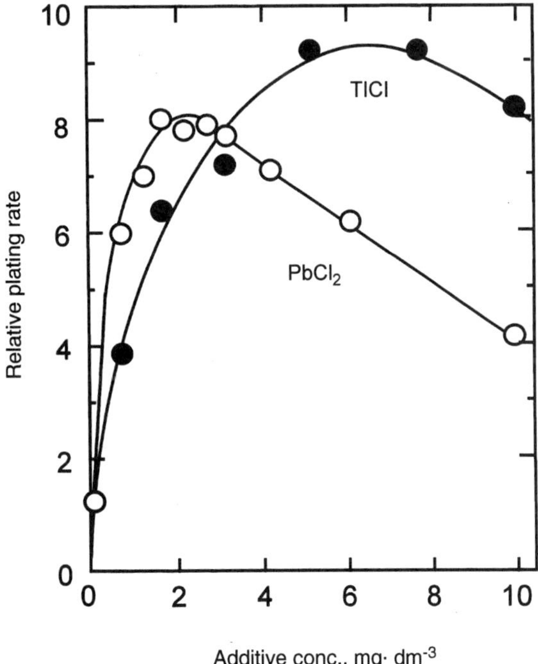

Figure 4 Effect of $PbCl_2$ and TlCl concentrations on plating rate of borohydride bath. (Matsuoka et al. [19])

Au(III) compound is that it allows the use of $KAuO_2$ or $KAu(OH)_4$ instead of the cyanide compound for the replenishment of gold, preventing the accumulation of free CN^- which is known to lower the plating rate. Corresponding Au(I) compounds do not exist in a stable form. Thus the Au(III) bath should be advantageous for maintaining the plating rate at a constant level for an extended period of bath usage.

Ohtsuka et al. [24] made a detailed investigation of their DMAB bath containing cyanoaurate(III) as the source of gold. They showed that the Au(III) species is reduced immediately by borohydride to Au(I) at the plating temperature of 70° to 75°C, and therefore the ordinary Au(I) cyanide salt, $KAu(CN)_2$, can be used for initial bath makeup while a solution prepared by dissolving $Au^{III}(OH)_3$ in dilute KOH is used for replenishment.

Simon [25] described the use of an "accelerator" to suppress the deleterious effect of the accumulation of free CN^- in the DMAB bath. The chemical composition of the accelerator has not been disclosed, however.

4. Adding an Accelerator for Anodic Oxidation of the Reducing Agent If the anodic oxidation of the reducing agent can be accelerated by adding a suitable catalyst in the bath, then the potential for this reaction should shift in the negative direction. From the general principle of autocatalytic metal deposition, it follows that the negative shift of the potential for the oxidation of reducing agent should increase the plating rate. Iacovangelo [26] found that carbonate ions and triethanolamine accelerate the oxidation of DMAB. The mechanism of the accelerating effect has not been clarified to the author's knowledge.

3.2 Methods of Improving Bath Stability

Quite generally, conditions favoring an increase in plating rate decrease bath stability in all autocatalytic systems, except when the rate increase under mass transport limiting conditions is brought about by raising the degree of bath agitation or the speed of substrate movement. The bath instability resulting from the use of conditions providing an excessive activity of the bath may be brought under control by adding a stabilizer such as one of the compounds mentioned in Section 3.1.

On the other hand, it should be realized that bath instability can be triggered by the introduction of certain impurities. As briefly touched upon in Section 1, the borohydride and DMAB baths are extremely sensitive to contamination with Ni^{2+}, Co^{2+}, or Fe^{2+}, while they are unaffected by Cu^{2+}. Figure 5 compares the effect of the addition of Ni^{2+} on the plating rate with that of Cu^{2+}, where the plating rate in the absence of the contaminating ions was the same as the constant rate obtained at the various concentrations of Cu^{2+}. It is seen that the presence of only 10^{-5} M of Ni^{2+} causes the plating rate to decline, and bath decomposition sets in at 10^{-3} M. Ions of Co^{2+} and Fe^{2+} exert similar effects. An electrochemical study of the effect of Ni^{2+} showed that the anodic oxidation of the reducing species, BH_3OH^-, is greatly inhibited by the preferential adsorption of $[Ni(CN)_4]^{2-}$ on the surface of gold [4]. As a method of preventing bath decomposition resulting from the presence of ions of the transition metals, Ali and Christie [27] added EDTA and ethanolamine, which form highly stable complex ions with the metals and suppress their reactions with the reducing agent (Table 3).

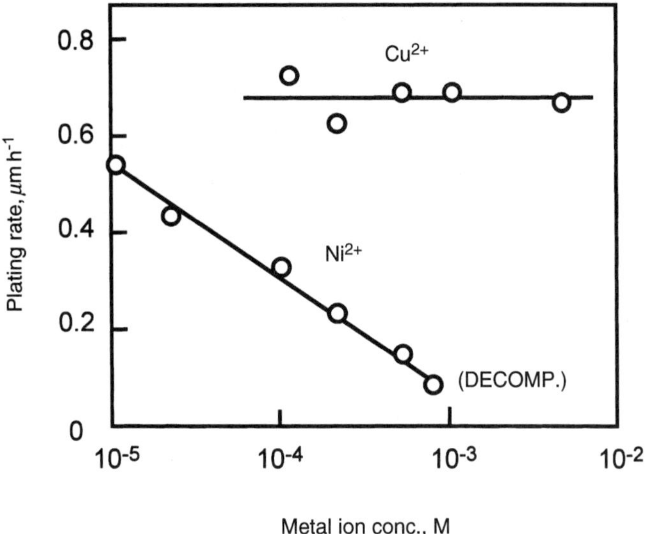

Figure 5 Effect of bath contamination with Ni^{2+} and Cu^{2+} ions on plating rate of borohydride bath. (Okinaka et al. [5])

TABLE 3 Borohydride Bath with EDTA and Ethanolamine

$KAu(CN)_2$	0.005 M
KCN	0.17
KOH	0.2
KBH_4	0.2
Na_2EDTA	5 g liter^{-1}
Monoethanolamine	50 cm^3 liter^{-1}
Temperature	72°C
Plating rate	1.5 μm h^{-1}

Source: Ali and Christie [27].

Decomposition of the borohydride and DMAB baths can also occur when they are contaminated with certain organics such as polyethylene [5]. One should be aware of possibilities of contamination with trace amounts of organic materials from storage bottles and vessels as well as from the water used to prepare the solutions.

3.3 Methods of Plating on Nickel Substrates

For the reasons described above, the borohydride or DMAB bath is not suitable for plating gold directly on nickel metal. It is a general practice to coat the nickel substrate with a thin layer of immersion (galvanic displacement) gold prior to plating electroless gold. However, because a complete coverage of the nickel substrate with immersion gold is virtually impossible, and because the bath is highly sensitive to an extremely small amount of dissolved nickel ions, the above approach cannot be expected to provide completely satisfactory solution to the nickel contamination problem.

TABLE 4 Bath with Two Reducing Agents (DMAB and Hydrazine)

$KAu(CN)_2$	0.005 M
KCN	0.035
KOH	0.8
K_2CO_3	0.45
Pb acetate	15 ppm
DMAB	0.05 M
N_2H_4	0.25
Temperature	80°C
Plating rate	
Initial (on Ni)	2.6 μm h^{-1}
On Au	7.8 μm h^{-1}

Source: Iacovangelo [26].

More recently Iacovangelo [26] proposed a new method of plating electroless gold on nickel without using displacement gold. The bath used in this method contains two different reducing agents: DMAB and hydrazine. The bath composition is given in Table 4. Lead acetate and potassium carbonate are added to accelerate partial cathodic and partial anodic reactions, respectively, leading to an increased plating rate. Upon immersion of a metallic nickel substrate in this bath, the initial deposition of gold takes place via the catalytic oxidation of hydrazine on the nickel coupled with the reduction of Au(I) to Au. After the nickel is completely covered with gold, DMAB serves as the sole reducing agent to deposit gold continuously on gold via the autocatalytic mechanism. In this method the dissolution of nickel by galvanic displacement can be virtually eliminated by selecting a proper bath composition. From the anodic polarization curves for the oxidation of hydrazine and DMAB on Ni and Au electrodes, Iacovangelo showed that in the potential range of interest, hydrazine is oxidized on Ni but not on Au (see Fig. 6), whereas DMAB is oxidized nearly 100 times faster on Au than on Ni (Fig. 7). The idea used in formulating this bath is based on the clever utilization of the difference in catalytic activity between Ni and Au for the oxidation of the two reducing agents. This concept should be of general virtue when substrate metal needs to be protected from its galvanic displacement reaction with a bath component.

In another development Iacovangelo and Zarnoch [28] found that the substrate-catalyzed deposition of gold on nickel with hydrazine as the sole reducing agent can be made to proceed to obtain a gold layer which is sufficiently thick for the purpose of bonding electronic devices and components. They demonstrated that the maximum thickness of gold obtained by the substrate-catalyzed reaction depends on the concentration of free CN^- ions in their bath (Fig. 8). Thus the maximum thickness of gold obtained from this bath is determined by the free CN^- concentration rather than the plating time as in conventional autocatalytic processes.

4 CYANIDE BATHS WITH OTHER REDUCING AGENTS

There are many descriptions, mostly in the patent literature, of cyanide-based electroless gold plating baths containing reducing agents other than borohydride or

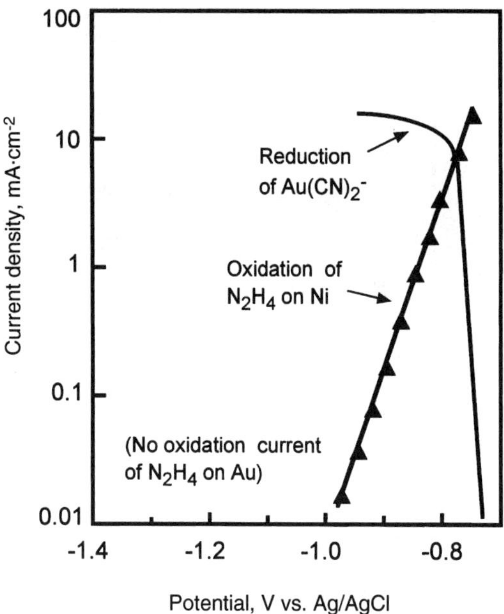

Figure 6 Polarization curves for the reduction of $Au(CN)_2^-$ and for the oxidation of hydrazine on nickel electrode: 0.8 M KOH, 0.035 M KCN, 0.05 M N_2H_4, 80°C. (Iacovangelo [26])

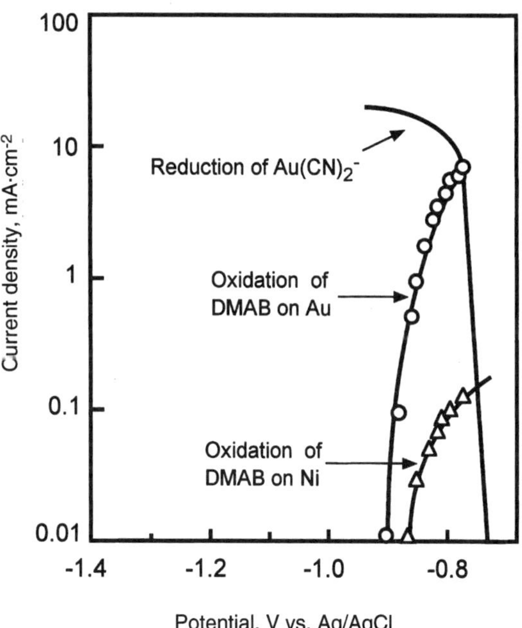

Figure 7 Polarization curves for the reduction of $Au(CN)_2^-$ and for the oxidation of DMAB on nickel and gold electrodes. (Iacovangelo [20])

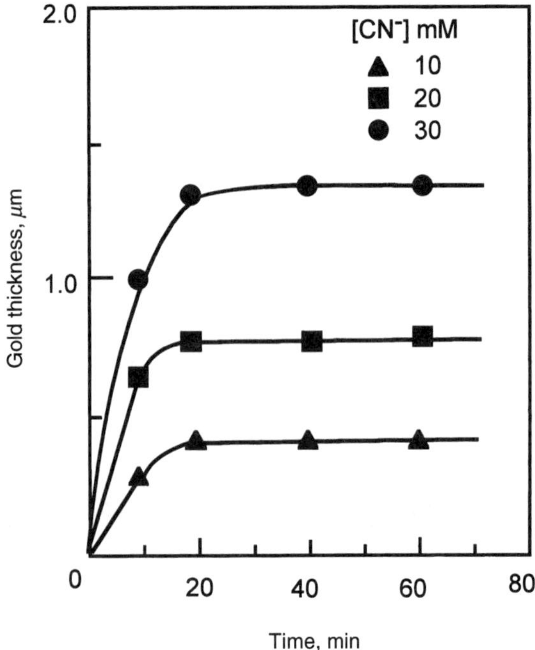

Figure 8 Gold plating thickness vs. time curves at various cyanide concentrations for substrate-catalyzed plating with hydrazine as reducing agent. (Iacovangelo [28])

TABLE 5 Cyanide Baths with Reducing Agents Other Than Borohydride or DMAB

Reducing Agent	pH	Temperature (°C)	References
Hypophosphite	7–7.5	93	Swan-Gostin [31]
	7.5–13.5	96	Brookshire [32]
	3–4	70–80	Ezawa-Ito [33]
Hydrazine	7–7.5	92–95	Gostin-Swan [34]
	5.8–5.9	95	Moskvichev et al. [35, 36]
	11.7	95	Wundt et al. [37]
Hydroxylamine	2.3–2.8	70–85	Dettke-Stein [38]
Cyanoborohydride	3.5	90	Bellis [39]
Hydrazinoborane	> 13	60	Efimov-Gerish [40]
Thiourea	6.5–7.0	83–90	Oda-Hayashi [41]
	4.2	80–85	Gesemann et al. [42]
Ascorbic acid	8	63	Andrascek et al. [43]
Titanium trichloride	5	75	Nishiyama et al. [44]

DMAB. One of the present authors already published extensive reviews of such baths [29, 30], and they will not be described in detail in this chapter. Table 5 lists those baths only with reducing agents used, pH, temperature, and references. It should be noted that the list includes baths operated at low pH values (e.g., those with hydroxylamine, titanium (III) trichloride, or thiourea) and those which do not generate gaseous hydrogen (e.g., the bath with titanium (III) trichloride). Because of such unusual features, those baths may find applications with special requirements. It should be cautioned, however, that some baths included in the list have been shown to be nonautocatalytic, and hence the maximum gold thickness that can be obtained with those baths is rather limited.

5 NONCYANIDE BATHS

The use of cyanide is undesirable for a few reasons other than its toxicity. In applications requiring gold deposition on a substrate with a circuit pattern delineated with a conventional positive photoresist, cyanide ions are known to attack the interface between the photoresist and the substrate, lifting the photoresist and causing extraneous gold deposition under the photoresist. The high alkalinity of the cyanide-based baths is also incompatible with the positive photoresist. In the borohydride or DMAB bath, the accumulation of cyanide ions with bath usage lowers the plating rate. For these reasons noncyanide baths are currently in demand.

Gold salts other than cyanide complexes used for formulating electroless gold baths include chloroaurate (III), Au(I)-sulfite complex, Au(I)-thiosulfate complex, and Au(I)-thiomalate. The noncyanide electroless gold baths developed prior to 1980 have been reviewed by Mathe [45] and Ganu et al. [46]; the reader is referred to these works. More recently a large number of patents and papers have appeared on electroless gold baths based on Au(I)-sulfite and/or Au(I)-thiosulfate complex and operated at near-neutral pH. Those baths have been studied in detail for the reaction mechanism, process control methodology with replenishment of bath constituents, and deposit characteristics. Here only the noncyanide baths developed and investigated after 1980 will be reviewed.

5.1 Baths with Au(I)-Sulfite Complex

Gold(I)-sulfite complex has been in use for many years as the source of gold in commercial noncyanide, *electrolytic* gold plating baths. Those baths contain proprietary stabilizing agent(s) because Au(I)-sulfite complex undergoes a disproportionation reaction in aqueous solution and decomposes spontaneously on standing to produce Au(III) and metallic Au(0).

For electroless gold plating, Richter et al. and Gesemann et al. [47–50] developed baths containing 1,2-diaminoethane and KBr as stabilizers in combination with hypophosphite, formaldehyde, hydrazine, borohydride, or DMAB as the reducing agent. Also described by Richter et al. [51] is a sulfite-based bath containing ethylenediamine or EDTA as the stabilizing agent, formaldehyde or hydrazine derivatives as the reducing agent, and an arsenic compound, which is reported to yield hard gold deposits.

Sato et al. [52], using Au(I)-sulfite-ethylenediamine complex as the source of gold, investigated the possibility of employing thiourea or its derivatives as the reducing agent. They found that the performance of the bath with methyl thiourea or acetyl thiourea was strongly pH dependent, and that the deposit obtained from the former bath was porous, whereas good deposits were obtained from the bath containing both thiourea and methyl thiourea at the concentration ratio of three.

Shindo and Honma [53] found that the stability of the sulfite bath can be improved by adding triethanolamine, nitrilo-triacetic acid, or sodium thiosulfate.

Using ascorbic acid as the reducing agent, Kato et al. [54, 55] had previously shown that Au(I)-sulfite complex can be reduced to gold metal autocatalytically, but the deposition rate is very low. They found that the deposition rate can be increased greatly by adding thiosulfate as the second complexing agent. The mixed ligand system will be described further in Section 5.3. Baths containing thiosulfate as the sole ligand will be described first in the following section.

5.2 Baths with Au(I)-Thiosulfate Complex

The Au(I)-thiosulfate complex, $[Au(S_2O_3)_2]^{3-}$, has been known since the 1910s. This complex ion has a large stability constant, 10^{26} [56], ranking next to that of $[Au(CN)_2^-]$, which is equal to 10^{39}. For comparison, the stability constant of the sulfite complex, $[Au(SO_3)_2]^{3-}$, is only 10^{10} [57]. It is thus natural that the use of the thiosulfate complex had been proposed for formulating electrolytic gold plating baths [58]. In contrast, the use of the thiosulfate complex for electroless gold plating is only a recent development. Sullivan et al. [59] made an extensive cyclic voltammetric study of the cathodic reduction of $[Au(S_2O_3)_2]^{3-}$ and the anodic oxidation of various reducing agents. They showed that among a number of reducing agents investigated, ascorbic acid is the only compound that is electrocatalytically oxidized at a reasonable rate within the potential range of the reduction of $[Au(S_2O_3)_2]^{3-}$ at a pH between 6.4 and 9.2 and at room temperature. They thus formulated the electroless gold plating bath with the composition listed in Table 6. An interesting observation made by the above authors is that the gold deposition rate is greater in air than in nitrogen atmosphere. This effect is attributed to the formation of H_2O_2 as a by-product of air oxidation of ascorbic acid, which reacts with excess free $S_2O_3^{2-}$ produced as a result of the reduction of the Au(I) thiosulfate complex. The accumulation of free $S_2O_3^{2-}$ ions has been shown to decelerate the gold deposition. It has also been shown that the air oxidation of ascorbic acid is catalyzed by the presence of $[Au(S_2O_3)_2]^{3-}$ ions. Sullivan and Kohl [60] extended the above study further to obtain an understanding of the reactions involved in this plating system by using a rotating ring-disk electrode. They showed that hydrogen peroxide does not significantly react with the Au(I) thiosulfate complex or ascorbic acid but reacts with free $S_2O_3^{2-}$ ions to form trithionate ($S_3O_6^{2-}$) and sulfate (SO_4^{2-}) ions. Utilizing the latter reaction to suppress the accumulation of free thiosulfate ions during the plating, they recommend periodic addition of H_2O_2 into the bath to maintain a constant plating rate. It should be noted that because this bath does not contain any stabilizing additive, the bath life is relatively short ($< 2\,h$), which needs to be improved for practical purposes.

TABLE 6 Au(I) Thiosulfate-Ascorbic Acid Bath

$Na_3\ Au(S_2O_3)_2$	0.03 M
Na L-ascorbate	0.05
Citric acid	0.4
pH (KOH)	6.4
Temperature	30°C
Plating rate	$0.76\ \mu m\ h^{-1}$

Source: Sullivan, Patel, and Kohl [59].

5.3 Baths Containing Both Sulfite and Thiosulfate

The noncyanide baths described above containing either sulfite or thiosulfate as a sole complexing agent appear to be of limited use because of insufficient stability of the systems. Baths containing both sulfite and thiosulfate are more stable, and those containing thiourea, ascorbic acid, hypophosphite, or hydrazine as the reducing agent have been developed. More recently baths containing both sulfite and thiosulfate but no additional reducing agent have been patented. Below, those baths will be described separately.

*1. **Thiourea Bath*** The thiourea bath was developed and subsequently improved by a group of investigators at Hitachi, Ltd. [61–73]. An improved version of the bath composition and operating conditions is shown in Table 7 [69]. It is seen that the bath is operated under a set of mild conditions. In this system thiourea has been shown to undergo complex chemical reactions through the formation of a radical intermediate, $(NH)(NH_2)CS\cdot$, to form final products including urea, a major product, and dicyandiamide [69]. This radical intermediate is believed to react with dissolved oxygen in the bath to form formamidine sulfinic acid, $(NH_2)_2CSO_2$, which appears to be responsible for bath instability. Hydroquinone appearing in Table 7 as an additional component of the bath reacts quickly with the radical intermediate before it produces the undesirable compound. It is also significant that the reaction between hydroquinone and the radical intermediate regenerates thiourea [72]. Thus hydroquinone acts as a stabilizer as well as a recycling agent for thiourea.

*2. **Ascorbic Acid Bath*** One of the present authors (M.K.) and his coworkers developed the ascorbic acid bath, which is operated at a near-neutral pH and at a mildly elevated temperature. The reason why ascorbic acid was chosen among a large number of reducing agents tested was explained based on the comparison of anodic polarization curves at a gold electrode. A similar approach was used by Sullivan et al. [59] in selecting ascorbic acid as the reducing agent for their room temperature, pure thiosulfate bath described in the preceding section. The original bath [54, 55] has since been improved considerably in terms of bath stability, plating rate, and substrate compatibility [74–77], and a comprehensive review of the improvements has been presented recently [76]. An example of the improved bath composition and operating conditions is given in Table 8.

The improvements made were primarily in bath stability, selectivity in plating patterned substrates, and in plating rate. The addition of a minute amount of a

TABLE 7 Sulfite-Thiosulfate-Thiourea Bath

$NaAuCl_4$	0.005–0.025 M
Na_2SO_3	0.04–0.2
$Na_2S_2O_3$	0.2–0.6
$Na_2B_4O_7$	0.066–0.13
Thiourea	0.0013–0.013
Hydroquinone	0.0018–0.018
pH	7.5–8.5
Temperature	60–90°C

Source: Inoue et al. [69].

TABLE 8 Sulfite-Thiosulfate-Ascorbic Acid Bath

$NaAuCl_4$	0.01 M
Na_2SO_3	0.08–0.32
$Na_2S_2O_3$	0.08–0.32
Na_2HPO_4	0.05–0.20
Na L-ascorbate	0.05–0.20
2-Mercaptobenzothiazol	Trace
pH	7.5
Temperature	60°C

Source: Kato, Yazawa, and Okinaka [76].

heterocyclic mercapto compound such as 2-mercaptobenzothiazole (MBT), 2-mercaptobenzoimidazole (MBI), and 6-ethoxy-2-mercaptobenzothiazole (EMBT) stabilizes the bath considerably while maintaining the plating rate at an acceptable level. The bath life of only 3 hours in the absence of stabilizers can be increased to more than 35 hours by the addition of a suitable amount of these stabilizers [76]. It has also been shown that increasing the sulfite concentration improves bath stability, and that the addition of excess sulfite helps prevent extraneous gold deposition from occurring on the surface of ceramic substrates with a circuit pattern delineated by using a conventional technique.

The plating rate of the improved bath containing a stabilizer such as MBT is of the order of $1\,\mu m\,h^{-1}$. It has been shown that certain additives increase the plating rate significantly. For example, the addition of 0.005 M of ethylenediamine increases the rate by a factor of three. The thallous ion is also an effective accelerator. The addition of 1 ppm of Tl^+ doubles the rate. The acceleration effect of Tl^+ has been shown to be due to its depolarization effect on the partial cathodic reaction brought about by the underpotential deposition (UPD) of Tl on gold. The effect is similar to what is known for the cyanide system with borohydride as the reducing agent.

More recently Honma and coworkers [78, 79] investigated the ascorbic acid bath containing both sulfite and thiosulfate as complexing agents. They found that the addition of nitrilotriacetic acid (NTA) improves the stability of the Au(I) sulfite complex through the formation of a complex which is more stable with respect to the disproportionation of Au^+. However, the bath stability was still insuficient, and effects of other additives were examined. They reported that $K_4Fe(CN)_6$, $K_2Ni(CN)_4$, 2,2′-bipyridyl, and cupferron were effective in the concentration range of 0.1 to 100 ppm. It was found that the bond strength between gold wire and the deposit was much greater when the bath with $K_2Ni(CN)_4$ or cupferron was used than that with 2,2′-bipyridyl. This difference was attributed, not to impurity content or deposit surface morphology, but to crystal orientation. Namely the deposits with good bondability had (220) and (311) preferred orientations. The above authors also found that aeration helps to stabilize the bath by suppressing the disproportionation reaction of Au^+. However, it should be noted that dissolved oxygen will react with free sulfite ions to form sulfate ions, which must be taken into consideration in controlling the concentrations of bath constituents. Another significant finding reported is that the addition of hydrazine as a second reducing agent increases the plating rate by a factor of up to two. Plating rates up to approximately $1.7\,\mu m\,h^{-1}$ were achieved at pH 6 and 60°C. (See the section below for hydrazine bath.)

*3. **Hypophosphite Bath*** Paunovic and Sambucetti [80] made an extensive investigation of non-cyanide electroless gold plating systems on the basis of polarization measurements. The results of their study using sulfite alone, thiosulfate alone, their mixture, phosphate, and pyrophosphate as complexing agents and hypophosphite, formaldehyde, DMAB, and TMAB (trimethylamine borane) as reducing agents, showed that the sulfite-thiosulfate mixed ligand system combined with hypophosphite gives the most satisfactoty performance in terms of bath stability and plating rate. Their baths were prepared by mixing equal parts of two solutions, *A* and *B*, followed by the addition of a reducing agent. The compositions of the two solutions were as follows: Solution *A* consisting of 0.005 M $NaAuCl_4$, 0.16 M boric acid, NaOH to adjust pH, and solution *B* consisting of 0.1 M each of Na_2SO_3 and $Na_2S_2O_3$, 0.16 M boric acid, and NaOH to adjust pH. The concentration of Na_2HPO_2 in the bath was 0.075 M. The above authors found that the addition of citrate at the concentration of 0.5 M increases the plating rate considerably. They state that citrate appears to act as an additional reducing agent. The plating rate achieved at pH 7.5 and 70°C was 0.9 $\mu m\,h^{-1}$. The bath stability was about 10 hours, but the authors state that it can be extended for much longer periods of time by the addition of a stabilizer such as SCN^-. The RBS and Auger spectrometric analyses of the gold deposit showed that sulfur is not incorporated in the deposit. Microprobe analysis showed none or less than 200 ppm of sulfur. Good bondability of gold wire to the deposit was confirmed.

*4. **Hydrazine Bath*** Shiokawa et al. [81] used hydrazine as the reducing agent in their sulfite-thiosulfate bath. An Au(I)-sulfite complex salt was used as the source of gold. They found that the presence of both sulfite and thiosulfate is necessary for bath stability. The following compounds were also incorporated for the reasons stated: EDTA or a similar strong complexing agent to mask metallic impurities such as Cu^{2+} and Ni^{2+}; an amine such as triethanolamine, ethylenediamine, or dimethylamine to suppress the formation of "gas pits"; a quinoline derivative such as 2-chloroquinoline, 2-chloromethylquinoline, etc., to prevent extraneous deposition on substrates patterned with photoresist; a salt of hydroxy polycarboxylic acid such as sodium or potassium citrate or tartrate which helps to stabilize the bath; and a minute amount of As^{3+}, Tl^+, or Pb^{2+} to improve plating uniformity. The bath is operated at pH 6 to 7 and 50° to 75°C, and it yields a plating rate in the range of 0.4 to 1.5 $\mu m\,h^{-1}$. The bath was operated successfully with replenishment up to five turnovers. An example of bath composition and operating conditions given in the patent disclosure is as follows: $(NH_4)_3Au(SO_3)_2$ 4 g liter^{-1}, Na_2EDTA 100 g liter^{-1}, dimethylamine 3 g liter^{-1}, $Na_2S_2O_3$ 10 g liter^{-1}, 2-chloromethylquinoline 10 mg liter^{-1}, K-tartrate 30 g liter^{-1}, Tl_2SO_4 3 mg liter^{-1}, hydrazine monohydrochloride 25 g liter^{-1}; pH 6.5; and temperature 60°C. This bath plated 0.93 μm in one hour.

In a later investigation by the same inventors, it was found that when the hydrazine bath was used to plate gold on copper clad printed circuit boards with fine circuit patterns, the copper tended to dissolve into the bath, causing abnormal gold deposition in the areas where copper was attacked. The accumulation of dissolved copper also caused the bath to decompose. To prevent these phenomena from occurring, the addition of benzotriazole (or its derivatives), a well-known corrosion inhibitor for copper metal, at a concentration in the range of 3 to 10 g liter^{-1} was found to be effective [82].

5. Reaction Mechanisms To prepare the baths described above, either a salt of Au(I) sulfite complex or sodium chloroaurate(III) can be used as the source of gold. When a solution of the latter compound is mixed with that containing sulfite and/or thiosulfate, Au(III) is reduced to form Au(I) complexes. Two Au(I) complex species can form according to the following reactions:

$$\mathrm{Au(III)} + \mathrm{SO_3^{2-}} + 2\mathrm{S_2O_3^{2-}} + \mathrm{H_2O} \Leftrightarrow [\mathrm{Au(S_2O_3)_2}]^{3-} + \mathrm{SO_4^{2-}} + 2\mathrm{H^+} \quad (14)$$

$$\mathrm{Au(III)} + 3\mathrm{SO_3^{2-}} + \mathrm{H_2O} \Leftrightarrow [\mathrm{Au(SO_3)_2}]^{3-} + \mathrm{SO_4^{2-}} + 2\mathrm{H^+} \quad (15)$$

Thus both Au(I) complex species can be present in the bath.

Kato et al. [55] determined cathodic polarization curves for the deposition of gold in solutions containing, respectively, sulfite alone, thiosulfate alone, and both sulfite and thiosulfate to obtain information as to which Au(I) complex is reduced from the mixed ligand system. The polarization curves obtained with equal concentrations of sulfite and thiosulfate (0.1 or 0.4 M) are shown in Figure 9. It is seen that the addition of thiosulfate ions to the solution of Au(I) containing sulfite alone shifts the

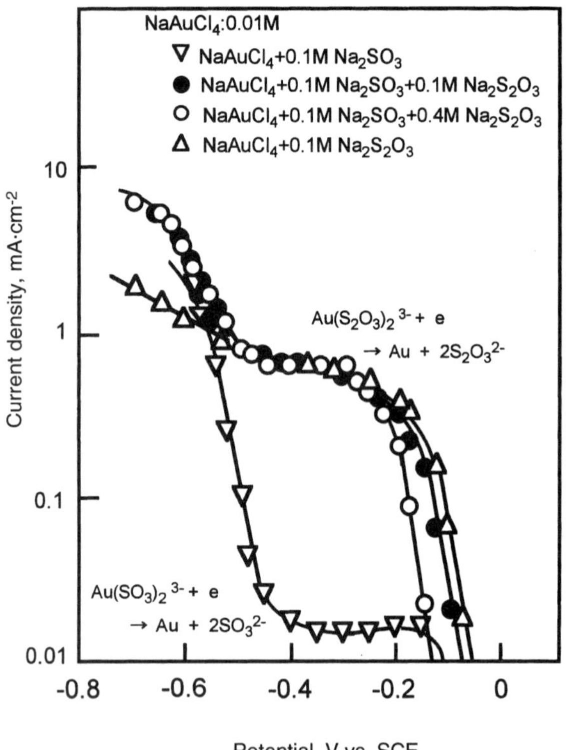

Figure 9 Polarization curves for gold deposition from various mixtures of $NaAuCl_4$, Na_2SO_3, and $Na_2S_2O_3$. Concentrations of SO_3^{2-} and $S_2O_3^{2-}$ were equal to each other in mixtures containing both anions. (Kato et al. [55])

polarization curve for the reduction of the Au(I)-sulfite complex to the potential range where the reduction of the Au(I)-thiosulfate complex takes place. This result indicates that the gold deposition takes place from the Au(I)-thiosulfate complex. This conclusion seemed reasonable in view of the fact that $[Au(S_2O_3)_2]^{3-}$ is much more stable than $[Au(SO_3)]_2^{3-}$ as indicated by their stability constants mentioned in the preceding section. In a later study [76], however, it was found that when sulfite was added in large excess over thiosulfate, such as 0.32 M of SO_3^{2-} and 0.08 M of $S_2O_3^{2-}$, the situation was completely different. Namely the polarization curve for the solution containing both sulfite and thiosulfate was closer to that for the solution containing sulfite alone rather than thiosulfate alone, as shown in Figure 10. Apparently, in the presence of a large excess sulfite, the gold deposition does not take place from $Au(S_2O_3)_2^{3-}$ as it does in the solution containing 0.1 M each of sulfite and thiosulfate. Okudaira and Takehara [73] made an ion chromatographic study of gold complexes formed in the sulfite-thiosulfate system, and concluded that the mixed ligand complex, $[Au(SO_3)(S_2O_3)]^{3-}$, can also form depending on the concentration ratio of the components.

In conjunction with the generally known anodic oxidation reactions of thiourea, ascorbic acid, hypophosphite, and hydrazine, major cathodic and anodic partial

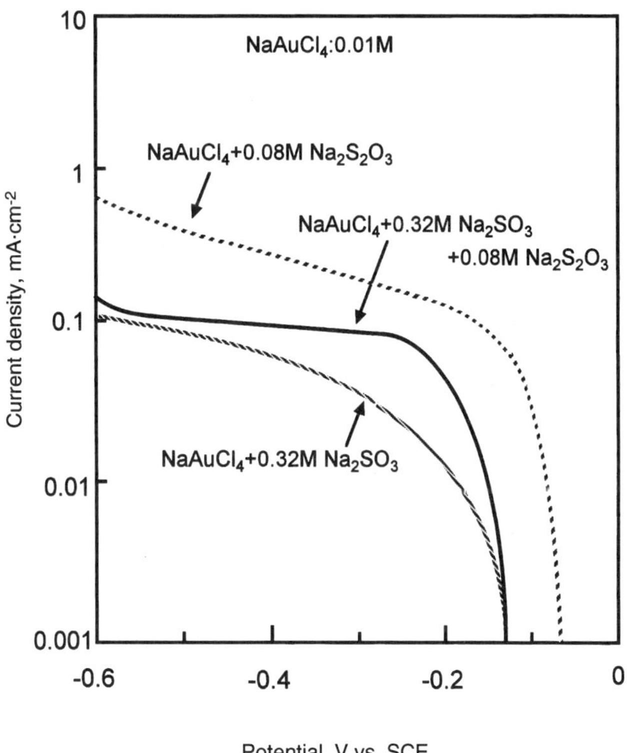

Figure 10 Polarization curves for gold deposition from various mixtures of $NaAuCl_4$, Na_2SO_3, and $Na_2S_2O_3$ with large excess of SO_3^{2-} over $S_2O_3^{2-}$. (Kato et al. [76])

reactions comprising the overall electroless gold deposition reaction from sulfite-thiosulfate mixtures can be written as follows:

Cathodic partial reactions:

$$[Au(SO_3)_2]^{3-} + e^- \longrightarrow Au + 2SO_3^{2-} \tag{16}$$

$$[Au(S_2O_3)_2]^{3-} + e^- \longrightarrow Au + 2S_2O_3^{2-} \tag{17}$$

$$[Au(SO_3)(S_2O_3)]^{3-} + e^- \longrightarrow Au + SO_3^{2-} + S_2O_3^{2-} \tag{18}$$

Anodic partial reactions:

Thiourea bath,

$$CS(NH_2)_2 + 5H_2O \longrightarrow \underset{\text{(urea)}}{CO(NH_2)_2} + H_2SO_4 + 8H^+ + 8e^- \tag{19}$$

Ascorbic acid bath,

$$C_6H_8O_6 \longrightarrow \underset{\left(\begin{array}{c}\text{dehydro-}\\ \text{ascorbic}\end{array}\right)}{C_6H_6O_6} + 2H^+ + 2e^- \tag{20}$$

Hypophosphite bath,

$$2H_2PO_2^- + 2OH^- \longrightarrow 2H_2PO_3^- + H_2 + 2e^- \tag{21}$$

Hydrazine bath,

$$N_2H_4 + 4OH^- \longrightarrow N_2 + 4H_2O + 4e^- \tag{22}$$

6. Baths Containing No Additional Reducing Agent Recently, Krulik and Mandich [83] discovered that the Au(I) sulfite-thiosulfate system functions as an autocatalytic gold plating bath without adding any conventional reducing agent. Such bath was found to plate 0.03 to 0.3 μm of gold in 15 min at pH 6.5 to 9.0 and 55° to 75°C directly on electroless nickel deposited on copper clad epoxy glass laminated printed circuit board. The gold deposition was confirmed to occur on a layer of immersion (galvanic displacement) gold deposited on the electroless nickel, indicating the autocatalytic nature of the gold deposition process. The inventors believe that the sulfite-thiosulfate mixture itself is a reducing agent system, and that especially sulfite functions as the main reducing agent in this bath. Because sulfite is oxidized by air and the oxidation catalyzed by impurities such as Cu^{2+} and Ni^{2+} ions, the addition of a strong complexing agent such as EDTA and NTA is recommended to reduce the effect of such contaminants. The bath is said to be completely stable for more than ten replenishment cycles over the period of many months.

In another patent the same inventors [84] describe that the plating rate of the above system can be increased by adding an amino acid such as glycine, alanine, glutamine, leucine, lysine, and valine. They obtained 0.39 to 1.0 μm of gold in 10 min using baths containing one of the amino acids or mixtures of two or more of those compounds.

5.4 Bath Containing Sulfite and Thiocyanate

Kawashima and Nakao [85] developed a bath containing sulfite and thiocyanate in place of thiosulfate as mixed ligands. Ascorbic acid was used as the reducing agent. The bath contained 0.01 M $HAuCl_4$, 0.1 M Na_2SO_3, 0.1 M KSCN, and 0.1 M ascorbic acid, and it was operated at pH 6.0 and 60°C. The plating rate was approximately 0.5 $\mu m\,h^{-1}$ on a copper or brass plate with successive layers of electroless nickel and displacement gold. The plating rate decreased with increasing concentration of Na_2SO_3 and increased with increasing KSCN concentration. The bath exhibited good stability during two to three temperature cycles consisting of heating at 60°C for 6 hours and room temperature stand for 18 hours, including the period for natural cooling. Good wire bondability is reported on a 0.5 μm thick film plated in this bath.

5.5 Bath with Au(III)-Phosphate Complex

Based on polarization measurements for the reduction of chloroaurate (III) in solutions of various complexing agents (phosphate, pyrophosphate, and glycine) combined with those for the oxidatin of various reducing agents (hydrazine, borohydride, and DMAB), Ohno and coworkers [86] selected a combination of phosphate and DMAB. Their typical bath composition is as follows: 0.005 M $NaAuCl_4$, 0.2 M Na_2HPO_4, 0.01 M DMAB, and 0.5 to 1.0×10^{-5} M MBT (mercaptobenzothiazole). The bath is operated at pH 11.0 and 60°C. The system is of interest because of the simplicity in bath composition. However, the plating rate reported is only 0.16 $\mu m\,h^{-1}$, which is too low for most applications. A further study is necessary to make the bath more practical.

Paunovic and Sambucetti [80] also evaluated an Au(III) phosphate bath using formaldehyde as the reducing agent. They found that the bath was stable for only 2 to 3 hours with a plating rate of 0.9 to 1.5 $\mu m\,h^{-1}$. It was stated that stabilizer addition must be investigated to improve this system.

6 CONCLUSION

It has been almost 30 years since the invention of the first autocatalytic gold plating process. Much improvement has been made to the original process during those years, and new baths containing no cyanide have been developed to meet requirements of the electronics industry. As a matter of fact, noncyanide baths, which can be operated continuously on a large scale in manufacturing environment, are now commercially available. As a result of the recent trend of accelerated miniaturization and densification of electronic circuits requiring gold deposition in small isolated areas, there are many instances in which the circuit fabrication process can be simplified considerably by using an electoless gold process. It is believed that electroless gold deposition will soon become an essential part of electronic circuit fabrication processes.

Finally, it should be mentioned that this review was not intended to cover all electroless gold plating processes described in the literature. Furthermore no mention was made of electroless plating of gold alloys. The interested reader is referred to the recent review article written by one of the present authors [30].

REFERENCES

1. Y. Okinaka, *Plating*, **57**, 914 (1970).
2. Y. Okinaka and C. Wolowodiuk, ibid., **58**, 1080 (1971).
3. R. Sard, Y. Okinaka, and J. R. Rushton, ibid., **58**, 893 (1971).
4. Y. Okinaka, *J. Electrochem. Soc.*, **120**, 739 (1973).
5. Y. Okinaka, R. Sard, C. Wolowodiuk, W. H. Craft, and T. F. Retajczyk, ibid., **121**, 56 (1974).
6. R. Sard, Y. Okinaka, and H. A. Waggener, ibid., **121**, 62 (1974).
7. L. A. D'Asaro, S. Nakahara, and Y. Okinaka, ibid., **127**, 1935 (1980).
8. D. Lamouche, P. Clechet, J. R. Martin, G. Haroutiounian, and J. P. Sandino, *Surf. Sci.*, **161**, L554 (1985).
9. L. M. Schiavone, *J. Electrochem. Soc.*, **125**, 851 (1978).
10. Y. Takakura, *Jitsumu Hyomen Gijutsu* (Metal Finishing Practice), **27** (1), 31 (1980).
11. Y. Takakura, Kokai Tokkyo Koho (Japanese Patent Disclosure), **269**, (1981).
12. J. A. Gardiner and J. W. Collat, *J. Am. Chem. Soc.*, **86**, 3165 (1964); ibid., **87**, 1692 (1965); *Inorg. Chem.*, **4**, 1208 (1965).
13. E. A. Efimov, T. V. Gerish, and I. G. Erusalimchik, *Zaschita Metallov.*, **11** (3), 383 (1975).
14. E. A. Efimov, T. V. Gerish, and I. G. Erusalimchik, ibid., **12** (6), 724 (1976).
15. D. M. MacArthur, *J. Electrochem. Soc.*, **119**, 672 (1972).
16. J. A. Harrison and J. Thompson, *J. Electroanal. Chem.*, **40**, 113 (1972).
17. T. Sawaguchi, T. Yamada, Y. Okinaka, and K. Itaya, *J. Phys. Chem.*, **99**, 14149 (1995).
18. J. D. E. McIntyre and W. F. Peck, *J. Electrochem. Soc.*, **123**, 1800 (1976).
19. M. Matsuoka, S. Imanishi, M. Sahara, and T. Hayashi, *Plating Surf. Finish.*, **75** (5), 102 (1988).
20. S. Sasaki, Kokai Tokkyo Koho (Japanese Patent Disclosure), 59-124428 (1984).
21. N. Honma, ibid., 59-229478 (1984).
22. W. Ott, European Patent Application, EP 281804 (1988).
23. M. F. El-Shazley and K. D. Baker, U.S. Patent 4,337,091 (1982).
24. K. Ohtsuka, K. Okuno, N. Hattori, and E. Torikai, Printed Circuit World Convention, Paper No. C 1/2, Scotland (1980).
25. F. Simon, *Gold Bull.*, **26** (1), 14 (1993).
26. C. D. Iacovangelo, *J. Electrochem. Soc.*, **138**, 976 (1991).
27. H. O. Ali and I. R. A. Christie, *Gold Bull.*, **17** (4), 118 (1984).
28. C. D. Iacovangelo and K. P. Zarnoch, *J. Electrochem. Soc.*, **138**, 983 (1991).
29. Y. Okinaka, in *Gold Plating Technology*, F. H. Reid and W. Goldie, eds., Electrochemical Publications Ltd., Ayr, Scotland, 1974, ch. 11.
30. Y. Okinaka, in *Electroless Plating: Fundamentals and Applications*, G. O. Mallory and J. B. Haydu, eds., American Electroplaters and Surface Finishers Society, 1990, ch. 15.
31. S. D. Swan and E. L. Gostin, *Met. Finish.*, **59** (4), 52 (1961).
32. R. R. Brookshire, U.S. Patent 2,976,181 (1961).
33. T. Ezawa and H. Ito, Kokai Tokkyo Koho (Japanese Patent Disclosure), 40-1081 (1962).
34. E. L. Gostin and S. D. Swan, U.S. Patent 3,032,436 (1962).
35. A. N. Moskvichev, G. A. Kurnoskin, and V. N. Flerov, *Zhur. Prikl. Khim.*, **54** (9), 2150 (1981).

36. A. N. Moskvichev, G. A. Kurnoskin, V. N. Flerov, and Z. P. Gerasimova, *Izv. Vyssh. Uchebn. Zaved., Khim. Tekhnol.*, **25** (9), 1104 (1982).
37. K. Wundt, B. Mankau, J. Schaad, and H. Meyer, German Patent DE 3614090 (1987); *Gold Patent Digest*, **5** (3), 14 (1987).
38. M. Dettke and L. Stein, German Patent DE 3029785 (1982); *Gold Patent Digest*, Pilot issue, 8 (1982).
39. H. E. Bellis, U.S. Patent 3,697,296 (1972).
40. H. A. Efimov and T. V. Gerish, *Zaschita Metallov.*, **15** (2), 240 (1979).
41. T. Oda and K. Hayashi, U.S. Patent 3,506,462 (1970).
42. R. Gesemann, J. Spindler, A. Eysert, R. Broulik, U. Heinzig, and H.-U. Galgon, German Patent DD 263307; *Gold Bull.*, **22** (3), 75 (1989).
43. E. Andrascek, H. Hadersbeck, and F. Wollenhorst, German Patent DE 3237394 (1984).
44. Y. Nishiyama, S. Wakabayashi, and N. Wakabayashi, Kokai Tokkyo Koho (Japanese Patent Disclosure) 60-125379 (1985).
45. Z. Mathe, *Met. Finish.*, **90** (1), 33 (1992).
46. G. M. Ganu and S. Mahapatra, *J. Sci. Ind. Res.*, **46** (4), 154 (1987).
47. F. Richter, R. Gesemann, L. Gierth, and E. Hoyer, German Patent DD 150762 (1981).
48. R. Gesemann, F. Richter, L. Gierth, U. Bechtloff, and E. Hoyer, ibid., 160283 (1983).
49. R. Gesemann, F. Richter, L. Gierth, E. Hoyer, and J. Hartung, ibid., 160284 (1983).
50. F. Richter, R. Gesemann, L. Gierth, and E. Hoyer, ibid., 240915 (1986).
51. F. Richter, R. Gesemann, L. Gierth, and E. Hoyer, ibid., 268484 (1989).
52. Y. Sato, T. Osawa, K. Kaieda, and K. Kobayakawa, *Plating Surf. Finish.*, **81** (9), 74 (1994).
53. H. Shindo and H. Honma, *Proc. 84th Conf. of Surface Finishing Society of Japan*, 163 (1991).
54. M. Kato, S. Hoshino, and I. Ohno, Kokai Tokkyo Koho (Japanese Patent Disclosure), 1-191782 (1989).
55. M. Kato, K. Niikura, S. Hoshino, and I. Ohno, *Proc. 82nd Conf. of Surface Finishing Society of Japan*, 109 (1990); *Hyomen Gijutsu* (J. Surf. Finish. Soc. Japan), **42**, 729 (1991).
56. J. Pouradier and M. C. Gadet, *J. Chim. Phys.*, **66**, 109 (1969).
57. P. Wilkinson, *Gold Bull.*, **19** (3), 21 (1986).
58. W. S. Rapson and T. Groenewald, *Gold Usage*, Academic Press, London, 1978, ch. 10.
59. A. Sullivan, A. Patel, and P. A. Kohl, *Proc. 81st AESF Technical Conf.*, American Electroplaters and Surface Finishers Society, 595 (1994).
60. A. Sullivan and P. A. Kohl, *J. Electrochem. Soc.*, **142**, 2250 (1995).
61. J. Ushio, O. Miyazawa, A. Tomizawa, A. Matsuura, and H. Yokono, Kokai Tokkyo Koho (Japanese Patent Disclosure), 87-247081 (1987).
62. J. Ushio, O. Miyazawa, H. Yokono, and A. Tomizawa, ibid., 87-86171 (1987).
63. O. Miyazawa, J. Ushio, A. Tomizawa, N. Matsuura, and H. Yokono, ibid., 88-79976 (1988).
64. J. Ushio, O. Miyazawa, A. Tomizawa, H. Yokono, N. Kanda, N. Matsuura, S. Ando, H. Okudaira, and K. Mori, ibid., 89-268876 (1989).
65. J. Ushio, O. Miyazawa, H. Yokono, and A. Tomizawa, U.S. Patent 4,804,559 (1989).
66. J. Ushio, O. Miyazawa, H. Yokono, and A. Tomizawa, ibid., 4,880,464 (1989).
67. J. Ushio, O. Miyazawa, A. Tomizawa, H. Yokono, N. Kanda, N. Matsuura, S. Ando, and H. Okudaira, ibid., 4,963,974 (1990).

68. H. Takehara, ibid., 94-73554 (1994).
69. T. Inoue, S. Ando, H. Okudaira, J. Ushio, A. Tomizawa, H. Takehara, T. Shimazaki, H. Yamamoto, and H. Yokono, *Proc. 45th IEEE Electronic Components Technology Conf.*, 1059 (1995).
70. S. Ando, T. Inoue, J. Ushio, H. Okudaira, A. Tomizawa, H. Takehara, T. Shimazaki, H. Yamamoto, and H. Yokono, *Proc. '95 Asian Conf. on Electrochemistry*, 408 (1995).
71. H. Okudaira, S. Ando, T. Inoue, J. Ushio, H. Takehara, A. Tomizawa, and H. Yokono, *Proc. 89th Conf. of Surface Finishing Society of Japan*, 29 (1994).
72. H. Okudaira, T. Inoue, S. Ando, T. Shimazaki, H. Yamamoto, and H. Yokono, ibid., 32 (1994).
73. H. Okudaira and H. Takehara, ibid., 30 (1994).
74. M. Kato, Y. Yazawa, and Y. Okinaka, *Proc. 90th Conf. of Surface Finishing Society of Japan*, 146 (1994).
75. M. Kato, Y. Yazawa, and Y. Okinaka, ibid., 148 (1994).
76. M. Kato, Y. Yazawa, and Y. Okinaka, *Proc. AESF Technical Conf.*, SUR/FIN '95, American Electroplaters and Surface Finishers Society, pp. 805–813 (1995).
77. M. Kato, Y. Yazawa, and S. Hoshino, U.S. Patent 5,470,381 (1995).
78. H. Honma, A. Hasegawa, S. Hotta, and K. Hagiwara, *Plating Surf. Finish.*, **82** (4), 89 (1995).
79. Y. Nishiwaki, H. Honma, and K. Hagiwara, *Proc. 9th JIPC Conf.*, Japan Institute of Printed Circuits, 45 (1995).
80. M. Paunovic and C. Sambucetti, *Proc. Symp. on Electrochemically Deposited Thin Films*, vol. 31, Electrochemical Society, Pennington, NJ, 1994, p. 34.
81. K. Shiokawa, T. Kudo, and N. Asaoka, Kokai Tokkyo Koho (Japanese Patent Disclosure), 3-215677 (1991).
82. K. Shiokawa, T. Kudo, and N. Asaoka, ibid., 4-314871 (1992).
83. G. A. Krulik and N. V. Mandich, U.S. Patent 5,232,492 (1993).
84. G. A. Krulik and N. V. Mandich, ibid., 5,318,621 (1994).
85. S. Kawashima and H. Nakao, *Proc. 10th JIPC Conf.*, Japan Institute of Printed Circuits, 45 (1996).
86. I. Ohno, H. Yajima, and H. Numata, *Proc. Symp. on Electrochemically Deposited Thin Films*, vol. 26, Electrochemical Society, Pennington, NJ, 1993, p. 130.

22 Electroless Deposition of Alloys

IZUMI OHNO

BACKGROUND

Electroless nickel and cobalt deposits usually contain phosphorus or boron as an alloy constituent, depending on the nature of reducing agent. Some of these alloy deposits show superior physicochemical characteristics, such as corrosion resistance, as-plated hardness, thermal stability, solderbility, and special electrical and magnetic properties.

Brenner and Riddel [1] were the first to produce electroless nickel-cobalt-phosphorus ternary alloy. Thereafter, many studies have been done on the electroless deposition of ternary and quaternary alloys. At present, electroless deposition of ternary and quaternary alloys is a practical and effective method for better control of the physicochemical properties of nickel- and cobalt-phosphorus deposits, thereby extending the range of application of electroless deposition. Most work (papers and patents) on electroless alloy deposition has been concerned with the composition and properties of the deposits plated at different conditions. A recent tendency in the study of electroless alloy deposition is toward applications in the electronics industry, such as film formation on semiconductors or multi-layer film formation.

Despite an increasing number of publications concerning electroless alloy deposition, the detailed mechanisms involved are not clear as yet. Basically, electroless alloy deposition consists of electrochemical partial reactions, such as cathodic reduction of the respective alloy components and anodic oxidation of reducing agent. However, these partial reactions are interdependent. Moreover, since the catalytic activity for anodic oxidation of the reducing agent depends on the nature of the metals [2], the rates of electroless deposition of the respective alloy components depend on the nature and composition of the depositing alloy in a rather complicated way.

1 ELECTROLESS ALLOY DEPOSITS AND PLATING BATHS

1.1 Nickel-Based Alloys

Electroless nickel-based ternary and quaternary alloys have been produced by adding cations or anionic complexes of the alloying elements into the electroless

Modern Electroplating, Fourth Edition, Edited by Mordechay Schlesinger and Milan Paunovic.
ISBN 0-471-16824-6

nickel bath [3]. Thus electroless nickel-phosphorus alloys containing iron [4–9], rhenium [10–17], molybdenum [18–23], tungsten [24–29], zinc [11, 30], tin [11, 20, 31], and copper [32–34] have been produced. Typical bath compositions for electroless nickel-based alloys are listed in Table 1. Electroless nickel-based alloy containing cobalt will be discussed in Section 1.2 on cobalt-based alloys.

Nickel-Iron Alloys Electroless nickel-iron-phosphorus alloys have been produced that contain up to 30% iron and approximately only 1% phosphorus [5]. In replacing the reducing agent hypophosphite by dimethylamine borane in the bath, nickel-iron-boron alloys were obtained [6]. Sodium potassium tartrate in the nickel-iron-phosphorus bath in Table 1 [5] can be replaced by the other complexing agents, such as tartrate-glycine [8] or tartrate-citrate [9].

Nickel-Rhenium Alloys Electroless nickel-rhenium-phosphorus alloys have been reported by several workers [10–17]. The alloys deposited contain a large amount of rhenium compared with the third metals in the other electroless nickel-base ternary alloys. Several electroless ternary alloys (nickel-rhenium-phosphorus, nickel-tungsten-phosphorus, nickel-zinc-phosphorus, nickel-tin-phosphorus) and quaternary alloys (55.5%nickel-32.7%rhenium-5.1%phosphorus-6.6%zinc, 53.0%nickel-44.1% rhenium-1.4%phosphorus-1.1%tin, 55.7%nickel-31.5%rhenium-2.6%phosphorus-10.1%tungsten) were reported by Pearlstein and Weightman [11]. Electroless baths reported later concerning nickel-rhenium alloys are based mostly on those in reference [11].

Nickel-Molybdenum Alloys Electroless nickel-molybdenum-phosphorus alloys as well as nickel-based alloys containing tungsten or zinc may be deposited from an electroless nickel bath containing oxy anionic complexes of molybdenum, tungsten and tin, respectively, by using hypophosphite [18–23]. In replacing hypophosphite by dimethylamine borane in the baths, nickel-molybdenum-boron alloys are obtained [18, 19]. The nickel-molybdenum alloys obtained were nickel-7.0%molybdenum-1.8%phosphorus and nickel-31.0%molybdenum-3.2%boron, respectively [19].

Nickel-Tungsten Alloys Electroless deposition of ternary nickel-tungsten-phosphorus alloy was reported first by Pearlstein, Weightman, and Wick [24] in 1963. The electroless plating baths are similar to those for electroless nickel-tungsten and nickel-molybdenum alloys, which contain citrate. Many publications reported about electroless nickel-tungsten alloys containing either phosphorus or boron [11, 20, 24–29]. The consumption efficiency of dimethylamine borane in electroless nickel-tungsten-boron is about twice that reported in binary nickel-boron alloys.

Nickel-Zinc Alloys Electroless nickel-42at%zinc-26at%phosphorus alloys [30] were produced by modifying an electroless cobalt-zinc-phosphorus bath [58, 59].

Nickel-Tin Alloys Electroless nickel-tin-phosphorus alloys containing up to 3%tin and 10%phosphorus may be deposited from a hydroxy acetate bath using sodium hypophosphate as reducing agent. In replacing hypophosphite by dimethylamine borane in the bath, nickel-tin-boron alloys containing up to 44% tin and 3.4% boron are deposited [18]. Addition of cupric salt to this bath results in quaternary nickel-tin-copper-phosphorus or poly alloy, nickel-2%copper-8%tin-16%phosphorus [20].

TABLE 1 Composition of Baths for Electroless Nickel-Based Alloys

Bath Constituents (g liter^{-1})	Ni–Fe–P	Ni–Re–P	Ni–Mo–P	Ni–W–P	Ni–Zn–P
Nickel sulfate, $NiSO_4 \cdot 6H_2O$	56 mM liter^{-1}	35	0.10 M	7	—
Nickel chloride, $NiCl_2 \cdot 6H_2O$	—	—	—	—	7.5
Sodium potassium tartrate, $KNaC_4H_4O_6 \cdot 4H_2O$	100–350 mM liter^{-1}	—	—	—	—
Sodium citrate, $Na_3C_6H_5O_7 \cdot 2H_2O$	—	85	—	40	—
Citric acid, $H_3C_6H_5O_7 \cdot H_2O$	—	—	—	—	19.8
Hydroxy acetic acid, $CH_2(OH)COOH$	—	—	0.30 M	—	—
Sodium hypophosphite, $NaH_2PO_2 \cdot 2H_2O$	10	10	0.28 M	10	3.5
Ammonium hydroxide(28%), NH_4OH	3.6 ml liter^{-1}	50 ml liter^{-1}	—	50 ml liter^{-1}	—
Ferrous ammonium sulfate, $Fe(NH_4)_2(SO_4)_2 \cdot 6H_2O$	20 ml liter^{-1}	—	—	—	—
Potassium perrhenate, $KReO_4$	—	0.2	—	—	—
Sodium molybdate, $NaMoO_4 \cdot 2H_2O$ (as oxy anionic complex)	—	—	0.05 M	—	—
Sodium tungstate, $NaWO_4 \cdot 2H_2O$	—	—	—	35	—
Zinc chloride, $ZnCl_2 \cdot 7H_2O$	—	—	—	—	5
pH	11.2	8.8–9.2	9	8.2	6–12
Temperature (°C)	75	98	87	90	75
Deposits (wt%)	25% Fe 1–0.5% P	46% Re	7.0% Mo 1.8% P	20% W 10% P	4–12at% Zn 2–26at% P
Reference	[5]	[11]	[20]	[11]	[30]

Nickel-Copper Alloys Amorphous nickel-(1–4)%copper-(1–16)% phosphorus alloys [20] were produced by Schwartz and Mallory, and the thermal stability of the alloys was reported by other workers [32, 33]. Copper-nickel alloys with a wide range of composition may be deposited in electroless nickel baths containing copper sulfate, hypophosphite, and boric acid. Addition of ferrous sulfate to the bath results in ternary copper-(6–10)%nickel-(2–4)%iron [34].

Electroless nickel-palladium- [35], nickel-vanadium- [36], nickel-platinum- [37] and nickel-chromium- [38], and nickel-boron-phosphorus [39–41] have also been reported.

1.2 Cobalt-Based Alloys

Similar to electroless nickel-based alloys, it is possible to produce electroless cobalt-phosphorus and cobalt-boron alloys containing one or more additional metals such as nickel [42–53], iron [54–57], rhenium [48], tungsten [48], zinc [58, 59], and silver [60].

Nickel-Cobalt Alloys It is possible to deposit nickel-cobalt-phosphorus alloys with a complete range of nickel-cobalt compositions from alkaline baths, since each metal is capable, independently, of electroless deposition. Alloy deposits containing cobalt are, however, not obtained from acid baths containing hypophosphite. Typical electroless plating baths for cobalt-nickel-phosphorus alloys are listed in Table 2. Addition of sodium tungstate dihydrate or potassium perrhenate to bath 3 in Table 2, instead of nickel sulfate hexahydrate, results in cobalt-phosphorus deposits containing about 9% tungsten or 30% rhenium, respectively [48]. Bath 5 in Table 2 has been developed especially for selective deposition of cobalt-nickel used for electric contact of CMOS (complementary-metal-oxide-silicon) device [53].

Cobalt-Iron Alloys Electroless cobalt deposits containing up to 45% iron [54] or 4% zinc [58, 59] have also been produced. Some of the alloys provided an improved resistance to tarnishing in salt environment [48] or superior magnetic properties [43–48, 54].

Other Alloys There have been several reports on electroless gold-based alloys [61–65]. Copper-selenium [66, 67] and zinc-arsenic [68] alloys.

Gold-Copper Alloys Electroless gold-copper alloys with a wide range of compositions may be deposited from alkaline baths containing copper sulfate, tetrasodium salt of ethylene diaminetetraacetic acid, potassium gold cyanide and formaldehyde [61]. Electrochemical polarization measurements were made on electroless gold-copper alloy deposits of various compositions. It is suggested that the competitive adsorption of formaldehyde and cyanide on the alloy surface determine the catalytic properties for electroless deposition [61, 62]. The rate of oxidation of formaldehyde on gold-copper alloys is lower than on gold-rich alloys in the cyanide baths [62].

Copper selenium has been formed by electroless deposition from a selenic bath containing copper sulfate and indium sulfate. Subsequently indium was incorporated into the film by cathodic polarization [66, 67].

Some quaternary alloys are obtained by addition of cations or anionic complex of different alloying elements to the plating beths of ternary alloys [64, 69–72].

TABLE 2 Composition of Baths for Electroless Cobalt-Nickel-Phosphorus Alloys

Bath Constituents (g liter^{-1})	1	2	3	4	5
Cobalt chloride, $CoCl_2 \cdot H_2O$	30	—	30	—	—
Cobalt sulfate, $CoSO_4 \cdot 7H_2O$	—	10	—	30	60
Nickel chloride, $NiCl_2 \cdot 6H_2O$	15	—	—	—	—
Nickel sulfate, $NiSO_4 \cdot 6H_2O$	—	15	30	30	2
Sodium citrate, $Na_3C_6H_5O_7 \cdot 2H_2O$	100	84	84.5	—	—
Ammonium citrate, $(NH_4)_2HC_6H_5O_7 \cdot 2H_2O$	—	—	—	—	55
Sodium potassium tartrate, $KNaC_4H_4O_6 \cdot 4H_2O$	—	—	—	200	—
Ammonium chloride, NH_4Cl	50	—	50	—	—
Ammonium sulfate, $(NH_4)_2SO_4$	—	42	—	30	40
Ammonium hypophosphite, $NH_4H_2PO_2 \cdot 2H_2O$	20	—	—	—	—
Sodium hypophosphite, $NaH_2PO_2 \cdot 2H_2O$	—	—	—	20	—
Hypophosphorous acid, H_3PO_2	—	8 ml liter^{-1}	—	—	8 ml liter^{-1}
Ammonium hydroxide(29%), NH_4OH	—	13.2 ml liter^{-1}	—	—	—
pH	9.5	8.5	8.9	8–11	8.5
Temperature (°C)	95	90	98	85	90
Deposits (wt%)	57–70%Ni 5.5–6.9%P 23–37%Co	50%Ni 5%P	30–75%Ni 6%P	65%Ni	2at%Ni 3at%P
Reference	[42]	[43]	[48]	[51]	[53]

1.3 Deposit Properties

Electroless alloy films have been used as conductors, contacts, resistors magnetic memory devices, and underlayers in floppy disk in electronics industry.

Magnetic Properties Initially the magnetic properties of electroless alloys of iron group metals have been of considerable interest. Electroless alloy films exhibit soft or hard magnetic property depending on the amount of nickel in nickel-cobalt and nickel-iron alloys. It was reported [42] that electroless cobalt-nickel alloys deposited from acid and alkaline baths have different magnetic properties. Electroless nickel-iron films have magnetic characteristics useful for computer memory [43]. Magnetic properties of nickel-molybdenum-phosphorus alloys depend on the molybdenum content and heat-treatment, and nickel-17%molybdenum-0.2%phosphorus does not exhibit any sign of ferromagnetism even after aging up to 400°C [19]. The magnetic properties of cobalt-nickel-phosphorus films, which were deposited from bath 1 in Table 2, were $Hc = 0.14–4.5$ oersted, $Br = 300–3000$ gauss, and $Bm = 3000–6000$ gauss, depending on the concentration of nickel chloride in the plating bath. Many workers have reported the possible application of electroless cobalt-nickel-phosphorus films for rapid memory devices. Heritage and Walker [43] reported that an electroless cobalt-nickel-5%phosphorus film exhibited a well-defined anisotropy ($Hc = 2$ oersted, $Hk = 6$ oersted) and switching constant of 0.15 μs. With good reproducibility, it was also reported [44] that superimposition of a magnetic field during electroless deposition has an appreciable effect on the anisotropy of the deposited film. Electroless alloy films were further applied to a perpendicular magnetic recording media [69–71]. Electroless multicomponent alloys (cobalt-nickel-manganese-phosphorus, cobalt-nickel-rhenium-manganese-phosphorus) for fabricating perpendicular recording media have been well reviewed [73].

Thermal Stabilities Thermal stability of electroless nickel-based alloys are enhanced by the codeposition of refractory metals, such as rhenium [16, 17], molybdenum [21–23], and tungsten [25–28]. Electroless nickel-44at%rhenium-phosphorus can be used for thin film resistor with high thermal stability [16]. A nickel-molybdenum-phosphorus alloy film was developed for a film resistor with low temperature coefficient of resistivity and high-thermal stability [21]. Electroless nickel-tungsten-phosphorus alloy films have been applied to the low-energy consumption thermal head for a heating resistor [28]. Depositing it on nickel-tungsten-phosphorus underlayer enhances crystallinity and smoothness of surface of cobalt alloy films and thus very high recording density is attained [27].

Structure Nucleation and growth of electrolessly deposited cobalt-65%nickel-phosphorus was discussed by Chow et al. [51], where it was claimed that codeposited phosphorus functions as an inhibitor for the growth of cobalt and nickel nuclei. The morphology and structure of a multi-layer component nickel-molybdenum-phosphorus/tin oxide/titan(Ni–Mo–P/SnO_2/Ti) was analyzed by Lo and Hwang [23]. The structure and properties of electroless cobalt-iron-boron film deposited on an amorphous nickel-based alloy ribbon has been studied as a candidate for perpendicular magnetic recording medium. Residual compressive stress in the deposited film was expected to enhance the perpendicular magnetic anisotropy during annealing [57]. Electroless nickel-boron-phosphorus films are used for an undercoat of magnetic films in many memory devices [39–41].

2 SUMMARY

Electroless alloy plating has many potential applications especially in the electronics industry. However, no systematic studies have been made on the process control of electroless alloy deposition, since there are too many interdependent factors affecting the composition and properties of the alloy. It is also not possible at this time to establish a relationship between the variations in the magnetic properties and the composition and structure of electroless alloy. Further basic studies are needed to develop the application of electroless alloy plating.

REFERENCES

1. A. Brenner and G. Riddell, *J. Res. Natl. Bur. Standards*, **37**, 31 (1946).
2. I. Ohno, O. Wakabayashi, and S. Haruyama, *J. Electrochem. Soc.*, **132**, 2323 (1985).
3. F. Pearlstein, in *Modern Electroplating*, F. Lowenheim, ed., Wiley, New York, 1974, p. 71.
4. P. H. Eisenberg, U.S. Patent 2,827,399 (1958).
5. A. F. Schmeckenbecher, *J. Electrochem. Soc.*, **113**, 778 (1966).
6. A. F. Schmeckenbecher, *Plating*, **58**, 905 (1971).
7. H. Matsubara, H. Mizutani, S. Mitamura, and T. Osaka, *Trans. IEEE Magn.*, **26** (3), 1210 (1990).
8. M. Matsuoka and T. Hayashi, *Plating Surf. Finish.*, **69** (12), 53 (1982).
9. D. Kim, H. Matsuda, K. Aoki, and O. Takano, *Plating Surf. Finish.*, **83** (2), 78 (1996).
10. K. M. Gorbunova, A. A. Nikiforova, and G. A. Sadakov, "Modern Problems of Metal Deposition by Reduction with Hypophosphite," *Electrochemistry*, Soviet Academy NAUK Moscow, M. M. Melikova, ed., 1966, p. 41; see reference [118].
11. F. Pearlstein and R. F. Weightman, *Electrochem. Technol.*, **6**, 427 (1968).
12. F. Pearlstein, U.S. Patent 3,485,597 (1969).
13. International Business Machines, British Patent 1,314,745 (1970).
14. F. Pearlstein and R. F. Weightman, U.S. Patent 3,751,939 (1973).
15. M. Gulla, U.S. Patent 3,764,352 (1973).
16. T. Osaka, T. Homma, M. Fukawa, H. Iwamoto, and J. Kawaguchi, *Denki Kagaku*, **59**, 723, (1991).
17. T. Osaka, M. Fukawa, and J. Kawaguchi, *Denki Kagaku*, **60** (6), (1992).
18. G. O. Mallory and T. R. Horhn, *Plating Surf. Finish.*, **66** (4), 40 (1979).
19. G. O. Mallory, *Plating Surf. Finish.*, **63** (6), 34 (1976).
20. M. Schwartz and G. O. Mallory, *J. Electrochem. Soc.*, **123**, 606 (1976).
21. I. Koiwa, M. Usuda, K. Yamada, and T. Osaka, *J. Electrochem. Soc.*, **135**, 718 (1988).
22. T. Osaka, H. Yamazaki, and I. Saito, *J. Electrochem. Soc.*, **136**, 3418 (1989).
23. Y. L. Lo and B. J. Hwang, *J. Electrochem. Soc.*, **143**, 2158 (1996).

24. F. Pearlstein, R. F. Weightman, and P. Wick, *Met. Finish.*, **61** (11), 77 (1963).
25. K. Aoki and O. Takano, *Plating Surf. Finish.*, **73** (5), 136 (1986).
26. K. Aoki and O. Takano, *Plating Surf. Finish.*, **77** (3), 48 (1990).
27. I. Koiwa, M. Usuda, and T. Osaka, *J. Electrochem. Soc.*, **135**, 1222 (1988).
28. H. Sawai, T. Kanamori, I. Koiwa, S. Shibata, and K. Nihei, *J. Electrochem. Soc.*, **137**, 3653 (1990).
29. J. L. X. Hu and D. Wang, *Plating Surf. Finish.*, **83** (8), 62 (1996).
30. M. Schlesinger and X. Meng, *J. Electrochem. Soc.*, **138**, 406 (1991).
31. H. Shimauchi, S. Ozawa, K. Tamura, and T. Osaka, *J. Electrochem. Soc.*, **141**, 1471 (1994).
32. A. M. Lunyatskas, *Zashchita Metallov*, **4** (3), 315 (1968).
33. N. Krasteva, V. Fotty, and S. Armyanov, *J. Electrochem. Soc.*, **141**, 2864 (1994).
34. A. Hung, P. C. Hung, and I. Ohno, *Plating Surf. Finish.*, **76** (12), 60 (1989).
35. F. Pearlstein and R. F. Weightman, *Plating*, **56**, 1158 (1969).
36. P. H. Eisenberg and D. O. Raleigh, U.S. Patent 2,828,227 (1958).
37. K. Aoki, O. Takano, and D.-H. Kim, *J. Surf. Finish. Soc. Jpn.*, **46**, 79 (1995).
38. K. Iwamatsu, *Proc. 74th AESF Ann. Tech. Conf.*, B4-1 (1987).
39. D. W. Baudrand, *Plating Surf. Finish.*, **66** (11), 18 (1979).
40. G. O. Mallory, "Recent Advances in Electroless Nickel Plating of Aluminum Connectors," Presented at the 17th Annual Connectors and Interconnection Technology Symp., Anaheim, CA, Sept. 19–21, 1984.
41. D. W. Baudrand and M. Malik, *Met. Finish.*, **84** (3), 15 (1986).
42. K. M. Gorbunova and A. A. Nikiforova, *Fiziko-khimicheskie osnovy protsessa khimicheskogo nikelirovaniya* (Physicochemical Principles of Chemical Nickel Plating), -Moskva, AN SSSR. 1960 (RZhKhim, No. 21 : 85470, 1960).
43. R. J. Heritage and M. T. Walker, *J. Electron. Control*, **7**, 542 (1960).
44. J. Bagrowski and M. Lauriente, *J. Electrochem. Soc.*, **109**, 987 (1962).
45. J. C. Hendy, H. D. Richards, and A. W. Simpson, *J. Mater. Sci.*, **1**, 127 (1966).
46. M. Johnson, *Proc. 3rd Plating in the Electronics Industry Symp.*, Feb. 1971, Am. Electroplaters' Soc., East Orange, NJ, p. 286.
47. G. W. Lawless and R. D. Fisher, *Plating*, **54**, 709 (1967).
48. F. Pearlstein and R. F. Weightman, *Plating*, **54**, 714 (1967).
49. R. D. Fisher and W. H. Chilton, *Plating*, **54**, 537 (1967).
50. J. H. Kefalas, *Plating*, **54** (5), 543 (1967).
51. S. L. Chow, N. E. Hedgecock, M. Schlesinger, and J. Rezek, *J. Electrochem. Soc.*, **119**, 1614 (1972).
52. T. Homma, K. Inoue, H. Asai, K. Ohrui, and T. Osaka, *IEEE Trans. Magn.*, **27** (6), (1991).
53. G. E. Georgiou, F. A. Baiocchi, H. S. Luftman, T. T. Sheng, M. J. Vasile, and R. V. Knoell, *J. Electrochem. Soc.*, **138**, 2061 (1991).
54. J. R. DePew and D. E. Speliotis, *Plating*, **54**, 705 (1967).
55. T. Osaka, N. Kasai, I. Koiwa, F. Goto, and Y. Sugiyama, *J. Electrochem. Soc.*, **130**, 568 (1983).
56. T. Osaka, T. Homma, K. Saito, A. Takekoshi, Y. Yamazaki, and T. Namikawa, *J. Electrochem. Soc.*, **139**, 1311 (1992).

57. K. Maruyama, T. Sato, I. Ohno, and H. Numata, in *Proc. Int. Conf. on Microstructures and Functions of Material*, ICMFM 96, N. Igata, Y. Hiki, I. Yoshida, and S. Sato, eds., p. 185 (1996).
58. R. D. Fisher, *IEEE Trans. Magn.*, **2**, 681 (1966).
59. M. Soraya, *Plating*, **54**, 549 (1967).
60. T. Miyabayashi, H. Wakayama, M. Miyata, and K. Kobayashi, in *Proc. Symp. on Electroless Deposition of Metals and Alloys*, vol. 12, M. Paunovic and I. Ohno, eds., Electrochemical Society, Penington, NJ, 1988, p. 196.
61. A. Molenaar, *J. Electrochem. Soc.*, **129**, 1917 (1982).
62. A. Molenaar and B. C. M. Meenderick, *J. Electrochem. Soc.*, **132**, 574 (1985).
63. G. Tomkyavichyus et al., *Zaschita Meallov.*, **24**, 849 (1988).
64. M. F. el-Shazly and K. Baker, *AES Electroless Plating Symp.*, vol. 20 (1982).
65. G. Stremsdoerfer, H. Perrot, J. R. Martin, and P. Clechet, *J. Electrochem. Soc.*, **135**, 2881 (1988).
66. R. N. O'Brien and K. S. V. Santhanam, *J. Electroanal. Chem.*, **260**, 231 (1989).
67. R. N. O'Brien and K. S. V. Santhanam, *J. Electrochem. Soc.*, **139**, 434 (1992).
68. H. J. West, *Met. Finish.*, **55** (1), 56 (1957).
69. I. Koiwa, M. Toda, and T. Osaka, *J. Electrochem. Soc.*, **133**, 597 (1986).
70. T. Osaka, T. Homma, K. Noda, T. Watanabe, and F. Goto, *IEEE Trans. Magn.*, **27** (6), (1991).
71. T. Osaka, N. Kasai, I. Koiwa, F. Goto, and Y. Sugiyama, *J. Electrochem. Soc.*, **130**, 568 (1983).
72. D.-H. Kim, K. Aoki, and O. Takano, *J. Surf. Finish. Soc. Jpn.*, **46**, 81 (1995).
73. T. Osaka, *Electrochimica Acta*, **37**, 989 (1992).

23 Preparation for Deposition

DEXTER D. SNYDER

1 PRINCIPLES

Success of electroplating or surface conversion depends on removing contaminants and films from the substrate. Organic and nonmetallic films interfere with bonding to cause poor adhesion and even prevent deposition. This surface contamination can be extrinsic, comprised of organic debris and mineral dust from the environment or preceding processes. It can also be intrinsic, one example being a native oxide layer. Cleaning methods are designed to minimize substrate damage, while removing the film or debris. If a metal's surface chemistry and processing history are known, one can anticipate cleaning needs and strategies.

Extrinsic organic and inorganic soils originate with processing of the substrate before plating, as well as from the general environment. Specific residues include lubricants, phosphate coatings, quenching oils, rust proofing oils, drawing compounds, and stamping lubricants. The mixture of potential contaminants to which a part is exposed is typically complex.

All metals form oxide and inorganic films with environmental gases and chemicals to a degree. Some of these films are protective against continuing attack, such as the hydrous aluminum oxide formed on aluminum alloys. Some are nonprotective such as ferrous/ferric oxides on steel. Some of these films can even be plated directly, with nickel on aluminum oxide over aluminum being an example. The cleaning and activation steps must account for the fact that surface oxide reforms at different rates on the different metals. For iron and nickel, the oxide reforms slowly enough that the parts can be transferred from a cleaning solution to a plating bath at a normal rate. For aluminum and magnesium, the oxide reforms extremely fast, so special processing steps are necessary to preserve the metal surface during transit into electroplating.

Porous substrates, such as cast iron, are a major challenge. Organic and inorganic residues penetrate the pores and must be removed to avoid their interfering with subsequent operations. Left in the pores, these residues decompose or vaporize during electroplating or when the component is used, causing blisters. To remove residues, the cleaner must penetrate the pores, solubilize the residue, then itself be removed.

Metals react with atmospheric constituents, such as water and oxygen, to form intrinsic films, which also must be removed through cleaning before plating.

Modern Electroplating, Fourth Edition, Edited by Mordechay Schlesinger and Milan Paunovic.
ISBN 0-471-16824-6

Pourbaix diagrams [1, 2], which are graphical renderings of metal/reactant equilibria, help one understand how to remove these films. Iron has a relatively large passivation range and shows immunity to corrosion at moderate cathodic potential [3], as shown in Figure 1. Copper, by contrast, is attacked at both high and low pH [3], as seen in Figure 2, so cleaners are designed for this window of passivation. *E*-pH diagrams are available for all common materials and plating systems [1]. Diagrams useful to evaluate cleaning approaches can be constructed for systems containing complexing agents and other constituents.

Cleaning processes are based on two approaches. In physical cleaning, kinetic energy is introduced to release both extrinsic and intrinsic contaminants from the metal surface. Examples are glass bead peening, ultrasonic agitation, and brush abrasion. In chemical cleaning, contaminant films are removed by active materials dissolved or emulsified in the cleaning solution. Extrinsic contaminants are removed with surface-active chemicals, and the chemical energies involved are modest. Intrinsic films are removed with aggressive chemicals that dissolve the contaminant and often react with the metal itself, and the energies involved are significant.

The substrate metal properties must be considered in balancing cleaning power with aggressiveness. Ferrous alloys are attacked by acids, withstand very alkaline cleaners, and have a nonprotective oxide [4–6]. It is important to protect iron alloys from oxidizing between plating steps. Iron oxide is nonprotective and must be dissolved by mixtures of acid and iron salts. Normal alkaline cleaners will not remove this film.

In air, aluminum, and its alloys are coated with a thin native oxide only a few monolayers thick, which forms very quickly on the metal [7]. Fortunately, this film

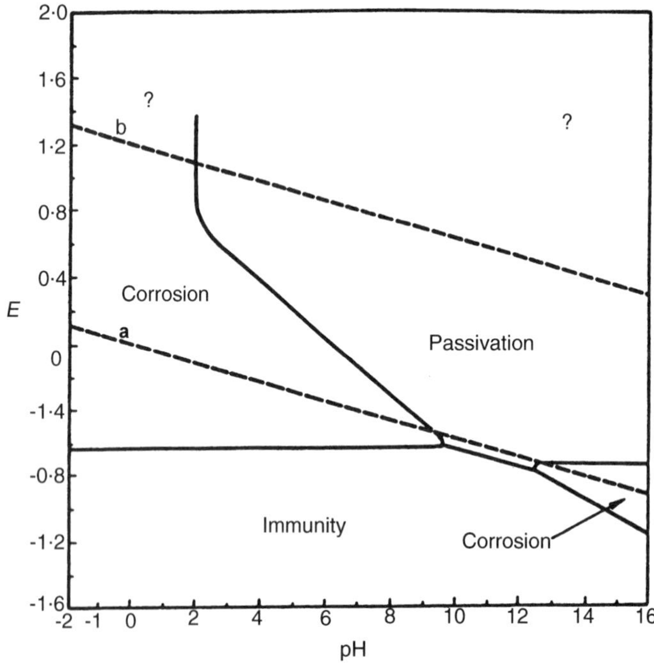

Figure 1 Simplified Pourbaix diagram for iron.

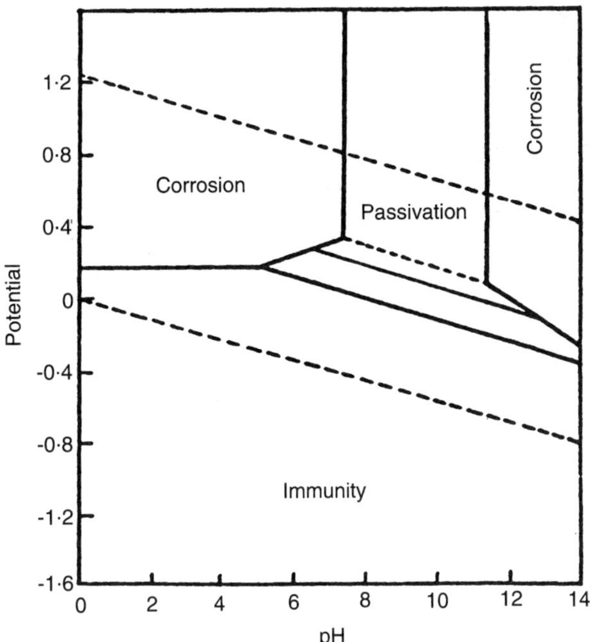

Figure 2 Simplified Pourbaix diagram for copper.

is protective in many environments, and one effective protective technology for aluminum alloys is to anodize, or purposely thicken the native oxide layer to μm thickness. Aluminum oxide layers can be dissolved in alkalis. Both strong acids and strong alkalis attack aluminum alloys.

Copper and its alloys require a milder cleaner than many other metals [8]. They are attacked by oxidizing acids, such as nitric, but they withstand alkalis. A copper cleaner can be built on a reducing acid, with complexing and wetting agents, and a corrosion inhibitor. Zinc is more active than copper, it is attacked by acids and alkalis at $pH > 10$ [9], and it requires a mild alkaline cleaner.

Cleaning of plastics and polymers before plating involves removing extrinsic organics and also etching the surface to provide for mechanical interlocking, where possible. The polymers most commonly electroplated include fluorocarbons, polycarbonates and acrylonitrilebutadienes.

The literature of metal cleaning for electroplating is extensive [10–13]. What has changed in recent decades are as follows:

- The growing requirement for efficient cleaning processes with no environmental impact
- Databases to recommend best cleaning alternatives for a given substrate, surface condition and subsequent plating
- Analysis tools to determine the actual degree of cleanliness

The principles of cleaning are well known, and the key is in knowing which combinations of metals and processes are effective in specific situations.

Mechanical blast cleaning works for nonprecision parts that are capable of withstanding the friction and force from the cleaning media [10a, 14]. While one avoids the need to deal with solvents, there is a requirement for high capital costs and large facilities. Mechanical surface preparation is expected to increase in use over traditional chemical surface preparation.

For extrinsic contaminants, solvents are preferred cleaners because they evaporate quickly and leave little to no residue on the metallic object [10b]. The cleaning ability of an organic solvent depends strongly on its normal boiling temperature. This follows due to the large dependence of the contaminant's viscosity on temperature. Solvent cleaners have long been a mainstay for vapor degreasing, and common chemicals have been tri- and tetrachloroethylene. However, solvents are being actively replaced wherever possible with environmentally benign alternatives [15–16]. One should select a solvent-based cleaner at this time, only if there is no technical alternative. In that case the process must be designed for compliance with environmental and worksite regulations, and the added cost can be significant.

Aqueous cleaning is a function of the chemistry, temperature, agitation and processing time [17]. The goal to maintaining the process is to remove oil and suspended solids from the medium that can be recycled. Aqueous systems tend to emulsify the oil, which then becomes part of the cleaner and eventually degrades its performance. In working out the optimum balance, one should team with the cleaner chemical suppliers to have all wisdom and alternatives available. One trend is to choose a cleaner that does not build a stable emulsion with the waste oil, but rather holds it loosely in suspension, then ejects it as a floating layer. This makes possible a steady-state performance of the cleaning system. Alkaline systems are effective in removing soils, and tend not to attack the substrate except at high pH. Acidic systems normally do attack the substrate, so this type of cleaning (called also pickling or activation, in different applications) must be carefully controlled by concentration and exposure time.

For some special cases, no cleaning is required before electrodeposition. For example, in the electrodeposition of copper conductors in integrated circuit (IC) fabrication, no cleaning is necessary when copper is electrodeposited on a PVD (physical vapor deposited) copper seed layer.

2 CONVENTIONAL PROCESSES

2.1 Organic Solvent Cleaners

While 1,1,1-trichloroethane has been banned for its role as an atmospheric ozone depleter, solvent cleaning is still possible within the current environmental and safety/health regulations [16]. Three replacements without ozone-depletion capability are methylene chloride, perchloroethylene, and trichloroethylene. Hydrofluoroethers are also available as alternatives to CFCs (chlorofluorohydrocarbons). These solvents must still be kept out of liquid waste disposal systems, and below predetermined thresholds in the workplace. All solvent based cleaning must now be closed-loop and zero-discharge. Solvents must be kept relatively clean, for both vapor degreasing and soak degreasing. When the solvent is dirty, the contaminants can be redeposited on the part.

2.2 Aqueous Cleaners

Aqueous cleaners normally contain surfactants, builders—such as alkali for neutralizing acid—to assist in cleaning, binders to hold the ingredients together, and buffers to help neutralize the surface activity of the parts [17]. Surfactants work by having the hydrophobic tail seek the oil. The negatively charged metal repels the negatively charged end, and the oil is gradually lifted from the surface. Sodium ions enhance the function of a surfactant.

Alkaline cleaners [10c] are largely intended to neutralize acidic species and soften the water, to prevent formation of insoluble calcium and magnesium soaps. Sodium hydroxide is the most important alkali in cleaners, but sodium carbonate is also used. Phosphates are used primarily for the water softening. Pyrophosphates are used for sequestering and chelating of zinc, copper, and magnesium, but they are not effective for calcium. The most effective all-around version is sodium pyrophosphate. Silicates are good buffers, and they are included in many alkaline cleaners.

Acid cleaners attack the metal substrate to remove the intrinsic films, and they also can facilitate removal of extrinsic soils [10d]. More care must be taken with acid cleaning, however, to avoid damaging the substrates. Oxide scale is removed from ferrous alloys using inorganic acids or acid salts with combinations of inhibitors, solvents and surfactants. Organic acids and acid salts have the added advantage of sequestering some contaminant species, after chemical reaction, to remove them from the surface. For electrolytic cleaning, sulfuric acid is the main choice. Acid treatment is used for aluminum alloys mainly to desmut, and exposure is normally a few minutes only. Copper cleaners are normally a mixture of acid, such as sulfuric, with an oxidizing agent, such as nitric acid. Aluminum is active in nonoxidizing acids, but inactive in oxidizing solutions. Zinc can be deoxidized with cold, dilute nitric acid containing an oxidizer.

In aqueous cleaning, problems still exist in the areas of rinsing and drying [11]. Rinsing is best addressed in closed-loop systems where the contaminant level can be monitored and controlled. Drying technology offer considerable advantages, and the contributors are to be found with air blow-offs and vacuum drying. Another major area for improving cleaning is to design products to be easy to clean. Better to reduce the problem as much as possible, for example, by avoiding blind holes and screw configurations.

Parts washers are often used with the aqueous systems used to replace solvent cleaners. Here, one uses a cleaning cabinet with high-pressure streams applied to the parts. These are basically industrial dishwashing machines. The detergent used is typically biodegradable.

In ultrasonic cleaning [10e, 18], sound waves dissipate energy which evaporates water into small vapor bubbles. Most of this energy dissipation occurs at nucleation sites on the metal surface and when the bubbles collapse. When the bubbles collapse, the implosion tends to dislodge the oil. Ultrasonic energy is generally added at 20 to 40 KHz. The higher the frequency, the smaller the bubbles. There is some thought that large bubbles are preferred because the larger bubbles generate more energy when they implode. Another school of thought says smaller bubbles are desired because they penetrate more easily into crevices and pores. There is no definitive evidence to decide this matter.

Electrocleaning is used as the final step, for such tasks as plating precision steel parts [11]. The actual cleaning comes from generation of tiny bubbles that lift and remove soils. Anodic cleaning, called *reverse*, involves production of oxygen bubbles. Cathodic cleaning, called *direct*, involves production of hydrogen bubbles.

2.3 Semiaqueous Cleaners

Semiaqueous cleaners are an alternative to solvent cleaning [10f]. These cleaners are made of synthetic organic solvents, surfactants, corrosion inhibitors, and other additives. Water is used in some portion. For water-immiscible cleaners, the active ingredients are held as an emulsion. For water-miscible cleaners, the water is actually filler to reduce volatile organic compound emissions. This class of cleaners still has some potential for volatile organics, aquatic toxicity, flammability, and human health effects.

2.4 Rinsing

Cleaning solutions can be inactivated by carryover from previous steps. This requires they be dumped prematurely. It can periodically reduce their cleaning effectiveness and the extra dumping increases disposal costs. Furthermore, contaminated cleaners can be carried over to degrade the plating solutions, compounding the cost. Effective rinsing is an integral part of cost-effective plating systems.

Rinsing system design is well covered in several texts [11]. Water requirements are reduced dramatically with multiple stage rinsing, including countercurrent flow where feasible. Water use efficiency is increased with such mixing tools as pressure spray, heating, mechanical agitation, and ultrasonic irradiation. With the trend to closed-loop operation, used rinse water can be reclaimed with combinations of filtration, evaporation, and ion exchange. In any event, maintaining a predetermined rinse water quality is critical to ensuring consistent rinsing results.

3 EMERGING TECHNOLOGY

There will be a steady increase in the number of environmentally alternatives to conventional solvent and water-based metal cleaning. As an example, carbon dioxide is being considered as a cleaning medium because it removes the threat of pollution.

Cold Jet CO_2 (dry ice) blasting, commercialized in the 1980s, has great potential to be used more in electroplating cleaning [19–20]. Carbon dioxide granules are produced and blasted onto the workplace with special nozzles, similar to blasting with sand, plastic beads, or soda. However, the particles sublimate on impact to yield a rapidly expanding gas and to release energy that creates microshocks to aid with the cleaning. CO_2 particles are nonabrasive, but one must tailor the process to avoid damage to the substrate.

Supercritical fluids (SCF) are another innovative route to cleaning metal surfaces [26]. This medium possesses liquidlike density and solvency, combined with gaslike viscosity and diffusivity. SCF carbon dioxide has a lower surface tension than liquid carbon dioxide, and it spreads more easily over a surface than does the liquid. This means SCF can penetrate substrates and interstices to dislodge and remove contaminants, then itself be easily and completely removed. This is a closed-loop, batch process operating at moderate temperatures and pressures on the order of 2000 psig. Cleaning times are 10 to 30 minutes, and the surface contaminants are the sole waste. Operating costs are low, but capital costs are high. This process does require pressure vessels and high-pressure controls.

4 MEASURING DEGREE OF CLEANLINESS

Increasingly surface analysis and characterization are accompanying cleaning decisions so one can detect when a cleanliness goal is reached [22–24]. The objective is to present to the metallizing process a surface free of foreign matter or films which interfere with adhesion. In the past the proof of cleaning success was in the plated part performing as required. This approach works as long as the substrate condition, the surface contaminants and the cleaning baths do not vary. When any element drifts, the degree of cleaning likely changes. Many failures would be apparent only after extended used as product. With affordable, modern surface, and solution analysis systems, the situation has changed. Unwanted surface species can be detected quickly and quantitatively to guide the decision to clean further or to metallize.

The final condition of a surface is the product of cleaning. Therefore surface analysis to monitor cleaning adds value by enabling one to find and maintain the optimum trade-off among several factors:

Substrate being used
Contaminants being removed in cleaning
Sensitivity of plating/surface conversion to state of cleanliness
Cost of plating/surface conversion per part
Budget

The surface condition before and after cleaning can be linked to requirements for subsequent surface conversion and plating steps. In some cases a cleaned surface can be validated before a particularly sensitive plating or surface conversion step in a high-value-added product. Where simple, inexpensive monitors are not yet available, one can still define the required cleanliness with laboratory analysis systems, and then check the substrate cleanliness periodically as parts exit cleaning.

Direct techniques analyze a part without need for removing any surface contamination. In some cases these methods are usable within the process itself. In-plant analysis options include the following [25]:

Magnified visual inspection
Fluorescence
Water break test
Contact angle
Gravimetric weighing of parts
Optically stimulated electron emission (OSEE)

Direct oxidation carbon coulometry (DOCC)
X-ray photoelectron spectroscopy (XPS)

Indirect methods require a solvent to dissolve residual contaminants for subsequent analysis by the following techniques [25]:

Gravimetric analysis
Ultraviolet analysis
Optical particle counter

Other laboratory/reference analysis options include the following [25]:

Scanning electron microscopy (SEM)
Energy dispersive electron analysis (EDAX)
Auger spectroscopy
Laser ablation spectroscopy
Raman spectroscopy
Total organic carbon analysis
Ion chromatography
Electron spectroscopy (ESCA)
Atomic tunneling microscopy (AFM)
Scanning tunneling microscopy (STM)
Cyclic voltammetry (CV)
Electrochemical impedance spectroscopy (EIS)
Fourier transform spectroscopy

5 STRATEGIES FOR CLEANING

Cleaning processes are designed to have maximum possible life, minimal maintenance, low environmental impact, and the ability to recycle fluids. This design for the environment approach is a major shift from the past when cleaning performance was the only criterion [26]. The extrinsic dirt can be dealt with through physical and chemical methods, and the part often must be shielded from the atmosphere between cleaning and metallization. Intrinsic residues more often involve intimate bonding with the substrate, so chemical cleaning is dictated. Important in this step is to document all prior processing for the metal part up to plating. This identifies compounds and surface modifications to which the metal has been subjected, and points to likely cleaning needs.

Cleaning technology has been traditionally the product of cut and try experience, and suppliers have closely held the knowledge. Therefore it is prudent to work in partnership with the effective suppliers—those with a track record of success—to find robust cleaners and pretreatment processes with a company having the knowledge to maintain them. With the Internet and computer databases, one can tap into the accumulated knowledge of commercial cleaning processes. Electroplaters will be increasingly able to identify and work with the cleaning suppliers having the most relevant products and experience.

Computerized mathematical models are being improved steadily to predict the outcome of cleaning and plating operations. These models take a variety of forms:

Full mathematical simulations of the cleaning and plating operations, including chemical, geometric and physical aspects

Expert systems that synthesize recommendations from question and answer, based on logical relationships programmed into the package by human experts

Databases with boolean lookup packages

Matching cleaning processes with the particular materials and electroplated coatings becomes efficient if one optimizes the control parameters with designed experiments. Software to select a design matrix and analyze the data is available for PC.

Finally, the choice and control of a cleaning process is a trade-off of all factors governing the total metal plating and surface conversion effectiveness:

Cost including waste/recycle

Substrates that can be handled

Toxics

Dirts that can be handled

Safety

Maintenance (life, contamination)

Analyses and control

What specific dirts must be removed

To what level must specific dirts be reduced

To the extent cleaning is effective, robust, and controlled, there are many low-risk options to control plating outcomes.

REFERENCES

1. M. Pourbaix, *Atlas of Electrochemical Equilibria in Aqueous Solutions*, Pergamon Press, Oxford, 1966.
2. M. Pourbaix, *Lectures on Electrochemical Corrosion*, Plenum Press, New York, 1973.
3. U. Evans, *An Introduction to Metallic Corrosion*, Edward Arnold Ltd., and American Society for Metals, 1981.
4. ASTM, B-183, Preparation of Low-Carbon Steel for Electroplating.
5. ASTM, B-242, Preparation of High-Carbon Steel for Electroplating.
6. ASTM, B-254, Preparation of and Electroplating on Stainless Steel.
7. ASTM, B-253, Preparation of Aluminum Alloys for Electroplating.
8. ASTM, B281, Preparation of Copper and Copper Alloys for Electroplating and Conversion Coating.

9. ASTM, B-252, Preparation of Zinc Alloy Die Castings for Electroplating and Conversion Coatings.
10. "Surface Engineering", *ASM Handbook Volume 5*, ASM International, Metals Park, OH, 1994. (a) "Mechanical Blast Systems," T. Kostilnik, pp. 55–66. (b) "Solvent Cold Cleaning and Vapor Degreasing," V. L. Rupp and K. Surprenant, pp. 22–32. (c) "Alkaline Cleaning," G. J. Cormier, pp. 18–20. (d) "Acid Cleaning," K. J. Hacias, pp. 48–54. (e) "Ultrasonic Cleaning," J. Hancock, pp. 44–47. (f) "Emulsion Cleaning," S. M. Nourie, pp. 33–38.
11. D. R. Gabe, *Principles of Metal Surface Treatment and Protection*, 2nd ed., Pergamon Press, Oxford, 1978.
12. L. J. Durney, ed., *Electroplating Engineering Handbook* 4th ed., Van Nostrand Reinhold, New York, 1984.
13. ASTM, B322, Cleaning Metals Prior to Electroplating.
14. See Reference [10a].
15. SAGE Solvent Alternatives Guide, Surface Cleaning Program of the Research Triangle Institute (1998), maintained on-line at http://clean.rti.org.
16. Products Finishing On-line, "Solvent Cleaning Is Alive and Well," Nov. 1998.
17. Products Finishing On-line, "Second Generation Aqueous Cleaning—An Update in Conventional Wisdom," Aug. 1997.
18. R. Walker, "Ultrasonic Agitation in Metal Finishing," in *Advances in Sonochemistry*, vol. 3, T. J. Mason, ed., JAI Press, Greenwich, CT, 1993, pp. 125–144.
19. R. Sherman, "Dry Cleaning Using CO_2 Snow," Soc. of Vacuum Coaters, *34th Annual Technical Conf. Proc.*, pp. 378–380 (1991).
20. L. C. Archibald and D. Lloyd, "The Cold Jet Process—An Environmentally Sound Alternative for Particles Removal from Advanced Substrates," in *Particles on Surfaces*, vol. 3, K. L. Mittal, ed., Plenum Press, New York, 1991, pp. 257–269.
21. Battelle Pacific Northwest Laboratories, "Supercritical Carbon Dioxide Cleaning Technology Review," July 1996.
22. L. E. Cohen, "How Clean Is Your Surface," *Plating Surf. Finish.*, 56–61 (1987); "Advanced Characterization: An Indispensable Tool for Understanding Ultraclean," ibid.
23. W. Vandervorst, H. Bender, W. Storm, M. Heyns, C. Polleunis, and P. Bertrand, "Processing," *Microelectron. Eng.*, **28**, 27–34 (1995).
24. T. Ohmi, "Ultra-clean Processing for ULSI," *Microelectron. J.*, **26**, 595–619 (1995); Battelle Pacific Northwest Laboratories, "Cleanliness Measurement Review," July 1997.
25. "Waste Minimization in Metal Parts Cleaning," U.S. Environmental Protection Agency, Office of Solid Waste, Aug. 1989.

24 Manufacturing Technologies

DENNIS R. TURNER

INTRODUCTION

Commercial electroplating began about 160 years ago. The manufacturing technology consisted of what we consider the minimum elements needed to deposit metal: open tanks, basic solutions for cleaning, rinsing, and electrodeposition, soluble or insoluble anodes, and a power source. There was virtually no ventilation or proper waste disposal. The entire operation was manual. Electroplating quality depended greatly on the experience and skill of the operator. Certain physical properties of deposits were improved after plating by heat treating or buffing. Over the years the industry grew as an art. Improved processes and techniques were discovered along the way and many were kept secret until others heard about them or stumbled on the technology themselves. Eventually platers began filing patents to protect their rights of discoveries. About the same time entrepreneurs emerged to make a business of developing and selling new ideas for better plating solutions, techniques, and machinery.

Following Edison's success with his research laboratory, many industrial companies established their own research and development laboratories, which included electroplating sections supporting manufacturing operations. Electroplating manufacturing technology was developed as needed: (1) to solve production problems, (2) to reduce cost, (3) to increase production rates, (4) to improve quality of the product, and (5) to meet new standards set by product designers and waste disposal regulations. Modern electroplating technologies developed for manufacturing include solutions with the best chemistry available, machines that automatically transport the product through all stages of processing, computerized monitoring, and control systems that also store and process data, on-line and off-line inspection facility, rapid chemical analysis systems, and an efficient waste treatment system. Many technologies are developed as manufacturing cost reductions. It can involve faster production rates with the same equipment, utilize cheaper materials, or eliminate unnecessary process steps or equipment. Also new technologies originally developed for nonelectroplating manufacture have been adopted for use for more advanced electroplating. Software for machine automation, data logging, and data processing are good examples.

Fully automatic electroplating machines can be justified only when large volumes of the same or similar product are processed. Modern automatic machines can be

Modern Electroplating, Fourth Edition, Edited by Mordechay Schlesinger and Milan Paunovic.
ISBN 0-471-16824-6

operated with fewer operators, but these people often must be more skilled and knowledgeable about instrumentation, sensors, and computers as well as electroplating.

1 ELECTROPLATING MACHINES

Plating machines were invented to do more cost-effective plating through increased worker productivity, improved quality and reproducibility, and higher plating speeds. Machine design and development improved as the need arose. Better materials especially special plastics minimizes solution contamination from the machine. Stronger and more corrosion resistant construction materials extends machine life while operating at higher production rates. Modern electronics and sensor technology have made possible more fully automatic plating machines.

Modern electroplating manufacturing technology is concerned with mass production, automation, quality, and flexibility. Mass production is important for low cost manufacturing. It requires machines that will process parts at a high rate while meeting all the specification requirements [1]. Automation helps to minimize labor costs and produce a uniform product. Quality is required to meet market acceptability and product life. Flexibility provides the ability of plating machines to process a variety of parts without extensive machine modification. It can significantly reduce the capital cost of more plating equipment. Plating machine types include barrel, vibratory, rack, edge board, strip and wire platers, and wafer platers.

2 BARREL PLATERS

Barrel plating began in the post–Civil War years. Barrels were the first plating device to dramatically increase productivity [2]. There is no racking and unracking of parts, and the operation corresponds well with other bulk treatment operations such as barrel polishing, burnishing, de-burring, phosphating, and oxide coating. Barrel plating machines usually require less space, and they are easily automated minimizing operator attention.

Plating barrels come in many sizes and designs. There two basic types: horizontal and oblique. Horizontal barrels are usually hexagonal with perforated cylinder walls and have a removable panel through which parts are loaded and unloaded (Fig. 1). They may be partially or entirely immersed and rotated via plastic gears. Oblique, or 45-degree, barrels may have solid walls open at the top and contain plating solution, anodes, and the parts being plated, or they may have perforated walls and partially or totally immersed in a tank with outside anodes. A cross section of an automatic barrel plating machine is shown in Figure 2.

Considerable work has gone into plating barrel design and evaluation of construction materials [2]. Construction of a typical barrel plating machine is concerned with (1) the barrel which is rotated so that parts inside are tumbled exposing different areas to plating, (2) a tank that holds the plating solution, anodes, support for the barrel assembly, and means for connecting the barrel to a current source, (3) a gearing system transforming motor power to barrel rotation, and (4) a means of making electrical contact to parts with flexible probes called *danglers* fed in through the rotating bearings or with metal studs fastened to the inside barrel walls.

Figure 1 Rotating plating barrel fixture.

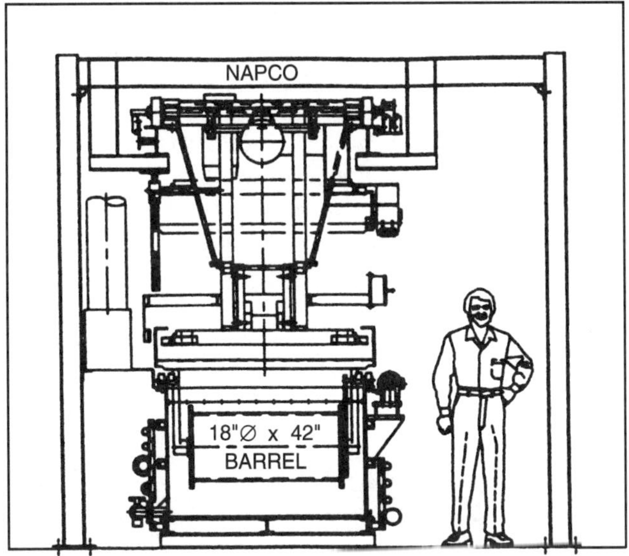

Figure 2 Cross section of an automated barrel plating machine.

To achieve the optimum in barrel plating performance in plating speed and deposit distribution, careful consideration must be given to part size, weight, shape, and type of plating to be applied [3]. Deposit thickness is made more uniform by the following processes:

1. Decreasing the deposit thickness
2. Increasing the plating time
3. Increasing the rotation speed of the barrel
4. Increasing the amount of barrel periphery open to plating solution

5. Decreasing the size of the load
6. Decreasing the size of the barrel

Plating current densities are usually low—1 to 2 amperes per square foot (ASF). Since the number of piece parts plated in a barrel is very large, the productivity of plating is high even though the plating rate is slow. Barrel plating machines are automated using lift arm or elevator-type barrel carriers. They are operated by indexing or stop-and-go movement. In more recent developments of automatic barrels, the cylinder and super structure assembly is mounted in a carriage, permitting up and down movement and a motor drive to rotate and elevate the cylinder. The carriage moves from tank to tank along rails either overhead or on the tanks.

3 VIBRATORY PLATERS

Vibratory plating is one of the newer developments in plating machines. It plates small parts in bulk like barrels, but instead of rotating, it relies on vibrational energy applied to the bottom of the processing cell to move parts about [4]. More uniform deposits are claimed for the vibratory plater compared to barrel plating. The technique is able to handle delicate fragile parts without distortion or damage. At present there are two types of vibratory platers: (1) where all processing is done in one place with solutions moved in an out for the plating sequence (Fig. 3) and (2) where the processing moves a basket with a vibrator attached from tank to tank in a sequence like a conventional barrel line [5, 6]. Both vibration frequency and amplitude are adjustable. This is necessary to compensate for process solution viscosity differences to achieve optimum agitation conditions.

Figure 3 Vibratory plating unit where all processing is done in one place moving solutions in and out using different anodes.

4 RACK PLATERS

Rack plating consists of attaching parts to an insulated frame and either manually or automatically moving the racks of parts through all plating steps [7]. Manual rack plating requires an operator to transfer racks usually on a time basis. Automatic plating produces more reproducible plating and is almost always used for large-scale production. Figure 4 is a cross section of a typical rack plating hoist assembly for printed circuit boards.

Good rack design and construction is important to successful rack plating. The following factors must be considered:

1. Size and shape of parts to be plated
2. Size and shape of rack based on tank size and other equipment
3. Method of supporting and making electrical contact to parts
4. Insulation of metal parts of rack except contacts
5. Dielectric shielding to control plating distribution on parts

Rack plating machines are of two types: straight line or return. Straight-line machines may be loaded at one end and unloaded at the other, or the process controller can be programmed to bring the rack carrier to the starting point where the same operator can unload and reload the carrier. Return machines are designed to have the carriers always move in the same direction and the tanks are laid out so that racks circle back and the unload station is next to or the same as the load position.

Rack carrier movement through the machine can be step and repeat or continuous. Step and repeat machines move carriers a fixed distance on a preset time cycle. When racks are moved from one tank to another, they are raised before

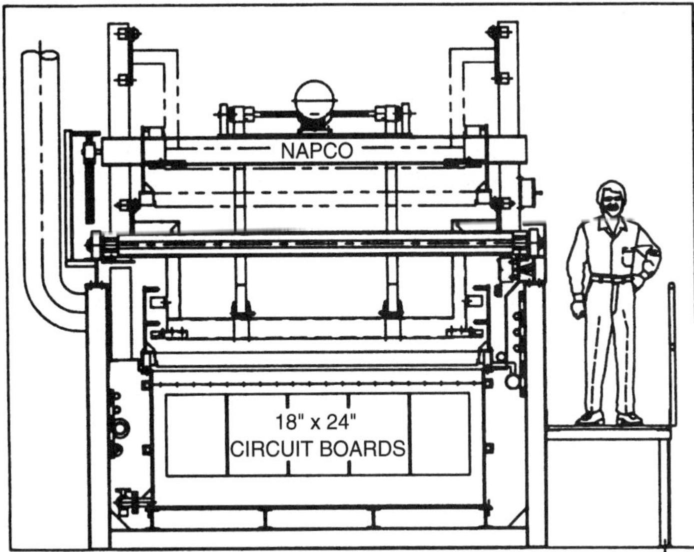

Figure 4 Cross section of an automated printed circuit board plating machine.

carrier movement then lowered. If racks are in a long plating tank, they are not raised until they reach the end of the tank. Plating thickness is determined by cathode current density and dwell time in the tank. Machines with continuous rack movement automatically raise and lower racks to move from tank to tank. As with step and repeat, dwell time at each process step is controlled by line speed.

Electroplating machines are designed to deposit metal as uniformly as possible. Side dielectric current shielding baffles, anode size, and placement are techniques used to improve plating distribution. Electroless plating is a relatively slow deposition process but it deposits metal more uniformly.

The slow plating rate can be compensated by plating many parts at one time. A plating machine for electroless plating is similar to one electroplating but different in several important respects. There is no power supply. Dwell time in the plating tank long, often 24 hours for a "full build" of copper on printed circuit boards. Continuous solution filtration is essential to remove particulates that could cause spontaneous metal deposition depleting metal ions from solution. The manufacture of PCBs using fully electroless copper is a viable technology to produce fine line conductors and high-aspect-ratio through-holes [8].

5 STRIP PLATERS

Most strip platers can be divided into two groups: (1) electrotinning and electrogalvanizing mill steel platers and (2) strip platers for electronic parts. Both types of strip platers are usually operated reel to reel continuously with splicing "on the fly."

6 ELECTROTINNING

Prior to 1937 all commercial tin plate was manufactured by the hot-dipping process. Electrolytic tin plate on steel became more economical than the hot-dipping process when continuous cold reduction mills came into operation in the early 1930s [9, 10]. By 1935 small experimental plating machines were designed and built capable of electroplating tin on steel strip at high speeds. Electrolytic tin plate became a commercial item in 1937. Electrotinning could produce a thinner more uniform tin coating than the hot-dipping process.

Three plating processes were developed: (1) alkaline stannate, (2) ferrostan (polysulfonate), and (3) Halogen (fluoride and chloride).

Plating baths for electrotinning are designed for very high speed tin deposition. Most commercial tin plating lines in the world use the ferrostan process which is based on a sulphonic acid electrolyte. Figure 5 shows schematic arrangements of three types of handling and processing machines. These are big engineering marvel machines designed for high-speed strip travel at over 11 m per second (~2000 ft per minute). The entry end "pays off" strip into a looping tower that stores a large amount of strip. When the "pay off" strip reaches near its end, it is stopped quickly spliced by welding to a new roll of strip. While the "pay off" speed is below normal line speed, strip in the looping tower keep it going until the "pay off" strip catches up and restores the volume of strip in the looping tower. A steady tension is put on the strip after the looping tower.

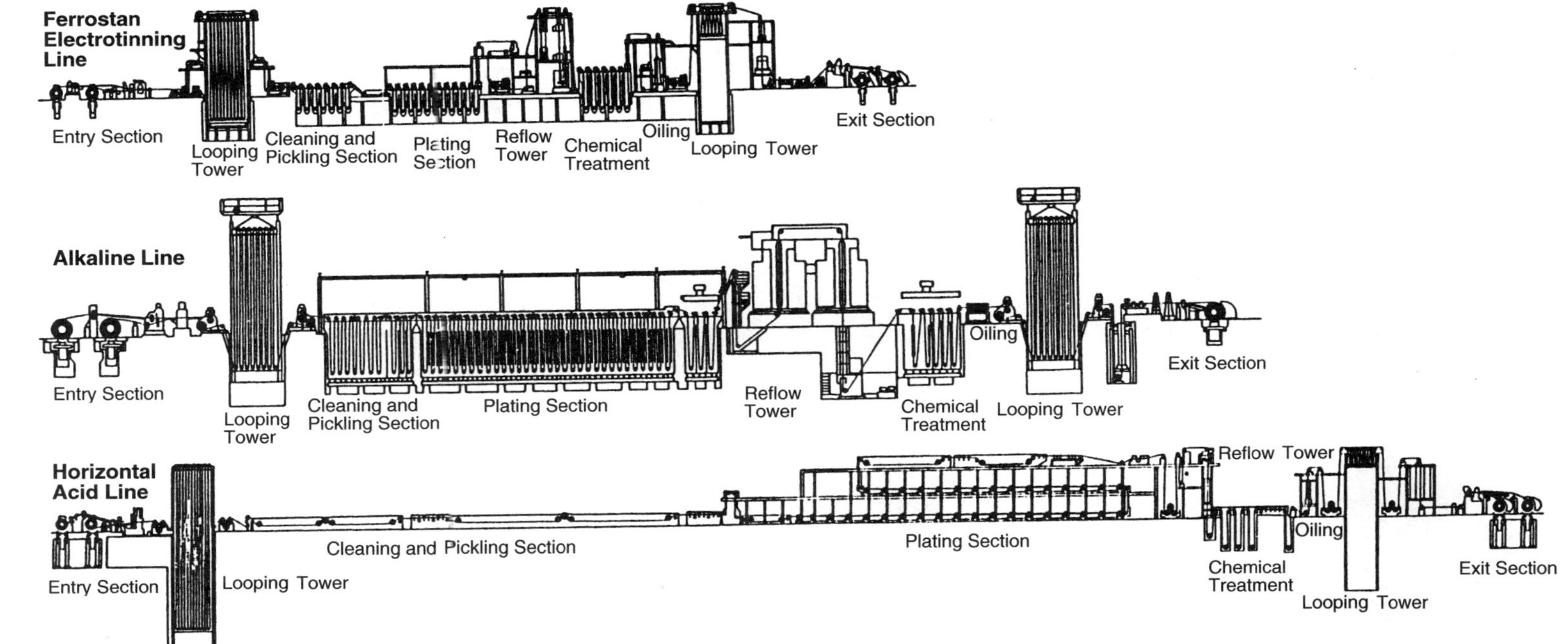

Figure 5 Schematic arrangement of the handling and processing units of three types of electrotinning lines.

Figure 6 Entry end of electrotinning line with one coil being paid off and another in reserve position—splice by welding.

Figure 7 Tin anodes being placed in tank. Note the size of the bus bars carrying current to cells.

Figure 6 shows the entry end of an electrolytic tinning line. Next the steel is cleaned and acid pickled before entering the plating section. Tin anodes need to be replaced often. The procedure is illustrated in Figure 7. Note the size of the bus bars that carry current to the cathode contact cylinders. High-speed strip travel requires very high currents to deposit the required tin. The typical electrical power capacity for a ferrostan line is 100,000 amp at 24 V.

Solution drag-out control is important since it can amount to 30 gallons per hour. After rinsing, the tin-plated strip passes a reflow tower that heats the strip above the melting point of tin. This gives the tin a brilliant luster typical of hotdipping. The fused and quenched tin is given a chemical treatment, rinsed, and dried. Finally, a lubricant oilfilm such as dioctyl sebacate is applied to improve its handling properties in succeeding operations. This is usually an electrostatic process using a high potential between strip and a fixed electrode while a mist of the oil is produced in between the surfaces.

The strip next enters the unit which supplies traction power to pull it through the machine. A tachometer generator is attached to the drive motor which produces the signals to regulate the plating and melting currents in relation to strip speed.

7 ELECTROGALVANIZING

Electrogalvanizing strip plating machines were developed after electrotinning machines [9]. They are similar in design as shown in the schematic arrangement of a line at United States Steel (Fig. 8). A looping unit is employed at both ends to enable splicing while the machine is running at full speed. Machines for plating on both sides use vertical cells like those used for electrotinning with either soluble or insoluble anodes and a zinc sulfate electrolyte.

The auto industry requires one-sided zinc-coated steel. United States Steel developed the radial cell system for zinc one side illustrated in Figure 9. The design achieves efficient plating power usage by reducing the distance between anodes and strip. CAROSEL is the acronym for consumable anode radial one-sided electrogalvanizing. The conductor rolls are 2.4 m (8 ft) in diameter. The total plating pass length is about 67 m (220 ft) for all 18 cells. The zinc for electrolytic replenishment is provided from cast zinc anodes supported on conducting bridges.

The anodes are moved across the bridges at a rate determined by the line control computer. Approximately one-half of the anode is consumed as it moves across the bridge. It is then removed and remelted for casting into new anodes. The plating current at each cell is 28,000 amp and a total of 1,008,000 amp for the 18 cells. The maximum plating voltage is 12 V. Strip widths on the line range from 91.4 to

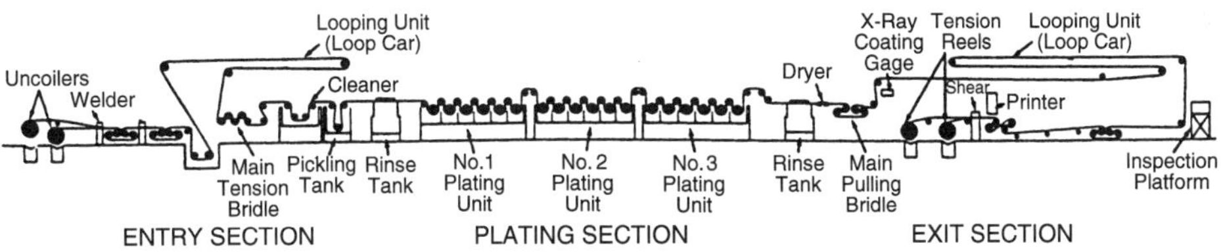

Figure 8 Schematic arrangement of a one-sided electrogalvanizing line.

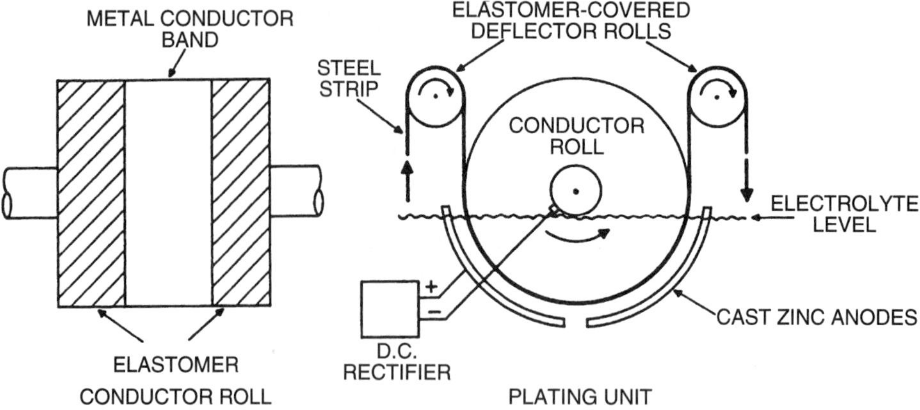

Figure 9 Schematic arrangement of conductor roll and plating unit CAROSEL for one-sided electrogalvanizing.

166.4 cm (36 to 65.5 in). Line speed is computer controlled to maximize production with the available current. Maximum line speed is 183 m per minute (600 ft min^{-1}).

After leaving the plating section, the strip is rinsed, dried, and moved to the main pulling station. An X-ray coating gauge continuously monitors the coating thickness and provides a feedback signal to the computer for line speed adjustments to maintain the specified zinc thickness.

Electrogalvanizing one side of strip up to 6 ft in width at 600 ft per minute is a commercial process. Machines for doing this are engineering marvels. They generally operate 24 hours a day 7 days a week and stop only for maintenance. Splices are made by welding "on the fly, and the plating thickness is continuously monitored by X-ray with the output signal fed to a plating current controller to maintain a constant deposit thickness. The large expense of these machines can only be justified if there is a large volume of product to be plated for several years.

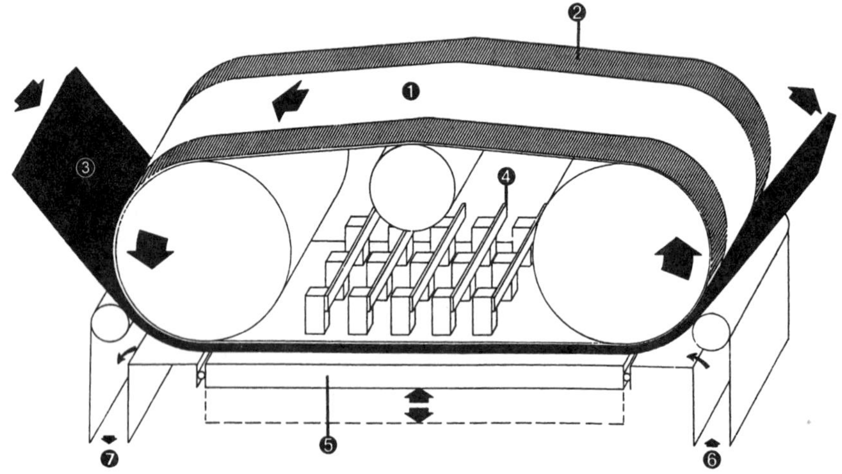

Figure 10 Schematic of the Belt cell for one-sided electrogalvanizing steel strip: (1) metal belt, (2) rubber bonded to top of metal belt, (3) continuous moving steel strip, (4) electrical contact to metal belt, (5) insoluble anode, (6) solution into cell, and (7) solution exit.

Another commercial electrogalvanizing plating cell design developed for depositing zinc on only one-side of sheet steel is shown schematically in Figure 10 [10, 11]. The "belt cell" flows solution between an insoluble anode and a cathode 6 mm apart at 4 m per second to achieve turbulent flow. Steel strip moves through the cell in the opposite direction to solution flow at speeds up to 200 m per minute. Dense coherent zinc deposits are obtained from sulfate solutions at high current densities up to 350 A dm^{-2} [12, 13]. Cathode efficiencies are as high as 95%, which is attributed to the high solution flow rate.

8 STRIP PLATING FOR ELECTRONICS

The explosive growth of the electronic industry began about 40 years ago, and it continues. It lead to a tremendous demand for contacts, lead frames, wire leads, chip carriers, and many other components requiring electroplating of tin, solder, nickel, silver, gold, palladium, or palladium-nickel. The huge number of components required make it necessary to develop electroplating technology capable of high speed and sometimes selective processing. Loose parts such as pins can be plated in high volume in barrels and vibratory machines. Gold and other precious metals often must be deposited selectively for economic reasons and strip platers are best suited. These are usually reel-to-reel plating machines designed to process one or more part shape. Modern technology has provided strip plating machine designers with a vast array of chemically stable materials, electrical and electronic devices, and subsystems. High-speed machines reel-to-reel platers can be divided into four segments: (1) strip transport system, (2) a table with small plating cells and larger solution reservoir tanks, (3) electrical power sources and controls, and (4) a display of plating conditions. An example of such a machine is shown in Figure 11. The machine was designed to electropolish, nickel plate, and gold plate telephone modular cord plug blades. All cells and reservoir tanks were covered to minimize contamination from the environment and make fume exhaust more efficient. It has four separate plating lines each running at 12 ft per minute. Figure 12 is a schematic of the strip plater.

Since the blade was formed by punching, the contact area had a shear/break surface and needed considerable electropolishing. Dwell time in the electropolishing cells required electropolishing selectively at a current density of about 4000 amp per square feet to achieve the desired finish. A high current to a strip can lead to several problems: (1) electrical arcing at the contacts can eventually causes a poor contact, (2) excessive heating of the strip, and (3) excessive heating of the solution. The arcing problem was solved by using tungsten/silver alloy contacting material.

Electropolishing quality degrades with solution heating. Strip heating is minimized by keeping the cell short and electrically contacting strip at both ends. Solution heating is controlled also by keeping anode-cathode spacing as close as possible and cooling the phosphoric acid electropolishing solution in the reservoir tank. Another problem occurs when electropolishing copper alloy strip. The solution becomes saturated with copper and eventually starts depositing at the cathode as a powder. The copper powder bridges the cell causing an electrical short so electropolishing stops. The powder is easily removed manually, but an automatic method is more economical. Two methods can be used: (1) a rotating cylindrical cathode with a scraper to remove the powder or (2) periodically vacuum off the solution with powder next to the cathode. The solution is returned to reservoir tank after the powder is separated out.

Figure 11 Four-line automatic strip plater for telephone modular cord blade plugs.

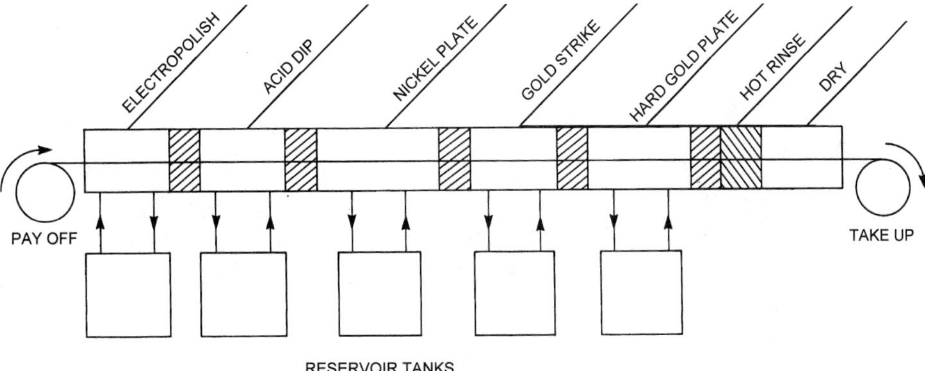

Figure 12 Schematic of the strip-plating facility. Each cell is a miniature chemical processing tank.

Good rinsing between chemical and electrochemical steps is very important in all electroplating systems. Spray rinsing is very effective provided nozzles stay unclogged. A filter in every spray rinse line is necessary. Nickel plating cells use small titanium baskets with bags and nickel pellets inside. The entire strip receives a gold strike followed by a selective hard gold plate just at the contact edge. Sensors detect when a payoff reel empties or a take-up reel becomes full. That particular line automatically stops until the operator changes reels and restarts the line.

Another type of strip plater is used in the printed circuit board industry to electroform a thin continuous sheet of copper which is bonded to a plastic dielectric. Copper is plated on the outside of a large rotating polished stainless steel drum. The drum is treated so the copper is removed easily in a continuous sheet. The outside surface of the sheet is purposely plated with slightly rough copper surface to improve adhesion when it is bonded to a plastic dielectric. This is called *vendor* copper and may be applied to one or both sides of a flexible or rigid dielectric.

A third type of strip plater manufactures flexible printed circuits. All chemical and electrochemical processing is done on wide roll to roll strip plating machines. A thin copper sheet is bonded to one or both sides of a flexible dielectric material such as mylar. A plating mask is applied to one or both sides. Rolls of strip, or *web* as it is called, then are passed through the plating machine to produce a printed circuit copper much thicker than the initial copper. Tin-lead solder may be plated over the copper for solderability and/or as an etching mask when the plating mask is removed and the exposed initial copper is etched away. The web moves over wide rollers on top and down into each tank in the processing line. The end of each roll is spliced to the next, so the machine sees a continuous web.

Narrow strip platers are used mainly to process parts for electronic applications. These are reel to reel platers and often deposit metal selectively. Considerable development has gone into these machines in recent years [16]. They are highly automated, plate at high speeds, and deposit metal with excellent precision. Typical applications are connector terminals and semiconductor lead frames. Gold was often used in these applications, but recently palladium and palladium-nickel alloy with a gold flash has replaced most of the gold [16]. Specially designed selective plating systems have been developed to minimize the usage of precious metals [18].

Figure 13 Control panel with processing display above.

Strip may be punched or unpunched. Plating unpunched strip is referred to as *stripe on strip*, with punching done afterwards. Most connector terminal plating is done on punched strip with terminals held together with tie bars, which are removed during connector assembly.

Development of high-speed strip platers began just at the time small low-cost microprocessors became available for machine control. This technique led to programmable logic controllers (PLCS) being used for overall process and machine control of strip platers. More recently small computers such as PCs have been used but not in direct machine control. PCs are very good for operator interface and reporting functions. PLCs are better for direct control [19, 20]. Often both electronic devices are used in combination for plating machine monitoring and control [21]. A main plant computer can be programmed to watch over plating operations and help to alert engineers of problems wherever they occur. The machine shown in Figure 12 utilizes a PLC for process monitoring and control with an interface PC data logging and analysis. Customers often request data on the processes used to electroplate their product and will pay extra for the information. A central console and process display panel (Fig. 13) allows the operator to quickly see the status of machine operation. Each process segment of the machine is designed to run in either the manual or automatic mode. This is desirable when a segment may not be functioning properly in the automatic mode.

9 BANDOLEER PLATING

Bandoleer plating involves attaching individual parts such as connector terminals to a carrier belt that moves through all processing steps like a strip plater. Parts that otherwise would be plated all over in a barrel or vibratory plater can be selectively plated on a bandoleer machine. A vibrator parts handler orients and moves the parts to the carrier belt where they are attached. At the end of the line, parts are detached and the belt circles back to the front of the machine.

9.1 Edge Board Platers

Edge board or tab platers are designed to plate precious metal contacts on printed circuit boards. These are relatively new machines in the plating industry and were developed to serve a specific need for rapid selective plating of gold on contact fingers. Boards are loaded automatically into the continuous motion transport system by indexing the board carrier at the proper time so that moving belts can grasp the board and start it through all processing steps similar to plating nickel and gold on connector contacts in a strip plater. Rubber belts also serve as plating masks for selective plating to eliminate taping. Transport and plating belts are specially designed to firmly contact boards with compliant rubber that will not wear excessively and not stretch. These properties can only be achieved with a compound belt made up of several layers. The belt may also incorporate metal inserts for electrical contact to the contact fingers.

Good solution agitation around the cathode is essential for high-speed plating and rapid movement of boards through the machine. Plating cell length is another factor that can determine machine speed. If line speed is limited by plating rate in a particular cell, any increase in that cell length can increase line speed. Improving solution agitation around the cathode area will also allow higher line speeds.

Good rinsing between chemical treatments is important. Belts also must be rinsed thoroughly before the belts return and contact new boards coming into the plating line.

Drying is done with a blast of hot air directed at the wet areas. Unloading is automatic with boards pushed into board carriers that are indexed to receive each board as it comes off the plating line. Most edge board platers are single lines as multiple are too complicated especially when belt transport and masking are used.

Some tab platers do not use belts. Masking is done with tape that can be done by machine, but this operation is more costly. In such machines contact to the fingers is made via a tie bar around the outer edge of the board which is cut off later. Electric current is brought to the tie bar either at the side or top. Board transport is by moving belts or an overhead trolley with fasteners holding boards.

10 WIRE PLATERS

Wire platers were the first high-speed platers with plating current densities up to 2000 ASF [9]. These high current densities are made possible by moving the wire at high speeds through plating solution causing rapid wire motion relative to solution not only parallel to the pull direction but also perpendicular as the result of wire vibration through the plating cell. Wire is an ideal cathode usually being round in cross section and continuously uniform in shape throughout the plating operation. Wire also lends itself to multiple line platers which is important in achieving high-volume production of plated wire needed in a variety of applications.

Studies on parameters influencing the speed and uniformity of electroplating wire include cell design and methods of forced fluid flow, as illustrated in Figure 14

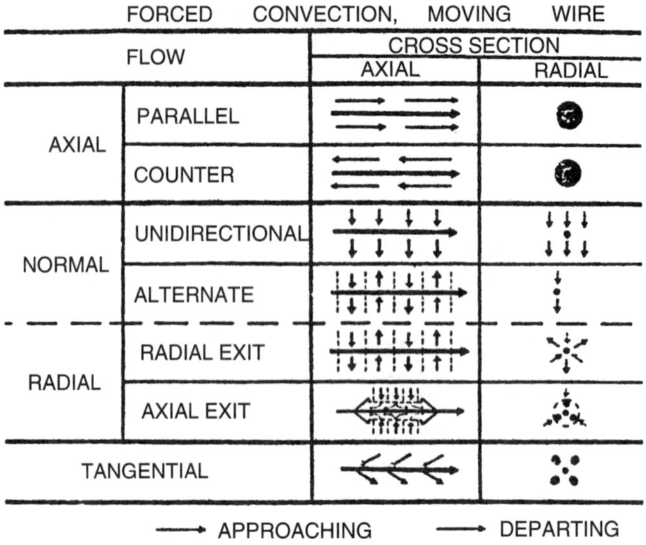

FORCED CONVECTION, MOVING WIRE

FLOW		CROSS SECTION	
		AXIAL	RADIAL
AXIAL	PARALLEL		
	COUNTER		
NORMAL	UNIDIRECTIONAL		
	ALTERNATE		
RADIAL	RADIAL EXIT		
	AXIAL EXIT		
TANGENTIAL			

→ APPROACHING → DEPARTING

Figure 14 Classification of cell designs on the basis of the direction of forced fluid flow.

[22, 23]. Classification of various cell designs is done on the basis of the direction of forced fluid flow which may be axial, normal/radial, or tangential. These are modified somewhat in practical cells [24].

Conditions which produce turbulent flow are favored for high speed plating. Tin and tin-lead alloys are plated on copper for electrical wire. Gold or palladium-gold is plated on copper wire for connector jacks, copper is plated on steel wire for communication systems, and zinc on steel for fencing and other outdoor applications.

11 DECORATIVE AND ENGINEERING PLATING

At one time a thin bright chromium deposit over bright nickel plating was used extensively on automobile and appliance parts. In recent times a combination of product styling changes and increased cost of the Ni–Cr finish due to tougher environmental regulations has reduced its use mainly to just trim. A significant improvement in the cost of hexivalent chromium disposal or recovery would help bring greater use of the attractive bright Ni–Cr electroplated finish.

Engineering electroplating technology generally involves improving mechanical wear or corrosion protection applying metal with special physical properties on very small or very large objects [25]. Outdoor objects such as statues and building domes are usually made decorative by plating but the process of doing the plating involves many engineering factors. The technique of brush plating is used extensively for applications where the part to be plated is too large or in a fixed position away from a plating shop. Special plating fixtures and solutions for substrate cleaning and depositing a variety of metals have been developed [26]. The technique is also used to repair worn or poorly machined parts. Electroforming of metal shapes can be another type of engineering plating [27]. Complicated forms impossible to machine are examples of ideal applications of electroforming. These can be very large objects or extremely tiny hairlike forms such as the copper coils used in computer memory disc drive heads [28, 29]. Another example is the manufacture of compact discs (CDs) where electroplated nickel stampers are electroformed with the precise image of the master audio or video recording [30].

12 LABORATORY TECHNOLOGY DEVELOPMENT

Electroplating manufacturing technologies are often developed first in a laboratory. An example is the reciprocating paddle plating cell which was developed by IBM for uniform and rapid deposition of NiFe, Cu, and Au on micro features for electronic and magnetic devices [31, 32]. Laboratory facilities are also important to help solve production problems when the plating process goes bad. Simulated conditions on a small scale can be set up to determine the cause of a problem and also how it may be corrected on the production line. Laboratories are also set up to do routine chemical analysis and plating tests such as the Hull cell or the new hydrodynamic controlled Hull cell which can solve manufacturing problems [33].

13 COMPUTER ROLE IN MANUFACTURING

Originally computers were designed for office use of word processing and spreadsheet applications. Their use has since expanded into many other areas including electroplating manufacturing technology. The development of more powerful hardware and software designed specifically for manufacturing technology with flexibility can monitor and control not only electroplating operations but also the business aspects of a plant. Computers make it possible through a system called *computer-integrated manufacturing* (CIM) [34].

The CIM concept aims to help management get the whole picture of the manufacturing operation. An efficient electroplating function requires adequate material resources for the production of products based on orders, status of product flow through plant operations, checks on quality, monitoring and control of instruments and processing devices, inventory of finished product, shipments made, and so on. CIM will also integrate the business aspects of the factory. The overall information helps a company know what its costs are in each segment of manufacturing. Computers are also helping to automate factory operations including electroplating.

Presently Programmable Logic Controllers (PLCS) are the key building blocks of industrial automation [35, 36]. They are designed from the ground up for machine monitoring and control. They are rugged and relatively inexpensive for automating machines. Furthermore they can be programmed in ladder logic diagrams which is similar to the electrical diagrams used by electricians that hard-wire a machine and are also called on to correct electrical problems. Programming is done using a cathode ray tube display that also pinpoints the location of an electrical problem. PLCs are becoming more advanced just as computers improve. They will probably continue to be the direct interface to electroplating machines but also interact with computers for information processing and transfer within a CIM system. Specially computer designers are developing hardware and software that will succeed in replacing PLCs in some machine monitoring and control applications.

14. SUMMARY

A wide variety of plating machines have been invented and developed to serve a specific need for electroplating parts with increased productivity, better deposit properties, and at a lower manufacturing cost. There is greater use of electronic controllers and sensors to increase the level of automation in the industry. Technology already exists to fully automate most plating machines. The utilization of this technology depends on its cost effectiveness and ability of plating engineers and operators to handle advanced technology such as computers. Plating processes continue to improve with better performance and knowledge about the best metal finish for various applications.

ACKNOWLEDGMENTS

Figures 2 and 4 were supplied by NAPCO, Inc. Figures 6 through 10 are reprinted with permission from *The Making, Shaping and Treating of Steel*, 10th ed., 1996 AISE.

REFERENCES

1. M. J. Moll and L. J. Durney, *Electroplating Engineering Handbook*, 4th ed. L. J. Durney, ed., Van Nostrand Reinhold, New York, 1984, p. 606.
2. W. H. Jackson, A. K. Graham, and R. K. Asher, *Electroplating Engineering Handbook*, 4th ed. L. J. Durney, ed., Van Nostrand Reinhold, New York, 1984, p. 573.
3. S. E. Craig Jr. and R. E. Harr, *Plating*, **60**, 617 (1973); **61**, 1101 (1974).
4. M. R. Kalantary, S. A. Amadi, and D. R. Gabe, *Electrochemical Engineering*, Inst. Chem. Symp. Ser. No. 112, Hemisphere, New York, 1989, pp. 199–206.
5. J. A. Lochet, Vanguard Research Assoc., Inc., company publication on vibratory plating.
6. P. A. Schopfer, MECO USA Inc. A company publication on Vibratory Plating.
7. H. Tilton, *Electroplating Engineering Handbook*, 4th ed., L. J. Durney, ed., Van Nostrand Reinhold, New York, 1984, p. 559.
8. H. Nakahara, *Printed Circuit Handbook*, C. F. Coombs, ed., McGraw-Hill, New York, 1995, ch. 20.
9. *The Making, Shaping and Treating of Steel*, 10th ed., AISE, 1996.
10. R. Winand, Belgium Patent No. 879,696 (Feb. 15, 1980).
11. R. Winard, *Oberflache-Surf.*, **25** (11), (1984).
12. A. Weymeersch, R. Winand, and L. Renard, *Plating Surf. Finish.*, **68** (4), 56 (1981).
13. A. Weymeersch, R. Winand, and L. Renard, *Plating Surf. Finish.*, **68** (5), 70 (1981).
14. A. K. Graham and S. S. Johnston, *Electroplating Engineering Handbook*, 4th ed. L. J. Durney, ed., Van Nostrand Reinhold Co., New York, 1984, p. 617.
15. D. R. Turner, W. W. Evarts, R. W. Duston, and G. L. Byars, *West. Electr. Eng.*, **22** (2), 26 (1978).
16. D. R. Turner, *Prod. Finish.*, **66** (2) (1987).
17. J. A. Abys, H. K. Strachil, I. Kadija, E. J. Kudrak, and J. Blee, *Met. Finish.*, **89** (7), 43 (1991).

18. D. R. Turner, *Met. Finish.*, **76** (10), 19 (1978).
19. B. W. Niebel, A. B. Draper, and R. A. Wysk, *Modern Manufacturing Process Engineering*, McGraw-Hill, New York, 1989, p. 805.
20. R. Mackiewicz, *Process/Industrial Instruments and Controls Handbook*, 4th ed., McGraw-Hill, New York, 1993, sec. 3.32.
21. M. Baab, *Control Eng.*, **40** (4), 46 (1993).
22. A. Tvarusko, *Plating*, **58**, 983 (1971).
23. R. C. Alkire and A. Tvarusko, *J. Electrochem. Soc.*, **119**, 340 (1972).
24. A. Tvarusko, *J. Electrochem. Soc.*, **120**, 87 (1973).
25. G. F. Bidmead, *Trans. Inst. Metal Finish.*, **59**, 129 (1981).
26. J. C. Norris, *Metal Finishing Guidebook*, M. Murphy, ed., 1992, pp. 301–320.
27. J. W. Dini and H. R. Johnson, *Plating Surf. Finish.*, **64** (8), 44 (1977).
28. D. A. Thompson and L. T. Romankiw, *IBM Disk Storage Tech.*, **3–5** (2) (1980).
29. L. T. Romankiw and T. A. Palumbo, "Electrodep. Tech., Theory and Practice," *Electrochem. Soc. Proc.*, vol. 17, 13 (1987).
30. H. V. Reynolds, *Plating Surf. Finish.*, **82** (5), 80 (1995).
31. J. V. Powers and L. T. Romankiw, U.S. Patent 3,652,442, (1972).
32. D. E. Rice, D. Sundstrum, M. F. McEachern, L. A. Klumb, and J. B. Talbot, *J. Electrochem. Soc.*, **135**, 2777 (1988).
33. I. Kadija, J. A. Abys, V. Chinchankar, and H. K. Straschil, *Plating Surf. Finish.*, **78** (7), 60 (1991).
34. J.-B. Waldner, *CIM Principles of Computer-Integrated Manufacturing*, Wiley, New York, 1992.
35. T. E. Kissell, *Understanding and Using Programmable Controllers*, Prentice-Hall, Englewood Cliffs, NJ, 1986.
36. E. Mandado, J. Marcos, and S. A. Perez, *Programmable Logic Devices and Logic Controllers*, Prentice-Hall, Englewood Cliffs, NJ, 1996.

25 Monitoring and Control

DENNIS R. TURNER

INTRODUCTION

Monitoring and control of electroplating processes is important to obtaining quality deposits. It is also essential for an automatic plating system designed to minimize operator attention of production. A skilled and knowledgeable operator can maintain a reasonable plating quality with little supporting equipment. However, trends to more automation to reduce costs with higher processing speeds and assurance of quality plating make it necessary to utilize many monitoring and control facilities.

1 ELECTROPLATING LINE

An electroplating line consists of a complex of equipment including motors, tanks, water rinses, product transport equipment, filters, heaters, power supplies, process monitoring devices, process control devices, and the like. Electroplating process conditions to be monitored and controlled in a typical line are as follows:

- Solution chemistry
- Solution temperature
- Solution level
- Solution flow
- pH when necessary
- Specific gravity when necessary
- Additive concentration
- Impurity levels when necessary
- Current at each electrochemical step
- Cell voltages
- Line speed
- Blow-off air pressure when necessary
- Exhaust vacuum

Modern Electroplating, Fourth Edition, Edited by Mordechay Schlesinger and Milan Paunovic.
ISBN 0-471-16824-6

2 SOLUTION CHEMISTRY

The composition of electroplating solutions has changed very little in recent years with the exception of high-speed plating where higher metal content and special additives are used. Wet chemical methods of analysis have been developed to determine the composition of most plating solutions [1, 2]. Instrumentation for rapid chemical analysis of inorganic and organic species has become highly automated providing accurate and rapid analysis.

Many industries utilize state-of-the-art electronic technology. Some instruments can be converted for on-line chemical analysis. The advantage of on-line monitoring and control of plating solutions is that deviations from optimum concentration ranges can be corrected as soon as set limits are exceeded.

Plating solution chemistry is usually monitored off-line by taking a solution sample and doing an analysis at the machine or in a chemical analysis laboratory. Solution adjustments are then made as necessary. Metal ion concentration, supporting electrolytes, additives, and pH are most often monitored. Metal ion concentrations are measured by colorimetry, polarography or with ion selective electrodes. Automated versions of these methods have been developed for on-line analyses of many metals including nickel, copper, lead-tin, and gold [4–6].

Fiber optics has made small colorimeters practical for automatic analysis of many plating solution species [7]. A light source and absorbance detector can be remote from the analysis cell via a bundle of flexible glass or plastic pipe fibers. The optimum wavelength of light to be used for a given species is best determined from a spectrographic scan over a wide range of wavelengths. One then selects a wavelength at or near a strong absorption peak and prepares a calibration curve. For example, copper (II) in electroless plating baths is determined at 620 nm wavelength. Cobalt (II) in hard gold baths is analyzed by first oxidizing it with hypochlorite to cobalt (III). Cobalt (III) forms a pink complex with EDTA which can be analyzed at 520 nm wavelength (8).

Polarography using a dropping mercury electrode (DME) has made considerable progress over the past 40 years as a laboratory technique for chemical analysis of plating solutions (Table 1), but it was not used in the manufacturing environment until about 20 years ago, mainly because of the sensitivity of the electrode to manual handling. When manual handling is eliminated by automation, the DME can be used as a reliable and reproducible sensor in automatic analyzes and controllers in industrial applications [4, 5].

Ion-selective electrodes have been developed which can monitor the concentration of a number of anions and cations [9–11]. A list of commercially available ion-selective electrodes is given in Table 2. The membrane can be glass as for a pH electrode which measures H^+ concentrations or liquid or solid material. Orion, one of the principal developers and suppliers of ion-selective electrodes, produced an automatic plating bath analyzer using their electrodes. Some ion-selective electrodes have limited lifetimes such as the CN^- electrode. Only periodic analysis is practical with a water soak in between applications. However, this problem can be minimized with a well designed sampling and flushing system.

A new wide strip electrotinning line built by British Steel in Wales incorporates a fully automatic analytical system to monitor solutions along the line [12]. Samples are taken automatically from cleaning, pickling, and plating tanks by taking a bleed

TABLE 1 Examples of Polarographic Analysis of Several Plating Baths

	Polarography can Determine These:		
In These Baths	Major Components	Trace Metals	Organic Additives
1. Zinc Sulfate	Zinc	Copper, Cadmium, Arsenic	o-Chlorobenzaldehyde
2. Palladium	Palladium, Chloride	Tin	Hydroquinone
3. Gold(I) Cyanide	Gold, Free Cyanide	Cadmium, Cobalt, Copper, Zinc, Iron, Tin, Chromium	
4. Watts' Nickel	Nickel, Chloride, Boric Acid		Saccharin, o-Benzaldehyde sulfonic acid
5. Electroless Copper	Copper, Formaldehyde		Mercaptobenzothiazole
6. Copper Sulfate	Copper		Thiourea and Thiourea derivatives
7. Solder Bath	Lead, Tin(II)	Tin(IV)	
8. Brass	Copper, Zinc	Lead, Arsenic	
9. Nickel-Cobalt	Nickel, Cobalt		

TABLE 2 Commercially Available Ion-Selective Electrodes

Ion	Membrane	Lower Detection Limit	Principal Interferences
H^+	glass	10^{-14} M	None below pH 13
Na^+	glass	10^{-6} M	H^+, Ag^+
K^+	glass	10^{-4} M	H^+, Na^+, Ag^+
K^+	liquid	10^{-4} M	NH_4^+, H^+
BF_4^-	liquid	10^{-4} M	NO_3^-, Br^-, ClO_4^-, I^-
NO_3^-	liquid	10^{-4} M	I^-, Br^-
Cl^-	liquid	10^{-4} M	I^-, NO_3^-, Br^-
Ca^{++}	liquid	10^{-4} M	Zn^{++}, Fe^{++}, Pb^{++}, Cu^{++}, I^-
Water hardness	liquid	10^{-3} M	Zn^{++}, Fe^{++}, Pb^{++}, Cu^{++}
F^-	solid	10^{-6} M	OH^- at high pH
Cl^-	solid	5×10^{-1} M	S^-, I^-, Br^-, CN^-
Br^-	solid	5×10^{-1} M	I^-, S^-, CN^-
I^-	solid	5×10^{-1} M	S^-
S^-	solid	10^{-17} M	None
CN^-	solid	10^{-6} M	S^-, I^-
Ag^+	solid	20^{-17} M	Hg^{++}
Cd^{++}	solid	10^{-7} M	Ag^+, Cu^{++}, Hg^{++}
Pb^{++}	solid	10^{-7} M	Ag^+, Cu^{++}, Hg^{++}
Cu^{++}	solid	10^{-8} M	Ag^+, Hg^{++}

from each tank and conveying it to a remote analytical station. There a robot prepares the samples for analysis by an inductively coupled plasma spectrometer. Results are fed back to the control station where the operator makes the necessary adjustments.

The chemistry of electroless plating baths change rapidly when operating. Solutions need to be analyzed frequently and chemicals replenished as needed. Manual bath sampling and analysis is often not practical. Machines have been developed which automatically take samples, analyze and reconstitute the bath chemistry [13]. These will be discussed later.

3 SOLUTION TEMPERATURE

Solution temperature can be monitored with a thermocouple, thermistor, RTD (resistance temperature detector), or silicon integrated circuit device. Many sensors for plating parameters are temperature sensitive and the output signal must be compensated for an accurate measurement. This is often done with an on-board silicon integrated device that senses temperature and corrected the parameter measurement [14].

4 SOLUTION LEVEL

The solution level in a tank can be monitored simply with a float containing a magnet that operates reed switches: one for the low-level position and one for the high-level. These switches are used to maintain solution level with replenishment water. Other more sophisticated devices utilize optics or sound to monitor and control solution levels. Plating tanks that have a high concentration of electrolyte can form salt deposits on a float causing it to become immobile. This is avoided by directing the replenishment water at the float area.

The exception to using automatic solution level control is for gold plating tanks. Platers can be concerned about the reliability of an automatic solution level system for a gold plating tank for fear of overflow and losing gold down the drain. Instead of automatic addition of water to gold and other precious metal plating tanks, an audio alarm can sound so the operator can adjust the level manually.

5 SOLUTION FLOW

Rapid solution flow around the cathode provides agitation for higher plating speeds. Large tanks for barrel or rack plating improve agitation at the cathode by moving the cathode or sparging solution or air from the bottom of the tank upward. In a strip plater that uses a line of small plating cells and reservoir tanks underneath, solution is pumped into the cells at rates sufficient to replenish the metal ion content but also, perhaps more important, to produce rapid agitation around the strip for high-speed plating. An on–off sensor using a magnet that operates a reed switch can signal if a pump malfunctions. For rate of flow, a paddle wheel magnet device, pressure differential, or electromagnetic sensor can be used.

6 SOLUTION pH

A pH electrode monitors hydrogen ion activity. Therefore it directly measures the acidity or alkalinity of a plating bath. All plating baths work best over a specific, sometimes narrow, pH range. The pH sensing and control systems are commercially available. The pH electrodes are also used to detect the concentration of certain compounds in plating baths by measuring solution before and after a reagent addition. For example, in electroless copper plating the amount of formaldehyde present can be determined by measuring the pH of a sample of solution before and after fixed amounts of H_2SO_4 and Na_2SO_3 are added. This method was used by Photocircuits and McDermid for automatic analysis and control instruments for electroless copper plating processes.

The standard glass electrode is reliable pH electrode provided it is kept wet. It cannot be immersed in strong alkali, fluoride or fluoborate solutions for very long before the glass is etched away. However, it can be used to measure pH in those solutions for a reasonable period of time if the electrode is rinsed and stored in distilled water after each quick pH measurement.

7 SPECIFIC GRAVITY

Specific gravity is monitored to maintain a concentration range for sufficient solution conductivity or to detect excessive build up of salts. A simple manually operated hydrometer can be used. A more elaborate automatic device uses a radio frequency oscillator with a loop probe exposed to the solution. The oscillator frequency shifts in proportion to the specific gravity.

8 ADDITIVES

Additives in plating baths are important to achieving the desired physical properties of many electrodeposits. Additive concentrations must be maintained within a specific range for best results. Techniques used for monitoring additive concentrations include voltametric stripping [15–21], polarography [22], and spectrophotometry [22]. The method used must be able to detect the active species in the presence of degraded nonactive material and/or impurities.

9 IMPURITIES

Solution impurity build up sometimes can be ignored. For example, when drag-out removes enough solution to keep the impurity level low or when the impurity has little or no effect on the deposit. Organic impurities can be removed by activated carbon treatment when necessary. However, it usually strips the bath of all organics so additives need to be replenished. Harmful impurities that are kown to build up in a plating solution should be monitored routinely. Often the first sign of an excessive impurity level is off-color plating. Water for rinsing often needs to be free organic and inorganic impurities [23]. A conductivity measurement is the best way to determine the presence or absence of inorganic impurities. A good strategy to minimize the amount of high-quality rinse water required and volume of wastewater produced is to use the good water first where impurities can be most harmful, for example, after nickel plating. When the rinse water conductivity reaches a set level, that water can be transferred back to the preceding rinses that do not require as high-quality water. This illustrated in Figure 1 for a strip plater involving electropolishing,

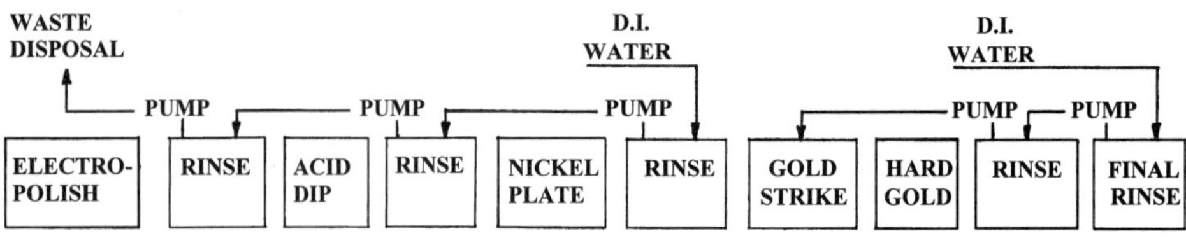

Figure 1 Water management system for a strip plater.

nickel plating, and gold plating. High-conductivity water is used only to maintain water levels in reservoir tank rinses after nickel and hard gold plating. When the conductivity in the nickel water rinse drops below a preset limit, water is pumped out of the rinse after electropolishing to waste disposal. That water is replaced by water from the rinse after acid pickling. Next water from the nickel rinse replenishes the acid rinse, and finally new water restores the rinse after nickel. The process continues until the conductivity of the nickel rinse water reaches a preset value. A similar feed back system is used in gold plating side. The gold strike solution is operated at about 65°C (150°F) and loses water rapidly through evaporation into the exhaust. A low-solution level activates a pump to transfer rinse water after hard gold to the gold strike reservoir. The final rinse water replenishes the preceding rinse. Fresh DI water restores the level in the final rinse. In addition to conserving water, this system keeps the gold in rinses moving back to the plating solutions. A gold reclaim plating cell in the rinse tank after hard gold plating can be used to reclaim gold.

10 PLATING CURRENT

The current at each electrochemical step can be displayed on an ammeter. It is also easily monitored by measuring the voltage drop across a resistor. That voltage can be used in an electronically controlled power supply to maintain the current at a constant value. An open circuit or power supply failure in the system can be detected by a Programmable Logic Controller (PLC) or computer to stop a plating line and sound an alarm.

11 CELL VOLTAGE

Monitoring electroplating cell voltages is necessary to detect cell electrical shorts between anode and cathode. This can occur in electropolishing cells of strip platers with close anode-cathode spacing when a copper alloy is the strip. Copper powder is produced, it collects at the cathode and eventually forms a conducting path between anode and cathode. Methods of automatically removing the copper powder are discussed in Chapter 24.

12 LINE SPEED

Line speed determines the dwell time of the product in each processing step. In plating cells it can determine the plating thickness. Big machines electrogalvanize steel strip monitor zinc thickness with an X-ray gauge that provides a feed back signal to a computer for line speed adjustment to maintain a specified zinc thickness [24]. Punched parts that can interrupt light periodically is used to control strip speed with a photo-optic system that produces an electrical pulse as each segment moves by. An electronic pulse counter is calibrated to display line speed and provide feedback to control line speed.

13 SOLUTION BLOW-OFF

Solution drag-out as parts leave a processing cell can be minimized by blowing air at the part in the right direction as parts leave the tank. Large mill plating machines for electrotinning steel strip can drag out 30 gallons of solution per hour unless special measures are taken to remove as much drag-out as possible [24]. Standard shop floor compressors should not be used for blow off air, since the air usually contains oil. A good roof air blower is preferred, its pressure should be monitored to ensure that adequate solution blow-off occurs.

14 FUME EXHAUST

A reliable fume exhaust vacuum system is important to the health of plating shop workers. A sensor placed in the exhaust plenum detects the level of fume exhaust. one type of vacuum sensor placed in the exhaust plenum works by monitoring the cooling of a heated leg of a Wheatstone bridge. If the exhaust air stops or deceases to an unacceptable level, the sensor can trigger an alarm.

15 MONITORING AND CONTROL EQUIPMENT

The designer of modern electroplating equipment has available an ever-growing array of devices for monitoring and controlling almost every aspect of the process. Plating machines can run automatically at high speeds, make adjustments, and signal an operator if help is needed. The capital investment and maintenance cost must be justified by an increased rate of production while maintaining or improving product quality. Equipment used in monitoring and control of plating include timers, ammeters, ampere-hour meters, sensors, transducers, programmable logic controllers, computers, and software.

16 TIMERS, AMMETERS, AND AMPERE-HOUR METERS

Timers with ammeters were the first means of monitoring the amount of metal electroplated. In many shops they are the only instruments needed to produce an acceptable product. The first automatic control device was the ampere-hour (or minute) meter that stops plating when a preset number of coulombs has passed through the cell. When high currents are involved, an intermediate relay may be necessary. Simple automatic plating machines rely on timers to advance product through processing steps.

17 SENSORS

Advanced sensors coupled with modern control technology revolutionized automatic manufacturing [25–27]. Sensors play a vital role in providing information about plating machines and processes that are necessary for automatic monitoring and control [28]. They have been described as "the eyes and ears" of electronic control systems on automatic plating machines. There are three basic sensor types: on–off, analog, and digital. The on–off type can be considered a form of digital sensor without quantitative information. For many situations, an on–off sensor is sufficient. For example, when a take-up reel of strip becomes full or a payoff reel becomes empty, a simple on–off signal is adequate. Analog sensors provide quantitative information that be read on a meter, but for automatic control it is usually necessary to convert the signal from analog to digital for electronic processing. Many sensors produce a digital signal directly. One example is a photoelectric sensor that generates pulses at a rate proportional to a plating line speed. Electronic processing then converts pulse rate to line speed which can be read on a digital voltmeter or fed into an electronic motor-speed controller.

An integrated system of sensors tied to an electronic controller is required to free an operator from machine watching and fine-tuning critical process variables. Sensor technology has developed rapidly in recent years, and we now have a vast array of

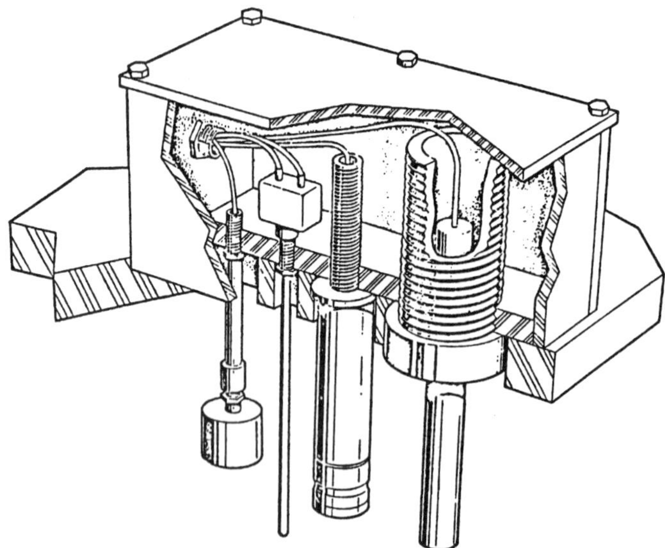

Figure 2 Typical sensor assembly on a plating tank cover. *Left to right*: Fluid level, temperature, pH, and conductivity.

TABLE 3 Sensor Types Useful for Electroplating

1. Chemical*	21. Motion*
2. Gas & vapor	22. Photoelectric*
3. Color	23. Position*
4. Displacement	24. Pressure*
5. Electrical properties	25. Proximity
6. Fiber optics*	26. Radiation
7. Flow*	27. Saw devices
8. Force	28. Smoke
9. Frequency*	29. Sound*
10. Friction	30. Sonar
11. Hall effect sensors*	31. Specific ion electrodes*
12. Humidity and moisture	32. Strain
13. Integrated circuit sensors*	33. Tactile
14. Laser beam*	34. Temperature*
15. Level*	35. Vacuum*
16. Light*	36. Velocity*
17. Load cells	37. Vibration*
18. Magnetic*	38. Video*
19. Mass	39. Volume
20. Mircrowave	40. Weight*
	41. Width

*Sensors already in use for monitoring and control of electroplating systems.

commercial sensors available and many are useful for plating applications. In fact the number of sensors available for some applications is so large that one must study the features of each sensor and make a choice. For example, a survey showed there were 144 manufacturers of level sensors making over 20 different types [29]. Table 3 lists 41 different sensor devices and phenomena useful for automatic electroplating. Those with an asterisk are used in automatic plating machines built today. Many others will be utilized when the specific need arises, and commercial devices are offered that are attractive to the machine designer.

There are several things to consider when choosing a sensor to monitor plating variables. The important items are accuracy, reliability, sensor life, cost, signal conditioning, and maintenance. The accuracy required depends on the process or machine part function. Solution temperatures controlled to $\pm 10°C$ are usually adequate. Cumulative variables such as plating current, plating efficiency, and line speed need to be controlled so the overall accuracy is within acceptable limits. For example, if each of the three parameters are allowed to vary as much as $+1\%$, then the plating thickness can vary as much as $\pm 3\%$. Plating machine sensors usually must operate in a hostile environment with corrosion, fumes, and electrical noise in abundance. For high reliability, the sensor design and choice of construction materials must be considered carefully. The sensor should hold its calibration as long as possible to minimize recalibration. The only assurance that a sensor is working properly is to either check it frequently or build in a self-checking system utilizing a sensor within sensor together with special electronics. These are called *smart sensors*, although sometimes there are other features included such having the sensor directly provide a control signal to the line. Smart sensors often can be reprogrammed by an

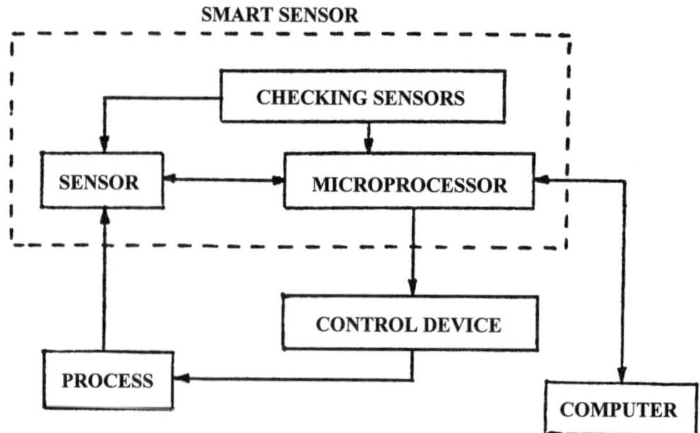

Figure 3 Smart sensor concept for monitoring with local control.

external computer. A schematic of a smart sensor and how it may interact in an automatic plating system is shown in Figure 3.

Sensor life depends not only on how well it is designed and made but also on how it is used and maintained. A glass pH electrode, for example, once put into use must be kept wet for reliable operation. If removed from plating solution, it should be stored in water or buffer solution until returning to the bath. A float level sensor in a plating solution can become inoperative if plating salts crystallize out on the float, causing it to stick. The author has seen this happen. The problem is avoided if water added to maintain the level is put in at the float. This removes tendency for salts to form at the float.

How much one should pay for a sensor depends on the value of the variable being sensed. Temperature sensors are relatively cheap and are used wherever necessary without much thought of expense. On-line plating thickness monitors are expensive, but they can save many times that annually by avoiding excessive gold plating or producing scrap due to underplating. Sensor signal conditioning involves converting the electrical signal into a form that can be displayed in engineering units such as grams per liter metal in solution. A raw signal may be conditioned using electronic hardware or computer software or a combination of the two. If only a few sensor systems are involved, hardware instead of software is usually the most cost effective if available. Software can be developed so that a computer can convert raw sensor signals to a form useful for process control, but this tends to be costly.

18 PROGRAMMABLE LOGIC CONTROLLERS AND COMPUTERS

Sensors are the "eyes and ears" of a modern automatic monitoring and control system but programmable logic controllers (PLCS), and computers are the "brains" that assimilate sensor information and with programmed instructions that initiate the control function.

The first automatic plating machines used timers to advance product through a line. Next a control system based on punched tape and photocells was developed that operated relays in the proper sequence to control a plating line. General Motors pioneered a wired relay logic system for machine control. Modification of that system by rewiring proved too time-consuming and costly. Electronics came to the rescue when the PLC was developed in 1969. Since then PLCs have become more sophisticated with computerlike features to monitor and control many machine operations including electroplating.

Initially computers were developed to perform office functions such as word processing and bookkeeping functions. They have since expanded their use in all aspects of modern life. While PLCs are still superior for direct machine monitoring and control, computers are better at processing data and communicating with other plant operations. Speciality computers are being developed for direct interface with machines for monitoring and control. Fuzzy logic control is a new technique that mimics human reasoning [30]. It is being used for control in many home appliances, cars, chemical processes, and the like, and promises to be used for manufacturing control in the future.

19 AUTOMATIC PLATING ON-LINE MONITORING AND CONTROL

Automatic electroplating utilizes on-line monitoring and control. The technology has developed rapidly with the availability of modern equipment including sensors, electrical and electronic devices, and better mechanical designs [31–34].

Figure 4 illustrates how the various parts of an automatic electroplating process monitoring and control interact. Sensors provide information about the plating process. After signal conditioning, information is sent to a PLC or a computer that is

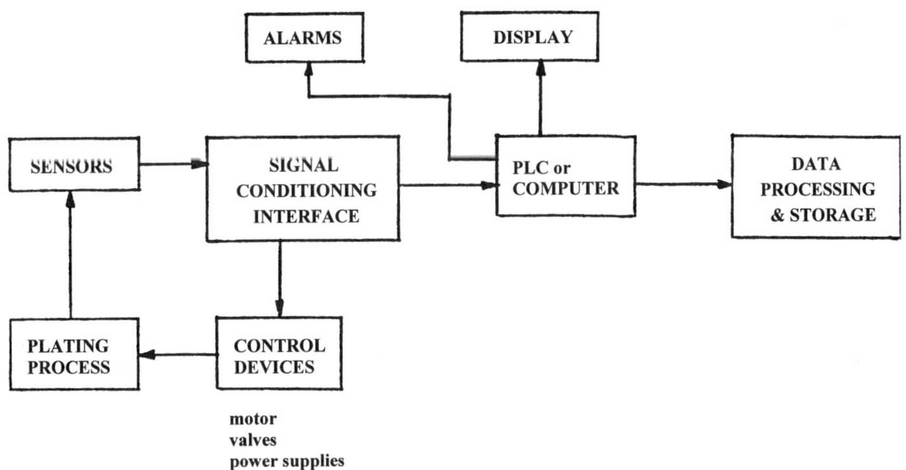

Figure 4 Schematic of an automated electroplating process monitoring and control system.

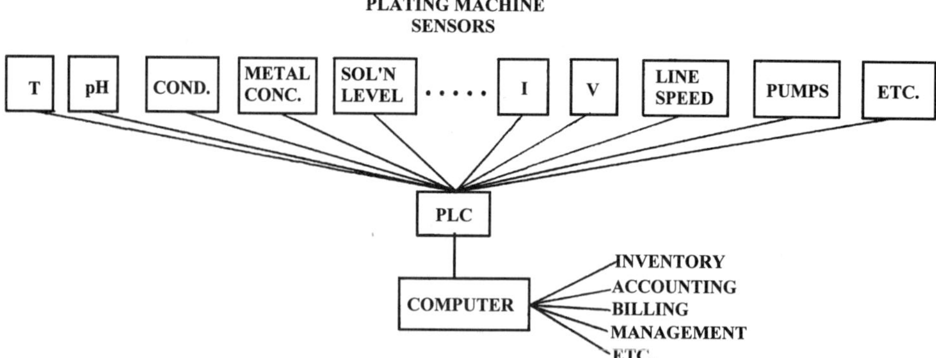

Figure 5 Schematic of a computer-integrated manufacturing (CIM) system for an electroplating shop.

programmed to control the system. Actual conditions can be data processed and stored and/or displayed. Deviations from the programmed condition limits will activate control devices such as motors, valves, and power supplies to bring the plating process back in control. If the operator must be involved, an audio alarm sounds. When there are multiple alarms, each condition has a different audio sound so the operator can quickly recognize the type of problem. Figure 5 illustrates how a central computer can integrate and control all parts of an electroplating manufacturing business.

20 CONTINUOUS SOLUTION FLOW ANALYSIS

Continuous flow analysis of electroplating solutions can be fully automated [35–37]. Palladium, tin, silver, iron, chromium acidity, persulfate, peroxide, and chlorate are among substances that have been determined automatically. A peristaltic pump withdraws a solution sample along with a reagent to form a color complex which is specific for the parameter to be analyzed. The light absorbance at a specific wavelength is monitored in a flow-through colorimeter, and with signal conditioning, the concentration is displayed on a recorder. With suitable electronics an automatic parameter control system can be established. Flow-through cells using ion-selective electrodes were used for continuous on-line monitoring of copper in electrowinning solutions [37].

21 ELECTROLESS PLATING SOLUTION MONITORING AND CONTROL

Electroless plating is an autocatalytic process initiated by a catalyst such as palladium and then continued by the deposited metal itself. Electrochemically the system is quasi-stable, as it must be to work. It requires a closer control of bath chemistry than electroplating solutions. Frequent electroless bath analysis and maintenance is necessary especially when fast plating is involved. Fast electroless metal deposition can result in spontaneous bath decomposition, causing metal particles to form and strip the solution of all metallic ions. Electroless copper plating parameters that need monitoring and control are temperature, pH, copper concentration, formaldehyde, and cyanide ion. Accurate temperature monitor/controllers are readily available. The pH value is measured with a standard glass electrode. Hydroxide solution is metered in as required to maintain pH. Copper concentration can be monitored colorimetrically or by polarography. Formaldehyde may be measured by adding 0.045 M sulfuric acid to a bath sample to reduce the pH to about 9.5. When a 0.05 M sodium sulfite solution (also at pH 9.5) is added, the resulting pH is proportional to the formaldehyde content. Polarography can be used to measure both formaldehyde and cyanide bath concentrations [4].

22 AUTOMATIC ELECTROLESS PLATING BATH ANALYZER/CONTROLLERS

Several on-line electroless copper plating analyzer/controllers have been developed for copper plating of printed circuit boards [4, 5, 13, 38]. They are used to monitor and control baths that deposit a thin layer of copper in through-holes prior to electroplating the thicker copper. Automatic analyzers or controllers are essential when the entire copper circuit is built up by electroless plating.

In 1971 Photocircuits developed their Mark VI automatic electroless copper plating bath analyzer to monitor and control their CC-4 electroless copper plating bath. Later McDermid developed a similar automatic bath analyzer which is shown schematically in Figure 6 [38]. A single peristaltic pump moves a sample and the reagents through the analyzer. The pH is adjusted first. Then the copper ion concentration is measured spectrophotometrically. A formaldehyde reagent is added next. The resulting pH is proportional to the formaldehyde concentration. The controller activates replenishment pumps to restore the copper ion and the formaldehyde concentrations to their optimum levels.

Bell Laboratories machines utilized polarography to analyze for Cu^{2+}, HCHO, and CN^-[4]. With the solution still, polarographic currents are measured at -1.0 V for Cu^{2+}, -1.8 V for formaldehyde, and -0.25 V for free cyanide. This is illustrated in Figure 7. The machine automatically averaged 10 measurements of each component for better accuracy. The block diagram, Figure 8, shows how the various parts interact. The actual machine, Figure 9, was built in two parts: one for wet chemistry on the left and the electronics the right. The two units could be separated by several hundred feet if necessary. One advantage of the separation was that the wet chemistry cabinet could be hosed down occasionally. Electroless copper solution flows through the flow gauge at the left continuously. Periodically a sample is taken into the analysis cell along with some NAOH reagent. The pH is measured, and dropping mercury polarograph measurements made. After electronic processing the results of analysis are displayed in digital units. If chemical additions are required, replenish pumps are turned on. A complete analysis can be made every 4 minutes.

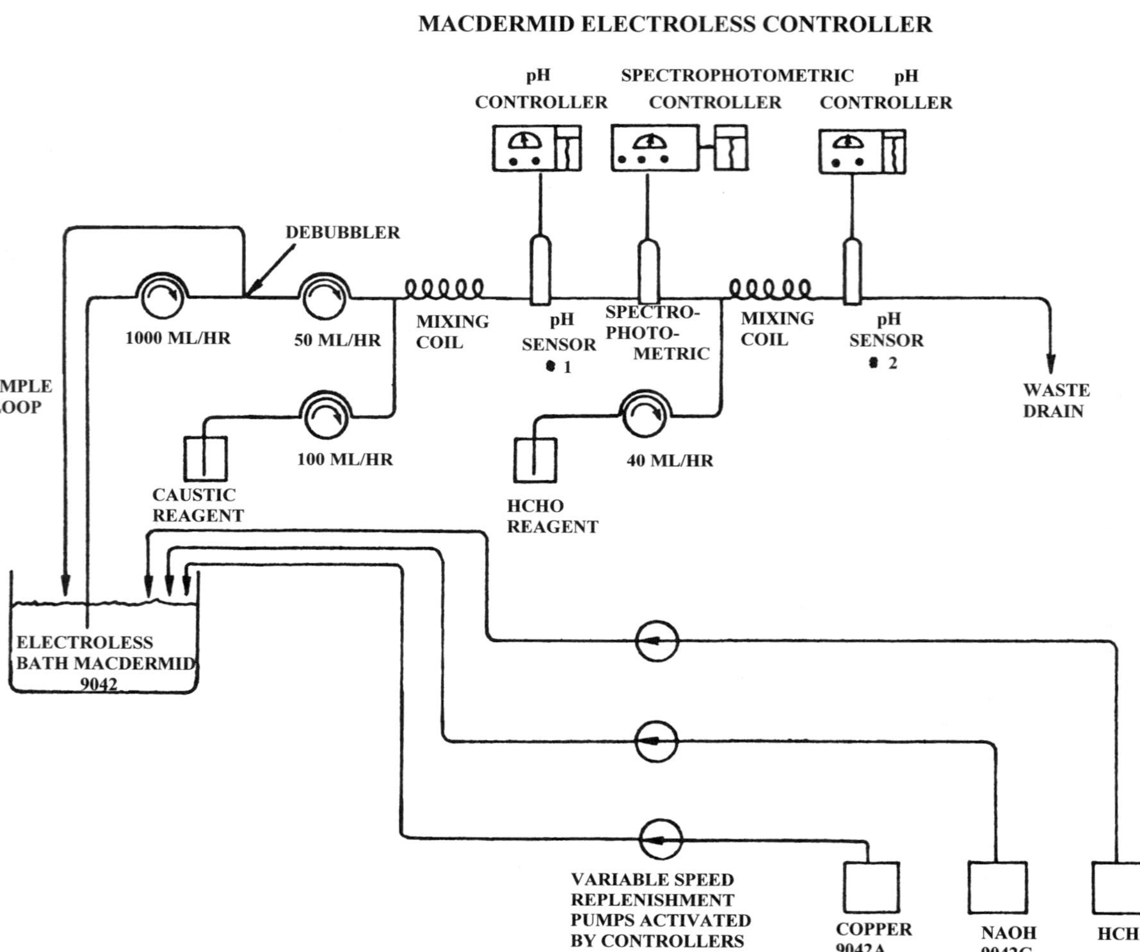

Figure 6 Schematic of McDermid electroless copper plating bath controller.

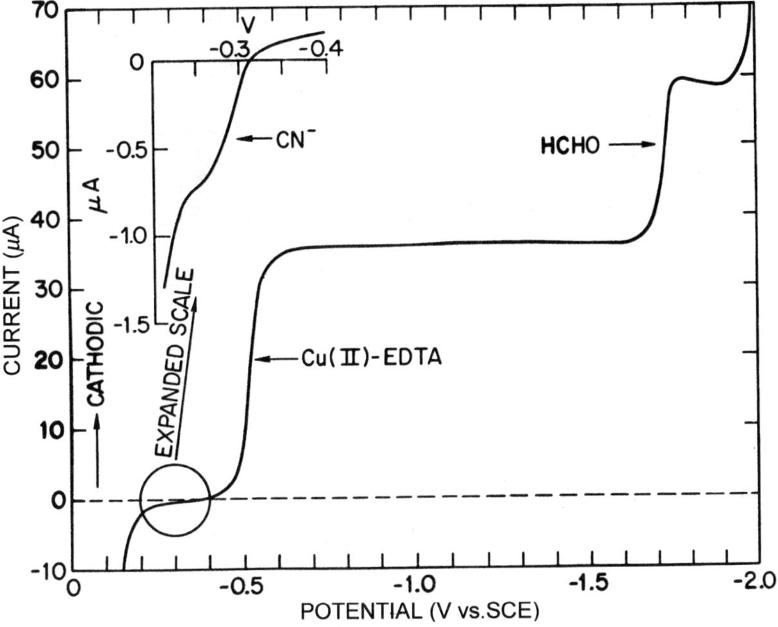

Figure 7 Polarographic analysis of electroless copper plating bath controller.

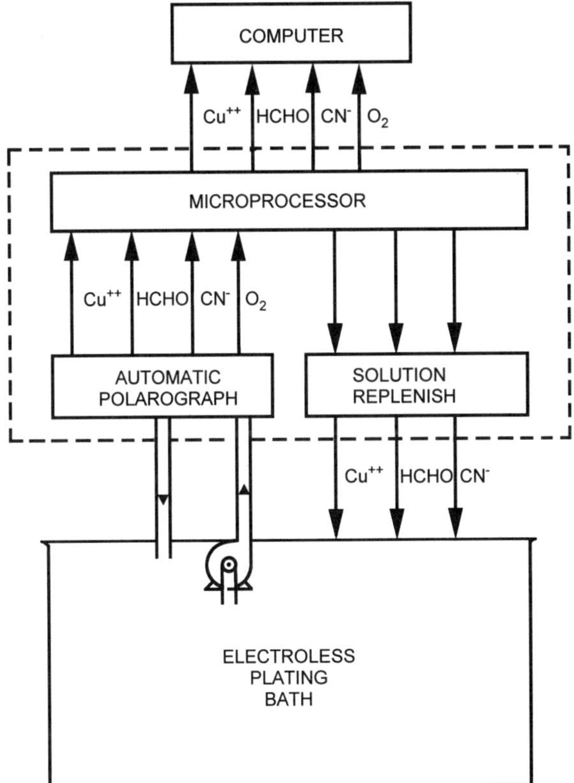

Figure 8 Electroless copper plating bath monitoring and control system.

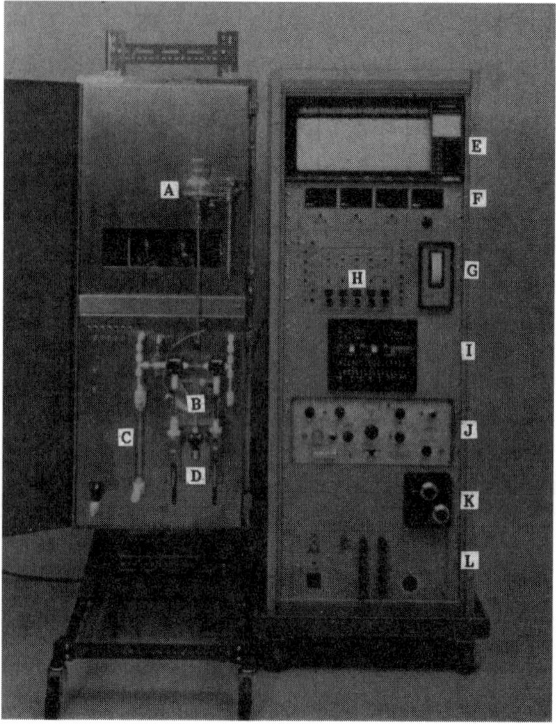

Figure 9 Electroless copper plating bath automatic analyzer and controller: (*A*) mercury reservoir, (*B*) analysis cell, (*C*) plating solution flow meter, (*D*) motorized syringes, (*E*) chart recorder, (*F*) analysis display meters, (*G*) pH monitor, (*H*) set point controls, (*I*) process controller, (*J*) polarograph, (*K*) viscosity monitor and (*L*) power source.

23 MONITORING AND CONTROL OF FINISHED PRODUCT

"Quality is not inspected in, it is built in." That quote is significant for several reasons: (1) inspection can be costly especially if done manually, (2) parts that are produced with variable quality can result in a costly scrap pile, and (3) if quality is built in, a more reproducible product is obtained. A well-engineered electroplating production line with good monitoring and control of important plating parameters is the best way to build in quality. Typical electrodeposit properties inspected are thickness, color, brightness, hardness, magnetic properties, and alloy composition. Automatic instruments are available for off-line and on-line measurements. Off-line inspection is usually done on a statistical sample for economic reasons. The advantage of on-line inspection is that every part is tested. If plating is out of specification, either plating conditions can be changed automatically or an operator can be alerted for a manual correction.

24 THICKNESS

Various wet chemical methods are available for measuring electrodeposit thickness: coulometric, dropping test and spot test [39]. A disadvantage of these tests is that the part used generally becomes scrap.

25 COULOMETRIC METHOD

The coulometric method is based on Faraday's law: One gram equivalent weight of metal is stripped away from an anode for every 96,500 coulombs of electricity passed through the cell. Four parameters must be controlled: surface area, amperage, time, and anode dissolution efficiency. At 100% anode efficiency, the deposit thickness is calculated from the formula

$$\text{Tks} = eit \times 10/Ad$$

where

e = electrochemical equivalent in grams per amp-s
i = constant current in amp
t = time in s
A = area in cm^2
d = metal density

It is necessary to use a specific electrolyte for each metal deposit and substrate [40]. The electrolyte must not chemically attack the plating. An automated instrument can detect the end point by sensing a voltage change when the substrate metal is exposed. The method is capable of an accuracy of +10% of the true value. One advantage of the method is its ability to measure combination deposits such as copper-nickel-chromium.

26 DROPPING TEST

The dropping test is very simple. It is performed by dropping chemical etch solution on a particular spot at a rate of about 100 drops per minute. The operator observes the time it takes to expose the base metal. The test is calibrated using a sample with a known thickness of plated metal. For consistent results, the drop size, temperature, and etch solution must be controlled. Operator skill is an important factor in achieving reproducible results with an accuracy of +15% [41].

27 SPOT TEST

This simple test was developed as a rapid and inexpensive test for chromium deposits on nickel or stainless steel. A drop of hydrochloric acid is placed inside a ring of wax on the part to be tested. Hydrochloric acid attacks chromium with the evolution of hydrogen gas bubbles. Gassing stops when all the chromium is dissolved. The time of dissolution is proportional to the chromium thickness. The test is calibrated with a known chromium-plated thickness specimen. An accuracy of +20% is obtained for deposits up to 1.2 μm thick [42].

28 NONDESTRUCTIVE TESTS

A nondestructive thickness test is usually preferred. There are several methods available, and most have been automated as stand alone instruments or incorporated

Table 4 Nondestructive Electroplated Deposit Thickness Tests

Method	Metals Tested
Beta Backscatter	Au, Ag, Cd, Cr, Cu, Ni, Zn, alloys
X-ray fluorescence	Au, Ag, Cd, Cr, Cu, Ni, Zn
Eddy Current	Cu, Zn, Cd, Ni, Cr

into on-line monitoring and control plating systems. The methods used include beta backscatter X-ray fluorescence, eddy current, and magnetic field measurements. Deposited metals tested by these methods are shown in Table 4.

28.1 Beta Backscatter

The radioactive decay of certain isotopes produce beta rays, which are fast moving electrons. When these are directed at an electrodeposit, some are slowed and change direction out of the deposit. The number of these backscattered electrons is proportional to the number of atoms per unit area and its atomic number. The backscatter is measured with a Geiger-Muller tube counter. The isotope used is based on its maximum beta ray energy and half-life. As the deposit thickness increases, the energy needed to penetrate the deposit increases. Metals with higher atomic numbers require higher energies for the same thickness. Instruments designed to measure metal deposit thickness utilize a Geiger-Muller tube electron counter and, with a built-in microprocessor, compute the thickness. The device is also capable of feedback current control of on-line plating [43].

28.2 X-Ray Fluorescence

This method is similar to beta backscatter in that a radiation is directed at the deposit and the energy emitted is measured. The method uses X-rays produced by an X-ray tube that is specific to the deposit metal. The radiation emitted from the deposit is proportional to the deposit thickness. A calibration requires an instrument to display the thickness and control the plating current on-line.

An X-ray fluorescence thickness monitor is used to control the thickness of zinc plating on steel mill strip by a feedback signal to a computer that adjusts the line speed in order to maintain a specified zinc thickness [24]. Emissions are specific for each metal, so alloy compositions can be determined. Metal deposit thickness in the range of 0.25 to 10 μm can be measured depending on the metal. Measurement accuracy is +10% with proper calibration [44].

28.3 Eddy Current

Eddy current thickness gauges are electromagnetic instruments designed to measure a change in impedance of a coil that induces an eddy current into the plated metal. The phenomena is based on the difference between electrical conductivity of the basis metal and the deposit that produces the change in impedance. The thickness test is performed with a specially designed probe. It is positioned perpendicular and in contact to the surface at the point of measurement. Thickness gauge instruments are available with digital display, memory, and computer prompting calibration procedure. The measurement accuracy is +10% to the true thickness in range of 5 to 50 μm. Factors that influence measurement accuracy are surface contour, surface roughness, and type of plating process that can influence deposit conductivity [45].

28.4 Magnetic Method

The magnetic method utilizes the magnetic influence of the electrodeposit for measurement. Two types of probes are in use: (1) a mechanical device that measures the influence of the plated metal on the attractive force between a magnet and the base material and (2) an electromagnetic probe that measures the influence of the plated metal on the reluctance of a magnetic flux through the deposit and base metal. Three types of deposits can be measured with the second probe: (1) nonmagnetic deposits on ferromagnetic base metals, (2) nickel deposits on ferromagnetic base metals, and (3) nickel deposits on nonmagnetic base materials.

Both probe types require perpendicular positioning on the surface. The magnet probe works by measuring the force required to pull it from the surface. The electromagnetic probe measures the reluctance difference with and without the plated metal. The use of both probes requires a calibration, and averaging several measurements improves the accuracy. Commercial instruments utilizing the electromagnetic probe are available with microprocessors that can automate the system by averaging data, display results, store or transmit information, and provide on-line monitoring and control [46, 47].

28.5 Color

The color of electrodeposits has an esthetic value particularly on jewelry, home appliances, autos, and so on. Even on parts that are hidden from view when assembled into an object, an attractive appearance is considered a quality factor. Off-color plating or variable plating color often is considered a sign of a process problem. Generally monitoring and control of color is done by visual inspection by a operator on the plating line. On-line automatic inspection of color is feasible for many plating lines. An optical fiber system was developed to monitor the color of gold plating on a moving strip shown schematically in Figure 10 [48]. An optical fiber bundle is positioned over a gold-plated strip. The bundle is divided into three parts. Light is directed into one arm that illuminates the moving strip. Reflected light from the gold surface passes into two separate optical fiber bundles. The light from one passes 700 to 800 μm through an optical filter that serves as a standard. The other light path passes 450 to 575 μm through the optical filter. An electronic comparator determines any deviation from the optimum color. The result can be displayed on a meter and/or fed into an automatic process control system to correct off-color plated gold.

28.6 Brightness

Bright plating for decorative applications usually requires closer monitoring and control of the bath chemistry. Additives that produce brightening, usually organic,

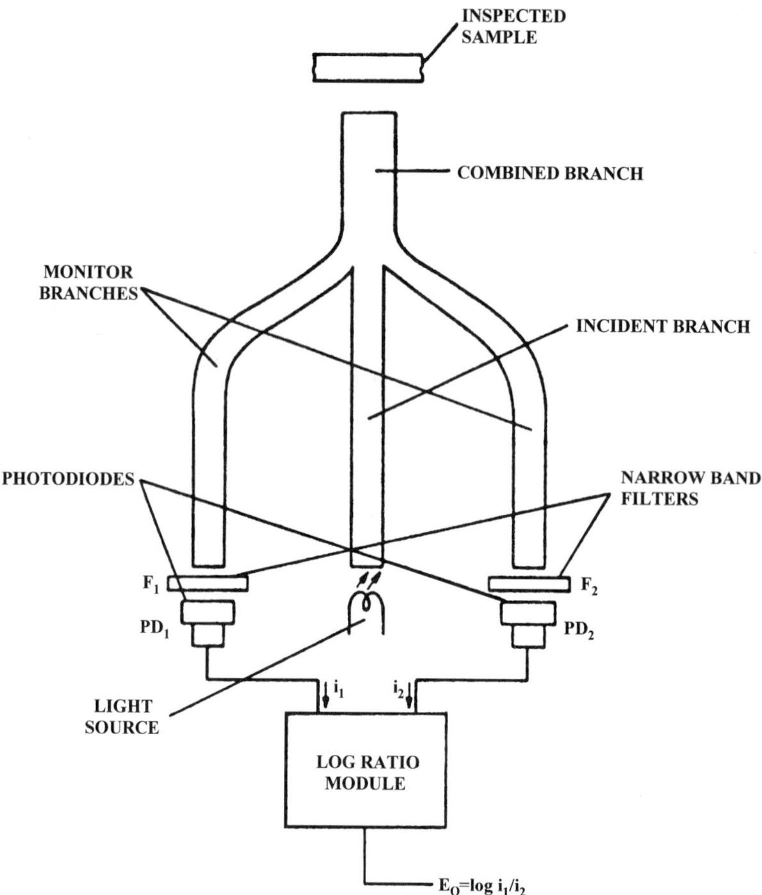

Figure 10 Schematic of an on-line gold plating color monitor.

can become depleted or degraded by oxidation at the anode. Additive depletion is easily corrected with more additive, and this can be done by metering it in on the basis of plating time. Brightener concentration also may be controlled by chemical analysis, plating tests (Hull cell), or cyclic voltammetric stripping (CVS) [15, 16].

CVS determines the concentration of brighteners or other additives such as levelers from the effect the additive exerts on the electrodeposition rate. The potential of an inert rotating platinum electrode is cycled at a constant rate in a bath sample so that a small amount of metal is deposited on the electrode and then stripped off by anodic dissolution. The charge required to strip the metal is related to the additive concentration. A computerized instrument has been developed that automatically determines additive concentration.

If too much brightener becomes chemically altered through oxidation, it can degrade bright plating. The only solution is to strip all the organics from solution with activated carbon and start over with fresh brightener.

28.7 Hardness

Electroplating hardness is a desirable property in applications where minimum wear is required. There are many test methods for hardness available but all are done off-line. Most testing is done with indentation instruments of somewhat different designs. The various types are Brinell, Rockwell, Vickers, Solersoope, Knoop, and Tukon.

29 SUMMARY

The electroplating industry has at its disposal an extensive amount of information and technology for monitoring and controlling all aspects of the plating process. Sensors have an important role in the monitoring the status of each important parameter that must be controlled for successful plating. As new phenomena and technology are converted into practical cost-saving devices for monitoring and controlling electroplating parameters and systems, the industry has adopted them. Electric devices and, in particular, computers have revolutionized the conversion of sensor signals to a useful form for monitoring and control. Further, plating information can be gathered and analyzed quickly to ensure that quality is maintained at all times. The design and engineering of individual sensor devices has improved greatly in recent years. Many incorporate a microprocessor for signal conditioning, and some now can also carry out control of certain plating conditions directly without relying on a PLC or computer. Automation of electroplating utilizing monitoring and control has resulted in lower manufacturing cost while improving quality and reproducibility of processing.

REFERENCES

1. J. Morico, *Electroplating Engineering Handbook*, 4th ed., L. J. Durney, ed., Van Nostrand Reinhold, New York, 1948, p. 289.
2. S. Hirsch and C. Rosenstein, *Metal Finishing Guidebook*, vol. 93, no. 1A, 482 (1995).
3. G. J. Shugar and J. T. Ballinger, *Chemical Technicians' Ready Reference Handbook*, 4th ed., McGraw-Hill, New York, 1996.
4. Y. Okinaka, D. W. Graham, C. Wolowodiuk, and T. M. Putvinski, *Western Elect. Eng.*, 22 (Apr. 1978).
5. S. Newman, *Prod. Finish.*, 92 (Nov. 1980).
6. Y. Okinaka, *Plating Surf. Finish.*, **72** (10), 34 (1985).
7. A. L. Harmer, *Proc. 2nd Int. Conf. on Optical Fiber Sensors*, Stuttgart, Sept. 17, 1984.
8. Y. Okinaka, *Plating Surf. Finish.*, **66**, 50 (1979).
9. M. S. Frant, *Plating*, **58** (7), 686 (1971).
10. H. Freiser, *Ion Selective Electrodes in Analytical Chemistry*, vols. 1 and 2, Plenum Press, New York, 1978.
11. P. L. Bailey, *Analysis with Ion-Selective Electrodes*, 2nd ed., Heyden, London, 1980.
12. D. Jones and R. Herbert, *Plating Surf. Finish.*, **83** (3), 22 (1996).
13. D. R. Turner and Y. Okinaka, Amer. Electroplaters' Soc. SUR/FIN Session M—Chemical Analysis, 1981.

14. C. Loughlin, *Sensors for Industrial Inspection*, Kluwer Academic, Boston, 1993, sec. 8.
15. D. M. Tench and C. A. Ogden, *J. Electrochem. Soc.*, **125**, 194 and 1218 (1978).
16. D. M. Tench and C. A. Ogden, U.S. Patent 4,132,605 (Jan. 2, 1979).
17. C. A. Ogden and D. M. Tench, *Plating Surf. Finish.*, **66** (9), 30 (1979).
18. R. Haak, C. A. Ogden, and D. M. Tench, *Plating Surf. Finish.*, **68**, 52 (1981).
19. R. Haak, C. A. Ogden, and D. M. Tench, *Plating Surf. Finish.*, **69**, 62 (1982).
20. P. Pinches and G. Bush, *Circuits Manufacturing*, July 1982, p. 36.
21. J. R. Smith et al., *Trans. Inst. Met. Fin.*, **73** (2), 72 (1995).
22. G. J. Shugar and J. T. Ballinger, *Chemical Technicians' Ready Reference Handbook*, 4th ed., McGraw-Hill, New York, 1996.
23. H. Thomson and Riley, *Phil. Trans. R. Soc., London*, **A302**, 327 (1981).
24. *The Making, Shaping and Treating of Steel*, 10th ed., AISE (1996).
25. T. Seiyama, ed. *Chemical Sensor Technology*, Elsevier, Amsterdam, 1988.
26. J. Janata, *Principles of Chemical Sensors*, Plenum Press, New York, 1989.
27. S. Soloman, *Sensors and Control Systems in Manufacturing*, McGraw-Hill, New York, 1994.
28. D. R. Turner, *Plating Surf. Finish.*, **73** (6), 30 (1985).
29. *Industrial and Process Control*, Oct. 15, 1985.
30. D. Drianko, H. Hellendoon, and M. Reinfrank, *An Introduction to Fuzzy Logic Control*, Springer-Verlang, New York, 1996.
31. A. Ivaska, *Proc. Anal. Div. Chem. Soc.*, **16**, 283 (1979).
32. D. R. Turner and L. T. Romankiw, *Electrochem. Soc. Proc.*, **17**, 719 (1987).
33. J. T. Y. Yeh, *Chem. Eng.*, **93** (2), 55 (1986).
34. C. G. Smith, *Electrochem. Soc. Symp. on Chemical Sensors Proc.*, Honolulu, 39 (Oct. 1987).
35. A. Aldo and J. DiLiddo, *Annu. Tech. Conf. Proc.—Am. Electroplat. Soc.*, Paper B-5 (1982).
36. J. DiLiddo and A. Conetta, *Am. Lab.*, **16** (4), 68 (1984).
37. F. C. Walsh and D. R. Gabe, *J. Applied Electrochem.*, **11**, 117 (1981).
38. T. A. Rau, MacDermid Electroless Copper Controller, Extended Abs., Electrochem. Soc. Meet., Las Vegas, 445 (Oct. 1979).
39. N. Sajdera, *Metals Finishing Guidebook*, M. Murphy, ed., 1992, pp. 552–570.
40. American Society for Testing and Materials Standards, vol. 02.05, B-504.
41. Ibid., B-555.
42. Ibid., B-556.
43. Ibid., B-567.
44. Ibid., 8-568.
45. Ibid., B-244.
46. Ibid., B-499.
47. Ibid., B-530.
48. F. W. Ostermayer Jr., U.S. Patent 4,278,353 (July 14, 1981).

26 Environmental Aspects of Electrodeposition

MICHA TOMKIEWICZ

INTRODUCTION

We address the environmental aspects of electrodeposition from the two complementary approaches that electrodeposition share with other environmentally sensitive technologies: prevention of environmental damage by management of the discharge and development of technologies that minimize toxic discharge. We discuss examples of zero-discharge electrodeposition technologies and some of the issues that require further development. We also include applications of electrodeposition and solar-induced electrodeposition in cathodic extraction of metal contaminants from waste streams.

The chapter relies heavily on the internet as an information source. We explore in detail the site of the Environmental Protection Agency (EPA) and mention relevant sites of professional societies.

1 GOVERNMENT INVOLVEMENT

Presently, in the United States, the leadership for controlling the environmental consequences of most technologies rests with the Environmental Protection Agency (EPA). The EPA was created in 1970 to be the federal watchdog for the environmental consequences of human activities through a series of laws and regulations that include:

- Clean Air Act, 1970
- Federal Water Pollution Act, 1972
- Safe Drinking Water Act, 1974
- Toxic Substance Control Act, 1976
- Resource Conservation and Recovery Act, 1976
- Comprehensive Environmental Response Compensation and Liability Act, 1980
- Superfund Reauthorization Act, 1986
- Occupational Safety and Health Act, 1970

Modern Electroplating, Fourth Edition, Edited by Mordechay Schlesinger and Milan Paunovic.
ISBN 0-471-16824-6

These laws and regulations are being constantly amended, and in addition, local governments issue their own regulations. In instances of conflict or differences, the most stringent regulation applies. With the passage of the Federal Water Pollution Control Act (FWPCA) amendments, Congress established limitations for direct discharges of specific categories of industrial waste. These "technology-based" limitations include Best Practicable Technology (BPT); Best Available Technology (BAT) to be achieved by specific dates (1977 for BPT and 1984 for BAT), and New Source Performance Standards (NSPS) to be achieved when a new source begins operation.

The electroplating and metal finishing control regulations are contained in the U.S. Code of Federal Regulations (CFR) Title 40, Parts 413 and 433. All sectors of the electroplating and metal finishing industry are covered by the metal finishing regulations except for existing job shops and independent printed circuit board manufacturers (IPCBM) facilities, which are indirect dischargers. Most countries carry and enforce their own set of regulations and incentives that are designed to protect the environment. These local legal restrictions are not the only one that today's business needs to address: International standards for Environmental Management Systems such as ISO 14001 are increasingly gaining in importance because of the global nature of many environmental issues that are reflected in present and future environmental and trade international treaties and because of the increasing global scope of the dischargers.

Fortunately, today the Internet offers a convenient information link that provides the opportunity for a continuous monitoring of EPA's regulations, publications, proposals, and so on, as well as technical help from companies that are willing and able to be subcontracted to handle the waste disposal aspects of the electroplater. The Internet became so useful that a significant portion of the reference material for this review came directly through the World Wide Web (WWW).

Figure 1 shows the Home-Page of the EPA [2]. This site is the main repository of practical information on environmental issues in the United States. Figure 1 lists the various audiences that the site tries to address. Figure 2 shows the first page of the Industry section, and Figure 3 shows the first page of the Small Business Information section. An obviously important component is the list of laws and regulations that applies to the industry. Figure 4 shows the first page of the Laws and Regulations section that can be accessed directly from the Home-Page and can also be accessed from within the site (see Fig. 2). This section directly connects the Federal Register. Opening the section on Compliance Assistance centers (Fig. 2), one gains quick access to industry specific compliance information. The section of most interest to the electroplating industry is the one addressing the National Metal Finishing Resource Center (NMFRC). Opening this section immediately makes available a list of relevant recent documents that can be viewed, downloaded, or ordered. Among the documents at the time of submission of this chapter (August 1999) one can find a recent consultant report on Environmental Management Systems: A Guide for Metal Finishers that adheres to the international ISO 14001 standard; an announcement about designating the Metal Services Industry as a Priority Sector; review of chrome plating emission test methods, alternative test methods for chrome plating compliance; application for the Metal Finishing Guidance Manual; among other things. The compliance of an institution in maintaining sound environmental practices is being helped by the transparency of the

U.S. Environmental Protection Agency

There is a Graphical version of this page.

Visit our Earth Day 2000 Web Site

There are two methods for ***you*** to get the information you need...

You can choose a ***user category*** to find selected resources geared to your interests:

Search - Search the EPA website
Search by Zip Code - Search for Environmental Information by Zip Code

Kids - Projects, Games, Art & Helpful Tips
Students - Environmental Concepts, Activities & Tips
Teachers - Basic Environmental Concepts & Teaching Aids
Children's Health - Tips for Protecting Children
Concerned Citizens - Tips for Home, Garden, Work & Beyond
Researchers & Scientists - Technical Documents, Research, Funding & More
Small Business - Assistance for Small Businesses
Industry - Assistance for Businesses & Entire Industries
State, Local, & Tribal Governments - Gateways to Funding, Partnerships, Regulations & More

You can choose a ***topic*** that covers the information you need from the EPA:

About EPA - Mission, Budget, People & Progress
News & Events - Press Releases, Speeches & Newsletters
Offices, Labs & Regions - Connect to EPA Across the Country
EPA Projects & Programs - By Natural Resource & Topic
Laws & Regulations - Find Federal Register Notices and More
Publications - Agency Catalog & Ordering Information
Money Matters - Contracts, Grants, & Environmental Financing
Other Resources - Hotlines, Clearinghouses, Libraries & More
Databases & Software - Find Environmental Data & Download Software

[EPA Home | Search | Browse | What's New | Comments | Español]

New home page and Web site features coming soon!

This page last updated on April 12, 2000
Privacy & Security Notice
EPA Server Information
http://www.epa.gov/epahome/text.htm

Figure 1 Home-Page of the EPA.

compliance to the local communities in which the facility resides. The mechanism for that is the entry for a search by a zip code shown in Figure 1. The first page of the Zip Code Search section is shown in Figure 5. One can search in four different data bases described in the figure. Taking the first category as an example and submitting the authors zip code, a list of EPA-regulated local facilities appear, the first page of which is shown in Figure 6. Among the information one can find for each regulated facility are the specific chemicals that the facility releases to air, underground, land surface, surface water, and the waste products that are transferred to other sites. One can also find the compliance status with regard to any of the discharges.

If we now go back to the Home-Page (Fig. 1) and click on publications, we will be provided with a description of the various EPA-related publications. If we now choose the National Publications Catalog and query for *electroplating*, we will be provided with a list of titles (Fig. 7) of recent publications that can be either inspected on-line or ordered (free of charge) from the EPA.

Document 40 CFR 413 of the Clean Water Act (recorded from the EPA site) describes the general provisions for the electroplating industry. Example for the requirements are shown in Table 1 (page 800).

Business & Industry

Industry Sector Notebooks Includes industry background, emission outputs, summaries of applicable regulations and references.	**Small Business Information** Access small business information including assistance, compliance & pollution prevention links.	**Programs & Assistance** Business initiatives, voluntary programs, and compliance assistance.
Compliance Assistance Centers Quick access to industry specific compliance information.	**Regulations & Legislation** Quick access to EPA regulations and environmental legislation.	**Contracts** Doing Business with EPA: Acquisition Resources, Policies & Contracts List
Publications Access to EPA technical documents and public information in paper and/or electronic format.	**Enforcement Data Systems & Models** Information about data systems & models used for analysis of enforcement related data.	**Envirofacts** A national information system that provides an integrated single point of access to data extracted from seven major EPA data bases.

Related Links

Year 2000 - EPA's Year 2000 strategy and other Y2K related issues.

EPA Information Resources - Exploring environmental issues through databases, tools, libraries, and other systems.

Small Business Administration EXIT EPA - Information on SBA programs, offices, electronic bulletin boards, loan and assistance services, and other Internet resources.

U.S. Business Advisor EXIT EPA - Provides one-stop access to federal government information, services, transactions, regulations and resources.

[EPA Home | Search | Browse | What's New | Comments]

http://www.epa.gov/epahome/business.htm
This page last modified: 04/17/2000 14:01:52
Privacy & Security Notice
EPA Server Information

Figure 2 First page of the Industry section of Home-Page of the EPA.

Since the limits are in concentrations, section 413.14 adds "No user introducing waste-water pollutants into a publicly owned treatment works under the provisions of this subpart shall augment the use of process wastewater or otherwise dilute the wastewater as a partial or total substitute for adequate treatment to achieve compliance with this standard." To put it differently, dilution is not a solution.

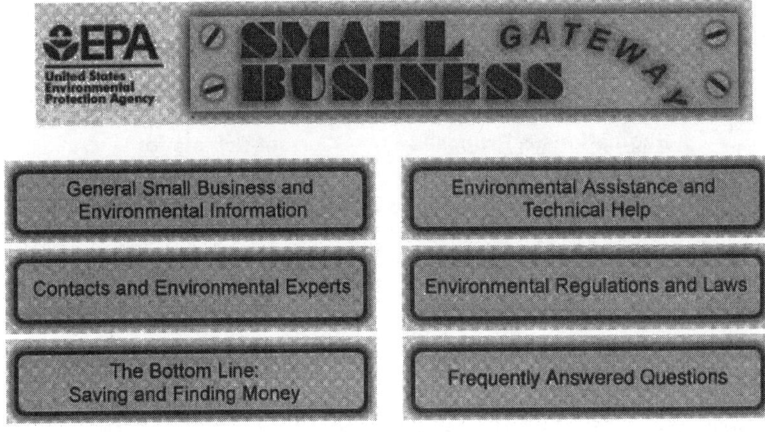

The U. S. Environmental Protection Agency (EPA) gateway to environment information and contacts for small businesses.

[EPA Home | Search EPA | Browse EPA | What's New | Comments]

Small Business Ombudsman Hotline: 1-800-368-5888

http://www.epa.gov/smallbusiness/
This site was last updated on May 10, 1999

Figure 3 First page of the Small-Business section of the Home-Page of the EPA.

All new facilities are subject to the same New Source Performance Standards (NSPS) regardless of size, type of facility, or type of discharge.

Publicly Owned Wastewater Treatment Works (POTW) are essentially designed to treat domestic sewage. Many industrial wastes are compatible with the treatment system but some are not. These either pass through the POTW untreated or interfere with the normal operation of the POTW. In both cases pretreatment regulations permit industry access to central wastewater treatment system, while at the same time protecting the quality of the receiving water body and the POTW.

Web sites of professional associations offer timely and specific information, particularly in terms of educational opportunities, research opportunities, and specific response to issues that are currently under legislative consideration. Most of these sites also offer chat-rooms and QA sessions that provide the opportunity for specific technical consultation. In almost all cases the chat-rooms cannot guarantee the competence of the participants. Almost all of these sites are interlinked and contain links to foreign organizations and to more specialized associations. Among the electroplating organizations one can find: The American Electroplaters and Surface Finishers Society [3]; The Home-Page of the Finishing Industry [4]; National Association of Metal Finishers (NAMF) [5], and the National Metal Finishing Resource Center [6]. For outsourcing waste management and cleanup services, one can contact a central site such as TDS Central [7]. The transparency of compliance to sound environmental practices is not limited to current practices. A new field is

Laws & Regulations

Additional Information:

Introduction to Laws and Regulations

Major Environmental Laws

Laws and regulations are a major tool in protecting the environment.
Find out about:

Regulations & Proposed Rules
New regulations, proposed rules, important notices and the regulatory agenda of future regulations.

Current Legislation
Current legislation before the U.S. Congress, Congressional Committees, and uncompiled Public Laws.

Codified Regulations
Federal regulations codified in the Code of Federal Regulations and additional material related to Title 40: Protection of Environment.

Laws
Public Laws passed by the U.S. Congress and codified in the U.S. Code.

Regulations and Proposed Rules

Federal Register - Environmental Documents
Full text of all Federal Register documents issued by EPA, and of selected documents issued by other Departments and Agencies. Notices, meetings, proposed rules, and regulations are divided into twelve topical categories for easy access (eg. air, water, pesticides, toxics, waste).

Federal Register database EXIT EPA
The Government Printing Office (GPO) maintains the official electronic database that covers Federal Register documents beginning in January, 1994. Unlike the EPA Environmental Documents service, it provides access to all Federal Register documents from all Departments and Agencies.

The Semiannual Regulatory Agenda
All agencies publish semiannual regulatory agendas describing regulatory actions they are developing or have recently completed. These agendas are published in the Federal Register, usually during April and October each year, as part of the Unified Agenda of Federal Regulatory and Deregulatory Actions.

- EPA Unified Agenda Preamble EXIT EPA Explains the rulemaking process, EPA's regulatory priorities and the contents of the Unified Agenda. Made available by the Regulatory Information Service Center of the U.S. General Services Administration (GSA).
- EPA Semiannual Regulatory Agenda Table of Contents EXIT EPA provided by the Regulatory Information Service Center.
- EPA's Statement of Regulatory and Deregulatory Priorities EXIT EPA Made available by the Regulatory Information Service Center.
- Search the Unified Agenda This site, maintained by the Government Printing Office, offers instructions and tips on how to search for Unified Agenda information and upcoming actions. The site searches the entire Unified Agenda.

Dockets
Dockets contain information and supporting documentation related to the rulemaking process. Public comments received on rules and proposed rules are also maintained in the appropriate Docket.

Electronic Enhancement of the Regulatory Process - This site is a comprehensive source for information on EPA's current electronic regulatory capabilities. Find information on electronic submission of public comments and data, documents, and the availability of public comments on the Internet.

Figure 4 First page of the Laws and Regulations section of the Home-Page of the EPA.

slowly developing that might be labeled Archeology of Environmental Contaminations. A good example that relates to the electroplating industry is the work by Spliethoff and Hemond [8] on the history of waterborne export of toxic metals from industrial/residential watershed.

ZIP Code Search

The zip code search requires a browser that supports JavaScript.
Non-JavaScript version

To search, enter your ZIP code, then select the type of information you want to find. Descriptions of EPA databases and the information they contain are listed below.

Enter your 5 digit ZIP Code:

Then select one of these databases:

- Pollution, hazardous waste sites, and other regulatory information -- through Envirofacts.
- A "live" Computer-generated map of regulated sites in your area -- through EnviroMapper.
- Environmental conditions and activities in your watershed -- through Surf Your Watershed.

More information about these databases:

These databases are maintained by EPA. A limited amount of supplementary information from other sources is also available here.

The **Envirofacts** Warehouse is a database that includes information on Superfund sites, drinking water, air pollution, toxic releases, hazardous waste, and water discharge permits. Through **Envirofacts,** you can get lists of which facilities in your neighborhood are releasing pollutants or are legally handling hazardous materials, where any Superfund sites are located and what their cleanup status is, and more. In many cases, you can link to more information about the chemicals involved at the listed sites, and find out whether they are potentially harmful.
[Back to ZIP Code Search] [Go to Envirofacts Homepage]

Through Envirofacts' **EnviroMapper** feature, you can customize a computer-generated map of your neighborhood to view the location of EPA regulated sites, schools, churches, streams, streets, and other geographic features.
[Back to Zip Code Search] [Go To EnviroMapper Homepage]

Watersheds divide the landscape into land areas that catch rain and snow and drain into groundwater or to rivers, streams, wetlands, estuaries, and oceans. **Surf Your Watershed** helps you locate a specific watershed and gives you a list

Figure 5 First page of the Zip-Code Search section of the Home-Page of the EPA.

2 ENVIRONMENTAL IMPACT OF ELECTROPLATING TECHNOLOGIES

Most of this book is concerned with detailed account of electrodeposition procedures for different metals. One can write an environmental impact statement for every metal and fill a separate volume in the process [9–10]. However, many of the issues are similar and can be easily extrapolated from one metal to the next. Because of the central role that the EPA plays in the regulatory environment, most of the

ENVIROFACTS Query Results

Page No. 1

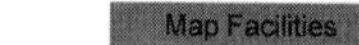

Please note that a maximum of 100 facilities will be mapped!

ZIP CODE: **11210**

LIST OF EPA-REGULATED FACILITIES IN ENVIROFACTS

To see a report on a facility click on the underlined Facility Name. Click on the underlined EPA FACILITY ID value to view EPA Facility information for the facility.

Go To Bottom Of The Page

FACILITY NAME/ADDRESS	EPA FACILITY ID	Permitted Discharges to Water?	Toxic Releases Reported?	Hazardous Waste Handler?	Active or Archived Superfund Report?	Air Releases Reported
1927 FLATBUSH CO % DAVID MADEL 1927 FLATBUSH AVENUE BROOKLYN, NY 11210	NY0001458520	NO	NO	NO	NO	YES
A MONACO & SON INC 2580 NOSTRAND AVE BROOKLYN, NY 11210	NYD981486012	NO	NO	YES	NO	NO
ABLE ANODIZING CORP. 984 E. 35TH ST. BROOKLYN, NY 11210	NYD041581463	NO	YES	YES	NO	YES
ACURA OF BROOKLYN 2729 NOSTRAND AVE BROOKLYN, NY 112105324	NY0002287233	NO	NO	YES	NO	NO

Figure 6 First page of the result of the Zip-Code search for zip-code 11210.

publications that cover the environmental aspects of the technology heavily rely on EPA's publications. Since the access to this information is now more convenient then it has ever been, the need for duplication of effort is considerably reduced.

A schematic outline of an electroplating process is demonstrated in Figure 8 [9]. A detailed account of the pretreatment process is shown in Figure 9. A central issue that we hope will be reflected throughout this chapter is the ability to include environmental considerations into the decision-making process of the applied technology.

To use copper plating as an example, one can use different baths that are summarized in Table 2 (page 801) [1].

The posttreatment process can include surface finishing such as chromating and coloring through selective oxidation of the surface. A schematic outline of the plating and the posttreatment processes is shown in Figure 10 and will be further discussed later. Each of the processes involve environmental consequences and costs associated with disposal. A schematic flowchart of the decision-making process is shown in Figure 11. The economic aspects of waste disposal will be discussed separately. However, environmental consequences are often reduced into political dimensions in which cost-benefit analysis is not always a solution. Local ordinances with zero tolerance to certain pollutants are not unknown.

Public Access Server Search Results

The "Highlight Search Terms" link displays the document with your search terms highlighted in red.
Click on the ">" to move to the next highlight or the "<" to move to the previous highlight.

Your query "**(electroplating)**" matched 2326 documents out of 316618. [Didn't Find What You Were Looking
Documents **1** through **10** are listed below in order of relevance. [Refine Search] [New Se

[Additional Help To Limit Your Search]

[Show Summaries] [Hide PDFs] [About PDF Fil

Rank **Title/Summary** **Fo**

1. **Unified Air Toxics Website - Rule and Implementation Information for Chromium Electroplating**
URL: http://www.epa.gov/ttnuatw1/chrome/chromepg.html Highlight Search Terms

2. **Policy and Guidance Record - Title V Operating Permits : Proposed Deferrals for halogenated solvent degreasing; perchloroethylene dry cleaning; ethylene oxide sterilization; chromium electroplating; a**
URL: http://www.epa.gov/ttncaaa1/t3/meta/m27732.html Highlight Search Terms

3. **The Electroplating Industry**
URL: http://www.epa.gov/ttbnrmrl/625/10-85/001.htm Highlight Search Terms

4. **electroplating**
URL: http://www.epa.gov/swertio1/osc/osc_register/outlines/electroplating.h tm Highlight Search Terms

5. **OCONOMOWOC ELECTROPLATING COMPANY, INC.**
URL: http://www.epa.gov/reg5sfun/sfd/npl/wisconsin/WID006100275.htm Highlight Search Terms

6. **W G Electroplating Site - menu**
URL: http://www.epa.gov/reg3hwmd/super/wgelectr/menu.htm Highlight Search Terms

7. **W G Electroplating Site, Administrative Record File Index, Removal 11/22/91**
URL: http://www.epa.gov/reg3hwmd/super/wgelectr/arrv0112291.htm Highlight Search Terms

8. **Matthews Electroplating - General Site Information**
URL: http://www.epa.gov/reg3hwmd/super/matthews/pad.htm Highlight Search Terms

9. **Matthews Electroplating Site - menu**
URL: http://www.epa.gov/reg3hwmd/super/matthews/menu.htm Highlight Search Terms

10. **Matthews Electroplating, Administrative Record File Index, Removal 07/20/93**
URL: http://www.epa.gov/reg3hwmd/super/matthews/arrv0072093.htm Highlight Search Terms

Pages: 1 2 3 4 5 6 7 8 9 10 11 12 13 14 15 16 17 18 19 20 21 22 23 24 25 26 27 28 29 30 31 32 33 34 36 37 38 39 40 41 42 43 44 45 46 47 48 49 50 [Next]

Figure 7 Results from running query on electroplating in the publication section of the EPA Home-Page.

Electroplaters have a number of options for cost-effectively bringing a facility into compliance with water pollution control regulations. Within the three general options—modifying the process, treating the wastewater, and becoming a direct discharger—there is a wide variety of possible combinations of treatment technologies to choose from.

TABLE 1a Discharging Limits for Facilities Discharging Less Than 38,000 Liters per Day

Pollutant	Maximum for any 1	Average daily values (in mg/L) for
CN	5.0	2.7
Pb	0.6	0.4
Cd	1.2	0.6
TTO (Total Toxic Organic)	4.57	

TABLE 1b Limits for Facilities Discharging 38,000 Liters or More per Day

Pollutant	Maximum for any 1 day	Average daily values (in mg/L) for 4 consecutive
CN	1.9	1.0
Cu	4.5	2.7
Ni	4.1	2.6
Cr	7.0	4.0
Zn	4.2	2.6
Pb	0.6	0.4
Cd	1.2	0.7
Total metals	10.5	6.6
TTO	2.13	

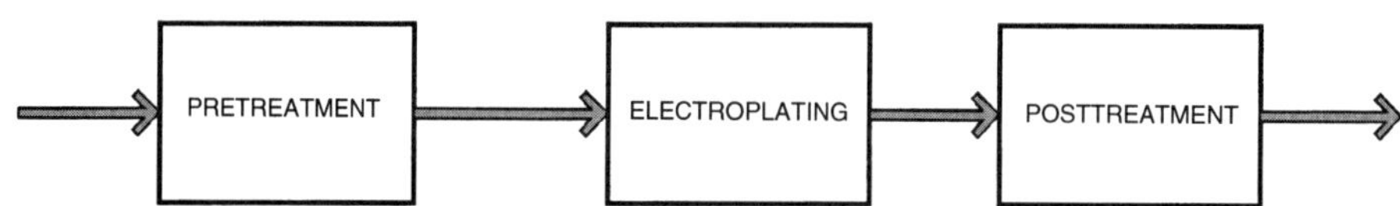

Figure 8 A schematic outline of an industrial electroplating process [9].

Conventional wastewater treatment technologies are shown in Figure 12 [10]. These treatments are based on precipitating the metals as hydroxide to be concentrated as a solid waste cake that can be separately disposed. Pollutants such as cyanide are destroyed through oxidation according to the following reaction:

$$2CN^- + 8OH^- \rightarrow 2CO_2 + N_2 + 4H_2O + 10e^- \quad (1)$$

This can be accomplished by using strong oxidants such as hypochlorite or ozone. The reaction with hypochlorite will take the following form:

$$2CN^- + 5OCl^- + H_2O \rightarrow 2CO_2 + N_2 + 5Cl^- + 2OH^- \quad (2)$$

Oxidized metals such as hexavalent chromium are reduced to a form suitable for the formation of an insoluble hydroxide.

In cases such as in the presence of chelating agents, the solubility of the metals in alkaline environments is too high to meet the standards and special techniques are required to reduce the metal concentrations. These techniques include precipitation as sulfides and ion-exchange technologies. If these alternative treatments are unsuccessful in precipitating the chelating complexes, one can dissociate the complexes by adjusting the pH to an extreme level while exchanging the complexing agents followed by neutralization of the solution.

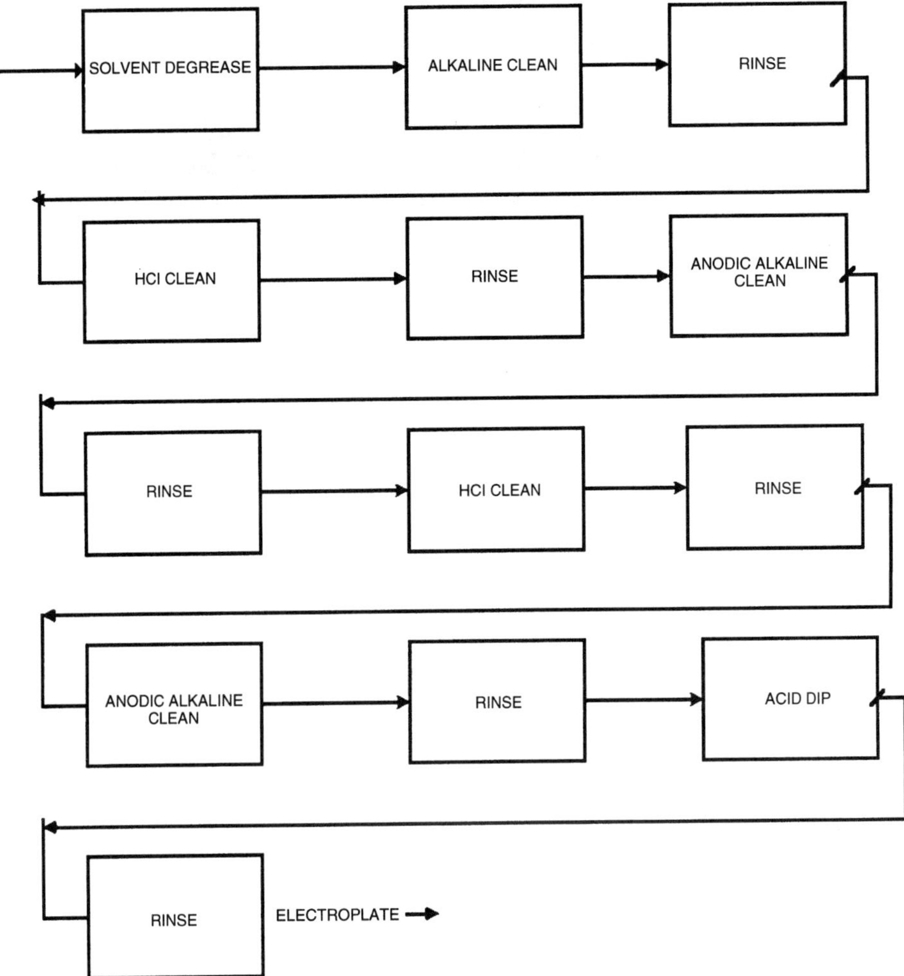

Figure 9 Schematic account of the pretreatment processes in electrodeposition [9].

TABLE 2 Different Baths for Cu Electrodeposition

Bath 1	Bath 2	Bath 3	Bath 4
CuCN	Copper sulfate	Copper fluoroborate	Copper pyrophosphate
NaCN (or KCN)	Sulfuric acid	Fluoroboric acid	Nitrate
NaOH (or KOH)		Boric acid	Ammonia
Rochelle salt			Orthophosphathe

The treatment of electroplating wastewater as mandated by the national pretreatment standards requirements will result in two streams: an effluent that must comply with regulations for acceptable pollutant discharge, and residue (sludge) containing a high concentration of the substances that the wastewater regulations prohibit discharging. Most electroplating sludges contain high concentrations of toxic heavy metals and are considered hazardous.

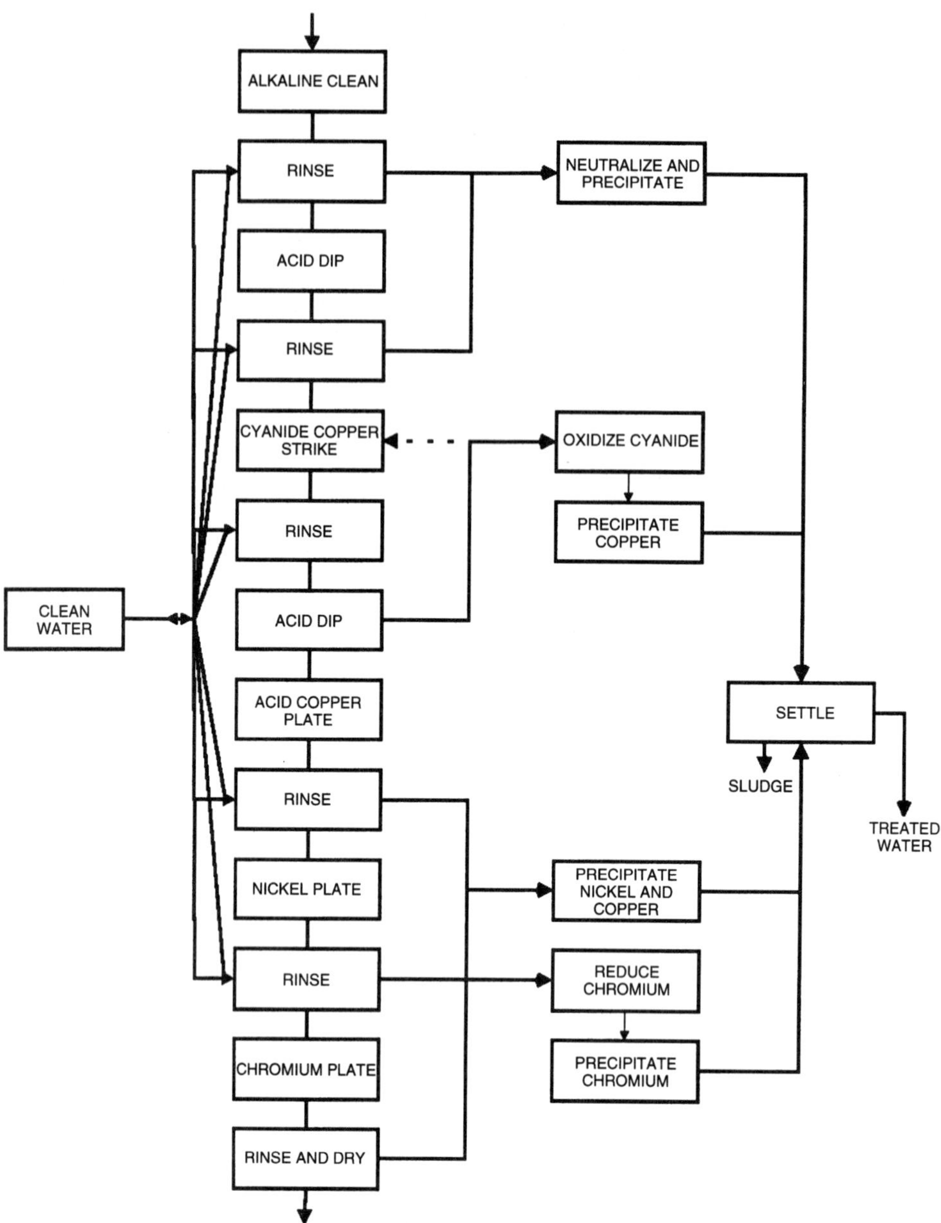

Figure 10 Schematic outline of the conventional on-site waste disposal procedures taken in industrial electrodeposition setting.

The high cost of sludge disposal demands that the design of wastewater treatment systems consider the relative volumes of the sludge generated by the different treatments. For example, insoluble sulfide precipitation can reduce the metal concentration in many waste streams to lower levels than can hydroxide precipitation. However, because this process uses ferrous sulfide as the source of the sulfide ions,

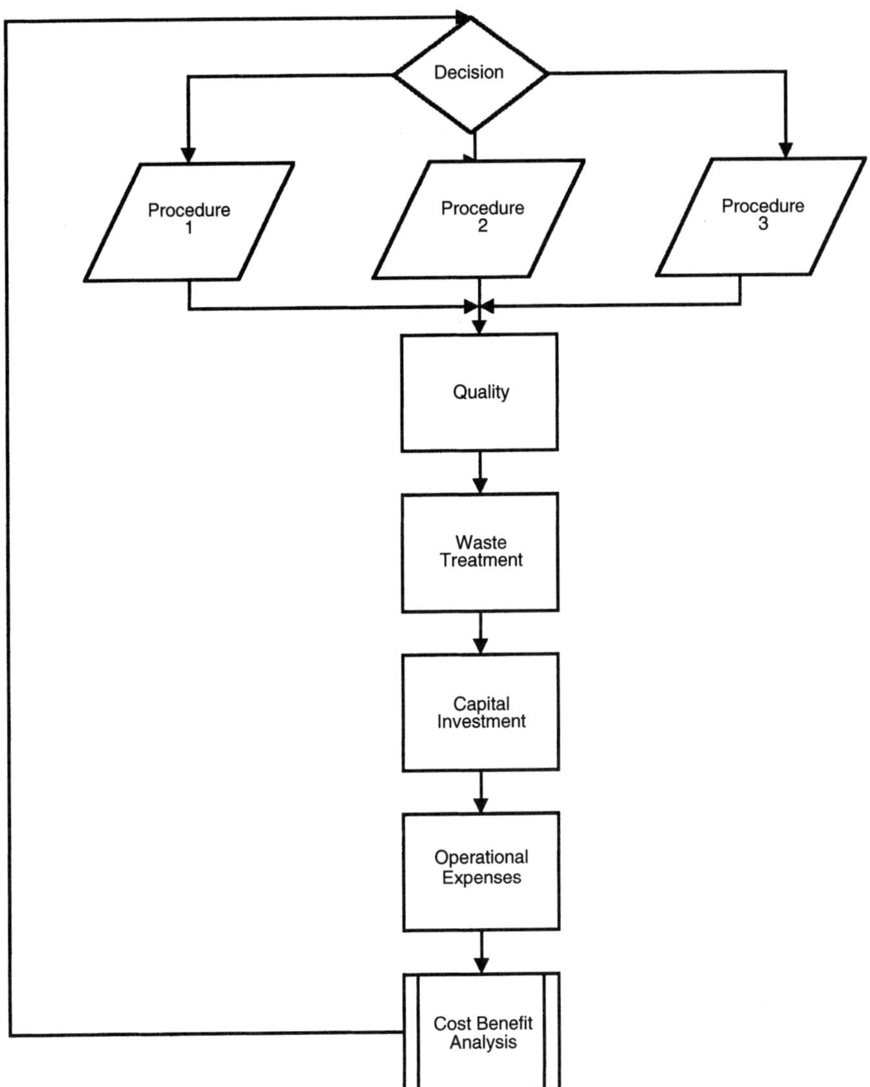

Figure 11 An outline of the flowchart of a decision making process in choosing a cost-effective procedure for copper plating.

ferrous ions are liberated in the reaction and converted to ferrous hydroxide that adds considerably to the sludge volume. Efficient dewatering of the sludge can also achieve significant reduction in the volume of the sludge.

An efficient way to minimize disposal costs is to minimize the total volume of water that is needed for the electroplating process. Useful techniques include simple steps such as preventing leaks, installing antisiphon devices equipped with self-closing valves on water inlet lines, using multiple counterflow rinse tanks to substantially reduce rinse water volume, using spray rinse, preventing unnecessary dilutions, recycling of wastewater, and using dry cleanup when possible.

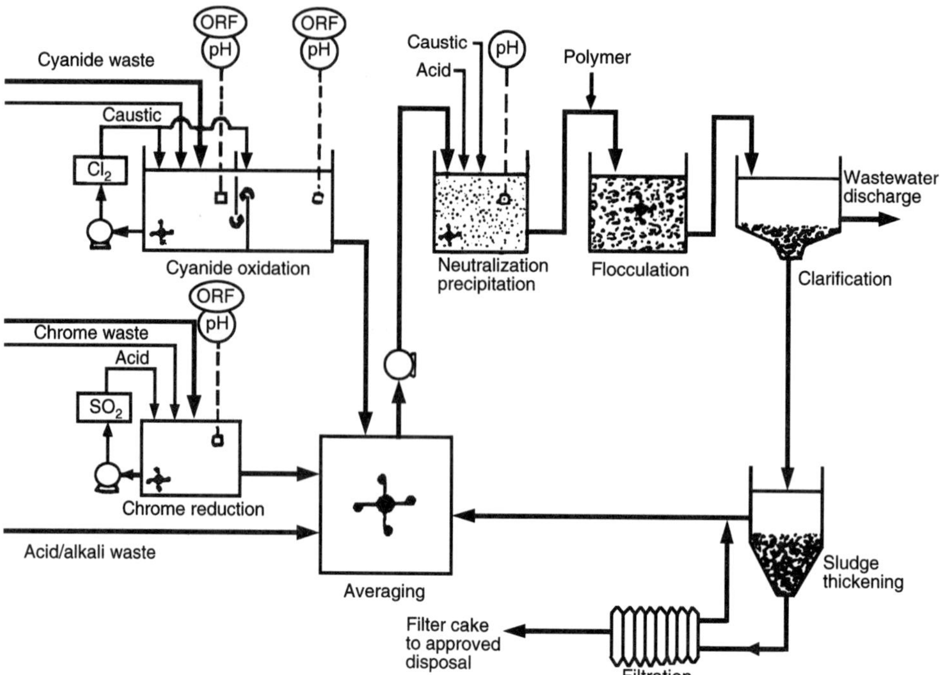

Figure 12 Conventional waste water treatment system for electroplating [10].

2.1 Impact Minimalization and Zero Discharge

The trend, where possible, to substitute modifications of the production process that result in waste minimization for waste treatment and environmental regulations is dominant in most discussions on environmental impact. The policy aspects of economic incentives will be discussed in a later section. Here we will concentrate on the technical aspects. Environmentally "acceptable" technological alternatives can often be found with varying degree of details in the EPA home-page section of Envirosense.

An EPA document [11] provides a comprehensive review of technologies that are targeted to eliminate, or significantly reduce the amount of any hazardous substance, pollutant, or contaminant released to the environment. The main target for such attempts is to eliminate or reduce cyanide in the electroplating procedures. Sodium and potassium cyanide are used in electroplating bath formulations for the deposition of copper, zinc, cadmium, silver, gold, and alloys such as brass, bronze, and alballoy.

Alternative technologies that are described in detail in the report include alkaline noncyanide copper plating, which was discussed before (in Table 2), Zinc-alloy electroplating, plasma deposition and thermal coating and work on new, environment-friendly alloys.

Zinc alloys can be used to replace cadmium coatings in a variety of applications [12]. They offer corrosion protection and lubricity. The most promising among these are zinc-nickel and zinc-cobalt. Zinc alone can provide corrosion protection equivalent to cadmium at thicknesses above 1 mil. Zinc cannot match cadmium when lubricity is desirable. EPA report [13] describes substituting of cadmium cyanide electroplating with zinc chloride electroplating. Product quality was defined in terms of corrosion resistance. Corrosion resistance was determined by salt-spray tests in accordance with ASTM B117-90.

Zinc-nickel alloys are being introduced in Japan and Germany in the automotive industry for fuel lines and rails, fasteners, air-conditioning components, cooling systems, and the like. Zinc-nickel coatings have also being introduced as cadmium replacements on fasteners for electrical transmission structures and on television coaxial cable connectors. Another alternative is to substitute electroplating with a dry process such as plasma deposition and thermal coating. The cost is significantly higher and severe restrictions for plating on irregularly shaped substrates exist. Nickel-tungsten-silicone-carbide coating was suggested by Takada [14] as a replacement for hard (functional) chromium. Nickel-tungsten-boron alloys were suggested as replacement for chromium. A family of these alloys is patented under the name of AMLATE.

So-called Zero-discharge processes are cyclical processes in which the water quality of the discharge is within regulatory specifications to be recycled into the public water system. Technology to achieve zero discharge is available. Major impediments are cost and drag-in.

Zero-discharge technologies are important not only in fulfilling the dream of eliminating adverse environmental consequences but in the more practical issue of establishing a baseline for charging the approximate price for pollution that the cleanup will cost. This "green tax" can be incorporated into the product's price

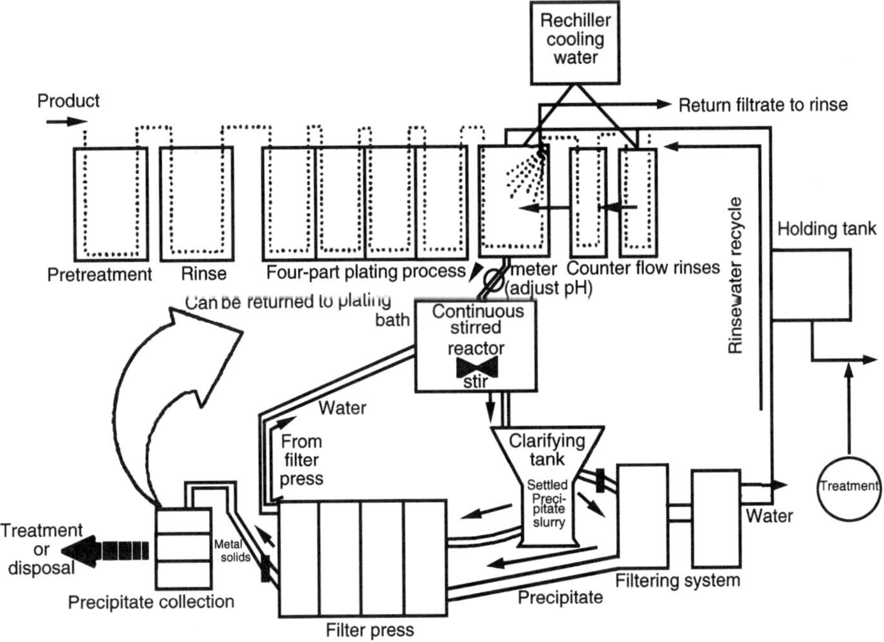

Figure 13 Schematics of the experimental P & H Plating Co. noncyanide plating line with the recycling system [15].

structure, thus forcing manufacturers to include environmental effects into the optimization of the manufacturing processes that will maximize the profitability of the product. A good example of such a process is a cooperative project between the EPA and the Illinois Hazardous Waste Research and Information Center (HWRIC) [15]. The demonstration was on a zinc plating line at a local Chicago facility (P&H Plating Co.). The first step was to convert existing cyanide based line to a noncyanide line. A schematic of the cyclical plating process is shown in Figure 13. Technical details are sketchy. The figure shows the usual parts of pretreatment, followed by the plating process, spray rinse, and submersion into two counterflow rinse tanks. Although the cleaning requirement of the ANC plating is considerably more stringent than the CN-based process, the process itself was similar. The hydroxide is precipitated by adjusting the pH, collected on filters, and dewatered using filter press. The hydroxide was stored and reused. Plating quality criteria were corrosion resistance and thickness. Quantitative comparisons were not provided, except to state that both methods produced satisfactory results.

3 APPLICATIONS OF ELECTRODEPOSITION IN ENVIRONMENTAL REMEDIATION

Waste streams similar in composition to the one described are not limited to the metal finishing industry. Electrochemical deposition of metals from waste streams is a technology that is being developed both for removal and for recycling. The industrial applications are still limited mainly because of relatively low cost of disposal of solid waste. Detailed discussion can be found in [16–20].

Similar to electrodeposition, the cathodic removal of metals may be represented by an electrowinning reaction:

$$M^{z+} + ze^- \rightarrow M \tag{3}$$

The soluble species may be simple a hydrated cation such as Cu^{2+} (aq) or a complex such as $[CuCl_4]^{3-}$ or $[Ag(S_2O_3)_2]^{3-}$. Desired cathodic reactions such as

$$Cu^{2+} + 2e^- \rightarrow Cu \tag{4}$$

may be accompanied by undesired solvent decomposition reactions such as hydrogen evolution, which in acidic medium will take the form

$$2H^+ + 2e^- \rightarrow H_2 \qquad E = +0.00\ \mathrm{V} \tag{5}$$

and in neutral or alkaline medium

$$H_2O + 2e^- \rightarrow 2OH^- + H_2 \tag{6}$$

Oxygen reduction will also often compete:

$$O_2 + 4H^+ + 4e^- \rightarrow 2H_2O \tag{7}$$

On the anodic side oxygen evolution will dominate when inert electrodes are used:

$$2H_2O + 4e^- \rightarrow O_2 + 4H^+ \qquad E = +1.23\ \mathrm{V} \tag{8}$$

Standard potentials of three of the metals often found in waste streams are shown below:

$$Cu^{2+} + 2e^{-} \rightarrow Cu \qquad E_0 = +0.34\,V$$

$$Ni^{2+} + 2e^{-} \rightarrow Ni \qquad E_0 = -0.25\,V$$

$$Zn^{2+} + 2e^{-} \rightarrow Zn \qquad E_0 = -0.76\,V$$

Cu is easy to electrowin because the standard potential is more positive than hydrogen. Nickel is more difficult, but raising the pH shifts the hydrogen potential to a more negative value. Zinc electrowinning is possible only due to the high overvoltage for hydrogen evolution on a zinc cathode. This requires a very pure Zn solution because deposition of any impurities will lower considerably the overvoltage for hydrogen evolution.

An electrowinning cell design that was used by Eco-Tec for waste recovery applications is described by Brown et al. [21]. The cell is based on Kennecott Copper High Efficiency Air Agitation (HEAA) technology, with the cell design optimized for recovery from the waste streams. Welsh et al. [22] describe a rotating cylinder electrode (RCE) cell for the removal of metals *via* cathodic depositions. References for other cell designs are also mentioned there.

Clarke et al. [23] describe proprietary EBONEX ceramic electrodes that are made of Magnelli phase suboxides of titanium dioxide. They find applications as cathodes in metal recovery systems from the waste stream, supporting anodes in electrowinning of Zn and Cu and as anodes for cyanide and organic waste destruction.

For an effective recovery, it is often required that the metals be electrodeposited in acceptable levels of purity from waste streams that contain more than one metal. This is often done by a judicial selection of the deposition potential. Armstrong et al. [24] discuss selective electrodeposition of cadmium, cobalt and nickel in pure form from binary mixtures. Raats et al. [25] discuss Cu–Ni and Zn–Cu mixtures, and Polcaro et al. [26] discuss deposition from a Pb–Cu mixture.

Electrowinning has traditionally been applied to metal recovery from concentrated solutions. Although attempts are being made to extend electrowinning to more dilute solutions [16], ion exchange is considered as a more effective tool for metal recovery. Brown et al. [21] describe a waste treatment system that combines the two techniques for total removal of metal contaminants. Cation exchangers exchange metallic cations for hydrogen ions:

$$RH + CuSO_4 \rightarrow R_2Cu + H_2SO_4 \tag{9}$$

Upon exhaustion, the resin must be regenerated by strong acids such as hydrochloric acid or by sulfuric acids, which will yield concentrated acidic copper solutions. This way metals can be concentrated from few parts per million to 30,000 ppm. By proper choice of regenerating acid, one can adjust for a desirable anions that will make the regenerated solution suitable for electrowinning and recycling. Unlike other separation processes such as evaporation and reverse osmosis which remove water from the waste, cation exchangers and electrowinning remove the metals from the waste.

Ion-exchange processes have not only the ability to concentrate the pollutants but also to separate and purify them by the differential affinities of the cations to the ion exchanger. Increased degree of metal separation can be achieved with application of chelating resins such as iminodiacetate chelating resins. These chelating resins can have strong enough affinity to bind metals from plating solutions that contain chelating agents such as quadrol, citrates, and EDTA. The authors describe the performance of a commercial ion-exchange system [27].

Three case studies in which the combinations of ion exchanger and electrowinning are used for metal recycling, are described in [22]. An excellent example is the one used by Hughes Aircraft in Orangeberg, South Carolina. Copper-containing waste concentrates from various processes, including electroless copper plating, nitric acid rack strip, and numerous spent acid baths, are segregated and collected in batch holding baths. The contents of these tanks are then bled into an equalization tank feeding the recovery system. After pH adjustment it passes through filters to the chelating ion-exchange units (IX). The effluent from these units is monitored by a colorimeter copper analyzer that initiates regeneration at a breakthrough level of $1\,mg\,liter^{-1}$. The ion exchange units are then processed in a batch-type Recowin electrowinning system where the copper is reduced from 10 to $20\,g\,liter^{-1}$ to about $1\,g\,liter^{-1}$. The spent electrowinning electrolyte is fed back through the ion-exchange units to recover residual copper.

3.1 Light-Induced Electrodeposition

The possibility of using solar energy as an energy source for environmental remediation has attracted considerable interest [28]. The pollutants might either be reduced or oxidized in an electrochemical cell in which at least one of the electrodes is a photoactive semiconductor. One can use an n-type semiconductor that will drive photogenerated holes to the semiconductor/electrolyte interface, or a p-type semiconductor in which photogenerated electrons will be driven to the interface. Presently most of the effort in this area is directed at photooxidation of organic pollutants [30–32].

Within the context of environmental aspects of electrodeposition, the most interesting processes are reductive processes. The reductive processes can involve either reduction of the metal ions to metallic form similar to the reduction processes that were discussed before or reduction of problematic metal ions such as Cr(VI) to more treatable ions such as Cr(III).

An example of a band diagram of a photoelectrochemical system designed for metal deposition is shown in Figure 14. The band diagram shown in Figure 14 is appropriate for large crystallites that can sustain a space-charge layer and drive light-induced carriers to the semiconductor-electrolyte interface. This example shows an n-type semiconductor in which light-induced holes will be driven to the surface, while the electrons will be driven to the counterelectrode. Under normal conditions (semiconductor in depletion and no current doubling), this configuration sustains oxidation at the semiconductor-electrolyte interface and reduction at the counter-electrode. The electron source can be water, as shown in Figure 14, or a sacrificial donor. If the equilibrium potential of the semiconductor is adjusted to match the counterelectrode, the system can operate under short bias conditions. Under these conditions one can replace the two-electrode configuration with a semiconducting powder in which both the reductive reaction and the oxidation reaction take place on the same electrode. However, when the particle size in such powders is reduced

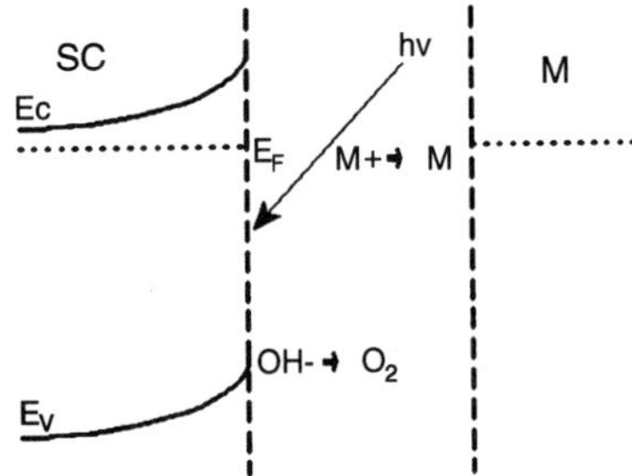

Figure 14 Energy band diagram of a photoelectrochemical system designed to photoreduce dissolved metal cations.

to nanometer scale, the particles can no longer sustain a space-charge layer and dopant-independent diffusion of light-induced carriers to the surface will take place. Semiconductor-electrolyte systems were described for gold recycling [31], silver recycling [33], mercury recycling [34], and platinum recycling [35]. In some cases water is the electron donor, and other cases organic solutes such as isopropyl alcohol are used as donors.

4 ECONOMIC CONSIDERATIONS

4.1 Accounting

It is not easy to arrive at cost numbers that will clearly define the costs that are associated with disposal. Typical numbers for the industry run at about 3% of total expenditure with a standard deviation of about 2%. Table 3 provides the cost data that were associated with experimental alternatives for the cyanide plating of zinc, with alternatives similar to one that were previously discussed for copper. The data in Table 3 can be interpreted as requiring between 60% and 87.5% of costs dedicated to waste disposal. The cost accounting in Table 3 is provided in support of the

TABLE 3 Comparison of Annual Operational Costs and Capital Investment for CN Process with Alkaline Noncyanide(ANC) and ANC with Recovery/Recycle (R/R*) System [8]

Process Operation	CN Costs ($)	ANC Costs ($)	ANC + R/R* costs ($)
Bath makeup	1771	1860	1860
Bath maintenance	22325	21225	19425
Water usage			
1. Use @ $7.56/7480 gal	1213	1213	364
2. Sewering @ $5.59/7480 gal	897	897	269
Waste-water treatment			
1. Cyanide oxidation	14000	0	0
2. Metal precipitation	69000	69000	20700
3. Labor @ $15/hr	7500	7500	2250
Sludge disposal @ $209/yd^3	2600	2600	1820
Total Operating Expenses	119306	104295	46688
Capital Investment		36000	87822

claim that the most environmentally friendly processing is also the most cost effective.

4.2 Policy

Most developed countries spend between 1% and 2% of their gross domestic product on environmental protection, and that proportion is likely to rise. In the United States it is estimated that the 1990 effort was 1.9% of GNP, and that is expected to grow to 2.7% by the year 2000 [36].

Regulation is by far the most common tool of environmental policy. What is often referred to as "command and control" policy is almost universally believed not to be the most efficient policy, both in terms of environmental protection and economic impact. The deficiencies of the regulations are discussed in [37] as follows:

- Encouragement of governments to do what they do worst—second-guessing companies about what is the best technology to achieve a particular goal
- Disguised true cost
- Unequal loading of costs
- Regulations that set both a floor and a ceiling on discharge (no incentive to discharge less than the regulation allows)
- Best results with big corporations but much less effective with dispersed sources

Market-driven economic environmental instruments fall into two categories [29]:

- "Green taxes" that put a price on pollutants that in principle reflect the cost they impose on society. The extent to which pollution will be reduced will depend on government's use of the fees. If government transfers the income into the general revenue, the effect on the pollution level will be difficult to predict.
- "Marketable permits" that are based on absolute quantity of pollution that is to be allowed, and then give or sell polluters rights to pollute up to that given level. Polluters can trade these rights with each other.

These permits make it easy to predict the total amount of pollution but very difficult to determine what the "effective" tax on polluters will be. They also make sense in environments that "share" pollution. It makes very little sense to exchange pollution permits between two aquatic reservoirs that do not mix effectively with each other so that, by cleaning one, one pollutes the other. The whole field of environmental justice is largely based on such practices. Green taxes set a price and allow the quantity to fluctuate, while permits set a fixed quantity while allowing the price to fluctuate. A comprehensive system analysis that models the impact of nine source reduction methodologies is given in [38]. An organization theory perspective of pollution prevention in the Chinese electroplating industry is presented in [39].

The search for effective economic incentives to tackle pollution is accelerating [36]. Many countries charge flat rate on water pollution. In many instances green taxes are being presented as "sin taxes," which puts them in the same category as taxes on cigarettes and alcohol.

4.3 Advocacy

The common desire of legislating the economy to operate under "automatic pilot" in order to reduce environmental impact is appealing. The cumbersome diversity of a mix of regulations, jaw-boning, and green taxes is no one's favorite. The mix is different in different countries. Despite this diversity of policy tools, one must grudgingly admit that considerable progress in integrating environmental considerations into the production facilities has taken place since 1974. One relevant quantitative indicator in support of this is that in the year 2000 the water consumption by manufacturing industries in the United States is expected to be only one-third of what it was in 1977, mainly because of the increased costs of disposing of wastewater.

An important aspect in which progress is limited is the incorporation of environmental considerations as an important element in the decision making. One particular aspect that limits progress in this area is the limited ability to quantify product quality and correlate it with commercial value. The two most common applications of the metal finishing industry are functional applications, such as corrosion protection and ornamental applications. Some of the considerations that are involved in quantifying product quality were previously discussed in the context of adopting cyanide-free deposition procedures and substituting toxic metals such as cadmium with less toxic ones such as zinc. A typical corrosion test can follow acceptable standards and is not very difficult to perform. However, ornamental applications are almost impossible to quantify. Public education and jaw-boning programs—similar to the ones that are being applied in the consumer-product industry ("green coating")—might be effective.

Figure 15 shows schematically a price-quality relationship that, if it existed, might facilitate cost-benefit analysis of alternative deposition technologies. Perhaps as important as the ability of a manufacturer to perform a cost-benefit analysis that will incorporate environmental considerations is the ability of society to funnel research support to technologies that offer the most potential benefits. Quantification of the quality of the surface treatments in terms of more sophisticated concepts than the prevailing ones is most urgently needed. Application of chemometric principles to

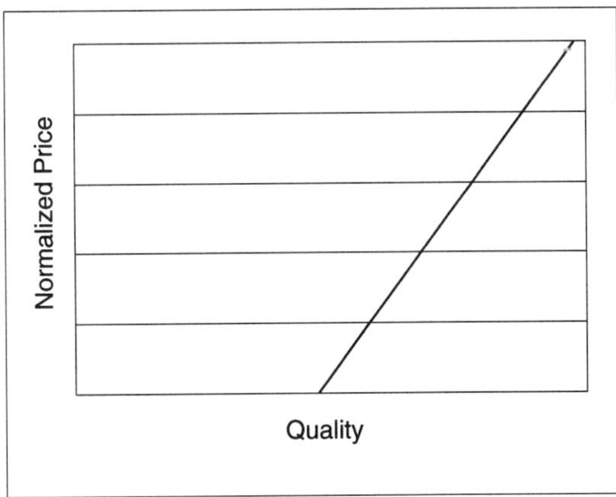

Figure 15 Schematics of normalized price-quality of a metal finished product.

this issue is likely to have a major impact on research and development that will result in better environmentally friendly products.

ACKNOWLEDGMENT

I acknowledge the partial support of the New York State Science and Technology Foundation through its Center for Advanced Technology Program and the NASA IRA program. I want to thank the following individuals and institutions that did their best to expose and educate me: Prof. Yehuda Klein, Economics Department, Brooklyn College; Richard Belgrave and Aldo Orlando, the Environmental Safety officers at Brooklyn College; Peter Dipietrantonio, the Environmental Compliance manager at LeaRonal Co. Of Freeport, NY; and Doron Almog, the Urban League for Environmental Protection, Israel.

REFERENCES

1. F. A. Lowenheim ed. *Modern Electroplating*, Electrochemical Society, Pennington, NJ, 1974.
2. *Http://www.epa.gov.*
3. Http://www.finishing.com.
4. Http://www.namf.org
5. Http://www.nmfrc.org.
6. Http://www.tsdcentral.com.
7. See, for example, *Electroplating and Related Metal Finishing—Pollutant and Toxic Materials Control*, Marshall Sitting, Noyes Data Co., 1978.
8. H. M. Spliethoff and H. F. Hemond, *Env. Sci. Tech.*, **30**, 121 (1996).
9. EPA/625/10-85/001, *Environmental Regulations and Technology* (1985).
10. Http://www.roskill.co.uk.
11. EPA/625/R-94/007, *A Guide to Cleaner Technologies and Alternative Metal Finishes.*
12. D. A. Schario, M. A. Klingenburg, and E. W. Brooman; ECS Meeting Abstract, MA96-2,309 (1996).
13. EPA/600/SR-94/074, *Substituting Cadmium Cyanide Electroplating with Zinc Chloride Electroplating*, B. C. Kim, P. R. Web, J. A. Gurklis, and R. K. Smith (May 1994).
14. Takada, U.S. Patent 4,892,627 (1990).
15. EPA/600/SR-94/148, *Alkaline Noncyanide Zinc Plating and Reuse of Recovered Chemicals*, Jacqueline M. Peden (1994).
16. F. C. Walsh in *Electrochemistry for Cleaner Environment*, J. D. Genders and N. L. Weinberg eds., Electrosynthesis Company, New York, 1992, ch. 4.
17. T. L. Hatfield, T. L. Kleven, and D. T. Pierce, *J. Appl. Electrochem.*, **26**, 567 (1996).
18. C. A. Hodges, *Science*, **268**, 1305 (1995).
19. W. J. Eilback and G. Mattock, *Chemical in Waste Water Treatment*, Wiley, New York, 1986.
20. K. E. Boumidel and R. Salhi, ECS Meeting Abstract, MA97-2, 661 (1997).
21. C. J. Brown, in *Electrochemistry for Cleaner Environment*, ed. J. D. Genders and N. L. Weinberg, eds., Electrosynthesis Company, New York, 1992, ch. 7.

22. F. C. Walsh and G. W. Reade, in *Environmental Oriented Electrochemistry*, C. A. C. Sequeira, ed., Elsevier, 1994, ch. 1; F. C. Walsh, in *Electrochemistry for Cleaner Environment*, J. Amsterdam, D. Genders and N. L. Weinberg, eds., Electrosynthesis Company, New York, 1992, ch. 4.
23. R. Clarke and R. Pardoe, in *Electrochemistry for Cleaner Environment*, J. D. Genders and N. L. Weinberg, eds., Electrosynthesis Company, New York, 1992, ch. 18.
24. R. D. Armstrong, M. Todd, J. W. Atkinson, and K. Scott, *J. Appl. Electrochem.*, **26**, 379 (1996).
25. C. M. S. Raats, H. F. Boon, and G. van der Heiden, *Chem. Ind.*, 465 (1978).
26. A. M. Polcaro and S. Palmas, in *Electrochemical Engineering and the Environment*, Ichem Symp. Series, **127**, 85 (1992).
27. C. P. Jones, M. D. Neville, and A. D. Turner, in *Electrochemistry for Cleaner Environment* J. D. Genders and N. L. Weinberg, eds., Electrosynthesis Company, New York, 1992, ch. 8.
28. For a recent review, see A. J. Nozik and R. Memming, *J. Chem. Phys.*, **100**, 13061 (1996).
29. For a recent review, see K. Rajeshwar, *J. Appl. Electrochem.*, **25**, 1067 (1995).
30. See, for example, D. F. Ollis, E. Pelizzetti, and N. Serpone, *Photocatalysis—Fundamentals and Applications*, Wiley, New York, 1989, ch. 18.
31. N. Serpone, E. Borgarello, and M. Barbeni, *J. Photochem.*, **36**, 373 (1987).
32. D. M. Blake, in NREL/TP-473-20300, *Bibliography of Work on Heterogeneous Photocatalytic Removal of Hazardous Compounds from Water and Air* (1995).
33. K. Ito and A. Fujishima, *J. Amer. Chem. Soc.*, **110**, 6267 (1988).
34. J. Domenech, M. Andres, and J. Munoz, *J. Electrochim. Acta*, **32**, 773 (1987).
35. B. Kraeutler and A. J. Bard, *J. Amer. Chem. Soc.*, **100**, 5985 (1978).
36. EPA-230-R-92-001, *The United State Experience with Economic Incentives to Control Environmental Pollution*, Alan Carlin (1994).
37. Frances Cairncross, *in Green, Inc.*, Earthscan Pub., 1995.
38. W. Glaze, Dissertation No. AAT 8919913 (1989).
39. K. A. Warren, Dissertation No. AAT 9630410 (1996).

Appendix

This Appendix contains tables of useful data miscellaneous information for use by electroplaters. It has been prepared by Arch B. Tripler, Jr., Battelle Memorial Institute, Columbus Laboratories, Columbus, Ohio.

TABLE 1 Physical Constants of Selected Elements

Element	Symbol	Atomic No.	Atomic Weight 1971	Density[a] g/cm^3, 20°C	Melting Point,[a] °C	Linear Coefficient Thermal Expansion near 20°C μm/°C
Aluminum	Al	13	26.9815	2.699	660.1	0.607[e]
Antimony	Sb	51	121.75	6.62	630.5	0.22–0.27[f]
Arsenic	As	33	74.9216	5.73	(820)	0.12
Beryllium	Be	4	9.0122	1.85	1350	0.31[h]
Bismuth	Bi	83	208.980	9.78	271.3	0.34
Boron	B	5	10.81	2.3	2300 ± 300	0.21[i]
Bromine	Br	35	79.904	3.12	−7.2	—
Cadmium	Cd	48	112.40	8.65	320.9	0.757
Calcium	Ca	20	40.08	1.55	810	0.56
Carbon (graphite)	C	6	12.011	2.25	3700 ± 100	0.015–0.11[e]
Cerium	Ce	58	140.12	6.9	600 ± 50	—
Chlorine	Cl	17	35.453	—	−101.6	—
Chromium	Cr	24	51.996	7.14	1890 ± 10	0.16
Cobalt	Co	27	58.9332	8.71	1492	0.312
Copper	Cu	29	63.546	8.933	1083	0.419
Fluorine	F	9	18.9984	—	−223	—
Gallium	Ga	31	69.72	5.46	29.7	0.46
Germanium	Ge	32	72.59	5.36	958.5	—
Gold	Au	79	196.967	19.32	1063	0.361
Hydrogen	H	1	1.0079	0.08375×10^{-3}	−259.14	—
Indium	In	49	114.82	7.28	155.0	0.84
Iodine	I	53	126.9045	4.93	113.5	2.36
Iridium	Ir	77	192.2	22.42	2443	0.17
Iron	Fe	26	55.847	7.86	1533	0.297
Lead	Pb	82	207.2	11.34	327.4	0.744[e]
Lithium	Li	3	6.941	0.534	186	1.42
Magnesium	Mg	12	24.305	1.74	651	0.66[j]
Manganese	Mn	25	54.9380	7.3	1260	0.56
Mercury	Hg	80	200.59	13.55	−38.87	—
Molybdenum	Mo	42	95.94	10.2	2620	0.12[e]
Nickel	Ni	28	58.71	8.8	1453	0.338[k]
Niobium (Columbium)	Nb (Cb)	41	92.9064	8.4	2500	0.18
Nitrogen	N	7	14.0067	1.1649×10^{-3}	−209.86	—
Osmium	Os	76	190.2	22.5	2700	0.12
Oxygen	O	8	15.9994	1.3318×10^{-3}	−218.4	—
Palladium	Pd	46	106.4	12.16	1552	0.30
Phosphorous (yellow)	P	15	30.9738	1.82	44.1	3.18
Platinum	Pt	78	195.09	21.37	1769	0.23
Potassium	K	19	39.098	0.87	62.3	2.1
Rhenium	Re	75	186.2	21.04	3180	0.18
Rhodium	Rh	45	102.9055	12.44	1960	0.21[e]

TABLE 1 *(Continued)*

Electrical Resistivity, microhm-cm	Electrochemical Equivalents mg/C for valence of 1[b]	Crystal Structure[c]	Lattice Constants at 20°C kX Units[d] a	b	c or Axial Angle or Temperature	Closest Approach of Atoms
2.655(20°C)	0.2796	fcc	4.0408	—	—	2.856
39.0(0°C)	1.262	rh	4.4974	—	57°6.5′	2.898
35(0°C)	0.7765	rh[g]	4.151	—	53°49′	2.50
5.9(0°C)	0.09340	cph[g]	2.2810	—	3.5771	2.221
106.8(0°C)	2.166	rh	4.7361	—	57°14.2′	3.105
1.8×10^{12}(0°C)	0.1120	or(?)	17.86	8.93	10.13	—
—	0.8281	or	4.48	6.67	8.72(−150°C)	2.27
6.83(0°C)	1.165	cph	2.9727	—	5.606	2.972
3.43(0°C)	0.4154	fcc[g]	5.56	—	—	3.93
1375(0°C)	0.1245	hex[g]	2.4564	—	6.6906	1.42
78(20°C)	1.452	fcc[g]	5.143	—	—	3.64
—	0.3674	tet	8.56	—	6.12(−185°C)	1.81
13(28°C)	0.5389	bcc[g]	2.8787	—	—	2.493
6.24(20°C)	0.6108	cph[g]	2.502	—	4.061	2.501
1.673(20°C)	0.6586	fcc	3.6080	—	—	2.551
—	0.1969	—	—	—	—	—
53.4(0°C)	0.7226	1fcor	4.517	4.511	7.645	2.437
8.9×10^{4}(0°C)	0.7523	dc	5.647	—	—	2.445
2.19(0°C)	2.041	fcc	4.0701	—	—	2.878
—	0.01045	hex	3.75	—	6.12(−271°C)	—
8.37(0°C)	1.190	fct	4.585	—	4.941	3.24
1.3×10^{15}(20°C)	1.315	or	4.777	7.251	9.773	2.70
5.3(20°C)	1.992	fcc	3.8312	—	—	2.709
9.71(20°C)	0.5788	bcc[g]	2.8606	—	—	2.476
20.65(20°C)	2.147	fcc	4.9395	—	—	3.493
8.55(0°C)	0.07191	bcc	3.5019	—	—	3.033
4.46(20°C)	0.2519	cph	3.2028	—	5.1998	3.190
185(20°C)	0.5694	cc[g]	8.894	—	—	2.24
94.1(0°C)	2.079	rh	2.999	—	70°31.7′ (−46°C)	2.999
5.17(0°C)	0.9943	bcc	3.140	—	—	2.720
6.84(20°C)	0.6085	fcc	3.5167	—	—	2.486
13.1(18°C)	0.9629	bcc	3.2941	—	—	2.853
—	0.1452	hex[g]	4.03	—	6.59(−234°C)	—
9.5(20°C)	1.971	cph	2.7298	—	4.3104	2.670
—	0.1658	c[g]	6.83	—	(−225°C)	—
10.8(20°C)	1.103	fcc	3.8824	—	—	2.745
10^{17}(11°C)	0.3210	c[g]	7.17	—	(−35°C)	—
9.83(0°C)	2.022	fcc	3.9158	—	—	2.769
6.15(0°C)	0.4052	bcc	5.333	—	—	4.618
19.3(20°C)	1.930	cph	2.7553	—	4.4493	2.734
4.5(20°C)	1.066	fcc[g]	3.7957	—	—	2.684

TABLE 1 *(Continued)*

Element	Symbol	Atomic No.	Atomic Weight 1971	Density[a] g/cm^3, 20°C	Melting Point,[a] °C	Linear Coefficient Thermal Expansion near 20°C µm/°C
Selenium	Se	34	78.96	4.81	220	0.94
Silicon	Si	14	28.086	2.42	1420	0.07–0.19
Silver	Ag	47	107.868	10.49	960.8	0.50[k]
Sodium	Na	11	22.9898	0.97	97.5	1.8
Sulfur (yellow)	S	16	32.06	2.07	113–119	1.6
Tantalum	Ta	73	180.947	16.6	3005	0.17
Tellurium	Te	52	127.60	6.25	452	0.423
Thallium	Tl	81	204.37	11.86	303.5	0.71
Tin	Sn	50	118.69	7.298	231.9	0.58
Titanium	Ti	22	47.90	4.54	1820	0.22
Tungsten (Wolfram)	W	74	183.85	19.3	3380	0.11
Uranium	U	92	238.029	18.7	1130	—
Vanadium	V	23	50.941	5.87	1735	0.20
Zinc	Zn	30	65.38	7.133(25°C)	419.47	1.01[l]
Zirconium	Zr	40	91.22	6.5	1750	0.13

[a]Densities and melting points taken from *Smithsonian Physical Tables*, Smithsonian Institution, Washington, D.C., 1954, 9th rev. ed.

[b]Calculated from 1967 atomic weights and value of faraday as 96,489 ± 2 C(NBS 1960, as given in *Handbook of Chemistry and Physics*, 50th ed., Chemical Rubber Company, Cleveland, Ohio, 1970, p. F78); for valences other than 1, divide value given by valence.

[c]The abbreviations used to denote crystal structure as follows:

fcc	face-centered cubic	1fcor	one face-centered orthorhombic
rh	rhombohedral	dc	diamond cubic
bcc	body-centered cubic	cc	cubic (complex)
cph	close-packed hexagonal	c	cubic
or	orthorhombic	fcor	face-centered orthorhombic
hex	hexagonal	bct	body-centered tetragonal
tet	tetragonal	fct	face-centered tetragonal

[d]The kX unit (1000 X units), used in all crystal structure measurements, differs slightly from an angstrom (10^{-8} cm); the tabulated lattice dimensions should be multiplied by 1.00202 to put them in angstroms.

TABLE 1 *(Continued)*

Electrical Resistivity, microhm-cm	Electro-chemical Equivalents mg/C for valence of 1[b]	Crystal Structure[c]	Lattice Constants at 20°C kX Units[d]			
			a	*b*	*c* or Axial Angle or Temperature	Closest Approach of Atoms
12(20°C)	0.8183	hex[g]	4.3552	—	4.9494	2.32
10^5(0°C)	0.2911	dc	5.4173	—	—	2.346
1.59(20°C)	1.118	fcc	4.0774	—	—	2.882
4.2(0°C)	0.2383	bcc	4.2820	—	—	3.708
2×10^{23}(20°C)	0.3323	fcor[g]	10.48	12.92	24.55	2.12
12.4(18°C)	1.875	bcc	3.2959	—	—	2.854
2×10^5(19.6°C)	1.322	hex	4.4469	—	5.9149	2.86
18(0°C)	2.118	cph[g]	3.450	—	5.514	3.401
11.5(20°C)	1.230	bct[g]	5.8194	—	3.1753	3.016
80(0°C)	0.4964	cph[g]	2.953	—	4.729	2.91
5.5(20°C)	1.905	bcc[g]	3.1585	—	—	2.734
60(18°C)	2.467	or[g]	2.852	5.865	4.945	2.76
26(20°C)	0.5280	bcc	3.033	—	—	2.627
5.916(20°C)[m]	0.6775	cph	2.659	—	4.935	2.659
41.0(0°C)	0.9454	cph[g]	3.223	—	5.123	3.16

[e]From 20 to 100°C.
[f]From 20 to 60°C.
[g]Ordinary form; other modifications known or probable.
[h]From 20 to 200°C.
[i]From 20 to 750°C.
[j]At 40°C.
[k]From 0 to 100°C.
[l]For polycrystalline zinc; in single crystals, varies from 61.5 (parallel to hexagonal axis) to 15 (perpendicular to hexagonal axis).
[m]For polycrystalline zinc; in single crystals, varies from 6.16 (parallel to hexagonal axis) to 5.89 (perpendicular to hexagonal axis)

Note: Many of the data in this table were taken from a similar table in *The Metals Handbook* published by the American Society for Metals. Permission to use these data has been received.

TABLE 2 Electrolytic Potentials of the Metals. Ions at Unit Molal Activity[a]

Metal/Ion Couple	Standard Electrode Potential, $E^0(V)$, 25°C	Metal/Ion Couple	Standard Electrode Potential, $E^0(V)$, 25°C
Li/Li^{+}	−3.09	Cd/Cd^{2+}	−0.403
Rb/Rb^{+}	−2.925	In/In^{3+}	−0.342
K/K^{+}	−2.925	Ti/Ti^{+}	−0.336
Cs/Cs^{+}	−2.923	Co/Co^{2+}	−0.277
As/As^{+}	−2.923	Ni/Ni^{2+}	−0.250
Sr/Sr^{2+}	−2.89	Mo/Mo^{3+}	ca. −0.2
Ca/Ca^{2+}	−2.87	Sn/Sn^{2+}	−0.136
Na/Na^{+}	−2.714	Pb/Pb^{2+}	−0.126
Mg/Mg^{2+}	−2.37	Fe/Fe^{3+}	−0.036
Pu/Pu^{3+}	−2.07	$H_2/2H^{+}$	0.000
Be/Be^{2+}	−1.85	Bi/Bi^{3+}	0.2
U/U^{3+}	−1.80	Sb/Sb^{3+}	0.2
Al/Al^{3+}	−1.66	As/As^{3+}	0.3
Ti/Ti^{2+}	−1.63	Cu/Cu^{2+}	0.337
Zr/Zr^{4+}	−1.53	$2Hg/Hg^{2+}$	0.789
Mn/Mn^{2+}	−1.18	Ag/Ag^{+}	0.7991
V/V^{2+}	ca. −1.18	Hg/Hg^{2+}	0.854
Nb/Nb^{3+}	ca. −1.1	Pd/Pd^{2+}	0.987
Zn/Zn^{2+}	−0.763	Pt/Pt^{2+}	ca. 1.2
Cr/Cr^{3+}	−0.74	Au/Au^{3+}	1.50
Ga/Ga^{3+}	−0.52	Au/Au^{+}	ca. 1.68
Fe/Fe^{2+}	−0.440		

[a]The numerical values of the standard electrode potentials have been taken from *The Oxidation States of the Elements and Their Potentials in Aqueous Solutions*, by Wendell M. Latimer, Prentice-Hall, Englewood Cliffs, N.J., 1952, 2nd ed., with three exceptions. The values for trivalent bismuth, antimony, and arsenic were taken from *Theoretical and Applied Electrochemistry*, by Maurice deKay Thompson, Macmillan, New York, 1939, 3rd ed. The sign convention used above is that of The Electrochemical Society.

TABLE 3 Standard Reference Electrodes

Half-Cell	$E_{25°C}$, (V)
0.1 *N* Calomel	0.334
1.0 *N* Calomel	0.281
Saturated calomel	0.242
Silver-silver chloride	
Chloride ion unit activity	0.222
0.1 *N* KCl	0.288

TABLE 4 Equilibrium Constants of Interest in Electroplating[a]

$AgBr = Ag^+ + Br^-$	5.0×10^{13}
$AgCl = Ag^+ + Cl^-$	2.8×10^{-10}
$AgCN = Ag^+ + CN^-$	1.6×10^{-14}
$Ag(CN)_2^- = Ag^+ + 2CN^-$	1.8×10^{-19}
$Ag_4Fe(CN)_6 = 4Ag^+ + Fe(CN)_6^{4-}$	1.55×10^{-41}
$AgI = Ag^+ + I^-$	8.5×10^{-17}
$Ag(NH_3)_2^+ = Ag^+ + 2NH_3(aq.)$	5.9×10^{-8}
$AuCl_4^- = Au^{3+} + 4Cl^-$	5×10^{-22}
$Au(CN)_2^- = Au^+ + 2CN^-$	ca. 5×10^{-39}
$Cd(CN)_4 = Cd^{2+} + 4CN^-$	1.4×10^{-19}
$CdI_4^{2-} = Cd^{2+} + 4I^-$	ca. 5×10^{-7}
$Cd(NH_3)_4^{2+} = Cd^{2+} + 4NH_3(aq.)$	7.5×10^{-8}
$Co(NH_3)_6^{2+} = Co^{2+} + 6NH_3(aq.)$	1.25×10^{-5}
$Co(NH_3)_6^+ = Co^{3+} + 6NH_3(aq.)$	2.2×10^{-34}
$Co(NH_3)_5 \cdot H_2O^{3+} = Co^{3+} + 5NH_3(aq.) + H_2O$	1.6×10^{-35}
$CrCl_2^+ = Cr^{3+} + 2Cl^-$	1.26×10^{-2}
$HCrO_4 = H^+ + CrO_4^{2-}$	3.2×10^{-7}
$Cr_2O_7^{2-} + H_2O = 2HCrO_4^-$	2.3×10^{-2}
$Cu + Cu^{2+} = 2Cu^+$	6.3×10^{-7}
$Cu(CN)_2^- = Cu^+ + 2CN^-$	1×10^{-16}
$Cu(CN)_3^- = Cu^+ + 3CN^-$	5.6×10^{-28} [b]
$Cu(CNS) = Cu^+ + CNS^-$	4×10^{-14}
$Cu(NH_3)_2^+ = Cu^+ + 2NH_3(aq.)$	1.35×10^{-11}
$Cu(NH_3)_5^{2+} = Cu(NH_3)_4^{2+} + NH_3(aq.)$	2.8
$Cu(NH_3)_4^{2+} = Cu^{2+} + 4NH_3(aq.)$	4.7×10^{-15}
$Fe(CN)_6^{4-} = Fe^{2+} + 6CN^-$	ca. 10^{-35}
$Fe(CN)_6^{3-} = Fe^{3+} + 6CN^-$	ca. 10^{-42}
$Fe_2(C_4H_4O_6)_3(aq.) = 2Fe^{3+} + 3C_4H_4O_6^{3-}$	10^{-39}
$In^{3+} + H_2O = In(OH)^{2+} + H^+$	2×10^{-4}
$Mn(OH)_2 = Mn^{2+} + 2OH^-$	2×10^{-13}
$Ni(CN)_4^{2-} = Ni^{2+} + 4CN^-$	1×10^{-22}
$Ni(NH_3)_4^{2+} = Ni^{2+} + 4NH_3(aq.)$	1×10^{-8}
$Ni(NH_3)_6^{2+} = Ni^{2+} + 6NH_3(aq.)$	1.8×10^{-9}
$PbBr_2 = Pb^{2+} + 2Br^-$	4.6×10^{-6}
$PbBr^+ = Pb^{2+} + Br^-$	7.1×10^{-2}
$PbCl_2 = Pb^{2+} + 2Cl^-$	1.6×10^{-5}
$PbF_2 = Pb^{2+} + 2F^-$	4×10^{-8}
$PbI_2 = Pb^{2+} + 2I^-$	8.3×10^{-9}
$PbI^+ = Pb^{2+} + I^-$	3.45×10^{-2}
$PdCl_4^{2-} = Pd^{2+} + 4Cl^-$	5×10^{-13}
$PtCl_4^{2-} = Pt^{2+} + 4Cl^-$	ca. 1×10^{-16}
$PtBr_4^{2-} = Pt^{2+} + 4Br^-$	ca. 3×10^{-21}
$Sn(OH)_4 = Sn^{4+} + 4OH^-$	ca. 1×10^{-57}
$Zn(CN)_4^{2-} = Zn^{2+} + 4CN^-$	1.3×10^{-17}
$Zn(NH_3)_4^{2+} = Zn^{2+} + 4NH_3(aq.)$	3.4×10^{-10}

[a]These values have been taken from *The Oxidation States of the Elements and Their Potentials in Aqueous Solutions*, by Wendell M. Latimer, Prentice-Hall, 1952, 2nd ed.
[b]Ernst and Mann, *Trans. Electrochem. Soc.*, **61**, 363 (1952).

TABLE 5 Electrochemical Equivalents and Related Data for the Commonly Electrodeposited Metals[a]

Element	Symbol	Atomic Weight (1971)	Valence	Density, g/cm^3 20°C	mg/C	g/A-hr	g/m^2 for 1 μm	A-hr to deposit 1 μm·m^2
Aluminum	Al	26.9815	3	2.7	0.0932	0.3355	2.702	8.078
Antimony	Sb	121.75	5	6.62	0.2524	0.9086	6.618	7.294
			3		0.4207	1.515		4.372
Arsenic	As	74.9216	5	5.73	0.1553	0.5591	5.729	10.26
			3		0.2588	0.9317		6.142
Bismuth	Bi	208.980	5	9.78	0.4332	1.560	9.788	6.278
			3		0.7220	2.599		3.766
Cadmium	Cd	112.40	2	8.65	0.5825	2.097	8.659	4.126
Chromium	Cr	51.996	6	7.14	0.08981	0.3234	7.146	22.11
			3		0.1796	0.6466		11.06
Cobalt	Co	58.9332	2	8.71	0.3054	1.099	8.719	7.926
Copper	Cu	63.546	2	8.93	0.3293	1.185	8.935	7.540
			1		0.6586	2.371		3.770
Gold	Au	196.967	3	19.32	0.6804	2.449	19.32	7.887
			2		1.021	3.676		5.257
			1		2.041	7.348		2.631
Hydrogen	H	1.0079	1	0.08375×10^{-3}	0.01045	0.03762		
Indium	In	114.82	3	7.28	0.3967	1.428	7.278	5.092
Iron	Fe	55.847	3	7.86	0.1929	0.6944	7.867	11.32
			2		0.2894	1.042		7.540
Lead	Pb	207.2	4	11.34	0.5368	1.932	11.35	5.880
			2		1.074	3.866		2.936
Manganese	Mn	54.9380	2	7.3	0.2847	1.025	7.302	7.116

Molybdenum	Mo	95.94	6	10.2	0.1657	0.5965	10.21	17.11
Nickel	Ni	58.70	2	8.8	0.3043	1.095	8.803	8.044
Palladium	Pd	106.4	4	12.16	0.2758	0.9929	12.17	12.26
			2		0.5515	1.985		6.129
Platinum	Pt	195.09	4	21.37	0.5055	1.820	21.38	11.75
			2		1.011	3.640		5.871
Rhodium	Rh	102.905	4	12.44	0.2666	0.9598	12.44	12.95
			3		0.3555	1.280		9.730
			2		0.5332	1.920		6.481
Silver	Ag	107.868	1	10.49	1.118	4.025	10.50	2.605
Tin	Sn	118.69	4	7.3	0.3075	1.107	7.302	6.604
			2		0.6150	2.214		3.300
Zinc	Zn	65.38	2	7.14	0.3387	1.219	7.146	5.863

[a]All values are based on 100% current efficiency and absolute electrical units.

TABLE 6 Values of Hydrogen Overvoltage, Direct Method[a]

		Overvoltage V at $16 \pm 1°C$ in 2 N H_2SO_4			
Material	C.D., A/cm^2	10^{-3}	10^{-2}	10^{-1}	1
Cadmium		0.98	1.13	1.22	1.25
Mercury		0.90	1.04	1.07	1.12
Tin		0.86	1.08	1.22	1.23
Bismuth		0.78	1.05	1.14	1.23
Zinc		0.72	0.75	1.06	1.23
Graphite		0.60	0.78	0.98	1.22
Aluminum		0.56	0.83	1.0	1.29
Nickel		0.56	0.75	1.05	1.21
Lead		0.52	1.09	1.18	1.26
Brass		0.50	0.65	0.91	1.25
Copper		0.48	0.58	0.8	1.25
Silver		0.47	0.76	0.88	1.09
Iron		0.40	0.56	0.82	1.29
Monel		0.28	0.38	0.62	1.07
Gold		0.24	0.39	0.59	0.80
Duriron		0.20	0.29	0.61	1.02
Palladium		0.12	0.3	0.7	1.0
Platinum		0.024	0.07	0.29	0.68
Platinum (platinized)		0.015	0.03	0.04	0.05

[a]The values listed were abstracted from the table prepared by M. Knobel, *International Critical Tables*, **6**, 329. Overvoltages obtained by the direct method necessarily include the IR drop and concentration polarization.

TABLE 7 Percentage Metal in Common Plating Salts (Based on 100% Purity)

Name	Formula	Molecular Weight	% Metal
Antimony trichloride	$SbCl_3$	228.1	53.4
Antimony oxide	Sb_2O_3	291.5	83.5
Bismuth oxide	Bi_2O_3	466.0	89.7
Cadmium cyanide	$Cd(CN)_2$	164.4	68.4
Cadmium oxide	CdO	128.4	87.5
Chromic acid	CrO_3	100.0	52.0
Cobalt ammonium sulfate	$CoSO_4 \cdot (NH_4)_2SO_4 \cdot 6H_2O$	395.2	14.9
Cobalt chloride	$CoCl_2$	129.8	45.4
Cobalt sulfate	$CoSO_4$	155.0	38.0
Cobalt sulfate	$CoSO_4 \cdot 7H_2O$	281.1	21.0
Copper carbonate (basic)	$2CuCO_3 \cdot Cu(OH)_2$	344.7	55.3
Copper cyanide	$CuCN$	89.6	70.9
Copper fluoborate[a]	$Cu(BF_4)_2$	237.2	26.8
Copper sulfate	$CuSO_4 \cdot 5H_2O$	249.7	25.4
Ferrous chloride	$FeCl_2$	126.8	44.0
Ferrous ammonium sulfate	$FeSO_4 \cdot (NH_4)_2SO_4 \cdot 6H_2O$	392.2	14.2
Gold chloride (ic)	$AuCl_3$	303.6	65.0
Gold chloride (ic), crystal	$AuCl_3 \cdot 2H_2O$	339.6	58.1
Gold chloride (ous)	$AuCl$	232.7	84.7
Gold cyanide	$AuCN$	223.2	88.4
Gold potassium cyanide	$KAu(CN)_2$	288.3	68.4
Gold potassium cyanide, crystal	$KAu(CN)_2 \cdot 2H_2O$	324.4	60.8
Gold sodium cyanide	$NaAu(CN)_2$	272.2	72.4
Indium cyanide	$In(CN)_3$	192.8	59.5
Lead carbonate (basic)	$2PbCO_3 \cdot Pb(OH)_2$	775.7	80.1
Lead fluoborate[a]	$Pb(BF_4)_2$	380.9	54.4
Manganous chloride	$MnCl_2 \cdot 4H_2O$	197.9	27.7
Manganous sulfate	$MnSO_4 \cdot H_2O$	169.0	32.5
Nickel ammonium sulfate	$NiSO_4 \cdot (NH_4)_2SO_4 \cdot 6H_2O$	395.1	14.9
Nickel carbonate (basic)	$2NiCO_3 \cdot 3Ni(OH)_2 \cdot 4H_2O$	587.6	49.9
Nickel chloride	$NiCl_2 \cdot 6H_2O$	237.7	24.7
Nickel sulfate	$NiSO_4 \cdot 7H_2O$	280.9	20.9
Platinum chloride, crystal (Chloroplatinic acid)	$H_2PtCl_6 \cdot 6H_2O$	518.1	37.7
Potassium stannate	$K_2Sn(OH)_6$	298.9	39.7
Rhodium phosphate	$RhPO_4 \cdot 3H_2O$	251.9	40.8
Rhodium sulfate	$Rh_2(SO_4)_3 \cdot 12H_2O$	710.2	29.0
Silver chloride	$AgCl$	143.3	75.3
Silver cyanide	$AgCN$	133.9	80.6
Silver potassium cyanide	$KAg(CN)_2$	199.0	54.2
Silver sodium cyanide	$NaAg(CN)_2$	182.9	59.0
Silver nitrate	$AgNO_3$	169.9	63.5
Sodium stannate	$Na_2SnO_3 \cdot 3H_2O$	266.7	44.5
Tin chloride (ous), anhyd.	$SnCl_2$	189.6	62.6
Tin fluoborate[a]	$Sn(BF_4)_2$	292.3	40.6
Tin fluoborate[a]	$Sn(BF_4)_2 \cdot 6H_2O$	400.4	29.6

TABLE 7 *(Continued)*

Name	Formula	Molecular Weight	% Metal
Tin sulfate	$SnSO_4$	214.8	55.3
Tungstic acid	H_2WO_4	249.9	73.6
Tungstic oxide	WO_3	231.9	79.3
Zinc carbonate	$ZnCO_3$	125.4	52.2
Zinc cyanide	$Zn(CN)_2$	117.4	55.7
Zinc fluoborate[a]	$Zn(BF_4)_2$	239.0	27.4
Zinc fluoborate[a]	$Zn(BF_4)_2 \cdot 6H_2O$	347.1	18.8
Zinc oxide	ZnO	81.4	80.3
Zinc sulfate	$ZnSO_4 \cdot 7H_2O$	287.6	22.7

[a]Does not apply to solutions as usually available.

TABLE 8 Specific Gravity-Degrees Baumé Conversion Table

Specific Gravity	Degrees Baumé	Specific Gravity	Degrees Baumé	Specific Gravity	Degrees Baumé
1.00	0.00	1.34	36.79	1.68	58.69
1.01	1.44	1.35	37.59	1.69	59.20
1.02	2.84	1.36	38.38	1.70	59.71
1.03	4.22	1.37	39.16	1.71	60.20
1.04	5.58	1.38	39.93	1.72	60.70
1.05	6.91	1.39	40.68	1.73	61.18
1.06	8.21	1.40	41.43	1.74	61.67
1.07	9.49	1.41	42.16	1.75	62.14
1.08	10.78	1.42	42.89	1.76	62.61
1.09	11.97	1.43	43.60	1.77	63.08
1.10	13.18	1.44	44.31	1.78	63.54
1.11	14.37	1.45	45.00	1.79	63.99
1.12	15.54	1.46	45.68	1.80	64.44
1.13	16.68	1.47	46.36	1.81	64.89
1.14	17.81	1.48	47.03	1.82	65.31
1.15	18.91	1.49	47.68	1.83	65.77
1.16	20.00	1.50	48.33	1.84	66.20
1.17	21.09	1.51	48.97	1.85	66.62
1.18	22.12	1.52	49.60	1.86	67.04
1.19	23.15	1.53	50.23	1.87	67.46
1.20	24.17	1.54	50.84	1.88	67.87
1.21	25.16	1.55	51.45	1.89	68.28
1.22	26.15	1.56	52.05	1.90	68.68
1.23	27.11	1.57	52.64	1.91	69.08
1.24	28.06	1.58	53.23	1.92	69.48
1.25	29.00	1.59	53.80	1.93	69.87
1.26	29.92	1.60	54.38	1.94	70.26
1.27	30.83	1.61	54.94	1.95	70.64
1.28	31.72	1.62	55.49	1.96	71.02
1.29	32.60	1.63	56.04	1.97	71.40
1.30	33.46	1.64	56.58	1.98	71.77
1.31	34.31	1.65	57.12	1.99	72.14
1.32	35.15	1.66	57.65	2.00	72.50
1.33	35.98	1.67	58.17		

TABLE 9 Specific Gravity and Degrees Baumé of Aqueous Hydrochloric Acid Solutions

Specific Gravity, 20/4°C	Degrees Baumé	% HCl	Grams per Liter	Ounces per Gallon
1.0032	0.5	1	10.03	1.34
1.0082	1.2	2	20.16	2.68
1.0181	2.6	4	40.72	5.43
1.0279	3.9	6	61.67	8.23
1.0376	5.3	8	83.01	11.08
1.0474	6.6	10	104.7	13.95
1.0574	7.9	12	126.9	16.9
1.0675	9.2	14	149.5	19.9
1.0776	10.4	16	172.4	23.0
1.0878	11.7	18	195.8	26.1
1.0980	12.9	20	219.6	29.25
1.1083	14.2	22	243.8	32.5
1.1187	15.4	24	268.5	35.8
1.1290	16.6	26	293.5	39.1
1.1392	17.7	28	319.0	42.5
1.1493	18.8	30	344.8	46.0
1.1593	19.9	32	371.0	49.5
1.1691	21.0	34	397.5	53.0
1.1789	22.0	36	424.4	56.6
1.1885	23.0	38	451.6	60.2
1.1980	24.0	40	479.2	64.0

TABLE 10 Specific Gravity and Degrees Baumé of Aqueous Sulfuric Acid Solutions

Specific Gravity, 20/4°C	Degrees Baumé	% H_2SO_4	Grams per Liter	Ounces per Gallon
1.0051	0.7	1	10.05	1.4
1.0184	2.6	3	30.55	4.07
1.0317	4.5	5	51.59	6.88
1.0453	6.3	7	73.17	9.76
1.0591	8.1	9	95.32	12.7
1.0731	9.9	11	118.0	15.73
1.0874	11.7	13	141.4	18.85
1.1020	13.4	15	165.3	22.05
1.1168	15.2	17	189.9	25.3
1.1318	16.9	19	215.0	28.7
1.1471	18.6	21	240.9	32.1
1.1626	20.3	23	267.4	35.6
1.1783	21.9	25	294.6	39.3
1.1942	23.6	27	322.4	43.0
1.2104	25.2	29	351.0	46.8
1.2267	26.8	31	380.3	51.0
1.2432	28.4	33	410.3	54.7
1.2599	29.9	35	441.0	58.8
1.2769	31.4	37	472.5	62.9
1.2941	33.0	39	504.7	67.3
1.3116	34.5	41	537.8	71.6
1.3294	35.9	43	571.6	76.1
1.3476	37.4	45	606.4	80.8
1.3663	38.9	47	642.2	85.5
1.3854	40.3	49	678.8	90.5
1.4049	41.8	51	716.5	95.5
1.4248	43.2	53	755.1	100.8
1.4453	44.7	55	794.9	106.0
1.4662	46.1	57	835.7	111.5
1.4875	47.5	59	877.6	117.0
1.5091	48.9	61	920.6	122.7
1.5310	50.3	63	964.5	128.7
1.5533	51.7	65	1010	134.8
1.5760	53.0	67	1056	141.0
1.5989	54.3	69	1103	147.0
1.6221	55.6	71	1152	153.8
1.6456	56.9	73	1201	160.0
1.6692	58.1	75	1252	167.0
1.6927	59.3	77	1303	174.0
1.7158	60.5	79	1355	181.0
1.7383	61.6	81	1408	188.0
1.7594	62.6	83	1460	195.0
1.7786	63.5	85	1512	201.6
1.7951	64.2	87	1562	208.4
1.8087	64.8	89	1610	214.6
1.8195	65.3	91	1656	220.6

TABLE 10 *(Continued)*

Specific Gravity, 20/4°C	Degrees Baumé	% H_2SO_4	Grams per Liter	Ounces per Gallon
1.8279	65.7	93	1700	226.5
1.8337	65.9	95	1742	232.5
1.8364	66.0	97	1781	237.5
1.8342	65.9	99	1816	242.0
1.8305	65.8	100	1831	244.5

TABLE 11 Total Concentrations of Copper Sulfate Plus Sulfuric Acid in Solutions of Given Specific Gravity[a]

Specific Gravity, 25/4°C	Copper Sulfate plus Sulfuric Acid		Specific Gravity, 25/4°C	Copper Sulfate plus Sulfuric Acid	
	g/l	oz/gal		g/l	oz/gal
1.01	20	2.7	1.13	217	29.1
1.02	36	4.8	1.14	234	31.3
1.03	52	7.0	1.15	251	33.6
1.04	68	9.1	1.16	268	35.9
1.05	84	11.3	1.17	286	38.3
1.06	100	13.4	1.18	303	40.6
1.07	117	15.7	1.19	321	43.0
1.08	133	17.8	1.20	339	45.4
1.09	150	20.0	1.21	357	47.8
1.10	166	22.3	1.22	375	50.2
1.11	183	24.5	1.23	393	52.6
1.12	200	26.8			

[a]From *Principles of Electroplating and Electroforming*, W. Blum and G. B. Hogaboom, McGraw-Hill, New York, 1930, p. 410.

TABLE 12 Conversion Factors

Multiply	By	To Obtain
Amperes	1.04×10^{-5}	Faradays per second
Amperes per square foot	0.108	Amperes per square decimeter
Ampere-hours	3600	Coulombs
Ampere-hours	0.0373	Faradays
Angstrom units	3.94×10^{-9}	Inches
Angstrom units	1×10^{-8}	Centimeters
Angstrom units	10^{-1}	Nanometers
Centimeters	0.3937	Inches
Centimeters	393.7	Mils
Circular mils	7.85×10^{-7}	Square inches
Circular mils	5.07×10^{-6}	Square centimeters
Cubic centimeters	0.061	Cubic inches
Cubic feet	1728	Cubic inches
Cubic feet	7.48	Gallons
Cubic feet	29.9	Quarts (liquid, U.S.)
Cubic inches	16.39	Cubic centimeters
Drams (avoir.)	1.77	Grams
Drams (avoir.)	0.0625	Ounces (avoir.)
Drams (fluid)	3.7	Cubic centimeters
Drams (fluid)	0.125	Ounces (fluid)
Faradays	96,501.2	Coulombs (int.)
Feet	30.48	Centimeters
Gallons	4	Quarts (liquid)
Gallons	3.785	Liters
Gallons	128	Ounces (fluid)
Gallons	8	Pints (liquid)
Gallons	3.782	Kilograms water (62°F)
Gallons	8.34	Pounds (avoir.) water (62°F)
Gallons (U.S.)	0.833	Gallons (Imperial)
Gallons per minute	0.002228	Cubic feet per second
Gallons per minute	0.0631	Liters per second
Grams	0.564	Drams (avoir.)
Grams	15.43	Grains
Grams	0.03215	Ounces (troy)
Grams	0.0353	Ounces (avoir.)
Grams per liter	1000	Parts per million
Inches	2.54	Centimeters
Inches	1000	Mils
Kilograms	2.205	Pounds (avoir.)
Kilograms	2.679	Pounds (troy)
Liters	0.03532	Cubic feet
Liters	1.057	Quarts (liquid)
Micrometers	3.94×10^{-5}	Inches
Micrometers	0.001	Millimeters
Milliliters	1.0	Cubic centimeters
Mils	0.001	Inches
Mils	0.00254	Centimeters
Ounces (avoir.)	28.35	Grams

TABLE 12 *(Continued)*

Multiply	By	To Obtain
Ounces per gallon (avoir.)	7.5	Grams per liter
Ounces (troy)	31.1	Grams
Ounces (troy)	480	Grains
Ounces (troy)	20	Pennyweights
Ounces per gallon (fluid)	7.7	Milliliters per liter
Pennyweights	24	Grains
Pennyweights per gallon	0.41	Grams per liter
Pints (liquid, U.S.)	473.2	Cubic centimeters
Pints (liquid, U.S.)	16	Ounces (fluid)
Pounds (avoir.)	453.6	Grams
Pounds (avoir.)	16	Ounces (avoir.)
Pounds (avoir.)	1.215	Pounds (troy)
Pounds (troy)	373.24	Grams
Pounds (troy)	12	Ounces (troy)
Pounds per cubic foot	16.02	Grams per liter
Pounds per gallon (U.S.)	119.8	Grams per liter
Square centimeters	1.97×10^5	Circular mils
Square inches	10^6	Square mils
Square inches	1.27×10^6	Circular mils
Square inches	6.45	Square centimeters
Volts per inch	0.394	Volts per centimeter

TABLE 13 Temperature Conversion Table

0–38			39–76			77–212		
°C		°F	°C		°F	°C		°F
−17.8	0	32.0	3.89	39	102.2	25.0	77	170.6
−17.2	1	33.8	4.44	40	104.0	25.6	78	172.4
−16.7	2	35.6	5.00	41	105.8	26.1	79	174.2
−16.1	3	37.4	5.56	42	107.6	26.7	80	176.0
−15.6	4	39.2	6.11	43	109.4	27.2	81	177.8
−15.0	5	41.0	6.67	44	111.2	27.8	82	179.6
−14.4	6	42.8	7.22	45	113.0	28.3	83	181.4
−13.9	7	44.6	7.78	46	114.8	28.9	84	183.2
−13.3	8	46.4	8.33	47	116.6	29.4	85	185.0
−12.8	9	48.2	8.89	48	118.4	30.0	86	186.8
−12.2	10	50.0	9.44	49	120.2	30.6	87	188.6
−11.7	11	51.8	10.0	50	122.0	31.1	88	190.4
−11.1	12	53.6	10.6	51	123.8	31.7	89	192.2
−10.6	13	55.4	11.1	52	125.6	32.2	90	194.0
−10.0	14	57.2	11.7	53	127.4	32.8	91	195.8
−9.44	15	59.0	12.2	54	129.2	33.3	92	197.6
−8.89	16	60.8	12.8	55	131.0	33.9	93	199.4
−8.33	17	62.6	13.3	56	132.8	34.4	94	201.2
−7.78	18	64.4	13.9	57	134.6	35.0	95	203.0
−7.22	19	66.2	14.4	58	136.4	35.6	96	204.8

TABLE 13 *(Continued)*

0–38			39–76			77–212		
°C		°F	°C		°F	°C		°F
−6.67	20	68.0	15.0	59	138.2	36.1	97	206.6
−6.11	21	69.8	15.6	60	140.0	36.7	98	208.4
−5.56	22	71.6	16.1	61	141.8	37.2	99	210.2
−5.00	23	73.4	16.7	62	143.6	37.8	100	212.0
−4.44	24	75.2	17.2	63	145.4	43.0	110	230.0
−3.89	25	77.0	17.8	64	147.2	49.0	120	248.0
−3.33	26	78.8	18.3	65	149.0	54.0	130	266.0
−2.78	27	80.6	18.9	66	150.8	60.0	140	284.0
−2.22	28	82.4	19.4	67	152.6	66.0	150	302.0
−1.67	29	84.2	20.0	68	154.4	71.0	160	320.0
−1.11	30	86.0	20.6	69	156.2	77.0	170	338.0
−0.56	31	87.8	21.1	70	158.0	82.0	180	356.0
0.00	32	89.6	21.7	71	159.8	88.0	190	374.0
0.56	33	91.4	22.2	72	161.6	93.0	200	392.0
1.11	34	93.2	22.8	73	163.4	99.0	210	410.0
1.67	35	95.0	23.3	74	165.2	100.0	212	413.0
2.22	36	96.8	23.9	75	167.0			
2.78	37	98.6	24.4	76	168.8			
3.33	38	100.4						

The numbers in the center of each group refer to the temperature in degrees Centigrade or Fahrenheit which it is desired to convert into the other scale. If converting from Fahrenheit to Centigrade, the equivalent will be found in the left column; if converting from Centigrade to Fahrenheit, the answer will be found in the column on the right.

INDEX

Modern
electroplating.

$150.00

DATE			